ESPRIT '89

ESPRIT '89

Proceedings of the 6th Annual ESPRIT Conference,
Brussels, November 27 - December 1, 1989

Edited by

COMMISSION OF THE EUROPEAN COMMUNITIES
Directorate-General TELECOMMUNICATIONS,
INFORMATION INDUSTRIES and INNOVATION

KLUWER ACADEMIC PUBLISHERS
DORDRECHT / BOSTON / LONDON

ISBN-13: 978-94-010-6968-7   e-ISBN-13: 978-94-009-1063-8
DOI: 10.1007/978-94-009-1063-8

Publication arrangements by
Commission of the European Communities
Directorate-General Telecommunications, Information Industries and Innovation
Scientific and Technical Communications Service, Luxembourg

EUR 12512
© 1989 ECSC, EEC, EAEC, Brussels and Luxembourg
Softcover reprint of the hardcover 1st edition 1989
LEGAL NOTICE
Neither the Commission of the European Communities nor any person acting on behalf of the
Commission is responsible for the use which might be made of the following information.

Published by Kluwer Academic Publishers,
P.O. Box 17, 3300 AA Dordrecht, The Netherlands.

Kluwer Academic Publishers incorporates the publishing programmes of
D. Reidel, Martinus Nijhoff, Dr W. Junk and MTP Press.

Sold and distributed in the U.S.A. and Canada
by Kluwer Academic Publishers,
101 Philip Drive, Norwell, MA 02061, U.S.A.

In all other countries, sold and distributed
by Kluwer Academic Publishers Group,
P.O. Box 322, 3300 AH Dordrecht, The Netherlands.

Printed on acid-free paper

Foreword

The 6th ESPRIT Conference is being held in Brussels from the 27th November to the 1st December 1989. Well over 1500 participants from all over Europe are expected to attend the various events during the week. The Conference will offer the opportunity to be updated on the results of ongoing Esprit projects and to develop Europe-wide contacts with colleagues, both within a specific branch of Information Technology and across different branches.

The first three days of the week are devoted to presentations of Esprit I projects, structured into plenary and parallel sessions; this year there is special emphasis on panels and workshops where participants can exchange ideas and hold in-depth discussions on specific topics. The different areas of Esprit work are covered: Microelectronics, Information Processing Systems, Office and Business Systems, Computer Integrated Manufacturing, Basic Research and different aspects of the Information Exchange System.

During the IT Forum on Thursday 30th November, major European industrial and political decision-makers will address the audience in the morning. In the afternoon, different aspects of Technology Transfer will be discussed with the participation of outside experts, and presentations on the future plans for community R&D in IT will take place.

This year, more than 100 projects will display their major innovations and achievements in the Exhibition Area which has been significantly expanded to include facilities for informal meetings to discuss all aspects of the projects. The exhibition offers a unique opportunity to acquire in a short time, first-hand, complete and comprehensive knowledge about the results obtained in all areas of the Esprit programme.

I would like to congratulate and thank everyone who has contributed to the Conference: the authors and reviewers of papers and project reports; the chairmen, the speakers, the panelists and workshop participants at the Conference; the project teams which set up the demonstrations and the various organisations who have made equipment available. The success of this Conference is due to all their efforts.

J.M. Cadiou
Director
Information Technologies - ESPRIT

CONTENTS

Information Processing Systems

Computer Integrated Manufacturing

Office and Business Systems

Information Exchange System

Basic Research

Indexes

PLENARY SESSION

Project no 927/2518

WAFER AND EPILAYER IMPROVEMENT CORRELATED WITH DEVICE PERFORMANCES FOR InP BASED OPTOELECTRONICS

I. GRANT, *ICI, Milton Keynes, GB*
M. HEYEN, H. JÜRGENSEN and D. SCHMITZ, *AIXTRON, Aachen, FRG*
M. RENAUD, M. ERMAN, J. LE BRIS, *LEP, Paris, F*
F. SCHULTE and Ch. STEINBERGER, *Aachen, Technical University Aachen, FRG*

ABSTRACT.   The project described in this paper is aimed to develop a reproducible technology for the preparation of InP substrates and epitaxial layers based on this material. This development is supported by extensive material characterization and the feed back from device manufacturing.

1. INTRODUCTION

The present project E 2518 is aimed to demonstrate the reproducibility of MOCVD equipment and growth processes for epilayers (GaInAs and GaInAsP) suitable for optoelectronic integration. Hence, ·the problem of growing large area (2 inches) wafer, with uniform characteristics has to be addressed. At the beginning of this project, we have therefore devoted special efforts towards the growth and the characterization of the uniformity of InP ingots as well as GaInAs and GaInAsP epitaxial layers.

Growth of (100) semi−insulating and n^+ InP ingots have been performed by ICI. Additional InP substrates have also been bought from outside−suppliers for comparison. The wafers uniformity has been tested at LEP by scanning photoluminescence.

The MOCVD growth process, developed at AIXTRON for the preparation of GaInAs MISFETs structures during the project E 927, has been applied to grow on InP substrates from ICI. Growth conditions for uniform GaInAsP layers ($\lambda_g = 1.3$ μm) have also been developed on substrates with various conductivity types. The uniformity of bandgap wavelength has also been tested by scanning photoluminescence. In the next period, the transparency of the quaternary material will be determined using a simple waveguide structure fabricated at LEP and the results will be correlated with the

3

photoluminescence data.

In order to test the electrical performances of the GaInAs layers, RWTH has continued the fabrication and characterization of MISFETs devices. Good high frequency performances have been obtained. A mask with ring oscillators has been fabricated.

2. CRYSTAL PULLING OF InP SUBSTRATES

ICI is engaged in the production of substrates of semi–insulating iron doped and Sulfur doped n^+ InP. For both materials the initial objective was the improvement of crystal uniformity with respect to dopants and defects. In case of semi–insulating substrates the primary aim is the minimization of Fe doping concentration consistent with uniform high electrical resistivty ($> 10^7$ ohm cm). In the n^+ material a low dislocation density is required at a Sulfur doping concentration less than $8 \times 10^{18} cm^{-3}$.

Non–intentionally doped InP has an n–type conductivity with a Hall carrier concentration in the range $10^{15} - 10^{16}$ cm^{-3}. To obtain semi–insulating electrical behaviour these residual shallow donors must be compensated by an excess concentration of deep acceptor states. Fe, with an acceptor level at $E_C - 0.65 eV$ is the impurity most commonly used to obtain high resistivity InP in the range $10^7 - 10^8$ ohm.cm. The minimum Fe doping concentration is determined by the background donor impurity concentration. The maximum Fe doping content is limited by the formation of FeP_2 precipitates within the crystal with associated twinning. Below this level, it is desirable to minimize the dopant concentration to avoid impurity outdiffusion from the substrate to the grown epitaxial layer.

Optimization of the dopant concentration therefore requires characterization of the residual donor impurities of the undoped material together with a quantitative analysis of the Fe incorporation in the crystal under standard conditions used for growth of polycrystalline undoped ingots subsequently used as charge material for single crystal growth. The background donor concentration was less than $10^{16} cm^{-3}$ in all cases.

A series of crystals was grown with different initial Fe doping concentrat- ions in the melt. The threshold value of Fe content for semiinsulating behaviour at $1 \times 10^{16} cm^{-3}$ seems to be consistent with the maximum donor background level. The wafers supplied to the contract have average dislocat- ion densities of less than $5 \times 10^4 cm^{-2}$.

The substrates were single side polished with an etched back surface and laser mark identification on the major flat on the rear surface. The polishing process was optimized for wafer cleanliness. Extensive characterization of the surface quality has been undertaken in RACE project R1029. The stand–

ard direction of misorientation adopted was with the normal to the wafer–surface tilted 2° towards (101). The specified wafer thickness is 375 microns $\pm$ 25 microns. The front surface wafer flatness is measured automatically using a TROPEL autoselect flatness tester. Results on sample wafers indicate an average optically measured flatness around 5 microns across the 2 inch wafer. Heavily doped InP can be grown with low or 'zero' dislocation density by selection of an appropriate dopant. Sulfur is most commonly used, at a concentration above $2 \times 10^{18} cm^{-3}$ in the crystal. A process was developed to adjust doping concentration in the range $5 \times 10^{18} - 8 \times 10^{18} cm^{-3}$ with an average e.p.d. less than 5000 cm^{-2}.

The polished surface characteristics of these batches are representative of current state of the art polishing quality as assessed in RACE project R1029 in comparison with other leading suppliers' wafers.

3. MOVPE GROWTH OF GaInAs AND GaInAsP EPILAYERS

The structures planned to be used in this investigation include quaternary layers (λ = 1.3 µm) to be grown on various types of InP substrates. An extended growth study on doped substrates resulted in the fact that the optimum growth temperature for ternary and quaternary layers on highly n–doped materials is 640°C. This growth temperature has been used for all growth runs in the quaternary material system.

The adjustment of the gas phase composition in order to obtain lattice matched layers with exactly defined emission wavelength is very complex in a four material system. For lattice matching and wavelength adjustment most convenient species are AsH_3 and TMIn. By varying the vapour pressures of these materials the process for the growth of lattice matched GaInAsP layers with an exact emission wavelength of 1.3 µm has been developed.

The FWHM obtained by double crystal x–ray diffraction (DXCD) is definitely below 20" of arc, which indicates excellent vertical homogeneity of the GaInAsP.

The growth experiments were investigated on Sn– and Fe–doped substrates. Differences in the lattice mismatch were detected on different substrates using the same process for growth. Thus the role of the substrate dopant on the layer properties was investigated by growing on different substrates in one run. The fact that no influence on the emission wavelength was found but the DXCD–measurement showed differences led to the assumption that the dopant in the substrates influences the lattice constant. This could indeed be verified and has to be considered in the adjustment of parameters to obtain optimum material parameters. The background doping level was lower than 2×10^{15} cm^{-3} in all measured points. 77 K mobilities were around 25.000 cm^2/Vs.

The InP/GaInAs structure for the MISFET's and Ringoscillators has been developed in project E 927 and is continued to be used in the present program in studies of the Aachen Technical University. It turned out that a modification of the contact layer thickness and buffer layer doping could improve the device performance. New structures including these modificat-ions have been developed.

The activities started as cooperation with E 263 in E 927 have been con-tinued. Zn-doping profile studies in GaInAs evaluated by SIMS at HHI in Berlin have been performed. First promising results obtained from spike doping have been reproduced and additional experimental structures have been prepared. After determining a maximum level of Zn-doping, reasonable for growth, several structures were grown and evaluated.

The maximum level for Zn doping was about $2.5 \times 10^{19} \mathrm{cm}^{-3}$, limited by crystal degradation. Steep doping transition could be obtained even at con-centrations of $10^{19} \mathrm{cm}^{-3}$ in the various structures. This behaviour indicates, that Zn is a promising dopant e. g. even for the fabrication of Heterobipo-lartransistors.

4. ASSESSMENT OF UNIFORMITIES OF InP SUBSTRATES AND GaInAsP EPILAYERS

Scanning Photoluminescence has been extensively used to test the homo-geneity of InP substrates from various suppliers. Iron doped semi-insulating and Sulfur doped n^+ InP substrates have been analyzed and the charact-eristics of these two types of substrates have been compared between the suppliers. The spatially resolved photoluminescence (SPL) has also been used to make mappings of the gap wavelength of GaInAsP epilayers.

The SPL measurements have been performed at room temperature on two inch wafers with an optical resolution ranging from a few hundreds of μm to 20 μm. The PL is excited either with an Argon laser ($\lambda = 0.5145$ μm) for InP analysis or with a YAG laser ($\lambda = 1.06$ μm) for Ga InAsP characterizat-ion. The 15 μm large laser beam is focused onto the surface, scanned over the sample and either the total PL signal or the PL signal at a given wavelength is collected. The total description of the apparatus which has other capabilities can be found in reference (1).

In case of doped InP material, either with iron or Sulfur, the photolu-minescence mappings mainly reflect the dopings homogeneities. PL intensity increases with Sulfur concentration and decreases with iron concentration. In fact, sulfur introduces radiative centers while iron produces essentially non-radiative centers. The dislocations can also be revealed by PL depending on the measurement parameters.

Using 4 wafers from a s.i InP ingot from ICI, it has been seen that doping striations occur only at the seed end. Besides, iron concentration and dislocation density increase towards the tail and the values estimated from PL measurements are in good agreement with SIMS analysis for Fe and with chemical etching for EPD. Therefore the scanning photoluminescence is a powerful and non destructive tool to test not only qualitatively but also quantitatively the homogeneity of InP wafers.

The homogeneity of the bandgap wavelength of $Ga_xIn_{1-x}As_yP_{1-y}$ epitaxial layers is of particular interest for many optical devices such as lasers and optical waveguides. Using scanning photoluminescence mappings recorded at several energies, we have determined the mappings of the bandgap wavelength and of the PL intensity at this wavelength. First, we can observe that the two mappings are not correlated. Secondly, the bandgap wavelength is constant at 1.296 µm over the whole wafer if we exclude 5mm at the edges. The PL intensity variations could be due to variations of thickness or impurities concentrations and this will be checked in the future.

The analysis by scanning photoluminescence of a GaInAsP epilayer grown at AIXTRON has shown the capability of MOCVD to produce epitaxial layers with a very good homogeneity of bandgap wavelength.

5. DISCRETE GaInAs MISFETS AND INVERTERS

MISFETS and inverters have been fabricated on epitaxial layers grown in the period of ESPRIT 927. For the fabrication of MISFETS and inverters in one run the following technology steps developped in ESPRIT 927 apply:

1) Mesa etching
2) Ohmic metallization (and annealing)
3) Channel recess of depletion type load and enhancement type driver
4) Channel recess of enhancement type driver
5) SiO_2 deposition and RTA annealing (750°C, 1s)
6) Gate metallization (Al, 300 nm)
7) SiO_2 opening and interconnection

Microwave measurements have been accomplished by the cascade microwave prober up to 20 Ghz. Fitting $|h_{21}|^2$ to 0 dB the device with 1.5 µm gate length yields a cut off frequency f_T of 16 Ghz. From the fit of the uni-lateral gain a maximum oscillation frequency (f_{max}) of 14 Ghz is deduced. An increasement of f_{max} can be achieved by reducing the gate resistance which is mostly contributed to the input resistance of 25 Ω.

For time domain measurements this sample has been bonded. The test pulse is generated by a HP tunnel diode mount with a rise time of about 20 ps.

A response time of 70 ps is obtained.

MISFET inverters have been fabricated in n− GaInAs with enhancement type driver and depletion type load. For both transistors the gate length is 2.0 μm. The voltage gain is −3. A high level output of 2 V and a low level output of 0.2 V is observed. The current of the enhancement type driver at

V_G = 0 V is similar to that of the depletion type.

Masks for ring oscillators have been designed and fabricated. These masks allow using logic gates of inverters in n− GaInAs with enhancement mode driver and depletion type load or inverters in p−type GaInAs with inversion mode driver and depletion type load.

6. IMPACT ON EUROPEAN PROGRAMS

The availability of basic materials like substrates and epitaxial layers is an essential requirement for the successful fabrication of advanced devices for microwave digital and optoelectronic applications.

Unfortunately, the resources for InP wafers in the Western community are very limited. Most material is from major Japanese suppliers. The complete dependence on these Far East suppliers has certainly to be overcome. Therefore the attempts of ICI to grow InP bulk material needs highest support.

Also the availability of equipment for epitaxial growth can be considered as a basic requirement. Moreover, AIXTRON is not only a supplier of equip− ment but transfers complete technology of material preparation with its reactors. Most of the InP based process know how has been supported with assistance of the ESPRIT program. The availability of the InP technol− ogy makes AIXTRON competetive not only on the European market but also in the USA. First customers have also been attracted from Taiwan and Korea.

The development of wafers and layers is only possible with the support of the corresponding characterization techniques. It is of high importance, that LEP participates in this program by making their most advanced characterization technique available.

The final prove for material quality however, is the device. The Aachen Technical University participates in this program by manufacturing devices. The feedback from the devices further improves structures and processes.

7. REFERENCES

/1/ Erman J., Gillardin G., Le Bris J., Renaud M. and Tomzig E. (1989) 'Characterization of Fe doped semi-insulating InP by low and room temperature spatially resolved photoluminescence', Journal of Crystal Growth, 469 – 482.

/2/ Splettstößer J., Schulte F., Trasser A., Schmitz D. and Beneking H. (1989) 'High speed $Ga_{0.47}In_{0.53}As$ MISFETs grown by metal organic vapor phase epitaxy', Proceedings of ESSDERC in Berlin, to be published.

DEPTH AND MOTION ANALYSIS:
THE ESPRIT PROJECT P940

MUSSO GIORGIO
Elettronica San Giorgio
- ELSAG S.p.A.
via Puccini, 2
GENOVA - ITALY

Summary

Depth and Motion Analysis represents one of the most important issues in passive three-dimensional Computer Vision. The ESPRIT Project P940 is a remarkable research initiative dedicated to these problems, in which some of the main University and Industry Organizations, working in the Computer Vision field, are involved.

The aims of this project are those of investigating passive methods and algorithms for Depth and Motion Analysis, and of designing and building a hardware architecture, capable of implementing this analysis on line with a mobile vehicle, navigating in a structured environment, and with a robot arm for recognition and location purpouses.

In this paper the technical contents of the project as well as the project evolution are described.

1. Introduction

Computer Vision, as well as any other AIP area concerning the intelligent machine-environment interaction, represents a research challenge stimulated by its widespread applicability in almost all the automation fields.

Due to its strategical importance, Computer Vision is investigated in many industry and university labs all over the industrialized world, particularly in the U.S.A. and Japan.

Fortunately, the Computer Vision area has not been neglected in Europe, where, since about fifteen years, many advanced research activities have been going on in industries and universities.

Just in this context, the ESPRIT P940 was launched in 1986, and, as a consequence of past experiences of the proposers, it was particularly focused on 3D vision problems and their real-time solutions.

3D vision approach was selected because many industrial applications of intelligent sensoriality cannot be led to simpler bidimensional problems: for instance, robot arm manipulation, non contacting measuring, as well as vehicle guidance and obstacle avoidance are intrinsically three-dimensional.

In particular, a three-cameras approach was selected for stereometric purposes, where extracted tokens (line segments in our case) are fused in the multi-image matching process and reprojected in their 3D space locations.

As far as motion analysis is concerned, the extraction of motion features has been approached by time-tracking the same tokens used for the stereo process, resulting in a further process unification and getting interesting possibilities for stereo-motion cooperation.

Besides the need of looking for a methodological solution to this class of problems, the aim of getting results as close as possible to industrial exploitation brought in the project a remarkable attention to the real-time capabilities; this led to the definition and realization of a hardware architecture, capable of solving stereo and motion problems at a speed suitable for application in both robot-arm manipulation and vehicle guidance fields. As specific exploitation areas, where specific demonstrations will be implemented as project outcomes, it was selected the automatic 3D model reconstruction of mechanical parts and their recognition for sorting and assembly purpouses in the factory automation domain.

Obstacle avoidance, ego-motion estimation and routing inside structured environments, like offices or workshops, represent the exploitation fields concerning vehicle guidance objectives, which are planned to be exploited for transportation tasks in Industrial Automation.

Given this framework, and after a deep analysis on the state of the art, the project P940 was defined, having major interests in the following areas: the theoretical approaches to stereoscopic vision, motion analysis and motion-stereo cooperation, the realization of a modular and flexible computing architecture (DMA) for the real-time implementation of these processes, and the selection of demonstrators where the full capabilities of the system, as well as its performances on real applications, can be evaluated.

Particularly for the demonstration phases, the DMA architecture will be integrated both with mechanical actuators, like robot arm and vehicle controllers, and with the high level knowledge of the specific application for achieving a full real-time demonstration of the selected tasks.

The Consortium of ESPRIT Project P940 is formed as follows:
- CAMBRIDGE University (U.K.)
- ELSAG S.p.A. (I) prime contractor
- G.E.C. (U.K.)
- GENOA University (I)
- INRIA (F)
- ITMI (F)
- MATRA ESPACE (F)
- NOESIS (F)

This paper is aimed to give a general description of the project, summarizing its technical contents on both methodological and implementation sides, and describing the achievements obtained so far. In doing that, particular attention will also be paid in underlining organization aspects, as well as Consortium cooperation issues, which have been absolutely determinant for the successful conclusion of the project first phase.

2. Project Background

Computer Vision concerns the capability of designing and building electronic systems which implement some form of artificial visual perception. In other words, systems that, acquiring and analysing images of the external world, are able to understand the "physical meaning" of what they are looking at.

Besides the scientific interest in understanding how biological vision systems operate, the possibility of allowing electronic systems to interact with the physical world by visual perception capabilities is getting more and more attention from many application fields.

Indeed, advanced automation, applied to manufacturing industry, continuous process plants and services, require much more intelligent ways for adapting automatically the system behaviour to external situations, that cannot be exhaustively foreseen in the system design phases (1).

Images represent input data for a vision system. In many cases, only one image is sufficient for a very intelligent vision system, like the human brain, to get a deep understanding of the three-dimensional world it describes. This capability is due to the large use of a-priori knowledge the human brain makes for interpreting bidimensional representations of the physical world.

In artificial systems, where the amount of a-priori knowledge we can represent and process is still very limited, we need to acquire much more information about the three-dimensionality of the observed scene.

This is the reason why a great effort of research is continuously spent in investigating methods for acquiring 3D information: the class of these methods is indicated by "shape from X", where, depending on the approach selected, X stays for stereo, motion, shadow, etc. (2).

All these methods are based on the acquisition of a set of images obtained from multiple viewpoints or by multiple locations of illumination sources, in such a way that 3D information can be recovered by matching corresponding features in different images.

Besides these passive methods, so named because they do not use special forms of illumination, we find active techniques, which acquire 3D information using some ad hoc illumination sources. Laser and ultrasound range finder, structured light, etc., belong to this second class.

How to approach a 3D vision problem, i.e. the suitability of active against passive techniques, represents an ever present issue, whose solution depends on considerations not only limited to vision aspects, but generally extended to the whole automation problem we are dealing with.

On the other hand, passive approaches seem to be more general, either from the scientific or the industrial stand point, because they do not imply a strong environment cooperation.

The above cited issues summarize the methodological framework we have to take into account in conceiving advanced researches in the computer vision area. This is not all we have to consider: as far as computer vision applications are concerned, a lot of attention has to be paid to the speed performances of the vision system.

As said before, visual perception has to allow the whole system to react, in real time, to the various situations of the external world in which it operates.

This implies that the large amount of computation, required to grow up the understanding of the observed scene, has to be performed in such a time as to allow the system to find and actuate an action plan useful for reaching the desired objective in the due time.

Partners in P940 Consortium, at the time of the proposal, had a good experience in all the topics above summarized, particularly in stereo approaches to vision, in high speed hardware design and in industrial application of Computer Vision systems. So, the definition phase of the project, as well as the following work, represent a synergetic fusion of complementary experiences, that allowed to originate one of the most advanced effort in Computer Vision actually in progress in Europe.

3. Project Contents (3)

3.1. Project objectives

As already mentioned in the introduction, the Esprit Project P940, called Depth and Motion Analysis, is an European research project involving several industrial and academic partners from three different countries: France, Great Britain and Italy.

The project started in June 1986 with the following main objectives:

1) Develop as general as possible algorithmic solutions to problems of 3D scene reconstruction from stereo images and to problems of motion analysis from image time-sequences.
2) Implement in hardware a selected chain of these algorithms to construct an early vision machine, the DMA machine, with a throughput suitable for a wide range of Vision applications.
3) Demonstrate the project outcomes in two distinct application fields:
   a) 3D model reconstruction of mechanical parts and their recognition and location for sorting and assembly purposes in the factory automation domain;
   b) obstacle avoidance, ego-motion estimation and navigation in structured environment of a mobile robot.

3.2. Project principles

The basic ideas of the project, concerning the 3D reconstruction of physical scene from stereo process and the analysis of image time-sequences for motion analysis are descripted in fig. 1 and fig. 2.

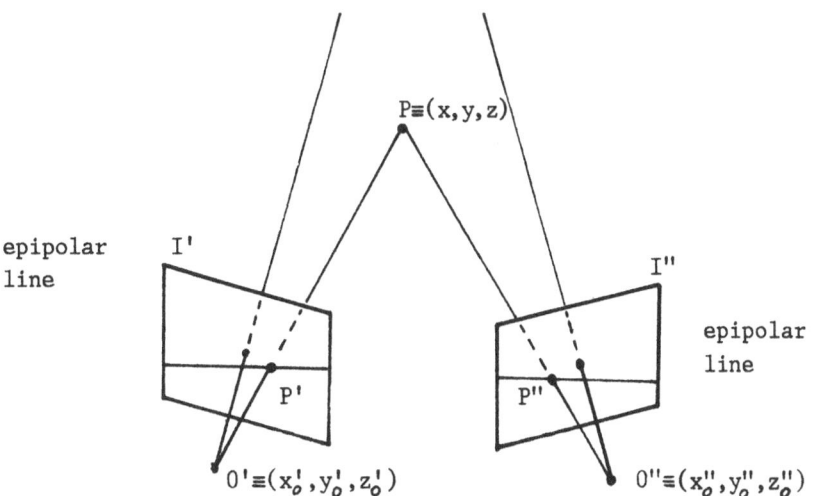

Fig. 1 : Scheme for stereo reconstruction

In fig. 1 the point P, located in three dimensional space at coordinates (x, y, z), originates the two projections P and P" in the two images acquired with optical centers located

14

respectively in $0 = (x, y, z)$ and $O (x, y, z)$. In principle given the two projections P and P", given that they correspond to the same 3D point, and knowing the locations and the optical parameters of the two imaging systems used, we can back-project in space P and P", determining in this way the three-dimensional location of the object point P (4), (5).

The main problem in this process concerns finding corresponding tokens in different images. This for two different reasons:

a) In dealing with complex scene, we obtain many projected tokens in images and finding corresponding tokens results in a complex matching process, that has to be designed as robust as possible. Indeed any mismatch gives rise to wrong 3D location estimates.

b) Blind trial and error matching processes have to be rejected to avoid computational complexity deriving from combinatorial explosion.

Apart from the possibility of structuring the data in the system memory to cut down the searching complexity, there exists a very natural way to reduce the searching time.

Indeed the point P" in image I", correspondent to P in image I , cannot be randomly located in the image. It will lay on the intersection of the epipolar plane, defined by the three points P, O , O', and the plane containing the image I".

So the searching process has to be executed only along this intersection line (epipolar line).

In fig. 2 the basic ideas used in the project for acquiring motion information are sketched (6), (7).

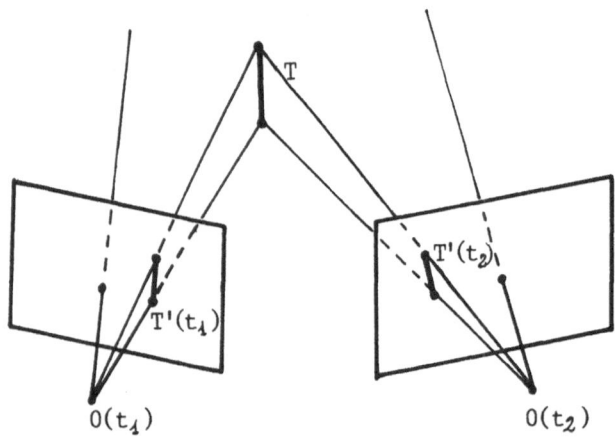

Fig. 2 : Scheme for acquiring motion information

In this case the two images are acquired by the same optical system at different time instants. If we track on the image the time history of projections of the same token, we obtain information useful for describing the relative motion between the optical center and the environment.

We may have two different cases, depending on the a-priori knowledge we have: if we know that the observed scene is fixed, and we know the scene, the tracking process executed on images allow to describe the observer ego-motion. On the other hand, if we know that the observer is fixed, we can get information about the movement of observed objects.

In addition, if we observe a fixed environment and we know the observer trajectory, we can obtain from motion analysis the three-dimensional structure of the observed objects (shape from motion). This case represents only a different way for implementing a stereo reconstruction process.

3.3. DMA Definitions

Given the project objectives (section 3.1) and the basic principles (section 3.2) the P940 research was referred to, the major effort was spent in looking for a computation chain capable of realizing stereo and tracking processes with advanced functional performances, and, at the same time, suitable to be implemented in a hardware machine for real time purpouses.

As a matter of fact, the two opposite aims, like 3D reconstruction accuracy and real time implementability in a reasonable hardware architecture, was even present during the initial project phases.

This work led to the following main definitions:

a) Trinocular stereovision cameras configuration (8).

Trinocular arrangement for cameras was selected to overcome the drawbacks characteristic of two-cameras stereo processes. In particular the two cameras approach does not allow to match points which belong to a segment laying on the same epipolar plane. Trinocular arrangement, with optical centers not aligned, allows to have in any case a couple of cameras whose epipolar plane does not contain a given segment.

Furthermore, the matching verification process can be made more robust: indeed the search for corresponding points between two images can now be reduced to a single verification at a precise location on the third image (see fig. 3).

In spite of the complexity added to the computing chain, that in this case has to process data coming out from one more camera, the trinocular arrangement results in much more robust and accurate matching output and in much less sparse range estimations.

b) Tokens to be extracted (9).

Tokens are the basic pictorial features that can be extracted from images, which the system makes reference to as primitive image descriptors. Selecting the kind of feature the system will be based on is always a critical design step.

In P940 we had a long debate on this topic.

In particular this discussion was aimed not only at finding the best solution from a merely computational point of view, but also to investigate if some kind of tokens can have influence on the set of possible applications, particularly in limiting them.

The basic idea of the selected approach is to avoid to work directly at the pixel level, but to build a simbolic token structure from each image, in such a way that tokens can be matched much more efficiently and safely.

We chose as tokens the linear segments coming from poligonal approximation of brightness edges.

They have the following important properties (9):

16

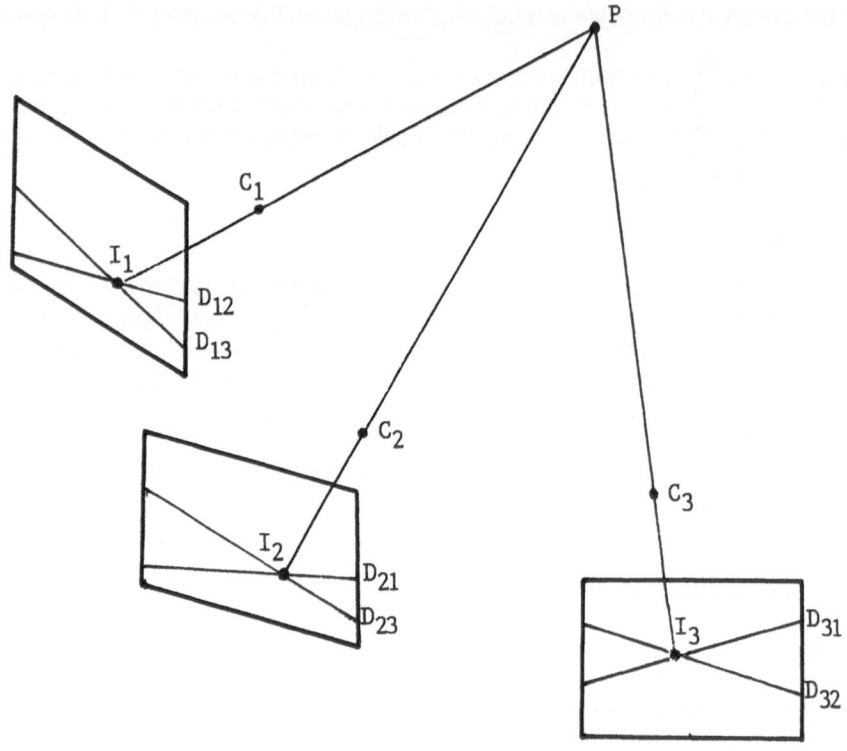

fig. 3 : Geometry of trinocular stereovision

1) They are compact, to allow for a concise description of images, and to reduce the complexity of the stereo-matching process.
2) They are intrinsic, in that they correspond to the projection of some physical object part.
3) They are robust to the acquisition noise.
4) They can be associated to a feature vector that can be used for the hypothesis verification in the matching process.
5) Their location in images can be precisely estimated, to allow accurate 3D reprojections.
6) Linear segments are built from edges, whose detection can be performed by very efficient and accurate algorithms.
7) Linear segments can be efficiently used for extracting motion information, avoiding more complex approaches based on the optical flow estimation.

As far as the application domains are concerned, an experimental work was done to verify that linear segments allow to describe structured environments for robot navigation aims (10), as well as small shape characteristics, always present in mechanical parts for recognition, location and handling purpouses (11).

c) Motion analysis.

Motion estimation, shape from motion methods and stereo-motion cooperation represent research areas much less mature than the stereo vision one.

For this reason, any definite decision was impossible at the project beginning, and two approaches to motion feature extraction have been maintained contemporaneously alive in the project.

One of them was based on tracking the same tokens used in the stereo process (line segments) (12), the second one was based on the extraction of the optical flow by differential methods (extended to the second derivative of image brightness to avoid the well known aperture problem) (13).

Though the second approach gave many interesting and original results, concerning the possibility of estimating the ego-motion from vanishing points movements and the object displacement and rotation by focus of expansion analysis, the line segment approach was selected, mainly for its lower computation complexity and for its large computing overlap with the system stereo chain.

The basic idea for motion extraction is to track line segments in an image sequence.

Each token is described by a parameter vector (center coordinates, length, inclination angle, etc.); a Kalman filter is used to aid tracking by providing reasonable estimates of the region where the matching process has to seek for a possible match in next image (9), (13), (14).

The output of this process is the trajectory each token had in the image plane, and this description can be used to describe the motion of observed objects or to describe the vehicle motion, depending on which a-priori knowledge we intend to use (see section 3.2).

An important feature of our approach has not to be neglected. Due to the fact that stereo and motion are based on the same image primitive description (line segments), very efficient methods for using stereo and motion cooperation can be identified.

Just as an example, in tracking corresponding tokens in different images, coming out from the three cameras, we can keep the correspondence information along the time evolution, avoiding in such a way a lot of work for searching and verifying matches in the stereo process.

This is an extremely interesting case where, in fusing data and information obtained from different and independent channels, we can achieve a synergetic overall advantage.

d) High level representation and process.

Project P940 was mainly conceived to deal with early vision problems, both on the methodological and implementation sides.

But, as far as we have to demonstrate meaningful applications, and we want to make evident possible exploitations of DMA machine, some extent of high level processes have to be defined, maintaining them as general as possible.

Even to this regard, line segments seem to be a quite flexible and efficient image primitive, suitable for representing shapes in the various aspects a three-dimensional problem can present.

We selected as basic 3D representation a modified version of Delaunay tetrahedization (15). This method allows to build the convex hull of a 3D environment starting from 3D located line-segments.

Furthermore this convex hull description gives rise to a description of visible surfaces in terms of triangular patches, whose space extention depends on the local density of detected tokens.

Finally, this kind of description allow to describe the visible object surface in terms of normal versors, with a space density "proportional" to the density of irregularities locally present on the object surfaces (16).

In the P940 project we use these properties both in the vehicle navigation and robot arm handling domains.

In the first case the 3D Delaunay reconstruction is used in computing the free space map, where the vehicle can move without crashing; of course, along its navigation, the vehicle has to update this map with parts that, in previous positions, were hidden (17).

In the second case, normals to object surfaces are used in a Hough transform derivate method, to recognize and locate quadric surfaces, which are then used for recognizing and locating complex mechanical objects (18).

The DMA machine does not implement any of the above mentioned high level processes. They are loaded in a host multiprocessor (like EMMA 2[1] developed by ELSAG or CAPITAN developed by MATRA) to which the early processing stages will be interfaced (19).

In general the high level vision is still an open problem. We think that results coming out from P940 can help Consortium's partners, or hopefully anybody else, in spending much more effort in investigating basic issues concerning high level vision.

3.4. DMA processing chain

The previous section was dedicated to basic considerations and statements taken to define the DMA elements, belonging either to algorithm or hardware implementation fields.

The results of definition phases, where a considerable experimental work and a remarkable level of cooperation and discussion took place inside the Consortium, gave rise to the following process chain.

step n. 1:   Cameras Calibration (20), (21).

This is an off-line process required for measuring the external coordinates of the three optical systems and the intrinsic camera parameters (focus length, photodetector array coordinate and scaling factors, etc.)

Several methods for calibrating optical systems have been developed in the project. The most remarkable ones are based on the use of known regular patterns which led to the solution of an algebric system of equation, whose unknowns are just the searched parameters values.

step n. 2:   Image Acquisition (9).

Image Acquisition and digitizer sub-systems were directly acquired by the available market.

The only feature to which we spent attention was the capability of using either RS170/CCIR standard or common non-standard cameras with progressive scanning (non-interlaced frames).

1   Registered Trade Mark

step n. 3:   Edge Detection (22), (23).

As edge detection algorithm, the image directional gradient was selected, based on a Canny operator followed by a non maxima suppressor to get edges with single pixel thickness.

step n. 4:   Edge linker.

Edge detected in step n. 3 are processed according to a contiguity predicate, for finding and making explicit the chain of adjacent edge pixels.
This algorithm is implemented in two subsequent sub-steps: creation of raw edge chain, and chain fusion.

step n. 5:   Polygonal Approximation.

Detected edge chains are approximated by polygonal approximations, where the approximation accuracy is continuously controlled. This algorithm was originally developed by Berthod (24).
The output is, for each image, a set of polygonals approximating the transformed edge image.

step n. 6:   Stereo Matching.

This process is dedicated to the token (line segments part of polygonals) matching process, where triplet of corresponding line segments are looked for in the three images acquired.
The process is mainly based on a Hypothesis Prediction and Verification approach, where an adjacency graph for segments is used to limit the search space in the system memory (this is a consequence of regularity hypothesis assumed for the external world, by which token adjacency is conserved in images deriving from slightly different view points) (25).
As far as the computing time is concerned, the matching algorithm represents one of the most critical points in the stereo chain. Three methods are used at the same time to avoid the combinatorial explosion that always takes place:

1) use of epipolar constraints to reduce to one dimension the searching space in the stereo images;
2) use of the third camera for speeding up the verification time;
3) use of a segment adjacency graph for reducing the searching time in the system memory.

The matching algorithm output is the set of segment triplets that correspond, in the three images, to the same physical feature of the observed scene.
Each segment triplet is described by the segment location in the respective image frame, in such a way that, given the calibration parameters, the 3D location can be easily computed.

step n. 7:   Token Tracking (12).

Image flow is described in terms of token trajectory on the image. Each token is described by means of a parameter vector, where parameters are represented by a value estimate, its variance, its temporal derivative, the variance of the temporal derivative and a covariance between the estimate and its derivative.

As line segments for each new image are made available the tokens already tracked find their match with the new ones. A Kalman filter approach is used to predict, for each tracked token, the image area where its new observation must be searched by token matching (9), (29).

3.5. Hardware Architecture

The above research clearly requires an application environment with real time capabilities, and within the project particular attention was devoted to the goal of achieving performances suitable for quite demanding applications, like a mobile vehicle moving in an indoor environment or a manipulating arm dealing with industrial scenes.

Although the project has covered both low and high level topics, the hardware implementation has been focused particularly on lower and intermediate levels of the process, so DMA may be defined as a front-end processor with number crunching capabilities in the range of the billion of operations per second. To achieve a full process implementation including also the higher levels, DMA is interfaced to a host multiprocessor, like EMMA2 (19), developed by ELSAG, or CAPITAN developed by MATRA.

For DMA development the basic starting hypothesis has been to define an "open" architecture, with the possibility of integrating also modules available from the market.

Other major hypothesis have been modularity and flexibility of configuration, to achieve efficiency also with different real-time requirements and to permit the implementation of additional improvements to the algorithmic chain.

In terms of performance, the reference target has been the TV video rate, at least as a medium term goal, that means an architecture able to support such throughputs, while for the processing module presently the range 5-25 frames/second has been considered as a good reference given the class of algorithms that have been investigated.

Table 1 summarizes the DMA main characteristics.

	INPUT	OUTPUT Worst case image	Task 3 Intermediate SW demonstration	Task 5 Hardware implement.
FIR filtering	Image 512x512x8	Gradient component Gx, Gy 2x(512x512x8)	4 68020 15 sec.	Dedicated HW 40 m sec.
Non Maxima suppression	Gradient components 2x(512x512x8)	Gradient module + binary edge image 512x512x8 Gx,Gy 2x(512x512x8)	4 68020 6 sec.	Dedicated HW 40 m sec.
Edge Linking	Gradient module + binary edge image + Gx+Gy 10% edge points	Linked chains 250 chains 10 K pixels	4 68020	4 ADSP 2100+ coprocessor (*) 120 m sec.
Polygonal approx.	Linked chains	Segments less then 1000	15 sec.	4 DSP 56000 200 m sec.(*)
Stereo matching	Segments	Matching validated hypothesis worst case 1000 normal case 200	5 68020 10 sec.	4 DSP 56000 200 m sec.(*)
Token Tracking	Segments	Tracked tokens max 250	SUN 3 8 sec.	ADSP2100 + coproc. 100 m sec.

(*) Expected computation time with typical configuration.
Larger and faster configurations are allowed.

Table 1 : DMA characteristics

3.5.1. General architecture

The hardware architecture is shown in fig. 4 for a configuration able to support 3 TV cameras; it is obvious that great attention has been devoted to the communication bandwidth, with three different link types:

– A general purpose bus: we chose VME for its high bandwidth associated to a worldwide diffusion, which implies a wide availability of commercial modules that may be included on the DMA; on the other hand this choice allows an easy integration of the DMA into several host environments.
Depending on the data rate requested by a specific application, the process can be spread on different VME buses.
– A video bus for high speed transfers of the images in raster scan format; here the choice has been for the compatibility with the Max-bus (developed by Datacube), that supports communications according to the RS170/CCIR standards as well as the transfers of portions of images (defined as Regions Of Interest).
– Some private inter-board links used for efficient local communications between identical modules that work in parallel.

With the above communication system an overall throughput in the range of hundreds of Mbytes/sec may be achieved.

Going now to a more detailed analysis of the DMA processing elements, a first consideration is that the dedicated hardware has been limited to edge detection and token tracking, while the intermediate processing sections make wide use of programmable Digital Signal Processors to permit a good flexibility in the algorithmic implementation and maintenance.

We now discuss the different modules.

3.5.2. Video Digitizer.

DMA uses video digitizers compatible with the above mentioned cameras and compliant to the Max-bus communication standard for image transfers.

The resolution is 8 bits/pixel (256 gray levels), and the acquisition chain includes features like input multiplexing, signal compensation, and direct implementation of point operations by means of look-up tables.

3.5.3. Multifunction Sequence Store.

This unit is not involved in the real time operation, but it is devoted to the diagnostics and used for development purposes.

This module is able to download synthetic or known images through the VME bus, transferring them to the video bus and acquiring, if requested, results in raster scan format; another fundamental feature is the capability of acquiring a sequence of images up to 1 sec. of duration, to debug motion algorithms.

3.5.4. Edge Detection.

This function is being implemented on a couple of boards, achieving a full TV rate operation dealing with a Canny operator, including a non maxima suppression to get edges with single pixel thickness.

For the Canny operator a FIR implementation has been chosen, due to the possibility of higher levels of integration with respect to HR alternatives.

The first board is a general purpose FIR architecture including 4 1D filters with length up to 64 taps, dealing with 8 bit data and 8 bit coefficients; data coming from the digitizer through the video bus are filtered row by row with two masks (derivative and smoothing components, operating in parallel) and the results stored into an intermediate memory buffer, addressable by rows and columns; then the second phase of the filtering is performed, applying the last two masks to the intermediate results read column by column.

The results are available both on the video bus, for real time operation, as well as on the VME bus, mainly for debug purposes and/or operation directly on standard work-stations. The output of the filter board is sent to the non-maxima suppression board, implemented at TV rate by means of a 3x3 neighbourhood analyzer devoted to the gradient maximum search.

3.5.5. Edge Linker.

The algorithm to be implemented is characterized by two major computational steps: edge chain creation and chain fusion. While the first phase is easily implementable in a parallel processing structure, the second one is essentially a global procedure, requiring a high performance of the processing elements (PE's).

The architectural choice has been oriented to a multi-DSP structure, to achieve at the same time high processing power, modularity of configuration and programming flexibility.

To speed-up the execution of specific primitives, each DSP takes advantage of a coprocessor, implemented as a piggy-back board using SMD technology.

The DSP selected for this architecture is the Analog Devices ADSP 2100, a 16 bit processor with both program and data buses extended off-chip able to carry out multi-function instructions with a cycle time down to 80 nsec.

For edge linking implementation the coprocessor analyzes in a single DSP cycle the 3x3 neighbourhood of a pixel, triggering if requested a specific routine on the DSP for the subsequent chain functions (open chain, close chain, etc...).

Each board includes 2 DSP units, with the possibility to build-up clusters of processors where a single PE is the cluster master and all other PE's are slaves.

Each DSP is equipped with 384 Kbytes of high speed data memory and 48 Kbytes of program memory.

The board has a flexible and high bandwith communication section, supporting 3 buses:

– VME bus
– Video bus
– Cluster bus (inter-PE link)

3.5.6. Polygonal Approximation and Stereo Matcher.

Both these functions have requirements for high dynamic range and accuracy that, together with good programming facilities have been the key targets in terms of implementation hypotheses.

These functions are quite demanding, but they may be decomposed into parallel processes with good efficiency, so the choice has been for a second multi-DSP architecture, based on the 24 bit processor Motorola DSP 56001.

This architecture is characterized by complementary features to the previous ones used for linking: in this case the possibility of coprocessors has been dropped, and we have preferred to increase the number of DSP modules on each board, 4 instead of 2.

In addition, the design has been carried out considering a future upgrade using the Motorola DSP 96001, which will add to the DMA machine the capability of floating point computations: this upgrade will take advantage of a software compatibility.

The module is designed for communication through the VME bus only, since it is devoted mainly to the processing of images in symbolic representation, and broadcast transfers are supported.

Each DSP has dual port memories (access from both DSP and VME), with up to 576 Kbytes of data memory and 192 Kbytes of program memory.

3.5.7. Token Tracker.

The token tracker works on symbolic data (tokens are line segments) and so it uses the VME bus as communication channel.

The processing structure has been defined according to the requested throughput of 10 frames/sec; this has forced us to choose a more dedicated approach, but even in this case the possibility of algorithmic improvement has been maintained.

The module has been implemented by a microcoded design, using a microcode memory addressed by a sequencer (Analog Devices AD1401) and providing as output the control of the arithmetic sections and of four address generators (Analog Devices AD1410) with related memory buffers.

The user(s) of the token tracker outputs depend on the global task that is implemented on the DMA machine; typical users are the stereo matcher (to simplify matching procedures) or an additional module computing the 3D structure directly from motion.

3.5.8. Supervision CPU.

This module consists of a general purpose CPU with the function of taking care of system start-up at power-on, while, during the normal operation, performing functions of data routing and process synchronization.

4. Demonstrators and Exploitation

All the Computer Vision Community has to be aware that, in spite of the remarkable advances achieved in understanding and implementing visual processes, Vision Systems are not so diffusely used in practice, as it was foreseen few years ago.

This is mainly due to the computing complexity implied by vision processes and to the unavailability of hardware machines, capable of implementing them in real time and reliably, as it is required by usual industrial standards.

The ESPRIT Project P940 tries find answers to these problems or, at least, to the more computation intensive phases of a three-dimensional vision process, namely the early vision stages.

To these ideas the demonstration tasks of our project were inspired. Three different kinds of demonstrations will be given in the project last phases:

25

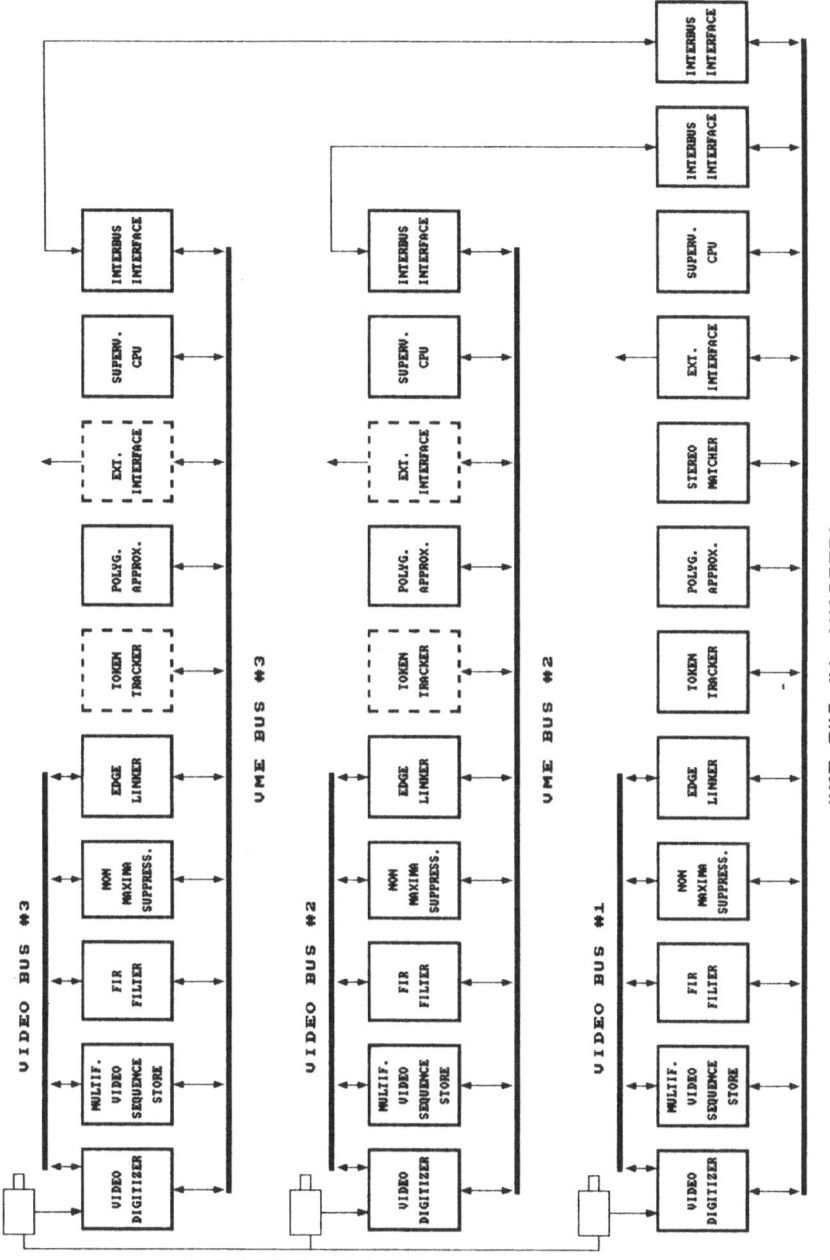

Fig. 4 : ESPRIT P940 - DMA Layout

a) Mobile Vehicle (3).

A mobile vehicle, equipped with the trinocular camera sub-system and radio-linked with the DMA machine hosted by a general purpose computing facility, will demonstrate a navigation application in a structured environment.

In this demonstration will be experimented the following vision functions:
- 3D reconstruction of the environment map and vehicle self-location by matching of this map with the known map (26);
- ego-motion estimation by time-tracking recognized environment features;
- free-space estimation by stereo processing and trajectory planning for reaching a predefined space location.

b) Robot arm manipulation (3), (27).

DMA machine will be integrated with an existing multiprocessor (EMMA2 developed by ELSAG) and with a robot arm controller.

The DMA stereo chain, together with high level running on EMMA2, will be used to demonstrate capabilities of recognizing and locating, in real time, mechanical parts observed from different viewpoints.

Besides the experimentation of the early processing stages implemented in the DMA architecture, this demonstrator will stress also the logical interface of data extracted from images with the high level representation of the application knowledge. Indeed for recognition and location purposes the object 3D reconstruction coming out from the stereo process has to be matched with the internal object representation existing in the system long term memory.

Two strategies for acquiring stereo data will be demonstrated: the use of the already described trinocular camera arrangement fixed just over the robot arm working space and the use of a single camera installed on the wrist and moved around by the robot arm.

c) Automatic 3D model reconstruction (3).

The third demonstrator is in some way connected with the robot arm manipulation task. Particularly in this case the capability of reconstructing and growing up a model of a 3D object only by its visual observation will be demonstrated.

A camera, moved around by a robot arm, will use the motion chain (based on token tracking algorithms) to create in the system memory a model of the visible surfaces of the observed object.

This demonstration could be considered in two ways:
the system evaluation in cases where some unknown object has to be manipulated (in this case we have no a-priori knowledge, about its shape) or the demonstration of a learning phase for the task described in the previous item b).

As far as the project exploitation is concerned, we have to remark that all companies involved in the project are well present on markets where intelligent sensoriality, and artificial vision in particular, are considered strategic achievements (all these companies are deeply involved in vision research since many years).

As an example we can mention ELSAG's applications on robotics and on measuring machines, space application already faced by MATRA ESPACE and robot arm manipulation for ITMI.

However, besides these remarkable application domains, we think that the DMA architecture, as a whole or only in part, can be used in much more extended fields where data fusion from multiple images or from time-sequences of images has to be applied.

5. Project Management and Consortium Cooperation (3), (28).

The project has been organized in eight tasks as depicted in the following Table II:
  T1 Computation of Passive Stero Vision
  T2 Motion Analysis
  T3 Integration of Stereo and Motion
  T4 Computation and Representation of 3D Shapes
  T5 Hardware Implementation
  T6 Mobile Vehicle Demonstrations
  T7 Robot Arm Demonstrations
  T8 Project Management

Table 2 : Project structure

In Table 3 the Participant's effort per Task, expressed in man year, is reported.

The project has been managed according to a matrix organization, where technical responsibilities were assigned to "key persons" (one per each task); they were charged of coordinating activities among all participants; decisional responsibilities were assigned to "project leaders" (one per each partner) as representatives of Consortium components.

Furthermore the project has been managed by a project manager, a deputy project manager, a technical coordinator and an administrative coordinator, all belonging to Prime Contractor Organization (ELSAG).

Besides the organizational and formal aspects, always very important for a large project as P940 certainly is, in our project a remarkable level of cooperation has been established by all people inside the Consortium.

A considerable exchange of experiences and know-how took place in all the project areas and very often people worked together exactly as they belonged to the same Company team.

Many researchers of one Organization spent some period working in some other Laboratory, and this reinforced the common knowledge of one another.

Standards for passing data, and software and hardware modules have been defined quite early in the project. This represented a remarkable benefit for the work progress and for the Consortium life.

6. Project Status and Conclusions

At present the first three tasks of the project are concluded, and the methodological aspects they were dealing with are completely defined.

The hardware design and realization of all the electronic modules is completed in its first release, and the Consortium is spending the major part of its present effort in integrating the DMA machine.

The next eighteen months before the project end will be dedicated to the second release of hardware modules, which will fulfill completely industrial quality standards, and to the

| | | | | T A S K | | | | | | |
	1	2	3	4	5	6	7	8	T O T A L	INITIAL BUDGET
Cambridge Univ.	1.1	=	=	2.8	=	=	=	=	3.9	4.0
Elsag	1.8	=	=	2.1	11.0	=	4.2	7.0	26.1	24.0
G.E.C.	=	2.0	1.0	4.0	=	=	=	=	7.0	6.0
Genoa Univ.	3.3	4.5	1.7	2.7	10.0	=	=	=	22.2	21.0
INRIA	1.3	1.6	2.9	4.0	6.1	8.0	=	=	23.9	23.0
ITMI	=	2.3	=	4.0	3.3	=	3.6	=	13.2	11.5
MATRA	1.6	1.3	3.0	=	10.7	=	=	=	16.6	15.0
NOESIS	=	=	=	=	=	11.3	=	=	11.3	8.0
T O T A L	9.1	11.7	8.6	19.6	41.1	19.3	7.8	7.0	124.2	112.5

Table 3 : Participants' Effort per Task, in man-years

YEARS 1, : Actual
YEARS 3, 4, 5 : Budget

realization of the demonstration facilities.

The hardware availability, the successful implementation of the algorithms developed and the good quality of work done by each participant are remarkable achievements obtained in this project.

Acknowledgements

The author wish to thank here all participants of ESPRIT Project P940 which have made writing this paper possible and which originate the great part of the work I described.

References

(1) MUSSO G., Applicazioni della visione artificiale nell'automazione industriale e dei servizi. La Visione delle macchine. Tecniche Nuove. Autunno 1989.

(2) ROSENFELD A., Computer Vision: basic principles. IEEE Proceedings. August 1988.

(3) ESPRIT P940 Technical Annex.

(4) AYACHE N. and LUSTMAN F. Fast and reliable passive trinocular stereovision. Proc. First International Conference on Computer Vision, pages 422-427, IEEE June 1987. London, U.K.

(5) AYACHE N. and LUSTMAN F. Trinocular stereovision, recent results. Proceeding of International Joint Conference on Artificial Intelligence. August 1987. Milano, Italy.

(6) FAUGERAS O.D., LUSTMAN F. and TOSCANI G., Motion and structure from point and lines matches. Proc. of First International Conference on Computer Vision, pages 25-34, IEEE, June 1987. London, U.K.

(7) Structure from Motion ESPRIT Project P940, Report R.2.5.1, June 1988

(8) Trinocular Stereovision ESPRIT Project P940, Report R.1.3.4, May 1988

(9) FAUGERAS O. et al. Depth and Motion Analysis: the machine being developed within ESPRIT Project P940. IAPR Workshop on Computer Vision. October 1988. Tokio

(10) AYACHE N., FAUGERAS O.D., LUSTMAN F. and ZHANG Z. Visual navigation of a mobile robot: recent steps. Proc. of Int. Symp. and Exposition on Robots. November 1988

(11) GARIBOTTO G. A real-time 3-D vision system for a variety of robotic applications. Proceedings 5th ESPRIT Conference 1988.

(12) Token Tracking ESPRIT Project P940, Report R.2.4.1., June 1988

(13) Differential Techniques for the computation of the Optical Flow ESPRIT Project P940, Report R.4.2.1, June 1988.

(14) JAZWINSKY A.M. Stochastic Processes and Filtering Theory. Academic Press, 1970.

(15) Delaunay Tetrahedrization ESPRIT Project P940, Report R.4.1.4, May 1988.

(16) Planes and Quadrics from 3D segments ESPRIT Project P940, Report R.4.1.6, May 1988.

(17) AYACHE N. and FAUGERAS O.D. Maintaining representations of the environment of a mobile robot. Robert Bolles and Bernard Roth, editors. Robotics Research: the Fourth International Symposium, pages 337-350, MIT Press, 1988.

(18) CASSOLINO C., MANGILI F. and VIANO G. Identification and Recognitionof 3D Surface Primitives-Internal Report, ELSAG, 1988

(19) APPIANI E., BARBAGELATA G., CONTERNO B. AND MANARA R., EMMA2: An Industry-Developed Hierarchical Multiprocessor for Very High Performance Signal Processing Applications. Published by the Proceedings of First International Conference on Supercomputing Systems, 1985. St.Petersburg, FLORIDA.

(20) FAUGERAS O.D. and TOSCANI G. The calibration problem for stereo, CVPR '86, pages 15-20, June 1986, Miami.

(21) Techniques for Camera Calibration in Stereovision ESPRIT Project P940, Report R.1.1.2, Nov. 1986.

(22) CANNY J. A Computational approach to edge detection. Published by IEEE Transactions on Pattern Analysis and Machine Intelligence, 8 No6 pages 679-698, 1986.

(23) DERICHE R. Using canny's criteria to derive an optimal edge detector recursively implemented. Published by The International Journal of Computer Vision, 2 April 1987.

(24) BERTHOD M. Approximation polygonale de chaines de contours. Programmes C, 1986. INRIA.

(25) MARR D., Vision. S.Francisco, CA: Freeman, 1982.

(26) AYACHE N. and FAUGERAS O.D. Building, registrating and fusing noisy visual maps. Published by Proc. First International Conference on Computer Vision, pages 73-82, IEEE, June 1987. London,U.K., also an INRIA Internal Report 596, 1986.

(27) Deliverables on Demonstrators Specifications.

(28) P940 Technical Annex, Last Revision, March 1989.

(29) CROWLEY J.L. Measuring Image Flow by Tracking Edge-Lines. Second International Conference on Computer Vision, September 1988.

LIST OF MEMBERS OF THE CONSORTIUM.

ELETTRONICA SAN GIORGIO-ELSAG S.p.A. - GENOVA (I)
UNIVERSITA' DEGLI STUDI DI GENOVA - GENOVA (I)
UNIVERSITY OF CAMBRIDGE - CAMBRIDGE (U.K.)
GEC-Research Laboratories - LONDON (U.K.)
ITMI - GRENOBLE (F)
INRIA - SOPHIA-ANTIPOLIS (F)
NOESIS - VERSAILLES (F)
MS2i (MATRA S.A. till Dec. '88) - PARIS (F)

CAD INTERFACES - A KEY TO COMPUTER INTEGRATED MANUFACTURING

INGWARD BEY
Kernforschungszentrum Karlsruhe GmbH
Dep. PFT
P.O. Box 36 40
D-7500 Karlsruhe 1

Abstract

CIM is a challenge for European industry for the next decade. The ESPRIT Programme with its part dedicated to CIM aims to create an environment in which multi-vendor systems can be implemented in a progressive manner and in which Community IT suppliers can compete effectively. To achieve this standard interfaces for open architectures and easy information exchange are necessary. The CAD*I project started in 1984 and terminated end of October 1989 contributed a lot to the goals of ESPRIT. A summary of results is given.

1. Introduction

The importance of integration is steadily growing: in electronics highly integrated circuits improve the performance of computers and controls, integrated machines perform many mechanical or technical operations formerly run on separate devices, companies (i.e. the automotive or aircraft industry and their suppliers) have closer market interrelations then ever before and - last but not least - our European nations are coming to an economical and political integration.

In this context efficient communications and information exchange play a very significant role for the success of European industry in world-wide competition. With the ESPRIT Programme, and especially the part of it dedicated to computer integrated manufacturing, Europe makes big efforts to have vendors able to supply modern CIM products which cover the needs of the users.

Although the realization of a CIM concept in a company requires not only to deal with technical but also strategic, economic, organizational, and human aspects (see i.e. ESPRIT Project 1199 "Human Centred CIM System" [25] from the users technical point of view the most important requirement refers to the availability of open, multi-vendor CIM systems and CIM components. Within ESPRIT-CIM two projects among others have to be mentioned here which made important contributions in this direction: ESPRIT 955 "Communication Networks for Manufacturing Applications" [26] and ESPRIT 322 "CAD Interfaces" (CAD*I). This paper concentrates on the goals and results of the CAD*I project.

2. Goals of the CAD*I Project

In mechanical industries, product design gives the starting point of product information. CAD is to be seen as the information generation activity, and interfaces in the CAD environment should allow for an efficient and secure data transfer to all other activities of the production process.

The impact of computer aided design (CAD) in industry has intensified the need for standard methods of data communication between CAD systems and from CAD to other areas of production, providing efficient storage, retrieval and exchange of information. Thus, the CAD*I project (ESPRIT Project No 322) aimed from the beginning to specify missing concepts, to develop the corresponding processors and to influence the international standardisation process in the field of CAD interfaces.

The CAD*I project was defined during 1984 with the primary objective of developing powerful, vendor independent interfaces for mechanical engineering applications as follows:
- CAD data exchange interface
- CAD data base interface
- interface to finite element analysis.

Another goal of the project was to develop a new level of communication in CAD/CAE systems which allows to replace formal geometric descriptions by nonformal descriptions like handwritten input or design by technical terms.

The project started officially in November 1984 and run for a period of 5 years. The research and development work was done by more than 50 people involved at 12 partner institutions participating in the project. They spent about 150 person-years of research and development effort.

Kernforschungszentrum Karlsruhe was responsible for the overall project management. The Commission of the European Communities provided over 5 million ECU funding for the project. The work was organized in working groups covering the different areas of research and development as follows:
- CAD data exchange interfaces for
  - wireframes
  - solids
  - surfaces
  - database systems
- networking
- advanced modelling
- FEM analysis interfaces to
  - FEM models
  - model optimisation
  - experimental dynamic model description.

3. Geometry Data Exchange

Numerous attempts to provide flexible, yet stable methods of CAD data transfer have resulted in several viable specifications [1]. Some of these became national standards, namely IGES [2] in the US, VDAFS [3] in Germany and SET [4] in France. However, all of them have had limitations in their scope and efficiency.

The CAD*I project started almost simultaneously with the US project to develop the Product Data Exchange Standard (PDES) [5]. CAD*I and PDES have many similarities: the approaches converge more than they compete. They are the basis for the future international "Standard for the Exchange of Product Model Data (STEP)", which has become a Draft Standard Proposal at ISO TC184 SC4 (see figure 1).

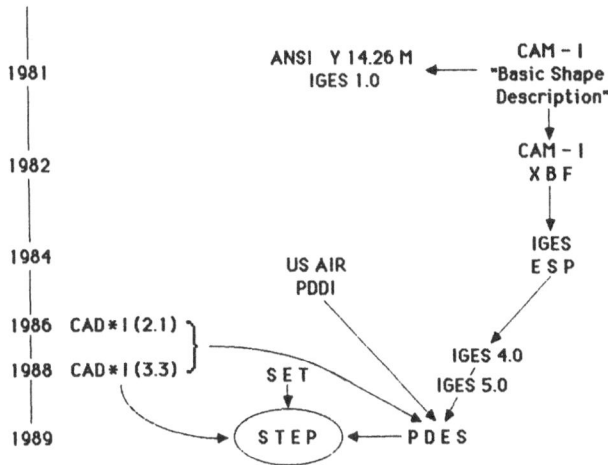

Figure 1: The development of solid model interfaces

3.1 Solid Model Transfer

CAD*I work covered geometry data exchange including wireframes, surfaces and solid models. Main emphasis was given to the problems of transferring solid models, both constructive solid geometry (CSG) and boundary representations (b-rep).

Sending system	Receiving system		
	CSG	B-rep	polyhedron

CSG
  Bravo3
  Euclid
  ICEM
	Bravo3
	Euclid
	ICEM
		Catia
		Proren
		Romulus
			Euclid

B-rep
  Catia
  Technovision
  Proren
  Geomod
  Romulus
	not possible	Technovision
		Proren
		Geomod
		Romulus
			-

polyhedron Euclid	not possible	Proren Euclid	Catia

Figure 2: Geometry data transfer between existing CAD systems using the CAD*I interface

A CSG model represents the object by a tree whose nodes are Boolean set operators and whose leaves are either bounded primitives or halfspaces. The Boolean operators are union, difference, and intersection. The set of primitives consists i.e. of block, cylinder, and cone. Halfspaces are defined by an infinite open surface that divides space into two distinct regions, solid and void.

CAD systems like i.e. Applicon BRAVO3, Control Data ICEM, or Mc Donnell Douglas UNISOLIDS are based on a CSG concept.

A boundary representation describes a solid object in terms of its enclosing surfaces. The B-rep structure is composed of a topological structure and associated geometry. The topological information consists of a graph of vertices and edges embedded in a surface. The sequence of edges that bounds a region of the surface, termed face, is called loop.

The CAD systems, i.e. Computervision/Prime MEDUSA, Isykon PROREN 2, Norsk Data TECHNOVISION, Hewlett Packard ME 30, or Siemens CADIS-3D are based on the boundary representation.

A special case of B-rep is the POLYHEDRON structure. It represents objects bounded by planar surfaces. Systems like Matra Datavision EUCLID and IBM CATIA make use of this kind of representation.

Some systems allow for certain combinations of the above mentioned concepts. It is evident that this variety of representation defines the need for a CAD system reference schema and the requirements for the specification of a neutral file for CAD geometry exchange.

Figure 2 shows the matrix of possible transformations covered by pre- and post-processors for the CAD*I neutral file format, which were developed in a prototype version during the project.

General solutions for data transfer must cover two main problems. They were taken into account within CAD*I:

a) specification of the neutral interface

b) development of pre- and post-processors for the specified interface.

3.2 Specification Work

The first step in developing a neutral file format consists in defining more precisely the information to be transferred and in analyzing in detail the different possible internal CAD system data structures characterized by their entities and properties. This last aspect led to a "reference schema". Only CAD systems whose data structures can be mapped onto this schema can be expected to send and receive solid model information properly via the neutral file. The principal entities of the CAD*I reference schema are described in detail in [14].

To define data to be transferred in an unambiguous way, a formal data specification language is needed. Within CAD*I the language named HDSL (High Level Data Specification Language) was developed to specify CAD data structures. HDSL is based on concepts of Pascal data structures. With the scope concept a block structure facility well-known in programming languages is introduced into data structure specifications.

All data structures of the CAD*I schema are described formally by HDSL. Semantics are defined by the sequence and the naming of the attributes in HDSL and informally by verbal descriptions (see figure 3).

3.3 Processor Development

The best interface specification has no practical importance if there are no commercial systems available for the customer which are based on it. Therefore, a main goal of the CAD*I project is to test the CAD*I specification for a neutral file for CAD geometry by implementing pre- and post-processors for a number of commercial CAD systems as shown in figure 4.

A series of test parts, both for CSG and B-rep models was developed to test the CAD*I specification with its special features. First data transfer (cycle and intersystem tests) were

geometric shape of a " BOX "	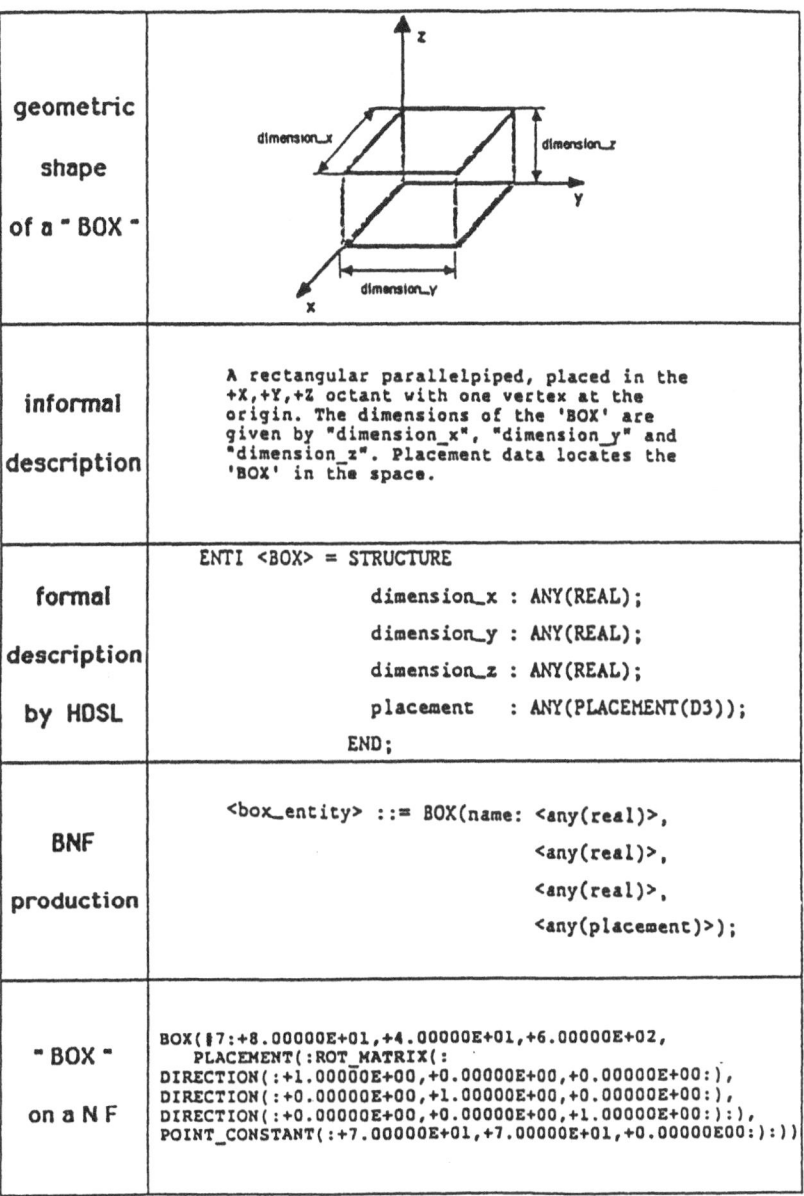
informal description	A rectangular parallelpiped, placed in the +X,+Y,+Z octant with one vertex at the origin. The dimensions of the 'BOX' are given by "dimension_x", "dimension_y" and "dimension_z". Placement data locates the 'BOX' in the space.

formal
 	
description
 	
by HDSL
	ENTI <BOX> = STRUCTURE
	
	        dimension_x : ANY(REAL);
	
	        dimension_y : ANY(REAL);
	
	        dimension_z : ANY(REAL);
	
	        placement : ANY(PLACEMENT(D3));
	
	    END;

BNF production	<box_entity> ::= BOX(name: <any(real)>, <any(real)>, <any(real)>, <any(placement)>);
| " BOX "<br><br>on a N F | ```
BOX(#7:+8.00000E+01,+4.00000E+01,+6.00000E+02,
    PLACEMENT(:ROT_MATRIX(:
DIRECTION(:+1.00000E+00,+0.00000E+00,+0.00000E+00:),
DIRECTION(:+0.00000E+00,+1.00000E+00,+0.00000E+00:),
DIRECTION(:+0.00000E+00,+0.00000E+00,+1.00000E+00:):),
POINT_CONSTANT(:+7.00000E+01,+7.00000E+01,+0.00000E00:):))
``` |

Figure 3: Example of the CSG primitive BOX in different description formats: geometry, informal description, HDSL, BNF, and CAD*I neutral file
```

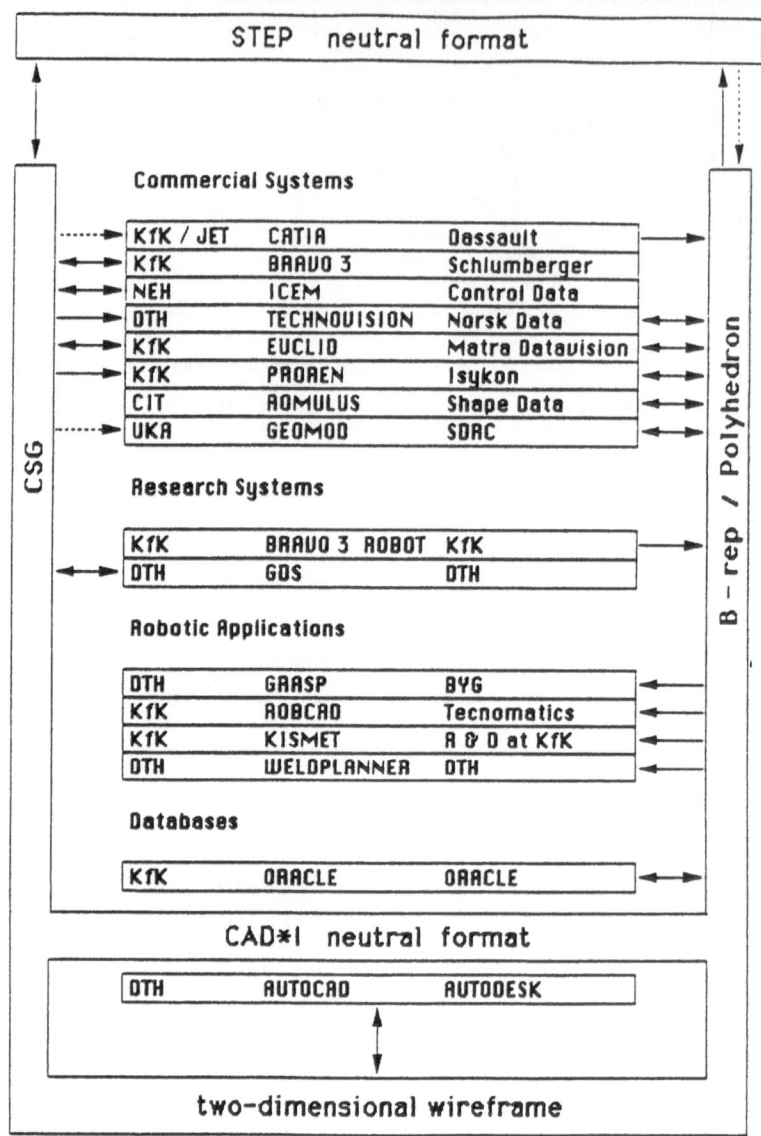

Figure 4: Available prototype CAD*I processors for commercial CAD systems

made at the end of 1986 and had been continued since then with very satisfactory results. A collection of test parts all of which have undergone successful transfer from one CAD system to at least one other is shown in figure 5.

It is worth mentioning that the very first transfer of a B-rep model between unlike CAD systems was performed early 1987 by transferring a test part via a CAD*I file generated on PROREN 2 implemented on a DEC Microvax II at Kernforschungszentrum Karlsruhe in

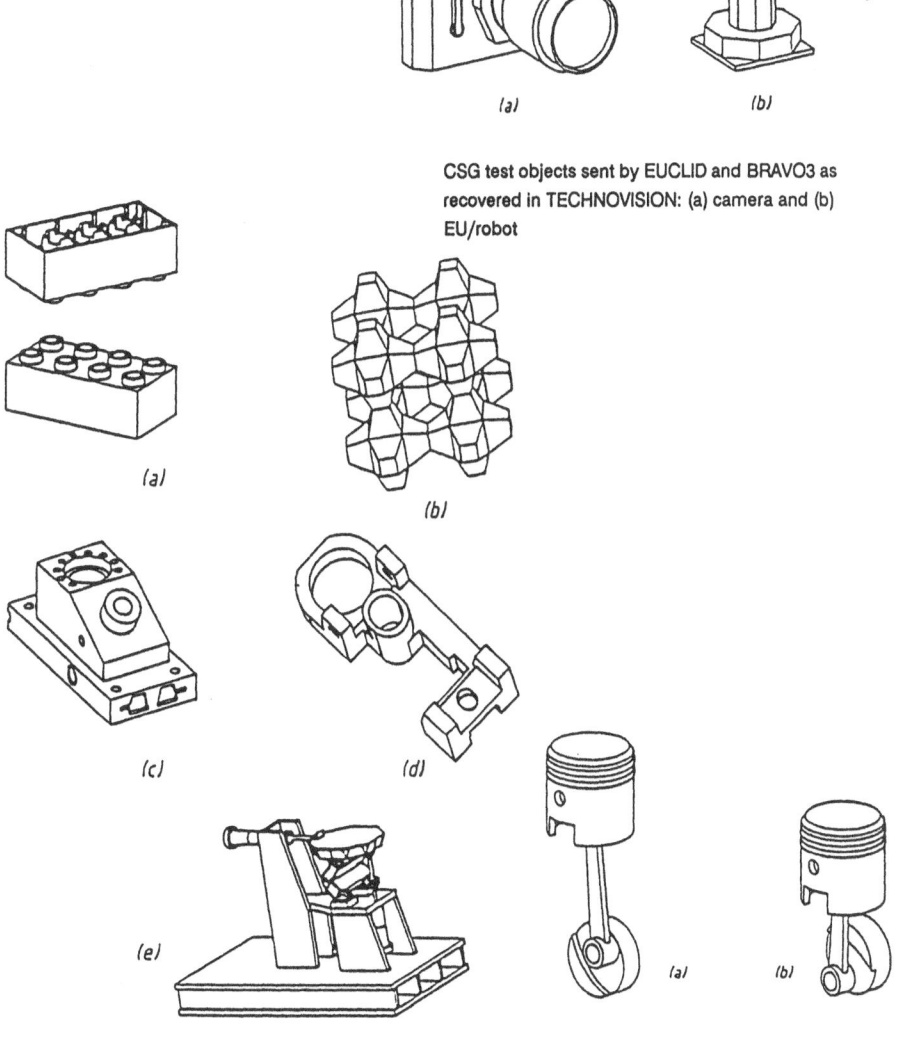

CSG test objects sent by EUCLID and BRAVO3 as recovered in TECHNOVISION: (a) camera and (b) EU/robot

(a)

(b)

(c)

(d)

(e)

(a)

(b)

B-rep test objects for cycle and intersystem transfer, as recovered in TECHNOVISION: (a) LEGO brick, (b) eggbox, (c) ANC 101, (d) MBB Gehaeuse, (e) vibrating platform

Kinematic model of a piston in GDS: (a) before and after a cycle test (b) parameter changed in the recovered model

Figure 5: Some of the test parts used to test the capabilities of the CAD*I neutral file and the prototype processors

38

Germany to TECHNOVISION on a Norsk Data computer installed at the Technical University of Denmark at Lyngby near Copenhagen.

During technical conferences and exhibitions, the CAD*I team has shown the capabilities of the specification and processors developed for available commercial systems on several opportunities: i.e. during the ESPRIT Technical Conferences of the last years, at the 3rd CAD*I International Workshop in Copenhagen 1988, at the EMO fair in Hannover 1989, and at the CIM-Europe Workshop in Karlsruhe in October 1989.

3.4 The Cad*i Drafting Model

An information model for the description of technical drawings has been developed. This CAD*I Drafting Model is a specification providing techniques to describe representations of product information (concerning i.e. product shape, physics, production method, assembly, quality) on a drawing or graphic form within a CAD*I neutral file. This information model represents actually the highest level of sophistication within the level concept of the Drafting Model of the STEP specification [23] (see figure 6).

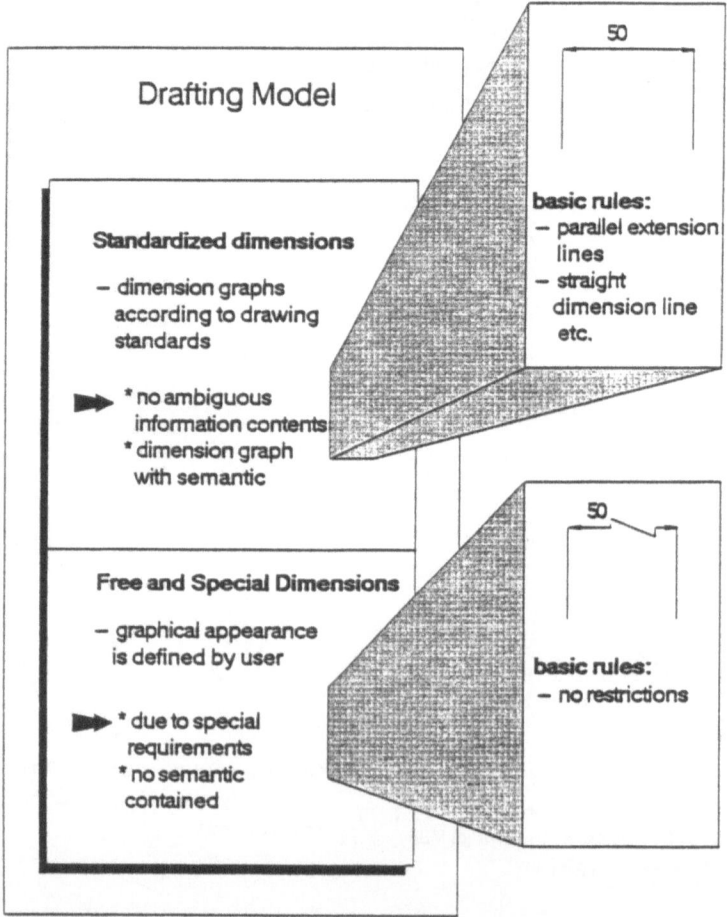

Figure 6: Basic principles of the CAD*I drafting model (source: BMW)

3.5 Surface Data Transfer

CAD systems use different description methods for defining curves and surfaces. To enhance transfer of curve or surface data from one CAD system to another it is not only necessary to have a powerful interface specification, but also approximation methods to adapt the different mathematical representations to the capabilities of the systems and the user required tolerance.

To be able to perform these transformations, a lot of work was spent within the CAD*I project to develop procedures for exact and approximate surface data transfer between compatible and incompatible formats [24]. An overview is given in figure 7. Tests were made to show their capabilities.

Source			Destination			
			low degree	high degree	rational	piecewise low degree
Form of math. description	neutral file format using it		CATIA	STRIM 100	INTERGRAPH	MEDUSA
Bezler Polynom. Coeff. B-Spline	CAD •I / STEP VDA IGES	low degree	EXACT	EXACT	EXACT	
Bezler Polynom. Coeff. B-Spline	CAD •I / STEP VDA IGES	high degree	CHEBYSHEV OR ORTHOGONAL	EXACT		CHEBYSHEV OR ORTHOGONAL
Polynomial Coeff.	SET	rational	ORTHOGONAL	ORTHOGONAL	EXACT	ORTHOGONAL
Bezler Polynom. Coeff. B-Spline	CAD •I / STEP VDA IGES	piecewise low degree		ORTHOGONAL		

Figure 7: Some solutions for curve and surface data conversion between incompatible CAD systems (source: C.I.T.)

4. Interface Between CAD and Data Base Management Systems

Companies produce various types of data (administrative, technological...) which are more and more often stored in a database. This enables any authorized user, i.e.:
- To know what data produced in the company is available to him
- to benefit of the results of the other users
- to be informed immediately of any modification done by any user
- to communicate better with users from other departments

However, these functionalities cannot be developed in the CAD area because of the difficulty to manage easily the isolated CAD data stored in independent and insufficiently flexible "files".

It is a nearly impossible job to create and maintain relations between pieces of information (entities) of one model stored in one "file" and pieces of information (entities) of an other model stored in an other "file".

The aim of CAD*I was to demonstrate that a data base system is the appropriate tool

40

to create, maintain and modify these complex relations. In that way, within CAD*I a standardised communication interface with a database system was specified and implemented. This communication interface consists of a set of standard subroutines for Fortran applications to write, to read, to modify, to delete, etc. CAD data in a CAD*I database.

It was decided to implement a subset of this package on the relational database system ORACLE supporting the standard SQL language.

To make a demonstration programme of the results obtained the interfaces shown in figure 8 were developed, including the corresponding test environment. More details are described in [17].

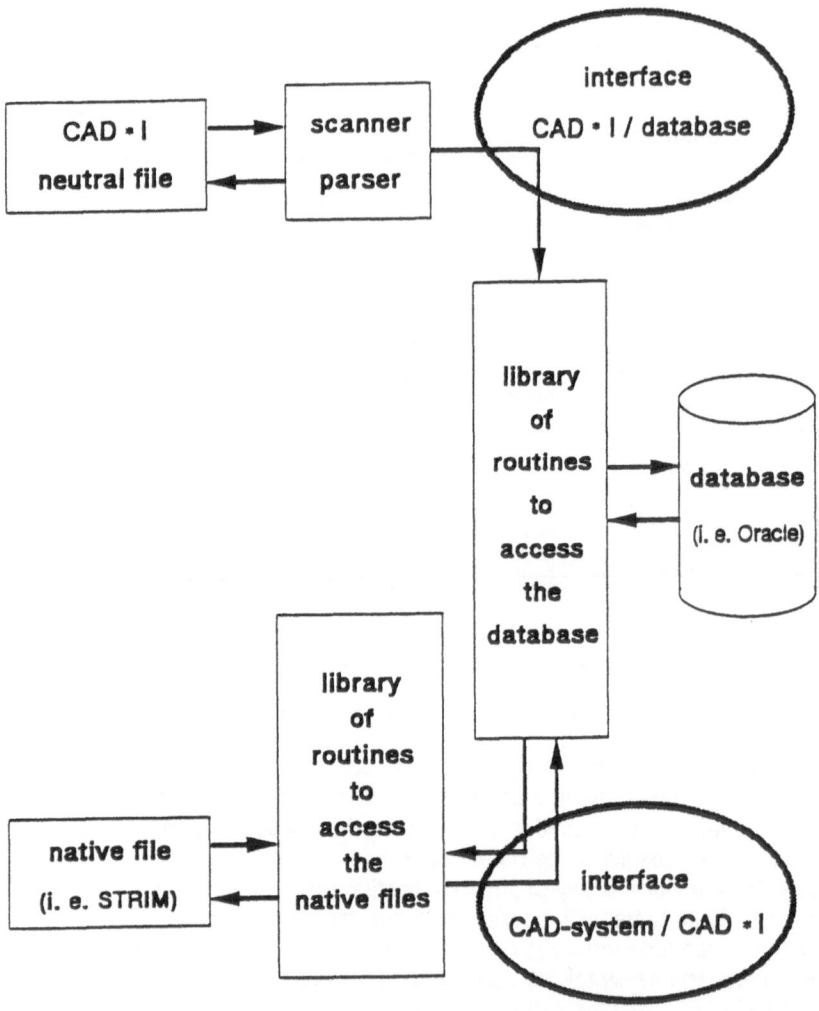

Figure 8: Access to CAD data via standard routines using a relational data base system and the CAD*I neutral format

5. Data Exchange Via Networks

One of the working groups set up within the CAD*I project investigated in the area of intertask communication (message exchange) via computer networks and made a feasibility study on exchange techniques for the exchange of CAD data in CAD systems implemented in a distributed manner, based on available computer network technology. Basic principles and upcoming standards for intertask communication were examined as well as the performance using existing network facilities to transfer CAD*I neutral files.

The following tasks were accomplished:
- Installation of network connections and communication software
- test of data transfer via Public Data Networks
- test of data transfer via available Local Area Networks
- test of data transfer via available Wide Area Networks
- report on file transfer experiments
- requirement definition for intertask communication
- study of intertask communication facilities
- feasibility study for implementing database interface routines on top of network drivers.
  A special report on this work is in preparation, others had been published [18].

6. Advanced Modelling

While present CAD systems are characterized by a rather primitive operational modelling interface (set operators, elementary geometric operators), future CAD systems are envisaged to require a high-level communication interface with pattern recognition and semantical scene analysis capabilities. Such interface techniques were developed in the project.

Following the global objectives of increasing the effectiveness and user orientation of communication interfaces of CAD/CAE systems (between system and system user) by improving the communication techniques actual CAD/CAE command languages are very formal and not really interactive and should be replaced by graphical/interactive and nonformal communication techniques based on AI. functions like pattern recognition and semantical scene analysis.

The working group was led by the University of Karlsruhe. The most important results concern:
- Analysis of needs and requirements of product and process design concerning the user-system communication,
- specification of a technical term dictionary and related technical semantics [12],
- analysing recent AI methods to be applied in CAD/CAE communication techniques (i.e. pattern recognition, semantical scene analysis),
- specification of new communication techniques using above results to gain interfaces which are able to serve also not well trained users and such who are not familiar with IT but carry high-level technical know-how,
- development of these (or a subset) techniques,
- implementation and testing of a prototype interface,
- verification of the prototype in a pilot installation.

A detailed description of the concepts of the developed Handsketching Input System was given in [8]. Further results including the DTT modeller (Design by Technical Terms) are reported in [17]. Prototype installations were demonstrated at the CIM-Europe Workshop in Karlsruhe, October 1989 and at the present ESPRIT Conference Week.

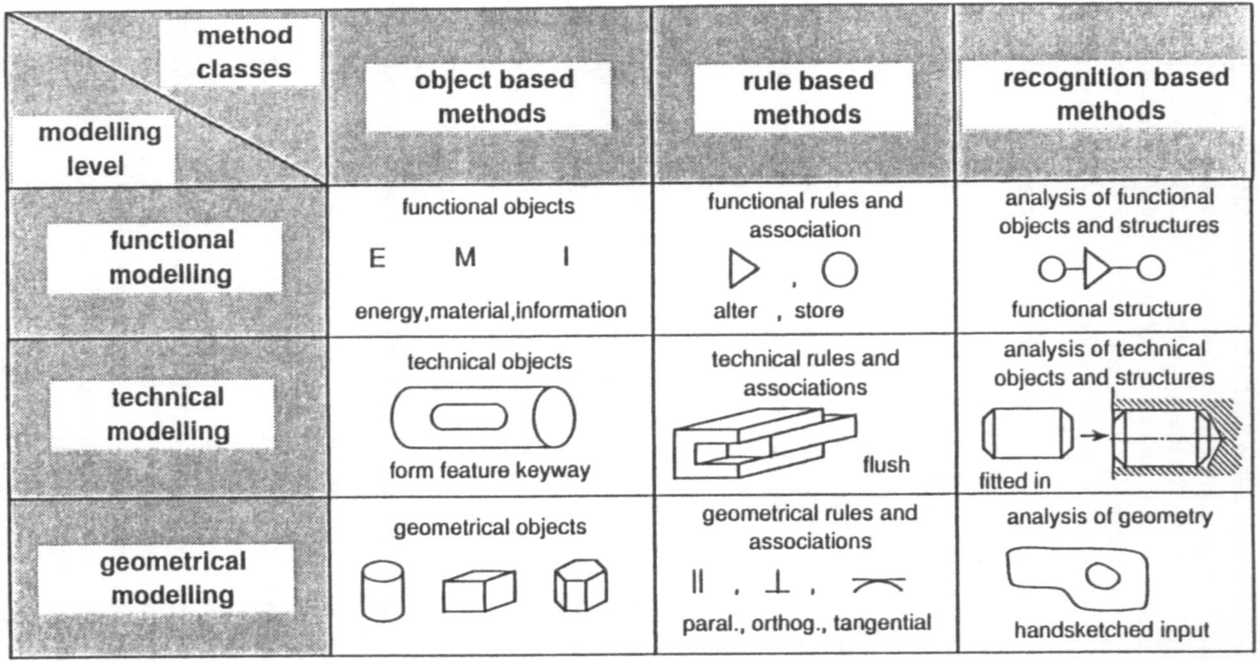

modelling level \ method classes	object based methods	rule based methods	recognition based methods
functional modelling	functional objects E M I energy,material,information	functional rules and association △ , ○ alter , store	analysis of functional objects and structures functional structure
technical modelling	technical objects form feature keyway	technical rules and associations flush	analysis of technical objects and structures fitted in
geometrical modelling	geometrical objects	geometrical rules and associations ‖ , ⊥ , ⟨ paral., orthog., tangential	analysis of geometry handsketched input

Figure 9: The concept for Advanced Modelling (source: University of Karlsruhe)

7. Interface to Finite Element Analysis

When the CAD*I project started, currently there did not exist standard integrations of FEA programmes in CAD systems. FEA models had to be derived from geometrical CAD models by manual operations. A generally usable standard interface between CAD and FEA programmes did not exist. Therefore, CAD*I aimed to develop these kind of interfaces, to provide processors for them and to introduce the developed concept into the international standardisation process.

The work in this direction was finished at the end of 1988. The specifications for the exchange of product analysis data are described in [19]. A syntax for representing the analysis data in a neutral file is presented; this syntax is formally described in Backus-Naur form. A reference model for the finite element (FE) analysis data is given which uses a formal data modelling language to define the fundamental data entities, their attributes and interrelations. Syntactical expressions (statements) for entities in these reference models are worked out. The overall internal structure of the neutral file, and the interaction between the neutral file and the computing environment via pre- and post-processors are presented. Figure 10 gives a look on the FEA systems on the market for which processors were implemented and tested.

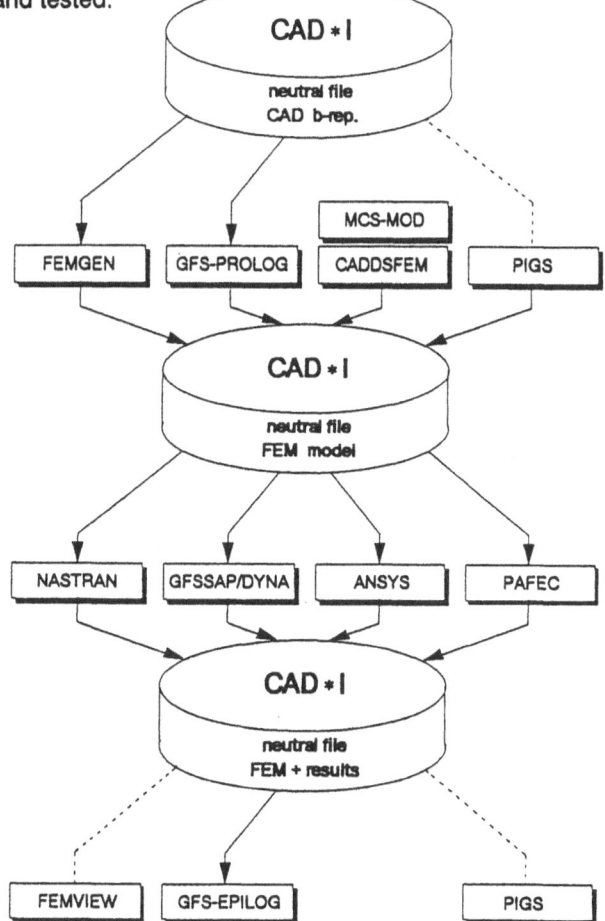

Figure 10: Use of CAD*I processors in the CAD-FEM domain (source: GfS)

8. Experimental Dynamic Structural Analysis and Model Optimization

In the same way as the work mentioned in the above chapter, activities for structural analysis and model optimization were finalized at the end of 1988.

The corresponding task groups of ESPRIT Project 322 CAD*I have studied methods for integrating FEM techniques and experimental testing methods for structural dynamic analysis, establishing a "link" between them. Results had been widely reported [20, 21, 22, 27].

One of the partners, LMS International - Belgium, the working group leader, has implemented several of the new developments in its new Computer Aided Dynamic Analysis system for UNIX workstations, the CADA-X system. The system is configured around a state of the art engineering computer workstation, and a modular front-end for data sampling and generation of excitation signals. The CADA-X system is actively marketed on a world wide basis.

9. Standardisation Activities

ESPRIT Project 322: CAD Interfaces (CAD*I) has been contributing to international standardisation efforts since the begin of project work to strengthen European influence in this area of vital importance for European industry.

A major goal is now reached with the elaboration of a draft proposal for the exchange of product model data, commonly known as "STEP", at ISO, the International Standards Organisation.

Important European contribution to the international standardisation efforts has been made by particular CAD*I contribution to the geometric specifications, the physical file format and the formal specification language used for STEP. The experimental geometric model exchanges between members of the CAD*I project using the closely related CAD*I neutral file specification have helped to validate some of the concepts before they were incorporated in STEP.

ISO has released to national standards bodies for review the draft proposal for a standard for the exchange of product model data. This proposal, commonly known as STEP, is the result of more than 4 years work by an ISO working group among them a group of CAD*I members who were specifically engaged with the solution of standardisation problems.

CAD*I work is going on to influence further international standardisation activities.

10. Outlook

With the end of the running time in October 1989 the CAD*I project has reached most of the originally planned goals. A lot of software has been written for tools, test programmes and processors. Prototype processors are available for several commercial CAD and FEM systems. They had been and can be purchased from the owning partners. CAD*I processors will be used as a basis for the development of STEP processors.

The interest and commitment of CAD and FEM vendors has steadily increased recently and there is a very good chance for vendors applying CAD*I results to be the first ones to have STEP processors available on the market.

One of the most important activities to follow the CAD*I/STEP line in CAD interfaces is the new ESPRIT Project 2195 (CADEX) which started in spring 1989. It has gathered more than 10 partners, mainly vendors of CAD and FEM systems and is led by GfS (Gesellschaft fuer Strukturanalyse GmbH) Aachen/FRG. This project will build upon the results of CAD*I to develop STEP processors and enable European companies to benefit from the developments in CAD*I.

The work and the results of CAD*I helped to be prepared in Europe for the CAD environment of tomorrow.

The "Conformance Testing Services" initiative from the C.E.C. is now supporting test centres in Europe to prepare themselves to certify STEP processors in the future. They will benefit also from the experiences gained inside the CAD*I project.

11. Acknowledgements

Now, after 5 years of intensive work, it is a pleasure to present our final results to the audience of the ESPRIT Conference 1989. As project manager for most of the running time of the CAD*I project I want to thank all participants in the project and the responsible persons at the C.E.C. for their cooperation and great personal efforts to reach the ambitious goals stated in 1984. It is very encouraging to see how ESPRIT has helped us to contribute in the direction of setting up cooperative research between industries and institutes in Europe and to come to results and products which support the perspective of the Common European Market of 1993.

References

[1] WILSON, P.R.: A Short History of CAD data Transfer Standards. IEEE, CG&A, Vol. 7, No. 6, pp.64-67, June 1987

[2] Digital Representation for Communication of Product Definition Data. ANSI standard Y 14.26 M, September 1981.

[3] DIN: VDA-Flaechenschnittstelle (VDAFS) Version 1.0. DIN 66301.Beuth Verlag, 1986.

[4] AFNOR: Automatisation Industrielle. Représentation externe des données de définition de produits. Spécification du standard d'échange et de transferts (SET). Version 85.08.Z68.300. AFNOR 85181, 1985.

[5] PDES File Structure Working Draft Version 3. PDES Chairman W.B. Gruttke. Physical File Structure/Formal Language Committee, Internal Notes, September 1985.

[6] BEY, I., LEURIDAN, J.: ESPRIT Project 322, CAD Interfaces: Interfaces in CAD Systems. In: ESPRIT '85. Status Report. pp 1273-1283. North-Holland. 1986.

[7] BEY, I., LEURIDAN, J.: ESPRIT Project 322: CAD*I. CAD Interfaces. Status Report 2. Kernforschungszentrum Karlsruhe. 1986.

[8] BEY, I., LEURIDAN, J.: ESPRIT Project 322: CAD*I. CAD Interfaces. Status Report 3. Kernforschungszentrum Karlsruhe. 1987.

[9] BEY, I., LEURIDAN, J.: ESPRIT Project 322: CAD*I. CAD Interfaces. Status Report 4. Kernforschungszentrum Karlsruhe. 1988.

[10] VAN MAANEN, J., THOMAS, D.: ESPRIT Project 322: CAD*I. Specification for Exchange of Product Analysis Data. Kernforschungszentrum Karlsruhe. 1987.

[11] HELPENSTEIN, H., LADEFOGED, T.: ESPRIT Project 322: CAD*I. Specification for Exchange of Finite Element Result Data. Kernforschungszentrum Karlsruhe. 1987.

[12] GRABOWSKI, H., ANDERL, R., RUDE, S.: ESPRIT Project 322: CAD*I. Technical Term Dictionary with Related Geometric and Technical Semantics. Kernforschungszentrum Karlsruhe. 1987.

[13] SCHLECHTENDAHL, I.E..: Specification of a CAD*I Neutral File for Solids. Version 2.1. Springer

46

Verlag Heidelberg. 1986

[14] SCHLECHTENDAHL, I.E..: Specification of a CAD*I Neutral File for CAD Geometry. Version 3.3. Springer Verlag Heidelberg. 1988.

[15] CAM-I: Geometric Modeling Project. Boundary File Design (XBF-2). Report No. R-81-GM-02.1. CAM-I Inc., Arlington/Texas. 1981.

[16] IGES, E.S.P.: Experimental Solids Proposal. 1984.

[17] BEY, I., LEURIDAN, J.: ESPRIT Project 322: CAD*I. CAD Interfaces. Status Report 5. Kernforschungszentrum Karlsruhe. 1989.

[18] TROSTMANN, E. et al: CAD data exchange via neutral interfaces. Enterprise Network Event (ENE '88). Baltimore, USA. 1988.

[19] D. THOMAS, J. VAN MAANEN, M. MEAD (Eds.): Specification for Exchange of Product Analysis Data. Version 3. Springer Verlag Heidelberg. 1989

[20] J. LEURIDAN
"The Use of Principal Inputs in Multiple Output Data Analysis", The International Journal of Analytical and Experimental Modal Analysis, Vol 1/3, pp. 1-8, Society of Experimental Mechanics Inc. 1986.

[21] D. OTTE, K. FYFE, P. SAS and J. LEURIDAN "Use of Principal Component Analysis for Dominant Noise Source Identification" Int. Conf. on Advances in the Control and Refinement of Vehicle Noise, 22-24 March 1988, Birmingham, UK, 8 pp. 1988.

[22] F.LEMBREGTS, J. LEURIDAN, J. LIPKENS and H. VAN DER AUWERAER "Comparison of Stepped-Sine and Broad Band Excitation to an Aircraft Frame", Proc. of the 6th Int. Modal Analysis Conf., 1-4 February 1988, Orlando, Florida, pp. 1706 - 1713. 1988.

[23] ISO: German Contribution to Drafting Model Standardization. Document No. N252, 1988.

[24] R.J. GOULT: Parametric Curve and Surface Approximation. IMA Conference on Mathematics of Surfaces, Oxford, September 1988.

[25] S. MURPHY: Human Centred CIM System. Proceedings of the 5th Annual ESPRIT Conference, Brussels, Nov 14-17, 1988. North-Holland, 1988, pp. 1615-1629.

[26] H. KREPPEL: CNMA - Progress in industrial communication. Proceedings of the 5th Annual ESPRIT Conference, Brussels, Nov 14-17, 1988. North-Holland, 1988, pp. 1571-1588.

[27] H.F. HELPENSTEIN: Interface to FE Analysis Data. Proceedings of the 5th Annual ESPRIT Conference, Brussels, Nov 14-17, 1988. North-Holland, 1988, pp. 1417-1429.

List of Partners in the CAD*I Project

Bayerische Motorenwerke AG/FRG
CISIGRAPH/France
Cranfield Institute of Technology /UK
Danmarks Tekniske Hoejskole/DK
ERDISA/Spain
Gesellschaft fuer Strukturanalyse mbH/FRG
Katholike Universiteit Leuven/Belgium
Kernforschungszentrum Karlsruhe GmbH/FRG
Leuven Measurement and Systems/Belgium
NEH Consulting Engineers ApS/DK
Rutherford Appleton Laboratory/UK
Universitaet Karlsruhe/FRG.

CONSTRUCTION AND MANAGEMENT
OF DISTRIBUTED OFFICE SYSTEMS
ACHIEVEMENTS and FUTURE TRENDS

R. BALTER
Bull Research Centre
2, rue Vignate
ZI de Mayencin
F-38610 GIERES

Abstract

The overall objective of the COMANDOS project (Construction and Management of Distributed Office Systems) is to identify and construct an integrated platform for programming and operating large-scale multi-vendor distributed systems. The project is planned over two main phases carried out both under ESPRIT-1 and ESPRIT-2 programmes. The ESPRIT-1 COMANDOS project was a three-year project (started in March 1986) which was mainly dedicated to the exploratory prototyping of this platform. Based on the results drawn from this project, a further consolidation and development phase, the ESPRIT-2 COMANDOS project running over a four-year period, should result in pre-industrial products for a number of system environments. This paper outlines the main achievements of the ESPRIT-1 COMANDOS project, and describes the intended consolidation and development work to be carried out in the framework of ESPRIT-2.

1. Project Objective, Significance and Timescale

The overall objective of the COMANDOS project is to identify and construct an integrated application support environment for programming and operating distributed applications which can manipulate persistent - i.e. long-lived - data. It is intended to provide such an environment in the framework of large-scale multi-vendor distributed systems.

Although the main intended application domain is "office systems", we expect that the COMANDOS platform may be valuable as a basis for integrated information systems in such application domains as CAD, software engineering and manufacturing administration.

COMANDOS itself does not provide end-user applications, but rather a basis for the development of these. The support environment includes infrastructure for:

- distributed concurrent computations in a loosely coupled LAN and WAN environment;
- storage and efficient retrieval of long lived persistent data;
- arbitrary programming languages;
- re-useable and extensible software modules;
- open systems, so that application builders can take components from independent sources and tailor them;
- secure and protected data, and access control;
- on-line management, monitoring and control of the distributed environment.

Further aspects of the project include tools to aid in office systems design and maintenance, and interworking with existing information systems.

This objective is extremely significant from an industrial viewpoint because the development of application software is currently a labour and cost-intensive proposition, particularly for distributed applications. Exploitation of data management systems is often tedious, particularly for applications which must handle large volumes of structured data (such as the multi-media document). The absence of a uniform mechanism for coupling independently coded software modules, possibly written in different programming languages, provides a major hinderance to software re-use and the development of an open market place for software. Finally the operational control of the distributed system is a further major consideration, for which current tools are limited, often restricted to system or hardware subsystem level, and mostly divorced from the executing applications.

Thus the COMANDOS project aims to provide an integrated platform for programming and operating distributed environments, particularly those requiring data management. This can be considered as a "long-term" objective, based on the development of innovative solutions, which require the integration of existing and emerging technologies (object-orientation, persistent storage, etc..).

The COMANDOS project is targeted for loosely coupled distributed systems of workstations, servers and processor pools, using internets based primarily on LANs. Connection to slower speed WANs will also be supported. There are many research and development teams worldwide building software support for such distributed environments. Basic kernels such as Mach at CMU [Jones 86], Amoeba at Amsterdam [Mullender 85], or Chorus [Zimmerman 84] offer relatively low-level and efficient distributed message passing support. The Apollo Domain system [Leach 83] provides a distributed store, and storage objects which can be accessed independently of their location. Distributed operating systems such as Locus [Walker 83] have included support for data consistency, and Argus has extended this by linguistic support for applications operating on long-lived data [Liskov 83]. Extensible DBMSs such as Exodus [Carey 86] and Starburst [Schwartz 86] allow application developers to tailor the data management support. The PS-Algol project at the University of Glasgow [Glasgow 86] provides a persistent storage tightly coupled with a linguistic support.

COMANDOS has borrowed many ideas from these systems and is attempting to integrate them into a unified platform for programming distributed applications which can handle persistent data. An object-oriented appproach appears to be beneficial for the reasons staded later below. Other projects are also taking such an approach, but few appear to have the embracing goal of COMANDOS. Emerald [Black 87] and Distributed Smalltalk [Decouchant 86] provide support for distributed programming in the context of a single language. Gemstone [Maier 86] and Trellis/Owl [OBrien 86] are integrating data management support into object oriented languages, but again are mono-lingual and currently have limited support for distribution. Galileo [Albano 85] provides a powerful data modelling system for persistent data but is ot yet distributed.

The goals of COMANDOS are thus extremely ambitious, but equally would be extremely beneficial and significant if they are achieved. There is currently no standard platform for programming distributed applications which can handle persistent data: therefore its availability would be highly advantageous.

Object orientation was adopted in COMANDOS for a number of reasons. Abstraction is not only an important programming concept, but it also aids distributed systems by allowing the implementation and configuration of common services to be hidden from clients (object orientation fits the classical "Client-Server" model). Bundling executable code and data together into a single entity also aids the construction of such systems via the modularity which results. Representing the same entity via different interfaces aids protection mechanisms. Code re-usage and incremental development aid programmer productivity but also allows implementors and administrators to tailor and extend a common system to their own requirements. Object oriented data models provide a rich framework in which to capture complex structures such as the multi- media documents found in office environments. Finally, the approach is appearing in international stand-ardisation activities such as ISO ODA for multi-media documents, the ECMA DASE proposal[Dase 86] and, to an extent, the ISO ODP work for distributed environments.

Further important considerations for the COMANDOS execution environment are integration both with existing environments and with management tools. In particular, co-existence with UNIX is crucial and it is essential that UNIX applications can be run in COMANDOS, and that distributed applications can be build by assembling of existing components already developed on top of UNIX. Management of a large scale distributed environment can be very complex, and is exarcerbated by distributed data management. Our view is that generally it may be beyond the ability of human administrators to control such complex system accurately, and machine assistance will be required. This problem, and that of providing feedback to office system designers, is also being studied by the COMANDOS consortium.

The ESPRIT-1 COMANDOS project was a three-year project which was mainly dedicated to the exploratory prototyping of this platform. However a number of significant achieve-ments heve been obtained in this first phase, which are described in section 2. Section 3 presents the intended exploitation of these results and the intended consolidation work planned in the framework of the ESPRIT-2 COMANDOS project. Finally the conclusion summarizes the overall ineterest of this project for the European IT industry.

2. Achievements to Date

Significant achievements of the project so far are :

- Definition of an application support interface which provides a persistent object-oriented view of the distributed environment. This application interface defines the COMANDOS virtual machine.
- Definition of an implementation architecture which supports the virtual machine, and which can be mapped onto various underlying systems (UNIX, native, low-level kernel, vendor-specific).
- Prototyping of this architecture, thus proving its viability on a number of underlying multi-vendor environments.
- Prototyping of the application support interface. This has been achieved in two ways: one by providing program libraries for existing languages (C and Modula-2), and the other in the form of a new programming language.
- Prototyping of on-line management facilities, including a distributed system obeserva-tion and control facility, security tools for authorization and risk management, and a distributed office system design tool for the design and integration of a distributed office system.
- Protyping of services for the co-existence with existing heterogeneous information systems.

2.1 Definition of an application support interface

The application support interface defined by Comandos is termed the **Comandos Virtual Machine**. This interface presents the functionality of a Comandos system to application programmers and system administrators. Technically it chiefly consists of two major components:

– A common and extensible **type model**, which allows data representation and implementation to be changed as necessary, while maintaining abstract structural and behavioural properties. Moreover this model intends to accomodate both the database and general programming language environments.

In COMANDOS, unlike some object-oriented systems, objects are typed. Further, **static type checking** is exploited as far as is possible, both to assist code generation for conventional hardware architectures and to detect programming ming errors as early as possible.

Each object has an associated type, which represents the visible properties and behaviour of the object. Each object also has an implementation which describes how the abstract interface represented by its type is implemented. A type may have several different implementations, perhaps reflecting different algorithms used to represent the type. There are various primitive types provided directly in the model, together with a number of conventional type constructors for building new types. COMANDOS also supports **inheritance, subtyping and genericity**, as these features all encourage code re-use.

– A **Computational model**, which allows distributed programs to be defined. This model provides the application designer with a multi-processor multi-node virtual machine, in which the parallelism is apparent and the distribution is hidden.

The computation unit (called activity) can be viewed as a sequence of synchronous operation invocations on objects. Objects invoked by an activity are dynamically bound in its virtual address space. A Job is a set of activities with share the same virtual address space. Communication and synchronisation between activities (within the same job or in different jobs) are carried out through the use of shared objects. The model provides facilities for the control of concurrent accesses.

Finally, **atomic transactions** and **exception handling** are combined to ensure distributed concurrency control and consistency management as well as aiding fault tolerance.

The Model is the basis for describing the interfaces between software modules in a programming language independent fashion, and is thus a mechanism to achieve software re-use. It is also the basis for describing the structure of data and information maintained on secondary storage, and is used during retrievals and query processing. It is abstract in the sense that it can be used to describe the functionality of software written in various current programming languages and is not overly dependent on any one of these.

Linguistic supports for both of these models have been prototyped (see 2.4). Further support provided by the COMANDOS virtual machine includes a security model, for secure and protected access, and a management model.

2.2. Definition of an implementation architecture for the Virtual Machine

The second major achievement of the project has been the design of an abstract architecture which can be the basis for implementing the Comandos Virtual Machine on a range of underlying operating environments. This architecture is abstract in that it is designed to be independent on any particular underlying host operating system or hardware. The architecture identifies and precisely defines a number of functional components, which together co-operate to provide the Virtual Machine, and which should be present in a Comandos implementation. In practice, attention has so far been focused on hosting the platform in a UNIX environment and directly on native hardware.

An important goal of the project is to provide a unified framework for the management of objects, which supports the viewpoints of both databases and general purpose programming languages. Usually, programming languages deal with volative objects, i.e. objects whose lifetime is limited to the execution of a program, while database languages deal whith persistent ones, i.e. objects whose lifetime is independent from that of the programs which use them. The COMANDOS architecture aims to support both mechanisms in an uniform way.

At the system level in COMANDOS therefore a two-level storage model is being proposed for the support of **persistent objects**. At the upper level, the Virtual Object Memory contains the objects that are used by the current activities. This includes persistent and volatile objects. At the lower level, the Storage Subsystem contains all persistent objects.

An important result has been the coupling of data management support with a persistent store, allowing efficient retrieval of information from secondary storage whether this data be requested directly (by name) or indirectly (by associative access based on properties and attributes of the data). A related result is that the precise format in which particular data is stored can be redefined and tailored to suit the requirements of particular applications, thus making the system open to new storage techniques and specialised algorithms.

The definition of a Virtual Machine and its conceptual models, and of an implementation architecture to support it, are both important because they aid the understanding of programming distributed, concurrent and/or data management oriented applications. They allow concepts to be clarified and a common vocabulary to be understood. They provide an evolution towards integrated thinking of application construction from re-usable components, and provide a measure of openness not otherwise readily achievable.

The basic features of the COMANDOS virtual machine and implementation architecture are described in [Horn 87]. Details of these two components are available in the CO-MANDOS architecture reference document [Comandos 87].

2.3. Prototyping of the COMANDOS implementation architecture

A strategic result of the project is the demonstration of various prototypes of the COMANDOS implementation architecture, thus proving its viability on a number of underlying environments. However these various implementations are based on a common strategy which consists to define a minimum system dependent distributed Kernel which, in cunjunction with a few essential system independent services, implement the virtual machine. This section describes the principle of the three kernel implementations carried out in the three-year first phase: GUIDE, IK, OISIN, Alves 88].

– The Bull/LGI prototype, called **GUIDE**, has been designed to provide a minimum basis for supporting distributed applications specified in terms of the COMANDOS model as quickly as possible, and to identify problems raised by the implementation of an object-oriented architecture above UNIX. This implementation runs as a "guest" layer on top of UNIX without any modification to the kernel. Moreover this approach allows to run existing UNIX applications on top of the COMANDOS platform (applications components being encapsulated within objects).

The current version of the GUIDE system is running on Bull DPX1000/2000 and Matra-Datasysteme MS-3. Both machines are based on 68020 processors and the host system is a version of UNIX System-V, plus socket communication features from BSD 4.2. [Decouchant 88].

– The INESC kernel, named **IK**, provides a single-user COMANDOS environment on a PC/AT computers (based on the INTEL iAPX286). The main goals of this prototype are:
  - to provide an implementation of the COMANDOS architecture on commercial, widespread and inexpensive machine.
  - to evaluate the suitability of a segmented memory management system, as provided by the iAPX286, to efficiently support object oriented systems.

The current implementation is running on a set of PC machines (Olivetti PE28 and ME28, and Sperry IT) interconnected on an Ethernet network.

– The main goal of the OISIN kernel was to provide an efficient implementation of the COMANDOS model on relatively sophisticated hardware. The chief features which distinguish Oisin from Guide and IK are as follows:
  - Kernel mode exploitation of a demand paged virtual memory environment,
  - use of clustering to reduce i/o operations and accelerate object invocations,
  - a multi-level i/o subsystem.
  - the absence of a UNIX style i/o buffer pool and instead exploiting all available physical memory effectively as an i/o cache.

The TCD implementation was primarily targeted at the NS32332 based Trinity Workstation, and is now being ported to Digital uVax-IIs.

It should be noted that these three prototypes are three implementations of the same architecture on various underlying system or hardware, thus exploring a number of implementation strategies in parallel [Comandos 89a]. A comparative study of the three prototypes has been carried out, in order to evaluate these mechanisms and the overhead induced by the underlying system (mainly for the UNIX implementation). This study is reported in [Adler 89].

2.4. Prototyping of the COMANDOS virtual machine interface

The programming environment provided by COMANDOS is intended to be multi-lingual, and a range of programming languages are expected to be used. To interact fully with, and thus exploit, the COMANDOS environment, a range of primitives are available via libraries (presntly for C and Modula-2). Existing programs are supported without requiring recoding, although relinking with standard environment libraries is necessary [Comandos 89b1]. This effort is currently extended to the support of C++.

However, consideration of our environment and virtual machine primitives have convinced us that it is useful to provide a language in which the concepts of the COMANDOS virtual machine are faithfully reflected. This language embodies the main features of the COMANDOS virtual machine. Some of its features may be regarded as "syntactic sugar", reducing the burden on the application programmer who is making extensively use of the COMANDOS program libraries. Moreover other features of the virtual machine such as typing and inheritance can only be expressed in linguistic terms. The resulting language, OSCAR, designed within the project, has been implemented in two ways: one provided as an extension of an existing language (preprocessing C), and a full compiler of a subset of the language. The language has been also used for programming some basic system services such as a name server and a type manager. The viability and usefulness of this new language have been demonstrated by programming basic system services (naming service, type manager), and a number of distributed applications [Comandos 89b2].

2.5. Definition and prototyping of Systems Services and on-line Management Tools

The COMANDOS Kernel is extended by a set of System Services, built as normal objects (using the COMANDOS language). Basic services provided so far include a Name Service for objects identification, a Type Manager for the management of user-defined types, and an Object Data Management System (ODMS). The overall objective of the ODMS is to provide advanced facilities for database handling (class management, query language embedded in the COMANDOS language, etc...). A complete design of the ODMS is described in [Comandos 89c], and a prototype implementation is under way.

The major contribution of the project in the area of system management is the definition of an adaptive approach to system design and management. This approach ensures that application systems are not rigid but can adapt to changing business needs and environments during their life time. In the adaptative approach the distributed information system supports three functions: identification of system models and descriptions, decisions about system changes for improvement based on these models, and implementing the changes by modifying the system.

Within the COMANDOS platform on-line management tools and facilities are provided which support the above three functions. They include a distributed, object-oriented system observation facility for system identification, supporting statistics collection and presentation, as well as alerting. The decision function is supported by performance modelling and configuration management tools, security tools for authorization and risk management, and a Distributed Office Sysem (DOS) design tool for the design and integration of a DOS. System modification is supported with a distributed object-oriented system control facility. It is important to note that these tools and facilities are applied to the COMANDOS platform as well as to applications.

The on-line management tools and facilities have been prototyped on top of UNIX (and thus should be portable to other UNIX environments). These prototypes are at various stages of development. The System Observation Facility and the System Control Facility are operational [Comandos 89d1]; the DOS Design tool [Comandos 89d2] and the Security tools [Comandos 89d3] are partially implemented. Some of the software packages developed in the project will result in products (see section 3).

In the consolidation year of COMANDOS results, it is intended to integrate these tools and facilities on top of the COMANDOS Virtual Machine.

2.6. Definition and Prototyping of the COMANDOS Integration System

The goal of the COMANDOS Integration System (CIS) is to support the interworking of COMANDOS applications with pre-existing and heterogeneous Information Systems by providing a uniform abstract interface to them. The advance of CIS is to consider the integration of generic information systems where a schema translation is usually not feasible, either because there is no schema at all (e.g. simple file system, CAD applications), or because the schema is not visible to the user (e.g. Public Data Banks).

The technical strategy adopted by CIS is to use the COMANDOS object-oriented model to build an uniform interface on top of the systems to be integrated. The use of an object-oriented model allows to overcome the limitations of the mapping between schema as it allows an "operational" mapping (achieved by a piece of code, embedded into the object definition, which performs the actual mapping).

The CIS prototype has been developped on a mixed MS-DOS and XENIX netword environment. It demonstrates the access to three kinds of application: a file-based application using either indexed and sequential access files; A CAD graphic application; a public data-bank [Comandos 89e].

3. Exploitation of COMANDOS-1 Results and Future Plans

The COMANDOS project started in March 86, and by September 87 has derived the conceptual model and underlying architecture of the COMANDOS platform. Prototyping of aspects of the platform then began, culminating in a public demonstration in November 88 at the 5th ESPRIT Conference exhibition. Since then, evaluation and consolidation of the prototypes has begun, including feedback on using the platform in practice.

This section describes the intended exploitation of COMANDOS-1 results. We distinguish two levels of exploitation, depending on whether we are considering "short-term" or "long-term" exploitation. Short-term exploitation includes the development of spin-off products and the transfer of know-how, while the so-called long-term exploitation refers to the continuation of the COMANDOS project , in the framework of ESPRIT-2, based on the results and experience of ESPRIT-1 COMANDOS.

3.1.Short-Term Exploitation of COMANDOS-1 Results

The final goal of COMANDOS is to provide an **innovative** support for distributed application development. Thus, expecting a short-term industrial product development based on the results of this first phase of the project is premature. However the acquisition of the object oriented technology is a result of primary importance in the light of further developments planned within the project (COMANDOS-2 proposal) or outside the scope of the project. The experience drawn from the project will be exploited in different ways, both within and outside of the consortium. We list below some of the intended experimentations:

– Design and prototype implementation of a Distributed Directory Service, in conformance with the ISO/ECMA standards.
– Development of an Hypermedia system.
– Development of a document handling system.

Moreover a number of software products and services based on COMANDOS-1

workpackages will be provided by two of the industrial partners.
a) A first product, based on the CIS prototype will allow to access the databank of "Corte di Cascazione" of Regione Toscana.
b) A set of services available in a UNIX System V.2 environment will be derived from the work in the "Tools" area.
- Performance modelling tool for distributed systems.
- Event Management Logger.
- Exploitation of conformance standards for alerting and statistics.

3.2. Future Plans - COMANDOS in ESPRIT-2

The COMANDOS project is now entering a second phase, which intends to consolidate and enhance the results of COMANDOS-1, in order to provide a **pre-industrial integrated application development environment**. Started in March 89 this development phase is planned over four years. Based on the results of COMANDOS-1, a further consolidation and integration phase will allow to provide an integrated prototype of the platform (called Release-1) at the end of the first year of COMANDOS-2. Following a period of evaluation and revision, a full pre-industrial prototype of the platform (called Release-2) will be provided at the end of the third year. In parallel Release-1 will be used as an experimental basis for the construction of extended facilities (mainly a distributed directory sService, and on-line management facilities) and for the development of a real-world application in the office area. The fourth year of the project will be mainly dedicated to consolidate and enhance the pre-industrial prototype, including the integration of various facilities already developed on top of Release-1. Main aspects of COMANDOS-2 may be summarized as follows :

– Revise, complete and fully implement the language defined within Esprit-1 COMANDOS. A full compiler will be developed, but the COMANDOS interface will also be made available as a set of libraries for some existing programming languages (namely Modula-2, C and C++).
– Provide a development environment for programming using the COMANDOS language. This environment includes facilities for symbolic debugging and syntactic editing.
– Revise the COMANDOS architecture, and implement the COMANDOS system (with full capabilities) both on a bare hardware and on a pre-existing low-level distributed kernel. Particular attention will be paid to efficiency and heterogeneity issues.
– Provide the compatibility with existing UNIX applications by extending the X/OPEN interface to encompass the COMANDOS features, and consequently to enhance UNIX towards the COMANDOS interface.
– Extend the functionalities of the kernel by a set of System Services, implemented as normal objects. The system services include a Distributed Directory Service and facilities for Object Data Management and Type Management.
– Refine the security architecture. Implement basic access-control mechanisms for the support of security tools (authorization, authentification, risk management).
– Provide on-line management facilities for the observation and the control of the distributed environment.
– Experiment the object-oriented technology for the support of powerful and flexible user interface management systems, based on X-Window facilities. - Finally provide a testbed application, both for checking the suitability of the COMANDOS model and language, and for evaluating internal mechanisms provided by the COMANDOS system. The

intended testbed application is a real-world application dealing with intelligent circulation of documents within an enterprise.

A major milestone in the timeframe of the ESPRIT-2 COMANDOS project is the delivery of Release-1 of the platform for further experimentations both inside and outside of the consortium. This can be viewed as an intermediate spin-off result which is expected to be use by some other ESPRIT-2 projects as a basis for the development of large distributed applications.

4. Conclusion

In summary, COMANDOS is conceptualising, designing and, most importantly **implementing a vendor-independent platform for distributed processing**, including management of long-lived data, programming language support and on-line administration. To date, no such vendor independent infrastructure exists. OSI may be used as a basis for interconnection of equipment from various vendors, but not as a common platform for the numerous interacting applications required in the integrated electronic organisation. The availability of such a vendor independent integrated platform for programming distributed applications, coupled with data management, operating in an environment of heterogeneous machines would be a significant advance from current exploitations of OSI. It would also advance the state of the art in distributed systems.

UNIX is an accepted vendor-independent system interface. Recent extensions to UNIX, such as the SUN Network File System or PCTE extend aspects of the UNIX interface to distributed environments. However none of them provide at the same time all of the facilities which characterise the COMANDOS programming interface: distributed objects considered in the SUN Network File System are basically limited to UNIX files [Sandberg 86]; the PCTE OMS (Object Management System) provides a simple object store but its type system and language support are rather weak [Bourguignon 85]. The standard UNIX interfaces (eg. X/OPEN and Posix) have yet to be transparently extended to include further aspects such as replicated data, transactional atomicity of fault tolerant behaviour. Rather than incrementally extend UNIX in a somewhat ad-hoc fashion, we propose to identify a sophisticated interface subsuming support for distributed processing, data management, security control, on-line administration and user interface generation, and then subsequently but during our project - to enhance UNIX towards this interface.

A major achievement of the project so far has been the definition of an Architecture of a Virtual Machine for general distributed processing. This architecture emphasizes the integration of technology from distributed operating systems, distributed programming languages and distributed databases. This integration objective is achieved through the use of an object-oriented approach. It should be noted that a similar approach is currently appearing in international standardisation activities such as the ECMA DASE proposal and the ISO ODP workitem. More specifically, these projects are considering a computational model for cooperative distributed processing which COMANDOS accomodates and for which COMANDOS can provide valuable input.

The industrial partners involved in the project are considering COMANDOS as a platform for the European IT industry to develop on the one hand easy portable distributed applications and to allow on the other hand the coexistence with old-style (UNIX oriented) applications. The results of the COMANDOS project will provide an increased level of harmonisation and consistency across the participating systems within the European market. Introducing the COMANDOS platform for the development of distributed applications means to significantly increase productivity and competitivity. This is especially valuable as the wide range of products within the various industrial companies lack respective functionality as well as homogeneity on the basic operating system level. Therefore the results of the COMANDOS project will strongly influence the future development of software products in a number of market segments.

The COMANDOS Consortium includes three leading European manufacturers (Bull, Nixdorf, Siemens), two software vendors (ARG, Chorus Syste'mes), two PTT laboratories (INESC, CNET-SEPT), and five Universities, who are located in different states of the European community.

The three main European manufacturers involved in COMANDOS are also members of the Open Software Foundation (OSF). It should be noted that the overall objective of the COMANDOS project is in conformance with the guiding principles of the OSF, and thus will be presented as a major contribution to the OSF Programme.

5. Acknowledgements

The author wishes to acknowledge all those who have contributed to the definition and prototype implementation of the COMANDOS platform.

Finally it is reminded that the ESPRIT-I COMANDOS consortium was composed as follows :

- Partners: Bull (F), ARG (I), ICL (UK), INESC (P), Nixdorf (G), Olivetti (I), Trinity College Dublin (IR).
- Sub-Contractors: IEI (I), Laboratoire de Genie Informatique-IMAG (F), Fraunhofer Institut Stuttgart (G), University of Glasgow (UK).

6. References

[Adler 89] Adler O. (February 1989), "The Comandos kernel implementations: object-oriented benchmarks, measurements and evaluation", Technical report D2-T3.2.1.4-890228.

[Albano 85] Albano A., Cardelli L. and Orsini R. (1985), "Galileo: a strongly typed interactive conceptual language", ACM TODS, Vol 10, No 2.

[Alves 88] Alves Marques J. and al. (November 88), "Implementing the Comandos Architecture", Proc. of 5th Esprit Conference, Brussels.

[Black 87] Black A. and al. (January 1987), "Distribution and Abstract Types in Emerald", IEEE Transactions on Software Engineering, Vol SE-13, No 1.

[Bourguignon 85] Bourguignon J.P., (September 1985), "Overview of PCTE: A Basis for a Portable Common Tool Environment" in Proceedings of ESPRIT Technical Week, September 1985.

[Carey 86] Carey M.J., and al. (August 1986), "The Architecture of the Exodus extensible DBMS". Proc. of 12th Int.Conf. on VLDB, Kyoto.

[Comandos 87] COMANDOS, (September 1987), "Object Oriented Architecture", Technical Report D2-T2.1-870904.

[Comandos 89a] COMANDOS, (March 1989), "Kernel implementation Report", Technical Report D1-T3.2.1.3-890306.

[Comandos 89b1] COMANDOS, (March 1989), "Language: pilot implementation report", Technical Report D1-T3.2.1.5-890219

[Comandos 89b2] COMANDOS, (June 1989), "Pilot conventional language interface", Technical Report D1-T3.2.4.1-890222

[Comandos 89c] COMANDOS, (April 1989), "ODMS: final design", Technical Report D1-T3.2.2-890401.

[Comandos 89d1] COMANDOS, (March 1989), "Prototypes of System Observation facility and System Control Facility for a Distributed office System", Technical Report D2-T3.3.15-880930.

[Comandos 89d2] COMANDOS, (March 1989), "DISDES: Distributed Information System Designer", Technical Report D1-T3.3.16-890406.

[Comandos 89d3] COMANDOS, (February 1989), "Security Architecture", Technical Report D1-T3.3.13-890216.

[Comandos 89e] COMANDOS, (June 1989), "Comandos Integration System: final report", Technical Report D1-T3.3.1-890601.

[Dase 86] ECMA, (October 1986), "DASE Model", ECMA TC32-TG2 Working Paper /86/69.

[Decouchant 86] Decouchant D. (September 1986), "Design of a distributed object manager for the Smalltalk-80 System", Proc. of ACM OOPSLA, Portland.

[Decouchant 88] Decouchant D. and al. (October 1988), "GUIDE : An implementation of the COMANDOS object-oriented distributed system architecture on UNIX", Proc. of EUUG Conference, Lisbon.

[Glasgow 86] Persistent Programming Research Group, (November 1986), "PS-algol Reference Manual", Persistent programming research report 12, Department of Computing Science, University of Glasgow and Department of Computational Science, University of St. Andrew.

[Horn 87] Horn C. and Krakowiak S. (1987), "Object Oriented Architecture for Distributed Office Systems" in ESPRIT '87: Achievements and Impact, North-Holland.

[Horn 88] Horn C., Ness A. and Reim F. (May 1988), "Construction and Management of Distributed Office Systems" Proceedings of EURINFO '88, Athens.

[Jones 86] Jones M.B., Rashid R.F. (September 1986), "Mach and Matchmaker: kernel and language support for object-oriented distributed systems, Proc. First ACM Conf. on Object-Oriented Programming Systems, Languages and Applications (OOPSLA), Portland, pp. 67-77.

[Leach 83] Leach P. and al. (November 1983), "The Architecture of an Integrated Local Network", IEEE Journal on Selected Areas in Communications, Vol SAC-1, No 5, pp843-857.

[Liskov 83] Liskov B. and al. (July 1983), "Guardians and Actions: Linguistic support for robust distributed programs", ACM TOPLAS, Vol 5, No 3.

[Maier 86] Maier D, and Stein J. (September 1986), "Development of an Object Oriented DBMS", Proc. of ACM OOPSLA, Portland.

[Mullender 85] Mullender S. (October 1985), "Principles of Distributed Operating System Design" Ph.D Thesis, Vrije Universiteit, Amsterdam.

[O'Brien 86] O'Brien P., Bullis B., Schaffert C. (September 86), "Persistent and shared objects in Treillis/Owl", Proc. of Object-Oriented database Systems, Asilomar Conference, Pacific Grove California.

[Sandberg 86] Sandberg R. (1986), "The Sun Network Filesystem: Design, Implementation and Experience" Proc. of Spring EUUG Conference.

[Schwarz 86] Schwarz P. and al. (September 1986), "Extensibility in the STARBURST Database System", Proc. of IEEE/CS, Asilomar.

[Walker 83] Walker B., Popek G., English R., Kline C., Thiel G. (1983), "The LOCUS Distributed Operating System", ACM Proc of the 9th SIGOPS.

[Zimmermann 84] Zimmermann H., Guillemont M., Morisset G., Banino J.S. (1984), "Chorus: a Communication and processing Architecture for Distributed Systems, RR 328, INRIA, Rocquencourt.

MICROELECTRONICS AND PERIPHERAL TECHNOLOGIES

DESIGN OF A SPEECH DECODER
FOR DIGITAL MOBILE RADIO:
HIGH LEVEL SYNTHESIS WITH CATHEDRAL II VERSUS
GENERAL PURPOSE DSP PROCESSORS

J. ADAMS, JJ. SCHMIT, M. VAN CAMP
ALCATEL-BELL
FRANCIS WELLESPLEIN 1
2018 ANTWERPEN
BELGIUM

Abstract

In this paper the implementation of a speech decoder algorithm for a Pan European Mobile Radio System with the Catedral II DSP silicon compiler, is described. A comparison is made between a general purpose DSP processor implemetation and the compiled Cathedral II implementation of the decoder algorithm.

1. Introduction

In implementing Digital Signal Processing (DSP) algorithms Cathedral II intends to be a cost effective solution, in terms of area usage and design time. This should be compared against the use of a general purpose DSP core combined with some glue logic, RAM and ROM. Cathedral II compiles the algorithms into a multiprocessor system on chip, starting from a high level description language (Silage). It is intended to be used by the system design engineer who need not to have the full detailed knowledge on designing circuits down to the silicon level. The content of this paper presents the design flow with Cathedral II in designing a Speech Decoder Chip for a mobile radio application.

The designed chip decodes a bit stream at 13 kbit/sec, encoded by a **Regular Pulse Excitation - Long Term Predictive Coder** scheme. Its output is a digital speech signal with 13 bit resolution and 8 kHz sample rate, according to the CEPT/GSM 06.10 specifications. [1]

The result of this project demonstrates the usefulness of such automatic synthesis system, featuring fast design turn around time with silicon design results at the top level design phase, and a cost efficient implementation of DSP algorithms on dedicated hardware architectures. It will also encourage the researchers in this domain to broaden the applicability of this kind of systems towards eventually other types of architectures.

In comparison with the final layout from the Piramid system, which is a production implementation of the Cathedral II concepts and design methodology, the "Edge" (CADENCE) Place & Route tools were used to compare the quality of the layout obtained from both systems.

The decoder (and also the encoder) algorithm will be implemented in a prototype system for the Cellular Radio by means of a commercially available DSP processor, the results of a comparison between the Cathedral II approach and the general purpose DSP implementation will be reported.

2. Algorithm description

The example chosen for this work is the speech decoder of the Pan-European Digital Mobile Radio (DMR) system. The procedure which has to be respected to implement this decoder is completely specified in the GSM recommendation 6.10 (Title: GSM Full Rate Speech Transcoding). A global blockdiagram of the decoder is given in appendix 1.

The input of the speech encoder is a 13 bit uniform PCM signal with a sampling rate of 8000 samples/s. The encoding procedure generates an encoded bit stream of 13 kbit/s. This encoded bit stream contains the parameters calculated by the encoding algorithm. These data are then processed by the channel coder and send to the receiver by the radio system. At the receiver side the bit stream is transmitted by the channel decoder to the speech decoder. The decoder delivers reconstructed speech samples. The coding procedure specified by the GSM recommendation is the Regular Pulse Excitation - Long Term Prediction - Linear Predictive Coder (RPE_LTP). [2]

3. Cathedral II design methodology [3,4]

3.1 Global Flowchart

Figure 3.1 gives a global overview of the design methodology which is used in the Cathedral II silicon compiler system. The methodology is based on the so-called "meet-in-the-middle" strategy. A clear distinction is made between the system design and the silicon design. The interface between both levels is situated at the building blocks (execution units EXU) of the target architecture.

Figure 3.2 shows the CAD toolbox of the Cathedral II system.

3.2 System Level Description

Within Cathedral II the Silage language is used as the design language. This language is optimized for high level description of signal processing algorithms.

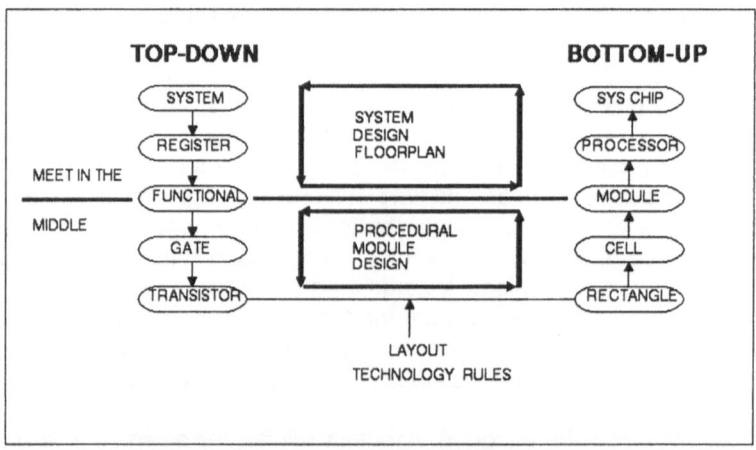

FIGURE 1 Meet-in-the-middle Design Methodology

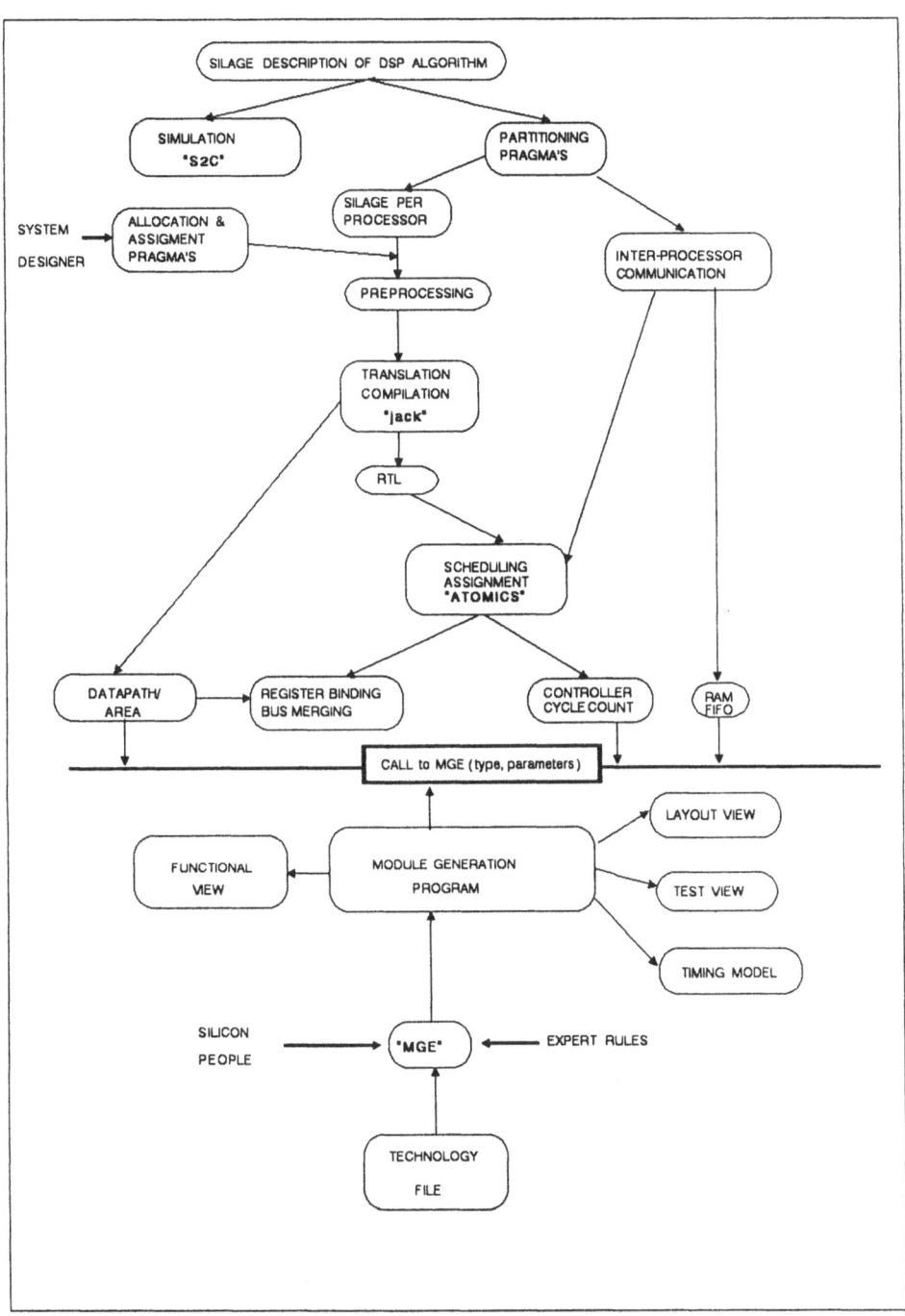

FIGURE 2 CAD toolbox for Cathedral II

Silage is an applicative language which is more adequate to describe hardware system than a procedural language. The main idea of Silage is to capture the signal flowgraph of a signal processing algorithm. It does not contain any structural or control information and does not enforce any degree of concurrency.

It is however possible to pass some structural hints to the compiler to guide the synthesis process by means of "pragma's".

The Silage simulator "s2c", which is based on compiled-code simulation techniques, allows the simulation of the implemented Silage code.

3.3 Architectural Synthesis

The basic architecture in Cathedral II is based on a multiprocessor configuration. Each processor is optimised for a particular part of the algorithm and consists of 2 major parts

- datapath: build with a limited set of parameterisable EXU
- controller: the implemented controller architecture is the multiple-branch controller which allows the implementation of a large variety of algorithms.

The algorithm is partitioned in seperate parts which will be executed by different processors. For each processor a seperate Silage code is generated. Starting from a Silage description for each processor the code is optimised with respect to the functions of the target architecture. The optimised code is then compiled by means of a rule based system, JACK, into a structure representation.

This structure is embedded in a number of unscheduled register transfer statements. (RTL)

The next step is to generate an optimal timing of the RT's. This is done by the scheduling program "ATOMICS". Atomics performs a microprogram scheduling and creates an optimal mapping of RT operations in the time domain. This tool takes into account all the implications of the actual controller structure such as the amount of pipelining, the type of controller logic as well as all the restrictions to prevent conflicts between RT's.

The result of this scheduling is a readable symbolic microcode, which can be processed by the remaining tools. Atomics gives also the cycle count. With this information the designer may change his implementation of the algorithm to achieve a lower cycle count and thus a more performant implementation.

3.4 Layout Synthesis

The symbolic microcode can be implemented on a variety of controller architectures, e.g. a single FSM or a ROM and PC based architecture. The efficiency depends upon the type of the considered algorithm. In Cathedral II a general multi-branch controller structure is used.

A controller generation environment (CGE) has been created which consists of a number of procedures that describe how an input description can be mapped into a particular controller architecture. These assembly procedures consist of a number of optimisations and minimisation steps such as state assignment, logic minimisation and layout generation.

The layout generation for each of the building blocks (EXU) is done in the "module generator environment" (MGE). This environment allows the designer to create, to generate and to adapt each module generator.

The generators are procedurally described and call symbolic descriptions of the leaf cells. By means of a compaction algorithm those leaf cells can be turned into real layout taking into account the specified constraints.

Besides the layout view a functional, timing and test view are also generated. These views are very important to check if the module fulfills the specifications such as timing conditions and to allow an evaluation of architectural trade-offs e.g parallel- or pipelined processors.

Once all the views are generated the designer can switch to the floorplanning phase. With the aid of an automatic place and route tool, different topologies can be generated and weighted versus the specifications. Eventually some iterations will be necessary to obtain an acceptable result.

4. Cathedral II usage

As already mentioned in the introduction the purpose of this paper is to discuss the results on the use of Cathedral II (Piramid) in designing a speech decoder chip for a mobile radio application.

The Piramid silicon compiler, developed by Philips Research lab, is an industrial implementation of the Cathedral II concepts. This tool comprises all the major tools from Cathedral II but also includes more enhanced features necesseray to use the system in a real design environment.

4.1 Silage Implementation of the Decoder

To enter the compiler, a high level description of the algorithm has to be provided by the system designer. This description is done in the Silage language. Since this is the major input to the whole system, the Silage code should be 100% correct and verified. The way an algorithm is implemented in Silage is strongly related to the available constructions in the language and to the available rules in the rule-based mapping system: "Jack".

For instance in the decoder algorithm different rates are used, which were hard to implement with the given set of constructions. This due to the fact that the Silage description can be considered as a program which is executed once for each set of input data. In more complex algorithms e.g GSM codec, some parts of the algorithm work on different speeds which should be taken into account by the Silage description.

4.2 Verification Of The Silage Code

The GSM provides also a set of testsequences, which allow the testing of the actual implementation. Only the input parameters, and the output signals (for the decoder) are given. With this limited data it is impossible to debug the Silage code. Therefor a C-implementation of the decoder algorithm was made. This C-code was easier to debug by means of a standard debugger tool than the Silage code which could only be executed. Once this implementation was done, we were able to view intermediate data on different nodes in the system. With this data we were able to retrace the problems in the Silage code.

The same C-code will be used as input for a TMS320 C-compiler to generate TMS assembler code. This will aid to make a comparison between a decoder implementation on a general purpose DSP machine versus the in Cathedral II promoted dedicated DSP machine.

5. Synthesis results

The compilation of the decoder chip was done in 2 phases. During the first phase only a part (10%) of the Silage description was taken and has been compiled by Piramid up to the layout phase. The results are given in table 1. Figure 3 gives the resultant layout.

# machine cycles		417 (41.7 usec) 100 ns/cycle
RAM	#	1
	size	83*16
ROM	#	1
	size	68*91
AREA	core	33.5 mm**2
	full	43.2 mm**2

TABLE 1 : Data for exercise 1

Figure 3 : Compiled Layout of Decoder

After the implementation of the additional rules, the complete Silage description of the decoder algorithm was processed by "Jack" and "Atomics". The results are given in table 2.

# machine cycles		4333 (433.3 usec) 100 ns/cycle
RAM	#	3
	size ram1 size ram2 size ram3	626*16 449*16 36*16
ROM	#	1
	size	307*146
AREA	core	94 mm**2
	full	121 mm**2

TABLE 2 : Data for exercise 2

The major difference in the resulting layout of the two exercises, is embedded in the number of RAM's that are used and in the size of the micro instruction ROM.

6. Floorplanning with "Edge" system

6.1 "Edge" System as Alternative for the Layout Part of Cathedralii

The "Edge" Place & Route system has proven its effectiveness for automatic routing of VLSI chips at Alcatel-Bell.

The router is gridless and based on a channel routing strategy. It allows 45 degree routing which results in a 10% save in routing area versus a Manhatten style routing. Critical nets such as power, ground and clock signals can be routed seperately from the other less-critical nets.

The placement can be done fully automatically e.g. first iteration or partially manual e.g. finalisation of the layout. The result is given in appendix 2.

Due to our experience with the "Edge" products, this system can be recommended as a valuable alternative for the current layout part of the Cathedral II Silicon Compiler.

6.2 Comparison with Piramid Results

The layout (routed floorplan), that is obtained with the "Edge" product is 10% smaller in size (7.33 * 5.33 mm2 = 39.08 mm^2) than the one obtained with the Piramid system (7.2 * 6.0 mm^2 = 43.2 mm^2). This is mainly due to the optimisations which were performed on the "Edge" system. The placement was obtained from an initial automatically placed version of the floorplan. After some (4..5) manually steered iterations and optimisations the final result was obtained. A maximum aspect ratio of 1.6 was allowed.

The layout produced by the Piramid system was obtained in a complete automatic way, without manual interference. We notice also that the module generators are producing a layout which is not fully optimal with regard to the floorplanning. This is due to the non-hierarchical floorplanning. Firstly the layout of the different modules are generated without taking into account the aspect-ratios of the other modules. After an optimal placement of the modules the routing of the signals can be performed. Because the router cannot route L-shaped channels the shapes of the modules are restricted to rectangles, which can result in wasted area.

7. Comparison with general purpose DSP

The Cathedral II system compiles a Silage description into a dedicaced hardware. It provides at the same time the microcode to implement in this hardware. This is an important advantage because no cumbersome assembler code has to be written and debugged. At the other hand there exists some tools for general purpose DSP processors e.g a C-compiler and code generator for TMS320C25.

An other advantage in this system is that the required executable units (ALU, ACU, multiplier, ...) are selected in an optimal way to execute the algorithm in the allocated time period with a minimal area. Also the number of busses can be specified. The width of the data bus, ALU, RAM and ROM are not fixed as in a commercial processor. The size of the memories are also dependent on the application. Hence there is no need to add off-chip external memory. A comparison with two commercial processors TMS320C25 and ST68930 is presented in table 3. For these processors the required memory and the number of instruction cycles are based on a detailed study.

	ST68930	TMS320C25	PIRAMID
transistors	120k	150k	-
area (mm**2)	57	56	121
technology	2u nmos	1.8u nmos	2u cmos
data busses	3	1	3
cycle (ns)	160	100	100
RAM	2*128*16	544*16	626*16 r1
			446*16 r2
			36*16 r3
ROM (coeff)	512*16	-	55*16
compiler	C	C	Silage
multiplier	16*16	16*16	16*16
accumulator	16	32	32
ROM (prog)	1280*32	4k*16	307146
DECODER IMPLEMENTATION			
instruction cycles	16120	31020	4333
execution time	2579 us	3102 us	433.3 us
external data mem.	220*16	220*16	-
internal data mem	2*128*16	544*16	1108*16
prog. mem.	220*32	260*16	-

TABLE 3 : Comparison Cathedral II with general purpose DSP

Some comments on the data in table 3.

- an important data, is the total number of instruction cycles and the associated execution time. It appears that the dedicated DSP implementation needs significant less cycles than the general purpose DSP machines.
- as indicated in the table for the 2 general purpose there is supplementary external RAM memory needed, which is not the case for the synthesised implementation.

– the resultant chip area of the synthesised version is significantly greater than the general purpose processors which are full custom layout chips. If the 2u cmos implementation is rescaled to a 1.8u cmos the obtained area is 98 mm2 which is 75% greater than a full custom approach. However one should notice that on the synthesised version the ROM and RAM sizes are greater than on the general purpose processors. There is also no external RAM needed in the synthesised version.

8. Conclusions

In this report we discussed the implementation of a speech decoder algorithm for the Pan European Mobile Radio system, with a silicon compiler based on the Cathedral II concepts. The different algorithms and concepts which are used, are very advanced in comparison with those implemeted in other commercial silicon compilers.

In compiling the decoder chip a lot of advantages in comparison with a general purpose DSP implementation did appear:

– The hardware is fully optimised versus the algorithm. e.g if no multiplier is needed or if we can afford the multiplication to be implemented with the slower implementation on an ALU no hardware is allocated for the multiplier. If we use a general purpose DSP the available hardware is fixed and can not be extended or diminished on chip level. It is for instance possible to add external RAM, which can influence the execution time to read or write from/to the RAM. In the Cathedral II system the RAM generator provides as much RAM as necessary on chip.
- The firmware is automatically embedded in the hardware. So there is no need to write assembler code or eventually C- code if a C-compiler exists.

As a result of this exercise an implementation of a speech decoder was made. Since it was only a first iteration some simplifications were made:

– the assumption was made that the hardware implementation of the arithmetic operations (add, subtract, multiply ..) behave as specified by the GSM specifications. Especially the potential saturation of the result of an arithmetic operation. The current module generators do not provide this hardware unit.
– Only 1 pass through the complete system was made, hence no other alternative implementations were studied such as:
- implementation without multiplier
- impact of the number of busses

Acknowledgements:

The authors wish to thank the people of the VSDM department of IMEC Heverlee and especially J. Van Meerbergen and O. McArdle from the Philips Research Labs in Eindhoven especially for their supporting effort in getting the Silage code compiled by the Piramid system.

References:

[1] GSM recommendations 06.01; 06.10;

[2] RPE: A novel approach to effective and efficient multipuls coding of speech. P. Kroon, E Deprettere, R Sluyter. IEEE Transactions on acoustics, speech and signal processing.

[3] Cathedral II: A silicon compiler for digital signal processing. H. De Man, J. Rabaey, P. Six and L. Claesen. IEEE Design & Test.

[4] Silicon compilation of DSP systems with Cathedral II. H.De Man, J.Rabaey, J.Van Meerbergen, J.Huisken. Esprit Project No. 97

RPE-LTP Codec

APPENDIX 1 : BLOCK DIAGRAM OF THE RPE-LTP CODEC CHIP

72

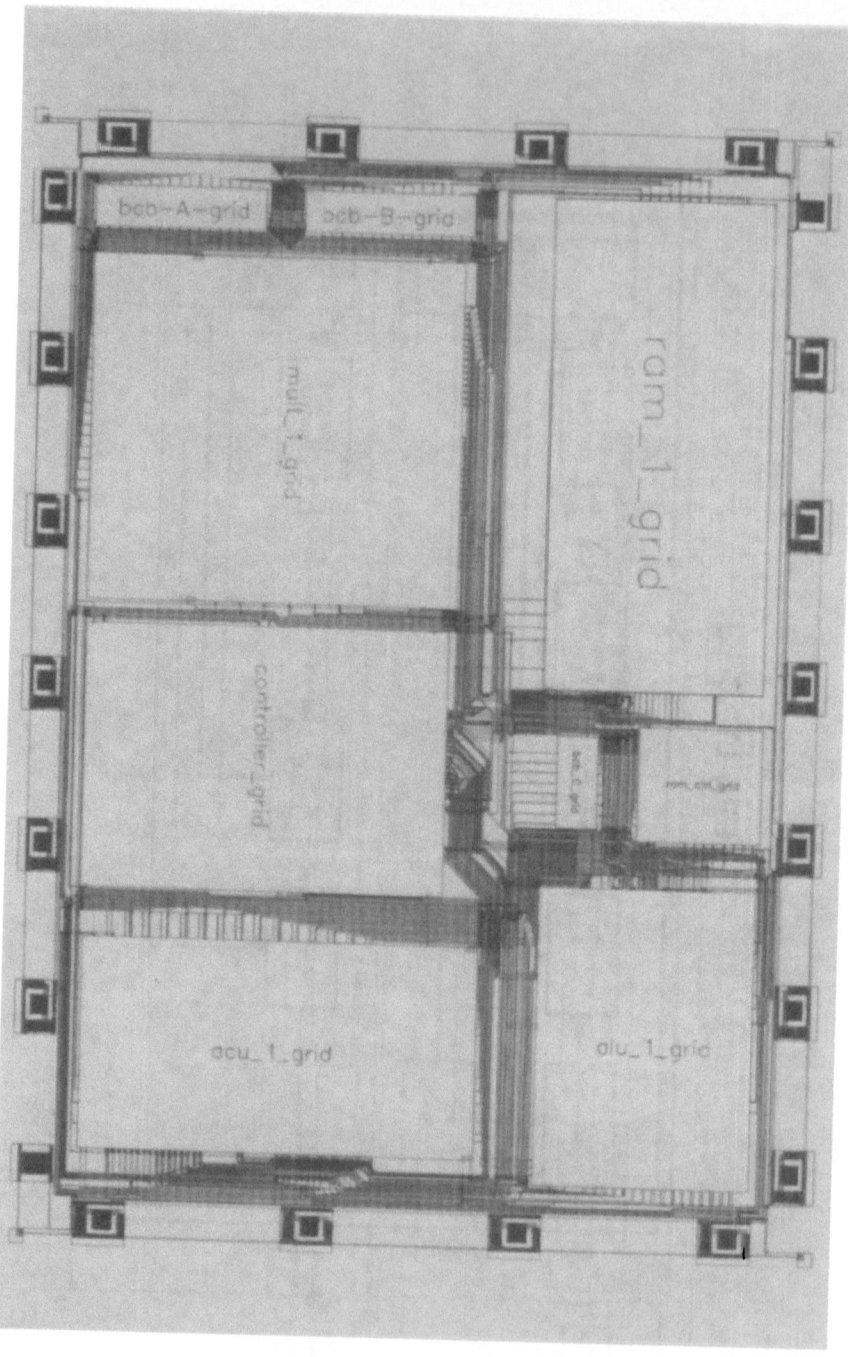

APPENDIX 2 : FLOORPLAN WITH "EDGE" SYSTEM FOR PHASE 2

Project No. 255

A SELF-CONSISTENT THERMAL SIMULATOR
OF MMIC MULTI-GATE GAAS ACTIVE DEVICES

GIOVANNI GHIONE
Dipartimento di Elettronica, Politecnico di Milano,
Piazza Leonardo da Vinci 32, Milano, Italy

CARLO U. NALDI
Dipartimento di Elettronica, Politecnico di Torino,
Corso Duca Abruzzi 24, Torino, Italy

ABSTRACT. The paper describes a CAD tool for the self-consistent thermal simulation of MESFET devices for MMIC's (MESS). The simulator was developed within the framework of ESPRIT Project No. 255 "CAD Methods for Analog GaAs Monolithic IC's". After introducing the physical problem, a *thermal resistance model* and a *self-consistent physical model* are discussed. The thermal resistance model, besides being an useful design tool *per se*, is needed to provide the self-consistent model with proper boundary conditions. Implementation details of the two-dimensional physical model are briefly reviewed, and results are presented.

1 Introduction

Thermal effects play a significant role in integrated active devices for logical or analog applications. Concerning GaAs MESFET's, the resistive heating within the active region affects both the DC and the AC performances. Since the mobility and saturation velocity of the material decrease with increasing temperature, heating reduces the saturation current and the device gain. In *logical circuits*, this means a larger gate delay and a reduced ability to drive output stages. Moreover, localized *hot spots* within the active region, which affect the lifetime and reliability of the device, are easily originated in GaAs MESFET's owing to the low thermal conductivity of GaAs [2] and to the spot-like distribution of heat sources [3]. Since the thermal behaviour is critical in MESFET design, an effective CAD tool should go beyond the simple electrical model and account for thermal effects.

The problem of simulating the thermal behaviour of Si and GaAs IC's has been widely addressed in the literature. The conventional approach is based upon the well-known concept of *thermal resistance*. The thermal resistance of a MESFET can be defined as the ratio between the power dissipated in the active region and the temperature difference between the active region and the heat sink. Actually, the temperature of the active region is not exactly uniform across the device, since heat dissipation is easier at the gate ends than in the middle of the gate fingers, and temperature differences arise for the same reason between parallel gates in multigate devices. The thermal resistance concept will be extended in the next sections so as to allow for temperature disuniformities; however, in well-designed multigate devices these are often negligible.

73

While the thermal resistance approach is simple and yields a fairly accurate estimate of the average device heating, it cannot describe the small-scale behaviour of concentrated heat sources, which lead to almost singular temperature distributions (hot spots). Moreover, the electro-thermal coupling is neglected. A *physical model* including heat diffusion together with carrier transport can directly provide a self-consistent thermal simulation giving high temperature resolution on the active region. Such models have been proposed in the past for silicon bipolar power devices [6,1] and for MOSFET's [12].

The coupled electro-thermal simulation is fraught with a basic difficulty: while electrical phenomena are restricted to a very small portion of the device cross section (i.e. the active region), heat flow extends far beyond and can have a three-dimensional (e.g. spherical) pattern. Under this respect, performing both the thermal and the electrical simulation on the whole device cross section (including neighbouring active regions in multigate MES-FET's) would dramatically increase the dimension of the problem and its computational intensity. However, since significant heating and thus strong electro-thermal coupling only occurs in the active region, while the rest of the device is almost source-free, the heat flow leaving the active region can be modelled by connecting the boundary of the simulated region to the heat sink through proper *thermal resistances*, which can be estimated by means of accurate, easy to compute analytical approximations. From a mathematical point of view, this amounts to imposing third-order boundary conditions to the heat equation on the periphery of the simulated region.

These ideas have been implemented in the current version of the MESS simulator, entirely developed within the framework of the ESPRIT project n.255, 'CAD Methods for Analog GaAs Monolithic IC's'. MESS is a FORTRAN 77 package for the DC and AC simulation of planar or recessed-gate MESFET's, initially written under VMS operating system (a UNIX version is also available).

The structure of the paper is as follows. Firstly, a new analytical thermal resistance model for multi-gate MESFET's will be discussed. The model not only is needed in connection with the self-consistent simulation, but can also be useful *per se* as an approximate design tool. Secondly, the self-consistent physical model and its computer implementation are discussed in detail. Finally, results are shown from the self-consistent simulation.

2 The thermal resistance model

2.1 INTRODUCTION

The concept of *thermal resistance* is based on an approximate assumption, namely that heat is injected into a well defined region of the device (e.g. the gate strip in a MESFET), which is *isothermal* and has *known shape and dimension*, while the rest of the device is free from heat sources.

Although none of these assumptions holds exactly true in MESFET's, a self-consistent two-dimensional analysis brings out that heat is generated almost only within a limited strip-shaped region of the active layer, placed immediately under the gate edge (see Fig.10). The temperature of the periphery of the hot strip is almost uniform. It is therefore clear that, at least from a macroscopic standpoint (i.e. neglecting local small-scale temperature peaks) the thermal resistance approximation applies fairly well to MESFET's.

Under these assumptions (i.e., heat is injected into an isothermal strip placed on the device surface, while in the rest of the device the temperature distribution is Laplacian) the (average) temperature rise of the active region with respect to the heat sink temperature is:

$$T - T_0 = R_\theta P \tag{1}$$

where $P = V_{DS}I_D$ is the total power dissipated in the active region, R_θ is the thermal resistance seen from the active region. The thermal resistance can be effectively approximated by means of the combined use of segmentation techniques and analytical (conformal mapping) methods, as discussed further on.

The idea that (1) defines an *average* temperature rise deserves some further comments. Let us call w_θ the width of the heat source. It has been shown (see e.g. [5]) that the best agreement with temperature measurements is found when $w_\theta = 8$ μm for a one-micron gate device, i.e. when the hot strip is approximately as wide as the source-drain spacing. The result can be surprising, since the self-consistent two-dimensional simulation clearly points out that the hot spot is narrower (e.g. 1-2 μm for a one-micron device). The discrepancy can be explained by taking into account that temperature measurement systems do not usually have a spatial resolution able to exactly resolve temperature variations occurring within the source-drain spacing. It is therefore clear that measurements tend to *underestimate* the peak temperature by averaging it with the temperature of the surrounding regions. As a consequence, the parameter w_θ which should physically correspond to the width of the hot spot (1-2 μm), actually describes a wider region, roughly encompassing the source-drain spacing.

The present research confirms that measurements can be matched by choosing for w_θ values in the range 6-8 μm. The same values are appropriate when computing the thermal resistance seen from the region where the self-consistent simulation is carried out. This region is approximately as wide as the drain-source spacing and thick as two or three times as the active layer thickness; w_θ can be taken as the overall width of the simulated region.

2.1.1 Treating the temperature dependence of thermal conductivity. The thermal conductivity K_θ of GaAs is an increasing function of temperature. Therefore, the heat equation $\nabla \cdot [K_\theta(T)\nabla T] = 0$ becomes *non-linear*, and the linear relationship between power and temperature rise within the active region ceases to hold. A classical linearization technique (Kirchhoff transformation) allows to turn the non-linear heat equation into a linear one ($\nabla \cdot \nabla \tau = 0$) *when the domain is homogeneous*, by means of the variable transformation (see e.g. [11]):

$$\tau(T) = T_0 + \frac{1}{K_\theta(T_0)} \int_{T_0}^{T} K_\theta(T') dT' \tag{2}$$

Notice that for $T = T_0$, $\tau = T_0$. By approximating the temperature dependence of $K_\theta(T)$, the Kirchhoff transformation can be explicitly solved and inverted so as to give the true temperature corresponding to a given τ. A suitable approximation for GaAs is [4]:

$$K_\theta(T) = 0.108 T^{-0.26} \text{ mW } \mu\text{m}^{-1}\text{C}^{-1} \tag{3}$$

where the temperature T is expressed in degrees centigrade. By inverting one gets:

$$T(\tau) \approx T_0 + [0.74(\tau - T_0)T_0^{-0.26} + T_0^{0.74}]^{1/0.74} \tag{4}$$

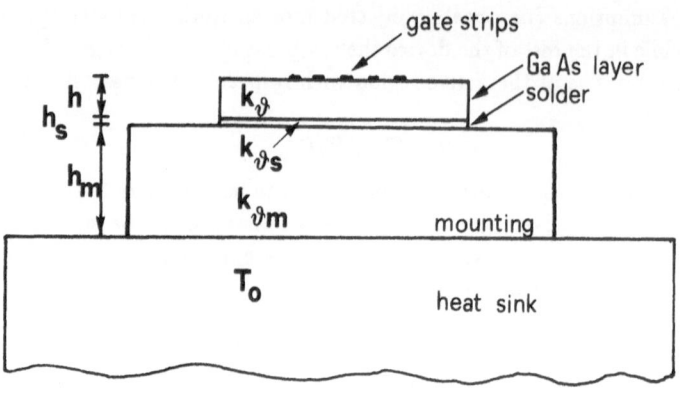

Figure 1: Multilayered MESFET structure with backside mounting.

Notice that T_0 is simply a reference temperature, not necessarily the heat sink temperature.

Now, thermal resistances can be used as follows. Given a reference temperature T_0, the appearent temperature τ is obtained as:

$$\tau = R_\theta(K_0)P + T_0 \tag{5}$$

where the thermal resistance is evaluated for $K_\theta = K_0 = K_\theta(T_0)$. The actual temperature T can the be obtained from (4) and finally the temperature-dependent thermal resistance can be defined as:

$$R_\theta(T, T_0) = (T - T_0)/P \tag{6}$$

Since the conductivity decreases with increasing temperature, the thermal resistance *increases* with increasing temperature.

2.1.2   Thermal resistance matrix of multi-gate devices.   Thermal resistance concepts can be readily extended to multi-gate devices made of several paralleled sections. Such devices tend to have *higher* thermal resistance than single-gate ones, owing to the proximity effect between neighbouring heat sources. If all active regions are supposed to be isothermal, an overall thermal resistance can be defined as the ratio between the temperature increase with respect to the reference temperature and the overall power generated within the device. However, the thermal resistance concept can be extended in a straightforward manner so as to deal with multi-gate structures having non-uniform temperature. With analogy to the *resistance matrix* of a multiport electrical structure, we can introduce the *thermal resistance matrix* $\mathbf{R}_\theta$ of a multi-gate structure, such as:

$$T_i = T_0 + \sum_{j=1}^{j=N} R_{\theta ij} P_j \tag{7}$$

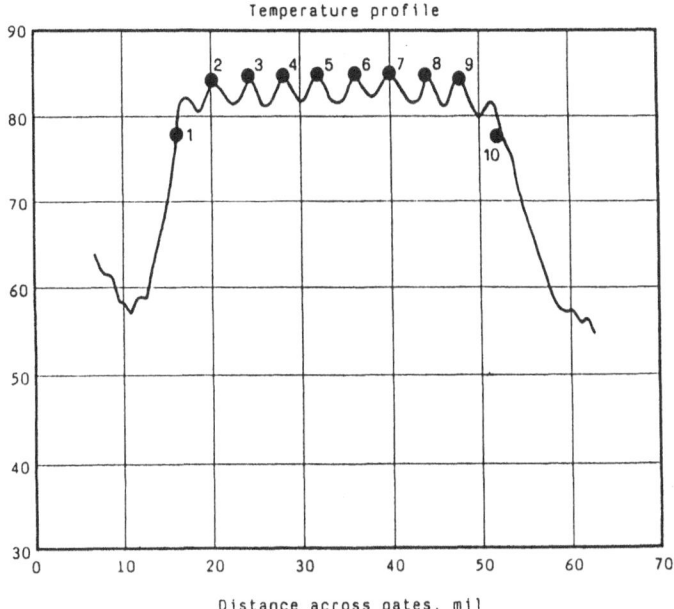

Figure 2: Temperature distribution in N-gate MESFET; dots are computed through the thermal resistance approach, the continuous curve is the measured distribution (TELET-TRA).

where T_i is the temperature of the i-th active region and P_j is the power dissipated in the j-th active region of the MESFET. If the *same* power P is dissipated in all active regions, one has:

$$T_i = T_0 + P \sum_{j=1}^{j=N} R_{\theta ij} \tag{8}$$

i.e. a temperature disuniformity arises between different active regions. Fig. 2 shows an example of temperature distribution on the cross section of a multi-gate device; external gates are cooler due to easier dissipation. The measured profile shows an oscillating behaviour due to the presence of the source regions, but fairly good agreement exist on the maximum (gate) temperatures. The thermal resistance matrix is approximated through conformal mapping techniques; the result is omitted for the sake of brevity (see [10]).

The temperature dependance of the thermal conductivity can be treated in much the same way as before. In fact, let us suppose that the power injected in each active region is known. Then, for each region a fictitious temperature τ_i can be defined, such as:

$$\tau_i = T_0 + \sum_{j=1}^{j=N} R_{\theta ij}(K_0) P_j \tag{9}$$

Then, (4) can be applied to *each* of the active regions, thereby yielding the true temperature T_i. Notice that in this case a temperature-dependent resistance matrix cannot rigorously be introduced, since the element $R_{\theta ij}$ is also influenced by the temperature of the other active regions. In other words, $R_{\theta ij} = R_{\theta ij}(T_1, ..., T_N)$.

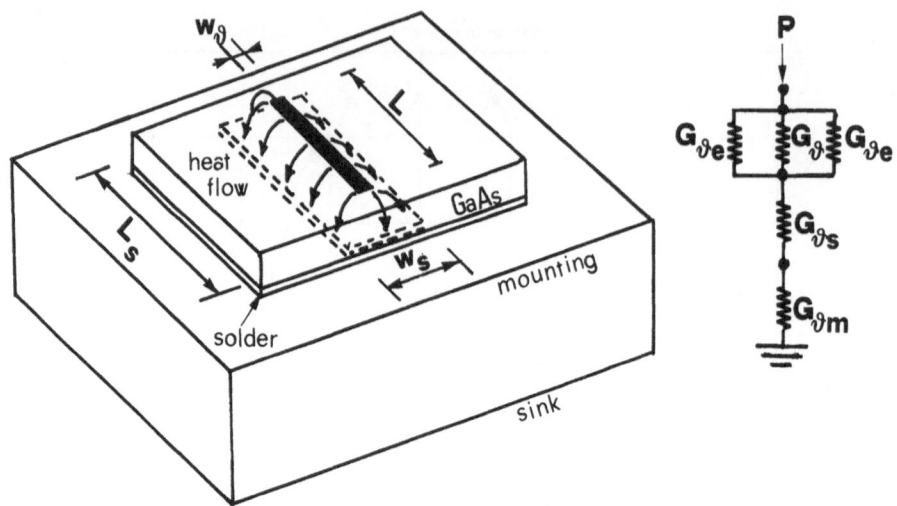

Figure 3: Segmentation of thermal resistance: left, device mounting; right, thermal equivalent circuit.

2.1.3 3D and mounting effects. From a thermal standpoint, MESFET's have a fairly complex structure (Fig. 2.1.1), in which the large-scale heat flow is three-dimensional. However, if the largest temperature drop takes place across the GaAs layer, the interfaces between such a layer and the solder and between the solder and the mounting can be approximately considered as isothermal. In this way, the overall thermal resistance can be segmented into the *series* of the mounting, solder and GaAs layer resistances, according to the electrical equivalent circuit shown in Fig. 3. The thermal resistances of solder and mounting can be separately evaluated according to heuristics well known in heat transfer analysis. Finally, Kirchoff transformation can be easily applied to the GaAs layer; in fact, given the dissipated power, the temperature of the bottom of the GaAs layer can be computed and taken as reference temperature for applying the Kirchhoff transformation. The analysis shall be confined to the so-called *backside mounting* (Fig. 2.1.1), which is normally used in IC's.

The analysis proceeds as follows. First, the system composed by the GaAs layer and mounting, plus the material (epoxy or solder) which connects them is analyzed through two-dimensional techniques, i.e. as a structure indefinite *along the gate fingers*. Efficent analytical approximations can be obtained trough conformal mapping techniques. Second, three-dimensional "end" effects are added in a way similar to the treatment of fringing capacitances in an electric problem. Third, the effect of the solder layer and of the mounting are separately evaluated and added, and finally the thermal resistance of the package (if any) is added to the whole structure.

The additional "fringing" thermal conductance due to the greater thermal dissipation taking place on the tip of the gate fingers can be evaluated on the basis of the fringing capacitance theory of symmetric striplines as follows:

$$G_{\theta e} \approx 2K_\theta \frac{K(k_1)}{K(k_1')} \frac{\delta + 2w_\theta}{4\delta + 2w_\theta} \delta \tag{10}$$

where:

$$k_1 = \tanh(\pi w_\theta / 4h) \tag{11}$$

$$\delta = 2h \log 2 / \pi \tag{12}$$

$K(k)$ is the complete elliptic integral of the first kind, while $k' = \sqrt{1 - k^2}$. The ratio $K(k)/K(k')$ can be expressed in terms of elementary functions with excellent approximation as:

$$\frac{K(k)}{K(k')} \approx \begin{cases} f(k) & 0.5 \le k^2 \le 1 \\ 1/f(k') & 0.0 \le k^2 \le 0.5 \end{cases} \tag{13}$$

where:

$$f(k) = \frac{1}{\pi} \log \left[2 \frac{1 + \sqrt{k}}{1 - \sqrt{k}} \right] \tag{14}$$

Notice that this is the thermal conductance for *one* gate end.

If we suppose that the heat flux at the *bottom* of the GaAs layer is approximately columnwise and that the equivalent flux area of a device of total width W and total length L is approximately $W_s = W + 2h$, $L_s = L + 2h$, we can approximate the solder or epoxy resistance through a simple plane parallel approach as follows:

$$G_{\theta s} \approx \frac{W_s L_s}{h_s} K_{\theta s} \tag{15}$$

This approximation is satisfactory since the thickness of the solder layer h_s is usually small with respect to the device dimensions.

Finally, the thermal conductance of *mounting* can be approximated as the capacitance of a metallic patch on a dielectric layer. The following expressions can be easily derived [15]:

$$G_{\theta m} \approx K_{\theta m} \left(2W_s \frac{K(k_L)}{K(k_L')} + 2L_s \frac{K(k_W)}{K(k_W')} - \frac{W_s L_s}{h_m} \right) \tag{16}$$

where:

$$k_L = \tanh(\pi L / 4h_m) \tag{17}$$

$$k_W = \tanh(\pi W / 4h_m) \tag{18}$$

The parameters $K_{\theta m}$ and $K_{\theta s}$ are the thermal conductivity of mounting and solder (epoxy), respectively; h_m is the thickness of mounting.

The thermal resistance of the package is rather difficult to evaluate *a priori*, since several geometries exist. The presence of cooling through extended surfaces like periodic fins makes the problem even more complex. Approximate expression for the thermal resistance of boxlike containers and for the effect of dissipators can be found in [13].

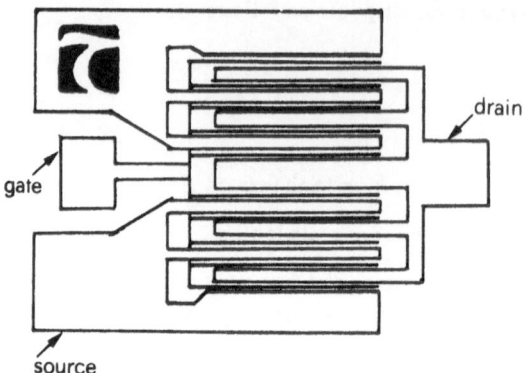

Figure 4: 10-gate MESFET (TELETTRA).

2.2 THERMAL RESISTANCE OF MOUNTED MULTIGATE STRUCTURES

Multi-gate MESFET's derive from the assembling of several lower-order cells. For example, the 10-gate MESFET shown in Fig.4 (Telettra) is the parallel of two 5-gate MESFETs with gate spacing of 20 μm, put at a distance of about 40 μm. In order to obtain a high-power 20- or 40-gate device, two or four 10-gate MESFETs can be put in parallel, at a distance of 100 μm. From the standpoint of thermal modelling, we can build up the thermal resistance model of the high-power MESFET by assembling a number of elementary cells (N_1-gate MESFETs, with N_1 "hot strips" of width w_θ, at a distance of l_1, see Fig.5) into higher-order blocks. For example, starting from a first-order 5-gate cell the 10-gate device is obtained by assembling $N_2 = 2$ first-order blocks put at a distance l_2; furthermore, the 40- (20-) gate high-power MESFET is obtained by assembling $N_3 = 4$ ($N_3 = 2$) second-order 10-gate cells put at a distance l_3.

Generally speaking, a high-power multi-gate MESFET can be interpreted as $M - th$-level structure made up of N_M ($M - 1)th$-level cells; those cells are in turn obtained by assembling N_{M-1} ($M - 2)th$ level cells, and so forth; the top-down procedure is stopped at the second-level cell, made of N_2 first-order structures. The first-order structure consists of N_1 *equispaced* hot strips. Let us define the spacing between the n-th order structure as l_n. Notice that such a spacing is defined with reference to the *sides*, not to the *centers* of the hot strips. The resulting structure can be easily analyzed by means of conformal mapping, thereby deriving the following result for the thermal conductance of the device:

$$G_{\theta 1} = LK_\theta \left\{ 2\left[2 - \prod_{i=1}^{i=M} N_i\right] \frac{K(k)}{K(k')} + 4 \sum_{i=1}^{i=M-1} (N_i - 1) \frac{K(k_{ei})}{K(k'_{ei})} \prod_{j=i+1}^{j=M} N_j + \right.$$
$$\left. + 4(N_M - 1)\frac{K(k_{eM})}{K(k'_{eM})} \right\} \tag{19}$$

where:

$$k = \tanh(\frac{\pi w_\theta}{4h}) \tag{20}$$

$$k_{ei} = \tanh(\frac{\pi w_\theta}{4h})\tanh(\frac{\pi(w_\theta + l_i)}{4h}) \tag{21}$$

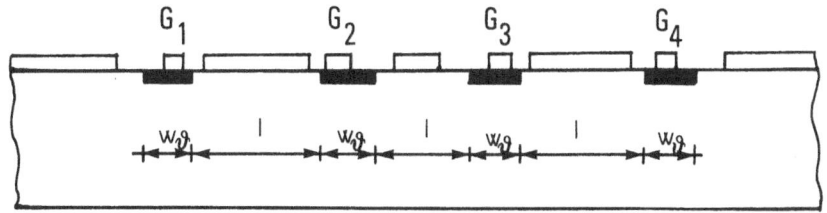

Figure 5: Characteristic thermal parameters of first-order MESFET cell.

in which: h is the substrate thickness, L is the length of each gate finger, K_θ is the conductivity of GaAs.

Given the thermal conductance of the GaAs layer G_θ, the overall thermal resistance of the device can be found according to the following steps:

1. Compute $R_1 = R_{\theta s} + R_{\theta m}$ (contribution of solder and mounting). For the solder, use (15) with parameters:

$$W_s \approx \prod_{i=2}^{i=M} N_i[N_1 w_\theta + (N_1 - 1)l_1 + 2h] \tag{22}$$

$$L_s \approx L + 2h \tag{23}$$

The conductance of the *mounting* can be expressed as:

$$G_{\theta m} = K_{\theta m} \left\{ L_s \left[2 \left(2 - \prod_{i=2}^{i=M} N_i \right) \frac{K(\kappa)}{K(\kappa')} + 4 \sum_{i=2}^{i=M} (N_i - 1) \frac{K(\kappa_{ei})}{K(\kappa'_{ei})} \prod_{j=i+1}^{j=M} N_j \right] + \right. $$
$$\left. + 2W_s \frac{K(\kappa_1)}{K(\kappa'_1)} - \frac{W_s L_s}{h_m} \right\} \tag{24}$$

where:

$$\kappa = \tanh(\frac{\pi W_s}{4h_m}) \tag{25}$$

$$\kappa_{ei} = \tanh(\frac{\pi W_s}{4h_m}) \tanh(\frac{\pi(W_s + l_i - 2h)}{4h_m}) \tag{26}$$

$$\kappa_1 = \tanh(\frac{\pi L_s}{4h_m}). \tag{27}$$

h is the substrate thickness, h_m is the mounting thickness, and $K_{\theta m}$ is the thermal conductivity of mounting. Such formulae are valid for $M \geq 2$. If $M = 1$ (i.e. a single first-order section is considered) one can put equivalently $M = 2$, $N_2 = 1$.

2. Given the heat sink temperature T_0, and supposing that the mounting is connected to the heat sink, compute $T_1 = T_0 + R_1 P$, where P is the power dissipated in the device.

3. Compute the device conductance G_θ from (19) and the "fringing" contribution $G_{\theta e}$ from 10. Then define:

$$R_\theta = 1/(G_\theta + G_{\theta e}) \qquad (28)$$

All thermal conductances must be evaluated taking for the conductivity $K(T_1)$.

4. Evaluate the apparent active region temperature τ_2 as: $\tau_2 = T_1 + R_\theta P$.

5. Now obtain the actual temperature of the active region as:

$$T_2 = T_1 + [0.74(\tau - T_1)T_1^{-0.26} + T_1^{0.74}]^{1/0.74} \qquad (29)$$

6. The temperature-dependent thermal resistance can now be computed as:

$$R_\theta(T_2, T_0) = (T_2 - T_0)/P \qquad (30)$$

2.3 COMPARISON WITH EXPERIMENTAL DATA

In order to assess the validity of the approach, the analytical expressions presented in the last section have been tested against experimental results obtained by Telettra. Such results concern the thermal resistance of two devices, a 20-gate 2-cell MESFET and a 40-gate 4-cell MESFET. Both devices have a substrate thickness of about 50 μm and have been measured in the following conditions: heat sink temperature, 30 °C; active region temperature, 50 °C. Both a wire-bound version and a version with source posts were analyzed. The results presented here concern the wire-bound structure. In such conditions, the thermal resistance measured were: 6.9 °C/W for the four-cell MESFET, 14.9 °C/W for the two-cell MESFET.

Both devices were simulated taking as geometrical parameters a gate-to-gate spacing of 20 μm (minimum) and 40 μm (maximum) and a spacing of 100 μm between neighbouring cells. In order to have a good match with experimental data, w_θ was varied between 4 and 8 μm (corresponding to the source-drain spacing).

The computed thermal resistances are plotted against power in Fig.6 for the 4-cell device and in Fig.7 for the 2-cell device. The central dot is the measured value, the upper and lower dots are thermal resistance obtained by supposing an uncertainity of $\pm$ 1 degree centigrade in the measured active region temperature. The best agreement is found for $w_\theta = 7$ μm (4-cell device) and for $w_\theta = 6$ μm (2-cell device). Notice that the contribution of solder and mounting to the overall resistance is almost negligible. For the solder (which is an alloy mainly made of tin) the conductivity of this material was used.

3 The self-consistent physical model

3.1 THE MODEL EQUATIONS

The steady-state physical model is based on the coupled solution of the diffusion-drift equations (Poisson and current continuity) and of the heat equation; a similar model was proposed in [6] for silicon bipolar devices. Hence, the unknowns are not only the charge density and electrical potential, but also the lattice temperature in the simulated region. No appreciable thermal dynamic effects occur in *small-signal* operation, since heat diffusion has slow time constants with respect to the microwave signal.

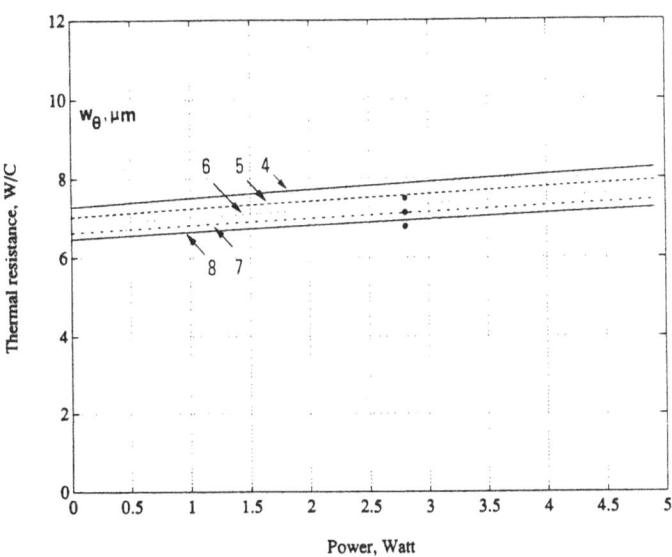

Figure 6: Thermal resistance of 4-cell (40-gate) power MESFET (see text).

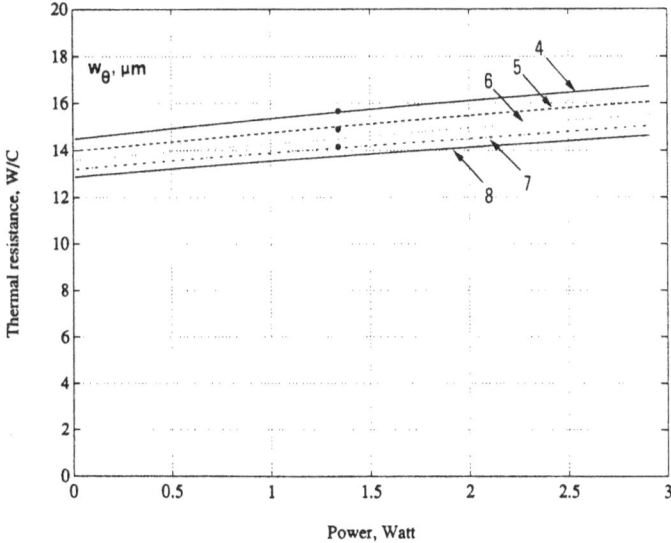

Figure 7: Thermal resistance of 2-cell (20-gate) power MESFET (see text).

Transport phenomena are described by the steady-state velocity field curve, which is approximated as in [7]; the diffusivity D is related to mobility through Einstein relation. The low-field carrier mobility μ_0, diffusivity D and saturation velocity v_{sat} depend on lattice temperature according to the approximate formulae [2, (120) and (122)]:

$$v_{sat}(T) \approx (1.28 - 0.0015T) \times 10^7 \quad \text{cm/s} \tag{31}$$

$$\mu_0(T) \approx \mu_0(300)(300/T)^{2.3} \quad \text{cm}^2/\text{Vs} \tag{32}$$

Expressions (31) and (32) are valid if T is not much lower than room temperature.

In steady-state conditions, the model equations are:

$$\nabla \cdot [n\mu(E)\nabla\Phi + D(E)\nabla n] = 0 \tag{33}$$

$$\nabla\Phi = -\frac{q}{\epsilon_0\epsilon_r}(N_D - n) \tag{34}$$

$$\nabla \cdot (K_\theta(T)\nabla T) = -\underline{E} \cdot \underline{J} \tag{35}$$

where n is the electron density, $\underline{E}$ the electric field, Φ the potential, $\underline{J}$ the current density, N_D the ionized donor density, T the lattice temperature, and $K_\theta(T)$ the thermal conductivity, which can be approximated as in (3) [4] on the range whereon devices usually operate. The other symbols have their usual meanings. The boundary conditions for (33) and (34) are conventional [14]; inhomogeneous Neumann conditions are imposed to the charge density on the upper free surface of the device in order to simulate the depletion layer due to surface band pinning.

3.1.1 Boundary conditions for heat equation.

As already recalled in the Introduction, the heat flow leaving the simulated region is accounted for through equivalent third-order boundary conditions. Experience over a number of cases shows that such a treatment of heat flow leads to fairly accurate results when the region whereon the coupled electro-thermal simulation is carried out has a horizontal extension of two-three times the source-drain spacing and includes a small portion of the buffer layer (e.g. $5 - 10$ times the thickness of the active region). Although this region is larger than those commonly considered in two-dimensional electrical MESFET simulation, the computational burden is still acceptable, above all if the simulation mesh is properly graded.

The simplified approach to outward heat flow treatment is able to approximately account for both interaction effects in two-dimensions and three-dimensional heat flow. Concerning the first point, the effect of neighbouring gates is automatically accounted for by the higher thermal resistance seen from each of them with respect to the case wherein they are isolated. In much the same way, three-dimensional temperature distributions can be investigated by dividing the active region of the device into subsections for which the thermal resistance is approximately constant, and carrying out the electrical simulation for each subsection with different thermal boundary conditions.

Following the above discussion, third-order boundary conditions are assumed to hold for (35), which read:

$$\nabla[K_\theta(T)T] \cdot \hat{n} = \frac{T - T_0}{\kappa_\theta} \tag{36}$$

where $\hat{n}$ is the outward normal unit vector, T_0 the reference temperature, and κ_θ is the distributed thermal resistance, which can be computed as $\kappa_\theta = R_\theta A$, where R_θ is the thermal resistance seen from a side of the simulated region and A is the surface of the side. Note that in principle $\kappa = \kappa(T, T_0)$, since the lattice conductivity depends on temperature ($K_\theta = K_\theta(T)$). For the same reason the heat equation (35) is weakly non-linear. The iterative technique used to solve the coupled heat and electrical equations automatically permits to account for such a dependence.

3.2 SOLUTION TECHNIQUE

Discretization is aimed at replacing the two-dimensional partial differential equations of the model by means of a system of linear or non-linear algebraic equations. The discretization of continuity equation is carried out according to the well-known Scharfetter-Gummel scheme on a triangular grid, while Finite Elements with linear basis functions and source term lumping are used for Poisson equation. The heat equation, which basically is a of non-linear Poisson equation wherein nonlinearity arises from the presence of the temperature-dependent thermal conductivity, is discretized again by means of linear basis functions on triangular elements and heat source lumping. This choice is consistent with the order of discretization chosen in the electrical part of the model, and surely grants an accurate enough resolution of the temperature profile in the critical part of the device. In fact, temperature is a smooth enough function on the device cross section, and temperature peaks only arise in a region which is already densely discretized (the active region).

Concerning *solution*, the electrical part is treated by means of a coupled Newton - Richardson scheme. This scheme allows better convergence properties and an increased program robustness when compared to the so-called decoupled (iterative) scheme. The heat equation has been solved in *decoupled* form. In other words, the initial temperature is set to a *constant* value, and the electrical part is solved; then, heat equation is solved setting the thermal conductivity to the value pertaining to the initial temperature. With the temperature distribution thereby obtained, the electrical parameters (mobility, diffusivity and saturation velocity) are newly computed and the electrical part is solved. The technique is iterated until convergence occurs, i.e. until the norm of the *difference* of the two last approximations of the temperature distribution is less than a prescribed value. Owing to the mild coupling existing between electrical and heat equations, and to the weak nonlinearity of the heat equation, convergence is always achieved in few (3-4) iterations.

3.3 RESULTS

A few results will be presented in this section to demonstrate the feasibility of the *electro-thermal* simulation. In Figs.8-10 the self-consistent electro-thermal simulation of a $1\mu m$ epitaxial backside-mounted MESFET is presented. The epilayer is 0.3 μm thick and the doping is 1.2×10^{17} cm^{-3}. As well known, the main effect of device heating amounts to a decrease of the saturation current; as a result, the VI curves tend to bend down for low gate bias. Such an effect is clearly seen in Fig.8, where the DC curves of the device are shown both in isothermal conditions (T=300 K) and accounting for device heating. In Figg.9-10 the potential distribution, electric field magnitude, temperature and generated heat distributions are shown for the working point $V_D = 9$ V, source and gate grounded. It is clear from Fig.10 that heat generation is confined to a very small circular region at the drain edge of the gate, which also corresponds to the maximum temperature. Significant temperature

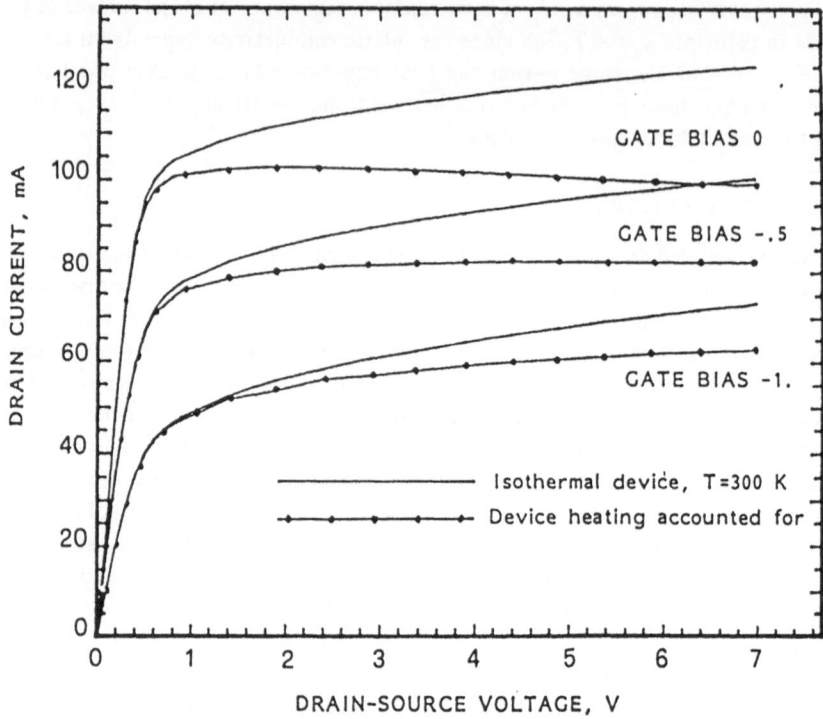

Figure 8: MESFET VI curves: effect of heating.

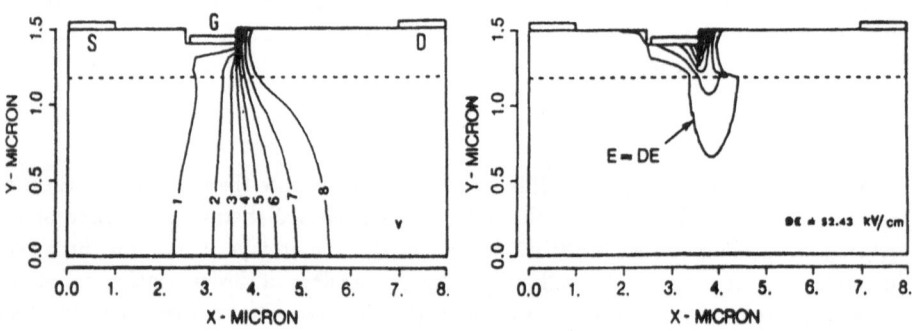

Figure 9: Potential (left) and electric field (right) from the self-consistent physical model.

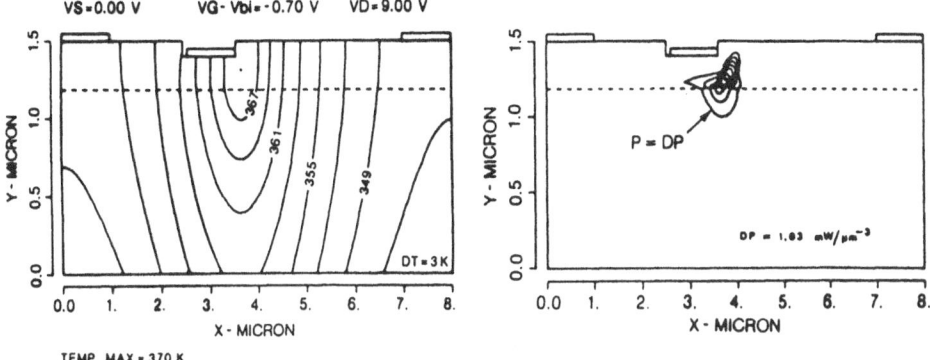

Figure 10: Temperature, K (left) and power density (right) from the self-consistent physical model.

gradients with a resolution of less than one micron can exist within the active region and on the device surface; such gradients are not expected to be observable through measurements, owing to the not high enough spatial resolution available from them. In Fig.11 the scattering parameters are shown for the same working point, in isothermal conditions and accounting for device heating. One clearly notices the effect of heating in reducing the device gain.

4   Conclusions

A self-consistent physical model for the thermal simulation of GaAs multi-gate MESFET devices has been presented. The model includes a new analytical thermal resistance model which has been tested against experiment and which can also provide useful design information *per se*.

Acknowledgments - The authors wish to thank Dr. Salvatore Iannazzo of the Components and Technologies Division of Telettra for effectively coordinating the project. The cooperation of the technical staff of the Components and Technologies Division of Telettra is also gratefully ackowledged.

References

[1] V.C.Alwin, D.H.Navon, L.J.Turgeon, "Time-Dependent Carrier Flow in a Transistor Structure Under Nonisothermal Conditions", IEEE Trans., vol. ED-24, No. 11, pp.1297-1304, November 1977.

[2] J.S.Blakemore, "Semiconducting and Other Major Properties of Gallium Arsenide", J. Appl. Phys., vol.53, no.10, pp.123-181, Oct. 1982.

[3] Harry F. Cooke, "FETs and Bipolars Differ When The Going Gets Hot", Microwaves, No. 2 (1978), pp.55-61, February 1978.

[4] Harry F. Cooke, "Precise technique finds FET thermal resistance", Microwave and RF, No. 8, August 1986, pp.85-87.

88

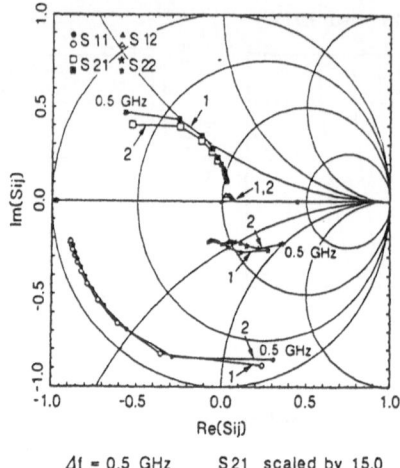

Δf = 0.5 GHz   S21 scaled by 15.0

Figure 11: MESFET scattering parameters: effect of device heating.

[5] J.V.DiLorenzo, D.D.Khandelwal, *GaAs FET Principles and Technology*, Artech House, 1982.

[6] S. P. Gaur, D. H. Navon, "Two-Dimensional Carrier Flow in a Transistor Structure Under Nonisothermal Conditions", IEEE Trans., vol. ED-23, No.1, pp.50-57, January 1976.

[7] G. Ghione, C. Naldi, F. Filicori, M. Cipelletti, G. Locatelli, "MESS - A Two-Dimensional Physical Device Simulator and its Application to the Development of C-band Power GaAs MESFET's", Alta Frequenza, Vol. LVII, N.7, Spetember 1988.

[8] G.Ghione, P.Golzio, C.Naldi, "Thermal Analysis of Power GaAs MESFET's", Proceedings of NASECODE V, pp.195-200, Boole Press, 1987.

[9] Ho C. Huang, F. N. Sechi, L. S. Napoli, "Measuring thermal Resistance in GaAs Power FETs", Microwave Systems News, Vol. 8, No. 10, pp.105-108, October 1978.

[10] S.Iannazzo, N.Fanelli, C.Naldi, G.Ghione, E.Pettenpaul, I.Wolff, Consolidated Interim Report, Esprit project No.255 "CAD methods for Analog GaAs Monolithic IC's", Period number 6, Luglio 1988.

[11] W.B.Joyce,"Thermal Resistance of Heat Sinks with Temperature-Dependent Conductivity", Solid State Electronics, 1975, Vol.18, pp.321-322.

[12] A.Schutz, S.Selberherr, H.W.Potzl, "Temperature Distribution and Power Dissipation in MOSFET's", Solid State Electronics, vol.27, no.4, pp.394-395, Apr. 1984.

[13] J.H.Seely, R.C.Chu, *Heat Transfer in microelectronic equipment*, Marcel Dekker, New York, 1972.

[14] S.Selberherr, *Analysis and Simulation of Semiconductor Devices*, Springer Verlag, Wien 1984.

[15] I.Wolff, N.Knoppik, "Rectangular and circular microstrip disk capacitors and resonators", IEEE Trans. MTT-22, No.10, pp.857-864, October 1974.

ADVICE PROJECT: FINAL BALANCE AND FUTURE PERSPECTIVES

M. Melgara
CSELT
Torino, Italy

I. Whyte
BTRL
Ipswich, UK

Y.J. Vernay
CNET
Grenoble, France

F. Boland
TCDU
Dublin, Ireland

B. Courtois
IMAG
Grenoble, France

Abstract

The increase of the complexity of present IC and the evolution of the technological design rules have enhanced the request of new techniques to evaluate the real behaviour of VLSI circuits. E-beam based techniques have shown to be a winning approach to examine the internal electrical and logical status of the circuit. The Scanning Electron beam Microscope can be really useful in IC debugging procedures if it is deeply integrated with the design environment. The ESPRIT Project 271 ADVICE[1] aimed to automate the electron beam tester, creating a link toward the design data. The foreground of the project is going to be a commercial product. Furthermore the basis for new researches have been established.

1. Introduction

The ADVICE project aimed to develop a methodology and an environment for automatic design error diagnosis using an electron beam. The project, partially funded by EEC under ESPRIT, saw the co-operation of three industrial partners (BTRL - UK, CNET - France, CSELT - Italy) and two Universities (IMAG - Grenoble, France, Trinity College - Dublin, Ireland).

The project started in December 1984 and it will end on November 1989. The total man power is 50 man/years and the expected total cost is 4 millions ECUs.

The complexity of present VLSI chips demands for powerful tools to detect possible design errors. This problem is made more difficult by the reduced amount of information available at the external pins. In addition, the physical dimensions of the internal lines make impossible a mechanical probing without both modifying the capacitance level and destroying the interconnections.

1. The research was performed within ESPRIT project 271 ADVICE (Automatic Design Validation of IC using E-BEam), partially funded by European Economic Community, being partners CSELT-Italy, BTRL-UK, CNET-France, IMAG-France, TCDU-Ireland.

The adoption of an E-beam allows to overcome these problems, increasing enormously the internal observability. However, the time required to position the beam and to acquire the measure, the debugging strategy to minimise the number of measures, the techniques to isolate the design problems represented key points to be solved in order to achieve real SEM automation.

The ADVICE project sought to provide the design/test engineer with an interactive environment, integrated with the design environment, to carry out all debugging procedure in a computer assisted/aided way.

A key point of the project was the integration between design and test environment. Design data are used to make the debugging process as automatic as possible: identification of physical co-ordinates on the chip starting both from layout information and line names used in high level description, layout pattern recognition to perform accurate positioning, comparison of physical measures against simulation results. The project has developed a user friendly working environment which puts under the designer finger tips all information related to the device under debug such as layout information, netlists, simulation/measure results, fault dictionary.

The ADVICE project was divided into two distinct but subsequent research periods. The first couple of years was mainly devoted to put the bases to the new marriage between SEM and Design environments: the equipment was assessed and all basical software tools were developed to achieve a sort of partial automation of the operating procedures.

The second period, starting from the third year, and lasting tree years, was intended to develop a more comprehensive automatic ADVICE system, deeply integrated with the CAD environment, taking advantages of the already obtained results.

2. Description of the ADVICE System

The ADVICE system is organised as a multi-window, multi-menu driven program (running on DEC VaxStationII/GPX), that will provide the user with an interactive, graphical environment [1].

The main characteristics of the system are:

1) The ADVICE system is a user-friendly environment in which a designer performs all operations related to the debug procedures, without leaving it, but having under his finger-tips all tools needed.
2) The ADVICE system is linked to the existing CAD and hardware world through standard interfaces (languages and procedures), that allows the development of a single nucleus, by all the co-operating partners.
3) The ADVICE system is running on a common workstation (Vax Station), integrated with commercial dedicated boards for image processing.
4) The ADVICE system automatically performs all trivial, repetitive and tedious operations, leaving the designer free of concentrating on decision tasks, eventually suggesting him basical strategies.

The block scheme of the hardware modules of the ADVICE system is provided in figure 1.

The VaxStationII/GPX screen is split in several windows, to provide the access to the different system tools. The following windows can be opened, on user request:

– SYSTEM DIALOGUE window;

– SEM control windows;

– LAYOUT window;

– ATE control window;

– IMAGE processor control window;

– WAVEFORM display window;

– METHODS (diagnostic tools) window.

The menus control each window and provide other commands useful for system operations (test pattern editing, background simulation, fault dictionary processing,...).

All textual dialogues (displayed messages, user specified commands and file names) are concentrated into the SYSTEM DIALOGUE window.

SEM windows are a set of interrelated windows that provide the remote control of all SEM parameters [2] (focus, brightness, astigmatism, zooming factor, beam current, etc.), of the device position (through the control of stepper motors) and of the electron beam position (deflection of the beam). The measurement procedures are also controlled by these windows. In particular, a direct control of the AVOSET equipment is provided [3]. The importance of the adoption of the AVOSET, a fast signal averager developed by BTRL within ADVICE, will be pointed out in next section.

The LAYOUT window is used to display the masks of the device. The IC layout can be described both in CIF or in GDS-II format. A hierarchical navigation in the layout description is provided, in order to cope with the problem of device complexity: sub-parts of the full device can be shown at mask level, while other parts are represented by the boundaries of the higher level cells. Measure point selection and location are mainly performed by the LAYOUT tool [4].

The ATE control window is composed by several menus to translate the simulation waveforms into the tester language, to load them into the tester and to control the test pattern generator.

The IMAGE control window allows to set image processor parameters and provides an entry to pattern matching procedures, used to determine the displacement between the SEM image taken around the measure point and the corresponding layout boxes, derived from the CAD data [5]. The WAVEFORM display window allows to handle both logical and electrical waveforms, either simulated or measured. Filtering and thresholding procedures are available to reduce measure noise and to obtain logic values [6]. Simulated and measured digital waveform can be automatically compared to highlight differences due to errors.

The METHODS window is a textual terminal which provides the user with a wide set of commands to load the logical description of the circuit, a file describing the simulated test sequences, a file of the correspondences between logical names and the associated measure point co-ordinates. Other commands are provided to navigate into the logical hierarchical data structure, to perform all the actions necessary to analyse the circuit behaviour and to diagnose all possible errors. Hilo hardware description language (supported by GenRad, Milpitas, USA), has been selected as a standard among the Partners to model the circuit at logical level.

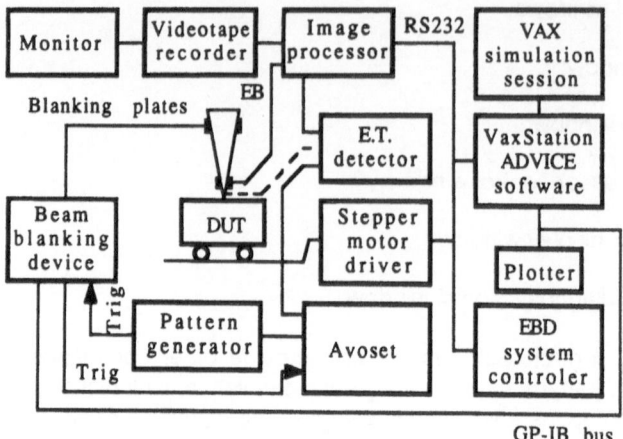

Figure 1: Computer controlled electron beam
debugging system

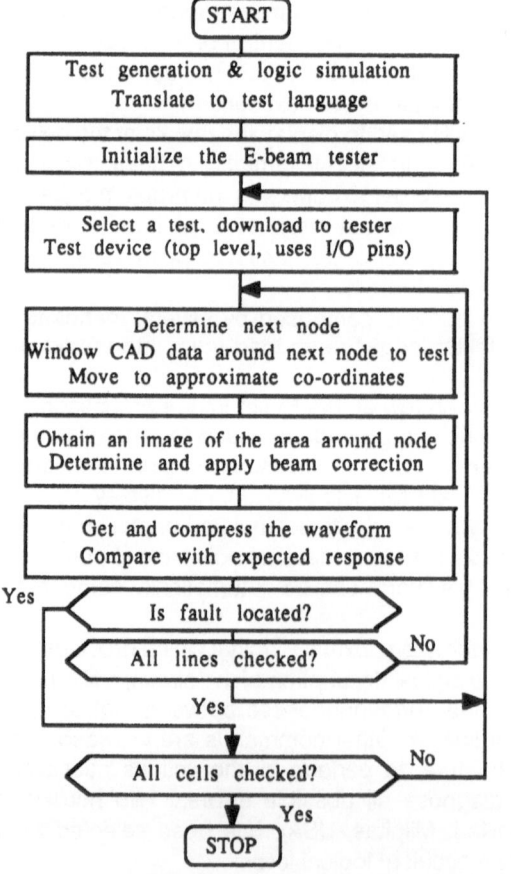

Figure 2: ADVICE System operation flow

A VMS process is associated to each set of window. A supervisor process, the ADVICE SHELL, controls the other processes. It receives and dispatches messages to the different processes. The communication mechanism is based on mailboxes: when a process needs a service, it sends a macromessage to the Shell; the Shell generates and forwards the relevant micromessages to the appropriate processes.

Each diagnostic session can be saved and restored both to work around possible machine failures and to interrupt very long circuit analysis procedures.

3. Usefulness of the ADVICE system

In order to understand the reasons that led research activities during the project, a classical flow of debugging operation, depicted in figure 2, will be described.

A first general problem to be faced was the definition of logical and topographical design rules to make a VLSI device electron beam testable [7].

Simulation sessions must be run to compute a sort of a-priori knowledge on the circuit. Ad hoc test pattern must be selected according to error location strategies and to respect EBT measurement constraints. The test sequence should be iterated several times to perform averaged measures. The AVOSET equipment and the SEM detection chain improvements had allowed to perform experiments with real test patterns of reasonable length. The AVOSET has allowed to use test sequences 1000 times longer than the one permitted by classical Box-Car averager; the real time measure frequence has been risen to 20 MHz, but if data are interleaved, the sampling rate can be increased up to 2 GHz.

Furthermore, theoretical researches on signal processing and the adoption of the AVOSET, providing the capability of acquiring several samples in stroboscopic mode during each waveform cycle (multisampling technique) has reduced of 2 to 3 orders of magnitude the time needed to acquire a measure.

One of the major problems encountered during the debugging session of an integrated circuit, is to exactly locate the beam on the appropriate tracks: users had to look for the points to be tested on the plot of the IC, then find again them on the SEM screen by moving the DUT. If we consider that a debugging session can involve thousands of measures and that several times the obtained SEM images are of rather poor quality (e.g. in stroboscopic mode) it comes out that this way of working (completely manual) is hardly feasible.

By using CAD information the beam positioning can be performed by an assisted three steps procedure:

– Identification of the points of interest on the layout representation;

– Positioning using stepper motors (coarse placement);

– Checking the correctness of the positioning and adjustment, if necessary (fine placement).

The first two steps can be easily obtained by extraction of coordinates from a layout representation (e.g. CIF format) and conversion into proper signals for stepper motors(figure 1).

The third step is a typical problem of pattern recognition and can be obtained in two different ways: manually by the operator, which compares the SEM image with the reference picture on a graphic terminal or, more powerfully, by an automatic pattern recognition tool.

In the ADVICE system the coarse placement is achieved by using a module which allows to explore, starting from the CAD data, a layout, to extract the coordinates of the measuring points and use them to drive the stepper motors [4]. This module allows a real time coarse positioning with an average accuracy +/- 3 microns.

The fine placement, unlike present commercial EBT systems, has been implemented by an automatic pattern recognition procedure, based on vector matching algorithm [5]. Commercial image processor boards have been inserted into the VaxStation frame.

Presently the time required to manipulate the SEM image and perform the matching with CAD data is about 14 seconds. The final goal is to reduce it to 10 seconds. It must be underlined that fully automatic positioning is mandatory if automatic debugging would be achieved, since all human operations to place the beam must be avoided.

The measure point can also be determined by selecting a logical name in the hierarchical netlist. Relevant measure points can be attached to the netlist, according to test point selection rules based on layout topographical considerations.

The problem of cross reference list building, to link logical names and physical co-ordinates, is solved in two ways: BTRL has implemented, starting from the ASTRA design system, a procedure that, given the logical name, extract the co-ordinates of the physical insertion of the wire in the specified cell; CSELT and CNET are facing the problem starting from the Dracula system design rule checker results: a multi-procedure system is under development to automatically derive the cross reference list from netlist information and Dracula outputs.

Once the point has been located, the measurement can be performed by using the AVOSET fast signal averager, and the obtained waveform is transferred to the WAVEFORM window on the VaxStation.

The compacted result of the comparison between the measured and the simulated waveforms can be stored into the hierarchical logical data structure of process METHODS, together with information about the status of the diagnostic session. In order to cut the number of measures to locate an error, the user can also be helped in the selection of next probing point by a pre-computed fault dictionary which summarises the behaviour of the circuit when a defined set of faults is injected in the device description.

Figure 3 provides a block scheme of the user assisted ADVICE system: the user, being in the middle of the system exploits all the facilities to control automatic measurements [8].

The present system has allowed to cut the time required to acquire 15 waveforms from about 2 hours, at the beginning of the project, to few minutes (figure 4).

A step further towards fully automated diagnostic procedures has been performed by the development within process METHODS of a set of diagnostic tools able to help the engineer in IC validation.

The diagnostic tasks (figure 5), sees the ADVICE interactive system as procedures able to interface them to the physical reality. The user can interact with the diagnostic tasks by the Diagnostic Task Shell, both issuing operative commands or monitoring the actions performed by them.

Starting from the knowledge about the logical model and the layout of the circuit, the simulation results and the status of the debugging procedures, the diagnostic tasks select next points to be probed, compare the measured and the simulated results, trying to locate the failing IC area. A probing algorithm [9], based on the circuit hierarchy analysis, its simulated functionality and its measured behaviour, has been developed to assist the user

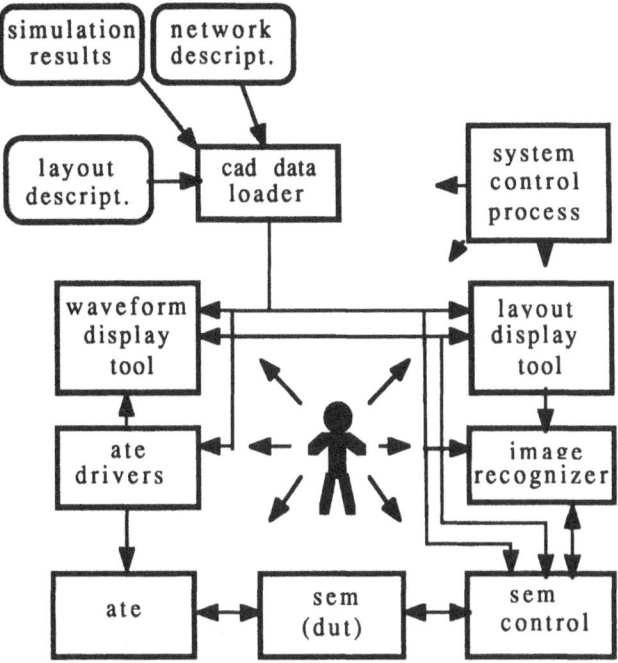

Figure 3: User assisted ADVICE system

Operation	Manually operated EBT	Automated EBT
EBT System set up (warm up & loading)	15	4
Selection of nodes on the layout	15	1
E-beam positioning	40	3
Measurement and acquisition	35	2
Hard copy,comparison with simulation	15	1
Average total time (minutes)	120	11

Figure 4: Average time required for 15 measures (minutes)

in the task of fault finding. Due to the complexity of automatic fault location process, the goal of this part of the project was mainly to demonstrate the feasibility of such an approach, rather than providing a real industrial tool.

4. Achievements, advances and perspectives

Five years ago, electron beam techniques were well established methodologies used by scientists. In the same period the ADVICE Project started its activities. The project was defined to solve problems that industrial Partners had to daily face in their labs.

At present the project is almost over. The final system is under integration. The ADVICE system has allowed to cut of some order of magnitude the time to validate IC's; it has made feasible VLSI circuit debugging by reducing the validation time from months to few weeks.

The major achievement of the project has been the transformation of an equipment for scientists into a system for engineers. This result could have been reached mainly thank to the synergic effect of the co-operation among the partners, a dynamically tuned goal definition (thank to the co-operation of EEC officers and reviewers), a gradual validation and adoption of partial results in every day work.

During development of the Advice system, several tools included into it have been used to validate several IC designed at the industrial Partner's premises. The extensive use has given two major advantages:

– the dramatic reduction of the debugging time;

– the positive feedback to Advice system designer, allowing them to improve the tools according to the real needs of industrial users.

In parallel with the last period of the project, industrial EBT equipments appeared on the market. The two most interesting systems available on the market are IDS 5000, developed by Schlumberger, San Jose', California, and ICT 9000, designed end produced by ICT, Munich, FRG. Those systems presently offer very powerful and reliable tools to IC test engineers. We hope this has been an effect of cross fertilisation of ideas generated by the dissemination of project results. At least we can assume that what was defined five years ago as a target for an advanced research reflects the development lines pursued, some year later, by the market.

A gap is still present between the ADVICE system and the commercial equipments, since more advanced facilities are included into it. The major advances of the Advice system with respect to commercially available systems are:

– fine beam placement is totally automated, thank to the pattern matching algorithm (the others automatically perform coarse placement in the area of interest and request the user intervention to place the beam in the measure point);

– test point selection can be made univocally by specifying the logical name, since test point location was done according to pre-defined test point selection rules, based on design for E-beam testability concepts (the others can highlight the full track corresponding to a given name into the netlist);

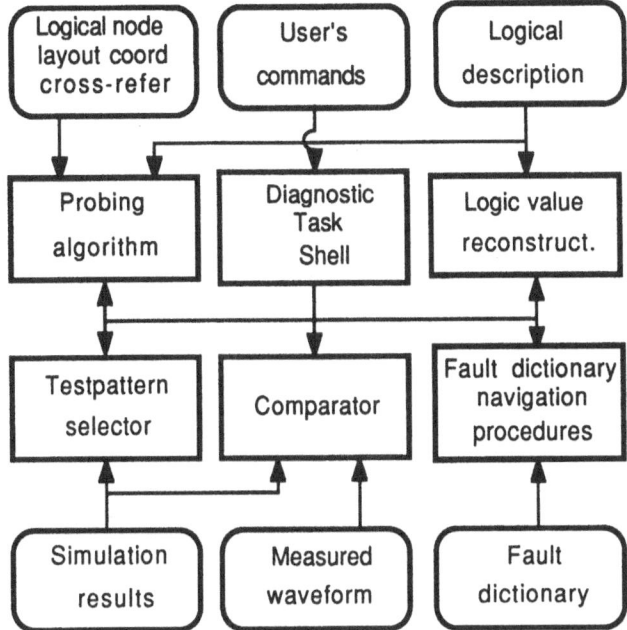

Figure 5: Diagnostic task unterconnection

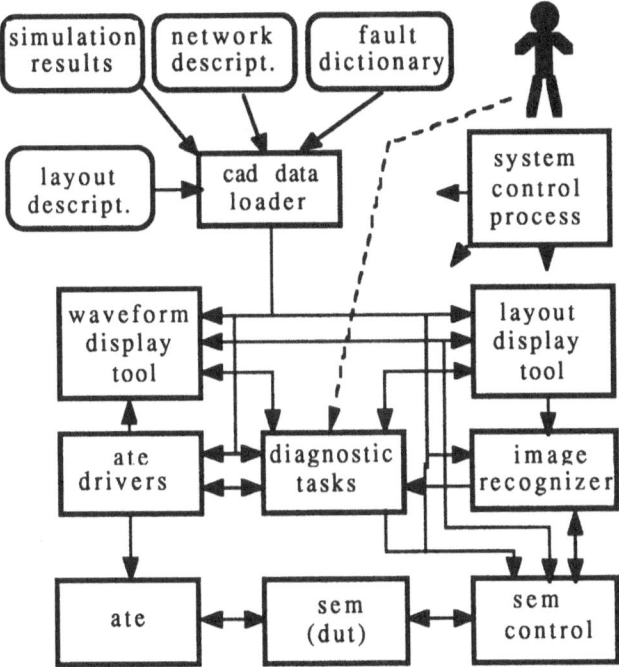

Figure 6: Fully automated ADVICE system

– test equipment control, loading and down-loading is performed directly within the Advice system by Process ATE (IDS 5000 has no direct link with the ATE: ATE control is performed independently from IDS processes; ICT 9030 SEM has been integrated into Advantest equipment: the SEM probe is seen as a pin of the ATE);

– diagnostic procedures have been implemented, able to navigate into the hierarchical logical description of the IC, to select the test pattern, to control the ATE, to retrieve simulation results, to help the user in the task of selecting probing point sequence to locate the error, to save into the hierarchical data structure information on the debugging procedure status (errors detected by the measures, assumption on defendant cells and wires) (the others have no diagnostic capability).

These differences have triggered the request from industrial EBT producers to get ADVICE project foreground and integrate it in their systems to come an industrial exploitation.

The research results of ADVICE have opened a new scenario towards very promising developments to improve the system and to be ready for the challenge of new technologies and the new requests of the designers.

A new EBT system should be targeted to wafer scale and ULSI devices: millions of transistors, with line width of 0.5 microns, on a 5 cm X 5 cm "die", working at hundreds of MHz.

Present system will require improvements of the SEM parts (gun, detector, spectrometer, ...). The mechanical stage to place chip will require wider and more accurate movements: new fine placement techniques must be developed.

The connection of hundreds of pins with the ATE will generate both mechanical and electrical problems. The integration between EBT and ATE should be improved, to exploit all the ATE facilities during the debug.

The link with the CAD environment should be moved from file based connections to a real integration between CAD Data Base and the EBT system, to exploit the inherent information contained in each description level of the IC design (design views).

This deeper integration will allow to perform a step further towards the implementation of higher level tools for fault location. Finally the problem of low yield of very large and complex circuits must be faced and solved by introducing in the devices reconfiguration capabilities, and develop new procedures able of correcting a failure, when this one has been located by previously mentioned high level tools.

The programme of a new research project, the follow up of the ADVICE in ESPRIT II, is under definition. The new proposal will include EBT system producers and VLSI producers, to be able to provide, at the end of the project, new equipments directly available on the market, validated through its massive use at the IC producers premises.

In summary it can be claimed the ADVICE project has met all the targets settled by the ESPRIT I programme.

A precompetitive, medium term research has contributed to push further the state of the art in the EBT field.

The foreground of the project will be industrially exploited into commercially available systems.

The project results have opened a new scenario for future research activities, aiming to design a new generation EBT system.

5. Conclusions

The ADVICE project has obtained very promising results [1-10] both on the technical and the international co-operation bases.

As a practical result it can be said that the time required to perform a measurement with the E-beam equipment has been reduced, respect to four years ago, to less than one tenth of previous one.

The development of the diagnostic tasks will allow the user to move, from the middle of the system, to a new position where he will act as a system supervisor (figure 6). Since more intelligence will be inherent to the system itself, user intervention should be limited to the verification/modification of the system actions.

At present, two companies producing E-beam systems are undertaking negotiations with the ADVICE project to commercially exploit the achieved results.

Acknowledgement

The authors would like to ackowledge all the people who cooperated to the Advice Project: G. Proctor, D. Ranasinghe, F. Stentiford, J. Dowe, D. Machin, T. Twell, S. Shaw, S. Matthews from BTRL, P. Rivoire, R. Mignone, J. Rouillard from CNET, M. Cocito, G. Bestente, P. Garino, M. Battu', G. Ghigo, F. Zanetti, M. Chiappone, M. Paolini, A. Di Janni from CSELT, J. Laurent, I. Guiguet, D. Micollet, M. Marzouki, R. Girka from IMAG, M. Whelan, I. O'Gorman, E. Lynch, J. Foley, J, Slevin, K. Hundertpfund, P. Fannin from TCDU.

Bibliography

[1] Y.J.Vernay, R.Mignone, P.Rivoire: "A SEM-based workstation for design validation", XII Europ. Solid State Circuit Conf. 1986, pp. 179-180.

[2] D.W.Ranasinghe, G.Proctor, M.Cocito, G.Bestente: "Computer control of Electron beam testing for design validation of VLSI circuits", XI Int. Congr. on Electron Microscopy, 1986, pp. 619-620.

[3] D.Machin, D.Ranasinghe, G.Proctor: "A high speed signal averager for electron beam test systems", 1st Europ. Conf. on Elect. and Opt. Beam Testing of IC, 1987, pp. 201-207.

[4] M.G.Battu', G.A.Bestente, P.G.Cremonese, A.B.Di Janni, P.A.Garino: "Automatic positioning for electron beam probing", XI Int. Congr. on Electron Microscopy, 1986, pp. 651-652.

[5] F.Stentiford, T.Twell: "Automatic registration of scanning electron microscope images", 1st Europ. Conf. on Elect. and Opt. Beam Testing of IC, 1987, pp. 215-221.

[6] E.R.Lynch, F.Boland: "Parameter extraction in E-beam testing", 1st Europ. Conf. on Elect. and Opt. Beam Testing of IC, 1987, pp. 195-199.

[7] M.Melgara, M.Battu', P.Garino, Y.J.Vernay, M.Marzouki, J.Dowe: "Design for E-beam debuggability", CAVE Workshop, Sintra, May

[8] M.Melgara, M.Battu', P.Garino, J.Dowe, M.Marzouki: "Fully automatic VLSI diagnosis in CAD-linked E-beam probing system", 1st Europ. Conf. on Elect. and Opt. Beam Testing of IC, 1987, pp. 283-296.

[9] M.Melgara, M.Battu', P.Garino, J.Dowe, Y.J.Vernay, M.Marzouki, F.Boland: "Automatic location of IC design errors using an e-beam system", Int. Test Conf. 1988, pp. 898-907

[10] I.Guiguet, D.Micollet, J.Laurent, B.Courtois: "Electron beam observability and controllability for the debugging of integrated circuits", XII Europ. Solid State Circuit Conf. 1986, pp. 181-183.

[11] D.Micollet, B.Courtois: "Design methods of electron beam sensitive devices in NMOS and CMOS technologies", 1st Europ. Conf. on Elect. and Opt. Beam Testing of IC, 1987, pp. 41-426. 1988

BIPOLAR CMOS ESPRIT PROJECT

Dr. P.A.H. Hart
Philips Research Laboratories
P.O. Box 80.000
5600 JA Eindhoven
The Netherlands

Dr. H. Klose
Siemens Research Laboratory
Otto-Hahn Ring 6
8000 Munich 83
Germany

1. Introduction

This paper describes the BICMOS projects 412 ESPRIT I and 2430 ESPRIT II.

Since the start of the latter (15/10/1988) both projects were merged to overcome a considerable imbalance of 3:1 in manpower between Philips and Siemens.

The merge had important consequences; much more emphasis will be given on the realization of circuits, a better balance has been established between partners (1.8:1) and two associate contractors have been added which are exclusively engaged in circuitry.

In this paper first some general statements will be made to outline the project, then the nature of the work and the progress will be described.

2. The Project

The BICMOS 412/2430 ESPRIT project is carried out by Philips and Siemens as the main partners. Philips is the prime contractor. The University of DUBLIN and INESC in Lissabon are associate contractors of Philips. The Entwicklungs Zentrum für Elektronik in Villach is associate contractor of Siemens and the University of Stuttgart is a subcontractor of Siemens.

An integrated circuit technology combining bipolar and CMOS technology is being developed at Philips and Siemens.

Additionally extensive physical and electronic studies and design are being undertaken in support of the technological work. In the matter of electronics design Philips and Siemens are being supported by their respective subcontractors.

The project now as BICMOS 412 has completed its fourth year and the merged project is in its first year.

The project planned is to end 15 November 1991, the present contract ends May 15th, 1990.

Because of their respective product backgrounds Philips and Siemens apply considerably different emphasis. The emphasis by Siemens is on digital systems, that one at Philips is on analog systems with a digital content. This gives their cooperation an interesting aspect. In order to have full benefit from the inhouse process development, design culture and CAD facilities the, -compulsory-, starting point has been a standard CMOS process at both Philips and Siemens. The work then essentially consists in the addition of the bipolar part. The basic principle of interaction between the companies is that we exchange process knowledge, agree on a common set of design rules for test vehicles, design circuits in each other's processes and then exchange these test circuits to do a common evaluation. Subordinate to all this is an exchange of modelling knowledge and measuring techniques and this includes surmounting differences in design and CAD culture. The role of the University of Stuttgart is to support Siemens with design optimization. The role of all associate contractors is to design a demonstrator circuit to be processed by one of the respective partners, this also entails suitable interface definition.

3. The Objectives

The objective of the merged project is to combine bipolar IC processing and CMOS processing into a combined bipolar CMOS process. This process should combine in an economic way the advantages of the bipolar and CMOS. Such a combination offers new ways of realizing combined analog-digital circuits. New circuit concepts with higher performance are possible. A visible trend is in the direction of increasing complexity. For instance for HDTV, gate arrays and microprocessor applications in the mid 1990's complexities of 100k equivalent gates at data rates of perhaps 600 Mbit/s and possibly even higher are needed.

Thus the objective of this project is :
– to realize this technology integration
– to realize adequate demonstrators to prove the maturity of the process and above all to show circuit and system advantages of this technology mix beyond pure device performance.

In detail this means, that :

– In the first three years a process named BICMOS 1 was to be developed by Philips and by Siemens. By Philips the feasibility of the process has been demonstrated by a circuit of 20 K transistor complexity level cf. (1, 20). Siemens has prepared the grounds for gate array IC's with on-chip memory function (2).

– In the following two years a process named BICMOS 2 is to be developed by Philips and by Siemens. By Philips it will be demonstrated (April 1990) a video AD converter of some 200k complexity, 432 MHz sampling rate, 13.5 MHz data rate and better than 10 bit accuracy. To this end BICMOS 2 will contain polysilicon emitter transistors and vertical pnp's.

Siemens will demonstrate the maturity of its BICMOS 2 process by a 16 k SRAM and gate array with 6k equivalent gates. These devices are the building blocks of high performance ECL gate arrays with on chip memory function.

The design activities of the University of Dublin and INESC are presently based on BICMOS 2P. They respectively will design and test a video FIR filter and a video A/D converter with added functionality (ready Nov. 1991). The Entwicklungs Zentrum für Elektronik will design and test an A/D converter for which presently two different options exist. These are a self calibrating very high resolution ADC with 16 bit accuracy at 200 kHz sampling frequency and data rate or a high speed 10 bit ADC for 100 MHz sampling frequency and data rate (ready Nov. 1991).

- In the last two years a process BICMOS 3 will be developed. This process will combine self aligned submicron poly emitter transistors and 0.7-0.8 μm CMOS. Suitable demonstrators showing the enhanced capability of the process will terminate the project. 18 months before the end of the project (Nov. 1991) these demonstrators will be finally fixed.

4. The Implementation of the Work

The first year was largely spent determining the proper requirements for the process and the circuits, and doing all the experiments needed to confirm views on technology and substantiate the many simulations done in the first year. Furthermore an existing Philips technology "BICMOS 0" was used to give a "flying" start with testing of electronic ideas and for calibration of the process simulation tools. Both Philips and Siemens had experimental circuits on these test masks (4). The floorplan has been described in (4).

The second year saw establishment of the process and harvesting the first measurements on elementary structures as well as on basic subcircuits. More test masks were generated and processed. A start was made with the concepts for BICMOS 2 and particularly for the demonstrator circuits cf. (3).

In the third year a more prodigious harvest of results of BICMOS I took place. The processes matured and were characterized. The design of the first large demonstrator was started. Numerous problems with CAD for the mixed analog-digital had to be surmounted. In reality two designs one in BICMOS 0 and one in BICMOS 1 were made and processed. The area occupied in the BICMOS 0 version was about 30 mm^2 (2 μm lithography) and 11 mm^2 for the BICMOS I version using 1.5 μm lithography. (1)

In the fourth year both in Philips and Siemens the emphasis switched from BICMOS 1 towards BICMOS 2. Work was started to implement all that is necessary to enable successful processing of the large demonstrators mentioned earlier. Of course a high priority was given to their design.

5. The Process

The bipolar and the CMOS processes have to be merged in such a way that the result is economically viable (i.e. comprise a minimum of steps) and still retains the advantages of both the bipolar and the CMOS discipline. A scheme how this is done has previously been presented (1-8). A cross-section of the overall result is shown in fig. 1. Results are given in Table I.

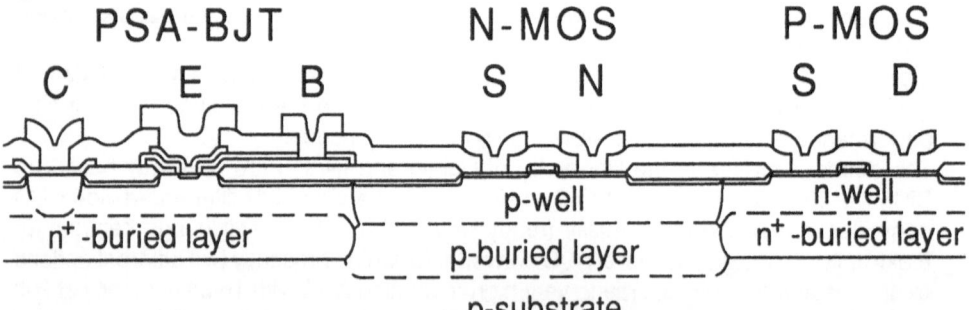

Figure 1. Cross-section of the BICMOS 2 devices.

		0P	1P	1S	2P	2S
NPN	HFe	200	100	55	100	100
	FT (GHz)	3	4	5.2	7	10
	V_{early} (V)	75	90	40	45	28
	$BVCEO$ (V)	16	12	11	10	6
PNP	Wb (μm)	3.5	2.0	-		-
	Hfe	100	126	-	150	-
	FT (GHz)	0.03		-	1	-
	V_{early} (V)	25	15	-	20	-
	$BVCEO$ (V)	20	17	-	15	-
NMOS	Vt (V)	0.75	0.9	0.76	0.9	0.82
	K (V)	0.44	0.50	1	0.50	0.9
	β (μA/V^2)	50	77	50	77	50
PMOS	Vt (V)	-0.80	-1.1	-0.77	-1.1	0.95
	K (V)	0.76	0.65	0.5	0.65	0.6
	β (μA/V^2)	17	23	18	23	18

TABLE 1. BICMOS

The present status of the BICMOS processes at Philips and Siemens is a 1.2/1.5 μm CMOS process together with a self aligned double poly bipolar transistor (Fig. 1).

As the market segments are different at which both companies aim the emphasis of additional devices and of the opimization directions are somewhat different too. For instance in the Philips processes PNP transistors are used while in the Siemens process fT for speed reasons is optimized, the incorporation of the self aligned NPN is common however.

As the goal of Philips is to realize analog/digital circuits several additional options are available and in some circuits necessary. The use of a separate N-well to enhance the characteristics of the NPN has previously been described (1). Other options are a vertical PNP and large capacitors for which the MOS thin oxide is used.

A true vertical pnp bipolar transistor is necessary to improve amplifier stages and thus to meet the high requirements put forward by DEMO 1. The inclusion of a vertical pnp is by no means a trivial matter, usually the process will become rather complex (10). We have succeeded to achieve inclusion by only two extra masksteps and judicious placement of buried layers and dopant choice. The cross section of the pnp is shown in fig. 2, data are given in table I with the 2P process (the 0P and 1P contain lateral pnp's only).

In order to avoid complications because of interaction of arsenic and boron, phosphorus was used for the buried isolating layer underneath the pnp. Two extra masks are needed to define respectively the buried phosphorus layer and the boron layer. No further additional steps for the pnp are needed, the pnp's isolation is the only modification to the process. For lateral isolation of the pnp the collector plug implantation of the npn is used. The pnp emitter uses the p MOS source-drain diffusion and the pnp's base is combined with the n-MOS LDD implantation. The complexity of the analog digital BICMOS 1P is 16 masks, the present complexity of BICMOS 2P including all options is 21. The latter figure can be reduced to about 18 by suitable tricks. This compares well what is for instance claimed by USA and Japanese makers.

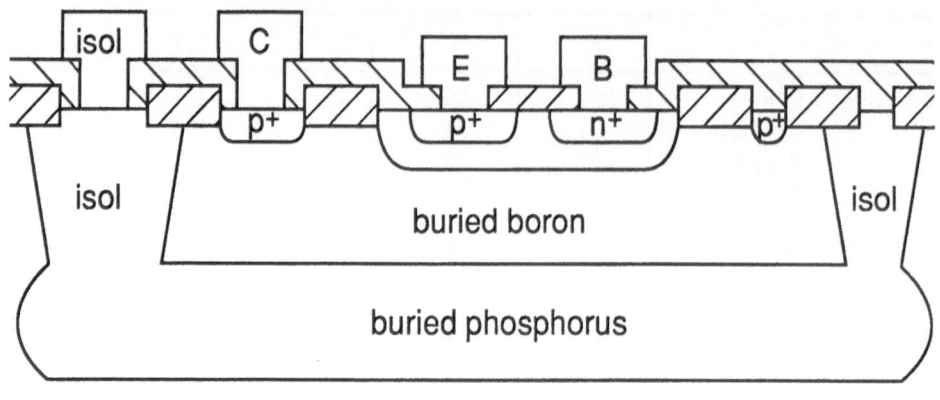

p-substrate

Figure 2. Cross-section of an isolated vertical pnp

As already stated BICMOS 2S is mainly aimed at digital applications like ECL-gate arrays with on chip memory function and micro-processor IC's. Technology and performance aspects are given in (11-13). This process allows for the realization of NMOSTs, PMOSTs and self-aligned npn bipolar transistor with an overhead of 5 mask levels compared with the underlying CMOS module. The minimal feature size is 1.2 μm. Despite the somewhat different application segments the BICMOS 2 processes at Philips and Siemens are identical in device structure (cf. Fig. 1). The main performance data of the BICMOS 2S process is summarized in Table I. As the bipolar transistor is optimized for high speed operation a cut-off frequency of 10 GHz was obtained. Typical ECL-gate delay times are 60 ps. To check out what the ultimate speed of this technology is, we used speed optimized pure bipolar circuits which were fabricated using the BICMOS 2S process (14). As an example the eye pattern of a 1:2 DeMUX is given in Fig. 3. The input data stream is 10 Gbit/s at a power dissipation of 675 mW. Both data rate of the DeMUX and the ECL gate delay time show that BICMOS 2S is among the BICMOS processes having the fastest devices worldwide (for comparison cf. (15).

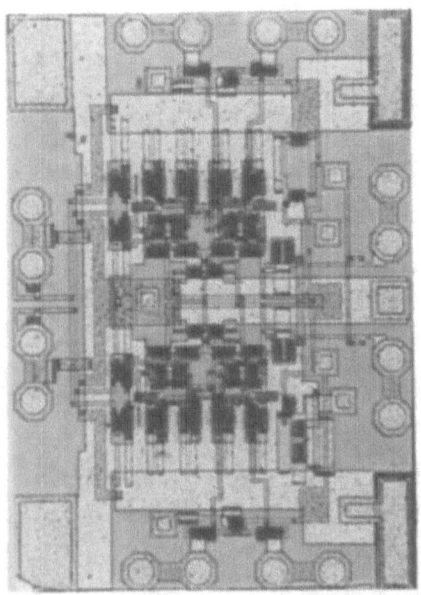

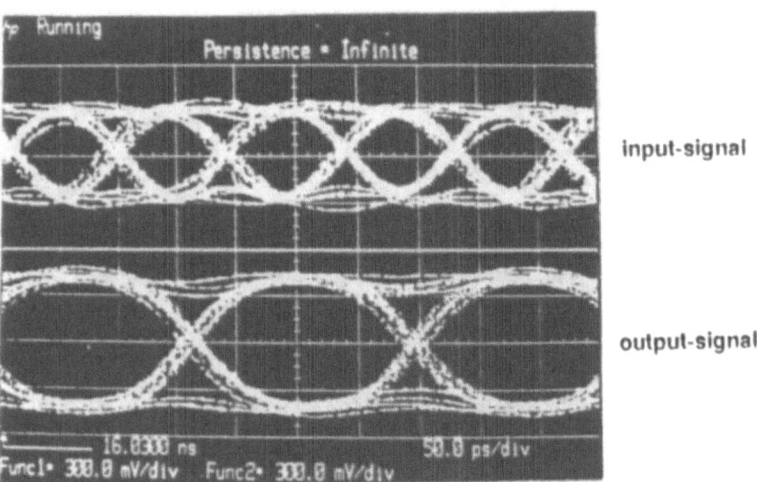

Figure 3. Chip-photo and eye-pattern of a 10 Gbit/s 1:2 DeMUX (cf (14))

6. DEMONSTRATORS

At Siemens so far emphasis was put mainly on basic circuitry and on the development of innovative circuit schemes like the "merged CMOS/bipolar logic" (MCSL) (16, 17). Using this circuit technique adder macros were realized (Fig. 4).

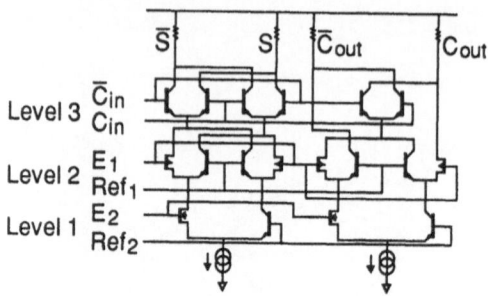

Figure 4. Adder cell in MSCL technique.

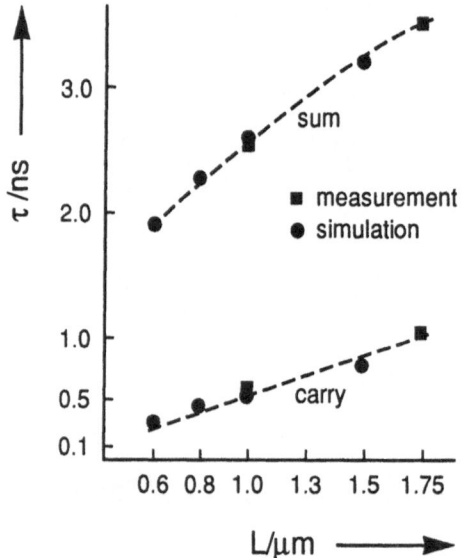

Figure 5. Carry and sum delay times vs. minimale gate length.

The inspection performance data is sketched in Fig. 5. Thus MSCL adders realized in 1.2 μm BICMOS have carry delay times of 100 ps and sum delay times of 700 ps. This is approximately a factor of 4 faster than pure CMOS.

The present memory circuit design activity is focussed on a 16 K SRAM which is intended to be used as a cache memory on ECL gate arrays (18). The address access time is approximately 4 ns (simulation). The block diagram and the performance data can be found in Fig. 6 and table II respectively. All main building blocks of the SRAM have been opimized by Monte Carlo methods developed and coded at the University Stuttgart (e.g. (19).

This activity is in its final state of design.

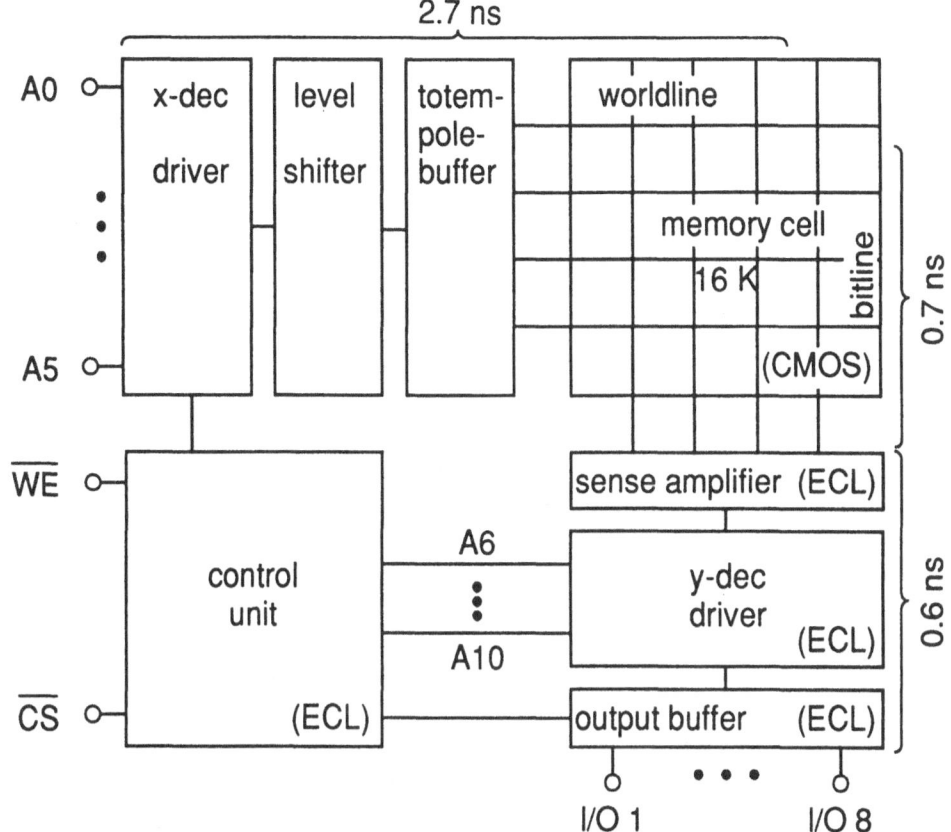

Figure 6. Block diagram of the 16 K SRAM.

Table II. BICMOS 2S 16k SRAM characteristics.

Organisation	2K word x 8 bit
Cell array	64 rows x 256 columns
Cell size	$432\ \mu^2$ (6 T)
Chip size	$24\ mm^2$
Address access time	4 ns (-4.5V, 60C)
Write pulse width	3 ns
Power dissipation	1.8 W
I/O level	ECL 100k
Sense amplifier voltage swing	50 mV
Package	64 pin Pinpack

The Philips demonstrator in the BICMOS 2P process is a combination of wide band, accurate analog circuitry and a VLSI signal processor. The analog circuits perform a double loop modulator with a sample rate of 432 MHz suited for A/D conversion of Video signals with 6.5 MHz bandwidth. The digital part, which consists of ECL circuitry running at 432 MHz and CMOS logic operating at 13.5 MHz data rate, has a complexity of some 200 k transistors and performs the decimation and low-pass filtering for the A/D conversion. The present status is that the blocs are being designed and first silicon is expected towards the end of 1989.

References

1. P.A.H. Hart and A.W. Wieder, Bipolar CMOS Esprit project, pp 95-108. Proceedings ESPRIT technical week 1988.

2. P.A.H. Hart and H. Klose, BICMOS versus CMOS and Bipolar p. 327-329. Proceedings ESPRIT technical week 1988.

3. P.A.H. Hart, A.W. Wieder, Bipolar CMOS project BICMOS 412. Proceedings ESPRIT Technical Week Brussels 1987.

4. P.A.H. Hart, K. Bürker and A.W. Wieder, BICMOS 412, Proceedings ESPRIT Technical Week Brussels 1986.

5. F. Rausch, H. Lindeman, W.J.M.J. Josquin, D. de Lang and P.J.W. Jochems, An analog BIMOS technology. Extended abstracts of the 18th Conference on Solid State Devices and materials. Tokyo 1986, pp. 65-68.

6. D. de Lang and W.J.M.J. Josquin, Optimization of a 1.5 m BICMOS process. BICMOS Symposion, Abstract no. 275. Electrochemical Soc. Philadelphia May 1987.

7. W. Josquin, D. de Lang, M. van Iersel, J. van Dijk, A. v.d. Goor and E. Bladt, The integration of double-polysilicon NPN transistors in an analog BICMOS process. Extended Abstracts of the 1988 SSDM, Tokyo. Paper no. 7, session A9.

8. H. Klose, T. Meister, B. Hoffmann, H. Kabza, J. Weng, Well-optimization for high speed BICMOS Technologies, ESSDERC, 1988.

9. D. de Lang, E. Bladt, A. v.d. Goor, W. Josquin, Integration of vertical pnp's in a double-polysilicon BICMOS process. Submitted to IEEE Bipolar conference, Sept. 1989.

10. Y. Kobayashi, C. Yamaguchi, Y. Amenuya and T. Sakai, High performance LSI process technology: SST CBI-CMOS p. 760-763, Proceedings IEDM, San Francisco 1988.

11. B. Hoffmann, H. Klose, T. Meister, A 10 GHz High Performance BICMOS Technology for Mixed CMOS/ECL ICs submitted to ESSDERC 1989.

12. H. Klose, T. Meister, B. Hoffmann, I. Kerner, P. Weger, 10 GHz BICMOS process for high performance CMOS/ECL integrated circuits, Submitted to IEEE Electron Devices Letters.

13. H. Klose, T. Meister, B. Hoffmann, B. Pfäffel, P. Weger, Low cost and high performance BICMOS processes : A comparison, Submitted to BCT Meeting 1989.

14. J. Hauenschild, H.M. Rein, P. Weger, H. Klose, A 10 Gbit/s monolithic integrated bipolar demultiplexer for optical fiber transmission systems fabricated in BICMOS Technology", submitted to Electronics Letters.

15. T. Chin, Non overlapping super-self-aligned BICMOS with 87 ps low power ECL, Proceedings IEDM, San Francisco 1988, pp. 752-755.

16. B. Zehner, H. Klose, D. Feige, A. Wieder, BICMOS, a technology for high-speed/high-density IC's, Siemens Forsch.- und Entwicklungsberichte, Ed. 17 (1988), Nr. 6, Springer Verlag pp. 178-283.

17. W. Heimsch, B. Hoffmann, R. Krebs, E. Müllner, B. Pfäffel, K. Ziemann, Merged CMOS/Bipolar Current Switch Logic, Proceedings of the ISSCC 1989, pp. 112-113.

18. W. Heimsch, R. Krebs, K. Ziemann, A 4ns 16k BICMOS SRAM, Symposium VLSI circuits, Kyoto, 1989.

19. W. Heimsch, R. Krebs, K. Ziemann, D. Moebus, Comparing CMOS and BICMOS NOR decoder structures using Monte Carlo optimization, Presented at ESSDERC 1988.

20. P. Nuijten and K. Hart, A digitally controlled twenty bit dynamic range BICMOS stereo-audio processor. Digest ISSCC Feb. 1989.

Project No. 802

The Layout Automation Tools in the CVS IC Design System

R. Airiau *, G. Arato **, J.M. Berge *, G. Bussolino **,
A.M. Fiammengo **, R. Manione **, V. Olive *, D. Rouquier *

* CENTRE NATIONAL D'ETUDES DES TELECOMMUNICATIONS
Chemin du vieux chêne. BP 98
38243 Meylan FRANCE +33 76.76.43.35
** CENTRO STUDI E LABORATORI TELECOMUNICAZIONI
Via Reiss Romoli, 274,
10148 Torino ITALY

ABSTRACT.

Layout automation of VLSI "custom cell-based" chips consists of two main tasks : the generation of necessary layout modules (cells) in a given target technology and the cells placement & interconnections routing. In the CVS system, developed whitin the ESPRIT 802 project, these tasks are carried out by the two tools illustrated in this paper : the Flexible Block Library (FBL) and the Floor-planner (ACCORDO). ACCORDO is presented in the first part of the paper, with some detail concerning the design flow of the floor-plan generation and the strategy adopted to solve each step in this flow. The FBL approach to guarantee technology independence, flexibility and maintainability in module generation is then discussed; significant examples are given for both the tools.

1 Introduction

The generation of efficient and technology independent layouts at chip level is one of the main targets of the CVS design system.

To reach the above ambitious goals an innovative system architecture and a set of specific tools has been developed in order to generate the layout of integrated circuits starting from their behavioural specification.

The present paper deals with the *layout generation end* of the design system whose two main components are the **Library of Process Independent Module Generators (FBL)** and the **Floor Planner (ACCORDO)** which exploits the results of the behavioural synthesis algorithms (see [1], [2]).

The ACCORDO Floor Planner is a layout automation tool that takes as input the Logic Hierarchical Description of an IC, obtains the desired Modules from the FBL and places & routes them in order to deliver the whole layout. User friendly interaction for Hierarchy Editing, Module parameters Specification, Slicing Placement Editing are among

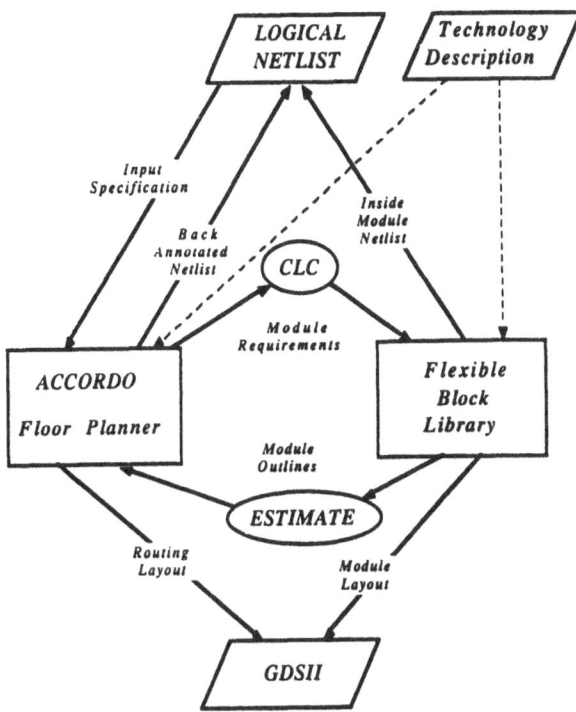

Figure 1: The CVS Physical Design Environment

the most innovative Floor Planning concepts implemented in ACCORDO (paragraph 3 will go into deeper details).

The Flexible Block Library (FBL) is in charge of supplying the layouts of the functional blocks according to parameters specified by the synthesis process and to the given target technology; this task is accomplished by a family of Module Generators implemented in the LOF4 environment, dealt with in paragraph 4.

2   The CVS Physical Design Environment

The Architecture of the Physical Design Environment of the CVS system can be sketched as in Figure 1 where the main input is the Logical Hierarchical Netlist (LHN) coming from the behavioural syntesis part.

Some remarks can be done upon the figure:

- during the behavioural synthesis process an interaction with the FBL (through the same CLC-Estimate protocol) already occurs in order to choose the best *module assignment* according to the user requirements of area and speed;

- ACCORDO enables the user to make further choices among the modules in the FBL as far as blocks with *geometric degrees-of-freedom* are used;

- The Chip Level Layout is generated by ACCORDO which delivers all the wiring geometries together with the instances of the Module Cells whose layout come from the FBL.

- Chip Level Verification of the Layout vs. Logical Netlist is possible thanks to the Back-Annotations of the *Inside-Module Sub-Netlists* from the FBL and of the wiring capacitances in the *Top Level Netlist* from ACCORDO

- Technology independence for both Wiring and Modules is obtained keeping the technology rules and parameters in a file.

3 The ACCORDO Floor Planner

The most suitable layout style for the implementation of **VLSI ASIC** chips is the well known (Macro)Cell methodology: each layout block is taken from a set of pre-designed (and characterized) cells, either fixed or procedural.

In the Macrocell methodology the entire circuit is partitioned into smaller (and homogeneous) units, each of them available from the Library of Module Generators; the so obtained units will further be Placed and Routed together with the I/O Pad ring in order to obtain the layout of the entire CHIP (see Figure 4).

The highly assisted interaction available in ACCORDO makes the user able to try different layout styles and module shapes starting from the same logic specification, even with incomplete designs (*Early FloorPlanning*).

The circuit partitioning (specified by means of the hierarchy) in the logic netlist taken from ACCORDO, due to its origin, obeys to behavioural synthesis criteria; in many cases this partitioning needs to be changed in order to meet layout requirements like minimum area, high speed, availability of a module generator capable of generating the given Macrocell, Power & GND considerations etc...; a Hierarchy Editor is avaliable that allows for such changes in the input Netlist partitioning.

Once the Macrocell layouts are obtained according to the desired parameters, they are Placed in a *generalized slicing* style; the slicing assumption is not a heavy constraint since any combination of Cut orientations is allowed; furthermore a Slicing-oriented Placement Editor is available to make local changes.

Signal nets are routed by an optimized Channel Router that always guarantees 100 % of routing; the output of the Topological Router is then compacted by a Contour Post Processor able to reduce the channel width of up to 30 %.

Power & Ground nets are propagated in a *planar way* according to a generalized comb fashion.

The Human Interface of ACCORDO is able to display a graphic representation of all the *cell views* during each phase of the process and to handle the user interaction in a unified way; the unified underlying graphic data structure is particularly suited for quick interaction even with very large amounts of data like the layout of an entire chip.

3.1 THE ACCORDO DESIGN FLOW

Figure 2 shows the ACCORDO design flow; on the left part of the picture the logic sequence of design phases is reported while on the right part, the main informations used/generated during the corresponding phase are depicted.

The main phases of the ACCORDO design flow can be so explained:

1. Logical FloorPlanning: in this phase the LHN of the input circuit is read and the Abstract Views (Physical Dimensions and Pin Positions) of the *Leaf Cells* are loaded.

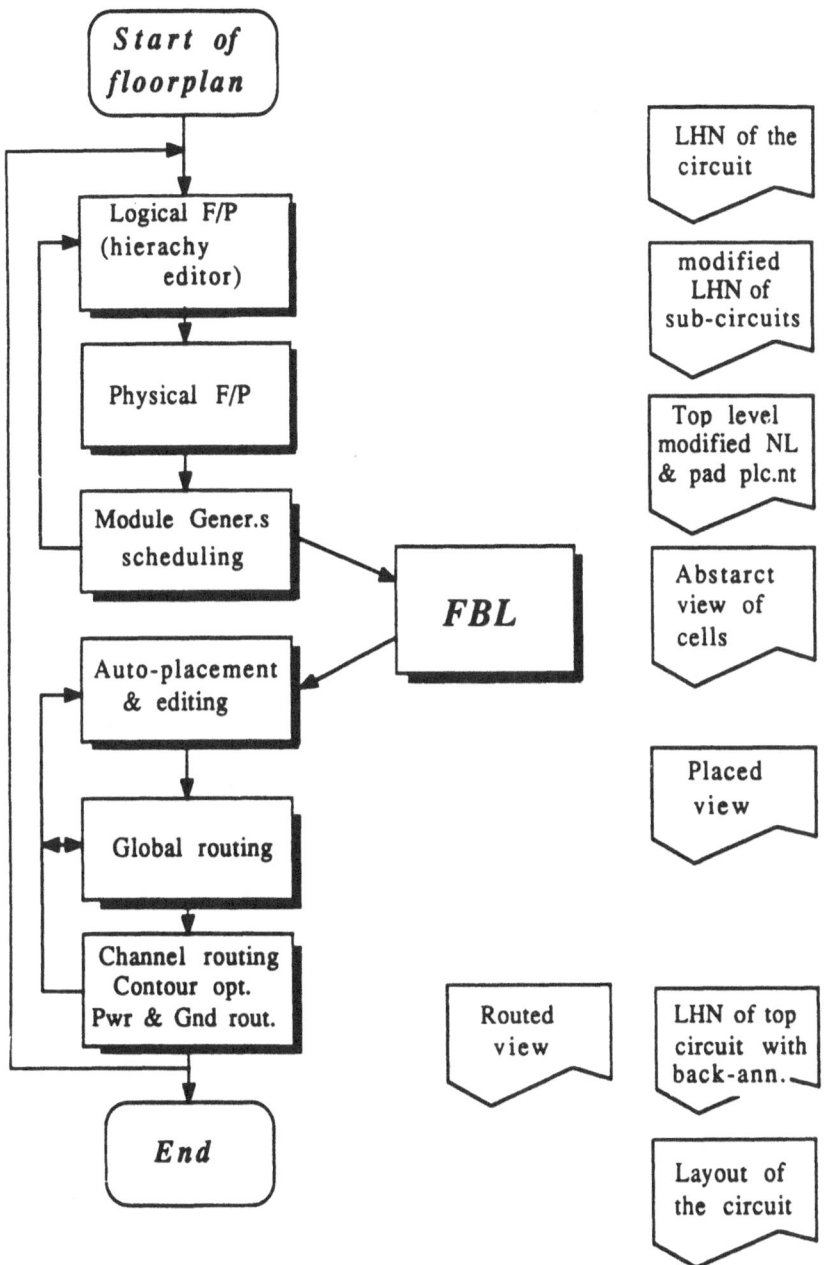

Figure 2: The ACCORDO Design Flow

(Leaf Cells are all the cells available from the Library that don't need a netlist to be specified; e.g. ROMs, RAMs, PLAs, ...)

One of the targets of the Logical FloorPlanning is to estimate the whole chip area as function of the input hierarchy; a graphical interpretation of these data is then presented on the screen in the style of *enclosing boxes* ; the *area overhead* due to the subcells routing is also estimated and shown on the screen.

This is the first time in the global IC design flow that the user has an almost precise idea of the real dimensions of the functional blocks instantiated into his circuit.

The above estimate can show that the input partitioning is not suitable in order to achieve a "good" layout, as already pointed out; in order to obtain a better "layoutable" hierarchy, some changes can be done in ACCORDO through the Hierarchy Editor.

Since The ACCORDO Place & Route phase will take the circuit Top Level Netlist, the heuristic that should guide the user during the Hierarchy Editing is that the Top Level SubCircuits should be homogeneous and have comparable area and "good" aspect ratios.

The most significant functions available in the Hierarchy Editor are: MAKE CELL that creates a new sub netlist that contains all the desired blocks; EXPLODE CELL that smashes a sub netlist and inserts its old components at the next upper level.

Before leaving the Logic FlorPlanning phase, the new circuit hierarchy can be saved (still in the LHN notation); no more changes unless backtracking are foreseen in the ACCORDO Design Flow.

2. Physical FloorPlanning: in this phase the alternatives in the geometrical parameters (if any) of the Leaf Cells at Top Level can be tried.

   A tight interaction with the FBL is performed; among the geometrical parameters that could be changed are: Area, Aspect Ratio, Port Positions, etc...

3. Module Generators Scheduling: in this phase, for each Top Level Subcircuit (either Leaf or Hierarchical) an Options File (CLC) is generated (options come from the possible topological/geometrical parameters); the syntax of the Options File is the one required by the specific Module Generator (FBL).

   In case of the Subcircuit represents an Hierarchical Module *(see [1])* (e.g. a standard cells block) its Flat Netlist and I/O port positions are also written.

4. Module Generators Run: in this phase, for each Top Level Subcircuit the proper (FBL) Module Generator is invoked and the corresponding MacroCell layout and Abstract View are read back.

5. Automatic Placement: starting from the Top Level Netlist, the Pads (or the Top Level Ports) Placement and the Abstract Views of all the Cells, a Slicing placement is built with a modified min-cut algorithm which tries to minimize the total net length.

   The Slicing Placement can be represented as a binary tree where each leaf node represents a cell an each internal node represent a channel that joins the two son cells; the cluster formed by the channel and the two associated leaf cells is grouped into a new cell (in the following named a Slice).

   The routing channels have their width automatically estimated according to the number of pins across the channel sides and the distance of the channel from the center of the chip.

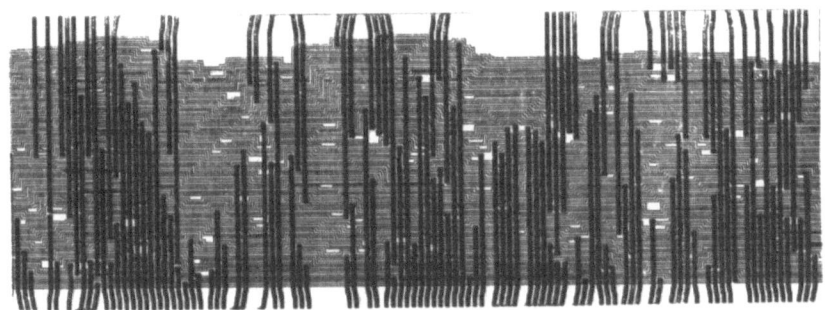

Figure 3: A Contour Router Example

6. Placement Editor: after the Automatic Placement step, some local adjustments can be done; in particular some cells or Slices *(see [1])* can be Rotated to best fit the available space or Mirrored to improve the routability.

   Cells can also be moved having the Slicing tree automatically updated; constraints like block alignment to a specific side or relative positioning of contiguous cells are handled as well; furthermore the channel width estimates can be modified.

7. Global Routing: starting from the final placement, the wiring geometries have to be assigned to the reserved area. In the case of the Macrocell design style the routing area is highly non regular and can be partitioned into regions in a number of different ways. Particularly useful for the routing purposes is the above defined Slicing Placements of the blocks. The Slicing placement leads to a partitioning of the routing area into regions that can be routed by a Channel Router (hence called Channels).

   The global routing purose is, trough a number of iterations, to establish which channels each interconnection will go through minimizing the net length and trying to obtain the desired withs for each channel. At the end of this step a net to channel assignment is available such as each routing channel has an associated list of incident nets. The global routing algorithm also takes into account the priority of the signal nets.

8. Channel + Power & Ground Routing: For each channel in the Slicing Tree the Channel Router and the Contour Optimizer are invoked. The scheduling of the channels follows the visiting order of the slicing tree in a depth-first way.

   The ACCORDO layering policy for the routing geometries specifies Metal_1 as the privileged Horizontal layer and Metal_2 as the privileged Vertical layer; this implies that horizonatal segments will run mostly on Metal_1 while vertical ones mostly on Metal_2.

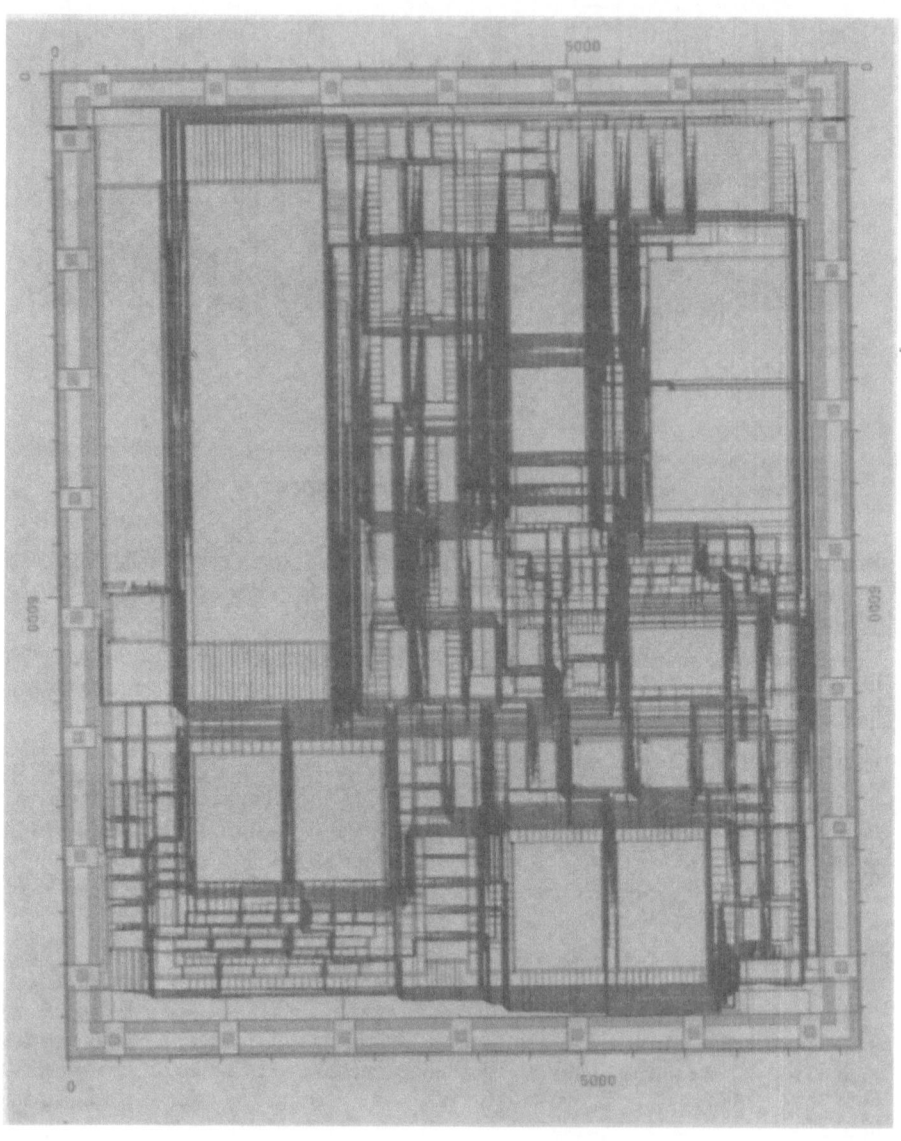

Figure 4: The whole layout of VITERBI

This convention is followed throughout all the ACCORDO layout, even in the routing of Power & Ground nets.

Figure 3 shows a part of a channel solved with the Contour Router.

When the routing process is terminated the layout of all the leaf cells and Macrocells is loaded and the layout of the entire chip can be generated.

After the layout generation, a delay back-annotation can be done writing the whole circuit LHN with delay times attached to the wires.

3.2 A DESIGN EXAMPLE

Some test circuits have been tried within the CVS system; a significant example is the VITERBI circuit, one of the official demo chips of the CVS project; Figure 4 shows the whole layout of an intermediate implementation of VITERBI.

4 Independence and Maintainability in Module Generation

Dependence on technology (or foundry), CAD environment (a given simulator or editor) or framework has to be overcome. Due to its cost, the life time of a complete library has to be compared to the average life time of CAD tools...

Maintainability is not the smallest problem in modern libraries with numerous complex generators (like RAM, PLA, ROM, Data-path, Control Unit, ...). In terms of the number of code lines to be maintained, the amount of software specific to the content of the library is greater than that of the tool itself. Maintainability thus becomes the complex problem that is so well-known in large software projects.

The tool LOF4, developed in CNET (French Telecommunications research center), handles these two aspects of silicon compilation.

4.1 LOF4 FEATURES

LOF4 is an independent module generation tool dealing with layout and logic descriptions of blocks. A set of primitives, including procedural description of layout, abutment operations, channel and "swith-box" routing, etc, is provided. Other primitives concern netlist description, graphic environment and interface to standard formats.

The input and output of this software are performed through widely used formats. LOF4 is intended for use in conjunction with any CAD system offering "classical" functions (schematic entry, layout editing, ...) but lacking the "silicon compilation" aspect.

Lof4 primitives allow :

- reading fixed "leaf-cells" using the GDS-II© and HILO© formats ;

- building leaf-cells by "procedural layout" techniques, using a technology file ;

- building the layout of regular or pseudo-regular blocks by procedural abutments and wirings ; cell placement operations are directed by the cell envelopes ; wiring operations are controlled by the port positions and by a technology file ; thus the building algorithm may easily be made completely technology-independent ; a channel router, switch-box router and river router are included ;

- building the netlist of blocks, either by extraction or by an independent algorithm ;

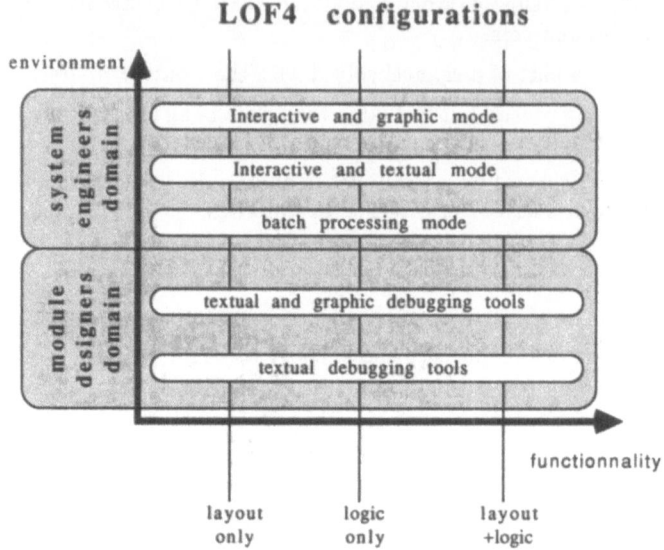

Figure 5:

- displaying the results through a read-only graphics editor adapted to GPX© workstations ;

- writing the resulting blocks (GDSII and HILO) ;

4.2 INDEPENDENCE

An independence strategy has been developed along three axes :

- Technological independence : the independence on technological updates due to scientific progress or foundry changes is given by the use of high level primitives (cell placement, routing, wiring, envelope and symbolic port management) which use technology-files for each piece of technology-dependent information. This independence covers usual technological updates (2.0 micron technology moving to 1.0 micron technology for example), it does not deal with major technological breakthrough (like new interconnection layers for example) which dramatically modify the entire design strategy of a module (in this example, "under-bus" design becomes possible).

- Software independence : LOF4 is a chain link which can be associated with other links from miscellaneous vendors (editors, simulators, data-bases,...) in order to constitute an efficient CAD system. Each of these links can, in an ideal situation, be the best of their category at a given date. To facilitate this linkage, interfaces to international standard formats (GDSII© HILO© , etc...) have been developed or are under study (EDIF, VHDL). An interface to the COSMIC data-base of CNET already exists. From the implementer point of view, this

ADA and VHDL

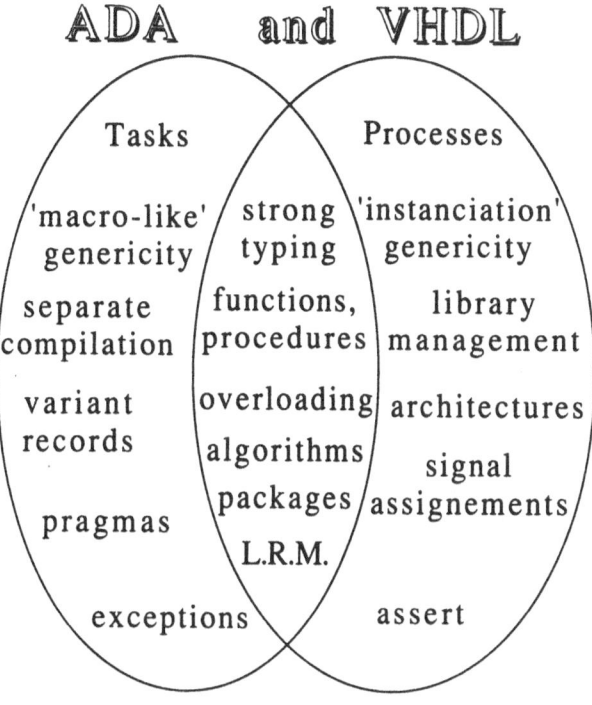

Figure 6:

software independence allows him to choose his development environment (any Ada programming environment fits). For the user, this characteristic means that he can freely change his CAD environment without losing his library. The library is Ada code and will be usable in the new environments. (This is not the case if the module had been written in a specific language proprietary of a given CAD vendor).

- Hardware independence : LOF4 has been written in Ada. This language allows a really efficient portability. The user graphic interface (GIX), based on the XWINDOW standard, can be easily separated from the LOF4 kernel, in order to use the library with classical textual terminals. Only the writing of new generators requires an Ada compiler licence (now available everywhere). This new generator will be easily portable (to others workstations for example) and its use does not require any Ada or LOF4 licences. A special effort has been made to provide a configuration adapted to each user's requirement. This feature will be shown later.

4.3 MAINTAINABILITY

As each generator is written in Ada, the LOF4 library has a good maintainability [4]. Considered by certain people as the "state of the art" in software engineering, this language, due to its numerous checks and high modularity, guarantees consistency and security of the system. These two characteristics are becoming increasingly necessary with the growing complexity of generators. This complexity implies a greater risk of software bugs appearing on layout. To make debugging easier, specific tools (graphic or textual) have been implemented, such as LOF4 basic primitives (browser, read-only editor,etc.).

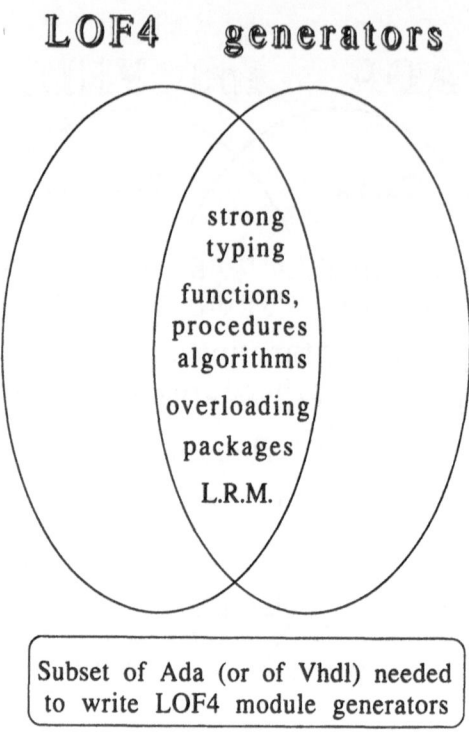

Figure 7:

4.4 CONFIGURABILITY

Another important advantage of the LOF4 system is its high modularity. Figure 5 shows the configurability of LOF4 to user requirements, depending on the desired functionality and target environment.

The two lower modes (module designers' domain) require the use of an Ada compiler. There is a strict dividing line between the configurations dedicated to system designers (where module design notions have been eliminated) and those dedicated to silicon designers. These people have to be familiar with a subset of Ada (specific Ada notions such as genericity or tasks are not necessary). When the silicon designer is getting ready to "speak" VHDL (behavioural description of his circuit), he will be pleasantly surprised by the ressemblance between it and Ada (as show in Figure 6).

VHDL and Ada are the only languages normalized before use. they share the same syntax and modern concepts such as strong typing, packages or overloading. However, there are many differences. For example, the concepts of time (and parallelism) in VHDL and Ada are quite different. Ada offers virtual parallelism in real time whereas VHDL offers real parallelism in virtual time. Nonetheless, the subset of Ada needed to write Lof4 modules is common with VHDL (Figure 7). Thus, an investment from a silicon designer to LOF4 will be useful to its knowledge of VHDL.

4.5 FEATURES

These features have been achieved using a 1.0 micron, two metal layer, N-well C-MOS technology (developed at CNET-Grenoble). Cycle time and other timing figures are taken from simulations ; they are TYPICAL figures ($25°C$, 5V). Length and height features correspond to the surrounding rectangle, and the actual area may be smaller.

4.5.1 Static RAM

The block is complete : address decoders and amplifiers, control and timing circuitry, three-state output buffers, power feed lines, ..., are included. The parameters (to be provided by the user) are :

- number of words (at least 4 ; maximum value : see max number of lines and columns , step 1) ;

- number of bits per word (at least 1 , max : 256 , step : 1) ;

- output bus load ;

- clock frequency (to derive width of power feeds, to eliminate solutions which are too slow) ;

features : (times in nS , L and H in microns)

Words	Bits	access time	read cycle time	L	H
64	4	7	21	463	184
		7,5	24	601	1190
		7,5	23	795	877
		8	25	1214	654
		8	25	1970	585
64	16	7,5	22	744	1843
		8	25	1172	1190
		8,5	25	1956	877
		8,5	26	3461	654
		9	29	6462	585
256	16	8,5	25	744	6182
		9	27	1172	3359
		8,5	26	1928	2006
		9,5	29	3432	1241
		9,5	31	6462	856
1024	32	10,5	34	3425	6345
		11	40	6427	3410
1024	64	11,5	42	6421	6345

4.5.2 ROM

The block is complete : address decoders and amplifiers, control and timing circuitry, three-state output buffers, ... , are included.

The parameters (to be provided by the user) are :

- number of words (at least 32 ; maximum value : see below) ;

- number of bits per word (at least 2, max : see below) ;

- file containing the "program" (textual format, information on request ; the file may be omitted to obtain an "empty" Rom) ;

- output bus load ;

- clock frequency (to derive width of power feeds, to eliminate solutions which are too slow) ;

Figures : (times in nS , L and H in microns)

Words	Bits	Lines	columns	access time	cycle time	L	H
64	4	16	16	5,2	25	410	540
256		16	8	512		4200	580
		16	256			2300	600
		32	128			1200	630
		64	64	7	25	760	800
512	8	8	512			4200	600
		16	256			2200	700
		32	128			1300	750
		64	64			800	900
		128	32			600	1200
		256	16	11,7	34,5	2160	600
4096	16	64	1024			8000	1100
		128	512			4200	1400
		256	256	12,5	38	2210	1980
		512	128			1300	3500
16384	16	256	1024	16,7	49	8000	2300
		512	512			4200	3800
		1024	256			2200	6700

4.5.3   Data-Path

features : The data-path generator is geared more towards Digital-Signal-Processing applications than CPU-like data-path generation. A classical "Bit-slice/Function slice" approach is used. A Data-Path is a collection of interconnected operators ; operators can be : Registers (numerous different types) ; Adders (normal or carry-look-ahead) ; Multiplexers ; Counters (up, down, up-down, reset, enable, load, ...) ; Bus-drivers ; ALU (complete set of operation) ;

input : the user gives a textual description containing :

- block name

- list of I/O (name, bit range, type, side)

- list of operators (name, type and subtype, bit range)

- list of connections

Each operator can have its own bit-width (example : two 16-bit registers feed an adder which in turn feeds a 17-bit register ...) ; each bit-slice can have its own connecting scheme (example : a 20-bit register receives its left bits from a 14-bit adder, and its right bits from some other oprerator) ; any "internal" point (such as "output not-Q of bit 12 of register A") can be routed to the periphery and used as a port of the data-path ; "shifting" connections (example : "output not-Q of bit 12 of register A" to "input D of bit 7 of register B") are allowed.

output :the layout is complete : decoders and amplifiers for control signal , power lines , ... , are included ; an optimal ordering of operators within the data-path can be computed by the data-path generator (minimize area / minimize total wire length), or the user may enforce a given ordering ;

figures :The following figures are taken from a test vehicle currently being fabricated. bit width : 32 ;

operator list : 3 registers (load-1 ; load-2 ; specific Reset pattern), 1 Inverter, 1 Nor, 1 Nand, 1 Xor 1 Alu, 1 Register (load-1, load-2, load-3, reset pattern, Q and NQ outputs), 1 Adder (normal), 1 Adder (carry look-ahead) , 1 Comparator, 1 Counter (up, reset pattern), 1 Counter (down), 1 counter (up-down), 1 Multiplexer, 1 Three-state Buffer, and 9 Switches.

Length : 2280
Height : 2475
Max. clock frequency : 38 MHz (because of the "normal" adder)

4.5.4 Glue Logic

The set of "standard cells" currently available is small (16 different types) but the place and route program is quite efficient : up to 1500 components in up to 20 rows ; ratio of total area / area of cells is around 1,8 .

user parameters :

- Hilo netlist

- Max length and/or max height and/or aspect ratio and/or number of rows

- I/O location (ordered list of ports for each border).

4.5.5 Delay Line

CNS has developed a digital delay line generator, for video applications. The parameters are :

- Length of the delay line, in number of clock cycles (between 1 and 2000)

- Number of bits to be shifted in parallel (between 1 and 12)

- Nominal clock frequency (up to 72MHz)

Several solutions, with different aspect ratios, are proposed by the generator, depending on the user parameters. features :

ncycle	nbits	freq	L	H
200	8	72	620	990
800	8	72	980	2580
1920	8	72	2650	2100
1920	12	72	3200	2760

At present a lot of LOF4 modules have now been developed. There are high flexibility modules like RAM, ROM, PLA, registers, multiplexers, adders, multipliers or analog blocks... but there are also large software programs such as irregular data-path generators, packaging tools or placement/routing for glue logic. In terms of number of code lines, the complexity of the set of modules is close to the complexity of the LOF4 tool itself (80 thousand lines), which nevertheless remains easily maintainable.

5   Two realistic examples of Ada source code

- This code builds a row of cells "RAAMLIAI", using the "He" function.

- Each instance is named RAAM#n (n varying from 2 to Nb_Raamliai).

- Each port VDDN is connected north to a wire of the same layer as the port, of computed length (Pas_max + Lmin are related to design rules and are read from a technology file), with the same width as the port (value -1, the value 0 could be used to specify the width to be minimal for a given layer/technology couple).

- At the end of this wire, a via is added if the original layer is Metal1.

- The resulting cell is stored in cell Core.

```
FOR I IN 2 .. Nb_Raamliai LOOP
    He(Core, Raamliai, "RAAM#" & Str(I));
    Pwire(Core, ( "RAAM#" & Str(I)) - "VDDN",
                    S(Undefined_Layer,Pas_Max + Lmin, North,  -1)
                  & S(Metal2, 0, North, Lmin)
          );
END LOOP;
```

- Depending on the value of Number Words Per line, a connection starts from the the port SAECRE or NSAECRE of cell Plan memoire to join the port EECRO of cell RAIMCOMI. A first segment (toward east) in Metal2 is drawn with a computed length (Ecart) and a minimal width (read in a technology file).

- The final port (EECRO) is then joined by two segment in metal2 layer, the first one being vertical. This wire is store in the cell Plan memoire.

```
Ecart := Ecart + Pas_Max;
IF Number_Words_Per_Line = 1 THEN
 Wire ( Plan_Memoire,                        -- receiving cell
          "RAALOCT" - "SAECRE",              -- starting point
          "RAIMCOMI" - "EECRO",              -- end point
          Vertical_First,                    -- for the last 2 segments
          S(Metal2,Ecart,East,metal2_min) -- first segment
        ) ;
ELSE
 Wire ( Plan_Memoire, "RAALOCT"-"NSAECRE", "RAIMCOMI"-"EECRO",
Vertical_First,
         S(Metal2, Ecart, East,  metal2_min));
 END IF;
```

6 Conclusions

The CVS Layout Automation subsystem has been discussed in the two complementary aspects of FloorPlanning and Library of Module Generators.

The main points of strength of the system are the ability to work on the circuit partitioning in order to make it efficiently *layoutable* , the availability of efficient Place&Route algorithms and a user friendly graphics interface which allows for manual changes in a number of checkpoints along the design flow; on the other hand the availability of a tight coupled Library of Parametric, Design Rules Independent and Characterized Module Generators gives the possibility to experiment with different design alternatives and technologies.

7 References

1. G. Arato, G. Bussolino, R. Manione: "ACCORDO. 2'nd Generation Floor Planning." CVS Rep. Sept. 1988.

2. G. Arato, G. Bussolino, A. Fiammengo, R. Manione: "The FloorPlanning Design Flow in ACCORDO." CVS Rep. Sept. 1988.

3. C. Jullien, A. Leblond and J.Lecourvoisier: "A database interface for an integrated CAD system" proceedings DAC,86 Las Vegas-Nevada, June 29-July 2, 1986 pp760-767

4. M. Loughzail, M. Côtè M. Aboulhamid and E. Cerny: "Experience with the VHDL Environment" proceedings DAC,88 Las Vegas-Nevada, June 29-July 2, 1986 pp760-767

5. Grady Booch: "Software engineering with Ada" (2th edition) Benjamin / Cumings publishing Company inc.

6. Norman H. Cohen: "Ada as a second language" McGRAW-HILL series in sofware technology

Project No. 824

RECONFIGURATION IN A MICROPROCESSOR: PRACTICAL RESULTS

R. Leveugle, M. Soueidan, G. Saucier
Institut National Polytechnique
de Grenoble / CSI
46, avenue Félix Viallet
38031 GRENOBLE CEDEX
FRANCE

N. Wehn, M. Glesner
Technische Hochschule
Darmstadt
Schloßgartenstrasse, 8
6100 DARMSTADT
F.R.G.

J. Trilhe
SGS-Thomson Microelectronics
Avenue des Martyrs
38019 GRENOBLE
FRANCE

ABSTRACT. This paper investigates microprocessor design techniques using redundancy in order to increase the yield by means of end-of-manufacturing defect tolerance. A lot of effort was done to obtain a regular design allowing the introduction of standby elements at an adequate level. The HYETI microprocessor chip is being manufactured within the ESPRIT 824 project (task C) and this paper reports on the practical results on silicon. In the HYETI chip, the area of the redundant elements was limited to less than 25% and a good compromise between area overhead and yield enhancement was achieved.

1. Introduction

The goal of the HYETI demonstrator of the ESPRIT 824 project (WSI) was to prove the feasibility of a high yield defect tolerant microprocessor. Such a device is supposed to become the core of an application specific microprocessor-based system integrated on a single chip. Very few studies deal with end-of-manufacturing defect tolerance for non regular circuits, namely microprocessors. The implementation of a correlator with redundancy was reported in [Teep 87] but only the reconfiguration of accumulators was considered and the circuit was realized in a bipolar technology. The introduction of redundancy to non regular CMOS circuits was considered in [Sume 86]; some figures were provided about area and yield but no practical information was given about the reconfiguration methodology. In fact, no paper was available in the literature, reporting the possibility to successfully use redundancy in a complex processing element in order to cope with some manufacturing defects.

In the ESPRIT 824 project, microprocessor design techniques using redundancy were investigated. A lot of effort was done to obtain a regular design allowing the introduction of standby elements at an adequate level. The reconfiguration strategy allows the tolerance of a large number of defects in order to increase the yield and to be able to use the processor in a WSI system. Nevertheless, all defects are not necessarily tolerated. The interconnection lines between the controller and the datapath are for example not reconfigurable. Some elements in the datapath are also not reconfigured because this would require a large area overhead. The area of the redundant elements was limited to less than 25% and a good compromise between area overhead and yield enhancement was achieved.

In this paper, we focus on the practical aspects of the HYETI implementation. The main characteristics of the demonstrator are given in section 2. Then, the design methodology is investigated. The design techniques used in the datapath and in the controller are described in section 3. Section 4 reports on the complete chip. In section 5, the main steps of the end-of-manufacturing reconfiguration are discussed. Finally, section 6 reports on the test results of the demonstrator.

2. General Presentation of the Demonstrator

HYETI is a 16-bit microprocessor with an addressing capability of 16 pages of 64 K words. The memory interface signals are compatible with those of Motorola family. Two hardware interrupt levels are provided, as well as a halt control signal. The circuit is implemented in the CMOS double metal 1.2μ technology of SGS-Thomson Microelectronics. The HYETI microprocessor is dedicated to highly dependable real time control systems and its characteristics have been determined in cooperation with industrial potential customers. Special emphasis has been given to on-line test. A compaction device allows to compact any information on the bus through a multiple-input LFSR with a programmable characteristic polynomial. The general architecture is shown in Figure 1.

The main part of the instruction set is quite classical (arithmetic and logical operations, transfer instructions, stack and interrupt instructions, ...). Masked instructions are also provided. These instructions allow the programmer to process selected bits in a 16-bit word. Finally, special instructions are implemented for on-line test purposes. Eight addressing modes are provided: relative, inherent, immediate, direct, register indirect, register indirect with post-modification and register indirect with pre-modification. For the two last modes, the indirection register is automatically modified by adding a programmable arithmetical offset. These modes were introduced for efficient matrix processing.

128

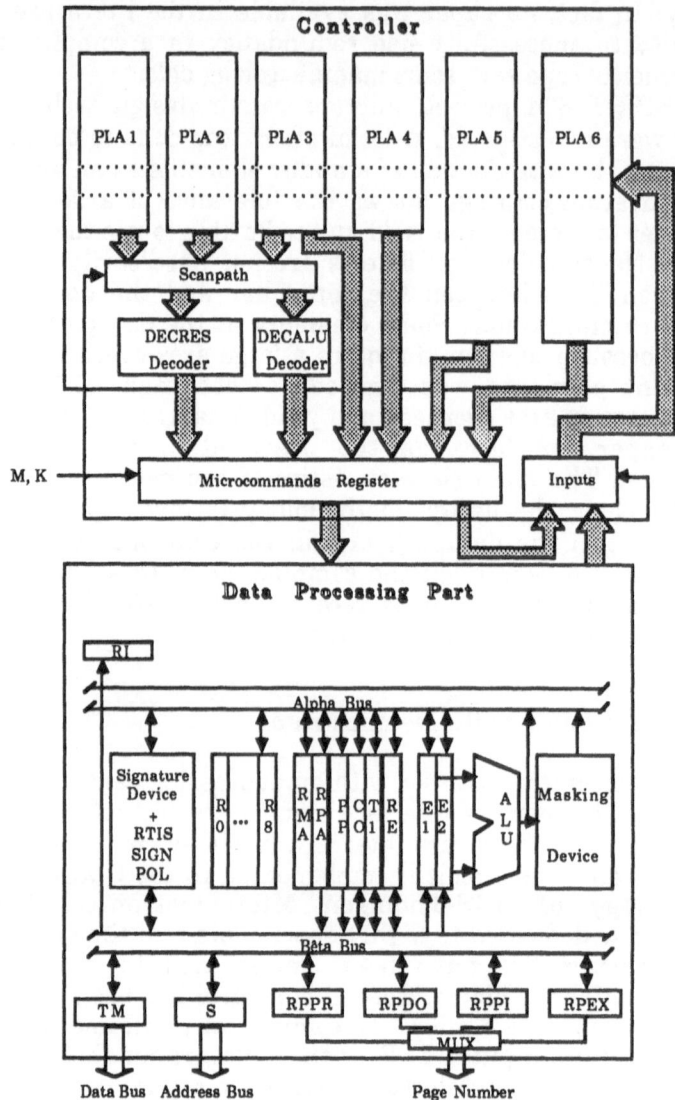

Figure 1: General architecture of the HYETI microprocessor

3. Design Methodology

As far as defect tolerance is concerned, the underlying strategy consists of
dividing the microprocessor into blocks where one defect can be corrected,
except some fatal ones on critical lines (for example, control signal lines,
which are not duplicated). This partitioning takes into account the general
architecture of the chip: the datapath and the controller are considered
separately. Furthermore, the definition of the blocks has to be closely

associated to the elements of the circuit. For example, it is impossible to consider that two PLAs are in the same block. They have obviously to be reconfigured separately. So, architectural and topological aspects of the design have to be considered during the partitioning.

Reconfiguration involves disconnection of defective elements and connection of spare ones. Several methods, software or hardware, can be used. Laser fuses and antifuses can be used when no reversibility is needed. Floating gate FETs can be used when several programming attempts are necessary. These three types of hardware switches were used in a first version of the microprocessor. Nevertheless, it is useful from an economical point of view to minimize the number of different switches, namely the number of steps and the number of machines necessary during the reconfiguration process. In consequence, the second prototype of HYETI only uses laser fuses, in spite of the difficult design problems which had to be overcome.

3.1. DATAPATH RECONFIGURATION

The partitioning of the datapath can be either functional or structural [Gene 86]. The functional partitioning consists in a separate reconfiguration of the different functional units. This partitioning may be necessary when there is no regular structure. The main drawback of this strategy is the complex connection network required between blocks in order to achieve the reconfiguration. In the HYETI datapath, we take advantage of the bit slice structure (structural partitioning). One spare slice is provided and can replace a defective one. The area of the datapath is about 7 mm^2 and this part of the circuit can be considered as only one block from the reconfiguration point of view.

A bit slice partitioning being chosen, two problems have to be overcome : the bus connections on the good slices and the bypass of the defective slice for the signals propagated between adjacent slices.

The first problem arises because some devices are not reconfigurable without a large area overhead. For example, in the case of the status register, each bit has a specific meaning and is associated with some random logic. The datapath is therefore composed of three parts, according to Figure 2. Depending on the slices used in the reconfigurable parts, the buses have to be reconfigured in order to correctly connect the non reconfigurable block. Furthermore, the bus structure is a complemented precharged and amplified one. Since two different buses are available in the datapath, four bus lines have to be reconfigured on each slice (two bus lines and the complementary lines).

The initial connections of the buses are made with laser fuses. After test, the reconfiguration consists in cutting some fuses and connecting new lines. These connections are made by means of transfer gates. A special reconfiguration signal Crec is generated for each slice. It indicates if the slice must be bypassed and its logic level is determined by laser fuses. On slice 0, the transfer gates are directly controlled by the Crec signal; on

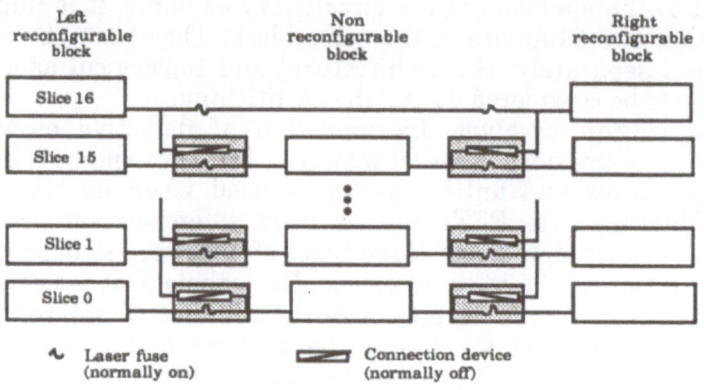

Figure 2: General floorplan of the datapath

the other slices (1 to 15), the transfer gates are controlled by a logic combination of the Crec signals: $C_i = Crec_i \cdot C_{i-1}$, $1 \leq i \leq 15$ with $C_0 = Crec_0$.

The reconfiguration of the signals propagating between slices is also made by means of transfer gates and controlled by the same signals Crec according to the scheme of Figure 3.

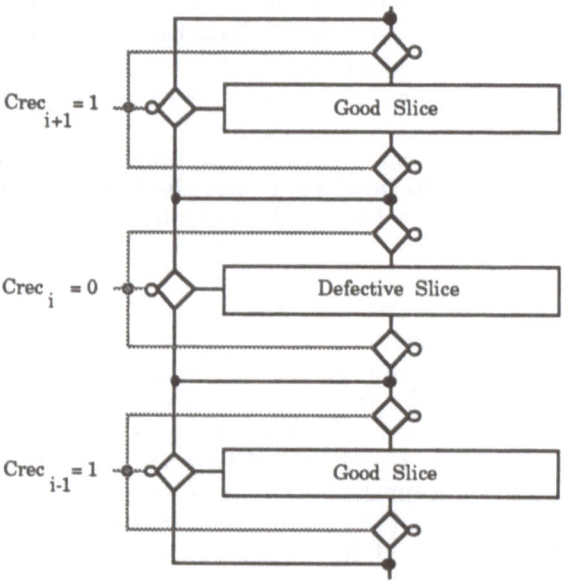

Figure 3: Reconfiguration of signals propagated between slices

A general study about defect-tolerant ALUs has also been carried out. Seven classical structures have been analyzed in order to establish if they are good candidates for integration in defect-tolerant processors. The four evaluation criteria were: complexity (area and number of transistors), performance, testability and reconfigurability. The iterative structure of an ALU was taken into account and the existence of a short pseudo-exhaustive functional test sequence (referred to as the C-testability property) was the testability requirement. The reconfigurability was evaluated according to the extra area (limited to 25%) and to the percentage of reconfigurable circuitry.

The "Carry Select", "Full Carry Look Ahead", "Ripple Within Groups, Look Ahead Between Groups" and " Look Ahead Within Groups, Ripple Between Groups" structures were discarded because of the area required for reconfiguration (more than 30%). The three other structures ("Ripple Carry", "Manchester" and "Carry-skip") were implemented. The comparative results for a reconfigurable 16-bit ALU with one spare slice are summarized in Table 1. A CMOS static ripple carry ALU was used in the present prototype of HYETI. The manchester ALU is faster with comparable area characteristics but the need of a precharge clock is a noticeable drawback. Finally, the carry-skip ALU is much faster but needs more area (34% more than the area of the ripple carry ALU). Furthermore, the test of the ALU is more complex and the non reconfigured part percentage is higher.

Table 1: comparative results for three ALU implementations.

Structure	Total area (mm2)	Maximal frequency (MHz)	Reconfigurability (% reconfigurable)
Ripple Carry	0.694	10	80
Manchester	0.687	15	80
Carry-skip	0.929	20	60

3.2. CONTROLLER RECONFIGURATION

A global lack of regularity can be noticed in the controller most commonly used structures. Several types of elements are necessary (ROM, PLAs, decoders and random logic) and the reconfiguration requires a lot of redundancy and several reconfiguration strategies. In the HYETI controller, only one type of element (PLAs) is used. An initial virtual PLA is partitioned into several ones, some of them implemented as decoders. A method has been defined in order to minimize the global size of the controller and the area needed for the interconnections [Leve 88]. The final controller is composed of six PLAs in the first stage and two decoders.

The initial PLA matrices, obtained after a complex synthesis process (optimized state assignment, minimization ...) with the ASYL system [Sauc 87], have been modified for test purpose in order to be able to localize a defect. The PLA testing strategy is based on the testing scheme presented by Bozorgui-Nesbat and McCluskey [Bozo 84] and improved later [Bozo 85], [Aart 86]. The basic idea of this scheme is the ability to activate an individual product term by adding some additional inputs to the AND-plane of the PLA. With this scheme, the test procedure requires only the application of some test vectors; no register is needed to control the activation of product or input lines. A more detailed presentation of the method can be found in [Wehn 88].

Each PLA is considered as a block from the reconfiguration point of view. Test inputs and redundant programmable product terms are provided. When taking into account the external interconnection overhead needed for redundant inputs and outputs, it is obvious that redundant product terms are a good compromise between area overhead and fault occurence. Input and output lines are designed on two layers with multiple connections. This technique ensures fault-tolerance with regards to open defects. On the other hand, decreased speed due to the enhanced capacitance must be accepted.

As mentioned earlier, the reconfiguration of the PLAs is made by means of spare programmable product terms which can replace the defective ones. The programmation uses laser fuses. A normal product term is implemented as shown in Figure 4a. A laser fuse allows to disconnect it from the OR array when it is proved to be defective. The programmable product terms are implemented as shown in Figure 4b. These product terms have, at the end of manufacturing, no influence since transistors are connected in the AND array on all the input and complemented input lines. Some of these transistors are disconnected from the AND array when the redundant product term has to be used. An example of reconfiguration is given in Figure 4c. The programmable product terms can also be disconnected from the OR array when necessary.

The main characteristics of the reconfigurable PLAs used in the HYETI controller are summarized in Table 2. The number of test inputs and programmable product terms, as well as the area overhead for reconfiguration, are listed in this table. Taking into account the routing overhead due to the test inputs of the PLAs, and even the scanpath register area, the complete area overhead in the controller is about 21%.

4. The Complete Chip

The chip sent to the foundry is shown in Figure 5. It incorporates about 50K transistors in a total area of 35 mm^2 (including the laser positioning patterns). The microcommands register can be loaded or read serially in a test mode. Furthermore, two scanpath registers are added onto the inputs of the controller and between the two stages of PLAs in the controller. The total silicon overhead, including both test and reconfiguration devices, can be evaluated about 8 mm^2.

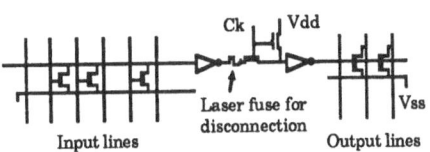

Figure 4a: Normal product term

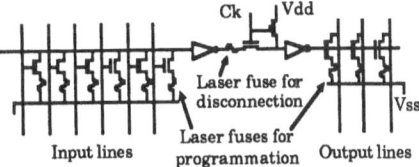

Figure 4b: Programmable product term

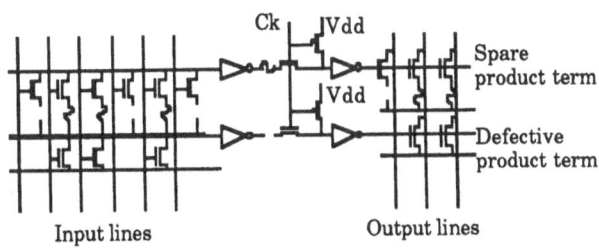

Figure 4c: Reconfiguration in a PLA

Table 2:
Characteristics of the reconfigurable PLAs used in the HYETI controller

PLA name	P1	P2	P3	P4	P5	P6	DECALU
No. inputs	15	15	12	26	26	26	4
No. outputs	7	5	29	23	3	3	10
No. product terms	96	94	101	108	166	178	13
No. test inputs	3	2	3	3	3	3	1
No. test product terms	1	1	2	1	2	2	1
No. prog. product terms	2	2	2	2	4	4	1
Complete size (mm ^{2})	0.704	0.662	0.976	1.206	1.380	1.502	0.226
Area overhead (%)	19	15	15	13	18	17	43

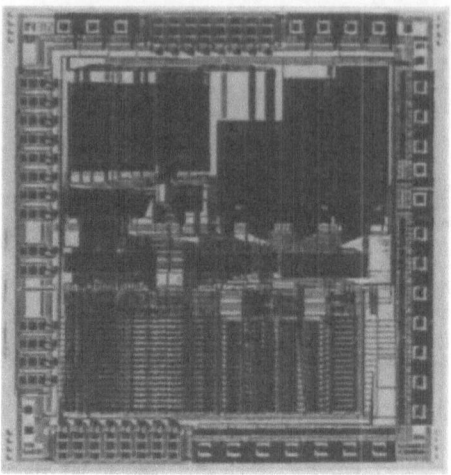

Figure 5: Layout of the HYETI chip

5. Off-line Test and Microprocessor Reconfiguration

At the end of manufacturing, a very efficient off-line test can be performed. This is due to the modifications introduced in the PLAs, to the special design of the ALU which has a C-testable structure and to the implementation of scanpath registers.

The test after manufacturing involves the following steps :
- Parametric test (general validation of the chip).
- Test of the three scanpath registers.
- Test of each PLA in the controller, using the scanpath registers.
- Test of the datapath elements, using the scanpath registers.
- Test of the whole microprocessor in working mode if all the other tests are successful.

When defects are detected in the reconfigurable parts of the processor, a reconfiguration phase takes place after this first test. In the different PLAs, the defective product terms are disconnected and replaced by spare ones which are programmed by laser cuts to implement the right function. In the datapath, the Crec signal of the defective slice D is set to 0 and all the fuses on the buses are cut on slices D ($0 \leq D \leq 15$) to 16. When no defect is detected or when the defective slice is the 17th one, only the fuses on this slice have to be cut.

When the reconfiguration is achieved, a second test is performed to verify the effectiveness of the corrections. Notice that multiple consecutive test and reconfiguration runs are not allowed. A defect can be repaired only if it can be localized during the first test with the general test patterns. This restriction is economically motivated, since test and reconfiguration of a chip is a time-consuming process.

6. Test Results

The HYETI microprocessor is now under test and the first results are very encouraging. At the moment, four main blocks have been validated: the scanpath registers, the PLAs, the static ripple-carry ALU and the signature device which has the basic structure of a Multiple-Input Linear Feedback Shift Register (MISR).

Figure 6 shows an example of activation of PLA P5 with a test pattern, using the scanpath registers. SCANIN is the serial input, T1 and T2 are the two non-overlapping clocks and the results for this particular test pattern are obtained on the two serial outputs SC65 and FETCH of the microcommands register. The SC65 output, which corresponds to the middle of the register, allows to greatly reduce the test time.

The delay measurement for DECALU is given in Figure 7. The photograph shows the rising time of an output during the evaluation phase (low level of the clock). Notice a non negligible part of the output capacitance was due to the test equipment itself. Therefore, the measured delay is an upper bound of the actual delay of the PLA in the chip.

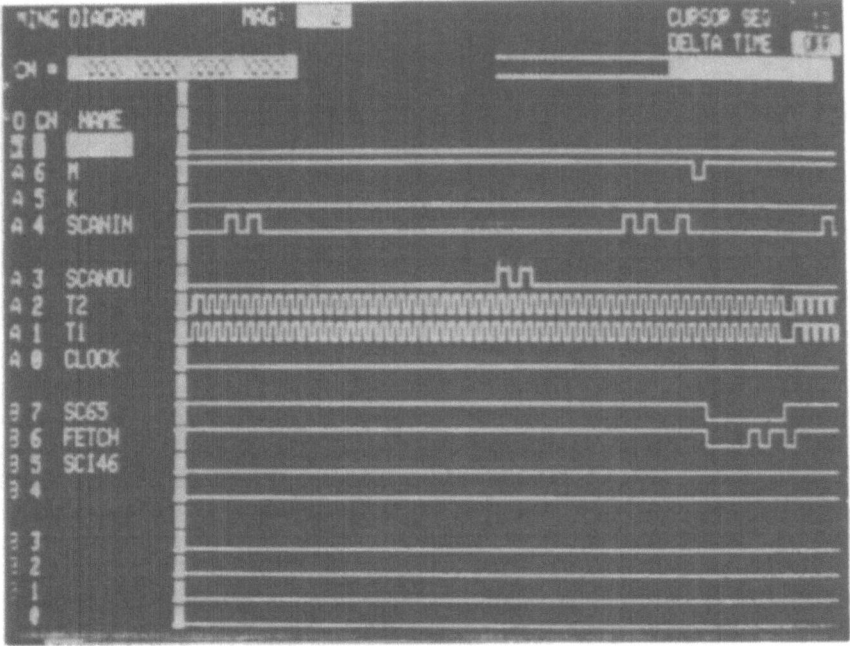

Figure 6: Example of activation of P5 using the scanpath registers

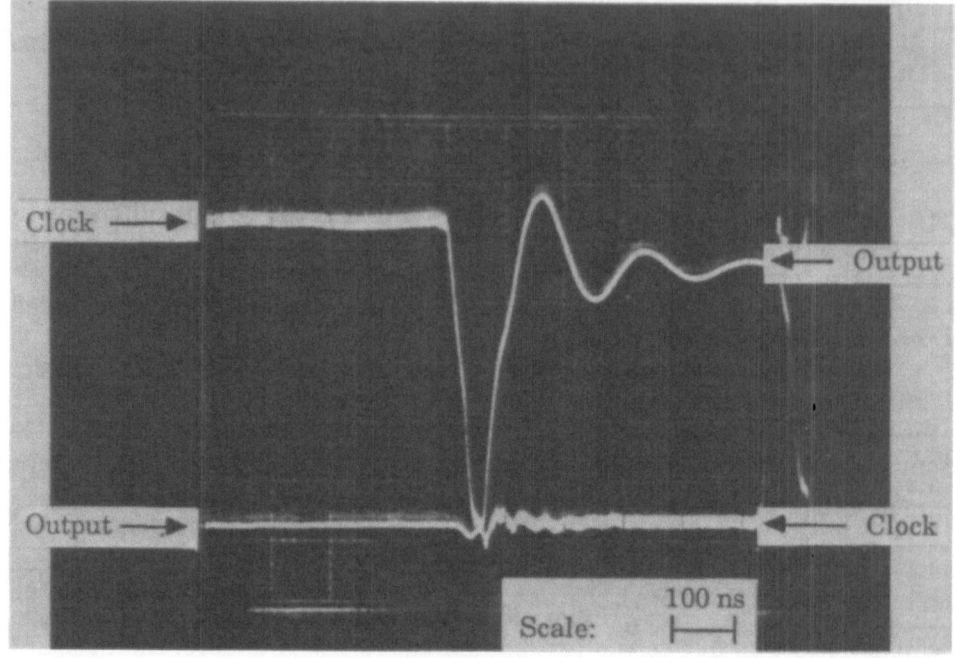

Figure 7: Delay measurement for DECALU

As already mentioned, two main blocks of the datapath have been validated. Experiences have also demonstrated the actual possibility of reconfiguration by using the spare slice.

The validation of the other blocks of the microprocessor is now going on. This validation phase will be followed by a quantitative evaluation of the effectiveness of the reconfiguration process. This evaluation will allow to obtain figures about the actual yield enhancement.

7. Conclusion

The techniques developped in the task C of the ESPRIT 824 project may be applied to any type of circuit with a bit slice datapath and a controller. The results now available on widely used blocks such as PLAs or ALUs are a major achievement in the end-of-manufacturing defect tolerance field. The method used in HYETI can be used to improve the productivity of VLSI circuits and may lead to the possible integration of whole systems on a single chip. Nevertheless, the necessary modifications of the circuitry have to be considered at an early stage of the design, especially for complex dedicated operators.

References

[Aart 86]   E. H. L. Aarts, F. P. M. Beenker, M. M. Ligthart, 'Design for testability of PLAs using statistical cooling', 23th Design Automation Conference, 1986

[Bozo 84]   S. Bozorgui-Nesbat, E. J. McCluskey, 'Lower overhead design for testability of PLAs', International Test Conference, 1984

[Bozo 85]   S. Bozorgui-Nesbat, J. Khakbaz, 'Minimizing extra hardware for fully testable PLA design', International Conference on Computer-Aided Design, 1985

[Gene 86]   P. Genestier, C. Jay, G. Saucier, 'A reconfigurable microprocessor for Wafer Scale Integration', in : "Wafer Scale Integration", G. Saucier and J. Trilhe, ed., Elsevier Science Publishers, Amsterdam, 1986

[Kuo 87]   S.-Y. Kuo, W. K. Fuchs, 'Fault diagnosis and spare allocation for yield enhancement in large reconfigurable PLAs', International Test Conference, 1987

[Leve 88]   R. Leveugle, M. Soueidan, 'Design of an application specific microprocessor', International Workshop on Logic and Architecture Synthesis for Silicon Compilers, Grenoble, France, May 1988

[Salu 83]   K. K. Saluja, K. Kinoshita, H. Fujiwara, 'An easily testable design of PLAs for multiple faults', IEEE trans. on Computers, vol. C-32, no. 11, November 1983

[Sauc 87]   G. Saucier, M. Crastes de Paulet, P. Sicard, 'ASYL: a rule-based system for controller synthesis', IEEE trans. on Computer-Aided Design, vol. CAD-6, no. 6, Novembre 1987

[Some 86]   F. Somenzi, S. Gai, 'Fault detection in Programmable Arrays', Proc. of the IEEE, vol. 74, no. 5, May 1986

[Sume 86]   G. W. Sumerling, G. E. Dixon, A. K. J. Stewart, 'An assessment of non-regular cell based architecture for ULSI and WSI', in : "Wafer Scale Integration", G. Saucier and J. Trilhe, ed., Elsevier Science Publishers, Amsterdam, 1986

[Teep 87]   G. H. Teepe, W. L. Engl, 'A bipolar correlator with redundancy', IEEE Journal of Solid-State Circuits, vol. SC-22, no. 6, December 1987

[Wehn 88]   N. Wehn, M. Glesner, K. Caesar, P. Mann, A. Roth, 'A defect-tolerant and fully testable PLA', 25th Design Automation Conference, 1988

[Wey 87]   C.-L. Wey, 'On the design of a redundant Programmable Logic Array', IEEE Journal of Solid-state Circuits, vol. SC-22, no. 1, 1987

[Wey 88]   C.-L. Wey, 'On yield consideration for the design of redundant Programmable Logic Arrays', IEEE trans. on Computer-Aided Design, vol. 7, no. 4, 1988

Hierarchical Design Rule Check for Full Custom Designs

W. MEIER
Siemens AG
Corporate Research and Development
Development Laboratory for
Process Technology
Otto-Hahn-Ring 6
D-8000 München 83

Abstract

In this paper we present the hierarchical design rule check programm HEXDRC, developed under the ESPRIT project AIDA. The main problem for a hierarchical layout verification tool is to handle interactions between different cells effectively. The kind of interactions however strongly depends on the actual design style, so that it will be impossible to find a solution which will work optimal for all layouts. To solve this problem, HEXDRC offers a lot of possibilities to modify the implemented basic algorithm by user instructions. We discuss the basic concept, describe how to adapt the program to a special design style and present first experimental results.

1. Introduction

The layout of an integrated circuit must obey the physical design rules required by the fabrication process. These rules specify the legal sizes of the layout structures, required spacings between different features and other constraints. If a layout does not fulfill these rules the produced chip will not function correctly. The task of a design rule check (DRC) program is to verify all these technology specific rules before starting the time consuming and expensive production of the masks for the chip fabrication.

Earlier design rule check programs always flattened out the chip layout ([3]). In this way the layout of a cell was rechecked for each instance of the cell. Errors found in a cell were reported many times without giving a hint which cell caused the errors. The amount of data to be checked grew linearly with the number of features in the fully instanciated layout.

For todays layouts, containing millions of transistors, it is impossible to verify a completely flattened chip in acceptable time. To overcome these problem hierarchical DRC programs have been developped during the last years ([1],[2],[4],[5],[6]). They try to take into account the original layout hierarchy and to verify the layout of each subcell, if possible, only once.

Hierarchical verification however is not an easy problem because the cells cannot be checked independently. The main problem is to handle interactions between different cells: Although a cell may be correct, a wire crossing an instance of that cell may cause a design rule violation. On the other hand a design rule violation found in a cell may disappear when the cell is considered together with its environment.

Existing hierarchical DRC programs mainly differ in handling interactions between cells. One solution is to look for instances of a cell with identical neighbourhoods and to check these situations only once ([1],[6]). This method needs a lot of bookkeeping and sometimes it is difficult to handle more than one level of hierarchy. Another approach is to replace each verified cell by an abstract, containing all information needed to check the cell against its environment on the next higher level ([4]). The abstract of a cell is independent of its environment, therefore this approach can be easily extended to implement incremental design rule checking. The abstract method is used in the presented hierarchical DRC too.

First hierarchical tools made restrictions concerning the layout style, for example they did not allow any features to overlap subcells ([5]). More recent DRC tools do not demand such restrictions about the layout, but although they are able to check any given layout, hierarchical verification of a bad designed layout may take even more time than a flat verification of the same layout.

Experiences made with existing DRC programs showed that besides the need of hierarchical tools it is also very important that designers accept a hierarchical design style. However this design style will depend on the kind of chips to be produced (full custom layout, standard cell chips) and therefore cannot be totally prescribed by the program.

In this paper we present the hierarchical DRC program HEXDRC, which has been developed at SIEMENS for the verification of full custom designs.

HEXDRC does not require any restrictions concerning the layout design, but it is assumed that the designers accept some kind of hierarchical design style. If this is true, the program will work most effectively. The design style however is not prescribed by the program. Although in standard mode it is assumed that subcells are not too much overlapped by surrounding data, the user may modify the basic algorithm. With these modifications the hierarchical algorithm can be adapted to a specific design style in order to verify a given layout in the most effective way. Especially for full custom chips the designers know in detail the hierarchical structure of the layout. Therefore it is very helpful that the user can tell the program which parts should be treated hierarchically and which parts should be flattened out.

Although HEXDRC is mainly intended for the verification of full custom designs it can be used to verify any kind of layout.

2. Basic Concept for Hierarchical Layout Verification

The basic idea of a hierarchical DRC is to check each cell in a given layout only once. In general however the cells may not be regarded totally independent because design rule violations may also result from interactions between different cells. These interactions mainly result from overlaps between different subcells or between primitives and subcells. In a hierarchical design such overlaps mostly appear in the boundary regions of the subcells.

Let us first assume that in a given layout no subcell is overlapped by primitives or other subcells by more than a maximum value MAX_OVERLAP and let MAX_DRC be the maximum design rule measure specified in the technology file. In this case layout structures of a cell which are involved in a design rule violation between a cell instance and its environment must have a distance less than MAX_OVERLAP + MAX_DRC to the cells boundary. Therefore, to verify a given cell it would suffice to check all subcells and then replace all subcells by their boundary parts before checking the given cell.

In general it will not be acceptable to restrict all overlaps in a layout to a maximum value. But obviously overlapping subcells by a great distance always will decrease the performance of a hierarchical DRC. Therefore it may be a good idea to distinguish between overlaps "near" the subcells boundary and other overlaps.

This leads to the following basic concept used in the program HEXDRC:

Besides the usual design rules the user may specify a maximum allowed overlap value MAX_OVERLAP in the technology file. If a subcell is overlapped by a primitive or by another subcell with an overlap less than MAX_OVERLAP this is defined to be a legal overlap otherwise it is called an illegal overlap (Figure 1).

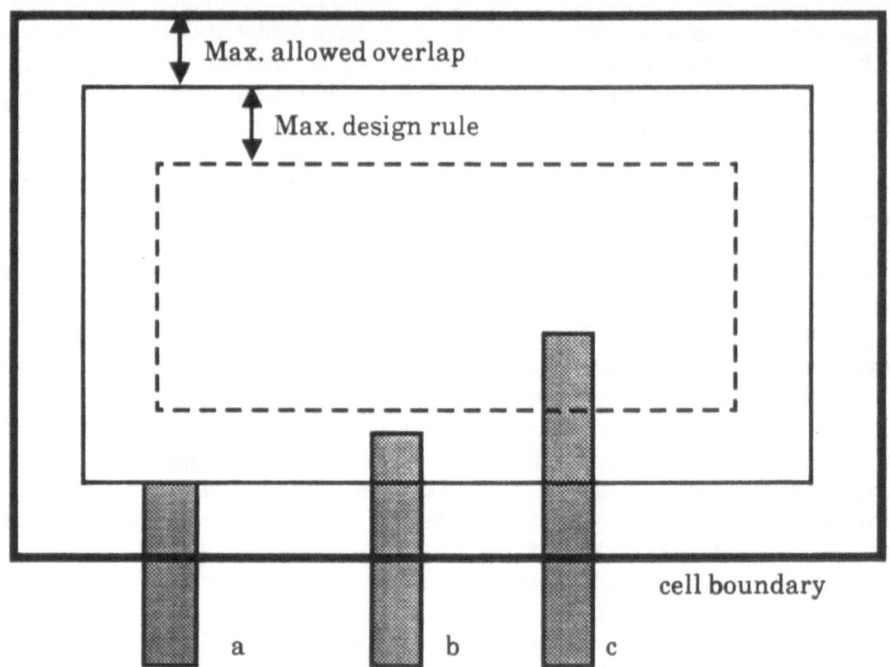

Figure 1: Legal and illegal overlaps.
       a : legal overlapping primitive.
      b,c : illegal overlapping primitives.

For each cell we define the abstract of the cell to consist of all layout structures of the cell which have a distance to the cells boundary which is less than MAX_OVERLAP + MAX_DRC (Figure 2). The boundary of the cell may be drawn on a special layer in the layout, otherwise it will be determined automatically by the program.

To detect all design rule violations, except those caused by illegal overlaps, it will suffice to verify all cells replacing the subcells by their abstracts.

Illegal overlaps between different subcell instances do not appear very often in a hierarchical design. To detect all errors caused by these overlaps, it will suffice to flatten out both instances.

There are two possibilities to handle illegal overlapping primitives: One possibility is to report these overlaps as errors. In this case the user has to check wether these overlaps

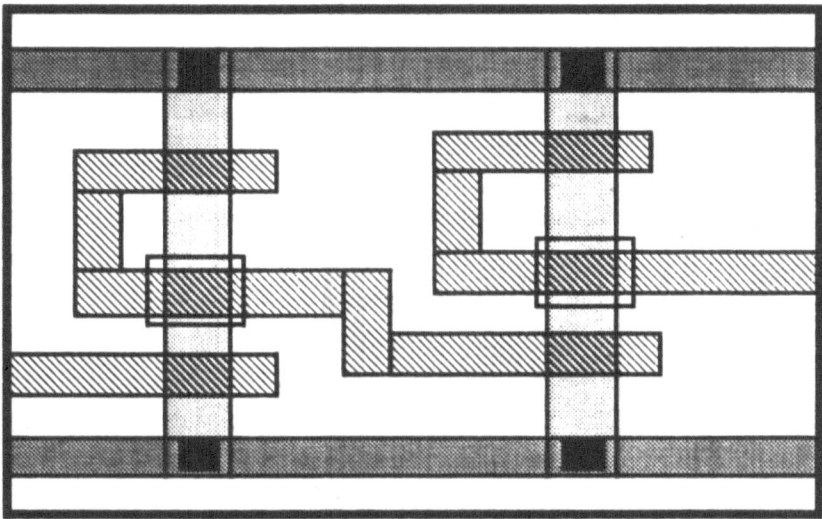

Full cell layout

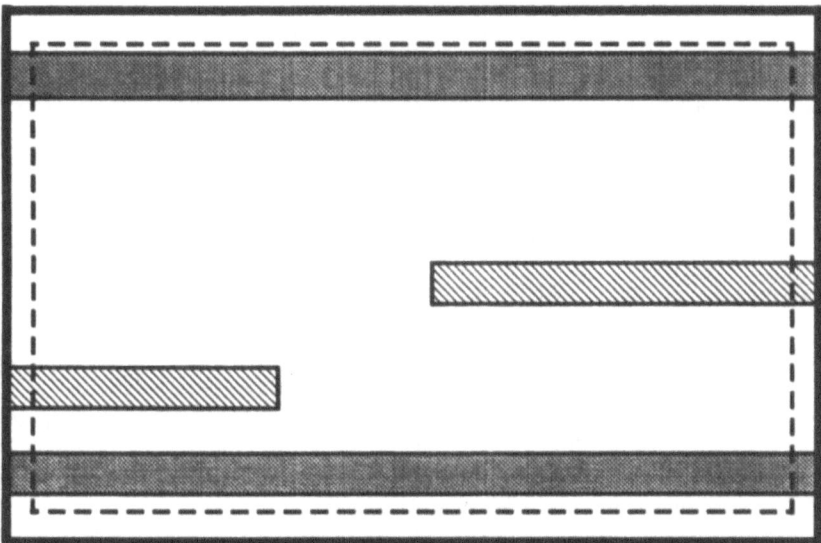

Cell abstract

Figure 2: Definition of abstract

cause design rule violations. The second possibility is to recheck overlapped cell instances together with the overlapping layout data in a separate step. This recheck will guarantee that all design rule violations caused by such overlaps will be detected by the program too (Figure 3).

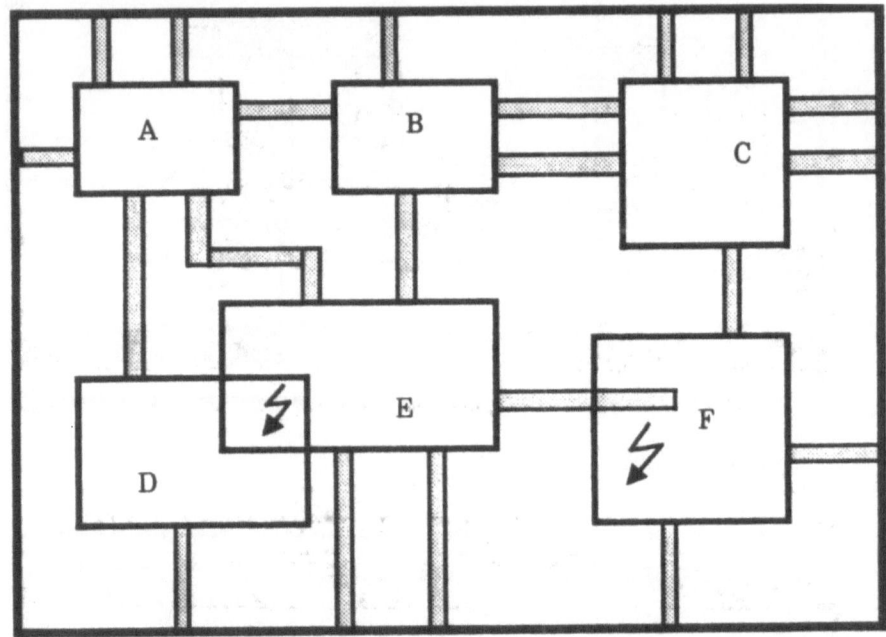

Figure 3: Handling for instances with illegal overlaps.
Instances D and E will be expanded.
Instance F will be rechecked in a separate step.

3. Main Algorithm for Hierarchical Drc

The user may select one or more cells from a given layout data base to be verified within one program run. Each of these cells may have an arbitrary hierarchical structure.

A selected cell and all subcells are verified recursively in an order corresponding to the layout hierarchy by the procedure CHECK_TREE (Figure 4):

For a given cell CHECK_TREE first determines illegal overlapping subcell instances. These instances are marked to be flattened out.

Before verifying the given cell, all subcells corresponding to hierarchically treated instances must be checked by recursively applying CHECK_TREE to these subcells. After this step for all hierarchical subcells the abstract consisting of all layout data in the boundary region of the subcell is availiable.

Finally a procedure CHECK_CELL is called to verify the actual cell and to generate the cells abstract.

```
procedure checktree ( root : cell );

begin { checktree }

  for all pairs (i1,i2) of instances in root do begin
    if illegaloverlapping ( i1,i2 ) then begin
      i1.expand := true;
      i2.expand := true;
    end;
  end; { for all pairs }

  for each instance in root do begin
    if not instance.expand then begin
      subcell := correspondingsubcell ( instance );
      if not subcell.checked then checktree ( subcell );
    end;
  end; { for each instance }

  checkcell ( root );
  root.checked := true;

end; { checktree }
```

Figure 4: Basic algorithm for checking a cell

The verification of a cell by CHECK_CELL is a flat DRC for the given cell together with the abstracts of the subcells. All layout data from the subcell abstracts and the complete layout of flattened subcell instances are regarded as belonging to the cell itself. CHECK_CELL verifies all design rules specified in the technology file. In addition during the cell check illegal overlaps between primitives and subcell instances are detected and stored in the internal data base for further analyzing. Finally the abstract of the cell is extracted.

Each cell verified by CHECK_CELL is marked as "checked" and will not be checked again during the program run.

4. Recheck for Cell Instances with Illegal Overlaps

During the verification of a cell illegal overlaps between primitives and subcell instances are detected. These overlaps may cause design rule violations, which could not be found when the subcells were replaced by their abstracts. Therefore these overlaps are reanalyzed in a separate step after all cells have been verified.

For each instance which is overlapped by primitives by more than the maximum allowed value the layout of the instance is rechecked together with all illegal overlapping primitives.

In general however it is not necessary to recheck the complete layout of the subcell instance and to recheck all design rules specified in the technology file:

In many cases only a small part of the subcell is overlapped by primitives. In this case it suffices to recheck the overlapped area (extended by the maximum design rule measure) together with the overlapping elements.

On the other hand illegal overlaps usually occur only in few layers. In this case it is not necessary to recheck all design rules. For example, if there are only overlaps in the metal layer, it suffices to recheck only rules which are concerned to metal. Other rules like dimension rules for gates cannot be violated.

To determine the set of design rules which have to be rechecked, HEXDRC assigns to each operation a set of concerned layout layers. In general this set is the union of the corresponding sets of all input layers needed for that operation. For original layout layers the corresponding set consists of the layers number, for derived layers the corresponding sets are recursively obtained from the input layers.

With this information it is easy to determine all operations needed for the recheck: A design rule must be rechecked only if the associated set of concerned layers and the set of layers containing overlapping primitives are not disjoint.

5. Modification of Basic Algorithm by User Instructions

For a given chip in the simplest case the technology files for a hierarchical DRC and for a flat DRC are nearly the same. The only additional information required for the hierarchical mode is the maximum allowed overlap to be specified by the user. Although the basic algorithm described above will always guarantee to detect all design rule violations, depending on the actual layout it will not always do it in the most effective way. Therefore the user has a lot of possibilities to modify the basic algorithm.

In general it is not the best way to handle all cells in a layout hierarchically. In some cases hierarchical verification may need even more time than a flat DRC. For example if the abstract of a cell contains nearly the complete layout of the cell, it will be better to flatten out all instances of that cell. The same is true for cells with many illegal overlapping elements. In this case the recheck would need the same time as the verification of the cell itself. Therefore, to improve hierarchical verification the user may specify in the runset which cells have to be treated hierarchically and which cells should be flattened out.

The basic algorithm assumes that there are not too much illegal overlaps in the layout. For a layout with wires crossing nearly all cells it will work very ineffectively. On the other hand layouts are often designed in a more restrictive design style. Using the basic algorithm the program however cannot take into account these additional informations.

To overcome these problems, the user can modify the abstract generation and specify its own hierarchy rules in the technology file. The most powerful option is the possibility that design rules may not only refer to the layout data of the actual cell but also to layers of (already checked) subcells. This for example allows to check that a layer in the actual cell does not overlap with a forbidden area in any subcell.

Example: Consider a standard cell block with METAL2 crossing all standard cells. In this case it would be very ineffective to recheck all cell instances together with the "illegal" overlapping METAL2.

Normally each cell already contains a special layer METAL2_KEEPOUT which defines those areas where no METAL2 may cross on the higher levels of hierarchy. These areas are for example needed by the router.

In this case the user first may specify, that illegal overlaps in METAL2 should be ignored and not reanalyzed. In addition he may define rules in the technology file which check

1. that no METAL2 of the actual cell overlaps the keepout areas of the subcells
2. that all METAL2 in the actual cell is inside the cells keepout area with a given distance

by the following commands:

```
CREATE      illegal = metal2 AND metal2_keepout(SUBCELL)
EXISTENCE   illegal -TITLE 'metal2 in keepout area of subcell'
DISTANCE    metal2 ONLY INSIDE metal2_keepout(CELL)
            -INSIDE 3.5 -TITLE 'metal2 not 3.5 inside protection area'
```

If there were no keepout areas defined in the cells, another possibility to check the minimum distance rule in METAL2 would be to import the complete METAL2 layer from the subcells into the root cell and then merge it with METAL2 of the root cell. This gives the possibility to make a flat check for this special design rule in the root cell.

6. Incremental Design Rule Checks

Incremental verification usuallay requires a suitable system en vironment. Although HEXDRC is a stand alone program using GDSII like layout formats as input and the program does not store any information gained during the verification of a cell some kind of incremental design rule check is already possible.

One possibility is to restrict the verification to all cells which have been changed since a given date. In this case the creation date of each cell contained in the layout file is evaluated. Not only cells with a creation date younger than the specified date will be checked, but also all cells which contain such changed cells as subcells.

A second possibility is to declare selected cells explicitly as "unchanged" by listing the cell names in the runset.

If a cell is unchanged, the program will not verify the cell but it will reconstruct the cells abstract. Replacing all instances of an unchanged cell by its abstract will guarantee that the instances will be checked completely against their neighbourhoods.

The reconstruction of a cells abtract usually needs very little time compared with a complete verification. Therefore the "unchanged" option can save a lot of time when a chip containing a correct block has to be verified.

7. Experimental Results

The program HEXDRC has been implemented on a SIEMENS mainframe. First experimental results were obtained by applying the program to existing layouts, which have not been designed with the existence of a hierarchical DRC tool in mind.

For a standard cell block with 520 instances of 25 different cells the CPU time was reduced from 9100 sec to 3100 sec. The chip did not contain any illegal overlaps. Rules concerning metal2 (crossing nearly all cell instances) were checked using the flattened layer in the root cell.

Another example is a logic block from a full custom memory chip. The block mainly contains 5 different subblocks which share a lot of small cells on the next deeper levels of hierarchy. Although the block is not very regular, the CPU time was reduced from 1517 sec to 1045 sec. The block contained a lot of illegal overlaps resulting from badly designed connections to subcell pins. After eliminating these overlaps in the design the runtime was further reduced to 970 sec.

The best result was obtained for a 4M DRAM, containing 8 identical layout blocks. In this case the design rule check could be improved by a factor 10. Further improvements of the algorithms will allow to use the internal hierarchy of these blocks even more effectively. After the implementation of these extensions we expect atleast a factor 20 compared with a flat design rule check.

Even if a layout is not very regular, the hierarchical DRC may improve the verification significantly: Checking all cells separately reduces the amount of data to be handled during a verification step. In this case the program keeps all information in memory and does not use temporary files which usually saves a lot of time.

If a layout contains many illegal overlaps a great part of the runtime is used for rechecking the cell instances. The area of the cell instances to be rechecked depends on the cells size. For small cells often more than 50 % of the subcells area must be rechecked. Illegal overlaps however often occur only in a small number of layers, therefore in many cases it suffices to recheck only 10 % or 15 % of all design rules.

One main advantage of the hierarchical DRC is the hierarchical error output. Together with each design rule violation the program reports the name of the cell in which the error was found. Errors in a cell are reported only once and not repeated for all instances of a cell. Even if hierarchical verification may not always save computer time it significantly reduces the time needed for evaluating the DRC results.

8. Future Work

In future the program will be improved to handle regular layout parts more effectively. For example, in case of a regular array of identical cells it is not necessary to replace each cell instance by its abstract. In this case it would suffice to verify that the cells fit together and then to treat the whole array as a single cell. This especially will improve the hierarchical DRC for highly regular memory chips. One solution to improve the DRC for regular arrays has been proposed in [2].

Another important point will be the extension to hierarchical netlist extraction. With the extracted netlist, for example, it could be verified that a layout is a correct implementation of the circuit specification. Netlist extraction is the most time consuming part of layout verification, therefore it will be very important, that it can be done hierarchically too.

9. Conclusions

Layout verification can be significantly improved by taking into account the layout hierarchy. The reduce in runtime however depends on the design style of the actual layout.

There will always be a tradeoff between a more restricted design style, allowing fast verification, and a more unrestricted design style, often needed to obtain a compact layout.

The presented program HEXDRC makes it possible to adapt the hierarchical design rule check to a given design style. This will help designers to accept some self defined restrictions and to learn how to use a hierarchical DRC effectively.

10. References

[1] Hannken-Illjes J., Golze U. (1986) 'A hierarchic incremental Designrule Checker' (in german), Informationstechnik, 3, 132-138

[2] Hedenstierna N., Jeppson K. (1987) 'New algorithms for increased efficiency in hierarchical design rule checking', INTEGRATION, the VLSI journal, 5, 319-336

[3] Perry S., Kalman S., Pilling D. (1985) 'Edge-Based Layout Verification', VLSI Systems Design, September 1985, 106-114

[4] Scheffer L., Soetarman R. (1986) 'Hierarchical Analysis of IC Artwork with User-Defined Rules', IEEE Design & Test, Feb 1986, pp. 66-74

[5] Wagner T. (1984) 'Hierarchical Layout Verification', Proc. 21st Design Automation Conference, 1984, pp. 484-489

[6] Yin M. T. (1985) 'Layout Verification of VLSI Designs', VLSI Design, July 1985, pp. 30-38

Keywords

Layout verification, design rule check, full custom design, hierarchy

Project no 888

Modular Testing supported by Block Environment Description

W. Roth, M. Johansson, W. Glunz
SIEMENS AG
Corporate Research and Development
Otto-Hahn-Ring 6
D-8000 Munich
FRG

Abstract

This paper presents an approach to generate tests for heterogeneous, modular circuits. The test patterns for the embedded blocks are transformed to be applicable and observable via the inputs and outputs of the embedding circuit. This transformation of test patterns is formally described by the *Block Environment Description (BED)*. This way it is possible to take advantage of the designer's knowledge about the circuit.

1. Introduction

Structured design for testability techniques like scan design [EiWi77, GeNe84] together with efficient algorithmic test pattern generation [Goel81, Joha83, Schu87] or built-in self-test are still sufficient to tackle test preparation for most of today's VLSI circuits. But these techniques allow only one universal solution for the whole circuit and do not consider the inherent function or global structure of the circuit, whereas recent designs are often composed of relatively large modules, which are provided in libraries, by module generators, or synthesis tools [DuKr89]. Often no gate level representation of these modules exists and therefore the traditional gate level test pattern generators can not be used.

Another fact is that today's design systems for integrated circuits support a modular and hierarchical design style as a mean to tackle high complexity. The advantage of using higher level primitives for test pattern generation was seen already some time ago [BeBr75, BrFr80], and there have been several attempts to use the modularity and hierarchy of VLSI designs for test pattern generation [BhHa85, SoGa85, ChPa87, HoGe87, Kris87, Marh87, MuHa88]. However, the circuits are getting more and more heterogeneous due to the integration of memories, generated or synthesized modules and complex macros and thesemodular, heterogeneous circuits demand also a heterogeneous test strategy, where dedicated tools or strategies can be used for the individual modules, and where the knowledge about the global function can be utilized also for test purposes.

Another well known observation is, that in spite of the complexity of the problem leading to the failure of automatic test generation programs, the human circuit designer often is able to find tests for large parts of his circuit. This implies that there is "high level information" available to the designer, which is not efficiently used by the tools. This information includes especially the knowledge about how to utilize the normal function of the circuit and inherent structures also for test purposes.

This research was partly supported by the European Community under contract ESPRIT 888

148

This paper describes a method that utilizes the knowledge about the accessability of embedded modules and makes it possible to combine test patterns, which already exist for individual modules, into test patterns for the circuit. The test patterns for the individual modules refer to the inputs and outputs of the module and must therefore be transformed to the inputs and outputs of the circuit. This can be done with the aid of the *Block Environment Description (BED)*, which describes how the individual signals at the module IOs can be applied and observed at the circuit IOs.

For circuits designed by humans the main goal is to give the designer the possibility to describe how to access the modules within his circuit. This is a way to make use of the designers knowledge about the circuit. The designer himself can choose the best test strategy or combination of test strategies for his design and he can try already in very early design stages whether his strategy will work. For circuits created by synthesis tools the goal is to make the information about the circuit available during the synthesis to be usable for the test generation. Further advantages are that the test patterns can already be generated for the modules without consideration of the embedding circuit and in any way desired. They are only subsequently transformed to test patterns for the entire circuit.

2. Testing Embedded Modules

2.1 Definitions and Preliminaries

We define an *embedded module* as a block of sequential or combinatorial circuitry for which test patterns already exist and which forms part of the whole circuit. Examples of such modules are parameterizable cells produced by module generators, synthesized blocks, or complex macros like microprocessor cores. Modules can also be created by partitioning of the circuit.

Figure 1 shows how the test stimuli can be applied to an embedded module and how the test responses can be observed through the embedding logic. The stimuli are propagated via the *control path* and the responses are propagated via the *observe path*; both paths have of course to be justified.

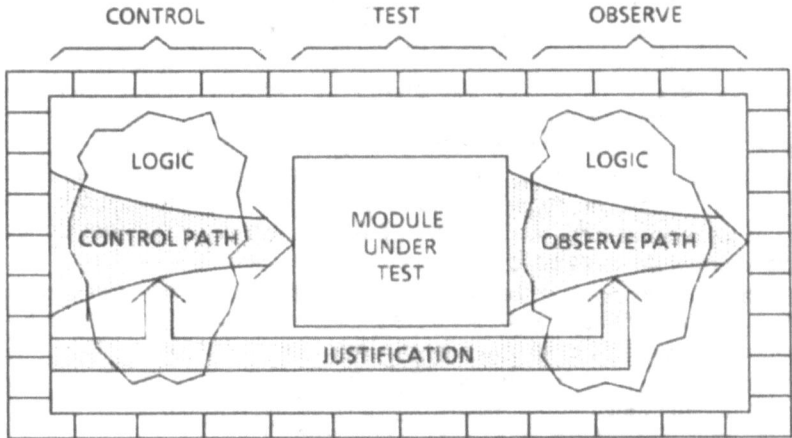

Fig. 1: Testing an embedded module

Since the desired propagations are not possible for all types of embedding, we have to restrict ourselves and decide what types we support and what to do otherwise. Therefore, we classify the paths:

The control (observe) path is *transparent*, if there is a one-to-one mapping of all inputs (outputs) of the module to inputs (outputs) of the circuit.

Beside simple examples of transparency like the path from one data input port of a multiplexer to its output port, also paths containing inversions, delays, or splitting into time slices, are transparent. An example of a delay is a register: the input data is mapped to the outputs after a clock cycle. Splitting arises e.g. if both input ports of a multiplexer shall be observed; first one of the inputs can be observed, then with another justification the other input is observed.

The control (observe) path is translucent, if there is an m-to-n mapping of all m inputs (outputs) of the module to n inputs (outputs) of the circuit, such that all possible module input (output) vectors are unambiguously mapped to circuit input (output) vectors.

For translucent paths n has to be greater than m; e.g. a decoder with 2 inputs and 4 outputs can be used as part of a translucent observe path, since all its input vectors are mapped to different output vectors. It can however not not be used as part of a translucent control path, since only 4 of the 16 possible output vectors can be provided.

2.2 Modular Testing Supported by Block Environment Descriptions

In our current version of the BED system we restrict the range of applicability to circuits where for each module the control path and the observe path are transparent. This way we can ensure that any given test pattern can be transformed without reducing the fault coverage; therefore we can produce the test patterns for the modules without restrictions inflicted by the embedding circuit. In many real circuits there are control and observe paths with the required properties for most modules. They make heavy use of existing bus structures and of the fact that modules like adders, multipliers, and ALUs have transparency modes. If transparent paths do not exist, they have to be achieved by appropriate DFT methods, like multiplexing, insertion of scan chains, and other methods taking advantage of the internal circuit structure. These constraints are not necessarily very hard, e.g. an area overhead of 5% [PoHJ88] or 3% [Bui88] respectively was reported, whereas a complete scan path would have required 10% or more.

Test patterns for the modules, i.e. the information as to how the isolated module is to be tested, must be present for each module of the circuit. Since the modules are of very different kinds and perhaps with different failure mechanisms to consider, the optimal methods to generate these tests will also be different.

For some modules, such as the ones produced by module generators, the test patterns are automatically generated along with the layout. One example is the parameterizable multiplier of [Beck85]. For this multiplier a small number of test pattern is generated automatically with a known high fault coverage (100 % single stuck faults). For other modules other test pattern generation strategies will be appropriate.

The test pattern transformation is done with the help of Block Environment Descriptions (BEDs). Figure 2 shows a small example of an embedded module with its BED (a real example is shown later). Assuming that the module under test has a dedicated set of test patterns, these have to be transformed to be applicable via the IOs of the circuit.

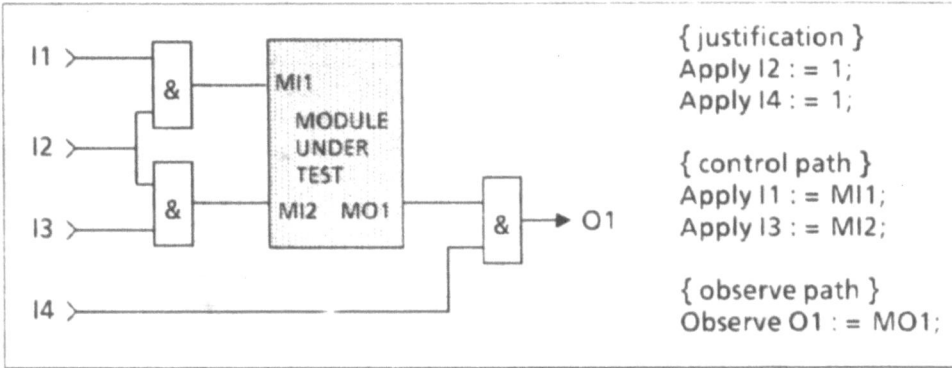

Fig. 2: Embedded module with its BED

2.3 Verifying the Block Environment Descriptions

As BEDs may be written or modified by a human designer, they may contain errors. To make sure that the BEDs are consistent with the circuit, there is a program which automatically checks the BEDs for consistency with the circuit. This verifies that the described transformations correspond to existing paths. This checking is done by symbolic simulation of the control, observe, and justification signals. Figure 3 shows a simple example: the justification signals at the circuit inputs I2 and I4 with their logical values of 1 and 0 respectively are propagated through the circuit; the control signals S1 and S2 at the circuit inputs I1 and I3 are propagated to the module inputs MI1 and MI2; but the observe signal S3 at module output MO1 cannot be propagated to the circuit output O1, as was stated in the BED for this module. This conflict is reported to the user, who has to correct his BED.

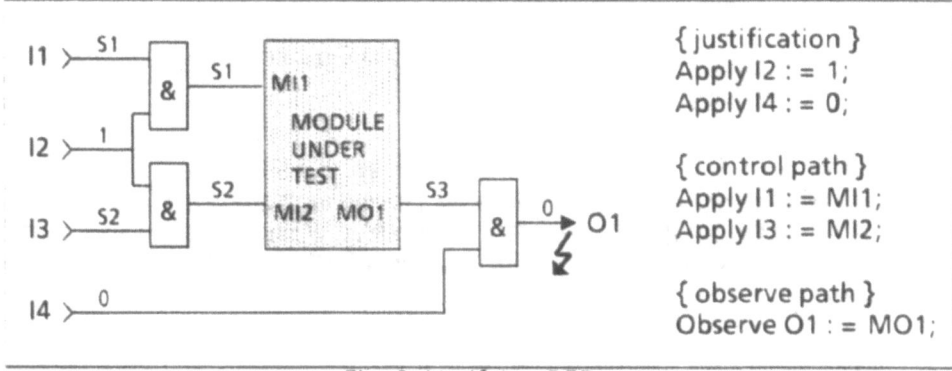

Fig. 3: Verifying BEDs

2.4 The Bed System

The method is being implemented in the BED-System (Fig. 4) in several software packages:

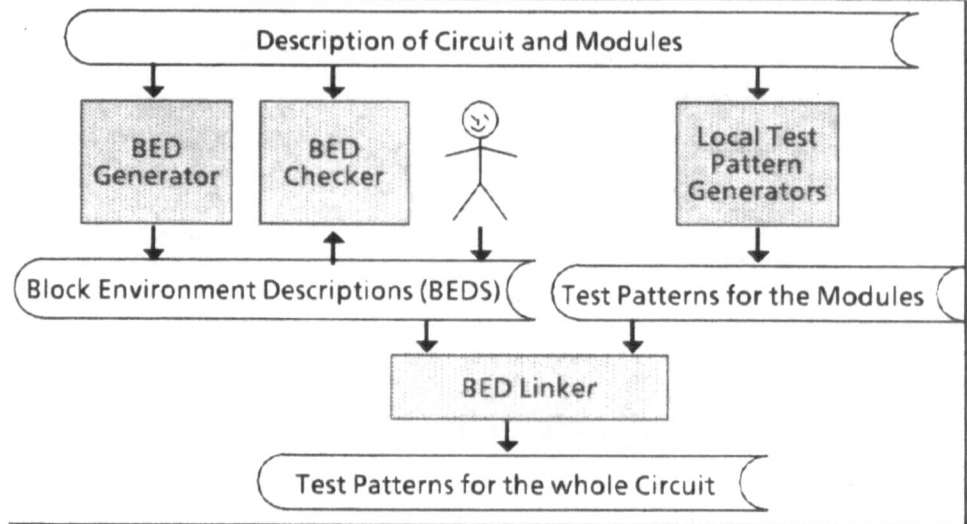

Fig. 4: BED-System

- The BED-Compiler reads the BEDs, does syntactical and semantical checks and produces an internal data-structure, which is then used by the BED-Linker and by the BED-Checker.

- The BED-Checker can be run optionally to check the compiled BEDs for compliance with the embedding circuitry.

- The BED-Linker reads the test patterns for the modules, transforms them and writes the transformed patterns for the whole circuit.

- The BED-Generator generates automatically parts of BEDs from the structural description of the embedding circuit.

3. Example

Figure 5 shows a 32-bit microprocessor based on the AMD 2901. It is built of several modules: *ALU, RAM, REGISTER, SHIFTER*s, and *MUX*es. Assuming that each of these modules has a dedicated set of test patterns, we can write BEDs for the transformation of the patterns.

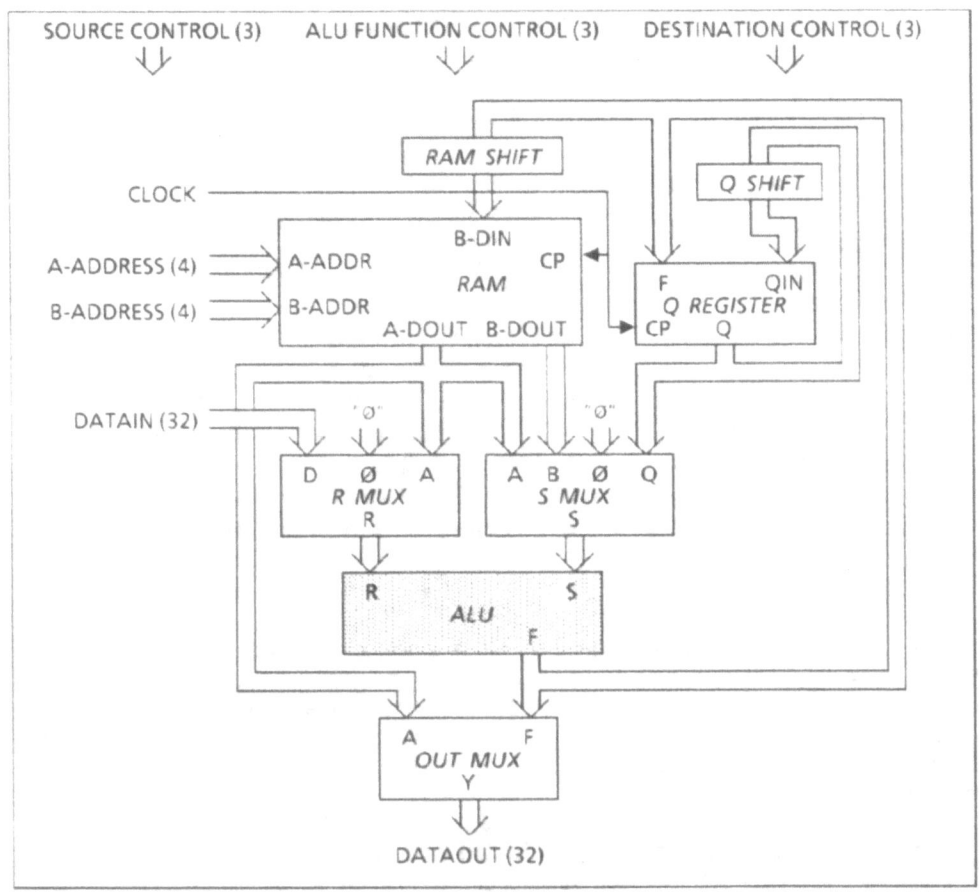

Fig. 5: Testing the ALU embedded in the microprocessor

Looking at the schematic it is obvious how to access the individual modules. E.g. to test the *ALU* we first apply the data for the S-input at DATAIN, propagate it through the *R MUX*, and the *ALU* by adding a 'zero', and load it into the *Q REGISTER*. In a second cycle we propagate this data from the *Q REGISTER* to the S input of the *ALU*, apply the data for the R input at DATAIN, and observe the responses through the *OUT MUX* at DATAOUT.

The BED for the *ALU* is following, and it shows that what was expressed with a couple of sentences in English can equally easy and short be expressed as a BED. To the right side of the BED there is an explanation of the statements.

```
BED ALU;                          Block Environment Description for ALU:
  Input  = R(32), S(32), FCT(3);  it has two 32 bit data inputs, a 3 bit
  Output = F(32);                 function input and a 32 bit data output;
  Initialize                      initialization describes fixed justifications:
  Cycle                           one pattern to set the destination control
   Apply DESTCTRL := "000";       so that F -> DATAOUT,  F -> Q
  For all pattern in ALU.TESTPATTERN  transform each testpattern of ALU into:
   Cycle Loop 1 CLOCK;            one test pattern with an active CLOCK
   Apply DATAIN   := S;           apply the stimuli for S at DATAIN
   Apply SRCCTRL  := "111";       DATAIN -> R, "0" -> S
   Apply ALUFCT   := "000";       add R+S; this cycle loads the stimuli
                                  for S into the Q register ;
   Cycle                          a second test pattern without clocking
   Apply DATAIN   := R;           apply the stimuli for R at DATAIN
   Apply SRCCTRL  := "110";       DATAIN -> R, Q -> S
   Apply ALUFCT   := FCT;         apply the stimuli for FCT to ALUFCT
   Observe DATAOUT := F;          observe responses of ALU at DATAOUT;
  EndFor;
EndBED;
```

Using conventional gate level ATPG for the example circuit would have severe disadvantages, because there is only one data input and one data output for the whole circuit, i.e. there is a high degree of reconvergent fanout. Additionally, the patterns have to be generated for a sequential circuit with feedback loops, with all the well known problems for ATPG. The register could of course be made scannable, but this still leaves the problem what to do with the RAM.

This example shows clearly the advantages of a modular test generation in connection with the usage of designer knowledge. For the individual modules the test patterns are available or can be generated easily. The designer can tell with relative ease how to access the single modules. It would in such a case be hard to convince the designer to spend extra hardware like scan path registers or additional multiplexers for test purposes.

4. Conclusion

The method for test generation presented in this paper utilizes the modularity of the circuit and the designers knowledge about his design. Test patterns for a design can be created in a very effective way from the module-specific test patterns.

The example showed that it is easy for a designer to write the BEDs for a circuit he is designing, and the resulting test patterns ensure a good fault coverage (provided the fault coverage of the test patterns for the modules was good). Very little CPU time is needed for the transformation of the test patterns, and if ATPG is used for all modules individually, this will be much faster than using ATPG for the whole circuit.

The BED system will be used by the end of the year to tackle a real design with a 32-bit architecture, containing ALUs, register files, a multiplication and division unit, and several other blocks.

5. Acknowledgements

We want to thank Dr. M. Marhoefer for his valuable inputs and suggestions.

6. References

[BeBr75] R. G. Bennetts, D. C. Brittle, A. C. Prior and J. L. Washington. "A Modular Approach to Test Sequence Generation for Large Digital Networks". Digital Processes, pp. 3-23, 1975.

[Beck85] B. Becker. "An Easily Testable Optimal-time VLSI Multiplier". Proc. Euromicro '85, pp. 401-409

[BhHa85] D. Bhattachary, J. P. Hayes. "High Level Test Generation Using Bus Faults". Proc. 15th Fault-Tolerant Computing Symposium, pp. 65-70, Ann Arbor, Michigan, June 1985.

[BrFr80] M. A. Breuer, A. D. Friedman. "Functional Level Primitives in Test Generation". IEEE Transactions on Computers, pp. 223-235, March 1980.

[Bui88] C. Bui. "Testability Using Random Access Test Register". Proc. 1988 IEEE International Test Conference, pp 994-995.

[ChPa87] S. J. Chandra, J. H. Patel. "A Hierarchical Approach to Test Vector Generation". Proc. 24th Design Automation Conference, pp. 495-501, 1987.

[DuKr89] P. Duzy, H. Kraemer, M. Neher, M. Pilsl, W. Rosenstiel, T. Wecker. "CALLAS - Conversion of Algorithms to Library Adaptable Structures". Proc. VLSI 89 Conference, pp. 197-208, 1989.

[EiWi77] E. B. Eichelberger, T. W. Williams. "A Logic Design Structure for VLSI Testability". Proc. 14th Design Automation Conference, pp 462-468, 1977.

[GeNe84] M. Gerner, H. Nertinger. "Scan Path in CMOS Semicustom LSI Chips". Proc. 1984 IEEE International Test Conference.

[Goel81] P. Goel. "An Implicit Enumeration Algorithm to Generate Tests for Combinational Logic Circuits". IEEE Transactions on Computers, pp. 215-222, March 1981.

[HoGe87] H. Hofestaedt, M. Gerner. "Qualitative Testability Analysis and Hierarchical Test Pattern Generation". Proc. 1987 IEEE International Test Conference.

[Joha83] M. Johansson. "The GENESYS-algorithm for ATPG without fault simulation". Proc. 1983 IEEE International Test Conference, pp 333-337.

[Kris87] B. Krishnamurthy. "Hierarchical Test Generation: Can AI Help?". Proc. 1987 IEEE International Test Conference, pp 694-700.

[Marh87] M. Marhoefer. "An Approach to Modular Test Generation based on Transparency of Modules". Proc. IEEE CompEuro Conference, pp. 403-406, 1987.

[MuHa88] B. Murray, J. Hayes. "Hierarchical Test Generation Using Precomputed Tests for Modules". Proc. 1988 IEEE International Test Conference, pp 221-229.

[PoHJ88] T. J. Powell, F. Hwang, B. Johnson. "Testability Features in the TM5370 Family of Microcomputers". Proc. 1988 IEEE International Test Conference, pp 153-160.

[RoJG89] W. Roth, M. Johansson, W. Glunz. "The BED Concept - A Method and a Language for Modular Test Generation". Proc. VLSI 89 Conference, pp. 143-152, 1989.

[Schu87] M. Schulz, E. Trischler, T. Sarfert. "SOCRATES A Highly Efficient Automatic Test Pattern Generation System". Proc. 1987 IEEE International Test Conference, pp 1016-1026.

[SoGa85] F. Somenzi, S. Gai, M. Mezzalama and P. Prinetto. "Testing Strategy and Technique for Macro-Based Circuits". IEEE Transactions on Computers, pp. 85-90, January 1985.

Project no 958

MULTICHIP PACKAGES IN THE ESPRIT PROGRAMME

G.Dehaine and K.Kurzweil
Bull S A
Les Clayes-Sous-Bois, France

R.Arrowsmith and D.Small
British Telecom Research Laboratories
Martlesham, United Kingdom

N.Chandler and S.Tyler
GEC-Marconi Research Center
Great Baddow, United Kingdom

Summary

A compatible set of advanced interconnect technologies has been developed and used in the construction of high density, high speed multichip modules. The two key interconnect technologies which have been developed are tape automated bonding (TAB) to 125 um pitch and >200 I/Os, and an additive multilayer board technology on a pcb base, with stripline signal layers suitable for GHz signals. All signal layers in this board technology have a 125 um pitch capability. Using the associated mechanical, electrical and thermal design tools that had been developed, these interconnect technologies have been proven in both high density (silicon to substrate ratio 36%) and high speed (>600 Mbit/s) multichip modules.

Introduction

In order to realise the intrinsic performance of ICs in a system, compatible performance in the interconnect technology is required to assemble the ICs. However, in 1985, it was clear that while the performance of ICs was continuing to evolve in both density and speed, the available interconnect technologies were already a significant limitation to high performance systems. It was also clear that the gap between intrinsic IC and interconnect performance was widening. Thus a project was proposed[1] to develop a compatible set of interconnect technologies which would enable the performance potential of advanced ICs to be realisable in systems in a three year time frame.

The project was established in the ESPRIT programme, a European Economic Community (EEC) initiative to promote the development of the IT industry in the EEC by encouraging trans-national collaborative projects. The ESPRIT programme was conceived as a ten year plan in two phases, with a projected budget in phase 1 of about $1.6 billion. The EEC were to contribute 50% to the costs of ESPRIT collaborative research projects, the remainder being resourced by the participating companies. This project, which was part of phase 1, was collaboration between Bull S A, France, and GEC-Marconi and British Telecom of the UK and was led by Bull S A.

In the ESPRIT collaborative project reported here, Bull S.A. was responsible for:

1. The definiton and evaluation of the demonstrator modules.
2. Bumped chip TAB development.
3. Overall project management for the EEC.

British Telecom was responsible for:

1. Bumped tape TAB development.
2. Provision of ICs for the test structure and demonstration modules.

GEC-Marconi was responsible for:

1. The high density substrate development.
2. Thermal management.
3. IC protection.
4. Repair and rework aspects of the modules.

In order to attempt to close the gap between IC and interconnect performances, the aims of the project were set to achieve an intercept between C and interconnection technology by 1989. In practice, all of these targets, as set out in Table 1, have been achieved and bettered during 1988. The performance of interconnection technology components has been proven in the demonstrators fabricated as part of the project.

TABLE 1
Interconnect Technology Features

	Available 1986	Project Target	Achieved
TAB leads	40-100	>200	364
pitch	200 um	125 um	100 um
Substrate track	125-175 um	50-80 um	65 um
pitch	300 um	125 um	125 um
vias	300 um	50-100 um	50 um
IC size	50 mm2	100 mm2	100 mm2
power density	1-5W/cm2	5-10W/cm2	20W/cm2

In this paper, results of the development and status are presented, firstly for the two key components of the interconnect technology, TAB and the high density board, and then the results of the application of the interconnect technology in the functional modules.

Tape Automated Bonding

The objectives set in the project were to develop and prove a TAB technology suitable for bonding ICs of 10mm x 10mm, with a lead count >200 and a pad pitch of 125 um. TAB had been selected for the chip to board interconnect because it appeared to offer a better capability to meet the project objectives and offered more scope for development, than the then two competing technologies, wire bonding and flip bumped chip. In 1986, automatic wire bonding machines were limited to wires no finer than 1 mil, which in turn limits the pad pitches on the chip to about 200 um. In addition, a 1 mil gold wire has a breaking strain of only 8 - 10 gms, which is inadequate for many applications. The alternative techniques of flip bump chip attachment requires that all the thermal stress differentials arising across the chip are absorbed in the bumps. Thus flip bump chip technology was not considered appropriate for this high power density application, particularly with the preferred edge pad arrangement on the ICs.

Two TAB techniques were evaluated in the project, bumped chip TAB in which Bull S A had considerable experience, and the novel BTAB (bumped tape TAB) process developed at BT. To date, bumped chip has proven to be the most appropriate TAB technology and the processing and results are given below. The development of BTAB and a further recent TAB technique, ball bumped TAB, which also requires no chip preparation, are presented. In addition, the future evolution of TAB is discussed.

High Density bumped chip TAB

a. Bump and Frame Technology

The basic techniques for 200 um pitch bumped chip TAB had been thoroughly assessed at Bull and the project plan was to evolve in steps from this base to the project objectives. Straight walled plated gold bumps were specified and have been developed in association with Micro-Electronic Marin of Switzerland to the dimensions in Figure 1.

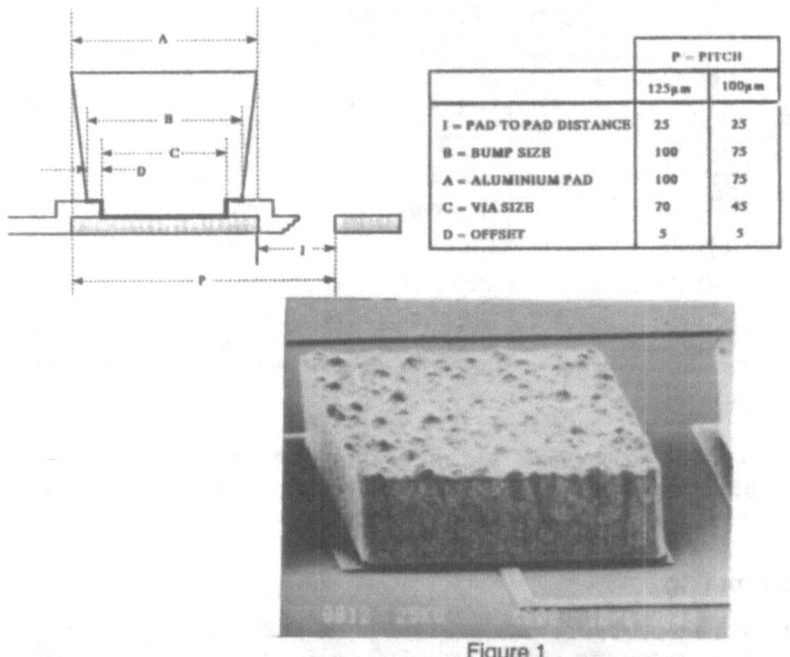

	P – PITCH	
	125μm	100μm
I – PAD TO PAD DISTANCE	25	25
B – BUMP SIZE	100	75
A – ALUMINIUM PAD	100	75
C – VIA SIZE	70	45
D – OFFSET	5	5

Figure 1

The bumps are 18 um high with a tolerance of 1.5 um and are formed by gold plating over a refractory metal deposited over the chip metallisation. The bump overlaps and seals the pad area, thus protecting the aluminium. Three layer tape was used for the frame. The final beam pattern consists of 35 um thick copper plated with tin. The tape was supplied by MCTS (part of Souriau, France). The tape design has straight short leads from the ILB (Inner Lead Bonding) to OLB (Outer Lead Bonding) sites, which then spread out to the frame probe test sites. This design has proven to be robust in operation, though it requires that the board has an identical pitch to the chip pads.

b. Assembly

ILB is by tin-gold alloying between the gold bump and tin plating on the lead, using a Farco F120 gang bonder. Careful control of the bonding cycle is required to avoid either oxidation of the tin or tin rich alloys forming. Similarly, the bonding operation must not induce cracking in the silicon or overlay oxide. Suitable parameters have been determined which produce good bonding without damage. After ILB processing, a gell type junction coating is applied to give additional mechanical protection to the IC during testing and assembly.

OLB is by tin lead soldering from an electrolytic deposit on the board to the tin plated beam using a Farco F120 OLB bonder, with a modified split field alignment microscope attachment. The OLB bonding uses a non-activated flux and also a film adhesive is used to retain the chip in place during OLB.

c. Results

ILB and OLB operations have been progressively assessed from 162 um pitch to 125 um and finally to 100 um(2). Examples of the bonding are shown in Figures 2 and 3.

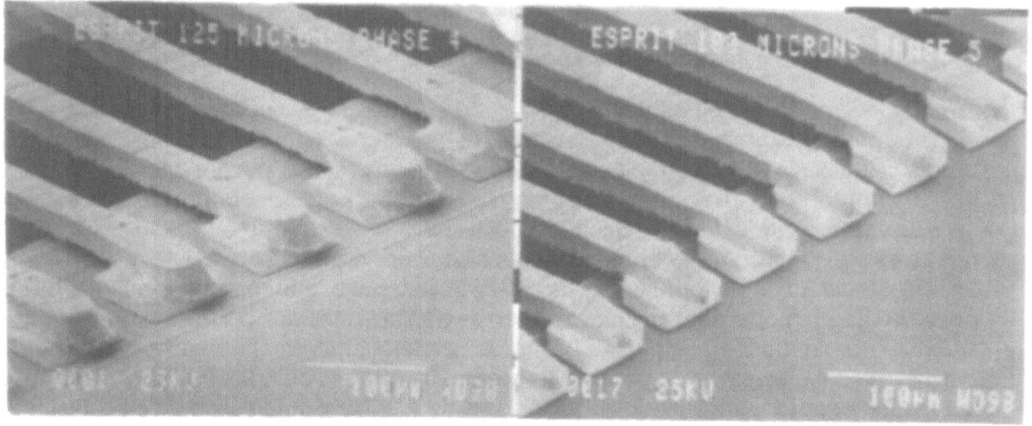

Figure 2   Figure 3

The project target of 125 um pitch was achieved in a demonstrator. A 100 um pitch process was also assessed, however, OLB alignment in particular, would need to be improved before this process could be considered for production.

BTAB and Ball Bumped TAB

BTAB and ball bumped TAB require no chip preparation for TAB bonding, but do require two or more stages of processing to produce the beam and bump. BTAB requires a two stage photolithography process on the tape. In the method developed at BT, the first step is to define the beam pattern and form a thin gold plate. Then, in a second photolithography process, the bump area is defined in thick photoresist and electroplated to about 25 um. The beam pattern is finally etched out using the gold as a mask and the frame is given a thin overall gold plating to prevent sticking to the thermode during inner lead bonding. Bumps are preferably smaller than the bonding area and have been defined to 50 x 50 fm. Figure 4a shows 70 x 70um bumps on beams. Currently available dry film resists have not yielded bumps within manufacturing tolerances and it has been found necessary to convert the processing to use a liquid resist, which appears to have overcome this problem.

Figure 4a   Figure 4b

Inner lead bonding using thermo compression bonding was successfully realised for chips having a gold metallisation. However, the process is not yet sufficiently reproduceable with aluminium based chip metallisation systems. It has been found that single point thermosonic bonding using a Hughes 2460 wire bonder provides considerably more satisfactory results with the aluminium based systems.

The same Hughes 2460 wire bonder was also used for producing ball-bumped TAB. This TAB technique is a four stage process. The first stage is to form conventional ball bonds on the chip. The second stage breaks the wire tail at the neck of the ball, as shown in Figure 4b. In the third stage, the ball tail is tamped flat for beam bonding. The TAB beams are then, in stage 4, bonded to the flattened gold bumps on the chip using thermocompression or thermosonic bonding. This TAB technique has given remarkably high yields of functional ICs and is also used in some of the modules. It has been successful down to 50 x 50 um pads on a 160 um pitch. Pure gold wire with a high percentage elongation was used in this assessment and although the tail length control was reduced, the soft gold balls have been found to provide improved compliance to absorb thermal stresses between IC and substrate.

OLB for both BTAB and ball bumped TAB is by lead tin solder, using the Farco 120 OLB bonder.

a. Future Trends

Of the TAB technologies evaluated in this project, only bumped chip TAB has reached the original targets and successful gang bonding has been achieved to 100 um pitch and with 364 leads. Further development of this technique will require improved registration control and would benefit from the relaxation in alignment if single point bonding with "nudge" capability becomes available. BTAB and ball bumped TAB do not yet offer equivalent performance in lead count and pitch. Ball bumped TAB in particular is considered to offer an easy entry into TAB using widely available auto-wire bonders and is suitable for all common IC metallisations.

High Density Substrate Technology

Ambitious objectives were set in he project for the substrate technology since the substrate density effectively determines the density capability of the whole module. The target pitch for the component attachment level was 125 um to match the TAB OLB. Similarly, the target pitch in two upper signal layers was set at 125 um. Further layers were to be provided on the PCB core for additional signal layers, power and ground planes. The substrate was to be capable of providing controlled impedance strip lines suitable for GHz signals. Module size was set at 10 x 10 cm.The technology selected by GEC-Marconi to achieve these targets was to assemble individually patterned copper clad laminates for the upper signal layers and the component attachment layer onto a conventional rigid multilayer PCB, which would provide the power, ground planes, and additional signal layers. The key developments required were in patterning the laminates, vias and laminate assembly. A cross section of the complete substrate, with two signal layers is shown in Figure 5.

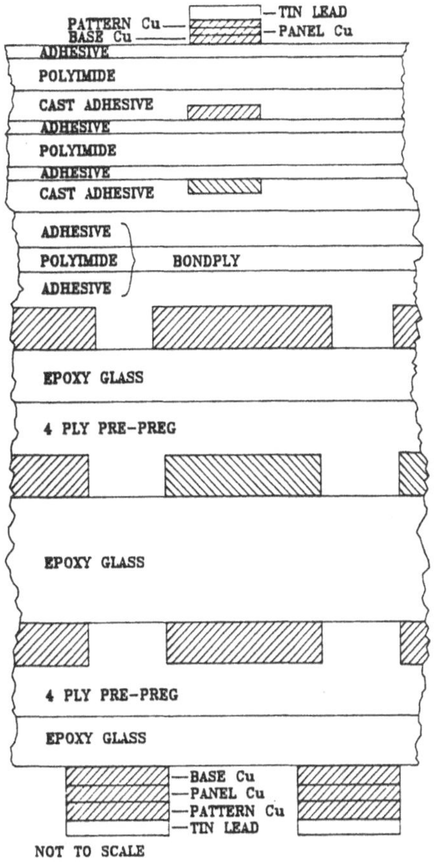

Figure 5

a. Laminate Processing

Various polyimide and PTFE copper coated laminates have been assessed. In the project, most of the work has used a 25 um polyimide based core with copper coating on both sides. The signal patterns are defined by photolithography, plating and etching in the copper layers over the 10 x 10 cm area. Lines and spaces simultaneously down to 40 um have been defined using these techniques. Allowing for the additional adhesive thicknesses between signal and ground plane, this laminate yields 75 ohm striplines with about 15 um copper thickness at a convenient track width, whilst retaining the 125 um pitch.

b. Vias

Vias < 50 um were needed to retain pitches of 125 um in the signal layers and these vias have been realised by laser drilling. Using a Nd-YAG laser, 50 um diameter holes have been obtained by unmasked piercing of the laminate. Larger holes can be made by moving the laminate. With an excimer laser, via holes have been defined using a metal mask which produces a clean hole as shown in Figures 6 and 7. These fine holes are mostly produced in the individual laminates before assembly, i.e. they form buried holes. Vias are plated to form the electrical connection between layers. Holes through the assembled multilayer structure have been drilled by laser and by fine mechanical drills and then plated.

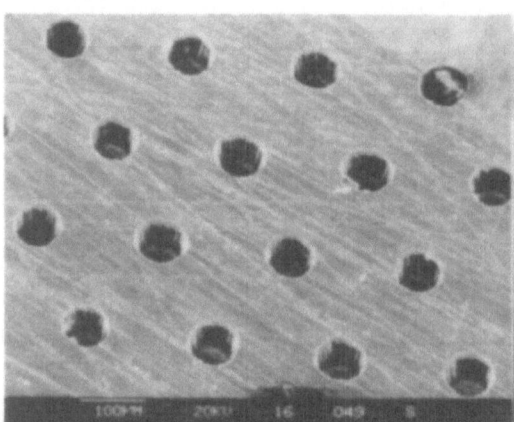 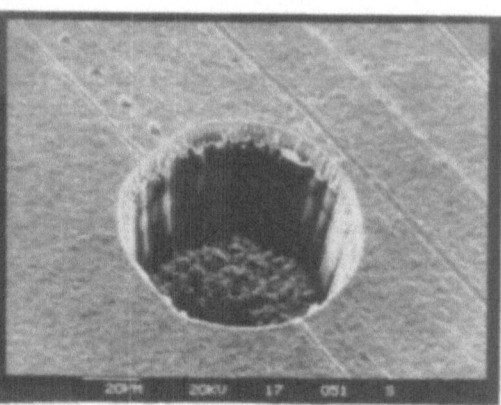

<u>Figure 6</u>   <u>Figure 7</u>

c. Substrate Assembly

Special tooling has been developed to achieve the very accurate registration of the track patterns on the laminates and to retain this registration during lamination of the whole substrate. Power vias at 350 um diameter are drilled after substrate assembly by conventional drilling. Finally, solder to the thickness optimised for TAB attachment is plated onto the top conductor layer, which primarily supports IC and passive component lands, plus vias to the signal and power layers. Solder of different thickness can be selectively plated if a range of TAB patterns is used in the module.

d. Future Trends

While polyimide laminates have been successful, it would be desirable to replace them with a material such as PTFE, which does not absorb moisture. This substrate technology has already demonstrated the capability of finer pitch tracking and low loss up to 3GHz. There is also potential for scaling up the size of the substrate and increasing the number of signal layers.

Technology Integration

Two multichip demonstrators have been produced during the project, the first primarily to assess mechanical and thermal aspects, and the second containing high speed data circuits to assess electrical performance.

The first board, shown in Figure 8, has a multilayer substrate with signal and attachment layers at 125 um pitch linked by 50 um diameter vias. There are sites for 36 TAB bonded ICs each with 284 leads at 125 um pitch. The ICs are 10 mm square and the board is 100 mm square. In this demonstrator, the ratio of silicon area to board area is extremely high, at 36%. This board demonstrates the good control of the bumped chip TAB at 125 um pitch ILB and OLB.

This board has primarily been used to assess the thermal management technique adopted by the project : immersion cooling in a fluorocarbon. First stage heat transfer by liquid avoids the very complicated contact cooling methods which have been described for other multichip modules, and appears to be compatible with all of the components used in the module. In this design, each TAB bonded IC can be powered to 10W and hence the whole board to 360W (3.6W/cm2). Immersion cooling has proved to be tolerant to a wide range of fluorocarbon boiling points and appears very suitable for even this high power density.

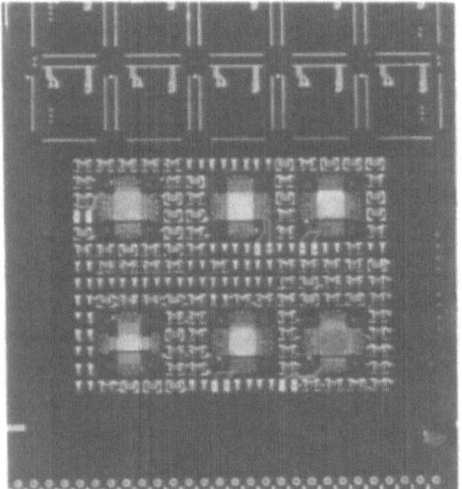

Figure 8   Figure 9

The second demonstrator module, shown in Figure 9, contains two high speed digital circuits with a maximum clock speed of 640 Mbit/s. One of the circuits, which is a 40 to 640 Mbit/s multiplexer, has been reduced in area by a factor of 10 times compared with the original circuit in leaded chip carriers.

The ICs for these test circuits are standard bipolar ECL arrays, with a maximum power dissipation of about 4.5W each. The chips have single or double level Al/1%Si/4%Cu metallisation and are TAB bonded, both using BTAB and ball bumped TAB. The module also contains ten power ICs to simulate a range of differential power distributions on the board.

These boards have only recently been assembled but early tests confirm that the 640 Mbit/s clock is transmitted through the board layers and that correct output signal levels are obtained. The second circuit on this module is a simple, 8 line 40Mbit/s bus driver to assess cross coupling in the high density stripline bus of the module.

For high frequency access, this module is assembled with discrete SMA connectors and is powered using a simple multi-pin through hole socket. No suitable high density multiple coaxial plug is currently available. For lower frequency tests, the board will be cut down to 80 mm square to fit directly into an available multi-pin socket.

Summary and Future Trends

This collaborative project set out in 1986 to assess and prove a set of interconnect technologies suitable for the assembly of high density, high speed multichip modules. The objectives set were considered ambitious, but all have been achieved. Bumped chip TAB to < 125 um pitch and > 284 leads, board technology with multilayer strip line interconnect and design, mechanical and thermal management tools, have been developed and proved in functional demonstrators.

Bumped chip TAB has proven the lead technology for high density TAB, though virtually all current standard ICs can now also be considered in TAB using BTAB or ball bumped TAB.

The module technology has shown considerable potential for further development in density and scale, though full exploitation of the high speed potential is currently limited by the lack of good high frequency module connectors.

The authors thank the respective department directors in Bull, GEC-Marconi and British Telecom for permission to publish this paper and also acknowledge the support given by the EEC, ESPRIT programme and particularly the EEC co-ordinator, Guy Doucet.

References

1. R.P.Arrowsmith, N.Chandler, G.Dehaine, K.Kurzweil, "High Performance VLSI Interconnection Systems", Proc. 4th Annual ESPRIT Conference, Brussels, September 1987.

2. G.Dehaine, K.Kurzweil, "Tape Automated Bonding Moving into Production", Solid State Technology, October 1985.

Project No. 962E-17

THREE-DIMENSIONAL SIMULATION OF VLSI DEVICES

P. CIAMPOLINI, A. PIERANTONI, M. RUDAN and G. BACCARANI
Università di Bologna
Dipartimento di Elettronica, Informatica e Sistemistica
viale Risorgimento 2, 40136 Bologna, Italia

ABSTRACT. Due to the increasing influence of three-dimensional effects in miniaturized devices, a strong need for efficient and reliable 3D device-simulation tools is forseeable in the next years. This paper describes some results obtained at the University of Bologna in the context of the ESPRIT Project 962 (EVEREST). A 3D on-state, one-carrier device simulator has been developed, namely HFIELDS-3D, which has been tested on a number of realistic devices. While the discretization techniques adopted in 2D can be straightforwardly extended to the 3D case, numerical efficiency has to be substantially improved, in order to cope with the huge number of non-linear equations involved. Geometrical accuracy is important as well, this making the 3D-mesh generation quite a critical problem. Our program adopts a prism-based discretization scheme, which allows for an effective trade-off between complexity and flexibility. Computational efficiency has been enhanced, by means of iterative linear solvers, through co-operative efforts in the Project framework. Some practical applications are also presented, concerning three-dimensional effects arising in EPROM cell arrays.

1. Introduction

Numerical simulation of semiconductor devices in two dimensions has become a well-established technique for the design of advanced electronic components and processes. As device miniaturization progresses toward submicron feature sizes, however, three-dimensional effects are getting more and more important due to the following reasons:

- It is increasingly difficult, and perhaps not even convenient, to rigorously scale both horizontal and vertical device dimensions by the same scaling factor. As a consequence, fringing effects become more and more important even for nominally-standard planar devices;

- Individual devices are being packed more and more densely to reduce the chip size. Thus, lateral coupling between elementary components and insufficient isolation may become a problem;

- New concepts and new device structures are being devised, such as the buried-electrode dynamic RAM cell, presently being used for high-capacity memory devices, and the floating-gate EPROM and E^2PROM cells, which are inherently three dimensional.

In order to keep up with the increasing requirements of advanced technology, it is therefore necessary to develop highly-efficient and reliable three-dimensional simulation tools, to assist the process designer in the difficult task of appropriately sizing device geometries and accurately predicting the achievable performance.

In recognition of the above necessity, the EEC approved in 1986 a 4-year project aiming at the investigation of efficient three-dimensional algorithms for device simulation, to be validated within one project code and/or individual codes developed by different partners. In this paper, we describe the results achieved thus far at the University of Bologna, where a three-dimensional device simulator called HFIELDS-3D has been written. Even though this code is representative of a small fraction of the overall effort devoted to the project, it must be pointed out that its development has taken advantage from the interaction among the various partners. More specifically, HFIELDS-3D employs two linear solvers – one of which suitable for symmetric matrices and the other suitable for asymmetric matrices – developed in the context of the Project by P. Mole and coworkers from STC.

From a conceptual viewpoint, 3-D modelling techniques should not significantly differ from 2-D techniques; however, a major obstacle to be overcome is represented by the enormous increase of the computational requirements. An accurate device description in three dimensions requires some thirty-to-fourthy times more grid points than the corresponding 2-D description. Thus, a two order of magnitude increase in CPU time should be expected, which requires a careful selection of the algorithms to be used for the solution of the linearized system equations. On the other hand, recent boosting of computer performance makes now affordable the huge amount of computation which is typically needed by the three-dimensional simulation of a realistic device.

An additional major problem is related to mesh generation, which turns out to far more difficult in three dimensions than in two. This problem has been tackled by selecting triangle-based prismatic elements as the fundamental building blocks of our meshes. So doing, we can easily accommodate non-planar geometries in two dimensions, while being restricted to a step-like modification of the device cross section in the third dimension.

The physical model is reviewed in section 2; the discretization techniques are described in section 3, along with some details on the mesh generation. Section 4 includes the description of numerical techniques and highlights the need of high-performance linear-algebra modules. Some application examples are then described in section 5, related to three-dimensional effects arising in an EPROM-cell array: first the influence of the narrow-channel effect is investigated; then, the coupling between adjacent devices is simulated. Results are compared, whenever possible, with 2D simulations. Finally, the conclusions are drawn in section 6.

2. Physical model

HFIELDS-3D solves the fundamental semiconductor equations in steady state

$$\text{div}\,(\epsilon_s \,\text{grad}\varphi) = -q\,(p - n + N_D^+ - N_A^-) \tag{2.1, a}$$

$$\text{div}(\mathbf{J}_n/q) = R - G \tag{2.1, b}$$

$$\text{div}(\mathbf{J}_p/q) = G - R \tag{2.1, c}$$

comprising Poisson's (2.1,a) and carrier-continuity equations for electrons and holes (2.1,b-c) on a three-dimensional arbitrary domain. In the above equations, the symbols are given the usual meaning: namely, φ is the electric potential, n and p are the electron and hole concentrations, respectively, $N_D^+ - N_A^-$ is the net ionized impurity concentration, $G - R$ is the net generation rate per unit volume, $\mathbf{J}_n$ and $\mathbf{J}_p$ are the electron and hole current densities, given by the classical drift-diffusion model

$$\mathbf{J}_n = -q\mu_n n \operatorname{grad} \varphi + q D_n \operatorname{grad} n \qquad (2.2,\mathrm{a})$$

$$\mathbf{J}_p = -q\mu_p p \operatorname{grad} \varphi - q D_p \operatorname{grad} p \qquad (2.2,\mathrm{b})$$

A number of different boundary conditions can be accounted for by HFIELDS-3D: in particular, since an EPROM cell will be simulated, a suitable boundary condition has to be expressed for the floating-gate electrode. This is done by specifying the amount of charge Q_{FG} injected into the insulated gate and applying the Gauss theorem to the gate surface G :

$$\int_G (\mathbf{D})_\perp \, dG = Q_{FG} . \qquad (2.3)$$

3. Discretization Techniques

For the numerical solution, the above equations are to be expressed first in a discrete form: whatever the discretization scheme, this also requires the discretization of the spatial domain, and thus the generation of a mesh approximating the real device.

Topologically regular meshes are easy to design and lead to highly-regular matrices (usually band matrices), thus lending themselves to highly-efficient solution algorithms. However, the geometrical flexibility of this approach is very poor. On the other hand, irregular meshes provide much greater flexibility, usually at the expense of non trivial mesh generation problems. Nevertheless, such a flexibility is essential for a general-purpose tool: therefore HFIELDS, in its 2D version, adopts a triangular-element mesh onto which the so-called Box Integration Method (BIM, [1]), is applied. As already mentioned, when moving from 2D to 3D, geometrical accuracy plays an even more important role, this suggesting to preserve most of the 2D features in the geometry management; hence, triangular grids should evolve toward tetrahedral element meshes. Unfortunately, the peculiar properties of the unknown functions which can vary by orders of magnitude over small regions, make most of the tetrahedral mesh-generation techniques used in different fields – such as structural analysis, or hydrodynamics – unsuitable here. In fact, the problem of automatically generating tetrahedral meshes over three-dimensional domains of arbitrary shape, while locally providing "good-quality" elements, is not fully solved yet.

As a trade-off between geometrical flexibility and mesh-generation feasibility, HFIELDS-3D adopts a triangle-based prismatic element mesh, whose generation can easily be accomplished in two steps. First, a planar triangular grid is generated, using a standard 2D tool and accounting for curved shapes; then, the same grid is replicated over many parallel planes, to distribute the nodes along the third direction. This results in the three-dimensional extension of the BIM technique, whose features are illustrated below in some detail. Fig. 1 illustrates the resulting mesh-generation technique, consisting in the development of a triangular grid on the front plane and of a rectangular grid on the lateral one. The two grids are then merged to get the final result. It is worth observing that such a strategy does not force any translational symmetry in the structure, but only requires a description of the geometrical changes along the third direction in a step-like fashion, thus allowing for the description of a complex geometry in a fairly realistic way. A further advantage of this mesh structure is that most of the database is shared with HFIELDS-2D[2], in such a way

168

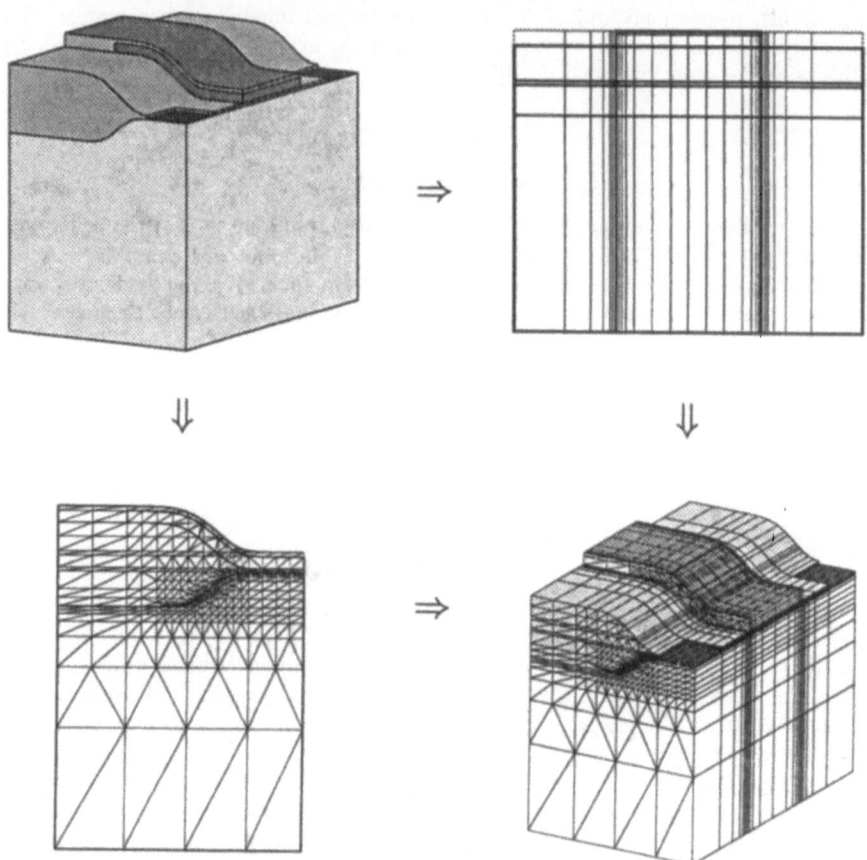

Figure 1: Construction of the 9407-node mesh of the simulated EPROM, resulting from a 409-node planar grid replicated on 23 successive planes. Due to symmetry, only half of the device is accounted for.

that the same user interface can be used. Once the grid has been obtained, equations (2.1) are integrated over each element, whose characteristic parameters are shown in Fig. 2. Each node is surrounded by several prisms (typ. 12): a "control volume" Ω_i, bounded by the surface Γ_i and made up by the shaded volumes of all the neighbouring prisms, is associated to the i-th node. Equations (2.1) then become:

$$\int_{\Gamma_i} \mathbf{D} \cdot \mathbf{i}_n \, d\Gamma_i = q \int_{\Omega_i} (p - n + N_D - N_A) \, d\Omega_i \tag{3.1,a}$$

$$\frac{1}{q} \int_{\Gamma_i} \mathbf{J}_n \cdot \mathbf{i}_n \, d\Gamma_i = \int_{\Omega_i} (R - G) \, d\Omega_i \tag{3.1,b}$$

$$\frac{1}{q} \int_{\Gamma_i} \mathbf{J}_p \cdot \mathbf{i}_n \, d\Gamma_i = \int_{\Omega_i} (G - R) \, d\Omega_i \tag{3.1,c}$$

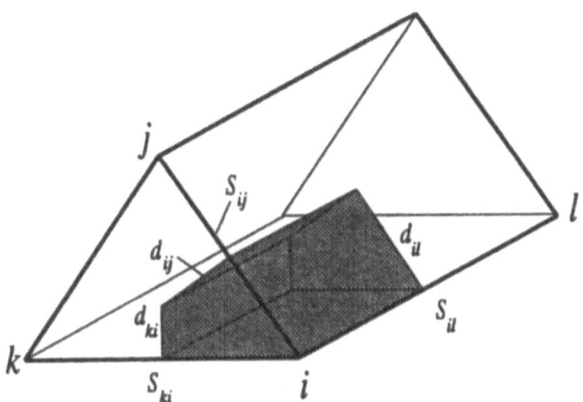

Figure 2: Construction of the elemental contribution to the box.

i_n being an outward–oriented normal versor. Poisson's equation (3.1,a) is discretized assuming a piecewise linear approximation for φ inside each element, while the discretization of the continuity equations (3.1,b) and (3.1,c) is carried out by means of a multi-dimensional generalization of the Scharfetter–Gummel method [3]. This leads to the following set of equations, related to the i-th node:

$$\epsilon_s \sum_{j \neq i} \frac{d_{ij}}{s_{ij}} (\varphi_i - \varphi_j) + q \, \Omega_i \, (n_i - p_i - N_i) = 0 \qquad (3.2,\mathrm{a})$$

$$D_n \sum_{j \neq i} \left[\frac{d_{ij}}{s_{ij}} (B_{ji} n_i - B_{ij} n_j) \right] + \Omega_i \, (R - G)_i = 0 \qquad (3.2,\mathrm{b})$$

$$D_p \sum_{j \neq i} \left[\frac{d_{ij}}{s_{ij}} (B_{ij} p_i - B_{ji} p_j) \right] + \Omega_i \, (R - G)_i = 0 \qquad (3.2,\mathrm{c})$$

being:

$$B_{ij} = \frac{q(\varphi_j - \varphi_i)/kT}{\exp\left[q(\varphi_j - \varphi_i)/kT\right] - 1}$$

the Bernoulli function. In (3.2) all the sums are extended to the nearest neighbours of the node i. Finally, as regards the boundary condition at the floating gate, the whole electrode, being assumed equipotential, acts as a single node and gives rise to an additional equation whose discrete form is the following:

$$\sum_j k_{fj} \cdot (\varphi_{FG} - \varphi_j) = Q_{FG} \qquad (3.3)$$

This equation is part of the Poisson system, the only practical difference consisting of a much larger number of non-zero terms in the f-th row, due to the larger number of adjacent nodes. The solution of Poisson's system then yields a self-consistent determination of the floating-gate potential φ_{FG}.

4. Numerical Techniques

The discretization transforms the original system, made of the three PDEs (2.1), into a set of $3 \times N$ non-linear algebraic equations – N being the number of nodes – whose linearized form has to be iteratively solved. For 3D applications, N usually ranges around some tens of thousands unknowns: to reduce the problem size, the Gummel successive procedure [4] may be used, whereby Poisson's and carrier-continuity equations are successively solved until an overall convergence is reached.

The solution of the linear system turns out to be the most demanding task as far as the computational resources are concerned: the direct solvers, commonly adopted for smaller size problems, perform very poorly as the number of unknowns increases: this is mainly due to the need of factorizing the system matrix, i.e., of decomposing it into two triangular (lower and upper) ones. This intrinsically expensive operation gives rise to the "fill-in" effect, which reduces the matrix sparsity and thus the efficiency of the algorithm.

By performing an approximate factorization and inhibiting the fill-in effect, the CPU time is significantly reduced. However, errors are introduced, which calls for iterative corrections. Methods based on this concept are called "iterative": among them, the Incomplete Cholesky Conjugate Gradient algorithm (ICCG [5]) is known to provide excellent results, and is applicable to symmetric systems (to Poisson's equation in this case). Therefore, we have compared the performances of the ICCG algorithm and of the SPARSPAK [6] direct solver adopted in the 2D version: Fig. 3 shows the behaviour of the two solvers as the number of unknowns increases. The break-even point occurs slightly above 2,500 nodes; for

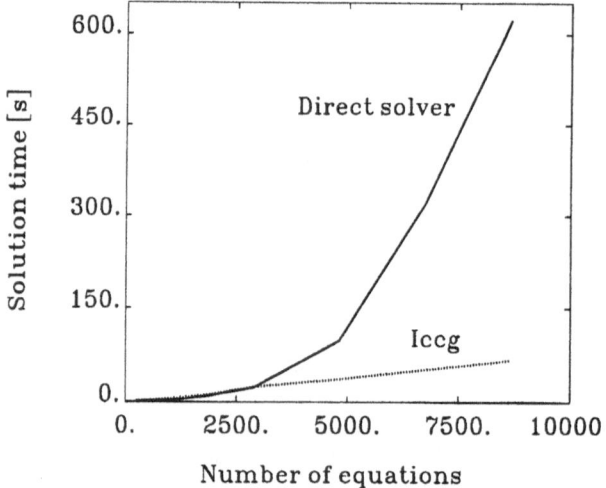

Figure 3: Solution time vs number of the unknowns for the linear system on a VaxStation 3200 ($\approx$ 3 MIPS). Comparison between ICCG and SPARSPAK.

higher values, the ICCG solver exhibits a remarkably slower raise.

Adopting the ICCG method to solve Poisson's equation, the related CPU time is reduced by over one order of magnitude for typical 3D-problem sizes. Similar improvements can be achieved with iterative solvers for non-symmetric matrices such as Conjugate Gradient Squatres (CGS) or Biconjugate Gradients (BG). These methods are necessary for the solution of the current-continuity equations.

The following table compares the performance of different solvers, reporting the total CPU times (on a Digital VS-3200 workstation) needed to simulate a $\sim 10,000$-node EPROM device in the on state. The data below are related to the whole simulation; hence, they include some set-up overhead:

Poisson	Continuity	CPU mins
direct	direct	$\sim 9,000$
iterative	direct	~ 360
iterative	iterative	~ 150

Thus, by adopting suitable linear solvers, CPU times, as well as memory requirements, can be strongly reduced, this making 3D simulations possible even without resorting to supercomputers.

5. Application examples

Thanks to VLSI technology, typical device size in large EPROM arrays can be scaled down to less than 1 μm. Such dimensions are small enough to influence both the short-channel and the narrow-channel effects, the prediction of which requires a three-dimensional analysis. To begin, a single cell has been simulated: Fig. 1 shows the 9,407-node discretization mesh which, taking advantage of the triangular-element flexibility, nicely accounts for the smooth transition between field and gate oxides and for the multi-layered gate structure. Due to symmetry, only half of the cell is drawn. The impurity profile has been expressed by means of analytical functions fitting experimental data.

Figures 4 and 5 show two different views of the computed potential pertaining to different section planes inside the cell. Fig. 4 refers to a vertical plane located along the channel symmetry axis. The "hole" above the channel region reflects the equipotential floating gate, where no equation is solved. Fig. 5 refers to a horizontal section located at the gate-oxide interface; it is worth noting the peculiar saddle shape of the channel region, responsible for the narrow-channel effect.

In order to investigate the behaviour of both the written and unwritten cells, two sets of turn-on characteristics have been simulated at different values of the charge injected into the floating gate. Each curve is compared in Fig. 6 with an homologous one computed on the corresponding 2D structure. The left curves refer to the unwritten cell (i.e., $Q_{FG} = 0$), while the right curves refer to the $Q_{FG} \neq 0$ case. As expected, injecting some charge results in a threshold-voltage shift. In both cases, the 2D currents stand well above the 3D ones: as shown in [7], this is mainly due to overestimating the channel width in the 2D case, where a constant width (i.e., the mask width) has been assumed. Fig. 7 shows the strong influence of the gate voltage over the effective channel width, which results in a modulation of both the device transconductance and the threshold voltage.

Another effect which may impose severe constraints on the design of the EPROM cell array is the so-called "drain turn-on" effect. Since all the drains of the cells pertaining to one row are connected to the same bit line, when a write operation is performed they are all raised to relatively high voltages (typ. 6-7 V), to induce hot-electron emission into

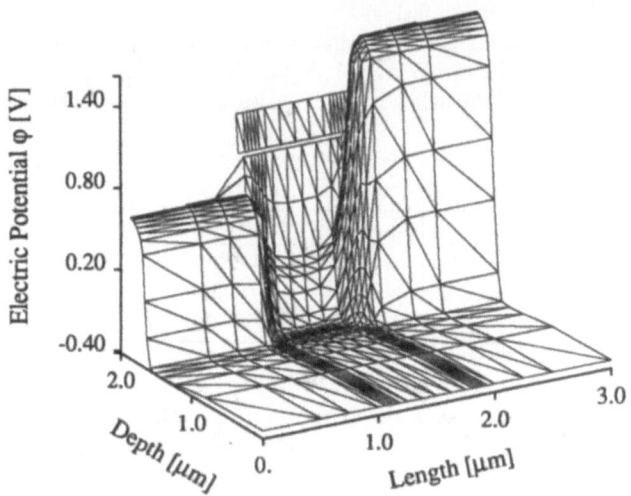

Figure 4: Perspective plot of the electric potential along a planar section, parallel to the current flow (i.e., vertical and normal to the front plane with respect to Fig. 1) and located at the middle of the channel: $V_{DS} = 1$ V, $V_{GS} = 1.6$ V, $Q_{FG} = 0$ C/cm^2.

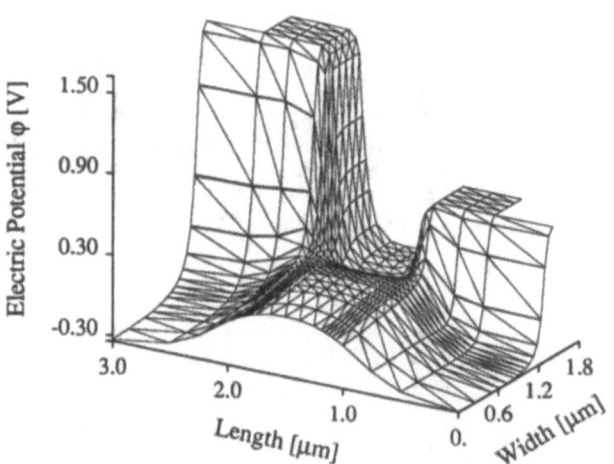

Figure 5: Perspective plot of the electric potential along a planar section, parallel to the channel interface (i.e., horizontal with respect to Fig. 1) and located at the inversion layer: $V_{DS} = 1$ V, $V_{GS} = 1.6$ V, $Q_{FG} = 0$ C/cm^2.

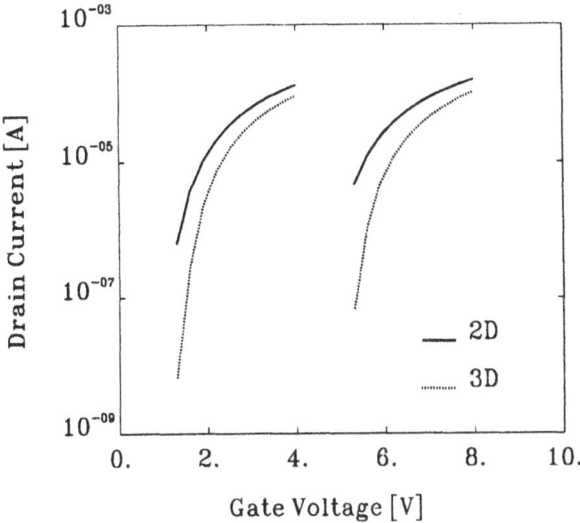

Figure 6: Transfer characteristics of the simulated cell. The curves on the left refer to "unwritten" cells ($Q_{FG} = 0$ C/cm^2) while the curves on the right refer to "written" cells ($Q_{FG} = -5.525 \times 10^{-15}$ C/cm^2).

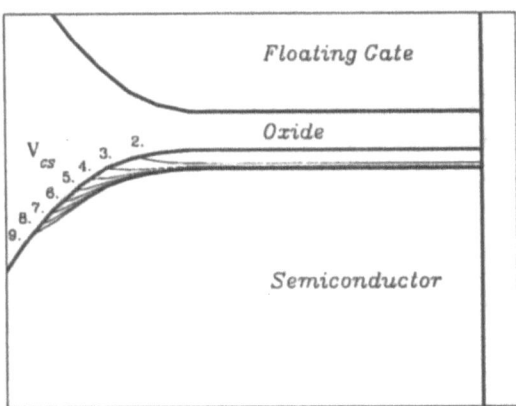

Figure 7: Expanded cross-sectional view of the channel region. The contour lines indicate the inversion layer edge, at different bias conditions: $V_{GS} = 2 \rightarrow 9$ V, $V_{DS} = 0$ V, $Q_{FG} = 0$ C/cm^2.

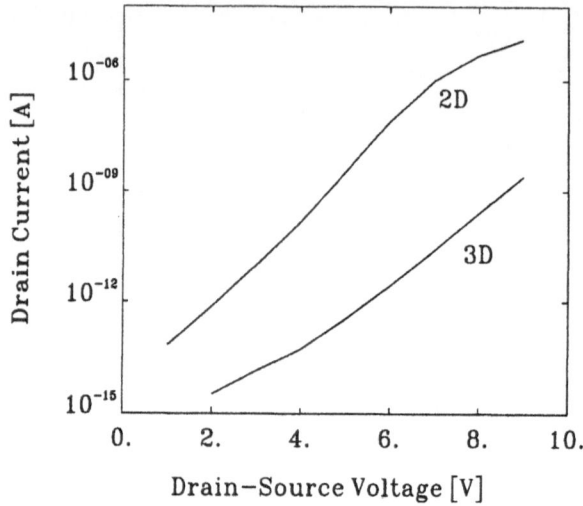

Figure 8: Drain turn-on characteristics of the simulated MOSFETs: $V_{GS} = 0$ V, $Q_{FG} = 0$ C/cm^2.

the selected floating-gate. Due to capacitive coupling between floating gates and drain regions, the floating gate potential for the unselected cells may significantly increase, which in turn induces larger leakage currents and possibly causes an incorrect write operation. The quantities driving this effects are the coupling capacitances among the second-level gate, the floating gate and drain region of each cell.

If a 2D analysis is performed, the region overlapping the bird's beak and the field oxide, which gives a significant contribution to this capacitive coupling, cannot be accounted for. This results in a reduced coupling between the floating gate and the second-level gate, which causes an unrealistic boosting of the drain turn-on effect, as shown in Fig. 8.

To further investigate the design constraints for the EPROM-cell array, a more complex structure has then been simulated, made up of two cells placed side by side. The 13,000-node discretization mesh is shown in Fig. 9. The region between the two cells acts as a parasitic MOSFET, which can be turned on by applying sufficiently-high voltages to the common gate electrode. To the purpose of artificially enhancing the parasitic effect, the oxide-thickness has been slightly undersized. In order to quantify the parasitic coupling, the turn-on characteristics have been simultaneously computed for one cell and for the parasitic MOSFET for different values of the floating-gate charge, as shown in Fig. 10.

Potential distributions at the non planar Si-SiO$_2$ interface are shown in Figs. 11, 12 and 13, giving evidence to the rise of a parasitic current flow under the field oxide, as well as to the depleting action of the charge stored in the floating gate. Finally, the evolution of the inversion layer edges at the same interface are shown in Figs. 14 and 15 for different values of the gate bias. The figures clearly show the influence of the floating-gate charge on the parasitic MOSFET.

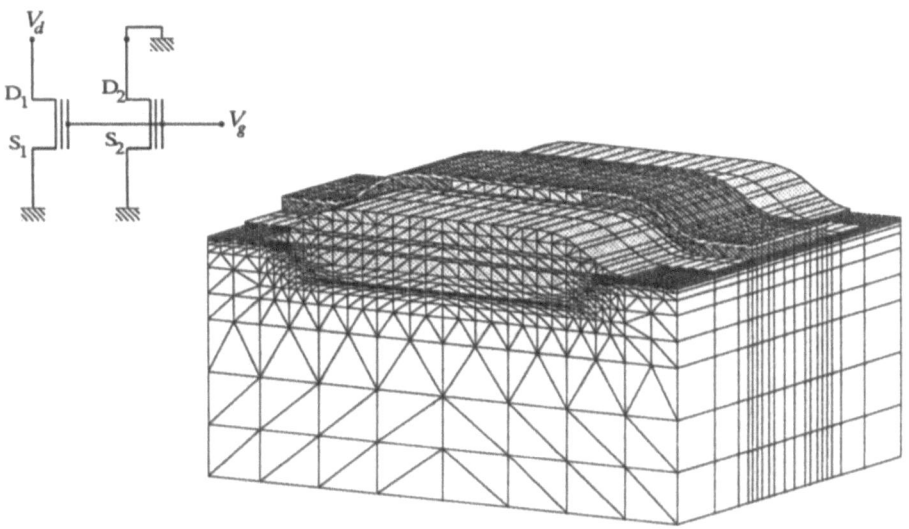

Figure 9: The 13,320-node mesh of the duel-EPROM structure, resulting from a 666-node planar grid replicated on 20 successive planes. Due to their symmetry, only half of each device is shown.

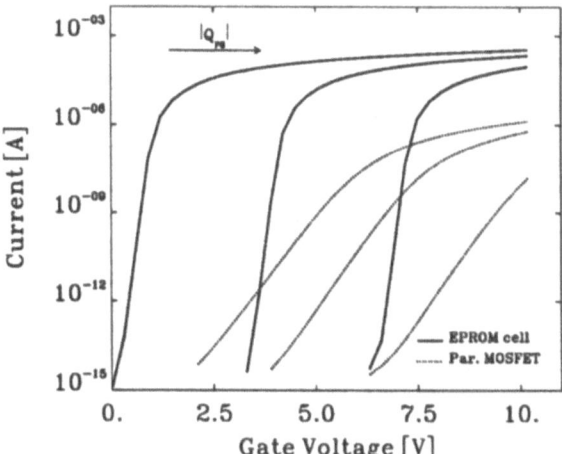

Figure 10: Computed turn-on characteristics: solid lines refer to the T1 EPROM cell, dashed lines refer to the parasitical MOSFET between T1 and T2: $V_{D1} = 1$ V, $V_{S1} = V_{S2} = V_{D2} = 0$ V.

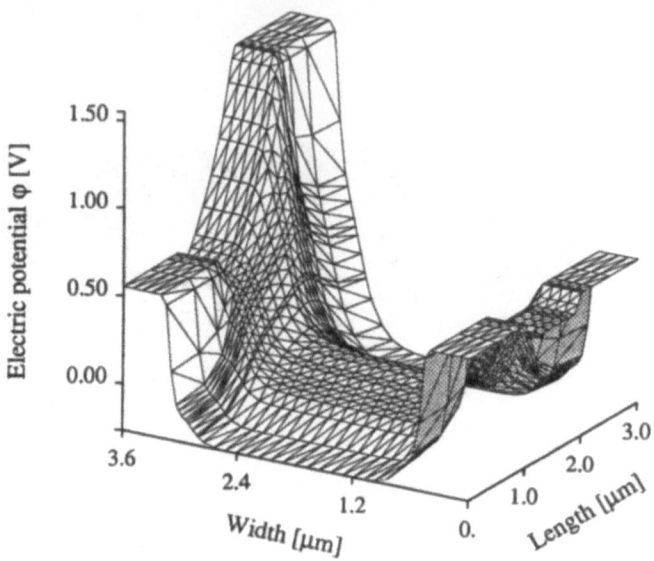

Figure 11: Potential distibution along the $Si - SiO_2$ interface: $V_{D1} = 1$ V, $V_{S1} = V_{S2} = V_{D2} = 0$ V, $V_G = 3$ V, $Q_{FG} = 0$ C/cm^2.

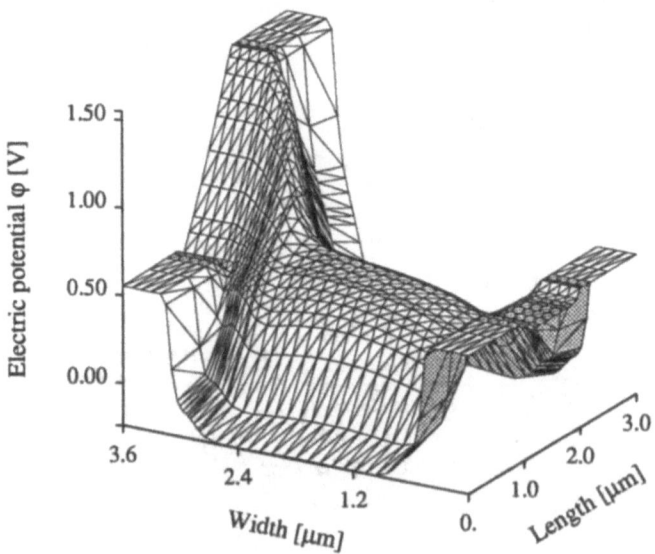

Figure 12: Potential distibution along the $Si - SiO_2$ interface: $V_{D1} = 1$ V, $V_{S1} = V_{S2} = V_{D2} = 0$ V, $V_G = 8$ V, $Q_{FG} = 0$ C/cm^2.

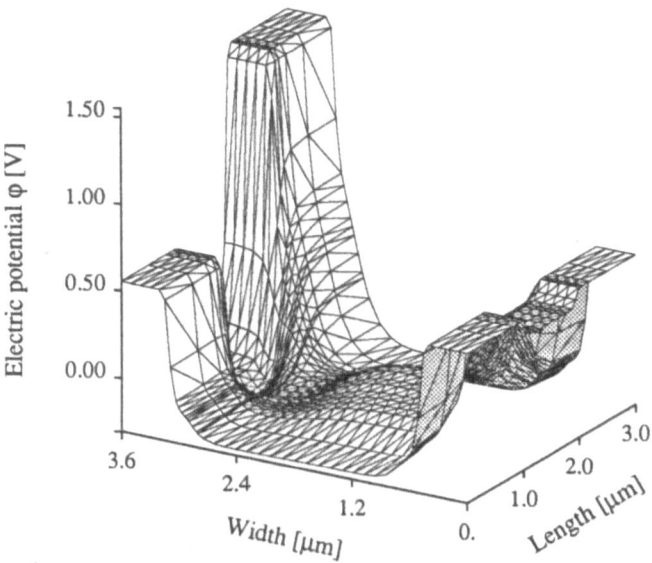

Figure 13: Potential distibution along the $Si - SiO_2$ interface: $V_{D1} = 1$ V, $V_{S1} = V_{S2} = V_{D2} = 0$ V, $V_G = 3$ V, $Q_{FG} = -5.525 \times 10^{-15}$ C/cm^2.

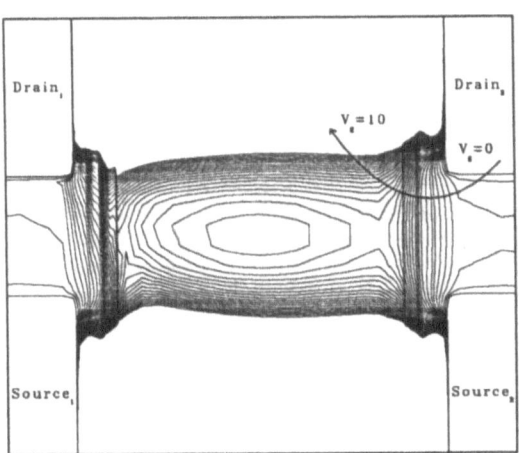

Figure 14: Contour lines, computed at the $Si - SiO_2$ interface, showing the inversion layer edges at different gate voltages: $V_{D1} = 1$ V, $V_{S1} = V_{S2} = V_{D2} = 0$ V, $Q_{FG} = 0$ C/cm^2.

178

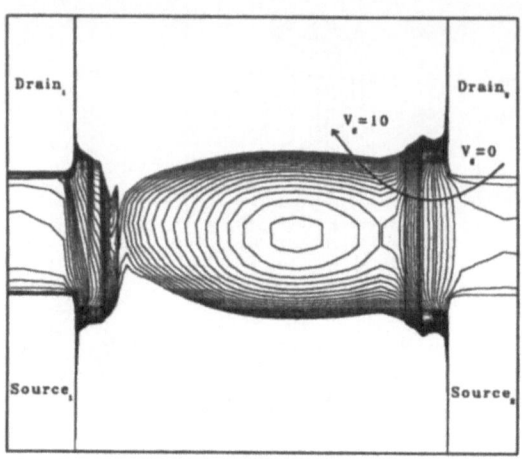

Figure 15: Contour lines, computed at the $Si - SiO_2$ interface, showing the inversion layer edges at different gate voltages: $V_{D1} = 1$ V, $V_{S1} = V_{S2} = V_{D2} = 0$ V, $Q_{FG} = -5.525 \times 10^{-15}$ C/cm^2.

6. Conclusions

In this paper, a three-dimensional device simulator has been presented, and the future need for efficient and flexible 3D tools is envisaged, following forseeable advances in VLSI technology. The practical success of such tools is probably conditioned by their capability to overcome some critical problems: improvements in numerical and computational efficiency should allow for 3D simulation on low-cost workstations. The rapidly-evolving scene in single-chip processor architectures is likely to help this process. Furthermore, the greatest geometrical flexibility has to be ensured while keeping mesh generation as easy as possible.

To partially fulfill these requirements, our program adopts a prismatic-element discretization mesh, which allows for considerable simplification of both mesh-generation and discretization problems, without introducing too-severe limitations on device geometry. Numerical efficiency has been enhanced by means of suitable iterative solvers, leading to a remarkable reduction in computational requirements.

The current status of the program is still largely incomplete: nevertheless, the simulation of some realistic EPROM cells allowed us to highlight the occurrence of phenomena which cannot be accounted for by a 2D device simulator. Such effects, on the other hand, may have a significant influence on the device behaviour and on the definition of the process design rules.

References

[1] R. S. Varga, *Matrix Iterative Analysis*, Englewood Cliffs: Prentice Hall, 1962

[2] G. Baccarani, R. Guerrieri, P. Ciampolini, M. Rudan, HFIELDS: a *Highly Flexible* 2-D *Semiconductor-Device Analysis Program*, Proc. NASECODE IV Conf., Dublin, 1985

[3] D. L. Scharfetter e H. K. Gummel: "Large-Signal Analysis of a Silicon Read Diode Oscillator", *IEEE Trans. on Electron Dev.*, vol. ED-16, pp. 64-77, 1969.

[4] H. K. Gummel, *A Self-Consistent Iterative Scheme for One-dimensional Steady State Transistor Calculations*, IEEE Trans. on Electron Dev., vol. ED-11, 1964.

[5] D.S. Kershaw, *The Incomplet Cholesky Conjugate Gradient Method for the Iterative Solution of Systems of Linear Equations*, J. of Computational Physics, n. 26, 1978

[6] A. George, J. Liu, E. Ng, *User Guide for SPARSPAK*, Research Report CS-78-30, Waterloo: University of Waterloo, 1980.

[7] P. Ciampolini, A. Gnudi, R. Guerrieri, M. Rudan, G. Baccarani, *Three-dimensional Simulation of a Narrow-Width MOSFET*, Proc. of ESSDERC 87, pp. 413–416, Bologna, 1987

Acknowledgement

The authors wish to acknowledge the contribution of P. Mole and his coworkers at STC, who made available their linear solvers based on ICCG and PCGS. Support from SGS-Thomson Microelectronics is also gratefully acknowledged.

Project No. 962E-17

Software for Modelling Semiconductor Devices in Three Dimensions

C. Greenough

Rutherford Appleton Laboratory
Chilton, Didcot
Oxfordshire OX11 0QX

D. Gunasekera, P.A. Mawby, M.S. Towers

University College, Swansea
Singleton Park
Swansea SA2 8PP

C.J. Fitzsimons

Numerical Analysis Group
Trinity College Dublin
Dublin 2

ABSTRACT: In recent years three-dimensional modelling of semiconductor devices has become increasingly important due to the continued miniaturisation of devices. There has been a corresponding increase in the research devoted to developing three-dimensional numerical models of devices. Here we discuss some of the work in the ESPRIT project EVEREST relating to this. We describe in detail the software implementation of the algorithmic techniques being developed in the project.

1 Introduction

The numerical simulation of semiconductor devices in two dimensions is well-established for the design of advanced electronic components and processes. As device miniaturization approaches submicron feature sizes, three-dimensional effects are becoming important even for nominally standard planar devices. In many devices key physical effects occur at corners where three-dimensional models are the only way to obtain an accurate soution. The need to simulate the essentially three-dimensional effects of threshold shift for small channel devices, channel narrowing and the accumulation of carriers at the channel edge is growing. These effects make two-dimensional simulation codes inadequate for device-performance prediction.

EVEREST is a four-year project supported by the European Community under the European Strategic Program for Research in Information Technology (ESPRIT), that is investigating suitable algorithms for the analysis of semiconductor devices in three dimensions, and developing software implementing the most effective of those algorithms. Both Japanese and USA industries have already been very active in this area of three-dimensional modelling and it is appropriate that a joint European effort be focused on the solution of such problems.

The code under development within the project provides Europe with an important set of analysis tools which will ensure that the European semiconductor industry is well placed in the race to develop new device structures.

2   Structure of the Project

The partners participating in this project are drawn from some of the major industrial and academic research and development teams in Europe. They are listed below.

- Analog Devices (Ireland)

- Philips (The Netherlands)

- SGS-Thompson (Italy)

- STC Technology Ltd (UK)

- Rutherford Appleton Laboratory (UK - Project Leader)

- NMRC (Ireland)

- IMEC (Belgium)

- University of Bologna (Italy)

- Trinity College Dublin (Ireland)

- University College Swansea (UK)

As can be seen, the list comprise four industrial partners, three large research laboratories and three universities. The Rutherford Appleton Laboratory is the overall project leader and prime contractor of the project.

Five main areas of activity have been identified and the project is divided into five workpackages, each with its own Workpackage Director. These five workpackages are:

- Physical Models and Validation (SGS)

- Discrete Problem Formulation (Philips)

- Mesh Generation and Refinement (UCS)

- Solution Procedures (Philips)

- Project Code (RAL)

The first four of these workpackages addresses one aspect of device modelling and the results produced are to be represented in the project code. The project will direct some 100 man-years toward developing solutions of the problems in simulating semiconductor devices.

2.1 Physical Modelling and Validation

Correct and reliable physical models are the key to establishing the validity and accuracy of numerical device simulation in two or three dimensions so as to account for the effects of mobility, heavy doping, etc. This workpackage is investigating the validity of the physical models being used within the project by the development of physical and analytical models, by parameter extraction to produce values to characterise the numerical models and by testing the research code against benchmark structures.

2.2 Discrete Problem Formulation

The objective of this workpackage is to investigate the different approaches to the discretisation of the conventional continuum device modelling equations. The extension of the standard Scharfetter-Gummel [1] scheme into two and three dimensions are being considered as well as the use of fitted and mixed finite element methods. A second strand to this workpackage is the development of error estimates to be used to control the propogation of errors in the solution as adpative mesh techniques are used.

2.3 Mesh Generation and Refinement

As three-dimensional simulations are very complex in nature it is important to develop effective techniques to generate the discrete-model mesh automatically. There is considerable activity in developing strategies to produce meshes in three-dimensions with the potential to be used adaptively. These techniques must the be coupled with the element selection algorithms to perform mesh refinement.

2.4 Solution Procedures

The most computationaly intensive part of majority of the numerical simulations is the linear algebra. Within the project there is a continual effort in improving the solution procedures being used. Currently the two most successful methods being used within the project are Incomplete Choleski Conjugate Gradients (ICCG) and the Conjugate Gradient Squared Algorithm (CGS). Research is continuing to improve these solution techqniues.

2.5 Project Software Suite

An important part of the project is the design and implementation of a complete simlutation system using the best consistent set of the algorithms developed by the project. This part of the project will be described in detail in the next sections.

3   Some Difficulties of Three-Dimensional Modelling

Although the implementation of a three-dimensional simulator is conceptually quite simple, the transition from two dimensions to three dimensions presents a significant number of problems in both the physical and numerical models. Some problems have arisen in the following areas during this transition:

- applicability of physical models in three dimensions,

- extension of discretisation schemes for the governing equations,

- geometric modelling of devices,

- mesh generation and refinement,

- impurity profile generation,

- solution of linear systems

- massive storage and *cpu* requirements.

Although, in general, the mathematical models describing the physical behaviour of semi-conductor devices currently used in two-dimensional simulators appear applicable in three-dimensional analyses, some care is necessary where models are dependent on vector quantities, such as the field-dependent mobility models.

From the numerical standpoint, one of the most important differences between two-dimensional and three-dimensional device-analysis programs stems from the massively larger number of mesh points necessary for an adequate description of a three-dimensional structure. In two dimensions between 1500 and 2500 grid points are sufficient for modelling many device types. If a similar density of grid points is assumed necessary in the third dimension, grids of some 30,000 points will be needed.

As the characterisation of the behaviour of a device requires the repeated solution of the sparse linear systems arising from the linearisation of the discretised equations, with the number of unknowns fast approaching 50,000 much attention must be given to the linear algebra used in three-dimensional simulators. This will require the optimisation of the basic algorithms and the techniques used to store the simulation data. As a matter of course the large *cpu* requiement will lead to the use of new parallel algorithms suitable vector or parallel processors.

Many of the other problems encountered in three dimensional simulations are not peculiar to device modelling. The areas of geometric modelling, mesh generation and display have been addressed in other fields such as computational fluid dynamics and electromagnetics, and we are able to draw on the experience in these areas.

4   Software Development

The design and implementation of a three-dimensional simulator is perhaps the most challenging task of the EVEREST Project, as it involves a co-operative effort between several partners and a clear definition of suitable interfaces between input pre-processor, mesh generator, numerical solver and display facilities.

In this section the software being developed within the EVEREST Project is described and, where appropriate, some details are given. The software consists of four main modules:

- A Simple Geometric Modeller and Mesh Generator

- An Impurity Profile Generator

- An On-State Transient Solver Module

- A Post-Processor

These elements of software are being developed at three sites as independent modules linked together by a common command parsing environment and data filing system.

4.1   The Geometric Modeller and Mesh Generator

These modules provide the project with a basic capability in geometric modelling and mesh generation. The geometric model is described in terms of a set of basic primitives: points, lines, surfaces and volumes. These can be used to construct a geometric model of most device structures. To enable the user to generate a geometric description a library of *standard* devices has been defined. This contains skeleton descriptions of a number of generic device structures.

The mesh generator then fills the model with tetrahedral elements and computes the subvolumes and subareas associated with the discretisation scheme.

4.2   The Impurity Profile Module

This module allows the user to implant the silicon device with either uniform or Gaussian-type impurity profiles. The uniform profiles are applied to defined geometric volumes and the Gaussian profiles are applied through associated implantation windows and simulate predeposition, drive-in diffusion and ion implantation. These simple tools provide a sufficient set to enable us to test the analysis code. Although the specification of impurity profiles in terms of uniform and Gaussian distributions is a very poor approximation to reality they provide an effective source of profiles in the development phase of the project.

To supplement these simple generations an interface to external profiles has been provided. It is possible to take a one-dimensional profile obtained from a process simulator or by experiment and distribute this over a doping window. This provides the simulator with a source of more realistic impurity profiles.

4.3   The Analysis Module

The analysis module is a fully-coupled three-dimensional on-state solver using the techniques developed in the rest of the project. The details of this module are described below. It provides the user with complete control over problem specification and the generation of results.

4.4   The Post Processor

The final module of the system and one of significant importance is the post-processor. This provides the user with the basic functionality to display the results of the analysis module. The module allows the user to display geometric, mesh and results data in a variety of ways: contour plotting and isometric projections, vector plots and one-dimensional graphs along sections.

5   The Numerical Model

5.1   The Governing Equations

The physical behaviour of a semiconductor device is described by a system of three partial differential equations subject to suitable boundary and initial conditions:

$$\epsilon . \nabla^2 \psi = -\rho \tag{1}$$

$$q\frac{\partial p}{\partial t} = -\nabla.\mathbf{J}_p - qR \tag{2}$$

$$q\frac{\partial n}{\partial t} = \nabla.\mathbf{J}_n - qR \tag{3}$$

where ψ denotes the electrostatic potential, $\mathbf{J}_n$ and $\mathbf{J}_p$ the electron and hole current concentrations respectively, R the net recombination rate, ρ the charge concentration and q the electronic charge. We refer the reader to [6, 7] for detailed discussions about (1-3) and to [14] for descriptions of the different boundary and initial conditions which apply.

The charge concentration ρ is given by

$$\rho = q(p - n - (N_A - N_D)) \tag{4}$$

where N_A and N_D are acceptor and donor atom concentrations, q the electron charge.

The expressions for the currents are derived from the Boltzmann transport equation. These are

$$\mathbf{J}_p = -q\mu_p(V_T\nabla p + p\nabla(\psi - V_T\log(n_i))) \tag{5}$$

$$\mathbf{J}_n = q\mu_n(V_T\nabla n - n\nabla(\psi - V_T\log(n_i))) \tag{6}$$

where n_i denotes the intrinsic concentration and V_T the thermal voltage. There are two types of boundary condition applied to the problem. As the contacts are assumed to be Ohmic both quasi-Fermi levels are equal to the applied bias.

$$\phi_p = \phi_n = V_{app} \tag{7}$$

Using charge neutrality

$$p - n - (N_A - N_D) = 0 \tag{8}$$

in conjunction with Boltzmann approximations for carrier concentrations gives Dirichlet boundary conditions for ψ, p and n.

On the other boundaries of the device Neumann boundary conditions are applied. These simply state that hole and electron current densities and electric field strength normal to the boundary vanish. This condition leads to

$$\nu.\mathbf{J}_n = \nu.\mathbf{J}_p = \nu.\nabla\psi = 0. \tag{9}$$

There are a number of other external and interface conditions that can be applied to devices. A more complete description some of these is found in [14]

The software incorporates the standard physical models for recombination and mobility. These are being updated as the project proceeds.

The recombination rate R is controlled by three mechanisms. At low and medium injection levels the dominant recombination mechanism is Shockley-Read-Hall [8], at high injection this supplemented by the Auger model[2]. The breakdown of a device due to the cumulative multiplication of free carriers under a strong electric field is modelled by the avalanche formula[15].

Another physical process that is modelled is the electron and hole mobility. There are a number of effects that contribute to the mobility models. The structure and temperature of the silicon lattice, the impurities present in the device, and the electric field present when voltages are applied. Expressions for μ_p and μ_n are given below; but we refered the reader to [7, 14] for a detailed discussion about mobility models. The lattice mobility is:

$$\mu_L = \mu_0(\frac{T}{300})^\alpha \tag{10}$$

The doping dependent carrier mobility model is

$$\mu(N) = \mu_{min} + \frac{\mu_{max} - \mu_{min}}{1 + (N_T/N_{ref})^\alpha} \tag{11}$$

The mobility for holes and electrons is specified when appropriate μ_0, μ_{min}, μ_{max}, N_{ref}, N_T, T and α is used, where

$$N_T = 0.34(N_A + N_D) + 0.66(p + n) \tag{12}$$

The doping dependent mobility model can be modified to include the effect of electric field.

$$\mu(N, E) = \mu(N) \cdot \frac{1}{[1 + (\mu(N)|E|/v_{max})^\beta]^{1/\beta}} \tag{13}$$

As devices become small the impurities present modify the intrinsic behaviour of the silicon. An effective intrinsic carrier concentration which models bandgap narrowing was introduced by Slotboom[9]. It is implemented according to the formula of Lanyon and Tuft [4].

$$n_{ie} = n_i \exp(\frac{3q^3}{32\pi(\epsilon kT)^{3/2}}\sqrt{N_A + N_D}) \tag{14}$$

5.2 The Choice of Dependent Variables

Though ψ, p and n are the logical choices for the dependent variables from the physics point of view, there are numerical reasons to choose other variables. Other common sets of dependent variables are: ψ, ϕ_p, ϕ_n and ψ, Φ_p, Φ_n.

$$\phi_p = V_T \log(p/n_i) + \psi \tag{15}$$

$$\phi_n = \psi - V_T \log(n/n_i) \tag{16}$$

$$\Phi_p = \exp(\phi_p/V_T) \tag{17}$$

$$\Phi_n = \exp(-\phi_p/V_T) \tag{18}$$

The variables ϕ_p and ϕ_n are the quasi-Fermi potentials. Φ_n and Φ_p are often called the Slotboom variables; they have no physical significance.

All three sets have advantages as well as disadvantages. There is a trade off between the exponential character of the unknown functions and the nonlinearity of the governing equations. The exponential character of an unkown determines the range of values it takes.

The set ψ, p, n yields the most linear operator, but there is a large variation in the magnitude of the variables and a Jacobian that is badly conditioned. ψ, ϕ_p, ϕ_n are of comparable magnitude, but they produce an operator which is highly nonlinear. ψ, Φ_p, Φ_n yield a symmetrical Jacobian but the variation in magnitude is too large for them to be expressed on most computers, thus rendering them useless in practice.

The EVEREST simulator uses ψ, ϕ_p, ϕ_n in the basic discretisation of the equations. Since it uses the correction transform technique in the nonlinear algebra [6], the nonlinear iterations are performed in the variables ψ, p and n.

5.3 Spatial Discretisation

The spatial discretisation is performed using a control region approximation [16]. This discretisation method for partial differential equations, cast in conservative form, has many properties making its use attractive in semiconductor device modelling. It also allows the convenient enforcement of natural boundary and interface conditions.

Control regions, or volumes, tessellate the whole of the solution domain and although they could be of any shape, they are simply connected convex polytopes surrounding a set of pre-defined points in the domain. A precise definition follows.

Let $\{p_i\}_{i=1,n}^n$ be a set of points in Ω, the domain of the problem; this set is called a mesh and the points nodes. With each node an associated control region is given by:

$$V_i = \{(x, y, z) \in \Omega : \|(x, y, z) - p_i\| < \|(x, y, z) - p_j\| \; \forall p_j, j \neq i\} \tag{19}$$

These regions are also known as Voronoi[10] polytopes and the collection of regions, $\{V_i\}_{i=1,n}^n$, tessellates Ω which is then known as a Dirichlet[12] tessellation. The dual of a Dirichlet tessellation of a three-dimensional region Ω is a tetrahedral mesh. It is from this tetrahedral finite element mesh that the Voronoi volumes are formed.

Let us further define cross-sectional subareas $\{a_i\}_{i=1}^6$, Voronoi subvolumes $\{v_i\}_{i=1}^4$ and pipe lengths $\{l_i\}_{i=1}^6$.

a_i = Area of the quadrilateral generated by connecting the midpoint of edge i, circumcentres of the two faces common to edge i and the circumcentre of tetrahedron.

v_i = Volume of the hexahedron generated by the closure of subareas of the edges common to node i and the faces common to node i.

l_i = Distance between nodes j and k.

Flux balances for control regions are performed by summing up the contributions from every associated tetrahedral element.

When discretising equation (1) , the contribution to the flux balance in control region i from tetrahedron k is

$$F_{ik} = -\sum_{j=1}^{3} \epsilon \frac{(\psi_j - \psi_i)}{l_j} a_j - q(p_i - n_i - (N_A - N_D)_i)v_i \tag{20}$$

where j refers to the three edges which are common to node i and to the nodes at their other end. The flux balance is

$$\sum_{k=1}^{m} F_{ik} = 0 \tag{21}$$

where m is the number of tetrahedra common to node i. We discuss the discretisation of the hole equation explicitly. An analogous discussion holds for the electron equation. The contribution to the flux balance of the hole continuity equation in control region i from tetrahedron k is given by

$$F_{ik} = \sum_{j=1}^{3} a_j J_{pj} + v_i R_i \tag{22}$$

where j refers to the three edges which are common to node i and to the nodes at their other end. The flux balance is given by

$$\sum_{k=1}^{m} F_{ik} = 0 \tag{23}$$

The current density between two nodes, approximated by a constant expressed in terms of nodal potentials (in a form which avoids cancellation problems) is derived below. This method was first presented in [6].

$$\mathbf{J}_p = -q\mu_p(V_T\nabla p + p\nabla(\psi - V_T\log(n_i))) \tag{24}$$

Let

$$V = \psi - V_T\log(n_i) \tag{25}$$

Using an integrating factor of $\exp(\frac{\nabla V x}{V_T})$ on (24)

$$\exp(\frac{\nabla V x}{V_T}).\mathbf{J}_p = -q\mu_p V_T \exp(\frac{\nabla V x}{V_T}).(\nabla p + p\frac{\nabla V}{V_T}) \tag{26}$$

Integrating (26) from node i to node j gives

$$\frac{V_T}{\nabla V}[\exp(\frac{\nabla V x}{V_T})]_0^l J_p = -q\mu_p V_T[p\exp(\frac{\nabla V x}{V_T})]_0^l \tag{27}$$

Rearranging this gives

$$J_p = -q\mu_p\frac{(V_j - V_i)}{l}\frac{[p_j\exp(\frac{V_j-V_i}{V_T}) - p_i]}{[\exp(\frac{V_j-V_i}{V_T}) - 1]} \tag{28}$$

Using

$$p = n_{ie}exp(\frac{\phi_p - \psi}{V_T}) \tag{29}$$

in (28) gives

$$J_p = -q\mu_p\frac{(V_j - V_i)}{l}p_i\frac{[\frac{n_{iej}}{n_{iei}}\exp(\frac{\phi_{pj}-\psi_j-\phi_{pi}+\psi_i}{V_T})\exp(\frac{V_j-V_i}{V_T}) - 1]}{[\exp(\frac{V_j-V_i}{V_T}) - 1]} \tag{30}$$

By using (24) this reduces to

$$J_p = -q\mu_p\frac{(V_j - V_i)}{l}p_i\frac{[\exp(\frac{\phi_{pj}-\phi_{pi}}{V_T}) - 1]}{[\exp(\frac{V_j-V_i}{V_T}) - 1]} \tag{31}$$

which can be rewritten as

$$J_p = -\frac{q\mu_p V_T p_i}{l}\delta_1(\frac{V_j - V_i}{V_T})\delta_2(\frac{\phi_{pj} - \phi_{pi}}{V_T}) \tag{32}$$

where $\delta_1(z) = z/(\exp(z) - 1)$ and $\delta_2(z) = \exp(z) - 1$.

The function $\delta_1(z)$ which must be coded with care to avoid zero division at $z = 0$, is implemented in the following manner.

$$\delta_1(z) = \begin{cases} z/(exp(z) - 1) & if|z| \geq 10^{-4} \\ 1 - z/2(1 + z/6)(1 - z^2/60) & if|z| < 10^{-4} \end{cases} \tag{33}$$

For completeness we present the expression for the electron current denisty

$$J_n = \frac{q\mu_n V_T n_i}{l}\delta_1(\frac{V_i - V_j}{V_T})\delta_2(\frac{\phi_{ni} - \phi_{nj}}{V_T}) \tag{34}$$

where δ_1 and δ_2 are as before. The derivatives of J_p and J_n are calculated from the derivatives of $\mu, p_i, n_i, \delta_1(z)$ and $\delta_2(z)$ using the chain rule.

6 The Solution Process

The solution of the nonlinear equation system arising from the discretisation the semiconductor equations is split into three stages: initial guess, nonlinear iteration scheme and linear solution. In this section we discuss the nonlinear scheme.

The initialisation scheme is a variation of Edwards *etal.* [13], which projects quasi-Fermi levels from one solution point to another by assuming zero change in the local divergence of current, and which works well for bipolar transistors and MOSFETs. Solution of (35) and (36) with constant up to date p and n, yields increments to quasi-Fermi levels. Hence new values of ϕ_p and ϕ_n can be calculated. Next, the electrostatic potential is determined by holding the majority carrier concentrations constant.

$$\nabla(\mu_n n \nabla(\delta\phi_n)) = 0 \tag{35}$$

$$\nabla(\mu_p p \nabla(\delta\phi_p)) = 0 \tag{36}$$

The nonlinear process consists of a Gummel iteration scheme and a fully coupled iteration scheme bound by a continuation method on the applied bias. In the current version of the solver the inner iterations of the Gummel scheme and the iterations of the fully coupled scheme are of damped Newton-Raphson type to which the correction transformation technique [6] is applied. Careful control of the error tolerances in the Gummel algorithm ensure that it is used from the initialisation to near the edge of the Newton convergence sphere of the fully coupled system[11]. The final solution is obtained with a fully coupled Newton-Raphson. The nature of the Gummel loops, correction transformation and damping can be control in the simulator by the user.

The linear systems which arise from the nonlinear discretisation are sparse. The linear solvers have been written specially to exploit this sparsity and to minimise storage requirements. The discretisation of the governing equations yield symmetric positive definite matrices which are solved by the conjugate gradient method with incomplete Choleski preconditioning[5]. The nonsymmetric systems which arise and from the fully coupled problem are solved by the conjugate gradient squared method with incomplete LU preconditioning[3]. The implementations [17] and [6] are used.

7 Numerical Results

We present results from the three-dimensional simulation of an MOSFET device taken from the EVEREST library of standard geometric models. The basic structure is shown in Figure 1.

Although this example is simple by nature it holds many of the necessary elements to test a three-dimensional simulator. The device is three-dimensional in structure and the results will have three-dimensional characteristrics. However the example does not attempt to simulate the serious three-dimensional problems associated with short and narrow channels.

The MOSFET is formed from a $4 \times 3 \times 2$ micron block with a channel length of about 2 microns. The gate oxide is 0.03 microns thick and the field oxide 0.5 microns. The device has been doped with a p-type background of $3 \times 10^{16} cm^{-3}$ and a n-type Gaussian implant under the source and drain giving a peak of about $1 \times 10^{18} cm^{-3}$.

The essential dimensions of the device, such as gate length (GL), are the only data provided by the user to the EVEREST pre-processor. The software then generates the appropriated geometric model of the device. Once the pre-processor has generated the tetrahedral mesh impurity profiles are introduced using the Doping Module. The output

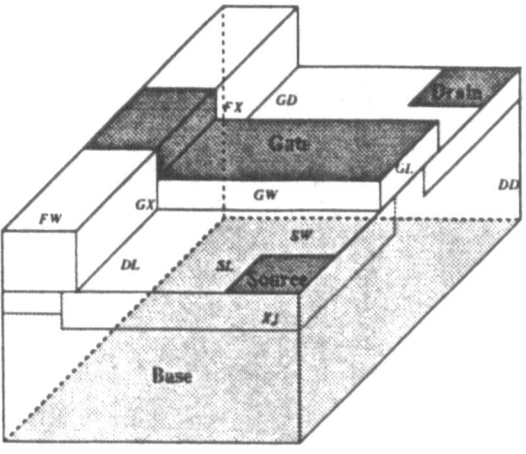

Figure 1: The MOSFET Geometry

of these two modules, which provides the basic information for the analysis module, is stored in the Neutral File System of the suite.

The mesh generated for this device has 1416 nodes and 5200 tetrahedral elements. Although this is quite a coarse mesh it is sufficient to demonstrate the system.

In the simulation presented here only a constant mobility model was used and no recombination of carriers was allowed. Since the MOSFET device is dominated by the flow of only one carrier, only the electron current has been computed.

8 Conclusion

In this project the problems of investigating the most efficient algorithms for three-dimensional simulation of semiconductor devices are being tackled. A number of discretisation schemes are being pursued and the results compared to identify the most successful methods. These are being incorporated into the project research code and the results achieved in both *off-state* and *on-state* solutions are very encouraging.

Although a three-dimensional *on − state* and transient solver have been implemented much work remains to be done in extending the work to use fully adaptive mesh generating schemes. The advances make by the project so far show greate promise. At the end of the project European industry will have access to an advanced three-dimensional design tool that will help them to successfully compete in international markets.

192

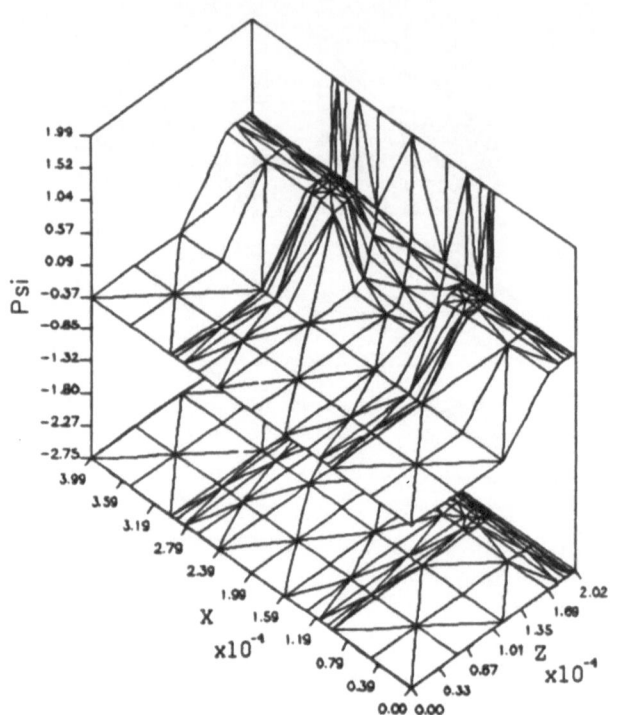

Figure 2: Potential plot with 3V gate bias

References

[1] D.L. Scharfetter and H.K. Gummel "
arge signal analysis of a Silicon Read diode oscillator", *IEEE Trans. Elect. Dev.*, ED–16, 64–77 (1969).

[2] J.O. Beck and R. Conradt, "Auger recombination in silicon", *Solid State Comm.*, vol.13 93–95 (1973).

[3] C. den Heijer, "Preconditioned iterative methods for nonsymmetric linear systems", Proc.Int.Conf. Simulation of Semiconductor Devices and Processes, Pineridge Press, Swansea, 276-285(1984).

[4] H.P.D. Lanyon and R.A. Tuft, "Bandgap narrowing in heavily doped silicon", *Proc IEDM*, pp 316–319 (1978).

[5] J.A. Meijerink and H.A. Van der Vorst, "Guidlines for the usage of incomplete decompositions in solving sets of linear equations as they occur in practical problems", *J. Comp. Phys.*, 44 134–155 (1981).

[6] S.J. Polak, C. Den Heijer, W.H.A. Schilders and P. Markowich, "Semiconductor device modelling from the numerical point of view", *Int. J. Num. Meth. Eng.* 24, 763–838 (1987).

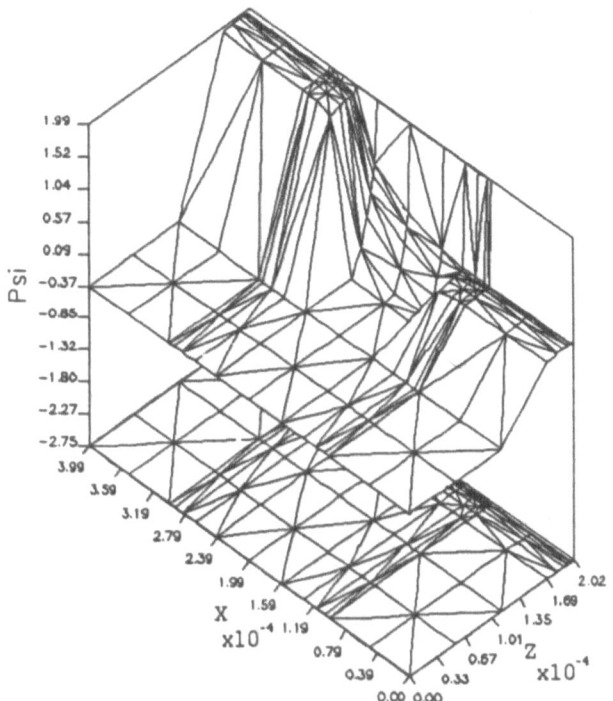

Figure 3: Potential plot with 3V gate and 2V drain bias

[7] S. Selberherr, "Analysis and simulation of semiconductor devices" Springer-Verlag, Wien-New York (1984).

[8] W. Shockley and W.T. Read Jr. "Statistics of the recombination of holes and electrons" *Phys. Rev.* 87, No.5, 835–842 (1952).

[9] J.W. Slotboom and H.C.De Graaf,"Bandgap narrowing in silicon bipolar transistors", *IEEE Trans. Elect. Dev.* 24, 1123–1125 (1977).

[10] G. Voronoi, "Nouvelles applications des parametres continus a la theorie des formes quadratiques", *J. Reine Angew. Math.* 134, No.4, 198–287 (1908).

[11] O.E. Akcasu, "Convergence of Newton's method for the solution of the semiconductor transport equations and hybrid solution techniques for multidimensional simulation of VLSI devices", *Solid State Elect. (GB)* 27, No. 4, 319–328 (1984).

[12] G.L. Dirichlet, "Uber die Reduction der positiven quadratischen Formen mit drei unbestimmten ganzen Zahlen", *J. Reine Angew. Math.* 40, No. 3, 209–227 (1850).

[13] S.P. Edwards, A.M. Howland and P.J. Mole, "Initial guess strategy and linear algebra techniques for a coupled two dimensional equation solver", NASECODE IV Proc, Dublin, Boole Press, 1985.

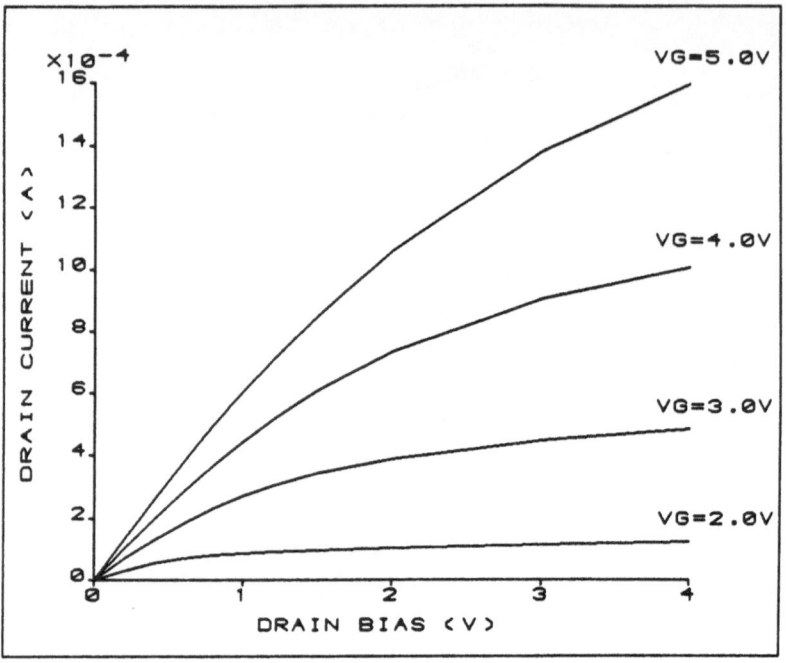

Figure 4: I-V characteristics for the MOSFET

[14] W.L. Engl, H.K. Dirks and B Meinzerhagen, "Device modeling",*Proc. IEEE* 71, No. 1, 10–33 (1983).

[15] J.G. Fossum and D.S. Lee,"A Physical model for the dependence of carrier lifetime on doping density in nondegenerate silicon" , *Solid State Electron*, Vol.25, No.8, 741–747 (1982).

[16] B.J. McCartin, "Discretisation of the semiconductor device equations", from *New Problems and New Solution for Device and Process Modelling* (ed. J.J.H. Miller) Boole Press, Dublin (1985)

[17] P.J. Mole, R. Debney and R. Van De Poel, "First issue release documentation for the linear solver, sparsity generation, Jacobian filling and machine constant routines", EC Microelectronic Project MR-02-RAL, Report GEC 5.3 (June 1987).

Project No. 1058

SPI: A Practical and Open Interface for Electronic CAD Tool Integration *

J. P. Schupp, J. Cockx [†], L. Claesen, H. De Man [‡]

IMEC Lab.[§], Leuven, Belgium, Phone: +32-16-281203

Abstract

This paper describes the SPI Interface as a practical and open interface to integrate electronic CAD tools. The goal of the interface is to provide a *direct communication* and *interactive feedback* between the primary design tools (schematics editors, symbolic layout editors, module generators etc.) and intelligent verification tools (electrical debugging, timing verification, simulation etc.).

The SPI Interface is the specification of a Structure Procedural Interface together with some utilities to standardise the *communication* among CAD Tools and to support their *integration*.

The data model of SPI is compatible with the ECIP [Eci 88a][Eci 88b] data model. The EDIF [Edi 87] terminology is used. Bindings do exist for C, Pascal, Lisp and Fortran, implementations do exist on a variety of Unix workstations.

The utilities are software *tools* and *libraries*. The software *tools* are a file dumping and a file restoring program. The software *libraries* include hierarchy and bus expansion, a browser, interprocess communication, merging of netlists from different producers into one netlist and tracing.

Because SPI offers an interface that supports, in an easy and efficient way, the integration of in-house as well as foreign CAD Tools, SPI will be made available and promoted in the European Electronic CAD Community.

1 Introduction

During the global design process of VLSI chips or VLSI modules, the verification of the correctness of the circuits takes a considerable amount of the design time. One bottleneck in the efficient application of this verification is the interactivity

---

*Research performed within the scope of the ESPRIT 1058 project: "Knowledge Based Design Assistant for Modular VLSI Design". Partners: IMEC Leuven Belgium, INESC Lisbon Portugal, Philips Eindhoven The Netherlands, Silvar-Lisco Leuven Belgium.

[†] Silvar-Lisco, Abdijstraat 34, B-3030 Leuven, Belgium, phone: +32-16-200016
[‡] Professor at K.U.Leuven
[§] Interuniversity Microelectronics Center

between the verification tools and the designer. The current design practice is that CAD tools are running *one at a time* and that *communication is via files and cross-reference lists*. This is extremely time consuming in a verification phase where the feedback between design definition and design verification is currently taking most of the designers time. Another disadvantage of the current CAD tools is that they often require different formats for representing design information, which necessitates the use of cross-reference lists and makes it harder for the designer to relate information from a verification tool to the original information.

To allow for a much faster feedback between verification tools and the designer, the SPI Interface is developed. The overall methodology and design philosophy of the SPI concept is described in more detail in [Ram 87] and [Cla 87].

In the area of CAD and Tool Integration in commercial applications and research of electronic design, a lot of effort is ongoing on DataBase Management Systems (data modelling, version-handling, maintenance, etc.) [Ram 87]. This could be considered to be *horizontal integration*, this is an *indirect interaction* among CAD Tools, see the horizontal double arrows in Figure 1. One can see that a tool only *interacts* with a user and a database. Interaction among tools must be performed along this database delaying the interaction with the designer.

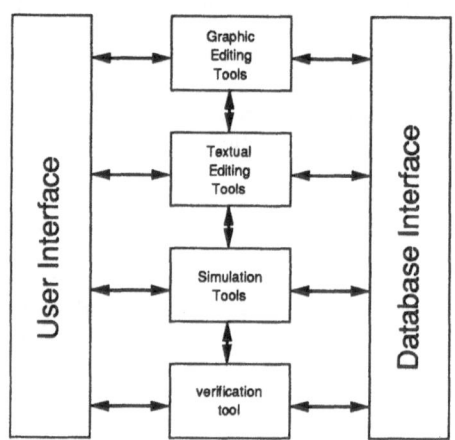

Figure 1: Horizontal and Vertical Integration of CAD Tools.

Instead of this horizontal integration the SPI Interface [Coc 89] concentrates on the *direct interaction* among CAD Tools. This could be pointed to as the *vertical integration*, see the vertical double arrows in Figure 1. Direct communication and interactive feedback is performed by the possibility to *highlight*, *select* and *back-annotate* (see subsection 3.2). The SPI Interface is not intended to be a DBMS to integrate horizontally, but to be complementary to existing frameworks and databases. Each CAD tool may have its own database but the use of some uniform database (and DBMS) is desirable.

The objective of the SPI Interface is briefly illustrated in Figure 2. A schematics editor contains a design called ALU with invertors and other gates. The invertor has been implemented as a set of rectangles in a layout editor. Both are visible in different windows, using different editors. A verification tool has been started in a

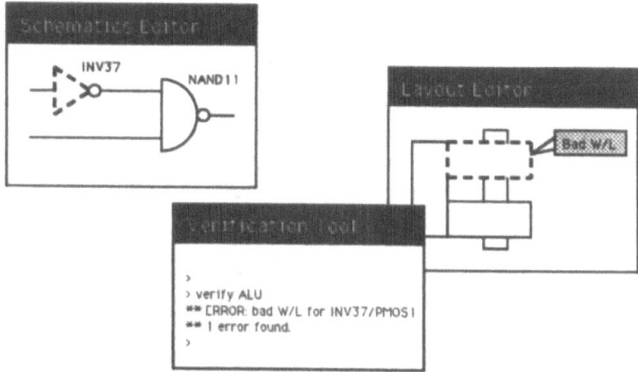

Figure 2: An example of tool integration with SPI.

third window. It issues an error message for one of the transistors of the invertor in ALU and highlights this invertor in both editors (see dashed lines). The schematics editor is for example an in-house editor integrated in some database environment. At the other side, the layout editor is a foreign tool with its own database. With SPI it is possible to communicate interactively from the verification tool to these two editors. These two editors can communicate with each other such that they produce *one* netlist to the verification tool. If there was no *direct communication*, the two netlists should have been merged in some way together with the generation and use of cross-references. If there was no *interactive feedback*, the interpretation of the error should have been very painful because then one had to interpret it using the cross-references and the two different netlists.

In the section 2 of this paper the Global System Architecture of the SPI Interface will be discussed. Section 3 is an introduction to the specification of the Structure Procedural Interface. In section 4 the integration of electronic CAD tools with the SPI Interface will be shown with an example and within the scope of the E-1058 project.

2   Global System Architecture

Figure 3 gives the major information flow (communication) in the architecture of the SPI Interface. The notational conventions used are indicated in the figure.

An implementation of the system can be done over one or more processes running on one or more machines. The functionality of the individual modules and the interfaces, depicted in Figure 3, will be explained in the following subsections.

2.1   UTM: Unix Tool Manager

The Unix Tool Manager is the main process controlling the global CAD activity. It is a very simple manager. It is no more than the possibility to execute additional commands under a Unix-shell. One can control the structure producers: e.g. to reset a process from a waiting state to an active state. Also from the shell one can

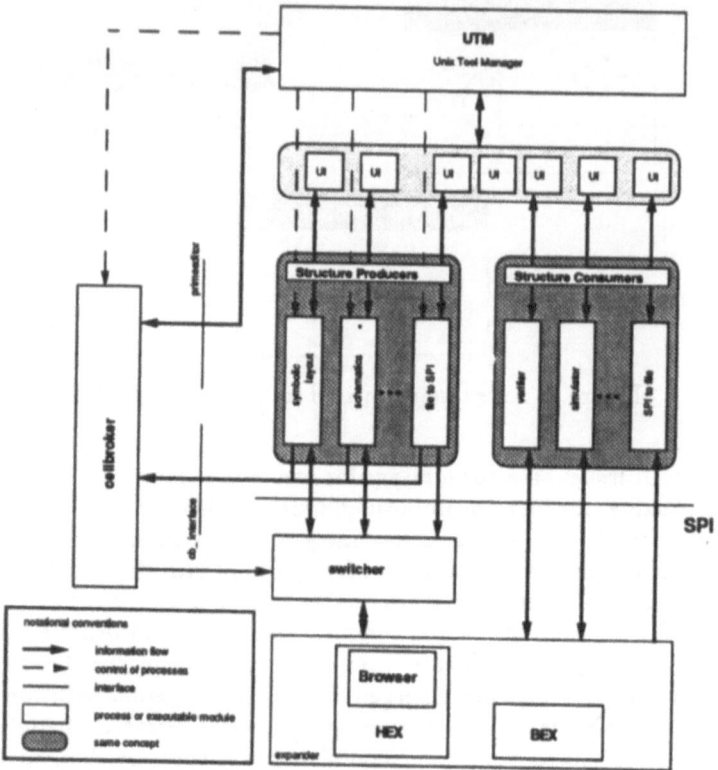

Figure 3: Global System Architecture.

start or delete the cellbroker or execute cellbroker commands. All the structure producers and consumers are started or ended in their own environment.

2.2 Application Programs

The application programs are the programs (the CAD Tools) which are to be integrated. The SPI Interface is built on the distinction between structure producers and structure consumers. A program can belong to the two classes. An example is a preprocessor program that takes structure and transforms it into another structure. With the SPI Interface the producers can run in *parallel*. The consumers must run *sequentially* unless a synchronising protocol is constructed among the consumers.

2.2.1 Structure Producers

The structure producer programs consist of application programs that have a direct graphical interaction with the user. These programs are for the most part editor-like and communicate in a graphical way (e.g. Schematics Editors). Sometimes a structure producer can be a textual editor with textual interaction (e.g. Structure Description Language Editors). The structure producer programs generate structural data that a structure consumer can read in via SPI calls.

2.2.2 Structure Consumers

The structure consumer programs are the programs that use structural data to perform new transformations on this data; for example a (timing) verification tool.

2.3 HEX: Hierarchy Expander

The hierarchy expander is a utility that provides a *flattened* circuit to a specific structure consumer from hierarchical structure producers using information obtained via SPI calls. With the use of the cellattribute *expansion_level* the depth of expansion can be controlled.

2.4 BEX: Bus Expander

The bus expander is similar to the HEX module. It is a utility that generates a circuit without busses (that is, all the busses are expanded) to a specific structure consumer.

2.5 Cellbroker

The cellbroker is a simple database that maintains and controls information about all cells defined in the different producer(s).

2.6 Switcher

The switcher can determine from the cellbroker where a certain cell is located and create a link from the consumer to the correct producer. It will also match and merge cells, instances and ports based on their (user)names.

2.7 Browser

This module is a software tool to browse through the design hierarchy during highlight and select actions.

2.8 UI: User Interface

Each CAD Tool may define its own user interface.

2.9 The Primeeditor Interface

With the so-called primeeditor interface a user can tell the cellbroker that a certain cell must be taken from a certain producer; e.g. when two producers define the same cell, a user must be able to indicate the producer that contains the *primary data*[1]. To allow the user to specify this, the *cell* command has been implemented. The *cell* command allows the user to set and list the prime producer for one or more cells.

2.10 The Cb_ Interface

With this interface one can tell the cellbroker to add or delete a structure producer or to add or delete a cell.

3 SPI: Structure Procedural Interface

The SPI Interface is a Structure Procedural Interface together with some utilities to standardise and to support:

- the *transfer* of structure *from* structure producers *to* structure consumers.

- the *transfer* of structure related information, called attributes, in a *bidirectional* way between structure producers and consumers using attribute- and backannotation operations.

- the *interactive communication* between structure producers and consumers using highlight and select operations.

Essentially the SPI Interface consists of two main concepts:

- The *(paper)* specification of the Structure Procedural Interface (including the data model).

- The Utilities including hierarchy and bus expansion, interprocess communication, merging of netlists from different producers into one netlist, tracing, browser[2] and file dumping and restoring; this is the only *software* of the SPI Interface. The utilities are designed and implemented such that they are *transparant* for the structure producers and consumers.

The SPI data model and the specification of the Structure Procedural Interface will be described in this section.

3.1 The SPI Data Model

The major foundation for CAD Tool Integration is *communication*. To facilitate successful communication a common *data model* is required. Together with this data model a *procedural interface* can be defined. The data on which the interface and the data model operate on, are *netlists*. Therefore *structure* producers and consumers are sometimes called *netlist* producers and consumers.

The SPI data model is illustrated in Figure 4. A netlist consists of cells. A cell can have ports, which define its external interface. A cell can contain instances of

---

[1]The *primary data* corresponds to the information that the designer wants to enter as the specification of the design.

[2]The browser is under development.

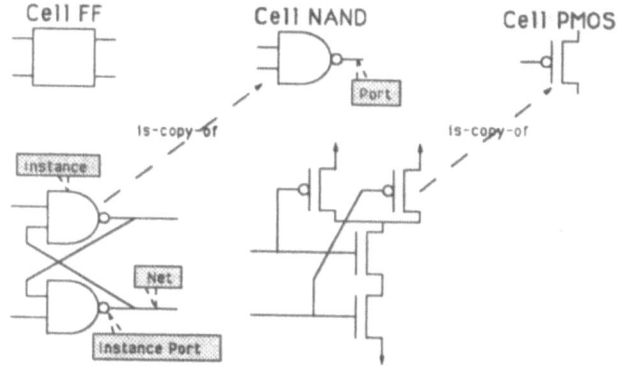

Figure 4: The SPI netlist data model: an example.

other cells and nets. When Cell NAND is instantiated in Cell FF, the ports of Cell
NAND are also instantiated in Cell FF; these instantiated ports are called instance
ports, see Figure 4. A net[3] connects zero or more instance ports and zero or more
cell ports. The fact that an instance port is connected to a net is called an internal
connection. The fact that a cell port is connected to a net is called an external
connection.

Cell, ports, nets, instances and instance ports can have attributes, which consist
of names and values.

The terminology used to describe netlists here reflects a similar definition from
EDIF [Edi 87]. The data model is compatible with the ECIP [Eci 88a][Eci 88b]
data model. An ECIP-style conceptual diagram of the SPI data model is given in
Figure 5[4]. Note that the terminology used in EDIF and ECIP differs. SPI uses

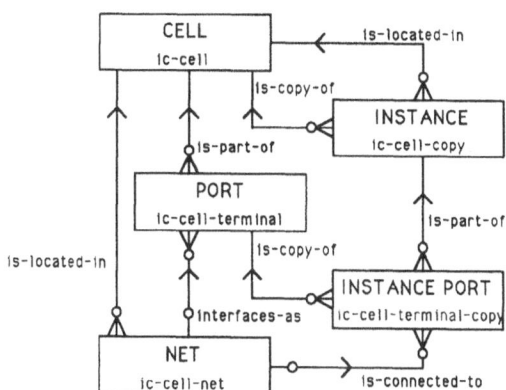

Figure 5: The SPI netlist conceptual model.

---

[3]One could also handle *busses* in the SPI Interface, but this will not be explained in this paper.
[4]The upper-case notations are SPI notations, the lower case ECIP.

the EDIF terminology, see the names between brackets. All the other names and notations are the ECIP conventions.

3.2 The Specification of the Structure Procedural Interface

The specification of the Structure Procedural Interface [Coc 89] is written using C syntax and contains the following sorts of *procedures*:

- control: for setting up the datastructure, initialization of the structure producer(s)

- request structure information: to get structure information from the producer

- request structure related information (*attributes*): to get attributes from the producer

- backannotation of structure related information: to create or modify structure attributes in the structure producer from a structure consumer.

- highlight structure objects: to highlight objects in graphical representation (e.g. schematics editor) or in textual form (e.g. textual editor)

- request selection of a structure object by the user: selection of an object by the user by pointing to it (e.g. with a mouse), or by referencing it by name (e.g. type name at keyboard)

- ask for usernames of structure objects: to pass the username of an object.

In the specification, only *long integers* and *character strings* are used. Therefore each language which has these two types (e.g. C, Lisp, Pascal, Fortran) can implement these specifications; CAD tools written in different languages can therefore be integrated together.

An example of a specification of a procedure for the request of structure information is

```
long SPIgetCell (name)
char *name;
```

parameters:

char *name	The name of a cell of which the netlist consumer would like to know the object handle.

returns: The object handle of the cell with that name, or zero.

usage: The netlist producer returns the object handle of the cell with the requested name, or zero if it does not know such a cell.

Remark: The (user)names of objects and the attributes (and values) are also standardised in the SPI Project [Sev 89][Coc 89].

4 Integration with SPI

The SPI Interface is supporting the task of integration. Figure 6 gives an example. The main aspects of SPI Interface helping the task of integration are

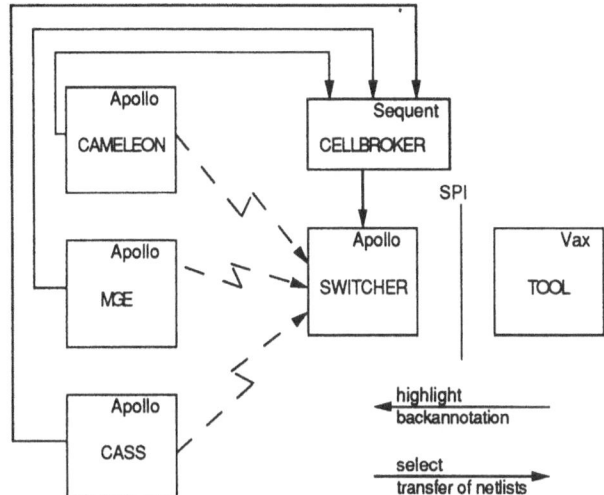

Figure 6: A possible session on different machines.

- **standardisation:** The standard procedural interface allows that the processes understand (can communicate with) each other. It also makes possible the implementation of utilities which can be shared among all the CAD Tools.

- **interprocess communication:** Allows editors to run on different Apollo workstations, cellbroker on a Sequent-machine, a verification tool on a VAX-Ultrix-machine.

- **netlistmerging:** The three editors, together defining a hierarchical design, produce 1 netlist to the consumer.

- **direct communication and interactive feedback:** transfer of netlists, highlight, selection and backannotation.

4.1 What do CAD Tools need to do to integrate with SPI?

Each CAD Tool can be developed independently and within its own environment (user interface and database). For this reason, together with the fact that different languages and different machines can be used, the SPI Interface is said to be *open*. Therefore making a coupling with SPI is not very complicated. What a structure producer and consumer have to do is described in the next 2 subsections. It is important to notice that this integration work can be done *after* the design and implementation of a CAD Tool.

4.1.1   The structure producers

A producer *has*

- to implement all the SPI Procedures as specified in [Coc 89]

- to tell the cellbroker (by calling cb_ interface routines) which cells it has in its own database.

The communication and usage of the SPI Utility (interprocess communication) is established by simple *linking* with the utility library.

4.1.2   The structure consumers

A consumer *can*

- call SPI Procedures.

The communication and usage of the SPI Utilities (interprocess communication, netlistmerging, hierarchy or bus expander, browser, tracing) are established by simple *linking* with the utility libraries.

Remark: The file dumping and restoring utilities are stand alone utilities which need not to be linked with the CAD Tools. Utility libraries only have to be *linked* because they are *transparant* software utilities (due to the standardisation).

4.2   How does the result of the integration of CAD Tools appear?

Assume that a design of the ALU cell has been made in MGE (a Module Generation Environment) using instances of the cells GFB & CARRY BYPASS. These cells are designed in both Cass (a Schematics Editor) and Cameleon (a Symbolic Layout and Compaction System). Suppose that the designer wants to do interactive timing verification using the cells as viewed in Figure 7.

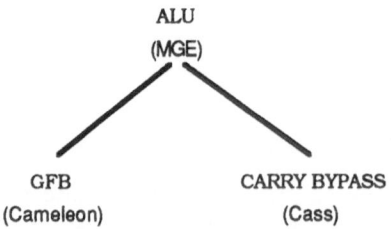

Figure 7: A view of the design of an ALU.

Description of a (artificial) session:

1. Take Figure 6 as the working environment. The user starts the 3 editors on 3 different Apollo workstations and the cellbroker on the Sequent-machine. The editors will tell the cellbroker which cells they have. In this case the cellbroker will known that MGE contains an ALU, Cameleon *and* Cass a GFB and a CARRY BYPASS.

2. To use only the cells as viewed in Figure 7 the user tells the cellbroker which cells are the primary cells. Here, the user indicate that the GFB cell has its primary data in Cameleon and the CARRY BYPASS cell in Cass.

3. Now, the user can start the timing verification tool on the VAX.

4. The tool wants to know which cell has to be verified. Therefore the select procedure is called. With the browser the selection is possible in all the editors. In this case the selection will be done in MGE.

5. After selection of the ALU in MGE the tool reads in the netlist. Because the timing verifier must have a flattened netlist without busses, it will use the hierarchy and bus expander. With the knowledge of the cellbroker, the switcher, in co-operation with the expander, can merge the netlist of the 3 cells in the 3 editors (ALU, GFB, CARRY BYPASS) into *one* flattened netlist.

6. The tool computes the longest path in the ALU and highlights this path with the highlight procedures. With the browser the user can browse through all the editors and cells to highlight the hierarchical path of his choice.

7. Then the tool and/or user will change the dimension attributes of one or more transistors in the longest path to speed up this path. The new attribute values will be backannotated in the right editor and cell by the use of the expander, the switcher and the knowledge of the cellbroker.

8. Then the user might restart the iteration (in point 5.) to reach a global optimum.

4.3   Integration in the Esprit-1058 Project

In the E-1058 the integration of CAD Tools has been achieved with the SPI Interface. Figure 8 shows the structure producers (above the SPI-line) and consumers (down the SPI-line) of the E-1058 project that have been integrated.

- The structure producers are

  - **Cass**, a Schematics Editor
  - **MGE**, a Module Generation Environment
  - **Cameleon-1**, a Symbolic Layout and Compaction System
  - **Hilarics-2**, a Structure Description Language
  - **SPICEtoSPI**, a SPICE Netlist to SPI Netlist Translator

- The structure consumers are

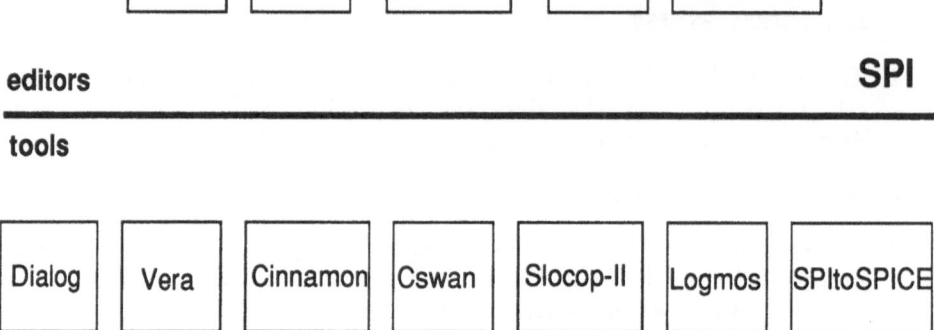

Figure 8: A very simple overview of the integration of the CAD-tools in E-1085.

- **Dialog**, a Knowledge Based Verification System
- **Vera**, a Knowledge Based Verification System
- **Cinnamon**, a Circuit Simulator
- **Cswan**, a (Parallel) Circuit Simulator
- **Slocop-II**, a New Timing Verification Tool
- **Logmos**, a Registor Transfer Simulator
- **SPItoSPICE**, a SPICE Netlist producer for SPICE-simulations

For the above CAD Tools the effort to couple a structure consumer to SPI is about 1 week, and 3 weeks for a structure producer. The implementors of these tools are satisfied with the *simple* and *easy* SPI Interface; *simple* and *easy* because SPI only handles structures (netlists). Therefore the SPI Interface is a *practical interface*.

5 Conclusion

In this paper, a *practical* and *open* Interface for interactive CAD tool integration has been presented. In addition to facilities provided by design management systems and DBMS, SPI provides a *direct communication* between structure producers and structure consumers. In this way the design cycle is shortened and verification tools can communicate more closely with the designer. An example is a timing verifier that can directly indicate the longest path in the schematics or symbolic layout by immediate highlighting without stopping either the timing verifier or the schematics or symbolic layout editor.

The SPI Interface is *open* because as few as possible constraints have been put on the tools themselves to be able to integrate already existing tools. Once a CAD Tool is integrated, a lot of other CAD Tools becomes available. It is also *practical* due

ESPRIT 1128

LARGE DIAMETER SEMI-INSULATING GaAs SUBSTRATES
SUITABLE FOR LSI CIRCUITS

G.M. MARTIN, P. DECONINCK, J. LE BRIS, J. MALUENDA
Laboratoires d'Electronique Philips
3, avenue Descartes, 94451 Limeil-Brévannes Cedex (France)

G. NAGEL, K. LOHNERT
Wacker Chemitronic
Postfach 1140 - D-8263 Burghausen (R.F.A)

M. CROCHET, P. NICODEME, F. DUPRE
Université Catholique de Louvain, Bâtiment Stévin
Place du Levant - 2B-1348 Louvain La Neuve (Belgique)

Summary

The goal of the work was to develop an industrial approach for large
diameter semi-insulating GaAs substrates suitable for LSI circuit applications. This goal has been reached to a large extent, since LSI circuit grade
material has been achieved using both In-alloyed and undoped GaAs grown with
special processes. This latter material prepared with a special thermal
treatment is even better than the In-alloyed one and this result can be
considered as a breakthrough in material development for GaAs digital IC's
applications. Progress has been verified until now on 2" material, but 3"
and, in particular, 4" materials already exhibit MSI grade quality indicating that these materials can also meet the LSI grade quality in near
future. These materials are now commercially available. Major progress has
also been achieved in the field of understanding of sources of inhomogeneities, discovered to be related to precipitate associated gettering
effects, and in the field of global LEC computer modelling. A computer code
is now available and is used in the material manufacturers premises for
simulation of both GaAs and Silicon preparation.

1. INTRODUCTION

The objective was to develop an industrial approach for the growth and
preparation of large diameter semi-insulating GaAs substrates (In-alloyed
and/or undoped) up to 4" by using improved LEC technology and well adapted
thermal treatments suited to prepare highly uniform active layers by ion
implantation for LSI circuit applications. Further on, reasonable characterization tools suitable for production as well as theoretical modelling of
the growth process for optimization of the thermal stress conditions were
expected to be developed. The aim of this work was to establish a source of
GaAs substrates with high quality and high reliability in Europe within the
frame of this project.

This objective has been reached to a large extent as will be described in this paper which aims to present the highlights and breakthroughs of the research during the three years of this very efficient collaborative program. Section 2 deals with the part of the research which can be considered as the basic support of the experimental work. This concerns the classification of sources of inhomogeneities in ingots on one hand and computer modelling of crystal growth treated as a global approach on the other hand. Section 3 presents the experimental results of improvement of material quality obtained along the main pathways defined using the deep knowledge of material behaviour during thermal processing, together with the results of scaling-up of a selected process for 3" and 4" diameter ingots. The last section summarizes the state of the art concerning the selection criteria for qualifying LSI grade wafers and the relevant tools and procedures to assess bare wafers.

2. BASIC SUPPORT OF CRYSTAL PREPARATION

2.1. Investigation of sources of inhomogeneities in substrates

During the project, a large effort has been devoted to understanding why, in certain cases, dislocations were seen to strongly influence the threshold voltage fluctuation of transistors made around them, while, in other cases, their dependence was only very vague. LEP has been able to show that precipitates (very likely As precipitates) play a major role in this matter. First of all, High Resolution Infra-red Tomography has been shown to be the best technique to visualize all precipitates in a material, their density, their size and their distribution. Then, it was demonstrated that, using A-B etching and, to a lesser extent, high resolution photoluminescence or cathodoluminescence, one can identify the only precipitates which are surrounded by impurities and active point defects. As shown in figure 1, it it clear that materials can be extensively different from the point of view of precipitate distribution ("milky-way" type distribution or precipitates aligned on the dislocation network, for instance). There also exist materials in which precipitates have gettered

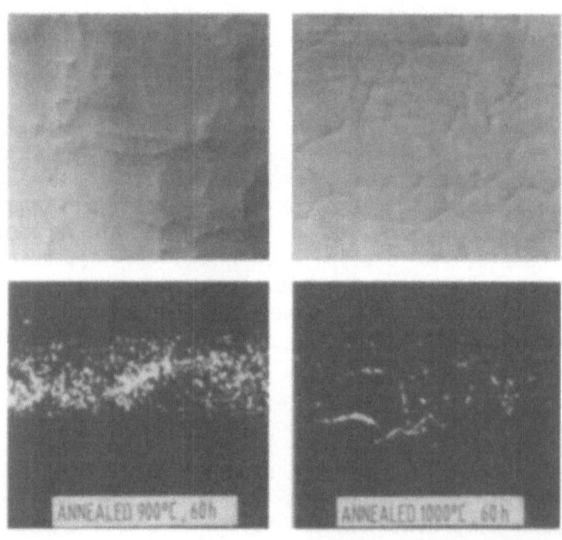

Figure 1 : A-B etching (top) and HRIT (bottom) pictures taken from the same area of two different materials

impurities and thus contain, or are surrounded by a large quantity of them. This investigation suggests that the importance of dislocations is monitored by the presence of precipitates. Very efficient impurity and/or active point defect gettering take place at temperatures around 900°C, while, at temperatures around 1000°C, those impurities are dissolved in the matrix.

Three classes of homogeneous materials have thus been identified[1]

- **Class G** , when all the impurities are gettered by precipitates, leaving a denuded and homogeneous matrix (except within a 1 or 2 μm wide gettered zone around each of them).
- **Class F** , when all the impurities are homogeneously distributed in the matrix, because the material is free from precipitates and dislocations.
- **Class D,** when all the impurities are diluted in the matrix after a very high (# 1000°C) temperature annealing.

Further investigation of the stability of properties of these 3 classes of materials has been studied after post implantation annealing at temperature around 850°C. It turns out that class D material degrades, which can be understood since gettering starts to take place at 850°C but is not completed, leading to rather large fluctuations of impurity concentration over large distances around precipitates. On the other hand, class G material are very stable since gettering effects were already completed. Of course, since no precipitates are present in class F material, they also remain quite stable.

As a conclusion, it becomes clear that besides In alloyed dislocation free (class F) materials, there also exist dislocated materials (class G materials) in which well controlled precipitate gettering effects allow the attainment of substrates with an homogeneity quite sufficient for LSI circuit applications.

The understanding of such mechanisms, as well as the definition of convenient assessment tools, have strongly helped defining the proper ingot annealing process.

2.2. Development and application of thermal modelling

A new approach has been successfully investigated for computer modelling of LEC growth. Instead of making simulations of isolated problems, like, for instance, convection in the melt or determination of stress in ingots, which require to be solved to make an hypothesis for the boundary conditions of the problem itself and which gives very limited information because the obtained solution is much too dependent of the hypothesis choosen at the initial boundary conditions, the idea has been to consider the problem as a global problem of simulation of the total system consisting of the puller, the melt, the boric oxide and the crystal. As such, the computer code is based on a finite element model which simulates all parts of the puller and all the liquid and solid parts of the III-V compounds. The boundary conditions do not correspond to any hypothesis, but precisely to only few imposed and measurable parameters such as the electrical power input, the melt charge, and the pulling speed. Simulation can actually provide information on the temperature of any point of that total system and, of course, especially on global exchange of heat between the crystal and its surroundings. Such a simulation is referred to as the current global modelling.

This global modelling was already started by UCLB before the current program for simpler cases than Czochralski growth of III-V compounds. In this latter case, two serious additional problems are encountered and have been solved : the problem of the liquid encapsulant (boric oxide) used to prevent As evaporation and the problem of gas and melt convection during growth. In particular, a new algorithm taking into account radiation through a semi-transparent body such as boric oxide has been developed, based on a simplified mathematical model of convective exchanges in the pressurized gas. The shape of the meniscus, which is present at the surface of the melt, has also been studied as well as its influence on the interface shape.

210

Furthermore, the modelling of the natural and forced convection in the melt was performed and a numerical method for calculating the thermal stresses in the crystal has been developed. As a result, it is now possible to impose the pulling rate of the crystal as a parameter and to be as close as possible to real conditions.

In a second step, the computer code has been tested by comparison with experimental temperature measurements. A first thermocouple is rigidly fixed at the top of the seed and dipped into the melt : it is located in such a way that during growth it is enclosed in the crystal. The experimental curve is shown as a solid line in figure 2. The vertical dotted line indicates when the material around the first thermocouple solidified. However, measurements show at that point a temperature of 1456 K while the melting temperature is 1511 K, thus demonstrating an incorrect calibration of the thermocouple. Thus the entire curve has been shifted by 55 K to get the dotted line in figure 2 as an

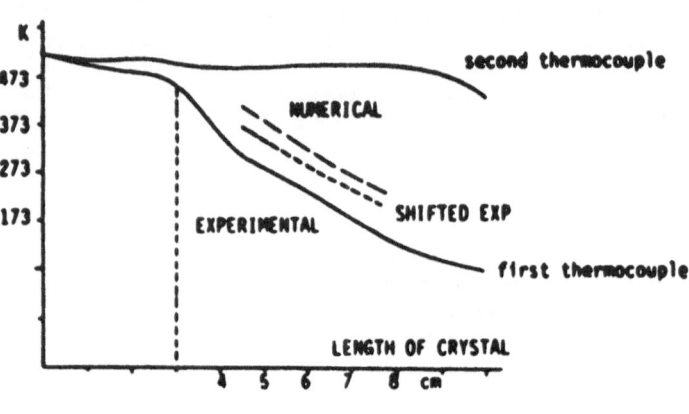

Figure 2

experimental reference for theoretical results. Four numerical simulations have been performed for different crystal lengths. One observes that the theoretical values are about 55 K higher than the experimental ones but the theoretical and experimental slopes are identical. This demonstrates that, at some distance from the interface, the code provides a correct calculation of the heat transfer between the crystal and its surroundings. The small remaining discrepancy of 35 K is most probably due to melt convection, neglected in these simulations and its effect upon the shape of the interface.

The computer code was then further optimized by taking into account the influence of the natural and forced convection on the temperature field and the shape of the interface

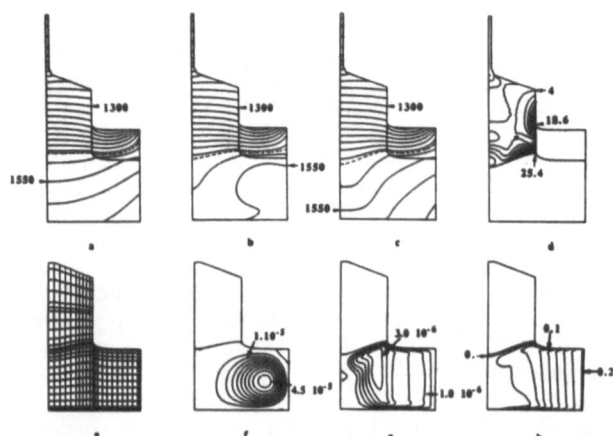

Figure 3 : Simulation of gallium arsenide growth :
a) isotherms in the melt and the crystal (in k) without melt convection
b) isotherms in the melt and the crystal with natural convection only
c) isotherms in the melt and crystal with natural and forced convection
d) isovalues of the von Mises invariant (MPa)
e) deformed mesh
f) streamlines with natural convection only (stream function is in m^3/s)
g) streamlines with natural and forced convection
h) contour lines of the azimuthal velocity (m/s)

In figure 3, the situations without convection, with natural convection and with mixed convection are compared. We observe that the temperature field in the melt and the shape of the interface are strongly influenced by the melt motion. One can note that this shape presents a double curvature as obtained in practical situations. Such a result obtained for the first time is considered to be both a proof of validity and a breakthrough in modelling.

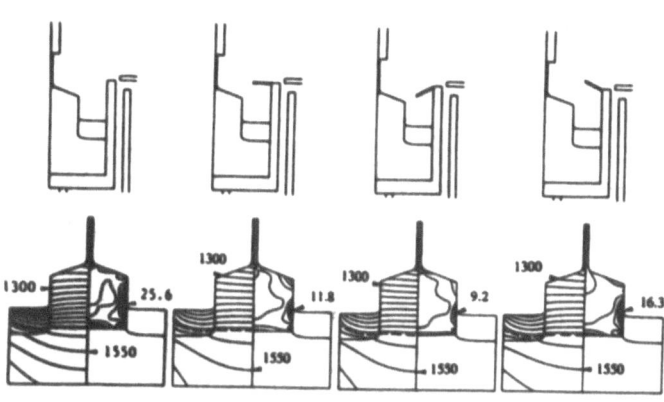

Figure 4 : Influence of a reflector on the temperature and the stress field for a short crystal

Following this, the power of the global UCLB model was applied to the simulation of new furnaces for obtaining decreased thermal stresses in crystals during growth. The influence of a molybdenum reflector, fixed at the top of the crucible, on the temperature and stress fields was studied. Four cases were considered : without reflector, with a horizontal reflector, with an inclined downwards and inclined upwards reflector respectively. The configurations and results are presented in figure 4 for a short crystal completely inside the crucible and in figure 5 for a longer crystal with the top emerging from the crucible. CLear conclusions can be drawn : firstly the presence of a reflector reduces the stress level by a factor of 2, mainly near the critical points, i.e.the interface and the junction of the crystal with the top of the layer. Secondly, the reflector also induces a stress increase near the top of the crystal, but this is less of a problem since the temperature is definitely lower at this point.

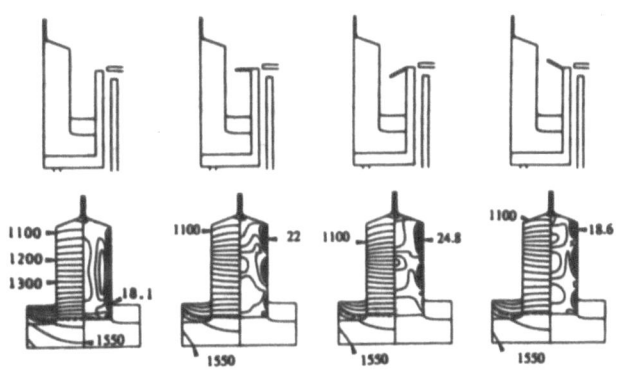

Figure 5 : Influence of a reflector on the temperature and the stress field for a long crystal

It is clear that the work performed during these three years has produced a powerful tool for designing new furnaces and growth conditions. It even makes possible the development of a dynamical global model for the evaluation of the stress history of the crystal which has crucial importance for the formation of dislocations and, even more important, as seen above, on the formation and gettering of precipitates.

3. CRYSTAL GROWTH AND CHARACTERIZATION OF LSI GRADE QUALITY GaAs

3.1.   Crystal growth

The current state of the art at the beginning of this project was the growth of 2" diameter (undoped, In-alloyed, Cr-doped) and first approaches to the growth of 3" (undoped) ingots by the low pressure LEC growth technology. In the frame of this project, the growth of 3" ingots (undoped, In-alloyed) has been further developed and in particular the growth of 4" (undoped) ingots has been developed. Figure 6 demonstrates the state of the art for the growth of 4" GaAs ingots and figure 7 shows a view of the low pressure LEC crystal growth equipment at Wacker.

Figure 6 : 4" undoped GaAs ingot   **Figure 7**   : View of the low pressure LEC equipment at WACKER

For improving both the substrate uniformity and the crystal yield, advanced LEC technologies such as magnetic field technique (vertical arrangement, field strength : up to 2500 Gauss) for stabilizing the melt convection, arsenic injection technology for controlling the melt stoichiometry and improved diameter control have been applied and investigated thoroughly. For further improving the substrate uniformity, optimized post growth thermal treatments have been developed and applied [2]. In order to reduce the dislocation density, different growth conditions and the application of special heat shields, developed by thermal modelling [3,4], have been investigated and applied successfully. The growth of completely dislocation-free GaAs has been performed by using In-alloying.

It should also be mentioned that improvements in the synthesis of the polycrystalline GaAs source material performed in a new developed separate high pressure equipment as well as the use of high purity raw materials (Ga,

As, B$_2$O$_3$) have made essential contributions for meeting the overall goal of this project.

As a result of these developments large diameter undoped GaAs substrates ranging up from 2" to 4" diameter with improved uniformity and high yield are now commercially available at Wacker.

3.2. Qualification of substrates for LSI GaAs circuits

LEP has extensively used the so-called microFET DRP (Dense Row Pattern) procedure [5] to probe the quality of substrates prepared by Wacker and LEP. This procedure was developed within another ESPRIT project (843) with Siemens. This procedure consists in testing 5 x 6 µm^2 microFET's densely packed in a 300 µm long row. Then the fluctuation of threshold voltage V$_T$ of microtransistors, in rows spread all over the wafer, is measured, leading to standard deviation σ (V$_T$) for each row. Cumulative curves of σ (V$_T$) for all rows are displayed in figure 8, and this provides the safest technique to compare materials. This is also today the only technique to qualify substrate for LSI circuit applications, since the corresponding criteria has been defined as P$_{15}$ 75 % (percentage of row on wafer with σ (V$_T$) below 15 mV) or M$_{50}$ 11.5 mV (M$_{50}$ being the median value of the curve).

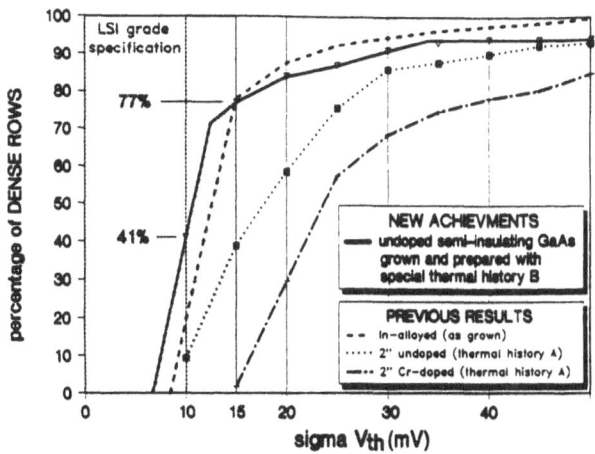

Main results on 2" Wacker materials are reported on figure 8. It is clear that undoped semi-insulating wafers from ingots prepared with special thermal history B (class G) are equivalent or even better than dislocation free, In alloyed wafers

Figure 8 : Cumulative curves giving the results of V$_{th}$ from each row of 30 microFET's all over 2" wafer for different types of material (full curve : recent results obtained with the special thermal treatment ; dashed lines : previous state of work)

(class F) and that they pass the LSI qualification test. This actually represent a breakthrough for LSI circuit applications and, in general, for high density circuit industrial applications, since previously dislocated materials (as grown, or annealed with thermal history process A) did not pass that qualification test.

DRP results recorded on 3" and 4" materials prepared by Wacker using similar special thermal history process B are also very impressive, as shown in table I. Even if those material do not pass yet the LSI qualification test, they are very close to the required quality.

Regarding the DRP results with respect to the threshold voltage standard deviations in figure 8 and Table I, one can make the following conclusions about the quality of different types of substrate materials :
a) Undoped GaAs substrates with special thermal treatments (WCT process B or LEP process L) exhibit the best DRP results among all the substrates and meet the LSI grade specification successfully. This progress has been

214

Type of substrate (thermal treatment)	Percentage of dense rows with V_{th} smaller than		σV_{th} (50 %)
	15 mV	10 mV	
2" Cr doped (Process A)	2 %	0	23.4 mV
2" undoped, low EPD (Process A)	38 %	10 %	18.5 mV
2" In-alloyed, dislocation-free (as grown)	77 %	18 %	11.8 mV
2" undoped (Process B)	77 %	41 %	10.4 mV
3" undoped (Process B)	21 %	4 %	20.4 mV
4" undoped (Process B)	46 %	12 %	15.9 mV

Table I : Micro FET DRP results on Wacker substrates

verified till now on 2" material. However recent investigations with 3" and in particular 4" material exhibit promising results indicating that this material can also meet LSI grade quality in the near future after further improvements. Both 3" and 4" wafers already exhibit excellent properties for MSI circuit applications. b) In-alloyed GaAs substrates exhibit the best DRP results among as grown process A thermally treated substrates and meet also the LSI grade specification, while corresponding undoped or Cr-doped Wacker grown GaAs substrates do not.
c) Progress have been achieved on Cr free ingots, with similar properties from seed to tail. This represents a good basis for GaAs industrialization which requires equivalent wafers down an ingot and from any ingot.

4. PROCEDURES FOR SELECTING BARE WAFERS

The DRP procedure is today the safest and only definite technique to qualify materials, but it is very time consuming. Thus, other tools have been used in order to get a faster feedback and to allow a first selection of high quality materials, directly at the suppliers premises. Many different techniques have been used, such as infrared absorption for the EL2 defect concentration mapping, infrared tomography for precipitates evaluation, etch pit density measurement, photoluminescence and bulk resistivity mapping. This paper presents the most relevant ones and especially those which are already used or are going to be used by Wacker as the GaAs wafer manufacturer.

Resistivity mapping has been extensively developed. Using that technique, the uniformity has been

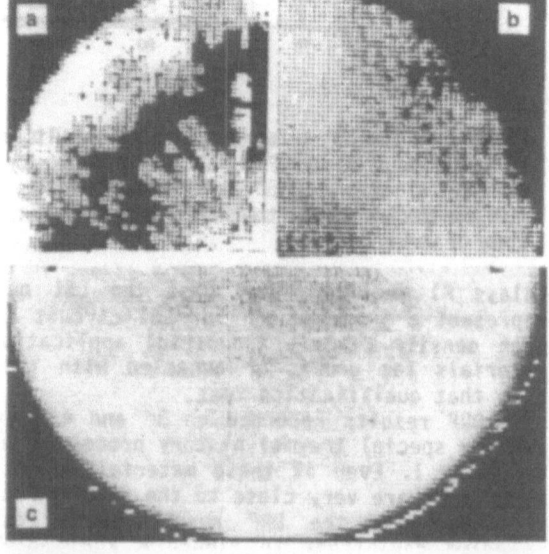

Figure 9 : Resistivity topographs on 2" undoped GaAs substrates in the as grown state (a), after the thermal treatment process A (b), and after the special thermal treatment process B (c).

shown to be improved significantly by applying process B as compared to process A. The typical fourfold symmetry, which is clearly observed in the as-grown state and still present in the process A annealed material, nearly disappeared in the improved material as shown in figure 9. The quantitative analysis (evaluation of linescan) of the uniformity in resistivity, as made by Wacker, on a series of 2"/3"/4" ingots, exhibits that process B treated samples have reproducibly lower standard deviations in resistivity ($\sigma \leqslant 15$ %) than process A treated samples ($\sigma \leqslant 30$ %). There is no significant difference in uniformity between seed and tail end.

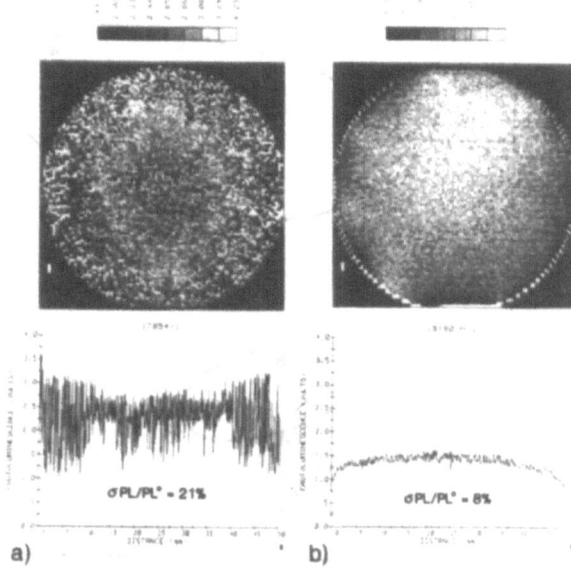

The uniformity of the photoluminescence (PL) mapping proved to be very useful. PL signal intensity could also be improved significantly and reproducibly with the special process B as compared to

Figure 10 : Topographs of the photoluminescence intensity on GaAs substrates grown and prepared with process A (a) and process B (b). The corresponding linescans are shown below. Measurements at 300 K, spot size : 20 um, spacings : topograph 500 um, linescan 100 um).

the original process A (see figure 10). The standard deviation (evaluation of linescans) is reduced nearly by a factor of three (process B : 6 % < σPL < 11 % ; process A : 20 % < σ PL < 33 %).

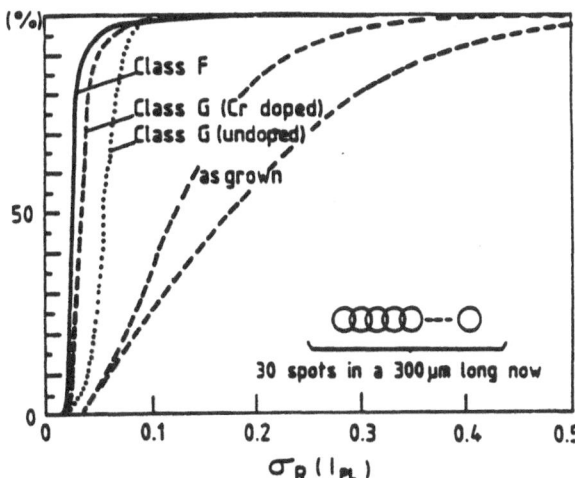

Figure 11 : Cumulative curves of σ(IpL measured from each DRPL row (see insert) spread over 2" wafers of different quality materials

An improved procedure for photoluminescence mapping has been developed by LEP. It involves the use of a high resolution spot (10 µm in diameter) and in probing 30 spots, 10 um from each other, along a 300 um long row. Then the PL intensity is recorded for each spot, leading to σ(IpL) in a row and to the relative σR(IpL) = σ(IpL)/IpL. The same statistical evaluation as for electrical DRP test is then made, leading to cumulative curves of σR(IpL) for a wafer, as seen in figure 11. Referring to the electrical DRP test, this optical technique is called the

216

DRPL technique (Dense Row Photoluminescence). Again, class G Wacker annealing process B) material is comparable to class F (In-alloyed) material, and is far better than other materials.

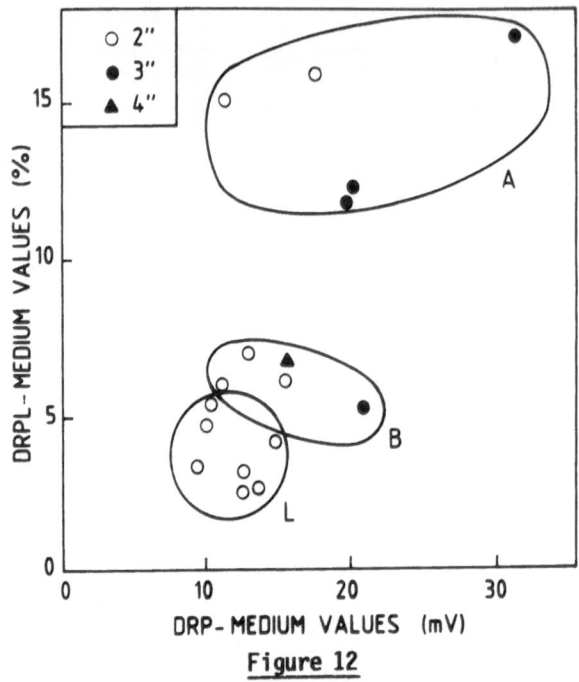

Figure 12

More statistical comparisons between electrical DRP and optical DRPL data have been made. M50 values have been measured for data recorded from cumulatives curves of DRP and DRPL measurements on adjacent wafers. The results are given in figure 12. It may be noted that the lower the M50 values, the better the materials. It is clear from that figure that the post growth ingot annealing process B (Wacker) or L (LEP) is far better than that given for process A (Wacker). Furthermore, in spite of some scattering in data, one may conclude from the same data, plotted in figure 13 with the LSI threshold acceptance limit, that high values in the optical DRPL tests correspond to poor materials, while materials exhibiting very low DRPL values have a high likelihood of meeting the LSI specifications. Even, if this optical DRPL assessment cannot yet be considered as a qualification test, it can be seen, today, as one of the best selective tests of bare wafers suitable for LSI circuit applications.

5. CONCLUSION

LSI circuit grade material has been achieved using both In-alloyed and undoped GaAs grown with special processes. Moreover, undoped material prepared with a special thermal treatment is even better than In-alloyed material and this result can be considered as a breakthrough in material development for GaAs digital IC applications. This improved undoped material is now commercially available.

From careful characterization of the substrates, a deeper understanding of sources of inhomogeneity in the material has been be obtained. Significant correlations between the formation of precipitates, native and extrinsic defects with the process conditions during crystal growth, post growth cooling and post growth annealing have been found. This has greatly helped to define the best annealing procedure needed to obtain the LSI quality undoped materials mentioned above. Also new tools for a very fast characterization and selection of LSI grade wafers have been developed and are, or soon will be applied by the wafer manufacturers to indicate the quality of wafers from 2 to 4" diameter.

The global approach for computer modelling has also proved to be extremely successful. As specific achievements, one can mention that it is now possible to obtain the shape of the growth interface and the temperature

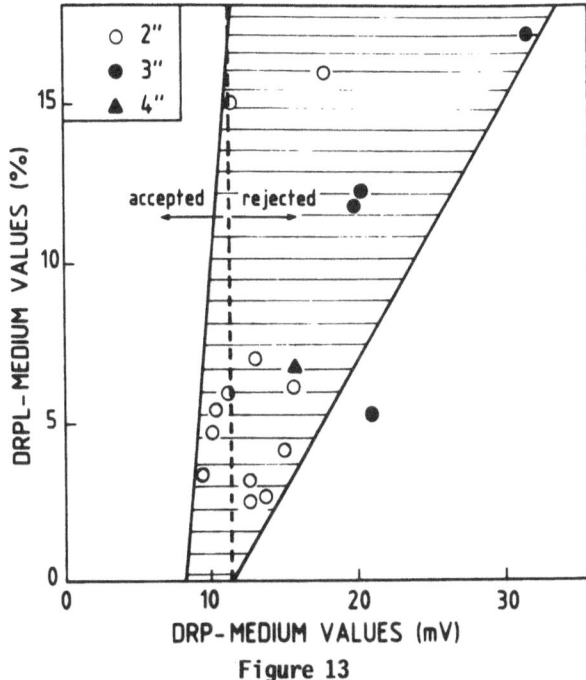

Figure 13

and stress field at any time. The former information is directly relevant to the single crystal material yield, while the other is essential to understand and control the dislocation density and, even more important, the formation and gettering of precipitates. This precipitate gettering effect is crucial for GaAs, as discovered in this program, but the importance of oxygen related precipitates is also well established in Si. The established computer code has been recognized as so useful that it has been implemented on Wacker's premises not only for simulations of GaAs growth but also for dealing with Si growth. This can be seen as a very good example of cross fertilisation between different areas of technology.

REFERENCES

(1) Suchet P., Duseaux M., Le Bris J., Deconinck P. and Martin G.M., Proc. 5th Int. Conference on Semi-insulating III-V Materials, Malmö (Sweden), (Adam Hilger, Bristol) 1988, p. 99

(2) Martin G.M., Deconinck P., Duseaux M., Maluenda J., Nagel G., Löhnert K., Crochet M.J., Dupret F. and Nicodème P., Proc. 4th Annual ESPRIT Conference, Brussels 1987 (North-Holland Amsterdam), p. 113

(3) Nicodème P., Dupret F., Crochet M.J., Farges J.P. and Nagel G., Proc. 5th Int. Conference on Semi-insulating III-V Materials, Malmö (Sweden), (Adam Hilger, Bristol) 1988, p. 471

(4) Nicodème P., Dupret F., Crochet M., Farges J.P. and Nagel G., Proc. 5th Annual ESPRIT Conference, Brussels (North-Holland, Amsterdam), 1988, p. 314

(5) Maluenda J., Martin G.M., Schink H., Packeiser G., Appl. Phys. Lett. 48 (11) (1986) 715

(6) Deconinck P., Farges J.P., Martin G.M., Nagel G. and Löhnert K., Proc. 5th Int. Conference on Semi-insulating III-V Materials, Malmö (Sweden), (Adam Hilger, Bristol) 1988, p. 505

PROJECT N° 1551

ADVANCED MANUFACTURING SYSTEM

A.M.S.

Jacques PINOT
SGS THOMSON MICROELECTRONICS
Z.I. ROUSSET - B.P. 2
13790 ROUSSET - France
Tel: (33) 42 25 87 29

André ROCHET
SGS THOMSON MICROELECTRONICS
GRENOBLE Plant - BP 217
38019 GRENOBLE CEDEX - France
Tel: (33) 76 49 36 00

KEYWORDS : C.A.M. System, Robotics, Clean Room, Automated Cell, Expert System, Recipe management, SECS protocol.

ABSTRACT : AMS is aiming at the definition of specifications for an ideal C.A.M. system covering all requested functionalities for integrated circuit automated lines. Automation concepts are experimented from software Equipment Interfacing with Protocol definition to robotized wafer handling in order to provide specifications and recommendations for Automation. The evolution and requirements for a CAM system are presented with Manufacturing Actors and Functional areas as variables. Then Equipment interfaces are considered; After a few years of evolution, SECS is now the standard carried by most of Equipment manufacturers but not always thoroughly or perfectly implemented. Some successful experiments using new software tools such as Recipe editor, Expert systems, Fab simulation are developed. Requirements for robots in clean environment are defined. Experiments on Diffusion and Photolithography areas are described and attractive results on particle contamination, yield improvement are given.

1 - SCOPE

AMS is a development program to explore the software requirements and the automation concepts to provide capability and competitiveness in the semiconductor industry.
Without supplying new products (software or equipments), AMS tests packages and develops experimental tools on real industrial situations to demonstrate the potential benefits and to back the European equipment, automation and software industry by defining specifications as for interfaces and communications protocols as for software functions.

The project include the following topics:

1. Production Information System
2. Networking, communications interfaces
3. Automation Island definition and experiment
4. Facility monitoring System
5. Material Handling Systems
6. Manufacturing line integration
7. Manufacturing requirements with emphasis on quality, service and production cost issues.

After nearly 3 years, different CAM systems have been experienced, and are running in production environment. Specifications for an ideal CAM system are well on the way, while experiments of automation are ongoing in Diffusion and Photolithography areas.

As a reminder, the project partners are :

SGS THOMSON MICROELECTRONICS S.A.	-	France
SGS THOMSON MICROELECTRONICS SpA	-	Italy
MARCONI ELECTRONIC DEVICES Ltd	-	United Kingdom

2 - OVERVIEW

2.1 - CAM system comparison considerations

The three partners of the AMS Project are now using three different CAM systems as ensivaged at the beginning of the project : SPN from H.P., COMETS from CONSILIUM and PROMIS from I.P. SHARP. During the course of the project, SPN disappeared as a commercial product and has been taken out of comparison. Nevertheless, the acquired experience is valuable for the specification of a new CAM system. CAM systems used are specific to the Integrated Circuit field; Comparing to general CAM, their main features are the following :

General CAM	Electronic CAM
. Operation range	. Data acquisition
. Requested quantity	. Lot tracking
. Starts linked to orders	. Equipment tracking
. Cost accounting	

The choices of PROMIS by MEDL and COMETS by ST Italy at the beginning of the project were due to the main following reasons : PROMIS is more suitable for ASIC manufacturing: merge, split, rework of lots are easier and PROMIS requests a lower level of resource both during implementation and in long-term maintenance. COMETS has greater functionality and flexibility; it uses more standard modules from DEC, such as DBMS, and is better adapted to mass production.

As a comparaison point, we can look at installation time :

Until lot tracking only	COMETS	times : 4 months	resources : 8 man months
Until lot and equipment tracking	PROMIS	times : 4 months	resources : 12 man months

These times and resources don't include process description and operator training depending on Fab specificity. Hardware can also be compared : COMETS and PROMIS are running on VAX and MicroVAX computers from DEC while SPN needs HP3000 and HP1000 computers.

Major weakness of these CAM system : Operator interface screens are not user friendly and require long time of training; Creation of new reports needs analyst intervention; Integration of environment parameters in lot history is not easy.

As a conclusion, technical comparison has to be followed continuously because product evolving. But the evolution of products depends not only on human resources of supplier but also on clear common improvement specifications defined by a large group of users.

2.2 - Requirements for an Advanced C.A.M. System

One main result of AMS project will be the definition of an advanced CAM system able to control future automated integrated circuit production lines.

From practical problems with current use of AMS partners' C.A.M systems, analysis of CAM user needs has led us to ten main functional areas which have then been investigated to build a complete C.A.M.

architecture. C.A.M. user needs were drawn from interviews of manufacturing Actors into main advanced plants of AMS partner companies.

Area investigations were made by interviewing both experts in specific domains such as Equipment maintenance Engineers or Process Engineering people, and expert in tools used in that domain, such as Engineer analysts.

This study has been conducted while keeping in mind 3 main objectives :

. User Orientation   . Integration   . Automation

The description of the model is covered by an ACTOR/AREA matrix. See matrix 2.2.

Manufacturing Actors	Functional Areas	
1.Operators	1. Manufacturing personnel Mngt	6. Material handling and tracking
2.Supervisors	2. Consumable management	7. Equipment management
3.Equipment Maintenance	3. Planning and scheduling	8. Facility monitoring
4.Product/Process Engineer	4. Specifications	9. Communication
5.Manager	5. Process control and Engineering	10. Quality control and acceptance

A sub-matrix gives the tasks covered by Cell Automation.

Each C.A.M. functionality covering one or more Matrix-cell is described in a C.A.M. specification which cannot be described in detail here. We can only highlight some important features not always taken into account in the current C.A.M. systems.

The first are Planning/scheduling tools which have to be specified from monthly planning to lot and task selection as mentioned hereafter in paragraph 2.4.5.

Due to the complexity of complete process flows, 200 to 300 operations, they ought to be described in two levels :

. Group of operations (performed successively in a cell) :50 to 70 per Process

. Process operation : Steps of process performed on one Equipment : 1 to 10 per group.

This process flow description is interesting for reports, planification/Scheduling, and number of lot transactions per process flow.

Before the end of the project, an important specification remains to be defined clearly: structure and functions of a CELL CONTROLLER included in a C.A.M. system. The software and hardware aspects have to be studied as well as sharing of the tasks between C.A.M. HOST and CELL CONTROLLER.

Extension and recommendations

From C.A.M. requirements and the associated C.A.M. specification set, we can define and ask the C.A.M. supplier for some improvement of current C.A.M. such as COMETS or PROMIS or others - which is today's approach - or develop a new C.A.M. System meeting all requested functionalities in New Software and Hardware generation - this is the trend for the near future.

2.3 - Equipment Interface

2.3.1 - Main aspects of SECS Protocol

The SECS protocol first appears as heavy and low. After, implementation of these point to point connection with more half-duplex than full-duplex use, the satisfaction is high. Integration in LAN via server is good and easy. Session init is dependent on hardware supplier and not on SECS protocol.

	PERSONNEL MANAGEMENT	CONSUMABLE MANAGEMENT	PLANNING SCHEDULING	SPECIFICATIONS	PROCESS CONTROL & ENGINEERING	MATERIAL HANDLING AND TRACKING	EQUIPMENT MANAGEMENT	FACILITY MONITORING	COMMUNICATION	QUALITY CONTROL AND ACCEPTANCE
MANUFACTURING MANAGEMENT	WORKED HOURS REPORTS BY AREA SHIFT	CONSUMABLE COST	PLANNING SCHEDULING SIMULATION			REVIEW PRODUCTION REPORTS	BOTTLENECK ANALYSIS	REVIEW FACILITY REPORTS		
PRODUCT/PROCESS ENGINEER	TRAINING	SETTING OF STANDARD CONSUMPTION		SPECIFICATIONS MANAGEMENT	CORRELATIONS ANALYSIS	SCRAP/HOLD REWORK DECISIONS	EQUIPMENT CHARACTERIZATION AND VALIDATION	FACILITY/PROCESS DATA CORRELATIONS		
EQUIPMENT ENGINEER				WRITE/USE MAINTENANCE SPECIFICATIONS	EQUIPMENT CONTROL DATA ANALYSIS		EOT REPAIR REPAIR PLANNING & SCHEDULING EOT PERFORMANCE REPORTING SPARE PARTS MANAGEMENT DIAGNOSIS			
MANUFACTURING SUPERVISOR	OPERATOR MANAGEMENT ACCORDING TO QUALIFICATION	CONSUMABLE COST	SCHEDULING	INQUIRIES ABOUT SPECIFICATIONS IN EXCEPTION CASES INFORMATION ABOUT NEW SPEC.	REVIEW CONTROL CHARTS	VIEW SHIFT PERFORMANCE REPORTS / SCRAPHOLD REWORK DECISIONS	CRITICAL EQUIPMENT FOLLOW-UP EQUIPMENT / STATUS CHANGE EOT START-UP			
MANUFACTURING OPERATOR		USE OF CONSUMABLE	SCHEDULING	RECEIVE OPERATING INSTRUCTIONS RECIPE	VIS. INSPECT. PERFORM MEASUREMENT RECORD DATA / REVIEW CONTROL CHARTS	WAFER HANDLING RECORD WIP MOVE VIEW SHIFT / PERFORMANCE REPORTS	OPERATE EOT EOT STATUS CHANGE RECIPE DOWNLOADING PREVENTIVE MAINTENANCE (1ST LEVEL)	FACILITY DATA COLLECTION		

TASKS COVERED BY CELL AUTOMATION

Matrix 2.2 : Actor/Area

The same is true for transport level and network level. The redundancy between network protocol header and SECS header is low and is mainly limited to identification and vertical check which is considered by us as a good thing. If we agree that the header and structure description can be important in regards to data volume, it works well and facilitates exchange for Supplier-Customer meeting.

The acknowledge both at application level and transmission level avoids the regular mixing and respects the OSI principle of level independence. For us also the few missing application acknowledge could be added (for variable table to use or for recipe comments to enable for example). The stream definition (standard) is good and covers the customer need allowing easy sequence programmation for each functionality.

SECS protocol finally appears to be a consistent protocol well embraced by engineering people in Front end. In test the volume of data is high and then the part SECS 1 with RS232 norm and 236 byte blocks could be improved (According to our experience in EWS test SECS 1 is sufficient. For parametric test SECS standard is not used today).

2.3.2 - SECS protocol and equipments

If we can observe many improvements during the last two years, the suppliers have not always respected the SEMI standard, and the customer needs. Our experience is the following :

- SECS 1 problems or non conformity
- Set up of the transmission parameter very limited (time out, retry number), and associated documentation
- SECS identification defined by the supplier
- System byte non conformity
- SECS 2 problems or non conformity
- Lack of minimal information before purchase
- Too much user defined stream versus SEMI standard streams
- Set of streams too poor. The minimum could be :
  . Reset, Alarms and downloading
  . Status, Remote control and data acquisition
  . others
- Unconformity in data format in regards to SEMI standards
- Complete recipe downloading not possible
- Documentation quality
  . change of software release quicker than documentation updating
  . sequence example for main functionality
  . description using SEMI structure
- Remote stream corresponding to local touch not available (for example a table of variables)
- Duplicated message not detected
- Multistate message not available.

Moreover, a test made with SECSIM (Std SECS Simulator) by the supplier and given before purchase would be a plus.

2.3.3 - Results from AMS Program

The work undertaken by the AMS team results in :
- a definition of needs for a generic interface
- a generic interface experimental tool
- experiments and knowledge exchange.

● Generic Interface requirement definition

The major requirement is the guideline "no programmation". The team defines a set of transactions between host and equipment allowing the major functions needed and a principle to make a parametric generic interface.

● Generic Interface experimental tool

The team retains for the experimental tool the already common principle of equivalence tables. After analysis of PC 10 (SPN), VSAP (DEC), PAM (COMETS) and SCOPE (PROMIS) the final tables were defined and the tool built up. The user only has to set up the data of the table to manage sequences of transactions and to assure with a unique interface the conversion of the stream from the standard definition to the equipment definition (ident., structure, format). Through this tool, transmission level and application level are well separated and the user can manage sequences of streams to realize transactions or complete functions.

The generic tool presents some limits today :

      a) custom calculations, if other questions continue to need programming

      b) the tool is only interfaced with COMETS. For example communication is made by COMETS
      mailbox and the tool uses primitives from COMETS.

An industrial product for the market could have equivalence table on the C.A.M. side and a powerful human interface to generate specific custom calculation without programming (a better human interface to set up the tables, both communication structure (mailbox)1 and resource optimisation.

● Knowledge exchange and experiments

To improve the cooperation between the partners and inside each firm, an AMS common support to describe the sequence of streams in SEMI rules, the specificities of each equipment (user defined stream, format, rules in streams chaining) has been defined. The first use is in progress and the consistency is yet unknown.

The tool generic is used to interface the THERMCO 10000. For this first experiment, it is difficult to separate time for debugging from time for interfacing but is appears clearly that the goal of 2 weeks is reached (equipment SECS analysis, table set up, test).

2.4 - Miscellaneous

2.4.1 - Advanced Recipe Technology (ART)

An important customer requirement is the Recipe Management including editing functions, dowloading and security control. ART is a generic recipe editor providing a single global method for creating and modifying equipment recipes irrespective of the particular type of equipment concerned. Editing operations are carried out on customized screen masks specially designed for each different type of equipment and each function used.

Internal format and external format can handle the same recipe information. A compiler using an equipment model converts the recipe from a format to the other.

The following stages are involved in recipe editing :

1. Identify recipe	4. Modify internal format
2. Retrieve file containing external recipe	5. Translate back to external format
3. Translate to internal format	6. Store external format in file

As a conclusion, the major objective of ART is to improve the reliability of recipe editing and minimize the risk of recipe errors. ART will therefore check modifications to recipes against the user making them and the privileges this user possesses.

This software module, running in production, is compatible with COMETS environment and could be adapted to other CAM systems.

2.4.2 - Operator Interfaces

Human interfaces of current C.A.M. systems are very poor. The study of a user friendly operator terminal based on Macintosh microcomputer and icon principle is in progress. The goals are to minimize training time for operator and decrease transaction time by a factor of 2 to 3. This approch is less heavy than X Window and X Term and does not allow central generation or maintenance.

The main advantage is the limited cost to the terminals (very low) without more CAM CPU requirements and extra software cost. Another advantage is the possibility to have screens not built by Analyst people but former operator.

The first results are very attractive. Integration with the CAM system and generalization to any CAM system are the main recommendation for this topic.

2.4.3 - Expert Systems

The use of EXPERT SYSTEMS is engaged by the AMS partners. The goals are a help for operators and less maintenance people intervention on everyday problems. The first application is the help to set up the steppers in case of alignment or prealignment problems. The results obtained are very attractive. They are mainly :

. Better and faster training of operators increasing people flexibility

. Reduction in Maintenance intervention

. Improvement of steppers throughput by downtime reduction.

A second experiment in development phase will treat a more complex case : analysis of electrical measurement results in order to identify origin of problems.

Extension.

For the first experiment, the set up of Equipment will be extended to coater and developer tracks.

For the second experiment, a highliht will be put on a user friendly interface for utilisation by non-specialist programming.

Criteria of subject selection for expert system applications.

Due to development and adjust time and consequently Return on Investment, subjects have to be selected under the following criteria :

. the lifetime must be 4 to 5 years comprising numerous updates

. the knowledge must really exist

. the Expert must be able to formulate it.

Taking into account these elements, applications developed for a wafer fab will be profitable if knowledge depth is limited between 2 to 6 inference levels.

Recommendation

EXPERT SYSTEM must be usable from any C.A.M. terminal in order to facilitate access by operator and decrease operator movement in very clean areas (class 10 to 1). That means Expert Syst must be a functionality of the C.A.M. system and ought to be implemented on the controller of this CELL.

Expert system development have to be limited to process aspect. Development dedicated to equipment itself (maintenance, trouble shooting) remains under supplier responsability.

2.4.4 - Photolithography cell controller

The sharpness of photolithography process and its repetition in the complete process flow are the main reasons for installation of a cell controller in this area.

Benefits that can be reached by its implementation are the following : Process stability and Quality improvement; Better equipment utilisation; Misprocessing error reduction; Recipe management; Real time equipment control and tracking; Automated data collection and SPC; Internal scheduling; Simple Man/Machine interfaces; Dependence product equipment support.

Achievement

Some difficulties have been found during the implementation of level 0. Equipment manufacturer did not implement SECS protocol in the right way and we therefore had to ask them for patches and new software releases. It took a long time to fix SECS problems. The track system in the first SECS software version did not provide full physical tracking of wafer but new version allows operation synchronization.

As a conclusion, the results of the first level of implementation are summarized in the following items :

  . 1 Mbit EPROM pre-production
  . Fully automated data collection : lot/equipment history; SPC
  . Simplified operation activity
  . High level of process flexibility
  . Process Quality.

Cell Controller Features

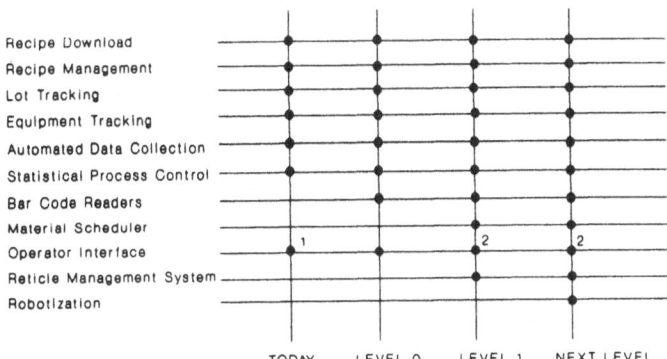

1. From Comets
2. Cell Controller Operator Interface

Extension

The functionnalities of a CELL CONTROLER can be classified in two categories : Passive ones and rather standard such as tracking of lots or wafers, tracking of equipment, management of alarms, recipe management etc... The active ones are more specific to the cell such as coordination of equipment, selection of lots, equipment parameters refinement after measurement etc...

In the AMS project, emphasis was put on the active and control functionnalities rather than on standard and passive ones.

The first implementations of photolithography cell controller allow better knowledge of physical problems of production. A lot of work is needed to define an accurate specification of the upper levels of a cell controller taking into account all general and specific requirements as shown in the upper Matrix

2.4.5 - Simulation - planning/schedulling

From the experience of "Knowledge System Group" of Reading University in "Scheduling in Job-Shop situations" using Reactive Constraint Scheduling Techniques, a partner of AMS project (MEDL) has undertaken a collaborative project with this group. The immediate object of this exercise is to provide suitable tools of simulation. This simulator will be linked with PROMIS data base in order to have a direct transfer of requested data : process description, WIP (Work In Process) status.

In particular it makes it possible to :
. Perform capability analysis
. Identify bottleneck
. Evaluate equipment utilization
. Study the effect of production mix on cycle time and throughput
. Evaluate scheduling policies
. Analyse "what if scenarios"

At first, a MODEL of the line has to be defined from the manufacturing process data. Five types of parameters, for which the impact on productivity and turn-around time had to be investigated, were identified :
. Type and number of equipment
. Skill and number of operators
. Availability of Equipment and operators
. Type, size and number of lots
. Type and frequency of reworks.

These five types of parameters determine the required level of detail of the model of the WAFER FAB which is broken in four parts: Factory model, Process model, Initial state model and Transaction model.
The definition of the model is a heavy task and a systematic approach to translating the data collected during a trial period into a usable plant model. A set of tools has been developed referred to as the MODEL BUILDER which will significantly improve future simulations.
Then EXPERIMENTS can be defined. An experiment uses a number of existing model files and input queries from the user in order to generate replacement model files reflecting the requirements of the query. At last the SIMULATOR is used to validate the factory model.

Ultimately, this work should lead to the creation of a Reactive Scheduler implemented on SUN station linked into Wafer and Equipment tracking of C.A.M. systems such as PROMIS or COMETS.
It would respond to the many non-planable events which occur in a Wafer Fab and reschedule accordingly in real time (1 min.).

In the future planning and scheduling tools have to be processed in one run, for operating people and directly on CAM data in real time.
These planning and scheduling tools will be grouped in different sets : monthly planning, weekly planning, daily scheduling, shift scheduling, start scheduling, lot and task selection.
It has been shown that driving starts with a smart tool and then using simple rules to select the lot at each equipment, even at each cell, could have a significant impact on the management of the line. Then, the first needs are shift and start scheduling with following features :

● Shift scheduling
Give operator/supervisor wip and capacity feasible targets at the beginning of the shift and provide feed-back on-line during the shift and at the end.

● Start scheduling

Give day by day, or shift by shift, the lots to start to meet demand, avoid long queues in front of equipment and avoid bottleneck starvation. A lot should be started if, and only if, it meets these 3 constraints.

These last two sets of tools will allow short cycle time and are necessary to attain full automation. They appear today to be key points for manufacturing of tomorow.

3 - ROBOTICS IN CLEAN ROOM

3.1 - Requirements for a Front-End

Introduction :   This chapter summarizes the needs and the applications for integrated circuits manufacturing with robotics.

Reminder :   - a Front End is a clean room in which bare wafers (silicium disk most often) are processed in order to obtain the same wafers with some hundreds identical products (memories, computer unit, logic or custom design).
- the flow in a Front-End is cyclic with a total of 200 to 300 operations.
- the main activities are photolithography, etching, diffusion, implant and deposition.
- Processes like photo or diffusion are performed from 8 to 16 times during a flow.
- An activity includes from 1 to 8 operations.
- An activity has the same requirements and often its operations are made in a same area.

3.1.1 - Needs by area

The principal needs are given in table 3.1.1
These data are not absolute but orientation.
They depend on the physical layout and the process choices.
All other robotics characteristics are very important (from the weight capacity to the offline programming or the user interface). For semiconductor industry reliability, suppliers quality cleanliness, communication interpolation and smoothness, the important criteria are . Weight, volume and Axis O are more layout dependent, morphology also at the lower level. Power of the language, process command, impact on the product (physical) are less important. Some others are not yet defined sensors/environment accuracy but what is clear F.E. needs industrial robots.

3.1.2 - Automation and Manufacturing Management (Planning - Scheduling)

Automation is not innocent with regards to planning, scheduling and general management. It depends also on the types of the automated cells :
Multiprocess or line : several successive equipments for the process flow are grouped inside a cell.
Monoprocess or workshop : the equipments of one process step are grouped inside a cell.
Only the simulation is able to supply data about cycle time, load rate, production in case of failure, productivity, parametric sensibility (mixt, volume), or transitory line balancing. Results will be obtained only if the model is a good simplification, the bottlenecks well identified and the objectives well defined (simulation is not a tool to do everything).

3.1.3 - Applications : see tables 3.1.3 a and 3.1.3 b

area Requirements	Photo	Diffusion	Etching	Implant	Deposition	Comments
Cleanliness	XX	XX	X	X	X	X:very high XX:critical
Velocity	1	2	2	3	-	1: very low 2:very high
Trajectory complexity	40 for stepper/ tracks	200/400	100/150	Not analysed	Not analysed	Physical points
Need for calibration	O	O.R.E.	O	O	O	. (O).Offset . (R).Robot calibration (E).Etalonnage site
Accuracy and incremental programming	Y-N	Y	Y	-	-	Local programming *
Tool	1	1/3	1	1	1/2	number of tools needed
Restart facility	XX	XX	X	X	Not analyzed	depends on cell complexity
Process or handling oriented	H	H	H	H	H	Robotics for handling
Sensors	N	Y	Y-N	N	N	Intelligent sensors like vision or 3D.
Morphology	open but low space	cartesian preferably	open	Not analysed	Not analysed	

* programming in object references and offline programming.

Table 3.1.1 : Needs by process area.

Area	Application	W	L	H	E	Prio-rity
General	1. Software interfaces for equipments (SECS II)				X	1
	2. Hardware interfaces for equipments				X	1
	3. Status information for equipments				X	1-5
	4. Equipment for box identification, cassette extraction and cassette to cassette transfer				X	3
	5. WIP storage with std mechanical interface				X	3
Photolitho-graphy	1. Coupling directly (interface) Stepper and tracks		X	X		1
	2. Coupling with flexibility tracks stepper, wet process and control		X			2
	3. handling for a workshop of tracks	X				5
Diffusion	For horizontal and vertical furnaces: 1. Loading and unloading carriers for diffusion or deposition					
	* quartz boat, intermediate carrier				X	1
	* PECVD carrier				X	1
	2. Transfer carriers to the cantilever					
	* Atmoscan				X	2
	* Paddle				X	2
	3. Loading directly wafer to paddle				X	5
	4. Coupling with cleaning (Cell)					3-4
Etching	- Handling for several etchers, stripper, control stations, and oven (cell between line and workshop)	X	X			4
Implant	- Handling for several implanters and coupling with post cleaning operation	X	X			4
Deposition	1. Handling for several sputters with pre cleaning operation	X	X			5
	2. Handling for deposition equipments with dangerous gases (APCVD - W.J. - Anicon)	X	X		5	
Wet Processing	1. Automation of the bench (transfers and times control)				X	3
	2. Coupling with equipment or cell integration			X		3
Control and test	- Standard interface for handling system					3

Table 3.1.3 a : Inside area applications

Application	Priority
- Low cost super clean (Class 10) AVG's with set up software For integration in CAM systems	3-5
- Super clean conveyors (Class 10) for box and cassette transfers	3-5
- Mobile like mail or documentation transport system	3-5

Legend :	1 and 2	today operation (1 to 2 years)	W : Workshop
	3	next automation step (3 years)	L : Line
	4	limited interest dependent on low payback	H : Handling
	5	interesting but without production demand or with remaining doubt.	E : Equipment

Table 3.1.3 b : Application between Areas

3.1.4 - Status of robotics - Profitability of investment

Technical stakes

Yield : Yield is improved by avoiding

Particule contamination

- defects or contamination

- miss processing

- breaks

The gain is = nb of wafers gained x price per wafer

Source	Gain	% actual
Air		5 - 10%
Operator	X	30 - 40%
Equipment		20 - 30%
Process	X	20 - 30%

Cycle time: By reducing the cycle time :
● the time to put a new product in production allows to be sooner on the market : higher market price, higher market share

● the capacity to answer to a customer and to change the mixt of the manufacturing is better.

The gain is dependent on internal rules.
Reminder :
Cycle time = queue time + pilot and monitor time + process time + control time

Productivity and reliability	Productivity and reliability allow higher volume of production. ● If the equipment is a bottleneck, The gain is equal to : 1. The number of wafers gained x price per wafer x % of use of this capacity during the time of payback 2. or the price of a new equipment x % defined by internal rules ● If the equipment is not a bottleneck, the gain is equal to : 1. 0 for equipment installed 2. or the price of equipments saved in case of new layout or increase of volume.
Direct labor	● For operator directly attached to the operations Gain = cost of an operator x number of entire operations gained ● For operator working partly at the operation Gain = 0 if no reorganisation is possible to concrete it
Indirect labor	Gain is = Cost of maintenance x % maintenance gained

Major stakes

Feasibility of a product	As the critical dimensions are smaller robotics could be the only possibility to produce with acceptable yields. In this case the profitability is more complex and the rules will be specific.
Automation Roadmap achievement	Some partial automation has no gain in itself. For these investments, the major criteria are the profitability of the total investment and directly the level of investment
Worldwide Competitiveness	For memory
Strategical constituant for equipments of the Future	For memory

3.2 - Industrial achievement of robotic projects

3.2.1 - Robotics in diffusion

3.2.1.1 - Studies

The team made an analysis on products and experiments.
- ASM : PECVD with one Robot one lift and cantilevers.
- SGS-THOMSON: Diffusion with one Robot and cantilevers.
- SEMCO : PECVD loading station and Mass transfer diffusion system.
- Bruce : Fully automated furnace with loading/unloading.
Future has been taken into account with concept like SMIF and vertical furnaces.

3.2.1.2 - PECVD achievements (status : beginning of production)
See Table 3.2.1.2

1. Analysis of the carrier mechanical definition and deformation, principle of use, robot or system precision and need of sensor.
2. Definition of a specification
3. Selection for the loading station only (main payback)

Item	Value	Comment
ROI	1 year	
Investment	150 KE	External expenses
Start up time	2.5 months	Delay for mechanical design and carrier reassembly procedure.
Integration in CAM SECS 2 protocol	Waiting for delivery	supplier specification match with user need
Remarks		. very poor documentation . No robotics language and closed application

Table 3.2.1.2 : Characteristic of PECVD loader

Item	Value	Comment
ROI	to be defined	
Investment	to be defined	
Robot		Replacement of a 6 axis robot by a robot for loading and a lift for the transfer in order to simplify the system to stick to the morphology of the site, to reduce wait time.
Integration in CAM SECS 2 Protocol		Many discussions to obtain the set of SECS transactions judged necessary by SGS-THOMSON and the conformity to SECS 2 standard
Remarks		Integration of sub-system is not an easy job.

Table 3.2.1.3 : Characteristic of diffusion furnaces loader.

3.2.1.3 - Diffusion (status : manufacturing in progress) See Table 3.2.1.3

This system will be a fully automatic cell with one robot to load and unload boats and a 3 axes lift to load and unload the cantilevers.

Item	Mechanics	Robot	Comments
ROI	1 year	1,5 year	
Investment	60 KE	120 KE	For the first equip.
Functionality	. Standard handling . Simple down- graded modes	. Standard handling . reject distinction . simple downgraded modes . lot interrupt . simplest interface with stepper	
Start up time	2 weeks	1 month	
Remarks	Very compact	No layout change.	

Table 3.2.2.1 : Characteristic of stepper and tracks coupling

Item	Value	Comment
ROI	2 years	
Investment	200 KE	
Functionnalities		. Wafer loading (tracks, stepper, tester) . Accept external input output for down graded mode. . Sofware Functionalities * Recipe down loading * Data acquisition * tracking (lot, equipment) * operator interface * scheduling and equipment coordination
Remarks		The main objective is to obtain a flexible cell.

Table 3.2.2.2 : Characteristic of link between stepper, tracks and tester.

3.2.2 - Robotics in photolithography darkroom

3.2.2.1 Coupling stepper and tracks

Comparison between mechanical interface and standard robot to link coating track, development track and stepper. The three equipment are today like a unique machine. See Table 3.2.2.1

3.2.2.2 - *Photolithography cell*

This project integrate a robotic interface between stepper - tracks and tester to constitute an automatic cell (called level O). See Table 3.2.2.2

3.2.3 - Quantification of results

3.2.3.1 - Cleanliness

Measures were made to quantify the contamination from a robot vs from a human. The main results are :

1.Results depend on test environnment (particle counter, point of measure) vary with the manufacturing of the robot (arm with air depression, type of belt, cable type) and quality could be improved by small modifications.

2. Low cost robots give different results for the arm and the hand.

	Arm			Hand		
Part. size	<0.3	0.3 ->1	>1	<0.3	0.3 ->1	>1
Robot	200	16	0	100	5	0
Human	50	30	15	300	250	50

3. Clean robots are very low particle generator.

	Appli 1			Appli 2			Appli 3		
Part. size	.5	0.5 ->1.	>1	.5	0.5 ->1.	>1	.5	0.5 1.	>1
Robot	1.5	4.	1.	82.	11.	2.		15	
Human	1.	10.	6.					50	

Remark

	PECVD		
Part. size	<0.4	0.4 ->11	>1.2
Robot	?	5.5	1.
30' waiting under VLF	?	6.	1.

* Part size in μm

3.2.3.2 - Results

	Photolitography			PECVD
	Mechanical interface	Fixed robot	Mobile Robot	
Yield	2,5%	2,5%		0,5%
Throughput	2%	2% -> 5% (remark : -1%)		0%
Cycle time	3%	3%		−ε % (masked time)
Disponibility	60%	99,5%		92%
Remark	. Stepper and interface together . Too early to have average value	. Robot alone . average value	no data available today	
Production management	. Efficient but without enough Flexibility (break down, different resist used	. Waiting for the functionalities extension . Flexibility limits . Preference for a unique solution		. Improvment for carrier cleaning needed. . Very satisfied
	Compact simple	. No dependant on layout. . Could be a standard.		

4 - FUTURE AND COOPERATION

4.1 - Strategy for global robotic implementation

4.1.1 - Guide lines

Apply robotics where :
    1. Equipments are reliable
    2. SECS 2 protocol are available
    3. Quick duplication is possible
    4. Key factors are improved (Yield, throughput, cycle time, availability and management constraints)
    5. Solution is consistent (specificities are sure to be taken into account and have manual or automatic solution)

4.1.2 - Robotic characteristics

1. Robot and application together
2. Market common product
3. Teach in, absolute and relative programmation
4. Langage oriented robotics and handling
5. Gripper and layout.

4.1.3 - Way

1. Reliability of equipments
2. Recipe down loading and data collection
3. In parallel :

Subsystem integration without cell controller	Cell controller definition	Extra cell transport
Cell integration with cell controller	Prototypes with level by level implementation	
Mono and multiprocess cell		

4. Process integration
5. Full material handling
6. Full remote control

4.2 - Future for robotics

4.2.1 - Extra cells transport

This robotics application is not fixed today. Many solutions are possible :
- AGVS (like Veeco, Daifuku or CEE product)
- conveyor (dedicated to clean room or clean standard mail dispatch system)

More that the transport function itself, what is researched concerns :
- the matching the scheduling at area level
- the full integration in CAM system
- the regard for timing and priority
- the allowance of clear lots storage.

4.2.2 - Diffusion

Horizontal furnace : loading robot and cartesian transfert to paddle
Vertical furnace : not defined.

4.2.3 - Photolithography

Two trends are followed up today :
1. Rigid interface between equipment (minimum investment, very simple)
2. Flexible mobile robot.

4.2.4 - Others areas

Not defined. As for photo two ways are seen :
- cell with cleaning process and control
- cell of process equipments.

5 - CONCLUSION

This paper presented in a first part an overview of the AMS project which covered the software part of the project : CAM system and some more specific modules such as Expert system application, Fab simulation, CELL CONTROLLER implementation. In second part Robotics in clean room was analyzed under requirements and results aspect.

During next year, the use of COMETS will be extended to the recent lines of SGS THOMSON MICROELECTRONICS with a set of common utilization rules.

PROMIS at MEDL will control the whole Wafer Fab, Assembly and Final test line .

Numerous Equipments (Process and Measurement) will be connected to CAM systems and SECS problems will be fixed. Experimental CELL-CONTROLLERS in Photolithography areas will be running in production.

All Island of automation in Diffusion areas as well as in Photolithography will be operational and comparison between experimented solutions will be possible.

Link between Facility monitoring system and CAM system will then be in experimentation.

Moreover, all software modules, new operator terminals, wafer handling interfaces, transport systems and wafer identifier will be ongoing on experimentation in lines.

Rich in experiment results in Production, R and D and ASIC lines, we will be able to define all requirements for :
. an advanced and flexible manufacturing system comprising the specification of an advanced CAM system
. a generic CELL CONTROLLER definition
. Recommendations about handling and robotics in cleanroom as defined in Chapter 4.

At the end of the project, the following phase ought to be the development by European Industry of the software packages and tools defined in the course of the AMS project.

As far as software, the main points could be a new generation CAM system with associated tools and general CELL CONTROLLER linkable to any CAM systems.

INFORMATION PROCESSING SYSTEMS

ESPRIT PROJECT 125 GRASPIN: ACHIEVEMENTS AND EXPERIENCES

W. D. ITZFELDT
GMD
Institut für Systemtechnik
Schloß Birlinghoven
D-5205 Sankt Augustin 1
West Germany

ABSTRACT. The ESPRIT 1 project GRASPIN has developed prototypes of a personal software engineering environment to support the construction and verification of distributed and non-sequential software systems. The GRASPIN environment provides a flexible framework with extensive facilities for the incorporation of new methods and tools, and for the customization to a variety of languages, applications, and target systems. Prototypes are implemented on both Lisp systems and PCTE-based machines.

This paper presents the main R&D results of the project, which ended in September 1989 after a lifetime of more than 6 years. Moreover, it reports first experiences gained by pilot users and discusses the lessons learnt in collaborating in an international project jointly performed by eigth partners from industry and academia.

1. Introduction

As a response to the "software crisis" already identified in the late 60s, the development of software engineering environments is under way all over the world. It is the aim of such activities to better control software development processes and thus to improve the quality of software products substantially and to reduce the development and maintenance cost drastically by means of well-engineered and integrated methods and tools.

The spectrum of software engineering environments reaches from mere tool kits to more or less integrated project support environments (IPSEs). On international level, R&D activities are directed towards so-called 3rd generation IPSEs, which incorporate knowledge bases, where the progression of IPSE generations shall lead to comprehensive Software or Information System Factories. It is widely agreed that software development environments of the future will fully be based on methods of both software engineering and artificial intelligence.

However, which methods and tools are necessary to form a 'complete' environment and how they should be integrated is still subject of extensive reasearch and explorative investigations. Before current tool kits in industrial software production can be gradually replaced by advanced IPSEs, a variety of experimental implementations will be necessary to clarify fundamental questions of systematic method support and tool integration.

Therefore, in order to successfully move to the 3rd generation of IPSEs, current software technology must be better understood and improved.

2. Project Goals

As one of the first ESPRIT projects in the area of Software Technology, the GRASPIN project was concerned with improving current software development approaches to encourage their use on a wider industrial scale.

The full title of the project, which also explains the acronym GRASPIN, is "Personal Environment for Incremental Graphical Specification and Formal Implementation of Non-Sequential Systems."

The GRASPIN project aimed at both, research and development goals:

Research activities were directed to improve current software development approaches and, where necessary, to develop new methods concentrating on the most critical areas of the software life cycle, namely the early phases of software development as well as validation and verification. A further goal was to combine these methods in a coherent way.

Development activities aimed to systematically support the methods by appropriate tools. The tools should be integrated into a personal software engineering environment prototype. As a 2nd generation type IPSE, the GRASPIN environment should provide a flexible framework with extensive facilities for the incorporation of new tools and methods, and for the customization to a variety of languages, applications, and target systems.

The environment should be particularly dedicated to the incremental development of distributed software systems. It should provide a coherent methodological support for specification and development of complex and reliable systems. Concepts, methods, and tools should cover as many technical activities in the software life cycle as possible, and reflect the cyclic nature of software construction with the countercurring validation processes.

Methodological improvements were expected e.g. from combining Petri net theory with algebraic specification and from supporting different languages for different 'phases' of the software life cycle. Technical improvements were expected e.g. from syntax directed editing techniques and from object oriented programming techniques. Great attention should further be given to possibilities of combining formal and informal specifications, and to cope with incomplete specifications, designs, and programs.

Prototypes of the GRASPIN environment were to be implemented on both, Lisp and PCTE/Unix machines.

3. Project History

The GRASPIN project started in September 1983 and ended in September 1989. It was conducted in four phases:
1. in a 1-year 'pilot phase', the development methodology was preliminarily defined and the requirements of the kernel system were analysed;
2. in the subsequent 2 1/2-years 'initial prototype phase', first prototypes of the kernel system were implemented;
3. a 2-years 'enhancement phase' was dedicated to
   • common specifications of kernel system interfaces in order to integrate different tools;
   • re-implementation/adaptation of the kernel system interfaces to the common specifications;
   • improvements and extensions to the method support of the environment;
4. in the last 7-months 'pre-exploitation phase', basically major project results were prepared for post-project exploitation, e.g. by pilot installations.

In the last two project phases, GRASPIN was jointly performed by eigth to nine academic and industrial partners from West-Germany, Italy, France and Greece. The full partners in the GRASPIN consortium were

- GMD, Sankt Augustin (D),
- Olivetti, Ivrea & Pisa (I),
- Siemens, München-Perlach (D),
- Tecsiel, Pisa & Roma (I),

GMD being the main contractor. Tecsiel, who was subcontractor until the end of the second project phase, joined the consortium as full partner in March 1987.

The following institutions acted as subcontractors of GMD, Olivetti, and Siemens respectively: Computer Technology Institute, Patras (GR), Epsilon, Berlin (D), Etnoteam, Milano (I), Kaiserslautern University, Fachbereich Informatik (D), and SLIGOS, Paris (F).

In early project phases, two other subcontractors were also involved, namely Language and Programming Systems (LPS), Torino (I), and Software Sistemi S.p.A. (SSS), Bari (I).

In the average, more than 40 people worked in the project, with an average manpower effort of about 25 person years per year. The total cost were approx. 15 MECU.

4. Project Results

The project has developed prototypes of a personal software engineering environment to support the construction and verification of distributed and non-sequential software systems. Each of these prototypes provides a flexible framework with extensive facilities for the incorporation of new methods and tools, and for the customization to a variety of languages, applications, and target systems.

4.1. THE ENVIRONMENT

The GRASPIN environment is a language directed environment of the second generation of integrated software engineering environments: Tool integration is reached by a common database for the sharing and exchanging of data between the tools. The structures of the data objects are specified by means of a declarative language. From these specifications, access operations are automatically generated; they constitute the common tool interface to the database through which all the tools access data objects.

The environment consists of a language independent Kernel System, a number of language dependent tools for validation and verification, and installation dependent interfaces to the host system [1].

The development of the GRASPIN environment has been largely influenced by concepts of current generators for structure editors, such as Mentor [2] and the Synthesizer Generator [3], and the Gandalf Environment [4]. However, the lack of flexibility of such generators has been overcome with a more powerful generation formalism. This formalism is provided by the declarative meta-language ASDL (Abstract Syntax Definition Language), being the core of the so-called Language Definition System of the Kernel System. The ASDL offers comprehensive features for embedding new methods and tools in the existing environment and for adopting it to specific project requirements of software developers. In this respect, GRASPIN is an open environment. Further, with ASDL integrated software engineering environments can be specified as specialisations or extensions of the GRASPIN Kernel System [5]. ASDL combines an object oriented type system with syntax directed translation schemes and a target language interface.

The GRASPIN environment is particularly developed for the construction and verification of distributed and non-sequential software systems.

4.2. METHODS AND TOOLS

As a 'general environment' [6] the GRASPIN prototype contains basic language independent tools like structure editors that support manipulation of both textual and graphical objects in all phases of the life cycle. However, main emphasis is placed on language specific methods and tools, which support the early phases of requirements definition and formal specification as well as validation and verification.

Informal and formal methods are equally provided: Structured Analysis (SA) and Entity Relationship (ER) diagrams are used for requirements definition, the semi-graphical language *SEGRAS* for formal specification and design, and ADA for programming.

The GRASPIN specification language *SEGRAS*, developed in GMD, combines the concept of abstract data types with the theory of high-level Petri nets. A software system is abstractly described in terms of types, operations, conditional equations and behaviour. Dynamic system behaviour (processes, synchronisation) is specified using high-level Petri nets (predicate-transition nets and predicate-event nets). The invariant system structure and data structures are specified algebraically. These relatively abstract and problem oriented descriptions can be interactively analysed, also in parts, e.g. with respect to type correctness, consistency and completeness. Incremental development is thus supported already in the design stage.

For static analysis of requirements definitions, algebraic specifications, and programs, a number of analysis tools are available. For example, SA and ER diagrams can be checked for their syntactic correctness; semantic checks concern, e.g., the consistency of connections among nodes in a graph. A Type Checker supports separate and incremental analysis of specification modules written in *SEGRAS*, by identifying all symbols occurring in a specification module and finding an appropriate typing so as to restrict the range of values which expressions may take during runtime. When performing static program analyses, data flow and control flow are analysed; cross references point to labels, variables, and datatype operations incorrectly used, etc.

Most of the GRASPIN verification tools make use of term rewriting techniques, and are embraced in the so-called Rewrite-Rule-Laboratory (RRLab). The RRLab supports three verification methods:

MISOP   method for the check of important specification object properties (confluence, consistency, completeness, termination, totality)

MCAI   method for the check of the correctness of an algebraic implementation step

MASN   method for the algebraic specification of finite nets with finite capacities

With the Net Analyser, lifeness and safeness properties can be analysed for a limited class of nets.

The Net Simulator provides for performing the common 'brute-force' simulations of PrT nets (single and multiple random step, user-selected step) and to test behavioural characteristics such as reachability of states interactively.

The Testbench supports program validation. It is constituted by automatic and semi-automatic components, both being integrated. According to users' needs, either semi-automatic or automatic tools can be used. Nevertheless, the automatic Testbench tools can be used as a stand-alone testing environment. They substantially support the generation of check lists and test cases.

All the tools are accessible via a uniform graphical user interface with multiple windows and menus. Therefore also graphical representations can be clearly visualized and directly manipulated. User interaction with the system is supported by choice via commands or function keys, menus and mouse.

4.3. IMPLEMENTATIONS

Several customizable prototypes of the GRASPIN environment are available.

The prototypes developed by the Italian partners Olivetti and Tecsiel run on Sun's Unix 4.2 BSD and are implemented in C language. These prototypes are particularly dedicated to Requirements Engineering, supporting subsets of Structured Analysis (SA) and Entity Relationship (ER) languages, and the Testbench [7].

The Lisp prototypes developed by the German partners, together with their sub-contractors, primarily support formal specification and verification. They are implemented in Common Lisp and run on Symbolics 36xx (with Genera 7.2), on Sun workstations with CLOS (Common Lisp Object System), CLUE (Common Lisp User Interface Environment) and CLX (Common Lisp X-Windows), and on the Macintosh II Ivory (with Genera 7.4i).

The very low performance on the Sun, however, makes this prototype unusable in practice. The results of a portation study show that selected tools of the environment can easily be ported onto a Macintosh II PC (with Allegro Common Lisp). Due to the degree of memory and computational power required by the whole environment a complete portation to this machine seems not to be reasonable. On the other hand, the MacII seems to be well suited as a front-end (also exploiting the Mac user interface) for a multi-user GRASPIN environment with only one Lisp machine host.

5. Exploitation and Impact

A number of collaboration agreements with both, organizations from industry and academia, have been established by the project
- to promote the concepts and methods supported by the GRASPIN environment, e.g. by means of special courses held at Universities;
- to evaluate the usability of specific tools and to get early feedback for improvements;
- to further develop and customize environment tools for specific languages, such as LOTOS [7].

For preparing the industrial exploitation of project results, amongst other activities in the last project phase, pilot installations of the GRASPIN environment were established to evaluate the feasibility of the GRASPIN methods and tools for realworld application problems, and to get feed-back for further developments.

SLIGOS, the 4th software company in Europe, whose main activities are in the area of distributed systems, acted as the first pilot user exploring the feasibility of the net-based tools of the Lisp prototype (on Symbolics workstations) when applied to specification and verification tasks of monetics applications, i.e. electronic payment systems. In the very early stage of this collaboration it already became evident, that
- knowledge of and experience with the (formal) methods is a pre-requisite for any pilot user - fortunately this prerequisite was given;
- lack of technical documentation did not have much impact on getting familiar with the tools;
- knowledge of the Symbolics user interface is a pre-requisite, which however did not have any consequence on the use of the tools;
- the experiment was successful even before the Sun3 version was available;

- performance aspects were not given enough emphasis when constructing the tools.

At the end, it could be shown, amongst others, that [8]

- the tools allow the analysis of communication aspects in monetics applications (modelled in term labelled systems);
- the tools provide a complete analysis of the associated condition-event nets of reasonable size;
- the modularity of the environment would allow easy introduction of algorithms;
- the simulation gives a visual representation of the behaviour of the designed payment system;
- algebraic specifications give a high level of abstraction of modelling;
- the advantages of specifying distributed monetics applications with the GRASPIN methodology compared with other approaches have been evaluated.

At the University of Kaiserslautern, the final version of the RRLab is currently used in two manners. First, students are instructed in practical application of algebraic specification and verification theory; second, students are going to enhance the facilities of the RRLab so that the laboratory is able to deal with cyclic rule systems without running into trouble with infinite computations. Since this educational work is done on Macintosh II PCs, equipped with Allegro CommonLisp and CLOS, the RRLab was installed on that type of machine. As spin off for GRASPIN, the team gained some first experience with respect to the portability of the RRLab, especially its applicability on non-Lisp machines. Apart from reimplementation of the LowLisp module, no further work had to be done. Consequently it is expected that portation of the system to other machines, such as Apollo workstations, will not raise problems.

Further, some of the project results, namely PCTE-based environment tools, have been used by the SFINX project to be incorporated in their Software Factory. In addition, a couple of ESPRIT 2 projects plan to use some of the GRASPIN tools based on formal methods.

6. Teamwork and Synergy

None of the partners had worked in a comparable joint international project before. For each team member, including the managerial staff, the need for collaboration was therefore perceived as a major challenge.

During the whole lifetime more than 100 people were involved in the project, each with different professional background and experience, and some only working for a few months. Quite naturally, learning was an ongoing task for most of them. Although knowledge distribution was a valuable side-effect, the high fluctuation rates in some teams sometimes made it impossible to reach milestones in due time. In one team also the project leader changed, in the average, annually.

Division of labour worked much better between industrial partners than between partners from research institutions and industry, provided they had the same goals. For the project as a whole, the benefits from cooperation were much less than expected, due to the lack of division of labour. Major obstacles to achieve a more effective division of labour in prototype development were

- different methodological approaches (GANDALF vs. MENTOR),
- different implementation strategies, mainly induced by company-specific policies,
- different hardware equipments of the teams.

Some partners even argue that they would have achieved better results in the same time if they would have done the work for their own. In fact, the price for cooperation between several partners from different countries was very high. On the other hand, of course, for the majority of the team members the experiences gained from this cooperation are judged very positively, and the parallel developments have led to a variety of (partially) complementary prototypes, which have already received much interest in both the academic and the industrial software community.

7. Conclusions

Within the last six years, the ESPRIT 1 project GRASPIN has developed prototypes of a personal software engineering environment to support the construction and verification of distributed and non-sequential software systems. The GRASPIN environment provides a flexible framework with extensive facilities for the incorporation of new methods and tools, and for the customization to a variety of languages, applications, and target systems. Prototypes are implemented on both Lisp systems and PCTE-based machines.

Industrial experience gained so far suggests that from a methodological point of view the GRASPIN environment can be seen as a significant step towards the next generation of CASE environments. Exploitation on a wider industrial scale, however, still requires additional technical consolidation of the results for adapting the underlying formal methods to specific industrial needs and to integrate the GRASPIN methods and tools with current industrial technology and standards.

8. Acknowledgements

I would like to thank all team members of the GRASPIN project, who have contributed to the results described herein. I am particularly indepted to Bernd Krämer and Heinz-W. Schmidt from the GMD team for their valuable hints to improve this paper.

9. References

[1]   R. Endres (ed.): Architecture of the GRASPIN Environment. GRASPIN Technical Paper GRA80/3, Sankt Augustin: GMD, May 1988
[2]   Donzeau-Gouge, V. et al.: Programming Environments Based on Structured Editors: The Mentor Experience. In: J.R. Barstow et al. (eds): Interactive Programming Environments, McGraw-Hill, New York, 1984
[3]   Reps, T. and Teitelbaum, T.: The Synthesizer Generator: Reference Manual. Technical Report, Computer Science Dept., Cornell University, Ithaca, N.Y., 1985
[4]   Habermann, H. and Notkin, d.: The GANDALF Software Development Environment, CMU Techn. Report, 1982
[5]   Krämer, B. and Schmidt, H.-W.: Developing Integrated Environments with ASDL. IEEE Software, January 1989, pp. 98-107
[6]   Houghton, R.C., Jr.: Characteristics and Functions of Software Engineering Environments: An Overview. ACM SIGSOFT, Software Engineering Notes, vol.12, no.1, January 1987
[7]   Mannucci, S. et al.: The Kernel of a Software Development Environment for Graphic Languages. Proc. Hawaii Int. Conf. on System Sciences, January 1989
[8]   Project Team: Final Project Report of Esprit Project 125 - The GRASPIN Environment. GRASPIN Technical Paper GRA 125/1, Sankt Augustin: GMD, September 1989

Project no 280
EUROHELP

Performance Interpretation in an Intelligent Help System

Maud Stehouwer
Jan van Bruggen
Courseware Europe bv
Ebbehout 1
1507 EA Zaandam
The Netherlands

Abstract.

Intelligent Help Systems have to keep track of the user's actions in order to monitor his performance. The help system monitors the performance of the user to detect whether the user is performing up to his level of proficiency with the information processing system. If the user makes errors or operates inefficiently, the help system will diagnose the user actions in order to remedy user misconceptions, or to introduce more efficient ways of performing tasks in the information processing system. The paper discusses performance interpretation in Eurohelp (Esprit project P280).

1. Introduction

One of the most outstanding characteristics of Intelligent Help Systems (IHS) is their ability to provide active and passive help that is sensitive to both the state of the information processing system (context-sensitive help), and the knowledge of the user (Fischer, Lemke & Schwab 1985). Moreover, intelligent help systems try to expand the knowledge and skills of the users in order to promote their efficient use of the full functionality of the information processing system. Especially the active help offered should thus be relevant to the current task of the user, and it should be in reach of the user's current knowledge and skills. In order to provide context- and user-sensitive help, the help system maintains models of the current task of the user, as well as a model of the user's knowledge of the information processing system, and how this knowledge can be further expanded.

In Eurohelp the task of performance interpretation primarily involves observation of the user's behaviour while he/she is working with the application, and identification of problems the user has with the application. Whenever a problem is identified, the performance interpretation modules analyse whether the IHS should intervene and identify the information that is relevant to feedback. In Eurohelp the performance interpretation modules serve several functions to allow the intelligent help system to provide its assistance to the user:
- recognition of the current plan of the user
- evaluation of the efficiency of the current plan
- diagnosis of inefficient or incorrect user behaviour.

Performance interpretation is not special to a help system: many systems, especially teaching systems, monitor and interpret the user's behaviour. Performance interpretation in an intelligent help system has to cope with two additional problems however:

248

In the first place the task of the user is in general not known to the IHS; only the user's behaviour can be observed. The IHS should be able to decide whether the user needs help, and determine the sort of help on the basis of this observed behaviour.

Secondly, diagnoses have to be set in a dynamic environment without interaction with the user. When a help situation occurs, the IHS is due to give feedback immediately. However the drawback here is, that there is no opportunity to gather further evidence from the user's behaviour, while an extensive interaction with the user is to be avoided.

A third problem for Eurohelp's performance interpretation is specific to Eurohelp's approach. Since Eurohelp aims at defining a shell for the development of intelligent help systems, the approach to performance interpretation has to be generic, rather then specific for a certain information processing system.

2. The Eurohelp Approach

2.1. General Description of the Approach

In Eurohelp performance interpretation can be seen as a cyclic process of plan recognition, planning, plan comparison and diagnosis. The plan recognizer tries to 'explain' user behaviour in terms of one of its known plans. Whenever such a plan is recognized, a planner is called to compute an alternative plan for the same task. If this planner returns a different plan, both plans are handed over to a diagnosing module that tries to explain why the user's behaviour was different from that expected on the basis of the planner's plan.

All modules make extensive use of Eurohelp's Application Model and student model. We will only touch upon these models here. The reader is referred to Breuker et al. (1989) for detailed information.

2.2. The Application Model as a Basis

The basis for performance interpretation is Eurohelp's Application Model (AM) on whose structure the performance interpretation operates. The AM holds explicit representations of tasks, plans and the conditions under which the plans apply. The task-plan hierarchy consists of hierarchical planning trees for the application tasks with all correct plans represented explicitly. At the top of this representation abstract tasks are situated, at the leaves simple interaction tasks, like 'Type-rmdir' are found. The representation supports plan recognition by its hierarchical structure which also supports planning by plan differentiation. Explicit effect descriptions are available for remediation purposes.

2.3. User Modelling Information

Eurohelp's user model is kept as an overlay on the task-plan hierarchy. The model holds the systems beliefs about the current knowledge (mostly procedural) of the user. The user model is used by the plan recognizer, planner and diagnoser modules (see below).

2.4. Plan Recognition as a Basis for Performance Interpretation

The plan recognition process is the basis of performance interpretation. Its purpose is twofold: in the first place understanding of user behaviour. Here correct behaviour (in terms of known plans) as well as alternative interpretations are involved. The second purpose is prediction of behaviour, which is used to find alternative interpretations as well. The Eurohelp plan recognizer matches incoming user actions against a pre-stored task-plan hierarchy of increasing abstraction.

2.5. Planning to Find the Most Efficient Plan

Whenever the plan recognizer recognizes a plan the Eurohelp's selection planner is consulted for a possible alternative plan. This selection planner uses plan rule-sets to select the plan from the task-plan hierarchy, that is the most efficient, given the state of the information processing system and the user's knowledge and skills. This plan is seen as the most appropriate plan for the user, and functions as an initial norm in the diagnosing process. Plan recognizer and planner may subsequently be called again from the diagnosing modules.

2.6. A Coaching Perspective on Diagnosis

Because the diagnoser has to operate in an environment with dynamic and incomplete knowledge, the character of the diagnostic process needs adjustment to this situation. The IHS's primary concern is to support the user in using the information processing system, and in learning about its possibilities. Analyzing the user's conceptions about the information processing system is of secondary interest. Stated differently: diagnosis serves coaching purposes first.

The coaching perspective implies that the diagnoser obeys two principles. The first principle is that of 'immediate assistance': whenever the user runs into problems, the help system should offer some support. The second principle is that of 'user-specific assistance': the assistance should be focused on knowledge that is applicable by the user in the current situation.

Working along the lines of the principle of immediate assistance creates special problems for a help system. On the one hand a detailed diagnosis of deficiencies will support remediation that is tailored to the user's problems. Such a detailed diagnosis, however, often requires further interaction with the user in which the system can ask for the user's intentions, or can probe its hypotheses by presenting questions or problems. Neither option is available to an IHS, which therefore has to operate with the user's behaviour as the only input. The principle of immediate assistance forces the help system to construct a diagnosis even if the supporting evidence is weak. Remember that in the systems we are discussing the user determines the topic of discourse by invoking the functionalities of the information processing system. A 'wait-and-see' strategy by the help system might therefore lead to a long delay before a second opportunity for intervention on the topic occurs.

Systems might only evaluate the situation of the information processing system for the user and let the user infer the correct knowledge himself. In Eurohelp however, the user is given as much support on the latter as possible. This, of course, is in line with the second principle mentioned above. We have pointed out the relevance of feedback that is given when it is needed and that is made user specific. By implication the diagnosing process should search for deficiencies in those knowledge topics that are relevant in the current situation for this specific user. This should be done in an environment where no deep diagnosing is possible.

The solution proposed here is based on two assumptions regarding the development of knowledge by the user. The first assumption states that user knowledge develops from more basic operational knowledge (commands), through more detailed operational knowledge (pre-conditions for commands) to support knowledge (effects of commands, related concepts). The second assumption states that a user does not have misconceptions about topics that he has little or no knowledge about. Working on these assumptions the diagnoser can determine the knowledge level that is relevant to the user, and direct the search for diagnoses toward this level.

<u>2.7. Performance Interpretation Process in Eurohelp</u>

The performance interpretation process tries to *understand* the user actions. Following Schank (1986) understanding is different from a deep explanation of the user's actions. More or less superficial checks on user behaviour are sufficient for general understanding. If user actions violate general expectations (i.e. an anomaly occurs), a deeper explanation process is required however.

The expectations concerning user behaviour in an information processing system imply that the use of the system is relevant to the user's purposes; that the user knows what he is doing when giving commands to the information processing system, and that the user wants to obtain his goals as efficiently as possible within his capabilities. The IHS expects the user to strive for some goal G (that makes sense in the application and is known to the IHS) by means of a plan P (a valid plan in the application and also known to the IHS). Further the IHS expects that plan P is a plan known to the user (its commands, preconditions and effects), and that plan P is also (one of) the most efficient way(s) known to the user to accomplish G.

Whenever these regular expectations about the user's behaviour are violated, for example when the user applies an unknown command, the performance interpreter constructs a deeper explanation for this behaviour in the user knowledge.

Two modules are responsible for this understanding process. The first is a monitoring module that checks whether the user action matches the expectations of the IHS and that detects anomalies. The second module, triggered when an anomaly is detected, diagnoses the anomaly by constructing an explanation for the unexpected user action from the user's knowledge.

The monitoring module receives its main input from the plan recognizer. We will discuss Eurohelp's plan recognizer first, and then describe the performance interpretation process in greater detail.

3. Plan Recognition in Eurohelp

<u>3.1. Basic Approach</u>

Eurohelp's plan recognizer is a rather straightforward plan matcher: received user actions are matched against the leaves of known plans. The plan recognizer selects one of the plans that start with the user action as its hypothetical plan and expands this plan to predict the next user action. As long as the actual next user action is in accordance with the expectation, the plan recognizer sticks to its hypothesis and predicts the next action in the hypothesized plan. As soon as a complete plan is recognized, the plan recognizer moves up its focus to higher level plans under which the recognized plan may be subsumed. Again a most likely candidate is selected, and this plan is expanded (the focus moves down) until the next user action can be predicted. This moving up and down of the plan focus resembles the FITS-2 plan recognizer (Woodroffe, 1988).

3.2. Focussing the Search

All plan recognizers face the problem of a potentially very large search space that has to be kept manageable by focussing the search. The Eurohelp plan recognizer uses two, rather crude, heuristics to focus its search: the strength of the user's knowledge of the plan, according to the user model, and the frequency of use of the plan: best-known plans that are often used, are hypothesized first. Although more application specific heuristic could be used, as in POISE (Carver, Lesser and McCue, 1984), at the moment the Eurohelp plan recognizer only uses the two heuristics mentioned. The focussing strategy is used both in bottom-up hypothesis formation as well as in top-down expansion of hypothesized plans.

3.3. Moving through the Task-plan Hierarchy

Eurohelp's Application Model has several features that allow the plan recognizer to quickly go up and down the task-plan hierarchy. Special relations in the Eurohelp Query Language link tasks and plans, and cater for partial orderings of steps in a plan, and mapping of arguments from one level in the task-plan hierarchy to another. The has-decomposition relation can be used to retrieve candidate plans for the user action, the other relations are used to test for constraints on ordering and arguments.

Whenever the plan recognizer moves up the task-plan hierarchy it may find that it can only partially or hypothetically instantiate the higher level plan. An example: a delete-character task may be part of a plan to delete a word character by character. This delete-word plan can only be instantiated in a hypothetical manner (an object of type word exists with first letter 'char'). The type of help system is important here: if the help system is pure 'add-on', it will have no knowledge of the type of char deleted, and so only conditional feedback can be given to the user (deleting a series of characters plus a space means to delete the rest of a word). If the help system has more knowledge about the state of the information processing system, it can collect individual deleted characters and match them against a grammar to recognize a 'word' object. If the help system and the application are fully integrated the help system can inspect the object hierarchy to find the 'word' object that is affected by the delete-char task.

3.4. Subplans

Subplans can interact with the plan in which they occur. In Eurohelp the task-plan hierarchy explicitly acknowledges possible inserted plans by grouping them in a set of tasks that can occur in a plan. Of course these are plans that have no negative interaction with the surrounding plan (for instance in MS-DOS the 'dir' command can be inserted in most other plans without interfering with that plan). Other plans do however interact by setting or removing pre-conditions for other plan steps. In Eurohelp positively interacting plans are made part of the task-plan hierarchy, i.e. they occur in the plan-decomposition of a higher level plan (cf the approach in the SINIX consultant, Hecking et al. 1988). An example: to delete all the files in a MS-DOS directory, there will be a plan that boils down to the command 'del pathspec\wildcard', and another plan with two steps: 1) move to the directory and 2) issue 'del *.*' command.

3.5. Backtracking

As stated before, the plan recognizer focuses its search on the plans that are best known according to the student model and most frequently used. This focussing strategy will result in several candidate plans (ordered by the heuristics mentioned above) and one hypothetical plan at each of the levels in the task-plan hierarchy on which the plan recognizer currently works. Whenever a user action cannot be matched against one of the expected actions under the current hypothetical plan, the plan recognizer backtracks by moving up the focus and reconsidering the unexplored candidate plans.

3.6. Limitations of the Approach

Eurohelp's plan recognizer is limited to matching user actions against the plans in the task-plan hierarchy. Plans that achieve the goal of a task, but that are not in the task-plan hierarchy will not be recognized. However, since the task-plan hierarchy contains both expert and novice plans, the plan recognizer is not limited to an expert's view on the information processing system. The plan recognizer assumes that the user has a correct view of the information processing system. If the user actions cannot be matched against one the known plans, for instance because the user confuses two commands, the plan recognizer gives up and the diagnoser has to deal with the partially recognized plan.

4. Monitoring: an Anomaly Detecting Process

In Eurohelp the user interaction with the system is intercepted by the help system, which can subsequently send it as input to be executed by the information processing system. The monitor performs some checks on the user actions before execution (syntax checks and checks on potential catastrophes), and other checks after execution by the information processing system. Pre-execution checks are based on single user actions, post-execution checks preferably on recognized plans. If no existing plan can be recognized, the interaction tasks are passed to the monitor.

Every user action is evaluated by applying a set of anomaly checking questions, based on the assumptions described earlier. These questions are handled sequentially and independently, each evaluating a specific aspect of the action. The anomaly checking questions use several sources of context information.

The first source of context information available to the IHS is knowledge about the available functionality of the system (commands) and knowledge about the state of the application (conditions set at a certain moment). This information is used in the following two anomaly checking questions:

1. Is the user action a legal action, i.e. can the command be executed? If the command cannot be executed in the application, the action is illegal, at least in the current context, and the IHS does not expect the user to apply the command now.
2. Is the user action a useful action, or is it redundant? If the user action does not lead to a change in the current state of the information processing system, then obviously the user action is point-less and is therefore not expected by the IHS.

The second source of context information available to the IHS is the knowledge state of the user as represented in the user model. The following two questions use this source:
3. Is the user action part of a plan known to the user, i.e. how plausible is it that the user knows what he is doing? If, according to the user model, the user does not know the

command, the IHS does not expect the user to use it.

4. Is the user action in accordance with his proficiency? If the user knows a more efficient way to perform the action the IHS expects the more efficient solution.

A third source of information deals with the semantics of the action and is in general not available to an IHS. To give an example, if a the user types a report and subsequently deletes the file with the report, there is no problem as far as the IHS is concerned, as long as the user's action are efficient. To signal catastrophes like these, a set of potential catastrophes can be defined for every information processing system. Using this set the IHS can warn the user for potential catastrophes. If the user indicates that the effect is not what he intended, an anomaly is detected. This leads to the fifth question:

5. Does the user action involve a potential catastrophe, and if so, is it probable that the user knows what he is doing?.

A fourth source of context information is the user himself. Whenever the effects of an action do not match the user's expectations, he might issue a help request. This request is handled by the passive help component of the IHS and will not be discussed here.

If one of the expectations handled by the anomaly checking questions is violated an explanation is desired. The diagnoser is called upon to search for a plausible explanation for the user's action in the user knowledge.

5. Diagnosing: an Explanation Seeking Process

The diagnoser is called when an expectation is violated and an explanation for an anomaly is needed. The explanations that are of interest to an IHS involve user knowledge: why did the user consider his action correct and efficient? If the diagnoser cannot find a plausible explanation in the user's knowledge, the behaviour is considered a slip, and the user should then be able to deal with the situation himself. The diagnosing process is decomposed in two main processes: norm generation and knowledge investigation. The first process infers the user's goal and generates a norm plan, the second process identifies the knowledge deficiency that explains the user's behaviour.

5.1. Norm Generation

To consider the relevant user knowledge, a norm plan is required to which the user's plan can be compared. For the generation of this norm plan, the goal of the user must be known of course. Preferably, from a performance interpretation point of view, the user is *asked* for his intention. However an IHS has to refrain from bothering the user as much as possible, and therefore, if it cannot ask the user, it has to infer the user's intention.

Contrary to most teaching systems, an IHS does not know beforehand the current task of the user. The plan recognizer is needed to infer the user's goal. As we have seen this intention inference process is in fact the basis for the signalling of anomalies. As long as the plan recognizer is able to recognize 'expected' plans and tasks, where 'expected' means 'no expectation violated', the intention of the user is hypothesized to correspond with the recognized tasks.

To hypothesize an intention several heuristic techniques are used, a.o. the detection of typing errors, the use of a bug library, manipulation of arguments, and syntactical manipulation of the interaction tasks of the command. Another important source for intention inference is the plan recognizer. Two techniques involve a query to the plan recognizer.

The first query checks whether an action might be a legal action in another context, in order to account for mode errors. This query can be applicable for executable as well as inexecutable commands: mode errors can occur when interaction tasks have different effects in different modes. The most notable example here is issuing commands to an editor while in insert mode.

The second query makes the plan recognizer return its expected actions. The task or tasks that are expected on the basis of the recognized plan(s), are hypothesized as intentions. This query is of course only powerful if the plan recognizer returns a limited number of expectations. Therefore this query will be most useful in applications with script-like sequences of tasks.

The other techniques mentioned are also more useful in one application than in another. For example, in an application with many one or two letter commands like vi, the detection of specific typing errors is less relevant. The knowledge engineer will have to decide which techniques are relevant to the information processing system. Which techniques are relevant in a specific situation is determined by the anomaly.

5.2. Investigating User Knowledge

Whenever an intention has been hypothesized, the selection planner is called to return a norm plan. Discrepancies between the norm plan and the user's plan trigger a diagnostic search. This search aims to find a cause in the user's knowledge that can explain the user's unexpected behaviour. The search is directed by a classification of the user behaviour, based on a syntactical comparison of user plan and norm plan. The discrepancy between the two plans is described in a so-called *deviance class*. Four main deviance classes are distinguished: redundant action, inefficient action, insufficient action and irrecognizable action.

Although deviance classes look like anomalies, there is an important difference: an anomaly concerns the detection of unexpected behaviour, the deviance classes imply an intention based description of behaviour. For example when the monitor detects a redundancy-anomaly, the inferred intention might be different from the one achieved by the redundant action, in which case the behaviour description will be insufficient or irrecognizable action, because the user did not accomplish the intended goal. Obviously a redundancy might also turn out to be just what it appears to be, a redundant action.

The knowledge deficiencies that can explain the user's behaviour are genericly described for every deviance class. For each class the set of relevant explaining hypotheses is defined along with a search strategy that obeys the coaching perspective principles described earlier.

Description Of Diagnostic Hypotheses. The diagnostic hypotheses are formulated as perturbations on correct knowledge topics. The following perturbations are distinguished: deletion of a topic (lack of knowledge), replacement of a correct topic by an incorrect topic (misconception) and addition of an irrelevant topic (overspecialization).

Examples of knowledge deficiency descriptions:
- lack of knowledge plan: the user does not know a command or combination of commands.
- misconception condition: the user thinks the plan is applicable under condition A, while in reality it is applicable under condition B only.
- overspecialization effect: the user thinks a plan (command) has an effect which it does not have in reality.

The Diagnostic Search. The search is aimed in the first place at finding a deficiency that gives a plausible and sufficient explanation for the user's behaviour. In the second place the diagnostic search is aimed at investigating whether the user has the knowledge that is required to solve the deficiency.

The Search For An Explanation. The set of potentially explaining deficiencies is defined genericly for every deviance class. In principal, in every situation, two kinds of explanations, equally relevant, are investigated. The first explains why the user did not apply the correct plan; the second why the user did apply the incorrect plan. In case of an inefficiency only the first explanation is relevant: the user's plan was a correct alternative. In case a redundancy was found, only the second explanation is relevant: why did the user do something where he should have done nothing? In case of an insufficient or irrecognizable action, both kinds of explanations are relevant.

For every deviance class one or two explanation sets are defined, and a generic description is given of how to obtain the specific hypotheses for the current situation. The sets identify the potential explaining diagnoses. The plausibility of these hypotheses is then tested against the user model.

The order in which the hypotheses are investigated is prescribed by a search strategy. The hypotheses are ordered in such a way that basic knowledge is investigated before more advanced knowledge. For example, knowledge about plans is investigated before knowledge about the necessary conditions for applying a plan. The search is continued until either a plausible hypothesis is found that gives sufficient explanation (according to explicit criteria), or the system runs out of hypotheses. The diagnostic search then shifts to the second aim and investigates the knowledge required to resolve the identified knowledge deficiency or deficiencies.

Investigating The Underlying Knowledge. The search for required underlying knowledge, is basically the same as the search for explaining knowledge deficiencies. For every deficiency a set of possible underlying knowledge deficiencies is formulated that contains deficiencies about related topics. All topics relevant to a correct understanding of the knowledge deficiency identified as an explanation are investigated. For example, when a lack of knowledge about a plan is found, the IHS investigates whether the knowledge about the subplans is available to the user. If the user has a misconception about a condition, the system investigates whether the user might have a deficiency concerning an effect, etc.

The Diagnosis Delivered To The Coach. The diagnosis delivered consists of a complete analysis of relevant knowledge. It contains all information that, in the current situation and considering the user's knowledge, might be relevant to remediation. Knowledge to resolve the misconception as well as knowledge to be able to apply the correct plan are implied.

All topics required for a good understanding of the current situation by the user are involved in the knowledge analysis. The analysis is independent of the coaching situation however. The decision to remediate on all topics, or only on parts of the topics, or on additional topics as well, is a tutorial one that is decided by the modules responsible for coaching.

6. Conclusions

A framework for performance interpretation is described, that enables an IHS to act according to the principles of 'immediate' and 'user specific assistance'. The framework provides a mechanism for diagnosis specification, in an environment where deep diagnosing is impaired by restrictions put forward by the principles mentioned. The performance interpretation process is able to come up with a diagnosis that is tailored to the knowledge level of the user. Although the diagnosis does not give a precise description of the user's knowledge, it gives a complete analysis of the knowledge state of the user with respect to the required knowledge in the current situation. This allows the IHS to give user specific feedback whenever this is required.

The generality of the approach sketched is, to a certain extent, an empirical matter. Of crucial importance here are the assumptions regarding the development of knowledge and misconceptions. It might be that a user, learning a new application, but bringing in knowledge from related applications, will develop knowledge, skills (and misconceptions) in a different manner than assumed by the current approach. In this case the performance interpretation would not provide the right focus for coaching.

7. References

Breuker, J., Duursma, C., Winkels, R., Smith M. (1989) *Knowledge representation in Eurohelp*: modelling operation and understanding of computer applications for help systems, pap. pres. Esprit Conference 1989.

Carver, N.F., Lesser, V.R., McCue, D.L. (1984) Focusing in Plan Recognition, in *Proceedings AAAI-84*, Un. of Mass., 42-48.

Fischer, G., Lemke, A., Schwab, Th. (1985) Knowledge-based Help Systems, in L. Borman and B. Curtis (Eds.) *Human Factors in Computing Systems*, CHI 85, April 14-18, San Fransisco.

Hecking, M., Kemke, C., Nessen, E., Dengler, D., Gutmann, M., Hector, G. (1988) *The Sinix consultant - a progress report*, Universität des Saarlandes, FB Informatik, KI-labor (SC-project) Memo Nr 28, August 1988.

Schank, R.C. (1986) *Explanation Patterns*: Understanding Mechanically and Creatively, Hillsdale N.J., Lawrence Erlbaum Associates.

Woodroffe, M.R. (1988) Plan recognition and intelligent tutoring systems, in J.Self (Ed), *Artificial Intelligence and Human Learning*; Intelligent Computer-Aided Instruction, London, Chapman and Hall.

Keywords

Diagnosis, Eurohelp, Intelligent Help Systems, Plan Recognition, User Modelling

Knowledge Representation in Eurohelp:

Modelling Operation and Understanding of Computer Applications for Help Systems.

Joost Breuker
University of Amsterdam
Dept. of Social Science Informatics
Herengracht 196
1016 BS Amsterdam
The Netherlands.
Tel: +31 20 525.2149
Email: breuker@swivax.UUCP

Radboud Winkels
University of Amsterdam
Dept. of Computer Science and Law
Kloveniersburgwal 72
1012 CZ Amsterdam
The Netherlands.
Tel: +31 20 525.3485
Email: winkels@uvalri.UUCP

Cuno Duursma
Courseware Europe BV
Ebbenhout 1
Zaandam
The Netherlands.
Tel: +31 75 172201
Email: duursma@coeur.UUCP

Mick Smith
ICL
Arndale House
Arndale Centre
Manchester M43AR U.K.

Abstract

EUROHELP is a shell for developing intelligent help systems. It is assumed that by specifying domain specific knowledge about an information processing system -e.g. Unix Mail or Vi- a help system for the particular application can be constructed. This leads to a number of requirements for the structure of the domain representation in EUROHELP. The application model should be generic -covering all kinds of applications- and multifunctional, i.e. it should serve different types of knowledge based modules. The various modules of the EUROHELP shell, like the plan recogniser, diagnoser, coach, etc. are specific interpreters of the representation of the target application (e.g. Unix-Vi).

The core of the representation consists of descriptions of system procedures ('commands') and structures of objects. The system procedures consist of actions which identify objects (object reference) and/or manipulate objects. This core can be viewed as a conceptual translation (model) of the target application. A more operational view of the functions of the target application is provided by a hierarchical task layer in which the mode concept plays an important role. The task layer bottoms out in system procedures. This task layer supports planning processes in a top-down way.

Although a principled approach has been taken to define the functionality of the EUROHELP system, the current design shows many pragmatic short cuts, which may make it also a practically feasible enterprise.

1. Introduction.

The research discussed here is part of the EUROHELP project[1]. This project is aimed at the construction of an environment for building Intelligent Help Systems (IHSs) for "Information Processing Systems" (IPSs, i.e. interactive computer programs). Core of this environment is a shell that contains all domain independent procedures and knowledge. The major task of a developer of a help system for some specific IPS will be to fill the shell with a representation of the domain concepts (commands, syntax, methods of object reference).

The EUROHELP shell and the tools for editing the knowledge base (Application Model) are implemented in Common Lisp running on a Xerox 1186 or SUN workstations.

In this paper we will focus on the generic representation of IPSs and the ways this representation can be used to fulfill the functions of Help Systems, but we will begin with a short description of what a full IHS consists of (see also Breuker et al., 1987).

1.1. The Architecture of an Intelligent Help System.

IHSs support users both in a passive and in an active way: The user may ask a question concerning the IPS, or the Help System may infer a need for information from the users performance. In the first case, the Question Interpreter will try to define the users need (preferably within the context of the performance), in the second case the Performance Interpreter will have to compare its ideas on the intentions of the user with the actual performance. This is achieved by a Plan Recognizer and a Planner, which cooperate (see section 4.1). Both Question and Performance Interpreter can consult the User Model to help them identify possible problems or needs for information.

Once a potential problem or occasion for expanding the user's knowledge has been identified, the Diagnoser will have to explain it in terms of either a lack of knowledge or a misconception on the part of the user. If it does find such an explanation, a local need is constructed: a description of the current user need in terms of (a) the immediate cause (a question and/or performance) and (b) the diagnosis. This local need is then sent to the Coach, who will plan a didactic discourse strategy to handle this user problem. Such a strategy is a hierarchical structure of which the terminal nodes are called tactics, i.e. communication acts (cf. Winkels et al., 1988). This tactic structure will then be transformed to natural language or some other means of representation by the Utterance Generator.

Figure 1 summarizes the described architecture.

1.2. Outline of the Paper.

After this short overview of the main functions of EUROHELP, we will now focus on the generic components of the (AM) and principles that guided the structuring of a generic AM. In section 4 the several interpreters (modules) that use this AM are discussed. In section 5 we present the practical conclusions about the principles as discussed in the next section.

---

1   The research is partially funded by the ESPRIT Program of the European community under contract P280. The project encompasses an effort of about 100 man-years over a 5 year period. Partners in the project are: AXION (Denmark), ICL, University of Leeds (U.K.), Courseware Europ, University of Amsterdam (The Netherlands).

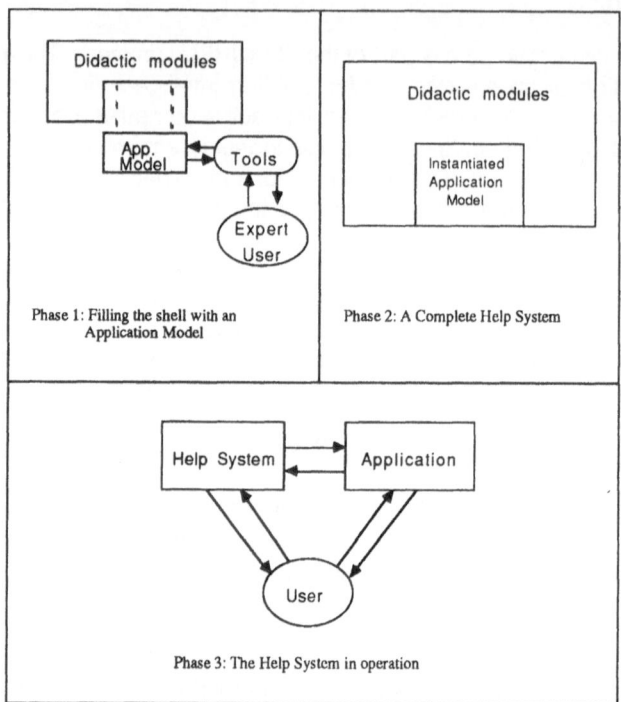

Figure 1: The three phases in building an Intelligent Help system by the use of the Eurohelp Shell.

2. Principles of Application Modelling for EUROHELP.

In selecting and designing the formalisms for Application Model (AM) in EUROHELP two important requirements were identified from the start.

Generic structures. The first one is that these formalisms should be both generic and specific for IPS, i.e. for a wide range of applications like editors, mail, spreadsheets, communication packages, etc.). The formalisms and structures of the AM should be sufficiently general to cover this large variety. On the other hand they should be more specific than a straightforward knowledge representation language, because it is assumed that the world of computer applications (IPS) is more specific than any world. This means that we have defined a limited set of types of objects, relations and actions that models this IPS world.

Multifunctional formalisms. The EUROHELP shell is a distributed knowledge based system, consisting of a number of cooperating modules. Normally each of these modules would have its own specific knowledge base, tailored to the functions of the module. This is advisable, both from an efficient processing point of view and from more fundamental considerations (Levesque, 1985). However, such solution is not very practical in EUROHELP. The description of an object or system procedure would be spread all over the IHS, with large amounts of overlap, and no way to ensure consistency and coherence. This means that some uniform rather than a hybrid formalism should do the job.

The requirements of multi-functionality and generality lead to the same principle: the specification of the structuring elements that constitute the world of IPSs. From the generality point of view it is hypothesised that this world is a rather 'narrow' one, as is the case with most 'artefactual' systems relative to 'natural' ones. From the multifunctional point of view the AM should be sufficiently 'deep' as to allow interpretations that drive planning, diagnosing, etc.

2.1. Operational and Support Knowledge.

The notion that 'deep' models support a larger variety of reasoning strategies has been defended by many researchers (e.g. Clancey, 1983; Davis, 1983; Wielinga & Breuker, 1986; Steels, 1987), but there is little evidence how wide this range is. Aside from the scope and 'depth' of knowledge, there is little doubt that declarative formalisms, which separate the specification of the knowledge from its use, are to be preferred to procedural ones, because it is easier for editing and supports a wider inference capability (Genesereth & Nilsson, 1987).

The distinction between declarative and procedural knowledge is analogous to operational vs support knowledge (Clancey, 1983). Support knowledge consists of descriptive models of a particular domain which allow one to infer what and why it is the case in a domain. The operational knowledge can be conceived of as a particular restructuring or compiling-out of this knowledge by experience in particular types of tasks, which can be conceived of as instructions of how to accomplish these tasks. Operational knowledge is in general represented as rules and plans; support knowledge as frames.

In principle support knowledge is sufficient to describe a system and derive all potential behaviour of such a system: it constitutes the deep, 'qualitative' model of a domain. However, for IPS applications one needs also an operational view on the knowledge for the following reasons:

– There is potentially far too much behaviour that can be derived. In an operational view, only that behaviour is of interest that is functional.
– The same behaviour (effect of a system procedure) may have multiple intentions. For instance, one can delete a file to prevent version confusion, or to annoy another user. In an operational view one can specify plans for 'version control' or for 'annoying ones colleagues'. Inferring the intentions of the user is of utmost importance in performance interpretation and question understanding.

From the example it is clear that there can be an endless variety of intentions and plans. In fact, these intentions are the link between the IPS-world and the world of the user, which may only partially overlap. The pragmatic way out is to combine the support knowledge with the operational knowledge, and to restrict the latter to standard plans, i.e. there can be plans for 'version control', it is very unlikely that 'annoying colleagues' will be available as a plan.

3. Description of Generic Structures.

In this section we will discuss the generic structures that have been developed to represent an application. The EUROHELP application model has to support two different kinds of functions (see section 2):

- Support the understanding of the application (support knowledge).
- Support the use of the application (operational knowledge).

Each of these two functions provides different requirements to the model. Therefore two different representations have been developed, that complete together the EURO-HELP application model. The support knowledge is represented by conceptual structure consisting of actions and objects; the operational knowledge is represented by a task structure. The two representations are connected at the level where use meets function-ality; the system procedure level.

3.1. Representation of Support Knowledge - Objects and Actions.

The functioning of the IPS has been modeled by representing both the objects available within an IPS and the way in which objects are changed by actions explicitly (see also Breuker \& de Greef, 1985; Duursma, 1988a,b).

The term 'object' denotes all pieces of information an expert handles during the performance of tasks with an IPS. Thus a window is considered an object, the cursor, but also a line and a file. Objects are characterized by their attributes and their relations with other objects. The "isa" relation and the "part-of"/"parts" relation are used to express the structure of the objects, the so-called "object-hierarchy". This object hierarchy reflects the structure of information within the system.

User interact with an IPS by invoking its commands. Two aspects of a command can be distinguished; the procedure that changes the state of the application and the interaction pattern that invokes this procedure. The changes made by system procedures to variables that are of any interest to the help system are reflected by changes in the object hierarchies.

To fulfill the requirements of both user-model and didactic goal generator, the effects of system procedures are described in abstraction hierarchies. At the top of the hierarchy the effect of an action can be expressed in terms of general action descriptions, providing an abstract description of the effect of the action. A few generic top-level classes of action abstractions have been discussed in Breuker and de Greef (85) and Duursma et al. (86). This set includes create, delete, show, hide, select, reject and change object actions. The first test of these primitives has been carried out lately in our prototype help system. The other relations in the representation of actions (inversion and analogy) will possibly be inferred from the abstraction. It must still be investigated whether this can be done automatically.

The abstraction hierarchy is built by describing the effect of the system procedure at increasing level of detail by the use of (combinations) of action primitives. The effects are not just decomposed to handle the complexity of the description, they represent a learning path of a novice user becoming an expert. At the top of the hierarchy a few action primitives are sufficient to represent the general effect of the system procedure. Descriptions at this level support explanations like "dd; deletes the current-line, remove $<file - name>$; removes the file $<file - name>$ from the current directory". To represent the effects of system procedures in more detail the primitives are specialized by adding before and after methods containing other action primitives.

3.2. Representation of the Use of an Application - From Intention to Interaction.

The representation of the operational knowledge is clustered around a set of tasks. These tasks represent user intentions; tasks users intend to perform by the use of the application. How these tasks can actually be performed is represented by plans containing sub-tasks in a multi level model. Basically the representation formalism is an AND/OR tree with added control and parameter passing constructions.

The plans represent the method to perform a task. Plans consist of a sequence sub-tasks, which can either belong to the same level as the plan, or to one level below the level of the plan. High level tasks are linked to high level plans, which contain sub-tasks that are more IPS specific. This structure is repeated, until the plans contain interactions only.

Some of the levels of this model are analogous to the levels in Moran's Command Language Grammar (CLG) [Moran 81]. In the CLG four levels are identified: task level, semantic level, syntactic level and an interaction level. The structure of the CLG has been taken as a starting point for the representation of operational knowledge in the EUROHELP application model, and it has been enhanced to make this model also suitable for planning and plan-selection.

The EUROHELP model contains tasks at three levels (see figure 2):

- User task level.
- System procedure task level.
- Interaction task level.

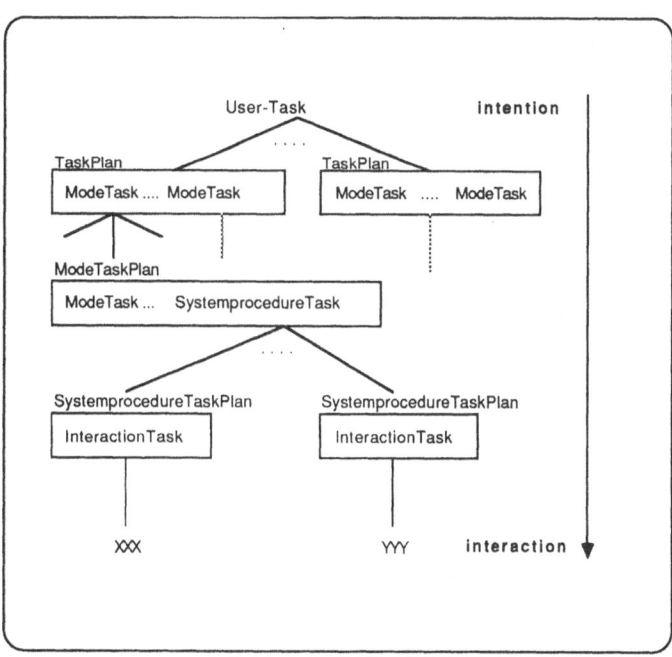

Figure 2: An overview of the task-plan hierarchy.

The three levels can informally be justified by describing the process the user carries out when she performs a task by the use of the application. This task performance consists of splitting up the task into a sequence of steps, that are collected in a plan. This decomposition is repeated until the steps match directly onto the procedures available in the application. At this point a specific effect of a command should be invoked to initiate a function of the application. In EUROHELP a specific effect of a command is represented by the concept 'system procedure'. The task to invoke such a piece of functionality is represented by the system-procedure tasks. How this task can be performed is represented by the system procedure task plans. These plans contain sequences of interaction tasks. The interaction tasks represent the interactions between the user and the application.

At each of the levels possibly several alternative plans can be used to perform a task (the OR branches of the tree). These alternatives are not always equally applicable at the same moment, their usefulness depends on the state the application is in. To determine the best applicable plan a rule is defined for each task. Each rule checks the current state to find out whether its plan should be applied within the current situation. In the CLG the plans to perform a task are represented as methods, for instance as 'semantic methods' at the semantical level. There is no way in the CLG to select the most efficient method.

At each of the levels a plan consists of a sequence of tasks (the AND branches of the tree). Because the tasks that are part of this plan need not always to be performed sequentially, a step structure has been designed. This step structure represents the ordering of the tasks by classifying their temporal inter-relations. Classes are e.g. 'immediately next' and 'sometimes next'. A second feature of the step structure is the representation of the control over the tasks. Though the plan represents the general method for performing the task, the step structure is needed to compute the precise form of the plan in the specific situation. At the moment the steps are computed by small 'black box' experts connected to each step in the step structure.

Finally there is a construction added to the task plan hierarchy to pass objects downwards and upwards through the hierarchy. Arguments of tasks always map directly onto the arguments of their plans. Within the plans arguments are split (or combined, going upwards) and passed to the sub tasks of the plan. Currently the argument mappings are also defined by 'black box' experts.

Task level

At the task level stereotypical user tasks for a class of applications (e.g text editors) are mapped onto application specific tasks (e.g. Vi). These stereotypical tasks are also called intentions. Thus 'intentions' are tasks a user wants to perform with the application and the application is supposed to support, but not necessarily directly. At the bottom of this hierarchy at task level, those tasks are represented that are directly supported by the application. The latter are also called mode-tasks, because these are the tasks that can be carried out within one mode. The top and the bottom of this hierarchy are connected by plans and, if necessary, sub-tasks. In this way a user-task is decomposed into a plan, of which the the tasks are again linked to other plans until eventually the mode-tasks are reached.

System procedure task level

At the system procedure level the mode-tasks are mapped onto a sequence of system procedures tasks. The system procedure tasks represent the task to invoke a system procedure. Plans at this level represent the way this piece of system functionality should be invoked. They consist of a partly ordered set of steps. The steps consist of interaction tasks. This level is in the CLG represented by the syntactical level. The semantical level of the CLG is not included in the task-plan hierarchy of the EUROHELP Application Model, but has been represented outside of this hierarchy as support knowledge.

Interaction level

At the interaction level the system procedures are mapped onto a sequence of interaction tasks. Interaction tasks represent the physical actions the user can perform on the system. This level describes pressing keys and clicking mouse buttons. It has a similar meaning as the interaction level of the CLG.

4. Functions: Interpreters in EUROHELP.

4.1. Performance Interpretation.

An Intelligent Help system built by the use of the development tools of EUROHELP will support the user in both an active and passive way [Breuker et al. 85]. The user can ask the help system a question, but the help system can also take the initiative to offer the user new information. In the first case the need of new information will be defined by the module that interprets the user questions. In the second case the performance interpretation module will recognize and compare user intentions with the new behavior and send a signal to the coaching component that the user needs more information.

The performance interpretation module consists of a few different parts:

– An expert-planner: The planner makes an ideal decomposition of a task, it produces a structure that represents an efficient solution path for the task [Duursma 89].
– A plan-recognizer: This module recognizes user tasks from a sequence of interaction tasks, it produces an instantiated user task [van Bruggen 89].
– A diagnoser: This module infers a knowledge need from the user's performance. It compares the recognized plan with the optimal plan, which results into a list of deviations. These deviations are categorized. Examples of categories are: inefficient, redundant, incomplete and erroneous. After this a reason for this behavior is searched for. During this search process the pre-conditions of the commands, the user model and a library of known bugs are consulted [Stehouwer at al. 89]

The process of performance interpretation is given in figure 3.

4.2. Didactic Goal Generation.

A fourth interpreter of the Domain Representation is the Didactic Goal Generator. It is part of the Coach and produces Didactic Goals: a didactic view of the domain. The major function of Didactic Goals is the specification of topics which can be be acquired by a user/student by applying some learning principle. The learning principles will be specified in the form of types of relations between concepts. In this sense, Didactic Goals have a similar representation format as genetic graphs (Goldstein, 1979). For a particular user the Didactic Goals consist of relations between knowledge the user already has and new topics in the domain.

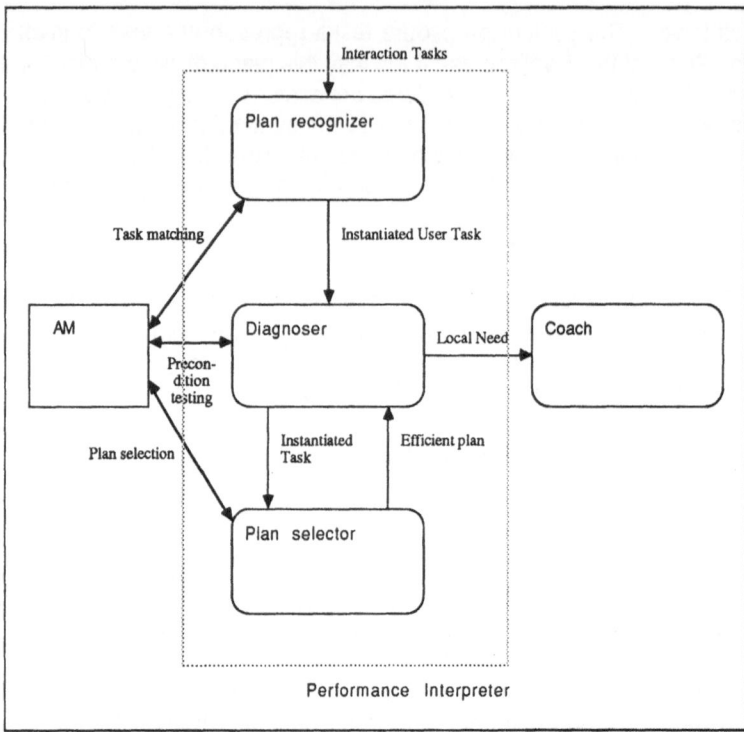

Figure 3: Data flow in the performance interpreter. The plan-recognizer follows the behavior of the user, when a sequence of interactions is found that completes a higher level task, it is passed to the Diagnoser. The diagnoser analyzes the recognized task and and when it is not erroneous it invokes the plan-selector to find an efficient alternative. This alternative is compared wih the recognized plan and an analysis is made. The result of this analysis is a local need, which is passed to the coach.

We distinguish four types of didactic relations:

1. **Generalisation - Specification**: Some concepts are more general than others. They share attributes, but the more specific one has extra attributes. All the shared ones can be transferred from the general concept to the more specific one if the Coach explicitly mentions this specification relation. This facilitates acquisition.

2. **Abstraction - Concretion**: When certain system procedures are reasonably well practiced by the user (skill), it is time to explain some underlying, hidden objects and models, i.e. to introduce some support knowledge (insight): abstraction. Also, it may be desirable to illustrate an explanation of how part of a system works (support knowledge) by telling the user how he can use it (operational knowledge), thereby specifying a concretion relation.

3. **Inversion**: Instead of transferring shared attributes, inversion allows one to transfer the inverse of attributes, i.e. "X is the opposite of Y".

4. **Analogy**: Explaining domain concepts by the use of analogies and metaphors. Here the idea is to have the user transfer a structure from one domain to the target domain (i.e. "X here is like Y there").

The Didactic Goal Generator interprets the Domain Representation to identify these didactic relations. For instance, some specification relations can be deduced from and through the isa and part-of hierarchies (see section 3.1).[2]

The generalisation-specification, inversion and some analogy relations[3] can be generated as soon as the Domain Representation for a specific IPS is ready. They are fixed, user independent didactic goals, and resemble genetic graphs. The abstraction-concretion relation however is user dependent and will have to be generated 'on the spot'. The need for abstraction will have to be inferred from the operational knowledge the user is supposed to have acquired (as reflected in the User Model). This will mainly be the case for the (hidden) objects and models the user has used or been referring to without (explicitly) knowing so. For instance, in the Unix Vi domain, the user has been issuing the commands "delete" and "put" several times, without knowing anything about "buffers" (the common hidden object of these commands). Then the concept of a "buffer" will become a Didactic Goal. Which in turn might trigger telling about how to address these buffers, so giving some new operational knowledge (concretion).

4.3. Topicalization.

Whenever the Coach is talking to the user, it is talking about something, about one or more topics. These topics will concern the target IPS most of the time; sometimes they will be about past coaching events or user performance. Topics concerning the IPS will have to be represented in the domain somehow. One could think of hardwiring them, e.g. much like the Bite-Size Architecture (Bonar e.a., 1986), but that would place the burden of deciding what is in a topic on the shoulders of the future IHS builder. We might first try to define some general principles with which to partition a domain representation into several topics, especially since the topics in a Help System tend to be rather small.

These principles are defined for system procedures, because they constitute the main part of the Domain Representation, and they are the entities the user is most confronted with.

A topic contains at least the name of the system procedure plus the intended effect, i.e. the reason or goal the Coach has to discuss it. Thus, if the user asked how to save the changes she has made in a file using Vi, the topic of the local need will at least be the name of the system procedure that accomplishes this, for example the 'ZZ' command, plus the effect that it saves the changes to the current file. The topic will also include the effect that one will leave Vi-mode. Besides the fact that "ZZ" leaves Vi, there is the condition that this command can only be executed from command mode. If the user is in command mode, this condition will not be part of the topic (though it may become a topic itself later on). If however, the user is not, this condition has to be part of the topic. This is desirable, because otherwise the command or plan will fail, which is an undesirable situation.

Another example that may throw some light on the issue of what is in a topic, comes from the Unix operating system domain. Suppose a user asks how she can rename a file "foo" to "file1". The topic will probably be the "mv" command, and the Performance Planner

2   For more detail, see Breuker et al. (1986).
3   Analogies with different domains, e.g. real world analogies like in the desktop metaphor, are difficult and likely pose new problems. Discussion of these kind of metaphors is outside the scope of this paper. Analogies within the domain of IPSs however are possible, e.g. between different modes, programs or systems ("ls in UNIX is like dir in MS-DOS").

might come up with the specific plan: "mv foo file1". There is no additional effect of leaving a mode, but there is however the potential effect that if file "file1" already exists, it will be destroyed. This obviously has to be mentioned. People should be informed or warned for any additional effects that delete something, because one of the general goals of users will be to preserve what they have got, unless explicitly stated differently (see also Luria, 1987).

Finally, effects of system procedures that are visible (e.g. on the screen; primitive actions "show" and "hide") should be mentioned in some situations. In general these effects will be self-evident, operationalisations of effects already described. E.g. a delete effect will be accompanied most of the time by the disappearance of an object from the screen. It will not be necessary to mention such. Other visual effects may already be covered by giving state feedback and will in that case be the intended effect of the topic. Sometimes however the visual effect may be contra-intuitive. In Vi for example a '$' sign will appear in the text on the screen when the user enters replace mode. This '$' is not really in the text, though the user might think so. Contra-intuitive visual effects will have to be explicitly mentioned in the domain representation by the Help System developer, since there probably will be no general way of inferring them from the Domain Representation.

So a more elaborate description of a topic is: The topic for every system procedure consists of its name, its intended effect, all additional and potential effects that apply in the current situation (if any), concerning the deletion of objects , visible changes of the state of the system that are contra intuitive, and a change of mode of operation, and all unfulfilled conditions, necessary to execute the procedure (see also: Winkels & Sandberg, 1987).

4.4. Natural Language Generation.

From the point of view of interpreting the AM, there is a cascade of interpretations before the Utterance Generator transforms a tactic structure into natural language output. The utterance generator contains a number of functions for generating pronouns, ellipsis, etc., but the major function with respect to the AM consists of transforming the components of a topic into lexical/syntactical elements.

5. Conclusions.

Finally we will evaluate the AM considering the following requirements: Generality, funcionality, and feasibility.

5.1. Generality.

Tasks and plans are used in every application. Also the levels that have been identified in the AM appear in every IPS. In the EUROHELP report 'the generality test' (vd Baaren et al. 1987) tasks and plans of different IPSs have been identified and represented.

The generality of the representation of the effects and objects has been tested for some text editors and for the Unix mail system. In such IPS the primitives have a clear and well defined meaning. These primitives can then be used to represent the meaning of more complex commands. This is different with applications like a database management system. The primitives that can be used in this case are well defined, but their definition does not have a clear, intuitive meaning. In some cases the operation does not even exist in the 'real world'. In this case the primitives must first be adapted, and the meaning of the primitives must be included as a didactic goal within the help system.

5.2. Functionality.

The functionality of some of the functional interpreters has been moved to the AM. The planner and plan-recognizer have been reduced to plan-selectors and plan-matchers; the plan-task hierarchy provides a complete task-plan structure, which can be compared directly with the actions of the user, and which also produces expert plans.

The diagnoser uses the task-plan hierarchy only indirectly, via the planner and the plan-recognizer. Because of the fact that expert plans and the recognized plans have exactly the same structure, deviations can easily be traced. The deviation can than be categorized by comparing the steps within the plans. A problem is still to trace the reasons of the differences between the steps by consulting the pre-conditions and effects of the system procedures.

The Didactic Goal Generator interprets the abstraction hierarchies that describe the effects of system procedures to generate a 'curriculum' for the given AM. For coaching, a topicalization process uses the same effect descriptions to decide what exactly to tell a user about a domain concept.

Both performance interpretation modules, and the DGG consult a user model, which is based on the 'overlay principle'. Because tasks and plans have been enumerated, the overlay method can also be applied to the operational knowledge.

5.3. Feasibility.

Because of the fact that both tasks and plans have been enumerated within the AM, the model does to require lots of computation. The technical feasibility only depends on the tools that need to be used to build and support the AM structure. Therefore a set of AM tools is being created at the moment. These tools will support a knowledge engineer to built and maintain the EUROHELP application model.

6. References.

1. Baaren, J. van der, Duursma, C., Hijne, H., Romeyn, T. & Vlugt, D. van der. (1987). Generality Test. COE/EUROHELP/022.

2. Bonar, J., Cunningham, R. & Schultz, J. (1986). An Object-Oriented Architecture for Intelligent Tutoring Systems. Proceedings of the first annual conference on Object Oriented Programming, Systems, Languages and Applications.

3. Breuker, J.A., (1988). Coaching in Help Systems. In: J. Self (ed). Artificial Intelligence and Human Learning. Intelligent Computer-Aided Instruction. London: Chapman & Hall, 1988.

4. Breuker, J.A., Winkels, R.G.F. & Sandberg, J.A.C. (1986). Didactic Goal Generator. Deliverable 2.2.3 of the ESPRIT Project P280, 'EUROHELP'. University of Amsterdam.

5. Breuker, J.A., Winkels, R.G.F. & Sandberg, J.A.C. (1987). A Shell for Intelligent Help Systems. Proceedings of the 10th International Joint Conference on Artificial Intelligence, 1, pp. 167-173.

6. Breuker, J.A. & Greef, P. de (1985). Information Processing Systems and Teaching & Coaching in HELP Systems. Deliverable 12.1 of the ESPRIT Project EUROHELP.

7. Bruggen, J. van (1989). Functional Specification of the Plan Recognizer. Report COE-EUROHELP-030. Courseware Europe BV.

8. Clancey, W.J. (1983). The epistemology of a rule based system -a framework for explanation.

Artificial Intelligence, 20, pp. 215-251.

9. Davis, R. (1983). Reasoning from first principles in electronic trouble shooting. International Journal of Man Machine Studies, 19, pp. 403-423.

10. Duursma, C.M. (1988a). Representing effects of commands in terms of changes in a structure of objects. Report COE-EUROHELP-025.

11. Duursma, C.M. (1988b). Description of the objects and effect description in a functional restricted editor. Report COE-UAM-EUROHELP/026.

12. Duursma, C.M. (1989). Functional Specification of a Selection Planner. Report COE-UAM-EURO-HELP/032. Courseware Europe & University of Amsterdam.

13. Duursma, C.M. & Maas, S.M. (1986). A report on a development of a domain representation for the EUROHELP sytem. COE-EUROHELP/016.

14. Genesereth, M.R. & Nilsson, N.J. (1987). Logical Foundations of Artificial Intelligence. Morgan Kaufmann, Los Altos, California.

15. Goldstein, I.P., (1979). The Genetic Graph: a representation for the evolution of procedural knowledge. International Journal of Man-Machine Studies , 11, pp 51-77.

16. Levesque, H.J. & Brachman, R.J. (1985). A Fundamental Tradeoff in Knowledge Representation and Reasoning. In: Brachman, R.J. & Levesque, H.J. (eds). Readings in Knowledge Representation. Morgan Kaufmann, pp. 41-70.

17. Luria, M. (1987). Goal Conflict Concerns. Proceedings of the tenth International Joint Conference on Artificial Intelligence, 2, pp. 1025-1031.

18. Moran, T.P. (1981) The command language grammar: a represenation for the user interface of interactive computer systems. International Journal of Man-Machine Studies , 15, 3-50.

19. Steels, L. (1987). The Deepening of Expert Systems. AI Communications, 0, pp. 9-16.

20. Stehouwer, M. & Berkum, J. van (1989). Functional Specification of the Diagnoser. Report COE-EUROHELP-031. Courseware Europe BV.

21. Wielinga, B.J. & Breuker, J.A. (1986). Models of Expertise. In: Proceedings ECAI'86, pp. 306-318.

22. Winkels, R.G.F. & Sandberg, J.A.C. (1987). The Eurohelp Coach: A Progress Report. Deliverable 2.5.1. of the ESPRIT ProjectP280, 'EUROHELP'. University of Amsterdam.

23. Winkels, R.G.F., Breuker, J.A. & Sandberg, J.A.C. (1988). Didactic Discourse in Intelligent Help Systems. Proceedings of the International Conference on Intelligent Tutoring Systems, Montreal, pp. 279-285.

Reliability and Quality in European Software Technology (REQUEST)

S.G. Linkman and G.H. Browton
STC Technology Ltd.
Copthall House
Nelson Place
Newcastle Under Lyme
United Kingdom
+44 782 662 211

Summary

This document gives a description of the achievements of the REQUEST project to date. It will describe the significant capabilities developed by the project in all of its areas of research and the way in which this technology is being incorporated into tools and the software engineering process in the partners of the consortium and to some outside bodies.

Main

The REQUEST project is divided into three sub-projects. The first deals with quality measurement and prediction, the second with reliability measurement and prediction and the third data collection and storage in support of the other sub-projects, but also investigating the issues related to those subjects.

This paper will outline the major research results, achievements and technical innovations of the sub-projects individually and then describe how the results blend together to create the products and processes which are suitable for industrial exploitation.

Sub-project 1

Sub-project 1 of REQUEST is concerned with quality measurement and prediction, and it has created the three part COnstructive QUality MOdel (COQUAMO)

The areas dealt with are:

1. the prediction of quality factors such as reliability which are likely to be achieved, assuming that the correct process is use. And further to provide help and advice on which quality drivers to adjust to maximise the probability of achieving the desired results.
2. the monitoring of software development to detect those components or systems which are likely to cause problems and provide advice on the causes and best corrective action. In addition standard reports will be constructed for each stage.
3. the assessment of the values of the quality factors actually achieved and the prediction of maintenance costs.

The predictive component relates to the impact of the non-tools related aspects of the

271

software production process, such as the company attitudes to Quality, personnel experience etc. This innovative approach is recognised by a number of authorities as being unique and holds great promise, not only in the prediction and control of quality, but in providing a mechanism for significant improvement on the current generation of cost models.

The monitoring component is designed to produce reports based around the process in use, and also to detect anomalous components which require managerial attention and provide advice on the correct managerial action. This component is capable of matching the analytical capability, for third generation type software, of some of the best analysts worldwide. This has been demonstrated as part of the extensive V & V of the REQUEST tools and models.

The final component of COQUAMO, the assessment component, is designed to provide assessment of the achieved levels of quality factor at the earliest possible stage to enable any corrective action in terms of either additional resources in the support process or changes to the product produced. In support of this, modelling of the support process is also provided.

In detail this part of REQUEST has achieved the following

In the planning component:

The capability to predict the likely achievement of the Quality Factors Reliability, Maintainability, Usability and Reusability.
Should the prediction indicate that the values are not adequate then the feasibility and advice components will provide the mechanisms for the user to adjust the values of the quality drivers to maximise the levels of the quality factors with minimum overall disruption to the development process.

In the monitoring component:

The capability to generate sufficient low level target values of software metrics to enable the detailed tracking with regard to cost, effort faults rates etc. and to detect components with a high proportion of unrealised targets.
The capability to detect those components, subsystems or systems which behave in an anomalous manner, or are following an anomalous trend, and are therefore likely to need corrective action if the overall project is to achieve its goals.
The capability to generate in a flexible manner standard reports on the progress of the software development. This mechanism being accessible to the user if required.
The provision of a general analysis tool to enable the skilled user to investigate the data themselves.
The provision of advice on the likely causes of deviations from the expected values and on the best corrective actions to take.
The monitoring component is able to achieve results comparable to skilled analysts, and provide advice based on good management.

In the assessment component:

The capability to predict the Reliability, Usability, Maintainability and Extendibility that the software will achieve in mature use, either in system test or early life.
The capability to predict the likely support costs associated with the values of the quality

factors and the support policy chosen.

The capability to assess the value of the actual quality drivers in the project, and hence provide the feedback to achieve improved prediction and control in the future.

Overall

In conjunction with the work of the TSQM project, the REQUEST models and tools provide a cohesive mechanism for quality planning and specification, process and product monitoring and quality assessment.

Sub-project 2

The work of sub-project 2 is concerned with the area of reliability and fault tolerance and has a number of distinct areas, these are:

- Improvement of software reliability modelling using proportional hazards modelling.
- The creation of a software reliability model based on structural concepts, taking into account information available from the component.
- The study of methods to combine the predictive results from a number of software reliability models into a single estimate of present and future failure behaviour.
- The extension of Scott's work in fault tolerant systems, removing a number of deficiencies and limitations of the original work.
- The provision of mechanisms for the assessment of the degree of diversity achieved in a system, and the assessment of the adjudication mechanisms in use.
- The creation of a mechanism for the establishment of the functional diversity.
- The creation of a tool to support the work described above and to make an integrated environment for reliability prediction

The above were the aims of the sub-project. The actual achievements were limited in some areas and extended in others as one would expect in any research activity.

This part of REQUEST has achieved the following:

Production of an integrated tool for the reliability growth model use, providing mechanisms for the selection of the most appropriate models for the observed behaviour. This tool(Perfide) incorporates the latest models in most areas together with graphical display of the relevant curves and plots. This provides the mechanisms for the project manager to make use of the work carried out in software reliability in a practical and useful manner.

Significant extensions to the current theories and practice in fault tolerant software in the areas of voter granularity and extensions to the work of Scott. The adjudication mechanisms have also been investigated, considering the impact of the mechanisms used to choose the course of action when faults occur.

The fault tolerant investigations have led to the implementation of tools to estimate diverse system reliability for two common types and the creation of guidelines for establishment and assessment of fault tolerant systems. The guidelines deal with areas such as correctness and coarseness of the acceptance tests, and the impact on the reliability of having intermediate checkpoints in the software which cause reduced diversity because the intermediate results must be generated.

The remaining areas of proportional hazards modelling extensions and the investigation of structural software reliability models did not make the early progress that would have allowed their inclusion in the tools developed. However significant late progress has been made and reports dealing with these issues will be available at the end of the project.

Overall

Significant progress has been made in the areas of fault tolerant software and the transfer of software reliability growth models from an academic subject into tools of practical application by a project or quality manager. The progress in the other areas has by enlarge come too late for inclusion into the tools developed by the project. To a great extent Perfide provides a tool for the specialist, or the more knowledgable project manager to use reliability models in the control of his project. Perfide providing an ideal partner in the reliability area in the same way that specilaist performance tools might.

Sub-project 3

Sub-project 3 of REQUEST is concerned with the problems of data collection and the provision of data in a suitable form for analysis by the researchers in the other sub-projects. However the activities required to support the goals outlined above led to a number of significant advances in the areas of data collection processes, data modelling and data storage and extraction.

In the area of data actually collected, however, the targets for the amount of data collected were not met, mainly due to the lack of data available in general that companies were willing to give without material recompense or the power of some central agency.

The work in the data modelling area has pioneered new techniques which bring many of the attributes of the growing discipline of process engineering and object oriented design to the areas of database design and data collection.

The process modelling concepts enable the data collection process to be fitted to the normal operational activities of staff, and hence have both a logic and a relevance associated with them. In the area of database design effectively an object oriented database has been created, together with the mechanisms which allow it to be built using current database technology and for the data to be entered extracted and manipulated in an object oriented way.

The results of this part of REQUEST are:

The design of data storage systems which now form a basis for future storage of software engineering data and will create the mechanisms to meet the future data needs of the Esprit projects. In addition the same technology provides the mechanisms for commercial and community organisations to establish internal databases for their own use in a way which allows the public domain data to be used.

The provision of more general methods and mechanisms whose use in the data collection and storage have allow the consistent customisation of the data model, and hence the collection process and the database to the needs of an individual organisation.

In the area of data collection achieved by the project, a number of datasets have been collected, however these tend to be concentrated in particular areas, for example in the fault data, and also cover a limited subset of development processes and phases. This has limited the usefulness of the data to the other sub-projects, who have tended to concentrate on the links with the database and made use of the data for validation of models rather than development of them.

Overall

The sub-project has succeeded in establishing the mechanisms that should be used in the establishment of the data collection, extraction, verification and storage of data relating to software engineering. It has developed methods which allow the creation of customised data, while still being able to ensure that the data collected is compatible with that collected over different companies, or in the same company as the process changes over time. The sub-project has succeeded also in raising the awareness within the industry of the usefulness of such data collection and analyses.

The sub-project did not meet the original aims in the amount of data to be collected, for a wide variety of reasons, nor in the timescales originally planned, however data did become available in time to be useful in verification and validation of the models and tools developed by REQUEST Synergy between the Research Themes

Synergy between the Research Themes

When the original REQUEST consortium was created it was a number of companies with widely divergent research and commercial requirements. This inevitably leads to a many difficulties in both management and technical direction. However through the Co-Operation and willingness of the partners much useful synergy has been achieved. Indeed each of the sub-projects would not have succeeded as well as it has if each had been alone.

The synergy has taken many forms, ranging from the interaction of ideas from the divergent backgrounds of the contributors to the actual use of one tool as part of another.

Particular examples of this synergy in use can be seen in the following examples, which are only the most obvious:

1. The COQUAMO III tool makes use of the knowledge and tools developed in sub-project 2 to enable the assessment of reliability and usability and security to occur at an early stage when corrective actions or resource rescheduling are still possible.
2. The Perfide tool developed by sub-project 2 makes use of the database developed by sub-project 3 to enable the storage of data and its comparison across many diverse projects.
3. The wide ranging requirements of the work being carried out in sub-project 1 have driven the creation of the capabilities created in sub-project 3, in particular much of the early work on the data model and the project report specification was carried out in sub-project 1

In addition the cross fertilisation of ideas between the project personnel has been promoted by the Technical Fora which occur three times a year and at which researchers and developers present their latest thoughts and research results for discussion amongst the project members

Future Developments

The exploitation of the work by the consortium members will clearly take divergent routes due to the different commercial and technical needs of the companies, for example some companies will concentrate on the technology transfer of the work carried out by the REQUEST partners in their own country, while others will take specific parts of the work to develop further. In other cases the work will be carried forward into future research, or included into tools and processes to be used as commercial or internal tools to help the evolution of the overall software engineering discipline, into one which can produce quality software with required reliability, with adequate data collection and process control to enable conformance to the relevant European standards.

It is to be hoped that the success of the REQUEST project, both in scientific technical and commercial terms and in the co-operation between the partners will provide an example for future ESPRIT projects.

Keywords

Quality, Reliability, Quality Management , Data Collection , Project Management , Project Control, Data Presentation, Data Analysis

X L F L : Extended Logical Form Language

S. Groß, A. Wehrmeyer, H. Rösner
Nixdorf Computer AG, DC-T21
Berliner Str. 95
D-8000 München 40

Abstract

The main goal of P311-ADKMS, Phase II, is the product oriented development of a multilingual natural language access (German and Italian) to a KL-ONE knowledge representation system which in turn automatically accesses a relational database via an extended query language for efficient recursive queries.

The following paper presents one of the major achievements of ESPRIT P311-ADKMS, Phase II, the joint development of a logic based Common Semantic Representation. The two partners involved (Nixdorf Computer and Olivetti) committed themselves to output this Common Semantic Representation in their respective Natural Language Handlers so that the subsequent processing can continue with **ONE** common mapping component.

The chapters "Introduction" and "Overall Architecture" were provided by Mrs. Dr. H. Roesner. The rest of the paper is the result of the common work between Nixdorf and Olivetti. The responsibility for this part lies with S. Groß.

Abbreviations

ADKMS: Advanced Data and Knowledge Management System
BACK : Berlin Advanced Computational Knowledge Representation
FOPL : First Order Predicate Logic
NLH : Natural Language Handler
NLH-G: Natural Language Handler for German
NLH-I: Natural Language Handler for Italian
SQL : Structured Query Language
XLFL : Extended Logical Form Language
XSQL : Extended Structured Query Language

1. Introduction: General Background

The general background and development of P311-ADKMS, phase I, has been described in [3].
Since November 1988, ESPRIT P311-ADKMS has entered its second phase. The main goal of P311-ADKMS, Phase II, is the product oriented development of the following system:
– a multilingual natural language access (German and Italian; responsibility: Nixdorf Computer, Olivetti) to
– a KL-ONE knowledge representation system (BACK: Berlin Advanced Computational

Knowledge representation; responsibility: TU Berlin, Nixdorf Computer) with
- automatic transfer to a relational database (responsibility: Nixdorf Computer, TU Berlin) with
- XSQL query language (extended SQL query language for efficient recursive queries; responsibility: TU Hildesheim, Nixdorf Computer),
- advanced Tools for the knowledge engineer and linguistic engineer (responsibility: Datamont, Quinary, Nixdorf Computer),
- an application which justifies the elaborate design (social structure of a holding; responsibility: Datamont, Quinary).

In the period preceding the first milestone of phase II (May 1989) the following tasks had to be fulfilled:
- Redesign and reimplementation of the BACK-System taking into account industrial requirements,
- Agreement of the partners on one Common Semantic Representation for representing the natural language queries. Agreement on ONE common mapping component that translates the Common Semantic Representation into a BACK query.
- Further overall integration of the system's parts.

The overall architecture of the system as aimed for in Phase II is the following:

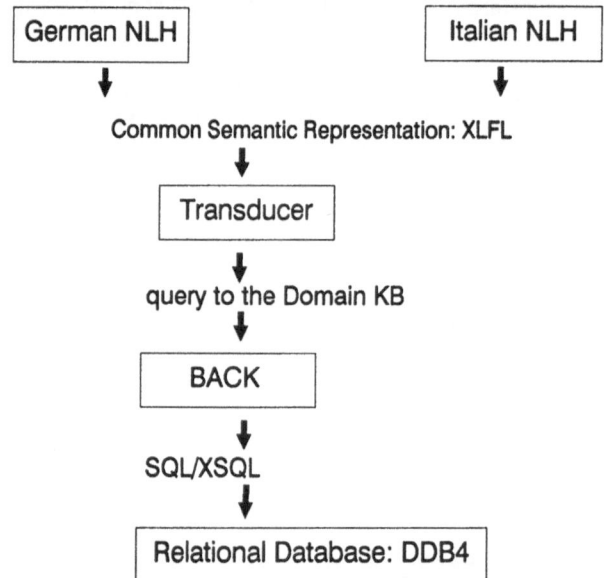

fig1: Overall architecture of ADKMS, phase II

This paper concentrates on the presentation of the Common Semantic Representation.
The Common Semantic Representation is the result of the cooperative work between Nixdorf and Olivetti. The need for a common semantic representation sprang up from the desire to simplify the architecture of the whole system and to have fuller integration of the components already developed by the individual partners.
From the architecture outlined above it can be seen that it is possible to work with ONE integrated system from the Common Semantic Representation downwards. Only the two

natural language handlers are different, which is natural since the natural language handlers have to cope with two different languages, namely Italian and German.

The Common Semantic Representation as it will be presented in the following can be seen as a real step forward in the conceptual and physical integration of the work done in ESPRIT P311.

2. Overall Architecture of the P311-ADKMS Natural Language Handlers

The overall architecture of the Natural Language Handlers consists of two main components: a linguistic component which independently of the knowledge represented in the knowledge representation system analyzes the input sentence and generates a logic based semantic representation of the input sentence, a mapping component which takes the semantic representation generated by the linguistic component, maps it to the knowledge representation system and generates the appropriate query. The linguistic component finds one (or more) logic based representations of a sentence. However, it is not concerned with interpreting it with respect to knowledge about the world or the knowledge in the knowledge base or database respectively. The mapping component has access to all the knowledge of the world of the knowledge base and also has knowledge of how to map the linguistic representation onto that world. Ambiguities which can only be resolved by knowledge about the world are no longer part of linguistic analysis. If the knowledge base allows only one interpretation the mapping component will find it and thus solve the ambiguities that might have remained after the processing in the linguistic component. Otherwise the mapping component will initialize a procedure to ask the user for the correct meaning. At the end of phase I of P311-ADKMS it was decided that both the NLH-I and NLH-G should have the same output. This means that the partners involved in the NLH-tasks agreed on a Common Semantic Representation for the natural language expressions that are analyzed by the NLH-I and NLH-G. The partners therefore developed a Common Semantic Representation, called XLFL (EXtended Logical Form Language). This Common Semantic Representation can be understood as a language independent (at least with respect to Italian and German) representation of the meaning of a natural language expression. The Common Semantic Representation is the interface that connects the language dependent parts of the general system, i.e. the NLH-I and the NLH-G with the language independent parts of the system, i.e. the transducer, the knowledge representation system and the database. It turned out to be useful to distinguish between two levels of semantic representation, namely the Common Semantic Representation and the tell- and query- languages for the knowledge representation system BACK. Like that it is possible to maintain the modularity of the system. The NLH-I, NLH-G and Common Semantic Representation constitute the linguistic part of the whole system. The main concern of the Common Semantic Representation is to represent the meaning of natural language expressions as a native speaker intuitively sees them. XLFL is independent of any particular application and it is therefore comparatively easy to plug the NLHs into new and improved versions of BACK or even a different knowledge representation system altogether. On the other hand, the BACK semantic representation languages, i.e. its tell- and query- languages, are mainly motivated by the current capabilities of the knowledge representation system. The conceptual seperation of a linguistic level of semantic representation, i.e. XLFL and a knowledge representation level of semantic representation, i.e the BACK languages leads to the problem that the much more powerful language XLFL has to be translated into the less powerful languages of the BACK system. This is not a trivial task, that has to be solved by the transducer. However, the seperation still proved to be useful, when the improved version of the BACK system appeared. Because of the distinction between two levels of semantic representation only the transducer had to be modified, but the NLHs could stay as they were.

3. XLFL: Common Semantic Representation

The following article contains a short outline of the specification of the common semantic representation XLFL as it will be used by the NLH-I and NLH-G in ESPRIT Project 311, Phase II. A more detailed version is available in Groß 89 [2]. On the following pages we will mainly be concerned with the representational adequacy of XLFL. However, XLFL was defined in such a way that each well formed formula of XLFL can be proved under Prolog directly, provided that the appropriate Prolog knowledge base exists. This was done in order to have a ready made test environment during the development of XLFL. For the reader who is acquainted with Prolog and who is interested in the Prolog proof rules for well formed formulas of XLFL we give these proof rules in the appendix. However, in the actual ADKMS of ESPRIT P311 we use BACK as the knowledge representation system so that the Prolog proof rules won't be used in there. Since the lingua franca of ESPRIT is English, functors and names in XLFL are English based. The representation language specified on the following pages owes a lot to ideas put forward by M. McCord [4]. The main part of the specification was done by Mr. Stefan Groß. The part on prepositions and adverbials in general was developed together with Mrs. A. Wehrmeyer. Mr. F. Orilia and Mrs. C. Dascanio from Olivetti made valuable comments on the treatment of copula constructions, coordinations and imperative sentences.

The article is divided into five parts. The first part states the motivation for choosing a semantic representation that is based on first order logic (FOPL). The second part gives a short outline of XLFL. The third part gives examples for XLFL representations. The fourth part gives proof rules for proving well formed formulas of XLFL in Prolog. The fifth part contains references.

3.1 Motivation for a FOPL Based Semantic Representation

A semantic representation that is to catch the meaning of a natural language expression has to fulfill two tasks: First it should be adequate with respect to the intuitive meaning of a natural language expression. We want to call this the "representational adequacy" of a semantic representation. This means that a semantic representation should be able to represent the meaning of a natural language expression in such a way that all the information contained in this expression is represented in a non-ambiguous way to the desired level of depth without changing the original meaning of the expression. It is important to note that a semantic representation can only be properly specified if we have a clear notion of the depth of the representation. For example, natural language express-ions might contain information on intonation and speech acts so that we have to decide beforehand whether or not we want to represent this information. The second, and more difficult task, is to specify truth conditions for this semantic representation so that it is possible to decide, whether a particular semantic representation of a sentence is true with respect to a particular model. Additionally, a semantic representation should be able to state explicit rules for valid inferences from given semantic representations. We want to call this "provability". Both of these aspects of a semantic representation are very well understood in FOPL. FOPL allows us to correctly represent a big subset of natural language expressions although there are sentences that cause problems in a FOPL representation, as, for example, sentences with belief contexts. The "representational adequacy" is probably not the strongest side of FOPL as a semantic representation for natural language expressions, although it does fairly well. However, if we come to "provability", FOPL becomes a very strong candidate for a good semantic representation. In FOPL we can exactly state the truth conditions for interpreting a well formed formula (wff) of FOPL and give sound and explicit rules for valid inferences. After what has been said it might seem best to take pure FOPL as a semantic representation. However, XLFL doesn't use pure FOPL. XLFL has some advantages over FOPL, because it can handle more natural language expressions than FOPL can. In particular, XLFL can handle all types of questions, imperative sentences and natural language quantifiers like 'few' and 'most', all of which FOPL doesn't deal with. Some modifications and extensions to McCord's LFL will be made (therefore EXTENDED logical form language) in order to capture certain natural language constructions not treated by McCord.

3.2 Outline of the Specification of XLFL

In the following we just state the most interesting features of XLFL and give a few examples. Readers who are interested in the details are referred to [2] and [3]. The definition of XLFL consists of the definition of predicates of XLFL and three formation rules to combine these predicates into well formed formulas of XLFL. The only logical connective that is allowed in the formation rules is the logical "&". Furthermore, the formation rules allow a kind of event variable that is not quantified over and that is used as an indexing operator. The predicates of XLFL are divided into two groups: lexical and nonlexical predicates. The lexical predicates in XLFL are generated by the morphologically realized words of a language via a lexical look-up that gets the XLFL-predicate for a word. The subclassification of the lexical predicates of XLFL results from the division of the words of a particular language (here: Italian and German) into word classes. The nonlexical predicates in XLFL are generated by inflectional information (e.g. genitive constructions) or by supra-morphological structures like word order (e.g. sentence type).

3.2.1 Lexical predicates in XLFL. The definition of a lexical predicate in XLFL consists of the definition of the functor of the predicate and the definition of the arity (number of arguments) of the predicate. The functor of a lexical predicate is in most cases the quotation form of the English translation of the word in the natural language sentence, except for prepositions, numbers and names. The following table shows a few examples of German, English and Italian words and the corresponding lexical predicate in XLFL.

CLASS	GERMAN	ITALIAN	ENGLISH	XLFL-PREDICATE
=======	========	=========	=========	================
det	der,die,das	il,lo,la,	the (sg)	the(BASE,FOCUS)
	ein,eine,ein	un,uno,una	a,an	ex(BASE,FOCUS)
	alle	tutti	all	all(BASE,FOCUS)
	einige	qualche	some	some(BASE,FOCUS)
	wenige	pochi	few	few(BASE,FOCUS)
	viele	molti	many	many(BASE,FOCUS)
noun	computer	computer	computer	computer(X)
	vater von	padre di	father of	father_of(X,Y)
name	nixdorf	nixdorf	nixdorf	name(X,nixdorf)
verb	schlafen	dormire	sleep	sleep(X) :E
	lieben	amare	love	love(X,Y):E
	geben	dare	give	give(X,Y,Z) :E
adj,e	rot	rosso	red	red(X)
,i	angeblich	supposto	alleged	alleged(NP)
,r	gross	grande	big	big(BASE,FOCUS)
prep		in	en	in loc(X,in(Y))
	waehrend	durante	during	temp(X, during(Y))
	von	di	of	modal(X,poss(Y))
wh-q	wer	chi	who	wh_q(X,person(X) & P).
	was	che cosa	what	wh_q(X,thing(X) & P).
	wo	donde	where	wh_q(X,loc(E,X) & P).
	wann	cuando	when	wh_q(X,temp(E,X) & P).
	welche	che	which	wh_q(X, P).

table 1: Representation of lexical predicates for German, Italian, and English words in XLFL

Some of the representations of word classes show interesting features.

1. Representation of DETERMINERS (det). In XLFL the syntactic word class DETERMINER (words like 'the', 'a', 'few') is mapped onto a set of predicates that bears some resemblance to quantifiers in FOPL. However, in the representation of determiners we deviate from the standard representation of FOPL. The main difference to FOPL is the way scope is expressed. In FOPL a quantifier scopes over a proposition PROP as in quant(X,PROP), where "PROP" has the structure "P <connective> Q". In XLFL the scope is taken apart into a pair of the form BASE,FOCUS, rendering quantified formulas of the form quant(BASE,FOCUS), where the quantification is over all free variables appearing in the BASE of the quantifier. Intuitively, you can view BASE to be the correlative to P and FOCUS to be the correlative to Q in the standard FOPL formula quant(X,P <connective> Q). However, the XLFL representation is more than just a change in notation. First, you can have more than one free variable in BASE. Like that it is possible to correctly represent the so called donkey sentences (see [1]). Secondly, with this notation it is also possible to represent and prove non-FOPL quantifiers like 'few' and 'most' in a uniform manner. Informally spoken, this can be done because the representation quant(BASE,FOCUS) makes it possible to refer to the point of relation with respect to which the relational quantifiers 'few' or 'most' are true. This point of relation will be represented in the BASE of such quantified expressions. Example (1)

and the proof rule for 'few' illustrate this point.

2. Representation of NAMES. Names are NOT treated as atoms in XLFL but rather like a two-place noun of the meaning "name of X is <name>" One of the advantages of this treatment is that in this way it is possible to modify names (e.g with titles). If names are used without determiners in the natural language expression there are given the quantifier 'the' as a default in the semantic representation. Example (2) illustrates these points.

3. Representation of ADJECTIVES. XLFL distinguishes three types of ADJECTIVES, namely extensional (e), intensional (i) and relational (r). This distinction is part of the lexical entry of an adjective. Extensional adjectives are those kinds of adjectives, for which we can correctly conclude that if adj(X)_noun(X) is true then also adj(X) & noun(X) is true. An example for this can be found in (5). For intensional adjectives, this is not true. With these adjectives, the modified nouns must be in the scope of the adjective. An example for this can be found in (6). Relational adjectives are represented via a two-place predicate. This has to be done, because a relational adjective can only be interpreted correctly, if we can state a point of relation with respect to which the relational adjective is true. Relational adjectives are somehow similar to quantifiers like 'few' or 'most'. The representation of relational adjectives is illustrated in (7).

4. Representation of PREPOSITIONS (prep). Prepositions are represented as a two-place predicate, where the functor is a semantic deepadverbial, the first argument the variable that ensures the proper attachment of a prepositional phrase and the second argument is composed of a functor that represents a sub-deepadverbial and a variable that is bound by the noun within the prepositional phrase. The word "deepadverbial" was coined to show the similarity between this word and the linguistic term "deepcase". Deepadverbials devide adverbial modifiers into a defined set of semantic classes. The distinction between deepadverbials and sub-deepadverbials in the representation of prepositions has several advantages. First, it is easier to classify the prepositions in a structured manner. For example, we only need the four deepadverbials "loc(ation)", "temp(oral)", "modal", and "depend" to capture all the German and Italian prepositions. Second, like that it is possible to map prepositions, many of the conjunctions, adverbs, and features of some wh-question words onto the same set of predicates. This is highly desirable because these wordclasses have semantically a lot in common. Third, it makes a natural link to wh-questions like 'where' and 'when', that require a prepositional phrase as an answer. An example for the representation of prepositions can be found in (8).

5. Representation of WH-QUESTIONS WORDS (wh-q). The representation of wh-q questions is based on the observation, that, logically spoken, the answer to such a question is the instantiation of a free variable in a logical formula. The representation $wh\_q(X,P)$ thus has a lot in common with the lambda expression lambda X. P. However, the representation shown above makes it possible to represent a semantic feature with the free variable in the representation of the wh-question word (as with "who" and "what"). Since the representation of the word "where" gets the semantic feature $loc(E,X)$, the answer to such a question can be any instantiation of a predicate $loc(E,X)$. In particular, X can be a function in itself so that for example $loc(E,in(pisa))$ and $loc(E,near(cologne))$, could be instantiation of $loc(E,X)$. Note how the semantic feature of the question word

"where" (and similarly "when") is unifiable with the representation of the preposition that is expected in answers to where- or when- questions.

3.2.2 Nonlexical predicates in XLFL. The nonlexical predicates in XLFL are mainly the sentence type markers decl(P) for declarative sentences, yn_q(P) for yes-no questions, imper(P) for imperative sentences. The sentence type marker wh_q(X,P) is a lexical predicate and has already been treated above.

Another nonlexical predicate in XLFL is generated by the genitive case. For the time being, we assume that genitive constructions express a possession relation so that the noun in the genitive construction possesses the object that is named in the nominal phrase modified by the genitive construction. Therefore genitive constructions are mapped onto a two-place predicate of the form modal(X,poss(Y)). Like that, natural language expressions like "Nixdorf's employees" and "the employees of Nixdorf" get exactly the same semantic representation. There are other ways to express possession relations in natural languages, as for example with possessive pronouns. We suggest that these are represented similarly to genitive constructions. However, we won't go into much detail here because this leads too far into the treatment of anaphoric constructions.

The sentence type markers decl(P) and yn_q(P) are simply one-place predicates, where P contains the XLFL representation of the natural language expression.

The sentence type marker imper(P) needs some comment. The interpretation of imperatives is a highly problematic thing if it is to be done in a general way. This is because an imperative is typically used to induce an action in the hearer. Only in some cases is such an action a verbal or written answer. However, in our application the action required by an imperative is always an answer. Since we always know the type of action required as a response to an imperative sentence we can actually treat this action as a default. This suggests that we consider phrases like "nenne mir .."/ "dimmi ..." (tell me ..) simply as imperative markers with no special verb meaning. We therefore say that in XLFL these phrases generate a predicate imper(P), where P contains the representation of that part of the sentence that follows the imperative marker. Example (12) illustrates this. This approach creates the problem that we are left with a sentence that is deprived of its verb. Since we assumed that in our application the verb has a default meaning it is possible to provide a dummy verb which will be called "system_list_value(X)". This predicate simply means to output all the solutions found to the query to the user.

3.3 Examples of XLFL Representations

In the following we give some interesting example sentences and their representation in XLFL. Some examples have a short comment. The reader who is interested in more detail is again referred to [2].

(1)   Few cars are pink
(1XLFL)   decl(few(car(X),pink(X)))
          Note the representation of non-FOPL quantifier

(2)   Mrs. Roesner sleeps
(2XLFL)   decl(the(name(X,roesner) & mrs(X), sleep(X):E))
          Note the representation of names and the possibility
          to modify them

(3)
(3XLFL)
	Peter works in Pisa
	decl(the(name(X,peter),
	  the(name(Y,pisa),
	    work(X):E & loc(E,in(Y)))))).
	Note how use is made of the E variable

(4) (4XLFL)	Paris is in France the(name(X,paris), the(france(Y), loc(X, in(Y))))). Note how the copula is represented
(5) (5XLFL)	the red car ex(car(X) & red(X), F)
(6) (6XLFL)	an alleged murderer ex(alleged(murderer(X)), F)
(7) (7XLFL)	Is Nixdorf a small company ? yn_q(the(name(X,nixdorf), ex(small(company(Y),X=Y)))) Note how extensional (5) intensional (6) and relational (7) adjectives are distinguished

(8)
(8XLFL)
	The man with the telescope saw the girl
	the(man(X) & the(telescope(Y),
	                    modal(X,possession(Y))),
	                    the(girl(Z), see(X,Z):E)).

(8XLFL')	the(man(X) & the(telescope(Y),
	                    modal(E,instrument(Y))),
	                    the(girl(Z), see(X,Z):E)).
	Note how the ambiguity in the prepositional phrase
	attachment is dealt with

(9) (9XLFL)	The man sold the car and bought a bicycle decl(and(the(man(X),the(car(Y),sell(X,Y))), ex(bicycle(Z),bought(X,Z))))). Note how coordination is dealt with.
(10) (10XLFL)	which man works wh_q(X,man(X) & work(X):E). Note how wh-questions are represented
(11) (11XLFL)	Who works in the headoffice wh_q(X,person(X) & the(headoffice(Y), loc(E,in(Y)) & work(X):E)) Note that we must understand "person(X)" in the representation of "who" as covering both persons and human organizations (like school, agency).

(12)   Tell me all employees who work at Datamont
(12XLFL)   imper(all(employee(X) & the(name(Y,datamont),
                            work(X):E & loc(E,at(Y))),
                                  system_list_value(X))).
                          Note how imperatives are represented

3.4 Appendix: Prolog Proof Rules

```
%for indexing operator
P : P:- P.

%for logical connective '&'
 &(P,Q):- P,Q.

%for existential quantifier
ex(P,Q):- P,Q.

%for 'the' quantifier, without uniqueness condition
the(P,Q):- P,Q.

%for 'the' quantifier, with uniqueness condition
the(P,Q):- card(P,1),
        P, Q.

    card(P,N):- set_of(P,P,S), length(S,N).
    length([],0).
    length([H|List],L):-
       length(List,L_minus_1),
      L is L_minus_1 + 1.

%for all quantifier
all(P, Q):- neg(P &  neg(Q)).

    neg(P):- P,!,fail.
    neg(P).

%for 'few' quantifier
few(P, Q):-
    card(P,N1),
    card(P & Q, N2),
    small(N2,N1).
```

 small can be defined for example as less than 10 per cent

```
    small(N2,N1):-
                    0.1>N2/N1.

%for 'most' quantifier
most(P, Q):-
    card(P,N1),
```

```
        card(P & Q, N2),
        large(N2,N1).
```

large can be defined for example as more than 80 per cent

```
            large(N2,N1):-
            0.8 < N2/N1.
```

```
%for decl sentence type
decl(P):- P.
```

```
%for yn_questions, answer 'yes'
yn_q(P):-
        P,!,nl,nl,
        write('The answer to your questions is:  '),
        write('YES').
```

```
%for yn_questions, answer 'no'
yn_q(P):-        nl,nl,
        write('The answer to your questions is:  '),
        write('NO').
```

```
%for wh_questions
wh_q(X,P):-
        findall(X,P,List),
        write('The answer to your questions is:  '),
        write(List).
```

```
%for imperative sentences
imper(all(P,Q)):-
        wh_q(X,P & Q).
        system_list_value(X):- write(X).
```

3.5 References

[1] Geach, P.(1972): Reference and generality.- Cornell University Press. Ithaca.

[2] Groß, S.(1989): Suggestion for a Common semantic representation.- Internal paper of ESPRIT P311-ADKMS, phase II, April 1989. Doc-Id: NCAG/P311/Groß/7: Common Semantic Representation/1.2

[3] Nixdorf Computer AG, Univ. of Hildesheim, TU Berlin, Datamont S.p.A., Quinary, S.p.A., Olivetti, D.O.R (1989): Deliverable 1, ESPRIT - P311: Advanced Data and Knowledge Management System, Phase II: November 88 - October 90, May 1989.

[4] Walker, Adrian, Michael McCord et al.(1987): Knowledge Systems and Prolog: A Logical Approach to Expert Systems and Natural Language Processing.- IBM T.J.Watson Research Center. Addison-Wesley, Readung/Mass.

Advice-Giving Dialogue: An Integrated System

Didier Bronisz, Thomas Grossi, Francois Jean-Marie
Cap Sesa Innovation,
Chemin du Vieux Chêne
Grenoble,
France

Abstract.

In this paper we present the implementation of an *advice-giving system* for financial investment for the final phase of the project Esteam-316. (This work was supported in part by the Commission of the European Communities as Esprit project 316 Esteam-316.) This system integrates multiple agents in a single architecture allowing cooperation between a natural language dialoguer, "intelligent" data base access modules, and a problem solver in the financial domain. Using a user model, this system adapts the mixed initiative dialogue during both the formulation of the problem and its resolution by the expert. A novice user has access to expert knowledge despite the weakness of his own knowledge.

After presenting the project Esteam-316 and the communicative structure of the implemented system, we give an example of a dialogue. With this example and its analysis we present the internal structure of the dialoguer and the details of its user model approach.

1. Problems of Advice – Giving Dialogue

Human—human dialogue seems easy to do but, like many human behaviors, it requires numerous and elaborate mechanisms. In a two-person dialogue, both participants build, according to their own intentions the structure of the discussion. Some topics are more important than others for one person, whereas they may be less so for the other. Ordinarily, the dialogue proceeds with many digressions like explanations or reminders of information. The topics discussed are ordered according to the preferences of each dialoguer, and the participants may often come back to a previous topic or begin another one before the current one is finished. Moreover, there is great deal of contextual information during the discussion. The typical Question/Answer interchange refers to previously mentioned information.

Advice giving corresponds to the class of situations where a consultant helps a client to solve his problem. The client has goals but cannot alone build a plan to achieve them. The consultant helps the client to express his goals and to build or complete his plan. Besides being an expert in the domain, the consultant the advice-giver must have a good understanding of the needs of the user: he has to explain what the client asks for, or to follow his preferences in order to build a good solution for the client's problem.

The preceding definition covers a wide range of cases depending on various factors. The client may be more or less familiar with the problem domain; the goals may be imprecise; the plan may already be partially defined at the beginning of the advice giving session. The performance of the consultant depends on the degree of expertise needed to solve the problem, and also on the amount and type of information at his disposal.

In the context of the project Esteam-316, we have limited our study to a specific type of advice giving situation: we are developing an advice-giving system for financial investment. In this context, the consultation consists of two main phases: problem formulation and presentation of a solution. During problem formulation the central task of dialogue management is to transform the user's goals into a suitable form for the problem solver (this form is referred to as a *basic investment situation* or *BIS*). The solution consists of a detailed plan that contains concrete means (savings account, securities) for achieving the initial goals. During such a dialogue the consultant builds and maintains a model of the whole plan of the client.

From the communicative point of view, we concentrate on both local and global aspects of cooperative behavior. In our view, the role of a consultant is to guide the consultation dialogue. However, this must be done in a flexible way in order to allow the client to express himself. A major problem of dialogue management is to handle the different interruptions resulting from the evolution of the user's intentions and adapt the remainder of the dialogue in a coherent way.

1.1 A Sample Dialogue

Before describing the Esteam-316 project development, we will motivate that presentation with a sample dialogue illustrating the above points. This dialogue example will be referred to in later discussions.

1 *System*: Hello, tell me about your investment plan.
2 *User*: I'd like to buy a car in two years.
3 *System*: Ok. Normally, an investment plan comprises three subparts: an *emergency fund*, for unforeseen expenses; a *fixed-term investment* when you know you'll need a certain sum at a certain date, as is the case with your car; and a *long-term investment* for an indefinite amount of time which can be aimed at either capital growth or regular income. How much do you want to invest?
4 *User*: $20,000.
5 *System*: Let's begin with the emergency fund. How much do you want to put into an emergency fund?
6 *User*: Let's talk about my car first.
7 *System*: Ok. How much money do you want to put into the fixed-term investment?
8 *User*: Oh, let's say $7,500, I guess.
9 *System*: I recommend that you put the money into a liquid security. Would you like me to propose one?
10 *User*: Yes.
11 *System*: Put your money into a 4.5 % savings account at Sofibanque. Let's return to the emergency fund, now. How much do you want to put in it?
12 *User*: $5,000.
13 *System*: Tell me also the delay.
14 *User*: 2 days.
15 *System*: Ok. I recommend that you put the money into a liquid security. Would you like me to propose one?
16 *User*: Are there any savings accounts with more than 4.5 % interest?

17	*System*:	Barclay's has a savings account at 5 % interest, but it is in England
18	*User*:	Ok, I'll take Barclay's.
19	*System*:	So, that leaves you with $7,500 for your long-term investment, is this ok?
20	*User*:	I wanted $10,000.
21	*System*:	If you want an investment plan for $20,000 with an emergency fund of $5,000 and a fixed-term investment of $7,500, it only leaves $7,500 for a long term investment. If you want to put $10,000, then you should modify something else in your plan.
22	*User*:	Could I put $2,500 into the emergency fund?
23	*System*:	Ok. I recommend for the emergency fund a bank account at Sofibanque earning 4.5%, you can put $10,000 into the long-term investment. What risk-level would you like?
24	*User*:	What is the risk-level of a long-term investment?
25	*System*:	The risk-level is represented by an integer between 1 to 10. It shows the ris you want to have for your portfolio. The higher the risk is, the more you might earn, but the more you could lose as well. So, what risk do you choose?
26	*User*:	What if I put a risk-level of 5?
27	*System*:	In this case, I recommend...Do you accept?
28	*User*:	And if the risk-level is 2?
29	*System*:	In this case, I recommend Which one do you prefer?
30	*User*:	The first one
31	*System*:	Ok. So, to sum up, you want to invest $20,000, with $2,500 in an emergency fund with a delay of 2 days, $7,500 in a fixed-term investment to buy a car in two years, and $10,000 in an long-term investment. Is that right?
32	*User*:	That's fine
33	*System*:	Ok. Goodbye!!

1.2 General analysis

From this dialogue, we can easily demonstrate the different points we mentioned above.

– In sentence 6 the user prefers to begin the discussion with the fixed-term investment, while the system began the discussion with the emergency fund. The system changes the order of the different topics according to the will of the user and the feasibility of such an exchange.

– In sentence 12, the user does not say "The amount of the emergency fund is $5000" but only "$5000". The system must understand such an elliptical response. In the same way, the system does not ask "Tell me the delay of the emergency fund of your Investment plan?" but only "Tell me the delay?". Normally, the two participants hide a lot of information. They just give sufficient information to have an unambiguous dialogue. But having a partial knowledge necessitates being able to confirm sometimes what has been understood. In sentence confirm, the system asks the user for a confirmation of the value of the fixed-term investment (2 years) which was previously given by him in the sentence 2. The system has understood that buying a car is

equivalent to having a fixed-term investment, but it asks him to verify this supposition.
- In sentence 24 the user begins a digression in order to have an explanation about a new term introduced by the system in asking for a value. The system has to recognize this new user's intention, cope with (it may be longer than a two-turn dialogue (User, system) as in the discussion of a given solution) and come back to the previous dialogue.
- In sentence 10 the user accepts the system's offer to come up with a detailed investment plan, while in sentence CA he decides to find out about other possibilities.
- In sentences 26 — 30 the system and the user explore the implications of a modification of one of the parameters.

All these considerations appear in any discussions, independently of the topics. There is implicit information (abbreviations, speaking manner) used in a discussion in a given domain. The dialoguer we are building does not handle such implicit information. We focus our attention on the domain-independent aspect of the dialogue organization. However, the advice-giving system must be able to explain what it does and how it does it, and also what the other components it is interfaced with do. We added, therefore, these domain-functionalities in order to implement a system able to help a *novice* user as well as an *experienced* one.

The system has to adapt its utterances and its explanations according to its perception of the user's knowledge. For instance, when the system presents a portfolio, it hides irrelevant information for a *novice* user while it shows it for the *experienced* one. In the same way, the system tries to use the user's vocabulary. In the example given above, the system uses the word car rather than *fixed-term investment* in order to help the user's understanding.

2. Esteam-316

2.1 Project Objectives

Automated advice-giving (that is, providing "expert" solutions to human "novices" by computer) is a complex task requiring the integration of knowledge and data from a variety of sources. Esteam-316 is engaged in the study of such *Advice-Giving Expert Systems* (AGES). The overall AGES problem consists of performing intelligently a single task, namely advice-giving and finding ways of representing and reasoning about diverse matters and of integrating the various system components. The main AGES architectural problem is providing the support and management for the cooperative functioning of several distinct, independent sources of knowledge or expertise. Each of these knowledge sources represents, of course, another more narrowly focused problem. The AGES human-interface problem is that of managing a dialogue with the user, as described above.

The Esteam-316 project's objective, then, is to study the design and construction of mixed-initiative, advice-giving expert systems which are general, distributed and cooperative:

- mixed-initiative in the sense that, although the system has a general plan for conducting the dialogue, it can accommodate requests for clarifications, change of subject, etc., from the user;
- advice-giving in the sense that it provides "expert" solutions to the "novice" user's problems;

- general in the sense that the system architecture and design are essentially domain independent, allowing the use of self-contained modules or *agents* that encapsulate knowledge or expertise;
- distributed in the sense that the system may be composed of multiple, logically independent software components which work together to respond to the user's request; and
- cooperative in the sense that the system actively helps the user to formulate his request and provides responses adapted to the user's expectations.

This objective presents a number of interesting and difficult research problems in the fields of user-system dialogue management, knowledge representation, database access and overall system architecture. The development, use and evaluation of possible approaches often requires the construction of prototype systems or sub-systems. (These prototypes also provide an opportunity to demonstrate of the project's results).

2.2 Make-Up of the Project

The Esteam-316 consortium comprises four partners and one sub-contractor: Cap Sesa Innovation (Prime Contractor) in Grenoble, France; ONERA—CERT in Toulouse, France; Philips Research Lab in Brussels, Belgium; CSELT in Turin, Italy; and the University of Turin on contract from CSELT. The project began in January, 1985, and will end in December 1989. It has a budgeted manpower of 106 person/years.

The project was broken down into five major tasks: an architecture for cooperation between agents; knowledge representation and acquisition; dialogue management; data and knowledge bases; and a final demonstrator incorporating as much as possible of the research results of the other tasks. This demonstrator and its individual components will be discussed in detail in the next section.

3. The Demonstrator

Our advice-giving system integrates various sub-systems, for example, a natural-language and intention recognition module, an expression generation module and a dialogue planning and management module. Great advances have been made in each of these domains but at present we are dealing only with the tip of the iceberg. In the project Esteam-316 we chose not to address specific issues such as explanation in great detail (there are many researchers already addressing these problems), but to build a general integrated system aimed at recognizing the user's intentions and answering him in an understandable way with expert knowledge. We will show how this system is able to incorporate modules which are more specialized in certain domains (problem solver or cooperative data base access).

Currently the project is oriented towards the development of a demonstrator or prototype, incorporating Esteam-316 research results and showing the feasibility of an AGES. The Cooperation Architecture is a conceptual framework for AGES design and a set of mechanisms to support implementation of that design. It is a cooperation architecture because it supports the active cooperation of independent components or modules of the AGES; it is thus the means for integration. Furthermore, it supports the integration of heterogeneous modules through encapsulation of modules as agents, and by providing module—module cooperation using any of three standard interaction models.

The general architecture and agents are shown in figure 1, following which we will present a description of the different agents included in the demonstrator and the communication between these agents.

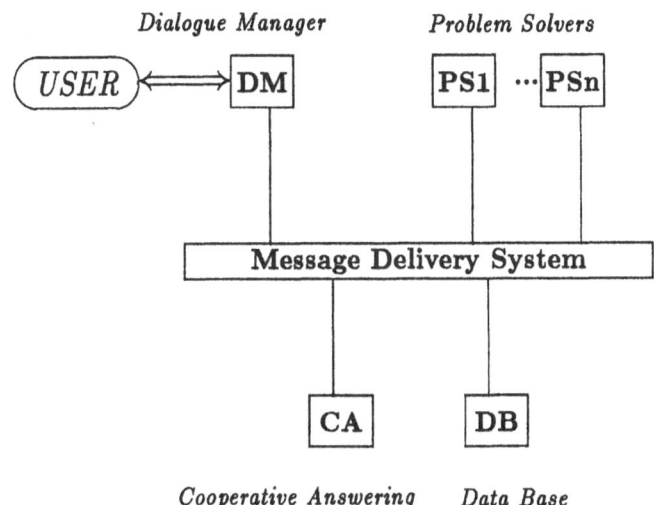

Figure 1. Esteam-316 Overall system organization

3.1 The Cooperation Architecture — CSELT

This section describes the architectural concepts and primitives for defining modular Agents i.e. programs with given functionalities, cooperating in the solution of a common application problem.

3.1.1 Motivation

The goals in formulating the architecture have been to customize:

- support of heterogeneity,

- control mechanism for coordinating the interactions,

- efficient communication

as required by AGES components of the demonstrator. In other words, we have set out to demonstrate that modules developed independently and using different languages or representations, can be made to communicate with each other and thus can together constitute a complex, advice—giving system. The architecture enforces the design and implementation of applications in terms of modular Agents that communicate across the frontiers of heterogeneous implementations via a set of communication primitives. More extensive descriptions of AGES Architecture concepts may be found in Esteam-316

Deliverable 5 "Design and Experimental Implementation of a Cooperation Mechanism".

An Agent is a modular component supported by the architecture for cooperation. A first characterization of an Agent is that of a medium/coarse grained computational entity; it is any part of a larger system able to perform almost autonomous operations, that is spending less time interacting with other Agents, than in its own computation.

From outside an Agent appears as an Object-like entity capable of holding information and providing operations on it, each such operation correspond to a functionality provided by the Agent to other Agents. An Agent can be requested, by some other Agent, to execute one of its functionalities via the interaction primitives of the architecture.

From inside the Agent appears as a structured entity split into two functionally distinct parts: the Agent Control part (AC) and the Agent Domain part (AD). The AD supports the computational capabilities of the Agent in its own domain of competence; the AD is constituted of a number of procedures each implementing an Agent functionality. These procedures may share information which is local to the Agent. This information is encapsulated within the AD and plays the role of Agent's status information. Each functionality is triggered by the Agent AC upon reception of an invocation, and may in turn invoke the local AC requesting the execution of some other Agent's functionality.

The AC is the homogeneous Agent front-end toward other Agents. Its role is twofold:
- first it implements a homogeneous interface to the Agent's own functionality; this interface makes functionalities accessible to other Agents in a standard way, despite the particular implementation of the Agent Domain part;
- then it implements the required communication protocols with other Agents allowing the Agent to access other Agents functionalities.

From an implementation point of view each Agent runs as a separate process. This feature has been introduced in order to accommodate problems coming from heterogeneity of subsystems modeled as Agents; they may thus be implemented in different formalisms. Since Agents run in a multiprocess virtually concurrent environment, a synchronization mechanism and a communication facility are required to allow them to interact meaningfully. The interaction mechanism provided by the Architecture provides them both.

This mechanism, named Direct Call, provides a procedure-like interaction paradigm and a Common Communication Language (CCL) for exchanging data. This mechanism forces Agents to run in a sequential way, it is implemented on top of a raw message-passing Inter Process Communication mechanism provided by the underlying Operating System and allows Agents to call each other's functionalities recursively, handling interactions in a LIFO style.

Agents in the demonstrator then cooperate requesting of each other the execution of some operation, via the Direct Call mechanism. In the context of AGES the operation requested has usually the meaning of solving a problem. The mechanism supports cooperation in the following fashion: while solving a received problem an Agent may request further information from its caller (e.g. in order to obtain additional data required at run time, and not contained in the original request).

3.1.2 The Interaction Mechanism

The Direct Call (DC) is a remote-procedure-call-like mechanism which allows Agents to transfer information and control to each other. Information is transferred in the form of parameter and return values; control is transferred in the procedure call like manner, suspending the caller and automatically resuming its execution upon return of the callee.

Recursive calls between Agents are allowed in by the Direct Call mechanism; the DC mechanism allows an Agent waiting for a Direct Call answer to receive and serve a further functionality execution request. This feature allows Agents to call back each other, cooperating via nested requests.

The Direct Call mechanism is divided into three major layers, the Transport Layer, the Protocol Layer, plus an Interface Layer on top. The lowest Transport Layer is completely transparent and is not supposed to be accessed directly by the user. The layered approach allows the definition of a hierarchy of protocols at different levels of concern, confining within each layer a number of problems peculiar to that level.

The Transport Layer provides a synchronized message passing facility with automatic logical-name to physical-address translation. It deals with the peculiarities of Interprocess Communication and Synchronization Primitives which are provided by the underlying Operating System.

The Protocol Layer implements the remote procedure call mechanism and handles recursive calls to the Agent. It implements the Direct Call mechanism providing just a raw, easy to port interface to it.

The Interface Level wraps the Protocol Layer adapting the raw interface to requirements of the different formalisms from which it is accessed. This layer then provides a formalism-tailored access method to the architecture. User-defined End-to-End interfaces may be added to this Level.

3.2 The Problem Solver — Philips

The Problem Solver (PS) [11] incorporates the results of two major research efforts: logic-based modeling and encoding, and explanation generation based on proof trees.

3.2.1 A Logic-Based Approach to Knowledge Representation

To provide a sound framework for knowledge representation, it must be possible to reason about the contents of the knowledge base (KB) and to specify the relationship between an answer produced by the PS and the knowledge that was used for producing it. Practically, this reasoning is only feasible if the KB and its relationships with a PS answer are described in a precise declarative manner, that is, without references to the execution strategy of the inference engine of the PS. First order logic has been studied for a long time as a formalism for modeling sound reasoning. It is therefore natural to use it as a framework to handle knowledge. The way to do that is to associate to every KB, a *logical theory* that is a precise characterization of its contents. The *axioms* of the theory express the declarative meaning of the KB rules. The evaluation of a query by the inference engine can be considered as a *formal deduction* from the axioms. The answer computed for the query is then interpreted as a *theorem* of the theory.

The interpretation of the execution of the PS inference engine as a formal deduction process requires that this inference engine be compatible with the laws of the logic. A logic-based knowledge representation formalism has been defined; it is a variant of traditional predicate logic formalism, augmented with some well-chosen higher-order features. The main departures from a first-order predicate formalism are:

- predicate names are terms of the formalism; this allows, for instance, to use structured names instead of atomic names, to write axiom schemas by using variables as or in predicate names, and to have relations be parameters of other relations, therefore

providing a uniform representation of knowledge and meta-knowledge.
- a special kind of terms, lambda-terms, are used to represent parameterized formulae; they are useful, for instance, to model dynamic user constraints, to represent sets or to attach attribute definitions to classes.
- a functional notation is available for binary relations which encode mappings, its use avoids the introduction of a large number of auxiliary literals and variables and therefore improves the readability and the structure of KB rules.

The meaning of knowledge bases written in the formalism is defined by the associated completed theories, following the approach of Clark [7,18]. To avoid consistency problems, completed theories may be given semantics based on 3-valued models [14].

The formalism has already been used in several applications [6] where its higher-order features have been put to work with success, namely:

- the definition of taxonomies where primitive and defined categories can be represented. Knowledge is explicitly structured around classes (primitive categories) and types (defined categories) which represent relevant abstractions of the application domain; entities of the application domain may be classified into these categories. Classes and types are organized in a hierarchy according to a partial order which corresponds to a specialization/generalization relation.
- the definition of various strategies for non-monotonic inheritance of attributes (binary relations between individuals and values) in hierarchies of classes. Attribute values or definitions bound to classes are transmitted through their subclasses (the classes below them in the hierarchy) and inherited by individuals belonging to them, according to axiom schemas defining the inheritance strategies.
- the definition of inference methods for plausible reasoning. The problem is to define a method to draw, for a given individual, all plausible conclusions, if any, from a set of rules, most of which admit exceptions. Those rules are encoded as links among nodes representing individuals and properties.

For any of these applications, a logical theory which models the application has been encoded using the formalism, therefore providing a declarative setting in which the generation of explanations is possible.

3.2.2 Explanation Facilities

Explanations are an essential part of expert systems and their systematic generation is an active research area. In the framework of logic-based knowledge representation formalisms, derivation trees built by the query evaluation process form the raw material from which explanations are generated. Currently, *how* explanations justify system answers by tracing rule applications, *why* explanations justify system questions by displaying a stack of goals, and *why-not* explanations justify system failures by enumerating failure branches of the evaluation tree.

When negation as failure is allowed in rules and queries, the fact that successes and failures of query evaluation are not handled uniformly is annoying since failure of a query is equivalent to success of its negation. Besides, enumerating failure branches is not as convincing as traversing a proof tree because completion axioms do not appear explicitly.

We therefore designed an approach that generates actual proofs for both how- and why-not-expla-tions. Our treatment of negation is based on the so-called Clark completed database [7] as the logical theory associated with the KB. Our inference system converts any finite evaluation tree of the Prolog-like inference engine into an equivalent natural deduction proof in the theory defining the logical contents of the KB. We also designed the corresponding interpreter to build a proof tree for any query. Successes and failures of query evaluation can then be treated more symmetrically, in that both can be justified by a proof (tree) as theorems of the theory [2,4].

Actual proof trees are important for logic-based representation formalisms, even if it is now recognized that a mere trace of the inferences drawn during the problem solving phase of the expert system only describes the system behaviour and does not justify it. Indeed, there is nothing to prevent from tracing inferences drawn during an explanation phase of the expert system. Hence, part of the knowledge base can be dedicated to the so-called deep explanations by encoding how general concepts and principles of some causal model of the application domain apply to the solution previously computed by the expert system.

It is also interesting to cater to explanations not directly related to some problem solving query [3,5]. Such explanations require dedicated domain knowledge, which may be about the terminology, the conceptual structure, or the basics of the domain. In order to be available, such knowledge has to be encoded in the KB so that it can then be directly queried, just like problem solving knowledge. The border between knowledge for problem solving and knowledge for explanation generation then becomes fuzzy and unessential. Generating more conceptual explanations becomes itself a type of problem to be solved and quite naturally requires corresponding knowledge. Knowledge for conceptual explanation is usually meta-knowledge, i.e. knowledge about knowledge. Since the formalism is "higher-order", knowledge and meta-knowledge are uniformly represented; entities that encode meta-knowledge are allowed to refer to other entities that encode object-knowledge.

3.3 The Cooperative Answering Module — *CERT*

In the context of traditional applications devoted to company management, like payroll computation, people or programs who have to access data in a Database have a very precise definition of the data they want to access. There are many other applications where people want to access data in order to make a decision, or to solve a problem whose solution cannot be found applying a simple algorithm. An important feature of this context, from the point of view of data retrieval, is that users don't have a precise idea of the data which can help them to solve a problem, or to make a decision. The objective of the Cooperative Answering module is to simulate the behavior of a person who wants to help as much as possible an interlocutor who asks them a question. That is, to try to understand why this interlocutor asks this question, and to determine what additional interesting information, not explicitly requested, could be provided in addition to the answer [9].

The function of Cooperative Answering is to transform the query to provide additional information. These transformations can be presented more easily using the concepts of Entity, and Attribute of entity, as in the Entity-Relationship Model [8]. In general the answer to a query provides some attribute values of entities satisfying some conditions. So a query can be decomposed into three parts:

– The type of entities which are expected in the answer. This part is called the "Entity".

- The condition entities of this type have to satisfy. This part is called the "Condition".

- The attribute of these entities whose values are expected in the answer. This part is called "Retrieved Attributes".

Then the additional information provided in the answers falls into the corresponding categories:

- Additional entities having a different type than those requested in the query.

- Additional entities satisfying "neighborhood" conditions.

- Additional attribute values.

Without going into further detail about the Cooperative Answering, this module accepts a query from the Dialogue Manager and then transforms it into several queries which are designed to elicit both a direct response to the user's original question and also responses providing complementary information that might be interesting to the user. The transformed queries are expressed in Predicate Calculus. An interface turns the predicate calculus queries into SQL queries capable of extracting a small subset of the Data Base (Oracle) from which one can compute the responses to the transformed queries. Examples:

- To identify the currency of the requested bond if the country of this one is different of the country of the user.

- To propose a stock with a very low risk in response to a query about bonds if there is no appropriate choice in the set of bonds.

3.4 The Dialogue Manager — *CSInn*

In the final integrated AGES demonstrator, the main problems are controlling the cooperative functioning of several distinct sources of knowledge, finding ways of representing and reasoning about diverse matters, and integrating the different contributions to perform intelligently a single task, namely advice-giving. The role of the Dialogue Manager is to provide a "friendly", comprehensible user interface that makes available to the user all the expertise of the various modules of the system in a flexible, uniform fashion. We present now how we are implementing the dialogue manager in order to provide such functionalities. The Dialogue Manager architecture is depicted in figure 2.

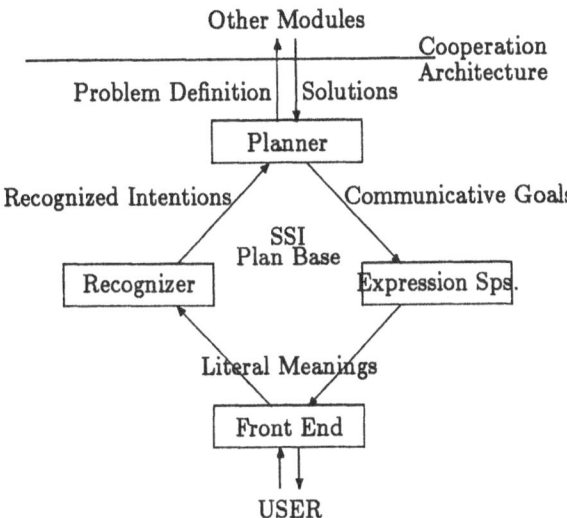

Figure 2. Architecture of the Dialogue Manager.

The main role of the *Recognizer* is to find the possible interpretation of the user's input (his intentions and, of course, what he is speaking about) [15]. The recognizer has to understanding the meaning of each of the user's sentences. To do this it uses the current focus of the dialogue [16,17] to manage recognition of about what the user speaks. Starting with low—level speech-acts such as *Acknowledge, Request, Inform, ...*[1], it tries to identify the indirect speech-acts (in sentence 2, the user expresses implicitly the fact that he needs or he wants an Investment Plan) and generates the list of the user's intentions. The use of focus also allows it to recognize the elliptical sentence given by the user. This parse allows the system to have a precise view of what the user wants to do although he might present his thought in greatly different ways.

The planner [10,12,13] receives the user's intentions from the recognizer. It organizes the dialogue according to its own intentions and the user's intentions regarding the current state of the dialogue. We explicitly distinguish two levels of work: the task-level plan and the communicative-level plan.

The *task-plan* describes the task-level problem the system has to solve for the user. In it we store all the information given by the user (the value of the amount of the investment plan ...) and by the problem solver (the different solutions) in a structured way. It allows the planner to decompose the whole problem in a coherent manner.

The planner formulates its reaction according to the user's input and the current state of the dialogue. This reaction includes all the modifications in the Task-Plan, the modifications in the User Model and the possible new goals introduced by the user. Of course, it checks the relevance of his sentence regarding the current state of the dialogue.

The expression specialists transform the communicative goals provided by the planner into a literal meaning [1]. When the user asks the system to obtain the explanation of a term or the role of an object as in sentence ~explain, The expression specialists select the appropriate words in order to give the most relevant answer to the user.

4. Conclusion

The integrated system elaborated in the project Esteam-316 is a demonstrator incorporating as much of the technology developed within Esteam-316 as possible. This experience comes from research in the areas of Knowledge Representation, Cooperative Answering, Intentional Answering, Deductive Databases and Mixed-Initiative interfaces (also see the deliverables and the prototypes issuing from the project). The imitation of human behavior in the domain of the advice-giving remains very delicate and elaborate but we believe we have made a significant contribution.

5. Acknowledgements

Some of the material for this paper was drawn from Esteam-316 deliverables, notably Del. 5 for the Architecture and Del. 13 for the Dialogue Manager. For their help and cooperation, we would like to thank the technical directors of our partners' Esteam-316 groups: Mauro Bert (CSELT), Robert Demolombe (CERT), and Alain Pirotte (Philips).

References

[1] J.F. Allen and C.R. Perrault, "Analyzing intentions in utterances". Artificial Intelligence, 3(15):143-178, 1980.

[2] A. Bruffaerts and E. Henin, Proof Trees for Negation as Failure: Yet Another Prolog Meta-Interpreter, in: Logic Programming, Proc. of the Fifth International Conference and Symposium, Seattle, August 15-19, 1988.

[3] A. Bruffaerts and E. Henin, Some Claims about Effective Explanation Generation in Expert Systems, in: Proc. of the AAAI'88 Workshop on Explanation, Saint Paul, August 22, 1988, 83-86.

[4] A. Bruffaerts and E. Henin, Negation as Failure: Proofs, Inference Rules and Meta-interpreters, Workshop on Meta-Programming in Logic Programming (META-88), H. Abramson and M. Rogers (eds.), Bristol, June 22-24 , 1988, The M.I.T. Press (in preparation).

[5] A. Bruffaerts, E. Henin and A. Pirotte, A Sound Basis for the Generation of Explanations in Expert Systems, Philips Technical Review, Vol. 44, no 8/9/10, pp. 287-295, May 1989.

[6] A. Bruffaerts and E. Henin, La construction de bases de connaissance pour les systèmes experts, in: Approche logique de l'intelligence artificielle, A. Thayse (Ed.), Dunod (in preparation).

[7] K.L. Clark, Negation as Failure, in: Logic and Databases, H. Gallaire and J. Minker (Eds), Plenum Press, New-York, 1978, 293-322.

[8] P. P. Chen, "The Entity / Relationship Model: toward a unified view of data." ACM TODS, Vol 1, Num 1 March 1976.

[9] F.Cuppens, R.Demolombe. Cooperative Answering : a methodology to to provide intelligent access to a Database. Proceedings of 2d Int.Conf. on Expert Database Systems. The Benjamin/Cummings Publishing Company. 1988.

[10] P. Decitre, T. Grossi, C. Jullien, J.-P. Solvay, "Planning for Problem Formulation in Advice-Giving Dialogue." Proceedings of the Association of Computational Linguistics, 1987.

[11] M. Hanet, The Problem Solver, in: Deliverable no. 16, ESTEAM-316 Esprit Project, July 1989.

[12] C. Jullien, J.P. Solvay, "Person-Machine Dialogue for Expert Systems : The Advice-Giving Case." Proc. 7th Int. Workshop Expert Systems & their applications, Avignon, 1987.

[13] C. Jullien, J.C. Marty, "Plan Revision in Person-Machine Dialogue.", 4th conf. of the European Chapter of the Association of Computational Linguistics, Manchester, April 1989.

[14] K. Kunen, Signed Data Dependencies in Logic Programs, Computer Sciences Technical Report #719, University of Wisconsin, Madison, October 1987.

[15] D.J. Litman and J.F. Allen, "A Plan Recognition Model for Subdialogues in Conversations." *Technical Report TR 141, Computer Science Dpt.*, University of Rochester, November 84.

[16] Barbara J. Grosz and Candace L. Sidner, "Attention, intentions and the structure of dicourse." *Computational Linguistics*, 12(3):175-205, 1986.

[17] Barbara J. Grosz, "Discourse Stucture and the Proper Treatment of Interruptions." *Proceedings of the IXth IJCAI*, Los Angeles (USA), 1985.

[18] J.W. Lloyd, *Foundations of Logic Programming*, Second, Extended Edition, Springer-Verlag, 1987.

PROJECT NO. 390

ALGEBRAIC SPECIFICATION WITH FUNCTIONALS IN PROGRAM DEVELOPMENT BY TRANSFORMATION

B. KRIEG-BRÜCKNER

FB3 Mathematik und Informatik
Universität Bremen
Postfach 330 440
D 2800 Bremen 33
FR Germany
usenet: bkb@informatik.uni-Bremen.de

ABSTRACT. The methodology of PROgram development by SPECification and TRAnsformation is described. Formal requirement specifications are the basis for constructing *correct* and efficient programs by gradual transformation. The power of compact development methods using the transformational approach, as supported by the PROSPECTRA system, is illustrated by an example. The algebraic specification language is then described, focussing on its extension by higher order functions. The functional programming paradigm leads to a considerably higher degree of abstraction and avoids much repetitive development effort, in particular through the use of homomorphic extension functionals. The combination with algebraic specification not only allows reasoning about correctness but also permits direct optimisation transformations.

1. Introduction

The ESPRIT project PROSPECTRA ("PROgram devlopment by SPECification and TRAnsformation") aims to provide a rigorous methodology for developing *correct* software and a comprehensive support system. It is a cooperative project between Universität Bremen, Universität Dortmund, Universität Passau, Universität des Saarlandes (all D), University of Strathclyde (GB), SYSECA Logiciel (F), Computer Resources International (DK), and Alcatel Standard Electrica S.A. (E) (cf. [1-3]).

The Methodology integrates program construction and verification during the development process. User and implementor start with a formal specification, the interface or "contract". This initial specification is then gradually transformed into an optimised machine-oriented executable program. The final version is obtained by stepwise application of transformation rules. These are carried out by the system, with interactive guidance by the implementor, or automatically by compact transformation scripts. Transformations form the nucleus of an extendible knowledge base.

The System completely supports the strict methodology of Program Development by Transformation (based on the CIP approach, see e.g. [4-6]). Any kind of activity is conceptually and technically regarded as a transformation of a "program" in one of the system components. This provides for a uniform user interface, reduces system complexity, and allows the construction of system components in a highly generative way.

The Objectives of the PROSPECTRA methodology, its development model, algebraic specification and transformational program development are briefly summarised in chapter 2.

Algebraic Specification with Functionals is a particular extension to classical algebraic specification (cf. [7]). Chapters 3 and 4 describe the combined advantages of functional programming and algebraic specification: a considerably higher degree of abstraction, avoiding much repetitive development effort, the use of homomorphic extension functionals as "program generators", and algebraic properties for reasoning about correctness and optimisation.

2. PROgram development by SPECification and TRAnsformation

2.1. OBJECTIVES

Current software developments are characterised by ad-hoc techniques, chronic failure to meet deadlines because of inability to manage complexity, and unreliability of software products. The major objective of the PROSPECTRA project is to provide a technological basis for developing *correct* programs. This is achieved by a methodology that starts from a formal specification and integrates verification into the development process.

The initial *formal requirement specification* is the starting point of the methodology. It is sufficiently rigorous, on a solid formal basis, to allow verification of correctness during the complete development process thereafter. The methodology is deemed to be more realistic than the conventional style of *a posteriori* verification: the construction process and the verification process are broken down into managable steps; both are coordinated and integrated into an implementation process by *stepwise transformation* that guarantees *a priori* correctness with respect to the original specification. Programs need no further debugging; they are correct by construction with respect to the initial specification. Testing is performed as early as possible by *validation* of the formal specification against the informal requirements (e.g. using a prototyping tool).

Complexity is managed by abstraction, modularisation and stepwise transformation. Efficiency considerations and machine-oriented implementation detail come in by conscious design decisions from the implementor when applying pre-conceived transformation rules. A long-term research aim is the incorporation of goal orientation into the development process. In particular, the crucial selection in large libraries of rules has to reflect the reasoning process in the development.

Engineering Discipline for Correct SW. The PROSPECTRA project aims at making software development an engineering discipline. In the development process, ad hoc techniques are replaced by the proposed uniform and coherent methodology, covering the complete development cycle. Programming knowledge and expertise are formalised as transformation rules and methods with the same rigour as engineering calculus and construction methods, on a solid theoretical basis.

Individual transformation *rules*, compact automated transformation *scripts* and advanced transformation *methods* are developed to form the kernel of an extendible knowledge base, the Method Bank, analogously to a handbook of physics. Transformation rules in the method bank are proved to be correct and thus allow a high degree of confidence. Since the methodology completely controls the system, reliability is significantly improved and higher quality can be expected.

Specification. Formal specification is the foundation of the development to enable the use of formal methods. High-level development of specifications and abstract implementations (a variation of "logic programming") is seen as the central "programming" activity in the future. In particular, the development of methods for "program synthesis", the derivation of constructive design specifications from non-constructive requirement specifications, is a present focus of research.

The abstract formal (algebraic) specification of requirements, interfaces and abstract designs (including concurrency) relieves the programmer from unnecessary detail at an early stage. Detail comes in by gradual optimising transformation, but only where necessary for efficiency reasons. Specifications are the basis for adaptations in evolving systems, with possible replay of the implementation from development histories that have been stored automatically.

Programming Language Spectrum: Ada and Anna. Program Development by Transformation [4-6] receives increased attention world-wide, see [8]. However, it has mostly been applied to research languages. Targetting the general methodology and the support system to Ada [9] (with Anna as its complement for formal specification, see [10]) make it realistic for systems development including concurrency aspects. $PA^{nn}dA$, the PROSPECTRA Anna/Ada subset, covers the complete spectrum of language levels from formal specifications and applicative implementations to imperative and machine-dependent representations. Uniformity of the language enables uniformity of the transformation methodology and its formal basis.

Stepwise transformations synthesise Ada programs such that many detailed language rules necessary to achieve reliability in direct Ada programming are obeyed *by construction* and need not concern the program developer. In this respect, the PROSPECTRA methodology may make an important contribution to managing the complexity of Ada.

Research Consolidation and Technology Transfer. The PROSPECTRA project aims at contributing to the technology transfer from academia to industry by consolidating converging research in formal methods, specification and non-imperative "logic" programming, stepwise verification, formalised implementation techniques, transformation systems, and human interfaces.

Industry of Software Components. The portability of Ada allows pre-fabrication of software components. This is explicitly supported by the methodology. A component is catalogued on the basis of its interface. Formal specification gives the semantics as required by and made visible to the user; the implementation is hidden and remains a (company) secret.

The methodology emphasises the *pre-fabrication* of generic, universally *(re-)usable* components that can be instantiated according to need. This will invariably cut down production costs by avoiding duplicate efforts. The production of perhaps small but universally marketable (Ada) components on a common technology base can also assist smaller companies in Europe.

Tool Environment. Emphasis on the development of a comprehensive support system is mandatory to make the methodology realistic. The system can be seen as an integrated set of advanced tools based on a minimal support environment, e.g. the ESPRIT Portable Common Tool Environment (PCTE). Because of the generative nature of system components, adaptation to future languages is comparatively easy.

The support of correct and efficient transformations is seen as a major advance in programming environment technology. The central concept of system activity is the application of transformations to trees. Generator components are employed to construct transformers for individual transformation rules and to incorporate the hierarchical approach of PA^{nn}dA (PROSPECTRA Anna/Ada), TrafoLa (the language of transformation descriptions), and ControLa (the command language); in fact, these turn out to be all sublanguages of the same language, for user program, transformation and system development. This integration and uniformity is seen as one of the major (unforeseen) results of the PROSPECTRA project (cf. chapters 4, 6 and [11-13]). Generators, in particular the Cornell Synthesizer Generator (cf. [14, 15]), increase flexibility and avoid duplication of efforts; thus the overall systems complexity is significantly reduced.

2.2. THE DEVELOPMENT MODEL

Consider a simple model of the major development activities in the life of a program:

Requirements Analysis

- Informal Problem Analysis
- Informal Requirement Specification

Development

- **Formal** Requirement Specification   ⇑ *Validation*
- **Formal** Design Specification   ⇑ *Verification*
- **Formal** Construction *by Transformation*   ⇑ *Verification*

Evolution

- Changes in Requirements ⇒ **Re-Development** ⇑

The *informal requirements analysis* phase precedes the phases of the *development* proper, at the level of formal specifications and by transformation into and at the level(s) of a conventional programming language such as Ada.

After the program has been installed at the client, no maintenance in the sense of conventional testing needs to be done; "testing" is perfomed *before* a program is constructed, at the very early stages, by validation of the formal requirement specification against the informal requirements.

The *evolution* of a program system over its lifetime, however, is likely to economically outweigh the original development by an order of magnitude. Changes in the informal requirements lead to re-development, starting with changes in the requirement specification. This requires re-design, possibly by *replay* of the original development (which has been archived by the system) and adaptation of previous designs or re-consideration of previously discarded design variants.

2.3. SPECIFICATION

A requirement specification defines *what* a program should do, a design specification *how* it does it. The motivations and reasons for design decisions, the *why's*, are recorded along with the developments.

Requirement specifications are, in general, non-constructive; there may be no clue for an algorithmic solution of the problem or for a mapping of abstract to concrete (i.e. predefined) data types. It is essential that the requirement specification should not define more than the *necessary* properties of a program to leave room for design decisions. It is intentionally vague or *loose* in areas where the further specification of detail is irrelevant or impossible. In this sense, loose specification replaces non-determinacy, for example to specify an unreliable transmission medium in a concurrent, distributed situation [16, 17].

From an economic point of view, overspecification may lead to substantial increase in development costs and inefficiency of execution of the program since easier solutions are not admissable. If the requirement specification is taken as the formal *contract* between client and software developer, then there should perhaps be a new profession of an independent *software notary* who negotiates the contract, advises the client on consequences by answering questions, checks for inconsistencies, resolves unintentional ambiguities, but guards against overspecification in the interest of both, client and developer. The answer of questions about properties of the formal requirement specification correspond to a *validation* of the informal requirement specification using a prototyping tool.

(2.3-1) Requirement Specification: Lists

```
generic
  type ITEM is private;
  "<": ITEM —> ITEM —> BOOLEAN :: for all x, y, z: ITEM => ¬ (x < x),   x < y ∧ y < z →x < z;
package LISTS is
  type LIST is private;
  empty:  LIST;
  cons:   ITEM —>  LIST —>  LIST;
```
```
  "&":    LIST —>  LIST —>  LIST;
  single: ITEM —>  LIST;
axiom for all  x, y, z: LIST =>
  x & empty = x,   empty & x = x,   (x & y) & z = x & (y & z);
```
```
  isEmpty: LIST —>                     BOOLEAN;
  head:   (x: LIST :: ¬ isEmpty x) —> ITEM;
  tail:   (x: LIST :: ¬ isEmpty x) —> LIST;
axiom for all e: ITEM; l: LIST =>
  isEmpty empty = true,     isEmpty (cons e l) = false,
  head (cons e l) = e,       tail (cons e l) = l,        (single e) & l = cons e l;
```
```
  "<=":      LIST —> LIST —> BOOLEAN;          - - lexicographic order
axiom for all l, l1, l2: LIST ; x1, x2: ITEM  =>
  l <= l & l2,      x1 < x2→ l & (single x1) & l1 <= l & (single x2) & l2;
end LISTS;
```

As an example consider the specification of lists in (2.3-1). This particular specification of lists is already more specific than perhaps necessary since it requires a strict order relation "<" as parameter, to be able to define the lexicographic order relation "<=" on lists. Its definition is a typical example of a non-constructive requirement specification: it is concise but does not give a clue for or overspecify an algorithmic solution.

For the purpose of exhibiting the use of functionals, a notation with explicit Curry-ing to allow partial parameterisation is used. During later stages of the development process, the system allows a switch from a more symbolic style to a more conventional Ada oriented notation using, for example, **return** instead of —> to denote a function(al type), see 2.4.

Note that we may have two views of lists: either constructed by empty and cons or by empty, "&" and single. Depending on the view, lists have different algebraic properties; in the second case those of a monoid (associativity and neutral element empty). Such strong properties become important for reasoning about optimisations, as we will see below. The definition of the selectors head and tail insures uniqueness of models up to isomorphism.

(2.3-2) Requirement Specification: Sets

```
generic
   type ITEM is private;
package SETS is
   type SET is private;
   empty:      SET;
   singleton:  ITEM —>   SET;
   "∪":        SET —>   SET —->   SET;
   "∈":        ITEM —>   SET —>   BOOLEAN;
axiom for all a, b: ITEM; x, y, z: SET; s: SET:: ¬ isEmpty s =>
   x ∪ empty = x,  x ∪ (y ∪ z) = (x ∪ y) ∪ z,  x ∪ y = y ∪ x,
   a ∈ empty = false,    a ∈ (singleton a),      a ∈ (x ∪ y) = (a ∈ x) ∨ (a ∈ y);
end SETS;
```

(2.3-3) Requirement Specification: (Partially) Ordered Sets

```
generic
   type ITEM is private;
   "<=": ITEM —> ITEM —> BOOLEAN::
   for all x, y, z: ITEM=> x <= x,  x <= y ∧ y <= x → x = y,  x <= y ∧ y <= z → x <= z,  x <= y ∨ y <= x;
package ORD_SETS is
   type SET is private;
   ... as for SETS above
   isEmpty:  SET —>                    BOOLEAN;
   min:      (s: SET:: ¬ isEmpty s) —>  ITEM;
   rest:     (s: SET:: ¬ isEmpty s) —>  SET;
axiom for all a: ITEM; x: SET; s: SET:: ¬ isEmpty s =>
   isEmpty empty,         a ∈ x → ¬ isEmpty x,        (min s) ∈ s,      a ∈ s → (min s) <= a,
   a ∈ (rest s) → a ∈ s,    a ∈ s ∧ ¬ a <= (min s) → a ∈ (rest s);
end ORD_SETS;
```

The specification of sets and ordered sets given in (2.3-2, 3) is an even better example of a non-constructive requirement specification. It is also an example of a loose specification. If several elements in a sequence of ∪'s are equivalent, these elements may or may not be multiply represented; sets and multisets ("bags") are allowed as models. This may be desirable: the semantics of some algorithm using sets may not depend on the existence of multiple elements. The user may know that multiple elements in a set are unlikely for the application and would not matter. In this case, the implementation freedom and potential efficiency should not be unnecessarily restricted. For example, in an automatically produced ordered list of titles we may not care for multiple occurrences if there are only few (cf. chapter 5). If we want to insist on multiple occurrences, we must add some function to be able to distinguish, such as a counting of elements; for true sets, we may add an axiom $x \cup x = x$ that forces absorption. (Omitting the commutativity and absorption axioms in ordered sets with the design specification of (2.3-4) yields priority queues, cf. [3]).

Design specifications specify abstract implementations. They are constructive, both in terms of more basic specifications and in the algorithmic sense. For a loose requirement specification, the design specification will usually restrict the set of models, eventually to one. (2.3-4, 5) show parts of a design specification for sets to multisets and true sets, resp.

(2.3-4) Design Specification for Ordered Multisets (partial)

```
                    min (singleton a )      = a,
   ¬ (min s <= a) →  min (s ∪ singleton a)  = a,
     min s <= a →    min (s ∪ singleton a)  = min s,
                    rest (singleton a )     = empty,
   ¬ (min s <= a) →  rest (s ∪ singleton a) = s,
     min s <= a →    rest (s ∪ singleton a) = rest s ∪ singleton a;   - - only one is removed
```

(2.3-5) Design Specification for Ordered Sets (partial)

```
x ∪ x = x,                      min as above
   a <= min s ∧ ¬ a ∈ s →    rest (s ∪ singleton a)   = s,
   a <= min s ∧   a ∈ s →    rest (s ∪ singleton a)   = rest s,
   ¬ (a <= min s) →            rest (s ∪ singleton a)   = rest s ∪ singleton a;
```

Implementation: What remains for abstract specifications is a mapping onto some suitable specification at a lower level of the system hierarchy, that is a standard one or one that has already been implemented. Certain abstract type (schemata) that correspond to predefined Ada types (constructors, selectors, other auxiliary functions and their algebraic specification), for example **record**, or the usual recursive variant (or union) types (free term constructions for lists, trees etc.) are standard in PA^{nn}dA (see [28]). They are turned into an Ada text automatically as an alternative (standard Ada) notation for the package defining the abstract type. We assume that a standard Ada implementation using access types (pointers) and allocators is still considered to be "applicative" at this level of abstraction and that side-effects of allocation will be eliminated during the development process by explicit storage allocation if required.

Consider sets implemented as lists (see [3]). In one implementation, all terms building one set using ∪ are represented by distinct lists:

x∪y and y∪x or x∪x and x

have distinct representations and a search is performed when ∈ is applied. In the other extreme, all such terms are represented by a list without duplicates; elements are ordered canonically and a search has to be made to eliminate elements from y that already occur in x upon ∪. Betwen the two extremes lie other admissible implementations, for example ordering but not eliminating multiple occurrences (or vice-versa), or one using a hash table. It should also be emphasised that not only list-like implementations are possible, of course. An implementation of an ordered set using, for example, a binary or balanced tree representation is also admissable. Similarly, any search algorithm, for example binary search, will do. Note the analogy between binary search and a binary search tree: the same idea is once represented in the algorithm and once in the data structure.

Integration of Construction and Verification: Not only is the program construction process formalised and structured into individual mechanisable steps, but the verification process is structured as well and becomes more manageable. If transformation rules are correctness-preserving, then only the applicability of each individual rule has to be verified at each step. Thus a major part of the verification, the verification of the correctness of each rule, need not be repeated. Verification then reduces to verification of the applicability of a rule, and program versions are correct by construction (with respect to the original requirement specification). This interactive, stepwise proof, aided by the system, is expected to be much easier than a corresponding proof of the final version.

As an alternative to proving the applicability conditions as they arise, the system can keep track of the verification conditions generated and accumulate them till the (successful) end of the development. This way, no proofs are necessary for "blind alleys", with the danger that the supposedly correct development sequence leading to the final version turns out to be a "blind alley" itself, if the proof fails. But even if we consider all proofs required from the developer (with assistance from the system) together, they are still much less complicated than a monolithic proof of the final version.

Transformations for verification (simplification of verification conditions to true) and an interactive Proof Editor are incorporated in the PROSPECTRA system. They use the PROSPECTRA paradigm for their implementation: inference rules are regarded as transformation rules, and proof strategies can be added by the user in analogy to transformation strategies (see [18, 19]). The Conditional Equational Completion subsystem ([20, 21]) can also be used for verification; its major use is a kind of "program synthesis" for deriving construcive design specifications.

2.4. TRANSFORMATIONAL PROGRAM DEVELOPMENT

Each transition from one program version to another can be regarded as a transformation in an abstract sense. It has a more technical meaning here: a transformation is a development step producing a new program version by application of an individual transformation rule, a compact transformation script, or, more generally, a transformation method invoking these. Before we come to the latter two, the basic approach will be described in terms of the transformation rule concept.

A transformation rule is a schema for an atomic development step that has been pre-conceived and is universally trusted, analogously to a theorem in mathematics. It embodies a grain of expertise that can be transferred to a new development. Its application realises this transfer and formalises the development process.

Transformations preserve correctness and therefore maintain a tighter and more formalised relationship to prior versions. Their classical application is the construction of optimised implementations by transformation of an initial design that has been proved correct against the formal requirement specification. Further design activity then consists in the selection of an appropriate rule, oriented by development goals, for example machine-oriented optimisation criteria.

Language Levels: We can distinguish various language levels at which the program is developed or into which versions are transformed, corresponding to phases of the development:

- formal requirement specification: loose equational or predicative specifications
- formal design specification: specification of abstract implementation
- applicative implementation: recursive functions
- imperative implementation: variables, procedures, iteration by loops
- machine-oriented implementation: machine operations, registers, array of words

All these language levels are covered by PA^{nn}dA, cf. [24, 26-29].

The (interactive) deduction of constructive design specifications from non-constructive requirement specifications, a kind of *program synthesis,* is supported by complex transformation strategies and tools. The enhancement of the Knuth-Bendix completion technique, for example, receives major attention in the PROSPECTRA project (see [20, 21]). Current research focusses on the development activities at the specification level, demanding most creativity from the developer. This language level is perhaps the most important programming language of the future.

Many developments at lower levels can also be expressed at the specification level, for example "recursion removal" methods transforming into tail-recursive functions [29, 30, 13]. As an example, consider the special case that $\oplus$ and n form a Monoid, where $\oplus$ is an arbitrary associative operation (cf. [29, 30]), expressed here at the specification level. The rule in (2.4-1) could be generalised and adapted further to apply to equations with constructors on the left instead of selectors on the right-hand sides. Assume that we want to derive a body for length. It is not in tail-recursive form: the addition of 1 still has to be made upon return from the recursion. By applying the transformation rule in (2.4-1), however, we can embed it into a function len that is tail-recursive, see (2.4-2). len can thus be transformed into a local loop, see (2.4-3). Note that the applicability condition, namely that + is an associative operation with neutral element 0, has to be proved with the aid of the system; but see also section 3.2 below.

Transformation Rules, Scripts, and Methods. Individual transformation rules are generalised to transformation *scripts*: sets of transformations rules applied together, possibly with local tactics that increase the efficiency. The long term research goal is to develop transformation *methods* that relieve the programmer from considerations about individual rules to concentrate on the goal oriented design activity. A transformation method is thus a set of rules or scripts with a global application strategy.

(2.4-1) Transformation Rule: Linear to Tail Recursion: Associative Operation

$f: S \longrightarrow M;$ **axiom for all** $x: S \Rightarrow$ $\neg\, \mathbb{B}\,x \rightarrow\quad f\,x = f\,(\mathcal{H}\,x) \oplus \mathcal{K}\,x,$ $\mathbb{B}\,x \rightarrow\quad f\,x = \mathcal{T}\,x;$
such that f does not occur in $\mathcal{T}, \mathcal{H}, \mathcal{K}$ **axiom for all** $x, y, z: M \Rightarrow$ $(x \oplus y) \oplus z = x \oplus (y \oplus z),\quad x \oplus n = x;$

$f:\quad S \longrightarrow M;$ $g:\quad S \longrightarrow M \longrightarrow M;$ **axiom for all** $x: S \Rightarrow$ $\qquad\qquad f\,x = g\,x\,n,$ $\neg\, \mathbb{B}\,x \rightarrow\quad g\,x\,y = g\,(\mathcal{H}\,x)\,((\mathcal{K}\,x) \oplus y),$ $\mathbb{B}\,x \rightarrow\quad g\,x\,y = (\mathcal{T}\,x) \oplus y;$

(2.4-2) Transformation: Linear Recursion to Tail Recursion: length

$length:\ LIST \longrightarrow\ INTEGER;$ **axiom for all** $x: LIST \Rightarrow$ $isEmpty\ x\ \rightarrow\quad length\ x = 0,$ $\neg\ isEmpty\ x\ \rightarrow\quad length\ x = length\ (tail\ x) + 1;$

$length: LIST \longrightarrow\ INTEGER;$ $len:\quad LIST \longrightarrow INTEGER \longrightarrow INTEGER;$ **axiom for all** $x: LIST\ ;\ r: INTEGER \Rightarrow$ $\qquad\qquad length\ x = len\ x\ 0,$ $isEmpty\ x\ \rightarrow\quad len\ x\ r = 0 + r,$ $\neg\ isEmpty\ x\ \rightarrow\quad len\ x\ r = len\ (tail\ x)\ (1 + r);$

(2.4-3) Ada Program: Applicative and Imperative Body of length (with Unfold of len)

| ```
function LENGTH(X: LIST) return INTEGER
is
begin
  If IS_EMPTY (X) then
    return 0;
  else
    return LENGTH(TAIL(X)) + 1;
  end if;
end LENGTH;
``` |

| ```
function LENGTH(X: LIST) return INTEGER
is
  V: LIST:= x;  R: INTEGER := 0;
begin
  while not IS_EMPTY(V) loop
    V := TAIL(V);   R := 1+R;
  end loop;
  return R;
end LENGTH;
``` |

Catalogues of Transformations: Some catalogues of transformation rules have been assembled for various high-level languages. Of particular interest is the structured approach of the CIP group. The program development language CIP-L is formally defined by transformational semantics (see [5, 6]), mapping all constructs in the wide spectrum of the language to a language kernel. Here, the kernel is PA^{nn}dA-S.

The PROSPECTRA catalogue is compiled in [30]. Various transformation scripts and methods such as fold-unfold, instantiation, variable and functional abstraction, recursion elimination, function composition, embedding, finite differencing, rewriting, narrowing and unification with a set of equations interpreted as rewrite rules, etc., have been implemented and are being complemented by other strategies and methods (cf. [30]).

3. Functionals

3.1. METHODOLOGICAL ADVANTAGES

Functionals, i. e. higher order functions with functions as parameters and/or results (cf. e. g. [37-41, 7, 23, 35]), allow a substantial reduction of re-development effort, in the early specifications and all subsequent developments. This aspect of functional abstraction is in analogy to parameterised data type specifications such as generics in Ada.

It is an interesting observation that many if not most definitions of functionals have a restricted form: the functional argument is unchanged in recursive calls. A functional together with its (fixed) functional parameters can then always be explicitly expanded by transformation (instantiation), cf. [11]. In the presence of overloading, a functional that is locally defined to a parameterised specification has an analogous effect to a polymorphic functional. The addition of implicit polymorphism to the specification language as in most functional languages (cf. e.g. [37]) is presently being considered.

Thus the major advantage of functionals appears, at first glance, to be "merely" one of abbreviation. In contrast to generics, tedious explicit instantiation is avoided for functional parameters, in particular for partial parameterisation ("Curry'ing"). However, working with functionals quickly leads to a new style of programming (i. e. specification and development) at a considerably higher degree of abstraction. As we shall see below, much repetitive development can be reduced to the application of homomorphic extension functionals; these can be considered as a kind of "program generators".

It is this aspect, that many functions should have the property of being homomorphisms, that goes beyond the correctness properties expressible in standard functional programming (in Miranda, for example). There, one tends to think only in terms of free term algebras (lists, trees etc.). Here, we have the whole power of algebraic specification available to state, for example, that the properties of a monoid hold and are preserved by a (homomorphic) function, indeed by a functional for a whole class of applications. Development (optimising transformations etc.) need be made only once for the functional. In fact, the recursion schema of homomorphic extension (see [42]) provides a program development strategy ("divide and conquer", cf. [43]) and an induction schema for proofs.

In meta-programming, these homomorphic extension functionals are important for the concise definition of program development tactics (see [11-13] and chapter 6). As we shall see in chapter 5, the algebraic properties of functionals allow a high level of reasoning *about* functional programs (postulated in [37-39]) that is supported by the PROSPECTRA system.

3.2. HOMOMORPHISMS AND HOMOMORPHIC EXTENSION FUNCTIONALS

The functionals Map, Filter of [37] and others are special cases of a more general homomorphic extension functional, see (3.2-1). Hom corresponds to the Monoid view of list construction (cf. section 2.3 above) and thus to a program development strategy by (binary) partitioning. An analogous homomorphic extension functional LinHom corresponds to the linear view and thus to a linear "divide and conquer" strategy, cf. [13]. Map can be defined as an automorphism (i. e. a homomorphism to the same structure); in fact it can be defined more generally to map between two lists of different component types.

(3.2-1) Homomorphic Extension Functionals over Lists

```
... inside LISTS
generic
  type M is private;
package LIST_MONOID_HOM is
    Hom:    (n: M) —> (op: M —> M —> M) —> (ITEM —> M) —> LIST —> M ::
            for all x, y, z: M =>  op x n = x,  op n x = x,  op (op x y) z = op x (op y z);
    axiom for all  n: M; op: M —> M —> M; h: ITEM —> M; e: ITEM; x, y: LIST =>
    Hom n op h empty       = n,
    Hom n op h (x & y)     = op (Hom n op h x) (Hom n op h y),
    Hom n op h (single e)  = h e;
end LIST_MONOID_HOM;
```

```
package AUTO_M is new LIST_MONOID_HOM (LIST); use AUTO_M;
Map:    (ITEM —> ITEM) —>        LIST —>   LIST;
Filter: (ITEM —> BOOLEAN) —> LIST —>   LIST;
Filt:   (ITEM —> BOOLEAN) —> ITEM —>   LIST;
axiom for all f: ITEM —> ITEM; p: ITEM —> BOOLEAN; x: ITEM =>
Map f = Hom empty "&" (single • f),
Filter p = Hom empty "&" (Filt p),   p x → Filt p x = single x, ¬ p x → Filt p x = empty;
```

Note that Hom requires, that the target algebraic structure has the properties of a monoid (actually, an injection function corresponding to single is combined with some f in the function h, now from ITEM to M; the function composition operator • is assumed to be universally defined in this paper). In this case we can transform Hom using the monoid properties of lists and employ the recursion removal transformation of (2.5-1) that is only applicable, if op and n form a monoid (cf. [37, 29, 13]), see (3.2-2).

In functional programming, such a global optimisation is not possible since we could not be sure that the binary operation is associative in general; there is no way to state such a requirement in a standard functional programming language. In conventional programming or algebraic specification without functionals we would have to separately prove the property and optimise for each case (each instance of the functional).

(3.2-2) Optimisation using Recursion Removal Transformation

```
... inside LIST_MONOID_HOM
    H2:  (n: M) —> (op: M —> M —> M) —> (ITEM —> M) —> M —> LIST —> M
         :: for all x, y, z: M =>  op n x = x,   op (op x y) z = op x (op y z);
    axiom for all  n: M; op: M —> M —> M; h: ITEM —> M; e: ITEM; y, z: LIST; r: M =>
                Hom n op h empty         = n,              - - linearize by substitution
                Hom n op h ((single e) & y)  = op (h e) (Hom n op h y),

    isEmpty z → Hom n op h z          = n,              - - eliminate constructors
  ¬ isEmpty z → Hom n op h z          = op (h (head z)) (Hom n op h (tail z)),

                Hom n op h z   = H2 n op h n z,          - - eliminate recursion
    isEmpty z → H2  n op h r z  = r,
  ¬ isEmpty z → H2  n op h r z  = H2 n op h (op (h (head z)) r) (tail z);
```

(3.2-3) Application of Homomorphic Extension Functionals over Lists

```
package toINT_HOM is new LIST_MONOID_HOM (INTEGER); use toINT_HOM ;
length:      LIST —>    INTEGER;
one:         ITEM —>    INTEGER;
axiom for all x: ITEM =>  one x = 1,     length = Hom 0 "+" one
```

```
package toBOOL_HOM is new LIST_MONOID_HOM (BOOLEAN); use toBOOL_HOM ;
Exist:       (ITEM —> BOOLEAN) —> LIST —>    BOOLEAN;
ForAll:      (ITEM —> BOOLEAN) —> LIST —>    BOOLEAN;
eq:          ITEM —>  ITEM —>   BOOLEAN;
isElem:      ITEM —>  LIST —>   BOOLEAN;
axiom for all x: ITEM; a, b: LIST =>
Exist = Hom false "or",           ForAll = Hom true "and",
eq x y = not (x < y) and not (y < x),      isElem x = Exist (eq x);
```

As an example for the instantiation of a homomorphic extension functional, see (3.2-3) for length (cf. (2.5-2, 3) above). Similarly, existential and universal quantification of a predicate over a list can be defined by homomorphic extension of the predicate over lists, using the algebraic properties of Booleans. isElem can be defined using isElem, for example. Note the use of partial parameterisation for eq in isElem.

Homomorphisms over sets are defined in an analogous way; there are, however, additional restrictions on Hom guaranteeing the preservation of the algebraic properties of sets (cf. (2.3-5)).

4. An Example of Transformational Development with Functionals

4.1. INFORMAL PROBLEM SPECIFICATION: KEYWORD-IN-CONTEXT

Consider a program that generates a KWIC Index (KeyWord-In-Context), informally specified as follows (taken from [44]):

> A *title* is a list of words which are either *significant* or *non-significant*. A *rotation* of a list is a cyclic shift of words in the list, and a *significant rotation* is a rotation in which the first word is significant. Given a set of *titles* and a set of *non-significant words*, the program should produce an *alphabetically sorted list* of the significant rotations of titles.

4.2. PROBLEM SPECIFICATION USING FUNCTIONALS

Let us disregard problems of layout, scanning of text etc. here and assume that a title has already been recognised as a list of words. The requirement specification is non-constructive, in a predicative style, stating the required properties (paraphrased underneath).

Note that the "alphabetically sorted list" has been modelled as an *ordered* set of titles, where the order (lexicographic order of list of words) is used as a parameter to the type schema. This is a good example for the advantages of the approach of algebraic specification over more model-oriented approaches: rather than using a few standard types (or schemata) such as lists, sets, maps etc., more specific and more problem oriented types can be defined (and re-used). It is a good idea to define the type of ordered sets separate from sets since it allows a number of different (and supposedly more efficient) implementations. In our example, we have not made a decision whether we sort elements upon adding to the set or whether we search the set when looking for an element (or printing it in order, corresponding to a posteriori sorting), cf. section 2.3. Apart from the conceptual advantage, it is thus better to encapsulate such a type separately rather than defining the notion of order for sets on its own; it may be difficult later on to recognise ordered sets and make a transition to a special implementation.

(4.2-1) KeyWords In Context

```
package KWIC is
    package W_LISTS is new LISTS (WORD, "<");   subtype WORD_LIST is W_LISTS.LIST;
    subtype TITLE is (t: WORD_LIST:: ¬ isEmpty t);                  - - A title is a list of words

    package W_SETS is new SETS (WORD);    subtype WORD_SET is W_SETS.SET;
    package T_OSETS is new ORD_SETS (TITLE, "<=");                  - - lexicographic order
    subtype TITLE_SET is T_OSETS.SET;              - -an "alphabetically sorted list" ... of titles

    rotate:                        TITLE —> NAT —>      TITLE;
    significant:  WORD_SET —> TITLE —>                  BOOLEAN;
    sigRotAll:    WORD_SET —> TITLE —>                  TITLE_SET;
    kwic:         WORD_SET —> TITLE_SET —>      TITLE_SET;
axiom for all nsig: WORD_SET; x: WORD; t, t1, t2: TITLE; i: NAT; ts: TITLE_SET =>
    select (t1 & (single x) & t2) i = x → rotate (t1 & (single x) & t2) i = (single x) & t2 & t1,
    - - a rotation of a list is a cyclic shift of words in the list
    significant nsig t  =  ¬ (head t ∈ nsig),
    - - a significant rotation t is a rotation in which the first word is significant,
    - - i. e. not in the given set nsig of non-significant words
    (0 <= i ∧ i <= (length t)-1 ) ∧ (significant nsig t) <-> (rotate t i) ∈ (sigRotAll nsig t),
    - - all significant rotations of one title
    t ∈ ts ∧ t1 ∈ (sigRotAll nsig t) <-> t1 ∈ (kwic nsig ts),
    - - all significant rotations of all titles from the given set of titles ;

private
    interval:    NAT —>    NAT —>     NAT_LIST;
    rotAll:               TITLE —>    TITLE_SET;
axiom for all nsig: WORD_SET; t: TITLE; i: NAT =>
    rotate t i       = drop i t & take i t,
    m <= i ∧ i <= n  <-> isElem i (interval m n),
    rotAll t         = Map (rotate t) (interval 0 ((length t)–1)),
    - - all rotations of one title
    sigRotAll nsig   = (Filter (significant nsig)) • rotAll,
    - - all significant rotations of one title
    kwic nsig        = joinAll • Map (sigRotAll nsig);
    - - all significant rotations of all titles from the given set of titles
end KWIC;
```

A design specification using functionals is then given in in the private part of (4.2-1). The same names for homomorphic extension functionals etc. have been used on sets for the analogous functions as on lists above. A few auxiliary functions are not defined here: joinAll concatenates all elements, for a list of lists, or unites all elements for a set of sets, or combinations. select selects an element of a list (counting from 0), take yields an initial segment, drop a final segment (cf. [37]).

4.3. OPTIMISATION USING ALGEBRAIC REASONING

Apart from the notion of an interval (a "generator" for a list of natural numbers), all definitions are already operational. However, we would like to improve efficiency by transformation in the system: in this case mostly fold-unfold, instantiation, algebraic manipulations of functionals, etc. For this reason let us consider some useful properties first. Most of the properties in (4.3-1) have been taken (and generalised) from [37]; they hold not only for automorphisms (from LIST to LIST) but for n, op of arbitrary type. Analogous properties hold for set homomorphisms, or combinations of the two.

As a typical example of abstract reasoning, (4.3-2) shows the derivation ("proof by transformation") of a distributivity property of homomorphic extension: it does not (semantically) matter, whether the filtering for significant title rotations is done before or after extension. We assume, that it is more efficient to do as soon as possible; therefore it is left innermost, in fact coalesced with the rest.

(4.3-3 to -6) give a series of derivations for the example using the properties of (4.3-1). Note the merge of several functionals in the classic style to one instance of Hom in (4.3-3) and (4.3-4) and its optimisation using the transformation of linear recursion to tail recursion that we had already shown for Hom in (2.5-1). We end up with two nested tail-recursions (loops), one in kwic2 and one in sigRotAll3. The rotation could also be optimised further.

(4.3-1) Algebraic Properties of Homomorphic Extension Functionals

(1)	(Map f) • (Map g)	= Map (f • g),
(2)	(Filter p) • joinAll	= joinAll • Map (Filter p),
(3)	joinAll • Map f	= Hom empty "&" f,
(4)	(Filter p) • (Map f)	= Hom empty "&" ((Filt p) • f);

(4.3-2) Distributivity of Filtering

axiom for all nsig: WORD_SET; x: WORD; t: TITLE; i: NAT =>

kwic nsig	= joinAll • Map (Filter (significant nsig) • rotAll)	-- *unfold sigRotAll*
	= joinAll • Map (Filter (significant nsig)) • Map rotAll	-- *(1)*
	= (Filter (significant nsig)) • joinAll • Map rotAll	-- *(2)*

(4.3-3) Homomorphic Definition and Unfold of kwic

kwic nsig	= Hom empty "∪" (sigRotAll nsig)	-- *(3)*
	= H2 empty "∪" (sigRotAll nsig) empty	-- *optimize Hom*
	= kwic2 nsig empty	-- *fold*
isEmpty ts → kwic2 nsig r ts	= r,	-- *unfold H2 in definition of kwic2*
¬ isEmpty ts → kwic2 nsig r ts	= kwic2 nsig ((sigRotAll nsig (min ts)) ∪ r) (rest ts);	

(4.3-4) Homomorphic Definition of sigRotAll

```
sigRotAll nsig t
= ( (Filter (significant nsig)) • (Map (rotate t)) )          (interval 0 (length t)-1) -- unfold
= ( Hom empty "∪" (Filt (significant nsig) • rotate t) )      (interval 0 (length t)-1) -- (4)
= Hom empty "∪" (sigRot nsig t)                               (interval 0 (length t)-1) -- fold sigRot
= H2 empty "∪" (sigRot nsig t) empty                         (interval 0 (length t)-1) -- optimize Hom
= sigRotAll2 nsig t empty                                     (interval 0 (length t)-1) -- fold
```

(4.3-5) Optimisation of sigRot

```
sigRot nsig t i    = Filt (significant nsig) (rotate t i),
                          -- (the singleton set of) a significant rotation , or empty
¬ (head (rotate t i) ∈ nsig) → sigRot nsig t i = singleton (rotate t i),          -- unfold
  head (rotate t i) ∈ nsig → sigRot nsig t i = empty,
```

```
                          -- simplify with head (rotate t i) = head (drop i t & take i t) = select t i,
¬ ((select t i) ∈ nsig) →    sigRot nsig t i = singleton (rotate t i),
  (select t i) ∈ nsig →      sigRot nsig t i = empty,
```

(4.3-6) Optimisation of sigRotAll

```
  isEmpty z →  sigRotAll2 nsig t r z = r,                    -- unfold H2 in definition of sigRotAll2
¬ isEmpty z →  sigRotAll2 nsig t r z = sigRotAll2 nsig t ((sigRot nsig t (head z)) ∪ r) (tail z);
```

```
                          -- instantiate  z = interval i j = cons i (interval (i+1) j)
¬ (i <= j) →  sigRotAll2 nsig t r (interval i j) = r,
  i <= j →    sigRotAll2 nsig t r (interval i j) =
                          sigRotAll2 nsig t ((sigRot nsig t i) ∪ r) (interval (i+1) j);
```

```
                          -- substitute (interval i j) by the pair i j
¬ (i <= j) →  sigRotAll3 nsig t r i j    = r,
  i <= j →    sigRotAll3 nsig t r i j    = sigRotAll3 nsig t ((sigRot nsig t i) ∪ r) (i+1) j,
```

```
                                                            -- unfold sigRot
          ¬ (i <= j) →  sigRotAll3 nsig t r i j = r,
i <= j ∧  ((select t i) ∈ nsig) →  sigRotAll3 nsig t r i j = sigRotAll3 nsig t r (i+1) j,
i <= j ∧ ¬ ((select t i) ∈ nsig) →  sigRotAll3 nsig t r i j =
                          sigRotAll3 nsig t ((singleton (rotate t i)) ∪ r) (i+1) j,
```

5. Conclusion: Meta-Program Development

The PROSPECTRA methodology and its program development system (see also (6-1) have been described. In particular, the importance of loose, property-oriented requirement specifications to avoid unnecessary over-specification and to leave room for design decisions has been emphasized.

The power of compact development methods using the transformational approach has been illustrated by an example. Current research and implementation focusses on methods for the early stages of development to aid the finding of problem solutions and the synthesis of operational versions.

The importance of the combination of algebraic specification with higher order functions has been stressed. The functional programming paradigm leads to a considerably higher degree of abstraction and avoids much repetitive development effort, in particular through the use of homomorphic extension functionals. The combination with algebraic specification only allows reasoning about correctness. For example, the statement of properties for parameters of a functional (such as those of a monoid) are not possible in conventional functional programming languages The (first order) algebraic properties of the types and also the (higher order) algebraic properties of the functionals permit general and powerful optimisations.

One important aspect of the PROSPECTRA approach has not been described in this paper: ist use for meta-programming and formalising developments. Various authors have stressed the need for a formalisation of the software development process: the need for an automatically generated transcript of a development "history" to allow re-play upon re-development when requirements have changed, containing goals of the development, design decisions taken, and alternatives discarded but relevant for re-development. A *development script* is thus a formal object that does not only represent a documentation of the past but is a plan for future developments. It can be used to abstract from a particular development to a class of similar developments, a *development method*, incorporating a certain strategy.

The approach taken in [11-13] is to regard transformation rules as equations in an algebra of programs, to derive basic transformation operations from these rules, to allow composition and functional abstraction, and to regard development scripts as (compositions of) such transformation operations. Using all the results from program development based on algebraic specifications and functionals we can then reason about the development of meta-programs, i. e. transformation programs or development scripts, in the same way as about programs: we can define requirement specifications (development goals) and implement them by various design strategies, and we can simplify ("optimise") a development or development method before it is first applied or re-played; in short, we can develop *correct*, efficient, complex transformation operations from elementary rules stated as algebraic equations.

The abstraction from concrete developments to development methods, incorporating formalised development tactics and strategies, and the formalisation of programming knowledge as transformation rules + development methods will be a challenge for the future.

There is a close analogy to the development of efficient proof strategies for given inference rules (transformation rules in the algebra of proofs). This is the basis for the development of the Proof Editor in PROSPECTRA [18, 19]. Perhaps the approach can be used to formalise rules and inference tactics in knowledge based systems.

Since every manipulation in a program development system can be regarded as a transformation of some "program" (for example in the command language), the whole system interaction can be formalised this way and the approach leads to a uniform treatment of programming language, program manipulation and transformation language, and command language. The uniform approach to program and meta-program development is perhaps the most important conceptual and methodological result of the PROSPECTRA project. But it also has had some major practical consequences. It has been exploited in the PROSPECTRA system (cf. 6-1) yielding a significant reduction of parallel work. The specification language PAnndA-S of programs (with Ada as a target) is also used as the transformation specification language TrafoLa-S. In this case, an abstract type schema to define Abstract Syntax is predefined, and translation to an internal applicative tree manipulation language is automatic. Work on a more powerful target language with higher order matching and functionals is nearly completed [45]. ControLa is a subset of TrafoLa.

Acknowledgements

I wish to thank B. Möller, and B. Gersdorf, J. v. Holten, S. Kahrs, J. Liu, D. Plump, R. Seifert, and Z. Qian for helpful comments.

References

[1] Krieg-Brückner, B., Hoffmann, B., Ganzinger, H., Broy, M., Wilhelm, R., Möncke, U., Weisgerber, B., McGettrick, A.D., Campbell, I.G., Winterstein, G.: PROgram Development by SPECification and TRAnsformation. in: Rogers, M. W. (ed.): Results and Achievements, Proc. ESPRIT Conf. '86. North Holland (1987) 301-312.

[2] Krieg-Brückner, B.: Integration of Program Construction and Verification: the PROSPECTRA Project. in: Habermann, N., Montanari, U. (eds.): Innovative Software Factories and Ada. Proc. CRAI Int'l Spring Conf. '86. LNCS 275 (1987) 173-194.

[3] Krieg-Brückner, B.: The PROSPECTRA Methodology of Program Development. in: Zalewski (ed.): Proc. IFIP/IFAC Working Conf. on HW and SW for Real Time Process Control (Warsaw). North Holland (1988) 257-271.

[4] Bauer, F.L., Möller, B., Partsch, H., Pepper, P.: Formal Program Construction by Stepwise Transformations - Computer-Aided Intuition-Guided Programming.IEEE Trans. on SW Eng. 15:2 (1989) 165-180.

[5] Bauer, F.L., Berghammer, R., Broy, M., Dosch, W., Gnatz, R., Geiselbrechtinger, F., Hangel, E., Hesse, W., Krieg.-Brückner, B., Laut, A., Matzner, T.A., Möller, B., Nickl, F., Partsch, H., Pepper, P., Samelson, K., Wirsing, M., Wössner, H.: The Munich Project CIP, Vol. 1: The Wide Spectrum Language CIP-L. LNCS 183, Springer 1985.

[6] Bauer, F.L., Ehler, H., Horsch, B., Möller, B., Partsch, H., Paukner, O., Pepper, P.,: The Munich Project CIP, Vol. 2: The Transformation System CIP-S. LNCS 292, Springer 1987.

[7] Möller, B.: Algebraic Specification with Higher Order Operators. in: Meertens, L.G.T.L. (ed.): Program Specification and Transformation, Proc. IFIP TC2 Working Conf. (Tölz '86). North Holland (1987) 367-398.

[8] Partsch, H., Steinbrüggen, R.: Program Transformation Systems. ACM Computing Surveys 15 (1983) 199-236.

[9] Reference Manual for the Ada Programming Language. ANSI/MIL.STD 1815A. US Government Printing Office, 1983. Also in: Rogers, M. W. (ed.): Ada: Language, compilers and Bibliography. Ada Companion Series, Cambridge University Press, 1984.

[10] Luckham, D.C., von Henke, F.W., Krieg-Brückner, B., Owe, O.: Anna, a Language for Annotating Ada Programs, Reference Manual. LNCS 260, Springer (1987).

[11] Krieg-Brückner, B.: Formalisation of Developments: An Algebraic Approach. in: Rogers, M. W. (ed.): Achievements and Impact. Proc. ESPRIT Conf. 87. North Holland (1987) 491-501.

[12] Krieg-Brückner, B.: Algebraic Formalisation of Program Development by Transformation. in: Proc. European Symposium On Programming '88, LNCS 300 (1988) 34-48.

[13] Krieg-Brückner, B.: Algebraic Specification and Functionals for Transformational Program and Meta Program Development. in Díaz, J., Orejas, F. (eds.): Proc. TAPSOFT '89 (Barcelona), Vol. 2. LNCS 352 (1989) 36-59.

[14] Reps., Teitelbaum: The Cornell Synthesizer Generator. Springer Verlag, 1988.

[15] Reps., Teitelbaum: The Cornell Synthesizer Generator; Reference Manual. Springer Verlag, 1988.

[16] Broy, M.: Predicative Specification for Functional Programs Describing Communicating Networks. Information Processing Letters 25:2 (1987) 93-101.

[17] Broy, M.: An Example for the Design of Distributed Systems in a Formal Setting: The Lift Problem. Universität Passau, Tech. Rep. MIP 8802 (1988).

320

[18] Traynor, O.: The PROSPECTRA Proof Editor. PROSPECTRA Study Note S.3.4.-SN-15.2, University of Strathclyde, 1989.
[19] Traynor, O.: The Methodology of Verification in PROSPECTRA. PROSPECTRA Study Note S.3.4.-SN-19.0, University of Strathclyde, 1989.
[20] Ganzinger, H.: Ground Term Confluence in Parametric Conditional Equational Specifications. in: Brandenburg, F.J., Vidal-Naquet, G., Wirsing, M.(eds.): Proc. 4th Annual Symp. on Theoretical Aspects of Comp. Sci., Passau '87. *LNCS 247* (1987) 286-298.
[21] Ganzinger, H.: A Completion Procedure for Conditional Equations. Techn. Bericht No. 243, Fachbereich Informatik, Universität Dortmund, 1987 (to appear in *J. Symb. Comp.*)
[22] Broy, M., Wirsing, M.: Partial Abstract Types. *Acta Informatica 18* (1982) 47-64.
[23] Broy, M.: Equational Specification of Partial Higher Order Algebras. *in:* Broy, M. (ed.): *Logic of Programming and Calculi of Discrete Design.* NATO ASI Series, Vol. F36, Springer (1987) 185-241.
[24] Breu, M., Broy, M., Grünler, Th., Nickl, F.: PAnndA-S Semantics. PROSPECTRA Study Note M.2.1.S1-SN-1.3, Universität Passau, 1988.
[25] Owe. O.: An Approach to Program Reasoning Based on a First Order Logic for Partial Functions. Research Report No. 89, Institute of Informatics, University of Oslo, 1985.
[26] Krieg-Brückner, B.: Systematic Transformation of Interface Specifications. *in:* Meertens, L.G.T.L. (ed.): *Program Specification and Transformation,* Proc. IFIP TC2 Working Conf. (Tölz '86). North Holland (1987) 269-291.
[27] Broy, M., Möller, B., Pepper, P., Wirsing, M.: Algebraic Implementations Preserve Program Correctness. *Science of Computer Programming 7* (1986) 35-53.
[28] Kahrs, S.: PAnndA-S Standard Types. PROSPECTRA Study Note M.1.1.S1-SN-11.2, Universität Bremen, 1986.
[29] Bauer, F.L., Wössner, H.: *Algorithmic Language and Program Development.* Springer 1982.
[30] Krieg-Brückner, B. (ed.): PROgram development by SPECification and TRAnsformation, Part I: The Methodology. PROSPECTRA Report M.1.1.S3-R-55.0. Universität Bremen. *(in preparation).*
[31] Dijkstra, E. W.: *A Discipline of Programming.* Prentice Hall 1976.
[32] Gries, D.: *The Science of Programming.* Springer 1981.
[33] Sharir, M.: Some Observations Concerning Formal Differentiation of Set Theoretic Expressions. *ACM TOPLAS 4: 2* (1982) 196-226.
[34] Paige, R., Koenig, S.: Finite Differencing of Computable Expressions. *ACM TOPLAS 4: 4* (1982) 402-454.
[35] Broy, M., Pepper, P.: Combinig Algebraic and Algorithmic Reasoning: An Approach to the Schorr-Waite Algorithm. *ACM TOPLAS 4:3* (1982) 362-381.
[36] Liu, J., Krieg-Brückner, B.: Transformational Development of Completion. PROSPECTRA Study Note M.3.1.S1-SN-8.0. Universität Bremen. *(in preparation).*
[37] Bird, R., Wadler, Ph.: *Introduction to Functional Programming.* Prentice Hall, 1988.
[38] Bird, R.S.: Transformational Programming and the Paragraph Problem. *Science of Computer Programming 6* (1986) 159-189.
[39] Bird, R.: Lectures on Constructive Functional Programming. in: (1989) 0-65.
[40] Karlsen, E., Joergensen, J., Krieg-Brückner, B.: Functionals in PAnndA-S. PROSPECTRA Study Note S.3.1.C1-SN-10.0, Dansk Datamatic Center, 1988.
[41] Nickl, F., Broy, M., Breu, M., Dederichs, F., Grünler, Th.: Towards a Semantics of Higher Order Specifications in PAnndA-S. PROSPECTRA Study Note M.2.1.S1-SN-2.0, Universität Passau, 1988.
[42] von Henke, F.W.: An Algebraic Approach to Data Types, Program Verification and Program Synthesis. *in:* Mazurkiewicz, A. (ed.): Mathematical Foundations of Computer Science 1976. *LNCS 45* (1976) 330-336.
[43] Smith, D.R.: Top-Down Synthesis of Divide-and-Conquer Algorithms. *Artificial Intelligence 27:1* (1985) 43-95.
[44] Bjørner, D.: Towards the Meaning of M in VDM. *in* Díaz, J., Orejas, F. (eds.): Proc. TAPSOFT '89 (Barcelona), Vol. 2. *LNCS 352* (1989) 1-35.
[45] Heckmann, R.: A Functional Language for the Specification of Complex Tree Transformations. *in:* Proc. European Symposium On Programming '88, *LNCS 300* (1988) .

KNOWLEDGE-BASED GRAPHICAL DIALOGUE :
A STRATEGY AND ARCHITECTURE

JOHN LEE and BOB KEMP
EdCAAD
Department of Architecture
University of Edinburgh
20 Chambers Street
Edinburgh EH1 1JZ
Scotland, UK.

TRAUDE MANZ
Fraunhofer Gesellschaft
Institut für Arbeitswirtschaft und Organisation (IAO/452)
Holzgartenstrasse 17
D-7000 Stuttgart 1, FRG.

Abstract.

The nature and role of graphical interaction in the project ESPRIT P393 (ACORD) are outlined. The objective of combining natural language and graphics as an interface to a knowledge-base system gives a particularly interesting slant to the treatment of semantics in relation to graphical depictions and interactions. An approach is introduced which takes semantic representation to be an integral feature of a graphics processor that operates by parsing a stream of interaction events according to a grammar. The semantic mapping function is determined by a visualisation system in a dialogue manager, and this leads to an idea of a very dynamic interaction which adapts naturally to the requirements of the individual user. Several correlative issues in dialogue managing are discussed and a prototype implementation is described.

1. Introduction

ACORD is a project with an ambitious goal: to integrate natural language (NL) text and graphics into a system that provides its user with the opportunity to engage in a dialogue uniquely combining power, flexibility and naturalness. The main themes of this project have been outlined by Klein (1987); here we concentrate on the demands made of graphics, and also dialogue management, in a system of this kind.

2. Communication and Semantics

One might say that any attempt to communicate involves the origination of a *message*. In NL, for example, a message might be a sentence. The aim of a system designed to communicate with a user must be to extract the *content* from the user's messages - to find out what the user means - and to respond with messages that the user can understand. In the ACORD project, we have subsystems that can accept NL messages in any of three different languages (English, French, German) and then generate appropriate responses in these languages. The first stage is to parse an input sentence in order to determine its syntactic structure, but also, using information stored in the parser's lexicon relating to the words in the sentence, to arrive at a representation of its *semantic structure*. This structure, independently of the original input language, encodes information specifying what entities are referred to in the sentence, and what is said about them (i.e. what properties and relations are asserted or questioned to hold of or between them). The representation language, somewhat like an augmented first-order predicate logic, is known as InL (Klein 1987, and references therein).

The InL expression derived from an input sentence is taken by the Dialogue Manager (DM), where it is resolved (e.g. to find referents for anaphoric expressions such as pronouns) and can then be used to query or update the Knowledge Base (KB), in which a hierarchical inheritance network contains detailed information about the objects represented. Responses given by the KB can be used to construct suitable InL expressions for input to one of the text generators.

When we turn our attention to graphics, we see that the same arrangement cannot quite apply. In principle, perhaps, one could use graphic images similarly to sentences, but in general (and especially in the context of a computer system) one finds that there are important differences. Pictures are always drawn on some surface, and persist there for some time. Like text, they can be used to communicate quite effectively in this static fashion, as for instance on posters. But in interaction, they tend to become the focus of a dialogue. The discourse revolves around the image, as a depiction of some domain of interest; the image is pointed at, touched, changed, in order more effectively to communicate information about the domain. Unlike text sentences (let alone spoken ones), which normally follow one another in sequence then disappear, the picture forms a continuing concrete context for the interaction.

This is a feature which can be of immense value to a dialogue manager. In NL discourse, topic and focus are more abstract and require constant inferencing for their determination. Description, ellipsis and the use of pronouns commonly assume such knowledge, but always implicitly; underdetermined references often have to be resolved across several sentences. A picture is a stable element that can be linked to NL by the natural use of *deictic* (demonstrative) expressions, such as 'this', 'that', 'here', 'there', etc. But the very possibility of such exploitation points to the fact that the picture is itself a representation of the dialogue context, and hence necessarily more structured than a set of NL sentences.

Actually, what we should rather say is that the picture is *based on* a representation of the context. The screen image itself may be fairly abstract: a map might be depicted as a set of nodes with links between them, which could equally be intended to convey information about any domain with a network-like structure. For it to serve properly in dialogue, it must be understood that the nodes represent *cities*, the links *routes*, and perhaps that certain little movable icons on these represent *trucks*, and if one of them is located by a node then that truck is *in* that city. The abstract image must have a *semantic interpretation*, given in terms of what entities it represents, and what properties and relations hold of and between them. Only then can it be used in communication about the domain in question (in this case, and in the case of the ACORD prototype implementation, the activities of a transport company).

There are two aspects to consider, of communication in graphical interaction. On the one hand, there is the picture itself. Given some arbitrary, but preferably obvious[1]

1   In the interests of brevity, a number of substantial questions are begged here, as elsewhere in the text.

decomposition, each graphical part of the picture must have an interpretation in terms of the domain. Some of the relations between the parts will also have interpretations, but indefinitely many of these could be picked out and only certain more obvious ones are likely to be used in practice. The picture communicates to the user, and sets a context. On the other hand, there are the interactive operations performed upon the picture by the user. These have a number of roles. The user might, while typing 'this truck goes to Paris', point to one of the little truck-icons (by clicking on it with the mouse). The operation here is the click, and the representation of the picture allows it to be resolved as a reference to some particular truck (say truck42). The effect of this should be to convey to the DM just what would have been conveyed if the user had typed 'truck42 goes to Paris'; a resolved InL expression is arrived at, which asserts that truck42 goes to Paris. But equally, the user might simply 'edit' the picture, by dragging the truck-icon to the node representing Paris. The goal to achieve is that this has exactly the same effect in the DM: an expression is created asserting that truck42 goes to Paris. Here, we are looking at the semantics of the editing operation considered as a communicative act.

3. Domain Dependence in the Semantic Mapping

For the system sketched above to work properly, there has to be a mapping established between the graphical objects, their properties and relations, and the objects with their various properties and relations in the KB. The main requirement of this mapping, as a semantic mapping, is that every meaningful aspect of the graphical depiction is mapped to something appropriate (not necessarily the other way round), and this does not rule out certain aspects of the depiction being treated as simply meaning*less*. It is clearly undesirable, however, that this mapping be hard-wired into the graphics system. Graphics in normal usage is a highly flexible means of communication. A given picture can represent many things (especially if it is a simple picture) and what it is taken to represent (how it is understood) may vary dynamically as a dialogue progresses. These are some of the most powerful aspects of the use of graphics, and we wish to preserve them as far as possible. It follows that, in many respects, one wants one's graphical representation system to be *domain-independent*. But this conflicts with the requirements of the semantic mapping.

Our answer is that the mapping must be variable, under the control of the DM. It must be possible to set up the mapping, essentially in a quite arbitrary way, so that it can later be used to interpret the user's actions in a domain-dependent way, and so that it can later be changed if desired in order to allow the interpretations of actions to vary through time. The full exploitation of such a facility would clearly require a highly sophisticated approach to dialogue management, and is far more than can be achieved given the current state of that art. A well-defined approach to the visualisation and graphic presentation of diverse types of information is yet lacking (although there are pointers in the literature - *cf.* Gnanamgari 1981, Mackinlay 1986). Gnanamgari Mackinlay It is nonetheless important to design systems with an eye to future developments and, more important still, in ways that reflect the theoretical structure upon which they are based. For the present, it suffices to rely upon simple heuristic techniques for determining the DM's decisions about visualisation.

4. A Strategy for Graphical Communication

It seems convenient and well-motivated within the terms of the above discussion, to regard the flow of information between the graphics system (GS) and the DM as having two logically separate functions:

(a) the function of setting up and maintaining the semantic mapping whereby pictures are interpretable;
(b) the function of causing pictures, with that interpretation understood, to appear and to change, and of registering the user's interactions with respect to a picture.

Of these, (b) is properly what was termed above *communication* with the user: there is a content to each message that passes between the user and the system, and it is just the handling of such messages that is at issue, making sense of them with reference to the domain knowledge. However, (a) is a *'metalinguistic'* function, specifying *with reference to the messages* how they should be interpreted. Such a metalinguistic function is not found in the NL subsystems. The lexica there are fixed and the interpretations of words are neither established by the DM nor variable at runtime. This is, of course, a limitation, in the sense that such metalinguistic functions are in fact a significant part of NL interaction between people, but nonetheless it is possible to get away with ignoring them because the conventions subserving the interpretation of text expressions are relatively stable and consensual. In the case of graphics, pictures are used in widely divergent ways, and one usually has to use a dialogue context to indicate the appropriate mode of interpretation.

The basis for such a dialogue depends to some extent on the intentions for the system. If the system were, say, a CAD system, where the user was using drawings as input to the system, for describing objects and their relations etc., then one would like to set it up so that the metalinguistic functions were integrated into the discourse, allowing the user to specify, through NL with deixis, what interpretations were appropriate for his drawings. ('*This* is a truck', '*this* is its cab', etc.) Such a line has been recently investigated in Pineda's GRAFLOG system (Pineda *et al.* 1988, Pineda 1988). Pineda Klein Lee Pineda Compositional In the context of ACORD, it seems preferable to place the semantic mapping initially in the hands of the DM, since the main function of pictures (as opposed to edits of them) is to communicate information to the user, and if the user is to update the KB, he will in any case have to do this in terms compatible with what already exists there. However, our goal will be in the future to extend to the user the ability to define and redefine modes of depiction and interaction more suited either to his own preferences or to new kinds of information which he may introduce to the KB. In the present implementation, we limit the subject of the user discourse to the represented domain, rather than the mode of its presentation. The user can understand the pictures because of the context in which they are produced, the use of labels, etc.

As observed above, the central vehicle for the semantic representation of domain information in the project is InL. Unresolved InL has a comple information in the project is InL. Unresolved InL has a complexity of structure unnecessary for the purposes of the graphics system, where there are no phenomena such as anaphora, and a simplified form resembling fully-resolved InL has accordingly been adopted. Here we shall call it the Graphical Representation Language (GRL). The graphics system has to interpret GRL expressions describing some event or state of affairs (say 'truck42 goes to Paris'), and cause appropriate updates to the screen to convey the information. Similarly, if the user interacts with the picture in some way, the GS must arrive at an interpretation of his action, and express it in GRL to the DM. This all depends on a prior setting of the mapping, which, along with any subsequent changes to the mapping, should not be confused with the main user dialogue.

5. Implications for Dialogue Management

We shall now look at some of the implications of the above strategy for the task of managing dialogue in the system. The central concerns here are how application knowledge is visualised, and how interaction with the user should be handled.

5.1. The Visualisation of Application Knowledge

Application knowledge in the ACORD system is either already part of the KB or it will be entered into the KB via natural language (perhaps including graphical deixis).

In order to present knowledge expressed in NL in a graphical form (to visualise it) and in order to interpret graphical actions by the user in terms of the application, a clear mapping between the application and its graphical presentation is needed. This mapping concerns those parts of the KB contents (a) which are presentable in a graphical form and (b) which it is sensible to visualise in the context of the application and the user dialogue

For the ACORD Demonstrator System this means that the set of visualisable objects has been restricted to vehicles, their loads, some cities and roads on the European Continent, the city depots and cities and roads contents. Introduction of new vehicles, changes of the locations of vehicles, changes of their loads, establishment of new depots and changes of their contents, are all topics for updates of the graphical presentation as well as for graphical user manipulations.

In order to have a continually updated graphical presentation, i.e. one which is consistent with the knowledge in the KB, the DM notifies the KB when starting up the system about those parts of the domain knowledge for which the DM wants to be notified whenever they are to be changed. After each NL input to the KB has been accepted, the KB tells the DM about any updates of the kinds the DM has initially specified. Whenever the user finishes a graphical manipulation the DM receives a GRL expression which corresponds to the one describing an equivalent NL input sentence giving the same information. The DM then transforms this into a KB update of identical form, and consequently the KB itself does not have to distinguish between NL and graphical updates. For efficient and effective updating, the DM itself stores all visualised information and updates this local knowledge-base whenever communicating with the graphical part of the ACORD system (the GS).

5.2. The Mapping Between Application and Graphics

In the current version of the ACORD system, the contents of the mapping between application and graphics was decided upon by the designer of the DM. That is to say, the designer created a set of rules describing what to present, how and when. Our intention is that in future versions, after an initialisation of the mapping by the designer (in order not to overload the user with presentation work), the user will be able to change the presentation and graphical interpretation behaviour of the DM by instructing it himself (using NL, probably with graphical deixis) as to how to present particular kinds of information, and in what circumstances. This requires a special selector and interpreter for semantic representations of the user's NL (plus deictic) input, before those data are passed along to the KB. Whatever is selected as display instructions by the DM would not be passed on since it would have no impact on stored information about the application domain. Rather, it has to be transformed into additional rules for the part of the DM that controls visualisation and the interpretation of graphical manipulations.

At present, the visualising component maps application and graphics as follows (*cf.* Figure 1):

Vehicles: For each concrete vehicle, i.e. each vehicle mentioned in the NL dialogue, a corresponding movable graphical object is created. Its shape (the icon used) is determined by the type of the vehicle (e.g. as soon as it is known to be a tanker a tanker shape is used, otherwise it is the shape of a truck). Its position is determined by the location of the vehicle, which has to be by a city or on a road. As soon as the vehicle is in a certain city, its graphical presentation is moved to the corresponding graphical city.

Cities: The graphical positions of the cities are determined by the geographical locations of the cities. The graphical shape of a city is determined by whether that city has a depot or not.

Roads: Roads are presented as connections between cities.

Loads: Loads of vehicles are presented in bar charts. Every presented vehicle has an accompanying chart describing its load. There is one bar for each sort of enumerable goods, e.g. for printers, screens, etc.

Stores: For every city which has a depot the contents of that depot are visualised on a bar chart similar to those used for vehicles.

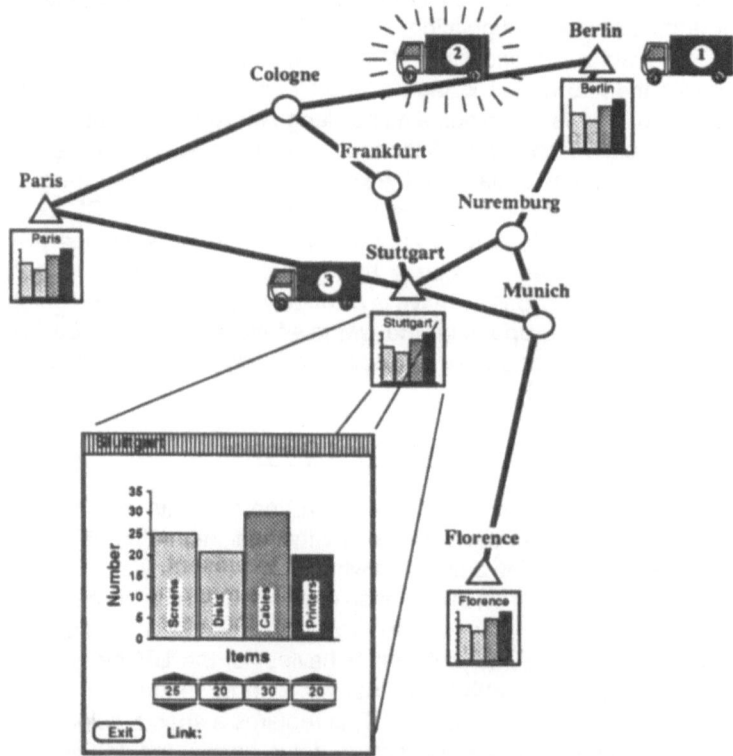

Figure 1. The map used in the ACORD interaction. Icons representing charts for depots can be opened as indicated. Not shown are the charts also associated with each truck.

All application-dependent data of the graphical presentation are stored and maintained by the DM visualising component. Whenever the KB notifies it about changes to those data it checks whether that update is already present in its visualisation base and if not then it collects all data needed for an appropriate graphical update, asks the GS to perform that update and stores the updated data in its visualisation base. In this way considerable redundancy due to the KB change messages can be avoided. (E.g.: the KB notifies two updates: 1. a new vehicle truck78 is known; 2. the location of truck78 is Stuttgart. Processing these update messages one by one causes the visualising component to retrieve from the KB the data which are also given in the second notification, before that notification is processed   it tries to find them as soon as it detects the new vehicle. Processing the second notificn as it detects the new vehicle. Processing the second notification would now be unnecessary because the data are already visualised, so the visualisation base prohibits a second attempt to update the graphical presentation which is already up to date with respect to those data.)

5.3. The Interpretation of Graphical Manipulations

In the current system there is a set of basic actions which the user can perform in order to change the graphical presentation and so to cause a respective change in the KB. For example, the user can see how a bar changes when clicking at an arrow in the bar chart: this manipulation is interpreted as a change in the data for the corresponding truck load or depot content. The other m for the corresponding truck load or depot content. The other manipulation possibility at present is to change the location of a truck by clicking and dragging it to another city.

In the envisaged version with a user-programmable mapping from application knowledge to graphics, user actions on the graphical side should also be changeable in their interpretation, and perhaps even in the way they are performed. Here the user will have the possibility of introducing a new interpretation for a certain action by typing in a natural language command such as *Interpret this action as: 'the vehicle goes to the city'* , which will then be accompanied by a graphical deictic action, being (say) a manipulative action e.g. clicking on the vehicle and dragging it to a city instead of a pointing action as already used for identifying an object. This possibly complex manipulative action will be finished by the user with a predefined 'ending action' telling the GS where to stop collecting the user's actions for the described manipulative action. The interpreter in the DM then will have to resolve *this action*, in the sentence, to the complex manipulative action described. It will have to identify *the vehicle* with the vehicle which is part of the manipulative action and to create a variable which will only be restricted to stand for an arbitrary vehicle not, e.g., for truck25 which is the vehicle represented by the one actually used in the action. The same identification action will have to take place for *the city*.

The GS will thus have to describe the user's complete action as a sequence of basic user actions which form the complex manipulative action. Then the DM will have to create the new interpretation for that action sequence. If the action sequence is new, i.e. there is no interpretation defined before the user introduces the new one, which means that the graphics component does not know how to react to that sequence, the DM will have to ask the user for appropriate graphical actions (in terms of predefined basic graphical actions such as echoing a ghost image) and tell the graphics component about the new complex user action and the reactions needed.

6. An Architecture for Realising the Strategy

We need a number of structural elements to realise the described functionality. These are depicted in Figure 2.

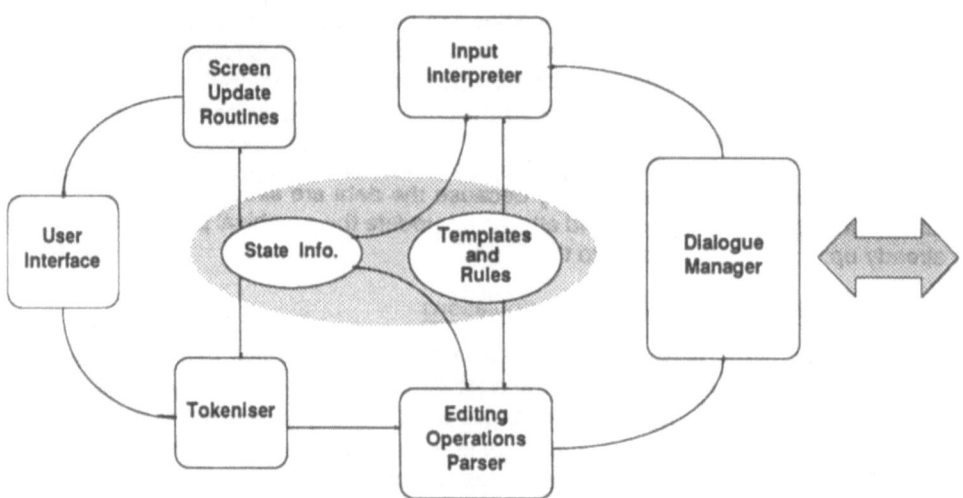

Figure 2. The architecture of the ACORD graphics system. Processing modules surround the information storage area.

At the bottom level, the GS provides a primitive set of graphical objects and editing operations. Anything drawn on the screen must be composed of those objects, and can be manipulated only by using compositions of those operations. At the stage of defining the semantic mapping, the DM can create specific such compositions of objects as *templates* , and for each template it will be specified what type of thing it depicts. Editing operations will similarly be composed into *rules*, relative to the templates, so that only certain actions may be specified for each one, and each rule will have a specified interpretation. These are then available for use in handling dialogue GRL expressions.

The GS can be thought of as five related areas:

1. The low-level graphics routines for updating the screen and returning 'tokens' representing user actions: tokeniser and screen update routines in the drawing. A token might be *button_down(Button, Obj, Loc)* representing button *Button* (e.g. 'left') on the mouse being pressed when the cursor is on object *Obj* at location *Loc*.
2. The individual object descriptions supplied by the DM. These provide information about the behaviour of an object, such as whether it can be moved.
3. The rules that mediate between the user's actions, GRL from the DM and the screen display. Some rules might be :

    rule 1:
    grab_object([source(Location)], Object) <—>
        button_down(left, Object, Location),
        constraint Location \= lost_zone,

```
        constraint movable(Object),
        drag_out(Object),
        action ghost_follows_cursor(Object).

    rule 2:
    grab_object([], Object) <—>
        button_down(left, Object, lost_zone),
        drag_out(Object),
        action ghost_follows_cursor(Object).

    rule 3:
    move([move(Object), Source, dest(Dest)]) <—>
        grab_object(Source, Object),
        drag_in(Dest),
        button_up(_, _, Dest),
        action move_object(Object, Dest),
        action exorcise_ghost(Object).
```

The rule bodies are composed of four types of goals: constraints, actions, tokens and references to other rules. Rules are used in two different ways by the Editing Operations Parser (EOP) and the GRL interpreter. The EOP uses the rules to produce screen changes from user actions and generates GRL as a by-product. The GRL interpreter produces a derivation for the given GRL expression, and executes the associated actions to produce screen changes.

4. The Editing Operations Parser (EOP) attempts to match a sequence of (tokenised) user actions against its rules by breadth-first search. For example with the sequence:

```
    button_down(left, truck42, paris), drag_out(truck42), ...
```

it would try to match with action. The first 'goal' in rule 3 is *grab_object* which is defined by rules 1 and 2. Rules 1 and 2 are both considered and the matching continues with their bodies. The first 'goal' in each body is *button_down* but only rule 1 matches and then the constraints following *button_down* are tested. Subsequently the second token arrives and the parse progresses, executing the action *ghost_follows_cursor* and returning to rule 3. The rule matching is done breadth-first because actions must be performed before all tokens for a derivation have been seen. In this way an abstraction of the user's actions are assigned an interpretation in terms of screen changes produced by the actions. The instantiated GRL in the head of each rule of the derivation is passed to the DM to communicate the changes made.

5. The GRL interpreter also searches the rules but it attempts to find a derivation for a given GRL formula. The tokens and constraints in the rules are ignored, only the GRL and the actions are used. Thus for the (simplified) GRL :

```
    [ move(truck42), source(stuttgart), dest(paris) ]
```

the derivation would use rules 1 and 3, with actions

```
    ghost_follows_cursor(truck42)
    move_object(truck42, paris)
```

exorcise_ghost(truck42)

having the effect of moving truck42 to Paris.

6.1. An Example

Given a GRL expression, the GRL Interpreter has to determine its effect relative to the current State Info. Suppose that we are at the stage where a map is on the screen, showing at least Paris and Stuttgart connected by a road. In this case, the State Info will contain instances of the template for *city*, named 'Paris' and 'Stuttgart', and an instance of the template for *road*, defined to link the two cities.

If an expression now states that truck42 goes from Paris to Stuttgart, the GRL Interpreter must first of all see whether or not truck42 is currently depicted. If it is not, then an instance of the template for *truck* must be created, named 'truck42', and entered into the State Info. The screen shows the map, but also an area where objects are placed that have no known location the 'Lost Zone'. Truck42 is initially a candidate for placement here. However, the movement of a truck from A to B is defined by one of the rules, as involving the truck first appearing at A then moving along to B; accordingly, the object (e.g. a truck-icon) depicting truck42 is placed on the node depicting Paris then moved, in a manner specified by the rule, along to the node depicting Stuttgart. For trucks, the rule may state that they move only along roads. In all these cases, the Screen Update routines must know how to realise the State Info on the screen at any instant, and may have built-in subroutines to execute certain actions of short duration.

If the user now depresses the mouse button with the cursor over the truck-icon, and moves it outside the boundary of the object, a couple of object-relative interaction events will be identified by the Tokeniser. Specifically, there is a button-down on the truck, followed by a dragging movement out of the truck's area. The EOP will interpret these as 'picking up' the truck, and start to move a ghost image of it with the cursor. Suppose the user moves the cursor along (sufficiently close to) the road object, until it reaches the Stuttgart node, and then releases the button. This tokenised sequence of actions is parsed until it is found to match a rule. In this case, it matches the same rule (move object from A to B) as was used in interpreting the incoming GRL expression described above. This rule can now be used in reverse to construct an analogous GRL expression stating that truck42 goes from Paris to Stuttgart. The expression is sent to the DM, which can treat it very much as though it were the result of parsing an NL sentence. At the same time, the State Info is updated appropriately (by deleting the truck at Paris and replacing it at Stuttgart) and the screen display is brought into equilibrium. Should the user infringe the constraints in the rule e.g. by not moving the cursor along a road then if no other rule can be matched the parse will fail; the screen will be restored to its original state and no GRL message forwarded to the DM.

Even such a simple example can give some feeling of the power of this approach. Clearly the rules and templates may eventually become extremely complex. In the ACORD Demonstrator System we augment the map with pop-up windows at the nodes, containing charts depicting various kinds of free storage capacity, etc. These charts can be attached both to depots (at nodes) and to trucks, and linked when a truck is at a depot so that, for example, a truck can be loaded from a depot by graphical interaction alone but the resulting KB state can be queried in NL, or further updated with a consequent change in the depicted charts (see Figure 3).

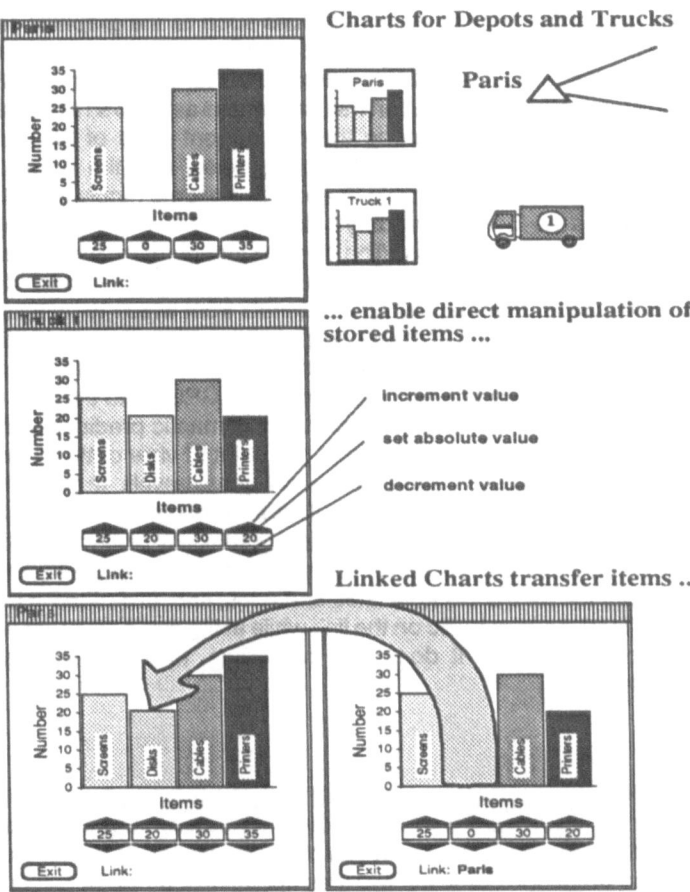

Figure 3. Update of the KB by manipulation of charts. The result can be queried or further updated using NL if desired.

(To preserve the efficiency advantage in graphics of packing data into tables, and the like, some of these features employ GRL expressions with special predicates that diverge from what one might expect to find in the corresponding InL expressions used in processing NL; but the principle remains the same.) Taken together, these rules and templates form a complete specification for the graphical interface provided by the system, given a suitable understanding of the primitive objects and actions they invoke. Indeed, the formalism in which they are couched can be regarded as an interface specification language.

7. Implementation

The overall ACORD architecture exploits the multi-processing facilities of UNIX by locating the major components of the system in separate processes. Most of these are instances of the C-Prolog interpreter with a few specialised, dynamically-loaded C functions to handle low-level operations or specialised activities such as the graphics screen operations and window-manager connection. Central interprocess communication is multiplexed by the DM, and the graphics system controls a further process that handles all screen I/O. In such an architecture, the question naturally arises as to how responsibility for various functions should be assigned among the component parts of the system. Issues concerning the weight of traffic on the connecting pipes, and the access one process has or doesn't have to information stored in another, become highly important.

Our current approach places the logical boundary between the GS as described above and the DM somewhere within the DM process. The domain-independent templates and rules, in particular, and parts of the GRL interpreter, can be regarded as being located in the DM. This means, for instance, that the abstract materials from which templates and rules are constructed in order to set up the semantic mapping, are directly available to the DMthe DM routines, and easier to integrate with the information from the KB that they must be mapped onto. It also allows for a certain simplification of the GRL fed to the GS. GRL, because of its relationship to InL, can in general contain any semantic predicates known anywhere in the system, e.g. in the NL lexica; but from the point of view of the GS many of these may fall into equivalence classes. Whether truck42 goes, travels, moves, drives, etc., from Paris to Stuttgart, may make no difference to its graphical representation. Such 'synonymy' can be filtered out in the DM, so that only a few graphically distinct predicates need to be recognised in the GS itself. Thus, the DM can in principle construct a complete domain-dependent configuration for the GS templates and rules, and 'download' it before interaction begins. This will reduce later traffic on the link while at the same time preserving the logical prerogatives of the design; it does not, of course, preclude subsequent alteration of the configuration.

8. Conclusion

In examining the conditions for graphical communication (especially, but not exclusively, in the context of integration with NL), we have shown the crucial importance of the semantic basis for interaction. Typical current systems do not respect this consideration. Rarely does one find a case where the complete two-way update link between screen and represented data or knowledge is supported at all, and even when it is it is usually in an *ad hoc*, hard-wired fashion.

A futuristic vision is of a system integrating graphics with NL, the latter using speech input and output. The user would be able to extract information both graphically and verbally, and himself inform the system using drawings of his own design for which he concurrently specified the method of interpretation. This vision is of course still a very distant one (and somewhat indistinct). Parts of it have been ambitiously worked upon already, especially the deictic links to NL (Bolt 1980, Hanne *et al.* 1986). Relatively little has been done on the integration of this work with the use of interactive alteration and construction of pictures for querying and update, but a start has now been made, and we feel that we have provided some useful pointers in the direction of its eventual realisation.

Acknowledgements

This work is made possible by the constant collaboration and encouragement of our ACORD colleagues. In NL: the Edinburgh Centre for Cognitive Science; the Institut für Maschinelle Sprachverarbeitung at the University of Stuttgart; the Section de Linguistique at the University of Clermont-Ferrand; the Laboratoires de Marcoussis. In Knowledge

Representation and Inferencing: TA (Triumph-Adler) Research, Nürnberg; BULL-SA, Louveciennes. From EdCAAD itself, we wish to acknowledge the contributions of Chris Tweed, Aart Bijl and Peter Szalapaj; and from the Fraunhofer Gesellschaft IAO the contribution of Toine van Hoof.

References

Bolt, R A, "'Put-That-There': Voice and Gesture at the Graphics Interface", *Computer Graphics* (ACM SIGGRAPH), Vol. 14, (3) pp. 262-270 (1980).

Gnanamgari, S, "Information Presentation through Automatic Graphic Displays", pp. 433-445 in *Computer Graphics 81*, Online (1981).

Hanne, K-H, K-P Faehnrich, and J Hoepelman, "Combined Graphics-Natural Language Interfaces to Knowledge Based Systems", in *AI-Europa Conference*, TCM-Publications, Liphook (1986).

Klein, Ewan, "Dialogues with Language, Graphics and Logic", pp. 867-873 in *ESPRIT '87: Achievements and Impact*, ed. CEC - DG XIII, North-Holland, Amsterdam (1987).

Mackinlay, Jock, "Automating the Design of Graphical Presentations of Relational Information", *ACM Transactions on Graphics*, Vol. 5, (2) pp. 110-141 (1986).

Pineda, L A, E Klein, and J R Lee, "GRAFLOG: Understanding Drawings through Natural Language", *Computer Graphics Forum*, Vol. 7, (2)(1988).

Pineda, L A, "A Compositional Semantics for Graphics", in *Proc. Eurographics '88*, North-Holland (1988).

ASPIS: A KNOWLEDGE-BASED APPROACH TO SYSTEMS DEVELOPMENT

M.J. ASLETT
GEC-Marconi Software Systems
Elstree Way, Borehamwood
Hertfordshire WD6 1RX
England

D. MELLGREN, Y.F. YAN
CAP SESA Innovation
33, Chemin du Vieux Chêne
38240 Meylan
France

F. PIETRI
Tecsiel SpA
Via Santa Maria, 19
56100 Pisa
Italy

Abstract.

The ASPIS project is building tools to support the early stages of the software life-cycle. The tools are based on artificial intelligence techniques and exploit knowledge about the development methodology as well as the application domain. This paper describes how this knowledge is used by the tools developed within the project.

1. Introduction

Because systems development is a knowledge-intensive task [Barstow87], CASE tools incorporating knowledge should yield significant gains in productivity. The main objective of the ASPIS project is to support the specification and design phases of the application life-cycle through the use of knowledge-based techniques. Two types of knowledge have been captured:

- *method knowledge*: knowledge about the methodology followed by the CASE tools.
- *domain knowledge*: knowledge about the application domain, which is critical to software developers [Adelson85].

The ASPIS tools follow the *assistant* philosophy. The knowledge-based tools provide a new form of support to users. The basic approach is that an assistant monitors the progress which a user is making using standard editing tools for the chosen representations. It is then able to make suggestions and perform checks on the current state of development using its knowledge. However the user is still in firm control of the overall development: any advice generated may be ignored.

There are currently four assistants within the ASPIS environment. The Analysis and Design Assistants are the main development assistants, supported by the Reuse and Prototyping Assistants.

Using its method and domain knowledge, the **Analysis Assistant** helps the user understand the problem and specify it correctly. The method used is based on Structured Analysis (SA) [Ross77] and the Entity-Relationship (ER) model [Chen76]. The requirement specification resulting from the analysis consists of SA diagrams describing the functions of the system, ER schemata representing the data, and structured texts representing non-functional constraints such as cost, response time, reliability or size.

The **Design Assistant** helps the user find a solution to the specified problem. It reads the requirements produced by the Analysis Assistant and helps the designer transform them into initial software and hardware architectures described in a design language. Domain knowledge is used in this step. The generated design documents are then transformed and refined through editing. The Design Assistant communicates with the Analysis Assistant when it finds incompleteness or incompatibility in the requirements. It also communicates with the Reuse Assistant to find reusable components.

The **Reuse Assistant** maintains a database of reusable analysis and design components, and helps the user in retrieving them. It observes the activities of the Analysis and Design Assistants in a passive manner to reduce the search space and uses the semantics and the structure of components to make the retrieval more efficient and precise. When a user decides that a retrieved reusable component is applicable, it is automatically inserted into the current document. It can then be modified and thus partial reuse is supported.

The **Prototyping Assistant** offers facilities to translate SADT diagrams into a logic-based formal specifications language (Reasoning Support Logic) which can be interpreted for animation. The result of animation can be printed as text or shown graphically by highlighting the data paths on the SADT diagrams. The Prototyping Assistant allows both the user and the analyst to validate the application specifications before design.

More information on all of the assistants can be found in [Hughes88], [Puncello88], and [Pietri87]. In particular, the architecture of the ASPIS prototype is discussed in section 5 of [Hughes88].

This paper first describes the analysis and design methods, and how knowledge about the method is used by the assistants. Then, it explains how domain knowledge is exploited within the context of the methods. Next, our knowledge acquisition tool is described. Finally there is a discussion of the knowledge-based approach to systems development.

2. Methods and Methodical Knowledge

In the ASPIS prototype, it has been necessary to limit the scope of methods being supported. Thus, the Analysis Assistant follows a single analysis method, and the Design Assistant a single design method. However, we have studied other methods, and we have remained quite generic in our approach. Future exploitation of ASPIS will show the same techniques and tools being applied under new methods.

2.1 Analysis Method

The current analysis method supported by ASPIS is based on two different languages: Structured Analysis (SA) for the functional specification [Ross77] and Entity Relationship (ER) for the conceptual data specification [Chen76]. Moreover, a set of Non Functional Requirements are captured using an ad-hoc formalism.

The method is composed of several stages to be performed in parallel from different points of view. This idea, derived from [Feather89], allows the complexity of the system specification to be broken down and the analysis to be carried out from different perspectives thus leading to a deeper insight. Besides, as a number of descriptions are produced for common aspects, consistency and completeness checks can be performed on the intermediate results.

The stages of the method comprise a Preliminary System Description and Viewpoint Detection, the Functional and Data analyses, using the SA and ER languages, from each viewpoint, the Viewpoint Specification i.e. the comparison of the SA and ER system descriptions, and, finally, the System Specification to integrate the various viewpoint results and obtain the final analysis documents that will be exploited by the designer.

At present, the Analysis Assistant (AA) supports the analyst in the particular case of one viewpoint. Both Functional and Data analyses can be carried out under the control of the AA which provides guidance and help on the criteria and rules to be observed. The Viewpoint Specification phase, is also supported by the AA to help the analyst compare the system data descriptions given in SA and ER. Through the exploitation of heuristics to derive information from the structure of the data included in the SA and ER documents, possible inconsistencies can be found by the AA between the two descriptions.

2.2 Design Method

The Design Assistant supports high-level hardware design as well as global and detailed software design. The design process has been defined in order to allow the linking of the analysis and design information in a partly automatic manner and to support the development of real-time systems. Starting from the functional and non-functional requirements produced during the analysis phase, the Design Assistant allows the user to

- select a hardware configuration that will support the software. The generated architecture is a collection of interlinked components described in HDL, a Hardware Description Language developed by the project.
- transform the requirements into a software architecture. The software components are described in a language called AMPHI (Abstract Machine and Process Hierarchies). Processes communicate via events and are grouped together into modules. Machines provide services and manage resources (data). The whole approach is based around abstract machines as in MACH [Galinier85], but with the addition of real-time facilities. Software metrics such as complexity and modularity measures can be used to help produce better designs.
- map the software onto the hardware, for instance it is possible to specify on which processor(s) a given process should run. This also allows the assistant to verify that the hardware configuration is consistent with the software.
- describe in detail all processes and operations in machines with pseudocode.

We also recognise that specification and design are not sequential processes in the real world. A protocol has been defined for communicating changes between analysts and designers. This is done using objects known as *marks*. So far we have defined four different types of marks. *Inconsistent* and *incomplete* marks describe problems with analysis documents. *Change* marks record the changes in analysis documents which have been made in order to solve the problems described in an *inconsistent* or *incomplete* mark. *Used* marks are set on the analysis documents from which design documents are derived.

2.3 Method Knowledge

The ASPIS development assistants contain knowledge about the methods that they support, in the following ways:

– *Steps*: The assistants know what the results of each analysis or design step should be and keep track of whether or not the step is finished. As subsequent steps typically depend on the results of earlier ones, they know which steps can be accomplished at any given moment. For instance, to perform the software-hardware mapping step of design, both hardware and software architectures must exist.

– *Languages*: The assistants know about the formalisms and languages used:

- In analysis, SA and ER diagrams

- In design, the AMPHI and HDL languages

Thus, the editors provided by the assistants can provide help that go beyond a purely syntactical level. Knowledge about the semantics of the languages is also useful in the implemented consistency and quality checks on the analysis or design documents. Syntactic correctness clearly does not guarantee a good specification or design.

3. Domain Knowledge

Knowledge related to an application domain (domain knowledge) is at the heart of the ASPIS philosophy of systems development. Indeed studies of the software design process have identified this as being a critical area in which tool support is required [Curtis88].

In order to capture knowledge of this form, we had to select a domain in which to work. The domain we chose was physical access-control systems. These systems regulate access to a site via the use of monitoring devices such as card readers and alarm sensors. This domain was chosen because

(a) it is a real-life example and therefore easy to visualise for non real-time experts.

(b) experts were available in the specification and design of this type of system.

(c) access-control systems are a type of embedded system which also include some data processing aspects.

(d) elements of complexity can be added as required, such as a complex man- machine interface or the selection and identification of hardware.

During many knowledge acquisition sessions in this domain we extracted and classified a large amount of knowledge. This domain knowledge relates to:

– Functional and data decomposition,
– Capturing design constraints,
– General hardware design,
– General software design,

– Detailed software design.

The rest of this section describes how domain knowledge is used and represented for functional and data decomposition and general hardware design.

3.1 Domain Knowledge for Functional and Data Decomposition

3.1.1 Scope

Having performed knowledge acquisition sessions for the analysis of requirements in our example domain, the following forms of knowledge were identified as being useful for supporting this process.

– Refinements of functions,
– Refinements of data,
– Relationships and criteria about functions and data.

The current prototype of the Analysis Assistant exploits all of the above sources of knowledge. Thus, a developer can use the domain knowledge to create a new functional specification in the current application domain, e.g. describe a function in more detail by using the alternative refinements stored in the domain knowledge. Appropriate selections may be made by using the comments associated with the overall refinements and the individual functions included in each refinement.

This use of domain knowledge is supported in a very flexible way. The developer is free to use only those items of domain knowledge which he/she thinks are suitable in the given context. The results from the domain knowledge may then be modified as desired.

In addition to using Domain Knowledge to provide suggestions on how to complete or build analysis documents, the Analysis Assistant exploits the concepts included in the knowledge base to monitor the user's intermediate results. Such observations are provided whenever the user's documents are not consistent with the knowledge coded in the Assistant. As for suggestions, the warnings produced may or may not be taken into account by the user; they just aim at advising the developer about the canonical contents of systems in the selected application field.

Finally, Domain Knowledge is used by the Classification Mechanism of the Analysis Assistant. Whenever the assistant is asked for suggestions about components of a document that are not known by the system (e.g. an SA box with a certain label added with the editor provided by the environment), the Classification Mechanism can be called upon to attempt an association of the components with the concepts known by the system. If the classification succeeds, the assistant will behave as if the suggestions were asked for the associated concepts. This process is currently driven by two principles: Comparison of the components' labels against a synonym dictionary and the exploitation of the domain concepts and relationships used for the suggestions as patterns to recognise typical situations.

3.1.2 Representation

Templates are defined for all of the objects which provide application-specific support. These templates are specified using an object-oriented Knowledge Representation System (KRS) developed within the project. Descriptions of this KRS may be found in [Pietri87] or [Hughes88].

For each of the templates, there are general routines which interpret the declarative representation and compare it with the current development. Thus any advice which is produced is consistent with the current state of the specification. Templates are provided for the following forms of objects: functions; refinements of functions; data; refinements of data; connections between functions and data; and domain-specific criteria. These are described below.

A template for a function contains the function's name and a textual description of what it does. A template for a refinement of a function contains the parent and child functions, a textual description of the overall refinement and an integer which represents the applicability of the refinement (according to a domain expert). The higher the value of this integer, the more likely an expert thinks the refinement is applicable in the current domain. The templates for data and data refinement are structured similarly.

Example: a possible refinement of the function 'administer security':

```
ref1_adm_sec : function_refinement
     suggest_rf: 'A possible refinement of ADMINISTER SECURITY is the following:'
     heuristic_rf: 1
     list_rf: ['classify people','classify places','classify alarms',
               'analyse log_of events','classify report formats']
```

A template for a connection between a function and a data item contains the names of the connected items along with the form of connection which can be input, control or output. These connections are ones that are very likely to appear in all specifications which include those data and function. For instance, the function 'administer system' usually maintains the data 'security rules'. Thus in the domain knowledge this connection would be recorded as 'administer system, outputs, security rules'.

Templates for domain checks consist of constraints which should be satisfied in the current application domain, along with descriptions of the consequences if they are not. For each check there are general routines which display these consequences if the constraints are not satisfied in the specification. Currently the following forms of checks are supported:

– *External Interfaces*. A template for this check contains an external data flow along with a description of the effects if it is not included in the specification.

– *Data Completeness*. Contains groups of data which should all be present together along with a description of the effects if they are not all present in the specification.

– *Data Flow*. Contains pairs of functions which should be connected together along with a description of the effects if there is no data flow connecting the functions in the specification.

Example: data flow check. If there is no data flow from the function 'administer system' to the function 'check access' then the security controllers can not modify the way in which access requests are validated.

3.2 Domain Knowledge for General Hardware Design

3.2.1 Scope

From knowledge acquisition sessions, we have developed a general hardware design process which is a synthesis of the various ad-hoc methods that were used by the experts we studied. This process is supported by a large amount of knowledge in the access-control domain. Although our actual general hardware design process is clearly a methodical process, it has resulted from domain specific studies. Thus an outline of this method is presented here, especially as there do not appear to be any standard methods in this area. It is also necessary to describe this design process in order to explain how domain knowledge is used.

Our general hardware design process starts by considering the non-functional design constraints in order to select a hardware configuration which satisfies all, or the most critical ones. The constraints have been specified by the Analysis Assistant.

The basic strategy for selecting a hardware configuration is to eliminate known configurations until only one is left. However as well as eliminating configurations, enhancements can be made to improve their rating with respect to a given constraint. The problem, as we shall see, is that every enhancement also has some drawbacks. These are presented to the user before allowing him/her to accept the improvement.

As a starting point for selecting a basic configuration, we determine whether the size of the system is large, medium or small. This is done using some classification rules related to the number of components in the system. Then we determine its distribution according to the maximum distance between components, which is a measure of the physical size of the system. This is again done using classification rules. The limiting values in the classification rules have been identified through knowledge acquisition sessions. As they affect the subsequent design process, we have ensured that it is easy to change their values.

Having classified the physical aspects of the system, we use previously developed rules to perform a first screening of some basic configurations, i.e. given the physical facts about the system, which of the basic configurations are most likely to be applicable? So far we have identified seven basic hardware configurations. These range from direct communication to a single processor to multiple processors connected via a network.

The result of this process is a shortened list of basic configurations. If there are several completely autonomous sites, each one of them should be considered as a subsystem and configured according to the strategy described below.

Next, we aim to rule out as many as possible of the remaining basic configurations by applying the design constraints in their order of priority (as stated in the specification). This is when domain knowledge is used in earnest. For each design constraint, domain knowledge is used to:

- display general comments related to the current constraint for every configuration which is still under consideration.
- identify improvements to each configuration for the current constraint. However as well as presenting the improvement, any drawbacks are presented, such as the additional cost.
- identify configurations which are definitely worse than the other configurations from the point of view of the current design constraint. Reasons are given as to why this is the case, but it is up to the user whether to accept or reject this advice.

– identify pairs of configurations which have conflicting advantages with respect to the current constraint. The relative advantages of both of the configurations are displayed and it is then up to the user whether to eliminate none, one or both of the configurations.

There are also domain-dependent scoring features. These present an overall desired value for each design constraint, based on the actual values in the specification, and calculated values for the chosen configuration. They are not essential to the system's reasoning, but are likely to be useful in practice. These ratings could help the user make up his or her mind when there are still several possibilities at the end of this process. They are also used to give the user an idea of how well the requirements are satisfied.

The current prototype of the Design Assistant supports all of the features which have been described in this section. As in the Analysis Assistant, a flexible approach has been adopted, in that all suggestions generated from the domain knowledge have to be accepted by the designer before being included in the current configuration. Once this stage is complete, the designer can freely edit the hardware architecture.

Currently the domain knowledge present in the Design Assistant covers the following forms of non-functional design constraints: size; physical distribution; cost; ergonomics; availability; reliability; security; response time; throughput; compatibility and safety. Although the knowledge is at a fairly high level of granularity, it has already shown a tremendous potential for this form of support in systems development. For example, the Design Assistant presents a number of possible solutions and then explains the advantages and disadvantages of the various alternatives. This is in contrast to many designers who seem to home in very quickly on one solution (often similar to their previous project!) without giving full consideration to alternative solutions.

3.2.2 Representation

Domain knowledge is represented by several types of relationships between the constraints and the basic configurations or enhancements. The form of each of these relationships is described below.

The 'general comments' relationship contains comments on the advantages and disadvantages of a certain configuration under the light of a given design constraint. An experienced designer may not see the value of these general comments, but they can provide useful guidance to a less experienced designer. Example: for a multiprocessor configuration,

conf3 : basic_configuration

    ...
    [security,
    "Security is distributed around each processor. There are more devices to protect",
    "than with a single processor, but less information is available at each one."]

The 'enhancements' relationship represents possible improvements to the basic configurations with respect to given design constraints. For instance, availability can be improved by adding a back-up processor. The explanation is obvious. On the other hand, the cost of this enhancement may be prohibitive:

```
imp13 : enhancement

      ...
      description: "Backup processor"
      implications: [[cost,
             negative,
             "The cost of the extra processor may be high"],
             [availability,
             positive,
             "It would be very unlikely that both processors go down at the
             same time."]]
```

This example shows that awareness of the drawbacks of any choice are essential. These drawbacks formally represent the inter-dependence of the different constraints. Thus they are recorded in relationships between the configurations and the possible improvements (the 'drawbacks' relationship).

The basic comparison relationship between two configurations from the point of view of a given constraint is called 'worse than'. This relation contains the reasons why one configuration is considered worse than another one from a certain design constraint. It is possible that there are factors against both configurations. The only restriction we impose is that all of the factors stating that A is worse than B are together. An example is that the cabling costs are lower when using multiplexers than with direct communication to a processor.

As for the score of a configuration under a given set of non-functional constraints, it is the sum of the ratings for each constraint, weighted by their importance factor. The scores are represented internally by integers. This allows us to give an absolute rating to each configuration, whereas the rest of the system generally uses a system of relative comparisons. In order to improve the clarity of the explanations for the user, there is a mapping from these integers to terms appropriate for each design constraint. For instance, a numerical score of 3 for reliability might be translated as "high". The scores for each design constraint are determined taking into account the basic configuration and any enhancements that are relevant to that constraint.

4. Domain Knowledge Editor

The process of knowledge acquisition is the traditional bottleneck in the development of knowledge-based systems. Indeed we experienced difficulties during our knowledge acquisition sessions. In particular, domain experts found it difficult to understand the knowledge after it had been encoded in our Knowledge Representation System. Therefore we developed a knowledge editor which allows an expert to input the knowledge directly.

Currently the knowledge editor allows an expert to define the functional and data decomposition (including checks) of a standard application, without having to know anything about their internal representation. Each item is described using terms defined by the expert, i.e. the ones the expert is familiar with.

This editor also allows experts to classify items into domains. The domains themselves are constructed in terms of hierarchies. Each domain is a specialisation of its parent domain. In this way the domain knowledge consists of a tree of domains where the root is the most general and the terminal nodes are the most specialised. If a domain has more than one parent then it contains specialised concepts of all its parents. In this way, a developer can decide on what level of domain support is required for the current application, by selecting the appropriate node in this hierarchy.

This tool has proved so successful that we plan to extend it to cover all areas of domain knowledge. For example, in acquiring knowledge for general hardware design, the knowledge editor will allow a user to add, delete or edit the various forms of relationships. It will set back-pointers automatically. It will also verify the transitivity of the comparisons between the basic configurations (worse-than relation).

5. Discussion

The ASPIS assistants only cover the early stages of the software life-cycle (up to detailed design). However, they could be combined with automatic programming techniques to cover the implementation phase as well. A well-known example is the MIT's Programmer's Apprentice [Waters82], capable of transforming *cliches* (designs) into Ada code. In ASPIS, the detailed design produced with the Design Assistant could be transformed into code, expressed in a programming language, in a similar fashion.

We believe that ASPIS has demonstrated that knowledge-based techniques provide an effective means of tackling the task of systems development. There are many advantages to the application of domain knowledge. It enables less experienced users to learn how the domain experts work by using the knowledge. At the same time, it allows the experts to make better use of their time so that they can consider new problems which always arise in the course of developing systems. The ASPIS assistants provide help that goes beyond the purely syntactic level (as seen, for instance, in syntax-oriented editors). Because of their method and domain knowledge, they are able to grasp the semantics of the documents on which the developer is working.

Effective tools with domain knowledge mean productivity gains. In order to evaluate the practical use of the ASPIS prototype, a software engineer, having no experience with either the ASPIS method or access-control systems, used the assistants to perform the analysis and design of an access-control system. He accomplished the analysis and design of a large system within a month. The translation from requirements into design documents, the domain- dependent suggestions, and the intelligent graphical editing were found to be of significant value.

The drawback of our ability to tailor the environment to specific needs is the initial cost of acquiring and incorporating knowledge. As seen in the previous section, ASPIS incorporates a knowledge editor that can help with this task. Still, the expertise must be available at least initially. The initial cost of adapting ASPIS has to be recovered. We believe that this can be done provided that a sufficient number of applications are to be built within the same domain and using the same analysis and design methods.

From a company's point of view, these domain knowledge bases will become commercially valuable commodities. To some extent they will free them from the problems of losing valuable personnel, as part of the knowledge will remain with the company. For the foreseeable future we can not envisage replacing human developers. However, we believe that a new generation of intelligent CASE tools will contribute substantially to resolving the "software crisis".

References

[Adelson85] Adelson, B. and Soloway, E.: "The role of Domain Experience in Software Design", IEEE Transactions on Software Engineering, special issue on Artificial Intelligence and Software Engineering, vol SE-11, Nov. 85.

[Barstow87] D. Barstow, "Artificial Intelligence and Software Engineering", Ninth International Conference on Software Engineering, April 1987.

[Chen76] Chen, P.: "The Entity-Relationship Model: Toward a Unified View of Data", ACM Transactions on Data Base Systems, vol. 1, no. 1, Mar. 76.

[Curtis88] B. Curtis, H. Krasner, N. Iscoe, "A field study of the software design process for large systems", Communications of the ACM, Nov. 1988.

[Feather89] M. S. Feather, "Constructing Specifications by Combining Parallel Elaborations", IEEE Transaction on Software Engineering, Vol. 15, No. 2, pp. 198-208, February 1989

[Galinier, 85] Galinier, M. and Mathis, A.: "Guide du concepteur MACH", Thomson-CSF, DSE and IGL Technology, 1985.

[Hughes88] A. Hughes, P. Puncello, F. Pietri, "Intelligent Systems Development in the ASPIS Environment", ESPRIT 88 Putting the technology to use, North-Holland 1988.

[Pietri87] F. Pietri et al., "ASPIS: A knowledge-based environment for software development", ESPRIT 87 Achievements and Impact, North-Holland 1987.

[Puncello88] P. Puncello et al., "ASPIS: A Knowledge Based CASE Environment", IEEE Software, March 1988.

[Ross77] D. Ross, "Structured Analysis (SA): A language for communicating ideas", IEEE Trans. on SE, SE-3 Jan. 1977.

[Waters, 82] Waters, Richard: "The programmer's apprentice: knowledge-based program editing", IEEE Transactions on Software Engineering, no. 1, 1982.

Project No. 415.

Parallel Computers for Advanced Information Processing:
The achievements of ESPRIT Project 415

Eddy A.M. Odijk [1]
Philips Research Laboratories
P.O. Box 80.000
5600 JA Eindhoven
The Netherlands

ABSTRACT. The employment of parallelism to extend the performance range of computer systems and the proper programming methods and languages to exploit this parallelism are considered to be key technologies for the IT industry of the '90s. Project 415, entitled "Parallel Architectures and Languages for Advanced Information Processing- a VLSI directed approach", has investigated a number of approaches towards high performance computer systems in particular for symbolic applications. Each of its six subprojects has adopted one programming model, advanced its theory and designed an architecture and language for its execution. Working groups have provided a disciplinary platform and ensured cross fertilisation between the subproject teams.

This paper reports on the aims of the project, the results that have been obtained and the intended way of further exploitation.

1 Introduction

Although hard to imagine in retrospection, parallel computing for other than numeric applications was in its infancy in 1984, when the plans for P415 were created. Numeric algorithms, characterized by regular datastructures and uniform operations on them, could benefit from the parallel execution means of vector computers and shared memory multiprocessor systems, supported by parallelizing compilers and careful tuning by the programmer. The latter category of systems, however, would not scale beyond some 20 - 30 processors because of its inherent communication bottleneck.

The new category of symbolic applications, partly overlapping with the general denominator of Artificial Intelligence, was shown to require programming styles and languages with a better expression power as well as more dynamic and flexible execution mechanisms. Additionally, high level concepts of communication and synchronisation that resulted from research in the 70s, needed to be integrated in such languages to achieve effective programming for and efficient exploitation of parallelism. These new concepts and the desire for a scalability of at least two orders of magnitude of parallelism necessitated the research to have a broad scope: the simultaneous and integral design of parallel architectures and

---

[1] The author has made extensive use of contributions provided by the subproject leaders and working group chairmen

languages for symbolic applications. Representative instances of the latter would proof the expressiveness of the language and the achieved performance of the parallel system. Finally, it was recognized that a better understanding of the theoretical foundations of the parallel programming styles and concurrency in general would have its impact on the quality of language and program design.

The start of Japanese Fifth Generation Computer System (FGCS) program in 1982 urged Europe and the United States to formulate their own large scale research programs. The FGCS program had immediately opted for one approach, based on logic programming. However, a number of programming models was at that time considered potentially feasible but neither the validity or even superiority of one of them had been established.

In 1984, P415 started as the first Esprit project in this area with its charter defined as "Parallel Architectures and Languages for Advanced Information Processing- a VLSI directed approach".

The general objective is to provide the European IT industry with the technology to reduce execution times for a wide range of applications through the use of concurrency, and to do so at a favourable price-performance ratio through proper architectures and exploitation of VLSI technology. Novel languages, incorporating modern programming concepts, reduce the cost of software design.

To reach this objective, the project aims at investigating the feasibility of six distinct programming styles and propose architectures and languages for their exploitation. In its aims, the project covers the understanding of parallelism, from the analysis and formulation of an application, through the semantics of concurrency in programming languages, to the principles upon which concurrent architectures should be based.

The intended results are the demonstrated feasibility of one or more of the six programming styles and designs on which further development towards industrial systems can be based. According to the Technical Annex, simulations would fullfill this purpose. Over and above, in most of the subprojects, real prototypes in hardware and system software have been constructed.

With P415 now being in its final year, this paper presents, the approaches that have been followed and the results that have been obtained. In section 2, the project organisation and partnership will be surveyed. Section 3 describes the technical goals and contents of each of the subprojects. Section 4 provides the project results. The next section shows that a number of plans and successor projects exist to exploit the outcome. Conclusions will be drawn in the final section.

2 Project Organisation

In accordance with the aim of investigating a number of distinct parallel programming styles, six subprojects were defined, each to be executed by one partner and its subcontractors (see Table 1).

At the conception of the project it was evident that the research and subsequent industrialisation of parallel processing would be important to the major European IT industries and need their support to have impact on the European technological innovation. Therefore, the full partnership was reserved to industrial partners with the commitment to execute one of

Sub	partners	subcontractors	Contributions of the partners
A	Philips	CWI, Amsterdam	Object-Oriented machine
		Techn.Univ. Eindhoven	
		Univ. Oldenburg	
	AEG	Techn.Univ. Berlin	VLSI simulator on DOOM
B	GEC	University College,London	lazy functional languages,
		Imperial College,London	reduction machine
C	BULL	LITP,Universite Paris	logic database machine
		ESIEE, Noisy le Grand	
D	CSELT	Universita di Pisa	mixed logic functional mach
E	Nixdorf	Stollmann,Hamburg	dataflow machine
F	Nixdorf	LIFIA-IMAG,Grenoble	connection method
		Techn. Un. Muenchen	logic machine

Table 1: Partners, subcontractors and approaches

the subprojects. These partners also managed the project through its Project Coordination Committee.

As Table 1 shows, many excellent universities and research institutes have entered the project as subcontractors and used their expertise to advance the state of art in theory and system design.

To facilitate a proper platform for the presentation and discussion of mutually important topics and advance the theory of concepts involved, project-wide working groups have been formed in the areas of Architecture and Applications, Semantics, and Proof Theory and Verification.

The project comprises some 280 manyears, 20 % of which were dedicated to the working groups.

3 Technical issues

Before describing the goals and technical directions of the subprojects, the issues at stake are presented in a general manner first.

In a language first approach, the feasibility of the selected programming model forms the hypothesis for each of the approaches. An abstraction of the corresponding implementation is the execution model. Table 2 provides a survey of these and links a number of attributes.

Logic and functional programming models belong to the class of socalled declarative systems while object-oriented programming falls within imperative programming. Declarative languages are considered higher level, more expressive, and concise and allow formal assessment due to their strong mathemathical basis. Uuntil recently however, their efficient implementation has not been well understood.

Two categories of parallelism exist: explicit parallelism, and implicit parallelism where the implementation (compiler and operating system) extracts an amount of parallelism from the program. In the first, e.g. in object-oriented programming, the programmer is responsible for the partitioning of a program into objects. To avoid burdening the programmer, the language should offer high level, "natural", facilities for communication and

sub	language style operational model	Parall.	Architecture
A	Object-Oriented control-flow	Expl	distrib. memory packet switch network
B	functional graph reduction	Impl	emulation on transputer
C	logic(many data) delta-driven	Impl	distrib. memory bus-based
D	logic+functional SLD resolution	Expl	shared memory MIN
E	single assignment dataflow	Impl	shared memory bus-based
F	logic(many rules) connection method	Impl	shared memory bus-based

Table 2: Execution models and attributes for subproject systems

synchronisation. The latter category of parallelism is in principle present in the execution model of declarative languages. Thus, programmers would be able to concentrate on the algorithm per se. The burden of extracting enough parallelism with a coarse enough granularity is put here on the compiler.

The prefered parallel architecture is related to the programming model and the required size and scalability of the parallelism. A general choice is that between shared memory and distributed memory architectures, and (mostly coinciding) between communication and synchronisation through shared variables or message passing. Other choices relate to the amount of hardware support for specific features of the execution model. An important unifying concept in the project is that of the homogeneous parallel machine, which consists of a number of identical so-called nodes. This concept provides for well extensible systems and allows an implementation taking advantage of the relatively low replication costs of VLSI based node realizations.

A common characteristic in the implementation of the various language models is the use of a dynamically varying number of processes, usually more than one per processing element, mostly with dynamically changing patterns of communication and interaction. This characteristic requires an operating system (kernel) to be resident on each of the nodes to manage the various resource at the node and the system level.

Thus, a number of common and related problems are posed, while their solution may differ with the language model. The subprojects have each directed their efforts at solving these problems in the contexts of the specific programming models. It is beyond the scope of this paper to provide the details of the solutions. References are provided to further information.

The project has put an emphasis on education and exchange of scientific results in this area. To this end P415 has organized two summerschools [19,20] and two PARLE conferences [3,4]. The latter proceedings include reports on the P415 subprojects.

4 Goals and results of subprojects and working groups

4.1 A: THE OBJECT-ORIENTED APPROACH

The essence of object-oriented programming is the subdivision of a system into *objects*, i.e. integrated units of data and procedures. Objects communicate by passing messages, which must be interpreted as a request to the object to execute a certain procedure (termed a *method*). To combine object-oriented languages with parallelism, we have chosen to associate with every object a process of its own. The model is intuitively appealing, with message passing as the only facility for communication between objects.

The systems' architecture consists of many, functionally identical and control-flow structured, computers connected via a direct message passing network. The computers contain a central processor, private memory and a dedicated communication unit to perform the message passing without interference of the cpu. The concept allows systems of more than 1000 of such computers.

The major goals of this subproject, derived from the chosen approach, can be stated as follows:

- A Parallel Object-Oriented Language, POOL, in which significant application programs can be programmed. The language provides the user with control of parallelism and granularity. It is important for the language to have clear semantics. Support for the verification of programs is desirable, and becomes even more important in a parallel environment.

- A prototype Decentralized Object-Oriented Machine (DOOM), consisting of some 100 identical self-contained computers, each having a powerful 32 bit processor, local memory and communications means, which are connected in a direct packet switching network. A DOOM node is able to execute many (presumably some 10 to 100) objects. The processor architecture must therefore support multi-processing. Each computer, called a node of the system, has a copy of the operating system kernel. This kernel performs local resource management, and cooperates with the other kernels for global operating system tasks. The prototype DOOM system is connected, as a satellite, to a host computer, where the programming environment resides. The prototype is based on existing technology, and will offer facilities for experimenting and for evaluation of performance aspects.

- Three significant applications in the area of symbolic processing that demonstrate the performance increase through parallelism on DOOM. The first of these is a parallel theorem prover. The second is a parallel version of the analytical component of the Rosetta natural language translation system. Rosetta is currently being designed in the Computer Science department at Philips Research.
  The third application is a multi-level VLSI circuit simulator designed at AEG (next section).

Results

The above goals have been obtained. Currently, a 100 node DOOM system is operational based on the 68020 processor and a proprietary *Communication Processor*. The latter performs, in hardware, complete end-to-end routing for fixed size packets, using alternative paths to avoid congestion and with a built in routing mechanism that has been proved to be deadlock free. This provides a powerful and efficient mechanism for higher layers of the system, and avoids interruption of the intermediate data processors.

POOL2 has been designed as the programming language for DOOM. Later on POOL-X, a super set of POOL2, was designed to program the database system of the PRISMA project, also at Philips Research. In the course of the project, both operational and denotational semantics for POOL have been obtained and proven equivalent. Furthermore a proof system has been designed. The POOL implemention on DOOM consists of an operating system kernel and a compiler. Beside this one, two other implementations have been obtained for both languages: *plx* is an interpreter system, running on seqential systems. *Parplx*, moreover provides an emulation of POOL on a parallel environment, and provides means for debugging and monitoring. All compilers have been implemented, using a newly designed compiler and code generator technology, based on lazy evaluation of attribute grammars (called Elegant).

Designs and implementations of a number of applications have been obtained. Among these are the above mentioned theorem prover and analytical component for natural language translation. First evaluations show promising speed ups. The VLSI simulator application will now be described more extensively.

4.2 A: THE MULTI-LEVEL VLSI SIMULATOR IN POOL

The main goal of the subproject is a new Parallel Multi-Level VLSI Simulator (PMLS) [7,8]. PMLS is running on general purpose parallel machines and combines multi-level simulation and exploitation of parallelism to achieve optimal performance. The initial implementation of the simulator is in the object-oriented language POOL for the parallel DOOM machine. Powerful simulators are a key element in today's and future digital circuit design environments and a significant application for future parallel machines.

PMLS focuses on the logic design levels, i.e. register transfer, functional gate and switch levels, with provisions for including the programming and the electrical levels. The simulator uses one simulation concept for all levels ("broadband concept") characterised by distributed discrete event simulation and the partitioning of the circuit into subcircuits which are simulated in parallel by subsimulator processes. Each subcircuit may contain elements at different abstraction levels.

Subsimulators execute asynchronously using their own local time, communicate by exchanging event messages and synchronise by using the so called 'time warp' approach, i.e. a subsimulator always runs forward at full speed; in case of an event message being received carrying a time stamp less than the local simulation time, the subsimulator rolls back to the earlier time. The concept delivers sufficient grain size to gain significant speed up on a parallel general purpose machine like DOOM. The implementation is object-oriented, allowing highly dynamic features like dynamic object creation and dynamic object interconnection, for dynamic simulation (e. g. incremental simulation, dynamic change of the abstraction level(zooming))), and providing a high flexibility.

Results
The above goal has been obtained. Currently, PMLS is in the final implementation stage:

The main parts of the simulator have been implemented in POOL, work continues on the advanced features (incremental simulation,zooming) and on tuning of the simulator.

The execution behaviour of the simulator is currently measured using the POOL interpreter/DOOM simulator as well as the 12-node DOOM prototype. First evaluations show promising speed ups.

The prototype version of PMLS is partially integrated into the existing environment of AEGs simulation system DISIM, a commercial VLSI simulation system.

We therefore have a complete system, going from a parallel object-oriented program implemented on a parallel machine and partially integrated into the environment of a commercial simulation system. This goed beyond our original goal.

A further result of the project is a Parallel Fault Simulator (PFS). Fault sets are dynamically split into subsets which are then processed as parallel jobs in the nodes of a LAN. PFS achieves a near linear speed up.

Summarising, VLSI simulation - as a special case of discrete event simulation - is definitely a promising application of parallel general purpose machines like DOOM.

4.3 B: THE FUNCTIONAL APPROACH

Our aim was to take a lazy functional language, which had no parallelism annotations, and implement it efficiently on a parallel architecture. At the beginning of the project, implementations of functional languages were generally interpretive, sequential, and slow. It seemed that significant increases in performance could be obtained firstly by designing specialised hardware to support their execution, and secondly by designing a parallel implementation which exploited their theoretically implicit parallelism.

Lazy functional languages were chosen because they seemed to be a very powerful programming paradigm, which is important for designing reliable and large systems. Furthermore, the parallelism was theoretically implicit in the execution mechanism of the language and thus invisible to the programmers.

We approached the problem in a 'language-first' manner. Firstly we developed a parallel evaluation model which retained the semantics of the language, and then investigated how this drove the architectural design. The original project description specified that we only had to take the process as far as designing a parallel architecture and simulating it.

Results

The subproject has advanced the state of the art in implementing lazy functional languages in a number of ways.

We defined a parallel model which is a natural extension of the sequential lazy evaluation, and which preserves the semantics of programs. The parallelism information for the model can be determined in the compiler using an analysis technique called *abstract interpretation*. It has been proved that if the information obtained from the abstract interpretation is used, then the semantics of the program is retained. The analysis technique has been implemented in another project led by one of our consultants.

The parallel execution model is based strongly on lazy evaluation, with some extra features to create subprocesses to do some of the computation. Therefore, the design of a parallel machine fell into two fairly distinct halves: the design of a machine to support the sequential core of the language, and the design of a parallel harness which supports the specifically parallel parts of the evaluation model. We designed a sequential abstract machine, an improvement on extant abstract machines, and then added the features determined in [9] to turn it into an abstract distributed memory architecture, showing how to compile code for it.

It was established that most features of functional languages can be implemented on current hardware, thus avoiding the need to design any specialised hardware. A demonstration implementation has been prepared on a transputer network. The parallel abstract machine code is translated into transputer machine code.

Functional languages, logic and object-oriented languages need automatic storage allocation and deallocation (*garbage collection*). We developed an award-winning garbage collection algorithm [10], and have subsequently improved it.

We therefore have a complete system, going from a functional program with no parallel constructs, to an implementation on a parallel machine. This exceeds our original goal.

4.4 C: THE LOGIC DATABASE APPROACH

The main objective is to design a parallel deductive database machine called DDC: the Delta Driven Computer [11]. The DDC is specialized to execute deductive requests in parallel by using relational operations. All relations are in main memory.

The execution model is based on an original technique called the Alexander method [12] which was designed to merge recursive views in queries. The advantage of this merge operation is that it produces a set of rules which can be executed in a forward-chaining strategy without computing useless information. This method is based on a compilation technique rather than other techniques based on interpretation.

The advantage is that following the execution model, the set of rules can further be compiled into a low level language where the parallelism is explicit. This language can directly be executed in each node of the DDC machine. Two intermediate languages (VIM and DDCL) are employed.

The DDC machine organisation is composed of between 4 to 256 identical nodes connected by a network. It is a "shared nothing architecture" in the sense that there is no shared memory. All communications between the nodes are done by message passing. Each node of DDC contains a general purpose processor, a special coprocessor called MUSIC, a large memory space and a communication device. The MUSIC coprocessor is designed to speed-up relational operations (partially financed by ESPRIT project 956 and a national project).

Results

DDC is currently simulated on a four processor UNIX machine (SPS7-70). It demonstrates that the compiled approach permits to exhibit parallel code from declarative languages (SQL and VIM). This simulation is very close to the real prototype (eight processors with no shared memory) on which the implementation of the DDC software has now started. The caracteristics of both the prototype and the simulation permit to deduce the performance on the future DDC machine by modelization.

The challenge in a parallel "shared nothing" architecture is to be able to produce an efficient parallel code. In DDC, this code is efficient if the grain of the parallelism is coarse (more that 1000 intructions between two communications).

4.5 D: THE LOGIC+FUNCTIONAL APPROACH

Differently from the other subprojects, the subproject aimed to evaluate the possibility and the advantage of integrating the dominant declarative programming styles, logic(L) and function(F).

Two languages have been designed: a first-order L+F integration which is both semantically well-defined and efficiently implementable by a single computational mechanism for sequential and parallel implementations. The corresponding language K-LEAF, based on Horn Clause Logic with Equality, is an extension of pure Prolog in order to express non-terminating conditional term rewriting systems with constructors.

IDEAL (an Ideal DEductive and Applicative Language), the first compiled higher-order L+F language was defined as a user language of very high level offering. Besides the usual Prolog-like capabilities, it offers the most distinctive features, now present in modern functional languages, namely lazy evaluation, type inference and higher order constructs.

Next, effort has been devoted to the efficient implementation of the integrated computational model on generic sequential processors and on parallel architectures. A conservative extension of the WAM (Warren Abstract Machine), K-WAM, was desgined to efficiently implement outermost resolution, based on the extensive experience available for sequential and parallel Prolog compilation, and suited to incorporate all the new incoming optimizations. Extensions to support the dynamic resolution strategy required by K-LEAF thus allowed the first compiled implementation of a sound L+F language.

A strong synergy with ESPRIT Project 26, where the architectural aspects were covered, enabled the implementation of a L+F language on a physical parallel machine. The machine embodies three main concepts: physically distributed/logically shared memory, very fast context switching built-in in the PE, and support of efficient packet switched non local communications. The machine is an ensemble of up to 128 Transputers fully interconnected by a low latency (2 mu-sec. transit time), high throughput (10Mbytes/s on each port), "cut-through" packet switching, buffered Delta network, implemented by means of a custom VLSI switching element circuit of about 30000 CMOS devices.

Results
Parallel execution models for L+F languages are viable on distributed memory architectures. For efficiency reasons, the programmer can control the granularity of parallelism, through special annotations and set-like constructs in a disciplined way,

A (deterministic AND)-OR parallel model reduction has been prototyped and a estriction of it, namely (independent AND)-OR parallelism, has been finally ported to our parallel machine, by means of an original mapping of AND to OR parallelism, as, independently, experimented with success in the GigaLips Project.

Most of the above mentioned results are assembled in a working prototype of a small configuration (16 PE's) machine able to run conventional benchmarks (e.g. n-queens, fibonacci numbers, etc.) and a few more realistic applications (e.g. grammar based image recognition, logic simulation/fault finding) written in AND-OR Parallel K-LEAF, showing quasi-linear speed-up's in the range 75%-90% of the ideal efficiency.

The measured performance of the mapping sequential K-WAM into C, on standard benchmarks is 80%-300% of Quintus 2.2, 50%-200% of G-Machine-based Lazy ML, and 20%-80% of C used with all of its imperative features. This experience suggests that a reasonably good (e.g. 160Klips on Sun3/280) performance can be obtained [14].

4.6 E: THE DATAFLOW APPROACH

The use of the dataflow principle for parallel execution has a long history where, since the early seventies, a large number of dataflow computer architectures have been proposed and some of them even built, e. g. the Manchester Data Flow Machine at Manchester University. But most of these systems are designed to exploit fine grain parallelism and to run programs in the area of number crunching thus leading to expensive special purpose hardware. In contrast, this subproject at Stollmann GmbH intends to build a dynamic dataflow machine suited to run commercial applications and to exploit coarse and coarsest granularity. The chosen application is in the database area, especially the parallel evaluation of database queries. Because of our concept of variable granularity, it was possible to deviate from a highly specialized hardware approach and to start by building a demonstrator with off-the-shelf hardware. The dataflow control mechanisms are lifted to the software level including distributed parallel firing. The basic principles of the approach are load distribution mechanisms aiming at exploitation of locality and dynamic load balancing mechanisms as e.g. task attraction.

We have designed an abstract architecture model, the Stollmann Dataflow Machine (SDFM). The SDFM units are implemented as software processes. The execution model consists of three different units: The execution units (EU) execute the fired dataflow nodes, that is, the executable instructions and produce output operands. The firing control units (FC) assign the produced output operands to the corresponding instructions and fire the executable instructions. The administration control unit (AC) has a special functionality as it monitors the system, starts and terminates programs and executes I/O instructions. These units are communicating asynchronously via buckets, a special kind of queues with a sort of intelligence and access strategies. The buckets are used for distribution strategies to exploit locality. That means the work has to be done by that unit where the corresponding data are stored. As well we have dynamic load balancing to avoid that exploitation of locality leads to non-optimal load balancing. In the case that a unit has run out of work it attracts work from other units. For the programming of the SDFM we have implemented the languages BLASS and CLAN. A program written in BLASS can be seen as the textual representation of a dataflow graph. CLAN is a functional, single assignment language similar to SISAL. To support coarse grain dataflow we have introduced the concept of user-defined instructions into both languages.

Results

The SDFM model has been implemented on our demonstrator, a multiprocessor configuration with four processor boards connected by a VMEbus. For the first implementation we have chosen 68020 processor boards equipped with dual ported RAM, thus admitting a global address space. One processor board is running UNIX V.3 and is used as a host. The other boards are running SRTX, a realtime operating system kernel with multitasking and dynamic memory management. SMOCS, a common layer to all processor boards provides a basic set of operating system functions for a multiprocessor system, especially global queues and management of the local, but shared memory.

Our application, the parallel evaluation of database queries, is realized in CLAN, by implementing the relational algebra operations as user-defined instructions in CLAN. These operations processing complete relations are splitted into parallel package versions to introduce more parallelism. Our performance measurements have shown that speedup depends largely on the grainsize. Anacceptable speedup can be gained by a grainsize of 0.1 seconds. Though there is no single bottleneck optimization of the system to reach an optimal grainsize of 0.05 seconds is possible and even finer grains by introducing special hardware.

4.7   F: THE CONNECTION APPROACH TO LOGIC PROGRAMMING

The main goal of the subproject is an inference machine based on the parallel automated theorem prover PARTHEO [17] for full first order predicate logic. The underlying proof calculus is Model Elimination, a specialisation of the Connection Method which has been developed by Prof. W. Bibel. The input language LOP for PARTHEO allows a straight forward declarative style of logic programming without being restricted to Horn Clause logic as in PROLOG.

The second goal of P415-F is to improve and implement the functional parallel programming language FP2. FP2 provides parallel processes based on term rewriting and communication via unification. It is used as a high level specification tool for the parallel algorithms used in PARTHEO.

LIFIA at Grenoble developed the language FP2 and provide the sequential implementation of FP2. The group at Technische Universität München worked on design and implementation of PARTHEO and LOP. The Nixdorf part of the project covered the parallel implementation of FP2, some contributions to the work on PARTHEO and the exploitation of project results.

Results
SETHEO, the Sequential Theorem Prover is an extension of the Warren Abstract Prolog Machine to full first order predicate logic. SETHEO is implemented in C and yields a performance of about 120 KLIPS on a SUN 4 machine. Running on every UNIX-machine SETHEO is one of the fastest existing high performance theorem provers as well as an efficient LOP interpreter.

PARTHEO, the Parallel Theorem Prover: A network of 16 transputers is used for solving independent parts of the search tree in parallel. Together with the fast abstract machine of SETHEO the parallel prover increases the performance of the inference machine. A user-friendly graphical user interface facilitates development and test of LOP programs.

Sequential FP2: Running on SUN workstations the sequential FP2 interpreter represents a powerful high level language for specification, verification and test of parallel algorithms.

Parallel FP2: A high performance FP2 interpreter [18] was implemented on a parallel VME-bus machine (Stollmann test machine built by subproject E).

4.8   WORKING GROUP ON ARCHITECTURES AND APPLICATIONS

The Working Group on Architecture and Applications (WgArch) is concerned with drawing together the results of the six Subprojects.

As described above, basically each subproject is responsible for investigating a particular class of parallel computer. The WgArch activities are concerned with identifying the strengths and weaknesses of these individual approaches, and drawing the results together to identify the concepts of future parallel computers.

The work of the WgArch divides into three major themes:

- comparing Subprojects: identifying the strengths and weaknesses of each of the six parallel computers and their associated programming styles.

- topics of common interest: studying important topics (e.g. programming styles, processor architecture, network architecture, memory management etc.) for all Subprojects and identifying a possible definitive approach.

- dissemination of results: continuously monitoring the progress and results of parallel architecture and applications projects elsewhere in the United States, Japan and Europe. In addition, Project 415 results were presented at the PARLE Conferences and two Summer Schools.

The WgArch holds 4 three day meetings each year, together with 2 one day meetings held in conjunction with the Project's general meetings. Each meeting typically comprises a tutorial/survey on an important topic by a WgArch member, common analysis/comparison work performed by each Subproject, a review of Subproject progress and one guest speaker.

Results

WgArch results mainly concern the sharing of results, and the development of common understandings, of the approaches of each Subproject. These results are documented in a series of: Survey papers, Position papers from each Subproject, and Discussion documents, on the important topics of parallel computing.

In addition, the WgArch on behalf of Project 415 and ESPRIT, has helped to organise a summer school in 1986 entitled: "Future Parallel Computers"; together with the two PARLE conferences [3,4].

As a finale to the work of Project 415, the WgArch has prepared a comprehensive overview of the project in a book [2]

4.9 WORKING GROUP ON SEMANTICS AND PROOF TECHNIQUES

For the full duration of the project a Working Group on Semantics (year 1) or on Semantics and Proof Techniques (years 2-5) has been in operation within project 415. Its primary purpose was to perform and discuss investigations in the area of semantics, proof techniques and specification techniques for parallel languages and systems. Its activities have in particular been directed towards a better understanding of the theoretical foundations which underly the programming styles at the basis of the respective subprojects. In this way, the feasibility of partial or full integration of the various computational models could be investigated and, when full integration turned out to be too ambitious, the range of commonalities in the mathematical treatments could be explored. In addition, the Working Group has made an effort at keeping abreast of the developments in the theory of parallelism in general, in order to enhance the awareness of the researchers in project 415 of the state of the art in this field.

The Group has held bimonthly meetings, about thirty altogether. where in-depth presentations were given on a broad scale of topics within the scope of the group both by its members and by guests. The presentations were devoted primarily to research reports on semantic and proof theoretic investigations stemming from the subprojects, and, to a lesser extent, to tutorial overviews of relevant developments elsewhere.

The Working Group has produced five deliverables of altogether about 2000 pages. In deliverable 5, a complete list is included of all papers within the scope of the Group written over the five years of the project. Part of the work of the Group was organized in subgroups which concentrated on one or more themes of specific mutual interest for the subprojects.

A substantial part of the work is also reported in the book [1] Besides in the reports of the subgroups already mentioned, intersubproject work was described.

Finally, we mention here the contributions of the Working Group to the educational efforts of project 415 such as the Advanced School on Current Trends in Concurency [19] and the PARLE conferences ([3,4]). Looking back over the past five years, we feel that the Working Group has been successful in its integrative role, in particular directed at the promotion and crossfertilization of foundational research in project 415.

5   Exploitation

The results of P415 have played a major role in the definition of two Esprit II technical integration projects (TIPs) in the area of parallel computer systems.

TROPICS, carried out by a consortium of Nixdorf, Olivetti, Thomson and Cap Sesa, led by Philips, aims at developing a high performance parallel computer system for office automation markets. A parallel data base server, compute and communication server as well as a parallel workstation will be realized based on the object-oriented technology of subproject A and the parallel database technology of the PRISMA project (Philips).

The goal of EDS (European Declarative System) is to design a massive parallel machine supporting declarative languages (LISP, ML, PROLOG) and database applications. Two different execution models for the database applications: the Extended Graph Rewriting model, proposed by ICL, and the Graph Saturation model, proposed by Bull, will be merged into a more powerful model.

Beside these major impacts of P415 on the future capabilities of the European industry, direct spinn-offs have also been obtained. These will be described per partner.

The exploitation by Philips will mainly be through the results from *TROPICS*. The transfer of expertise to the Datasystems Division has already begun. Furthermore, the Elegant compiler technology is now in use at various divisions within the company. Researchers at a number of Philips Research Laboratories have already started or plan to use the DOOM system to speed up their experiments.

AEG is aiming at the development of the next generation VLSI simulation tools, based on the utilisation of parallel processing on general-purpose parallel machines. The prototype versions of PMLS and PFS are the basis for future parallel simulation kernels in AEGs commercial VLSI simulation system DISIM. The intended exploitation of the project results aim at a drastic performance increase and cost effectiveness of the DISIM system.

Most of the GEC's results are independent of the target architecture, and so are useful for implementation on any parallel MIMD-style architecture. Furthermore, some things, like the garbage collection algorithm and our method for developing parallel architectures for languages with implicit parallelism, are applicable to other programming language styles. GEC envisages that the power of functional programming to develop systems rapidly will receive the most immediate application within their company.

CSELT's results on parallel models for declarative languages, both from P26 and P415, are almost directly exploitable on multiprocessor machines(EMMA2) of Elsag, belonging to the STET Group, as well on commercial multiprocessors. The low level compilation technique is already exploited in an industrial development jointly supported by RSE (Raggruppamento Selenia Elsag) and SIP (the Italian telecom operating company), for em bedded knowledge-based systems.

For Stollmann's SDFM it was decided to use a machine with standard components as a testbed with more general distributed data processing capacities. This machine called GLOBUS will be available as a development and testbed environment for multiprocessor concepts, and marketed accordingly. Transfer to RISC processors (88000) and inclusion of X-Window as comfortable user interface is already planned. To exploit dataflow principles in "coarse grain" application servers (database or communication), an integration of distribution and dataflow support into ASICs and incorporation of fault-tolerance mechanisms is currently being planned at Stollmann.

PARTHEO is considered as inference machine for Artificial Intelligence systems at Nixdorf. Esprit Project 311 and the TWAICE expert system project (design for assembly) at Nixdorf are investigating the use of SETHEO/PARTHEO. Besides that, LOP is an interesting alternative to PROLOG. Thus exploitation of LOP as logic programming languague is being studied at Nixdorf. Last, but not least many research institutions all over the world already bought a licence for SETHEO and are using the prover.

6   Conclusions

Six different language and execution models for parallel computer systems have been investigated and designs for corresponding languages and their implementations made. Beyond the original scope, hardware prototypes have been constructed, ranging from some 4 computers (DDC, SDFM) to 100 (DOOM). These prototypes allow for a good evaluation of the approaches, which will be undertaken in the final year.

Acknowledgement
The author is very grateful to the subproject leaders and chairmen of the working groups of P415 for providing draft sections and for carefully reading and commenting the paper: W. Bronnenberg, [2] F. Lohnert, [3] G. Burn, [4] B. Bergsten [5] P.Bosco [6] E. Glück-Hiltrop [7]

---

[2] Philips Research Laboratories, Eindhoven, The Netherlands
[3] AEG Research Institute, Berlin, FRG
[4] GEC Hirst Reserach Centre, Wembley, United Kingdom
[5] Bull Centre de Recherche, Louveciennes, France
[6] CSELT, Torino, Italy
[7] Stollmann GmbH, Hamburg, FRG

W. Ertel [8] J. de Bakker [9] P. Treleaven [10]

References

[1] de Bakker,J.W.(ed.) (1989) Languages for Parallel Architectures: Design, Semantics, Implementation Models, Wiley.

[2] Treleaven,Ph.(ed.) (1989) Parallel Computers: Object-Oriented, Functional and Logic, Wiley.

[3] de Bakker J.W, Nijman A.J, Treleaven P.C.(eds.) (1987) Proceedings PARLE Parallel Architectures and Languages Europe, Vol 1: Parallel Architectures, Vol. 2: Parallel Languages, LNCS 258, 259, Springer.

[4] Odijk,E.A.M. , Rem,M., Syre, J-C. (eds)(1989) Proceedings PARLE Parallel Architectures and Languages Europe, Vol 1: Parallel Architectures, Vol. 2: Parallel Languages, LNCS 365, 366, Springer.

[5] Bronnenberg W., Nijman L., Odijk E., van Twist,R (1987) 'DOOM: a Decentralized Object-Oriented MAchine', IEEE Micro, Vol. 7, no. 6, oct.1987, pp. 52-69.

[6] Odijk E, Bronnenberg, W.,(1988) 'Parallel Computing: the Object-Oriented Approach', Proceedings of CONPAR 88, 12-16 September 1988, Manchester, United Kingdom, Cambridge University Press.

[7] Aposporidis,E., Lohnert,F., Mehring,P., Hoppe,F., Post,H.-U.(1989) 'A Parallel Multi-Level Simulator for VLSI', Proc. of the Workshop in this Conference: "ESPRIT Project 415 - Parallel Architectures and Languages for Advanced Information Processing - A VLSI-Directed Approach"

[8] Aposporidis,E., Lohnert,F.(1989) 'Parallel Multi-level VLSI Simulator - An Object-Oriented Approach', Proc. ESM'89, Rome, June 7-9,1989

[9] Burn, G.L.(1988) 'Developing a Distributed Memory Architecture for Parallel Graph Reduction' ,Proceedings of CONPAR 88, 12-16 September 1988, Manchester, United Kingdom, Cambridge University Press.

[10] Bevan, D.I.(1987) 'Distributed Garbage Collection using Reference Counting', Best Paper Award, PARLE 1987, June 1987 Eindhoven, The Netherlands, Springer Verlag LNCS 259, pp.176-187.

[11] Bergsten B., Gonzalez-Rubio R., Kerherve B., Rohmer J.(1988) 'An Advanced Database Accelerator', IEEE Micro, Oct. 1988.

[12] Rohmer J., Lescoeur R., Kerisit J.M.(1986) 'The Alexander Method. A technique for the processing of recursive axioms in deductive Database', New Generation Computing, 4, 1986.

[13] P.G. Bosco, E. Giovannetti and C. Moiso(1988) 'Narrowing vs. SLD-resolution', Journal of Theoretical Computer Science, Vol. 59, no. 1-2 (North-Holland, 1988), pp. 3-23.

[14] P.G. Bosco, C. Cecchi and C. Moiso: *An extension of WAM for K-LEAF: a WAM-based compilation of conditional narrowing*, in Proc. 6th Conf. on Logic Programming, Lisboa, (MIT Press, 1989).

[15] Glueck-Hiltrop,E., Joehnk,M., Schuerfeld, U.(1988) 'The Stollmann Data Flow Machine', in Proceedings of the Conpar Conference. Jesshope,C.R. (Ed.). September 1988.

[8]Nixdorf Computer, Munich, FRG
[9]CWI, Amsterdam, The Netherlands
[10]University College, London, United Kingdom

[16] Zucker, W.(1986) 'The Stollmann DF Design Principles', ESPRIT Summerschool on Future Parallel Computers, June 1986.

[17] Ertel W., Kurfeß F., Pandolfi X., Schumann J.(1989) 'PARTHEO, A Parallel Inference Machine', Proc. of PARLE Conference 1989, Springer 1989.

[18] Rogé S.: 'A Parallel FP2 Interpreter', ESPRIT Conference 1988, North Holland 1988.

[19] de Bakker J.W. , de Roever W.P. , Rozenberg G.(eds.)(1986) Current Trends in Concurrency, Proc. LPC/ESPRIT Advanced School, LNCS 224, Springer, 1986.

[20] Treleaven P., Vanneschi M.(eds)(1986) 'Future Parallel Computers', Proc. ESPRIT Summerschool on Future Parallel Computers, Pisa, June 1986, LNCS 272, Springer, 1987.

KNOWLEDGE BASE DEVELOPMENT IN A STANDARD FRAMEWORK[1]

GIUSEPPE ATTARDI and MARIA SIMI
DELPHI S.p.A.
Via della Vetraia, 11
I-55049 Viareggio LU
Italy

Summary

In the ESPRIT Project P440 (Message Passing Architecture and Description Systems) several aspects of knowledge representation have been explored: formalisms, languages, techniques. The most significant techniques developed within the project are: taxonomic reasoning, meta level strategies and concurrent object oriented programming. These techniques are embedded in the systems developed in the project: Omega, KRS and DELPHI Common LISP. As a generalization of our experience in building knowledge based systems, we present a framework which provides a range of tools for building knowledge base applications. The framework is conceived so that industrial standards are selected where appropriate and areas where standard are missing are highlighted.

1. Introduction

The ESPRIT Project P440 (Message Passing Architecture and Description Systems) aims at exploring techniques to push the limits of knowledge base systems in terms of performance and size.

We worked on the assumption that complex knowledge bases could only be managed if suitable structuring mechanism are provided so that the search and deductions could be focused and properly directed. In addition, concurrency would be used to speed up the search both through the various paths in the knowldege base and by applying alternative strategies.

The project has gone through the following phases:

1. design of suitable formalisms for knowledge representation
2. development of techniques for reasoning on the knowledge base
3. verification of the techniques
4. generalization of the results

The formalisms developed are Omega, a logical calculus of descriptions, and KRS, a programming language for knowledge representation.

The techniques that have been developed are:

1   This work has been partially supported by ESPRIT Project P440 (Message Passing Architectures and Description Systems) and is a contribution to Cost13 n.21 (Advanced Issues in Knowledge Representation).

1. taxonomic reasoning, a form of deduction which explores a taxonomy of concepts
2. meta level strategies, which allow to taylor the reasoning strategies to each specific task
3. parallel constructs for spreading the deduction across the nodes of the taxonomy

The sperimentation phase has produced various software systems, which have been made available also outside of the project: the Omega Knowledge Base Development Environment, the KRS language and an implementation of Common LISP (DELPHI Common LISP) which includes a Multithread facility for concurrency (3) and the Common LISP Object System for objectoriented programming.

The generalization of the experience has led to the design of a framework for building knowledge base applications which is based on industrial standards, and which incorporates as "generic" tools the essential building blocks that were used in the construction of Omega and which can be helpful in a variety of knowledge base applications.

In the following sections we will put our approach into the perspective of knowledge representation research, we will sketch the techniques developed in the project and finally we will illustrate the structure of the knowledge base development framework which constitutes a generalization of our experiences.

2. Perspective on Knowledge Representation

Historically the research in knowledge representation started investigating the problems of formulation, i.e. how to translate into a precise formalism the expressions of natural language, the same problems that philosophers and linguists had been debating for long time: reference, co-reference, definite or indefinite descriptions, referential transparency and so on.

The issues addressed were of the kind: "How one would represent statements like 'the present king of France is bold' or 'Kepler did not know that the number of planets is 9'?".

Experiments with deductive systems that were being conducted at the same time, demonstrated however that even for problems whose formulation in predicate logic was straightforward, the computational demands for even the simplest deductions were overwhelming.

So while research is still continuing in problems of formalization, and new issues like default reasoning, nonmonotonic reasoning, reasoning about beliefs, about time, etc. have been added to the list of unsolved problems, an alternative line of investigation has emerged.

This research addresses the problem of developing formalisms with a proper balance between the overall expressive power and the computational complexity of the basic constructs. A simpler formalism is expected to be more amenable to algorithmic treatment, for instance as the basis for an automated deduction system.

Two approaches have been followed, a "programming language" approach and a "logic formalism" approach.

In the first approach, the basic formalisms were programming languages which were designed or extended to provide constructs useful to represent knowledge (6),(19). Typically new constructs were introduced to define objects and classes of objects, usually related by some kind of inheritance hierarchy. These capabilities were useful to model the individual objects of the domain of discourse, and programs could be written to manipulate those objects. Beyond these basic descriptive capabilities, the bulk of the semantic knowledge on the domain, would be expressed by procedures associated to the objects, or sometimes attached to the slots of some object. For instance in LOOPS one could supply a procedure to be triggered when the value of a slot of an object changes. This approach can be characterized by:

- quite limited descriptive capability
- ability to handle inheritance and defaults
- no mechanism to formulate general assertions
- full reliance on procedural knowledge

An attempt to enrich the descriptive capabilities of these systems was to combine them with a rule base mechanism. A rule consists of a condition and a sequence of actions to be executed when the condition is fulfilled. Since the condition can be expressed with a pattern containing variables, this provides a way to represent general statements. This gave rise to the so called "hybrid" knowledge representation facilities whose most significant example is KEE (11).

In a hybrid system knowledge is still represented mostly in a procedural way, since rules are still procedural rather than assertional. Even though the rules can be attached to certain classes of objects, the triggering of rules happens externally to the class system, since conditions may refer to more than one object and may consist of combinations by means of logical connectives. Hybrid systems have been heavily promoted, even though building an application using such a system consists largely of programming in the underlying programming language.

Hybrid systems nonetheless do not achieve the expressive power that is often requested to model many situations. For instance in KEE it is not possible to express that "all divisions of a company are located in New York" or that "a subsidiary of a company is owned by that company". Such limitations are due to the fact that the declarative facilities of the object-oriented language do not provide logic variables nor universal abstraction. Variables are allowed in the rule-based component of such systems, but this provides no assertional capability for either descriptive or prescriptive purposes.

An attempt to overcome the limitations of an objectoriented programming language from the point of view of knowledge representation has been made in our ESPRIT project by the team from the Vrije Universiteit in Brussels in developing the language KRS (20). KRS is an object oriented programming language where a naming mechanism is provided so that within the description of an object (called concept in KRS terminology) a reference can appear to some component of the same description. This naming/reference mechanism is supported in the implementation by a constraint propagation technique to maintain the co-reference of the components. This is still not a general assertional mechanism, but it provides a convenient alternative to the attachment of procedures triggered on slot access. In KRS no built-in rule mechanism is provided, but libraries with similar functionality can be programmed in KRS.

The second approach to knowledge representation looks for formalisms for expressing knowledge in a declarative way, with sufficiently rich assertional capabilities and with a deductive apparatus for deriving conclusions or answering queries on the knowledge base. How information obtained from interacting with the KB system could be further manipulated is often left as an interface problem between a programming language and the knowledge base system.

The computational tractability problem consists of finding a suitable compromise between the expressive power of the formalism and the computational complexity of the deductive system. An influential example of this approach is the KRYPTON system (9) where a split of the knowledge base is proposed between an assertional and a terminological component. The terminological component describes the structure of the objects in the domain and possibly relations between them (e.g. inheritance). Only punctual information can be placed in the terminological component. The assertional component uses a logical calculus by which general facts can be expressed. A specialized algorithm is envisaged for the terminological component, while some kind of theorem prover is needed to reason with facts on the assertional component.

There has been a lot of investigation on where to place the border line between the two components. Unfortunately, theoretical results show that in order to keep the complexity of the decision procedure within acceptable limits (polynomial time), one has to severely restrict the expressive power of the terminological component (16),(17).

On the other hand there has been criticism (16) about the benefits provided by an automatic classifier like the one present in NIKL (21), which is the foundation of the terminological reasoner.

Only a few systems have attempted an integrated approach for both the descriptive and assertive aspects of knowledge representation: for instance LOGIN (1) and Omega (2).

LOGIN extends a logic programming formalism with structured terms and a notation for subsumption between classes. The classical resolution algorithm is modified to take into account the subsumption relation during unification. The performance benefits of this solution seem to come from the interleaving of unification steps and subsumption tests, since this can often lead to prune the number of alternatives to be examined during a proof.

3. Omega

Omega is a logic for knowledge representation in which terms (called descriptions) represent sets of individual objects and the basic assertion is about inclusion between sets.

We interpret the theoretical complexity results mentioned in the previous section to mean that even in a apparently limited language, queries can be expressed which have the appearance of real puzzles. We don't expect a reasoning system to be good at solving puzzles, but rather to be able to find immediate answers to simple queries, like those for determing subsumption between two concepts.

Omega explores the idea of "taxonomic reasoning", that is, basing all reasoning on the traversal/instantiation of the lattice of descriptions. Omega is an integrated knowledge representation system, where all knowledge is represented in a single lattice: from factual knowledge, to general rules, to dependencies and constraints.

So for instance both the facts:

Henry-VII is (a Father (with child Henry-VIII))
Henry-VIII is (a Father (with child Mary-Stuart))

and the general assertions:

(a Father (with child =x)) is (an Ancestor (with descendant =x))
(an Ancestor (with descendant (a Father (with child =x))))
        is (an Ancestor (with descendant =x))

will be represented in a structure corresponding to the lattice of Figure 1, which is built by a "classification" algorithm.

Arrows in the picture correspond to "is" links stemming either from assertions (thick arrows) or induced during classification (thin arrows). Symbols like =x and =y denote universally quantified variables. A taxonomic reasoner can solve the query:

Is Henry-VII (an Ancestor (with descendant MaryStuart))?

by classifying the two descriptions:

Henry-VII
(an Ancestor (with descendant Mary-Stuart))

and then trying to find a path between the two. Classification will place the second description as indicated by the dashed arrow in Figure 1.

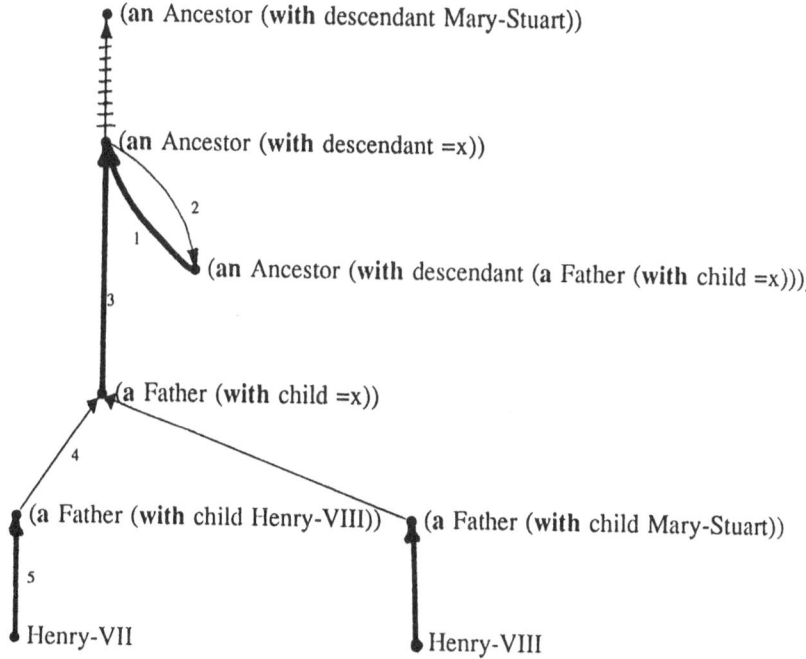

Figure 1.

To find a path, the link labeled 1 can be followed by binding =x to Mary-Stuart and obtaining

(an Ancestor (with descendant (a Father (with child Mary-Stuart))))

A subordinate search is started from the description (a Father (with child Mary-Stuart)) to find an instantiation, in this case Henry-VIII. We now have

(an Ancestor (with descendant Henry-VIII))

so we can continue through link number 2 binding =x this time to Henry-VIII. When link 3 is traversed, the description obtained is:

(a Father (with child Henry-VIII))

This description has two downward links, but only the link labeled 4 can be followed, given the present binding of =x to Henry-VIII. Continuing further the traversal through link 5, description Henry-VII is reached and the goal established.

The main advantage of the classification is that it creates in advance all the links that might be necessary during a search, so that deduction is guided through them. The structure of links is reminiscent of the technique of connection graphs in classical theorem proving.

The second technique which is supported by Omega is the ability to program deductive strategies by means of a meta level facility. Omega assertions about Omega itself are used to express which strategies to adopt in particular circumstances.

Finally, an object oriented concurrent programming language has been developed. The object oriented nature of the language has been useful in expressing the object oriented nature of the knowledge representation system, and concurrency has been used to explore various form of control during reasoning.

The original implementation of Omega was done on a LISP machine using its rich set of tools for object oriented and graphic programming.

After the system was completed, we were confronted with the problem to make our results available on a wider range of machines. Rather than doing a direct port, we decided to generalize our experience, by identifying a number of fundamental building blocks which we had used and which could constitute the elements also for other applications.

4. Architecture of a Standard Framework

The overall architecture of the knowledge base development framework is presented in Figure 2. This framework tries to cover many aspects of knowledge base applications, ranging from deductive support, to graphical presentation, to access to external data. We illustrate together both the general framework and our particular instantiation. So, in describing each component, we also indicate an appropriate standard that we have selected to fulfill that role. Many of these standards have already acquired a well recognized state. In a few areas standards are less established or even lacking. Nonetheless we believe that it is quite useful to have a reference model for knowledge base applications so that the overall picture becomes clear and the focus of the tool developers is oriented towards producing components that will fit a global design.

One fundamental design decision is the choice of an object-oriented approach for the higher layers of the framework. This provides the essential capabilities for allowing:

- composability
- reusability
- extensibility

In our framework, the tools are conceived as generic tools, in the sense that they operate on objects according to a determined set of generic functions, which constitutes the interface to these objects. Objects of various kinds can be handled by the tools provided that the generic functions required by the interface are defined for their classes. These classes can be defined from scratch, but most likely they will be just specializations of existing classes. Most of the generic functions of the interface would then just be inherited from the superclasses or be adapted with the technique of "method combination" (10).

For instance a tool for displaying tree structures in a graphical form would rely on generic functions like "get-sons" and "get-parents" on objects of class "display-node". In order to get this tool to work for either a tree representing a file system or a tree for a concept taxonomy, it would be sufficient for the nodes of these structures to incorporate by inheritance the class "display-node" and to specialize the "get-sons" and "get-parents" operations.

This metaphor of programming by refinement or specialization has been successfully explored in Smalltalk and other similar programming environments (13).

4.1. Physical Resources

At the basis of our construction there are some physical resources that the applications exploit. We broadly distinguish three kind of resources:

- processing resources (CPUs)
- data storage resources (disks)
- graphical resources (displays)

These resources have a physical location and specified capacities. We would like our applications to have access to these resources independently from their location.

4.2. Network Protocols

Suitable network protocols provide remote access independent of physical location to the class of resources that we are considering.

4.2.1. Network File System

The Network File System (NFS) developed by Sun Microsystems has become an industry standard for providing uniform access to files across a network.

4.2.2. Remote Procedure Call

The Remote Procedure Call is also a protocol present in various network implementations, whose aim is to provide uniform access to processing resources across a network. This is an area where further standardization effort is required, in particular on issues like dynamic software migration (23), which would allow one to migrate programs for execution on a remote target, rather then being limited to invoke those programs which have been previously installed on the target.

4.2.3. X Window Protocol

X Windows is a window system developed at MIT (12) which provides transparent network access to graphic display resources. Each display is managed by a "window

A STANDARD FRAMEWORK
FOR KB APPLICATIONS

TOOLS

OMEGA KB   DEDUCTION   PRESENTATION

OBJECT

PCLOS   CLOS   CLUE

(Persistent Object System)   (CL Object System)   (Graphic Toolkit)

LANGUAGE

OMS   COMMON   CLX

(Object Management System)   LISP   (CL X Windows)

NETWORK

NFS   RPC   X WINDOWS

(Network File System)   (Remote Procedure Call)   (X11 Protocol)

RESOURCES

Figure 2.

server" to which "client" applications connect to perform graphic operations. The dialog between server and clients happens trought a particular network protocol called X11 Window Protocol. As a network protocol, X11 is display independent and machine independent, and this allows client applications written in different languages to work in multivendor architecture. X11 is a de-facto standard endorsed by a large number of manufacturers and it is under consideration by the international standardization bodies.

4.3. Programming Language

The various facilities provided by the network layer get together within a programming language either as primitive constructs of the language, or as libraries or as binding for the language.

4.3.1. Common Lisp

Common LISP is our choice as a standard programming language, since it has gained acceptance by a wide community, it is being considered for standardization by ANSI and it also has a role in the ISO LISP standardization.

4.3.2. Oms

The Object Management System (OMS) is a component of the Portable Common Tool Environment (PCTE), a basic support layer for software engineering developed within the european ESPRIT programme. The OMS is a data management system of the entityrelationship family. It provides basic storage for objects consisting of attribute values and of links to other objects. A schema definition mechanism is used to introduce new types of objects.

4.3.3. Clx

The Common LISP X Window (CLX) is the language binding of X Windows for Common LISP. CLX is part of the X Window specification, and it is for LISP what Xlib is for C.

4.4. Object Oriented Layer

This level extends the facilities at the previous level and provides a higher degree of modularity and composability through the use of object oriented programming. These facilities are embodied in a collection of objects with associated operations. New objects can be created by composing previous objects. Operations on the new objects can be derived or specialized by means of an inheritance mechanism.

4.4.1. Clos

The Common LISP Object System (CLOS) is an extension to Common LISP, already under consideration by ANSI for inclusion in the Common LISP standard. CLOS is based on the notions of "instance", "class" and "metaclass". An instance is a primitive data structure of the language. Each instance refers to the class from which it originated. A class is itself an instance whose role is to describe the structure of its instances and to provide them with a common behaviour. A class can be defined as a composition of other classes by "multiple inheritance". A metaclass is a class whose instances are themselves classes, and so it describes how its instance classes should behave, for example when a new instance is created.

CLOS also introduces the notion of "generic function", i.e. of a function whose behavior can be specialized with respect to the types of its arguments. A generic function consists in a collection of "methods", each one specific for certain argument types. When a generic function is invoked, the suitable method must be selected according to the types of the actual parameters of the invocation. Because of inheritance, more than one method may be applicable, so they will be sorted and the most specific or a combination of the applicable methods is selected and applied to the parameters.

Each CLOS class represents a Common LISP type, and thepate class hierarchy if fully integrated with the Common LISP type hierarchy. Therefore the generic function mechanism works across the whole language, allowing the technique of programming by specialization to be exploited uniformly.

CLOS is defined and implemented in terms of itself, e.g. by a certain number of (meta)classes and generic functions on those. This is not just an interesting way to present the system, but it is essential to open up the system for extension while remaining within a single framework.

For example, the basic operations to create objects and to access slots of an object are implemented as generic functions on the kernel system metaclasses ("standard-class" and "standard-object"). These operations can be specialized to achieve a different behavior for objects which have a different internal representation. Despite the internal implementation difference, one would still apply the standard operations ("make-instance" and "slot-value") to these objects as well. In the next section we present a non trivial application of this property, other examples are reported in (5).

4.4.2. Persistent Clos

The OMS is used as an intermediate layer to provide persistency at the Common LISP object level. Two special classes are introduced to model persistent objects and their classes: the metaclass "persistent-class" and the class "persistentobject". The class "persistent-object" must be included by any class of persistent objects. It defines the essential structure of persistent objects used by the system to manage them. The metaclass "persistent-class" is the class from which classes of persistent objects are created. It provides a specialized method for the generic function "make-instance", so that when an instance of a persistent class is generated, an entity is actually created on the OMS to contain the data for the object and also an appropriate "object handle" is created in memory containing a reference to the allocated OMS entity.

Each OMS entity is also recorded in a list within the entity representing its class. This provides the basis for navigation through persistent objects, which is possible by means of the following construct:

```
(for-each (obj <persistent-class>)
<body>)
```

which executes <body> several times with the variable obj bound each time to a different element of the persistentclass.

The primitives for accessing slots of persistent objects are also specialized for "persistent-class" so that access is redirected through the object handle to the OMS entity.

4.4.3. Clue

The Common LISP User Interface Environment (CLUE) is an object-oriented graphic toolkit (18). It is written in CLOS on top of CLX and suport the event/callback model of user interaction, whereby the application defines the interaction events relevant for it and also which procedures (callback) should be invoked to handle those events. The control flow of the application becomes then totally "event driven", with events triggering the corresponding action.

4.5. Tool Layer

This layer consists in a set of tools which implement some of the fundamental functions of a knowledge base application. Their object-oriented nature allows the knowledge engineer to mix and match them together to create the final application.

4.5.1. Deduction Tools

Support of deductive activities is provided by tools which realize the appropriate control structures. For instance backward reasoning requires the management of a goal tree and various kinds of search require the saving and restoring of certain states in the computation.

A general goal tree management tool is built in terms of objects representing nodes. A goal tree has the structure of an "and/or" tree, where "and" nodes represent how a goal is split into subgoals and "or" nodes (called attempts) represent different ways ("strategies") to achieve a goal. The main operations on the goal tree nodes are done through the following generic functions:

solve	applies a certain strategy to a goal, creating an "attempt" node and the corr sponding goal subnodes
tackle	chooses one or more of the sobgoals in an attempt and tries to solve them.
succeeds	it is called when the goal has been fulfilled, so that the success can be prop gated further, possibly to higher nodes of the tree.
fails	it is applied to a goal to discard it and propagate the failure through the tree may happen either because there is no apparent way to achieve this goal or just because one wishes to stop the search for solutions which involve this goal.

Once a proof has been constructed, the goal tree can be explored for instance to generate an explanation of the solution.

Support for logic variables and unification is provided by a "variable" object class and a generic function "unify".

The generic function "unify" is specialized on the two arguments representing the terms to unify. During unification some variable might get bound in the current environment, and in such a case the binding is recorded in a trail structure for use in case of backtracking. The results of "unify" are a boolean value and a list of constraints that need to be fulfilled for the terms to unify.

This technique provides a smooth integration of logic programming and procedural programming, so that for instance the solutions of a query can be examined in turn and further processed.

4.5.2. Presentation Tools

The tools in this collection are helpful for the knowledge engineer to present information and interact with the user of the application. Some of the tools are specifically provided for displaying information about the knowledge base and the status or the progress of the activity of the inference engine.

For instance a generic grapher is a graphic tool useful to display the concept taxonomy in the knowledge base, or the status of the goal tree during a deduction.

A browser tool is used to navigate through the knowledge base. An interactive form design tool and a form filling interface provides the support for interacting with the knowledge base in a Query By Example style (22).

4.5.3. Query Tools

Some query operations are often useful, and so are part of our toolkit. The most obvious one is the query "whichis?", which retrieves all individuals which fullfill a given description. Another frequent operation is to collect all descriptions relating to a certain individual and fuse them whenever possible. For instance, if an individual Ind is described as being both:

(a Person (with-every child (a Male)))

and

(a Person (with-every child (a Female)))

the attribute child of these two descriptions can be fused and then would collapse to Nothing (the empty set). As a consequence the whole description would also collapse to Nothing , giving raise to a contradictory situation of a non existing individual Ind.

5. Implementation

The elements of the framework described in the paper have reached various stages of development, with a few of them already at the stage of fully supported products.

Significant amount of effort was spent in tuning the performance of the lowest layers to ensure a proper foundation for the overall framework. The Common LISP implementation consists of an interpreter written in C and a compiler which uses C as its intermediate language. Its performance, measured in terms of the standard Gabriel Benchmarks, is comparable to that of the best LISP implementations on the market.

The Common LISP kernel has been extended with special support for CLOS and logic programming. The most critical elements of the CLOS implementation are generic function invocation and slot access. Since invoking a generic function requires to determine the type of its arguments and to dispatch accordingly to the appropriate method, a generic function invocation can be considered equivalent to a normal function call followed by a case expression on the type of the arguments. In such comparison, the generic function call is approximately 4 times faster. Such performance was achieved by using a variety of techniques, and in particular by cashing methods in hash tables. Slot access optimization is done by the compiler whenever it can determine the class of the value of an expression. In such cases a slot access is turned into just an indirect memory reference, without involving any generic function invocation.

The CLX implementation has also been carefully tuned: in various benchmarks CLX programs take approximately 1.7 the time required by the corresponding C program.

The implementation of Common LISP consists of 50000 lines of C code, and 25000 lines of LISP code. CLOS consists of 6000 lines of LISP code, CLX of 17000 lines of LISP code. CLUE consists of 7000 lines of CLOS code. The graphic toolkit library contains currently a dozen of contact windows for a total of 9000 lines of CLUE code. The knowledge base tools consists of 7500 lines of CLOS code.

The size of the executable including CLOS, CLX, and CLUE is approximately 4.5 Megabytes. The system has been ported to various workstations (Sun, Apollo, HP, Unigraph), minis (DEC VAX), and Intel 386 based PCs under Unix.

The system has been in use both internally and at a number of customer sites since early 1989, for building or porting applications of significant complexity.

6. Conclusions

The ESPRIT Project P440 has been an opportunity to tackle many significant issues of knowledge representation and of advanced programming. We had to face challenging problems with respect to the state of the art of knowledge representation formalisms, but we also had to face significant software engineering problems in order to develop systems which could be practical, robust, portable, and open.

We solved most of these issues by adopting and adhering coherently to a small set of principles: having always a precise semantic model as a reference in the design and using object oriented programming techniques in the implementation.

Our experiments on knowledge representation systems will continue in further ESPRIT2 projects where we plan to verify our ideas on machines with a high degree of concurrency.

REFERENCES

(1) AIT-KACI, H. and R. NASR, "LOGIN: A Logic Programming Language with Built-in Inheritance", Journal of Logic Programming, 3(3), pp 187-215, 1986.

(2) ATTARDI, G. and M. SIMI, "Consistency and Completeness of Omega, a Logic for Knowledge Representation", Proc. of 7th IJCAI, Vancouver 1981.

(3) ATTARDI, G. and M. SIMI, "A Description Oriented Logic for Building Knowledge Bases", Proceedings of the IEEE, Vol 74, No 10, pp 1297-1472, October 1986.

(4) ATTARDI, G. and S. DIOMEDI, "Multithread Common LISP", ESPRIT MADS Tech. Rep. TR-87.1, DELPHI, 1987.

(5) ATTARDI, G. et al., "Metalevel Programming in CLOS", ECOOP 89, S. Cook (ed.), Cambridge University Press, 1989.

(6) BOBROW, D.G. and T. WINOGRAD, "An Overview of KRL, A Knowledge Representation Language", Cognitive Science, 1(1), pag. 3-46, 1977.

(7) BOBROW, D.G., et al., "Common LISP Object System Specification", X3J13 Standards Committee Document 88-003 (ANSI COMMON LISP), June 88.

(8) BRACHMAN, R.J. and J.G. SCHMOLZE, "An Overview of the KLONE Knowledge Representation System", Cognitive Science, 9(2), pag. 171-22, 1985.

(9) BRACHMAN, R.J., V. GILBERT PIGMAN, H.J. LEVESQUE, "An Essential Hybrid Reasoning System: Knowledge and Symbol Level Accounts of Krypton", Proc. of 9th IJCAI, Los Angeles, 1985.

(10) BOBROW, D.G. et al., "The Common LISP Object System", SIGPLAN Notices, 1988.

(11) FIKES, R. and T. KEHLER, "The Role of Frame-based Representatio in Reasoning", Comm. of the ACM, 28(9), pag. 904-920, 1985.

(12) GETTY, R. and R. SCHEIFLER, "The X Window System", Transactions on Graphics, 1987.

(13) GOLDBERG, A., "Smalltalk 80", Addison-Wesley, 1986.

(14) HAIMOWITZ, I.J., R.S. PATIL and P. SZOLOVITS, "Representing medical knowledge in a terminological language is difficult", Symposium on Computer Application in Medical Care, 1988.

(15) "KEE User's Manual", IntelliCorp, 1988.

(16) LEVESQUE, H.J. and BRACHMAN, R.J., "A Foundamental Tradeoff in Knowledge Representation and Reasoning", in "Readings in Knowledge Representation", Brachman and Levesque Eds, Morgan Kauffman, 1985.

(17) NEBEL, B., "Computational Complexity of Terminological Reasoning in BACK", Artificial Intelligence Journal, 34(3), April 1988.

(18) OREN LA MOTT and K. KIMBROUGH, "Common LISP User Interface Environment", Texas Instrument, 1988.

(19) ROBERTS, R.B. and I.P. GOLDSTEIN, "The FRL Manual", AI Memo 409, MIT AI-Lab, 1977.

(20) STEELS, L., "The KRS Concept System", Vrije Universiteit Brussel AI-Lab Tech Rep 86-1, Brussels, 1985.

(21) VILAIN, M.B. "The restricted language architecture of a hybrid representation system", in Proc. of the Ninth IJCAI, pag. 547-551, 1985.

(22) ZLOOF, P. "Query by Example", IBM Journal of Syst. and Devel., 1986.

(23) S.BELHASSEN, E.DUROCHER, I.FILOTTI, W.QING, "Process Migration in a heterogeneous network of operating systems", ESPRIT CHAMELEON Tech. Rep. TR-88-55, DELPHI, 1988

List of Members of the Consortium

DELPHI S.p.A. Via della Vetraia, 11 I-55049 Viareggio LU Italy

VUB AI Lab Pleinlaan, 2 B-1050 Brussels Belgium

LABORATOIRES DE MARCOUSSIS CGE Research Center Route de Nozay F-91460 Marcoussis France

An Intelligent CASE Tool for the Jackson System Development Method

Patrick Connolly
Seamus Kearney
David O'Neill
Generics Software Limited
Clonard House
Sandyford Road
Dublin 16
IRELAND
Phone: +353-1-954012
Fax: +353-1-954011
EMail: {connolly,kearney,oneill}@genrix.ie

Abstract

ESPRIT Project 510 ("ToolUse") is conducting research into *method-driven* approaches to software development. A particular area of interest is the development of tools to *actively* assist in the application of software development methods. This paper describes the *JSD Advisor*, a knowledge-based prototype tool supporting the Jackson System Development (JSD) method. Active support for the JSD method is made possible by the incorporation of both method and application domain knowledge in the Advisor knowledge base. The method knowledge takes the form of an explicit *JSD development rule set*. The application domain knowledge takes the form of an explicit *application domain model*. The role of application domain modelling within the software development process is discussed. An assessment of the JSD Advisor is given. This research led to the development of the commercial product *GenASSIST*, an intelligent CASE tool supporting JSD.

1. Introduction

Software development is a highly complex process. The effective use of methods can help to reduce complexity and improve the quality of the software produced. A software development method is an organised collection of notations, techniques and formal or semi-formal procedures for carrying out one or more of the major life cycle activities. A method identifies the deliverables and prescribes the form or the notation in which they will be produced [HJK87c].

Tool support is necessary to ease the user's task of applying a method, and to constrain users to adhere to the recommended approach. Many existing tools and environments tend to focus on the more routine activities, such as the entry, processing and management of the various deliverables produced during the software life cycle. ("Deliverables" should be taken here in a wide sense, including requirements, specifications, and programs). Much less support is available to *actively* assist in the application of a software development method, i.e. to provide decision support [Horgen 85, O'Neill 87]. However, a truly *method-driven* approach to software development requires the availability of such tools. This paper describes the *JSD Advisor*, a knowledge-based prototype tool supporting the

Jackson System Development (JSD) method. Active support for the JSD method is made possible by the incorporation of both method and application domain knowledge in the Advisor knowledge base.

The next section discusses ESPRIT Project 510, which provided the context for most of the work described here. The reasons for choosing to support JSD and a brief overview of the method follow. A detailed description of the JSD Advisor is given, including a discussion of the role of application domain modelling within the software development process. An assessment of the Advisor is presented. The *GenASSIST* tool, which is a commercial spin-off from this research, is also described.

2. Esprit Project 510 : "ToolUse"

ESPRIT Project 510 ("ToolUse") aims to produce "an advanced support environment for method-driven development and evolution of packaged software" [Horgen 85]. "Method-driven" means the ToolUse environment will embody generalisations and abstractions of a number of software development methods, which are represented in such a way that the environment can treat methods as objects. It was intended that the ToolUse environment would encompass techniques from a range of design methods represented in a completely formal development language [Connolly 89].

A number of software development experiments on various design methods were undertaken. The short-term objective was to evaluate and describe the particular methods [HJK87c]. The long-term objective was to generalise the experimental results into development rule sets which could be formalised and used as a basis for developing prototype *method advisor* tools. These tools would then be used to support the application of the associated method, to support future knowledge gathering experiments and to support knowledge validation through the replay of experiments. The methods studied included:

- JSD (Jackson System Development) [Cameron 86, Jackson 83]
- SARS (System for Application Oriented Requirements Specification) [Epple 85]
- VDM (Vienna Development Method) [Bjorner 82]
- OOD (Object Oriented Design) [Booch 87]
- HOOD (Hierarchical Object Oriented Design) [Heitz 87]
- ESTEREL [Berry 84]

In the case of JSD, the process was carried one stage further by adding application domain knowledge (in the form of an explicit application domain model) to the method knowledge encapsulated in the development rule set.

3. Jackson System Development

JSD was chosen as a method for research within the ToolUse project for the following reasons:

- The method is widely used in industry throughout Europe.
- The method has a wide domain of applicability, including data processing and real time applications.
- Unlike many other methods, JSD addresses most of the software life cycle, from requirements analysis through to implementation and subsequent system maintenance.

- The JSD method provides comprehensive guidance.

From a high level, the JSD method can be divided into two distinct phases, the *specification phase*, followed by the *implementation phase*.

The specification phase is concerned with the production of a (nearly formal) specification of the system to be implemented. The specification phase is divided into a *modelling stage* and a *network stage*. The modelling stage is concerned with the production of a model of those areas of the real world which are of relevance to the system. The model produced consists of a number of *entities* (objects in the real world), modelled as long-running processes (i.e. a time-ordered sequence of the *actions* that an *entity* performs and suffers). The system *functions* are added to the model in the network stage. Each function is also expressed as a *process*, and is connected to the *entity processes* and other *function processes*. The resulting JSD specification therefore consists of a network of communicating processes.

The implementation phase is concerned with fitting the JSD specification to the target machine. A number of formal, mechanisable transformations are available to assist in this task.

4. The JSD Advisor

The JSD Advisor is a prototype knowledge-based assistant for the JSD method. The Advisor is centred around a *JSD development rule set* (providing method knowledge) and an application domain model (providing application domain knowledge)[1].

The JSD development rule set is an attempt to crystallise the knowledge applied by an experienced JSD developer into a concise form (namely, a knowledge-base of development rules). The rule set was developed from the existing literature on JSD [Jackson 83, Cameron 83, Cameron 86], and by applying JSD to a set of case studies [PC86f]. The first draft of the rule set was expressed in English and structured hierarchically according to the phases and stages of JSD. Control was expressed by sequencing, iteration, alternative and backtracking constructs [PC86a]. The rule set was further refined in subsequent drafts [PC86c, PC86h]. Although the initial drafts of the rule set addressed the entire JSD method it was important to focus the development of a support tool on a specific phase given the time and resources available. Effort was concentrated on supporting the JSD specification phase (i.e. the modelling and network stages) as this phase was considered to be of most interest.

The idea of using rule bases for software development will be familiar to those acquainted with transformational programming [Burstall 77]. Transformational systems such as CHI and CIP derive a machine-oriented program from a high-level specification using sets of transformation rules [Green 81, Moeller 84]. The JSD Advisor differs from such systems in that it is method-specific and incorporates application domain knowledge to address the derivation of a specification from an application model [Connolly 89].

Application domain modelling (explained in detail below) was introduced into work on the Advisor with the objective of making more active assistance possible [Barstow 85]. The particular domain chosen for modelling was the *Server-Client* domain, which is a generalisation of the process control and resource management domains. JSD is a

---

1   The boundary between method-specific and application domain-specific knowledge in the
     JSD Advisor is not, in fact, easily drawn. The two areas are discussed separately here for
     clarity

particularly suitable context for this work, since JSD advocates a modelling approach to software development.

The Advisor is implemented using an expert system builder tool called KES (Knowledge Engineering System) [KES 86].

4.1 Application Domain Modelling

Software development may be regarded as the production of a series of models or descriptions, beginning with a completely world-oriented model and progressing toward models which are more and more machine-oriented [Greenspan 84, Mostow 85a]. These models, called *application models*, describe the *application* to be developed, i.e. the functional and non-functional requirements of the application, and all the information needed to implement the requirements. The initial world-oriented application model may be written in a natural language such as English, while the final application model, which is totally machine oriented, is written in machine code. Each application model may be derived from the previous application model through a series of correctness-preserving transformations (for example, a compiler performs a correctness-preserving transformation from an application model written in some "high level" computer language to an application model written in machine code.).

4.1.1 Application Models and Application Domains

When developing a new application, an experienced software developer remembers concepts applied in developing previous "similar" applications, and reuses those concepts where possible [Greenspan 85, Adelson 85][2]. To take a particular example, imagine the case of a developer who has previously implemented a Pascal compiler, and is now to implement a Modula-2 compiler. These two applications are clearly "similar". Most of the concepts applied in the development of the Pascal compiler could be reused in developing the Modula-2 compiler, even in the case where it was not possible to reuse any of the physical code.

It is therefore useful to introduce the idea of an *application domain*. Broadly speaking, an application domain is a set of "similar" or "related" applications, e.g. the domain of all text processing applications, or the domain of all process control applications. When developing a new application, it will generally be possible for a developer to reuse concepts applied in developing previous applications which fall within the same application domain.

For our purposes, it is not important to classify all possible applications into a set of exhaustive and mutually exclusive application domains (even if this were possible). It is only important to understand (1) that different applications can be in some way "similar", (2) that concepts applied in developing one application can be reused in developing other "similar" applications, and (3) that similarity between different applications can be expressed by conceptually grouping these applications into an application domain.

---

2   In this context, "concepts" should be taken in a broad sense, including more abstract concepts such as general specification and implementation principles, as well as more concrete concepts such as objects, functions, and other structures used in implementing previous applications

4.1.2 Application Domain Models and Software Development

The notion of an application domain gives rise to that of an *application domain model*. An application domain model is a model or description of a particular application domain. An application domain model therefore models the concepts which are common to the applications which fall within the application domain.

These ideas lead to a modelling approach to software development, pictured in Figure 1.

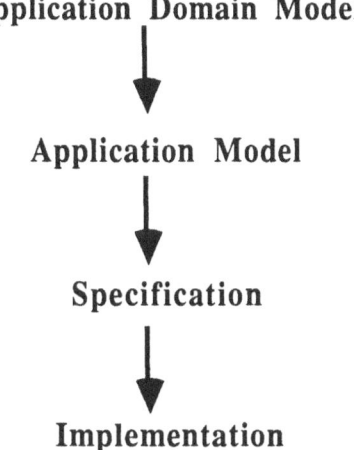

Figure 1: A Modelling Approach to Software Development

When developing an application, the concepts expressed in the application domain model are drawn upon to produce an initial (perhaps informal) application model. From this model a specification is produced, which is successively refined to produce an implementation. (Strictly speaking, specifications and implementations are themselves merely application models at different levels of abstraction, but it is useful to maintain the distinction). It is important to note that a software developer (perhaps unconsciously) follows this approach, even when only using pencil and paper, and even when not familiar with the modelling terminology used here. In the simplest case, a software developer may use only one application domain, namely the set of applications which the developer has previously implemented. The developer has a mental model of these applications, i.e. an in-built application domain model, and he draws from this model in developing new applications. Thus, an experienced software developer continuously enhances his in-built mental application domain model, and develops software by applying knowledge from this application domain model.

Summarising, an *application domain* is a (possibly infinite) set of *applications* having some common domain of interest. An *application domain model* is a description of the general concepts, structures and principles which apply to an *application domain*. This model contains the knowledge necessary to derive a particular application model, i.e. a description of an application at a suitable level of abstraction.

4.2 The Server-Client Application Domain Model

The Advisor's application domain model is based on a *Server-Client* paradigm. A

Server-Client paradigm is usually found in applications or application components which in some way involve resource management or process control. The Server-Client application domain model describes a class of applications where one object or person, the *Supplier*, provides services to other objects or persons called *Clients*. These services may include the provision of objects to Clients. These objects are called *Service Objects*. Client requests for Service Objects are processed by the *Server*.

The model is concerned with the management of Service Objects, and in particular with the scheduling of these objects in response to Client demands for their use. Applications which fall within the Server-Client domain include:

- A library, which provides books for its members, and which must keep track of its books at all times. In this case, the Supplier is the library, the Server is the librarian, the Service Objects are the books, and the Clients are the library members.
- A memory management system, which provides memory buffers to processes. In this case, the Service Objects are the memory buffers and the Clients are the processes. The Supplier and Server are in fact the same object, namely the memory management system.
- A car rental firm, which provides cars to its customers and which must be able to fulfil customer requests for cars and monitor the servicing of its cars.
- A printer scheduling system, which maintains queues of print jobs for a number of printers.
- A CPU time slice arbiter, which assigns the CPU of a computer system to executing processes for a certain time slice.
- A car repair service, where mechanics are assigned to cars.

4.3 Software Development With The JSD Advisor

The Advisor uses its in-built Server-Client application domain model to produce an application model based on an analytical questioning session with the user. The question texts use the terms supplied by the user from previous answers, and thus the Advisor's questions use terms familiar to the user. In addition, the user can ask the Advisor for an explanation of any question asked. A retraction facility is provided that allows the user to backtrack and withdraw an answer to a question, i.e. to change the application model.

From the application model, the Advisor infers a corresponding JSD Specification using the JSD development rule set stored in its knowledge-base. The user can ask for a justification of the Advisor's decisions (e.g. why a particular entity appears in the JSD specification). The justifications are given in English, again using the terms given by the user. This process is pictured in Figure 2.

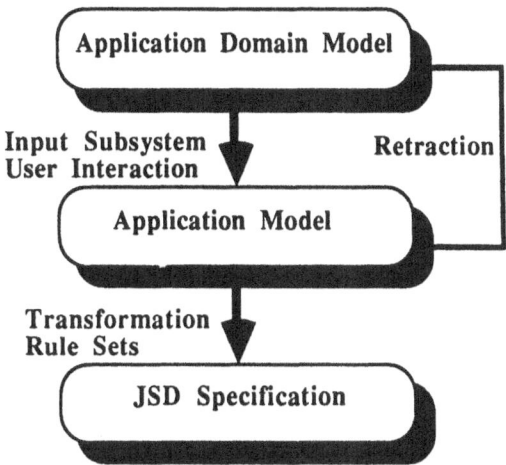

Figure 2: Software Development with the JSD Advisor

4.3.1 Derivation of the Application Model

The Advisor's application domain model contains the concepts needed to build an initial world-oriented application model, i.e. a description of the application using the terms of the real-world environment in which the software system is to function. The application model is built by the Advisor through analytical questioning of the user. For example, consider a library management system. A world-oriented application model for this system would discuss books, librarians, library members, book reservations, the action taken for overdue books, etc. These real-world terms are substituted for the generic terms in the Advisor's application domain model. For example, the Advisor's application domain model includes the generic term *Supplier*. The Advisor will ask the question:

What is the Supplier called?

The user might reply "Library". In all subsequent questions referring to the Supplier, the Advisor will substitute the term "Library". Another generic term used in the Advisor's application domain model is *Service Objects*. The Advisor will ask the question:

What are the objects supplied by the Library?

The user might reply "Books". Subsequently, the Advisor will substitute the term "Books" in all questions referring to Service Objects, and so on. The user can ask the Advisor to explain its questions and the possible answers in more detail. Thus, for the library example, when the Advisor asks:

Who are the Clients for Books?

... the user can request an explanation and the Advisor will explain exactly what is meant by "Clients".

The Advisor therefore derives a specific application model by *instantiating* the generic Server-Client application domain model.

4.3.2 Derivation of the JSD Specification

Once an application model has been built, the Advisor transforms this model into a JSD specification. The Advisor performs the transformation by applying the *JSD development rule set* stored in its knowledge base. The rules encapsulate the knowledge of an experienced JSD developer. They express both general knowledge of the JSD method, and knowledge on how to transform an application model into a JSD specification.

For example, while building the application model for a library management system, the user might tell the Advisor that a library member borrows books and may return them overdue, and that the system is to calculate the payment (or fine) due by library members when books are returned overdue. From this, the Advisor can deduce that *Book* must be an entity in the JSD specification, with entity actions *Borrow*, *Return*, and *Return Overdue*, and that *Payment Calculation* is an information function, triggered by the *Return Overdue* action of the *Book* entity.

The transformation from an application model to a JSD specification is performed in a stepwise fashion, corresponding to the steps of the JSD specification phase:

- Choose Entities.
- Choose Entity Actions.
- Structure Entities (not implemented at time of writing).
- Define Entity Data.
- Choose Functions.
- Define Data Stream and State Vector Connections.

The user can indicate which step assistance is required for, and the Advisor will ask for all information necessary to perform the step.

The user can also ask the Advisor to explain its conclusions. Thus, for the library example, when it is deduced that *Book* is an entity, or that *Borrow*, *Return* and *Return Overdue* are entity actions, the user can ask why these deductions were made.

5. Implementation of the JSD Advisor

The JSD Advisor is implemented on top of an expert system builder tool called KES [KES 86].

KES (Knowledge Engineering System) offers:

- Attributes, classes and inheritance.
- Backward chaining (by production rules) and forward chaining (by demons).
- Adequate user interface facilities.
- Truth maintenance.
- Embeddability: a KES knowledge-base may be embedded within a C program.

The implementation of the Advisor is divided into a knowledge-base and an embedded system:

    Knowledge Base < − −-> Embedded System < − −-> User

The knowledge base is implemented using the knowledge representation mechanisms provided by KES. It contains the application domain model, the inferred JSD specification, and the JSD development rule set. The rule set controls the instantiation of the application domain model to produce an application model, and infers a JSD specification from this application model.

The embedded system is interposed between the knowledge base and the user. The embedded system is implemented as a C program and provides:

- The user interface, including a command language to allow access to the application model and JSD specification stored in the knowledge-base.
- A JSD help facility.
- A session handling facility.
- Some consistency maintenance facilities which can not be implemented in KES.

5.1. Representation of the Application Domain Model and the JSD Specification

Both the application domain model and the JSD specification are represented by a set of KES *attributes* and *classes*. A brief edited extract from the Advisor knowledge base is reproduced in Example 1, to give a flavour of the approach adopted.

> **attributes**:
>
> Supplier: **str**
> {**question**: "What is the Supplier called?"}
> {**explain**: "This object, called the Supplier, supplies, i.e. sells,",
> "hires, or otherwise provides goods or services to Clients ..."}.
>
> \ ...
>
> **classes**:
>
> Service Objects:
>   **attributes**:
>   \ ... attributes of Service Objects here
>   %
> **endclass**
> {**question**: "What are the Service Objects supplied by the Supplier?"}
> {**explain**: "The definition of a Supplier is that it provides",
> "goods or services, called Service Objects, to Clients.",
> "You are being asked here to list these Service Objects ..."}.
>
> Example 1: Representation of the application domain model

It can be seen that the *Supplier* name is represented by a KES string *attribute*, whereas *Service Objects* are represented by a KES *class*. The texts enclosed in curly brackets are called *question* and *explain* attachments. Whenever a rule requires a value for an attribute, or a list of members for a class, the appropriate question attachment is displayed to prompt the user to input the value or class members. The user has the option of typing "explain" for more information on a question, in which case the appropriate explain attachment is displayed.

The process of instantiating the application domain model to form an application model is therefore a process of entering values for attributes and members for classes. The attribute values and class members are requested as they are required by the rules for building the JSD specification.

5.2.Representation of the JSD Rules

KES provides *production (if-then) rules* and a *backward chaining inference engine*. The antecedents of the JSD rules refer to the application model; the *consequents* express the information which can be inferred about the JSD specification. A simplified rule is reproduced below in Example 2.

> Rule I0a:
> SO : Service Objects
> **if** Server Type = Human Being **and**
>   SOClient Identifies = Must Always | May But Need Not **and**
>   SOServer Must Check Availability = yes **and**
>   SOAssistance Needed In Checking Availability = yes
> **then**
>   SOIs Entity = **true**.
>   SOAll Availability Actions Relevant = **true**.
>   SOHas Availability Function = **true**.
> **endif**
> {**explanation**: "If the Service Object can or must be identified, and",
> "the Server is to be assisted with the checking the availability of",
> "the Service Object, then the Service Object MUST be an Entity,",
> "with ALL the actions which determine its availability for use being",
> "relevant for the Entity's model process."}.

Example 2: A Simplified Rule

This rule states that if (1) the Server is human, and (2) when requesting Service Objects, Clients can request particular Service Objects, and (3) when a request for a particular Service Object is made, the Server must check whether the particular Service Object is available and (4) the system to be implemented is to assist in checking the availability of these objects, then (1) the Service Object is an entity in the JSD specification, (2) all actions dealing with the availability of the entity are relevant entity actions in the JSD specification, and (3) the JSD specification must include a function for checking the availability of particular Service Objects.

The text in curly brackets is an *explanation attachment*. Each rule may have an explanation attachment, which gives an English explanation of the rule. For every value inferred by rules, KES keeps a list of the rules which fired to infer that value. This list is displayed when the KES "justify" command is issued. In the case of the Advisor, explanation attachments and the justify command are used to explain the particular decisions taken by the Advisor, e.g. why a particular entity, entity action, function, etc., forms part of the JSD specification. Generic terms such as "Service Object" and "Entity" are substituted with the application-specific text entered by the user.

It should be noted that KES backward-chaining rules were not sufficient to represent the full JSD rule set, and that some of the JSD rules are in fact represented as KES *demons*.

6. Assessment of the JSD Advisor

The JSD Advisor attempts to capture, for reuse, some of the software engineer's expertise. Not surprisingly, it has proved difficult to isolate individual factors from among the myriad of (frequently implicit) decisions that constitute a development [Ryan 88]. Developers of early systems such as MYCIN were astonished at the difficulty involved in codifying even a small proportion of the knowledge that is retained and used by experienced professionals [Buchanan 84].

In order to assess the usability of the JSD Advisor (and to assist in validating the JSD development rule set), a case study evaluation was performed. A JSD specification was developed by hand and compared with the specification produced by the JSD Advisor. The application chosen was a computer sales and services company [FC89a]. The specifications were compared with respect to completeness and correctness [PC89a, PC89b].

As regards completeness, the Advisor covered approximately half of the application. It was necessary to enhance the JSD specification produced by the Advisor to cover areas not modelled by the tool. This result was as expected. Regarding correctness, all rules that fired were correct, apart from one rule which inferred a superfluous entity action and interactive function. The questions, explanations and justifications generated by the Advisor were meaningful and understandable, apart from a few cases at the "edges" of the Server-Client application domain model.

7. GenASSIST: An Intelligent CASE Tool

The JSD Advisor is a prototype tool which has proved very useful for experimentation within the ToolUse project. The Advisor is currently being enhanced and packaged independently as the basis of a commercial CASE tool called GenASSIST.

The ultimate aim of CASE (Computer-Aided Software Engineering) is to automate to the greatest extent possible all the activities within the traditional software life cycle from initial analysis and design through to code generation and maintenance.

The current generation of CASE tools tend to address the more mundane tasks of software development. For example, graphical editors eliminate the need for paper, pencils and erasers. Intelligent CASE tools will offer more active support in the use of a method. "Active support" means that the CASE tool not only constrains the developer, but also drives the development process, for example by taking the initiative in dialogues. Intelligent, or knowledge-based, support implies that non-trivial deductions are made by the system based on previously encoded knowledge [Connolly 89]. GenASSIST belongs to this emerging generation of intelligent CASE tools.

GenASSIST improves on the JSD Advisor by expanding the Server-Client model, by completing the rules to cover the model, by improving the questions, question explanations and design decision justifications and by expanding the embedded system. GenAS-SIST runs on VMS and Unix based systems and can be made to run beside existing tools for the JSD method (e.g. Speedbuilder and PDF). The structure of GenASSIST is shown in Figure 3.

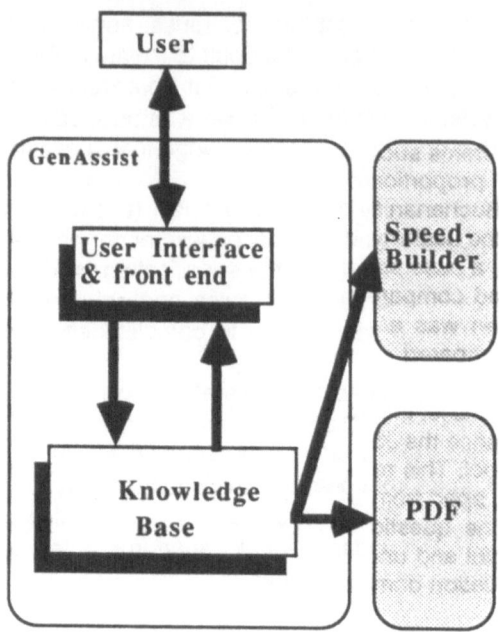

Figure 3: GenASSIST Architecture

GenASSIST can automatically produce a JSD specification for those areas of an application which fall within the boundary of the Server-Client application domain model. In addition, every development decision made can be justified by GenASSIST. The justifications relate the development decision to user input, thus providing requirements tracing.

GenASSIST illustrates some of the benefits of integrating software engineering and expert system technology [Mostow 85b]. In the particular case of GenASSIST:

- The approach is declarative: the implementation parallels the way we think about the problem.
- (Relatively) fast rule base development and maintenance is possible.
- Rapid prototyping of the user interface is possible.
- The KES "justify" command allows GenASSIST to explain its design decisions.
- KES allows access to the embedded system (i.e. a C program) for the more procedural aspects of GenASSIST.

To date, market reaction has been very positive. GenASSIST was a finalist in the 1988 RITA award. It is currently being evaluated by both Michael Jackson Systems Limited in London and Jackson System Corporation in the United States.

Acknowledgements

The work described here was carried out by various members of the ToolUse project. The project is made up of teams from software houses and research establishments from

Belgium, France, Germany, Ireland and the United Kingdom. The work reported here was partly funded by the CEC under the ESPRIT programme.

Generics is a registered trademark of Generics Software Limited.

GenASSIST is a trademark of Generics Software Limited.

JSD is a development of Michael Jackson Systems Ltd.

Speedbuilder is a trademark of Michael Jackson Systems Ltd.

PDF is a trademark of UKAEA Harwell.

KES is a trademark of Software Architecture & Engineering, Inc.

References

External Publications

[Adelson 85] B. Adelson, E. Soloway, "The Role of Domain Experience in Algorithm Design", IEEE Transactions on Software Engineering, Vol. SE-11 (11), November 1985.

[Barstow 85] D.R. Barstow, "Domain-Specific Automated Programming", IEEE Transactions on Software Engineering, Vol. SE-11 (11), November 1985.

[Berry 84] G. Berry, "The ESTEREL Synchronous Programming Language and its Mathematical Semantics", INRIA Technical Report, 1984.

[Booch 87] G. Booch, "Software Engineering with Ada", Benjamin Cummings, 1987.

[Bjorner 83] D. Bjorner, "Software Engineering Aspects of VDM", International Seminar on Software Engineering, Capri, North-Holland, 1983.

[Buchanan 84] B.G. Buchanan, E.H. Shortliffe, "Rule-Based Expert Programs: the MYCIN Experiments of the Stanford Heuristic Programming Project", Addison-Wesley, 1984.

[Burstall 77] R.M. Burstall and J. Darlington, "A Transformation System for Developing Recursive Programs", Journal of the ACM, Vol. 24 (1), January 1977.

[Cameron 83] J.R.Cameron "JSP & JSD : The Jackson Approach to Software Development", IEEE Tutorial, 1983.

[Cameron 86] J.R. Cameron, "An Overview of JSD", IEEE Transactions on Software Engineering, Vol. SE 12 (2), February 1986.

[Connolly 89] P. Connolly, D. O'Neill, K. Ryan, "Intelligent Support for the Jackson Method", CASE'89 Workshop, London, 1989.

[Epple 85] W. Epple, G. Koch, "SARS - A System For Application Oriented Requirements", Proc. IFAC Real Time Programming, North-Holland, 1985.

[Green 81] C. Green et al, "Research on Knowledge-Based Programming and Algorithm Design - 1981", Report Kes.U.81.2, Kestrel Institute, Palo Alto, California, 1981.

[Greenspan 84] S. Greenspan, "Requirements Modelling : A Knowledge Representation Approach to Software Requirements Definition", Technical report CSRG-155, University of Toronto, March 1984.

[Greenspan 85] S. Greenspan, "On The Role of Domain Knowledge in Knowledge-Based Approaches to Software Development", Proc. International Workshop on the Software Process and Software Environments, California, March 1985.

[Heitz 87] M. Heitz, "HOOD : A Hierarchical Object Oriented Method for Technical and Real Time Software", CISI Ingenierie, Toulouse (in French).

[Horgen 85] H. Horgen, "ToolUse: An Advanced Support Environment for Method-Driven Development and Evolution of Packaged Software", Proc. ESPRIT Technical Week '85, North-Holland, 1985.

[Jackson 83] M. Jackson, "System Development", Prentice-Hall, 1983.

[KES 86] KES, "Knowledge Base Author's Reference Manual", Software A&E Inc., Arlington, Virginia, USA.

[Moeller 84] B. Moeller (ed.), "A Survey of the Project CIP - Computer-Aided, Intuition-Guided Programming - Wide Spectrum Language and Program Transformations", Technische Universitaet Muenchen, Institut fuer Informatik Report TUM-I8406, 1984.

[Mostow 85a] J. Mostow, "Toward Better Models of the Design Process", The AI Magazine, Spring 1985.

[Mostow 85b] J. Mostow, "What is AI? And What Does It Have To Do With Software Engineering?", foreword to IEEE Transactions on Software Engineering, Vol. SE-11 (11), November 1985.

[O'Neill 87] David O'Neill, "Support Tools for the Abstract Data Typing Approach to Software Development", M.Sc. Thesis, Trinity College Dublin, April 1987.

[Ryan 88] K. Ryan, "An Experiment in Capturing and Classifying the Software Developer's Expertise", First International CASE Workshop, Boston, 1988.

Internal ToolUse Documents

[DO88d] Patrick Connolly, David O'Neill, "The JSD Advisor", Project ToolUse, Generics Software Ltd., Ref: DO88d, July 1988.

[FC89a] Fiona Clarke, Patrick Connolly, "A Description of the Computer Firm Case Study" Project ToolUse, Generics Software Ltd., Ref: FC89a, 1989.

[HJK87c] S. Kearney and H.-J. Kugler (eds.), "Derivation of Experimental Design Rules and Their Impact on DEVA", Task E1 Intermediate Report, Deliverable 25, Ref: HJK87c, December 1987.

[PC86f] Patrick Connolly, "The OBDH Case Study - A Solution", Project ToolUse, Vector Software Ltd., Ref: PC86f, 1986.

[PC86a,c,h] Patrick Connolly, "JSD Rule Set", Project ToolUse, Vector Software Ltd., Ref: PC86a/c/h, 1986.

[PC89a] Patrick Connolly, "The Computer Firm Case Study - A Solution", Project ToolUse, Generics Software Ltd., Ref: PC89a, 1989.

[PC89b] Patrick Connolly, "Validating the JSD Rule Set", Project ToolUse, Generics Software Ltd., Ref: PC89b, 1989.

PIMS, A PROJECT INTEGRATED MANAGEMENT SYSTEM

Jacques PARIS
Philippe VAUQUOIS
Cap Gemini Sogeti
Cap Sesa Innovation
33, Chemin du vieux Chêne - ZIRST
38240 - MEYLAN
FRANCE
Tel [33] 76 90 80 40
Fax [33] 76 41 06 29
Email: paris@capsogeti.fr, vauquois@capsogeti.fr

Abstract.

The PIMS project (Project Integrated Management System) is an ESPRIT I project that run from December 1985 to June 1989. The global aim is to analyse the software project management process and to build prototypes for supporting software project managers. The work consists of a theoretical study of project management and the implementation of a project management support system.

An important result of PIMS is the definition a theoretical framework and practical solutions for the parametrising of management methods. The theoretical study highlights the notion of Project Management ACtivity (PMAC) which represents the basic actions a project manager has to perform. These PMACs are organised into networks describing the different methods used for the management of software projects.

The final PIMS prototype has been delivered in June 1989. This prototype implements ideas developed in the theoretical study and provides support to project managers in two ways: first, it guides in the method application and second, it provides tools for performing the PMACs. This prototype consists of the infrastructure (object oriented database, user interface), conceptual schemas (the project data and the method description) and of the techniques for method support (Process Definition and Process Guide) and for support of basic management activities. PIMS proposes a set of standard techniques which support the basic management activities: project setup, planning, progress control and reporting.

Each PIMS consortium partner has defined an internal use of the PIMS results. The academic partners will publish a book. The industrial partners will mainly use the prototype for demonstrations, teaching and internal use. The solutions provided by PIMS have to be assessed and improved in a new project as a preparation for full scale industrialisation.

1. Introduction

The PIMS project was launched at the very end of December 1985 and finished in June 1989. The project is led by Cap Sesa Innovation (Cap Gemini Sogeti group) in Grenoble, France and regroups teams from BSO (Eindhoven, The Netherlands), PA Consulting (London, UK), Senter for Industriforskning (Oslo, Norway), London School of Economics (London, UK), London Business School (London, UK) and University of Amsterdam (Amsterdam, The Netherlands). It is a 3.5 years long project with a total effort of about 56 man-years.

The global aim is to analyse the software project management process and to build prototypes for supporting software project managers. Theoretical and practical work have been intertwined through the whole life of the project: theoretical aspects giving life to practical realisations and use of practical results feeding back the theoretical approach.

The PIMS project has been divided in three main phases: a preliminary phase that defined the first prototype (6 months), the development of the first prototype (18 months) and finally the development of a second, enhanced, prototype (18 months).

The prototypes have been developed on top of a specific infrastructure.This infrastructure proposes unified and powerful ways and mechanisms for representation and use of the project management knowledge. It also provides an attractive and convivial user interface which includes windows, graphical display, object-oriented dialogue.

This infrastructure allows the integration of different tools: *Process Definition* supports the definition of methods, *Process Guide* helps the project manager in the application of a defined method, and *standard tools* support the basic management activities (most important are work breakdown, resource allocation, scheduling, tracking, predicting, diagnosis and reporting).

The PIMS approach is given below, followed by a presentation of the results achieved by the project for both theoretical and practical aspects. Finally, the future use of PIMS results is presented.

2. PIMS Approach

This chapter first recalls the goals of PIMS, then defines the application domain it addresses and last evaluates the generalisation of the results obtained.

The PIMS acronym summarises the goals of the PIMS project:

P for Project: PIMS focuses on software projects
I for Integrated: integration has to be done both for the user interface and the functionalities. An integrated user interface means that all the interactions are built on the same model. Integration of functionalities is achieved if any functionality is available at any moment with the only restrictions coming from the method used (eg. the unavailability of prerequisite data).
M for Management: for helping in project management, it is not enough to provide a set of separate tools which support low level mechanical activities like planning creation or project tracking. An upper level has to be provided, helping the project manager to choose an activity according to the current state of the project and the method used.
S for System: we want to provide a System which helps project managers.

The domain of software project management is very large, and the PIMS project has chosen to support medium size projects in the field of business application. This corresponds to projects lasting one to two years and involving up to a dozen people. In such projects, the structure consists of the project manager and his team. We can then focus on the actual problems of project management without taking care of multilevel management.

On this application domain, PIMS aims at providing solutions for description and integration of the process model, support and guidance.

The result we have achieved in PIMS can be transposed to larger teams and larger projects because our domain of study embodies the basic structural hierarchy: project manager/team and senior manager/project manager. PIMS can also address other areas by eliciting knowledge specific to each domain: estimation parameters for real time applications, specific work breakdown models for artificial intelligence applications for example.

PIMS supports experienced and novice project managers. This support is proposed in a unified way, with flexible help at any moment. It proposes an aid through the whole lifeof the project development, starting before the appointment of the project manager and ending after the delivery of the system, since it also includes facilities for contract elaboration and seeks to extend the limit of system delivery to system maintenance, an important part of the software system life cycle.

In the two next chapters, the theoretical and the practical approaches are successively outlined. From the theoretical approach has emerged a good modelling of the software project management process. The practical approach has built prototypes based on ideas expressed in the theoretical work.

3. PIMS Theoretical Results

A basic result of the PIMS project is the PIMS Frame of Reference (PIMSFOR) [PIMS 88a]; it will be finalised by a book to be published at the beginning of 1990.Its goals are to increase the understanding of software project management and to provide material for the PIMS prototypes. It provides a Frame of Reference for the design of computer-based project management support systems. For all the results of the PIMSFOR, the reader has to refer directly to this document, the current paper will only give an overview of the results and their implication on the PIMS prototypes.

The analysis of the project management hasbeen directed by the refinement of the project manager goals. These topics are further developed below.

3.1. Entities within whe Frame of Reference

The figure 1 presents the main entities within the Frame of Reference and their relationships.

The **project environment** contains the contractor organisation (which employs the project manager), the client organisation (which has requirement and budget for a software development), the project (which includes the working team and the equipment) and the world of technology (current state of hardware, software).

Requirements are set from the project environment. These requirements act as constraints for the **Project Work System** (PWS). The PWS is the human and technical system the project manager uses to carry out his responsibility and produce the required software product. Work within this systemis defined in term of tasks assigned to the resources. The **Project Status Model** (PSM) is a model representing the PWS. It is built by looking at a project from several perspectives. These are mainlythe project function perspective, the activity perspective, the time and resource perspective, and the personal perspective. The PWS and the PSM represent static views of the objects the project manager is working on.

The **Project Management System** (PMS) includes all managerial operations on the PWS. It includes the plan creation, its control, its progress monitoring and, eventually its correction.The **Project Management Model** (PMM) is a model of the PMS defined in terms of a structure of PMACs linked through input and output predicates. This model is able to determine at any moment what activities must be carried out.

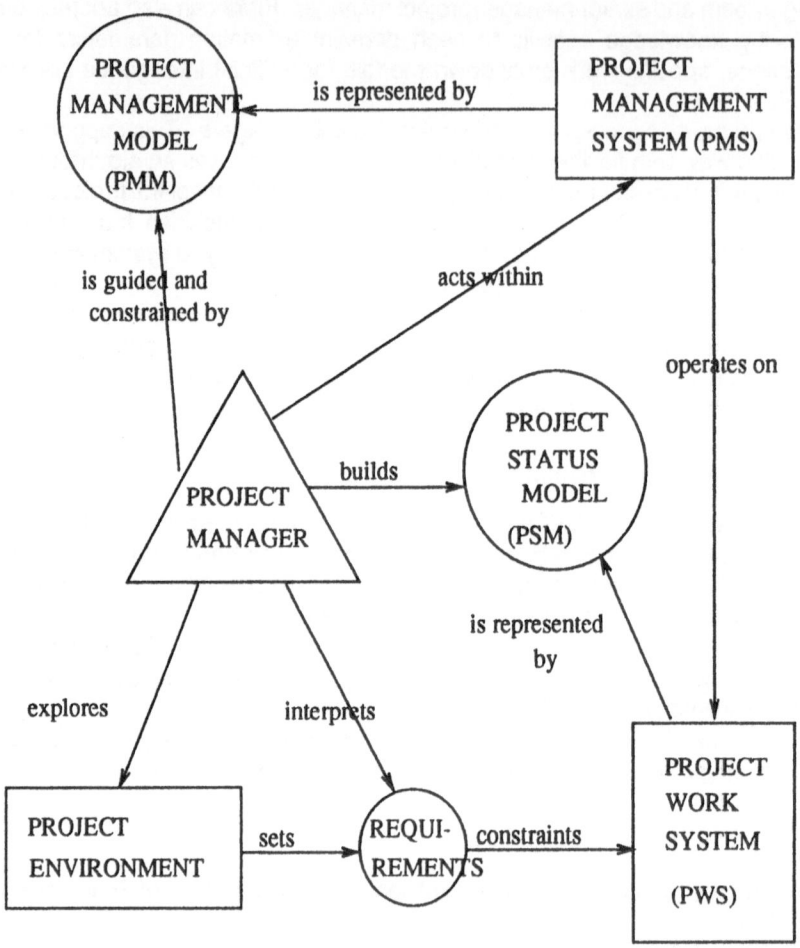

Global Project Management Model
- Figure 1 -

3.2. Goals of the Project Manager

PIMSFOR has first investigated the different goals of the project manager. The overall goal is to manage all his projects well while advancing his and his organisation's expertise. This concept of goal is used for understanding what motivates the activities of the project manager at different points during the project's lifetime. Progress towards the achievement of any particular goal can be evaluated at the conclusion of an activity in order to assess whether the activity has been successfully carried out or not.

In order to use goals in this way, they have to be detailed in such a way that we can measure their achievement. As an example, we give here the first level of refinement taken from PIMSFOR:

Manage all projects well

- Transact with project environment well
  - Clarify external requirements for the project (1)
  - Satisfy external stakeholders in the project (2)

- Manage this project well
  - Create and implement a project plan (3)
  - Maintain relationship between plan and reality (4)
  - Facilitate project work (5)

- Advance expertise (6)

These six numbered subgoals correspond to the first level of decomposition. Each one suggests a number of accurately describable tasks or task-clusters where valuable support can be provided; this level corresponds to what we call the Project Management Activities.

3.3. Project Management Activities (PMACs)

A Project Management Activity (PMAC) represents a part of the project manager job. A PMAC is defined by expected results, by the means to reach them (maybe a set of subPMACs), and by preconditions and postconditions. A PMAC can be activated only if its preconditions are verified. Preconditions can involve results of previous PMACs or temporal dependencies. The postconditions check the achievement of the PMAC and the progress towards achievements of subgoals.

These management activities are organised as a network, representing the project management method. Each node of the network is a PMAC. Links represent PMAC dependencies. More generally, a method is described by a hierarchy of networks because a PMAC may be composed of sub-PMACs.The global structure represents the overall project management process.

For instance, the goal "Manage Project Well" has produced the subgoal (4) "Maintain relationship between project plan and reality". This subgoal generates the following PMACs:

- tailor management process (TM): tailor the standards and procedures recommended by the organisation to the needs and requirements of the project
- quality assurance planning (QP): plan and control the quality of the project deliverables
- set-up monitoring procedures (SM): set up a reporting structure in such a way that current state can be compared with the planning
- identify, predict discrepancies (IP): identify and predict deviations from the plan given the results of monitoring
- diagnose,remedydiscrepancies (DR): assess significance of discrepancies, identify exceptions and choose a corrective course of action for exceptions
- case study (CS): create case studies for the purpose of future organisational use and learning

The PMACs are defined in the Project Management System and suggest objects to be modelised in the Project Status Model.

3.4. Impact of PIMSFOR on PIMS Prototypes

The PIMSFOR proposes 3 levels for the architecture of a Project Management Support System (PMSS):
- at the top level, a network of PMACs defines the method to be used for managing the project; this method is used for guiding the project manager in choosing the activity to be performed.
- at the intermediate level, the processing related to a PMACis defined; the basic part is realised by Techniques gathered in Environments
- at the lowest level, the data the project manager is working on are modelised in a PSM.

4. PIMS Prototypes

The PIMS project has produced two prototypes. These prototypes implement results of the theoretical work on top of a specific architecture. This chapter first describes the implementation choices, then the resulting implementation.

4.1. Implementation Choices

In the middle of 1986, we decided to develop a specific architecture for the first implementation of PIMS. The reasons were mainly the lack of appropriate products at this time and the need for integration facilities at the two levels of functionality and user interface. The prototypes have been implemented on Sun machines.

4.2. PIMS Architecture

Figure 2 gives a schema of the PIMS architecture. This architecture provides common facilities for

- OBJECT DEFINITION AND MANAGEMENT
  The **Information Manager** (IM) [PIMS 87] is used to define and manage objects of the PMCS and PSCS.

  The **Data Modelling Language** (DML) is an object definition language. It allows the description of classes and supports multiple inheritance mechanisms. A class contains attributes with several facets for typing (mono or multivalued, basic type or any other DML defined class) and reactive deamons (IfNeeded to calculate a value when accessing the attribute and AfterModify to propagate a modification). Constraints can be defined on attributes, objects or classes and within several classes; constraints are defined by predicates which are verified at the end of a transaction.

  An orthogonal set of functions allows to manipulate classes, instances and attributes. The method activation is performed by message passing. Other functionalities allow to define transactions, to manage return points, to check constraints and to manage historical values or versions.
  The IM has been implemented with Flavors in Allegro Franz Common Lisp [ACL 88] and runs on Sun machines.

- USER INTERFACE
  The Front End (FE) [PIMS 88c] consists of a Dialogue Manager (DM) and of visualisers.

The DM controls the logical dialogue with the user. A visualiser is an entity able to display objects of the IM and to forward to the DM user interactions on the displayed objects. Available visualisers are form visualiser (text, sliders, scrollable lists) and graphical visualiser (tree, network, table, calendar, Gantt, histogram). The FE is implemented partly in Lisp, using the IM facilities, and partly in NeWS [NeWS 87].

- TECHNIQUE INTEGRATION
A technique implements a set of functionalities provided to the PIMS user.

To facilitate technique integration, a technique is split up into procedural and user interface parts. A set of techniques is collected into an Environment.

The procedural part is a list of Lisp functions which define the processing of the technique.

Descriptive languages are used for describing the user interface: external user views, commands and attachment of commands to objects. A command description defines the external name of the command, the associated help, the associated internal function and the parameters (type, default value, help and precondition). A view description defines how to display an object on the screen; it is a reference to a visualiser and specific parameters for this visualiser. Each technique defines which commands have to be attached to an object; this set of commands is evaluated dynamically; if several techniques are active, one object will get the union of all the commands required by each technique. These descriptions are compiled or interpreted, and then used by the Dialogue Manager.

- REPORT PRODUCTION FACILITIES
PIMS uses the facilities of an external document editor for producing reports.

First the document editor is used to build the predefined structure of a PIMS document. This structure contains annotations defining accesses to the PIMS database or describing dynamical structure construction. For instance an access to the total cost of the project will be replaced by the value stored in the PSCS. A dynamical definition allows for example to create a chapter per task, or a table row per person.

The document creation is performed by a C process, automatically launched by PIMS, which expands all annotations of the document model by accessing the PIMS database. The result is then editable and printable by the document editor.

The external document editor used is GRIF [GRIF 86]. This work has been done in cooperation with the Eureka Software Factory project [ESF 88].

This architecture allows prototyping. It has been very helpful for technique integration. All the techniques can share the same set of data handled by the Information Manager. The resulting user interface, generated by the Technique compiler, is coherent for all the techniques.

5. Implementation Results

PIMS#1, the first prototype, was developed for validating the global PIMS approach: it

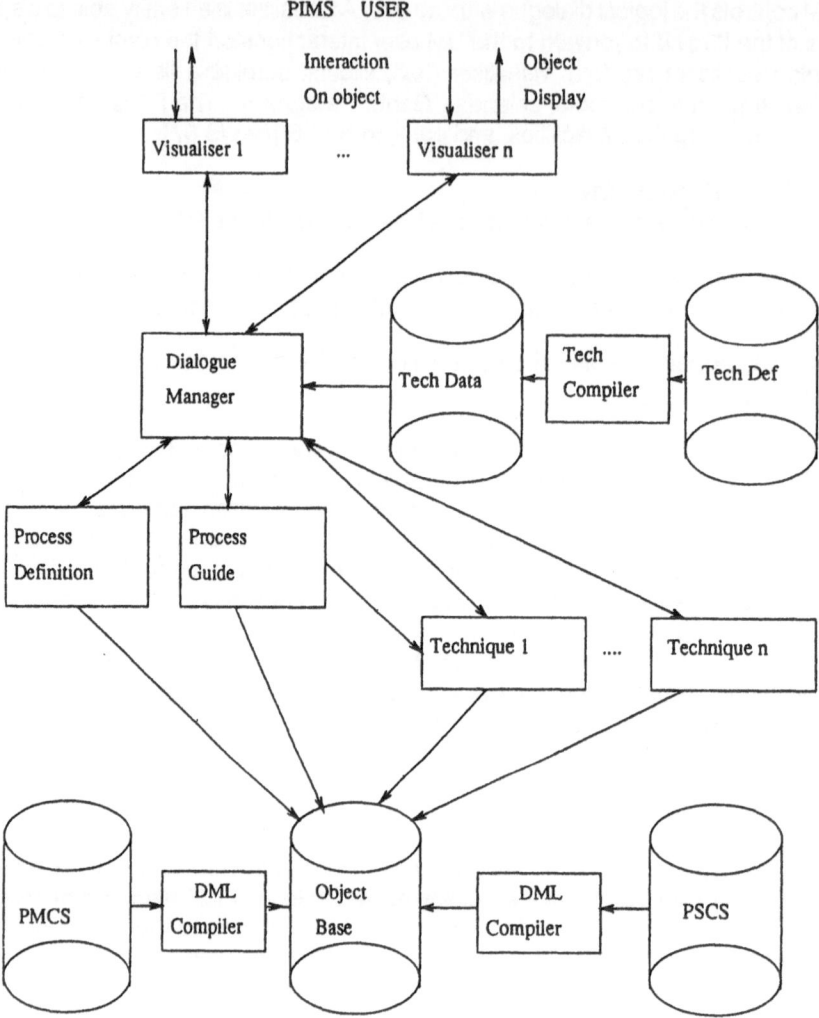

PIMS architecture
- Figure 2 -

consists of a set of tools organised around a common object-oriented database and provides a unified user interface.

The second prototype, PIMS#2 [PIMS 88b], is more polished and integrates feedback from PIMS#1 (comments and remarks given during demonstrations and experiments). PIMS#2 also expands the scope of PIMS#1 by providing a real integration of management methods, giving more and more powerful functionalities, and improving the performance of the infrastructure and user interface.

The results of PIMSFOR have been implemented in the prototypes: conceptual schemas have been defined for implementing the Project Management Model and the Project Status Model. The Project Management Conceptual Schema (PMCS) defines classes for PMACs definition. The Project Status Conceptual Schema (PSCS) defines all the classes for the project data (project, deliverables, tasks, resources, allocations, ...).

The users of PIMS are the Quality Assurance (QA) people and the project manager. The QA people is in charge of defining methods. The project manager has to use a predefined method for managing its project.

PIMS proposes to support the project managers by providing a set of environments dedicated to management activities. Each environment is oriented towards the specific capabilities which allow the achievement of particular goals; it is implemented as a set of techniques.

The overall system is controlled by a method which constitutes a guide environment. This method determines at any moment the current state of the project and the next actions whose the project manager has to perform. These actions are supported by environments: project setup, project planning and progress control.

Here is a short description of the environments PIMS proposes as they can be perceived by the user. Figure 3 presents how the techniques are gathered in PIMS environments.

METHOD DEFINITION

This environment allows method description and tuning; it contains the technique *Process Definition*.

In a company, the Quality Assurance (QA) people has the responsibility of proposing and promoting standards and methods for a better quality. He is the only person allowed to define and modify methods. The description of the method consists in the definition of a network of PMACs. To each PMAC are associated precondition, postcondition and a PIMS environment.

The method defines a set of activities to the project manager, meaning that parallel work is possible. Each compound activity of the network can be exploded for consulting its contents. Dependencies between activities are shown on the edges linking the activity "bubbles".These dependencies are either logical dependencies (this activity starts when the status of the previous one changes) or temporal dependencies (this activity has to be performed 10 days after the previous one is completed).

PIMS environment	used techniques
Method Definition Method guidance Project setup	Process definition Process guide Project initialisation Calendar definition Resource definition External risk analysis
Project Planning	Work breakdown Estimation Scheduling Resource allocation Planning assessment Calendar definition Resource definition
Progress control	Tracking Monitoring Predicting Diagnosis and repair Calendar definition Resource definition
Common facilities	Calendar definition Resource definition Report generator

- PIMS environments and techniques -
- Figure 3 -

It is recognised that the use of a method is sometimes too constraining. In some cases, the method, which has to be quite general, needs adjustments due to project size or domain.The QA people may give some freedom to the future user (a project manager) while describing the methodology: he may allow that some activities may be bypassed or deleted by the project managers.

Project managers then select one method among several and have the opportunity to make a final tuning before running it for a specific project. This tuning is limited by the possibilities set by the Quality Assurance people.

METHOD GUIDANCE

As soon as the project manager connects to the PIMS system, the *Process Guide* technique displays the current status of his project in term of positions in the PMAC's network of the method; this status is where he was at the end of the previous session. This display indicates which activities must be performed and which ones can be performed. For instance, it may be time to make a monthly report, to distribute the work description of a task to a team member or it is necessary to report a problem, to acknowledge a partial result in the project (put a stamp on a delivery) or to report the arrival of an external document.

With such a representation, the user is able to browse the method for learning more about it. He can have a look to the next activities and plan them appropriately. This approach allows a smooth use of a method.

PROJECT SETUP

When starting working on a project with the PIMS system, initialisation is necessary: every pertinent information about the project has to be entered using the *Project initialisation* technique.This includes information about the contract, the deliverables, the milestones, the client, the possible team members, the initial budget, the method which has to be followed and so on.

The *External Risk Analysis* calculates the risk of the project before building a plan for the project. The risk is rated through six categories (type of the client, nature of the contract, domain for instance) and about sixty criteria. The result is a percentage of risk, global and per category as well as corresponding advices which indicate how to cope with the resulting expected risk. For example, have very tight progress control periods if the project is very time-constrained.

PROJECT PLANNING

The planning environment provides techniques to build a plan which is then used as the reference for project progress control. The planning environment includes techniques for decomposing a project into a hierarchy of task, defining the products, estimating the workload and the cost, scheduling the project tasks, declaring the resources and allocating them properly. Planning must be done at the start of the project and may be updated according to the needs: a 3 months detailed planning has to be always up to date for instance, and when large discrepancies are detected during the progress control, the planning must be reevaluated.

PIMS makes the re-planning work easier by making full reuse of the information which is still pertinent. This means that for re-planning his project, the user only has to modify the data he judges wrong and PIMS uses this new data for computing the new plan. The *Workbreakdown* technique is domainoriented.It manages a library of predefined decomposition templates. The template may be attached either to the management standards or to the project domain. Rapid detailing facilities allow to build the workbreakdown as a cartesian product of a functional breakdown and of a life cycle method.

The *Task Scheduling* and *Resource Allocation* techniques support the project manager in expressing the different time, dependency, availability, competence, human constraints. PIMS starts from these constraints for providing a solution which satisfy all of them: if no solution exists the user is asked to relax one or several blocking constraints until an acceptable solution is reached.

The *Planning Assessment* technique evaluates the planning resulting from the previous operations: strengths and weaknesses are highlighted in a way similar to external risk analysis.

PROGRESS CONTROL

This environment handles data collection, discrepancies evaluation, prediction and diagnosis.

When the project manager wants to know about the project progress, he enters the progress control environment and gets information such as effort consumption, remaining effort, documents already produced, milestones reached, money spent. The up do date information is displayed through standard graphics (curves, histograms) and the user gets at one glance the current status of his project. Facilities for predicting and for problem diagnosis are also provided. This is particularly useful for making about re-planning.

The *Collect Data* technique allows to enter data describing the current state of the project: allocated people, started or completed tasks, effort spent, other expenses. This information is available at different levels: basic task, the finest grain in the task decomposition, and compound task including project level. In the last case, the information is obtained from depending sub-tasks.

Monitoring technique compares actual figures to planned ones. Discrepancies are pointed out. Therefore, it is possible to get at one glance the global state of the project as well as indications on specific parts.

Predicting provides different models which show the tendency ofthe project. Results are time and cost predicting on short, middle or long term.Prediction functionalities give general information about a possible future of the project: end date and final cost of a specific task or of the project as a whole. These results are computed from various predicting models. These models are based on propagation in the future of the current problems. For example, if it happens that all the specification tasks have required 10% extra effort, the prediction function assumes that each of the future specification tasks will have a slippage of 10%.

The *Diagnosis* technique determines the causes of the discrepancies.In some cases, PIMS can propose remedial actions. The system analyses its internal data and proposes real causes (task started late) and possible causes (productivity of person X decreased, X is no more motivated to work on the project). PIMS can even propose remedial actions for some problems: it may be the case that all the "design" tasks in the project are late. The system can propose some correcting factor for reevaluating the effort. The most common remedial action suggested is replanning.

OTHER FACILITIES

PIMS provides common facilities and techniques. They can be used used by the previous environments.

A common *help facility* is accessible at any moment within the PIMS system. Help is associated to each of the displayed elements (object or command). This is done at the management method level where the goals and the means are explained, for any activity the project manager is able to select. This is also true in the running of the management techniques: help can be obtained on any functionality which is proposed in the current environment.The help PIMS proposes is a sort of guidance for providing advice on what to do next.This is very important for novice project managers who can then learn with the system.

Each technique can use the *Report Production* facilities of the architecture for producing predefined documents. For example the progress control environment defines commands forproducing reports about the project; it contains information at the project level (effort spent, predicted end date and cost) and at task and deliverable level. For printing such a standard document, the user will only have to communicate the month name to obtain the monthly report. In this case the exact set of data is determined by the indication of the month.

A *Calendar Definition* technique allows to set a calendar for the project. It displays a calendar view in which dates can be selected. This technique is included in each environment.

A *Resource Definition* technique is used for human resource definition and browsing (name, skills, cost, availability, ...). This technique is included in each environment.

6. Future of PIMS

PIMS project has produced both theoretical (PIMSFOR) and practical (PIMS#2 proto-type) results. The project finishes in June 1989, but the work already done will be extended.

The *academic partners* (LSE, UVA) have planned to put together in a book all the results of the theoretical studies and of the knowledge acquisition. The potential readers of this book are both project managers and specialists of project management. The book will be published at the beginning of 1990.

Each *industrial partner* (BSO, CSI, PA, SI) has defined its own plan for using PIMS results. First the PIMS prototype will be directly used in showing demonstrations with some marketing goals; this prototype is a kind of runnable specification of an advanced project management support system. A second way to use PIMS in the company is to really integrate PIMS results in the organisation; for example, CSI is currently having a PC-based project management system written, built on the PIMS results.

Beside this PIMS usage by individual partners, new projects are defined for using PIMS results as a springboard for further enhancements. Some new areas need investigation and this could be tackled in the future:

– connection of management facilities developed in PIMS with a development support as design and code generation; the expected result is a complete CASE.
– implementation of simulation facilities
– packaging of the architecture to produce a kit for object oriented rapid prototyping
– use in education

More generally, the ideas proposed in the PIMS project could be viewed as the project management part of a Software Factory, meaning that more links with other activities could be developed.

7. Acknowledgements

We would like to thank all the people (the over 40 people who did are too many to be named here) who have worked and struggled on this project. This includes the current team in France, The Netherlands, Norway and United Kingdom, but also all the individuals who have contributed at some time, to make PIMS what it is.

8. References

[ACL 88] Allegro Common Lisp User Guide, Franz Incorporated, release 3.0, June 1988.

[ESF 88] ESF Technical Reference Guide, ESFinternal document, November 1988

[GRIF 86] V.Quint, I.Vatton - GRIF: an Interactive system for structured document manipulation,

Proceedings of the International Conference, J.C van Vliet, ed., Cambridge University Press 1986

[NeWS 87] NeWS 1.1 Manual, Sun Microsystems, 1987.

[PIMS 87] M.S.Doize, C.Fernstrom - User's manual for the PIMS#1 Information Manager, PIMS internal document, December 1987.

[PIMS 88a] D.Berkelay, J.Hawgood, R.deHoog, P.Humphreys, M.Vogler - A Frame Of Reference for software project management, PIMS internal document, September 1988.

[PIMS 88b] John Hawgood - PIMS#1 User Manual, PIMS internal document, May 1988.

[PIMS 88c] S.Konc, A.Leclerc, J.Paris, D.Ribot, Ph.Vauquois - PIMS#2 Front End, PIMS internal document, November 1988.

The internal PIMS documents are available from the authors of this paper.

Keywords.

management method, management method support, model for project, model for project data, object oriented database, object oriented user interface, project management, process model.

Project no 820

DIAGNOSTIC REASONING WITH THE QUIC TOOLKIT

ALBERTA BERTIN
GIORGIO TORNIELLI
CISE Tecnologie Innovative SpA
P.O. Box 12081
I-20134 Milan
Italy

BEATRICE DE RAVINEL
Framentec S.A.
Tour Fiat, cedex 16
92084 Paris la Defense
France

Abstract.

The QUIC (Qualitative Industrial Control) project (ESPRIT P820) is developing a composite framework, termed toolkit, for supporting the design and development of knowledge-based systems in the domain of industrial automation. One of the main issues arising in this application domain is the representation and use of deep knowledge about system behavior. This paper presents two different diagnostic systems based on a deep model of the system to diagnose, that have been developed from the toolkit for solving two real size diagnostic problems: the diagnosis of malfunctions in the steam condenser of a thermal power plant and the diagnosis of faults in a subsystem of the electrical power supply of a telecommunication satellite. In this paper the diagnostic strategies adopted by the two systems and how they have been implemented within the toolkit are described.

1. Introduction

The QUIC toolkit is a development environment for knowledge based systems in the area of Industrial Automation. One of the main motivations behind its design has been the recognition that representation and use of <u>knowledge about technical artifacts</u>, (e.g. formal models of these artifacts), is essential if effective, reliable and safe knowledge-based automation systems are to be constructed. The QUIC toolkit aims at providing a set of tools ranging from an appropriate language for describing a technical system to a set of inference mechanism able to reason on this description (for a comprehensive description of QUIC refer to Stefanini and Leitch [1988]).

Diagnostic systems have been the major focus of development for KBSs in industrial automation, and indeed represent a major motivation for this project. However, diagnostic systems can be implemented in two fundamentally different ways. <u>Hierarchical classification</u>, developed for medical applications by Gomez and Chandrasekaran [1981], implements diagnosis as one of identifying a malfunction (patient case description) as a node in a fault (disease) hierarchy. In this case the fault hierarchy is obtained by pre-compiled knowledge and represents a decision process. A similar approach, known as heuristic classification, has been developed by Clancey [1985] as a generalization of the Mycin expert system. An alternative approach to diagnostic systems receiving much attention, is <u>model based diagnosis</u>. In this case, an explicit model of the system to be diagnosed is represented and used to generate hypotheses which are matched against the current state until there is no discrepancy or, at least, the discrepancy is minimized. For instance DART [Genesereth, 1984] and GDE [de Kleer and Williams, 1987] represent two of the most mature attempts of transferring new modeling concepts in the realm of practical applications. Both focus on the same task, i.e. diagnosis of electronic circuitry. Compliant with the current way of representing and reasoning about electronic circuitry, they represent the system to diagnose as a set of interconnected components of known transfer function. The inference mechanism adopted is basically different, however, as Genesereth uses a linear input resolution algorithm on a set of first order clauses, while de Kleer and Williams couple constraint propagation with an Assumption-based Truth Maintenance System (ATMS) for managing different diagnostic hypotheses.

The unifying feature of GDE and DART is the kind of diagnostic problem they solve, namely, they aim at picking up the component(s) which do not behave correctly given a description of the correct system structure and behavior. Hereinafter, we will refer to this problem as one of *fault insulation*.

However, fault insulation alone may be not enough for many practical applications. For instance, a component usually presents many faulty states, e.g. a faulty resistor may be either equivalent to a short or an open circuit. When a component has been discovered to be the cause of a symptom, an interesting subproblem is to identify among a set of known possible fault causes the one which explain the observations.

One of the diagnostic systems we will present in this paper, named Ontological Diagnostic System for identifying Discontinuous Changes (ODS-DC) is based on an extension of the strategy adopted by GDE in that it is able to identify both the components which do not work properly and the actual explanation of the abnormal behavior. This diagnostic system has been applied to the diagnosis of faults in a subsystem of the electrical power supply of a telecommunication satellite.

Usually, a fault affecting an electronic component radically alters the system structure. This is the case, for instance, of a fuse which can be normally considered an ideal wire, while it becomes an open circuit when a failure occurs. Another cause of change in the behavior of a physical system is due to violations to some design assumptions which do not imply any modification in the system structure, like variations in the values of design parameter. Hereinafter, we will refer to variations in the values of design parameters as *malfunctions*. In monitoring of continuous processes, like the steam condenser of a power plant, the problem is just the one of early detection of malfunctions in order to avoid a subsequent drastic alteration in the physical structure of the system. The second diagnostic system we will present in this paper, named Ontological Diagnostic System for Parameter Identification (ODS-PI), has been designed with the objective of detecting malfunctions rather than performing fault insulation.

In this case, the system model consists of a set of equations, relating a set of observable state variables to a set of parameters, whose olutions correspond to equilibrium states of the system. In normal working condition the system is supposed to operate in a definite state, corresponding to nominal values of the design parameter set.

The diagnostic problem here consists of discovering which parameters have a value different from the nominal one when the observed system state is different from the expected. Furthermore, for each parameter the sign of the variation, that is if the value of a parameter is increased or decreased with respect to the nominal value, must be determined for activating the appropriate recovery procedure.

The paper is structured in the following way:
- section two gives a description of the ontological part of QUIC toolkit that has been used for implementing the two diagnostic systems.
- Section three gives a brief description of the methodology supported by the toolkit and adopted for implementing the two diagnostic systems.
- Section four and five give an explanation of the application domain and diagnostic strategy of the ODS-PI and ODS-DC.
- Finally, section six provides conclusive remark and a discussion about future developments.

2. The Ontological Components of the Quic Toolkit

The toolkit is a composite framework for supporting the design and development of knowledge-based systems in the domain of Industrial Automation developed within ESPRIT Project P820. It consists of a set of modular and tailorable tools plus a design methodology.

As outlined in the introduction, a significant aspect of the QUIC project is the recognition of the various sources of knowledge available to the designer of automation systems. In addition to the experiential knowledge of process operators and engineers, industrial designers utilize theoretical knowledge of the scientific principles underling the operation of a physical system. We term the former experiential knowledge *empirical*, and the latter *theoretical*.

A similar dichotomy has been recognized long time ago as one of "shallow" versus "deep" knowledge [Michie, 1982]. There is however no general agreement on how "deep" knowledge structures should be generated. The problem is that, as defined, deep knowledge is relative, i.e. it is deep only with respect to a particular shallow description and can be shallow with respect to a deeper model. This appears to be no significant characteristic of a representation that qualifies or the description "deep". In order to avoid this confusion, we prefer to focus on two different types of knowledge, empirical and theoretical.

This dichotomy reflects the way in which the knowledge is generated. Empirical knowledge is induced from observations whereas ontological knowledge is deduced from theoretical principles.

In the specific field of diagnostic systems, this dichotomy is reflexed in the distinction, mentioned in the introduction, between classification and model-based (also called "generative") systems. For sake of consistency with our project terminology, we will continue to refer hereinafter to empirical versus ontological distinction.

Each source of knowledge requires a representation language and an associated inference mechanism. We will focus in this section on ontological knowledge only.

The representation language adopted for representing ontological models of technical artifacts within the QUIC toolkit is the **Component Based Language** (CBL). The basic concept of the CBL is that a physical system can be represented by describing its internal structure as a set of interconnected components. A model is defined for each component from which the component's behavior can be inferred, and hence the behavior of the whole system. Thus the CBL provides a definition section, for defining abstract models of components, and a declaration section, for describing the actual system in term of a set of component instances, and their interconnections.

The component model is given in terms of a set of equations constraining the component variables, and of some component specific information. Equations may be interpreted either in *quantitative* or in *qualitative* domains. The former represents the domain of real numbers while the latter consists of a set of three values (inc, dec, std), which represent the possible qualitative variations of a variable with respect to a reference value[1]. The qualitative equations that can be defined by means of the CBL are similar to the *confluences* as defined in de Kleer and Brown [1984]. Each component description may be partitioned according to different views of the underlying physical process (e.g. the electrical and thermal view of a resistor) each of which may be considered as a model of the process in a specific situation.

The CBL declaration section allows the user to describe a particular physical system by declaring its components as instances of component types provided by means of the definition section, and to state the connections between component instances.

An example of component definition, taken from the description of an electronic circuit, is the ideal diod as shown in Figure 1 below.

```
COMPONENT_CLASS ideal_diode
TERMINALS t1 t2 OF_TYPE i_v

VIEW on
 RELATIONS
   t1.i + t2.i = 0 ;
 CONSTRAINTS
   t1.v - t2.v > 0 ;
END_VIEW

VIEW off
 RELATIONS
   t1.i = 0 ;
   t2.i = 0 ;
 CONSTRAINTS
   t1.v - t2.v < 0 ;
END_VIEW

END_COMPONENT_CLASS
```

Figure 1

The main inference mechanisms supporting the CBL that have been used for implementing the two diagnostic systems described in sections 4 and 5 are: the causal ordering system, the constraint propagator and the assumption-based truth maintenance system. These are briefly overviewed hereinafter (more specific details about them can be found in [Arlabosse *et al.*, 1988]).

The **Causal Ordering System** analyzes the CBL ontologic model and provides a partial ordering between the variables referred to in the set of equations [Iwasaki and Simon,

---

1   The qualitative domain has been used in an earlier diagnostic system developed with the QUIC toolkit and described in [Gallanti et al. , 1989]

1986]. This ordering reflects the dependencies between the variables of the model under a certain view. This is defined once one establishes which variables of the model are **exogenous**, i.e. their values are believed to be determined by the environment surrounding the system only. Once these have been identified, the causal ordering method allows to identify an ordering between the variables which appear in the equation set of the model. The exogenous variables are considered as the "primary causes" that drive the system. In summary, causal ordering permits a causal interpretation of a mathematical model to be determined.

Constraint Propagation [Sussman and Steele, 1980] is a simple way of interpreting the set of equations representing the component behavior. An equation is represented by a constraint net which consists of cells each of them representing a system variable; each cell participates in one or more constraint expressions, i.e. relations among the cell values. Each equation is represented by one or more constraint expressions. Each cell may have a value, which may come from the user or may be deduced from other values in the constraint net. When a cell is assigned a value, each constraint it participates in is considered to determine if enough information is available to deduce a value for another cell. Discovering a new value may thus allow many other values to be determined, thus "propagating the constraints". An advantage, with respect to techniques based on algebraic manipulation, is that constraint propagation may work when the operators do not define a mathematical field as it is the case for the operators used in Qualitative Physics [Bobrow, 1984].

The toolkit also provides an **Assumption-based Truth Maintenance System** (ATMS) [de Kleer, 1986]. Conventional, justification-based, TMSs such the one by Doyle [1979] associate a state of **in** (believed) or **out** (not believed) to each problem solver datum. The entire set of in data defines the current problem solver context. The task of the TMS is to keep consistent the data base, adding and removing data from the in and out lists at each inference made from the problem solver. In conventional TMSs the notion of context under which a datum is believed is implicit, as there is only one in list.

The ATMS removes this limitation of conventional TMSs, as each datum is labeled with the set of assumptions, i.e. the context, under which it is believed. The basic idea is that the assumptions are the primitive data from which all other data are derived, thus they can be manipulated more conveniently than the data sets they represent.

The functionality provided by an ATMS is general enough to envisage its use in a variety of tools. ATMS allows problem solving architectures where multiple potential solutions are explored simultaneously. Constraint Propagation applied to diagnostic reasoning where the variable values are computed under different diagnostic hypotheses, is one of these cases, so that the QUIC toolkit provides a constraint propagator coupled with an ATMS.

3. The Toolkit Use Methodology

The methodology provided by the toolkit for building an application in the domain of Industrial Automation is based on a classification of the tasks that are typical of this domain. Such classification is done by recognizing that we are dealing with dynamic systems that evolve over time. The adopted classification accords with the conventional classification of control theory, and consists of the following five primitive tasks:

– INTERPRETATION:
  - The transformation of observed data into the adopted state representation.

- EXECUTION:
  - The transformation of decisions in the system state into data suitable for action.
- DECISION:
  - The decision making process whereby the known or assumed present state is used to construct conclusions (hypotheses) at the same time.
- PREDICTION:
  - The generation of future states from known or assumed present states.
- IDENTIFICATION:
  - The determination of (unknown) past states from known or assumed present states.

Based on the preceding classification of 'primitive' industrial control tasks we can develop a range of *systems*, like diagnostic or monitoring systems, for industrial control. Each system is obtained as a combination of the primitive taks. In particular, a model-based diagnostic system, like the ones overviewed in the following sections, consists of three tasks: an *interpretation* task, an *identification* task, and a decision task according to Figure 2.

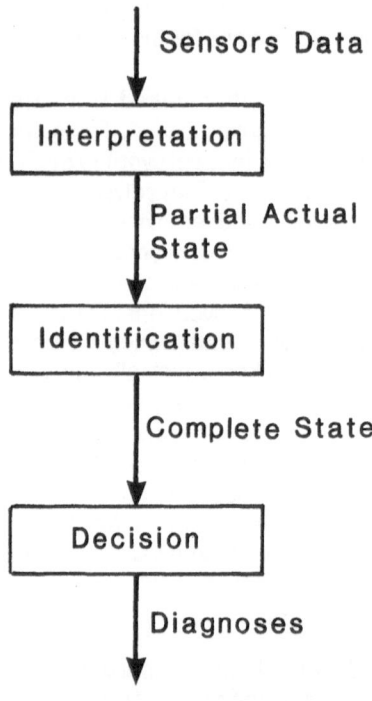

Figure 2

The former is designed to translate a set of input data (ideally taken from sensor readings) into the actual partial state, according to the given model represented with the CBL. The latter allows to completely identify the unknown state of the system, by reasoning on the actual partial state.

For supporting this methodology, the toolkit includes two classes of software packages: **tool components** that perform some well defined but basic functionality (the constraint propagator, the atms and the causal ordering reviewed in section 2 are example of tool components), and **tools** providing full support for performing a primitive task. The latter are ordinarily obtained by assembling specific tool components.

The methodology is supported by a development environment based on an object oriented paradigm so that each tool or tool component is represented as an object. For developing a specific application the user has to combine those objects which fulfill the application requirements.

The two diagnostic systems we will focus on in the following sections have been developed within the QUIC toolkit using the tool components presented in section 2. As a result of their development, two new identification tools have been inserted in the toolkit. These tools are at the core of the two diagnostic systems, and cope with the two different problems mentioned in the introduction: identification of discontinuous changes and identification of malfunctions. Both problems are clearly identified, and appear specific of important classes of applications (continuous processes, electronic circuits), whereas it is impossible to envisage diagnostic situations where both problems are to be faced. Therefore, it is reasonable to provide the toolkit with two distinct tools for the two problems - possibly as basic building blocks for generic diagnostic systems.

The interpretation tool developed for parameter identification was also added as a generic tool to the toolkit. On the contrary, no data interpretation tool was developed for the discontinuous changes problem, because the format, where the input data are available, was in that case the same one adopted for the state representation adopted for identification.

4. An Ontological Diagnostic System for Parameter Identification

4.1. Problem Description

The behavior of a dynamic system can be modeled by a set of differential equations relating its actual state to the input vector and to the previous state. Most industrial processes are designed to operate, however, in a stationary condition identified as a maximum of a defined function of merit. In these situations, we may assume that the observable stationary state solely depends on the values attributed to the set of input variables. Among these we can distinguish between proper input variables ($\underline{u}$) which are observable, and unknown disturbances that can modify some parameters $\underline{p}$ of the system, whose values are defined when the system is designed and should be constant in order to maintain the system at the desired state.

The state $\underline{s}$ can be expressed as a function of the input variables $\underline{u}$ and the parameters $\underline{p}$ that can be modified by disturbances:

$$\underline{s} = f(\underline{u}, \underline{p}) \quad (1)$$

In these system a diagnostic problem arises when the system state changes independently from a variation of the input vector $\underline{u}$, due to disturbances affecting one or more system parameters. Under well known conditions the system is stable and one must assume that, following parameter variations, the system recovers a stationary state $\underline{s}'$ such that:

$$\underline{s}' = f(\underline{u}, \underline{p}'). \quad (2)$$

and this is detected by continuously monitoring some measurable variables.

The diagnostic problem consists of determining which variations:

$$\Delta\underline{p} = \underline{p}' - \underline{p} \quad (3)$$

may have occurred to generate the displacement:

$$\Delta\underline{s} = \underline{s}' - \underline{s} \quad (4)$$

of the system state.

Let's stress again that we refer to system anomalies that are called malfunctions rather than faults, in that the structure of the system model remains unchanged.

In malfunction conditions it is often reasonable to assume that variation $\Delta\underline{p}$ is small enough that $\Delta\underline{s}$ linearly depends on $\Delta\underline{p}$:

$$\Delta\underline{s} = C \cdot \Delta\underline{p} \quad (5)$$

where the coefficients of the sensitivity matrix C depend on the nominal values of the system variables.

Ordinarily, the number of observable state variables (m) is smaller than the number of parameters (n) that can be affected by a malfunction and, therefore, C is a rectangular mxn matrix with n>m. In this case diagnosis consists in identifying, among the infinite solutions to system (5), the ones which comply with some minimality criterion like identifying the solutions involving the minimal number of parameters different from zero (i.e. the minimal number of parameters involved in a malfunction).

4.2. The Diagnostic Problem in the Steam Condenser

A case study where this kind of diagnostic problem arises is diagnosis of a steam condenser in a thermal power plant.

The role of the condenser in the water-steam cycle of a thermal power plant is to cool the steam coming from the turbine to bring it back to the liquid state.

Condensation is obtained using surface condensers (i.e., without mixing the cooling water with the steam), in order to keep the chemical characteristics of the cycle water unaffected. The cooling fluid is open-loop water taken from a river or from the sea. In order to increase cycle efficiency, fluid condensation occurs in depressurized conditions, i.e. the pressure inside the condenser is kept lower than the outside atmospheric pressure. The cooling water is circulated by a water pump by a tube bundle located inside the condenser, where steam condensation takes place.

A simplified schema of a steam condenser is reported in Figure 3.

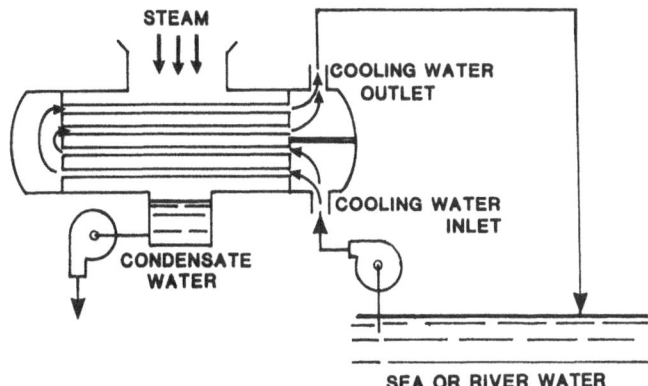

Figure 3

The two main processes taking place in the condenser are heat exchange between the two fluids, and cooling water circulation. Each of these two processes is ruled by a set of steady-state equations, like (1), identifying the stationary state, usually corresponding to a defined function of merit, in which the system has been designed to operate.

A malfunction is detected by continuously monitoring some measurable variables like the steam and coolant temperatures, or the load loss indifferent sections of the circuit.

The most likely malfunctions affecting the condenser modify parameters like the exchange surface and the cross-section of the tube bundle, the thermal exchange coefficient, the mechanical power supplied by the pump, the average thickness of the tubes. For instance, tube clogging affects cross section and exchange surface of the tube bundle; calcarean crusting or erosion affect tube thickness. Of course, the above mentioned parameters cannot be directly measured (in which case the diagnosis would be trivial), but the malfunctions can be ascertained through reasoning upon (5).

4.3. The Diagnostic Strategy

The strategy we introduce for solving our diagnostic problem makes use of models of the system to be diagnosed at different levels of abstraction, qualitative and quantitative[2]. In particular, our algorithm uses a quantitative, real-valued algebraic model like (5), and a qualitative causal model that can be easily derived from the former in an automated way by means of Causal Ordering. The causal model is intrinsically qualitative in that the value of the variables involved in the causal net range on a binary domain (i.e. variable values may be changed or not changed with respect to their reference value). Such model is used for the generation of diagnostic hypotheses. The qualitative causal model is very abstract so that conclusions drawn from it are quite generic, even if they may be generated in a very efficient way. The real-valued model is indeed used for validating or rejecting the diagnostic hypotheses drawn from dependency analysis.

In summary, the main steps of our diagnostic algorithm are:

2   When a qualitative model is derived from a quantitative one [Kuipers, 1984, de Kleer and Brown, 1984], the former is an abstraction of the latter in that qualitative variables range in a finite set of values vs. an infinite set

- given a model of a physical system to be diagnosed in form of linear equations like (5) entered by means of the CBL, a causal ordering between the variables belonging to such model is computed. The causal net obtained from a simplified model of the condenser is shown in Figure 4 below.

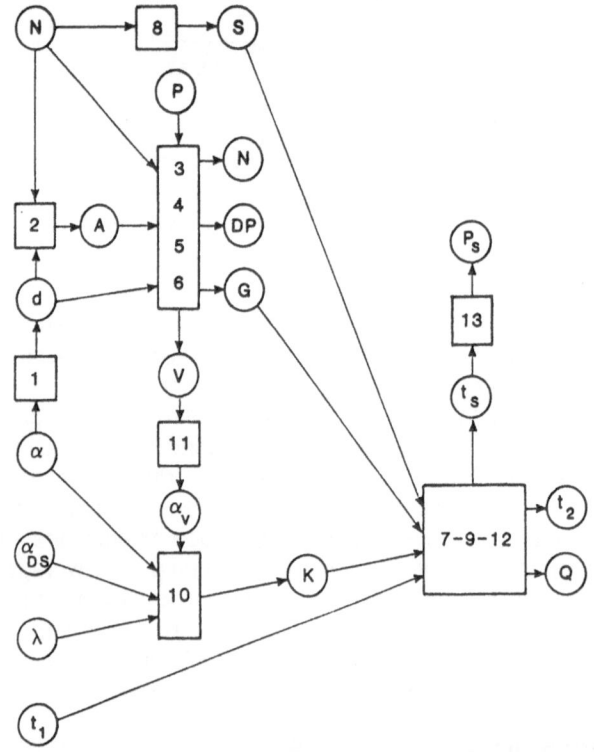

Figure 4

The causal net consists of two kind of nodes: nodes denoting variables (the circles) and nodes denoting equations (the rectangles). The causal net used for diagnostic purposes has been obtained considering exogenous the parameters that can be modified by a malfunction. These parameters are shown in the left side of the net.

- From an analysis of the dependencies between observables and parameters that can be affected by malfunctions, and between observables and other observables made explicit by the causal net, a set of diagnostic hypotheses, or explanations, explaining all the observations is obtained.
The computation of these diagnostic hypotheses is done by means of a logical calculus which implements reasonings like the following:
  - when the value of an observable which depends on parameters only is changed, then at least one of the parameters the observable depends on must be changed;
  - when the value of an observable which depends on parameters only is not changed, then either no parameter is changed or at least two of them are changed but their effects on the observable compensate each other;
  - when the value of an observable is changed and this variable depends on parameters

and on other observables whose value is not changed, at least one of the parameters must be changed.

Once the explanations for each single observable are computed, these are combined in order to obtain an explanation for the full set of observations.

In order to clarify the candidate generation phase, let us assume that from the causal net we obtain that the observable variable o_1 depends on the parameters p_1, p_2 and p_3, whilst the observable variable o_2 depends on the parameters p_3 and p_4:

$$\text{Depends } (o_1 \ (p_1 \ p_2 \ p_3))$$
$$\text{Depends } (o_2 \ (p_3 \ p_4))$$

Furthermore, let us assume that the value of o_1 is changed with respect to its nominal value, while the value of o_2 is not changed.

The explanation for the observation of o_1 is that at least one parameter among p_1, p_2 or p_3 must be changed. This is represented in logical terms by the formula:

$$CH(p_1) \text{ or } CH(p_2) \text{ or } CH(p_3)$$

where the predicate CH(x) means that the value of the parameters x is changed with respect to its nominal value.

The explanation for o_2 is that either p_3 and p_4 are not changed or both of them are changed. In logical form this is represented as:

$$(\neg CH(p_3) \text{ and } \neg CH(p_4)) \text{ or } (CH(p_3) \text{ and } CH(p_4))$$

The explanations for the two observations may be combined in order to obtain an explanation for both the observations by means of logical conjunction:

$$(CH(p_1) \text{ or } CH(p_2) \text{ or } CH(p_3)) \quad \text{and}$$
$$((\neg CH(p_3) \text{ and } \neg CH(p_4)) \text{ or } (CH(p_3) \text{ and } CH(p_4)))^{(6)}$$

The actual diagnostic hypotheses may be obtained by transforming this formula in the disjunctive normal form: each disjunct represents a diagnostic hypothesis able to explain all the observations. The diagnostic hypotheses for the previous example may be obtained by translating (6) into the disjunctive normal form:

$$(CH(p_1) \text{ and } \neg CH(p_3) \text{ and } \neg CH(p_4)) \text{ or}$$
$$(CH(p_2) \text{ and } \neg CH(p_3) \text{ and } \neg CH(p_4)) \text{ or}$$
$$(CH(p_3) \text{ and } CH(p_4)) \text{ or}$$
$$(CH(p_1) \text{ and } CH(p_3) \text{ and } CH(p_4)) \text{ or}$$
$$(CH(p_2) \text{ and } CH(p_3) \text{ and } CH(p_4))$$

- The last step consists of validating each diagnostic hypothesis on the quantitative model. This is equivalent to solve the linear system (5) with respect to the parameters listed in each of the hypotheses. When the system does not admit solution for a given

414

hypothesis, this is rejected and another hypothesis is validated: only the ones that survive quantitative validation are the actual diagnoses.

4.4. Implementing the Strategy in the Toolkit

The architecture implementing the ODS-PI diagnostic strategy is drawn in Figure 5. As mentioned in section 3, the ODS-PI is implemented by means of two tools: an interpretation tool and an identification tool. The former is in charge to compute the nominal values of the measurable variables and to detect when the value of a measurement is different from the nominal value. This information is given to the identification tool which is in charge to identify the complete actual state of the system, and therefore to discover which parameter variations may explain the observations.

The interpretation tool is implemented by means of the constraint propagator (CP) used for the computation of the nominal values of the variables, and by means of the CBL which is used for describing the system model. The identification tool uses the causal ordering (CO) for computing the causal model starting from the same CBL description.

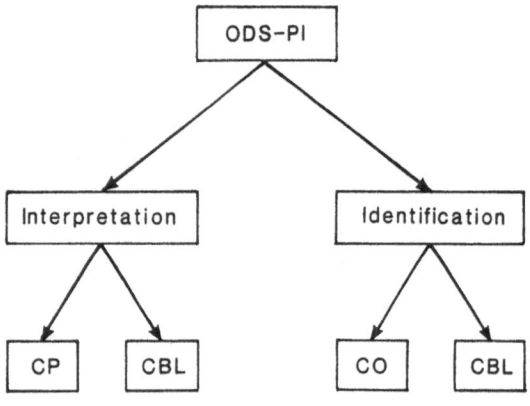

Figure 5

5. An Ontological Diagnostic System for Identifying Discontinuous Changes

5.1. Problem Description

The objective of the system is to diagnose problems within the electrical power supply of a geosynchronous telecommunication satellite. The satellite receives energy from two single axis sun-oriented solar array wings. Solar array power is directly transferred to two independent power distribution buses, with a cross-trap possibility between them. During launch and eclipse the power is supplied by two batteries. Every equipment needing power supply is connected to both buses. Selecting some equipments consists in sending some telecommands to switch some relays. One or more switches may be stuck during this operation. There also may be some blown fuses, short circuit,... Two kinds of telemetries are available to analyze the problems: status telemetries indicating the presence of voltage, and analogical telemetries for voltage and current. They are all available when the diagnosis starts. This means that no more telemetries will be available afterwards.

The circuit is a globally complex system rich in interaction paths between relatively simple components (relays, fuses, resistive fuses, resistors, diodes, ...) making it well adapted to a model-based approach. This approach offers the advantage that the symptoms are not related to the faults ; this allows to diagnose non anticipated faults ; furthermore if the structure of the circuit is modified or if some components are added or removed, it is not necessary to design a new diagnosis.

The behavior of the system is described by the behavior of each component. A component may have several functioning modes. A mode is described by a set of equations involving the internal parameters and the terminals of a component. The failure modes may also be described for each component. The diagnosis system is flexible enough to take into account the presence or absence of failure modes. Describing the failure modes answers to one of the objectives of the power supply diagnosis: the objectives are, given a set of symptoms (such as: some equipments have no power), to identify the components whose functioning present some anomalies and to identify the type of these anomalies, i.e. the failure mode of such components.

A reference mode is known for each component when the diagnosis starts. It is deduced from a telecommand, or is a normal operating mode, or is a mode which has been identified in a previous diagnosis session.

5.2. Diagnostic Strategy

The diagnosis strategy uses and extends the approach of the General Diagnostic Engine (GDE) of de Kleer and Williams [1987]. GDE produces a set of single and multiple suspects, called candidates, each one explaining all the observations. GDE assumes that each component is in one of two states: working or not working. We extend this approach considering that a component may have multiple working and failing states.

The diagnostic principle consists in comparing the reference state (predicted from the reference model) and the observed state (computed from the telemetries). A difference is due to an unexpected behavior of the system: the reference model does not explain the system behavior. Diagnose corresponds to identify a new model explaining the behavior.

The following steps are considered:

5.2.1. Generate conflict sets.

In this step the equations of the reference mode of each component are used to compute the reference state. The observed state is matched against the reference one: each difference creates a conflict set. A conflict set can be seen as a set of conflicting modes belonging to a path between two observations. Only a subset of the observations are matched against the reference state, since some observations may belong to paths which are not activated in the reference model. Let Sub-obs be this subset.

A conflict set is a set of modes which cannot work together. For example, considering the possible modes up and down for a relay, and nominal and blown for a fuse, and supposing that the reference modes are:

fuse1-nominal, fuse2-nominal, relay1-up, relay2-down, relay3-down

we have found the two following conflict sets:

(relay3-down relay1-up fuse2-nominal)

  (relay2-down relay1-up fuse2-nominal)

which are interpreted as:

  (not (relay3-down and relay1-up and fuse2-nominal))
  and
  (not (relay2-down and relay1-up and fuse2-nominal))

where relay3-down represents the assumption that relay3 is in the mode down.

5.2.2. Generate candidate sets.

  A candidate is a minimal set of modes which are all assumed not to work properly. For example there are three candidates obtained from the previous conflict sets, each one explains all the observations of Sub-obs.

  (not relay2-down and not relay3-down)
  or
  (not relay1-up)
  or
  (not fuse2-nominal)

  These are obtained by computing the normal disjunctive form of the nogoods, and by keeping only the minimal sets.

5.2.3. Generate fault assumptions.

  This step does not exist in GDE. Using the possible modes of the components involved in the candidate sets, and assuming that a failed mode is any mode different from the reference mode, the fault assumptions are generated. Each fault assumption is a minimal set of modes that explains all the observations contained in Sub-obs. From the previous candidates we obtain:

  (relay3-up and relay2-up)
  or
  (relay1-down)
  or
  (fuse2-blown)

5.2.4. Test the fault assumptions.

  A test is achieved first by using the equations of each failed mode contained in a fault assumption in order to modify the reference model, so that to incrementally modify the reference state, and second by matching with the observations.

  A test is necessary when two conditions are realized: first the set Sub-obs is different from the complete set of observations and second testing the fault assumption allows to match some of the new observations. Such a test may lead to suspect new components.

  Even if there are no new observations, a test may be useful in order to prune some fault assumptions which represent a fault mask. A fault mask is produced by multiple points of failure, when one broken component outputs an incorrect value but a second broken component masks the effects by producing the expected value. In the power supply case study some fault assumptions such as (relay4-down and fuse3-blown) representing a fault mask have to be tested in order to be pruned.

A test is useful when trying to identify among several failure modes which mode explains the observations.

A test is not necessary if a fault assumption contains a component mode which is the only remaining possible mode. For example a fault assumption like: fuse2-blown, is not tested. An exception to this remark concerns the relays: switching a relay connects two different parts of the circuit and may allow to match new observations.

If the new state explains all the observations then the fault assumption is accepted. Otherwise some new conflicts are detected. Using the new conflicts we are able to modify the set of candidates and the set of fault assumptions. This modification is not incremental: some fault assumptions may be discarded (in particular the fault assumption being tested), other may be increased, new ones can be found.

For example when testing (relay1-down) we find the new conflict set (not (fuse3-nominal and relay1-down)). The fault assumption (relay1-down) is discarded and the new fault assumption (relay1-down and fuse3-blown) is computed.

5.3. Implementing the diagnosis in the toolkit

The ODS-DC is implemented by mean of an Identification Tool which uses the CBL, the CP and the ATMS tool-components. The system is modelled in CBL and translated in a net of constraints implementing all the component modes.

The steps one and four of the ODS-DC uses the CP and the ATMS. The ODS-DC decides what modes to activate and when. It uses the ATMS for making assumptions on component modes. When a mode is activated an assumption is created and associated to each constraint which implements the mode. When a constraint is triggered an ATMS justification for the computed value is created: it involves the values which allow the computation and the assumption. This allows to record the modes used in computation.

Matching the telemetries with a state consists of declaring the telemetries as ATMS premises and to propagate this knowledge in the constraint net. The contradictions between a telemetry and a prediction or between two predictions are detected by the CP and recorded in the ATMS. The ATMS builds the conflict sets and maintains the consistency on the data base of computed data.

6. Conclusions

In this paper two diagnostic systems have been presented: the Ontological Diagnostic System for Parameter Identification (ODS-PI) and the Ontological Diagnostic System for identifying Discontinuous Changes (ODS-DC). The diagnostic problems solved by the two systems are different. Together, they cover a broad spectrum of industrial diagnostic problems. The ODS-DC performs fault insulation, i.e. it aims at identifying the components belonging to a complex system, in our case an electronic circuit, that do not work properly. The adopted strategy is an extension of the one adopted by GDE in that ODS-DC is able to identify the faulty components as well as to choose in a catalog of possible fault causes the actual cause of the misbehavior of each broken component. The ODS-PI assumes that the system structure does not change when a malfunction occurs, and tries to identify the parameters whose value is changed and that have caused the malfunction. This last system is specially useful when a continuous process is continuously monitored so that the diagnostic strategy may be activated as soon as a displacement of the working state is detected: this may prevent subsequent drastic alterations of the system structure.

With respect to other recent developments in the field, the ODS-DC shows similarities with parallel works which extend the GDE functionality, namely by De Kleer and Williams [1989] and Struss and Dressler [1989]. The latter work is one of the outcomes of a German project on qualitative modeling, TEX-B.

While applied to malfunctions, instead than structural fault identification, the ODS-PI adopts a candidate generation method similar to the diagnostic strategy due to Bakker *et al*. [1989]. This method explains the inferences that can be drawn from observations about variables that maintain their expected value. Parallel work of Raiman [1989] generalizes this approach by recognizing the alibi principle as a fundamental one for diagnosis.

Future work will try to focus better which features characterize the whole range of the different diagnostic scenarios (e.g. malfunctions vs. structural faults; availability of fault models; presence of diverse behaviors of components in different operational statuses) as a preliminary to the design of a general diagnostic framework. Through a comparison of previous approaches to model-based diagnosis, Struss [1989] provides a foundation to this work.

Acknowledgement

This paper describes developments partially undertaken within ESPRIT project P820, partly funded by the Commission of the European Communities within the ESPRIT programme. Project P820 consists of a consortium composed of CISE, Aerospatiale, Ansaldo, CAP Sesa Innovation, F.L.Smidth, Framentec and Heriot-Watt University. The authors want to acknowledge here the contribution of all the members of the project team to the ideas expressed in this paper, while taking full responsibility for the form these ideas are expressed.

References

[Arlabosse *et al*. , 1988] F. Arlabosse, B. Jean-Bart, N. Porte and B. de Ravinel. An efficient problem-solving architecture using ATMS tested on a non-toy case study. *AI communications* , North Holland, December 1988.

[Bakker *et al*. , 1989] R.R. Bakker, D.C. van Soest, P.A. Hogenhuis and N.J.I. Mars. Fault Models in Structural Diagnosis. *Proceedings of the International Workshop on Model-Based Diagnosis* Paris, France, July 1989.

[Bobrow, 1984] D. G. Bobrow, editor. Special Volume on Qualitative Reasoning about Physical systems. *Artificial Intelligence* 24 (1-3), 1984.

[Chandrasekaran and Mittal, 1983] B. Chandrasekaran, and S. Mittal. Conceptual Representation of Medical Knowledge for Diagnosis by Computer: MDX and Related Systems. In *Advances in Computer* Yovits, Marshall (Ed.) pages 217-293. Academic Press, 1983.

[Clancey, 1985] W.J. Clancey. Heuristic classification. *Artificial Intelligence* Vol. 27, pages 289-350, 1985.

[de Kleer and Brown, 1984] J. de Kleer, and J. S. Brown. A Qualitative Physics Based on Confluences. *Artificial Intelligence* Vol. 24, pages 7-83, 1984.

[de Kleer, 1986] J. de Kleer. An assumption based truth maintenance system. *Artificial Intelligence* , Vol. 28, pages 127-162, 1986.

[de Kleer and Williams, 1987] J. de Kleer, and B. C. Williams. Reasoning about multiple faults. *Artificial Intelligence* Vol. 32, pages 97-132, 1987.

[de Kleer and Williams, 1989] J. de Kleer and B.C. Williams. Diagnosis as Idenfying Consistent Models of Behavior. *Proceedings of the International Workshop on Model-Based Diagnosis* Paris, France, July 1989.

[Doyle, 1979] J. Doyle. A truth maintenance system. *Artificial Intelligence* . Vol. 12, pages 127-162, 1979.

[Gallanti *et al.* , 1989] M. Gallanti, A. Stefanini and L. Tomada. ODS: a diagnostic system based on qualitative modelling techniques. *Proceedings of the Conference on Artificial Intelligence Applications* , Miami Florida, March 1989.

[Genesereth, 1984] M. Genesereth. The use of design descriptions in automated diagnosis. *Artificial Intelligence* Vol. 24, pages 411-436, 1984.

[Gomez and Chandrasekaran, 1981] F. Gomez and B. Chandrasekaran. Knowledge Organization and Distribution for for Medical Diagnosis. *IEEE Transaction Systems Man and Cybernetics* Vol. SMC-11, pages 34-42, January 1981.

[Kuipers, 1984] B. Kuipers. Commonsense Reasoning about Causality: Deriving Behavior from Structure. *Artificial Intelligence* Vol. 24, pages 169-203, 1984.

[Iwasaki and Simon, 1986] Y. Iwasaki and H. A. Simon. Causality in Device Behavior. *Artificial Intelligence* Vol. 29, pages 3-32, 1986.

[Michie, 1982] D. Michie. High Road and Low Road Programs. *IA Magazine* Vol. 3, No. 1, pages 21-22, 1982.

[Raiman, 1989] O. Raiman. Diagnosis as a Trial: the Alibi Principle. *Proceedings of the International Workshop on Model-Based Diagnosis* Paris, France, July 1989.

[Reiter, 1987] A theory of diagnosis from first principles. *Artificial Intelligence* Vol. 32, pages 57-59, 1987.

[Stefanini and Leitch, 1988] A. Stefanini and R. Leitch. QUIC: a development environment for Knowledge Based Systems in industrial automation. *Proceedings of the Third Esprit Conference* , Bruxelles, North-Holland, pages 674-698.

[Struss, 1987] P. Struss. Multiple representation of structure and functions. In *Expert Systems in Computer Aided Design* , J. Gero editor, North-Holland, 1987.

[Struss, 1989] P. Struss. Diagnosis as a Process. *Proceedings of the International Workshop on Model-Based Diagnosis* Paris, France, July 1989.

[Struss and Dressler, 1989] P. Struss and O. Dressler. "Physical Negation". Integrating Fault Models into the General Diagnostic Engine. To appear on the *Proceedings of the Eleventh Joint Conference on Artificial Intelligence* Detroit, MI, Aug. 1989.

[Sussman and Steele, 1980] G.J. Sussman and G.L. Steele. CONSTRAINTS - A Language for Expressing Almost-Hierarchical Descriptions. *Artificial Intelligence* Vol. 14, pages 1-39, 1980.

QUIC Toolkit Demonstrator Applications.

A. Cavanna
Ansaldo SPA
V. Dei Pescatori, 35
I-16128 Genova
Italy

J.C. Chautard, C. Honnorat
Aerospatiale
100, BD Du Midi
F-06322 Cannes-la-Bocca
France

M. Levin, B. Klausen
F.L.Smidth & Co. A/S
Vigerslev Alle 77
DK-2500 Valby
Denmark

Abstract

The QUIC Toolkit can be characterised as an application environment specially designed to support process control applications. The QUIC Toolkit methodology and terminology has been validated on three different applications ranging from diagnosis of power plant condenser subsystems over monitoring and control of spacecrafts to control of vital parts in cement manufacturing plants. For each application a detailed system description is provided together with a task analysis which provide the mapping of the demonstrator requirements onto the QUIC Toolkit. Using the methodology and terminology of the QUIC Toolkit each application is then hierarchical decomposed into its basic building blocks. Finally, validation results are described and discussed.

1. Introduction

In this paper application of the QUIC Toolkit on three different demonstrators covering a broad range of process control application is presented.

The QUIC Toolkit, which is developed within ESPRIT Project 820: *Design and Experimentation of a Knowledge Based System Development Toolkit for Real-Time Process Control Applications*, has four main motivations as presented in ESPRIT '88 (Leitch & Stefanini, 1988):

- The development of a powerful but general QUIC Toolkit (**QU**alitative **I**ndustrial **C**ontrol) for an extended range of applications within the Industrial Automation.
- The recognition of the respective advantages of ontological (theoretical) knowledge and empirical (experimental) knowledge, and to support reasoning with both sources of knowledge within the QUIC Toolkit.
- The development and application of ideas and concepts emerging from recent work on Artificial Intelligence on Qualitative Modelling or Qualitative Physics (Bobrow and Hayes, 1984).
- The validation of the tools supported by the QUIC Toolkit on realistic industrial case studies provided by the three industrial partners of the consortium.

The project is now in its final stage and the kernel of the QUIC Toolkit has been designed, implemented and validated on various applications. The following main sections are included in this paper:

2. QUIC Toolkit Architecture giving a brief overview of the QUIC Toolkit architecture.
3. Description of the Demonstrators giving an overview of the three demonstrators.
4. Task Analysis of Demonstrators maps the control tasks of the demonstrators onto the primitive tasks defined for the QUIC Toolkit.
5. Description of the Toolkit Applications describes three QUIC Toolkit applications, one for each demonstrator, according to the methodology and terminology defined for the QUIC Toolkit.
6. Results of Validation each of the three demonstrators has applied tool/tool-components provided by the the QUIC Toolkit. The validation results achieved by each demonstrator is described.

2. QUIC Toolkit Architecture

A brief overview of the QUIC Toolkit architecture will be given in the following. A more detailed description of the QUIC Toolkit concept, architecture, knowledge representation languages and applications is included in ESPRIT '88 (Leitch & Stefanini, 1988).

One of the main motives of P820 has been to develop a toolkit providing a set of high level tools special tailored for industrial process control tasks. Five *primitive tasks* have been identified:

Decision
The decision making process whereby the known or assumed present state is used to construct conclusions (hypotheses) at the same time.
Prediction
The generation of future states from known or assumed present states.
Identification
The determination of (unknown) past states from known or assumed present states.
Interpretation
The transformation of observed data into the adopted state representation.
Execution
The transformation of system state into data suitable for action.

Within the QUIC Toolkit five conceptual levels have been defined: Strategic, Tactical, Teleological, Functional and Object. The Strategic and Tactical levels are composed from the QUIC Toolkit by the KBS builder for a given application. The Teleological, Functional and Object levels exist within the QUIC Toolkit. The boundary defined between the Teleological and Tactical level is believe to cover a wide range of industrial process control tasks. The levels will be described more detailed next:

Strategic Level
The goal of this level is to build *Target-KBS* 's satisfying the overall objective of a process control application. A *Target-KBS* provides control and coordination between the various *Systems* based on the QUIC Toolkit. This level represents the 'highest' level of the QUIC Toolkit.

Tactical Level
>   The goal of this level is to build *Systems* solving clearly defined sub-parts of the overall objective of a process control application. A *System* provides control and coordination between the various *Tools* provided by the QUIC Toolkit.

Teleological Level
>   The goal of this level is to build *Tools* able to solve tasks relevant for process control application. A *Tool* is defined by the primitive tasks and the basic problem description language. Besides, it is build on generic *Tool-components*. This level defines the interface between the QUIC Toolkit and applications.

Functional Level
>   The goal of this level is to define a set of generic *Tool-components* used to construct different *Tools* on the Teleological Level. This level is defined by the various programming languages defined by the Object Level.

Object Level
>   The goal of this final level is to implement the generic *Tool-components* on the *Functional Level* and to provide access to various programming languages and environments. In this level the major software development effort is expended.

3. Description of Demonstrators

This section gives a brief description of the three demonstrators. The three demonstrators provided by the industrial partners of the consortium are:

Demonstrator 1 (Ansaldo)
>   malfunction detection and diagnosis in the thermal cycle of a power plant

Demonstrator 2 (Aerospatiale)
>   data interpretation and control of a spacecraft, including misposition/attitude detection and diagnosis, and assistance to correction manoeuvres

Demonstrator 3 (F.L.Smidth)
>   start-up/shut-down, on-line monitoring, and control of the main subsystems of a cement manufacturing plant

The main objectives of those demonstrators in relation to the project have been:
– to identify, or to refine, the requirements for the QUIC Toolkit
– to validate the QUIC Toolkit architecture and the tools provided by the QUIC Toolkit on realistic demonstrators.

3.1. Description of Demonstrator 1

3.1.1. Conceptual architecture.

Demonstrator 1 has been implemented for performance diagnostic of a Power Plant condenser. Performance diagnostic is intended as the set of tasks (usually performed by a condenser expert), consisting of:
a) judging whether the measured values of process variables which characterise the operation of a condenser match with the values that the same variables should theoretically have

b) in case of mismatch, individuating the design parameters whose value is modified with respect to the normal operating ones.

Demonstrator 1 is characterised by two main objectives:

a) to incorporate tools of the toolkit in a way suitable for test and validation

b) to be an industrial prototype vz suitable for realistic applications and showing better performance than more traditional systems.

Demonstrator 1 includes two fundamental kinds of knowledge, ontological and empirical. The two types of knowledge are organised within two different Diagnostic Systems, the Heuristic Diagnostic System (HDS) and the Ontologic Diagnostic System (ODS). The latter, which is the core of the Demonstrator, has been built using the toolkit tools. The HDS and the ODS are integrated as a general Condenser Diagnostic System. Benefits related to the integration of ontologic and empiric knowledge in an industrial prototype capable of testing and validating the toolkit (i.e. the ODS) can be summarised as follows
− to obtain a significant test bed for the toolkit
− to limit the work area of the ODS
− to have a flexible testing tool
− to increase the industrial significance of the system both in terms of efficiency and of effectiveness.

3.1.2. Process description.

A Power Plant condensing system has the task of condensing the exhaust steam flowing from the discharge of the turbine at the lowest possible pressure, whose value is imposed by the characterististics of the available cooling medium. The condensing system may be considered as composed by subsystems
− the condensate extraction subsystem
− the non-condensible gasses extraction subsystem
− the cooling water subsystem.

Being Demonstrator 1 goal the diagnosis of malfunctions having an impact on system performance, the condensing system thermal-hydraulic process is of primary importance. A simplified description of this process follows.

The condenser is a shell and tube heat exchanger. Steam from the turbine is condensed is the shell side by means of cooling water flowing in the tube side. Condensed water is pumped (figure 3.1.2.a) via two electromechanical pumps, to another small shell and tube heath exchanger. The shell side of this exchanger is fed by the discharge section of two air ejectors, whose function is to extract from the condenser shell side the non-condensible gasses coming from various parts and processes of the Plant, and whose presence would be detrimental to condenser performance. From the outlet of the air ejectors heat exchanger, water is sent to an other exchanger , whose function is to condense steam from the turbine seal. From the turbine seal H/X water is sent to the first regenerative H/X. Inlet flange of this H/X is considered as one of the boundaries of the condensing system.

Cooling water for the condenser comes from the nearest available body of natural water, or from cooling towers, via intake structures and filters, and is pumped via two electromechanical pumps to the tube side of the condenser. Heated water is discharged from the condenser and sent back to the body of natural water of to the cooling towers.

424

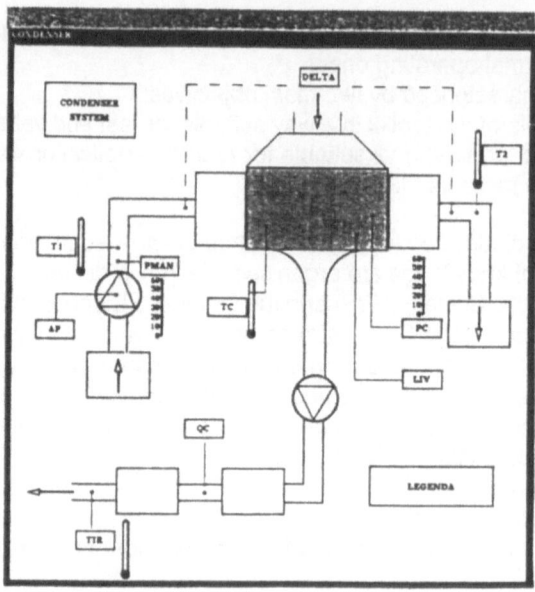

Figure 3.1.2.a.

Plant model

The main source of knowledge regarding the condenser system originates in the plant design phase, and can be expressed in the form of a physical-mathematical model. This plant model is made up by a set of thermal-hydraulic equations including semi-empirical relationships and involves process variables and design parameters. Some of the process variables are independent (exogenous), such as the quantity of heat to remove in the condenser, the cooling water temperature, the flow rate and thermodynamic characteristics of the motive steam to the air ejectors, the status of the cooling water pumps, etc. These variables define the boundaries of the condensing system model. The design parameters are also exogenous quantities, having values imposed by:
- design decisions
- malfunctions of the system causing modifications of the values of some design parameters with respect to the design decisions.

The plant model must contain all the design and process information necessary for performance diagnosis purposes. In this regard, the following considerations have to be done:
- the performance diagnosis is performed with the plant in stationary conditions
- the environmental variable cooling water temperature is sufficiency slow to induce only unsignificant transients
- the performance modifications that need to be detected are in general small with respect to the field of validity of the model equations.

A physical-mathematical model constituted by algebraic equations is therefore sufficient. These equations describe the impulse, mass and energy transport phenomena taking place in the condensing system circuits. The value of the design parameters entering these equations is the nominal one.

3.2. Description of Demonstrator 2

The application domain of Demonstrator 2 concerns geostationary telecommunication spacecraft monitoring, diagnosis and control from a ground station. The purpose of Demonstrator 2 is to help the human operators of a ground based station to detect, diagnose and correct anomalies of a satellite as well as making it perform manoeuvres of orbit position and attitude (i.e. platform orientation) correction. The ultimate goal of such a system is to keep the telecom satellite pointed towards the earth so that its mission is not interrupted.

This application domain proved to be rich and gave birth to three different case studies each of which corresponds to a basic functionality of the target control system:

1/ *Anomaly detection and diagnosis*
   case study: the electrical power supply which is the most representative subsystem for diagnosis problems.
2/ *Operation plans execution (in real time)*
   case study: the attitude and orbit control of the satellite (North/South orbit control)
3/ *Operation plans generation*
   case study: generation of operation plans for telecom payload management.

We made the difficult choice to define case studies of realistic industrial size, being fully aware of the complexity it introduces in the demonstrator. We have considered it uncircumventable for the sake of toolkit validation's quality.

3.2.1. Description of the electrical power supply.

The electrical power supply is the spacecraft's subsystem which is in charge of providing every equipment with the electrical power it needs, whatever the external conditions are (eclipse or not etc..). In our specific case study see (Chautard et all., 1988), the satellite receives electrical energy from two single axis sun oriented solar array wings. Solar array power is directly transferred to two independent power distribution busses, with a cross strap possibility between them. During launch and eclipses, the power is supplied by two Ni-Cd batteries. Every equipment needing electrical power is connected to both busses. It owns a DC/DC converter providing it with a correct voltage. Figure 3.2.1.a, extracted from the full case, gives an idea of the complexity of such a system.

In this particular case study any component may own multiple possible states, describing its different ways of functioning and its nominal, degraded or faulty states. A component also owns *modes*. The difference between modes and and states is illustrated in the following example: A relay with four pins has two states (up and down) which describe its functionality; it may be in the state "up" either in a nominal mode or in a "stuck-up" mode which is a faulty mode.

The diagnosis tool has as input the *reference state* of the system i.e. the state into which it is expected to be (this information may come from simulation) as well as the set of telemetries (TM) actually coming from the satellite. These TMs are of two kinds: status TMs indicating the presence of voltage or not, analogue TMs giving some current and voltage measures. The purpose of the diagnosis is given a set of symptoms, to identify the component(s) the behaviour of which proves to be abnormal and to identify the type of anomaly (i.e. the failure mode of such components).

426

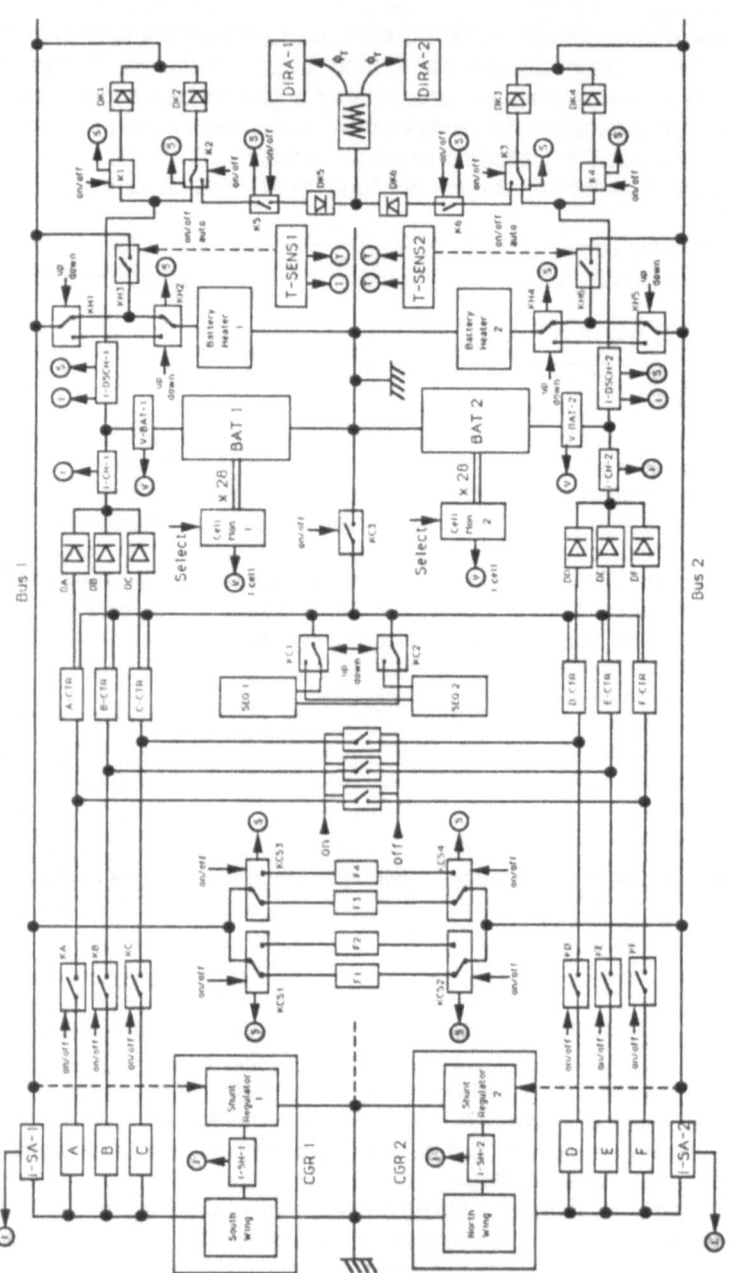

Figure 3.2.1.a.

3.2.2. Satellite attitude and orbit control.

a/ On-board attitude and orbit control

On-board attitude and orbit control is performed by the Attitude and Orbit Control Subsystem (AOCS). Figure 3.2.2.a depicts the AOCS as composed of sensors (earth or sun sensitive, or inertial), processing electronics (AOCE) and actuators (momentum wheels and thrusters). The AOCE performs attitude data processing and controls feedback between sensors and actuators according to a command law which depends on the operation phase and characterise the AOCE modes. To whatever operation phase is associated a set of possible sensors and possible actuators depending on the kind of work to be performed (accurate or not, translation or rotation etc...).

SENSORS   CALCULATOR   ACTUATORS
                 (command law)

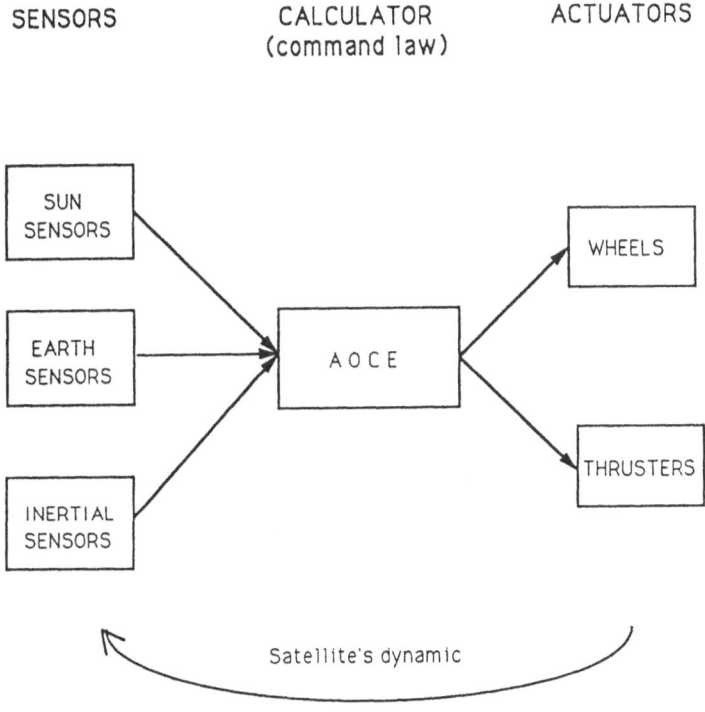

Figure 3.2.2.a.

b/ The North-South orbit correction procedure

Ideally, a geostationary satellite should stay on a single point of the geosynchronous orbit. This orbit is a circle which plane is the equator, which center is the earth's center and which radius is 36 000 km. Unfortunately, different kinds of phenomena (such as gravitational moon interferences) do perturb this orbit. The global resulting effect is an angular drift of the orbit's plane (see figure 3.2.2.b). Therefore, the human operator must regularly make the satellite perform a South to North orbit correction (called North-South station keeping or NSSK) while maintaining the mission uninterrupted (the satellite must be kept pointed towards the earth).

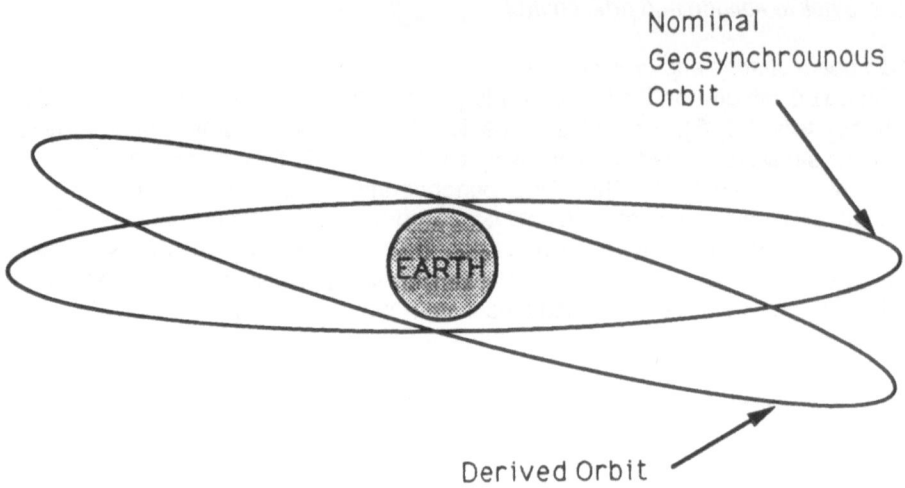

Figure 3.2.2.b.

This manoeuvre must begin at a given time and not last more than an allowed interval of time. It consists of three phases. First, the firing plan (expressed in terms of a list of dated thrusts) must be elaborated and the satellite equipments which are required by the NSSK command law must be selected and configured. Second, the actual thrusts are performed. Finally, the satellite must be taken back to its normal on orbit mode (NOO mode) after stabilisation. More than the complexity of the different equipment preparations itself, what makes this manoeuvre one of the most difficult is that it is operated in open loop, that is, there is no on-board control of the thrusts themselves, nor even any telemetries of the satellite's position.

c/ *State of the art in satellite control from the ground*

A satellite is monitored from the ground by human operators. In a control center, an operator knows about the current state of the satellite through telemetries (TM) and sends telecommands (TC) to control it (see figure 3.2.2.c). The knowledge telling him what TC he should send and when in order to perform various operations is structured into procedures , all of which are tied together into operation lists . The procedures are written in pseudo code for human use and are literally followed by the operators. Whenever an unforeseen event occurs, a station engineer or even an expert should be consulted in order to adapt or rebuild a procedure.

d/ *Requirements*

The goal of this case study is to demonstrate that one could automatically command a geosynchronous spacecraft, covering not only the task of a human operator but also expressing and exploiting satellite expertise. This requires, a formalism for modelling the satellite, a formalism for expressing procedural knowledge, and capabilities to operate on these knowledges. The real time control of the satellite needs the system to be reactive: given the environment, the purpose of the procedure and the state of the satellite, the system must react within limited time. Moreover, even if the system actually controls the satellite, it must stay attentive to other monitored events.

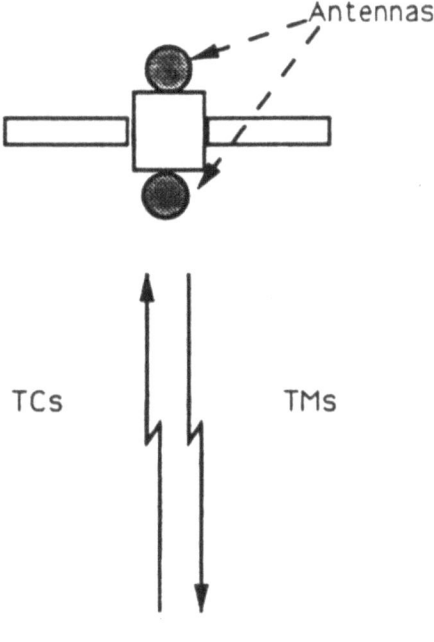

SATELLITE

Antennas

TCs   TMs

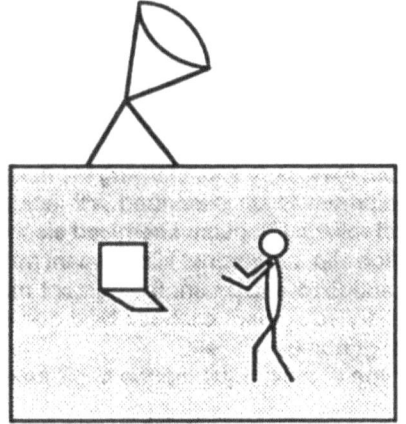

GROUND STATION

Figure 3.2.2c.

3.2.3. Telecom payload management.

The purpose of this case study is to provide a typical example of a satellite subsystem for which it is particularly interesting to have a tool for automatic generation of management procedures.

The spacecraft's subsystem into question is the payload of a TV transmission satellite. This payload consists in five channels for direct TV broadcasting. In this aim, it is composed of an antenna module with one reception and one transmission antenna, both controlled in fine pointing. The heart of this payload is the repeater, the task of which is to receive and transmit at high power up to five TV channels. It includes a power supply (RPS), a broad band section unit (BSU) for global preamplification and translation to 12GHz of the 17GHz uplink signals, a passive input multiplexer (IMUX) for "channelisation", five amplification chains and finally a passive output multiplexer (OMUX). Each amplification chain is itself composed of a channel amplifier (CAMP) which gain is selectable from ground and a high power stage using a travelling wave tube (TWT) with its specific power supply. One of these chains is equipped with a TWT redundancy.

Each of this electronic is controlled by ground telecommands.

From a deep description of this subsystem, the purpose of the operation plans generator is to build up, given a high level goal such as "the payload is configured for using the T1 T2 and T3 channels", a set of procedures in the Event Graph formalism (or close) that do implement the payload transformation via telecommands when executed.

3.3 Description of Demonstrator 3

In this sub-section a brief description of Demonstrator 3, the cement manufacturing plant demonstrator, will be given. The main subsystem in a cement manufacturing plant is depicted in figure 3.3.a.

3.3.1. Description of a cement manufacturing plant.

The following description covers a specific modern cement manufacturing plant. Details include the main processing equipment starting from raw material and ending with cement ready for delivery either by bulk dispatch or packed in sacks.

Starting from the quarry and following the material flow through the manufacturing plant the main equipment to be described will include: crusher, circular prehomogenisation stores, vertical roller mills, pulse-energised electrostatic precipitator, continuous raw meal homogenisation silo, precalciner kiln, cement mill, and automatic cement packer. Besides the above mentioned equipment the cement manufacturing plant also includes a multi-plicity of auxiliary equipment such as: heat generators, dust bag filters, various conveyor systems, coal grinding mills, etc.

For each type of main equipment a short functional description will now be given.

Crusher.
In the crusher irregular lumps of rocks with a longest dimension of approx. 2 meters are being crushed into pieces with a longest dimension of approx. 25 millimeters. The crusher will normally be located at a suitable place between the plant and the quarry.

431

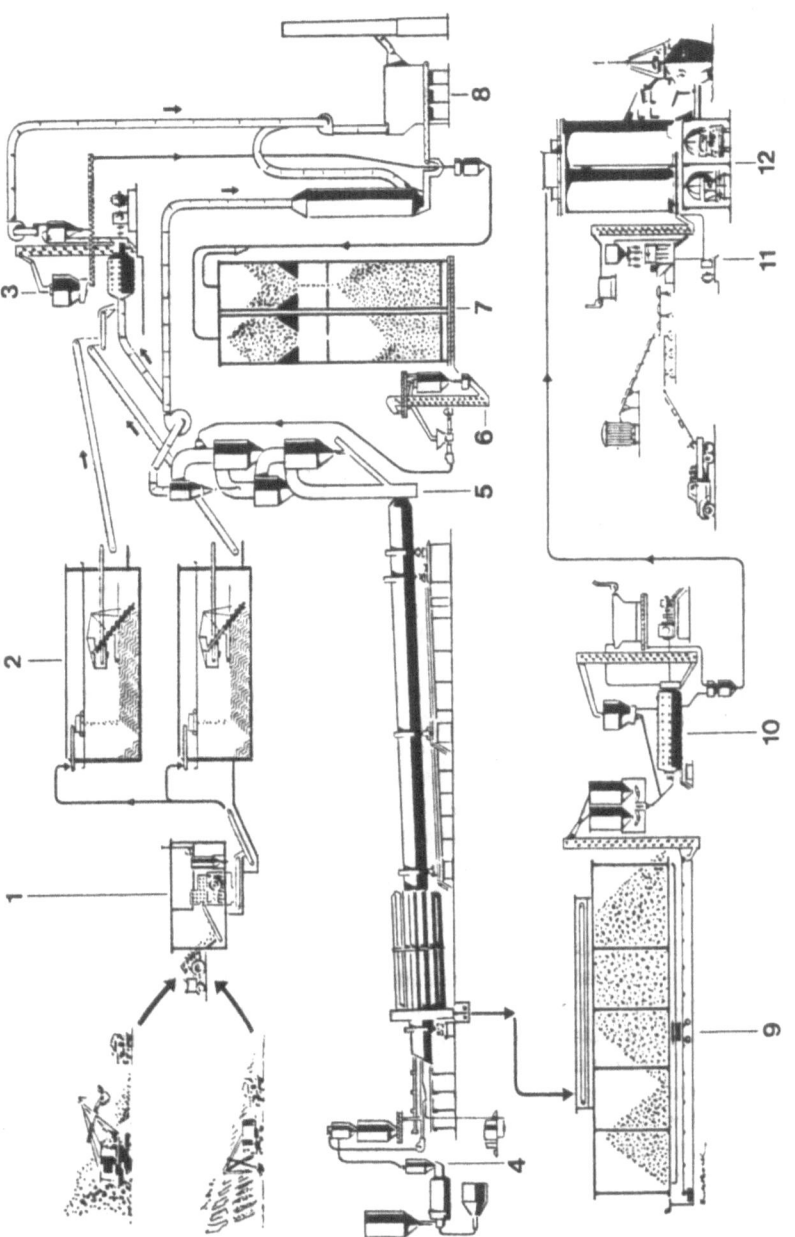

Figure 3.3.a.

Circular prehomogenising store.
In this store crushed material is added to one end of the circular pile while material to be further processed is removed from the other end of the pile. Homogenisation is due to the techniques used for adding new material and removing material from the pile.

Raw mill (Vertical roller mill).
Raw materials from different stores are mixed before it is ground in the roller mill. The mixing is done so that the output from the mill has approximately the correct chemical composition. The actual grinding of the raw materials to a fine raw meal is done in the roller mill. When the size of the particles in the raw meal is small enough they become suspended in the hot gasses led through the mill. The hot gasses are taken from the kiln subsystem. Part of the natural moisture in the raw material is evaporated by the hot gasses. The temperature of the gasses is hereby lowered making it possible to lead the gasses directly to the electrostatic precipitator without cooling.

Pulse-energised electrostatic precipitator.
In the precipitator the air suspended particles are precipitated and the resulting raw meal is carried to a silo. Samples of the raw meal are taken periodically and brought to a laboratory for chemical analysis. To avoid high-voltage problems (coronas, sparks) it is important that the temperature of the gasses in the precipitator is below the safety level and that the gasses do not contain a mixture of components such as carbon monoxide which is susceptible to explosion initiated by sparks.

Continuous raw meal homogenisation store.
From the precipitator the raw meal is led to the top of the store. As small samples of this raw meal are being analysed at regularly time intervals and as the retention time for raw meal in the storage by far exceeds the time needed for the laboratory analysis of the raw meal as well as the sampling intervals, the chemical composition of the material within the store is fairly well known. Due to this a considerable improvement in the correct chemical composition of the raw meal leaving the silo is possible by proper management of the outlet valves in the bottom of the silo.

Precalciner kiln.
The raw meal leaving the silo is fed into a preheater string where the temperature of the raw meal is raised. The preheated raw meal is then fed into the precalciner where part of the calcination process takes place (Decomposition of calcium carbonate into calcium oxide and carbon dioxide is called calcination. The process requires that energy is added, and it is reversible). The precalcined raw meal is now fed into the rotary kiln where the remaining calcination takes place and where also the raw meal is burned to clinker (During the burning process the clinker minerals called tricalcium silicate (alite), dicalcium silicate (belite), tricalcium aluminate and tetracalcium aluminoferrite (brownmillerit) are formed). After the burning the clinkers are cooled and carried to a store. Samples of the clinkers are taken periodically for various laboratory analysis to determine its quality (litre weight, free lime, alkali, etc.).

Cement mill.
In a final energy consuming process the clinkers with some additives, are finally ground into the powder known as cement. Gypsum is added to improve the strength and to delay the setting time of the cement. However, gypsum changes character when it is heated

and therefore the temperature of the material within the mill must be kept below a given level. This is accomplished by pouring water onto the clinker in the mill. Apart from gypsum, other types of additives may also be used to make cements. The cement is then finally stored in a silo.

Automatic cement packer.
From the silo cement is shipped either in bulk or packed in paper sacks. In the latter case filling of sacks is done automatically by a complex packing machinery. The following palletising of cement sacks is also done automatically.

From the description given it follows that the cement manufacturing plant is divided into a number of subsystems separated by stores. This means that the couplings between subsystems are relatively loose, the main exception being the close coupling between the kiln subsystem and the vertical roller mill the latter needing hot gasses from the kiln subsystem to operate.

3.3.2. The Control Objective of Demonstrator 3.

The control objective of Demonstrator 3 is Overall Plant Control of a cement manufacturing plant. The Overall Plant Control may be divided into four control activities as listed below:

Plant repair and maintenance
controls the repair and maintenance work that constantly is carried out in the plant. This activity is today mainly supervised by operators. The knowledge applied can basicly be classified as empirical procedural knowledge.
Plant start-up
controls the plant from a non-production state (cold or hot kiln system) to a normal production state. This activity is today mainly carried out by operators. The knowledge applied can basicly be classified as empirical procedural knowledge.
Normal production
controls the plant during normal production. This activity is today partly carried out by operators and partly by automatic controllers. The knowledge applied to carry out this activity can be classified as empirical declarative knowledge.
Plant shut-down
controls the plant from a normal production state to a non-production state. This activity is today basicly carried out by operators. The knowledge applied can basicly be classified as empirical procedural knowledge.

As seen from the above description, the control knowledge applied in Demonstrator 3 is basicly classified as empirical knowledge. Besides different knowledge representations are required: declarative and procedural. Except for the *Normal production* control activity, the activities are in general carried out manually by operators.

4. Task Analysis of Demonstrators

In this section the control tasks of the three demonstrators is identified. Those tasks are mapped onto the primitive tasks provided by the QUIC Toolkit.

4.1. Task Analysis of Demonstrator 1

4.1.1. General.

Demonstrator 1 is a model based diagnostic system that utilises ontological knowledge of a steam condenser plant, and which is augmented by specific empirically based information. The primitive tasks for this system are the following
– interpretation task
– decision task (empiric)
– identification task
Due to development reasons, two different interpretation task modules were developed, one serving the identification task, the second serving the decision (empiric) task. Interpretation and decision (empiric) task modules together constitute the Heuristic Diagnostic System, while interpretation and identification constitute the Ontologic Diagnostic System.

From a conceptual point of view however, a single interpretation task module is sufficient, giving rise to the general task structure represented in figure 4.1.1.a.

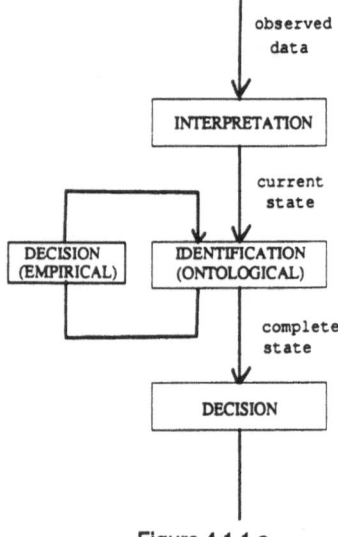

Figure 4.1.1.a.

4.1.2. Plant Model and Interpretation Task.

Plant model includes a description of the condensing system process as well as the way the parameters of the process equations are determined by the design parameters. On the bases of this description it is possible to define an interpretation task module capable of evaluating the deviations of measured variables with respect to their values computed on the bases of condenser design parameters. When the deviations go beyond a predefined threshold, a process of malfunction cause identification is started.

Performance of this task requires the availability of a language for representing ontologic models of physical systems. The language provided by the QUIC toolkit is the Component Based Language.

4.1.3. Decision Task (empiric).

The empiric decision task mainly relates to empirical information. The task consists of constructing hypotheses on the possible causes of malfunction, based on the actual state of the plant. This state is partly unknown, being based on a very limited set of measurements, those that are always available through the standard monitoring equipment of the plant. The goal of the task is to identify a sub-set of the complete list of possible malfunctions. This sub-set will be used as one of the inputs for the identification task.

4.1.4. Identification task.

The task goal is to identify the unknown state of the system, represented via the ontologic plant model, by comparing its actual state with the state variations caused by each one of the set of fault mechanisms. Identification is generally incomplete because the actual state of the plant is partly unknown. In fact, it is necessary to cope with situations where the measurements which can be taken have different cost and reliability. Thus it is necessary first to process the set of measurements having minimum cost and maximum reliability (in practice, the already mentioned set of measurements which are always available through the standard monitoring equipment). When these are not sufficient to single out the malfunction cause, a new measurement is selected, the actual state is expanded and a new identification process is started. The selection procedure mainly relies on the plant model, as the new measurement selected, among the ones having the same cost and reliability, is the one which has the higher discriminating power with respect to the likely faults. As for the case of interpretation task, the plant ontologic model is represented by using the Component Based Language.

4.2. Task Analysis of Demonstrator 2

Basically, this application includes in the tactical layer a diagnosis system and a command system.

The diagnosis system (known as the Ontological Diagnosis System or ODS) performs in the toolkit terminology the tasks of:
- *interpretation:* integration of telemetries into the satellite model knowledge base and propagation to infer the complete state of the system. This is done for the observed state as well as the reference state (the expected state).
- *identification:* comparison between the reference state and the observed state, isolation of discrepancies. From the discrepancies, a set of fault assumptions is generated and tested.

The diagnosis principle consists in comparing the reference state predicted from the reference model and the observed state computed from the telemetries. Any difference is due to an unexpected behaviour of the system. Diagnosing consists in identifying a new model explaining the actual behaviour of the electrical power supply.

The diagnosis system uses the Component Based Language (CBL) for modelling the system (ontological knowledge). The system's CBL description is translated into a net of constraints implementing all the component modes. The combined Constraint Propagator (CP) and Assumption based Truth Maintenance System (ATMS) tool-components are used for the *interpretation* and *identification* tasks.

The command system relies on the Procedural Actuation Tool. This tool uses the Event Graph formalism for expressing procedural knowledge and exploiting it in order to eventually perform actuations on the system (i.e. send TCs to the satellite). The Event Graph interpreter is used to perform the *decision* and *execution* task in the toolkit terminology, exploiting a model of the satellite (design knowledge expressed with the KEE representation language) and a base of procedures (coded within the Event Graphs formalism). Further more in case of procedure deficiency it will allow (this is on going work) to build out of the complete model of the satellite (close from design knowledge) new procedures within the Event Graph formalism (or close to it) using planning techniques involving the CP and the ATMS (this is demonstrated on the third case study).

4.3. Task Analysis of Demonstrator 3

The control activities of the Overall Plant Control were identified in section 3.3. Those activities can be mapped onto the QUIC Toolkit in order to automate them. The four activities may be mapped onto two generic control systems, which are:
Feedback Control System (FCS)
This generic system is intended for automation of the *Normal production* control activity.
Sequence Control System (SCS)
This generic system is intended for automation of the control activities: *Plant repair* and *maintenance*, *Plant start-up* and *Plant shut-down*.
Those two control system can be related to three primitive control tasks recognised by the QUIC Toolkit: *Interpretation Task, Decision Task and Execution Task*.
The three primitive tasks and the interrelations for the FCS and SCS are shown in figure 4.3.a.

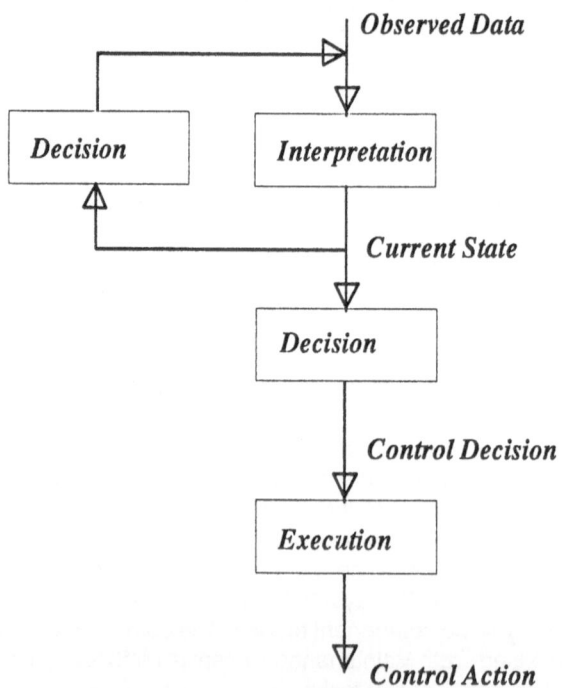

Figure 4.3.a. The primitive tasks of FCS and SCS and their interrelations.

The *Interpretation Task*
  generates a situation description from observable signals in the plant. The *Interpretation Task* co-operates with a *Decision Task* to provide for sensory signal validation prior to forming the current state.
A *Decision Task*
  generates conclusions based on the validated situation description achieved by the *Interpretation Task* and a *Decision Task*.
The *Execution Task*
  transforms the conclusions obtained by the Decision Task into signals driving the actuators of the plant.

From a control engineering point of view, it can be concluded, that the primitive task classification provided by the QUIC Toolkit is suitable to describe the FCS and SCS identified by Demonstrator 3.

5. Description of toolkit applications

In this section applications from the three demonstrators are decomposed according to the hierarchical structure of the QUIC Toolkit architecture.

5.1. Description of Demonstrator 1 Toolkit Application

5.1.1. General.

Reference is made to the layered toolkit architecture, which defines five conceptual layers or levels: Strategic, Tactical, Teleological, functional and Object.

At the STRATEGIC level, the goal of the application KBS is the minimisation of the Power Plant heat rate, by automatic and manual acquisition of data from the plant process and providing the Plant Operators with information regarding the reason of abnormal heat rate in terms of deviations of plant parameters whit respect to the design values.

The strategic goal has been achieved by using, at the TACTICAL level, a diagnostic system corresponding to the system previously defined as composition of primitive tasks. Figure 5.1.1.a. represents this system, as well as the general architecture and control strategy which can realise the strategic goal of demonstrator 1. A supervisor/controller module is fed by low cost and high reliability measurements (cost 0 measurements) from the Plant Data Acquisition System (the DAS is not part of Dem 1). Some of these measurements are used for condenser performance surveillance, i.e. for triggering the Heuristic Diagnostic System. Output of the HDS, together with cost 0 measurements, go to the ODS which manages the diagnostic problem and requires further (cost 1 and 2) measurements when necessary.

The ODS has been realised by the interconnection of the Ontological Interpretation Tool and the Ontological Identification Tool (TELEOLOGICAL level). ODS general structure is shown in figure 5.1.1.a. This figure, as well as the following description of the ontologic interpretation and identification tools, are taken from (Stefanini, 1988).

438

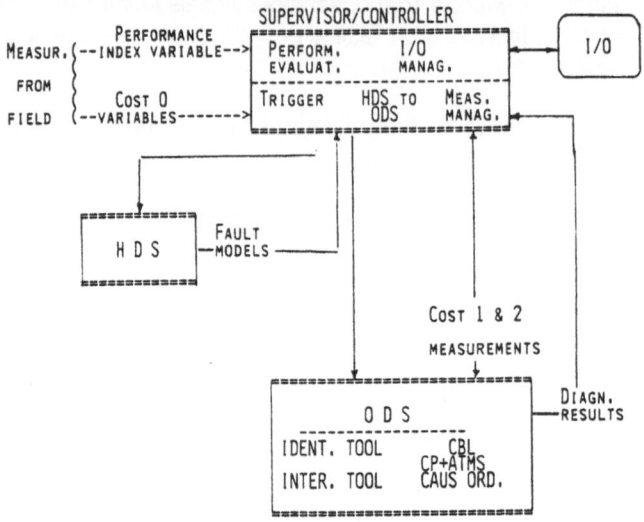

INTEGRATED SYSTEM GENERAL ARCHITECTURE

Figure 5.1.1.a.

5.1.2. Ontologic Interpretation Tool.

This tool is basically designed to translate a set of input data into the actual state, according to a given model represented by the Component Based Language, of the artifact which is being monitored. Thus the tool performs the interpretation primitive using an ontologic model. To this aim, the tool is built around a constraint propagator tailored for propagating real values. The tool also embodies:
– the CBL compiler, for translating the CBL source model into an internal representation suitable for processing by a causal and constraint net generator
– an ATMS
In addition to its basic function, the tool also computes and stores a reference state from given reference values of a subset of state variables; and computes the qualitative variation between the actual and reference state.

5.1.3. Ontologic Identification Tool.

This tool identifies an unknown state of of a system, whose ontologic model is represented with CBL, by comparing its actual state with the state variations caused by each one of a set of fault mechanisms. The state identification is actually done using a qualitative representation based on vectors that represent the qualitative variation of the actual state with respect to a reference state: this is compared with the qualitative variations which are induced by each fault mechanism. Fault mechanisms, in turn, are represented as the qualitative variation (with respect to its reference value) of a parameter being directly affected by the fault cause. For instance, a fault in a hydraulic circuit induced by crusting of the inner wall of a tube is represented as a qualitative decrease of the cross-section of the tube.

The tool uses the constraint propagator for propagating the effects of their fault mechanisms over the qualitative model of the system, and makes intensive use of the ATMS coupled with the constraint propagator for checking consistency of the qualitative variations associated with each fault with the state variation actually observed. In this way, the tool implements the primitive task of identification to establish the particular fault mechanism that is generating the observed misbehaviour.

The causal ordering system is also embodied within the tool, in order to discriminate between sources of "propagation stuck" conditions in the constraint propagator. Essentially, this allows the detection of "loops", due to simultaneous equations, from ambiguities generated by the indeterminacy of the qualitative algebra.

5.2. Description of Demonstrator 2 Toolkit Application

5.2.1. Description of the diagnosis application.

From the higher point of view, the diagnosis system is given :
– a set of observations under the form of telemetries.
– a set of reference initial values allowing to compute a reference state via propagation. It is known from a previous diagnosis session or is a normal operating mode.

It outputs a diagnosis consisting in a new reference state explaining the behaviour of the system.

The overall diagnosis strategy uses and extends the General Diagnosis Engine (GDE) of de Kleer, see (de Kleer & Williams, 1987), and follows the path hereunder:
a) generation of conflict sets (nogoods) The reference state is processed from the equations of the reference mode of each component. Then it is matched with the observed state. Each discrepancycreates a conflict set.
b) generation of candidate sets This step outputs from the conflict sets candidate modes which are all assumed not to work properly, into a disjunctive form.
c) generation of fault assumptions Each fault assumption is the minimal set of modes that explains all the observations. This step generates from the candidate sets, a disjunction of possible modes which could explain the actual observations.
d) tests of the fault assumptions One can test a fault assumption using the equations of each failed mode included in it by incrementally modifying the reference state and by matching the latter with the observed state. If the new state explains all the observations then the fault assumption is accepted.

5.2.2. Description of the command application.

The command system uses two sources of knowledge. The first is the satellite's state description (SSD) in terms of components. In particular it includes the telemetries coming down from the spacecraft. It must be considered as a photography of the system at a given time. The second is the satellite's set of operating procedures (SOP). These procedures express how to reach a goal by decomposing it into subgoals to be achieved and eventually by acting directly onto the system via telecommands. The main characteristic of this application is that it operates in *real time*.

From a high point of view, the application is decomposed into, first, a satellite simulator

440

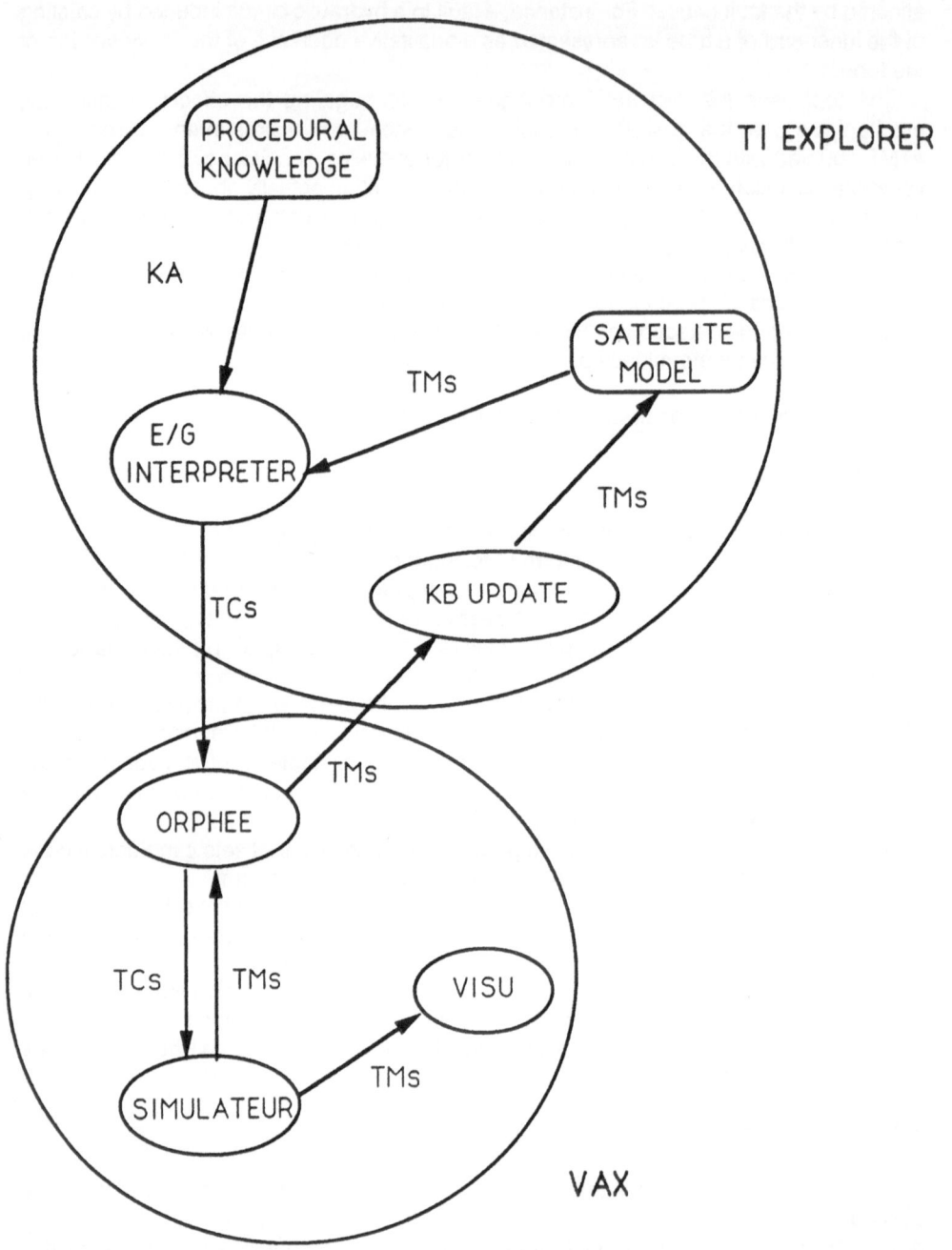

Figure 5.2.2.a.

which accepts a number of TCs and periodically outputs formats of TMs and second our command system which takes the TMs stream as periodic input and computes thus the observed state of the satellite upon which it can reason and generate the appropriate TCs given an initial goal.

As said above, the event graph interpreter is able to execute procedures coded as event graphs while monitoring a dynamic set of constraints and being sensitive to every new event.

In order to implement this we followed the scheme presented in figure 5.2.2.a.

ORPHEE is dedicated to manage the data exchanges, from a software point of view, between the satellite simulator and the command expert system, and, from a hardware point of view, between a VAX on which runs the simulator and a TI Explorer via an Ethernet link.

KBUPDATE is dedicated to periodically fetch the TMs coming from the simulator and inject them into the SSD providing the E/G interpreter with the most recent observed state of the satellite.

E/G INTERPRETER is given an initial goal to be reached and reactively executes the appropriate procedures.

VISU is a module whose task is to provide a visualisation of the main attitude parameters of the satellite (TMs) under the form of curves (angular positions and rates vs time). This allows the control center's engineers to supervise the manoeuvre being executed.

N.B.: The application corresponding to the third case study (the payload) is not described here because this is on-going work and all of the architecture choices are not definite yet. The interesting fact is that it is one of the best application from the toolkit point of view because it will use the CBL for knowledge representation, the CP and ATMS to build out procedures within the E/G formalism, to be executed by the E/G interpreter. This is a good example of a Knowledge Base System built out of P820 toolkit's bricks.

5.3. Description of Demonstrator 3 Toolkit Applications

Following the architecture defined in IR 2.10 (Leitch, 1987) the generic structure of the overall plant control can now be outlined as shown in fig. 5.3.a.

The Strategic Level.

This level defines the Target KBS: *Overall Plant Control*. The Target KBS controls and coordinates the four Systems: *Plant repair and maintenance, Plant start-up, Normal production, and Plant shut-down*. The control and coordination job is quite simple and can easily be implemented by a simple LISP programme.

The Tactical Level.

This level defines the four Systems: *Plant repair and maintenance, Plant start-up, Normal production and Plant shut-down*, constituting the *Overall Plant Control*. The Systems control and coordinate the Tools constituting the Systems. The control and coordination job is simple and can be implemented by a LISP programme. The systems may be described as follows:

Normal production system (FCS)

This system is build on the Tools: *Interpretation Tool (IT), Execution Tool (ET), Procedural Decision Tool (PDT)* and *Empirical Decision Tool (EDT)*. PDT is required to specify the control of the EDT (fuzzy rules) and the procedural knowledge dealing with normal production.

442

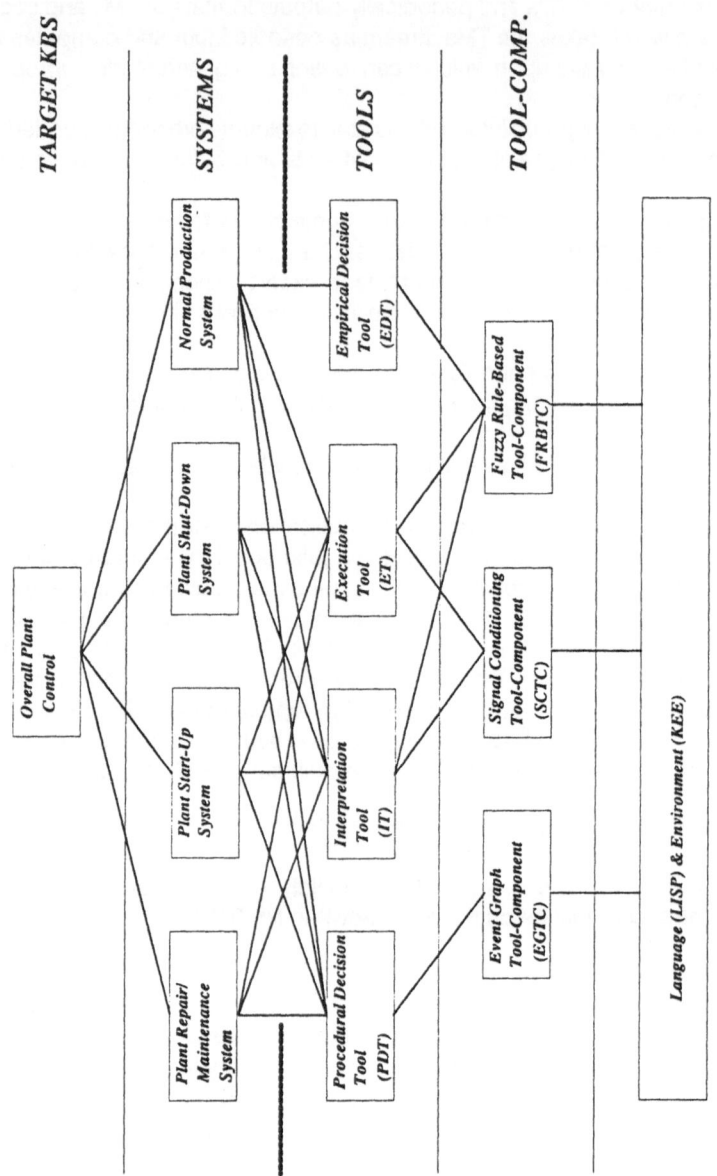

Fig 5.3.a. Overall plant control described by the QUIC Toolkit

Plant repair/maintenance, start-up and shut-down systems (SCS's)
All these systems are build on the Tools: *Interpretation Tool (IT), Execution Tool (ET)* and <u>Procedural Decision Tool (PDT)</u>. The PDT is required to specify the procedural knowledge (event graphs) and to control the execution of the procedural knowledge attached to the Systems.

The Teleological Level.
This level defines the Tools constituting the interface between the QUIC Toolkit and the application. Four generic Tools are required for this demonstrator:
Procedural Decision Tool (PDT)
This Tool is basicly build on the *Event-Graph Tool-Component (EGTC)*.
Interpretation Tool (IT)
This Tool is build on a general *Signal Conditioning Tool-Component (SCTC)* and a *Fuzzy Rule-Based Tool-Component (FRBTC)* extending the interpretation capabilities of the Tool.
Execution Tool (ET)
Uses the same tool-components as the *Interpretation Tool (IT)*.
Empirical Decision Tool (EDT)
This Tool is basicly build on the *Fuzzy Rule-Based Tool-Component (FRBTC)*.

Common for those Tools is that they primary are based on Tool-components defined on the Functional Level.

The Functional Level.
The functional level consists of general blocks called Tool-components. Tool-components are used to build the task dependent Tools on the Teleological Level. The need for three Tool-Components has been identified:

Event-Graph Tool-Component (EGTC)
This Tool-component supplies a functionality making it possible to perform procedural reasoning. This Tool-Component is provided by the QUIC Toolkit.

Signal Conditioning Tool-Component (SCTC)
This Tool-component supplies general functionality for signal conditioning. In regulatory feedback control applications signal conditioning is mandatory for various reasons. An important reason is to avoid unfortunate control actions due to spurious sensor behaviours. Another reason being the transformation between raw input/output data and the internal data representation used by the decision making tools.
Conditioning of raw input data in regulatory feedback control systems requires in general application of various kind of techniques including filtering (linear/non-linear), smoothing and prediction. The signal conditioning tools should therefore be built upon a signal conditioning tool-component (SCTC) which should have a functionality which resembles that found in various simulation packages like SIM-NON, Elmquist (1975). This Tool-component is currently not provided by the QUIC Toolkit.

Fuzzy Rule-Based Tool-Component (FRBTC)
This Tool-Component supplies a functionality making it possible to perform fuzzy rule-based reasoning. This Tool-component is provided by the QUIC Toolkit.

6. Results of Validation

As mentioned, one of the main objectives of the three demonstrators has been to validate the tools provided by the QUIC Toolkit. In this section a brief summary of the validation results achieved by the three demonstrators is described.

6.1. Results of Validation on Demonstrator 1

6.1.1. General.

As already said, the object of the validation activity was validation of the toolkit ontologic tools, interpretation and identification, and the corresponding tool components, i.e. the CBL, the Constraint Propagator plus ATMS, the Causal Ordering tool component.

The validation methodology consisted of two conceptual phases:
- a black box analysis of the ODS with reference to realistic test cases, in terms of effectiveness and efficiency. Effectiveness mainly refers to capability of identification of the necessary cost 1 and 2 measurements, and to discrimination capability.
- a tool functionality analysis

6.1.2. Black-box analysis.

A short description of the most meaningful test cases follows:

a) Partial clogging of condenser cooling water discharge structures
   This malfunction has the final consequence of increasing the pressure of the condensing steam, due to a loss of heat exchange effectiveness. The affected design parameter is the hydraulic resistance coefficient downstream the condenser. Cost 0 measurements are not sufficient for discrimination. In particular, by using only cost 0 measurements, it is impossible to discriminate between an increase of hydraulic resistance downstream and an increase upstream the condenser. A measurement of pressure anywhere in the cooling circuit outside of the condenser, which is available among the cost 1 measurements, allows discrimination. The HDS applies to the ODS a set of possible malfunctions corresponding to the two mentioned resistance increases. The ODS was fully successful in verifying the consistency of the proposed (by HDS) fault mechanism with the set of cost 0 measurements, in identifying the necessary cost 1 measurement, and in discriminating. Execution time was satisfactory.

b) Obstruction of a fraction of cooling water tubes of the condenser tube bundle
   This malfunction has the final consequence of increasing the pressure of the condensing steam, due to a loss of heat exchange effectiveness and a decrease of heat exchange surface. The affected design parameter is the number of tubes of the condenser tube bundle. Cost 0 measurements are not sufficient for discrimination. In particular it is impossible to discriminate among the following three malfunctions: 1) decrease of the number of tubes in the tube bundle, 2) increase of hydraulic resistance of the inlet or outlet sections of the condenser, 3) increase of condenser tube friction coefficient.
   To discriminate, a pressure measurement inside the condenser would be necessary. This measurement is not available. The ODS succeeded in confirming the set of possible malfunctions associated with cost 0 measurements and in identifying the unfeasibility of a discrimination. Execution time was long.

6.1.3. Tool functionality analysis.

The tool functionality analysis revealed a certain number of problems regarding a) the equation structure in the ontologic model, b) the constraint propagation in the qualitative domain and c) some limitations in the representation of qualitative relations:

a) equation structure
   Problems arose in the following two cases:
   1. a variable appears more than once in the same equation. A typical case is encountered in the computation of the mean logarithmic temperature difference in the condenser.
   2. simultaneous equations. A typical case is encountered when coupling a pump and its hydraulic circuit.
   Both these problems are being solved by symbolic manipulation.

b) constraint propagation in the quantitative domain
   Main problem in this field relates to numerical errors, with reference to computational errors and to instrumentation errors.
   Problems arise when it is necessary to compare two computed or measured (or combinations) values.
   A simple solution was the introduction of a global threshold, under which two numerical values are assumed to be coincident.
   Final solution regards error propagation according to the rules of numeric analysis.

c) limitations in the representation of qualitative relations
   Relations in the qualitative domain are obtained by total differentiation, disregarding the derivative coefficient.
   Problems arose in case of equations containing numerical coefficients with different order of magnitude. Solution of the problem is to take into account in the qualitative relations the order of magnitude of the derivative coefficient.

6.2. Results of Validation on Demonstrator 2

Validation on Demonstrator 2 led to two main results according to each of the case studies:
   Concerning diagnosis, Demonstrator 2's application led to some new requirements on the Constraint Propagator and the Component Based Language. Briefly speaking, the high complexity of the system being diagnosed in this case study introduced the necessity to have *hierarchical and view-oriented representation and reasoning*. This had some impact on the CP and the CBL. Due to the characteristics of the case study, it appeared through validation that some more features should be added to the CBL. In particular: dealing with components which have multiple modes and components which have multiple functioning regions; dynamical activation of views, modes and regions; and solving simultaneous equations by building compound components.
   For what concerns command, validation led to an important redefinition of the event graph language and interpreter so that it could now embed some of the main functionalities of what we called the Procedural Expert System (PES) following the terminology of (Georgeff, 1986) (see (Bailby et all., 1988) and (deKleer & Williams, 1987)). A prototype of the PES as well as a functional comparison between the PES and E/G drove to the idea

that each could take benefit of the other if we could build a tool embedding a common formalism and similar functionalities. Basically, what was missing in the former E/G tool was the possibility to express goals to be reached in order for an arc to be crossed. Reaching a goal would trigger another graph dedicated to the specific task of solving this goal. Thus, this led to confer to the E/G tool some planning aptitudes in that it is now able to decompose goals into subgoals, going down graphs considered as recursive transition networks until actual actuation (TCs).

6.3. Results of Validation on Demonstrator 3

As the control knowledge of Demonstrator 3 tend to be empirical, the main validation tasks of Demonstrator 3 has been to validate two empirical tool-components provided by the QUIC Toolkit:
- The Fuzzy Rule Based Tool-Component (FRBTC)
- The Event Graph Tool-Component (EGTC)

According to the architecture described in section 5.3 those tool-components are essential for the Overall Plant Control.

In the following the validation results will be summarised:

6.3.1. The Fuzzy Rule Based Tool-Component (FRBTC).

The FRBTC supports a declarative knowledge representation language based on fuzzy logic introduced by Zadeh, see e.g. Zadeh (1971) and Zadeh (1972). This knowledge representation language has been developed within Demonstrator 3 with special emphasis on ability to capture empirical operator knowledge used as basis for decision making in control of rotary cement kilns. Although the FRBTC is designed with a special purpose in mind it should be general enough to be used in other contexts. For a much broader scope on application of fuzzy logic a recent paper of Zadeh (1988) should be consulted. A more detailed description of the FRBTC can be found in Deliverable Report D2.3 (D2.3, 1988).

The validation of the FRBTC on Demonstrator 3 was done using three different approaches:

1. *Testing of FRBTC-based regulatory controllers against models of different complexity*
   These tests showed that the FRBTC was well suited for implementation of feedback controllers. Furthermore, it was demonstrated that that the controllers would have properties which are very attractive from a control engineering point of view.

2. *Analysis of FCL[1] fuzzy programmes in order to map FCL based controllers onto FRBTC based controllers (to check the sufficiency of the functionality provided by FRBTC)*
   These analysis of FCL programmes demonstrated that the substance of the programmes could easily be mapped onto the knowledge representation language of the FRBTC and in a much more expressive way.

3. *Evaluation through feedback from process engineers*
   With one exception the feedback from the process engineers did not provide requirements which could not easily be handled by the FRBTC either directly or through minor

1   FCL (Fuzzy Control Language) is a control tool developed by F.L.Smidth & Co. A/S and applied the last decade, especially in the field of automatic cement kiln control.

modifications/extensions of its functionality. The main problem were due to the fact that some control engineers did not find the separation between knowledge and control over this knowledge very appealing. However, as the criticism basically were related to on-site validation work the problem could be overcome by designing an presentation interface hiding the separation.

6.3.2. The Event Graph Tool-Component (EGTC).

The EGTC supports a procedural knowledge representation language sharing many features with the more known Petri-Net formalism, see (Reisig, 1982). A more detailed description of the EGTC may be found in Deliverable Report D2.3 (D2.3, 1988).

This validation has focused on the validation of the EGTC knowledge representation language. The basis for the validation has been knowledge from the application domain of Demonstrator 3.

The knowledge acquisition has not covered the complete sequence control knowledge of the application domain. Instead the knowledge acquisition has focused on relevant test cases. It is assumed, and believed, that the selected test cases cover the major part of the sequence control problems found in the application domain. It is thus our feeling, that the knowledge acquisition has given a good understanding of sequence control problems in the application domain.

The validation of the EGTC knowledge representation language has shown, that the formalism to a large extent is able to express the sequence control knowledge of the application domain in an explicit way. Besides, the formalism provides sufficient function-ality/features to implement large parts of a sequence controller in a prototyping environment. However, especially in the area of malfunction handling some additional management functionality/features are required. In the new version of the EGTC those requirements have, however, already been taken into account.

7. Conclusion

In this paper three different applications of the QUIC Toolkit have been presented. Although these applications cover a broad range of process control applications it has been shown that the control tasks can be defined in a uniform way using the terminology and methodology provided by the QUIC Toolkit. *The QUIC Toolkit thus seems to provide a very general terminology and methodology for process control tasks*.

In addition it has been possible to map the three applications onto tools/tool-components partly already supported by the QUIC Toolkit. *This shows, that the tools/tool-components provided by the QUIC Toolkit already cover a large class of applications*.

Finally, the sufficiency of the tools/tool-components included in the QUIC Toolkit has been validated against demonstrator applications. *This validation demonstrates that the tools/tool-components to a large extend are sufficient for the three demonstrators*.

In summary it may be concluded, that the QUIC Toolkit covers the major aspects of the tasks relevant for the three demonstrators.

Acknowledgement

This paper describes developments partially undertaken within ESPRIT project P820, funded by the Commission of the European Communities within the ESPRIT programme. Project P820 consists of a consortium composed of CISE, Aerospatiale, Ansaldo, CAP Sesa Innovation, F.L. Smidth, Framentec and Heriot-Watt University. The authors want to acknowledge here the contribution of all the members of the project team to the ideas expressed in this paper, while taking full responsibility for the form these ideas are expressed.

References

Bailby, G., Chautard, JC. and Honnorat, C. (1988) 'Functional Comparison of PES and the E/G tool', ESPRIT Project P820, Internal Report IR4.2, 1988.

Bobrow, D. G. and Hayes, P. J. (Eds.) (1984) 'Qualitative reasoning about physical systems', Artificial Intelligence Special Volume 24, (1-3).

Chautard, JC., Courtin, JP. and Honnorat, C. (1988) 'Demonstrator 2 requirements stemming from first phase', ESPRIT Project P820, Internal Report IR4.4, 1988.

D2.3. Many Authors. (1988), 'Kernel Toolkit User Manual', ESPRIT Project P820 Deliverable Report D2.3, June 1988.

Davis, R. (1984) 'Diagnostic reasoning based on structure and behaviour', Artificial Intelligence 24, 347-410.

de Kleer, Johan and Williams, Brian C. (1987) 'Reasoning about multiple faults', Proceedings of the national Conference on Artificial Intelligence 1987.

Georgeff, M. P. and Lansky, A. (1986) 'A system for reasoning in dynamic domains: Fault diagnosis on the space shuttle', Technical note 375, 1986 - SRI International.

Leitch, R. R. (1987) 'Task Classification for Knowledge Based Systems in Industrial Control', ESPRIT PROJECT 820 (QUIC). Internal Report 2.10, December 1987.

Leitch, R. R. and Stefanini, A. (1988) 'QUIC: a development environment for Knowledge Based Systems in industrial automation', ESPRIT '88, Putting the Technology to Use, Proceedings of the 5th Annual ESPRIT Conference Brussels, November 14-17, 1988 (Part 1), Elsevier Science Publications B.V., Netherlands.

Reisig, W. (1982) 'Petrinetze', Springer-Verlag, Berlin, Heidelberg, New York, 1982.

Stefanini, A. (1988) 'Task Dependent Tools for Intelligent Automation', ESPRIT PROJECT 820 (QUIC), Status Report on Esprit Project 820 "QUIC", March 1988.

Struss, P. (1987) 'Multiple representation of structure and function', in John Gero (ed.), Expert System in computer aided design, North Holland 1987.

Zadeh, L. A. (1971) 'Toward a Theory of Fuzzy Systems', in R. E. Kalman and N. de Claris (eds.), Aspects of Network and System-Theory, Holt, Rinehart and Winston Inc., 1971.

Zadeh, L. A. (1972) 'Outline of a New Approach to Analysis of complex Systems and Decision Processes', IEEE Transaction on systems, Man, and Cybernetics, vol. SNC-3, No.1, January 1973.

Zadeh, L. A. (1988) 'Fuzzy Logic', IEEE Computer, April 1988.

Project no 857

THE GRADIENT BACKBONE ARCHITECTURE

K.ZINSER, P.ELZER, H.W.BORCHERS, C.WEISANG
ABB Asea Brown Boveri AG
Corporate Research CRH/L2
P.O.Box 101332
D-6900 Heidelberg
F.R.G.

Abstract

This paper describes a communication architecture for the exchange of knowledge and data between expert systems. Knowledge can be structured and represented in various forms. The expert systems can be implemented in different hardware and software environments. The overall system, therefore, can be regarded as a distributed system. Missing functionality of a module is handled gracefully. The implementation of the communication protocol on top of the buslike communication backbone architecture follows the ISO/OSI layer model specification. An application example of an operator support system (GRADIENT), based on multiple expert systems, illustrates the concept and operation of the communication backbone.

Introduction

GRADIENT (GRAphical DIalogue environmENT, P857) is a project that investigates the possibility to support operators of industrial processes in their supervision and control task. This is achieved by a set of interacting expert systems for on-line, real-time diagnosis and graphical presentation of information. The modules of the GRADIENT system are shown in Figure 1 and briefly described below. More details can be found in /1/ /2/ /3/. The paper at hand presents a description of the architecture used for communication between the individual modules of the GRADIENT system. The advance of the backbone structure and the communication protocol is twofold: First, expert systems contain knowledge that is too highly structured for conventional (even the distributed) database managment systems or blackboard models. Second, modern operator support systems as GRADIENT are too specialised not to be distributed over several modules, that are usually hosted on dedicated hardware (e.g., algorithmic processing, symbolic reasoning, graphics). Therefore, an envisaged communication protocol must be independent from hardware, networks, and operating system software, and must allow for multiple data and knowledge representation formats. One possible communication architecture to fullfill these requirements is presented in this paper. For better understanding, the relevant aspects of the GRADIENT architecture shall be summarized as follows:

Most of the GRADIENT modules base their operation on an elaborate model of the technical process under consideration. The Quick Response Expert System (QRES) performs a fast diagnosis of incoming process values in order to abstract the system status. The Support Expert System (SES) provides deeper reasoning for assessing the impacts of failure states, state-based reasoning and other operator and dialogue support.

450

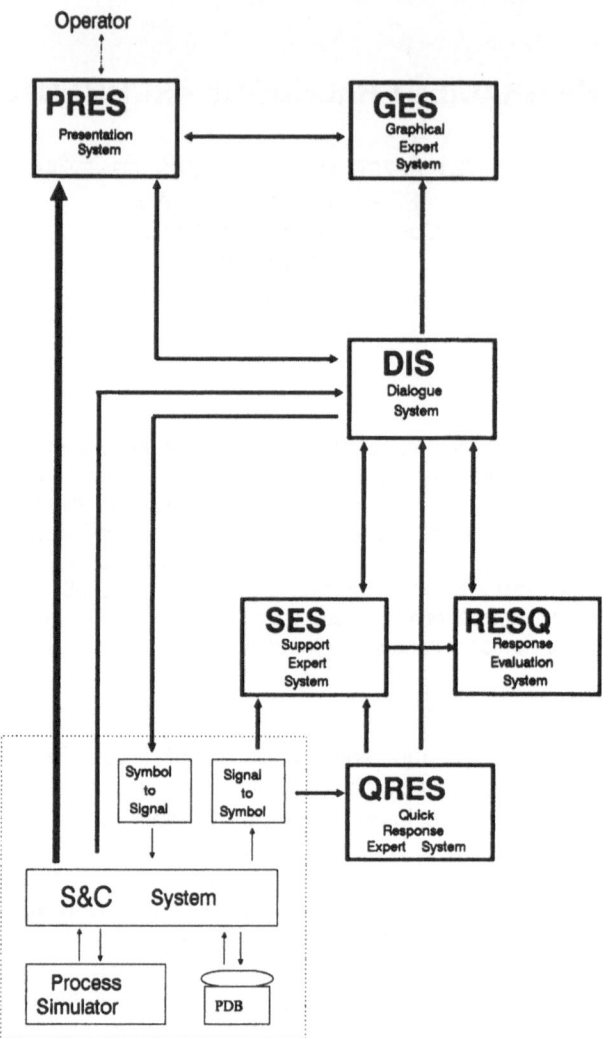

Figure 1: Overall GRADIENT Dataflow Diagram

Operator support in terms of process visualisation is provided by the Presentation System (PRES) and the Graphical Expert System (GES). The PRES features fast bit-mapped graphics and multiple windows on multiple screens, and user interaction by a mouse or similar input devices. Information is presented in form of the 'Picture Pyramid' that implements additive and alternative information zooming and panning. All aspects of the user interface (e.g., colors, shapes, information coding, information rates) are based upon a user model and are designed to match cognitive requirements of the operator tasks. The Graphical Expert System (GES) further supports on-line picture generation and display in accordance with the state of the process and operator dialogues as conceived by the other modules, especially SES and DIS. The aim is to present the operator with the proper information at the right time in the most suitable representation and arrangement.

The Dialogue System (DIS) is central to GRADIENT and is based on knowledge of typical dialogue sequences in the respective domain. DIS is the supported by the Response Evaluation System (RESQ) for deeper analysis of the state of interactions of the operator with the technical process. DIS achieves a multi-channel asynchronous dialogue facility through the 'Channel Handler' (CH) /4/. The exchange of messages between the GRADIENT modules is controlled by the CH. Thus, the arrows in Figure 1 describe the functional information exchange between modules rather than the physical communication links.

In order to better illustrate the central idea of the 'backbone architecture', Fig. 1 was redrawn into a more general form, with CH being separated from DIS. The result is Figure 2

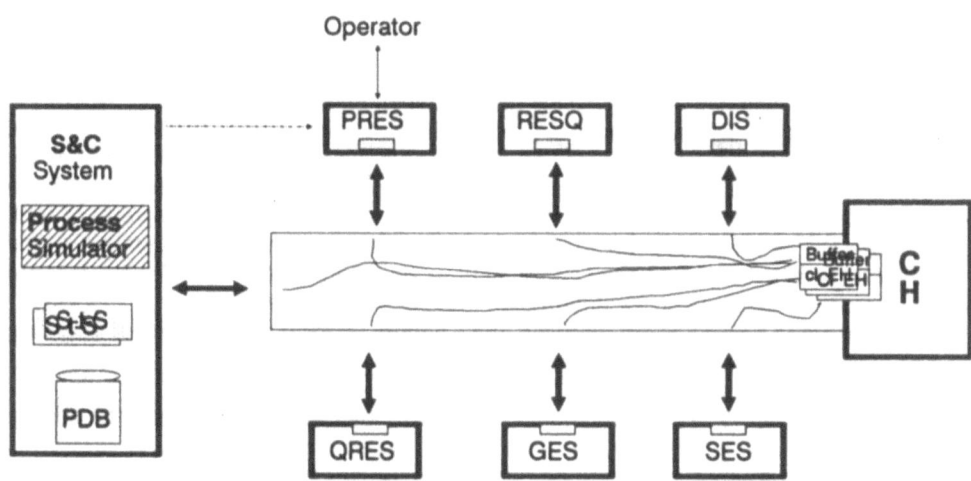

Figure 2 : GRADIENT Backbone Architecture

which provides a schematic of a bus-like architecture. The CH operates as a 'bus controller' coordinating all the channels of communication between the individual modules. This 'communication bus' is called the GRADIENT Backbone which interconnects all the modules. It comprises the following main components:
- Channel Handler
- GRADIENT modules
- Composite Objects.

Finally, as a module external to GRADIENT, the simulator combines and provides the following functionality: All process values need to be stored in a Process Data Base (PDB) from where they can be requested by all other modules, along with history, trends, etc. A direct link may exist between the simulator's Process Data Base (PDB) and PRES. The conversions from numeric values to symbolic information and the other way around are also provided by this simulator module. It is suggested here to route such communication through the DIS. This includes also information of the Signal-to-Symbol and the Symbol-to-Signal transformations. These transformations are essentially needed for input and output of Composite Objetcs as described in a later section. However, low-level data preprocessing and abstraction , e.g., sending of alarms according to some limits or the detection of trends, may also be performed at this stage.

The Backbone Manager

In order to achieve the desired communication for GRADIENT, the functionality required of the backbone manager was specified. The Channel Handler as part of the Dialogue System has been redesignated as the communication backbone manager, with the central function of coordinating all communication between the modules of the GRADIENT system. As described above, and illustrated in Figure 2, the Channel Handler serves to control the sequencing of individual communications from each of the other modules. The Channel Handler is interruptable by successive module communications, and asynchronously handles many messages in a timesharing manner.

The Client-Server Relationship

The Channel Handler manages the communication backbone on the basis of a client-server model. The server role is filled by the Channel Handler serving clients (the communicating GRADIENT modules). Thus, the server (i.e., the Channel Handler) is able to oversee the interprocess communication from modules which may be running on different machines, or as separate processes on a single machine. The format of such communication is described as the Composite Objects Intermodule Communication Protocol in detail below, in terms of information to the Channel Handler and also in terms of the protocol which is used for information transfer between modules.

With respect to serving client modules, the Channel Handler has the task of attending to and maintaining the following environment:

- a number of clients whose communication with other clients has to be ordered and relayed;
- the clients communication has to be handled according to their priorities;
- a clients priority may be represented as an attribute of the client or it may be conveyed by a message from the client to the Channel Handler (i.e., dynamic priority allocation).

The processing of successive client communications requires that at any moment of the Channel Handler operation a set of tasks is pending and ordered according to their priorities. The priority of client messages can be preset or may be changed dynamically by the clients themselves. This allows for the expected differences in priority across different GRADIENT modules. Thus, except process display updates, which may be routed directly to the Presentation System on a separate communication link (cf. Fig. 1), the operator warning systems, QRES and RESQ, will normally pass messages of higher priority than other modules. Of course, circumstances may dictate that this be changed for particular cases, which is provided for by the possibility of dynamic priority setting.

Channel Handler Architecture

At present, the Channel Handler exists as a set of 'C' procedures, which operate in a

UNIX environment. This has proven sufficiently flexible in concept and efficient in operation to pro cess the communication needs of the GRADIENT backbone. The functional architecture of the Channel Handler is illustrated in Figure 3. Further detail of the operation of the Channel Handler is given below.

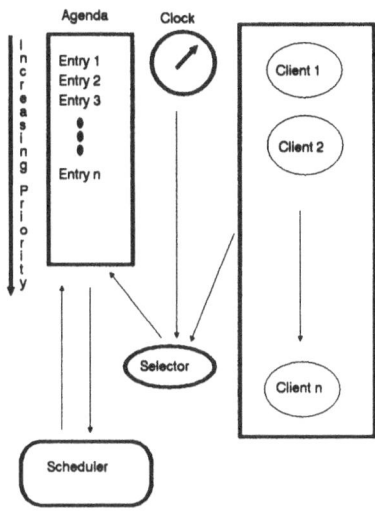

Figure 3 : Channel Handler Functional Diagram

Operation of the Channel Handler

Communication through the Channel Handler is initiated by interrupts (requests) from clients (other GRADIENT modules, or the operator through PRES). The Channel Handler is notified of the interrupt and the client's own event handler, which exists within the Channel Handler, is called. This event handler decides what to do with the message. Usually it is forwarded to the destination client (e.g., specified by a destination field in the message), although other operations are possible for the event handler, e.g., it could write a log entry. When Channel Handler forwards a message to its destination, the receiving client is required to handle the interrupt created by the arriving message. This is achieved for each individual module by a resident module handler (as described later). In addition, a module base within the Channel Handler has been provided that holds all client specific data (e.g., on what machine the client (module) runs, what protocol is used, default priority, if logging is required). This allows for the pre-specification of client types and protocol characteristics at Channel Handler startup time.

Error Handling

Since the Channel Handler is in a central position with regard to intermodule communi-cation, it is reasonable that it should also perform the role of error handling for any communication that violates the standard protocols. Although this is an appropriate task for the Channel Handler, the precise need for this facility, and the form that it might take, will only emerge when implementation of the integrated GRADIENT system will have

454

progressed further. However, the Channel Handler may actively adapt to accommodate the absence of particular GRADIENT modules, due to partial implementation of a module, or due to temporary module failure. In any case, graceful degradation is an absolute requirement.

Module Architecture and Communication

Overall Structure

The GRADIENT system provides its functionality by a number of interacting expert systems each of which is contained in a module. Thus, each GRADIENT module needs to convey the same architectural aspects with respect to interacting with the rest of the GRADIENT system. Figure 4 presents the overall structure of an arbitrary GRADIENT module, indicating components generic for all modules (as far as communication with the GRADIENT system is concerned), and components specific to the module . The terms generic and specific refer to the functionality of the respective module component.

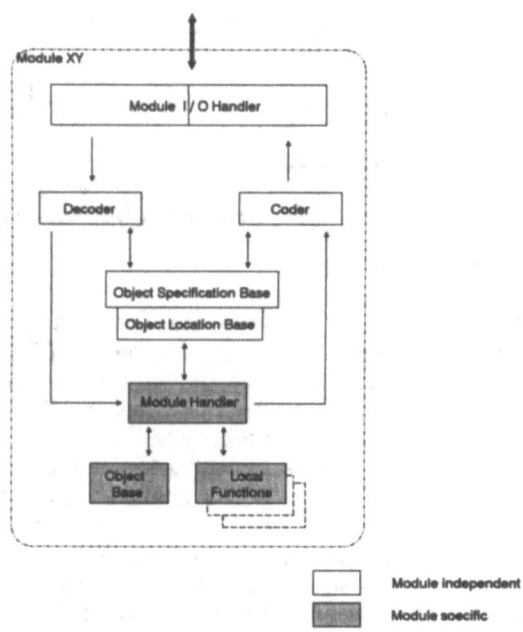

Figure 4: GRADIENT Module Architecture

When a message is received by the Module Input Handler, it is first relayed to the Decoder that has access to the module's Object Specification Base to interpret the incoming data. The decoded message is then forwarded to the internal Module Handler that decides when and how the module processes the message. A response message is then coded and sent out by the Module Output Handler. The next section provides more detail on the module-generic components, the module-specific components are desribed in Section 4 below, as far as the Composite Object Communication Protocol is concerned.

The module-generic I/O components are currently being implemented on the various hardware platforms as present in the GRADIENT project.

Generic Communication Components

The module-generic components represent the communication endpoints for inter-module communication. A module can either send out a message to another module, or receive a message from another module. A message to another module can be a request or just an information update, vice versa a message from another module can just be data (or an answer on a former request), or a request that needs to be answered. If the *Module I/O Handler* receives an interrupt, it relays the message to the Decoder. At this point, the message is just a byte stream as sent from another module via the Channel Handler. The Decoder's task is to interpret the incoming message into a form understandable to the module. The decoding is based upon the *Object Specification Base* that contains a global data dictionary of all the Composite Objects. The Object Specification Base is the same for all modules (in the same way data dictionaries are common in distributed data base systems.

The module-specific components can now access the message and decide on how to process it. Eventually, the module's *Local Functions* will produce a response message. A mesage (i.e., a request) is also generated when the module requires remote data for its own operational procedures. The Coder will translate an outgoing message from the module's internal representation to a byte stream that can be sent via the Channel Handler. The Coder will make use of the Object Specification Base for the translation, and, most importantly, of the *Object Location Base* to find out the destination module. The destination module could also be determined by a direct interaction of the module's local function with the Object Location Base. The message is finally relayed to the Channel Handler for delivery.

Composite Objects

Composite Objects Protocol

The Composite Objects protocol provides for communication between modules in GRADIENT. It consists of an *Object Specification Language*, an *Object Language* and primitives for intermodule communication. The protocol is described and specified in more detail in /5/.

The Object Specification Language provides a syntax for the definition of the Composite Object types that are stored in the Object Specification Base for reference whenever an instance of the type in question is encountered.The purpose Object Language provides a syntax for Composite Object instances and their manipulations. The language deals only with simple manipulations of objects like

- creation and deletion
- accessing and changing
- copying and composing.

The language does not support operations on simple objects, e.g., adding of integers. Neither does it support programming structures like sequences, repetitions or conditions. Like the Object Specification Language it is defined in BNF syntax and is easy to parse.

Intermodule Communication

As mentioned above it was suggested that each module should know the types of objects available as well as the location of specific objects. Accordingly, two types of communication, local and remote to the module, are supported. Both server functionalities are supported by the Module Handler (c.f., Figure 4).

The task of the local server is to give a module's Local Functions access to the contents of the Local Object Base. The server will evaluate any expression in the Object Language, and, if the server finds out that some part of an expression must be evaluated at another module (e.g., a request for data held at another module's Object Base), the server will send that part of the expression to the other module (via the Channel Handler), await the answer and then continue its evaluation. The task of a remote server is to help other modules to get access to the knowledge stored locally in the Object Base. The remote server is to some degree application dependent as there are some consistency problems. These include updating during reasoning, reading during local updates.

In order to handle these conflicts the local application (i.e., the Module Handler together with the module's Local Functions) has to decide when the remote server can be active.

Implementation of a Prototype

A prototype version of a Composite Objects Coder/Decoder has been implemented in accordance with the concepts sketched out above /6/. This prototype contains an Object Specification Base and an Object Location Base. For the time being, there are no parsers for the Object Specification Language and Object Language, as it was not needed for prototype's size of object bases.

The implementation of the protocol itself is designed in three layers. These three layers correspond to the ISO/OSI layers 5 (Session Layer), 6 (Presentation Layer), and 7 (Application Layer), respectively. Table 1 illustrates some of the commands provided by the Session Layer and the Presentation Layer.

Table 1 : Session Layer and Application Layer of Protocol

Lowest layer (network specific) : Session Layer (ISO/OSI)
init_communication, exit_communication
> (establish or exit the communication
> using a specified syntax or medium)

readstream, writestream
> (the actual data transfer on
> the selected stream and medium)

Second layer (protocol specific) : Presentation Layer (ISO/OSI)
co_code, co_decode
> (transform the representation of Composite Objects
> so that they can be used by readstream and writestream)

co_update
> (update the specified Composite Object Bases)

cmd_parse
> (interpret and evaluate the command
> to take appropriate action).

On the lowest layer calls for opening a connecting on a stream, closing down communication, reading and writing on streams are implemented in network specific procedures. These streams are known to the Channel Handler for its operation and managment as described above.

On the second layer, these calls are used for parsing commands from a stream, coding and decoding Composite Objects and reading from or writing to the respective Composite Object Bases. This layer interfaces to the application.

The highest layer represents the application where streams are actually opened, commands are generated and evaluated according to the functionality of the respective module, using the functionality provided by the layer(s) below.

Summary

A communication architecture for a system of communicating modules that contains knowledgebased systems was proposed. First, the GRADIENT system as the application example was described. Then, the backbone architecture was described consisting of the Channel Handler as the Backbone Manager and the GRADIENT modules as the communication partners, along with the specification of their architecture. Finally, a communication protocol based on 'Composite Objects' was presented and its implementation and operation was mapped on the ISO/OSI communication model. It was argued that, with respect to exchange of knowledge between expert systems, existing approaches to distributed database design and blackboard architectures have disadvantageous qualities that can be overcome by the proposed backbone architecture.

Acknowledgements

The work reported is partially supported by the Commission of the Eurpean Communities under project title P857 GRADIENT. The consortium partners are AXION, Denmark, ASEA Brown Boveri, FRG, University of Leuven, Belgium, University of Kassel (UKS), FRG, University of Strathclyde (UST), Scottland.

The Composite Object approach was developed by a special workshop consisting of members of all project partners. The Channel Handler concept and software were developed and implemented at Strathclyde University and the Scottish HCI Center. The backbone architecture was discussed and passed by a special GRADIENT working group whose formulations and impacts on this paper are hereby acknowledged, especially G: Johannsen (UKS) and G. Weir (UST).

Literature

/1/Alty, J.L., Elzer, P., Holst, O., Johannsen, G., Savory, Smart, G. (1986) Literature and User Survey of Issues Related to Man-Machine Interfaces for Supervision and Control Systems", ESPRIT'85, Status Report of Continuing, North Holland.

/2/Elzer, P., Borchers, H.W., Siebert, H., Weisang, C., Zinser, K., (1987) MARGRET - A Preprototype of an Intelligent Process Monitoring System", ESPRIT'87, Achievements and Impact, North Holland.

/3/Elzer, P., Siebert. H., Zinser, K., (1988) New Possibilities for the Presentation of Process Information in Industrial Control, Proceedings of the 1988 IFAC Conference on Analysis, Design and Evaluation of Man-Machine Systems, Oulu, Finland.

/4/Shahidi, A., Mullin, J., Weir, G. (1987) GRADIENT Dialogue Channel Handler, P857-UST-WP04-021.

/5/Nyholm, K., Zinser, K., Ravnholt, O. (1988) P857-CRI-WP03-074, Composite Object Reference Document, CRI.

/6/Elzer, P., Zinser, K. (1988) A Communication Protocol for Distributed Expert Systems, 8th IFAC Workshop on Distributed Computer Control Systems, Vitznau, Switzerland.

ESPRIT 892 (DAIDA)

THE DAIDA DEMONSTRATOR:
DEVELOPMENT ASSISTANCE FOR DATABASE APPLICATIONS

MATTHIAS JARKE, DAIDA TEAM *
Universität Passau, P.O. Box 2540, D-8390 Passau

ABSTRACT. The DAIDA project investigates comprehensive support for the development of database-intensive applications. The DAIDA methodology involves going from *concept-based specifications* via *process-oriented development* to *quality-assured database software*. A prototype DAIDA environment has been built that integrates knowledge-based development aids and graphical interaction tools under the control of a global knowledge base management system. An extended example illustrates the approach.

1 Review of DAIDA Project Goals

A large share of today´s software market is concentrated on data-intensive information systems applications; as databases extend their scope to new areas such as design, process control, or multimedia applications, this share may even grow further. Nevertheless, we are far from a good understanding or truly professional support for the production and maintenance of such systems. In the DAIDA project, a group of software houses, research centers, and universities from five European countries have collaborated to provide knowledge-based languages, methodologies, and tools as *Development Assistance for Interactive Database Applications*. DAIDA started in early 1986 and is currently in its fourth and final project year, intended to draw the individual project results together and contribute to industrial impact via the development of an integrated DAIDA demonstrator prototype.

A basic DAIDA methodology emerged during the project. Placatively, it can be summarized by the following strategic goals:

DAIDA Goals:

From	**Concept-Based**	Specifications
Via	**Process-Oriented**	Development
To	**Quality-Assured**	Database Software

---

* The DAIDA team consists of the software houses BIM, Everberg/Belgium (Raf Venken -- administrative manager; Eric Meirlaen, Jean-Marc Trinon, Irene de Zegher), GFI, Nanterre/France (Gerard Bonin, Alain Rouge), and SCS, Hamburg/Germany (Rainer Haidan, Ralph Meyer, Barbara Piza, Ingo Röpcke); the FORTH Computer Research Center, Iraklion/Greece (Manolis Marakakis, Alex Borgida, Panagiotis Katalagarianos, Lydia Kavraki, Michael Mertikas, John Mylopoulos, Yannis Vassiliou); and the universities of Frankfurt (Joachim Schmidt, Florian Matthes, Peter Niebergall, Jürgen Stuchly, Ingrid Wetzel, Ariane Ziegler) and Passau/Germany (Matthias Jarke -- technical manager; Michael Gocek, Manfred Jeusfeld, Eva Krüger, Hans Nissen, Thomas Rose, Martin Staudt).

The rationale behind these goals is as follows. Communication between developers and users in the requirements analysis and system specification phases has been notoriously difficult. *Concept modelling*, using object-oriented representations and hypertext-like user interface technologies with animation by prototypical examples, appears to be one of the few methods for achieving good requirements, the area where the most costly mistakes are made.

The life-cycle of many information systems goes beyond individual generations of hardware, system software, or software development teams. Up to 70% of information systems costs are spent in the maintenance phase. It is increasingly recognized that the quality of IS maintenance work is proportional to the degree that experience gained by the initial development team can be transfered to the maintenance task. Given the high turnover of software personnel, this implies the need for maintaining a record not only of the outcomes but also of the decisions and tool applications in the design process. Not accidentally, *software process modelling* has recently become a buzzword not only in DAIDA [OSTE87].

Often, information systems are used in time- or correctness-critical applications. Depending on the criticality of these non-functional requirements, a software environment should therefore provide a range of tools for producing and evaluating *quality-assured software*. Full software quality assurance (with respect to well-defined specifications) by appropriate formal development methods, verification and testing tools may be too expensive for certain applications; but at least, the decision not to use available quality assurance technology should be a conscious one. As an important side effect, we note that the use of formal software development tools may also facilitate development "replay" in the maintenance phase -- the initial costs may be recovered later on.

Although the above goals are ambitious, there should be little disagreement about their potential usefulness by now. The main problem in DAIDA, of course, was how to derive from these goals a coherent methodology; also, how to support this methodology with an environment which makes maximum use of available software engineering knowledge while allowing the later addition of emerging knowledge- based theories and tools. Elsewhere, we have analyzed these requirements, and related them to other proposals in the literature [DAIDA88]. In this paper, we report the resulting DAIDA methodology (section 2), architecture (section 3), and tools (section 4), and then illustrate them by an example (section 5). A summary of the main lessons learned in DAIDA concludes the paper.

2   The DAIDA Methodology

This section elaborates the three DAIDA goals into a comprehensive methodology for the development and maintenance of database applications. As a testbed for the general methodological issues, DAIDA has made specific choices for languages, techniques, and tools. In the discussion, we try to separate the general conclusions to be drawn about the methodology, from the evaluation of the specific choices made. This suggests a number of improvements for further work, e.g., in industrializing the methodology.

2.1 CONCEPT-BASED SPECIFICATIONS

The key point here is that conceptual languages allow us to work with adequate concepts when specifying the semantics of an application. In requirements modelling, we need extreme freedom of defining our own concepts and terminology. In contrast, the design phase needs a predefined but powerful set of constructs. Even at the database and program design level, we need a broad conceptual basis, offering a full spectrum of data types (orthogonal and persistent, including sets) and predicates (nested and quantified) to relieve the database programmer from tedious details.

As already indicated, at least three groups of activities must be supported in a comprehensive information systems development and maintenance environment such as DAIDA (cf. [DAIDA88]):

- capturing the *application's* requirements
- organizing the *system's* data objects and transitions
- producing high-quality database application *software*.

2.1.1 *Capturing the Application's Requirements*. In DAIDA, the requirements modelling phase precedes a formal system specification. It is not confined to describing the system's requirements alone but also takes into account the broader usage context of the system.

Therefore, according to the DAIDA methodology, requirements analysis involves at least two tasks: describing the subset of the world in which the information system is intended to function (*world model*), and describing the data and functionality requirements of the information system within that mini-world (*system model*). The basis for the information system's data model is laid by the description how the system model represents objects in the world model. Seeds for a functional and user interface specification are given by the description how the system model is embedded in the world model.

As a consequence, the requirements model should be managed as a knowledge base of application concepts; its development process should be viewed as *knowledge base construction* [BJM*87]. This requires a conceptual modelling language with particular features:

- Application areas for information systems vary widely; nevertheless, the knowledge representation language should provide a means of communication between designer and users (or their managers) as close to the application as possible, in order to permit active user involvement. The language should therefore offer *extensibility*, i.e., the capability to define application-specific sets of concepts on a case-by-case basis.
- Requirements should be visualized in user-friendly graphical or text-based interaction. A *hypertext style of interaction* [CONK87] in which different members of the analysis team can cooperate in the knowledge base construction, appears quite promising. The need for learning a formal syntax should be restricted to the support team and avoided for the users. Even for the support team, a lot of guidance should be provided.
- Furthermore, users often need *animation* to understand formal requirements analyses. It should therefore be possible to run examples through the requirements description, using derivation rules or similar approaches to simulate system behaviour.
- The requirements analysis does not just describe a static database design but also the dynamic behaviour of the information system. Moreover, both the world model and the requirements of the system (i.e., how the system model looks like and how it is embedded in the world model) evolve as new things are learned or as the requirements themselves change. The representation of *time* is therefore a very important issue in choosing a conceptual modelling language for requirements analysis. Another consequence of changing requirements is that it must be possible to manipulate the requirements model as a dynamic knowledge base, not just as a one-shot documentation. Dependencies among requirements should therefore be preserved in addition to the requirements themselves.
- As mentioned, requirements analysis can be a major cooperative task with contributions from various stakeholder and developer groups. The language should offer viewpoint or *modularization* concepts to model the evolution of individual opinions as well as their integration in a commonly agreed requirements model (the "contract" between system customer and system user).

Most current requirements modelling languages and tools are too inflexible to handle these requirements. They usually provide only nice graphical tools for the acquisition and documentation of requirements but the result is *not* maintained as a knowledge base that can be (a) transfered to the specification and design phase, (b) reused for subsequent requirements changes. Even advanced entity-relationship model extensions, such as the entity-relationship-activity-event (ERAE) model [HAGE88] which come closest to our requirements, do not offer sufficient extensibility.

DAIDA´s choice for a conceptual modelling language is the knowledge representation language CML/Telos [KMSB89] which was specifically designed with the above requirements in mind; it was extended in DAIDA to a version called SML that covers systems modelling aspects for software engineering in database-intensive information systems. The basic language integrates predicative assertions and an interval-based time calculus in a structurally object-oriented framework with built-in axioms for aggregation, generalization, and classification; it has by now been applied to several medium-scale analyses, including the modelling of the DAIDA environment itself [JJR89a]. With respect to the above-mentioned requirements, we have made positive experiences with the following SML features:

- Classification allows the partitioning of the knowledge base in an arbitrary number of *meta-levels* such that each level defines the sub-language in which objects of the level immediately below can be described. It is relatively easy to define domain-specific concepts at a metalevel which are then instantiated by actual requirements.
- The integration of structural, semantic network-like principles with more textually oriented frames and rules provides the basis for a hypertext-like interface in the SML environment. The predicative sub-language can be used as a basis for filtering, in order to make only relevant parts of the requirements model visible.
- The embedded time concept supports both history time to model the dynamics of world and system, and transaction time as a basis for version management.
- Predicative assertions can be interpreted either as integrity constraints to support the language extensibility, or as deduction rules that derive implicit information about objects in the knowledge base. The developer can use deduction rules for interactive animation of the requirements model by asking deductive queries about prototypical example objects. Collections of rules can also be interpreted as viewpoints which express the view of different team members; but this has not been explored in DAIDA.

One weakness of the current SML version is its lack of a modularization capability. Our experiments showed the need for better information hiding, for assigning knowledge base modules to development subgroups, and for knowledge base configuration operators that integrate versions of individual modules into versions of the complete model [RJ89]. A second problem is the optimal choice of predicative assertion languages; there exists a basic trade-off between expressiveness and computational tractability, when requirements models become large. A combination of assertion languages, a powerful and expensive one for analysis-in-the-small and a simpler and more efficient one for large-scale configuration management, may be the appropriate choice.

2.1.2 *Organizing the System Data Objects and Transitions.* While the requirements model focuses on knowledge about the application domain and its information needs, we now have to bring in concepts about the general domain of database-intensive information systems. The representation formalism for this level (called *conceptual design*) must enforce a clear understanding of the semantics of data objects and of events that create and change data in the intended information system. This activity still requires a semantically rich set of concepts but this set should be *fixed* as a uniform structuring mechanism for information systems.

The application-specific knowledge representation of the requirements model has then to be *mapped* to this structure; on the other hand, the final conceptual design then should constitute a safe starting point for the remaining, more formal step of quality software development. Again, this double role places stringent requirements on a good language for semantic information systems modelling.

Firstly, the design language should not be too different from the requirements language. To simplify the mapping, it should offer similar conceptual abstractions and surface syntax. Moreover, it should mirror standard semantics of information systems, such as data classes, short-term transactions, and longer-term interactions of system and users (scripts).

On the other hand, the conceptual design level should not be a simple elaboration of the system model. Rather, it should represent the results of distinct design decisions over and above the initial set of requirements [CKM*89]. For example, such choices include decisions about the satisfaction of temporal conditions by transactions, scripts, or the definition of time periods for which data should be kept (so-called relevance periods). Similarly, assertions at the requirements level can be satisfied by integrity constraints on data, by precondition tests on transactions, or by specification of structures and operations that satisfy assertions by design (e.g., error-preventing user interfaces). Even a re-grouping of system model data structures can be such a design decision; for instance, the designer may decide to organize generalization hierarchies of concepts by their temporal actuality rather than by content-oriented subclasses.

Finally, as a starting point for formal refinement methods leading to quality-assured software, the conceptual design must be formally consistent and complete with respect to a clearly defined semantics. A heuristic understanding of the specification, combined with partial formalization, is therefore insufficient at this level.

In contrast to the requirements level, a fair number of semantic data models exist as candidates for organizing conceptual designs [HK87]. However, many of them consider only the static aspects of database design and are therefore less suitable for a DAIDA-like methodology. DAIDA chose to adapt the TAXIS language [MBW80] to our needs. The revised version -- TDL [BMSW89] -- offers generalization hierarchies and set-oriented predicative assertions but neither CML's meta-level extensibility nor its built-in time concept.

In contrast to the history-oriented view of the world taken by SML, TDL thus takes a state-based view of computation like most programming languages. The management of data is organized as *entity classes* related by attributes; atomic state transitions are effected by *transactions*, while *scripts* describe the long-term pattern of global coordination and timing.

While these basic concepts appear reasonable, several features of TDL turned out to be hard to use, to formalize, or to implement; these include aspects of the assertion language, the scripts, and certain attribute categories for transactions. Revision work to better satisfy the goals set out for the conceptual design level above is still ongoing. Nevertheless, one of the lessons learned is that a direct transition from the requirements model to a formal specification of the system is problematic with large information systems: it seems important to reorganize the application knowledge gathered in the requirements phase from the viewpoint of how information about this application is to be managed in the information system from an overall, data dictionary-like perspective, beyond the specification of each individual database program. Whether this can be handled by a specialized sublanguage of CML, or needs own language constructs, is a question for further research.

2.1.3 *Producing High-Quality Database Application Software.* The development of correct and efficient database application software is neither simply a database design task nor a classical programming effort where most emphasis is on the optimization of each individual application, rather than of the IS as a whole. Instead, this task requires integrated concepts based on advanced database and systems programming technology.

464

Based on earlier research in database programming languages [SCHM77], DAIDA´s choice has been the language DBPL [SEM88]. DBPL fully integrates the concepts of a set-oriented (extended relational) model of data into those of the systems programming language, Modula-2, in a multi-user setting. By declaring variables of arbitrary types within a *database module*, values of these variables can be made persistent (i.e., their lifetime goes beyond that of a single application run). A type *relation* is introduced to deal with large sets of similarly structured data elements; predicative rules and constraints are supported using so-called *selectors* and *constructors*, giving DBPL the capabilities of a deductive database. *Transactions* are used as units of integrity and concurrency control. Moreover, DBPL does support modules as in Modula, an important feature we identified already as missing in the languages above. This currently forces the prototype DAIDA environment to make some of the modularization decisions quite late in the software development process, something the final version of the methodology will have to remedy.

2.1.4 *Summary.* To summarize the section on concept-based specifications, we should like to re-emphasize that the individual choices for the languages are less important than the groups of activities they represent in the DAIDA methodology. Fig. 1 illustrates these roles. At the requirements level, we try to *capture* knowledge about the role of systems in the world, seen from a world model perspective. At the design level, we *organize* the same systems from the specialized viewpoint of an integrated information system world. Finally, software-specific concepts serve as the basis for quality-assured program *production*.

At least as important as the conceptualization of these individual tasks is the conceptualization of their *interrelationships*. There are two aspects to this. First, we need an abstract conceptual model which describes all of the individual conceptualizations in a uniform manner. Second, we need a representation of the relationships between these abstract objects; these relationships cannot be deduced from looking at the objects alone but derive from knowledge about the process of development. This leads us to the next strategic goal of DAIDA: process orientation of the environment.

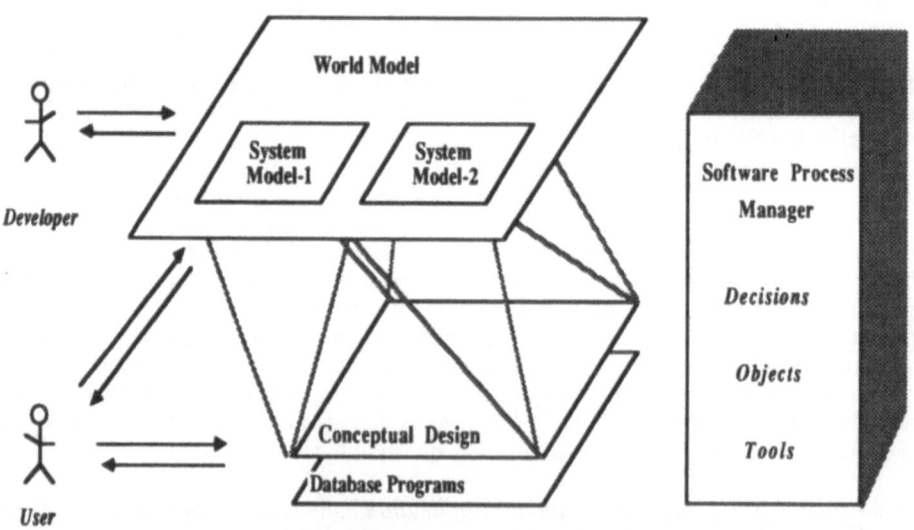

Fig. 1: Conceptualization tasks in DAIDA

2.2 PROCESS-ORIENTED DEVELOPMENT

The software development process works incrementally, and it runs through stages. Neither the stages nor the steps of that development process are predetermined. Instead, they are determined by the development tasks, available tools and the human development team.

The second strategic goal of DAIDA was to support such a process explicitly, and over long development periods. Conceptual languages of requirements, design, and database programs are not sufficient for this task; a conceptual model of the software development process itself must be maintained. DAIDA´s software process model, called the *D.O.T. (decision-object-tool) model* [JJR89a], represents states of a development process by documenting, as *objects*, relevant properties of results achieved in that state; we represent state transitions by documenting and justifying the decisions leading to the results; *decisions* can address refinement within a DAIDA level, mapping between levels, versioning to change previous decisions, or configuration to aggregate existing system components. Since the development environment may change over the life time of an information system, we also represent the *tools* that support the execution of decisions. All of this information is managed to transfer development experience to subsequent stages of a system´s life cycle.

DAIDA documents states and transitions by a Global Knowledge Base (GKB) accessible from all levels and languages. The conceptual modelling language SML proved suitable to represent the software process as well. The knowledge base provides information required by the development process and assures certain formal properties of it. This limited but broad control is achieved by using SML´s metaclass hierarchy to document:

- the basic software process model which defines the D.O.T. concepts
- a particular development environment described in that language
- concrete development projects within such an environment
- prototyping experiments for such a project.

Each instantiation level is controlled by the one above it. Level models may also evolve along version histories. The graphical view of SML serves as a basis for browsing in version histories, along development levels, along usage relationships, etc. while the formal view allows to restrict attention by predicative querying prior to graphical exploration.

In the DAIDA prototype, the software process metamodel is naturally instantiated by the three-level DAIDA methodology. However, another instantiation of the same model, even using the same three languages, may support quite different development paradigms. For example, DAIDA´s GKB models development tools as *reusable components* as if developed earlier by the DAIDA methodology itself. Taking this idea one step further, the GKB could *in general* be viewed as a repository of reusable development experience from which new applications can be developed largely by configuration of existing components rather than by rewriting software from scratch. An even looser coupling of GKB and development environment would just document the availability of software at a rough specification level.

2.3 QUALITY-ASSURED SOFTWARE

To provide a controlled degree of quality assurance in the software, it is not sufficient just to document the relationships between the three levels of representation and their changes over time. Supporting tools, in particular for the validation of requirements or designs and for mapping among the three DAIDA levels are needed. *Validation* aspects are covered by prototyping facilities using BIM-Prolog [MTV88]. For the mapping tasks, *knowledge-based assistants* should not just help the user satisfy the functional requirements of the intial system model but should also support non-functional goals (e.g., efficiency, accuracy, etc.).

In DAIDA, there are two main mapping tasks: from system model to conceptual design, and from conceptual design to database program. While some problems addressed by the former were mentioned in section 2.1.2, we now provide a glimpse at the latter.

The TDL-DBPL mapping process is more formalized than the SML-TDL mapping. Using Abrial´s notion of Abstract Machines as an intermediate representation [AGMS88], a specific group of DBPL application modules is derived from the TDL model as follows. The developer selects a coherent subset of TDL classes as the requirement for the intended program. This subset is then automatically translated into an Abstract Machine representation which is checked for consistency and formal completeness. From this initial abstract machine, refined machines are derived in part automatically, in part manually. Each refinement step generates a number of proof obligations. The developer can either just sign them off as supposedly satisfied; or he can carry out a formal, computer-assisted proof -- this provides the freedom of choosing various degrees of quality assurance. The last refinement result should be so close to a DBPL representation that an automated translation is possible.

In order to support automated refinement and correctness proofs, DAIDA is developing a number of special proof techniques (pre-proven lemmas and tactics how to use them) specifically devoted to the database programming task from object-oriented specifications such as provided by TDL. This is not an easy task, and we were fortunate to be able to start with at least a good kernel theorem-proving assistant (Abrial´s B tool [ABRI86]) and an initial set of theories useful for software development in general. The current state of this work is reported in [WSS89].

3 The DAIDA Architecture

Knowledge-based support for the DAIDA methodology is provided by the DAIDA environment. It consists of a set of dedicated toolboxes coordinated by a global KBMS. From a functional viewpoint, the architectural requirements can be summarized as follows:

- Several related languages with different constructs and usage patterns each require their own sub-environment. Nevertheless, common functionalities among these sub-environments should be supported in a uniform way to avoid confusion of the user.
- Mapping between the three levels is supported by specialized assistants. Since these incorporate development theories which may change, they are organized as extensible toolkits. In particular, for quality reasons, it must be possible to include theorem-provers for partially automated programming and verification of critical components.
- Information systems development is a continuous and usually cooperative process of analysis and re-analysis, design and re-design, programming and program re-organization. Information about this process should therefore be stored in a central repository of software process experience which we call the GKBMS. The GKBMS interacts with other tools by documenting and retrieving their results and the underlying decisions. But it should also model the evolution of the DAIDA environment itself, e.g. to explain or replay the execution of old design decisions.

From these requirements, the functional architecture shown in fig. 2 has been derived; its technical implementation follows a client-server approach. Note that fig. 2 does not imply a "waterfall" methodology although it looks like one. The specification and design assistants contain full animation and prototyping tools; due to the GKBMS, systems can be developed and maintained incrementally. We only insist that for each TDL design, an appropriate system model exists, and a TDL design for each DBPL program; for existing software, these models can be built by reverse engineering (this was in fact done for tools used in DAIDA).

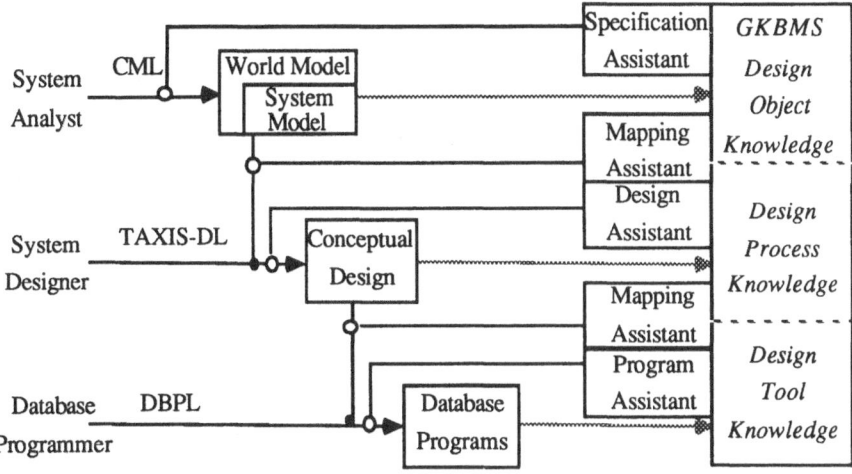

Fig. 2: Functional View of DAIDA Architecture [DAIDA88]

4 The DAIDA Tools

For a number of critical components of the above architecture, prototypical tools have been built. Together with existing components from our earlier work or from external sources, these tools form the basis for the integrated demonstrator prototype. Additionally, they were designed to have also value of their own; several are converted into commercial products.

4.1 GRAFIC

GraFlc is a graphical browser and editor which has been used as a graphical front-end for all of the other DAIDA tools mentioned below. Based in part on requirements by other DAIDA partners, GraFlc was designed and implemented in Prolog/SUNView by GFI [RB89]. It is able to represent an underlying semantic network structure (as in the SML or TDL languages), using graphical types (shape, customized icons, ...) for nodes and links to distinguish different classes of objects, and graphical status information (color) to indicate operations on the knowledge base (e.g., selected objects, inaccessible objects, etc.). Used as a browser, GraFlc provides a standard layout to represent, e.g., the result of a query to the knowledge base; this layout can be changed interactively by the user. When GraFlc is used as an editor, the developer can insert new nodes and links with menu choice of graphical types and possibility integrity checking by an underlying SML or TDL system.

4.2 SMLS

This component, designed by SCS [PR89] as an extension of earlier work in ESPRIT LOKI, is intended to support the requirements analysis phase. SMLS provides a kernel implementation that maps SML models to Prolog programs, and a window-based user interface. Formal support for knowledge acquisition is provided by a deductive integrity checking component. Animation and thus intuitive understanding is facilitated by querying properties of example instances in the model; the query interface is quite refined, allowing for example queries with more general negation than typical deductive databases [MH89]. To obtain an overview of the knowledge base, there are special model management classes, browsing tools and pop-up menus reflecting the context of work in which they are invoked.

468

4.3 TDL-MAP

An interactive graphics-oriented environment helps to map system models to corresponding conceptual designs. Since there are important user decisions involved here, only part of this process is automated. Currently, structural mappings including the mapping of generalization hierarchies are supported [KMMV89]. Extensions study mapping of temporal requirements, and of predicative assertions [CKM*89]. From the available knowledge we create a partially filled-in frame of the corresponding TDL code which is completed manually; the consistency control mechanism of TDL can further support this activity by error control. The status of each mapping process is indicated graphically. The assistant is constructed by FORTH-CCI in collaboration with GFI; fig. 3 gives an impression, using several instances of GraFIc.

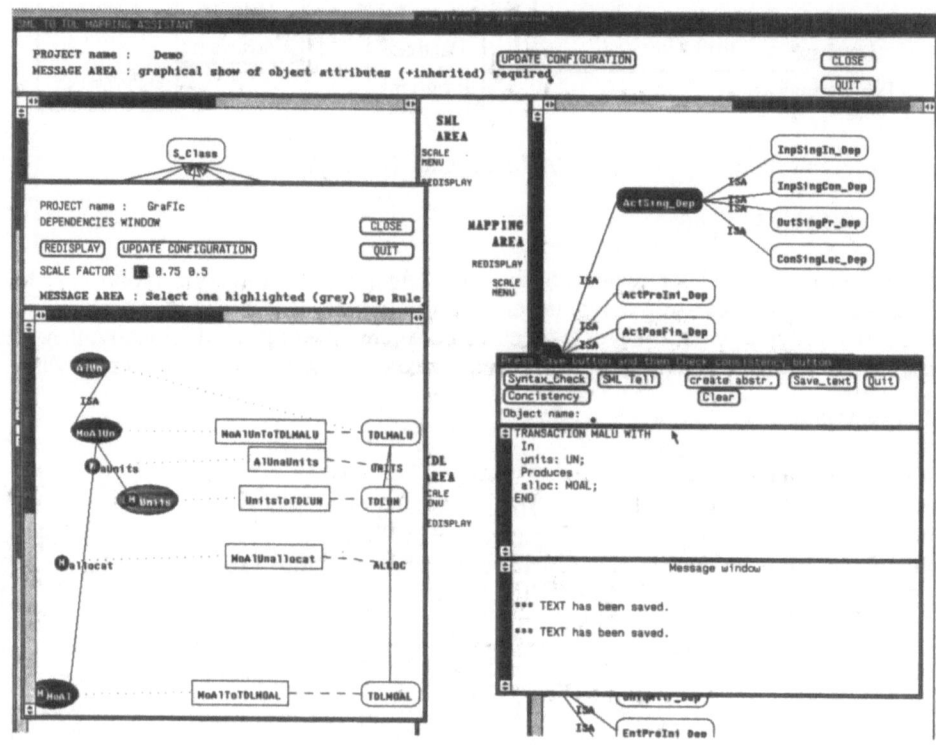

Fig. 3: TDL-MAP: The SML-TDL Mapping Assistant

4.4 PROBE

PROBE [ZMT89] -- built by DAIDA partner BIM -- is an object-oriented extension of BIM-Prolog which serves as a basis for the implementation of a TDL-based design and prototyping environment. TDL structures are mapped to PROBE predicates; moreover, PROBE supports structural and predicative integrity enforcement. Using the Prolog embedding of PROBE, the developer can add certain control information to TDL representations to make them executable as full functional prototypes; a library of standard functions for this purpose is being developed.

4.5 DBPL-MAP AND DBPL-USE

The TDL-DBPL mapping assistant, DBPL-MAP [BMSW89], bridges the gap between object-oriented system descriptions and set-oriented system implementation in a database programming framework. Due to the quality-assurance requirement, it should be possible to provide full verification and partial automation to this process. The mapping assistant therefore applies a mathematical theorem-proving assistant for software development, called the B-Tool [ABRI86], extended by a set of theories and proof tactics specifically dedicated to the tasks of (a) consistency checking of the input TDL model, and (b) verification of refinement steps towards a correct and efficient database program.

DAIDA extended the DBPL database programming environment, in particular by a language-sensitive editor, DBPL-USE [NS89], to simplify those program development and modification steps which are not formally constructed by DBPL-MAP. Additionally, DBPL itself was extended from a relational model to an orthogonally persistent type system. DBPL-MAP and DBPL-USE were built at the University of Frankfurt, partially in collaboration with the BP Research Centre (a DAIDA subcontractor) and Passau University.

4.6 CONCEPTBASE

ConceptBase, developed partially within DAIDA at the University of Passau [JJR88], is an SML-based kernel system and usage environment which serves as DAIDA´s Global KBMS component. The system is operational on SUN/UNIX and VAX/VMS workstations; the implementation language is BIM-Prolog, using SUNVIEW for the SUN version.

The tools mentioned under 4.2 to 4.5 document their activities in the GKBMS, and retrieve information about previous work of their own or of other tools. As a data model, ConceptBase provides the D.O.T. model of software processes discussed in section 2.2, i.e., it documents design objects, decisions, and tools used in executing these decisions. This model was also heavily applied in the implementation of the system itself [JJR89b]. For human users, there is a hypertext-style interface for browsing, filtering, and editing the knowledge base along dimensions of interest to software developers (e.g., development hierarchies, version histories, call relationships, etc.). For other DAIDA tools, ConceptBase offers a programming interface via a client-server protocol. The second prototype, to be completed in 1989, also provides version and configuration management services [RJ89].

5   A DAIDA Development Example

To give a flavor of the DAIDA methodology, tools, and environment, we give a detailed example of development and maintenance. The example is from an actual information systems project in a research company [MCNE89]. Although the description below is informal, it has been fully formalized and implemented for the DAIDA demonstrator.

The example is shown in fig. 4; four stages of the system (shown from right to left) are illustrated at the three levels of system model, conceptual design, and database program (from top to bottom). Consider a world with *persons* some of which are *employees* of research *companies*. In the initial requirements analysis, it is assumed that each employee only works on at most one *project*; this is indicated in SML by making the workson attribute instance of an attribute class *Single*. Similarly, uniqueness of employee names is given by attribute class *Unique*. Persons may turn into employees by *hireEmp* activities which instantiate the *belongsto* link between employee and company. As a specialization of this general hiring activity, persons may also be hired directly for specific projects (*hireEfP*). An SML integrity constraint restricts employees to work only on projects of their own company.

470

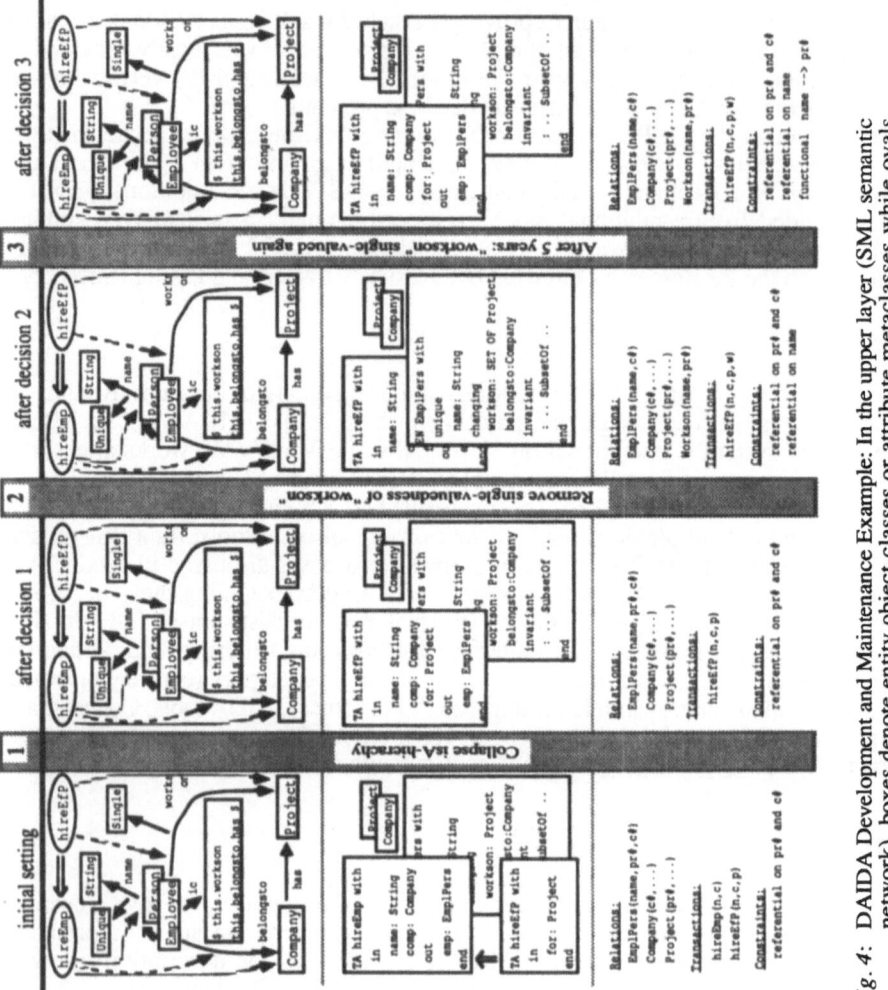

Fig. 4: DAIDA Development and Maintenance Example: In the upper layer (SML semantic network), boxes denote entity object classes or attribute metaclasses while ovals denote activities. Solid lines denote attribute and instantiation links, double arrows isa links, dotted lines activity input, and broken lines activity output. At the DBPL level only the rough schema of relations, constraints, and transactions is shown.

The developer decides to store only information about persons who are employees; therefore, the generalization hierarchy Employee-Person is mapped, by TDL-MAP, to a single TDL entity class, *EmplPers*, which gets the attributes of both its origin classes by inheritance. In contrast, each SML activity class is separately mapped into a TDL transaction. The same direct mapping applies for project and company objects. DBPL-MAP converts this structure into a relational database with a relation for each TDL entity class, artificial keys *c#* and *pr#* to ensure object identity, and a referential constraint that makes sure that employees work in existing companies and on existing projects. The implementation of the transaction specifications is more complicated: for the *hireEfP* (hire employee for project) transaction, DBPL-MAP has to take into account the inherited parameters and functionality of the hireEmp transaction, and it has to add a precondition that the integrity constraint of the world model (mapped to a *Subset* invariant of *EmplPers* in TDL) is satisfied, prior to executing the transaction code. The whole process is recorded in the GKBMS.

In evaluating this system concept, two major criticisms come up, resulting in decisions 1 and 2 (gray vertical bars) of fig. 4. First, prototyping shows that users are confused if they have to deal with two different kinds of transactions. As a consequence, the SML-TDL mapping is changed so that the *isa* hierarchy of transactions is collapsed, using inheritance. Interestingly, the DBPL code need not be changed at all, provided an artificial project like "general hiring" is introduced for those employees not hired for projects; then, the transaction program *hireEmp* can simply be discarded and *hireEfP* used alone.

In decision 2, it is noticed that employees may in fact work on more than one project, i.e., the instantiation link to the attribute class *Single* is removed in the SML model. The mapping of this change to TDL is quite simple: the attribute *workson* becomes set-valued. However, DBPL-MAP has now a much tougher job: if we want to retain a normalized relational database schema, the *deletion* of the *Single* constraint requires the *addition* of a new relation *Workson* which represents the many-to-many relationship between employees and projects. Together with a referential constraint, it ensures the appropriate construction of this relationship from existing projects and employees. On the other hand, the *pr#* attribute can be omitted from the *EmplPers* relation. Of course, this new database structure also implies substantial changes in the transaction, *hireEfP*, which we cannot describe here in detail (but which are covered by the demonstrator prototype).

After these revisions, the information system is implemented and filled with data; thus, the existence of the information system structure becomes a major factor in the world model which must be taken into account in subsequent requirements changes. When, after five years, the company decides that employees should better concentrate on one project again, one cannot simply return to the state after decision 1. Instead, decision 3 preserves the existing implementation, adding only an integrity constraint: the functional dependency that each employee can be assigned to at most one project in the *Workson* relation. Transaction programs need not be changed in this case since the DBPL integrity checker will verify the correctness of transaction results automatically; of course, an initial personnel reassignment has to assure that the constraint is satisfied in the beginning. Dependencies maintained by the GKBMS support this kind of reusing previous development experience.

6 Conclusion: Lessons Learned From DAIDA

The example indicates the scope of development and maintenance decisions we can support with the integrated DAIDA demonstrator prototype. Nevertheless, there are a lot of more complex mapping and re-mapping tasks which require further research and development in theories and supporting tools. Additionally, we need more experience with the practical usefulness and cost-benefit ratio of the DAIDA tools in real-world settings.

Instead of speculating on possible results of such future work, we conclude this paper by summarizing some of the lessons learned during the project. The discussion follows the same pattern of conceptualization, process orientation, and quality assurance that was emphasized before as the main goals of the DAIDA methodology.

Conceptualization: The project started with the standard waterfall-like model of specification, design, and implementation, only augmented by the idea of adding knowledge-based support and prototyping. The understanding obtained during the project looks similar only on the surface. Instead of just starting with formal specifications, the SML model emphasizes knowledge capturing for current and intended roles of the system in its environment (world), seen from a pure application perspective. The TDL model is no longer a vague concept of "design" but has a distinct role of drawing together the individual system perspectives in an integrated information system; the emphasis is on organizing computerized information, rather than describing an aspect of the world. In the TDL-DBPL mapping assistant, the limited but important role of formal methods in producing quality-assured database software from portions of such an integrated conceptual structure has been clarified, also the need for high-level database programming tools to facilitate this process. Finally, we observed the need of representing (the history of) the development process as an important object in the world, not just as a formal object: We have to know how the production of DBPL software was related to the choice of overall information systems organization, and from what world model the requirements for this organization were derived. More importantly, we have to record how requirements, designs, implementations, and their interrelationships change over time, due to maintenance and reusability requirements; and we have to support this change consistently, in-the-small, in-the-large, and in-the-many.

This brings us to the second goal, *process orientation*: Initially, we conjectured that automated formal methods are unavailable in the short run, and thus followed the well-known idea to provide assistants at least for those tasks that can be automated. It quickly became apparent that the two DAIDA mapping tasks are of a very different nature, one (DBPL-MAP) more amenable to formal methods than expected, the other (TDL-MAP) focusing on human choices and hardly explored in any previous work. Additionally, although DAIDA developed reasonable languages for world modeling and conceptual design, methodologies and development assistants for working with these languages -- for capturing requirements in a goal-oriented manner and for integrating views -- are still in their infancy.

Similarly, understanding the "passive" process support component took substantial work. We started with the empirically observed need [DJ85] to preserve process knowledge for maintenance and reusability in large information systems. Dependency modeling appeared as an easy way out since AI had some experience in this area. However, it was quite unclear how to obtain dependency information and we found that the initial model was by far not rich enough. To have meaningful dependencies that are not just uninterpreted hypertext links, we needed a uniform description of all objects relevant to the DAIDA process. Only after some experiments, we got the idea to view software environments as a special kind of world whose structure could be modeled with specialized SML metaclasses: the D.O.T. model. In this model, dependencies were conceptualized as being created by *decisions*; this, in turn, lead to the idea of providing a decision support methodology where decisions are modeled as being goal-driven, resulting from argumentation, and being controlled by methodologies. Moreover, advanced CASE environments such as DAIDA support the execution of decisions by *tools*; usage of these tools should also be recorded since it may be difficult to understand the results afterwards. Nevertheless, few other environments have a clear concept how to model (a) what tools are available with what capabilities, and (b) how they have actually been used (e.g., for purposes of replay in maintenance or reuse); here, our trick was to use the DAIDA methodology itself to model these tools. Meanwhile, usage of this model has migrated into the active DAIDA process tools, and into other projects as well.

Only in the later phases of the project it became clear that the DAIDA methodology *per se* might easily fall into the same trap as many previous projects, namely, that it could support only relatively small-scale projects. Within DAIDA and in follow-up projects there have therefore been substantial recent efforts to scale up the models to the cases of programming-in-the-large (esp. version and configuration management [RJ89]) and programming-in-the-many (project and documentation management [HJK*89]).

Finally, **quality assurance**: At the beginning of the project, this was not a major concern; where it was, we talked about integrity checking and prototyping. By understanding the needs of maintainability and by following the idea of system model embedding in a world model, new options emerged. With SML, we have now the possibility to formally represent and enact (by prototyping) the role systems may play in the world; thus, the main aspect of quality assurance in the SML-TDL mapping is validation. In contrast, for the TDL-DBPL mapping, we learned to appreciate the advantages of formal methods in terms of repeatability and verified correctness at least within a validated frame of reference. Lastly, the formal software process model can enforce some measure of process quality; it also provides documented flexibility in the way other quality assurance measures are actually applied.

Several of the individual tools discussed in section 4 are currently in various stages of industrial product development. Moreover, the experiences gained in DAIDA are exploited in further projects, among them several ESPRIT II projects such as the Technology Integration Projects ITHACA and MULTIWORKS and the Basic Research Actions FIDE and COMPULOG. ITHACA exploits DAIDA experience in developing a software information base, MULTIWORKS extends the methodology to co-authoring support in a multimedia environment, FIDE concentrates on further development of database programming languages, and COMPULOG investigates basic issues of realizing software environments by extended logic programming. In all of these projects, synergy of DAIDA results with experience and tools from other ESPRIT projects is being achieved.

References

[ABRI86]   Abrial, J.R. (1986). An informal introduction to B. Manuscript, Paris, France.

[AGMS88]   Abrial, J.R., Gardiner, P., Morgan, C., Spivey, M. (1988). Abstract machines, part 1 - 4. Unpublished manuscript, Oxford University, UK.

[BJM*87]   Borgida, A., Jarke, M., Mylopoulos, Y., Schmidt, J.W., Vassiliou, Y. (1987). The software development environment as a knowledge base management system. In Schmidt, J.W., Thanos, C. (eds.): *Foundations of Knowledge Base Management*. Springer-Verlag.

[BMSW89]   Borgida, A., Mylopoulos, J., Schmidt, J.W., Wetzel, I. (1989). Support for data-intensive applications: conceptual design and software development. *Proc. 2nd Workshop on Database Programming Languages*, Portland, Or.

[CJM*89]   Constantopoulos, P., Jarke, M., Mylopoulos, J., Pernici, B., Petra, E., Theodoriou, M., Vassiliou, Y. (1989). Scripting the ITHACA software information base. Report ITHACA.FORTH.89.E2.#1, Iraklion, Greece.

[CKM*89]   Chung, L., Katalargianos, P., Marakakis, M., Mertikas, M., Mylopoulos, J., Vassiliou, Y. (1989: Mapping advanced concepts: mapping of time and assertions. ESPRIT 892 (DAIDA), Iraklion/ Greece.

[CONK87]   Conklin, J. (1987). Hypertext: an introduction and survey. *IEEE Computer 20*, 9, 17-50.

[DAIDA88]   Jarke, M., DAIDA Team (1988). The DAIDA environment for knowledge-based information system development. In *ESPRIT '88: Putting the Technology to Use*, North-Holland, 405-422.

474

[DJ85]   Dhar, V., Jarke, M. (1985). Learning from prototypes. *Proc. 6th Intl. Conf. Information Systems*, Indianapolis, Ind, 114-133.

[HAGE88]   Hagelstein, J. (1988). Declarative approach to information systems requirements. *Knowledge-Based Systems 1*, 4, 211-220.

[HJK*89]   Hahn, U., Jarke, M., Kreplin, K., Farusi, M., Pimpinelli, F. (1989). CoAUTHOR: a hypermedia group authoring environment. *Proc. European Conf. Computer-Supported Cooperative Work*, Gatwick, UK.

[HK87]   Hull, R., King, R. (1987). Semantic database modeling: survey, applications, and research issues. *ACM Computing Surveys 19*, 3, 201-260.

[JJR88]   Jarke, M., Jeusfeld, M., Rose, T. (1988). A global KBMS for database software evolution: documentation of first ConceptBase prototype. Report MIP-8819, Universität Passau, FRG.

[JJR89a]   Jarke, M., Jeusfeld, M., Rose, T. (1989). A software process data model for knowledge engineering in information systems. *Information Systems 14*, 3.

[JJR89b]   Jarke, M., Jeusfeld, M., Rose, T. (1989). Software process modeling as a strategy for KBMS implementation. In Nishio, J., Kim, W., Nicolas, J.-M. (eds.): *Deductive and Object-Oriented Databases*, North-Holland, to appear.

[KMMV89]   Katalagarianos, P., Marakakis, M., Mertikas, M., Vassiliou, Y. (1989). SML-TDL mapping assistant: architecture and development. ESPRIT 892 (DAIDA), FORTH-CCI, Iraklion, Greece.

[KMSB89]   Koubarakis, M., Mylopoulos, J., Stanley, M., Borgida, A. (1989). Telos: features and formalization. Report KRR-89-04, University of Toronto, Ont.

[MBW80]   Mylopoulos, J., Bernstein, P.A., Wong, H.K.T. (1980). A language for designing interactive data-intensive applications. *ACM Trans. Database Systems 5*, 2, 185-207.

[MCNE89]   McNeil, I. (1989). Validating and verifying SML specifications - efficient prototypes and proofs. ESPRIT 892 (DAIDA), BP Res. Centre, Sunbury, UK.

[MH89]   Meyer, R., Haidan, R. (1989). Negation and metalevel reasoning in SML. ESPRIT 892 (DAIDA), SCS, Hamburg, F.R. Germany.

[MTV88]   Meirlaen, E., Trinon, J.M., Venken, R. (1988). An object-based prototyping workbench in Prolog. In *ESPRIT '88: Putting the Technology to Use*, North-Holland, 423-437.

[NS89]   Niebergall, P., Schmidt, J.W. (1989). DBPL-USE: a tool for language-sensitive database programming. ESPRIT 892 (DAIDA), Universität Frankfurt.

[OSTE877] Osterweil, L. (1987). Software processes are software too. *Proc. 9th Intl. Conf. Software Engineering*, Monterey, Ca, 2-13.

[PR89]   Piza, B., Röpcke, I. (1989). SMLS: Systems Modelling Language Support System on SUN. Manual, ESPRIT 892 (DAIDA), SCS, Hamburg, FRG.

[RB89]   Rouge, A., Bonin, G. (1989). GraFIc 2.0. Manual, ESPRIT 892 (DAIDA). GFI, Nanterre, France.

[RJ89]   Rose, T., Jarke, M. (1989). A decision-based configuration process model. Report, Universität Passau, FRG.

[SCHM77]   Schmidt, J.W. (1977). Some high-level language constructs for data of type relation. *ACM Trans. Database Systems 2*, 3, 247-261.

[SEM88]   Schmidt, J.W., Eckhardt, H., Matthes, F. (1988). Extensions to DBPL: towards a type-complete database programming language. ESPRIT 892 (DAIDA), Universität Frankfurt, FRG.

[WSS89]   Wetzel, I., Schmidt, J.W., Stuchly, J. (1989). A mapping assistant for database program development. ESPRIT 892 (DAIDA), Universität Frankfurt, FRG

[ZMT89]   de Zegher, I., Meirlaen, E., Trinon, J.-M. (1989). BIM-Prolog Object Extension: User Manual. ESPRIT 892 (DAIDA), BIM, Everberg, Belgium.

AMORE - OBJECT ORIENTED EXTENSIONS TO PROLOG FOR THE RUBRIC IMPLEMENTATION ENVIRONMENT

Z. PALASKAS & P. LOUCOPOULOS
Department of Computation
UMIST
P.O. Box 88, Manchester M60 1QD
U.K.
Tel: ++ 44-61-236 33 11

FRANS VAN ASSCHE
James Martin Associates
Rue de Geneve 10,
EVERE 1140 Brussels,
BELGIUM
Tel: ++ 32-2-241 88 00

Abstract.

The RUBRIC paradigm to developing information systems is based on the premise that the development process should be viewed as the task of constructing or augmenting an organisation knowledge base which is declaratively defined. Such an approach requires a number of features of its implementation environment. These features are provided by the AMORE language which is discussed in this paper. The primary objective of the AMORE language is to provide support for the information base of RUBRIC and to offer facilities for developing a set of tools which assist in the specification and maintenance of this information base. This language has been designed as an extension to Prolog offering Object Oriented concepts. The integration of the two approaches offers a flexible environment for the development of advanced information systems.

1. Introduction

Current practices in the development of information systems either rely on the traditional, ad-hoc approach or make use of one of the many development methods and associated CASE environments. The shortcomings of the former are well documented [Chapin, 1979; Lehman, 1980; Boehm, 1981; Bubenko, 1986]. The latter approach has produced better results if only because it imposes a design discipline in the entire development process. This is achieved through the use of a meta-model for system representation, a process by which to construct a specification and a set of computer-assisted tools to aid model development. Examples of contemporary methods are Information Engineering [MacDonald, 1986], JSD [Jackson, 1983], NIAM [Verheijen & van Bekkum, 1982], SASD [DeMarco, 1978]. However, despite the considerable wealth of experience and resources available to system developers, problems of unreliable and inflexible systems are still widely reported by end users [Morris, 1985]. These problems manifest themselves in computer systems which are often unmanageable, unreliable, inflexible and hence difficult to maintain [Maddison, 1983; Olle et al, 1983, 1986].

A major shortcoming of current practices is that the modelling of organisational policy within an information system is completely defined only at a low level of abstraction, namely programming code. The result is that whereas end users perceive and often define a business system in terms of policies or rules, such a view is not directly visible in the derived system specification. This low level representation of business rules results in computer systems which are difficult to maintain for whereas end users define require-ments changes in terms of the policy which dictates the behaviour of the business environment, maintenance staff need to translate these into program concepts. The intertwining of business policy with other programming constructs which deal with issues such as interfacing, file management and error detection and correction makes the task of system evolution a complex undertaking.

Balzer [1983] has argued for a new approach which recognises the need of addressing system evolution at a high level of abstraction. The RUBRIC (RUle Based Representation of Information-systems Constructs) paradigm [van Assche et al, 1988a; van Assche et al, 1988b] goes a long way in meeting this objective. This paradigm is based on the premise that organisational policy must be separated from the procedure which supports that policy and so allow a greater distinction between the what and the *how*. This approach conforms to the proposal by ISO [van Griethuysen et al, 1982] and is a logical extension of the well accepted practices of *data* and *interface* independence from application programs.

The RUBRIC paradigm states that development of an information system should be viewed as the task of developing or augmenting an organisation's knowledge base (Fjeldstad et al, 1979; Mathur, 1987; Greenspan, 1984). Within RUBRIC this knowledge base is concerned with the definition of the principle *objects, relationships, happenings* within the organisation and *constraints* on the operating of the organisation [van Assche et al, 1988].

The basic RUBRIC paradigm is that development of an information system should be viewed as the task of developing or augmenting the policy knowledge base of an organisation, which is used throughout the software development process, from require-ments specification through to the run-time environment of application programs. Within RUBRIC this knowledge base is concerned with the definition of the principle *objects, relationships, happenings* within the organisation and *constraints* on the operating of the organisation [van Assche et al, 1988]. The RUBRIC model which satisfies this objective consists of:

– **The structural component** which is expressed as a binary entity-relationship model [Chen,1976]. This model differs from the original E-R model in that it regards any association between objects in the unified form of a relationship thus avoiding the unnecessary distinction between attributes and relationships [Kent, 1983; Nijssen et al,1988]. The RUBRIC E-R model represents explicitly *entity types* and *value types*. For each relationship the RUBRIC model recognises two reference predicates which are used to make statements (e.g. "a PRODUCT is sold at a PRICE") and two referent functions which are used as a selection mechanism for entities or values (e.g."a PRODUCT ... sold at a PRICE..").
– **The static rules** which either represent integrity constraints on the E-R model or derivations on it. An example of an integrity constraint might be the involvement subset constraint "EMPLOYEE that has SALARY is_a_subset_of EMPLOYEE working for DE-PARTMENT". An example of a derivation rule would be "A supplier is cheapest supplier

for a product P if a supplier makes an offer for product P and the price of that offer is the minimum".
- **The dynamic component** which deals with the definition of transactions and operations. A transaction is a logical unit of work which consists of one or more operations. An operation acts on a database object and describes the behaviour of this object as a result of some stimulus. A transaction is regarded as having the properties of *consistency, atomicity* and *durability.* By consistency it is meant that an activity must obey some predefined rules. By atomicity and durability it is meant that either all actions are done and the effects of carrying out the activity persist from then on, or none of the effects of the activity survive because of some violation of a static rule. Each transaction and therefore the entire behaviour of an information system is determined by *dynamic action rules.*
- **The dynamic rules** which represent either dynamic constraints or control of operations. An example of a dynamic constraint might be "prices can only change overnight".

A dynamic action rule for the control of operations consists of three parts: the *trigger* which is an expression describing the conditions under which an operation should be considered for execution, e.g. the condition of 'stock level < reorder level' being true would be a trigger to the operation 'issue purchase order' (van Assche et al [1988] describes in more detail the full range of conditions for triggers); the *preconditions* which are expressions which must be true if an operation is to be executed, given the occurrence of the associated trigger; and finally the *action* part which represents the detailed operation. In an object-oriented context an operation represents a method.

The RUBRIC paradigm is supported by a number of facilities which assist in the specification and validation of a conceptual schema and its exploitation at run-time. The abstract architecture of the RUBRIC environment is shown in figure 1.

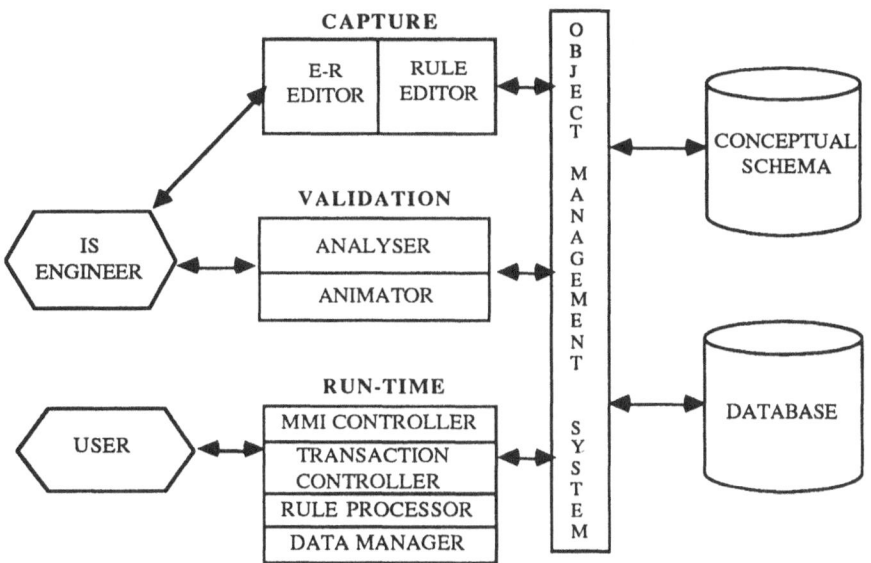

Figure 1. The RUBRIC implementation environment

The RUBRIC model determines certain requirements of its implementation language. Specifically:

- The structural component requires the definition of constructs in the AMORE language which correspond to constructs in the extended E-R model.
- The static rules require the definition of constraints on entity, relationship and value types during requirements specification and the handling of these constraints during prototyping or run-time.
- The dynamic component requires the specification and execution of operations, i.e. the detailed behaviour of an information system.
- The dynamic rules require the definition of the interaction between rules which determine the control of operations and a mechanism for executing these rules.

In order to realise these diverse requirements, an integration of database technology, rule-based knowledge representation techniques and object-oriented programming was necessary. At the implementation level, this integration manifests itself in the AMORE (A Method Object Rule Environment) language. This language provides object-oriented facilities as extensions to the Prolog programming language. The motivation behind the AMORE development is discussed in section 2. A detailed description of the AMORE language is given in section 3 in terms of the language's basic components.

2. Integrating Logic Programming and Object Oriented Concepts

The primary objective of the AMORE language is to provide support for the information base of RUBRIC and to offer the facilities for developing a set of tools which assist in the specification and maintenance of this information base. The implementation vehicle of AMORE is the Prolog language, chosen primarily for its flexible framework, clarity and high expressiveness of its semantics and most importantly its suitability for rapid prototyping [Venken, 1984]. Because AMORE is a flexible tool which offers facilities for the adaptation of the environment according to the application in hand, the language can also be viewed as a general formalism which extends Prolog to take advantage and exploit ideas given by the Object Oriented paradigm. The task of merging Logic Programming and Object Oriented concepts, can been seen as an attempt to maintain as much as possible the Prolog philosophy, and introduce a constrained set of new syntactic concepts upon the language, thus hiding from the user internal predicates and functions.

Previous work has demonstrated that Prolog can be extended to support many knowledge representation and computational formalisms [Newton, 1986] such as frames [Cuadrado, 1986], semantic networks [Malpas, 1987] and Object paradigm [Iline, 1987; Koseki, 1987; Fukunaga, 1986].

The motivation for integrating Logic Programming and Object Oriented concepts is in order to exploit features of each approach; for whereas Logic Programming is suitable for expressing reasoning, Object Programming provides facilities for describing complex structures. Iline [Iline, 1987] considers the Object Oriented paradigm as being suitable for structuring a domain, maintaining consistency of the modelled domain and activating general reasoning mechanisms. Recent work, such as PEACE [Koseki, 1987] and LAP [Iline, 1987], shows that this integration of programming paradigms results in flexible systems which can be adapted in practical applications of Artificial Intelligence, particularly Knowledge Based Systems. Commercial tools and systems for the creation of knowledge based applications, such as ESP frame engine [ESP, 1986] and Flex [Flex, 1988] and inference engines such as Class [Buis, 1988] have been developed by integrating frame technology with Prolog reasoning facilities. Similar multiple-paradigm languages and tools have been developed based upon Lisp, such as KEE [KEE, 1985], LOOPS [Bobrow 1986], FLAVORS [Moon 1986], FORK [Beckstein 1987].

AMORE consists of a compiler which translates class definitions into an executable Prolog specification, and a runtime support enviroment. The latter is implemented as a set of predicates which support the execution of messages and a set of tools and utilities which are built on top of the language and can be used by the user defined object classes. AMORE was designed to be as close as possible to Prolog in specification and execution, with the introduction of a minimum amount of externals, and a messaging mechanism to support backtracking inside the context of the objects.

3. The AMORE Language

3.1. Class Definition - an Overview

The use of the term *class* in AMORE is not different from its' use in other object-oriented languages such as Smalltalk and EIFEL [Meyer 1988] or object-oriented language extensions such as FLAVORS [Moon,1986] and Objective - C [Cox, 1986]. A class refers to a set of objects which share a set of common properties.

Classes form latice inheritance networks, that is, multiple inheritance is supported. Cyclic class definitions are also supported and precedences are resolved by applying a modified depth first algorithm [Stefik, 1986].

Class properties are defined in terms of *slots* and *methods*. Slots are viewed as the constructs which specify the static structure of objects whereas methods is the means by which objects describe operations in their domain.

Slots and methods are accessed by *messages*. Messages that refer to slots, can be queries or assertions. Queries usually return the values of the addressed slot which in turn can be stored or derived. Slot value assertions are under the control of the runtime constraint management system, which performs controlled constraint evaluation whenever a slot value is changed. Slots can also have associated *daemons* which are activated during updates of slot values. When messages refer to methods the object performs the actions as specified by the code for these methods.

3.2. Slot Definition

The objective of a slot definition is to determine an aspect of the structural properties of the objects. These structural properties are directly mapped from the RUBRIC structural components which is expressed as an extended E-R model. A slot corresponds to a relationship between an entity type and a value type. A slot description includes *type, cardinality, inheritance, default values* and *general constraints, inverse relationship* and *daemons*.

3.2.1. General Constraint Specification.

The type, cardinality, constraint and inheritance facets produce a combined constraint formula which is then stored as the aggregate constraint applying on the slot. These constraints are validated during run time when a slot value is updated. The specification of general purpose constraints is achieved through the **constraint** facet which permits any valid Prolog goal. The special variables of **it**, **super** and **self** can be used inside the context of the constraint facet. The following example demonstrates that the only possible values of the slot *sex* are *male* and *female*.

```
defclass person.
        slots.
                sex,
        constraint ( member(it,[male,female]),
                age,
        type integer(0,120).
endclass.
```

Constraints override constraints of superclasses or can be appended to produce an additive effect.

3.2.2. Inverse Relationship Specification.

In the definition of a conceptual schema in terms of the E-R model, a convenient construct is that of the inverse relationship. For example, in the E-R diagram of figure 2 the role 'occupied_by' is the inverse of 'lives_in'. Inverse relationships are specified through inverse slot facets.

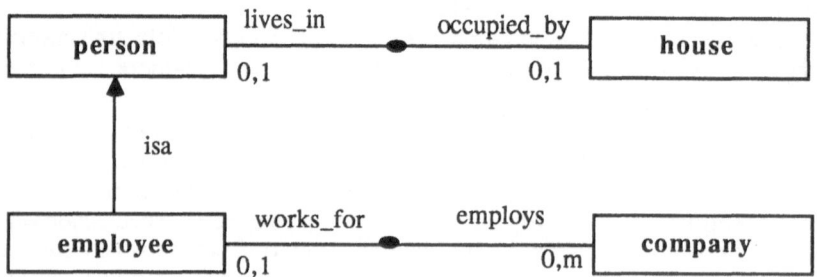

Figure 2: An example E-R diagram representing relationship roles and their inverse

The definition of some entity types, relationship types and their inverse for the E-R diagram of figure 2 are specified as follows:

```
defclass employee.                      defclass company.
        supers person.                          supers amore_object.
        slots.                                  slots.
        works_for,                              employs,
                type instance_of(company),              type instance_of
                cardinality(0,1),                               (employee),
                inverse_of employs.                     cardinality(0,many),
endclass.                                               default [],
                                                        inverse_of works_for.

                                        endclass.
```

Suppose, there are the instances of employees **e1**, and the instances of the class company **c1**. The statements:

e1 < -works_for : = c1

will automatically set the value of the slot **employs** of the instance **c1** to the value [e1]. It should be be noted that it is only permitted to update a foreign slot in the case that the slot has cardinality greater than one. In the case of single valued foreign slots, an update is permissible only when a relationship already exists with the owner of the foreign slot.

3.2.3. Daemon Specification.

There may be two categories of daemons associated with slots. Daemons which operate when the contents of slots are queried (**if_needed**), and daemons which operate when the values of the slots are updated (**before_changed, when_changed** and **after_changed**). Daemons of the first category may be used to effect a tight binding of class instances with a database facility; queries on the slots are transformed to queries to the database and can also be used to access derived rather than stored information. Daemons of the second category are invoked when the values of the slots they control are updated. There are three types of update daemons, one type operates before the slot value has been changed whereas the other two operate after the value of the slot has changed.

The following example demonstrates the use of daemons. The class definition for a product can be freely translated to the requirements that the stock on hand for the product can never fall below zero, and the sales price of the product can change only overnight e.g. after 5p.m. and before 8 a.m. The *if_needed* daemon associated with the slot cheapest_supplier derives the appropriate value forming a Prolog query. The special variable **self** refers to a particular instance of product and the variable **it** refers to that product's slot value cheapest_supplier which is derived and returned.

```
defclass product.
common.
category purchasing.
slots.
     stock_on_hand,
            constraint( it  0 ),
     cheapest_supplier,
            type instance_of(supplier),
            if_needed (cheapest_supplier(self, it, _)).
            category pricing.
slots.
     salesprice,
            type float(0.0,1E7),
            default 0.0,
            before_changed ( hour(_,_,_,_hour),
                                 (_hour  > 17 ;_hour < 8) ).
endclass.
```

3.3. Methods

Methods are used to describe the behaviour of an object in response to a message received by that object. A method definition is equivalent to a clause definition in Prolog and can have a variable number of arguments. Three types of methods are supported: **before, primary** and **after. Before** and **after** methods as analysed by Stefik and Bobrow [Stefic & Bobrow, 1986] *declaratively* add actions to the inherited behaviour of one object with respect to a message. An example of the use of methods inside class definitions follows:

```
defclass supplier.
common.
category offer_creation .
method before.
make_offers:-
        ask('make new offer?',_reply),
        _reply == no,!.
method primary.
make_offers:-
        self<- make_offer,
        self<- make_offers,!.
method primary.
make_offer:-
        ask('product?',_product),
        ask('price?',_price),
        offer <- insert([_product, self,_price]).
endclass.
```

Within the context of a class, methods can backtrack as can be seen by the definition of the make_offers method. The special variable self is unified with the instance of the class that receives the message.

The primary method overrides inherited primary methods. All before and after methods are done in an order determined by the precedence list of the class.

Messages sent to methods have the effect of firing these methods in a three-phase sequence. First, the *before* methods are executed and these methods are retrieved by classes in the precedence list which is searched upwards, from the most to the least specific class. Second, the primary method is fired. Third, all *after* methods traversing the precedence list in reverse order are fired.

For procedural method combination, AMORE provides the reserved variable **super** which in association with a message passes that message to a superclass of the class in which the method is declared.

3.4. Messages

Messages are used to access the private data and operations of objects. All messages involve one object and one selector. The selector might refer to either an object slot or an object method. Four operators are provided to support message passing. These are the *send message* '<-', the direct access '<+', and the two assignement operators ':=', '=>'. The send message operator is used to invoke methods of objects or to access slots. In this case proper evaluation of the constraints and daemon invokation is performed. The direct access operator is used for slots only and does not invoke daemons or perform any evaluation of slot constraints; it should be used with caution when efficiency is important. Below are given some examples on the use of messages:

1. supplier1 <- make_offers
2. offer<-insert([product1,supplier1,20000])
3. self<+daughter<+name := self<+wife<+daughter<-name
4. _address := supplier1 <- address
5. supplier1 <- address => _address

Example 1 sends a message make_offers to the instance supplier1. Example 2 sends the message insert with argument [product1, supplier1, 20000], to the instance offer. Example 3 demonstrates message chaining, and it translates as *set the value of the slot daughter of this instance to be the value of the slot daughter of the wife of this instance*. The access of the slot name of the instances' wife, will result in a daemon invokation. Examples 4 and 5 show the equivalence in the use of the assignment operators.

3.5 Class and Instances in AMORE

In a class definition, slots or method can be declared to belong to the class or to the instances of the class. Class slots and class methods, refer to own properties of the class, whereas instance slots and methods are inherited by all instances of the class. The concept of distinguishing class and instance slots and methods is similar with KEE as explained by Fikes [Fikes 1985]. In AMORE this concept is syntactically supported by the keywords of common and private, the former refering to all instances and the latter to the class itself.

The following example demonstrates the use of private and common slots. The property heaviest truck is defined as being a property of the class of trucks, whereas the properties weight and tyres are properties of all instances of trucks.

```
defclass truck.
private.
slots.
        heaviest.
common.
slots.
        weight,
        tyres.
endclass.
```

Messages to classes are treated in a similar way to messages sent to instances. The information about the firing of a message which is sent to an instance will be found in the class definition for that instance and in its superclasses.

A metaclass network facility, similar in functionality to Smalltalk metaclasses is provided. When a class is created a database holding all existing class names is updated. The metaclass name which is a system generated name, will be the name of the class appended by a '_class' postfix. In AMORE the name of a class must be unique. Classes inherit slots and methods by the metaclass network, in a similar way in which all instances of a class inherit methods and slots by that class and its' superclasses.

There is a set of enviroment classes which provide methods inherited by all user defined classes. Such a method is the method **new** defined in the class *class*. This method is used to create new instances and may be inherited by all classes via the metaclass network. System defined classes are the classes *metaclass, class, object, amore_object*. The *amore_object* class is usually made the superclass of the topmost user defined class to properly pass down a set of useful methods, to assist browsing, tracing, compilation, message profiling information and inspection of the classes.

Examples of an instance creation are:

1. supplier<-new(s1)
2. supplier<-new(_s)

Example 1 creates a new instance of supplier with identifier s1, and example 2 returns a system generated identifier which will be supplierN where N is selected so that the instance will be unique.

An instance stores the link to the class it belongs via the slot *class*. The fact instance_of(class,instance) is inserted to the database to assist queries on all instances of a particular class. AMORE provides the possibility of an instance to become instance of more than one class, by letting the class record of an instance to accept multiple values.

Instance manipulation methods such as persistent storage and retrieval of instances, deletion and creation are defined in the class class and are inherited from all AMORE classes. Specialisation of methods for the creation and deletion of instances may include definition of initial and final actions regarding these database operations and their functionality is similar to that found in other languages.

3.6. Temporary objects in AMORE

A temporary object is a garbage collectable object which has the same behaviour as a permanent object, without being stored in the Prolog database. An example of a temporary object definition is:

x = **object**(person,)

which will unify the variable '_x' with a temporary instance of the class person. Prolog does not support destructive assignement and it would normally be impossible to assign new values to the attributes of one temporary object. The technique which has been employed in order to overcome this obstacle, associates the object record with a list ending to a free term. Updates in slots of one object are appended at the end of that list. Therefore, the list associated with a temporary object carries the history of updates in the objects' lifetime. The slots and their values are inserted in this list in the form *att* : *value*. The use of temporary objects is better clarified with the following example which shows a sequence of statements that handle temporary objects:

1. _x=object(person,[name:george,age:23|_]),
2. _y=object(person,_),
3. _name := _x<+name,
4. _age := _y<- age ,
5. _y <- name := george.

The first and second statements unify the variables '_x', and '_y' with the respective temporary objects. The third and fourth statements demonstrate the use of the messages for accessing slot values in temporary objects. In the third statement the variable '_name' will be unified with the value george, where in the fourth statement the variable '_age' will be unified with the value **undef**. During the execution of this statement, the temporary object represented by the variable '_y' will be set to

   y = object(person,[age:undef|]) .

The fifth example will set the temporary object unified with the variable '_y' to

   y = object(person,[age:undef, name:george|]) .

This approach introduces overheads in the retrieval and the updating of slot values proportional to the number of the object updates. To overcome these, appropriate predicates have been defined in the AMORE which create copies of objects that retain the most recent values for the slots. Also, there exist methods which transform a temporary object to a persistent object and vice versa.

3.7. Categories

Slot and method categories definition, has been introduced in AMORE to assist the browsing of classes. Slots and methods can be grouped according to their specification to form categories. Method categorisation follows the Smalltalk model and is a way of indecing method names under similar functionality. AMORE extends this concept to also accommodate slots.

Category statements may be introduced at any point inside a class definition, and all slots and methods following a category statement will be grouped under that given group name. Methods and slots by default are grouped in the category miscellaneous. The management of categories is catered by the compiler. An example of a class including a category statement is the following:

```
defclass product.
        common.
        category selling.
        method primary.
        sell(_customer, _quantity):-
                _present_stock:=self<+stock_on_hand,
                _new_stock is _present_stock - _quantity,
                self<-stock_on_hand:=_new_stock.
endclass.
```

3.8. Transaction and Constraint Handling

Transactions are considered to be atomic updates of an object's attributes. At run-time it is possible to select protected mode of operation to ensure that failure in the execution of a message would leave the database in a consistent state. In order that the integrity of all objects is maintained, the kernel records the updates in a stack, so that in case of a failure or a constraint violation, it becomes possible to return to a previous consistent state.

When a message is issued inside the context of another message a new stack level is generated. When a message terminates succesfully all the updates of the lower stack level will become updates of the parent stack level. When the topmost level is reached again all the updates become permanent.

During a transaction, it is sometimes desirable to allow possible violations of constraints and evaluate them at the end. This is accommodated by switching the message execution to a mode of relaxed constraints operation. The default mode of operation evaluates the object constraints immediately.

3.9. Other Utilities

The language provides methods which can be used for instance and class inspection, browsing of the classes, tracing, performance optimisation by producing profiling information and caching message execution. Messages can be compiled so that runtime searching is avoided. AMORE provides two facilities to assist method compilation. If the first facility is used, a class will compile all messages sent to it. The second facility allows selective compilation of messages. Message compilation is based on the idea that when a message is sent to an instance of a class, constraints and methods applying to this message will allways be fired in the same place, once the classes have been permanently defined. Therefore, it is possible to produce and store the information which describes where and how a message would fire, e.g. which *before*, *primary* and after methods are executed, which *daemons* are invoked and which *constraints* apply for this message.

3.10. Implementation

AMORE is implemented using BIM Prolog on the SUN workstation under UNIX and LPA-Prolog on the APPLE Macintosh. The AMORE kernel comprises of approximately 250 predicate definitions and 3,000 lines of Prolog code. The compiler produces a set of properties for the compiled class and a set of predicates which reflect the code for the methods, slot constraints and daemons. The design of the compiler allows the incremental refinement of the class description. Thus, classes can be refined by incremental specification, i.e. it is not necessary to recompile the whole definition of a class. The compilation and code generation phases are separated, and the compiler will attempt to compile all correct classes in one file generating switchable warnings and errors during compilation.

Object data is stored in property lists, and indexed by the name of the object. The Macintosh implementation takes advantage of the existing property management system of LPA-Prolog, avoiding the inefficient asserts and retracts necessary for the updating of object records when these are stored in a Prolog database. AMORE offers rich specification, supports multiple inheritance, before and after methods, constraints and daemons on slots. However, a rich specification comes in expence of performance and therefore a number of techniques have been employed in order to optimise the language. One of the AMORE requirements is to support applications in real time. This requirement is addressed in the implementation as follows. Class information is separated into changing and unchanging parts where unchanging parts can be compiled by Prolog compilers and optimisers. Precedence information is cached and method and daemon precedence lists are filtered to avoid unnecessary searching. Messages inside method definitions are compiled and no overheads are introduced in interpreting them. AMORE internally supports the caching of messages in classes, so that frequently used messages in classes can be compiled to offer better performance.

AMORE has been used in developing the HCI part of RUBRIC which has resulted in a performance withing accepted limits. The language is currently used for the implementation of the rule editor and the run-time environment of RUBRIC.

4. Conclusions

Continuing challenges in the field of information systems development has encouraged researchers to seek solutions which better address the issues of software quality and productivity. Currently a number of different paradigms are under active development c.f. [Hagelstein, 1988; Jarke, 1988]. The RUBRIC paradigm complements these developments by adopting a rule-based approach to the development of information systems. The RUBRIC system architecture is being realised through the AMORE language has been developed by integrating the Logic Programming and Object-Oriented paradigms. The basic functions supported by AMORE are [van Assche et al, 1988]:

- the definition of information requirements
- the manipulation of information elements, including integrity checking and derivation of information elements.

The language has evolved into a flexible and general programming system which addresses the initial RUBRIC requirements but also offers the potential to be used in a variety of disciplines. It integrates all RUBRIC components, offering an enviroment where maintainability and structuring specification is enhanced. The possibility to produce fast models and prototypes was invaluable in a research project where new ideas need to be rapidly tested and evaluated.

The emphasis in the design of AMORE was to extend Prolog rather than form a self contained language, and this has been achieved by introducing a constrained set of new but consistent syntactic constructs to describe the semantics of the language. The design has retained the philosophy of Prolog and in the authors' experience it is possible to understand and effectively use the AMORE concepts very rapidly by someone familiar with the Prolog language.

Acknowledgements

The RUBRIC project is partly funded by the Commission of the European Communities under the ESPRIT R & D programme. The project collaborators are UMIST, James Martin Associates (Netherlands), BIM (Belgium) and IESB (Ireland).

The authors wish to acknowledge the contribution that Paul Layzell, Bill Karakostas and Chris Westrup (UMIST), Geert Speltincx (JMA), Raf Venken and Daniel Donner (BIM) have made to the RUBRIC project.

References

Van Assche, F., Layzell, P.J., Loucopoulos, P., Speltincx, G., (1988a), Information Systems Development : A Rule-Based Approach, Journal of Knowledge Based Systems, September, 1988, pp. 227-234.

van Assche, F., Loucopoulos, P., Speltincx, G., Venken, R., (1988b), Development of Information Systems: A Rule Based Approach, Proceedings IFIP TC2/TC8 Working Conference on 'The Role of

AI in Databases and Information Systems', Canton, China, July, 1988, Kung & Meersman (eds), North Holland, pp. 533-540.

Balzer, R., Cheatham, T.E, Green, C., (1983), Software Technology in the 1990's: Using a New Paradigm, Computer, November 1983, pp. 39-45.

Beckstein, C., Gorz, G., Tielemann, M., (1987), FORK A System for Object and Rule Oriented Programming, Proc. ECOOPS 1987, European Conference on Object Oriented Programming.

Buis, M., Hamer, J., Hosking, J.G., Mugridge, W.B., (1987), An Expert Advisory System for A Fire Safety Code, Applications of Expert Systems, Quinlan J.R. Proc. 2nd Australian Conference held in May 1986, Turing Institute Press, Sydney, 1987.

Bobrow, G.D., Kahn, K., Kiczales, G., Masinter, L., Stefik, M., Zdybel F., (1986), Common Loops: Merging Lisp and Object-Oriented Programming, ACM OOPSLA Conference, 1986.

Boehm, B.W., (1981), Software Engineering Economics, Prentice-Hall, 1981.

Bubenko, J., (1986), Information System Methodologies - A Research View, in 'Information Systems Design Methodologies: Improving the Practice', eds Olle, T.W., Sol, H.G., Verrijn-Stuart, A.A., North-Holland.

Cox, J.B., (1986), Object Oriented Programming : An evolutionary Approach, Addison Wesley Publishing Company 1986.

Chapin, N., (1979), Software Lifecycle, INFOTEC Conf in Structured Software Development.

Chen, P.P., (1976), The Entity-Relationship Model- Towards a Unified View of Data, Transaction on Database Systems, Vol. 1, No. 1, pp. 9-36, 1976.

Cuadrado, J., Cuadrado, C., (1986), AI in computer vision : Framing doors and Windows, BYTE, January 1986.

DeMarco, T. (1978), Structured Analysis and System Specification, Yourdon Press, 1978.

ESP frame engine, Users Guide Expert Systems International Ltd. 1986.

Fjeldstad, R.K. et al, (1979), Application program maintenance, in Parikh & Zveggintzov (1983), pp. 13-27.

Fikes, R., Kehler, T., (1985), The Role of Frame - Based Representation in Reasoning, Communications of the ACM, September 1985, Volume 28, Number 9.

FLEX expert system toolkit, Users guide, Logic Programming Associates Ltd.

Fukunaga K., (1986), An experience with a Prolog Based Object - Oriented - Language, Proc. OOPSALA-86, Portland Oregon, September 1986.

Goldberg, A., Robson, D., (1983), Smalltalk - 80: The language and its Implementation, Addisson-Wesley 1983.

Greenspan, S.J., (1984), Requirements Modelling : A knowledge representation approach to software requirements definition, Computer Systems Group, University of Toronto, Tech. Rep. CSRG-155.

Hagelstein, T., (1988), Declarative Approach to Information Systems Requirements, Knowledge Based Systems, Vol 1, No. 4, September 1988, pp. 211-220.

Iline, H., Kanoui, H., (1987), Extending logic programming to Object Programming: The system LAP, IJCAI 1987 Milano, pp.34-39.

Jackson, M., (1983) System Development, Prentice-Hall.

Jarke, M., Jeusfeld, M, Rose, T, (1988), Modelling Software Processes in a Knowledge Base: The Case of Information Systems, Knowledge Based Systems, Vol 1, No. 4, September 1988, pp. 197-210.

KEE software developement system, User Manual, Intellicorp 1985.

Kent, W., (1983), The Realities of Data: Basic Properties of Data Reconsidered, Proc of thre IFIP WG 2.6 Working Conference on Data Semantics (DS-1), Hasselt, Belgium, 7-11 January, 1985, North-Holland, 1986, pp. 175-188.

Lehman, M., (1980), Programs, Life Cycles and Laws of Software Evolution, Proc. of IEEE, Vol. 8, No. 9, September 1980, pp.1060-1076.

MacDonald, I, (1986), Information Engineering- An improved, automatable methodology for designing data sharing systems, in Olle (1986), pp. 173-224.

Malpas, J., (1987), PROLOG: A relational language and its applications, Prentice Hall International , 1987.

Maddison, R., (1983), Information System Methodologies, Wiley-Heyden.

Mathur, R.N., (1987), Methodology for Business Systems Development, IEEE Transactions on Software Engineering, Vol. SE-13, No. 5, May 1987, pp. 593-601.

Meyer, B., (1988), Object Oriented Software Construction, Prentice Hall International 1988.

Moon, D.A., (1986), Object programming with FLAVORS, in Object Oriented Programming Languages Conference, OOPSLA Proceedings, ACM SIGPLAN Notices (1986) 1-8

Morris, E.P., (1985), Strengths and Weaknesses in Current Large Scale Data Processing Systems, Alvey/BCS SGES workshop, January 1985.

Newton, S. Lee, (1986), Programming with P-Shell, IEEE Expert, Summer 1986.

Nijssen, G.M., Duke, D.J., Twine, S.M., (1988), The Entity-Relationship Data Model Considered Harmful, Proc 6th Symposium on Empirical Foundations of Information and Software Sciences, Atlanta, Georgia, USA, October 1988.

Olle, T.W. et al (eds) (1983), CRIS - Information System Design Methodologies: A Comparative Review, North-Holland.

Olle, T.W. et al (eds) (1986), CRIS3 - Improving the Practice, North-Holland.

Stefik, M., Bobrow, D.G., Mittal, S., Conway, L., (1983), Knowledge programming in LOOPS : Report on an experimental course, Artificial Intelligence 4, No. 3.

Stefik, M. and Bobrow, D., (1986), Object Oriented Programming: Themes and Variations, AI Magazine 6. no 4 1986

Sterling, L. and Shapiro, E., (1986), The Art of Prolog, MIT Press 1986.

Venken, R., Bruynooghe, M., (1984), Prolog Language for prototyping of Information Systems, in Approaches to Prototyping 1984.

Verheijen, G. & van Bekkum, J. (1983) NIAM : An Information Analysis Method, in Olle (1983)

Yoshiyuki, K., (1987), Amalgamating multiple programming paradigms in Prolog, IJCAI 1987 Milano, pp.76-82.

KEY WORDS

Object Oriented, Logic Programming, information systems development, conceptual schema, rule-based specifications, constraint handling.

Project no 1219(967)

PADMAVATI
PARALLEL ASSOCIATIVE DEVELOPMENT MACHINE AS A
VEHICLE FOR ARTIFICIAL INTELLIGENCE

Patricia GUICHARD-JARY
THOMSON-CSF Division CIMSA-SINTRA
160 boulevard de VALMY , B.P 82
92704 COLOMBES -FRANCE-
tel (33) 1.47.60.36.37

This ESPRIT Project, started in 1986, is a cooperative effort between THOMSON-CSF (prime contractor - France) , GEC (United Kingdom), and FIRST (Greece). PROLOGIA (France) and NSL (France) are also involved as sub-contractors.

Its objective is to develop a machine for symbolic and time critical applications, such as image recognition (THOMSON-CSF), natural language (GEC), and parallel expert system (FIRST).

PADMAVATI is a MIMD machine connected to a SUN host. It is composed of 16 Inmos Transputers T800, with a 16Mbyte DRAM. A Content Addressable Memory extends this memory capacity (up to 680kbytes), and is especially used by symbolic languages. A dynamic Network and a static ring connecting the 16 transputers provide non local communications.

As far as languages are concerned, domains investigated are parallelism and associativity in LISP, PROLOG, and C.

1. Introduction

The ESPRIT Project 1219(967) "PADMAVATI" focuses simultaneously on the hardware level, the parallel computational model, and the time critical application level.

At the hardware level, our research had to provide enough performance possibilities, keeping flexibility to support different explicit/implicit computational models.

So, our architecture is based on an array of Transputers (from 8 transputers up to 256 - but the prototype will be composed of 16 processors-), which are interconnected by :

– a dynamic network, of the DELTA topology,
– a static ring, using the tranputers links.

Associatives devices -Content Addressable Memories- have been added to this architecture, for symbolic languages such as LISP and PROLOG. Other associatives techniques (hash-coding for example) are also used in PROLOG. The speed-up obtained by theses features can then be added to that obtained by the parallelism provided by the architecture.

Several parallel computational models have been studied for the languages PROLOG and LISP : one is called "macro-parallelism" and define parallelism between static tasks, the other model is called "micro-parallelism", it is only used in Prolog and define parallelism between goals dynamically created during the resolution.

Applications studied in this project are: image recognition, natural language understanding, and parallel expert system.

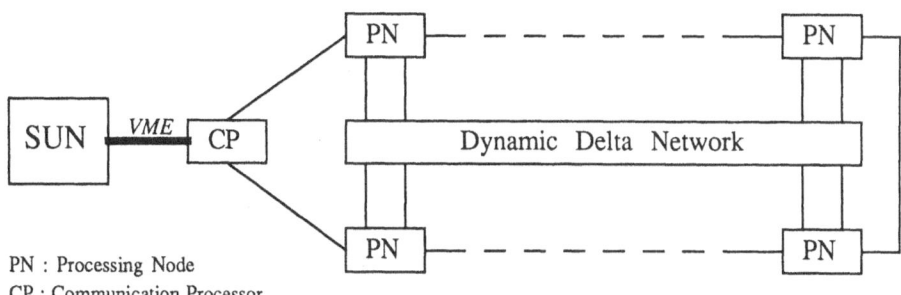

PN : Processing Node
CP : Communication Processor

Figure 1: PADMAVATI basic architecture

This paper will present our hardware architecture, and then the software one. However, the application parts will not be related.

2. Hardware Architecture

2.1. Processors.

PADMAVATI is a multi-processor machine with distributed memory, each processor providing fast context switching (about $1\mu s$) and synchronisation in order to support parallelism. Considering these features, the transputer is an adequate processor to build such an architecture. We chose the T800 transputer, as it is presently the higher performance transputer.

So, a PADMAVATI processor is composed of a transputer T800 with its 4Kbyte on-chip memory, and from 4M to 16Mbyte of DRAM. It is also possible to connect an associative Memory on the memory bus extension, developped in the project.

However, there are two particular processors on this machine. One is the "Communication Processor -CP-" used for communications between PADMAVATI and the host machine. It is also a transputer. The other one is obviously the host processor. Presently, this latter is a SUN with a 680x0 processor.

Henceforth, we will speak about Processing Node, instead of Processor. The SUN and each transputer of PADMAVATI are considered as processing node because they can run user's processes. Except the CP which only runs part of the system managing the communication between SUN and PADMAVATI.

492

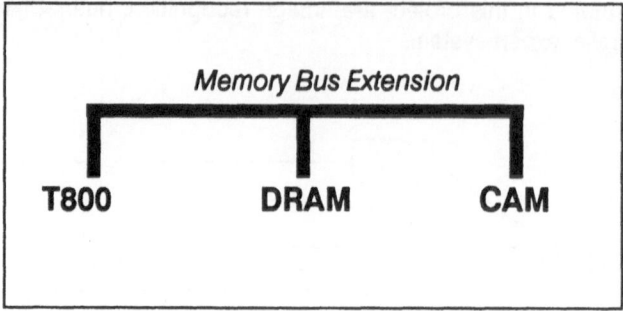

Figure 2: A PADMAVATI Processing Node

2.2. Processors Inter-connection.

Two different networks connect the nodes.
The first one is the simplest: a processor ring, using the transputer links. This network will be used for several purposes:

- to load the executive on each processor
- to load the application on the processors
- to debug applications
- to optimize communication between two neighbouring processors

The second one is a dynamic network belonging to the DELTA network family.

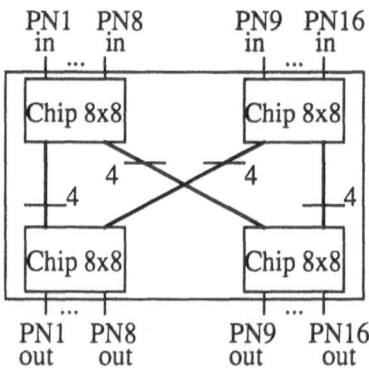

Figure 3: Dynamic DELTA network

Dynamic means that it is able to connect at run time every PADMAVATI node to every other, just using the number of the destination node. The connection is then like any other transputer link connection. The only difference is that the network does not get advantage of the bi-directionality of a link (for network complexity reasons) and uses two transputer links: one is used for input messages, and the other is used for output messages.

This network is extensible, connecting from 8 to 64 transputers. It is composed of 8x8 VLSI chips especially developed for the project.

Functionnally, it uses the path switching protocol with self contention resolution. This protocol does not have a big time overhead (a path needs about 50ns to be established), and does not need queue management which the packet switching protocol needs.

2.3. Associative Memory Extension

Content Addressable Memory (C.A.M) or Associative Memory [1] not only stores data, but can also access that data in parallel. Each bit of storage can compare the value it contains with a broadcast value (the search key). Bits are arranged in words and if the whole word matches (all its bits match the corresponding bits in the search key) then a flag bit associated with the word is set. The flags can then be used to select which words take part in subsequent operations. For instance it is possible to write the same value to all flagged words simultaneously.

Thus, CAM is accessed by specifying a value instead of a location. This would be of limited use if it was only possible to tell whether any data matched the search key or not, but it is also possible to retrieve data associated with matching data. This can be done with a partially specified search key where some bits are masked or "don't care". Alternatively, associated data can be stored in consecutive words thus allowing access by a combination of matching and relative location.

PADMAVATI CAM is organised hierarchically. A CAM array consists of a number (1-16) of cascaded CAM boards. Each CAM board contains a number (6-8) of CAM modules, and each CAM module contains (10-12) CAM chips. Each CAM chip contains 148 words, and so the maximum size of a CAM array is 227328 words [2].

Interest of such a memory is the seed-up of symbolic languages. The way of using the CAM in LISP and PROLOG will be described in the next chapter.

3. Software Architecture

Our objective is to provide the user with a system which can run parallel applications with parts written in different languages and loaded on different nodes that can be transputers or the host (see 2.1).

The PADMAVATI Run Time System (RTS) provides primitives for interprocess communications, and also for memory management and processes management.

3.1. A Model for Communicating.

Our model is based on communication by message passing. We define a message as: data (with variable size) + a priority + a tag. So, a message reception can be selective, depending on the given priority, and/or the given tag.

Communications may be synchronous or asynchronous.

The destination of a message is a port. A port is identified by the user with a name, and by the system with a (node number, local address) pair.

A port can be defined as a "Mail-Box", to which some processes can send messages, and from which others can read the messages. Thus, processes send and receive messages on named ports, and don't have to know the physical location of their communicating processes.

A centralised "Name Server" records the names of ports and their location on PADMA-VATI nodes. RTS (distributed on each node) manages creation of ports and, sending, reception of messages.

This model allows non-local as well as local communications.

The transputer communication model provides synchronous communications using channels. Our system also allows the use of transputer channels for local communications.

3.2. Memory Management.

The RTS also provides primitives for memory allocation (MALLOC, and FREE), and primitives for on-chip memory allocation. This latter possibility is very interesting for time critical applications.

3.3. Process Management.

Different aspects of process management are included in this part.

First of all is dynamic process creation, which is possible owing to dynamic memory allocation, and transputer instructions such as: STARTP, RUNP (to start a process) and ENDP, STOPP (to stop a process).

The other aspect is event management, such as semaphores, with adequate primitives to suspend and then wake-up processes.

4. Languages for A.I.

Languages for Artificial Intelligence developped in this project are PROLOG and LISP. These languages are studied according to the hardware components, the hardware architecture, and the system of the PADMAVATI machine:

- transputer
- associative memory (Content Addressable Memory)
- multi-processor machine with distributed memory
- message-passing
- synchronous, asynchronous communication
- dynamic process management

For each language, an interpreter and a compiler are implemented, running on the transputer node: interpreter to get an interactive interface, compiler producing transputer code to run applications fast.

The host been considered as a PADMAVATI processing node, particular implementations (semi-compiler, interpreteror simulator) for LISP and PROLOG must run on the SUN.

First, the PADMAVATI PROLOG system will be presented, and then the LISP one.

4.1. PROLOG

We propose to the user a PROLOG with BSI syntax and built-in predicates [3] [4].
Different ways to optimize the Prolog system on PADMAVATI have been studied:

4.1.1. Compilation

Objective is to take advantage as much as possible of the transputer by compiling high level languages into transputer code.

PADMAVATI's PROLOG compiler produces Warren Abstract Machine (WAM) code [5] [6], which is then either transformed into transputer code to run on PADMAVATI (whole compilation), or interpreted to run on the host (semi-compilation).

4.1.2. Parallelism

The second way to optimize Prolog is to use as well as possible the PADMAVATI architecture.

Parallelism in PROLOG has been investigated through two systems: one is based on message passing, the other one uses a fine grain parallelism.

PROLOG with message passing:

We can characterize this parallelism as a coarse grain parallelism: an application is divided into big tasks, that can run simultaneously, and sometimes exchange messages - data, or goals to solve -.

This model is built on top of the RTS model defined in chapter III. So, built-in predicates have been added for port manipulation (creation, message passing: message writting, message reading).

Division of application into big tasks implies a static process allocation for these tasks. But we are considering a dynamic process allocation (to solve a goal sent in a message for instance).

A fine grain parallel PROLOG :

This model proposes parallelism within a PROLOG clause, instead of big tasks [7].

On one hand, we have explicit AND-parallelism, annotated by "&&" construct, and allowing variable sharing.

example: P(...) :- Q(...) && R(...).

On the other hand is OR-parallelism, which consists of the resolution of the same goal with all the clauses defined for the corresponding predicate. It will be also an explicit parallelism, with annotated packets.

This dynamic parallelism implies also dynamic process creation, for each parallel sub-goal resolution.

First measurements of AND-parallel FIBONACCI program, running on a four transputer ring, (OR-parallelism is not yet implemented) shows that speed-up is almost linear with the number of processors.

As a conclusion on parallelism in PROLOG, we can say that these two systems are not incompatible and may constitute only one system at the end of the project: PROLOG concurrent tasks may use also a fine grain parallelism within their resolution.

4.1.3. Optimisation of clauses selection

In this way of optimizing Prolog, we especially use the associative features of PADMA-VATI.

Efficient clause selection can avoid useless unifications and some choice points creations. Two methods have been studied, one is a software associative method to access a clause (Hash-coding) [8], the other one uses the Content Addressable Memory (Filtering).

The clause selection mecanism is divided into two parts: one part is static (during clauses analysis) and consists of computing all clauses heads of a PROLOG program in order to put them in memory, and the second part is dynamic (during resolution) and is a direct access to a clause candidate for the resolution of the current goal.

Each clause selection method implies extraction of relevant information (a PROFILE) from a PROLOG term (clause head or goal). A profile summarizes the structure of a term and the nature of its arguments.

Example: term = P([1,2|X],a)
                profile = (P,(list,1),(atom,a))

According to statistics done on 40 PROLOG programs [9], it appears that most of head predicates have less than 3 arguments (average number is 2.35). So, our system takes into account only 3 arguments in a head to compute a profile (Measures done on several Prolog programs running on a mono-transputer Prolog interpreter, show that selection on three arguments is better, in term of avoided fails, but it is not always better in term of speed-up [10]).

HASH-CODING

In the static phase, the hash-coding method hashes clause head profiles into addresses, and then stores at the computed addresses the corresponding clauses (in fact, it is not the whole clause which is stored, but just its address).

During the dynamic phase, the profile of the current goal is extracted, then hashed (by the same function that hashed clause head profiles), and the first candidate clause is then accessed directly.

This method has been implemented in a PROLOG interpreter running on one transputer during the first two years of the project [11], and it will soon be implemented in the compiler. Results obtained are quite important: a total resolution runs 3 times to 17 times faster than the same interpreter without associative clause selection [10]. A drawback of this method is that, according to the number of arguments taken into account, memory overhead may be more than 40% !

FILTERING

This second method is based on filtering (comparison in parallel between profile of the current goal and all clauses profiles stored in the CAM). During the static phase, all profiles and addresses of the corresponding clauses are recorded in CAM. During the dynamic phase, the profile of the current goal is searched for simultaneously in all the CAM words, and all the matching entries are "flagged".

This method is also promising. A small example runs almost five times faster on an interpreter with CAM than on a classical one. Larger programs promise even greater speed-ups. [12]

We have also implemented filtering on RAM: clauses profiles are stored in RAM, and comparison between profile of the current goal and clauses profiles is sequential instead of parallel. We observed an average speed-up of 3.

To conclude on such associative methods for clause selection, we can say that the most important point is that they both provide constant access time, at the dynamic phase, whatever the number of accessed clauses. Thus, they are all increasingly interesting as the number of clauses for the same predicate in a PROLOG program increases.

4.1.4. Optimization of Unification

In addition to clause selection, CAM can be used by unification.

Unification is the kernel of a PROLOG system. It basically does matching between two terms, and binds variables when it is possible. Values bound to variables are often used during further unifications. So, it could be interesting to optimize this stage. The CAM will be used for this.

A variable binding is characterized by :

- binding time :resolution step at which the binding has been created
- variable number: composed of the variable number in the clause and the step number
  (= resolution step at which the clause containing the variable has been used).
- value

Our study is based on [13] [14], in which promising results are presented (average speed-up of 4).

4.1.5. Intelligent Back-tracking [7] [15]

This study is an enhancement of the Prolog language, and consequently does not exploit the underlying architecture.

The "intelligent back-tracking" method allows back-tracking to a choice point which is the most likely reason for failure (this reason for failure is determined by using a "dependence graph", which record dependences between variables and clauses in which these variables have been bound), instead of taking the last choice point as it is usually done. The resulting speed-up is important, but PROLOG semantic change a little because of side effects.

4.1.6. C Predicates and functions

It is sometimes difficult, and not efficient, to write predicates in PROLOG. So, our system offers the possibility to define predicates and functions in C, and then use them as "built-in" predicates.

4.2. LISP

LE_LISP, from INRIA [16] [17], has been choosen to be implemented on transputer, and extended with parallel primitives.

Domains especially investigated in LE_LISP are :

4.2.1. Compilation

The LE_LISP Compiler produces an intermediate virtual machine code, which is then transformed into transputer code to run on PADMAVATI.

4.2.2. Parallelism

The computational model is based on the RTS model, and uses a coarse grain parallelism with message passing. A LE_LISP application is then divided into concurrent LE_LISP processes, statically allocated on the processing nodes.

LE_LISP is then extended with primitives to manipulate ports (creation, message writing, message reading).

4.2.3. Use of CAM

Different ways to use the CAM in LE_LISP could be envisaged :

- storage of association lists, and property lists: obviously, an associative access will be better than a sequential one!
- memo-functions
- direct user access via special commands.

In the project, only the third possibilty will be implemented.

4.3. Conclusion on A.I. languages

C has not yet been discussed, but it will run on the machine. Access to message-passing primitives, will be possible through C librairies. Thus, a C program may be parallel. In fact, most of the system software is written in C and it is also used for the language implementations.

The parallel computational models of PROLOG, LE_LISP, and C are the same, based on message passing using ports. So, these three languages can communicate.

5. Conclusion

PADMAVATI is a multi-processor machine with many interesting features.
On the hardware level:

- The transputer is a powerful processor, dedicated to parallelism
- The dynamic Network allows all kinds of distributed applications
- Associative memory, is very usefull for symbolic languages
- The hardware architecture is very simple, and is extensible

On the software level:

- A.I. languages take advantage of the hardware components efficiently.
- Processes are independent of their location on processing node, this removes the need for the user to consider process placement during implementation.

The PADMAVATI prototype, composed of 16 Processing Nodes, will be available in mid 1990. Then, we shall be able to measure the exact performances of the machine and the software systems implemented on. Applications on image recognition, natural language understanding and partial differential equations processing, will be good tools for test.

References

[1] Kohonen T., "Content Addressable Memories" Springer series in Information sciences, 1980

[2] Howe D., "Padmavati CAM chip functional specification" Internal report REP-14, July 1989

[3] "Prolog syntax: draft 5.0 and comments" Report BSI PS/239 (or ISO N9), march 1988

[4] "Prolog built in predicates" Report ISO N28

[5] Warren D.H, "Implementing Prolog" D.A.I research reports n 39, 40, May 1977

[6] Warren D.H, "An abstract Prolog Instruction Set" Technical Note n309, SRI International, Menlo Park,1983

[7] Bodeveix J.P, "LOGARITHM: Un modèle de Prolog Parallèle. Son implémentation sur transputers" Thèse Docteur en Sciences. Université Paris-Sud. Janvier 1989

[8] Knott G.D., "Hashing Functions" The computer journal, vol 18, n1, pages 38-44, 1975

[9] Py M., "Static measures on Prolog II programs" Padmavati internal report, April 1987.

[10] Jary P., De Joybert X., "Selection de Clauses en Prolog" Actes du 8ème Séminaire Programmation en Logique, Trégastel,24-26 mai 1989

[11] De Joybert X., Jary P., Pressigout P., Battini F. "LOGICS : A prolog interpreter + Hash-coded memory" Annex 1 to the PADMAVATI 4th six month report, March 1988

[12] Howe D., "CAM for Prolog clause access" Padmavati internal presentation. September 1988

[13] Oldfield J.V, Storman C.D, Brule M.R "The application of VLSI Content-Addessable Memories to the acceleration of Logic Programming Systems" IEEE 1987

[14] Oldfield J.V, Storman C.D, Brule M.R, RIBEIRO J.C "An architecture based on Content Addressable memory for rapid execution of Prolog"

[15] Bruynooghe M., Pereira L.M "Deduction Revision by Intelligent Backtracking" Implementation of Prolog, ed. Campbell, 1984

[16] Chailloux J.,Devin M., Hullot J. "LeLisp a portable and efficient Lisp system" Proc. of the 1984 ACM symposium on Lisp and Functional Programming, Austin Taxas, August 1984

[17] Chailloux J., "Manuel de Reference LeLisp 15.2" 3ème Edition , Novembre 1986

Padmavati Consortium Members :

THOMSON-CSF Division Cimsa Sintra: 160 boulevard de Valmy B.P 82 92704 COLOMBES FRANCE -	Prime Contractor	PROLOGIA: 70 route Léon Lachamp Case 919, Luminy 13228 MARSEILLE cedex 09 FRANCE	Sub-Contractor
GEC hirst Research Centre: East Lane WEMBLEY Middlesex HA9 7PP - UK -	Partner	NSL: 57-59 rue Lhomond 75005 PARIS FRANCE	Sub-Contractor
FIRST Int. Ltd: P.O Box 1340 26110 PATRAS - GREECE -	Partner	THOMSON-CSF LER: Avenue de Belle Fontaine 35510 CESSON-SEVIGNE FRANCE	Sub-Contractor

KEYWORDS

Associative, Delta-Network, Distributed, Dynamic, LeLisp, Parallelism, MIMD, Port, Prolog, Transputer.

PROJECT NO. 1098: KBSM

THE KADS KNOWLEDGE BASED SYSTEM METHODOLOGY

Marco de Alberdi and John Cheesman
STC Technology Ltd
London Road
Harlow, Essex
CM17 9NA
U.K.   Tel: + 44 279 29531

ABSTRACT. The Knowledge Analysis and Design Structured (KADS) methodology produced by P1098 is a comprehensive, sound methodology for the commercial development of Knowledge Based Systems. This paper provides an overview of the project structure showing the objectives and interworking of the theoretical, consolidation and experimental streams of work, and focuses on some key results of the KADS process. Some of the development and user benefits of this rigorous modelling approach are highlighted, together with their effects on company strategies. Some early exploitations of the methodology are presented. We conclude that KADS provides a basis for developing complex knowledge based system applications enabling European industry to obtain competitive advantage from new information technology.

1. INTRODUCTION

P1098 started in late 1985 with the principal objective of producing a comprehensive, sound methodology for the commercial development of Knowledge Based Systems. Now four and a quarter years later, we believe that our results and achievements suggest we have made very substantial progress towards that goal, having prototyped and tested the KBS Analysis and Design Structured (KADS) methodology:

- We have developed a generic methodology with a firm theoretical basis, consisting of a range of methods and techniques to cover the Analysis and Design of Knowledge Based Systems

- We have developed a workbench incorporating the methods and techniques to support Analysis and Design activities

- We have tested and explored the methodology in a variety of situations, including the development of large and complex commercial systems which embed KBS

2. OVERVIEW OF THE PROJECT

There are six partners in the second phase of the P1098 consortium. These are STC Technology Ltd., University of Amsterdam, CAP SESA Innovation, NTE Neu Tech, SD Scicon and Touche Ross Management Consultants. (Partners in phase 1 of the consortium were STC IDEC Ltd, University of Amsterdam, CAP Sogeti Innovation, Scicon Ltd, The Polytechnic of the South Bank and SCS Organisationsberatung und Informationstechnik GmbH.). We have been working closely together on the KADS project and we have structured our work into three streams of activity:

The theoretical stream; taking results from software engineering and Artificial Intelligence research and deriving the methods and techniques.

The consolidation stream; packaging the methods and techniques and providing tools, including a software workbench.

The experimental stream; using the methods, techniques and tools to address KBS development in different contexts (countries, tasks, domains, and commercial market segments).

The development of the methodology is thus derived from a solid theoretical basis and is validated against real problems to assess feasibility and commercial viability. This sequentially staged model of methodology development is potentially extensive in terms of elapsed time, and runs the risk of the "knock on" effects of slippage. This problem has been largely avoided with monthly technical and management meetings and six monthly project technical weeks, where the whole project team meets for presentations, workshops and reviews. Lateral communication has been vital. We suggest that impact in this project on the methodology development has stronger positive correlations with the number and frequency of working papers than with the overall effort spent on a task.

The main project results are:

- A model based paradigm of the commercial development process for KBS

- Practical methodological support for analysis and design activities

- A Knowledge Engineering workbench known as Shelley.

3. AN OVERVIEW OF THE KADS DEVELOPMENT PROCESS

The KADS view of a Knowledge Based systems development is that of a sequence of modelling activities. The process involves translating the problems or current situation into an abstract form, mapping that abstract representation into a system environment and finally refining that model and reducing the level of abstraction, resulting in a physical artefact, that is, a real computer system. (See figure 1.).

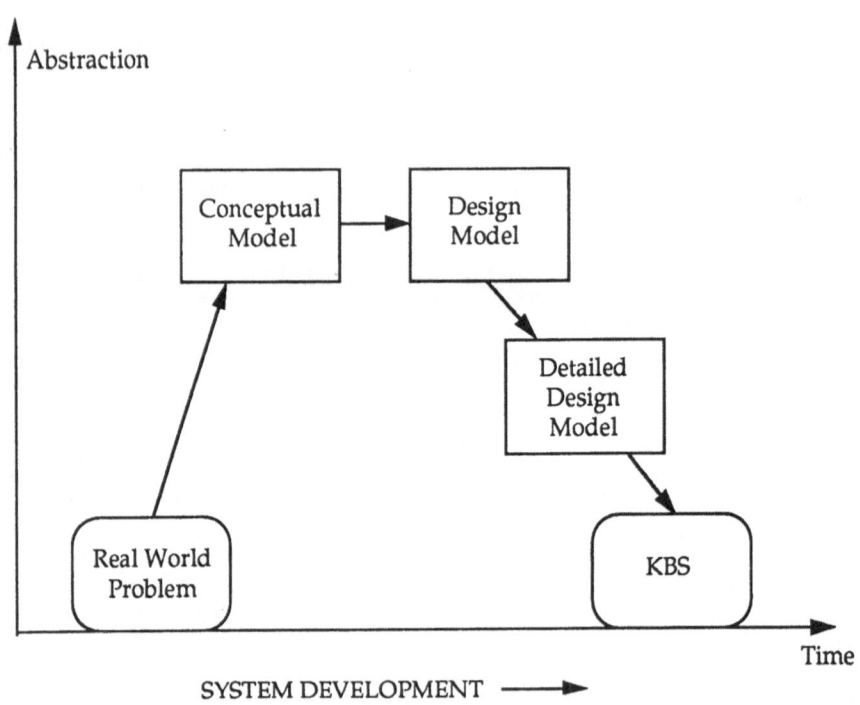

Figure 1: KBSM - The Modelling Paradigm

Three models are involved. The first of these is the model of expertise or conceptual model. This is an abstract description of the problem solving process as exhibited by experts in the real world. The conceptual model is the major outcome of the KBS stream of a requirements analysis, and is discussed in more detail in the next section. The emphasis of the conceptual model is on capturing and representing the expertise which exists in the real world and it is essentially neutral with respect to design and implementation considerations.

The second model is the design model. This is at the same level of abstraction as the conceptual model but it represents the structure and problem solving scope of the proposed system. This is usually a subset of the real world problem solving process, produced by considering the conceptual model together with the external requirements from the requirements analysis stage, such as user, system and environmental constraints. The design model effectively represents a high-level design specification for the final system.

The third and final model in the process is the detailed design model. It is a transformation of the design model to a lower level of abstraction giving rise to a physical design structure for the resultant system. This then acts as an input to implementation leading to actual code and a real system.

4. THE CONCEPTUAL MODEL

Much of our work in P1098 has been concerned with developing and experimenting with this four layer knowledge model, from it's creation from real world data to its transformation into code. This approach creates some interesting possibilities and opportunities when combined with a configurable methodology. The KADS methodology is essentially a framework and a set of techniques that can be configured into a set of activities and outputs to achieve a particular goal. This is in effect another way of saying that we can apply the four layer model approach to developing knowledge based systems in a number of ways.

The conceptual model of expertise [Wielinga & Breuker 1986] is a key component of the whole KADS process, being used to drive the original requirements study and acting as a framework for knowledge modelling in the subsequent design and implementation stages.

The conceptual model distinguishes knowledge at four distinct levels which are described below. Each level is explained using examples from the domain of strategic marketing as studied during an STC experiment which formed part of the KADS experimental stream [Campbell et al 1989].

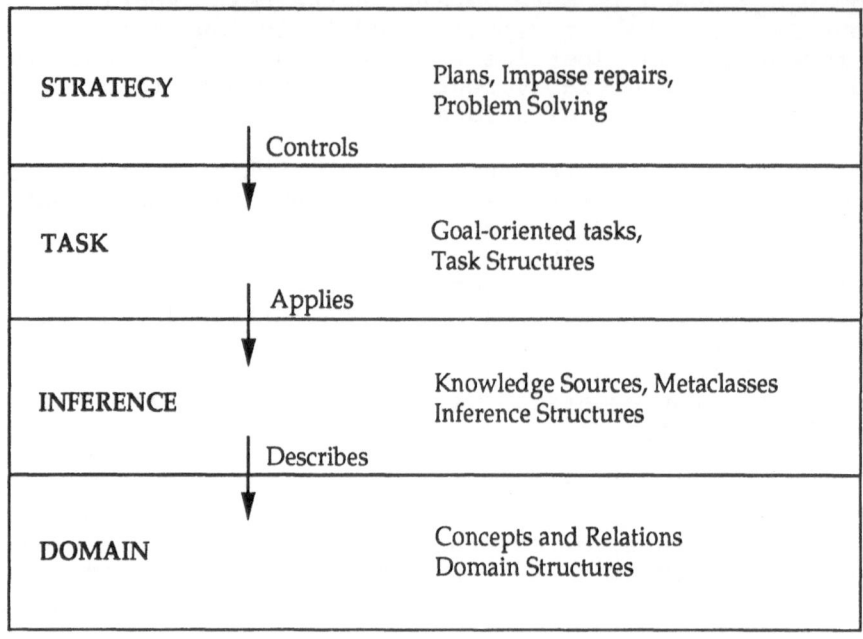

Figure 2: 4 Layer Conceptual Model Overview

4.1   Static Domain Layer

This category of knowledge details the domain dependent concepts which are used, how they are structured and what relationships exist between them. It is static in the sense that it is independent of the way in which it is used.

In marketing, concepts and relations may exist at a variety of levels of complexity. Relatively simple concepts like products, customers, outlets and competitors, and relations between these such as sales and profit can be combined into complex concepts such as outlet performance or outlet catchment areas. These can be expressed quantitatively in a variety of ways. For instance, outlet performance could be defined in terms of outlet profit/time, outlet turnover/ time or outlet turnover/outlet size.

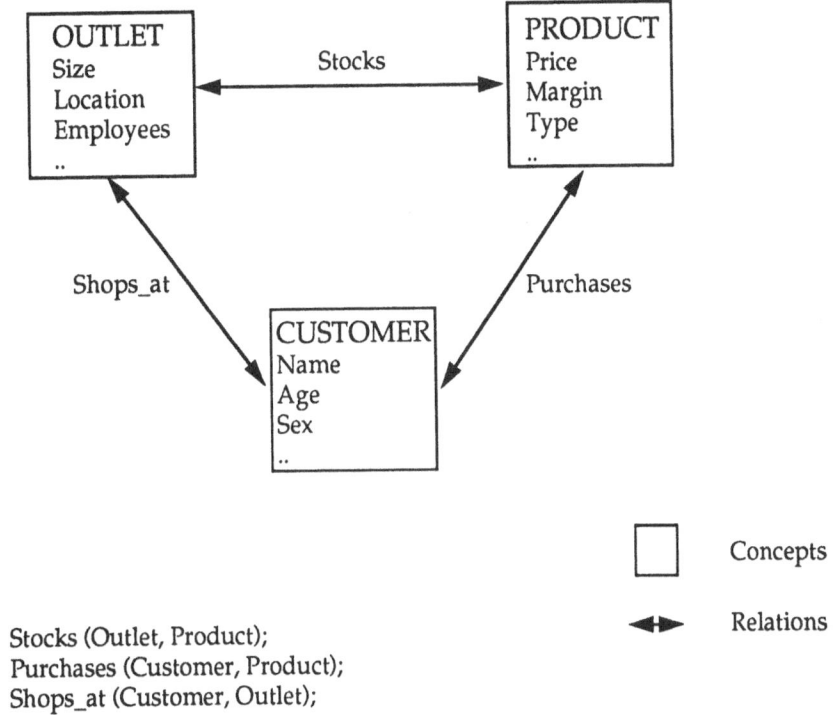

Stocks (Outlet, Product);
Purchases (Customer, Product);
Shops_at (Customer, Outlet);

Figure 3: Domain Layer extract

These domain specific concepts and relations are captured and represented separately and independently of any consideration of inference, process or tasks over them. This not only allows a clear separation of concerns but it also very conceptually clean.

4.2 Inference Layer

Knowledge in this layer consists of a description of the potential inferences which exist between groupings of entities in the domain layer. These groupings are referred to as metaclasses. They represent the role particular aspects of the domain layer play in an inference. In this sense, metaclasses can be seen as describing or classifying sections of the domain layer from the point of view of a particular inference. The way a particular inference is made is described in this layer in terms of a knowledge source which is essentially a transformation process from one or more metaclasses to another, single metaclass.

An example of a marketing inference would be the abstraction of some key attributes of a particular concept class, such as outlet to form a simpler performance indicant class in terms of those attributes.

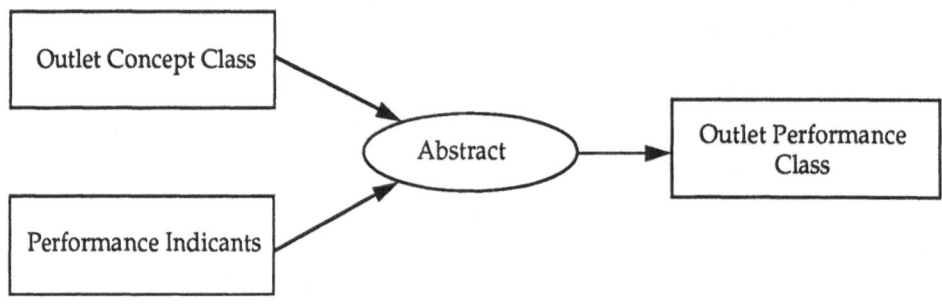

Figure 4:   Inference Example

This is effectively a focusing of attention by considering only some of the attributes as contributing to the concept of outlet performance, such as a sales, size, number of employees and so on.

In general, metaclasses are involved in a variety of inferences. By representing a set of metaclasses and inferences together in a single diagram a structure is formed. In KADS terminology this is referred to as an inference structure: (see Figure 5).

It is important to note that the arrows on the diagrams represent the direction of each inference. However, they do not indicate the ordering of inferences and hence the inference structure is only a representation of convenience.

An inference layer specification includes not only the diagrammatic inference structures but also a formal description of each of the knowledge sources and metaclasses, identifying both the inference method and the mapping onto the domain concepts respectively.

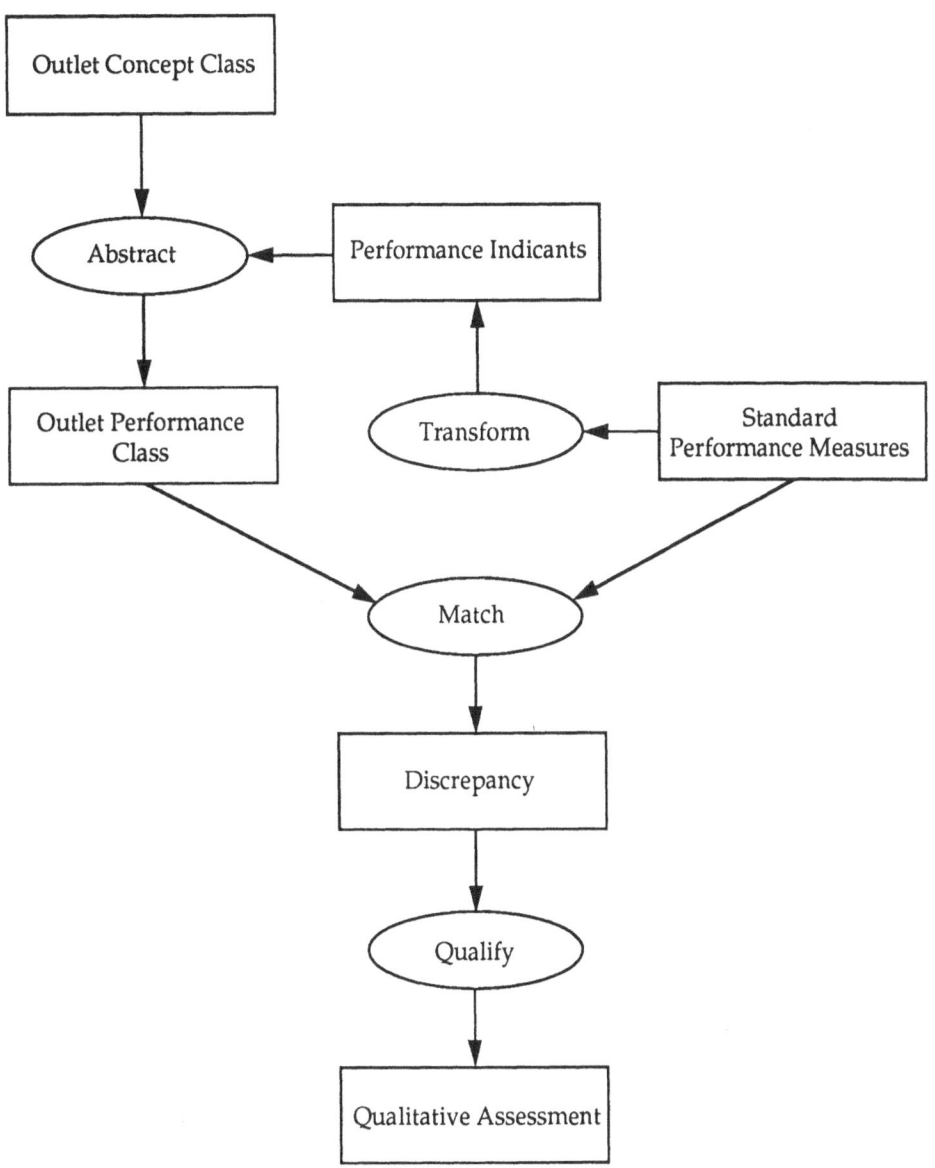

Figure 5:   Inference Structure

4.3 Task Layer

The task layer details how the inferences referenced in the inference layer may be ordered and sequenced to achieve a particular goal. In this way the task layer can be seen as applying the inference layer. Each individual ordering is referred to as a task structure. The knowledge in the task layer usually consists of a number of these task structures.

An example of a task involving inferences specified in the inference structure in figure 4 would be an 'Assess outlet performance" task. Such tasks may be expressed in a variety of ways, but formal process definition type specifications are usually used.

Task "Assess Outlet Performance":

```
Transform (Standard-Performance-Measures,
           Performance-indicants);
Abstract  (Outlet-Concept-class,
           Performance-indicants,
           Outlet-Performance-class);
Match     (Outlet-Performance-class,
           Standard-Performance-Measures,
           Discrepancy-class);
Qualify   (Discrepancy-class, Decision-class);
```

For example, an assessment of the performance of a particular outlet may begin with a turnover to square footage ratio indicating standard performance. The performance indicants would be the key attributes of outlet turnover and square footage which would then be abstracted from an actual outlet description and the ratio would be formed. The two ratios may then be matched to provide a discrepancy such as a percentage difference. This could then be qualified as being above average performance or "normal" for instance.

4.4 Strategic Layer

This layer corresponds to the capturing of control knowledge in terms of controlling the application and sequencing of specific tasks and is generally more loosely defined. This gives flexible movement between problem solving tasks allowing for the avoidance or repair of impasses and the use of the appropriate task at each stage in the problem solving process.

The strategic layer may often be omitted from the scope of system behaviour in the design model, leaving strategic knowledge still on the human side of the joint human/ computer problem solving process. The implication of this is that the human computer interface (HCI) design is strongly related to strategic layer issues. The mapping between HCI design and the strategic layer, combined with the flexible, controlling role of the strategic layer in the conceptual model, gives rise to open, flexible and interactive Knowledge Based systems.

5. SOME BENEFITS OF THE KADS METHODOLOGY.

In principle, a simple knowledge based system developed using KADS for a well formulated, fairly restricted domain will exhibit behaviour very similar, or even identical to a knowledge based system developed without KADS. Although, we would still argue that the KADS version would probably be easier to maintain and enhance, because of the clear separations in it's knowledge structure. In particular, domain specific knowledge is represented in the domain layer whereas the higher level layers are in some sense domain independent. This allows the same tasks and inferences to be applied to new domain by simply changing the domain layer. This is a major benefit of the layered approach.

More interesting, however, is the opportunity to experiment with KADS to explore the development of a complex system for a difficult domain. Knowledge Based Systems traditionally become very difficult to build when there is more than one expert, when no one expert's expertise seems to cover the whole of the problem area, when the problem area appears diffuse and non-formulated with seemingly open ended tasks, where there is a lot of ambiguity in the terminology of the problem area, and finally, when the system to be developed is generic, rather then bespoke or developed to suit a single organisation. Traditional wisdom has it that these are all good reasons for turning down a knowledge based systems project.

Unfortunately, many real-world problem domains appear to be difficult and unstructured and so traditional development wisdom rejects the opportunity to develop some potentially high value solutions. In one of our experiments we have built a generic model of the expertise of strategic marketing in the retail industry and translated this into a high level design that can be mapped onto a number of different system configurations and components. The KADS design decomposition of the reasoning or problem solving can be combined with other methods to develop a system with embedded knowledge. [Campbell et al. 1989].

This generic approach, addressing the marketing function across the retail market segment has a number of significant advantages deriving from the rigorous modelling approach. The generic conceptual model provides the reference framework for requirements analysis for each implementation. We know both what needs to be known and where that knowledge fits. The conceptual model, also, provides a framework to transfer that knowledge to the immediate project team and other teams. KADS also promises a high degree of design and software reusability when applied to generic problems. Indeed, we have found the conceptual model we developed for retailers to be applicable to other market segments. KADS has had a major impact on the product strategy of the company concerned. This generic approach has potential benefits for the user organisations. It promises to provide affordable information technology solutions to real and significant problems. In the specific instance of marketing systems the approach spreads scarce marketing planning and analysis expertise enabling more users to gain a competitive edge through marketing. A different approach to generic models is being pursued in another of our experiments [Chon & Streng 1989]. Here, a synthetic generic model of continuous process monitoring is being developed. This aims at re-using design components for a specific class of problem rather than a vertical market segment like retailing. Although, of course, selecting a particular class of problem is really another way of segmenting a market. This represents a different type of company market strategy.

We will have to wait to see whether the early promise of these generic applications of KADS meets the KADS criterion of "commercial development". This properly belongs outside the province of the P1098 project as a commercial exploitation issue. However, there are several early indications that there is much from the project that is exploitable.

6. EXPLOITATION OF RESULTS

Companies both inside and outside the consortium are taking up KADS or supplementing their own methodologies with KADS. These include a number of major consultancies and systems developers. Training courses in KADS have been run in Holland, Germany, Sweden and the U.K. Also, in the U.K., three of the consortium partners have been participating in the government sponsored GEMINI initiative, to develop a fully integrated system development method which will make use of KADS concepts and methods.

Consortium partners have also collaborated in licencing a project software spinoff, PCE, the Programmable Computing Environment. This is a user-interface management system used in P1098 both for user-system interface prototyping and as the user interface to the Shelley workbench that supports the KADS methodology. This was licenced to Quintus in the U.S., who productised it and released it as PRO-WINDOWS. Heads of agreement to licence PCE to other companies have also been signed.

There seems to be a growing awareness of KADS in the community and it is now very much in the public domain. The project has published a lot of its work both in books and conference papers and these have aroused a lot of interest. KADS is a young methodology and it is too soon to tell whether we have achieved the objective of producing a comprehensive, sound methodology for the commercial development of Knowledge Based Systems. However we believe that P1098 has made very significant progress, not least in extending our capability to deal with more complex problems in difficult domains. We hold that we have achieved the development of methodological support for knowledge based systems for the commercial environment and that this will enable European industries to achieve a competitive edge through the use of new information technology.

7. KEY WORD INDEX

KADS, Conceptual Model, Shelley, KBS Methodology, Software Engineering.

8. CURRENT CONSORTIUM MEMBERS

STC Technology Ltd.
University of Amsterdam
CAP SESA Innovation
NTE Neu Tech
SD Scison
Touche Ross Management Consultants

REFERENCES

Campbell, H.M., Cheesman, J.M., de Alberdi, M,. Hesketh, P, Lesan, H, Lewis, A.E. and Tansley, D.S.W. (1989). 'Strategic Marketing Information System' F8 System Documentation: ESPRIT project P1098, Deliverable E81, STC Technology Ltd.

Wielinga, B. and Breuker, J. (1986). 'Models of Expertise', ECAI '86 Proceedings, pp 306 - 318.

Chon, Y. and Streng, K. (1989). 'Reusable Analysis Model for Process Monitoring and Preventive Maintenance in Manufacturing Environments': F10 Working Paper: ESPRIT project P1098, NTE.

Project no 1219(1106)

Technical Diagnosis Based on Numerical Models Using PROLOG III

Thomas JOST
Robert Bosch GmbH,
Geschäftsbereich
Industrieausrüstung
Prüftechnik
Franz-Oechsle-Str. 4
Postfach 1129
7310 Plochingen
West-Germany
Phone: +49-7153/66-529

Reinhard SKUPPIN
FAW - Ulm
Helmholtzstr. 16
Postfach 2060
7900 Ulm
West-Germany
Phone: +49-731/501-413
earn: SKUPPIN@DULFAW1A

Summary

A technical diagnosis system is under development in order to validate the constraint logic programming language PROLOG III. Based on technical models already introduced earlier the paper outlines the formalisms that can be used to represent these models in PROLOG III. Since technical systems can be largely described by measurable quantities, emphasis is laid on the use of numerical constraints, one of the main features that extends PROLOG III over current "standard" PROLOG implementations. After evaluating experiences gained with a formal, but straightforward approach, the representation is refined to account for runtime requirements of an appropriate inference strategy.

1. Introduction

The use of constraint solving techniques is pursued by different research groups (see [2] or [4]) to enlarge the power of the PROLOG programming language. Admission of constraints on various data types enhances the expressive power of the language, improves run-time performance, and opens fields of application that could previously be treated in PROLOG only with difficulty, for example problems containing numerical data.

The main objective of the ESPRIT Project "Further Development of PROLOG and its Validation by KBS in Technical Areas" (P1219/1106) is the development of the constraint logic programming language PROLOG III (see [1] and [2]). The essential features that extend PROLOG III above current "standard" PROLOG implementations are:
- the use of linear numerical (in)equalities over rational numbers,
- the full treatment of Boolean Algebra,
- direct list concatenation.

This work is validated through the use of PROLOG III in an application oriented domain; we have chosen the field of vehicle engine diagnosis. In the first project phase, when PROLOG III was not yet available, a pilot system was developed in the PROLOG II language [5] that covers this field of application by means of a knowledge base which reflects the

1.The FAW - Institute at Ulm is a subcontractor of Daimler-Benz AG, 7000 Stuttgart, West-Germany

514

expertise of experienced service engineers. A suitable inference engine together with a rather simple user interface complete this system [7]. The experience gained with the development of this system lead to the conclusion that this kind of a 'surface reasoning system' would not be sufficient for a thorough validation of the new and extended features of PROLOG III. A model-based approach is now being tried which seems to be more appropriate.

Krautter and Steinert [6] gave an overview of the principal ideas and first results of research in this direction. They showed how technical systems (in this case a vehicle engine) can be modelled in terms of components with characteristic input- and output-properties. Then they outlined how these models can be transformed to a PROLOG III representation. Example queries together with their respective answers by the PROLOG interpreter were given. Our objective in this paper is to report on the progress of the ongoing research activities:

First we shall rediscuss the modelling principles in a formal way and then show how arbitrary   models can be transformed to a PROLOG III representation in a rather straight-forward manner. Based on experience gained with this approach we shall then improve the representation, especially with regard to possible inference concepts.

2. Knowledge Representation

2.1 Underlying Engine Models

There has been further work by the project team in the field of knowledge acquisition and modelling, but we shall restrict our discussion to the two engine models given in [6]:
In the first case (see Figure 1) the engine is decomposed into three main subsystems:

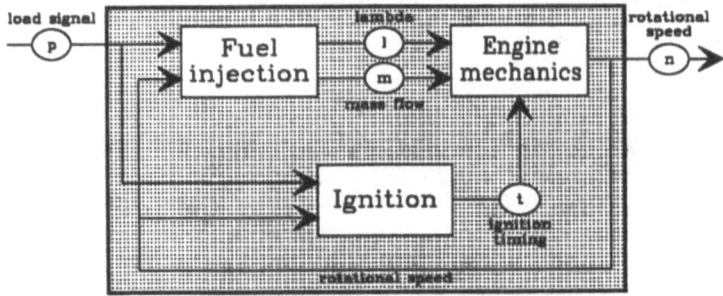

Figure 1: Simple Engine Model (3 Components)

The subsystems cooperate to produce a shaft rotation speed n from a given input load signal p. They mutually influence each other by the quantities they exchange, in this example there are only three such quantities (air/fuel ratio Lambda l, mass flow m, ignition timing t). Model I is an abstraction of a more detailed arrangement of the physical reality. Introducing a more detailed view we get the more elaborate Model II (see Figure 2):

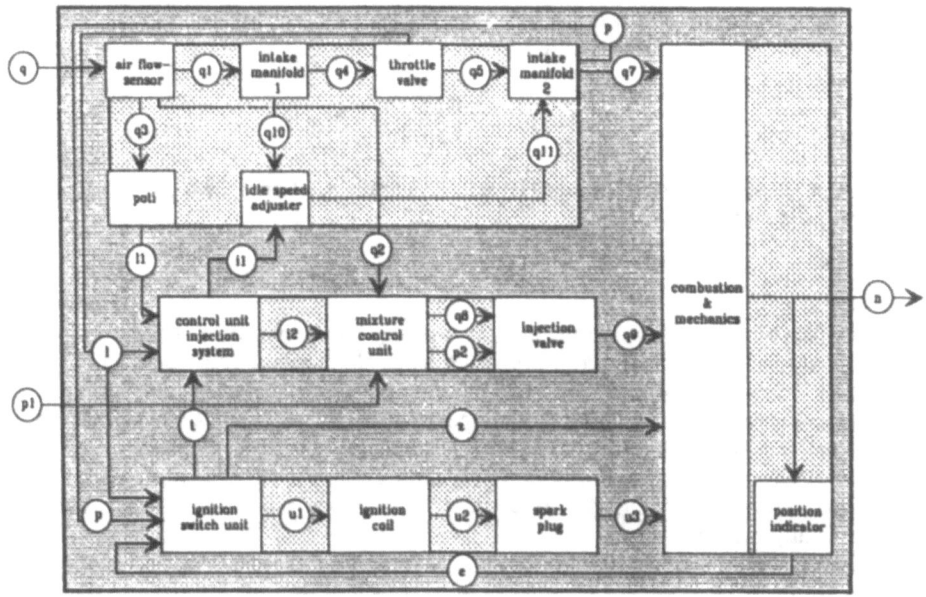

Figure 2: Extended Engine Model (14 Components)

There are two input quantities (air flow and load signal) and one output quantity (speed). In the first view there are 4 subsystems which exchange 11 quantities; in the second view there are 14 components and 22 quantities. If required, this step of deriving a more detailed model may be applied again to any desired level. How deep to go is merely a question of the specific interest.

2.2 Modelling Principles

As shown in section 2.1 a physical device may always be interpreted as a composition of smaller interacting subdevices. Let us call the components at the lowest level of detail atomic parts. In case of failure, those parts can either be adjusted, exchanged, or repaired. Interaction between components occurs by means of physical quantities. As an example of composition and interaction, consider again Model I as illustrated by Figure 1.

According to the philosophy outlined in [6], the components fuel injection, ignition and engine mechanics may be represented by mathematical relations; let us call them F, I and E respectively. These relations depend on the load signal p, the Lambda quotient l, the mass flow m, the ignition timing t, and the rotational speed n. The connections between components are modelled by the following equations:

$$\langle l,m \rangle = F(p,n) \tag{1}$$
$$\langle t \rangle = I(p,n) \tag{2}$$
$$\langle n \rangle = E(l,m,t) \tag{3}$$

Communication between components is established by the equality of names of their input/output quantities. For example, there is an influence of the ignition upon the engine mechanics due to their common quantity t. Equation (1) is a mapping from a 2-dimensional domain into a 2-dimensional range. This mapping may be decomposed into a pair of more simple mappings with 1-dimensional ranges only:

$$<l> \quad = \quad F_1(p,n) \qquad (1')$$
$$<m> \quad = \quad F_2(p,n) \qquad (1'')$$

By substituting equations (2) and (1') and (1'') into equation (3) we get a mathematical model of the overall system:

$$<n> \quad = \quad E(F_1(p,n),F_2(p,n),I(p,n)) \qquad (4)$$

Within this model all dependencies are expressed by means of the input quantity p and the output quantity n, whereas the quantity n also introduces circular feedback. We will call models like (4) compound models since they consist of compound mappings.

Compound models are in general nonlinear and not solvable symbolically. But in order to use them for diagnostic purposes there is need for symbolic manipulation. For that reason, models like (4) serve in general only as a source for approximate models which themselves can be manipulated symbolically. We will use piecewise linearization for the purpose of approximation.

We can derive an approximate model from the mathematical foundation of the domain given by equation (4) once it is specified. An arbitrary physical device may be viewed as a mapping from an m-dimensional domain into a k-dimensional range,

$$f: D^m\text{-}R^k \qquad (5)$$

where D^m denotes the domain and R^k the range of the mapping. As already pointed out, there is a necessity for linearization. Since linear relations always have ranges with dimension 1, they force a decomposition of the general case (5) into a set of mappings where the power of the set is given by the dimension of the range of (5):

$$\{ f_i: D^m\text{-}R_i, 1<i<k \} \qquad (6)$$

where the R_i denote the individual ranges. As an example of dimension 2, please see the equations (1), (1') and (1'').

For the sake of linearization, the domain of the mappings must be partitioned into intervals of dimension m. To each of these intervals a linear function is attached which is assumed to approximate the non-linear function locally. For r different intervals we have clearly r different linear approximations. Henceforth, we call m-dimensional intervals operation spheres and the linear approximation functions linear constraints or simply constraints. Notice that an operation sphere is defined by a set of inequalities (at most 2 inequalities for each dimension of the domain of the original mapping), where constraints are linear equations. Figure 3 illustrates the partitioning of a mapping with 1-dimensional domain and 1-dimensional range. In that case the operation spheres simply reduce to ordinary intervals.

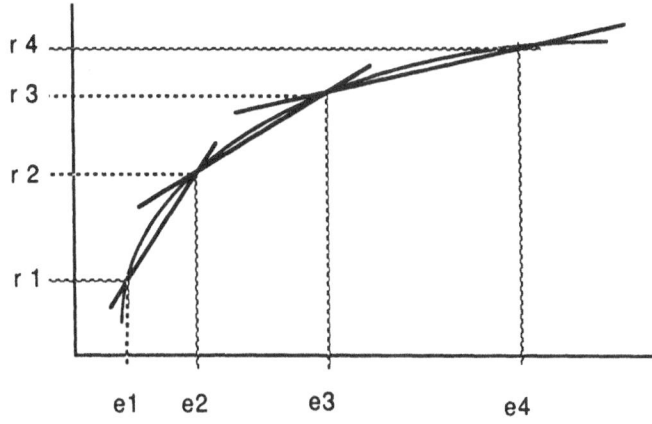

r 4

r 3

r 2

r 1

e1   e2   e3   e4

Figure 3: Piecewise linearization of a nonlinear function

Given that the dimension of the range of the original mapping is k and r different operation spheres are used for local approximation, the total number of constraints is clearly k * r, whereas only k constraints (namely those of the same operation sphere) are used to approximate locally the ok-state of the component of interest. If a constraint c is used for approximation within the operation sphere s, we say c is bound to s, and we write c ⌊ s.
All together, this leads to the following definitions:

Def. 1: Given a set $C = \{C_1, ..., C_m\}$ of constraint sets approximating a set of mappings with respect to a given set of operation spheres. Let S be a set of inequalities denoting an arbitrary but fixed operation sphere. Then, for all i, the pair ,C_i is called a local constraint set if C_i has the property: c C_i if and only if cS.

Def. 2: A linear numerical model of an analogue device is a set of local constraint sets such that their operation spheres cover the whole domain without intersection.

The non-intersection requirement in definition 2 assures that for a given set of input quantities of a physical device a unique constraint set may be identified which describes the expected ok-state of the physical device for the given set of input quantities.

2.3 A Formal Representation Approach

1) The first step towards formalizing the representation consists in mapping the system network onto a tree structure (decomposition tree); part of this representation tree for Model II is given in Figure 4:

518

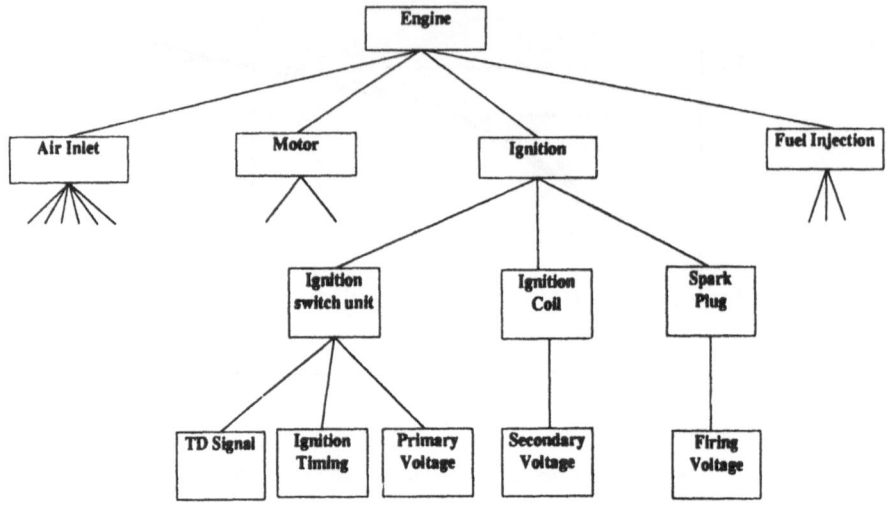

Figure 4: Tree representation of a technical system

The leaves of this tree are given by those quantities that appear somewhere as output of a given component, the other nodes denote the possible subsystems, individual components etc.

2) In the next step a PROLOG term of the form *node(b,i,o,q)* [*] is associated with each node of the tree:
   - the identifier *node* is the corresponding label,
   - *b* is a Boolean variable to describe the state of node (1' = true means *operative*, 0' = false means *defective*),
   - $i = <\_1, ..., i\_>n$, $o = <\_1, ..., o\_>m$, and $q = q<\_1, ..., q\_k>$ are the lists of input, output, and internal quantities, respectively.

Since leaves represent output quantities (see step 1) they have no internal quantities. In this case the lists *o* and *q* are used to distinguish between the actual (as determined by measurement) and the nominal (as required by correct operating conditions) output value; note also that these lists are of length 1.

3) In the third step the system interconnections are established by building PROLOG clauses: for any node, the head of the clause is given by the corresponding term, whereas the body is obtained from the terms of the immediate subnodes. The central mechanism here is to use PROLOG unification over the variables. An associated Boolean constraint states that correct operation at any node is assumed if correct operation at all subnodes is given.

In the case of a leaf, the body of the clause consists of a definition of the nominal (correct) behaviour of the quantity under consideration together with a tolerance condition to determine the state; in a realistic situation one must always account for some variation of a numerically measured value, due for example to the precision of the measurement.

As an illustration consider the following extract from the representation of Model II:

---

[*]   Throughout the text PROLOG III - Code will be typed in italics.

```
/* Definition of the Engine */

    Engine(b,<<q,pl>,<>>,<<n>,<>>,<<q9,i1,l1,q2,q7,t,z,u3,p,e>,<l>>)->
        Air_Inlet(b1,<<q,il>,<>>,<<ll,p,q2,q7>,<l>>,x1)
        Motor(b2,<<q7,q9,z,u3>,<>>,<<n,e>,<>>,x2)
        Ignition(b3,<<e,p>,<l>>,<<t,z,u3>,<>>,x3)
        Fuel_Injection(b4,<<t,l1,pl,q2>,<l>>,<<q9,i1>,<>>,x4)
        { b = b1 & b2 & b3 & b4 };


/* Definition of the Ignition Sub-System */

    Ignition(b3,<e,p>,<l>>,<<t,z,u3>,<>>,<<u1,u2>,<>>)->
        Ignition_Switch_Unit(b31,<<e,p>,<l>>,<<t,z,ul>,<>>,x31)
        Ignition_Coil(b32,<<ul>,<>>,<<u2>,<>>,x32)
        Spark_Plug(b33,<<u2>,<>>,<<u3>,<>>,x33)
        { b3 = b31 & b32 & b33 };


/* Definition of the Ignition_Switch_Unit */

    Ignition_Switch_Unit(b31,<<e,p>,<l>>,<<t,z,ul>,<>>,<<>,<>>)->
        TD_Signal(b311,<<e>,<>>,<<l>,<>>,<<l'>,<>>)
        Ignition_Timing(b312,<<e,p>,<l>>,<<z>,<>>,<<z'>,<>>)
        Primary_Voltage(b313,<<e,p>,<>>,<<ul>,<>>,<<ul'>,<>>)
        { b31 = b311 & b312 & b313, e >= 240, e <990 };


/* Transfer Function Ignition-Timing */

    Ignition_Timing(b312,<<e,p>,<l>>,<<z>,<>>,<<z'>,<>>)->
        nominal_Timing(z',<e,p,l>)
        within_tol(b312,z,z');


    nominal_Timing(z,<e,p,l>)->
        { l = l', e >= 240, e <480, e' = e - 240, z = 155/10 - 52/2400*e' };


    within_tol(l',x,x') -> { 95/100*x' < x <105/100*x' };                Program Example 1
```

2.4 Implementation Experiences

Implementing Model I according to these ideas with the presently available PROLOG III prototype interpreters produced a program that is a straightforward representation of the given technical system. As a confidence test, systematic test series were performed with parameters of correct and faulty system states that had been determined by a kind of classical simulation: the corresponding PROLOG III queries lead to the same results. The PROLOG III representation was also used to determine possible states of the engine with only part of the measurement parameters bound or to determine parameter dependencies under given state assumptions (see for example Figures 5 and 6):

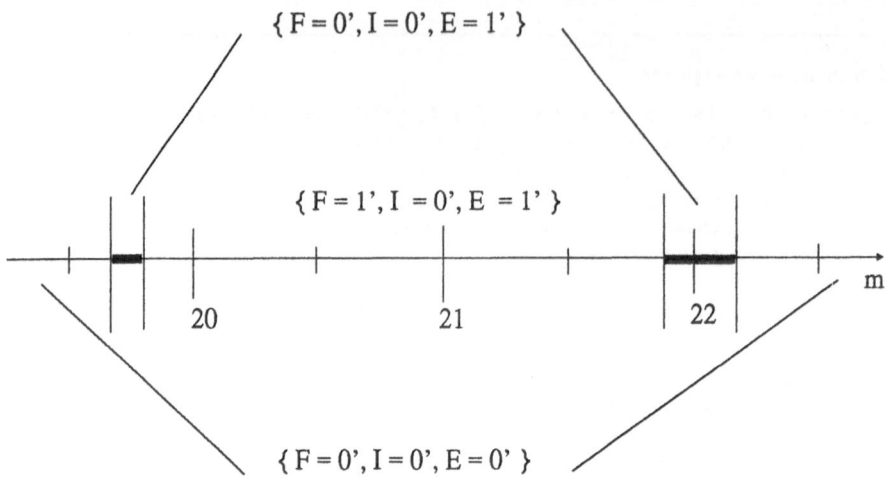

Figure 5: Different component states depending on the value of the parameter mass flow (all other parameters are supposed to have fixed values). For m [19.80,21.89] only component ignition is defective (single fault); for greater and smaller values of m, component fuel injection becomes faulty at first (double fault) and then all three components are out of order.

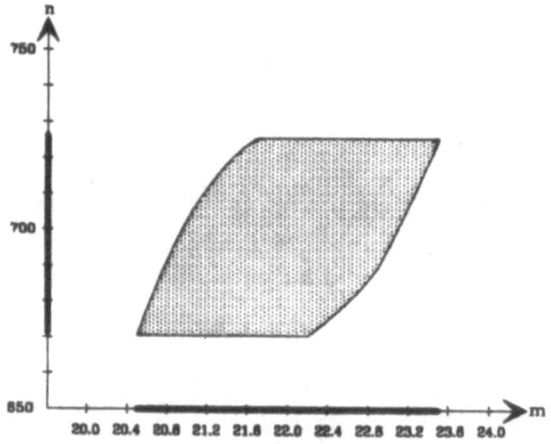

Figure 6: Relationship between speed n and mass flow m; the assumption is that the values of all other parameters are fixed and that all components are in ok-state. The shaded area indicates the admissible solutions.

These results underline the general philosophy of PROLOG as a declarative language, since queries with arbitrary variable bindings can occur and especially the use of arithmetic data in PROLOG III is an improvement over the traditional use of numerical data types in "standard" PROLOG implementations.

However, extending this way of straightforward representation to the more realistic, but still rather simple Model II revealed that the kind of simple queries cannot usefully be asked without further restrictions. Firstly it is clear that for freely chosen parameter settings the size of the solution space is governed by the order of magnitude of the power set of the leaves in the representation tree (2^4 for Model I, but 2^{23} in the case of Model II). Hence additional assumptions, expressed as supplementary constraints, have to be made: for instance, making the 'Single Fault Assumption' (i.e. only one output function is wrong at a given time) allows analysis in several example cases. However, situations can easily arise, where the evaluation time for one query still is in the order of minutes; this is too slow to be applicable in an interactive dialogue system.

We therefore have studied a different approach that will be discussed in the next section.

3. Improved Representation, Constraint Propagation and Diagnosis

3.1 Decomposition Trees and Corresponding PROLOG III Notation

Staying with the idea of mapping the network of a technical model onto a decomposition tree and because evaluating whether a set of constraints is satisfied or not is very fast in PROLOG III we are lead to a representation which associates the linearized model to each node of the tree. The model is supposed to represent the ok-state of exactly that device which is represented by the node. As we already stated in section 2.2, communication between various components is achieved by quantities common to the models of these components. Therefore, this communication is expressed by the interchange of the corresponding values between the local constraint sets. The question arises how to organize all the constraint sets to assure coordinated access to and communication between them as well. Therefore we must provide information about how and where communication occurs, that is, we attach in addition to each tree node a set of parameters, where a parameter is defined as follows:

Def. 3: A parameter of the decomposition tree is a pair $<x,i>$ where x holds the value and i is an identifier.

From the point of view of a component there might be parameters which carry out the communication with its environment as well as parameters which are only of internal importance. Let us call the last category of parameters internal parameters and the former ones input/output-parameters (i/o-parameters for short).

Def. 4: A node in the decomposition tree is a tuple **(node, p, m, s)** where the identifier **node** is the label, **p** the list of i/o-parameters, **m** a linear numerical model, and **s** a list of successor nodes of arbitrary length.

By definition 4 we get a hierarchy of tuples which is supposed to be an isomorphic structure first of the hierarchy of domain components, and second to the possible communication paths of the domain components, whereas the numerical models are supposed to be approximations of the correct behaviour of the respective component. The following figure illustrates the refined structure of such a hierarchy.

522

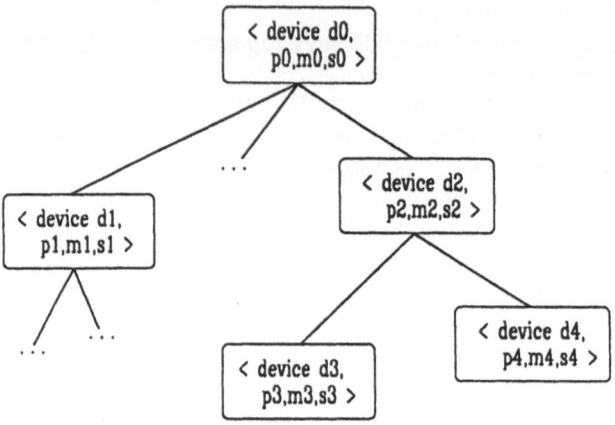

Figure 7: Decomposition Tree

An i/o-parameter of a given component may be an internal parameter defined in a higher level of the hierarchy. By definition, atomic parts of the domain have no internal parameters at all as they are never considered for further decomposition. Hence, we have the following proposition:

Prop. 1: For each parameter p of the decomposition tree there exists at least one node with p element of the list of i/o-parameters.

Communication throughout the entire domain is directed by the identifiers of the parameters. Based on definition 3 the basic communication principle is as follows:

– If the identifiers of two parameters are unifiable, their values must be unifiable.

Thus, unification between parameters serves to mimic the communication between different components. Whether two components represented by nodes in the composition tree share a common parameter is easy to determine by examining the identifiers of their parameter sets. If two different nodes share a parameter with the same identifier then their values must be the same.

Taking advantage of the typical properties provided by PROLOG III, linear numerical models as given by definition 2.2 are simply stated as PROLOG III facts with associated constraints whereas decomposition trees can be given by "standard" PROLOG facts. The general structure of the corresponding PROLOG expressions may, for example, look like:

$$node\ (<<i\_1,Input\_1>,...>,<<o\_1,Output\_1>,...>,<subnode\_1,...>)->;(6)$$
$$cset(node,<i\_1,...>,<o\_1,...>) -> \{\ c1(i,o),...\ \}\qquad\qquad (7)$$

Relation (6) denotes the decompostion of the tree node 'node' and relation (7) states a constraint set for a unique operation sphere associated with 'node'. Obviously, a linear numerical model of a node is made up of as many relations of the kind (7) as there are operation spheres. The separation of (6) and (7) allows access to the tree structure independently from constraint evaluation. This allows employment of all kinds of tree search strategies without having to evaluate constraint sets redundantly.

Clearly, in (6), if the list of subnodes is empty, the node represents an atomic part. Further, all nodes, whether atomic or not, own their specific linear numerical model. Commonly, for atomic parts it is relatively easy to get the mathematical information needed to develop numerical models. In case of compound parts things turn out to be more complicated. The question is, how can compound models be generated based on the numerical models of their atomic parts? The generation of compound models can be done by introducing a pre-modelling-phase based on the description defined in section 2.3. By definition, models like those in section 2.3 only use constraint sets for their leaf descriptions. If we want to generate, for example, the numerical model for the ignition (which is a compound part) and a node with name 'ignition' exists according to 2.3 we only have to state

$$make\_constraint\_set(ignition); \quad (8)$$

where the following PROLOG relation is supposed to be valid:

$$make\_constraint\_set(i\_d) \rightarrow$$
$$i\_d(1',i,o,q) \quad (9)$$
$$assert(cset(i\_d,i,o), <>);$$

As a result of query (8) PROLOG III will try to resolve the predicate $ignition(1',i,o,q)$ which in turn will generate the compound model (provided an ignition model exists according to section 2.3).

As an illustration consider the following extract from the represention of the Model II ignition system (see also Program Example 1):

```
Engine(<<q,air_flow>,<p1,load_signal>>,<<n,rotational_speed>>,
        <Air_Inlet,Motor,Ignition,Fuel_Injection>)->;

Ignition(<<p,under_pressure>,<e,electrical_speed_signal>,<l,idle_contact>>,
        <<z,ignition_timing>,<u3,firing_voltage>,<t,TD_signal>>,
        <Ignition_Switch_Unit,Ignition_Coil,Spark_Plug>)->;

Ignition_Switch_Unit(<<e,electrical_speed_signal>,<p,under_pressure>,
                <l,idle_contact>>,<<t,TD_signal>,<z,ignition_timing>,
                <u1,primary_voltage>>,<>)->;

Ignition_Coil(<<u1,primary_voltage>>,<<u2,secondary_voltage>>,<>)->;

Spark_Plug(<<u2,secondary_voltage>>,<<u3,firing_voltage>>,<>)->;

cset(Ignition,<S$1 +240,p'_10 +940,1'>,<S$1 +240, -(13/600)S$1
    +(31/2),(1/200)p'_10 +(1/500)S$1 +(759/50)>)->,

{S$1 >= 0,
 p'_10:num,
 S$2 >= 0,
  S$2 = +S$1,
  S$4 > 0,
  S$4 = 750 -S$1,
  S$3 > 0,
  S$3 = 240 -S$1};

  . . .                                            Program Example 2
```

3.2 Semantic Principles

We are now ready to give an architecture for an inference procedure based on linear numerical models. Given a device D, let CD denote its linear numerical model. If an operation sphere is given by a set S of inequalities, let D|S state that D is working in S. Finally, let C_S CD denote the local constraint set with all its constraints bound to S. Remember also, that CD is supposed to approximate the ok-state of D. This fact serves as a base for the semantic meaning of a constraint violation by data observed from D:

– if C_S is violated by data observed from D, and D is operating in S, there must be at least one fault in the domain of D.

If it is the case that D is an atomic part then we have arrived at a diagnostic goal, that is, D must be adjusted, exchanged or repaired. Otherwise, D is composed by further sub-devices $D_1, ..., D_n$ and the semantic interpretation above leads to the conclusion:

– if D has at least one fault and D is composed by $D_1, ..., D_n$ then, for some i, there is a D_i which has at least one fault

Hence, with T denoting the composition tree and $T|D$ the node of T corresponding to D we may formulate an inference scheme using local constraint sets as follows:

3.3 Inference Procedure

```
procedure inference:
    0)      initialize the set of observed data
            O := { operation sphere data }
    1)      P := set of i/o-parameters of T|D
    2)      C := CS, that is, select the appropriate local constraint
            set according to D|S
    3)      push C onto the stack of active constraints
    4)      take the set P, and the data previously observed; try to
            unify the parameters in P with the already observed data,
            according to the communication principle; exclude all
            unified parameters from P;
    5)      did any conflict arise due to the unification attempt?
    6)      YES, is D an atomic part?
            6.1)        YES, D malfunctions, display the result
            6.2)        NO, D' := set of sub-nodes of T|D; recurse on
                        each element of D' beginning with step 1 until a
                        malfunction within an atomic part is found
    7)      NO, is the set P empty?
            7.1)        YES, no faults within D, go to step 8
            7.2)        NO, ask the user for the remaining parameters in
                        P; store the observed data and try again to unify
                        the parameters in P; repeat step
    8)      STOP, no faults found within D;
```

Using the composition tree, the above inference scheme works in a depth first search. With respect to the isolation just mentioned, this does not necessarily need to be the case. Expanding the nodes of the composition tree with heuristic information (e. g. fault statistics), the tree can also serve for branch and bound or A* search strategies. The basic inference scheme above would require only little modification.

4. Conclusion

The approach to diagnostic reasoning presented in this paper overcomes the drawback of a purely rule based approach. This is due to the representation of the physics of the reality in a form that models the ok-state. In contrast to explicitly stated causality links in rule based approaches, we view fault as a divergence from the ok-state, whatever the divergence may look like in detail. Using the structures as we defined them, we are able to pursue the trace of misbehaviour to an arbitrary, but previously defined level of detail. That is, we find the causality links and thus the origin of misbehaviour.

In section 2 we have shown how to represent models of physical devices in PROLOG III. This representation makes extensive use of of 'numerical constraints', one of the distinguishing features of the new programming language. Since the treatment of numerical data is remarkably different in "classical" PROLOGs from the way it is handled in 'constraint logic languages' like PROLOG III, it would have been very difficult or even impossible to realize this model-based approach in a more traditional PROLOG. The code size of the application is of the order of some tens of Kbytes; therefore the suitability of hardware is determined by the memory space requirements of the PROLOG III interpreter. From the application point of view it would be possible to represent substantially larger models with some hundred Kbytes on rather small (personal) computers.

However, the implementation experience gained with the first approach had revealed that this is suitable in an interactive dialogue application only for relatively small models. Due to the increased complexity of larger models the evaluation times become unacceptably long. Therefore we had developed a revised way of representing the models as discussed in section 3. The experiences gained so far with this approach have shown that, with respect to time efficiency, it is a definite improvement compared to the first (straightforward) one. The results are promising in the sense that it should be possible to represent even larger models without affecting performance too seriously.

But we have to admit that the approach presented so far is not able to diagnose dynamic misbehaviour, as it is the aim of qualitative physics (see e.g. [3]). The inference scheme as discussed in 3.2 and 3.3 is grounded on fixed operation spheres. To search for causality links of dynamic misbehaviour would require repeated changes between operation spheres. At this stage it remains as an open question how it is possible to guard and to judge transitions between operation spheres.

Acknowledgements

We are indebted to J. Franco for critically reading earlier versions of this paper. He made a lot of valuable suggestions which helped to improve the manuscript. Our thanks also go to all the colleagues in the ESPRIT-Project for the spirit of fruitful cooperation. Finally we like to mention G. Ruck for his patience and help in drawing figures and setting up the final layout of the text.

References

[1] Colmerauer A. 1987a, Opening the PROLOG III Universe, BYTE-Magazine, August 1987, pp. 177 -182

[2] Colmerauer A. 1987b, Introduction to PROLOG III, Annual ESPRIT-Conference, Bruxelles September 1987, North-Holland, pp. 611 -629

[3] DeKleer J. and Brown J.S. 1984, A Qualitative Physics based on Confluences, Artificial Intelligence 24, pp. 7 - 83

[4] Dincbas M. et al. 1988, The constraint logic programming language CHIP, Proceedings of the International Conference on Fifth Generation Computer Systems, December 1988, ICOT, Tokyo, pp. 693 - 702

[5] Giannesini F. et al. 1986, PROLOG, Addison Wesley

[6] Krautter W. and Steinert M. 1988, A Knowledge Representation for Model-Based Reasoning using PROLOG III, Annual ESPRIT-Conference, Bruxelles November 1988, North-Holland, pp. 814 -825

[7] Skuppin R. and Weber R. 1987, PROMOTEX: Ein wissensbasiertes System in PROLOG für die

Kraftfahrzeug-Diagnose, 2. GI-Kongress, München Oktober 1987, Springer IFB 155, pp. 426 - 432

Key Words

model-based reasoning, knowledge representation, inference strategy, constraint logic programming, PROLOG III

PRESENTATION OF THE SEDOS ESTELLE DEMONSTRATOR PROJECT

J.M. AYACHE (1), J. BERROCAL (2), S. BUDKOWSKI (3), M. DIAZ (4), J. DUFAU (1),

A.M. DRUILHE (5), N. ECHEVARRIA (6), R. GROZ (7), M. HUYBRECHTS (8)

(1) VERILOG - Toulouse - France

(2) Escuela Tecnica Superior de Ingenieros de Telecomunicaciones Madrid - Spain

(3) BULL/DAS - Paris - France

(4) Laboratoire d'Automatique et d'Analyse des Systèmes Toulouse - France

(5) MARBEN INFORMATIQUE - Paris - France

(6) ENTEL - Madrid - Spain

(7) Centre National d'Etudes des Télécommunications - Lannion - France

(8) EXPERT SOFTWARE SYSTEMS Ghent - Belgium

1. Introduction

The SEDOS ESTELLE DEMONSTRATOR Esprit project 1265 is conducted by a consortium of three partners : **VERILOG** (prime contractor), **EXPERT SOFTWARE SYSTEMS, MARBEN,** and five subcontractors : **LAAS, CNET, ETSIT, ENTEL** and **BULL.**

SEDOS ESTELLE DEMONSTRATOR has been set up from the results of the ESTELLE task of the SEDOS Esprit project 410 in order to develop and integrate basic support tools and to evaluate on selected examples the Formal Description Technique ESTELLE [1] standardized by ISO.

As a consequence, the project has been divided into two main work packages :

- the development of an ESTELLE workstation (EWS),
- the evaluation of ESTELLE and of the EWS on real and significant applications.

The objectives of the SEDOS ESTELLE DEMONSTRATOR project are :

- to produce a basic set of **prototype tools** in order to increase efficiency and productivity in the design development phases : formal description, validation and implementation.
- to propose an **open environment** which architecture is designed to allow the integration of prototype tools from any origin through an intermediate form of the ESTELLE source (common data structure of all components of the environment);
- to produce a set of **validated** ESTELLE descriptions selected in four important application fields [2] :
  1. Computer Networks : ISO - TRANSPORT, SESSION, ACSE, ROSE, FTAM ;
  2. Industrial Systems : A flexible assembly cell, a power plant control system ;
  3. Telecommunications : protocols of layer 2 and 3 of the future paneuropean radiocellular network, a double call service in the ISDN framework ;
  4. Space communications : three layer 2 protocols COP0, COP1, COP2 from the CCSDS (Consultative Committee for Space Data Systems) ;

The first part of this paper is dedicated to the EWS description and the second part presents the interests and the expected results for each of the selected evaluation examples.

2. Description of ews

2.1 GENERAL ARCHITECTURE OF EWS

EWS consists in five interconnected components which functional relationships are depicted as follows (Figure 1) :

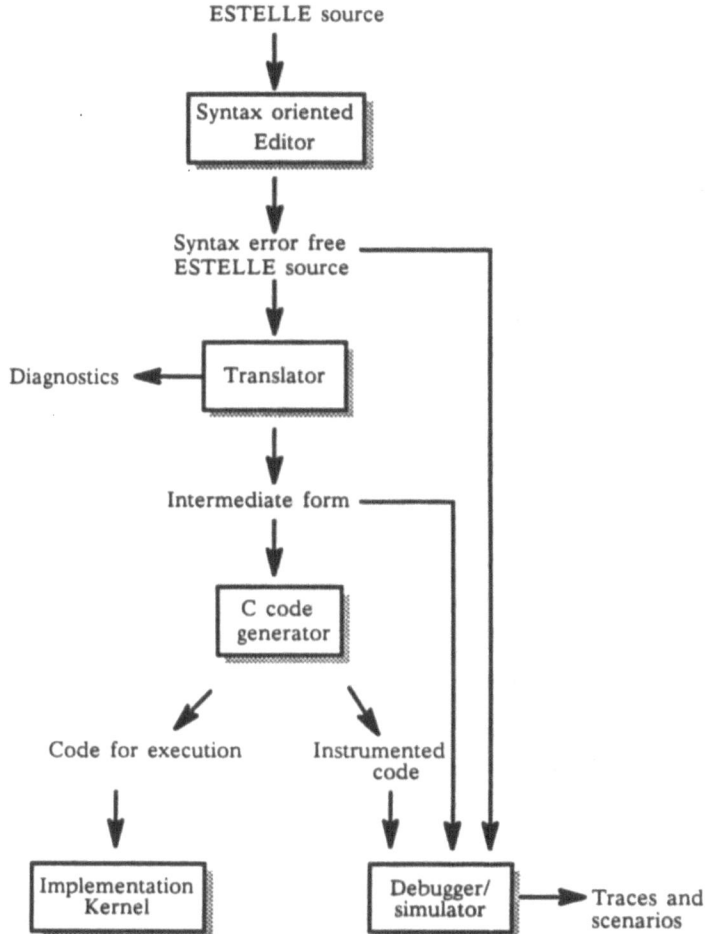

Figure 1 : General Architecture of EWS

1. the **syntax oriented editor** constitutes the entry point of the software production line ; it is able to take into account an existing ASCII file or to support the user from the very first line ; the editor produces a syntax error

free ESTELLE source on a ASCII file ;

2. the **translator** is in charge of computing an ESTELLE specification into an **intermediate form** file after semantics checks ;

3. the **C code generator** is in charge of producing C representations of all data structures and operations of the ESTELLE specification starting from the Intermediate Form informations ; the produced C code is available on ASCII files and is compiled using the **ESTELLE primitives library** ;

4. the **EWS simulator** allows interactive experimentation of the ESTELLE specification ; all objects previously produced for a given specification are involved to provide the use of an attractive interactive debugger :

. the **C generated code** is the basis for the specification execution (after compilation and biding with the existing simulator object code),

. the **intermediate form** provides symbolic informations to the user,

. the **ESTELLE ASCII source** is used by a "read only" version of the editor which manipulation is combined with simulation features.

5. the **implementation kernel** uses the C generated code and provides an ESTELLE runtime primitives library and extra features needed to run the ESTELLE specification on a target machine ;

All these EWS components are integrated within an **interactive monitor** which is in charge of verifying the consistency of the user's commands.

2.2. THE SYNTAX ORIENTED EDITOR

The syntax oriented editor is a very powerful editor for producing and maintaining ESTELLE source texts. It offers the advantages of an advanced screen oriented editor combined with the concept of syntax directed editing [3].

The EWS editor takes advantage of the available computer resources in using its knowledge of the programming language to support the programmer at every moment. The user can enter text in a linear manner, as with a normal full screen editor. However, optional menus, adapted to the syntactic context, assist the user during insertion. Furthermore, the editor will constantly check the validity of the entered tokens without any need of explicit calls to the checking mechanism.

The editing manipulations are based on lexical units grouped into basic syntactical units which in their turn are grouped into larger syntactical units. Lexical units are the most elementary components of ESTELLE. Examples of lexical units are names, numbers, operators, keywords ... The only elements that can be inserted, edited, selected, deleted, replaced, moved ... are lexical units and syntactical units. Amending of names (character level) is also possible. Removing of structures are made safe by using the clipboard concept.

Units selections are available through the use of a pointing device (mouse); manipulation activations are accessed by both mouse and keyboard ; moving through the text is accessed by pointing on the scroll bar.

Program texts are automatically formatted and can be displayed with different levels of detail according to viewing strategies. In the default configuration the full text is shown on the screen ; by command, the user can ask the editor to skip all comments, to focus on the ESTELLE syntactical constructions, the PASCAL constants, the types and variables declarations, the ESTELLE transitions parts ...

A view of the EWS editor screen is shown on Figure 2.

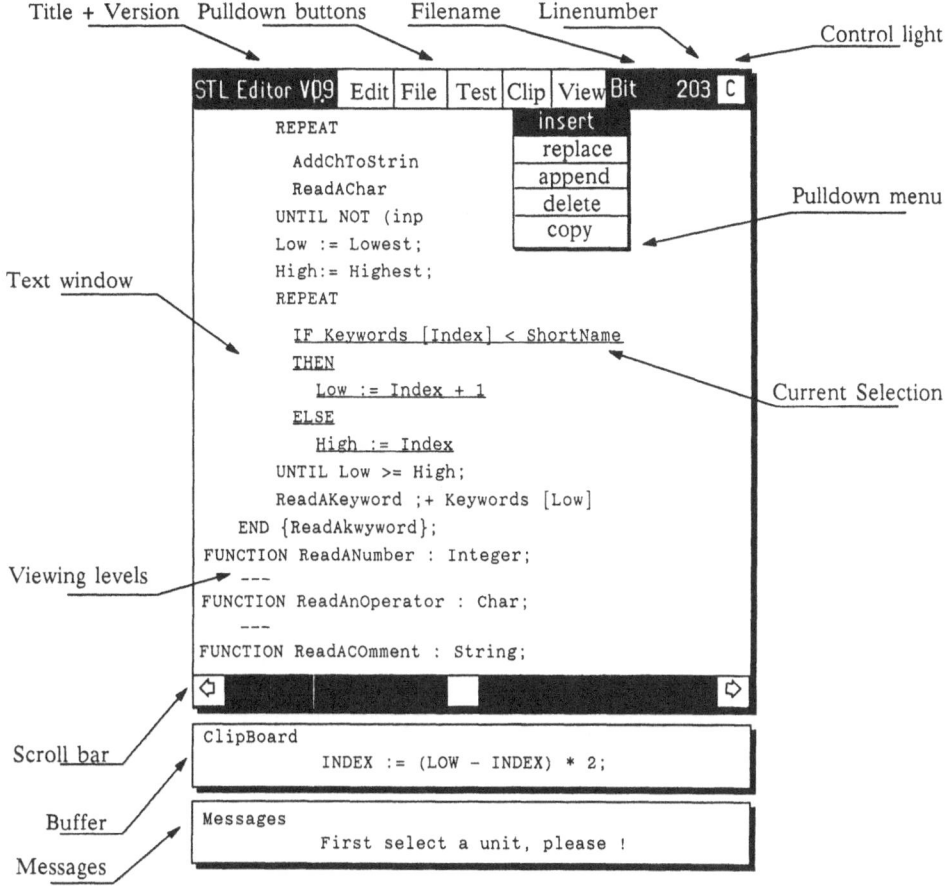

Figure 2 : EWS syntax oriented editor screen

The EWS editor combines the advantages of advanced text editors (full screen editor, use of a high-resolution display and use of mouse) and of syntactic directed editors (syntax error free editing, on-line help, automatic completing of the insertion when no non determinism is left, support of novice users).

As an advanced product, the EWS editor has been developed with the latest software technologies. In order to deliver a family of editors, it is constructed using the MIRA tool [4], offering flexibility in incorporating other languages. The internal data structure to represent the ESTELLE text is based on the abstract syntax tree concept. It involves a unique memory allocation scheme providing both fast edition and efficient garbage collection.

2.3. THE TRANSLATOR

The **translator** [5] has three functions :

. as a **source analyser**, it delivers cross references on dedicated files,

. as a **semantics checker**, it delivers detailed diagnostics to the user,

. as a **intermediate form generator**, it translates the ASCII ESTELLE representation produced by the editor into a data structure suitable for further processing : this data structure (called the intermediate form of the ESTELLE specification [6]) is the common input for any other component of the production process.

The intermediate form is produced as a result of the complete analysis (lexical,syntactical and semantical). Its contents preserves the semantic of the input specification (no loss of information) but its structure and access are more suitable for automatic processing. The intermediate form is the central object of EWS.

The entire translator is developed in portable C with the help of a compiler generation tool : SYNTAX [7].

2.4. THE C CODE GENERATOR

The purpose of the **code generator** is to translate an ESTELLE specification into a "C" program that can be executed according to ESTELLE semantics.

The same generated C programs are monitored by an implementation motor and a simulation motors : using the same generated code basis, simulation of the ESTELLE specification is fully coherent with its possible implementation.

The C generated code is closely dependent of the organization chosen for implementation and simulation [8]. From both implementation and simulation motors, ESTELLE module instances are considered as independent tasks for which a specific context is generated. This context consists in :

. a set of data corresponding to the internal variables of the ESTELLE body,

. a set of data corresponding to the external variables and interaction points of the ESTELLE module header,

. control data required by both simulation and implementation motors to invoke the tasks execution,

. a set of evaluation functions corresponding to all transition guards,

. a set of execution functions corresponding to all transition statement parts.

Note that the ESTELLE statements such as WHEN, OUTPUT, INIT ... are translated into a suitable sequence of calls to a generic library of C functions. All PASCAL parts of ESTELLE are translated into their corresponding C types or C control structures.

For simulation, extra instrumentation of the C code is processed in order to support debugging oriented features such as execution stop on breakpoints, read/write operations on internal variables of the ESTELLE module instances.

2.5. THE EWS SIMULATOR

The EWS **debugging oriented simulator** [10] allows an interactive execution of the ESTELLE specification.

Execution is possible according to two major modes :

– fully interactive i.e. transition by transition with the opportunity for the user to select which transition has to be fired next,

– fully automatic i.e. by a random selection of the transitions to fire during a simulation time denoted as simulation step.

When executing the transitions, traces to various specification objects can be made (variables, major state of module instances, contents of queues associated to interaction points) : any access to the specification objects is proposed on a symbolic mode (the user asks for objects with their ESTELLE identifier) while a set of intermediate form access functions ensures for consistency between ESTELLE symbolic objects and C real objects.

Execution can stop on user defined breakpoints inside transition statement parts or PASCAL procedures/functions.

Automatic backtracking on executed transitions can be performed as well as automatic replay. Executed transition sequences can be stored on files and replayed during further sessions.

All interactive features are fully supported by an advanced screen technology based on the use of pop-up menus, graphic displays, on-line commands interpretor. The EWS editor is used by the simulator in a read-only mode in order to support functions such as breakpoints selection, automatic pointing on last fired transition, advanced pattern research.

A view of the simulator screen is shown on Figure 3.

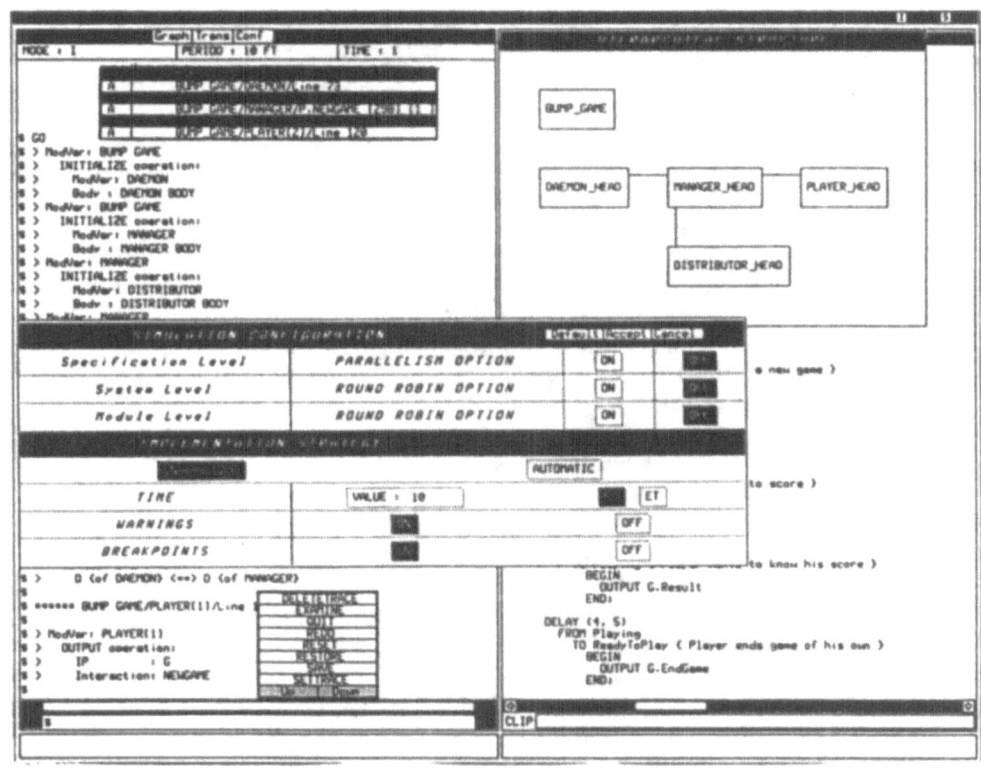

Figure 3 : EWS simulator screen

2.6. THE IMPLEMENTATION KERNEL

The **Implementation Kernel** constitutes the runtime library needed for the execution of generated C code in a real implementation environment. The primitives of this kernel enable the implementation of an ESTELLE system (systemactivity or systemprocess) within a single task of the target operating system. This kernel provides C functions implementing all basic ESTELLE operations such as ATTACH, OUTPUT, FORONE ... [9].

The implementation kernel includes the implementation motor which is in charge of the scheduling of the module instances. In addition, memory management routines are provided to allow efficient manipulation of data buffers (assembling, fragmenting ...). Moreover, primitives for communication with other tasks of the target operating system are provided to connect the generated code to the real implementation environment.

Implementation kernel (at first developed for Unix System V target machines) has been designed to be as much system independent as possible. When migrating from one implementation operating system to another, only a few clearly located code has to be rewritten (inter task communications and time clock management).

3. Presentation of the evaluation examples

3.1. GENERAL PURPOSE OF THE EVALUATION PHASE

This part presents the work conducted in ESTELLE SEDOS DEMONSTRATOR for evaluating both the ESTELLE Formal Description Technique and the tools developed within the project. The evaluation phase relies on developing examples selected in four important areas : Computer networks, space communications, telecommunication and industrial systems. For these examples, evaluation consists in :

- a **formal description phase** : starting from existing specifications either in formal language or in natural language, this phase allows to evaluate the suitability of ESTELLE, of the syntax oriented editor and of the translator;
- a **simulation phase** : in order to check the correctness of the dynamic behavior of the ESTELLE specifications, this phase permits to experiment the EWS simulator;
- when possible, an **implementation phase** : this work focuses on the integration of the generated code in a real operating environment; for some applications, the availability of implementations allows to measure generated code characteristics such as efficiency or capability of integration.

In addition, the evaluation phase is supported within EWS by a study of all important parameters inherent to a development phase : those parameters are related to the practical use of ESTELLE and to the behavior of software engineering tools users. All along the applications development phase, this evaluation work is made possible by the collection of data related both to the developed ESTELLE specifications and to the use of the tools. Some of those data are automatically collected, others are collected with dedicated interactive forms. These data are treated according to general evaluation criteria deduced from "classical" software tools (editors, compilers, pretty printers, symbolic debuggers ...) or to specific criteria derived from the intended goals of the project (interest of formal techniques for functional validation or for real implementation phases). Within EWS, all data collection is supported by a fully integrated tool called **data collector** which functionalities are : automatic collection handling, interactive forms handling, transport of the collected data from the collection sites (all partners dealing with application development) to the treatment site (VERILOG).

3.2 SELECTED EXAMPLES FROM COMPUTER NETWORKS AREA

Computer networks area can be considered as the major field of application of an ESTELLE workstation in the sense that the concepts developed in ESTELLE are really close to the software concepts generally needed in this field. Four protocols have been selected to be submitted to EWS :

. ISO Transport class 4,

. ISO Session,

. ISO ACSE and ROSE,

. ISO FTAM.

ISO Transport class 4 protocol: *description, simulation, implementation.*

Transport class 4 [11] has been chosen for its complexity and interest. The wide range of involved functions (such as multiplexing, segmentation, sequencing, concatenation, flow control, error recovery ...) constitutes a solid experimentation example for ESTELLE constructs.

The Transport class 4 specification which is used is the one developed in 1985 and revised in early 1986. The simulation task objective is to check the correctness of the updated specification. In order to be used in EWS, additional specification must be developed : Transport User and Network Service specifications. Most of the specification will not be changed and as the original version has been fairly tested, the simulation task is supposed not to present significant problems.

The proposed work will cover the EWS capabilities related to code generation tools and efficiency of the generated code. It will provide a working implementation of the class 4 Transport protocol in a Unix system and a set of performances measures over an Ethernet LAN.

This work is conducted by **ETSIT** (Spain).

ISO Session protocol : *description, simulation, implementation.*

This application basically consists in the simulation and the implementation of ISO Session protocol [12].

There are several aspects that make this protocol an interesting application to evaluate the ESTELLE workstation :

. the application is big enough to produce a complex specification to test performances and reliability of EWS,

. in addition to the Session protocol, a complete transport T.70 protocol has to be included in order to make possible the implementation of the Session layer on a real X.25 environment.

Among the wide variety of functionalities, the following functional units have been chosen to be coherent with T70 : kernel, half duplex, minor synchronisation, activity management, data exchange, exceptions and typed data.

The main objectives of this application are the description of Session and T70 layers (starting with a skeleton already available), their simulation and implementation on Unix system V.

This work is conducted by **ENTEL** (Spain).

ISO ACSE and ROSE protocols : *description, simulation.*

This study consists in the specification and simulation of two important Application Elements (ASEs) belonging to the OSI Application layer [13]. These application elements are the Associated Control Service Element (ACSE) and the Remote Operations Service Element (ROSE). They use the services of the Presentation layer and optionally, of the Reliable Transfer Service Element (RTSE).

When theses ASEs are available on one site, several Application layer configurations can be obtained according to the presence or absence of ACSE, ROSE and RTSE on the other site. This configuration is known after establishing the Application association, when the Application context has been negociated : the first point of interest of this application is to observe the ESTELLE Workstation capability to handle this kind of situations, in which a system is built in a dynamic manner.

The specific tasks to be developed with the ESTELLE Workstation are the following :

. Formal specification of the ACSE protocol,
. Formal specification of the ROSE protocol, assuming that the RTSE service is not included in the application context,
. Formal specification of the ROSE protocol, assuming that the RTSE service is included in the application context,
. Simulation of the above three protocols,
. Formal specification and simulation of the complete system : ACSE and the two options of ROSE (with and without RTSE).

This work is conducted by **ETSIT** (Spain).

ISO FTAM protocol : *description and simulation.*

The File Transfer Access and Management application is an ISO-standardized application protocol [14]. It allows the following services :

. transfer of data,
. data unit logical access belonging to different file structures,
. file attributes consult and in some case file attributes modification.

In 1983, twelve European constructors formed the SPAG (Standard Promotion and Application Group) and defined the functional profile concept, in order to facilitate the interoperability of their implementations [15]. These profile consist in making choices on standards and options which meet particular functional needs.

Due to the fact that there is a strong need for a basic file transfer, the SPAG has decided to define a profile, named A111, which only allows the transfer of whole files with text and/or binary data. No file access is needed for the basic profile and the file access structure is the simplest one. This profile is supposed to fulfill the requirements of many user groups where only simple files (in many cases of restricted length) are exchanged.

A restricted FTAM specification has been written in ESTELLE within the ESPRIT SEDOS ST410 project : it is composed of two parts, the initiator and the responder; they are a flat description of the protocol (no structuring). It is possible to strongly structure the specification with nested modules, each level refering to a specific regime within the protocol.

The work conducted in this study is the complete and structured description of A111 FTAM, including a simplified description of the Presentation service, and its simulation. As the size of this application is very large, it will present a great interest in measuring the performances of the EW. This work is conducted by **BULL** (France).

3.3 SELECTED EXAMPLES FROM SPACE COMMUNICATIONS AREA

The main interest of this application field stands in the emergence of the space domain in the protocol standardization area. With the technological progress and the space race, the space domain is opened to a strong development for the next years.

The protocols examined within this application are under standardization by the CCSDS (Consultative Committee for Space Data System). The CCSDS is an organization dedicated to the goal of defining recommended, internationally-agreed standard techniques for space mission data exchange. It is supported by all space agencies. This standardization came from the need of the different agencies (NASA, ESA, CNES ...) to access their spacecrafts via the same ground stations [16].

The structure required for a space application can be considered as several interconnected networks which are composed of ground stations and spacecrafts. Access to theses networks is made by ground stations connected to control centers.

As the ISO Reference Model, the architecture defined by the CCSDS is composed by seven types of services which are gathered into three global layers providing the following services:

. the Data Management Service includes the application process layer (interfacing with the telecommand system), the system management layer (translation of command directives into transportable telecommand application data units), and the packetization layer (providing end-to-end transport of application data on spacecraft);

. the Data Routing Service includes the segmentation layer (multiplexing of variable-length packetized data units), the data transfer layer (controlling the delivery of data units and offering a reliable sequential channel without loss or duplication);

. the Channel Service includes the coding layer (providing reliable delivery of data across physical medium) and the physical layer (providing a physical connection via radio signals from the ground station to the spacecraft).

For the Data Transfer from ground station to spacecraft, three Command Operation Procedure (COP) have been defined [17] :

. **COP0** : acceptance and retransmission of sequential frames without numbering; this procedure is adapted for spacecraft in low earth orbit ;

. **COP1** : Acceptance of frames in a strict sequential order, without loss or duplication and retransmission of rejected or missing frames from the first missing one; this procedure is adapted to spacecraft on orbits such as the data communication delay is less than 20 seconds ;

. **COP2** : selective acceptance of numbered frames by the spacecraft while the ground station only retransmits the rejected or missing frames; this procedure is recommended for distance beyond 10 E+06 kilometers.

The specific tasks to be developed with the ESTELLE workstation are the following ones :

. modeling of the three Command Operation Procedures defined in the Transfer Layer of the CCSDS: these specifications are available on natural language for data structure definition and architecture details (accepted as standard by CCSDS) and state matrices (under study by CCSDS) for the dynamic behaviors;

. modeling of the service provided by the lower layers (Channel service standard is available),

. simulation of the resulting specifications to validate each COP state matrix.

This work is conducted by **VERILOG** (France).

3.4 SELECTED EXAMPLES FROM TELECOMMUNICATIONS AREA

Two examples of application have been selected because of their present interest :

. a set of radio cellular network protocols (from the future pan-European digital cellular mobile communication system), the Link Access Protocol Data Mobile and a mobile-network signaling protocol,

. an ISDN (Integrated Service Digital Network) service located at the application level : a double call service.

The radio cellular network application : *description, simulation*

In 1982, the CEPT (Conférence Européenne des administrations des Postes et Télécommunications), in charge of the standardization in the field of postal and telecommunication services, decided to set up a new standardization group (the Groupe Spécial "systemes publics de communication avec les mobiles", GSM) in order to elaborate the specifications of the future pan-European digital cellular mobile communication network in the 900 MHz frequency band. One of the main attribute of this system is that it will allow international roaming mobile subscribers to access the telecommunication services provided by European operators in a fully compatible way.

All these protocols are currently under study within the scope of the layered model for Open Systems Interconnection (OSI) of ISO and CCITT (Comité Consultatif International pour le Télégraphe et le Téléphone), and are as far as possible based on standardized protocols by CEPT, CCITT and ISO, enhanced to cope with the pecularities of the radiocommunication cellular system : these form the ISDN protocols Q.920, Q.921, Q.930 and Q.931 [18] [19] [20].

One major interest of this work is to verify as deeply as possible all the protocols used on the radio path between a mobile station and a land-based system infrastructure. The protocols under specification specified are complex and new according to many aspects. Therefore, it is of tremendous importance that these protocols are specified without errors and ambiguities. Moreover, in order to insure the good operation of a mobile in all the concerned public

Land Mobile Networks of CEPT it will be necessary to certify the protocols developed by the manufacturers.

Taking account of the actual status of definition of the protocols, the following tasks are foreseen :

. validation of the link layer, close to Q.920 and Q.921,

. validation of the layer 3 protocol between mobile and network, close to Q.931 used to set up and release the communications.

This work is conducted by **CNET-PARIS** (France).

the ISDN double call application : *description and simulation.*

In the Telecommunications area, ISDN (Integrated Service Digital Network) is the backbone on which all future public services will rely in the next decades. The telecommunication world is currently switching massively from analogic to digital technology. Protocols that have been defined to standardize ISDN interfaces belong to the lower three layers of the OSI model [21].

The application studied for the EWS evaluation is the functional specification of a "double-call" facility, in the framework of the CERAME (CNET) project : CERAME is aimed at developing a prototype ISDN exchange integrating most of the ISDN features already defined. ISDN software is developed by CERAME according to four phases :

. functional analysis : formalizing the service and all its interrelations with other services on a purely functional basis (concept of modular decomposition),

. organic analysis : real design of the software by assigning functions defined to foreseen hardware and software elements,

. coding,

. testing.

The double-call service facility makes possible for a telephone caller to suspend a connection with another subscriber, without losing this connection, to call a third person, and then to be able to switch between the two connections as often as wished.

The work which is conducted within the project consists in :

. translation of the informal existing specification of the double-call service into ESTELLE; this work presents two interesting points :

. representation of data management,

. integration of data base interface;

. validation of the functional analysis by simulation.

This work is conducted by **CNET-LANNION** (France).

3.5 SELECTED EXAMPLES FROM INDUSTRIAL SYSTEMS AREA

Real time control systems constitute an application field where the geographical distribution and the specificity of the equipments are increasingly leading to distribute computers. Therefore, a great increase in the use of industrial Local Area Networks (LAN) is observed and the use of formal description techniques to describe distributed applications is expected.

Two examples have been selected :
- . one in the field of discrete manufacturing production: the control of a Flexible Assembly Cell (FAC),
- . the other in the field of continuous production : the monitoring of a field of heliostats of a solar power plant.

The FAC application : *description, simulation and implementation*

A Flexible Assembly Cell (FAC) can be defined as a subsystem in a manufacturing workshop. This part of the system is in charge of the manufacturing of a given set of equipments, further assembled together in bigger sets of equipments.The variety of the set of equipments that can be built together defines the cell flexibility [22].

The actual flexible assembly manufacturing systems have been designed and developed according to centralized implementation concepts. One interest of this application is to evaluate ESTELLE as a new approach for designing a distributed FAC.

The work conducted in this application consists in :
- . definition, structuration and description of such a distributed architecture in ESTELLE,
- . validation of the cell behavior using the simulator,
- . if possible, implementation of the ESTELLE specification with the following phases:
  - . code generation from the ESTELLE specification,
  - . development of non ESTELLE software to interface equipments such as robots, camera and controllers,
  - . integration of all software on a real experimentation site.

This work is conducted by MARBEN and LAAS on an existing assembly cell available at LAAS.

The plant monitoring application : *description and simulation*

The selected heliostat field of the THEMIS experimental plant is composed by about 200 focalising heliostats.The field is divided into 9 adjacent groups, each of them composed by 17 to 30 units [23]. Sun tracking is ensured by a open-loop control. The heliostat orientation is provided from calculated sun position data and target coordinates. The architecture is composed by functional units and security units. The kinds of functions required are :
- . operator interface,
- . central management including data acquisition, check and display of process state objects,
- . remote actions, control and changes,
- . local control and optimization.

As in most distributed process control applications, three main levels of hierarchy can be distinguished : central control, group control and unit control. The communication system is supported by two LANs, providing interfaces between the consecutive control levels.

This application addresses the genericity of this example by considering the first two levels of hierarchy. As a consequence, the use of ESTELLE and of the EWS in order to define high level functional descriptions of the complete system will be evaluated.

The work conducted in this application is the following :

. Three ESTELLE descriptions will be produced :

    . an user specification which gives the functional user requirements and fully reflects the problem complexity,

    . two different high level design descriptions providing two possible expressions of the answer to user needs :

        . the first one includes the use of static and dynamic modules instantiations,

        . the second one, more complete, contains the modeling of the communication system between the central control and the group control ;

. one description will be simulated using EWS simulator.

This application is conducted at LAAS.

8. Conclusion

EWS has been developed under UNIX System V and is available on Apollo and Sun graphic workstations. All graphic and interactive features are supported by an integrated dedicated package in charge of the window management, the graphic displays and the input devices.

With the concept of intermediate form, EWS constitutes an open environment on which additional tools such as verification tools, test sequences generator ... can be easily integrated to the existing toolset.

Sedos Estelle Demonstrator project is currently in the beginning of its second phase in which EWS is being evaluated with all selected applications.

The main advances represented by the project are :

- the degree of the tools integration to form a continuous production solution,
- the degree of workstation validation through experimentation on significant number of applications,
- the illustration of the productivity increase in using both ESTELLE formal description and a dedicated workstation on actual examples.

Reference documents

[1] :   "Estelle – a formal description technique based on an extended state transition model", ISO DIS 9074 July 2, 1987.

[2] :   "Example selection for ESTELLE Workstation evaluation", ESTELLE SEDOS Demonstrator, Esprit Project 1265.

[3] :   "ESTELLE Syntax Oriented Editor : user's guide", ESTELLE SEDOS Demonstrator, Esprit Project 1265.

[4] :   "MIRA : A computer-aided software engineering tool, user's manual" EXPERT SOFTWARE SYSTEMS, Ghent, Belgium.

542

[5] :   "Estelle Translator : Tutorial", ESTELLE SEDOS Demonstrator, Esprit Project 1265.

[6] :   "Estelle Translator : The Intermediate Form", ESTELLE SEDOS Demonstrator, Esprit Project 1265.

[7] :   "Syntax : manuel utilisateur" INRIA, France.

[8] :   "Requirements for ESTELLE code generator", ESTELLE SEDOS Demonstrator, Esprit Project 1265.

[9] :   "From ESTELLE formal semantics to ESTELLE Kernel implementation", ESTELLE SEDOS Demonstrator, Esprit Project 1265.

[10] :   "Requirements for ESTELLE simulator", ESTELLE SEDOS Demonstrator, Esprit Project 1265.

[11] :   "Connection Oriented Transport Protocol Specification" ISO 8073.

[12] :   "Basic Connection Oriented Session Protocol Specification" ISO 8327.

[13] :   "Application Layer Structure" ISO/TC97/SC21 N1494.

[14] :   "File transfer Management Access" ISO DIS 8571.

[15] :   "SPAG : Guide of the Use of Standards" ISBN : 0 444 701923 North-Holland.

[16] :   "Telecommand, Summary of Concept and Service" Green Book, Issue 5, February 1986.

[17] :   "Recommendation for space data system standard : Telecommand Part 2.1 : Command Operation procedures detailed specifications and state matrices", CCSDS Document, Red Book, Issue 2, April 1986.

[18] :   "ISDN User-Network interface data link layer specification, applcation of CCITT recommmendations Q.920 (I.440) and Q.921 (I.441)" Draft CEPT recommendation T/CS 46-20 revised (Nov. 1986).

[19] :   "ISDN User-Network interface layer 3 general aspects" CCITT recommendation Q.930 (I.450), red book, vol. VI, Fascicle VI.9.

[20] :   "ISDN User-Network interface layer 3 specification" CCITT recommendation Q.931 (I.451), red book, vol. VI, Fascicle VI.9.

[21] :   "ISDN Packed Switching : from Experimentation to System Integration" Le Roux A., Dauphin J.L, Laurens J., International Switching Symposium, proceedings, Phoenix, Arizona, March 1987.

[22] :   "NNS, a knowledge based on-line system for assembly workcell" H. CHOCHON, R. ALAMI, IEEE International conference on Robotics and Automation, 1986.

[23] :   "Commande d'un champ d'heliostats par microinformatique répartie sure de fonctionnement" M.D CABANNE, M. DIAZ, J.M. PONS, M. AGUERA, C. BOURDEAU, C. BURGAT, Congres AFCET, Toulouse 1986.

Project № 1520

THE ALF APPROACH TO PROCESS MODELLING

Ph. Griffiths [2], Ph. Jamart [3], A. Legait [1], D.E. Oldfield [2]

[1] GIE Emeraude   [2] ICL Defence Systems [3] Université Catholique
   c/o Research Centre   *Eskdale Road*   de Louvain
   BULL 58F23   *Winnersh*   *Unité d'Informatique*
   68, Route de Versailles *Wokingham*   *2, place Sainte Barbe*
   F-78430 Louveciennes   *Berkshire RG11 5TT*   *B-1348 Louvain-la-Neuve*
   FRANCE   *UNITED KINGDOM*   *BELGIUM*

ABSTRACT. The **ALF** (Accueil de Logiciel Futur) Project is concerned with building a third generation software engineering environment (i.e. a fully integrated environment using a rule-based control system). A key aspect of this system is its flexible approach to the support of software design methods, and this paper concentrates on the way such methods, or *Software Processes* are modelled in ALF.

1. Introduction

Process Modelling has recently come to the fore as an important research topic (c.f. [14]). ALF has its own way of modelling the software development process, providing a consistent user interface for all users of the system, from Project Manager to Programmer; Method Designer to Quality Auditor. The ALF concept of Process Modelling differs from other third generation software engineering environments in providing assistance to the entire software development team by guidance and explanation, and by taking initiatives when appropriate.

The remainder of this paper is structured as follows. First we introduce the general requirements of Process Modelling accepted by most researchers in this field. Then we present the general concepts of interest for Process Modelling. Next we introduce the ALF approach to Process Modelling. We also compare the ALF approach with those of other projects. Finally, we present the work organisation of the ALF project, and our plans for exploitation of the results.

2. Requirements for Process Modelling

The Software Crisis that is evident today, can only become more acute in the near future, as the number of engineers graduating from academia and going into industry will decrease, and as the importance of software will still be increasing in Information Systems.

Research in the field of Software Processes has shown the importance of Process Modelling, to provide help and assistance to the Software developer, and to improve software productivity.

We may consider the Software Process to be a description, with constraints, of a set of cooperating activities performed in order to create and evolve a software system. Using this definition of a Software Process it is clear that there is one Software Process for each software system that is developed. Several Software Processes may be based on similar principles, i.e. may refer to the same generic model for Software Processes. A *Process Model* describes features of a class of processes but not those features which are unique to each process. By adapting the model to project-specific (including organization/company-specific) requirements and by deriving processes from the adapted model, we obtain project-specific processes. Very general features of a set of Process Models are described by generic models. Obviously, a generalized description can be used more often than a specific solution. The benefits of describing general features of processes rather than describing each process individually are primarily those relating to the economy of reuseability, as discussed in [5] and [10].

We can distinguish two ways of describing Software Process Models: description by means of a formal notation, and narrative or informal description.

The widespread use of narrative descriptions is highlighted by the fact that organizations have recorded their standard operating procedures in this way. This widespread use shows that the idea of describing Software Processes is not a new one. Of course, operating procedures have been described for each software development. The problem with most of these descriptions was, and is, that they lack precision. They are given in the form of general guidelines and advice like "start with a requirements phase" or "test each module carefully". Therefore, they are the source of many misunderstandings and mistakes.

While the principle of describing Software Processes is not a new one, the idea of integrating a formal notation of Process Modelling into a Project Support Environments is more recent. These third generation *Integrated Project Support Environments (IPSEs)* are environments which are built around a knowledge base in which knowledge about the planned software process is stored. A Software Process Model is used to describe this knowledge. Besides other uses, this knowledge is exploited to help and to assist all kinds of users to facilitate their work and in order to yield lower costs of software development and improved productivity. This assistance ranges from control to explanation, and may be provided by the system at the user's request or on its own initiative. Another important feature of third generation IPSEs is that they enable the development of software systems in different application domains, where different life-cycle models are appropriate. Such an IPSE will support the modelling of existing methods, adaptations of existing methods, or new ones.

Bearing in mind these aspects of Software Process Modelling and the advantages of describing Software Process Models formally, the main requirements for a system supporting Process Models are:

- The system shall support the formal definition of Software Process Models.

- The system shall support the derivation of single process descriptions from a Software Process Model. Such descriptions form the basis for the development of one particular software system.

- The formal language for describing Software Process Models shall enable the description of various methods in a uniform way. As a method is a description of part of a Software Process life cycle and as a Software Process can be considered as a formal description of a method, the description language should enable a uniform description of both.

3. Process Modelling: its components

Having presented the motivations and the high level requirements for Software Process Modelling, we introduce the following key concepts of Process Modelling:

- Objects
  All Models of Software Process incorporate an object model. The most common features of such models are typing, inheritance of types and data, and the grouping of data at various degrees of granularity. The views which different users have of various parts of the object base are sometimes defined by *roles* or *schemas*.

- Activities
  All activities must always be described. The description may range from a model of tools with their syntax and logical conditions, to a mathematical description.

- Decomposition
  Decomposition is the division of tasks into sub-tasks. Even the simplest models can be hierarchically decomposed.

- Cooperation
  Activities do interact and they have to be synchronized. This synchronisation may be achieved using messages, or simply by object sharing.

- Control
  Control of objects and activitives may take three forms. Activities may have to be executed, at a given time, under given conditions; activities are imperatively executed. Activities may become executable; the domain of activities is only retricted. Finally, if activities are forbidden, or fail, recovery procedures under the initiative of the system may be provided. For the implementation of control, projects have variously used rule based mechanisms, algorithmic or triggering mechanisms.

- Incomplete knowledge
  Even if most projects use some kind of formal description for Process Models, these descriptions are often incomplete, and for some authors this seems to be part of the nature of Process Modelling. Human creativity is impossible to describe with a complete formal model.

- Adaptability
  It is recognised by most authors that the adaptability of a Software Process Model from one case to another is an important feature of the problem. Adaptation may be supported by version control, inheritance facilities or other mechanisms.

4. The ALF approach to Process Modelling

In section 3. above we introduced the essential elements required for process modelling. In this section we describe the formalism in ALF for process modelling, called *Model for Assisted Software Process (MASP)* and show how it deals with process modelling.

The MASP concept is a key part of ALF. The MASP is the formalism which ALF uses to describe different software process models in a uniform way. A MASP that describes a particular software process, for example SSADM (System Structured Analysis and Design Method), is not specific to a particular project until values for such things as team members, module names, tools, hardware resources, etc., are specified, or *instantiated*. This can be repeated, for any given MASP, once for each project or sub-project. An instantiated MASP, or *IMASP*, may then be executed by the ALF system to yield an *Assisted Software Process* (*ASP*). An ASP is uniquely identified by its run-time behaviour.

Each instantiation of a MASP produces an IMASP, which is a static process description and different from any other IMASP produced from the same MASP. The execution of an IMASP results in the execution of operators and interaction with the user. A consequence of this is that different executions of one IMASP may yield different sequences of operator executions and therefore different results. So a MASP describes a class of IMASPs and each IMASP represents a class of run-time behaviours.

Now we will see how the MASP concept address the concepts introduced in section **3**. The components of a MASP are:

> An object model
> A set of operator types
> A set of expressions
> A set of orderings
> A set of inference rules
> A set of characteristics

In what follows here, we give a brief account of these concepts, but a more complete description of the MASP can be found in [7] and [1].

Objects The objects of a MASP are described using the ERA[4] (Entity, Relationship, Attribute) approach of PCTE[3] (Portable Common Tools Environment). This is perhaps not the most advanced method of representing objects but it is currently available and at low risk. Furthermore the MASP itself contains enough functional richness to be able, if necessary, to implement more advanced models, for example an object oriented interface.

Activity Each ASP is an activity, but how are ASPs activated? There are three mechanism whereby things happen in ALF. The most obvious is through user actions. The other two are through *inference rules* and through *characteristics*. Inference rules cause forward progress, they use *expressions* to observe the MASP and when certain situations arise they activate *operators*. Characteristics are similar to ALF's inference rules but are backward looking, they activate operators in order to recover from erroneous situations.

Decomposition One important feature of ALF is that MASPs can be decomposed into sub-MASPs. The operators in the MASP are subdivided into two classes. The first, called *elementary*, are actual tools, for example a specific compiler or a particular structure editor. The second, called *non-elementary*, are themselves MASPs. Thus any method is not described by just one MASP but by a hierarchy of MASPs and, similarly, any project is described not by a single IMASP but by a set of hierarchies of IMASPs.

Cooperation Projects cooperate by sharing tools and objects (for example, software modules or users). MASPs and IMASPs can also share these resources. Projects can also reuse the MASPs themselves. Users are symbiotic with MASPs. Since ultimately all activities originate with some user action, IMASPs cannot proceed indefinitely without users; similarly all activity is controlled by IMASPs, even the production of MASPs. However, useful work can continue without the presence of users. So users are not tied to MASPs and MASPs are not tied to users. Users can change their context from one project, say, to another or within a project by temporarily changing their *role* perhaps from a software developer to a project manager or team leader.

Control Our philosophy regarding control is to restrict the sequencing of operators, which is done by the use of *orderings*. If no restriction is imposed by an ordering then all possible sequences of operator invocation are permitted. However just because an order permits an operator to be invoked does not necessarily mean that the operator is invoked. There are two ways of actually invoking an operator, the first is by a user action, the second by the firing of a rule (including the violation of a *characteristic*). So the ALF approach is much more flexible than a purely imperative approach. Recovery and initiative taking are achieved through characteristics and rules, as described above in the item on Activity. An ordering does not have to specify a unique path through a set of operators, although this capability exists in case a particular method requires it.

Incomplete knowledge Closely coupled with the MASP interpretation system, is a knowledge manipulation system. This understands *orderings* and can infer possible paths through a set of orderings to arrive at a given point. This is done by examining not just its orderings but all of the MASP's components; for example the pre and post conditions of the operators. See the item above on Activity for inference mechanisms used.

Adaptability The architecture of ALF allows one MASP to monitor the behaviour of another. This mechanism is used for the generation of *history* and for *feedback*. These two (*feedback* for short term results, *history* for the longer term) are used to modify the behaviour of the ALF system by making changes to the details of MASPs and therefore IMASPs.

Genericity MASPs are generic. The instantiation process associates actual objects with the object types referenced by a MASP. Similarly for operators, which can be either tools or MASPs, the instantiation process associates actual operators with the operator types described in the MASP being instantiated.

5. Comparison with other approaches

We present in this section the way other systems or projects related to Software Process Modelling deal with the key concepts introduced in section **3.**

The systems or projects discussed are Arcadia [13], Marvel [8], IPSE 2.5 [12], EPOS [6], Bisiani's Tool to coordinate tools [2] and Genesis [11].

Note that Genesis was not designed as a Software Process Environment but is related to Project Management Modelling.

Objects All authors recognize the need for supporting these three relations for objects: *classification/instantiation* (i.e. typing), *generalization/specialization* (i.e. inheritance) and *aggregation/decomposition*.

In most systems, the object structure is kept in a database, while the actual contents of the objects are stored in files. This has two drawbacks compared with the ALF/PCTE approach, where the database and the file system are integrated: firstly, it is more difficult to integrate existing tools (see *Modelling Activities* item); and secondly, the need for maintaining the consistency of both data systems generates extra overheads. However, the ALF approach also has a drawback: when the granularity of data is fine, each access to an atom of data may require a disk access; we are working on overcoming this by suggesting improvements to the implementation of PCTE.

In Arcadia and Genesis, historical links are automatically recorded: this means that when an activity produces an object using another object as input, the result is automatically linked to the input. Moreover, in Arcadia, all objects created by tools (i.e. atomic activities) are automatically stored in the database.

In most systems, predefined attributes are managed by the system itself.

Genesis proposes the concept of *information abstractor* (although *extractor* would perhaps have been a better name) to transform internal contents of objects (i.e. generally the contents of files) into external information (i.e. attributes, entities and links).

EPOS includes mechanisms for version and configuration management.

All theses functionalities may be supported by ALF, but they are not built-in: MASPs or tools must be written to support the functionality; this means that these functionalities may be user-defined to local requirements.

Modelling activities Activities are in general modelled either by atomic tools or by a description in a very high level language.

Linking between atomic tools of the underlying operating system and activity models is a classical problem. The tools have been designed for acting on a file system; if one wants to integrate them in the IPSE so that they act on the database, one must either modify them or call the tools indirectly by putting them into an envelope. This envelope can then update the database. For instance, in Marvel a tool description (i.e. the activity part of a rule) is a mapping to a command line acting on concrete files; Bisiani's Tool uses a similar approach. This problem does not occur in ALF, since the PCTE database and the file system are one and the same.

In Arcadia, a tool is a fragment of an activity; in other words, a tool performs very elementary activities; any substantial activity involves complex interaction between various fragments.

Decomposition Decomposition may be achieved either by defining activity models in terms of models of sub-activities, or by relating each activity to a goal, where goals are then structured into subgoals. This latter approach is followed by Bisiani's Tool. In IPSE 2.5, a Sofware Process is a set of *roles*. Each role is a network of atomic activities and of subroles which may communicate with each other. This seems to be quite similar to the ALF approach, but ALF uses only two concepts, roles and Software Processes, both being represented by MASPs; moreover, it is not necessary in ALF to know whether an operator type will be instantiated to a MASP or to an atomic tool.

Cooperation In IPSE 2.5, activities belonging to different roles may interact by exchanging information. In contrast, Arcadia supports communication by object sharing. Both complementary solutions are supported by ALF, using either the triggering mechanism defined for ALF or the communication mechanisms of PCTE.

Control There is a generally recognized need to be able to describe predictable development routes without being restrictive; i.e. imperative control alone is inadequate. Systems propose some mixing of positive (i.e. what the system can do by itself) and negative descriptions (i.e. what the system cannot do by itself and must not allow the user to do); positive descriptions are generally modelled either by rules triggered by predicates or by a goal orientated approach. In Genesis, control is represented by rules directly attached to the ERA model; this is an object-oriented approach.

We will distinguish, as in previous sections, three control principles: imperative control, restrictive control and recovery.

Imperative control In IPSE 2.5, activities in one role are connected by a Petri-like net indicating imperatively an ordering for these activities. Partial ordering is obtained for a full Software Process, since each role of the Software Process has its own net.

Restrictive control The ability to select an activity, even when it is involved in an imperative ordering, may, in general, be protected by preconditions. This way of representing knowledge is used by almost all systems mentioned. Such preconditions are *necessary*, because they do not allow the automatic activation of the operator. The main difference between those systems is that such preconditions may be attached either to the activity itself or to the Software Process using the activity.

Recovery and initiative taking In Marvel, when a tool is invoked and its precondition is not verified, this precondition is considered as a temporary goal for the system; opportunistic reasoning based on the pre and post conditions of rules is used to build a plan, i.e. a sequence of activities which leads to the precondition being verified. Opportunistic reasoning means that backward chaining from the target and forward chaining from the current state may be used.

Arcadia supports the *dynamic synthesis* of tools from fragments. A tool, i.e. a *plan*, is a network of *nodes* (object types) connected by *edges* (elementary tools). Thus planning in Arcadia relies on a monolithic typing mechanism.

In Genesis, the project is at any moment directed by a set of goals. The rules are used to automatically perform certain activities.

In IPSE 2.5, predicates are used to trigger activities composing roles.

In ALF, we distinguish clearly between preconditions attached to the operator type, which are *necessary* conditions, and rules attached to the Sofware Process, which are *sufficient* conditions. However, preconditions become temporarily sufficient conditions when a goal must be reached provided the action helps to reach the goal, this occurs when a characteristic has been violated. No other known system provides such a flexible approach.

Adaptability In most systems, it is possible to build big tools for describing substantial activities whose knowledge is complete. However, adapting a tool for use in a process description is not easy when the database and the file system are two distinct components.

The need for supporting refinement of the activity description is recognized by most authors. In EPOS, it is suggested to consider the problem of refinement as a version management problem. Another suggestion is to support specialization of activity types. In IPSE 2.5, the PML language supports inheritance of object types and activities are considered to be a special kind of object. We are currently designing an inheritance system for MASPs. We think also that the monolithic typing systems proposed in the literature are not adequate for managing real refinement of Software Processes.

Attaching *sufficient* preconditions to tools is not a flexible approach because any modification of the Software Process may involve the modification of these sufficient preconditions, i.e. in the case of the tool itself. So, preconditions attached to tools must be *necessary* preconditions and sufficient preconditions must be attached to the Software Process Model.

Adaptability also means being able to modify process descriptions on the fly, i.e. dynamically during the actual execution of the process. This seems to be supported by IPSE 2.5, but we have been unable to find published details of the mechanism.

Genericity The word *genericity* is widely used in the Sofware Process community to refer to the parameterization of processes and to the ability of a description to produce a class of concrete plans. We think the concept of genericity should be reserved for type parameterization, either by classical polymorphism or by type inheritance.

Of course, every system mentioned supports some kind of modelling which may, by an instantiation process, produce different plans, but real genericity is supported by the inheritance mechanism of IPSE 2.5.

The use of activity types instead of concrete activity identifiers in the Sofware Process descriptions is also generally considered to be a form of genericity. This is supported by EPOS which uses the version management mechanism to select concrete tools, since tools are a special kind of object.

Incomplete knowledge Most authors recognize the necessity of allowing partial descriptions of Software Processes, because the expert knowledge is itself incomplete. This means that at certain moments, the system must be allowed not to know what to do. This implies that users must be allowed to propose, by themselves, the activation of some tools.

In Marvel, the user may invoke tools explicitly when the system has no goal to reach. Nothing is said about the possible action of the user when the system is *unable* to reach a goal.

In Bisiani's Tool, the system is quite similar and the command which is generated may allow user interaction; but once again, what happens when the planner cannot derive a plan?

6. Project Summary

6.1. PAST ACHIEVEMENTS

The first year of the project was devoted to studying various topics and to organizing the consortium. The main result of this initial study phase was the set of requirements for Process Modelling presented in section 2 and 3 of this paper. This work led to the definition of the MASP concepts (section 4).

In the second year, we refined the various concepts of the MASP, and designed the user-level MASP language. This language is complex, and as a MASP designer is only concerned with a specific Software Process Model and does not want to worry about the syntax in which this model is to be expressed, the first tool that needed to be implemented during year two was a MASP editor. Another tool was specified and designed: a MASP consistency checker. This tool will check the consistency of each component of a particular MASP or sub-MASP.

6.2. CURRENT ACTIVITIES

The main internal components of the ALF system will be a *Piloting System* and an *Information System*. These are to be built on an *Executive System* which is based on the PCTE interfaces. The ALF system will be provided with a comprehensive and consistent user interface, which will be developed on top of X-Windows. The first two components will have to support the description, the instantiation and interpretation of the ALF MASP. We have specified these components during the second year of the project and will implement them during the third. The Piloting System will benefit from previous work in artificial intelligence; it will be implemented using Sepia [9] (a Prolog compiler developed by the

European Computer-Industry Research Centre (ECRC)) and XRete [16] (a real time expert system generator developed by the Central Reserach Laboratory of Thomson CSF). The Information System will benefit from previous work in databases as perfomed in the PACT [17] and PCTE+ [3] projects. Work on PCTE enhancements which has continued throughout the first two years will also continue within the project.

6.3. FUTURE WORK

MASPs will be written during the third year of the project. The methods which will be considered are VDM (Vienna Development Method) [15], configuration management, Hood [18], M2C (Méthode de Conception Cértifiée) [19], cost estimation and functional nets. There will be also a MASP for MASP designers; MASP design is one activity among others, and the ALF system will support it as well.

During the fourth year of the project, we will experiment with the MASPs written in the third year. An assessement of the models and of the system will follow.

6.4. EXPLOITATION OF RESULTS

There will be three levels of exploitation results. The first level of exploitation will be mainly academic, by publications, theses and lectures on the ALF approach to Process Modelling.

The second one will be proposals for PCTE enhancements, to define the next generation of Public Tool Interfaces. One of the aims of the project is to propose enhancements to the PCTE standard to make it fit for Process Modelling. Where requirements highlight deficiencies within the actual PCTE specification, new proposals may be formulated to help defining the next generation of Public Tool Interfaces compatible with PCTE.

The third level of exploitation will be the in house and commercial exploitation by the ALF partners of the prototypes developed in the third year. This commercial exploitation will probably require some work on the standardization of Process Modelling Languages.

7. Acknowledgments

The authors acknowledge the contribution to this paper from all the members of the ALF Consortium, who are: GIE Emeraude (France), CIG-INTERSYS (Belgium), Computer Technologies Co.-CTC (Greece), Grupo de Mecánica del Vuelo, S.A. (Spain), International Computers Limited (United Kingdom), University of Nancy-CRIN (France), University of Dortmund-Informatik $\underline{X}$ (Germany), Cerilor (France), Université Catholique de Louvain (Belgium) and University of Dijon-CRID (France). This work is partially sponsored by the Commission of the European Communities under the ESPRIT programme (Project Ref. N⁰ 1520).

8. References

[1] Benali K. et al. (1989) "Presentation of the ALF project", Proceedings of the International Conference on Software Development Environments & Factories, Berlin, May 1989.

[2] Bisiani R., Lecouat F. and Ambriola V.(1988) "A tool to coordinate tools", Expert Systems, November 1988.

[3] Boudier G., Gallo, F., Minot, R., Thomas, I. (1988) "An Overview of PCTE and PCTE+", in Proceedings of the 3rd ACM Symposium on Practical Software Development Environments, Boston, November 1988.

[4] Chen P. P. (1976) "The Entity-Relationship Model: Towards a unified view of data", ACM Transactions on Database Systems, Vol. 1, No. 1, March 1976.

[5] Chroust G. (1988) "Models and Instances", Software Engineering Notes Vol. 13, No. 3, July 88.

[6] Didricksen T. et al. (1989) "Change Oriented Versioning", in Proceedings of the ESEC'89 Conference, Warwick, September 1989.

[7] Griffiths Ph. et al. (1989) "ALF: its Process Model and its Implementation on PCTE", in Proceedings of the International Conference on Software Engineering Environments, Durham April 1989.

[8] Kaiser G.E., Feller P.H., and Popovitch S.S. (1988) "Intelligent Assistance for software development and maintenance", in IEEE Software, May 1988.

[9] Meier M., et al. (1988) "SEPIA Version 2.0 User Manual", ECRC Report: TR-LP-38, September 1988.

[10] Osterweil L. (1987) "Software Processes are Software Too", in Proceedings of the 9th International Conference on Software Engineering, Monterey, CA, March 1987.

[11] Ramamoorthy C.V., et al. (1985) "GENESIS: An Integrated Environment for Supporting Development and Evolution of Software", in Proceedings of COMSAC, 1985.

[12] Roberts C. (1988) "Describing and Acting Process Models with PML", in [14] Proceedings of the 4th International Software Process Workshop, Moretonhampstead, May 1988.

[13] Taylor R.N. et al. (1988) "Foundations for the Arcadia Environment Architecture", in Proceedings of the ACM SIGSOFT SIGPLAN Software Engineering Symposium on Practical Software Development Environments, Boston, MA, 1988.

[14] Tully C. (ed.) (1988) "Proceedings of the 4th International Software Process Workshop", ACM SIGSOFT Software Engineering Notices, Vol 14, No 4, Moretonhampstead, May 1988.

[15] Jones C.B. (1986) "Systematic Software Development Using VDM", Prentice/Hall International, ISBN 0-13-880717-5.

[16] (1989) "Specification du language", Thompson CSF/LCR, version 1.0.

[17] Thomas I. (1989) "Tool Integration in the PACT Environment", Proceedings of the 11th International Conference on Software Engineering.

[18] (1987) "The HOOD Manual", CRI-CISI INGENIERIE-MATRA, issue 2.0.

[19] Tankoano J. (1987) "Méthode de Conception Cértifiée", Thèse d'état Université de Nancy.

Project no 1550

The DRAGON project

Andrea Di Maio
TXT S.p.A.
Via Socrate 41
20128 Milano (Italy)

Ian Sommerville
University of Lancaster
Bailrigg
Lancaster LA1 4YR (U.K.)

Frank Bott
University College of Wales
King Street
Aberystwyth, Dyfed SY23
3BZ, Wales (U.K.)

Rami Bayan
GSI Tecsi
6 Cours Michelet
Paris La Defense (France)

Martin Wirsing
Universitaet Passau
Postfach 2540
D-8390 Passau (Deutschland)

Abstract

This paper presents the results of the DRAGON project, that aims at providing effective support to reuse in real-time, distributed Ada applications. The various aspects of the project are outlined, together with some hints about the success in partner cooperation.

1. Introduction

The DRAGON project was conceived within the context of real time, distributed, non stop, embedded systems, such as are found in applications in manufacturing industry and aerospace. Within this context, the original objectives of the project were:

– to develop methods and tools for designing reusable software components (*design for reuse*) and for designing specific software systems in an environment rich in reusable software components (*design with reuse*);
– to develop tools for partitioning Ada programs into units which can be mapped on to different physical processors, independent of the software design;
– to develop techniques for reconfiguring distributed systems without stopping execution, both to change the physical configuration and to replace software components;
– to validate the above work on a case study, by reworking the design and implementation of an existing system.

To meet these objectives in an integrated manner, a single paradigm for the structure of the systems is required. The participants came to the conclusion, early in the project, that this paradigm should derive from object oriented design; in other words, the object should constitute the fundamental unit of distribution and reconfiguration and object and classes should be the fundamental units for reuse and formal specification. This conclusion, however, raises as many questions as it answers.

Four very different views of object oriented design (OOD) are represented, for example, by Smalltalk (Goldberg & Robson 1983), Eiffel (Meyer 1988), HOOD (1987) and the work of Booch (Booch 1986). Smalltalk and Eiffel are both languages first and foremost. Despite their very considerable technical interest and the insights which their philosophies reveal, their limited availability and still experimental nature made them unacceptable as a basis for DRAGON. Furthermore, they have nothing to offer in the real time area nor in the area of distribution.

HOOD and Booch's software components are both firmly based in Ada and are therefore linguistically acceptable. However, precisely because they are based on Ada, they take a much narrower view of OOD than is taken by object oriented languages; in particular they do not provide for *inheritance* mechanisms. This lack was felt to be insuperable barrier to achieving software reuse on a large scale. However, HOOD is a 'design methodology' in a way which none of the others are and the discipline which this implies was felt to be essential if DRAGON was to be industrially viable.

The central concern of the project was therefore to develop an object oriented approach to the design and implementation of real time, distributed systems which:

- would be compatible with the use of Ada;
- would support distribution;
- would support reuse by taking a broad view of OOD;
- could be used as the basis for an industrially acceptable design method.

This approach would then form the unifying paradigm for the other aspects of the project.

The following sections of the paper describe the object oriented approach itself; the design method and the tools which support it; the more general reuse toolset; and the way in which formal methods are integrated into the whole. Finally some comments are offered on the level of integration and collaboration achieved within the project.

2. The Object Oriented Paradigm

At the very beginning of the project it was felt that the object oriented approach was an attractive way to provide a single framework within which to address the various issues with which DRAGON is concerned. Nevertheless, it was soon realised that just defining another object oriented design method or programming language was not enough for our purposes. We had to ensure that our approach could accommodate the requirements for reuse, distribution and dynamic configuration in as uniform a manner as possible.

At this stage Ada was felt to be an uncomfortable constraint: the importance of the language was evident, for reasons of standardisation and maintenance and because of the strong commitment to Ada and support for it on the part of certain key organisations; on the other hand, Ada lacks some of the features of object oriented languages which are most conducive to software reuse. Indeed, although the term 'object oriented' is often used with reference to Ada and many object oriented techniques have been developed that are targetted to Ada, Ada lacks the mechanisms for inheritance and polymorphism (except in the rather limited form of overloading) which are essential if the full potential for reuse which the object oriented approach offers is to be realised.

These considerations led us to the following set of requirements for the object oriented framework within which the project would operate:

- the programming paradigm should be as compatible with Ada as possible;
- a new language should be produced, capable of being used as a programming language or a program design language (PDL) and capable of being translated into fairly readable Ada automatically;
- the framework must accommodate the needs of both reuse and distribution;
- an object oriented design method should be produced which makes use of the language and which is as compatible as possible with existing methods, in particular with HOOD.

These requirements then led to the design of DRAGOON (Distributed Reusable Ada Generated from an Object Oriented Notation). We perceive DRAGOON more as a notation than a language. Its flavour is very similar to Ada but it introduces features such as classes and inheritance that are necessary for real object oriented design but which are missing from Ada. It should not be regarded as an extension or revision of Ada but rather as an additional layer which allows the programmer and designer to work at a higher level, using object oriented constructs which are better adapted to reuse and distribution.

There have been other attempts to add an extra, object oriented layer to Ada; examples are InnovAda (Simonian & Crone 1988) from Harris Corporation and Classic-Ada (SPS 1988) from Software Productivity Solutions. What distinguishes DRAGOON from such proposals is its scope. It is far more ambitious in that it addresses distribution, reuse and, at a certain extent, real time issue within a single, uniform framework. This also makes it more powerful than other object oriented languages such as Objective C (Cox 1986), C++ (Stroustrup 1986) and Eiffel (Meyer 1988), all of which were major influences on its design.

2.1. The Language

The most important novel features of DRAGOON (Di Maio et al. 1989) with respect to Ada are:

- classes
- abstraction
- multiple inheritance
- polymorphism
- behavioural inheritance
- full support to distributed execution.

A major improvement is the availability of **classes** as library units. Taking Meyer's view (Meyer 88) a class can be regarded as the implementation of an abstract data type: the class contains (and hides) a set of instance variables and provides a set of subprograms (called *methods*) that can be called by clients (i.e. users) from outside. Different instances can be created from a class that are called objects: each object has a state consisting of the values of the instance variables of the corresponding class (that can be in turn references to other objects).

In order to support incremental design of software components, DRAGOON allows classes to be defined in subsequent steps, with increasing details. Therefore **abstract** can be used which introduce some features whose implementations are *deferred* to a later stage (heir classes, see below).

Multiple **inheritance** allows a class to inherit the features of one or more *parent* classes:

such features can then be modified in various ways (redefined, renamed, restricted, etc.).

Polymorphism is a consequence of inheritance that overcomes the strong typing scheme of Ada. Although beneficial for safety purposes, strong type checking turns to be a constraint for incremental components to be developed. Polymorphism allows new components to be used by clients of the parent ones, which increases flexibility and extensibility. Moreover DRAGOON's polymorphism is fully compatible with Ada's strong type checking: for instance, the inheritance mechanism is static, thus permitting compile time checking and increasing efficiency.

In the domain of embedded real-time systems for which Ada is intended, support for concurrency is an important requirement. Traditional approaches to programming concurrent systems involve interprocess communication and synchronisation features using relatively low-level constructs, such as monitors, semaphores and the rendezvous (ALRM 1983). The direct inclusion of synchronisation solutions in this way is one of the greatest obstacles to the reuse of components because it requires early decisions about the detailed approach to be used for managing interactions between concurrent objects.

The idea introduced in the DRAGOON language is to separate the concerns of the programmer for the functional aspects of an object and the control of its use by others, by enhancing the conventional inheritance mechanism. The programmer is able to build concurrent and distributed software incrementally, superimposing the necessary synchronisation properties on existing classes in the library, at any time. Consideration of these matters can therefore be deferred until the latest possible stage of the system development, so improving the potential for reuse.

DRAGOON objects can be either passive or active: in the latter case an independent thread of control is started when the object is created.

Abstract *behaviours* (i.e. policies ruling the interleaving of different threads of control affecting an object) are defined by constructing *behavioural classes*, containing the abstract specification of a method interleaving policy. A concurrent (protected) class can be built by inheriting from the non-concurrent (unbehavioured) and from the appropriate behavioural class. This special kind of inheritance is called **behavioural inheritance**.

As a consequence, it is possible to reuse unbehavioured classes in different concurrent applications, just by inheriting from different behavioural classes. The properties of polymorphism, that enable the concurrent behaviour of a supplier class to be changed without its clients having to be modified, still hold, which greatly improves the prospects for reuse.

In an object oriented paradigm, the application is made of several independent objects which communicate by method calls, i.e. by exchanging messages.

2.2. Distribution

Distribution in DRAGOON is concerned with the **clientship** relation which allows the instance variables of an object to refer other objects (of different classes), so producing a logical network of objects.

Traditional approaches to distributed systems implement an application as a set of programs communicating through services provided by the underlying run time support (or operating system). In such a model the internal communication between modules belonging to the same program (procedures, packages, Ada tasks, ...) is viewed conceptually as being different from external communication between programs. The user is thus forced to decide from the early stages of design about the model of communication and the specific software allocation to choose and this prevents any future maintenance.

In DRAGOON an instance variable can refer an internal object (i.e. belonging to same address space of the client object) as well as an external one (i.e. belonging to a different address space than the client).

The decision about whether implementing an object as an internal or an external one depends on the following issues:

- degree of reusability desired for an object: when an object component is to be reused under executable binary form (this is the case of most commercial components) then it is *de-facto* in a different address space than its clients;
- moveability of objects: when an object needs to be replaced or migrated dynamically to a different hardware then it must be implemented in a separate address space;
- hardware diversity: when a client and a server execute on different hardware, they are in separate address spaces as a consequence;
- shareability of objects: when an object serves objects in different address spaces then it must be itself in a separate address space.

When an object is to run on a given machine, a directive has to be given to the DRAGOON compilation system. This is achieved again by a particular use of *inheritance*. The class from which the object is instantiated inherits an execution support class which offers additional services concerning the manipulation of objects such as downloading, migration, stop or others. These services are viewed exactly like any other service provided by the object.

Such an object in DRAGOON is called an **executable object**. Actions like reconfiguration and site failure detection are then based on manipulating these objects by using assignments of instance variables and method calls: distribution is then handled inside the language in a fairly elegant and effective way, allowing the designer to concentrate on the functional aspects of the application and to delay allocation decision until the definition of the executable objects mentioned above.

2.3. From DRAGOON to Ada

After the design of the language, a major effort was required by the definition of the corresponding Ada transformation. In fact DRAGOON classes are automatically translated into source Ada, to be compiled by (any) host compiler.

Several alternatives were investigated, all aiming at achieving an effective and readable transformation. The adopted solution was chosen according to requirements such as minimisation of recompilations, run-time overhead and readability. It is worth mentioning that the implementation of behavioural inheritance makes provisions for alternative behaviour implementations: in other words, it is possible to implement synchronisations using either Ada tasks or different (and perhaps more efficient) mechanisms, without any impact on the preprocessing tools.

3. The Dragon Design Method

While object oriented design is widely believed to be an excellent way of designing software systems, there have been few attempts to produce an industrially usable development method based on this technique and there is no consensus about the structure of the underlying model. A method like HOOD, which insists on strictly hierarchical decomposition of objects and which makes little use of classes and none of inheritance, is very far from the original ideas exemplified in Smalltalk; indeed, many practitioners would deny its right to be considered an object oriented method. On the other hand, the 'methods' described by Booch and Meyer lack the discipline necessary to turn a philosophy into a method.

It was originally intended that the Dragon design method, provisionally christened DEMON, should be an extension of HOOD. However, the underlying model in HOOD is so far away from the spirit of object oriented design that many of the benefits of that approach are lost. Although DEMON retains many of the characteristics of HOOD it is, in spirit, much closer to Meyer's work. It is an attempt to impose on Meyer's ideas the disciplines necessary to produce an industrially usable design method, while avoiding the defects evident in HOOD.

3.1. The DEMON Design Model

In contrast to HOOD, DEMON is firmly based on the notion of classes and of class inheritance. The designer does not design objects but classes. The idea of a component library, consisting largely of classes, is central to the concept of reuse in DEMON. The library is assumed to transcend individual projects. Any project can therefore expect to exploit an existing library and must expect to contribute to the library, for the benefit of future projects.

The contents of the component library are described in the component catalogue and the tools used for searching and maintaining that catalogue are to be regarded as part of the DEMON toolset, although, of course, they are capable of being used in a wider context.

There are two fundamental relationships which may subsist between a pair of classes: clientship and inheritance. Each of these induces the structure of an acyclic directed graph on the set of classes. (Note that this allows a class to inherit directly from, or be a direct client of, more than one other class; this is not possible if the induced structure is a strict hierarchy.) These relationships are essentially the classical ones as used in Smalltalk. Class A is a client of class B if the definition of A refers to an instance of an object of class B. Class A inherits from class B if it is defined by modifying the definition of B or adjoining new operations or attributes. Class definitions cannot be nested.

In contrast therefore to HOOD, DEMON permits the use relationship to exist between classes rather than objects, in the form of clientship, and does not permit the HOOD inclusion relationship to exist between classes; of course, a class definition may include the declaration of objects of other classes — this is just an example of clientship. HOOD has no equivalent of inheritance; classes are used only as a shorthand for defining several objects with identical characteristics.

Classes in DEMON, like objects in HOOD, may be active or passive. An instance of an active class has a single thread of control, which is specified in the class definition, and all instances of active classes are regarded as potentially executing in parallel; an active class is thus analogous to an Ada task type and its instances can be considered candidates for distribution. Instances of passive classes have no such thread of control and thus correspond to procedures or functions.

Much traditional reuse has been of a 'bottom-up' nature, i.e. reusable routines have formed part of the underlying virtual machine on top of which an application system is implemented. Particularly in the area of distributed embedded systems, it is important to allow for 'top-down' reuse, in which the upper levels of a system are reused but lower levels are reimplemented to accommodate, for example, new hardware. This is supported in DEMON by the use of fully abstract classes; these are class definitions which specify only the interface which the class presents to the outside world. High level classes can be defined as clients of these classes; specific implementations are classes which inherit their interface from the abstract class.

DEMON supports generic classes, i.e. classes parameterised by other classes. Many of the 'Booch-like' classes will be of this form, providing implementations of stacks, queues, sets, etc.

A system designed using DEMON will appear as a single object belonging to a class representing the complete system; this class may perhaps be parameterised. Thus a class representing central heating control systems might be parameterised by the number of rooms to be controlled.

3.2. The DEMON Design Process

Essentially the DEMON design process is a process of successive refinement, modified in some important ways. We present the process as a process for designing a class. When the process starts, the classes to be designed may have been identified as part of an initial design phase carried out using the designer's notepad or the starting point may be the class modelling the complete system. In either case, the design of the top level class or classes will result in a need for the design of other classes, to which the process is then applied recursively.

The design process for a class consists of the following steps:

1. Produce an informal description of the class.
2. Use the informal description to search the component catalogue for a suitable existing class.
3. If the search produces no class which, by itself, meets the informal description adequately, got to step (6).
4. If the search is successful, establish what parameterisation, if any, is necessary for each instance of the class and record the objects as being of that parameterised class.
5. Examine all the classes of which the class identified is a client, directly or indirectly, to establish whether their implementations are suitable for use in the class being designed. Suitability is likely to be assessed in terms of hardware dependency and performance. Apply the design process recursively to any classes which are found to be unsuitable. If the design method has been followed correctly when building up the class library, then the unsuitable classes will be descendants of abstract classes; the redesigned classes must be descendants of the same abstract classes. Once such classes, if any, have been redesigned, the design of the original class is complete.
6. Refine the informal description to produce a precise description of the interface provided by the class, to include its operations and attributes and the exceptions which it may raise. This description must include the signatures of all operations and the types or classes of the attributes; where called for in the quality plan, pre- and post-conditions and invariants should be included.
7. Construct a new abstract class, inheriting appropriate characteristics from suitable related classes identified in (2). The new class should meet the requirements of the object as identified in (1); it may (and usually should) be made more general than the object requires, through the use of parameterisation. The new class is then entered into the class library. Component catalogue information for the class is produced at this stage.
8. Construct a new class which implements the abstract class constructed in step (7).

Further component catalogue information may be produced at this stage.

Step (8) can be further broken down into a sequence of activities as follows:

- formulate an informal solution strategy;
- identify the objects required within the class;
- group these objects into classes and apply the above procedure to the classes;
- specify the operations, exception handlers and, in the case of an active class, the thread of control, using the DRAGOON as a PDL.

When the design process is applied initially to the whole system, steps (1) and (2) constitute the process of requirements definition and cannot be dismissed as straightforward steps in the design process. The output from existing methods of requirements analysis tends to be less than satisfactory as a starting point for object oriented design because the methods are oriented towards functional decomposition. It is hoped to address this point more fully in future work.

3.3. Tool Support for DEMON

A DEMON design is represented by the structure of the class to which the system object belongs, and its history; the definition of this class, as of any other class, will involve the declaration of objects belonging to other classes, classes of which the system object class is a client. The system object class, and all other classes, may also be descendants of other classes.

It follows from the above that tools to support design using DEMON must operate on the project copy of the component library. The library, in effect, constitutes the design database. This database records all the structure in the library. Essentially then the toolset consists of tools for updating, displaying and verifying the contents of this database. Conceptually, the database can be regarded very naturally as having two levels: one level is concerned with inter-class structure and the other with intra-class structure. It is important to understand that all information about the design is held in this database and it is this that drives the support tools. Thus the requirement to display information about the design may require display in various textual and graphical forms but these forms are to be regarded, both by the user and by the tools, as different representations of the same information.

Because of the complexity of the relationships which develop as the library evolves, the graphical display tool allows for considerable flexibility in choosing the area of interest to be displayed and the selection of features required on the display. It is possible, for example, to select a class and ask for it to be displayed along with its immediate clients and an indication of which of its operations its immediate clients use.

Since the DRAGOON is suitable for use as a PDL, the natural textual form for displaying the structure of the class is the DRAGOON, with annotation included to display, for example, the informal description. This means that the DRAGOON structure editor, which works from the class structure held in the database, can be regarded as the fundamental tool for recording the design.

4. The Reuse Toolset

An explicit aim of the DRAGON project is to develop a set of tools to support software reuse and one of the activities of the project was to identify the components of such a toolset. Logically, several different components can be identified:

- Tools to assist in recording and the retrieval of information about reuse products (software components).
- Tools which actively support reuse in the specification, design and implementation phases of software production.
- Tools which assist with the construction of generic reusable components.
- Tools which support system configuration from a library of reusable components.
- Tools which assist with the recording and retrieval of information about the reuse process.

Of course, the resources available for this project were such that we could not possibly contemplate providing all of these tools so it was decided to concentrate development on providing tools to provide information about reusable products and to carry out research in how reuse in the design process can be supported. To this end, our work has been concerned with the development of three major tools:

- *Designer's Notepad* (**DNP**)
- *Software Components Catalogue* (**SCC**)
- *Component Information Store* (**CIS**)

All of the reuse tools have been built as general-purpose tools which can be configured to specific requirements. Instantiations of all of these tools are currently being implemented which are explicitly tailored to the DRAGON method and DRAGOON. Thus, an object-oriented DNP where the user designs using class information is being built as is an object-oriented component retrieval system. At the time of writing, working prototypes of the DNP and the SCC are complete and DRAGOON-oriented versions of both of these systems are being implemented. An initial version of the Component Information Store is planned for completion by October 1989.

4.1. The Designer's Notepad

The Designer's Notepad (DNP) is a tool which is intended to support a design process where a designer produces a 'rough' design then refines this via a number of explorations and 'thought experiments' to produce a final design solution. We believe that reuse is a critical driver of this 'rough' design phase and that existing CASE tools do not provide effective support for the early stages of the design process.

Our model of design relies on the designer using the DNP to produce a 'rough' design using reusable components then refining that design using the DRAGON method or otherwise. It is intended to provide an explicit conversion program whereby designs produced using the DNP may be imported into the DRAGON method tools for further development.

The DNP is a highly interactive system which is intended to replace the working notebook of a software designer. A key requirement therefore is that it should offer a fast and effective user interface and that it should be very simple to use. To satisfy this requirements, we have built the system around two metaphors.

1) The user constructs his or her design on electronic paper. There is no limit on the size

of a piece of 'paper' and as many pieces of paper as are required may be used. Software design graphs consisting of boxes and lines may be created on this paper and the nodes and links may be annotated with text or other information.

2) Notes may be associated with any part of the design (nodes, links or other notes) and the manual analogue of these notes is the 'Post-it' sticky notes which can be attached to paper documents. Thus to associate a note with the design, the user fills in the note and sticks it to the appropriate design entity. A number of different colours of notepad are supported representing different types of note.

In designing the basic functionality of the DNP, our principal design driver was the need for a simple, quick-to-use interface. Thus, only a small number of functions have been provided but these are powerful, orthogonal and general-purpose. The functions which are currently supported are:

1. *Add an entity to a design graph*
The entity name is simply typed. The DNP automatically encloses the typed name in a box and positions it on the design graph. Of course, its position may be adjusted by the designer.

2. *Construct a relationship between design entities*
The user points at each of the entities in turn and these are linked. There is no requirement for entities to be displayed in the same window. When a link is drawn across windows, the nodes at each end of the link are visually tagged. Selecting this tag will cause the linked node to be displayed (if necessary) and highlighted.

3. *Add a note to the design*
The user fills in a note as required with text and sticks this onto the appropriate entity. To view notes associated with an entity, the user moves the cursor over the entity and chooses a menu item allowing notes to be displayed. A note can be added to any component of the design graph - a node, a relationship, another note or the overall design itself. Any number of notes may be associated with the graph.

4. *Initiate a catalogue retrieval*
The user fills in a retrieval note (a special kind of text note) and sticks it on to the appropriate entity. The effect of this action is to initiate a catalogue query and the results of the query are also 'stuck' to the design entity as notes. This facility means that a designer can be provided with reuse information as design possibilities are being explored and can therefore take into account the existence of reuse components when formulating his or her design.

5. *Examine a design in more detail*
A node in the design graph can be exploded and displayed as a graph.

6. *Construct or choose an alternative design*
To support the exploration of different possible designs, alternatives may be associated with nodes in the design graph.

7. *View a map of the design*
Different overall views of the design may be examined and used as a navigation aid.

4.2. The Software Component Catalogue

A Software Component Catalogue (SCC) is analogous to a catalogue in a library in that it should be easy to use with minimal training and should effectively narrow the user's search for material. It is unrealistic to expect a catalogue to return a single component in response to a user's query; indeed this is undesirable as a characteristic of software components is their modifiability. Components which are close to the user's requirement are potentially useful and should be brought to the user's attention.

Thus, the objective of the cataloguing system is to provide the user with a relatively small set of potentially reusable components in response to a particular query. The user then examines these components in more detail to determine if they may be reused as they are or if they may be reused with some modification. We do not require the user to develop a detailed component specification in order to use the catalogue and a simple one line description is usually adequate to develop a catalogue query.

The component catalogue is a separate tool which can be accessed from the DNP but which can also act as a component classification and cataloguing system in its own right. Both keyword-based retrieval and browsing facilities are offered. The basic ideas underlying the classification system are derived from natural language understanding systems and have been described elsewhere by (Wood & Sommerville 1988). In essence, a component is represented as a frame and matching a request against the component database depends on a frame matching process.

The system which has now been developed represents a significant extension over that described in Wood and Sommerville. As well as 'intelligent' retrieval facilities, based on a query language, the system also supports:

1. A powerful classification tool which allows components to be entered into the catalogue under a number of different classes.
2. A knowledge-base browser which assists the classifier in understanding and extending the classification system. We believe it is essential that any classification scheme for software components is user-extensible as no standard component taxonomy has yet been developed.
3. An interface to a component documentation browser which allows the user to access the detailed documentation associated with a component.
4. A catalogue browsing system which allows a user to browse available components in a relatively unstructured way in the hope of discovering reusable components. This system and the query system are tightly integrated so that the set of components resulting from a query may be browsed and the browser may use any component returned as a result of a query as a starting point for catalogue browsing.

4.3. The Component Information Store

The Component Information Store (CIS) is seen by the user as the repository for the actual material which is to be reused. Thus, it contains the code of reusable components (perhaps expressed in more than one language), specifications, designs, documentation, etc. We have decided to maintain this information through this tool for two principal reasons:

1. It is useful for a user to be able to access all of the information (specification, design, implementation, etc.) associated with a logical component such as a parser, a class

implementing a dictionary etc. In general, this information will not be available in one single place in an object-management system but may be spread over information belonging to several projects.

2. We required a portability interface whereby the component catalogue could be interfaced with different object-management systems such as PCTE, a Unix file store, etc. Thus, the software component catalogue references the information store and the information store then references objects whose contents are of interest.

As well as a component catalogue interface, the Component Information Store also supports a browsing interface so that once a component has been located, the user may browse the information store and find other related component information. For example, the user may query the SCC looking for components which implement menus and this may reveal the existence of an Ada menu package whose source code is available in the CIS. The CIS browser is entered from the SCC and, from the Ada source code, the user may move directly to user documentation for the package and also to related information such as an Ada window management package. In the same way as the SCC is analogous to a library catalogue, the CIS browser is analagous to a reader browsing the shelves of books in a library.

5. Formal Methods

Reusability means production of software components which may be used for many different applications. This requires both high reliability of the reused component and a checking procedure to establish whether a component retrieved from a library is reused correctly. Both correctness and correct reuse of components can be supported and ensured by using formal methods.

Therefore this is their primary role in the DRAGON project. In particular, the following topics are addressed:

- definition of formal semantics for DRAGOON;
- a formal approach to the systematic reuse of software components;
- object-oriented algebraic specifications.

One of the prerequisites for applying formal methods to programs is that the programming language itself has a formal mathematically well-defined semantic description. For the semantics of DRAGOON an algebraic approach has been chosen. This implies that the semantic expressions can be evaluated using prototyping tools for algebraic specifications, verification can be based on well-known algebraic methods and the equational axioms support automatic transformation and manipulation of programs and components.

Algebraic specifications provide a flexible tool for the formal description and development of software. Main advantages of this technique are abstractness, modularity and the availability of prototyping tools (supporting early testing) and of strong proof methods (supporting formal verification). Basically, abstract data types are described by giving names for the sets of data to be manipulated, names for the relevant operations and axioms describing the characteristic properties of the operations. Such an abstract, implementation independent description of data structures is suitable for supporting understandability and information hiding.

In the process of formal program development abstract specifications can be refined by making some design decisions, such as the choice of data representation or the choice of algorithms, leading to more concrete specifications of the problem. This shows that algebraic specifications are well suited for the description of software at different levels of abstraction. Based on clean formal semantics, these descriptions can be related or even transformed (Wirsing 1989).

The method of algebraic specifications is used to give a formal semantics to DRAGOON (Breu & Zucca 1989).

Each DRAGOON class is associated to a structured algebraic specification. Objects are described by values of a sort corresponding to the name of the class. Methods are described by operations which in the sequential case are state transition functions, whereas in the concurrent case they are, more generally, processes. The relations between classes are modelled by the corresponding operators on specifications; in particular, inheritance and clientship are represented by specific *enrichments* of algebraic specifications, and the interface of a class can be restricted and renamed by two other operators *Export* and *Rename*. The concurrent behaviour of the objects and methods of a class is modelled by (an algebraic specification of) transition systems which describe the dynamic evolution of processes in a state-oriented way (Astesiano et al. 1988).

The basic idea behind this approach is to model the evolution of an *object environment*. At each stage of its life the object environment describes the state of all the existing objects determined by the state of their attributes. Objects can evolve either performing their own actions if they are active, or as a consequence of the evolution of *processes* representing method calls performed by active client objects. A state transition relation describes how this dynamic system of objects and processes evolves. Thus, in this model methods of a class are described by processes, that can perform basic actions such as reading or updating of instance variables as well as creation of new objects.

In order to achieve systematic reuse, a software component cannot be provided only as a piece of code. To retrieve it, a more abstract description has to be given as well. For example in the approach of (Matsumoto 1984), a component consists of four parts at different levels of abstraction: a requirement specification, a design specification, a source program and the object code. Our approach (Wirsing 1988) (Wirsing et al. 1988) is based on the algebraic specification of abstract data types and is closely connected to the work on the semantics of DRAGOON.

Generalising the approach of Matsumuto, a reusable component consists of a tree of formal (algebraic) specifications where a specification SP1 is a child of another specification SP if it is an implementation (i.e. if the signature of SP is contained in that of SP1 and if all models of SP1 restricted to the signature of SP are models of SP). Hence the tree may also be seen as a 'top-down' program development tree where each node of the tree is an instance of a certain level of abstraction of the program, with the root being the most 'abstract' formal abstraction and the leaves representing the 'concrete' implementations. Every node of the tree is itself a structured specification. In contrast to other approaches these trees are considered as objects of the language and can be constructed and manipulated by the operators of the language, enforcing modifications of components that are consistent for all nodes of the tree.

Combining the object oriented language with this notion of reusable component, it can be proven that the inheritance relation between two classes corresponds exactly to the implementation relation between the two associated algebraic specifications in the semantic domain, i.e. the specification of a heir class is an implementation of its father class. This reflects exactly the intention of inheritance as pointed out in (Meyer 1988).

Moreover, the algebraic semantics of classes forms the basis for an object-oriented specification language which combines object oriented and algebraic principles in a coherent way. This language which is currently under development enriches the usual framework of algebraic specifications by features such as objects consisting of a reference and a state, object sharing, dynamic creation of objects. The structuring mechanisms for specifications are the same as for the algebraic language of the semantics of DRAGOON.

6. Cooperation

The success of the DRAGON project so far can largely be attributed to high levels of collaboration and cooperation achieved. As a result of the requirements analysis under-taken during the first six months, and the consequent decision to adopt an object oriented approach, it was necessary for all partners and subcontractors to modify their original plans and interests and to change substantially the scope of their work from that specified in the original technical annex. All responded enthusiastically and showed complete commitment to the new approach, despite difficulties caused by shortage of staff and the difficulties of recruiting new staff. The result should be a much more tightly integrated toolset than many similar research projects have produced.

One important factor in the success of the collaboration has been the regular project meetings to which all staff working on the project have been invited. These meetings have taken place about four times a year, rotating around the sites of the partners and subcontractors. Central technical issues have been discussed and decided at these meetings, with the result that everyone working on the project has been aware of the issues and has felt involved in the decisions taken. The social side of such occasions should not be undervalued; if people from different organisations are to work together effectively, it is important that they should develop satisfactory personal relationships. This is doubly important when they come from different countries and do not share a common cultural background. The project meetings have provided a valuable opportunity in this regard, particularly so because of the variety of venues in which they have taken place.

It is also appropriate to note here the contribution of Dornier Systems, the partner responsible for the case study. Staff from Dornier have participated actively in the definition of the language and the design method, giving valuable input from the world of real projects.

7. Acknowledgements

The DRAGON project is partially funded by the Commission of the European Communities under the Esprit programme. We would like to thank all the participants for their valuable contribution: Cinzia Cardigno and Stefano Genolini (TXT), Catherine Destombes and Francois Kaag (GSI-Tecsi), Wolfgang Kratschmer, Sigrid Kutzi and Siegfrid Raber (DORNIER Systems), Bob Gautier and Andy Ormsby (University College of Wales), John Mariani, Ron Thomson and Neil Haddley (University of Lancaster), Egidio Astesiano and Elena Zucca (Universita' di Genova), Stefano Crespi-Reghizzi, Dino Mandrioli and Flavio De Paoli (Politecnico di Milano), Martin Wirsing, Rolff Hennicker and Ruth Breu (Universi-taet Passau), Stephen Goldsack and Colin Atkinson (Imperial College).

References

ALRM (1983). The Ada Language Reference Manual. ANSI/MIL-STD 1815A.

Astesiano, E., Giovini,A. & Reggio, G. (1988). *Concurrency '88*. F.A. Vogt (ed.). Lecture Notes in

Computer Science. Springer, pp. 140-159.

Breu, R. & Zucca, E. (1989). An algebraic compositional semantics of an object oriented notation with concurrency. *Towards an Algebraic Compositional Semantics for DRAGOON*, Deliverable WP2.D2.T1 of the project DRAGON (Esprit 1550).

Booch, G. (1986). Software Engineering with Ada. Second edition. The Benjamin/Cummings Publishing Company.

Cox, B. (1986). Object Oriented Programming: an Evolutionary Approach. Addison-Wesley.

Di Maio, A., Cardigno, C., Bayan, R., Destombes, C. & Atkinson, C. (1989). DRAGOON: an Ada based Object Oriented Language for Concurrent, Real-Time, Distributed Systems. Proc. Ada Europe Conference, Madrid. Cambridge University Press.

Goldberg, A. & Robson, D. (1983). Smalltalk-80: The Language and its Implementation. Addison-Wesley.

HOOD Manual (1987). Issue 2.1.

Matsumoto, Y. (1984). Some experiences in promoting reusable software. IEEE Trans. on Soft. Eng. Vol SE-10, no. 5, pp. 502-513

Meyer, B. (1988). Object-Oriented Software Construction. Prentice Hall.

Simonian, R. & Crone, M. (1988). InnovAda: True Object-Oriented Programming in Ada. Journal of Object-Oriented Programming. Vol. 1, no. 4. pp. 14-21.

Strom, R. (1986). A Comparison of the Object-Oriented and Process Paradigms. SIGPLAN Notices. Vol. 21, no. 10

Stroustrup, B. (1986). The C++ Programming Language. Addison-Wesley.

SPS (1988). Classic-Ada: User Manual (draft). Software Productivity Solutions.

Wirsing, M. (1988). Algebraic description of reusable software components. *COMPEURO 88*. E. Milgrom, P. Wodon (eds.). Computer Society Press, 1988, 300-312.

Wirsing, M., Hennicker, R. & Breu R. (1988). Reusable specification components. *MFCS 88*, M. Chytil, L. Janiga, V. Koubek (eds.), Lecture Notes in Computer Science, Springer, pp. 121-137.

Wirsing, M. (1989). Algebraic Specification. *Handbook of Theoretical Computer Science*, J. van Leeuwen (ed.), Amsterdam, North-Holland, 1989, to appear.

Wood, M. & Sommerville, I. (1988). An information retrieval system for software components. IEE/BCS Software Engineering Journal. 3 (5), pp. 198-207.

REAL-TIME PERCEPTION ARCHITECTURES :
THE SKIDS PROJECT

André AYOUN, Christophe BUR, Robert HAVAS
Nicole TOUITOU, Jean-Michel VALADE
MS2i
3, Avenue du Centre - Les Quadrants
78182 Saint Quentin en Yvelines
France

ABSTRACT : SKIDS stands for Signal and Knowledge Integration with Decisional Control for multiSensory systems. The project aims at defining a generic architecture for multisensor perception. General concepts have been defined and the implementation of a demonstration has started.

Basic problems of multisensor fusion have been met: they are described in this paper. The demonstration is described along with the approaches to face the general problems of:

-control of attention, resource allocation, data consistency maintenance, uncertainty management.

1.   Introduction

SKIDS concentrates on the particular problem of multisensory perception. The perception process consists of several layers of information processing. This information comes from various sensors and the output is a high level representation of the world called the "perceived situation". The perception machine therefore contains a number of interpretation processes. The interpretation processes may be monosensorial or multisensorial - involving fusion of data coming from different sensors.

Algorithms and signal processing techniques exist for monosensorial interpretation. Multisensor fusion consists of combining the representations output by the low level signal processing layers to the high level interpretation procedure. The basic assumption of SKIDS is that a practical way to combine the multisensor information is to fuse high level - numeric and symbolic data rather than low - level representations . The main difficulties are therefore:

-the organization of the interpretation flow,

-the control, in real-time, of the interpretation process.

Both topics are relevant to the architectural domain rather than to the algorithmic one.

A Hierarchical/Distributed organization has been selected in SKIDS, in order to ensure a manageable design and real-time efficiency.

In particular, a distributed task management knowledge-based system has been defined. This organization raises the problems of the behaviour of a complex system having control loops at different layers: stability, transient behaviours and tuning are still research issues.

The practical implementation in the SKIDS demonstrator considers mainly three processing layers: a high-level situation-assessment layer, a medium-level multisensory-fusion layer and low-level logical-sensor processing layer. Five broad categories of perception functions have been defined: detection, identification, localization, tracking, characterization.

To achieve these functions, the hardware architecture is composed of interconnected processing clustered either dedicated (mainly to signal processing) and general purpose (for higher level functions).

The knowledge based control system is two-fold: it ensures that high level perception requests are serviced, by actuating the proper tasks; it ensures some optimization of the resource allocation (mainly the processing power of non-dedicated clusters).

The main result up to now, apart from the demonstrations, is a general formalism which ensures:
    -a consistent design of the functional architecture,
    -the implementation of a general control mechanism based on a distributed task management system.

The papers presents first the problem of an automatic multisensor perception system for general surveillance system. Then, the demonstration of SKIDS, illustrating an indoor factory-like surveillance application is described in more detail.

2 . Functions of a perception machine

In a general surveillance problems the need is:
    -to detect unexpected events,
    -to maintain a complete global situation representation.
When a human operator uses such a surveillance system, he wants to be warned only if something really important happens, and in this case he may need to look "deeply" in the situation as established by the machine. If we consider the situation as a big state vector, this state vector must contain very symbolic information at the top level and (possibly) raw data at the sensor level (such as for example video-tapes for identifying the gangsters in a bank hold-up).

This leads to identify 4 types of perception requests, which can be combined:
    -Detection:
    •   events are generally detected by a monosensorial process, the detection is then transmitted to higher levels, the consistency with other observations checked; if the alarm is serious enough the detected event is passed on upstream; the seriousness is evaluated differently depending on the level,
    •   abnormal situations are detected on a longer term by processes which permanently scan the situation and match it to model situations. Such detection processes are permanent (they could be initiated or inhibited by the high level control and their modalities can be controlled as well). The model situations are a priori defined and for the moment the learning of model situation is not considered.
    -Tracking: the tracking process keeps detected events or assigned objects in the field of attention of sensors. It does not necessarily maintain high level representations: there is generally no comparison to models in the tracking process. Basically the tracking process permits to minimize the combinatorial data association problem by using temporal continuity.
    -Characterization: for detected events or non recognized objects, a characterization request is emitted by the high level. It permits to associate new attributes to the objects or to the events.
    -Identification or recognition is a process which compares assigned objects to models. The process is on demand but the identity is then kept attached to the object if "tracking" continues.
    -Localization of objects is on demand. It has not to be confused with location maintenance or map-updating which is implicitely done for detecting abnormal situations.
Higher level perception requests are handled by predefined plans using the above mentioned requests. For example, the FIND request need to actuate first an IDENTIFICATION request followed by LOCALIZATION request.

3 . Architecture

The architecture is characterized by the organization of the interpretation functions (represented by the interpretation grap), by the organization of the data and by the organization of the control.

3.1 THE INTERPRETATION GRAPH

The interpretation graph is a graph whose nodes are the interpretation tasks and whose edges are the data representations. To satisfy a perception request from the high level, the machine-control selects a path in this graph. This path may be a-priori selected by a perception planning strategy or generated in real-time by the distributed control system. The interpretation does not contain information about control.

The graph is used to display the possible routes from low-level sensory data high level interpreted information (fig. 1).

It must be noted that the different pathes do not strictly give the same information: they differ by quality performance, response-time, ressource needed. The real time control selects the route with respect to these criteria.

3.2. DATA REPRESENTATIONS AND DATA BASES

These are basically two types of data: the data which convey the perceptual information called the perception data and the control data. In relation with the perceptual data, there are corresponding models to which the perception data are matched (for recognition operations).

3.2.1 Perception data
The perception data range from raw data-coming from the sensors to very "symbolic" data such as the situation. These data are inputs and outputs of the elementary perception tasks of the interpretation graph. They may be redundant, even duplicate, local or shared in a common data base.

The stucture of the perception data base follows a "formal frame". These are mainly two classes of data: the "objects" and the models. The objects are described in a hierarchical form i.e. objects are decomposed into subobjects whose relations are described into the object representation. For the sake of efficiency, only one or two levels of hierarchy have been implemented: the Situation is a "super-object" which contains perceived-objects (which themselves should correspond to physical objects of the real world).

The objects are of three types:
    -the map of the scene (a priori known but updated from the perception; see MOUTARLIER [13],
    -the objects of the environment and,
    -the sensors, which are perceived differently by the machine (the sensors are internally perceived and they have more elaborated models than the objects of the outer world).

The structure of an object representation is:
    -detection data,
    -identification data,
    -characterization (various features),
    -localization,
    -tracking.

The different items of the object representation contain datation elements
    -for detection: the date of detection corresponds to the first appearance of the object in the scene,
    -for localization and tracking: date of the last update.

The characterization data are split into geometric features (Region - Edge - Vertices description) and other global attributes (colour, noise activity,...).

The data are either numeric/quantitative (for example localization, dimensions) or symbolic/qualitative. Attached to these data, confidence measures are expressed as accuracy measurements (such as the error covariance matrix for a localization feature) or uncertainty measurements (such as probability or possibility-plausibility pairs).

A non-trivial problem is the defining of the procedure for interpreting quantitative data into more symbolic qualitative representations, using a tradable representation of the uncertainty.

The perception data contain redundant attributes in order to simplify further processing: it results in more voluminous representations. However, the representation is sometimes not directly usable by the interpretation processes: some projections (to generate "views") may be necessary.

The models of the objects contain a priori information necessary to recognize, track, localize… the objects: this information is mainly data but encapsulate rules.

3.2.2   Status and control parameters

The control of the sensing and processing modalities needs inputs and outputs for every perception task. Outputs are status and other performance values. Inputs are control parameters (focus-of-attention parameters, modality-selector,…). The control system needs execution models for planning and for checking the perception process: during the execution, the status and performance reported values are compared to model-status… These models are described in tables and rules.

3.2.3   The blackboard and data base

As the SKIDS machine is inherently distributed, the data base as well as the knowledge sources for interpretation are distributed.This means that some data may be duplicated in different location, or known only be one subsystem, or shared thanks to access control mechanisms.

The implementation in local or global databases is linked to the use of the data. For example, the pixels of the images, as well as raw data from the different sensors do not need to be transmitted. They are locally stored. On the other hand the location of the objects are used by all the subsystems, at least to keep track of the objects: the objects are therefore in the common data base. A blackboard architecture implements this distributed data base for symbolic information.

3.2.4   Records of successive values of the data

The successive values of the data may be needed to interpret a situation: rather than keeping track of various hypotheses, it is often more efficient to go back to past data if it is possible to store them. At low level, the raw data are too voluminous to be stored; on the contrary a state-vector - of a Markowian process - contains all the information needed for further prediction. The baseline in SKIDS is the storing of high level data only (history of the situation) and mainly for the needs of reporting to the human operator.

3.3   THE CONTROL

The control in a closed-loop automated system generally refers to the control of the plant. In a perception machine, the control refers to the control of the modalities of the sensors, of the sensory processing, and the resource (memories, processing) allocation. In fact, especially for the control of sensors such as a mobile sensory-head, the problems of multi-layered control are encountered: communication, time-constraints, stability…

However we mainly address here the control of the perception process itself which is composed of:

       -Perception plan generation.

       -Real-time control of the execution.

       -Distributed operating system for execution control and hardware ressource allocation, synchronization,…

Of course, these three aspects of the control are interdependent: the operating system layer must be capable of servicing task activation decided by the upper layer knowledge based real-time control.

3.3.1   Perception plan generation

The generation of perception plan before execution may be:

       -predetermined,

-based on a priori knowledge and assumptions about the data; it consists in this case in the application of rules to determine an "optimal" route in the perception graph.

The idea in SKIDS is that the a priori knowledge is distributed: at every layer, there are different sets of rules for determining the optimal selection. The rules are compiled before running, resulting in decision trees at every layer of the decisional control.

This solution suppresses the effective generation of plan and leaves most of the control to the execution control. In the SKIDS demonstration, the rule compiler is Kheops, developed by the LAAS.

3.3.2 Execution control

The task management is distributed. There are mainly three layers:

-the sensory-processing layer,

-the control of the fusion processing modalities for achieving the tasks of recognition, detection, identification, localization, characterization...

-the (higher level) perception strategy selection...

The distributed task management induces a non predictable behaviour of the control process: this is context and internal machine state dependent. The control structure is schematized on figure 2: at high level, the situation assessment process selects a perception goal such as "find the conveyor" which is decomposed into elementary perception goals: scan the scene (do-loop) while activating permanent detection tasks, when the object is found, localize it. The localization task manager, when activated, itself activates lower-level interpretation tasks, according to the precompiled rules. It may happen that the required data be already calculated and therefore avoid the activation of a perception task.

3.3.3 Distributed control and resource management

The executive control uses a set of control actions which permits the execution of tasks in different physical processing nodes. Synchronization features and message passing tools allow the correct functioning. Among them, the time-out and priority system are the main features which ensure the proper activation of task and cascaded extinction of given-up tasks.

A distributed clock is used in the demonstrator.

The functional architecture is mapped onto the hardware architecture in a partly dynamic way: the ressource allocation is dynamic, as far as the processing power is concerned; but the procedures are resident in the different processing clusters: in other words, there is no dynamic memory allocation.

Coherence of the data

This is the central problem of any distributed system, either database or processing distributed system. A perception machine is distributed on theses two respects. It is felt that the use of specialized and partitioned blackboards, with dated messages should help. However, the implementation of the full-size system is necessary to give more general answers to the practicability of various theoretical solutions.

Uncertainty and inaccuracy management

From a theoretical point of view, probabilistic approach and Bayes' rule seem to give a valid frame. However many practical problems are encountered:

-how to extract indices of confidence from the matching of model to perceived data?

-what is the meaning in terms of likelihood of this index of confidence?

-how to use them in an operational way.

Both accuracy to uncertainty and uncertainty management are still unsolved and is an important part of the research activity at LAAS and at OXFORD (see 3.2.1.).

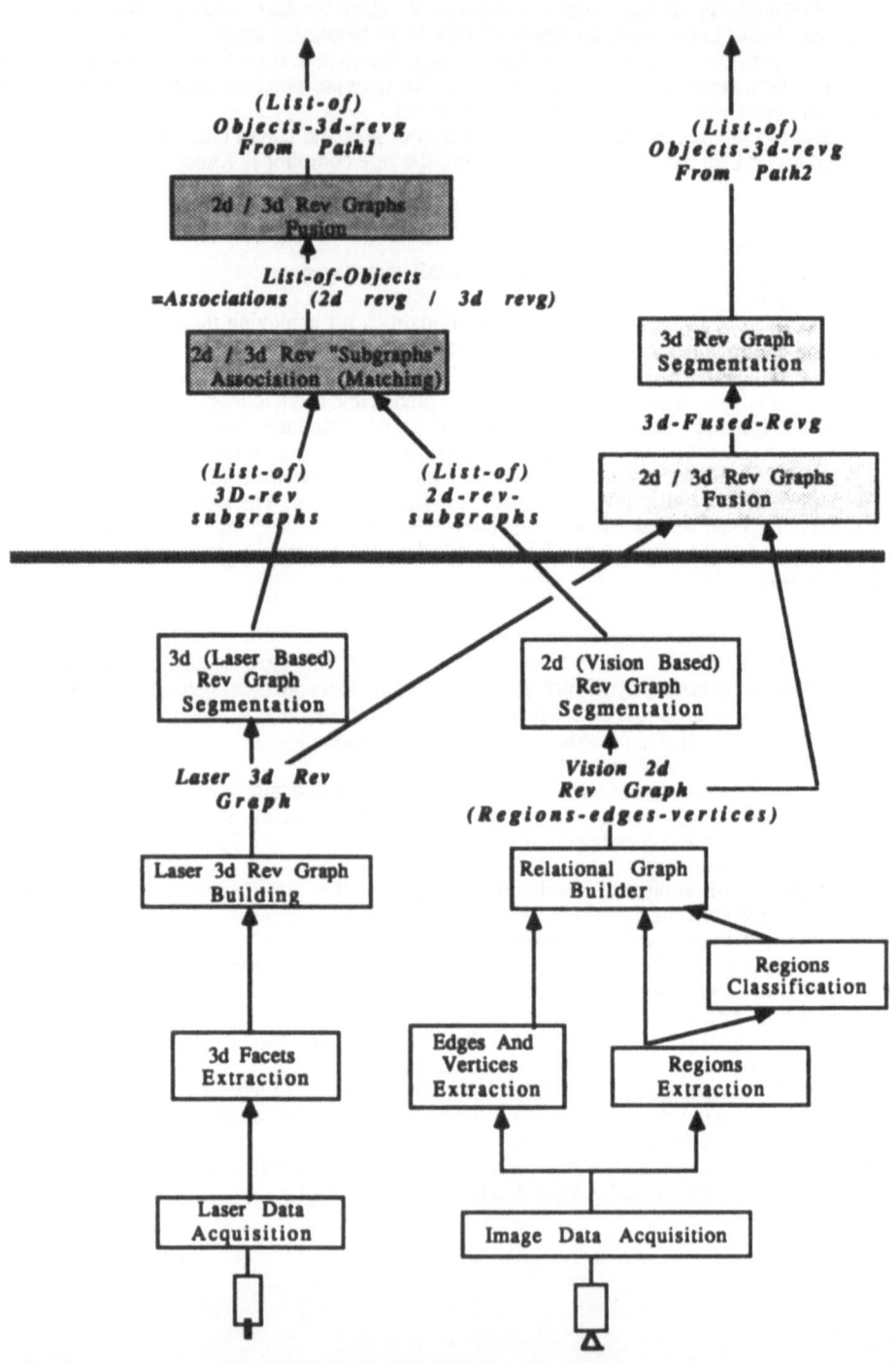

Figure 1
THE INTERPRETATION GRAPH FOR THE GEOMETRIC
CHARACTERIZATION OF AN OBJECT

575

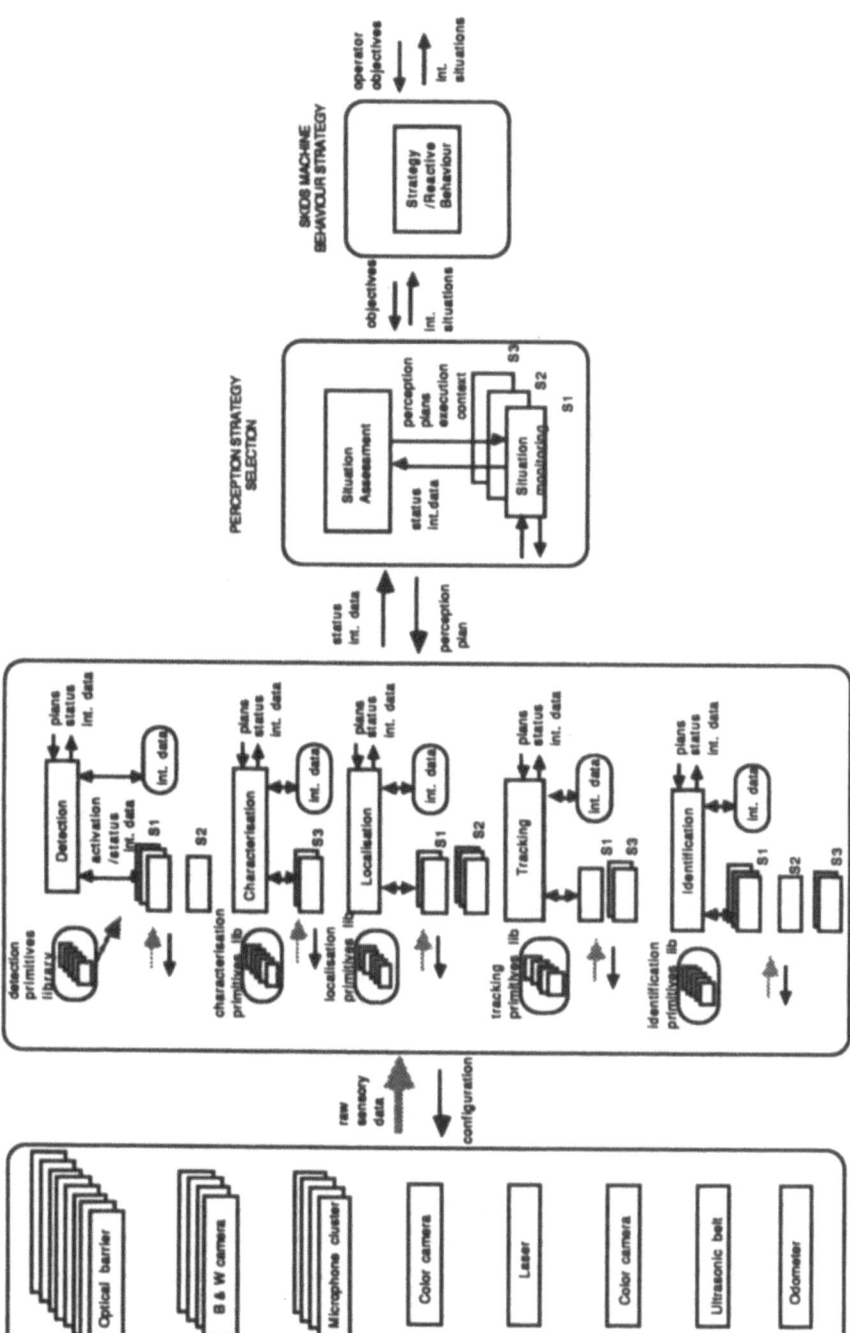

Figure 2
THE DISTRIBUTED CONTROL ARCHITECTURE

576

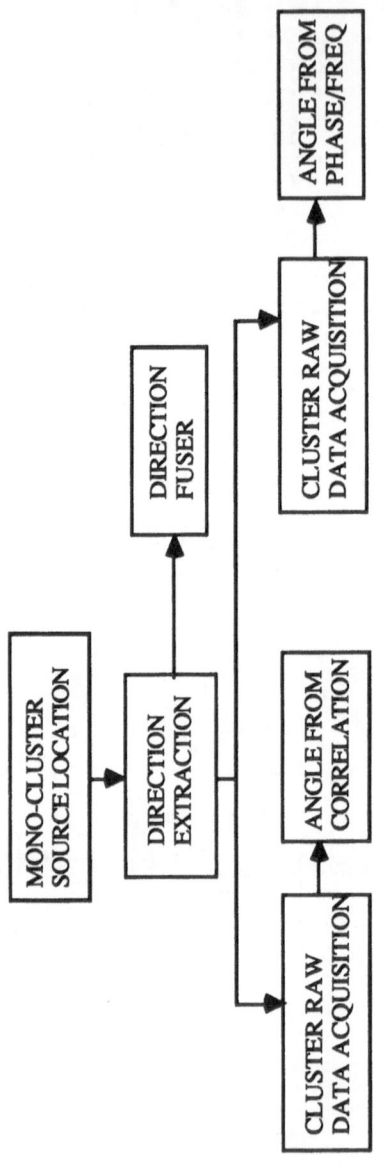

Figure 3
THE CONTROL GRAPH FOR THE PASSIVE ACOUSTIC SYSTEM

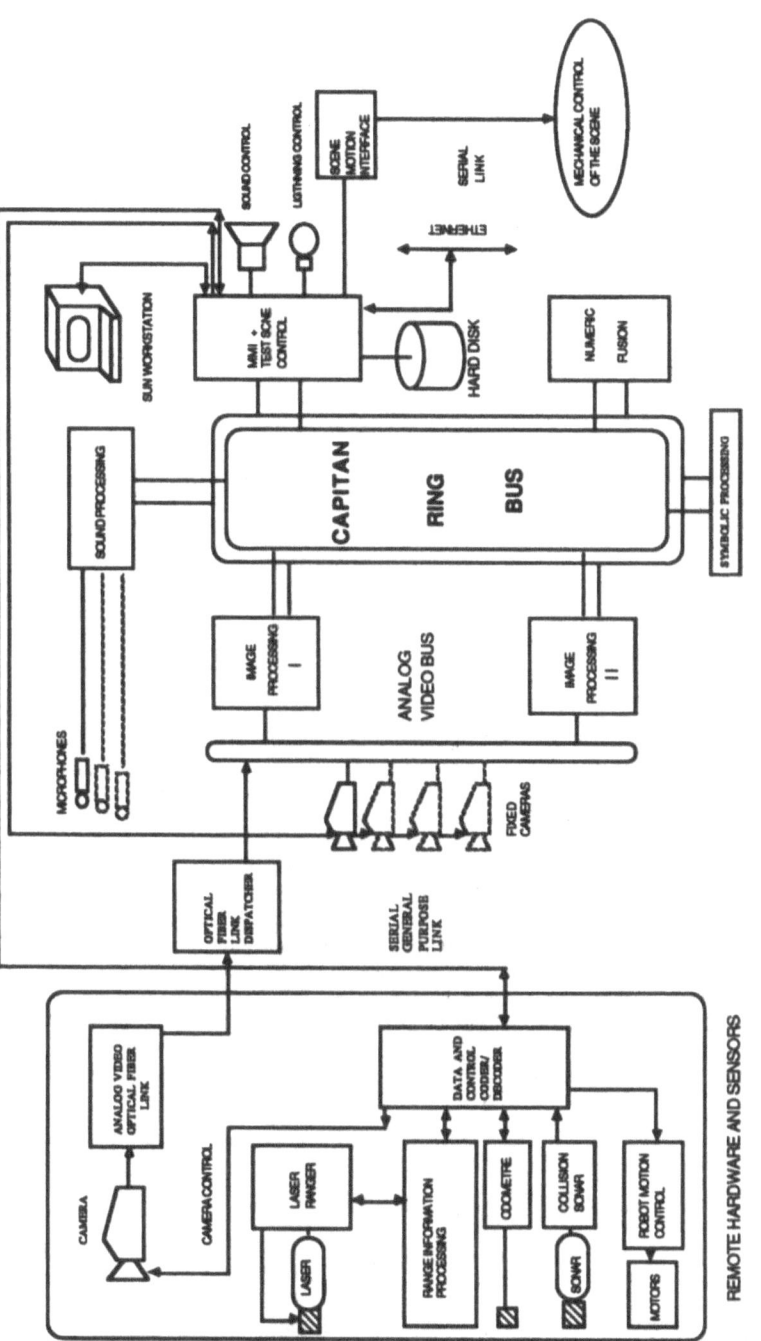

Figure 4
THE HARDWARE ARCHITECTURE

4. The demonstrator of SKIDS

The prototyping phase will consist in building a demonstration site, a demonstration machine and running a scenario in the site after having assigned perception goals to the machine.

The test site has been defined and is being set-up. Continuous integration of the results in this test site and real size experiments conducted by the partners aim at providing experimental support to the project and making the integration easier.

One of those experiments, conducted at LAAS, performed fusion between 3D data obtained by stereovision and by a laser range-finder. Another experiment has been performed by BAe to test detection-and-tracking of mobile objects using multi-camera vision. At last, MS2i is experimenting the "minimal surveillance" experiment based on optical barriers detections, odometric and ultrasound-proximity measurements from the mobile sensing unit. This experiment particularily aims at demonstrating the architectural principles in a very simple environment.

4.1 DEMONSTRATION SITE AND SCENARIOS

The demonstration site has been defined as a laboratory version of ta surveillance environment. It is a 10 x 15 meters room partitioned into a big room and a smaller one, to separate the fields of view of different sensors. A part-conveying platform (AGV) loops around the site, carrying boxes between a few workplaces set-up in the room.

The site is equipped with passive acoustic sensors (4 arrays of 5 microphones), 8 optical barriers, 4 fixed B&W cameras, 1 pan-and-tilt orientable colour camera. The mobile sensing unit consists in a general purpose robot platform carrying a mobile head consisting in an orientable colour camera and a 2D-scanning laser both mounted on a pan and tilt platform. This robot is also equipped with odometers (coders on the wheels) and an ultrasonic belt for proximity sensing.

Five basic scenarios have now been defined such that they face SKIDS with a complete set of perception goals.

1)   The AGV starts, drops a part and stops in front of it after having completed its loop.

2)   A part falls off a workplace during its manufacturing, on the AGV stopping point.

3)   The AGV arrives at its normal docking site but the fallen part prevents correct unloading operation, and the AGV remains stuck.

4)   A humain being enters the site (crossing one or more optical barriers), picks up a box on the floor, goes across the room, and puts it on a desk.

5)   An optical barrier is short-circuited, thus continuously indicating it is crossed, the AGV looses its track and stops after going at random for a while, a camera is turned off.

Each scenario has a particular purpose. For instance, scenario 4 intends to demonstrate detection capabilities, scenario 5 tests the reconfigurability of the machine.

The demonstration site have been defined to ensure that:

-Complexity of the environment can be tuned from a very simple block-world environment for early tests to almost real world operation for the final experiments,

-a good testing environment is provided, where experiments can be repeated in a noise-free environment and where events can be recorded for results analysis.

4.2 DATA REPRESENTATION AND INTERPRETATION FUNCTIONS

A part of the interpretation graph (fig. 1), and a part of the control graph (fig. 3) are presented in the case of the demonstration machine.

This figure represents a part of the interpretation graph for the characterization perception function. The characterization function is made of a set of elementary functions which are the nodes of the graph.

The goal of the characterization is to output a description of a perceived object and an appropriate representation of this object. In particular, a number of interpretation functions for characterization are dedicated to build a geometric description of the object, using the "rev graph" representation (i.e. Regions Edges Vertices).

Fgure 1 illustrates the different interpretation functions involved for building a "3D rev graph" of a perceived object. Two interpretation paths appear, depending on the level at which the segmentation (of the global rev graph) into distinct objects is performed:

-PATH 1: The segmentation into subgraphs is done separately on the two rev graphs from different sensors origin (2D vision rev graph and 3D laser rev graph).

The subgraphs from the two origins are then associated with a matching procedure, and then refused in order to build 3D rev graph representations of the objects involved in the "Field of View".

-PATH 2: The global 3D rev graph is first built by fusion of 2D and 3D global graphs (from vision and laser data) and then segmented into separate 3D subgraphs corresponding to distinct objects.

If a request for geometric characterization of an object is expressed, the decision between the two interpretation paths is to be done by the decisional control system on the basis of criteria such as:

-Computation resources available and response time required

(path 2 is more resources and time consuming than path 1).

-Required accuracy on geometric features

(path 2 allow more accurate geometry as the fusion is made directly on "raw" data, no segmentation error being introduced.

The control graph in figure 3 illustrates the hierarchical breakdown of tasks into more specialized ones (vertically) and the possible transitions between tasks. The transitions are: sequencing, choice between alternative agents and concurrent activation of parallel agents (not shown here). These rule-based transitions use control actions on the agents according to the defined local interpretation strategy to perform distributed control. The control actions are request to a global controller that manages conflicts. Control action are start, stop, repetitive start, wait for data, allocate a resource, change a modality (e.g. a physical parameter of a sensor). These actions are not OS-level, they are task-level control messages, but the OS-level actions that implement those control actions have been designed in the SKIDS Distributed eXecutive as very similar to the task-level ones.

The representations are organized into hierarchical classification of data types and structure. In the description of a task are included the input and output classes of representations, together with the a-priori models of its execution capabilities.

4.3 DEMONSTRATION MACHINE ARCHITECTURE

The figure 4 gives an overview of the demonstration machine VME clusters are linked by the Capitan ring bus. Some are dedicated (e.g. vision processing) and others perform general purpose numeric and symbolic processing. In addition, some information signal processing is done on-board the mobile sensory head acting as a remote VME cluster.

This architecture has been designed to:

-use off-the-shelf equipment and comply to industry standards when possible,

-be flexible to adapt to possible architecture changes during the course of the project (the ring structure can be changed by software to a pipe-line structure);

-offer fine-grain control on the hardware to allow implementation of features such as dynamic reconfiguration, load balancing, etc...

This required the implementation of a real-time distributed infrastructure the SDX (SKIDS Distributed eXecutive).

4.4 THE MINIMAL SURVEILLANCE EXPERIMENT

This experiment is based on a minimal set of sensors: only optic barriers and the odometry and ultrasound ring on-board the robot are used.

The perception goal is to monitor elementary events which can occur during the experiments such as entering of a man. This means that the machine shall be able to distinguish between itself (the robot) and external objects and that it shall be able to make decision such as: it is an obstacle or it is a known part of the world by comparison with the map. As the process is real time, the rules are compiled using KHEOPS.

The scenario only consists in a nominal motion of the robot. During its course, it encounters a fixed obstacle it should be capable to detect. A man enters the room, crossing several optical barriers: it should be detected.

The implementation has been performed in a way to validate the principles of decomposition into activable tasks with control parameters. Estimation on the robot location is performed using on-board measurements and comparison with a map. Along with the position estimation from the ultrasound measurements, obstacle detection is performed. However, the decision that this is an obstacle is made locally only after sufficient evidence is gathered. The high-level reasoning makes decisions such as it is a man or a failed sensor ...

4.5   RESULTS

For the moment only preliminary results have been obtained. They show that the process of decision-making may take place at various levels and that the optimization is rather a question of architecture. In fact it is preferable for efficiency of the implementation to have a sub-optimal system (in terms of false detection rates or of accuracy) than a theoretically optimum system which would need high-rate communications of complex data (such as conveyed temporal correlation or parameters of the multiple hypotheses tracked at a higher level). Though no quantitative result has yet been obtained, the experiment showed that many fundamental mechanisms (synchronization, real-time reasoning, coherence checking, fusion of localization data, sensor failure detection, hierarchy of interpretation levels, ...) may be tested on simple sensory configurations.

The next steps will consist in verifying the validity of the principles and the universality of the representations (such as object representations, geometry and uncertainty representation). The integration of the different components of the SKIDS demonstrator will also permit to evaluate how the complexity is a function of the number of sensors, if the communication can be managed by an integrated perception machine and how the distribution of the control functions should be implemented.

5 .   Conclusion

The study of the multi sensory perception problem permitted to structure a perception machine, with very general principles. The first implementation in the SKIDS demonstrator shows that the approach is very promising.

It should help to envisage real size integration in different applications such as:
-surveillance of large plants/installation,
-mobile systems situation perception and control.

A conclusion from the first two years activity is that the complexity forces to make compromises between efficiency, flexibility and quality of the perception. Loose interfaces between subsystems resulting from the adopted structure are helpful for integration and maintenance. But they yield performance degradation when high level data are fused.

We are now confident that the full implementation of the demonstrator will validate the principles and that scientific results will be important.

ACKNOWLEDGEMENT -
The work described had been using the collaboration of all the partners of SKIDS, : BAe (U.K.), CFR (DK), KAE (D), LAAS (F), MS2i (PRIME, F) , MAPS (SP), the Universities of BARCELONA, OXFORD and PATRAS

REFERENCES

[1]   **A. SHAFER S., STERTZ A. and E. THORPE C.**
AN ARCHITECTURE FOR SENSOR FUSION IN A MOBILE ROBOT
CMU Research report CMU - RI - TR - 86 - 9

[2]   **ALAMI R.**

NNS: A LISP - BASED ENVIRONMENT FOR THE INTEGRATION AND OPERATION OF COMPLEX ROBOTICES SYSTEMS.
IN PROC. IEEE INTERNATIONAL CONFERENCE ON ROBOTICS AND AUTOMATION, Pages 349-353. ATLANTA.

[3]   ATHANS M.
ARCHITECTURE AND CONTROL COMMAND AND CONTROL THEORY A CHALLENGE TO CONTROL SCIENCE IEEE TR ON AUTOMATIC CONTROL.
Vol AC-32, No. 4, APRIL 87.

[4]   BROOKS R.
SYMBOLIC REASONING AMONG 3D AND 2D IMAGES.
ARTIFICIAL INTELLIGENCE 17-285-349, 1981.

[5]   CHEN S.
MULTISENSOR FUSION AND NAVIGATION OF MOBILE ROBOTS.
INT. JOURNAL OF INTELLIGENT SYSTEMS, 1987.

[6]   CHIU SL., DJ. MORLEY and JF MARTIN
SENSOR DATA FUSION ON A PARALLEL PROCESSOR
IN PROCEEDINGS OF THE IEEE CONFERENCE ON ROBOTICS AND AUTOMATION.
Pages 1629-1633. SAN FRANCISCO, CA, APRIL, 1986.

[7]   CROWLEY J.
COORDINATION OF PERCEPTION AND ACTION IN A SURVEILLANCE ROBOT
IJCAI 87, MILAN, AUG. 87.

[8]   HAYES-ROTH B.
A BLACKBOARD ARCHITECTURE FOR CONTROL
ARTIFICIAL INTELLIGENCE VOL 26 (1985) P. 251-321.

[9]   H. LEVIS A.
ARCHITECTURES FOR COMMAND CONTROL SYSTEMS.

[10]   KOHL. C.
A GOAL-DIRECTED INTERMEDIATE LEVEL EXECUTIVE FOR IMAGE INTERPRETATION.
IJCAI 87, MILAN, AUGUST 1987.

[11]   LANUSSE A.
MODELES D'INTEGRATION ET DE RECONFIGURATION MULTISENSORIELLE EN ROBOTIQUE.
MARI COGNITIVA 87, PARIS, MAY 1987.

[12]   LEYTON M.
NESTED STRUCTURES IN CONTROL: AN INTUITIVE VIEW.
CVGIP 37, 20-53 (1987).

[13]   MOUTARLIER Ph. and CHATILA R.
STOCHASTIC MULTISENSORY DATA FUSION FOR MOBILE LOCATION AND ENVIRONMENT MODELLING.
TO BE PRESENTED IN 5th INTERNATIONAL SYMPOSIUM OF ROBOTICS RESEARCH (1989).

[14]   THORPE C., KANADE T.
VISION AND NAVIGATION FOR THE CMU NAVLAB
SPIE VOL 727 MOBILE ROBOTS (1986).

Project no 1609

SYSTEM MEASUREMENT AND ARCHITECTURE TECHNIQUES (SMART)

M. K. CROWE
Dept of Computing Science
Paisley College of Technology
High Street
Paisley PA1 2BE
Scotland

J. A. CARRASCO
Departament d'Enginyeria Electronica
Universitat Politecnica de Catalunya
Diagonal 647
08028-Barcelona
Spain

Abstract

The SMART project aims to assist the designers of real-time fault-tolerant (RT/FT) systems by providing a number of specially-developed tools within an open environment. The tools developed within the SMART environment support dependability evaluation, fault tree analysis, and simulation of real-time architectures. The SMART environment is open so that designers of RT/FT systems can add their own tools to it.

1. Introduction

The SMART project set out to develop an environment which formalized the evaluation of performance of fault-tolerant systems and to give technical support for their optimisation. In the early stages of the project it was decided to restrict its scope to real-time fault-tolerant (RT/FT) systems.

It used as a main starting point METFAC, a Markovian modelling tool for dependability evaluation developed at the Universitat Politecnica de Catalunya, Barcelona, Spain. Other partners are MATRA-Espace, who are contractors for the data management systems of the main European space applications (Columbus, Hermes, Ariane); CEA, the French nuclear energy authority; CISI-Ingenierie, Paris, France; CRI, Copenhagen, Denmark; and CCS-SCYT, Madrid, Spain, who brought their expertise and approaches to the design of RT/FT systems; while Paisley's contribution has been in tool building.

In common with most ESPRIT projects, the SMART project contributes to the development of European research and its exploitation in industry. Its specific objectives, in relation to fault-tolerant real-time systems, place it at the edge of what is achievable, and the development of research in this area has ruled out as infeasible (within the project timescale), or not useful, a number of approaches that seemed promising at the outset of the project. This article attempts to indicate some of these aspects in addition to documenting what has been achieved.

The SMART project began in May 1987 and is due to end in late 1989. This article gives a view of the project as of September 1989, and so some of the content of this paper deals with work in progress. All the research aspects of the project have been completed, and the remainder of the project has been carefully planned out to be completed on schedule. The development part of the project has resulted in an environment which offers three tools built upon a frame server specially developed in the project. The tools address the areas of dependability evaluation, fault tree analysis, and simulation of real-time architectures. This environment is being demonstrated as part of the ESPRIT Technical Week.

This paper includes synopses of a number of papers internal to the project, which are individually acknowledged. It is a pleasure to thank our colleagues on the project, especially Pascal Paulet (MATRA) and Hans-Kurt Johansen (CRI), for their contributions to the content of this paper; and the anonymous reviewers for their helpful comments.

2. Outline of the SMART project

2.1. Objectives

The SMART project originally intended [1] to assist the design of fault-tolerant systems using a three-dimensional metric system with product, development, and management as axes. An environment was envisaged as supporting data collection, measurement analysis and modelling, to assist system managers and designers in the development of fault-tolerant systems having to meet constraining dependability requirements. It was hoped that the project would be able to use techniques and tools developed in other ESPRIT projects, and that the project would lead to recommendations for the next generation of fault tolerant systems.

It was noted that in nuclear, avionics, and space applications, industry faces a trade-off between high reliability and productivity, with a corresponding need for evaluation and modelling tools. During the early stages of the SMART project, attention was restricted to these areas, which have an important real-time aspect in addition to fault-tolerance, and a data model based on them was constructed.

2.2. Metrics

Metrics, in the sense of numbers associated with systems or components, have an obvious value in classifying or characterising systems. However, in the sense of something that can be calculated automatically from source code, occurrence of faults, or management history, metrics were not found to be helpful in the RT/FT field, which is characterised by low rates of observed faults, small size of software components, and one-off management techniques. For similar reasons, the use of proportional hazards functions and time series, which had been intended alongside METFAC, were found not to be appropriate [2].

A further problem in relation to the original objectives of the SMART project was that the focus on RT/FT systems also made most models for the development environment and the management process break down. For such systems, the development cost is normally a secondary consideration to safety and running costs, and the system designer generally has little interest is management aspects.

2.3. Amended Objectives

At the same time that these negative results were being assimilated, however, it had become clear that important advances could be made by an appropriate redefinition of the objectives of the SMART project. This redefinition of objectives was based on a theoretical study of RT/FT systems, and an assessment of the needs of industry in the RT/FT field. This led to some aspects being selected for completion within the SMART project, and others being highlighted for further work elsewhere.

It was noted that in order to cover the application areas of interest in the SMART project, including avionics, space missions, and nuclear safety, both non-repairable and repairable systems should be considered. For some systems, design support can be provided for the average user on a safe theoretical basis. For others, the dependability models are so complex that the tools can only be put at the disposal of an expert user. The following section describes the theoretical principles underlying this classification [3].

3. General Structure of Real-Time Fault-Tolerant Systems

3.1. Dependability and Coverage

In considering RT/FT systems, the key concepts are dependability and coverage[4]. Dependability deals with the factors that justify faith in the system's ability to perform according to expectations, and includes such attributes as reliability, safety and availability. Coverage is the probability that the system will remain operational following a given exceptional condition, such as component failure.

In many application areas, the normal way to obtain a high dependability system from hardware components is to arrange them in a parallel processing configuration with redundancy. Typically, the computer system receives input data from the process to be monitored using N-fold redundant sensors and produces a set of corresponding analog or digital output signals. These N signals enter a majority voting device which produces the final value output by the process. In cases of unacceptable disagreements between corresponding values a fault/error handling mechanism is activated. This mechanism is responsible for identifying and isolating faulty components and reconfiguring the system in a gracefully degraded mode of operation.

3.2. Numerical Difficulties

In RT/FT systems the execution of application tasks is organised in control loops with a cycle time of the order of tens of milliseconds. A task that fails to complete in that time is considered to have failed, and any output from it is ignored. The fault/error handling mechanisms mentioned above therefore have to be fast since the operation of the system cannot be stopped for recovery. This implies that recovery actions are typically several orders of magnitude faster than failure and repair actions.

Systems of the type considered in the SMART project need to be ultra-reliable. For example, international airflight regulations require that a failure of the system must occur with a rate smaller than 108-99 per hour for a 10 hour aircraft flight. When failure rates have to be as small as this, it is impossible to determine their values by testing, and mathematical methods are required instead. Because the applications are in safety-critical systems, the objective of dependability analysis is to prove that the dependability calculation is based on conservative models of the system, and known conservative approximation methods are used to model fault/error behaviour and handling (see, for instance 5). High precision computations on large Markov models are not necessarily required, particularly since the transition rates are not known precisely. On the other hand, a sensitivity analysis on input data such as component failure rates is required. Getting good values for the model parameters is costly, and sensitivity analysis can be used to establish acceptable error margins for the various parameters, in relation to the size of their contribution to the metric of interest.

3.3. Decomposition Techniques

In fact, the difference of orders of magnitude between the durations associated to recovery actions and failure/repair actions, which is typical of RT/FT systems, allows the use of the behavioural decomposition technique[6], which has been proved to give conservative estimates, and fortunately, resolves many of the numerical difficulties mentioned earlier[7]. In this technique, the behaviour of the system is modelled using a fault occurrence and repair model (FORM) and a number of fault/error handling models (FEHM). The FEHMs are used to estimate fault coverages, which are incorporated in the FORM as instantaneous switching probabilities following fault actions.

The development of tools for specification and solution of FEHMs has been relegated in SMART to further work. Nevertheless, the simulation tools provided by the SMART project for interactively analysing the mechanisms for dealing with errors are of considerable value within the design and management teams of a project.

The tool dedicated to FORM specification and solution in the SMART environment is an enhanced version of METFAC. A hierarchical high-level language for describing the behaviour of fault-tolerant repairable systems has been developed as an alternative to production rules, and a FORM generator based on that language has been developed and integrated with the original version of the tool. This language is described in the next section.

4. Dependability Evaluation

4.1. Metfac

Dependability evaluation is supported in the SMART environment by an enhanced version of METFAC[8]. METFAC is a tool for specification and solution of Markovian dependability models using homogeneous continuous-time Markov chains. For such models, METFAC computes a number of dependability metrics including reliability, availability, mean time to first failure, as well as performance and cost-related metrics. The main enhancement to the tool implemented in the context of the SMART project has been a new more user-friendly front end for model specification based on a high-level language [9], covering occurrence, propagation and repair of software and hardware faults in hierarchical systems. This high-level language can be used for a wide variety of fault-tolerant repairable systems, but as certain aspects of the modelling process have been built in to the language, thus restricting the class of models for which the high-level language is appropriate, the original front end, based on production rules, has been retained as an alternative.

4.2. The Form Language

The high-level language takes the SAVE [10] modelling language as its starting point but adds new features which expand significantly both the user-friendliness and modelling power of the language used in SAVE. The main features added are: support for hierarchical model specification through the cluster concept, parameterisation of component and cluster classes, and support for the specification of systems with several operational configurations which can be used successively as resources fail. The introduction of the cluster concept not only allows concise specifications of complex systems, but supports the reduction of the state space through use of aggregation techniques.

Experiments using the language have shown its power and facility in describing fault-tolerant computer systems. For example, the Ariane system [11], a case-study proposed by MATRA, contains 54 components which would have to be considered different in a flat, component-level description, whereas using the hierarchical language it is sufficient to describe only two cluster classes and four component classes.

Rather than give the full syntax of the language, its flavour can perhaps best be illustrated by an extract from this example (Figure 1). The extract shows of the system specification, which follows the definition of some system parameters and constants (such as the failure rate "mcufr" below), and is followed by the specification of cluster classes, component classes, and repair teams.

The language also allows the modelling of error propagation, and of repair strategies assigning priorities to repairing failed components. The language was developed at UPC together with the specification of the FORM generation algorithms. Paisley developed a special-purpose window-based editor and preprocessor for the language, and SCYT developed the FORM generator.

5. Safety Analysis

For the purposes of the SMART project, safety is defined as a measure of the time to catastrophic failure. Safety related specifications are generally expressed using requirements such as Fail Operational (FO) or Fail Safe (FS), or indicating fail-safe only on the second fault (e.g. FO/FS).

The use of timed Petri nets [12] is unpractical owing to the size of the Petri nets involved. Two approaches are available:
 (a) failure mode effect analysis (FMEA), which is qualitative in nature, and helps to identify critical components;
 (b) fault tree analyser, which obtains estimates for the unreliability of the system associated to each minimal cut in the fault tree.

Tools for performing the fault tree analysis exist. The contribution of the SMART project will be to provide means of combining the fault trees of components to enable the fault tree analysis of the system to be done without first generating the expanded fault tree of the system.

FIGURE 1 ARIANE

```
System
    Made of
        1 Mt of TC(mcufr, mcuc)
        1 S1t of TC(s1cufr,s1cuc)
        2 S2t of TC(s2cufr,1)
        5 S3t of TC(s3cufr,1)
        1 Bus1 of BusC
        1 Bus2 of BusC
        1 Recovery of RecoveryC
    Resource Attributes
        Can_by_B1: Bus1[1] and Mt.Can_by_B1[1] and S1t.Can_by_B1[1]
                and S2t.Can_by_B1[2] and S3t.Can_by_B1[5]
        Can_by_B2: Bus2[1] and Mt.Can_by_B2[1] and S1t.Can_by_B2[1]
                and S2t.Can_by_B2[2] and S3t.Can_by_B2[5]
        Can_by_B1_and_B2: Bus1[1] and Bus2[1] and
                Mt.Can_by_B1_and_B2[1] and
                S1t.Can_by_B1_or_B2[1] and
                S2t.Can_by_B1_or_B2[2] and
                S3t.Can_by_B1_or_B2[5]
    Operational Requirements
        Recovery[1] and (Can_by_B1 or Can_by_B2 or Can_by_B1_and_B2)
    Configuration
        Requirements: Can_by_B1
        Clusters: TC
            Using_B1: all
        Components: BusC, Recovery
            Operational: all
    Configuration
        Requirements: Can_by_B2
            Clusters: TC
        Using_B2: all
            Components: BusC, Recovery
        Operational: all
    Configuration
        Requirements: Can_by_B1_and_B2
        Clusters: Mt
            Using_B1_and_B2: 1
        Clusters: S1t, S2t, S3t
            Using_B1_or_B2: all
        Components: BusC, Recovery
            Operational: all
```

6. Real-Time Architecture Simulation

The tool described in this section is being developed as part of the SMART project to assist the designers and validators of RT/FTS. Interaction with the display and the user is by means of the X Window System [13].

6.1. Architecture Description

The real-time architecture simulator (RTAS) provides a method of system description which takes into account the two complementary aspects of a system: architectural and behavioural. The architectural aspect describes the construction of the system from components (modular objects), which may be hardware or software. The behavioural aspect describes the evolution of the system in time from one state to the next. In principle the modelling can take place at any level of design detail, but in practice it is most useful on models of the global design of the system.

The architectural description includes both logical and graphical information. The logical information describes the relationships (connections) between components of the architecture, while the graphical information describes how the model can be displayed for the user (creation of a block diagram). The RTAS can be commanded to display a synoptic of the system from the views of the different components during the simulation. A graphical editor is provided to enable the user to modify the appearance of a displayed object.

The behavioural description can deal with error states and error handling in addition to the nominal behaviour of the system: a timed transition high level Petri net formalism is used. Graphical and textual editors are provided for such Petri nets.

6.2. Using The Simulator

Interaction with the RTAS allows an initial state to be set up, and the system behaviour set in motion for a given number of steps or time interval. System evolution is shown on the display by using highlighting, updating variables, etc. No performance is measured, but the real-time functionality can be checked using tracing facilities, animated block diagrams and chronograms.

In this way the system designer can show the nominal behaviour of the system, and can intervene to inject events into the system, by setting variables tested by transition predicates in the Petri nets. Consequently, the FEHM behaviour can be explored, enhanced, and to some extent validated: both hardware and software faults can be analysed, depending on the level of detail provided by the model.

Because fault scenarios can be complex, tracing and some other monitoring facilities are available. This helps when an interesting behaviour mode needs further analysis.

6.3. Implementation

The architecture of the SMART environment is such that the RTAS is implemented as a set of processes co-operating through the medium of the X server and the SMART frame server, which is described in the next section.

A components library for grouping the user models is supported, with a pictorial high level language and synoptic binding in addition to description of operation using a programming language (C). This allows modelling by simple Petri nets with complicated operations, even referring to external C code, as an important alternative to detailed Petri nets.

Systems such as RTAS are frequently implemented using Smalltalk [14]. However, the disadvantage of Smalltalk is that it does not integrate well with Unix [15], and it was considered important to develop an open system in the SMART project, that could be integrated at a later stage with other developments such as IPSEs [16], the PCTE [17] etc.

7. The Frame Server

This section describes the SMART frame server [18], which has been implemented as part of the SMART project.

The frame server provides the infrastructure of the SMART environment. The concept of frame used in the SMART project is rather different from that of Minsky [19], but similarly provides an extremely general method of describing objects and classes. For example, a model description in the FORM language, as described above, is a single frame, as illustrated below. The FORM language defines what attributes should be expected in such a frame.

The server manages a set of independent contexts on behalf of client applications. In each context, a set of frames forms a semantic network which can be navigated, consulted or updated by one or more clients. The use of the frame server minimises access to the disk, and allows a client to be notified if an item of information of interest to that client is updated.

7.1. Frames

Frames have types, but the semantics of the frame type is defined by another frame (usually a Sublanguage-specification frame). The semantics of a frame type is context-dependent, since it depends on which sublanguages are defined in the context's semantic network, and it is possible for a given frame to appear to have different contents in different contexts. There are a number of predefined frame types, including Sublanguage-specification, Context, and Window-data: a window-data frame can be used to describe the contents of a window on the display.

Frames contain attribute/value lists: by default an attribute value consists of text. Repeating attributes are supported, and attributes may have sub-attributes.

For example, representing sub-attributes by indentation, the FORM fragment given earlier corresponds to the frame portion in Figure 2.

7.2. Semantics

Attributes may have associated hook functions, which leads to a kind of access-oriented programming. An important hook function is make-link, which means that the attribute value is the name of a frame to be installed at that point in the semantic network. Methods associated with a frame type allow the user to define new hook functions: and frame types can specialise other frame types, inheriting their methods and top-level attributes, so that object-oriented programming is supported too.

A transaction mechanism supports graceful error recovery and preserves consistency during complex updates to the semantic network. Syntactic and lexical definitions allow the frame server to select portions of attribute values.

Frames on disk can be used to provide a starting state for the semantic network. Window-based browsing and editing tools are provided. A memory-resident frame for each context allows access to internal information about that context, such as the identities of any frames that have been modified but not written to disk.

It follows from the above description of the frame server that the data model in SMART is open. New attributes can be added to existing frame types without any change being needed to existing software. New views of existing frames can be provided simply by using a modified sublanguage-specification. This makes development of the model a simple matter.

Figure 2 A Frame Example

Attribute	Value
System	
Made_of	
Pieces	1 Mt of TC(mcufr, mcuc)
Pieces	1 S1t of TC(s1cufr,s1cuc)
Pieces	2 S2t of TC(s2cufr,1)
Pieces	5 S3t of TC(s3cufr,1)
Pieces	1 Bus1 of BusC
Pieces	1 Bus2 of BusC
Pieces	1 Recovery of RecoveryC
Resource_Attribute	Can_by_B1: Bus1[1] and Mt.Can_by_B1[1] and S1t.Can_by_B1[1] and S2t.Can_by_B1[2] and S3t.Can_by_B1[5]
Resource_Attribute	Can_by_B2: Bus2[1] and Mt.Can_by_B2[1] and S1t.Can_by_B2[1] and S2t.Can_by_B2[2] and S3t.Can_by_B2[5]
Resource_Attribute	Can_by_B1_and_B2: Bus1[1] and Bus2[1] and Mt.Can_by_B1_and_B2[1] and S1t.Can_by_B1_or_B2[1] and S2t.Can_by_B1_or_B2[2] and S3t.Can_by_B1_or_B2[5]
Operational_Requirements	Recovery[1] and (Can_by_B1 or Can_by_B2 or Can_by_B1_and_B2)
Configuration	
Requirements	Can_by_B1
Clusters	TC
Mode	Using_B1: all
Components	BusC, Recovery
Mode	Operational: all
Configuration	
Requirements	Can_by_B2
Clusters	TC
Mode	Using_B2: all
Components	BusC, Recovery
Mode	Operational: all
Configuration	
Requirements	Can_by_B1_and_B2
Clusters	Mt
Mode	Using_B1_and_B2: 1
Clusters	S1t, S2t, S3t
Mode	Using_B1_or_B2: all
Components	BusC, Recovery
Mode	Operational: all

7.3. Implementation

Frames on disk are implemented as text files with the names of attributes stored along with their values, and the hierarchical structure being controlled by special characters. This is an efficient use of disk space since a single frame can contain a substantial amount of information. When a frame is read into a context, the frame type controls the building of efficient data structures in the server's memory which facilitate access to the frame contents. Linked frames are installed in these structures so that their contents appear as sub-attributes of the links.

Clients of the frame server issue commands and queries to the server: just as for data bases, the queries include the names of attributes requested. The corresponding retrieval is retained by the server to be the subject of further requests by the client (e.g. update, insert, delete).

The server-client communication is built on the reliable transport layer of the Unix inter-process communication system, and is similar to that of the X Window System.

8. Conclusions and suggestions for further work

The SMART project has been successful in devising an environment of use to the designers of real-time fault-tolerant systems. It has brought a number of theoretical results on dependability into the mainstream of industrial exploitation.

8.1. Further Research

Further work in the dependability area is required into

8.1.1. FEHM. The study of general exit time distributions for FEHM, possibly leading to a kind of importance sampling technique in the simulation algorithm;

8.1.2. Metric evaluation. Following up suggestions in the literature for evaluation of dependability and related metrics [20,21,22];

8.1.3. Sensitivity analysis. Multi-parameter sensitivy analysis using eigenvector analysis of a locally linearised matrix of partial derivatives;

8.1.4. Multiphased systems. Development of numerical methods for the evaluation of metrics for multiphased systems;

8.1.5. Approximation Techniques. Development of approximation techniques for FORM's with provable error bounds.

8.2. Tool Enhancements

8.2.1. METFAC.

The SMART project is making the METFAC dependability evaluation tool more accessible to industrial developers. A number of proposed enhancements to METFAC would develop from the research areas mentioned above: application to multiphased systems, and approximation methods to reduce the state space.

8.2.2. RTAS.

The SMART real-time architecture simulator supports an integrated approach to system simulation and validation of fault/error handling methods. It would be convenient to add the standard verification algorithms for Petri nets (cyclicity, liveness, boundedness) and facilities for verifying invariants. Other ways in which the RTAS could be developed would be to generalise the user interaction and display methods so that quite general systems could be animated.

8.2.3. Frame Server.

The frame server concept appears to have a usefulness in a number of other areas, and its potential use is currently being explored in a number of research projects. Further interfaces with databases and windowing systems could be developed so that it could be used in other contexts, and a number of optimisations would be desirable, such as more efficient use of virtual memory.

8.3. Evaluation

Some validation studies are already in progress as part of the SMART project, and further evaluation studies, possibly coupled with some of the above suggestions for further work, may find a place in the ESPRIT programme. It is expected, however, that an industrially-significant toolset to help with the design of RT/FT systems will emerge from the SMART project, and therein will lie its success.

References

1. Kuntzmann, A. (1986) 'SMART: System Measurement and Architecture Techniques', ESPRIT project proposal.

2. Musa, J. D. (1987) 'Software Reliability Measures: Guiding Software Development for Cost-Effective System Quality', ESEC '87 1st Europoean Software Engineering Conference (Tutorial 2).

3. Johansen, H. K. (1988) 'Dependability Evaluation of Real-Time Fault-Tolerant Systems', SMART working paper, CRI Copenhagen.

4. Laprie, J. C. (1985) 'Dependable Computing and Fault-Tolerance: Concepts and Terminology', Proc. 15th Int Symp on Fault-Tolerant Computing, pp. 2-11.

5. Wensley, J. H. (1978) 'SIFT: Design and Analysis of a Fault-tolerant Computer for Aircraft Control', Proceedings of the IEEE, vol. 66, no. 10, pp. 1240-1255.

6. Geist, R. M. and Trivedi, K. S. (1983) 'Decomposition in reliability analysis of fault-tolerant systems', IEEE Transactions on Reliability, vol. R-32, pp. 463-468.

7. McGough, J., Smotherman, M., and Trivedi, K. S. (1985) 'The conservativeness of reliability estimates based on instantaneous coverage', IEEE Transactions on Computers, vol. C-34, pp. 602-609.

8. Carrasco, J. A. and Figueras, J. (1986) 'METFAC: Design and Implementation of a Software Tool for Modelling and Evaluation of Complex Fault-Tolerant Computing Systems', Proc 16th Int Symp on Fault-Tolerant Computing, pp. 424-429.

9. Carrasco, J. A. (1989) 'A high-level modelling language for fault-tolerant computer systems', SMART working paper, UPC Barcelona.

10. Goyal, A., Carter, W. C., de Souza e Silva, E., Lavenberg, S. S., and Trivedi, K. S. (1986) 'The

System Availability Estimator', Proc 16th Int Symp on Fault-Tolerant Computing, pp. 84-89.

11. Sotta, J. P. (1988) 'Modelling the Ariane System', SMART working paper.

12. Leveson, N. G. and Stolzy, J. L. (1985) 'Safety Analysis using Petri Nets', Proc 15th Int Symp on Fault-Tolerant Computing.

13. Fulton, J. (1988) X Window System, Version 11, X Consortium, MIT Laboratory for Computer Science.

14. Goldberg, A. (1983) 'The Influence of an Object-Oriented Language on the Programming Environment', ACM Computer Science Conference, pp. 35-44, Orlando, Florida.

15. Ritchie, D. M. and Thompson, K. (1978) 'The Unix Time Sharing System', The Bell System Technical Journal, vol. 56, no.6, pp. 1905-1929. Unix is a trademark of AT&T Bell Laboratories.

16. US Dept of Defense (1980) 'Requirements for Ada Programming Support Environments', STONE-MAN.

17. ESPRIT (1985) PCTE: A Basis for a Portable Common Tool Environment. (Functional Specification)

18. Crowe, M. K. and Oram, J. W. (1989) 'An Applicationlevel Frame Server', SMART Technical report, Paisley College of Technology.

19. Minsky, M. (1980) 'A framework for representing knowledge', in Mind design: Philosophy, psychology, and artificial intelligence, ed. J Haugeland, MIT Press, Cambridge, Mass.

20. Heidelberg, P. and Goyal, A. (1987) 'Sensitivity Analysis of Continuous Time Markov Chains using Uniformization', Proc of the 2nd Int Workshop on Applied Mathematics and Performance/Reliability Models of Computer/Communication Systems, Rome.

21. Singh, C., Billington, R., and Lee, S. Y. (1977) 'The Method of Stages for Non-Markov Models', IEEE Transactions on Reliability.

22. Conway, A. E. and Goyal, A. (1987) 'Monte Carlo Simulation of Computer System Availability/Reliability Models', Proc 17th Int Symp on Fault-Tolerant Computing, pp. 230-235.

THE KIWI(S) PROJECTS: PAST AND FUTURE

The KIWIs Team [1]

Abstract

Knowledge base management systems play a crucial role in advanced information processing. Nevertheless, new sophisticated applications using large amounts of data and knowledge, possibly stored on multimedia devices and distributed systems, call for novel more reliable, more efficient, more powerful and, at the same time, more friendly systems. In order to (at least, partially) reply to such requests, a number of strategic programs and projects are being carried out. In particular, the European ESPRIT program has funded and is funding a number of projects involved with knowledge base issues. Among them, the KIWI projects 641 and 1117 have designed and developed a prototype of an advanced knowledge base system; the new KIWI project 2424 is now using and further extending and improving the results of the previous projects in the development of an industrial prototype new-generation knowledge base management system.

1. Introduction

A cornerstone for advanced information processing is the availability of knowledge base systems that support sophisticated applications requiring complex operations on large amounts of data and knowledge, possibly stored on multimedia devices and distributed systems. Such an ambitious goal has motivated the proposal and the performance of the KIWI(s) project within the framework of the ESPRIT program.

The preliminary definition of the KIWI knowledge base system was given in the ESPRIT project 641 (*Knowledge-Based User-Friendly Interfaces for the Utilization of Information Bases*) from December 1984 to November 1985 by the following consortium: CRAI (Italy, prime contractor), Dansk Datamatik Center (Denmark), ENIDATA (Italy), INRIA (France), Philips Intl B.V. (The Netherlands), University of Antwerp (Belgium) and University of Rome (Italy) [S1]. The actual design and implementation of the KIWI system was carried out by the same consortium within the ESPRIT project 1117 from February 1986 to December 1988 [KT]. The latter project has achieved a number of interesting results, particularly in the area of object-oriented languages for knowledge representation, combination of logic programming and databases and graphical user interfaces to knowledge bases. Nevertheless, a tight integration of object-oriented programming, logic programming and database system has not been pursued since the project intended to deeply investigate each of them and therefor only achieved a loose combination of them, suitable for a

---

1   The KIWIs team is presently composed by: A. Mecchia, C. Pizzuti, G. Rossi, P. Rullo (*CRAI, Rende Italy*), C. Del Gracco, G. Germano, P. Naggar (*ENIDATA, Rome, Italy*), E. Laenens, J.J. Snijders, F. Staes, R. Van Dam (*PHILIPS Application & Software Services AIT, Eindhoven, the Netherlands*), M. Ahlsen, J.A. Bubenko, M. Norrie (*SISU, Kista, Sweden*), B. Verdonk, D. Vermeir (*University of Antwerp UIA, Wilrijk, Belgium*), D. Saccà, A. Volpentesta (*University of Calabria, Rende, Italy*), S. Christodoulakis (*University of Crete, Crete, Greece*), A. D'Atri, L. Tarantino (*University of L'Aquila, L'Aquila, Italy*).

short-term industrial exploitation. For long-term exploitation, the strength of the project results on each of the above issues has prepared the ground for future research activities focused on knowledge base systems of great power and versatility where various techniques and programming styles are tightly combined and gracefully harmonized.

As follow-up to the project 1117, a new project within the ESPRIT program has started on January 1989 and will last till December 1991: the project 2424 - *KIWIS, Advanced Knowledge-Base Environment for Large Database Systems*. The consortium is now composed by: CRAI (Italy), ENIDATA (Italy), Philips Application and Software Services AIT (The Netherlands, prime contractor), SISU (Sweden), University of Antwerp (Belgium) University of Calabria (Italy), University of Crete (Greece), and University of L'Aquila (Italy). The aim of the new project is to build a system that provides a tight integration between advanced knowledge-based paradigms (notably, object-oriented and logic programming) and database techniques. Such an integration is necessary to arrive at the performance level required for a novel industrial product and at the expressive power required by sophisticated knowledge-intensive applications. Moreover, the new system will support links with existing databases as well as with other similar systems (thus, from one KIWI system to a *federation* of KIWIS - as suggested by the project acronym) so as to allow exchange of information without prior awareness of the knowledge or data stored in these systems.

The aim of this paper is two-fold: (a) to present the main results of the projects 1117 (Section 2) and (b) to describe the main goals and the preliminary results of the new project 2424 (Section 3).

2. Main results of the project 1117

2.1 GENERAL

The KIWI system consists of three layers:

(a) the *User Interface* (**UI**) assists users in the interaction with the KIWI system;
(b) the *Knowledge Handler* (**KH**) implements the knowledge representation language OOPS, which is basically object-oriented but enriched with other powerful paradigms such as rule-based programming, monitors and logic programming;
(c) the *Advanced Database Environment* (**ADE**) supports the combination of relational database and logic programming technologies to efficiently implement the OOPS concepts.

Prototypes of the three layers of the KIWI System have been realized and integrated even though full integration was outside the scope of the project. In particular, the integration of KH and ADE is not a tight integration but a loose one. This means that the KH is responsible for the implementation of many of its methods, whereas the ADE provides the support to manage "regular" knowledge. In spite of the abovementioned limitations, the loose integration has been proved to be effective and examples have been developed which involve all three layers. In addition, the KIWI project has added interesting results to the current state of the art: in fact, within the project, many novel issues concerning languages, knowledge base and database systems have been deeply analyzed. In particular four interesting results have been achieved:

a) the definition of an object-oriented language (OOPS+) with many advanced features;

b) the extension of relational database systems to include logics;

c) the design of a graphical interface to the overall system.

d) the implementation of three prototypes, running on SUN workstations, one for the KH, one for the ADE, and one for the UI; the three prototypes provide an efficient environment for developing advanced data-intensive knowledge-based applications of great versatility.

2.2 The Knowledge-handler and OOPS+.

The Knowledge Representation Language OOPS+ was developed as an object-oriented database programming language to be used in the KIWI system as an interface to databases via an advanced deductive database environment [LV, VL]. The language aims to provide a suitable tool for knowledge-based applications programming, integrating concepts from semantic modeling, databases and logic and object-oriented programming in a simple and orthogonal fashion.

The basic concept in OOPS+ is the notion of (persistent) objects. Several kinds of objects are recognized, the most important of which are

(a) Primitive objects such as *str* and *int* (the set of all strings and all integers respectively), integers, strings, *nil* and * (the object that has all objects as an instance).

(b) Records, which are labeled tuples of references to other objects. For example,

$$fred = (name = \text{``fred''}; age = 5)$$

defines a record object with two fields, labeled name and age.

(c) Sets and lists of (references to) objects such as *persons* = { *john, joe, fred, jane*} and *ranks* = < {*fred, jane*} , *john*>.

(d) Function objects. Functions may be written using a mixture of imperative and logic programming (for querying) constructs. They may also be set up for automatic execution whenever certain events occur. This trigger mechanism is useful, e.g., to enforce constraints.

Between objects, an "instance-of" relation, denoted as ':' is defined which can be intentional (structural similarity) or extensional (set membership). For example, we have that

$$fred : (name = str)$$

because the *name* field of *fred* refers to an instance of *str*. On the other hand, we also have that

$$fred : persons$$

since fred is a member of the *persons* set object.

OOPS+ is strongly (but not statically) typed. Type constraints are specified by requiring that a record field always refers to an instance of an associated (type) object. For example,

$$fred = (str\ name = \text{``fred''}; persons\ father = joe)$$

requires fred's name to be a string, his age must be an integer and his father must be an existing person. Any object can be used as a type, and thus types are first class objects which may evolve, e.g. when using a set as a type.

Functions are also (polymorphically) typed since the actual parameter (record) object must be an instance of the formal parameter object.

An OOPS+ program then is just a definition of a 'top level' record object, called the program object. The program below specifies a small knowledge base.

```
student =
       (
       name = str;
       enroll = [enrollment] default {};
       grade = () - int default .. ;
       );
enrollment =
       (
       course = courses;
       score = int;
       );
courses = { cs100, ... };
```

In the example, [enrollment] denotes a power object, the instances of which are all finite sets of enrollment instances.

A record object may specify default instances for its fields. Together with the operation of type casting, this provides the usual inheritance facilities. For instance, the, expression

```
(student) (name="fred")
```

will equip fred with the default (empty) set of enrollments and (a reference to) the default grading function.

Finally, we mention the possibility of creating the union or intersection (denoted a|b and a&b) of two objects. Such objects have as instances the union (resp. intersection) of the sets of instances of their components. Union and intersection objects are useful as types, as in

```
person = male | female
student = person & (grade = int)
```

Implementation

The full OOPS+ language has been implemented as a stand-alone prototype system on a SUN workstation. The system consists of a compiler which translates OOPS+ source into a knowledge base population and a run-time system that supports function (method) execution. For system development, a graphical browser interface is provided which gives direct access to all features of the language for both update and retrieval. In addition, an application development tool is available which allows for easy customization of the graphical representation of objects, as well as the linkage of input/output events to OOPS+ functions.

2.3 The Advanced Database Environment

The main feature of ADE [S2] is the combination of logic programming and relational databases; the logic programming language supported is a simple extension of DATALOG (i.e., logic programming without function symbols) where base predicates correspond to objects rather than simple relational tuples. Actually, an object is vertically partitioned, thus it is represented as multiple tuples that are spread amongst several relations (a surrogate in every tuple is used to recompose the object).

To implement DATALOG programs, ADE uses a Prolog system coupled with relational databases for handling facts of the knowledge base and for retrieving additional facts from external databases. However, since DATALOG is based on the pure semantics of definite Horn clause queries (minimum model and fixpoint computation) that is not realized by the particular execution model of Prolog (SLD resolution with leftmost goal expansion), ADE modifies DATALOG programs into Prolog programs where the fixpoint computation is enforced. In addition, while rewriting a logic program, ADE selects a safe and efficient implementation, thus the programmer is relieved from the burden of ordering rules and goals.

Following a "database approach" the intensional information is compiled into efficient code that is applied at run time on the extensional database. The strategy used to answer a query is the query/subquery approach. Whenever it is possible, the fixpoint computation is made in two steps; in the first step initial bindings are propagated to restrict the actual answer computation of the second step.

The compilation methods used by ADE are the *Minimagic method* with the differential technique and the *Magic-Counting method* for (sub)queries with linear recursion [SZ1-SZ4]. The output of a query compilation is a Prolog program where unification is removed and the only operations are handling facts (assert or retrieve) and nested-loop joins. Theoretical complexity analysis [MPS1, MPS2] and experiments performed on fact bases with a mean size of 500 facts show that prototypes can be rather efficient since compiled queries run much faster (up to 20 times) than direct Prolog execution.

ADE is also responsible for handling the KIWI Knowledge Base. To this end, each object of the KB is considered to consist of a name (or surrogate) and of relationships to other objects; it can be an instance of other objects; besides, it may have instances and, therefore, be a class; it may be a subclass of another object (ISA relationship), thus its instances are also instances of the other object; finally, an object may have properties (attributes) and values for those attributes.

A synthetic description of the main relationships among objects is kept in central memory in order to have a scheme of the KB available for fast manipulation. This allows to transform the object-oriented requests into the tuple-oriented ones (possibly involving many relations) only by consulting the KB scheme stored in central memory.

Implementation

The ADE has been implemented as a stand-alone prototype on a SUN workstation and consists of two parts: KNOW-ADE, which interfaces the KH layer and maps its requests into tuple oriented operations, and ADE-MACHINE, which i) manages the knowledge base, stored as a relational database, ii) provides efficient implementation of logic queries using rule rewriting methods and C-Prolog as execution environment and iii) interfaces with external relational databases managed by INGRES.

2.4 The User Interface

KIWI is a system oriented to different classes of users with different needs and requirements, according to their skill and expertise with the system. The aim of the KIWI User Interface (UI) is to support non expert users in their interaction with the OOPS+ knowledge base [DDM]. The friendliness of the UI is obtained through a graphical environment which both allows for an easy dialogue and multiple presentation frames and supplies information on the knowledge base in a more natural and friendly way for the naive user.

The UI uses two models to represent and display information in a friendly way [DLP]: an *external* model for the actual interaction with the user and an internal model to interface to the knowledge base. The internal model, which is not visible to the users, defines structures and operations for the external model starting from the objects described in OOPS+. The model is a particular kind of semantic network based on the formalism of directed multi-graphs with labeled nodes and unlabeled edges. The external model (also called *display* model) uses semantic trees as basic structures for the user interaction. A semantic tree for a given object shows the relationships of the given object with all other objects and is constructed using the semantic network of the internal model. A graphical model (called *box_in_box* technique) is used to display a portion of a semantic tree.

In order to retrieve information from the KIWI system, the User Interface supplies the user with simple, intuitive commands and techniques to perform an analysis (*navigation*) of the knowledge base. Three kinds of techniques have been studied, implemented, validated and fully integrated in the KIWI system, making use of OOPS+:

1) *Elementary browsing*, to retrieve information by examining the "neighborhood" of an object and by moving the visibility portion of its associated semantic tree (by panning and zooming operations).
2) A more sophisticated style of browsing (*synchronized browsing*) has been investigated to speed up repetitive searches starting from the results of previous elementary browsing.
3) A third technique (*view manipulation*) provides structures and operations for gathering browsing results. It allows the user operating on existing semantic trees, to create new (virtual) semantic trees.

Implementation

The prototype of UI, running on a SUN workstation, is composed by two parts: the actual user interface (KIWI-UI) and the KIVIEW system [MDT]. Both use the SunView libraries, which provide tools for a friendly interaction (such as pop-up menus, panels, pop-up windows, and so on). As for the KIWI-UI, an ad hoc graphical kernel has been implemented (according to the GKS standard) which supports a multi-window environment and handles complex graphical hierarchical structures. KIVIEW is a prototype whose underlying knowledge representation formalism is simpler than the one of OOPS+ and it was used to experiment with the formalism for displaying semantic networks.

2.5 The KIWI Methodology

The KIWI methodology for information retrieval is the specification of a way to assist information consumers in searching for and accessing data in relational and bibliographic databases that are external to the KIWI system [La1, La2]. The KIWI Knowledge-based Information Retrieval Assistant (KIRA) is a prototype software system, which implements the methodology.

The methodology and the KIRA system focus on those users of the KIWI system, who are professional experts in their own subject area, but are not expected to have knowledge of OOPS+, query languages, nor database descriptions. Another important group of users are information retrieval experts, who are responsible for describing subject areas and external databases to the system.

The methodology can be seen as a *specification* of how to support these users, while the KIRA system can be seen as a *demonstration* and *test vehicle* for it.

The following *principles* of information retrieval assistance have been identified as basic elements of the methodology:

(a) The user's *subject terminology* must be known to the assistant and coupled with descriptions of the relevant databases.
(b) Users must be interviewed in order to help them find out how their information needs can be best described or specified.
(c) The assistant must be able
  - to find a set of interpretations of the user's need specification and rank them,
  - to construct correct database queries for the selected interpretation(s),
  - to rank the obtained answers.
(d) The user must give feedback about the relevance to his information need of the answers obtained from the databases.
(e) Users must be allowed to choose a greater or lesser degree of control over the assistance and retrieval processes.

These principles have been further developed and specified in a formal way using OOPS+ giving a requirements specification for the software development of an assistant program.

The KIRA system is a prototype assistant program, which implements the KIWI methodology based on the OOPS+ specification. It has been used as a demonstration of the ideas in the methodology and as a vehicle for experimentation and "thinking-aloud" tests with users.

Implementation

The KIRA system is programmed in Smalltalk/V for the MS-DOS operating system, and part of it has been ported to Smalltalk-80 for the UNIX operating system. Smalltalk was selected because of its close relationship to OOPS+. In fact, it has turned out to be easier to translate between the two, where OOPS+ is the more high-level and easy to overview and read.

2.6 Conclusion

The project has delivered 20 technical reports, a number of papers published in international conferences and journals and the following prototypes:

− KH - the Knowledge Handler

- ADE - the Advanced Database Environment
- UI - the User Interface,

which have been loosely integrated. Further on, the KIWI methodology for information retrieval has been experimented through the KIRA prototype.

3. Goals and Preliminary Results of the Project 2424

The new project is a natural outgrowth of the previous one; however, while the previous project was pre-competitive research-oriented, the new project aims to use and improve the results of the previous one in the development of an industrial prototype new-generation knowledge-base management system, co-ordinated with other similar systems.

The new KIWI system can be used both as a sophisticated stand-alone "personal knowledge machine" that supports knowledge-based applications, as well as a "window on the world" that provides a seamless integration of information coming from a wide variety of other sources with the local knowledge bases. The features of the new KIWI system include:

(a) An advanced graphical user interface design system that provides a smooth transition between default (generic) and special purpose (application) knowledge base usage.
(b) A new knowledge representation and manipulation language, called LOCO, which is based on a tight integration between the object-oriented and the logic programming paradigm. In addition, the language supports such features as defeasible and default reasoning, making it suitable also for AI-flavored applications, e.g., expert systems.
(c) Efficient query evaluation algorithms, extending the state-of-the-art in deductive database technology.
(d) Representation and manipulation of a large variety of objects including texts with the usual textual information retrieval operations, simple images (bitmaps) together with a number of operations such as display, entry, clip, past, scale, zoom, etc., and arbitrary objects (called UFOs) together with their operations and access methods, that are supported by "foreign" systems.
(e) An efficient underlying main-memory resident database management system that is tuned to the requirements of the efficient manipulation of large numbers of complex objects and deductive queries.

Integration with external information sources is possible in two ways:

(a) To provide a tight and transparent read-only interface to a variety of external information sources, including traditional databases and text or picture ones. These databases may reside on the same or remote systems, using traditional or 'new' devices (such as CD-ROM). The interface module of the KIWI system will be extendible so that the above list is not exhaustive.
(b) to allow for a federation of KIWI nodes for information sharing and cooperation without a commitment to a centrally maintained global schema.

In both cases the imported information is completely integrated with the local knowledge base.

Presently, a preliminary description of the new KIWI system as well as of its interfaces to external information sources has been completed and a first release of its architecture definition is available. In order to develop the system, the approach of fast prototyping will be adopted instead of giving a detailed specification before implementation. A minimal system should be implemented soon, having all the major components of the final system, although each component with limited functionality. This minimal system will be useful to clarify requirements, specifications, interfaces and coordination aspects. Research will be pursued in parallel to expand and improve the various components; the expanded functionalities will be implemented in later versions of the system. Finally, a case study will be started from the very beginning of the project in order to (a) have a consistent example on which all system components can be tested, (b) verify the capability to support real knowledge-based applications and (c) communicate the insights and the capabilities of the systems to non-technical persons in order to receive useful feedbacks and to promote industrial exploitation.

Acknowledgement

Many persons were working on the implementation of the KIWI system. We are grateful to all of them; moreover, we wish to thank all the colleagues from Dansk Datamatik Center, INRIA and the University of Rome who gave a great contribution to the performance of the previous KIWI projects.

References

[DDM] D'Atri, A., Di Felice, P., Moscarini, M., "Dynamic Query Interpretation in Relational Databases", *Proc. of 6th ACM SIGACT-SIGMOD-SIGART Symp. on Principles of Database Systems*, San Diego, March 1987, pp. 70-78.

[DLP] D'Atri, A., Laenens, E., Paoluzzi, A., Snijders, J., Tarantino, L., "A User-Friendly Graphical Interface to Knowledge and Databases", 1988, submitted for publication.

[KT] KIWI Team, "A System for Managing Data and Knowledge Bases", in *ESPRIT '88: Status Report of Continuing Work,* The Commission of European Communities (Eds), Elsevier Publishers B.V. (North-Holland), 1988.

[La1] Larsen, H.L., "Knowledge Representation in IRIS, an Information Retrieval Intermediary System" *Proc. of the 7th International Workshop on Expert Systems and their Applications*, Avignon, May 1987.

[La2] Larsen, H.L., "KIWI: knowledge-based user-friendly system for the utilization of information bases", in *Knowledge Engineering, Expert Systems and Information Retrieval, I*, Wormell (ed.), Taylor Graham, 1987, pp. 113-126.

[LV] Laenens, E., Vermeir, D., "A Language for Object Oriented Database Programming", *Proc. ECOOP'88*, Oslo, Springer-Verlag, 1988.

[MDT] Motro, A., D'Atri, A., Tarantino, L., "The Design of KIVIEW: An Object Oriented Browser", *Proc. 2nd Intl. Conf. on Expert Database Systems*, April 1988.

[MPS1] Marchetti-Spaccamela, A., Pelaggi, A., Sacca, D., "Worst-case complexity analysis of methods for logic query implementation", *Proc. ACM SIGMOD-SIGACT Symp. on Principles of Database Systems*, 1987, pp. 294-301.

[MPS2] Marchetti-Spaccamela, A., Pelaggi, A., Sacca, D., "Comparison of methods for logic query implementation" *Journal of Logic programming*, to appear.

[S1] Sacca, D., Vermeir, D., D'Atri, A., Liso, A., Pedersen, S.G., Snijders, J.J., Spyratos, N., "Description of the overall architecture of the KIWI System", in *ESPRIT '85: Status Report of Continuing Work*, The Commission of European Communities (Eds), Elsevier Publishers B.V.

(North-Holland), 1986, pp. 685-700.

[S2] Sacca, D., Dispinzeri, M., Mecchia, A., Pizzuti, C., Del Gracco, C., Naggar P., "The Advanced Database Environment of the KIWI System", in [DE], pp. 20-27.

[SZ1] Sacca, D., Zaniolo, C., "On the implementation of a simple class of logic queries for databases", *Proc. 5th ACM SIGMOD-SIGACT Symp. on Principles of Database Systems*, 1986, pp. 16-23.

[SZ2] Sacca, D., Zaniolo, C., "Implementation of recursive queries for a data language based on pure Horn clauses", *Proc. 4th Int. Conf. on Logic Programming*, Melbourne, 1987, pp. 104-135.

[SZ3] Sacca, D., Zaniolo, C., "The generalized counting method for recursive logic queries", *Proc. 1st International Conf. on Database Theory*, Rome, 1986, Theoretical Computer Science, November 1988.

[SZ4] Sacca, D., Zaniolo, C., "Magic counting methods", *Proc. ACM SIGMOD Conf.*, San Francisco, 1987, pp. 49-59.

[VL] Vermeir, D., Laenens, E., "OOPS+, an Object Oriented Database Programming Language", *Journal of Object Oriented Programming*, to appear.

Project no 2424

LOCO, a Logic-based Language for Complex Objects

E. LAENENS
Philips Applications & Software Services
Advanced Information Technology
Building HCM5
P.O. BOX 218
5600 MD Eindhoven, The Netherlands

D. VERMEIR
B. VERDONK
Dept. of Math. and Computer Science
University of Antwerp, U.I.A.
Universiteitsplein 1
B2610 Wilrijk, Belgium

Abstract

Both object-oriented programming and logic programming have received increased attention over the last decade, due in part to their applicability to a wide variety of areas. LOCO
(LOgic for Complex Objects) is a database programming language that combines the declarative elegance and power of logic programming with the advantages of object-oriented
systems: object identity, inheritance, default reasoning, encapsulation etc. A LOCO program
describes a knowledge base (schema and initial population) as a set of interrelated objects.
We take the view that an object can be identified by its properties, i.e. relationships to other
objects. The properties of an object are described using a logic program, hence, to LOCO,
an object is just a set of clauses. However, the clauses in an object's definition do not constitute
the entire knowledge about that object. A specificity relation defined on the objects allows for
the introduction of some rules for knowledge flow between them. This specificity relation (also
called subobject relation)[1,2] is sufficiently general and powerful to be useful to model e.g.
delegation[1,2] classification and/or generalization hierarchies, etc. Therefore, the language
does not enforce a particular modeling paradigm. The core of the language is based on a
nonmonotonic logic[3,4] called Inheritance Logic[5] which has been developed to provide a
formal model for object-oriented concepts, within a logical framework.

1 Objects

1.1 Simple Objects

In LOCO, the properties of an object are described using a logic program. Thus an **object**
is just a set of clauses. Simple objects are made up of a set of base properties (i.e. facts):

```
nautilus =
       {
       name("Nautilus").
       nrOfRooms(267).
       address(city="Bajamar",street="av. de las Piscinas 2").
       telephone(540500).
       rooms(category=1,nr=106).
       rooms(nr=161,category=2).
       };
```

The above declaration describes a hotel called "Nautilus" as a set of facts which are presumed to be true about this object: it has 267 guest rooms, an address, a telephone number and rooms. It should be noted that integers and strings are predefined objects in LOCO.

It is possible to specify facts using named arguments, as in address and rooms, so that the position of the arguments has no meaning and the arity is not fixed. The above fragment expresses the fact that there are 106 rooms of the first category and 161 rooms of the second category in the hotel. So it is possible to have many clauses for the same predicate in a single object.

1.2 Complex Objects

Complex objects extend the notion of objects in two ways: one is that it is possible to specify derived properties using rules, the other allows for the usage of simple or complex objects in the definition of properties.

Derived properties have a syntax that is similar to prolog: in particular, variables start with upper case.

```
nautilus =
       {
       ..
       street("av. de las Piscinas 2").
       city("Bajamar").
       address(Street,City) :- street(Street), city(City).
       };
```

When asked for the address of the hotel, LOCO will use the available rules and facts about address to compute all solutions.

The following person is the owner of the hotel nautilus and can therefore be defined in terms of the previously defined object (refered to by) *nautilus*.

```
gonzales =
              {
              ..
              profession("hotelkeeper").
       owns(nautilus).
       };
```

2 .The specificity relation

The specificity relation, also called **instance-of** relation, is defined on objects and is used in the definition of information flow between these objects (see sections on inheritance, overruling and defeating).

If X instance-of Y, then X is called the instance or subobject of Y, while Y is called the class or superobject of X. We also say that Y is more general than X while X is more specific than Y.

In LOCO, the instance-of relation on simple objects is predefined. E.g. *"boy"* and 7 are instances of *str* and *int* respectively.

Consider the following declaration of a class. It is a description of features that are common to all hotels: they have a name which is a string (str), they have an address which consists of 2 strings and an integer-valued telephone.

```
hotel =
        {
        name(str).
        nrOfRooms(int).
        address(street=str,city=str).
        telephone(int).
        };
```

One can then make nautilus an instance of hotel by writing

```
(hotel) nautilus =
        {
        ..
        };
```

The instance-of relation can be specified with the superobject, with the subobject or both, as is illustrated below:

```
nautilus = {..};
        neptuno = {..};
        hotel =
                {
                ..
                _instance(nautilus).
                _instance(neptuno).
                };
```

The built-in _*instance* predicate establishes the specificity relation between its argument and the object of which it is a property. Note the ' _ ' in the definition of the instance property. This makes instance an **own** property (see below).

The last example is equivalent to

```
hotel = {..};
        neptuno = {..};
        (hotel) nautilus = {..};
        (hotel) neptuno;
```

and to

```
        nautilus = {..};
        neptuno = {..};
        hotel =
                {
                ..
                _instance(nautilus).
                };
        (hotel) neptuno;
```

The built-in _instance predicate adds much power to the language. Indeed, the _instance property of an object need not be base, it can as well be derived. Moreover, it gives the ability to query the specificity relation.

The following example illustrates this:

```
(hotel) romanHotel =
        {
        _instance(H) :-
                hotel._instance(H),
                H.city(rome).
        };
```

where '.' denotes the selection operator: object.propertyName(arguments).

In order to properly support inheritance in this case, the instance relation's extension would have to be maintained, e.g. using the trigger mechanism (see below).

Note that objects that are instances may themselves have instances. The relationship between objects is therefore hierarchical, but it is a directed acyclic graph rather than a tree, for an an object may be instance of many other objects.

```
(restaurant hotel) nautilus =
        {
        ..
        };
```

The above expression defines nautilus to be both a hotel and a restaurant.

There is a built-in predicate class which associates an object o with all the objects of which it is an instance. Class can be thought of as defined by the rule

```
o =
    {
    ..
    class(X) :- X._instance(o)
    }
```

In the next sections, we discuss the effect of instance-of declarations.

2.1 Monotic Inheritance

Our first knowledge flow rule defines the notion of inheritance: information (i.e. base and derived properties) filters down from an object to its instances.

Consider the following example where sheratonStockholm is declared as an instance of sheratonHotel which in turn is an instance of hotel. Hotel is the most general of the three objects while sheratonStockholm is the most specific.

```
hotel =
        {
        name(str).
        nrOfRooms(int).
        street(str).
        city(str).
        address(street=Street,city=City):- street(Street), city(City).
        telephone(int).
        };
(hotel) sheratonHotel =
        {
        name("Sheraton").
        category(5star).
        not cheap.
        };
(sheratonHotel) sheratonStockholm =
        {
        nrOfRooms(416).
        street("Tegelbacken 6").
        city("Stockholm").
        telephone(142600).
        };
```

According to the inheritance rule, the *address* property as defined at *hotel* will be available at *sheratonHotel* and consequently at *sheratonStockholm*. The same holds for the other properties of *hotel*. Similarly, the properties *name*, *category* and *not cheap* of *sheratonHotel*, are inherited by *sheratonStockholm*.

However, this information flow from a general object to more specific objects may be blocked in several ways. One is by introducing **own** properties. These are properties that are owned by the object in which they are defined, in that they do not pass through to its instances. As an example, recall the own predicate *_instance*. This property will not be visible in objects more specific than its owner, e.g. *_instance(nautilus)* will not be visible in *neptuno*.

The other ways of information blocking, overrulers and defeaters, are related to nonmonotonic reasoning and are discussed in the respective sections.

2.2 Nonmonotonic Reasoning

It may be the case that inheritance yields conflicting or inconsistent information for a certain object. In this section we indicate how LOCO deals with such situations. First we need a definition of conflicting and inconsistent information.

Inconsistent information

A positive and a negative fact $p(a)$ and not $p(b)$ are said to be inconsistent if $a = b$ or if a is an instance of b. For example (assume that *europeanCountry* is an object of which all European countries are instances), *like(italy)* and *not like(europeanCountry)* are inconsistent while *like(brazil)* and *not like(europeanCountry)* are not.

Conflicting information

Let $p(a)$ be a fact about some object. We say that $p(b)$ is conflicting with $p(a)$ if (1) a is no instance of b and (2) b is no instance of a and (3) a and b have no common instances. A negative fact *not p(c)* is conflicting with $p(a)$ if a is an instance of c or c itself, so any inconsistent information is conflicting. As an example, *checkoutTime(12)* is conflicting with *checkoutTime(14)* but not with *checkoutTime(int)*. Also, *not allowed(mammals)* is conflicting with *allowed(snoopy)*, but not with *allowed(wanda)* where *wanda* is a fish.

2.2.1 Overruling

Consider the following example:

```
hotel =
       {
       ..
       facility(restaurant).
       };
(hotel) inn =
       {
       ..
       not facility(restaurant).
       };
(inn) youthInn =
       {
       ..
       };
```

In this case, the result of inheritance is unreasonable as we find inconsistent information at the object *youthInn*: *not facility(restaurant)* and *facility(restaurant)*. Intuitively, it is clear that *youthInn* does not have a restaurant. Hence the next rule: information at a particular object gets **overruled** at a more specific object *o* by conflicting information at *o*. Thus only the more specific information passes through. In our example, *facility(restaurant)* is overruled by *not facility(restaurant)* at the object *inn*; only *not facility(restaurant)* is inherited by *youthInn*. Note that the overruled information remains valid at some higher level in the object hierarchy, e.g. *facility(restaurant)* is true at *hotel*.

In the next example, the conflicting information is not inconsistent:

```
hotel =
        {
        ..
        checkoutTime(12).
        };
(hotel) nautilus =
        {
        ..
        };
(hotel) neptuno =
        {
        ..
        checkoutTime(14).
        };
```

LOCO supports the notion of **defaults** through the combination of inheritance and overruling. Here, we clearly expect *checkoutTime(12)* to hold at the nautilus hotel, since this is the default inherited from the class object *hotel*. However, for the neptuno hotel, we have more specific information available, namely that the checkout time is 14. Hence, the default *checkoutTime(12)* is overruled by the fact *checkoutTime(14)* at neptuno.

In general, LOCO will always give priority to the most specific information available and ignore any inherited (less specific) conflicting information. The idea is that default information is canceled by more specific information.

However, in some case this may lead to unintuitive results, as in the following example:

```
hotel =
        {
        facility(tv) :-
                    category(3star).
        };
nautilus =
        {
        category(3star).
        facility(sauna).
        };
```

Intuitively, in the above example we would like for both *facility(sauna)* and *facility(tv)* to be true. However, *facility(sauna)* is conflicting with *facility(tv)* (we assume that *tv* is not an instance of *sauna*, nor the reverse). This results in the overruling of *facility(tv)*, in favor of *facility(sauna)*. This can be fixed by introducing two new predicates: *defaultFacility* and *extraFacility*:

```
hotel =
        {
        defaultFacility(tv) :- category(3star).
        facility(X) :- defaultFacility(X).
        facility(X) :- extraFacility(X).
        };
nautilus =
        {
        category(3star).
        extraFacility(sauna).
        };
```

Now both *facility(tv)* and *facility(sauna)* will hold at *nautilus*. A more fundamental solution is to allow the explicit specification of what constitutes compatibility on a per-predicate basis. In practice this would boil down to the optional declaration of a predicate as "cumulative" ("set-valued") denoted *propertyName{arguments}*, or "single-valued" denoted *propertyName(arguments)*:

```
hotel =
        {
        ..
        facility{tv} :- category(3star).
        };
nautilus =
        {
        category(3star).
        facility{sauna}.
        };
```

For cumulative predicates, the definition of conflicting information would coincide with the definition of inconsistent information. Therefore, *facility{sauna}* does not overrule *facility{tv}* as they are not inconsistent (and consequently not conflicting).

2.2.2 Defeating

Another case is when there is multiple inheritance (an object being an instance of several other objects) and the information available from the various classes is inconsistent. For example, suppose we accept that, by default, hotels located in Rome, do not feature a pool. On the other hand, Sheraton hotels have, by default, a pool. What about the Sheraton hotel in Rome?

```
romanHotel =
        {
        ..
        not facility{pool}.
        };
sheratonHotel =
        {
        ..
```

> *facility{pool}.*
> *};*
> *(romanHotel sheratonHotel) s;*

In this case, two inconsistent pieces of information, namely *facility{pool}* and *not facility{pool}* are available at *s*, and neither of them is more specific than the other. In this case, LOCO refuses to draw any conclusion and *facility{pool}* will be neither true nor false. In general, information *i* **defeats**, at an object *o*, information *j* if *i* and *j* would yield inconsistent conclusions at *o*. Note that if *i* defeats *j*, then *j* defeats *i*.

2.3 Delegation

Assume that *pedro* inherits the hotels owned by *gonzales*, his grandfather. This can be expressed in LOCO as:

> *gonzales =*
> *{*
> *name("gonzales").*
> *profession("hotelkeeper").*
> *owns{nautilus}.*
> *owns{neptuno}.*
> *};*
> *(gonzales) pedro =*
> *{*
> *name("pedro").*
> *owns{pedroInn}.*
> *};*

pedro inherits all properties of *gonzales* except his *name* which is overruled at the object *pedro*. This example illustrates that we deal with delegation, which is a general class-independent term for dynamic hierarchical resource sharing. Thus, '*pedro* is an instance of *gonzales*' should be interpreted in a broader (class-independent) context, e.g. '*pedro* is like *gonzales*'.

It turns out that LOCO supports the concept of delegation which is much more powerful and flexible than inheritance. Inheritance is a specialization of delegation in which the entities that inherit are classes. In other words, LOCO supports plain inheritance (i.e. multiple inheritance, defaults, overruling, etc.) in the special case where the specificity relation reflects the subclass-superclass hierarchy.

3. Queries, updates and triggers

3.1 Queries

The following query yields all hotels that have more than 30 rooms:

> *? hotel.instance(X), X.nrOfRooms(N), N > 30.*

In general a query of the form

? *object.propertyName(X).*

will yield all X's such that propertyName(X) holds at object, whether this property has been explicitly declared at object or inherited by object from any of the objects of which it is an instance. Therefore, in the following example

> *hotel =*
> > *{*
> > *address(street=str, city=str).*
> > *};*
> *(hotel) crest =*
> > *{*
> > *address(street="5th avenue",city="antwerp").*
> > *};*

the query

> ? *crest.address(street=X,city=Y).*

will result in *X=str, Y=str* and in *X="5th avenue", Y="antwerp"*.

This mechanism can be put to good use, e.g. in the definition of symmetric relationships:

> *person =*
> > *{*
> > *parent(person).*
> > *child(C) :- C.parent(self)*
> > *};*

The rule in the object *person* above says that C is a child of a particular person p if p is a parent of C. Note that the keyword $self$ always refers to the object in which the clause is "evaluated". At first sight, rules like the one above make LOCO second-order, since C.parent can be regarded as a predicate variable. It is clear, however, that such expressions can be easily mapped into their first-order equivalents, e.g. by adding an extra "owner" argument to each predicate, which would transform *C.parent(P)* to *parentOf(C,P)*.

3.2 Updates

LOCO will support the addition and removal of clauses to objects, as well as the creation or destruction of the objects themselves and the manipulation of the partial order relation. The latter is accomplished by applying the former to the built-in _instance predicate. The problem of specifying updates in a declarative language is that, in many cases, it is necessary to specify control information, e.g. by giving a meaning to the order of the literals in the body of a clause. One of the proposals in the literature[6] suggests adding control primitives to the language. Another approach, which is the one taken in the present prototype implementation, creates for each update new more specific versions of existing objects in which the update clause is inserted. Using this approach, the update

> ? *replace[crest.address(street="grand avenue",city="antwerp")].*

to the database

```
(hotel) crest =
        {
        address(street="5th avenue",city="antwerp").
        telephone(3238302760).
        };
```

has the following effect

```
(hotel) crest_1 =
            {
            address(street="5th avenue",city="antwerp").
            telephone(3238302760).
            };
(crest_1) crest =
            {
            address(street="grand avenue",city="antwerp").
            };
```

LOCO automatically keeps track of which is the latest version of an object and the rules of inheritance logic make sure that the semantics of the update (which is really the semantics of the object after an update has been applied), conform to our intuition. Therefore the query

$$? \; crest.address(street=X,city=Y), \; crest.telephone(Z).$$

will correctly yield X="grand avenue", Y="antwerp" and $Z=3238302760$.

3.3 Triggers

Closely related to the problem of updating is triggering. Triggering implies that some additional actions are performed when an event occurs, i.e. when an object is updated. Although the issue of triggers is still under investigation in LOCO, here are some hints of the possibilities. Consider the situation where, whenever a hotel gets an extra facility, its ranking is upgraded. This could be expressed in LOCO as follows.

```
hotel =
        {
        ..
        add[facility(X)] - delete[ranking(X)], add[ranking(X+1)].
        };
(hotel) crest =
        {
        facility(sauna).
        ranking(2).
        };
```

The "clause" above is not a logical rule but an event handler. Its left hand specifies is the update operation which triggers the actions on the right hand side. Event handlers are inherited by other objects just like any other clauses. The major difference between event handlers and clauses is that event handlers are constructed from update operations, not from predicates. Updating the above database

```
? add[crest.facility(jacuzzi)].
```

will not only cause the execution of the update operation, but will also activate the event handler, resulting in

```
(hotel) crest_1 =
       {
       facility(sauna).
       ranking(2).
       };
(crest_1) crest =
       {
       facility(sauna).
       facility(jacuzzi).
       ranking(3).
       };
```

4.Theoretical background

LOCO is based on a logical formalism, called inheritance logic (IL)[5]. IL models the most important aspects of object-oriented programming languages, such as object identity, multiple inheritance and defaults. The logic is based on a partially ordered structure of objects, each of which has an associated theory. The proof theory takes into account the precedence structure between rules as implied by the partial order. The semantics of an IL theory are reflexive in that the domain of an interpretation is confined to the set of objects of the theory. Although the logic is naturally nonmonotonic, soundness and completeness hold under certain natural restrictions. Fixed point semantics for this logic are currently under development.

5. Other Issues

The following issues are still under investigation: typing, encapsulation and extendibility, external methods.

Typing

The effect of declaring an object as an instance of another object can also imply that the instance object's properties are constrained by the higher object's properties. This is closely related to the issue of types and requires further investigation.

Encapsulation

The scope rules for LOCO should be defined, as well as some encapsulation mechanism which will facilitate the specification of very complex knowledge bases.

Extendibility and External methods

In order to make LOCO an extendible language, it is sufficient to provide for the inclusion of new "built-in" methods (rules) and objects (classes) that may be implemented in other languages.

6. Conclusion

Through the LOCO language, we achieved a tight integration between the object-oriented and logic programming paradigms. LOCO is based on inheritance logic which gives clear definitions and formal semantics of the important notions of these paradigms. We introduced defeasible and default reasoning in LOCO rather than monotonic reasoning as the latter has proven to be too restrictive for AI-flavored applications (e.g. expert systems). In addition, updating is based on the ideas of version control, while a trigger mechanism is used to support constraint checking. Moreover, the language holds the promise of efficient implementation.

References

1. H. Lieberman, "Using Prototypical Objects to Implement Shared Behavior in Object-Oriented Systems," *OOPSLA'86*, pp. 214-223, 1986.

2. L. Stein, "Delegation is Inheritance," *OOPSLA '87*, pp.138-146, 1987.

3. D. McDermott and J. Doyle, "Non-monotonic logic I," in *Artificial Intelligence*, vol. 13, pp. 41-72, 1980. Also in 'Readings in nonmonotonic reasoning', M.L. Ginsberg

4. D. Nute, "Defeasible reasoning and decision support systems," *Decision support systems*, vol. 4, pp. 97-110, 1988.

5. E. Laenens, D. Vermeir , B. Verdonk , and A. Cuyt, *A logic for objects and inheritance*, Submitted in 1989.

6. S. Naqvi and R. Krishnamurthy, *Database Updates in Logic Programming*, 1988.

COMPUTER INTEGRATED MANUFACTURING

Project no 179

THE GEC MATRIX COPROCESSOR: A NEW ARCHITECTURE FOR REAL TIME CONTROL APPLICATIONS

P.R. DOREE
GEC Hirst Research Centre
East Lane
Wembley, Middlesex
HA9 7PP, England

Abstract.

There is an increasing requirement for control systems to be responsive in real time to the changes normally found in real world environments. This is particularly the case where the control system is expected to work faster and under more difficult conditions. Such conditions can only be overcome by the development of novel VLSI processor architectures for real time control. The GEC Matrix Coprocessor integrated circuit is the result of a number of years research effort into new types of processor architecture for real time control and was part funded under ESPRIT 179.

1. Introduction

The GEC Matrix Coprocessor (MCP) architecture was defined with a number of prime objectives in mind. To produce a processor with a simple but yet powerful instruction set so that both linear and non-linear control algorithms could be described relatively easily, to allow microinstructions to execute within a single clock cycle, and to minimise the amount of data traffic between the coprocessor and local memory. The first objective is essential in facilitating easy use of the device. The second objective is essential in providing high performance, and the third objective is essential for interfacing the device to other microprocessors acting as input/output processors which can be sited within the coprocessor system.

2. The GEC Matrix Coprocessor Architecture

The GEC Matrix Coprocessor chip is a high unique performance integrated circuit dedicated to matrix/vector arithmetic and co-ordinate transformations. As a peripheral device to a host microprocessor, the Matrix Coprocessor allows computationally intensive tasks to be computed at high speed with minimal host interference except for data setup and retrieval. Thus the Matrix Coprocessor provides a means of accelerating crucial computations whilst the host processor is performing systems management functions such as data acquisition. The Matrix Coprocessor system is illustrated in Figure 1.

The Matrix Coprocessor functions either as a matrix/vector solver, or as a co-ordinate transformer and provides a simple but powerful set of matrix/vector instruction primitives (ADD/SUBTRACT, COPY, MULTIPLY, INVERT, BUILD, CORDIC, CROSS PRODUCT) which support a wide range of matrix and vector based algorithms. An algorithm is assembled into a sequence of matrix primitives by the MCP Compiler.

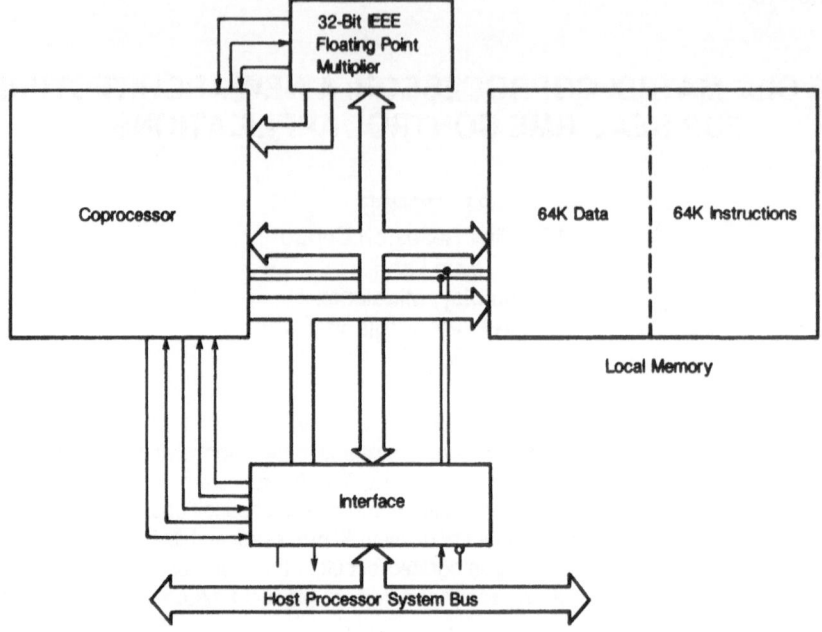

FIGURE 1. MCP SYSTEM ARCHITECTURE.

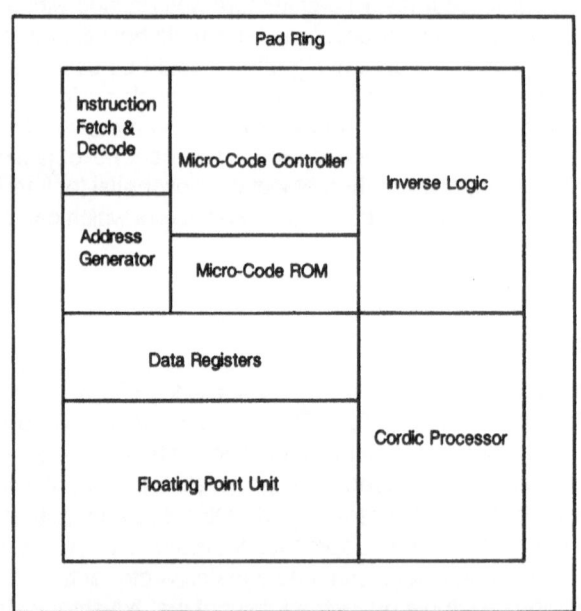

96 K Transistors

1 μm CMOS

FIGURE 2. GEC MATRIX COPROCESSOR CHIP ARCHITECTURE.

The Matrix Coprocessor is a load/store architecture. The principle behind this type of architecture is that the MCP instructions address data elements in local memory only, and the actual arithmetic computation is hidden to the user in the data input/output. To add together two matrices, the user would just specify the addresses of the two source matrices, and the address of the result. This means that the user has no control over which internal registers are used by any instruction except the Status registers which can be read at the end of an instruction sequence. There are no separate load and store instructions provided to the user, the loading and storing of data operands is implicit in the instruction.

The instruction format is generally a three address format in which the two sources and destinations addresses in local memory are specified, except two instructions which use only one source address.

The on-chip floating point unit supports single precision 32-bit floating point arithmetic which is partially compatible with the IEEE P754 Rev 10 standard, and any denormalised data operands are internally treated as zero. An external 32-bit IEEE compatible multiplier integrated circuit must be used in the Matrix Coprocessor system to support the multiply function.

Figure 2 shows the internal Matrix Coprocessor architecture. In terms of performance at 4 MHz a (4x4) matrix multiply takes approximately 23 μs.

3. The Matrix Coprocessor Operating Protocol

A simple strategy has been adopted for controlling the Matrix Coprocessor by the host processor. On power up, the GEC Matrix Coprocessor adopts a wait state with all buses and addresses disabled. The host processor then downloads instructions and data into the coprocessor local memory (which is a block of RAM, or for fixed applications, fast EPROM with the instructions already loaded). The address of the first instruction in the algorithm sequence is then written to the Matrix Coprocessor Program Counter by the host processor. A start signal is then asserted by the host processor which causes the Matrix Coprocessor to fetch, decode, and execute each instruction sequentially until either an error is detected or the end of a block of instructions is reached. When either the end of a block of instructions is reached or an error, the Matrix Coprocessor generates an interrupt, upon which the host processor would be actioned by the user to read the contents of the Matrix Coprocessor's internal Status registers to determine the appropriate course of action. Such an action would be either to initiate another instruction sequence or investigate the error. Such error conditions can be invoked by requesting the Matrix Coprocessor to invert a non-invertible matrix.

4. The Matrix Coprocessor Instruction Set

The GEC Matrix Coprocessor instruction set is designed for the high level solving of general matrix-matrix, matrix-vector, and vector-vector equations on operand sizes up to (32x32), and also for co-ordinate transformations and cross products on (1x3) vectors. These instructions are:

- Matrix ADD/SUBTRACT
- Matrix MULTIPLY
- Matrix INVERSE
- Matrix COPY
- Matrix BUILD

- Vector CORDIC
- Vector CROSS PRODUCT

Vectors are a special case of matrices with unity in one dimension, and are automatically accommodated by the Matrix Coprocessor. The functions provided by these instructions are briefly reviewed as follows:

4.1 Matrix Add/Subtract

The operation performed is C = A +(-) B. The instruction supports transposing of either of the source or destination operands.

4.2 Matrix Multiply

The operation performed is C = A * B. The instruction supports transposing of either of the source of destination operands.

4.3 Matrix Inverse

The operation performed is B = INV(A). The Inverse instruction is iterative, and the number of iterations (k) depends on the number of rows in the matrix to be inverted. The instruction also employs partial pivoting to improved the numerical robustness. The matrix operands is assumed to be square, however a rectangular matrix can be inverted when solving a linear system of equations (Ax = B). Inverse as implemented on the GEC Matrix Coprocessor is a very powerful instruction, and a number of results are generated besides the inverse. One of the by-products of Inverse is computations of the matrix Determinant. Inverse will also indicate whether the matrix A is of type "Positive Semi-Definitive" and indirectly provides the matrix rank.

The GEC Matrix Coprocessor will also invert any sparse matrix A, when an inverse exists, provided that there is at least one non-zero element on every row in A. If the matrix A to be inverted is singular, i.e. when the determinant is zero, then the Matrix Coprocessor will invert those linearly independent rows or columns and provide an inverse with rank m, where m is smaller than the original matrix declaration, and suspend processing by generating an interrupt to the host processor. The user would then read the MCP Status register contents, and decide what appropriate cause of action is necessary.

4.4 Matrix Copy

The operations performed is B = A. The instruction supports transposing of either of the source of destination operands.

4.5 Matrix Build

This instruction allows a matrix to be formed from data elements which are distributed anywhere in coprocessor data space. This allows for the construction of matrices, and vectors in real time.

4.6 Vector Cordic

Cordic is a vector based operation which uses (1x3) vectors in Matrix Coprocessor data space. The operation performed is 3-dimensiqonal for rotation, and 1-dimensional for vectoring. Appropriate choice of control fields allows transcendental functions to be created.

4.7 Vector Cross Product

Cross product is a vector based operation which uses two (1x3) vectors of elements from which the cross product is computed.

4.8 Matrix Trace

Although not provided as an explicit Matrix Coprocessor instruction, the matrix Trace is computed using a combination of Build and Multiply primitives. The Build instruction is used to construct a (1xn) row vector containing data elements from the matrix leading diagonal. This vector is then multiplied by a (nx1) column vector of 'ones' to obtain the matrix trace. Matrix trace is an example of an operation that can be formed from kernel instruction primitives.

5. Development Environment

For a new microprocessor to be usable by systems designers it must be supported by appropriate development tools. Certain tools are available for the GEC Matrix Coprocessor. These include an MCP Compiler, a software design kit for the remote development of applications, and an evaluation card which will be available in late 1989. Most of the software has been written in the C programming language, which means that the software design kit is currently hosted on a IBM PC-AT or compatible PC running MS-DOS. The evaluation card is AT compatible.

6. GEC Matrix Coprocessor Applications

The GEC Matrix Coprocessor is intended to have wide applications in the fields of real time control and graphics. The following example illustrates the simplicity of real time Kalman Filtering with a processor with a powerful instruction set such as the MCP.

6.1 Kalman Filtering

A typical linear control algorithm is the Kalman Filter which is used for estimation of the system state in the presence of noise. Kalman Filtering is essentially a technique for computing the optimal estimate of the state of a system by combining a system measurement with "a priori knowledge" of the system. Optimal means that estimate with the minimised mean square error or variance of each signal component.

The essence of the algorithm is the iterative solution of the matrix equation of the form:

$$x(k) = Ax(k-1) + (I-HAx(k-1))ky(k-1)$$

where the Kalman Gain is given by:

Figure 3: Kalman filter algorithm written in MCP Matrix Vector Language

The character '#' in the example below represents a comment character. Comments are added for the sake of clarity.

MATRIX

\#   p[4,4] Estimate Covariance Matrix

\#   System Model Parameters
\#   a[4,4] = System Matrix
\#   h[2,4] = Observation Matrix
\#   y[2,1] = Actual Measurement Matrix

\#   Statistics of the random processes q,r
\#   q[4,4] = System Noise Covariance Matrix
\#   r[2,2] = Measurement Noise Covariance Matrix
\#   k[4,2] = Predictor Gain

\#   Matrix Declarations

p[4,4] c[4,4] a[4,4] h[2,4] q[4,4] i[4,4] r[2,2] y[2,1] k[4,2] x[4,1]

BEGIN

\#   Compute Predicted Covariance Matrix (Prediction Step)
      p = (a * p * TRANS(a)) * (c * q * TRANS(c));

\#   Compute the Predicted Gain
      k = p * TRANS(h) * INV(h * p * TRANS(h) + r);

\#   Compute the Estimated Covariance Matrix (Filtering Step)
      p = (i - (k * h)) * p;

\#   Compute the Current State Estimate
      x = (a * x) + (k * (y - (h * (a * x))));

END

$$K = PH^T(HPH^T+R)^{-1}$$

the demanding part of this algorithm is the inversion of the (nxn) matrix $(HPH^T+R)^{-1}$. This inversion presents no problem to the GEC Matrix Coprocessor for matrices up to (32x32). Larger matrices can be processed by matrix partitioning. The algorithm essentially is composed of four processing steps for each time point which are listed as follows:

- Step 1. Compute the Predicted Covariance Matrix
- Step 2. Compute the Predictor (Kalman) Gain
- Step 3. Compute the Estimated Covariance Matrix
- Step 4. Compute the Current State Estimate

Each of these steps can be performed in real time for single order systems using the Matrix Coprocessor. The filter operates in a 'predict correct' fashion in that a correction term is added to the predicted estimate to obtain the filtered estimate. This set of equations is easily written in MVL (Matrix Coprocessor "Matrix Vector Language") and is shown in Figure 3. A Kalman algorithm of the form illustrated compiles into 22 MCP instructions, and takes less than 500 ms to execute, which is fast enough for real time applications. The only interaction between the host processor and the Matrix Coprocessor would be the transfer of time varying data sets and a final interrupt to the host processor after computation of the current state estimate. Note that the Kalman gain does not need to be computed "off-line" and then used when required. Using the GEC Matrix Coprocessor, the Gain in Step 3 can be computed in real time. This could be of particular importance where the gain is time invariant.

7. Availability of the GEC Matrix Coprocessor

The GEC Matrix Coprocessor chip together with the Evaluation card will be available in late 1989. The Matrix Coprocessor chip, fabricated in 1μm CMOS will be packaged in a 144 PGA.

8. Conclusion

The GEC Matrix Coprocessor integrated circuit is the result of a number of years research effort into new types of processor architecture for real time control and offers high performance for low system cost. The part funding from ESPRIT has provided an opportunity to explore a new IC architecture for real time control systems that has application in other areas.

PRODUCTION ACTIVITY CONTROL (PAC) :
PILOT IMPLEMENTATION and PROJECT EVALUATION

R. Trentin
SESAM S.p.A.
Cso. Svizzera 185
10149 Torino
Italy

Abstract.

EP477 Control System for Integrated Manufacturing has now reached its fifth year and the relevant software packages have been implemented. These are currently being installed and tested in three real live environments (COMAU in Grugliasco, DEC in Clonmel, Renault in Cleon).

This paper is an update of the "state of the art" of the Project, now reaching the final phases. It is mainly concerned with the progress from the Production Activity Control Implementation model to the Pilot Implementation taking place on the Project's partners sites, validating the PAC concepts in a test environment.

A number of significant feed backs allows the project team to better understand the significance of the work performed until now and to make proposals for future developments and enhancements.

1. Introduction

The main purposes of the COSIMA (Control System for Integrated Manufacturing) Project are:

- Define a generic Production Activity Control (PAC) architecture for discrete parts manufacturing, in order to ensure that the production orders for each work cell are met, providing an analysis mechanism to evaluate the effectiveness of the manufacturing activities.

- Develop decision, design and integration tools to support engineers in the understanding, design and development of PAC system. Typically these tools are:

  - a manufacturing profile, which collects all the significant characteristics of the plant (machines, setup times, tools, ...),

  - a library with a set of software building blocks performing activities of monitoring, dispatching, moving, etc., depending on the plant profile,

  - a selection mechanism which assists in choosing the right software building blocks for the real manufacturing environment.

– Implement PAC system in live environments: the goal is to move from a simulation model to an implementation model, to fully test and validate the PAC concepts.

Figure 1 shows the Decision Making Hierarchy in a manufacturing environment specifying the context where the PAC system is positioned.

DECISION MAKING HIERARCHY

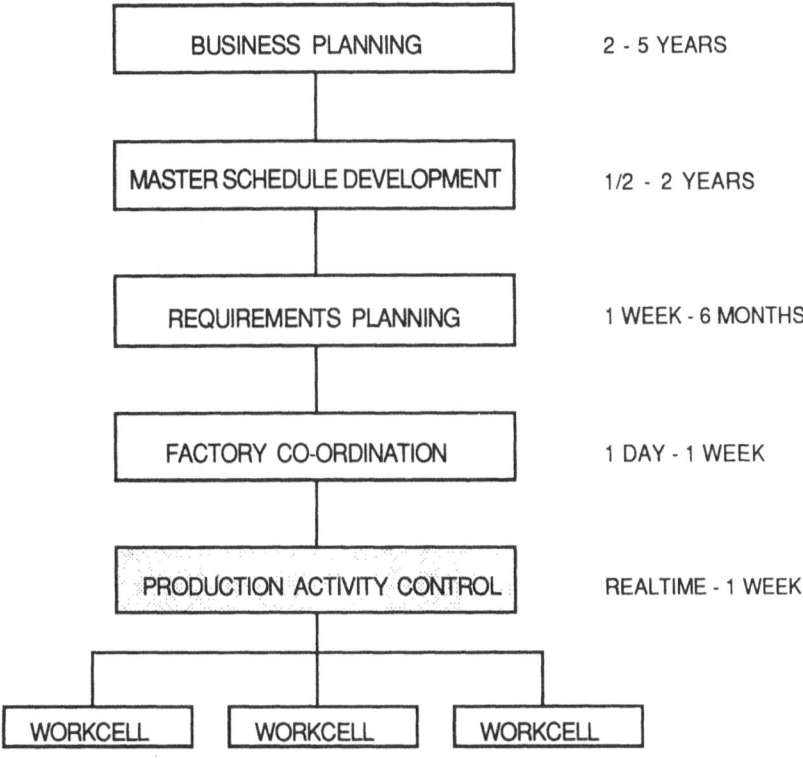

Fig. 1: Diagram showing the Decision Making Hierarchy

This paper is in six sections:

– The COSIMA Consortium
– A description of the PAC Architecture
– Particulars of the Clonmels test site

- Particulars of the Comau test site
- COSIMA Implementation in Renault
- Enhancements & Conclusions

2. The Cosima Consortium

The COSIMA Consortium brought together the following industrial partners and institutions :

- COMAU SpA, Torino, Italy
- Digital Equipment GmbH, Munich, West Germany
- Digital Equipment International B.V., Clonmel, Ireland
- Renault Automobiles, Paris, France
- SESAM SpA, Torino, Italy
- University College Galway, Galway, Ireland

3. Production Activity Control Architecture

The COSIMA project has modelled Production Activity Control (PAC) systemsased on an architecture with five fundamental building blocks (see [1]), as shown in figure 2, whose description follows:

- Scheduler -

It performs the function of scheduling production activities on the shop floor by specifying the timing of operations in order to comply with due dates, priorities, availability of resources, etc. Driven from MRP Planned Orders, the Scheduler provides a time sequenced operation plan to the Dispatcher building block.

- Dispatcher -

It executes in real time the sequenced operation plan provided by the Scheduler. It does this by assigning sequenced production orders to Producers and sending appropriate commands to Movers to ensure the availability of materials, tools and fixtures (where present) in the right place at the right time.

- Mover -

It co-ordinates the transport of material in the manufacturing plant, controlling the shop floor transport devices as carousels, robots, automated guided vehicles (AGVs) and manual transporters.

- Producer -

It controls specific types of production equipment such as CNC machines and robots through standard protocols. The producer isolates the physical level of production devices from the control level by translating general instructions from the Dispatcher into specific device instructions. It also communicates performance data from the plant to the Monitor.

- Monitor -

It performs the real time feedback function, collecting data on equipment utilization, materials, stock status and quality management and reports them back to the appropriate building blocks within the PAC system or the PAC user interface, thus supporting the decision making process of the PAC system.

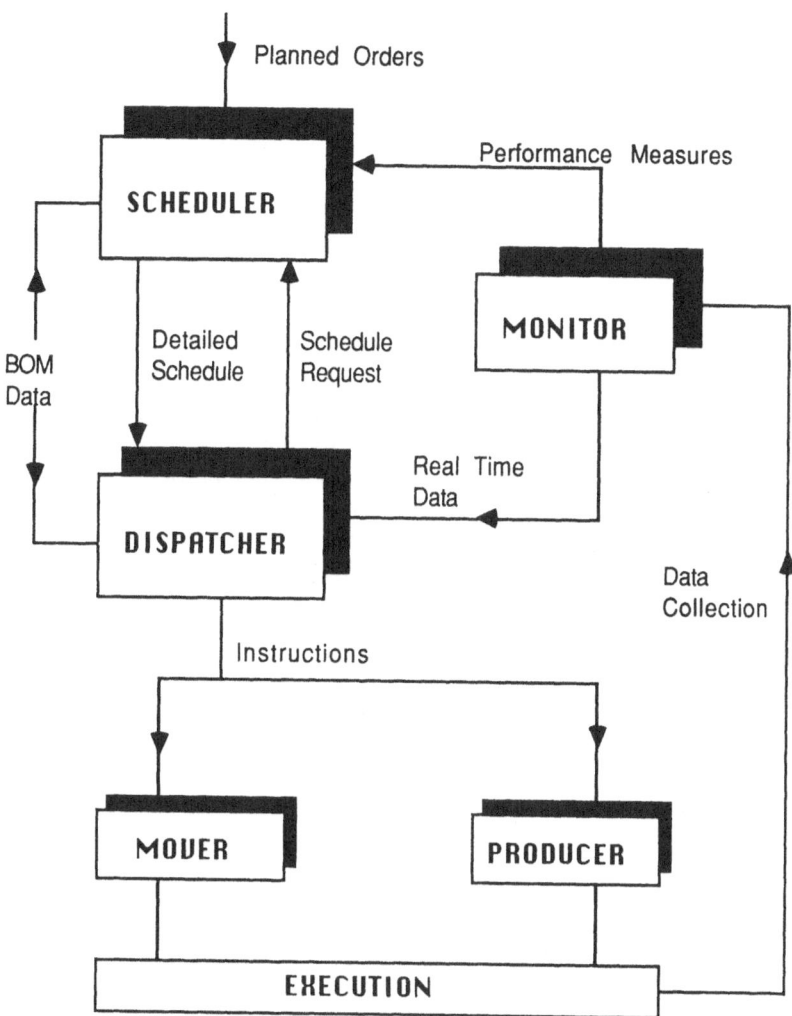

Fig. 2: Data Inputs and Hierarchy of PAC

When a plant control system is to be defined, or refined, the PAC Simulation Model, developed by the Consortium before the Pilot Implementation, allows the manufacturing engineer to design and test a PAC system based on the characteristics of a particular production setup. This is composed by applications which refer to each building block of the PAC architecture and communicates with each other via a software bus, the Application Network.

In this way the user can experiment with different control strategies, defining thus the manufacturing system, always and only using the unique PAC architecture. Finally, the implementation of a pilot PAC system in the Test Site environment can follow. This is done by replacing one of the above applications and the Shop Floor Emulator, that simulates all the events that occur on the shop floor, by Producer and Mover building blocks to achieve a Production Activity Control Implementation model. This model can then be fully tested in the Test Site Environment.

The advantage of the unique COSIMA architecture is that whether a building block has to be substituted, as the Scheduler, Dispatcher or Monitor, the PAC frame remains the same and the new module will be receiving the same information from the shop floor through the Producers and Movers.

4. Clonmel Test Site

The first Test Site Environment in wich the Pac Implementation model has been tested (March 1989) is the Digital manufacturing plant in Clonmel, involving a significant part of the electronic assembly environment, at a workcenter level (see[2]).

The PAC architecture applied to the Clonmel environment consists of the Scheduler, Dispatcher and Monitor building blocks which interact with the workcells on the shop floor.

The implementation model is shown in figure 3, where are clear the interactions between the software buildings blocks through the Application Network.

This COSIMA PAC modules have been designed to integrate with existing systems on the shop floor and their main functions are:

- On line scheduler: facilitates real time requests from the shop floor.
- Dispatcher: allows the supervisor to request a schedule and then send it to the relevant workcell.
- Monitor: compares the schedule plan to what actually is happening on the shop floor, and through a user interface allows the supervisor to track specific parts at each workcells.
- Application Network: provides the message passing facility for the live PAC system. Its main benefit is that it ensures:
  - flexibility: different building blocks can reside on different nodes in a local area network;
  - modularity: more buildig blocks can be added on to the PAC system without difficulty.
- Workcell reporter: this facilitate the viewing of workcell schedules at the workcell and also allows the operator to enter the current status of a job.

The demonstration of Clonmel was in two parts:

- Off-line implementation: this showed the use of PAC as an off line planning tool and the following software was demonstrated:

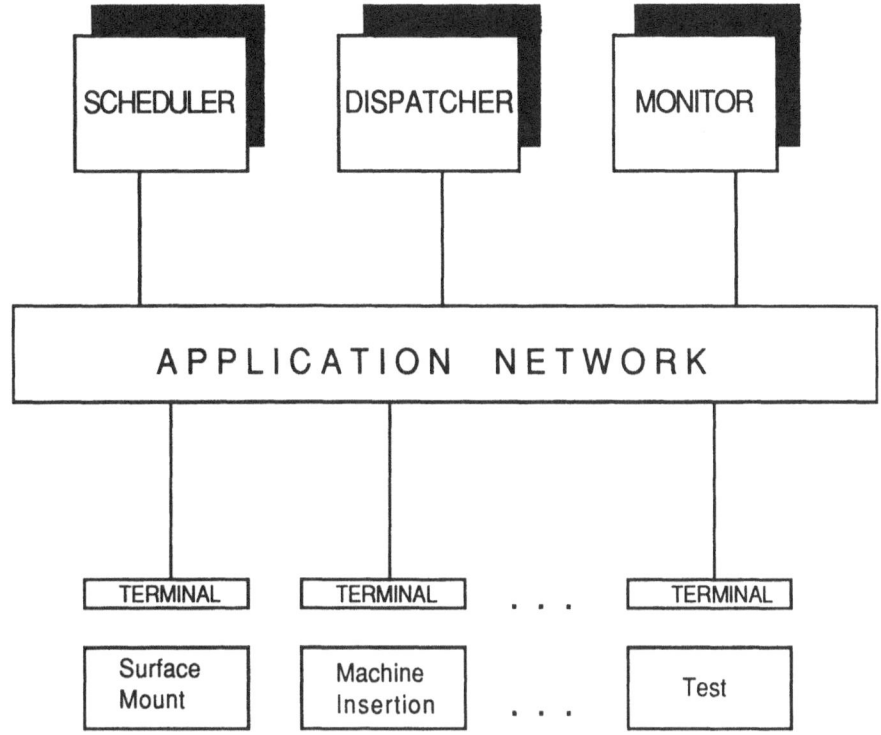

Fig. 3: PAC Implementation Model in the Clonmel Environment

- the manufacturing data_base which holds the relevant manufacturing data for the PAC modules to function, and the data_base interface which is a user friendly application allowing the data to be updated and changed;

- the Plant level scheduler and the plant simulator, both of which are used to develop and test schedules before their implementation on the shop floor.

– On-line implementation: the modules of PAC which are used in real time by the production planning personnel were shown.

5. Comau Test Site

The next Pilot Implementation Test Site will be in the Comau manufacturing plant in Grugliasco, in October 1989 (see [3]). The PAC architecture for it, in a context of mechanical working, where pieces of robots are produced, is shown in figure 4.

The PAC modules developed for the Comau's FMS are (see[4]):

– Scheduler : produces a sequence of operations for each machine in the system based on the weekly production plan from the Methods department.

632

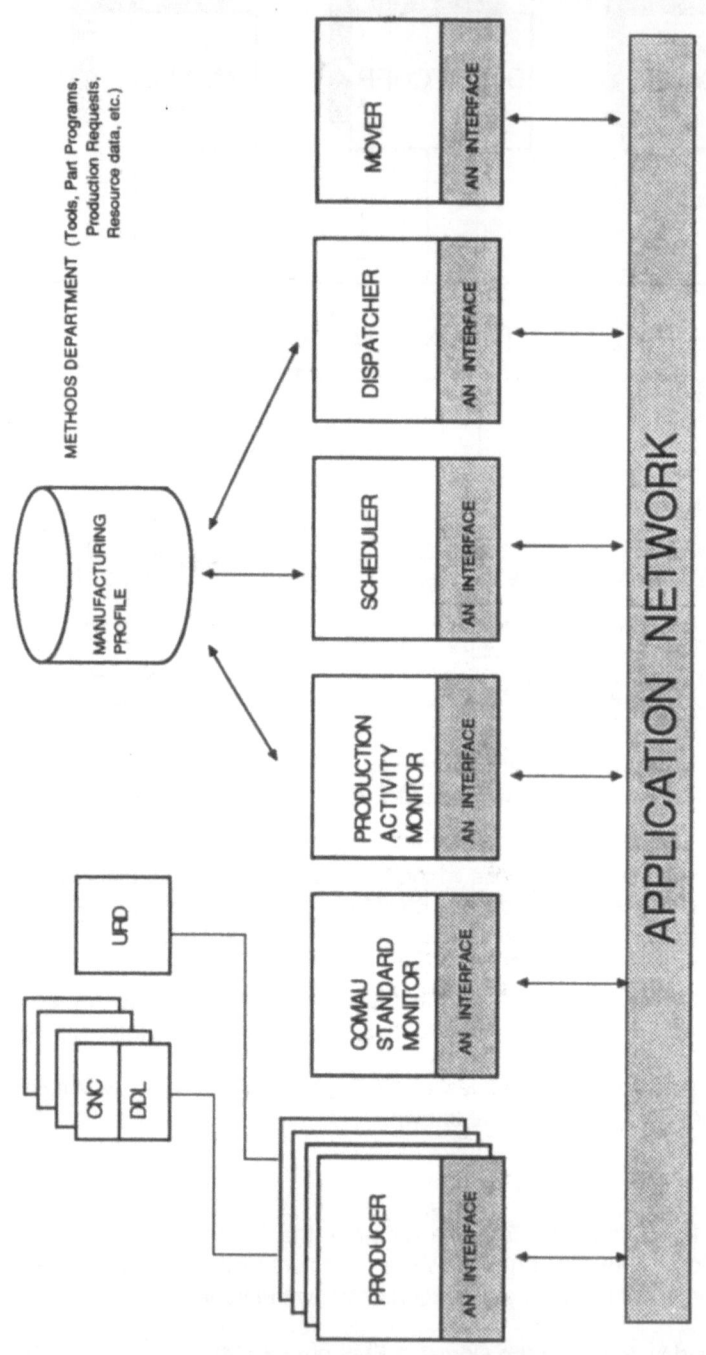

Fig. 4 : COSIMA Software Architecture at the Grugliasco Plant

– Dispatcher : is a decision support system for the plant supervisor, the real-time scheduler for the system. The correct following of the instructions issued by the dispatcher will lead to the implementation of the schedule. These instructions take the form of 'DISPATCH' and 'MOVE' messages.

– Monitor : consists of both the COMAU Standard Monitor functions, HERMES, and Production Activity Monitor. The Production Activity Monitor for the FMS will act as supplementary monitoring function to reside alongside the existing HERMES monitoring functions. The COMAU Standard Monitor is a software environment which comprise both diagnostic information, i.e., a schematic layout of the FMS showing the machines with differents colors relating to the different machine states, and monitoring information. This information is stored in a database and it is accessible through a report generator or by any application that needs the data. The Production Activity Monitor consists of collection of the messages, analysis and reporting.

– Mover : is a very limited system for the Grugliasco FMS because the actual moving system consists of a forklift and a driver. Thus the Mover performs a conceptual rather than a functional purpose. It demonstrates however, the possiblity of having a fully automated system being controlled by the instructions from the Dispatcher, thus, if an automatic transportation system was installed in the plant then the Mover process could be enhanced so that intructions from the Dispatcher would be translated into device specific orders which the transportation could comprehend and perform.

– Producer : it is formed by different layers as follows:

a) System (AN) Interface, which handles the communication with the operator and the other facility for the live PAC system. blocks. It analyzes the meaning and source of received messages from the operator and then calls lower level services. These services are represented by a peripheral interface layer: URD, DDL. The Data Collection Unit (URD) is a COMAU property board used to count the number of parts machined well or scrapped on the machining centers, interfaced with the FMS operator by a keyboard and with all the machining centers. The Dnc Data Link (DDL) is a microcomputer based intelligent unit developed at COMAU, used to upload, store and download part programs from/to CNC and to handle tools attributes.

b) File Tranfer Functions, which receives request for file transfer and generates a sequence of calls to lower level service to execute it.

c) Tool Parameter Handling Function, for an "new tool loaded" event, detected by the event generator.

d) Event Generator, is called synchronously with the polling function, calling internal functions or enabling the Producer to send messages to other building blocks on significant changing of states.

e) Machine Status Polling checks the DDL periodically to retrieve the machine status, which is then passed to other building blocks.

f) URD and DDL Interface implements the URD and DDL connections through standard communication protocols.

Possible future developments of this implementation could be related to addition of extra machines, use of the automatic moving systems, extension of the part family and variety of the production mix, involving in that way further development of the building blocks, which does not necessarily mean that other building blocks need to be adapted or changed. This because of the modularity and flexibility of the PAC architecture, which also allows for addition of extra HW/SW.

6. COSIMA Implementation at RENAULT

The Renault test site is Cleon, a manufacturing plant located 100 kilometers West from Paris where many different types of car engines are machined and assembled.

Renault's priorityis to produce the right product at the right time with the right quality. Workshops therefore concentrate on improving the utilization of their manufacturing resources and, in particular, carefully scheduling their production plan.

In this context, where a very wide variety of pieces have to be handled, distinguished and mounted according to the production plans, a scheduler and a monitor building blocks have been developed.

Since January 1989, this software, implementing a Scheduling and a Monitoring building blocks, has been running on the shopfloor.

By October, a list of suitable modifications will be issued and a refined version of the software delivered to the plant.

In this way, Renault will have provided the COSIMA project team with a real live implementation of two main PAC concepts (Scheduler and Monitor) and an interesting feedback from the manufacturing teams.

7. Enhacements & Conclusions

At this stage of the COSIMA Project is going to be supported by three different Pilot Implementations in three European countries, so, it is been demonstrating the feasibility of implementing the PAC system based on the architecture and simulation model previously developed.

In this way, it has been established reference sites of PAC installations with the result of a proven base technology applicable across various industries, e.g. electronic, and automotive industry.

In that respect, major other companies have declared their interest on these applications, leading to the possibility for standardization of business protocol.

Beside we must highlight other positive aspects of the Project such as the high level of the Partners involvement with a good exchange of experiences in a technologically advanced context. Success assessment follows PAC Pilot Implementation; this should focus on the potential for implementing PAC in the Test Sites environment based on the lessons learned from the pilot implementation.

The feed backs from the team have allowed some enhancements and new proposals:

– A new version of the AN has been released

– COSIMA PACK-Kernel-in ObjEcTS (PACKETS) : an architecture and software system for modelling, describing and implementing single-level PAC systems. It describes the techinques for creating a reference model, integrating a PAC applications (Building Blocks) based on that model. The PACKETS project is aimed to prove not only the

PAC architecture but also the advantages of an object oriented approach, including reusability, extendibility, and the flexible applicability of objects.

– The COSIMA Partners are involved in the analysis of an exploitation plan of the Project which is really influencing their standard way of working. In that respect, the Scheduler building block built for the Pilot Implementation in the Grugliasco plant satisfies the needs of the shopfloor, so it will be extended to cover the scheduling of fifteen machines. Some future developments in the application of the PAC concepts are planned with the use of AGV (Automatic Guided Vehicles) in other major FMS, outside Grugliasco plant.

The Simulation Model has been proved to be an important tool for analysis in the developing of new plants. It is worthwhile to stress, though, its aspect of standardization and adaptability, as explained in the COMAU Implementation description, regarding the integration of the Production Activity Monitor building block and the COMAU Standard Monitoring functions.

A major exploitation of the PAC concepts can be achieved analysing the Production Management System Market trends, such as the increase of sophistication of the applications, the key role played by the system integrators, the increased requirements of the networking capabilities and the need of standards demanded by the users, among others. An approach to that exploitation could be summarized in the following steps:

- Define a market
- Develop a business plan
- Find independent Software Vendor
- Define selling model
- Develop a system

References

1) Browne J.andDuggan J., 'Esprit Project 477: Production Activity Control Design and Implementation' Proceedings of the 5th Annual ESPRIT Conference.

2) Lyons g. and Duggan J. 'Functional Specification for Production Activity Control Pilot Implementation at Digital Equipment Corporation, Clonmel, Ireland', Final Implementation Report, January 1989

3) COSIMA Project Team, 'Preliminary Implementation Description at Grugliasco Plant', 5th Year Interim Report, July 1989

4) Higgins P., Copas C. and Trentin R., 'Grugliasco PAC Software Description', Internal Document, January 1989

Keywords

Production Activity Control (PAC); Monitor; Dispatcher; Mover; Scheduler; Producer; Application Network (AN); Test Site Environments

Project No. 623

INTEGRATED PLANNING AND OFF-LINE PROGRAMMING SYSTEM FOR ROBOTIZED WORK CELLS

R. Bernhardt

Fraunhofer-Institute for Production Systems and Design
Technology; Head: o.Prof. Dr.-Ing. Drs. h.c. G. Spur
Department of Automation, Director: o.Prof. Dr.-Ing.
G. Duelen
Pascalstr. 8-9, D-1000 Berlin 10

ABSTRACT.   The paper gives a short survey of the ESPRIT 623
project: Operational Control for Robot System Integration
into CIM, Systems Planning, Implicit and Explicit Programming
in which twelve companies, universities and research
institutions are involved. About three years after the
project's commence work has started to combine software
modules developed by different partners to integrated systems
with increased functionality. To accomplish this four working
groups have been installed which are working on different
aspects, e.g. access to different data bases or direct usage
for industrial applications. These realized system reflect
also the specific interests of the involved companies. In
this paper the realized integrated planning and explicit off-
line programming system is presented in more detail and some
applications reaching from off-shore over the automotive
industry up to space technology are described.

KEYWORDS.   CIM, CAP, Layout Planning, Assembly, Off-line
Programming, Simulation, Industrial Robots, Information
System, Telemanipulation, A&R in Space, Off-Shore.

1. INTRODUCTION

Robot systems are important components for flexible automation. Planning and programming of such systems is a very complex and time consuming task and requires computer aided supporting means which are themselves integrated parts of CIM systems. The overall objective of the ESPRIT 623 project is the realization and industrial application of computer aided manufacturing system planning and off-line programming for robot integrated CIM systems /1/.

The partners involved in the project are IPK (Berlin, Germany) as the prime contractor, the companies KUKA (Augsburg, Germany), Renault Automation (Paris, France), FIAR (Milano, Italy), PSI (Berlin, Germany) and the universities/ research institutions UCG (Galway, Eire), UKA (Karlsruhe, Germany), UPM (Madrid, Spain), UNL (Lisboa, Portugal), PM (Milano, Italy), LADSEB-CNR (Padova, Italy) and UA (Amsterdam, Netherlands).

After the project's start, three subgroups were installed to realize subsystems and components in the areas manufacturing system planning, explicit and implicit off-line programming of industrial robots.

After this project phase the realized modules have been interfaced and combined to demonstrate the increased functionality of integrated planning and programming systems. This has done by the definition and realization of four demonstrator systems. Three of them are shortly described in chapter 2, while the structure, functionality and applications of the forth is presented in detail from chapter 3 onwards.

2. BRIEF DESCRIPTION OF REALIZED DEMONSTRATOR SYSTEMS

The main objective for realizing demonstrator systems by
integrating modules of different partners is to show the
increased functionality and efficiency of an integrated
planning and programming procedure for robotized manufacturing cells.

Thereby the different principles like automatic or
interactive planning functions and explicit or implicit
programming procedures are considered. Additional realistic
industrial applications or well-known benchmark tests have
been selected. A further important criterium for the
selection of applications is to cover a broad spectrum of
problems and to show their solutions. Examples are to select
applications

-   with parts of very different size requiring gripper
    exchanges,
-   parts of different shapes requiring multifunctional
    grippers, or
-   assembly operations with different mating directions
    requiring complex fixtures and robots with six degrees of
    freedom.

The realized demonstrator systems or - depending on the
production task - parts of them are also used by the project
partners for a variety of industrial applications.

In this chapter three demonstrators are briefly
described. Detailed descriptions have already been published
for the first one, and are currently being prepared for the
others. For all demonstrators unified reference models have
been elaborated (fig. 1 to 4). The rectangular blocks on the
left side of these models show the activities, i.e. which
functions have to be executed. Also the partner who realized
the module is indicated by the abbreviation used in the
introduction. The eliptic blocks on the right represent the
information required and/or produced by the activities.

A demonstrator has been realized in cooperation of
Renault Automation and the University of Karlsruhe /2/. It is
oriented to a variety of manufacturing tasks like welding,
glueing, laser-cutting, assembly. The main objective of this
demonstrator is to show the transfer of execution planning
information between different data bases. After performing
the execution planning, the trajectory information are stored
in a semantic form, independent from the physical data base
structure. This information then has to be transformed into
specific robot trajectories which are simulated for test
purposes and executed by the real robot. The reference model
of this demonstrator is shown in fig. 1.

For the validation of the demonstrator an assembly task
has been chosen at UKA. First the task execution planning is

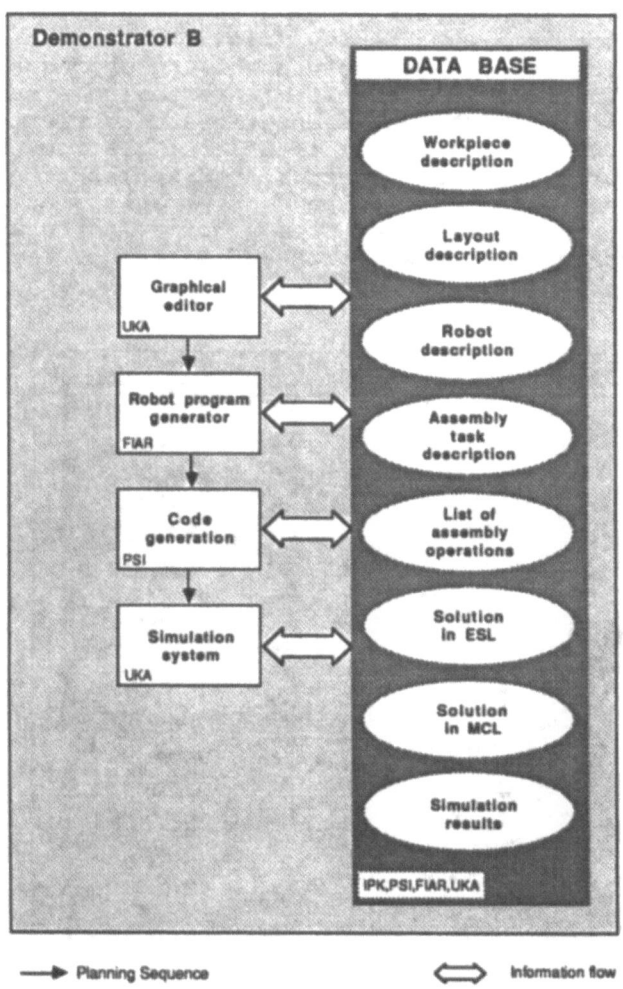

Fig. 1: Reference Model Interactive Robot Programming

performed at UKA and stored in a DEC/RDB data base, then it
is transfered to the ORACLE data base at Paris, transformed
into robot trajectories, simulated and executed. This procedure was also followed vice versa for other manufacturing
tasks planned at Renault, Paris, transfered to the data base
at Karlsruhe and simulated. Within the demonstrator system,
two explicit programming procedures have been used: a
language-based from UKA and an interactive one from Renault.
    The demonstrator system has alredy been installed and
will be presented in October 1989.

640

A further demonstrator system deals with advanced concepts in implicit programming (fig. 2). The results achieved by the implicit robot programming approach can be used for application in flexible manufacturing systems. The assembly application chosen is generic and complex enough to be realistic. The Demonstrator is able to realize the following representative scenario: At a workstation, supported by a comfortable graphic interface, a user interactively describes a work cell. Normally the work cell components are standardized parts and therefore may be selected from a predefined library. Then the user graphically

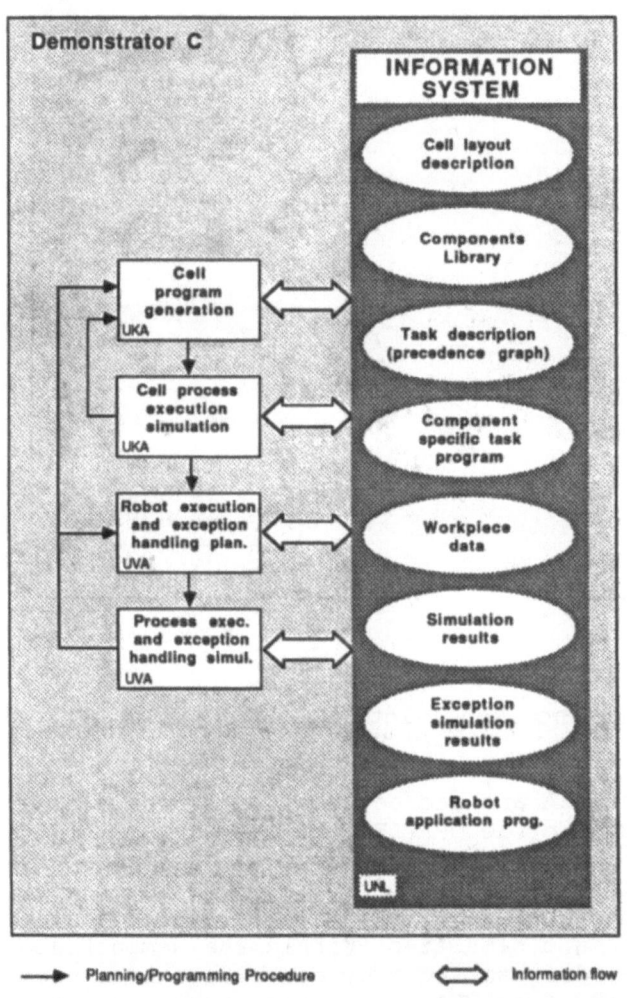

Fig. 2: Reference Model Task Level Robot Programming

specifies the assembly task to be performed and starts the planning system. The system automatically generates a sequence of safe robot actions/motion necessary for performing the task. It translates these robot instructions to executable robot commands specific for the the robot type and passes them on to the simulation system which visualizes the robot motions found for the specified assembly task. Therefore the actual presence of the real robot or of the work cell is not needed.

The first version of the Task Level Robot Programming System was installed and presented in April 1989 in Lisbon.

The objective of a further demonstrator "High Level Interpreter" is to demonstrate the feasibility of a flexible control system for a robot production cell (fig. 3).

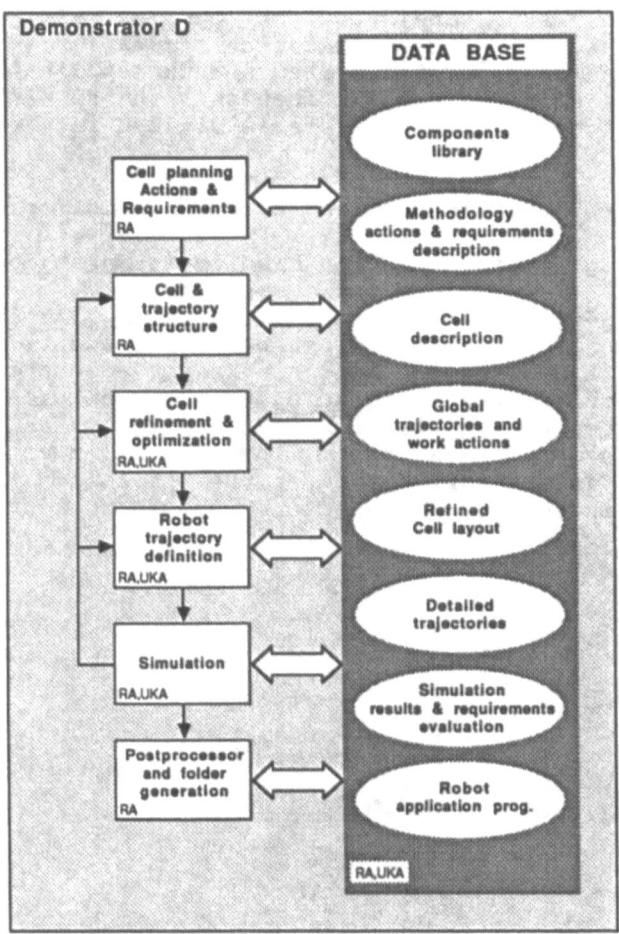

Fig. 3: Reference Model High Level Interpreter

Flexibility requires exception handling capabilities of the control system at cell level and at the individual robot level. The required decision making capabilities of the High Level Interpreter (HLI) include cell level scheduling, task scheduling, monitoring and exception diagnosis, recovery planning and motion planning.

The decision making behaviour of the HLI needs to be planned before execution of the robot cell's program. Therefore the HLI imposes additional requirements on the programming environment. We contribute to the off-line programming environment by offering tools for generating multi-robot cell control programs and tools for specifying the exception handling mechanisms. Also the simulation environment is extended to perform test and validation of the generated programs and exception handling mechanisms.

The demonstrator consists of the modules cell program generation and cell process execution simulation for generating the cell control system and the application programs for individual cell components. The specification of the exception handling mechanisms is divided into two modules:

-   robot program execution and exeption handling planning and
-   robot program and exception handling simulation.

To test the generated programs and exception handling mechanisms, a strong interaction with the active agents must be established. A downloading approach is not adequate for this purpose and therefore a simulation environment will be used which simulates the interaction with the robots and the sensors during program exection. The generation of robot programs suitable for execution on the simulation system is carried out in the module robot program generation.

A first version of this demonstrator was presented in April 1989 in Lisbon.

3. INTEGRATED PLANNING AND INTERACTIVE OFF-LINE PROGRAMMING

3.1. Objectives

The objective for realizing this demonstrator system, involving the Systems Planning and Explicit Programming groups, is to achieve a first industrial implementation of components and tools developed by the various partners within the two groups. The first version of the integrated planning and explicit programming system for assembly tasks was installed and presented in October 1988.

The principle tasks of the integrated system are the

- detailed planning of an assembly process,
- selection of all components and their arrangement,
- generation of robot application programs and
- simulation and test of the assembly task execution within the planned work cell.

The integration of components and tools is achieved primarily through information exchange and management, performed via a relational data base system. This enables a safe data handling and leads to a most flexible system structure related to the integration of functional system units and a quick adaption to user needs.

3.2. System Functionality

The integrated planning and explicit programming procedure involves a number of steps realized with the developed components and tools. In each step specific planning activites have to be executed using information generated in the previous steps and producing information for subsequent ones. Due to the detailing and growing of information in each step the procedure cannot be purely sequential: feedback loops are required. A common data base for information management and access is shared by all. In fig. 4 a reference model of the demonstrator system is presented.

The left side represents the activities of the planning/ programming procedure and their interrelations. The right side shows the required and produced information. To distinguish the planning / programming procedure and the information access and exchange, different graphical representations of arrows have been used.

The integrated procedure starts with the generation of an assembly sequence plan which includes the operation sequence planning, the precedence analysis, the preselection of robots which are suitable for the different assembly steps and the selection of available feeders and grippers. Based on the assembly sequence and the suggested devices the components are selected and roughly arranged. During an interactive and

644

iterative procedure, this layout is improved and verified
using supporting tools for e.g. cycle time estimation and
material flow simulation. In the next step the generated
planning data are transformed into a structure suitable for
the off-line programming procedure. Besides the layout
description the assembly sequence is described with a
semiformal (similar to PSL) containing also geometrical,
technological and control specific data. The planning phase
is completed with a layout optimization procedure considering

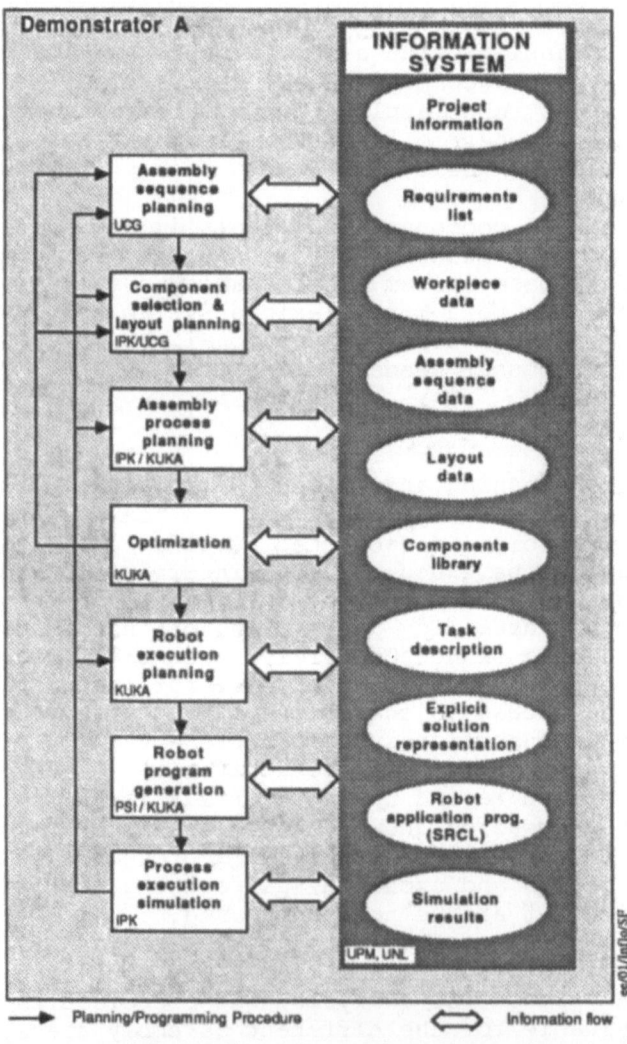

Fig. 4: Reference Model of the Integrated Planning and
Explicit Off-line Programming System

type variants and locations of robots, tools and peripheral devices and their influence on important parameters reaching from cycle times up to the total cost of the work cell.

Based on the generated description of the assembly task and the optimized layout, the off-line programming procedure for the robot is performed. As a first step the execution of the task by the robot has to be planned. This is related to the determination of the motion types and parameters, the conversion and integration of technology information as well as to the scheduling of the task and the explicit solution representation (ESR) which is a data structure containing all necessary information to generate robot application programs in a specific language. In the next step out of the ESR a robot program is generated automatically. For the first realization a subset of the Siemens robot language SRCL is used. The executability of the off-line created application programs has to be tested, i.e. to verify whether or not the assembly task is fulfilled. Due to the complexity of the necessary decision making process required for verification, the human operator has to be provided with a presentation of the task execution as realistic as possible. This is achieved via the simulation system by visualizing the movements of the robot as well as the actions of peripheral devices of the assembly cell. Additionally robot controler specific cycle times and joint parameters are measured.

As to be seen from the reference model the planning/ programming procedure is not a sequential one. To denote its iterative nature, feedback loops at specific points are necessary for the alteration or refinement of previous steps.

The elaborated reference model shows the main functions of the system and the planning/programming procedure but not the data flow in between. Therefore the system has been analyzed and a functional model has been elaborated showing the functions as well as the information linkage. For the representation SADT (Structured Analysis and Design Technique) has been applied. The system was hierarchically decomposed into subfunctions resp. activities. At the highest level shown in fig. 5, the information relations between the planning/programming system and its environment are specified globally. In general it is distinguished between

- data consumed by the function (project information, requirement list, workpiece data),
- data produced by the function (robot application program),
- the control flow structure which are data controlling the function (manager, component library) and
- constraints and realization aspects (mechanisms) which realize or support the function (user).

646

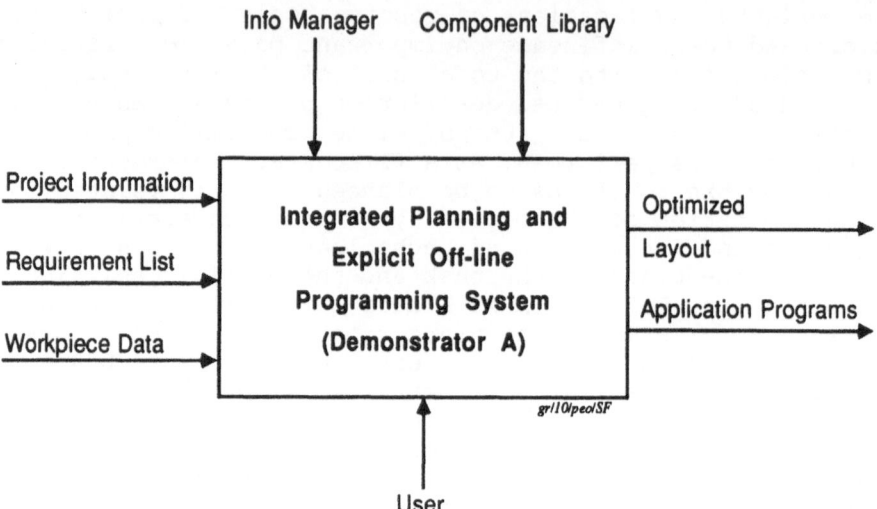

Info Manager   Component Library

Project Information → **Integrated Planning and Explicit Off-line Programming System (Demonstrator A)** → Optimized Layout

Requirement List →

Workpiece Data →

Application Programs →

grl10lpeolSF

User

Fig.5:   Overall SADT-Diagram of Demonstrator A

This method has been applied to generate the next lower level of the functional model of the Demonstrator A. In fig. 6 the SADT diagram of this level is presented showing the consumed/produced information of each functional module as well as the control information and mechanisms.

3.3.   Production Task

For the first version of the Integrated Planning and Interactive Off-line Programming System the selected production task was a subassembly of a dot matrix printer. The parts to be assembled are

- the plate of the printer chassis,
- the stepper motor,
- the gear,
- the rocker,
- the lock washer and
- two screws.

In fig. 7 the parts of matrix printer line feeder subassembly are shown.
    To assemble this drive unit of the printer's line feeder two interlocking drive gears must be positioned. Before the stepper motor is screwed to the plate of the chassis the gears of the motor pinion and the rocker must be positioned and interlocked. This requires an oblong hole in the chassis's plate limiting the motors range of motion. After the motor gear and rocker have been correctly joined, the motor is screwed to the plate.

647

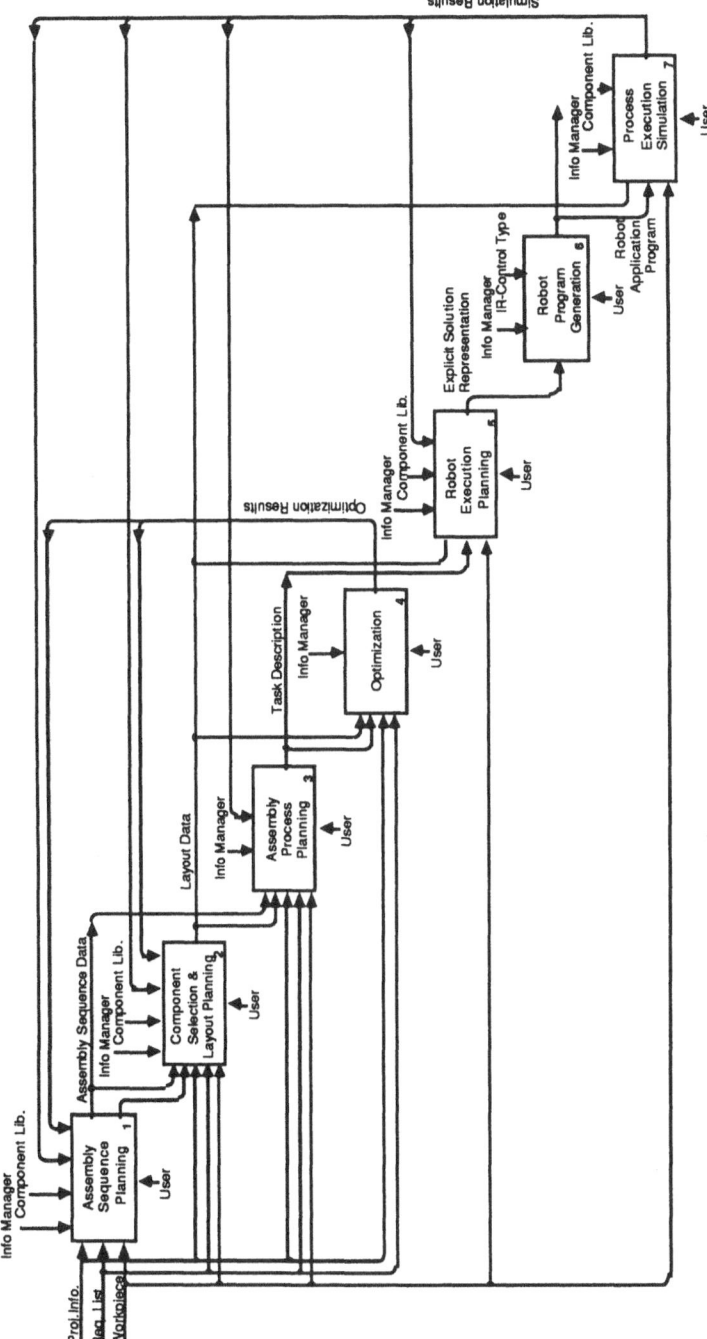

Fig. 6: SADT-Diagram with Main Functional Modules

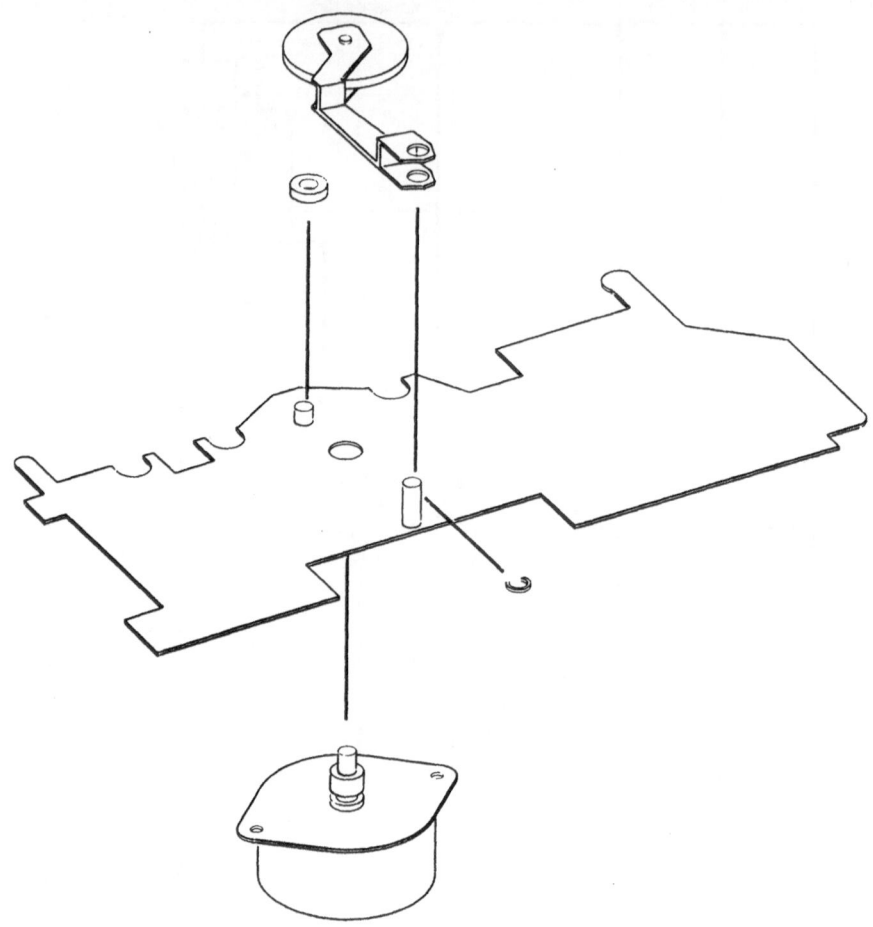

Fig. 7: Parts to be Assembled

Fig. 8 shows an example of a robotized cell containing an KUKA IR 162/15 robot, different part presentation devices, feeders, grippers and magazines which has been planned and programmed by applying the system described above.

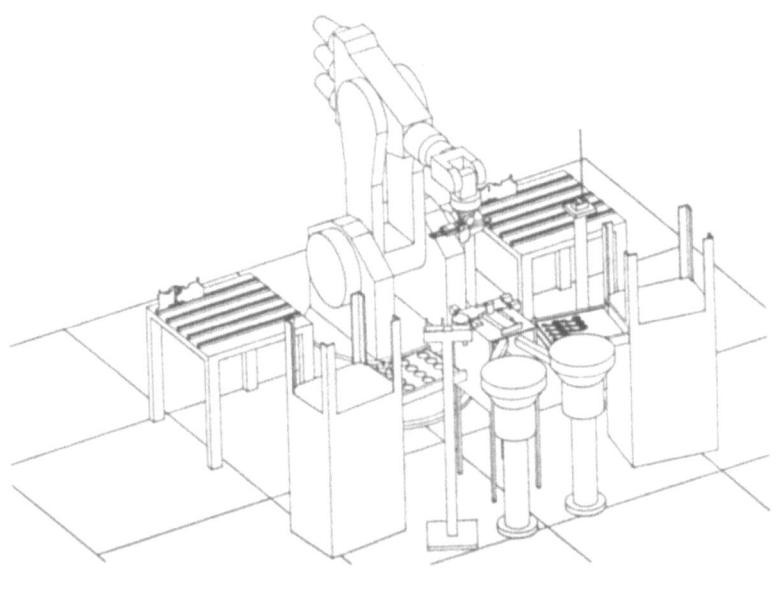

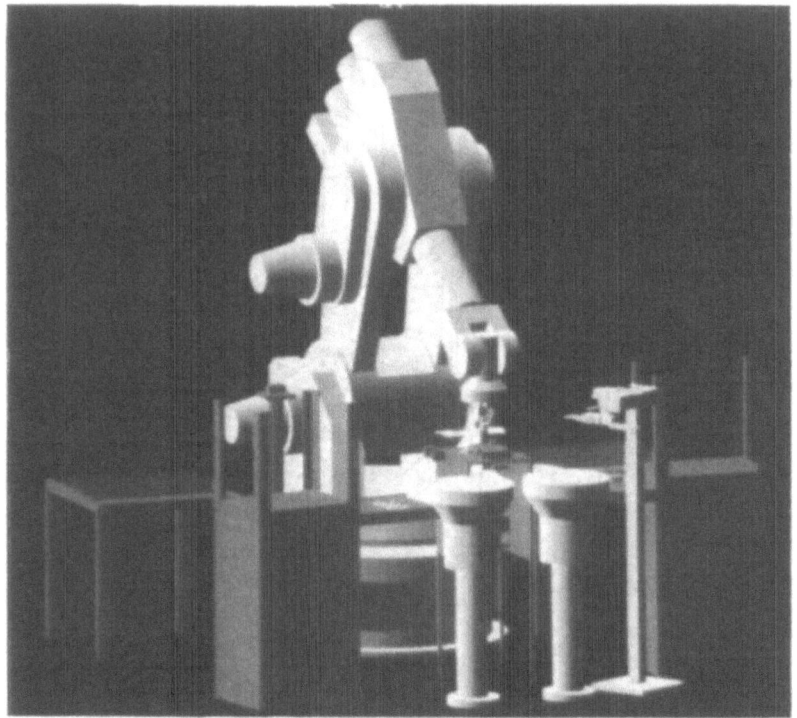

Fig. 8   Example of a Robotized Assembly Cell

4. APPLICATIONS

The Integrated Planning and Interactive Off-line Programming
System has been used by the involved partners for a broad
spectrum of applications. This is possible because of the
modular system structure allowing the exchange of user and
task specific modules.

In the following subchapters some examples are presented
to point out the broad application spectrum. The selected
applications are all related to finished or running
industrial projects including real components.

4.1. Complex Operations in Restricted Spatial Conditions

An important area of use for planning systems are robot
applications, in which extremely complex operations within
restricted spatial conditions have to be performed /3/. An
example is given by fig. 9 (simulation) and fig. 10 showing

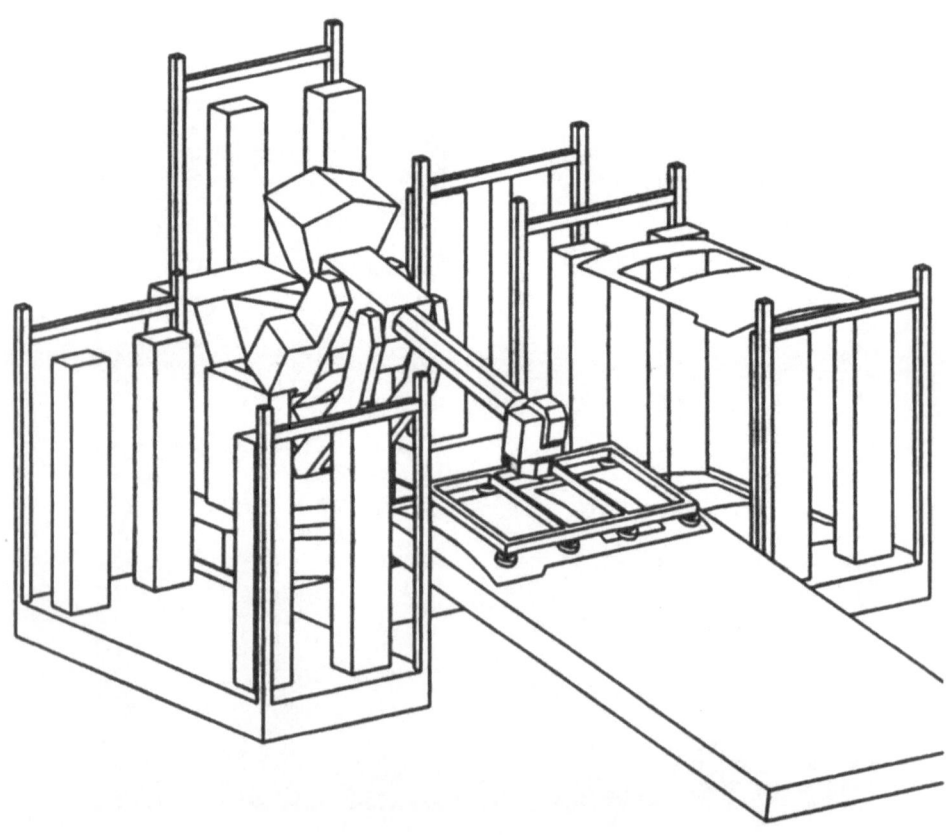

Fig 9: Complex Robot Application (KUKA Simulation)

Fig. 10:   Realized Work Cell (KUKA)

the realized work cell. A KUKA robot IR 662 equipped with a multi-purpose end effector has to serve for two different types of operation, gripping and spot welding. Additionally the operation has to obey very critical time specifications. It is a very difficult task to meet all these different kinds of requirements. Therefore the application of computer-based optimization techniques was considered. The main objective was to get an approximate layout definition for all components which are involved. As a multi-purpose tool for gripping and spot welding has to be used, the mounting position of the tool at the flange has to be evaluated carefully. This is essential in order to avoid singular configurations of the robot arm. Therefore this optimization aspect has to be included, too.

4.2. Inspection of Underwater Structures

The overall objective of the so called OSIRIS project is to make available an underwater working robot in connection with a qualified carrier /4/. The main tasks are cleaning and inspection of seam welds at underwater structures without diver assistance. This requires the planning and programming of the submersible vehicle's motion as well as the task

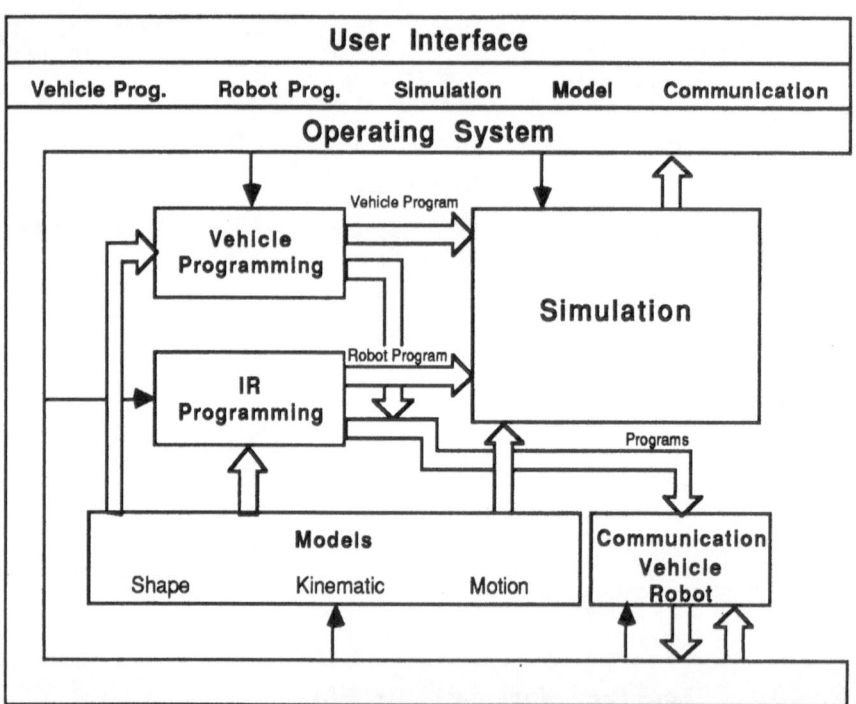

Fig. 11:   Principle Functional Structure of the OSIRIS
            Programming and Simulation System

execution of the robot itself. Therefore the existing system has been enlarged by the additional module "Vehicle Programming" as shown in fig. 11. As the first step of the programming procedure,the application program for the submersible vehicle is generated. The motion execution is tested via the simulation system (fig. 12). In the next step the same procedure is applied to generate and test the robot program. In fig. 13 a simulation of the modeled environment is shown. Additionally a "work cell" was built in a laboratory consisting of a part of the industrial robot. Off-line generated and tested robot programs were transfered to the robot control and executed by the robot. Fig. 14 shows a photo of the experiment set-up. The project is conducted by the partners Interatom, GKSS and IPK-Berlin.

Fig. 12:   Simulation of the Submersible´s Motion

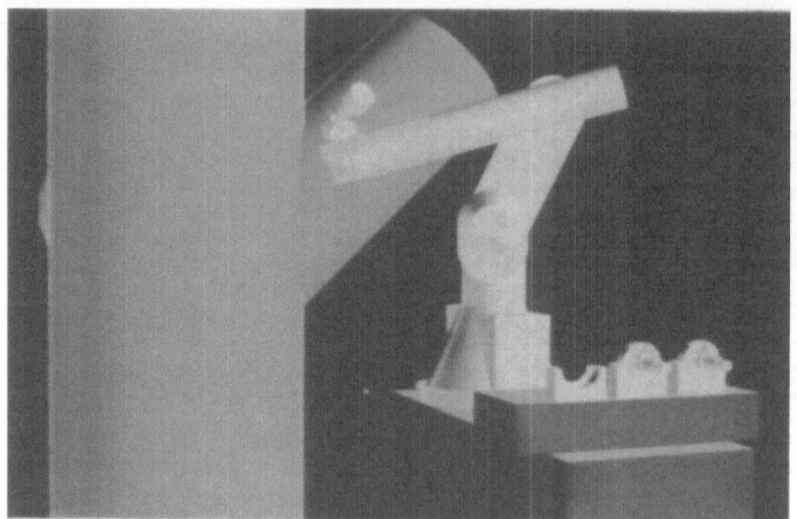

Fig. 13:   Inspection Robot with Tool Exchange System and
            Underwater Structure (partly visible)

Fig. 14:   Realized Laboratory Experiment

4.3.   Automation and Robotics in Space

Robotics is to be seen as a key technology for the automation
in and commercial use of space in future. Concerning
production technologies, automation and teleoperation concepts have to be developed and tested. In this context also
the results achieved withing the ESPRIT 623 project are
helpful or can even be directly used for space applications.
In the following subchapters two examples are given.

4.3.1. Robot Technology Experiment (ROTEX)

For the German Space-lab mission evisaged for 1991 a robot
technology experiment is in development. Thereby a robot in
the experiment box (fig. 15) has to fulfill different tasks
after a caliberation procedure /5/. In fig. 16 details of the
experiment cell are shown. The concerning robot programs have
to be planned, generated and tested by simulation in the
ground station and transfered for execution to the orbit. The
principle system structure is shown in fig. 17. The project
is financed by the German Ministry of Research and Technology
with Dornier as prime contractor and IPK-Berlin as a
subcontractor of Dornier.

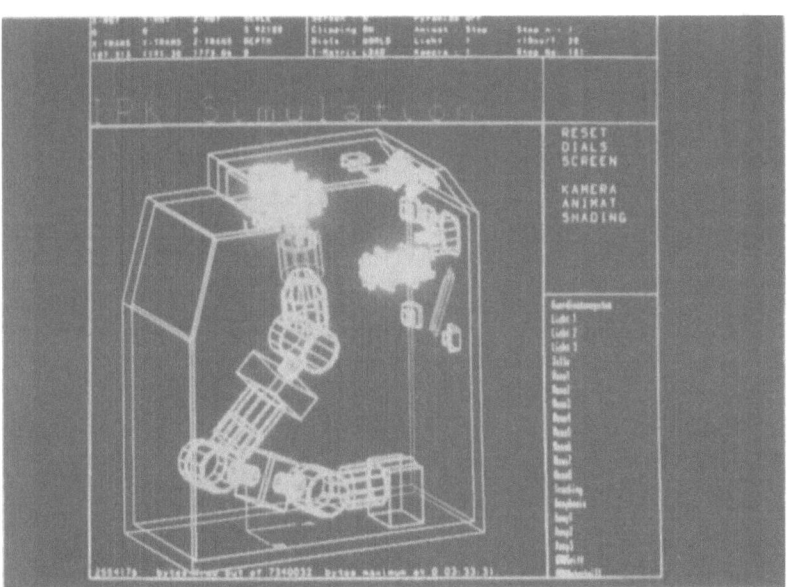

Fig. 15:   Preliminary ROTEX Experiment Simulation

656

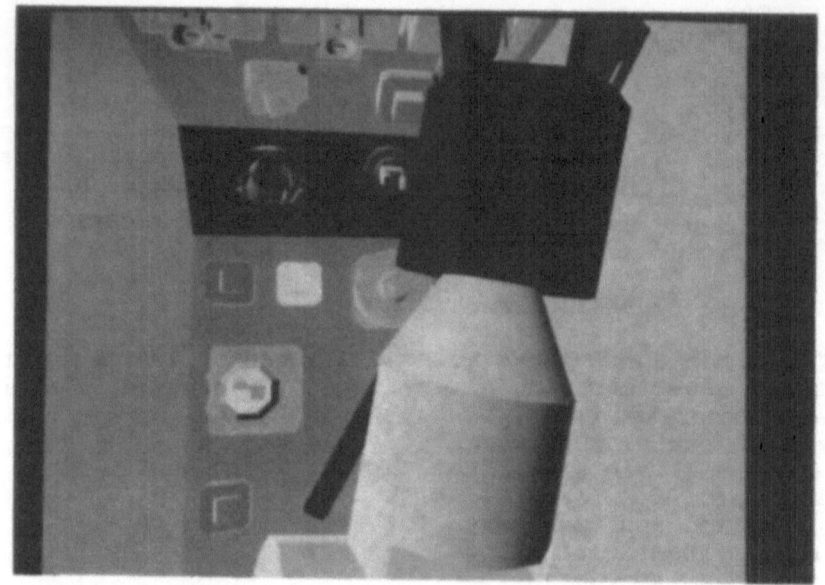

Fig. 16:   Details of the ROTEX Simulation

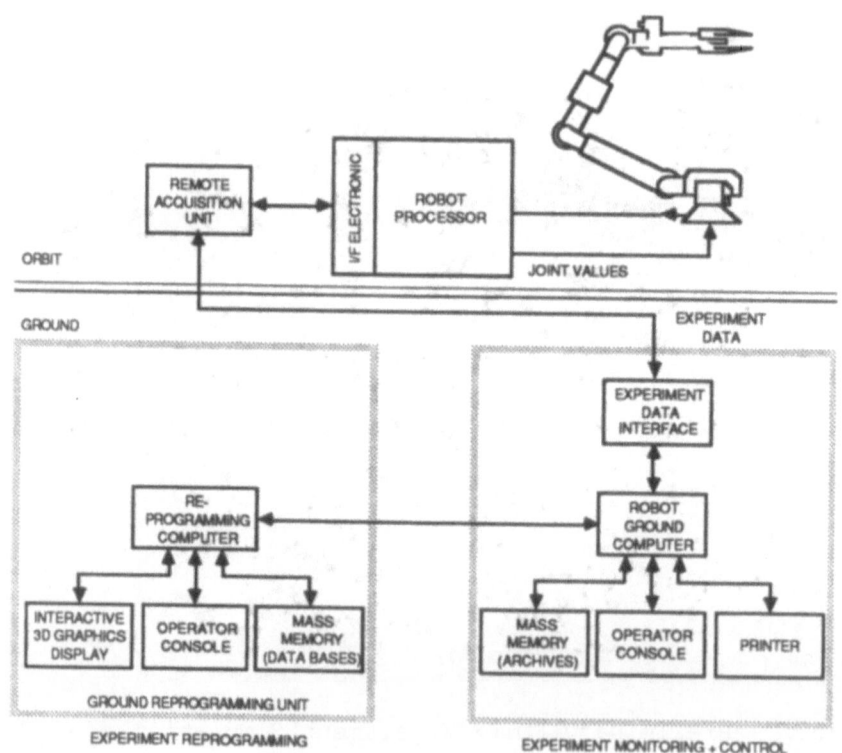

Fig. 17:   System Structure of ROTEX

4.3.2. Lab for In-Orbit Automation (LORA)

At the IPK-Berlin a lab for automation and robotics (A&R) in
space has been built up which provides the development environment for automation procedures and components for space

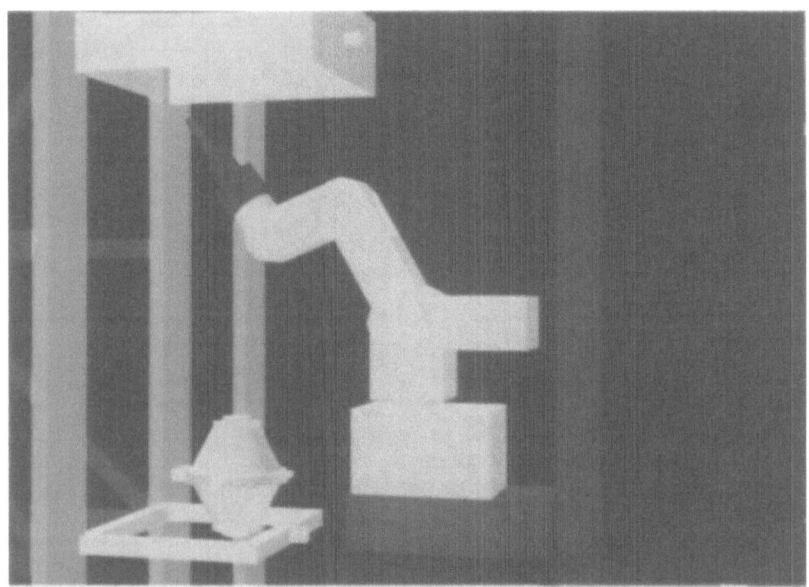

Fig. 18:   Simulation of the Task Execution (LORA)

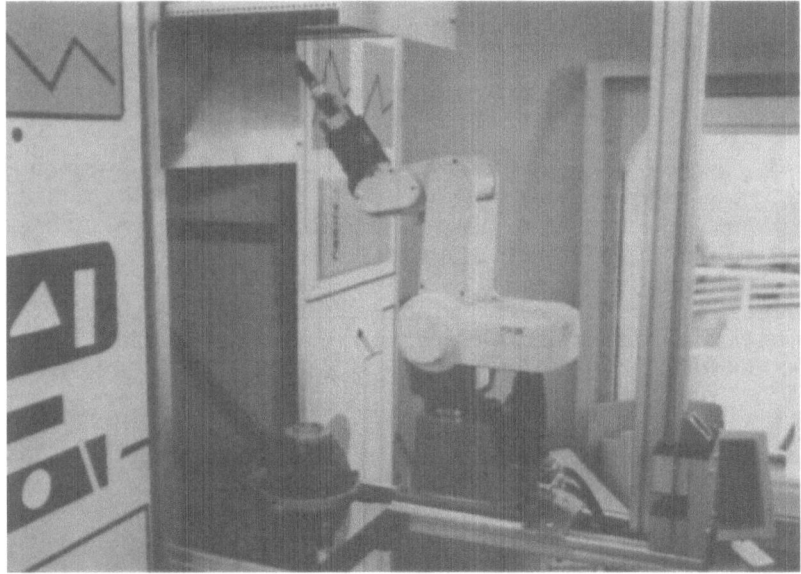

Fig. 19:   Realized Work Cell (LORA)

applications. This lab allows the development and test of automation and experiments under realistic conditions. The project is partly supported by the Senate of Berlin and a number of research institutions and Berlin SME´s participate in it. As a first experiment, a robotized work cell (5-axes robot mounted on two external linear axes) for the exchange of samples from a melting stove has been realized. Thereby the task execution is planned and tested by the simulation system (fig.18). A photography of the real work cell is shown in fig. 19.

5. CONCLUSION

Robots are important components for flexible automation. With the improvement of their features the related to

-   robot control functionality,
-   sensor processing,
-   off-line programming and
-   the availability of computer-aided planning means,

the application area for robots will be more and more enlarged. This is valid in manufacturing industry as well as for other industries like off-shore and space where they can lead to a new dimension in executing tasks which are dangerous for humans.

For the manufacturing industry which is mainly reflected in the ESPRIT 623 project it can be pointed out that the developed and industrially applied systems brought advantages which can be summarized as follows:

-   Costs for design, optimization programming and workshop test are significantly reduced.
-   At a very early stage in a project, highly accurate information is available which leads to getting a manufacturing system faster in operation.
-   Several alternative solutions can be analyzed and compared which results in an improvement of the quality of the solution.

Finally it can be stated that all planning means are to be integral parts of the CIM system. This requires a standardization of their interfaces to ensure their cooperation via an adequate information system.

References

1   G. Spur et. al.: Planning and Programming of Robot
    Integrated Production Cells.
    4th ESPRIT Conference and Exhibition.
    Brussels, Sept 28-30, 1987.

2   B. Duffau: Interactive Off-line Robot Programming,
    Industrial Applications.
    CIM-Europe Conference.
    Athens, May 1989.

3   G. Stark: Operational Control for Robot Integration into
    CIM and its Applications.
    IFAC Conference "SYmposium on RObot COntrol" (SYROCO).
    Karlsruhe, Oct 5-7, 1988.

4   B. Schubert; G.F. Schultheiss; G. Duelen; U. Kirchhoff;
    F.L. Krause; R. Rieger: Advanced Development of Automated
    Remote Handling.
    In: Proceedings of the 2nd Workshop on Manipulators,
    Sensors and Steps toward Mobility.
    Manchester, Oct 24-26, 1988.

5   G. Duelen; U. Kirchhoff; R. Bernhardt: Ground Based
    Supervisory and Programming System for Automatic
    Experiment Execution in Orbit.
    First European In-orbit Operations Technology Symposium.
    Darmstadt, Sept 7-9, 1987.

Acknowledgement

The Integrated Planning and Off-line Programming System
presented in detail has been realized in the frame of the
ESPRIT 623 as a specific demonstrator project conducted by
the following partners:

IPK, Berlin: Dr. R. Bernhardt, A. Deutschländer
                       G. Schreck, V. Katschinski, V. Gleue,
                       St. Krüger

UNL, Lisbon: Prof. A. Steiger, Dr. L. Camarinha,
                       A. Barbosa, D. Ferreira, L. Correira,
                       J. Moura-Pires

UPM, Madrid: Prof. E. A. Puente, Prof. A. Jimenez,
                        Dr. F. Sastron, Dr. F. Prieto,
                        Dr. C. Cerrada, L. M. Fletes

PSI, Berlin: W. Jakob, H. Nase

KUKA, Augsburg: Dr. H. Wörn, G. Stark,
                        Dr. K. Schwendinger, W. Miosga,
                        G. Schratzenstaller, E. Postenrieder

UCG, Galway: Dr. J. Brown, K. Tierney, R. Bowden,
                        S. Wadhwa, P. O´Gorman

The author in his capacity of demonstrator project manager
reports on the work done by the above mentioned researchers.

Project No. 688

CIM-OSA - ITS GOALS, SCOPE, CONTENTS AND ACHIEVEMENTS

KURT KOSANKE
AMICE Consortium
489 Avenue Louise
B1050 Brussels
Belgium

JAKOB VLIETSTRA
APT Nederland BV
Larenseweg 50
1200 BD Hilversum
The Netherlands

KEYWORDS: architecture, open system, information technology, models, communications, services, business process, integration, manufacturing enterprise

ABSTRACT: With the advent of Computer Technologies, and supported by major breakthroughs in electronic engineering and informatics, Computer aided Design (CAD) and Computer Aided Manufacturing (CAM) penetrated our design, engineering and production activities. The result has been continuous improvement in productivity. The major drawback was a gradual deterioration of information infra-structures in our enterprises because of the fragmented solutions offered by CAD and CAM. Solutions to this drawback were sought in integrating the various enterprise functions, and this resulted in Computer Integrated Manufacturing (CIM). This article will highlight a new dimension in the application of CIM and Information Technology: the Open System approach to CIM. The article contains a summary description of the various aspects of CIM-OSA: Computer Integrated Manufacturing - Open System Architecture, and its technical specifications. CIM-OSA is developed in an Information Technology project and carried out by AMICE, a consortium of 21 European organizations whose work is sponsored by the European Economic Community within the ESPRIT programme.

1 Introduction

World-wide fast changing markets of goods, capital and know-how lead to increasing global competition. To succeed and grow in these markets requires an easily adaptable and efficient enterprise operating on a product innovation, cost and organizational basis. It requires quick responses as well as adapting to changes in market demands, in product and manufacturing technology, and in the social environment. Active management of change is the most significant future requirement for an enterprise operation.

Understanding a large complex system such as an enterprise is a challenging task. Only with abstraction, aggregation and structuring can this task be achieved. However, for people, this understanding exists in a static way, and accommodations to system changes are usually rather slow.

Information technology can provide the means for information abstraction, aggregation and structuring thus mastering complexity in an efficient and flexible way. A reference architecture is necessary for creating individual solutions that are consistently applicable across enterprises.
Information technology plays an important role in efficiency improvement. This role is still marginal in terms of enterprise needs because of organizational "fences" as well as our limited intellectual capabilities. Therefore, the required reference architecture must also these means for information abstraction, and aggregation of information.

A reference architecture applicable to different enterprises and different industries should be sufficiently generic. At the same time, this reference architecture must be specific enough to be applicable for the end user: the business professional. (Business users or business professionals are all people with responsibilities for particular business processes. Business users or business professionals are: technical staff, administrative staff, and all levels of management). Therefore, the architecture has to provide a computer supported descriptive language, or a set of generic building blocks that express and specify the business requirements. These requirement specifications must be directly transferable (computer supported) into solutions that can be implemented as enterprise system components.

2 AMICE and the CIM-OSA Goals

CIM-OSA is developed in an ESPRIT project carried out by the AMICE consortium, an association of 21 European firms. The consortium is composed of Information Technology vendor industries, manufacturing enterprises, software houses and an educational institute:
AEG (D), Aerospatiale (F), Alcatal (B), APT (NL), British Aerospace (UK),
Bull (F), CAP Sogeti Innovation (F), PROCOS (DK), Digital (D),
Dornier (D), FIAT (I), GEC (UK), HP (F), IBM (D), ICL (UK),
Italsiel (I), Philips (B), SEIAF (I), Siemens (D), Volkswagen (D), WZL (D).

The goal of CIM-OSA is to guide the user in specifying, procuring, implementing, and using system components in his enterprise operation. It will enable the enterprise to conduct its business in a real time adaptive mode. Therefore, CIM-OSA will support operational flexibility as well as multi-disciplinary information integration.

This is being achieved by providing a descriptive rather than prescriptive methodology. Therefore, CIM-OSA furnishes a reference architecture from which particular architectures can be derived for the individual enterprises. Using CIM-OSA results in a complete description of the enterprise. This description is stored and manipulated by the relevant information technology base of the enterprise during the build time phase, and is being used to control and monitor the enterprise operation during the run time phase.

CIM-OSA does not consider system design as a one time affair. Therefore, CIM-OSA will provide sufficient support for updating, modifying and extending the current system. It will allow for evolutionary design enabling the enterprise to start using CIM-OSA methodologies and components in areas of the enterprise rather than requiring a complete system description. CIM-OSA guides the vendor in conceiving, designing and marketing system components that provide consistent inter-working capabilities.

The reference architecture consists of sets of generic building blocks and partial models that will contain the base definition for marketable products in the manufacturing and information technology area.
CIM-OSA places emphasis on the open system approach. An integrating infrastructure will allow consistent enterprise operation in heterogeneous manufacturing and information technology environments. To enable system consistency, vendors are encouraged to provide the relevant services of this infrastructure as part of their respective products.

3 CIM-OSA Scope

CIM-OSA is an architecture that aims to improve the operation of the enterprise. As such, CIM-OSA embraces the complete enterprise operation, and describes it as an integral system.
Only if the total operation is explicitly described can the consequences of internal and external changes be made visible in real time. This means that all internal processes of the enterprise as well as all its external relations to vendors, customers, governmental and other agencies have to be part of the system description.
Therefore, the scope of CIM-OSA provides guidance for design and execution of the whole enterprise operation with all its aspects (development, production, marketing, financing, administration, etc.). However, the project is currently focussing its work on the major areas of CIM only (development, planning and production). Further work is required to develop the concepts beyond the CIM area of the enterprise.

4 CIM-OSA Contents

Enterprise integration is a very complex task. The project has identified three levels of integration to guide the development of CIM-OSA (see figure 1.)

Physical System Integration is mainly concerned with inter-system communication: the communication among parts of the system. This level of integration is currently provided by present information technology concepts and standards (e.g. OSI, CNMA, MAP/TOP). CIM-OSA therefore, is mainly concerned with the other two levels of integration:

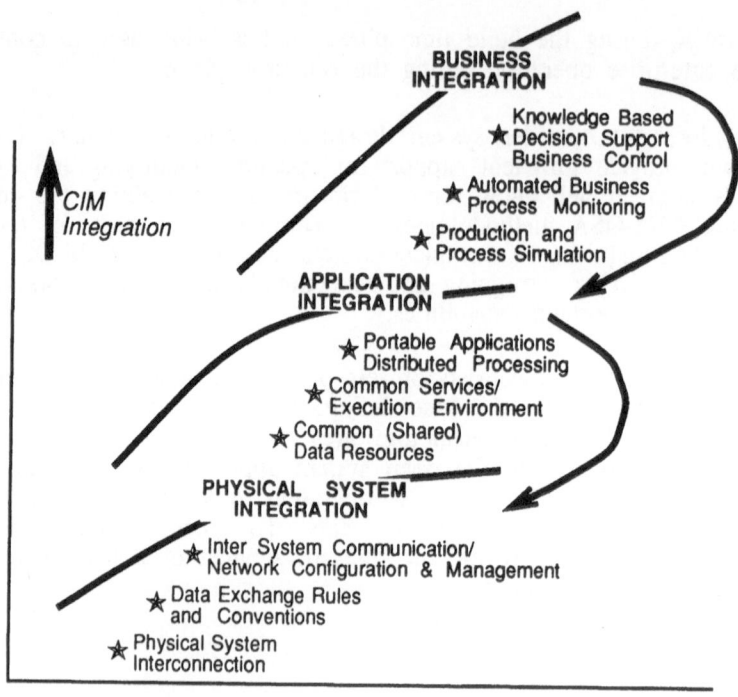

Figure 1: The Levels of Enterprise Integration

Application Integration and Business Integration.

CIM-OSA provides general solutions for intra-system communication: communication among different systems (Application Integration), for enterprise requirement definition, and CIM system design and operation (Business Integration). These solutions are forwarded to various standardization bodies for their comments and evaluations, and thus contributing to the processes that lead to future (industrial) standards.

CIM-OSA will make use of the available standards (OSI, others); will consider emerging standards (ODP, others) when ever applicable, and will strongly promote those areas in which a need for standards is recognized.

4.1 CIM-OSA CONCEPT

The project has developed two main concepts that are complementary to each other. The CIM-OSA Integrating Infrastructure (IIS) provides for application integration, and the CIM-OSA Modelling Framework supporting Business Integration.

Intra-system communication is achieved through the CIM-OSA Integrating Infrastructure. The IIS provides a structured set of common System Wide

Services that will avoid redundant functions in the system. By offering these services system wide in a uniform manner, the basis for required integration is created. The role of the IIS is to provide the Computer Integrated part of CIM; that is, the integration is achieved through Information Technology, i.e. by computers. (See figure 2.)

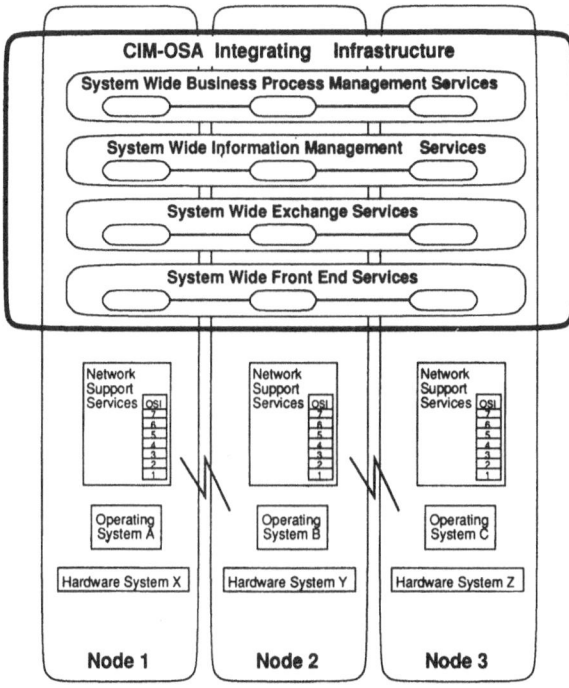

Figure 2: CIM-OSA Integrated InfraStructure

The IIS provides a set of specified services to control and integrate the definitions in the Particular Implementation Model and the resources (humans, machines, application systems) performing these functions. The services will:

* isolate the application programs from the data processing and business environments

* operate systems wide and not require the application modules using these services to have any knowledge of their location and/or distribution.

* provide common protocols allowing humans, machines and application programs to interact with the IIS in an uniform way.

The System Wide Services appear to the user of the service as a single service across all nodes of the system. The user doesn't have to know how and where the service is provided.

Within the services of the IIS we distinguish four sets of services. Function related services include all those services related to enterprise functions. These services will manage the control and execution of the enterprise activities. They provide the Integration of Functions, and contain: Business Process Control, Activity Control, and Resource Management.

Information related services include services that support the information processing activities of the enterprise. They deal with locating, accessing, storing and maintaining the consistency of information.
These services provide the Integration of Information with System Wide Data and Data Management

Communication related services include the services required for the control of Intra and Inter System Communication. They provide the Integration of Communication (Communication Management and System Wide Exchange).

Front End related services include the services required for the control of Communication with the Human, Machines, and Applications. They provide the Integration of Front Ends consisting of Human Front End, Machine Front End, and Application Front End.

The IIS provides the business process services for processing CIM-OSA compliant system descriptions as well as the traditional areas of information technology (communication, information and front-end services).
The modelling framework consists of several modelling levels for requirements, system optimization and system implementation. Different aspects (views) of the enterprise operation such as function, information, resources and organization may be modelled separately.

The framework of CIM-OSA provides **three modelling levels** (see figure 3):

At the **Requirement Definition** Level, the business requirements of the enterprise are identified. These requirements are represented in terms of business processes, their inputs and results, the procedural rules and the enterprise activities within the business processes, describing WHAT has to be done in the enterprise.

The **Design Specification** Level is used for the design of the business processes and enterprise activities describing HOW they are performed. On this level, volumes will be added, parameters will be specified and additional constraints will be considered.

The **Implementation Description** Level provides the MEANS of executing the model by selecting the physical entities like programs, machines and humans

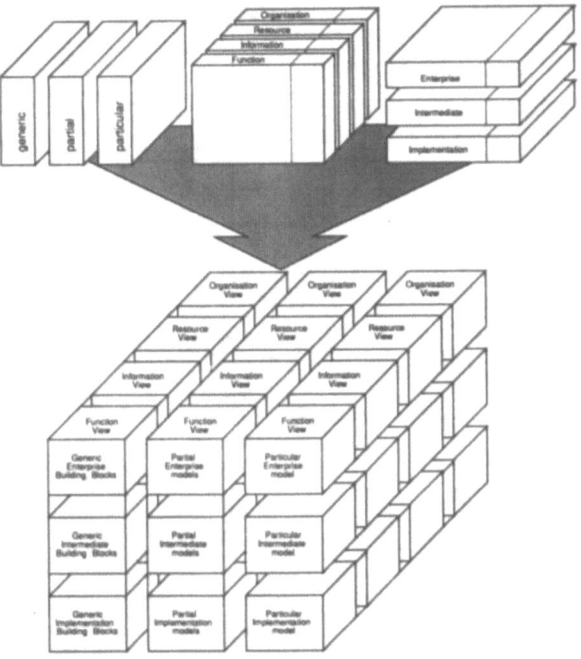

Figure 3: Overview of CIM-OSA Architectural Framework

needed to run the business process defined at the requirement level. These entities representing the components of a CIM system will be selected according to the specifications of the Design Modelling Level from a component catalogue offered by various vendors. Components required, but not available on the market, must be developed by the particular enterprise. In the framework, the process of Stepwise Derivation from the requirement level to the implementation level will be supported by guide-lines or computer programs.

Four different complementary **enterprise views** have been defined that allow the modelling of the major aspects of the enterprise (See also figure 3).

The Function View is a representation of the enterprise operation in terms of a structured hierarchy of business processes. Each business process is constrained by the declarative rules, defined by its triggering events, the results it produces and by its control flow description which is called the procedural rule set. The internal structure of a business process may be lower level business processes or enterprise activities. Enterprise Activities represent the lowest level of decomposition (See figure 4).

The Information View gathers all information defined and contained in the enterprise. The information is structured in information classes and enterprise information objects; with their object views and object editions within domains to be defined by the enterprise model designer. All information is composed of information elements, the smallest addressable atomic information unit.

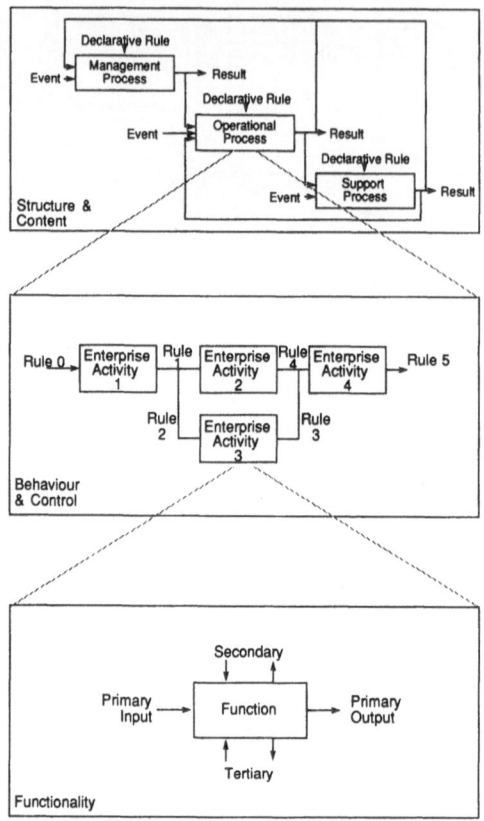

Figure 4: The Enterprise Model

The Resource View contains all the information about the enterprise's resources. The view is structured by using a hierarchical concept of cells for grouping resources according to the enterprise requirements.

The Organization View contains all relevant information about the responsibilities within the enterprise, and allows for structuring the different responsibilities in the enterprise for function, information and resources. The enterprise views will be generated one after the other, and this process which is called stepwise generation, will be supported by a set of programs guiding the enterprise designer.

Enterprise systems must be implemented by employing People and using Manufacturing and Information Technology. Figure 5 shows the implementation of the CIM-OSA models that contain all the enterprise requirements and resulting designs of the real world of people and technology components. The models and the real world comprise the CIM system of the enterprise.

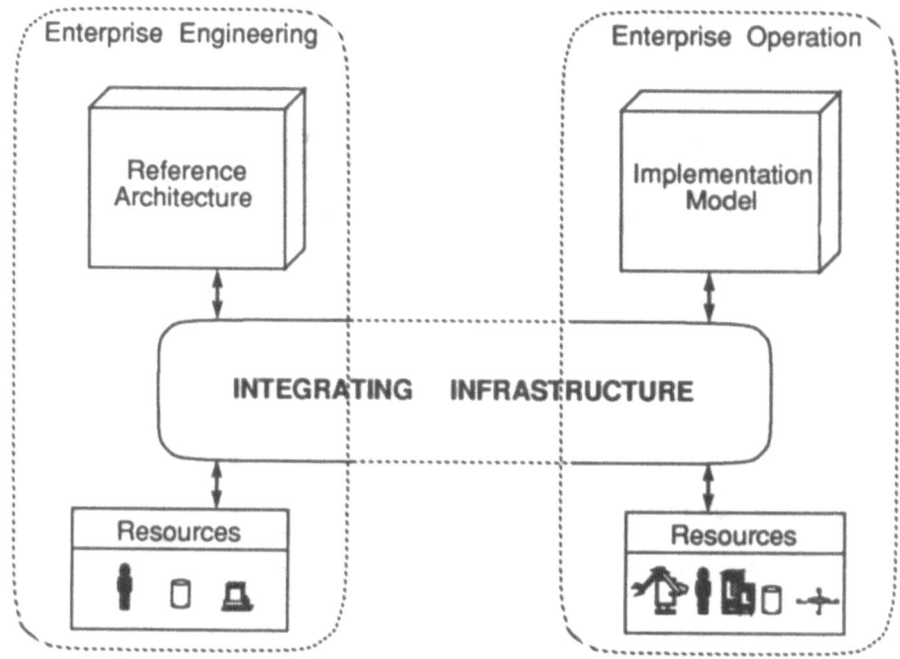

Figure 5:   Implementation of CIM-OSA

4.2 THE CIM-OSA CREATION PROCESS

CIM-OSA provides a set of guide-lines for creating the contents of the particular architecture from the reference architecture. To facilitate the definitions of these guidelines, the creating process has been decomposed into its 3 dimensions. Instantiation, derivation and generation are applied to all the levels and views of the architecture. The creation process inside the reference architecture is controlled by the owners of the reference architecture. The process of creating a particular architecture is done by the user of CIM-OSA. This process will be guided and supported significantly by information technology to assure the creation of consistent CIM system descriptions.

The integrating infrastructure consists of a set of services aimed at consistent operation in heterogeneous environments. CIM-OSA provides a build time and run time support environment for the design, implementation and execution of enterprise operations. Both environments use the integrating infrastructure (IIS) that provides specific information technology services for the enterprise operation, but more important, provides for vendor independence and application portability.

To achieve the required Business Integration, Application Integration and Physical Integration, CIM-OSA provides for an integrated environment divided into two parts:

1) In the **Integrated Enterprise Engineering Environment** the released implementation model contains the CIM-OSA defined model of the business processes and enterprise activities required to implement the CIM-OSA Guidelines. The guidelines are for the specification of the requirements, design, implementation and release of the enterprise system (and changes to it). The model also includes the related information, resource and organization views.

2) In the **Integrated Enterprise Operation Environment** the released implementation model is the released particular implementation model. It contains a model of the business processes and enterprise activities required for the operation of the enterprise, and the related information, resource and organization views.

5 Achievements

The CIM-OSA concepts developed in the project have been verified, to some extent, internally in the project. Promotion is done by presentations and demonstrations at European and International events as well as within the Consortium. National, European and International standardization bodies have made use of our input, and and further proposals are in progress.

5.1 RESULT OF DEVELOPMENT WORK

Both the modelling concepts and integrating infrastructure have been validated by either case studies in different environments, or through scenarios developed for special areas of the enterprise. In addition, an interactive demo has been developed showing the aspects of build time and run time phase for CIM-OSA particular architectures.

The modelling applications are currently hampered by lack of real computerized support for design, and optimization of particular architectures. First versions of such tools will be available and demonstrated at this conference. Nevertheless, current results have been used by partner companies to model particular areas of their operations.

Infrastructure related work is currently conducted to show the feasibility of the different services. The results will once more be demonstrated at this conference.

5.2 STATE-OF-THE-ART

The CIM-OSA modelling concept is unique in several aspects compared with the state of the art: ICAM, CAM-I (DPMM), NBS (AMRF), ISO TC 184/SC5/WG1, ESPRIT Project 34, ALVEY (ANSA). This is shown in the following table:

	State-of-the-Art	CIM-OSA
Scope	architecture/model for particular and unique industrial environments	Reference Architecture for all industrial environments and guide lines for CIM product development
Environment	discrete parts manufacturing with emphasis on mechanical parts and aerospace manufacturing	discrete part manufacturing (electrical, electronic, mechanical) and CIM vendors (equipment, machine tool and IT industry)
Functions	focus on manufacturing functions and complementary functions taken into account	all functions in a manufacturing enterprise
Models	hierarchical/top down decomposition of functions and data (control structure and management)	reference catalogue for enterprise, intermediate and implementation modelling
Methodology	free form textual description and graphs/flow charts DVM: Data Vector Modelling IDEF: ICAM Definition based on SADT, ERA	System design using finite set of generic building blocks for computer supported modelling and computer supported enterprise operation

5.3 PROMOTION

During the course of its development work the CIM-OSA project has established the CIM-OSA concepts and published them in a number of project internal and external documents. [1] [2] [3].

5.4 STANDARDIZATION

The CIM-OSA Modelling Framework has been introduced into the standardization bodies (national: DIN, European: CEN/CENELEC, international: ISO). The submission into ISO by DIN has been accepted and has generated a

672

new work item in ISO TC 184/SC5/WG1. The framework is also considered as an ENV (European Prenorm) by CEN/CENELEC/WGARCH.
A similar proposal is currently developed for the IIS by the CIM-OSA Project.

6 Summary

To improve enterprise efficiency and competitiveness requires efficient cooperation and information accessibility across and beyond the enterprise. The use of information technology will significantly support this process, provided that the functional and organizational structures of the enterprise are adapted to changing environments and the requirements of efficient enterprise operation.

CIM-OSA provides a concept for the future needs of industry; and is presently validating the content of the architecture. Figure 6 illustrates this concept and provides a general view of the way how CIM-OSA will be used to design, implement, operate and maintain future CIM systems. Such systems will provide industry with opportunities to streamline production flows, to reduce lead times, and to increase overall quality while adapting the enterprise fully to the needs of the market.

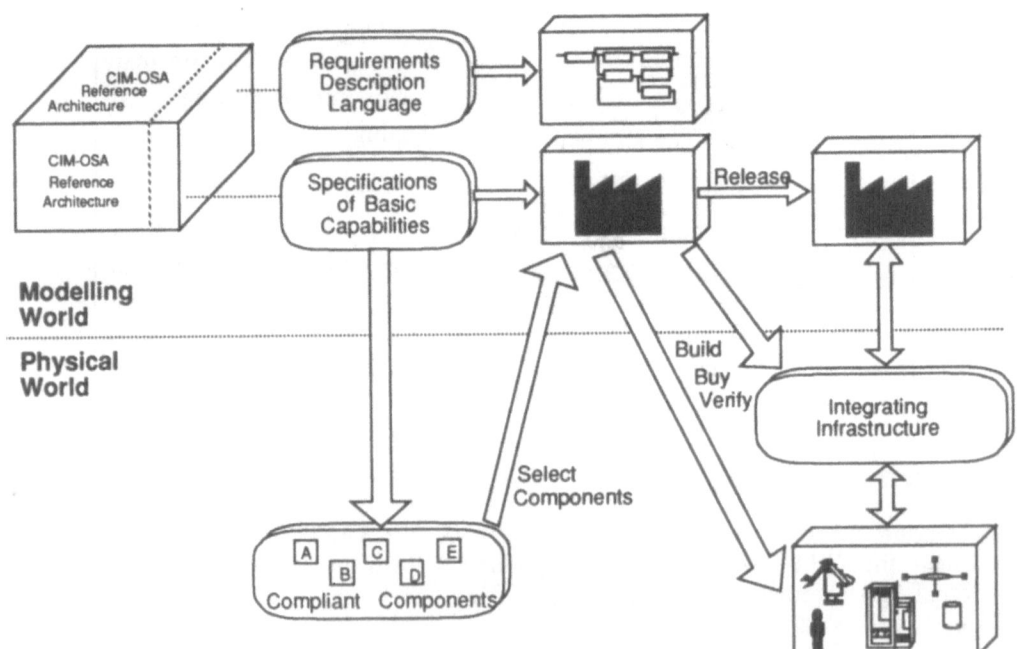

Figure 6: CIM-OSA Information Technology Environments

The enterprise modelling approach will provide a descriptive language for the enterprise requirements. Intermediate and implementation models will support consistent, optimized system design and implementation (including system qualification and release). Both design and operation will be supported by the system wide integrating infrastructure. This will provide inter- and intra-system communication across multi vendor environments.

On-going promotion and standardization activities will provide an industry wide acceptance and support for CIM-OSA. The AMICE consortium itself will contribute to this acceptance by actively pursuing CIM-OSA in their own organizations.

References

[1]   AMICE Project Team.
        CIM-OSA: Reference Architecture Specification.
        CIM-OSA Esprit Project, 489 Avenue Louise, Brussels, Belgium, 1988.

[2]   AMICE Project Team.
        CIM-OSA: A Primer on Key Concepts and Purpose.
        CIM-OSA Esprit Project, 489 Avenue Louise, Brussels, Belgium, 1987.

[3]   AMICE Project Team.
        CIM-OSA: Strategic Management and Design Issues.
        CIM-OSA Esprit Project, 489 Avenue Louise, Brussels, Belgium, 1987.

Project No. 809

Integration of Dynamic Expert Scheduling in Production Control

H. DE SWAAN ARONS[1]

D. RIEWE[2]

ABSTRACT. The paper deals with the integration of a dynamic expert scheduler in a Flexible Manufacturing System. The dynamic expert scheduler is a combination of an expert system and a scheduler. The expert system has to cope with unpredicted events during manufacturing, to decide whether or not these give rise to the necessity to reschedule, to determine appropriate dispatching strategies and to evaluate generated workplans. The scheduler generates workplans based on the current cell status and the advised dispatching strategies generated by the expert system, both for the cell and the machines in the cell. The dynamic expert scheduler has to carry out its on-line task in a production control system. This work has been carried out as a part of Esprit project 809 *Advanced Control Systems and Concepts for Small Batch Manufacturing*[3].

KEYWORDS. Dynamic Scheduling, Expert System, Knowledge-based Decision Making, Real-Time Processing, Flexible Manufacturing

1. Introduction

Esprit project 809 *Advanced Control Systems and Concepts in Small Batch Manufacturing* aims to provide an advanced production control system for the manufacturing of mechanical parts in the metal cutting industry in small to medium batches. Production will be considerably improved by increasing the flexibility of the production control system. Developing such a system has been the main aim of the project. One of the means to achieve this is to make use of decision support, which is especially needed in the scheduling area. Therefore, as a part of such a production control system a dynamic expert scheduler has been considered to improve the quality of decisions with respect to rescheduling, generation and evaluation of workplans. For this purpose, in the beginning of the project the Consortium partners Delft University of Technology (DUT) and Krupp Atlas Datensys-

---

[1] Delft University of Technology, Dept. of Technical Mathematics and Informatics, Dept. of Technical Mathematics and Informatics, Julianalaan 132, 2628 BL Delft, The Netherlands

[2] Krupp Atlas Datensysteme, Altendorfer Strasse 104, D - 4300 Essen 1, Federal Republic of Germany

[3] This project has started March 1986 and is carried out by prime contractor Dextralog and ICL (both GB), Krupp Atlas Datensysteme (FRG), Delft University of Technology and University of Twente (both NL)

teme (KAD) started a feasibility study of an dynamic expert scheduler. This work was carried out in Work Package 2 (WP2) and it was reported of in [Hei 87].

The results were most encouraging. They demonstrated that a combination of an expert system and a scheduler/simulator providing on-line assistance with respect to event handling, decisions to reschedule, selection of dispatching strategies, generation and evaluation of workplans, could considerably speed up the generation of new workplans and improve their quality. The feasibility study also identified items for further research in order to have full advantage of a dynamic expert scheduler. Although the WP2 prototype appeared successful, it was only meant to demonstrate that better workplans could be generated and much faster.

Partly based on the results of WP2 a dynamic expert scheduler is under development that will be incorporated in the production control system of a Flexible Manufacturing System. This dynamic expert scheduler is the subject of this paper which discusses its design principles, how it is integrated in the production control system being developed within the Esprit project 809 as a whole, the kind of events it has to react on, and how expert system and scheduler/simulator cooperate in order to generate acceptable workplans.

In the following section a brief description is given of the goal of Esprit project 809. In section 2 the scheduler/simulator developed by KAD is discussed in more detail. It explains how it has been designed, how it transforms the inputs into the workplan and how its outputs are used by either the expert system or the operator. The expert system developed by DUT is described in section 3. It is discussed how it handles events of various types (shopfloor, cell etc) based on knowledge incorporated in a rule base. In section 4 a few words are spent to the integration of the dynamic expert scheduler in the production control system as a whole.

Some conclusions and results are discussed in section 5.

2. Overall project Esprit 809

The main goal of Esprit project 809 is to improve production control for small batch part manufacturing systems, in order to achieve both a more flexible use of resources and a decrease of the size of economic batches [Tie 88], [Tie 89]. For this reason an integrated production control system has been built for a manufacturing system in the metal cutting industry. This system is intended to improve the control of a manufacturing system by integrating scheduling, production control and monitoring functions. Therefore, the production control system consists of several modules. For better understanding of tasks and position of the dynamic expert scheduler in the production control system its various modules are are briefly explained with the help of figure 1.

The production control system will start its activities by activating the Scheduler when production orders come in from system level. The components Scheduler and Expert System together form the dynamic expert scheduler. Although this paper is mainly concerned with the actions that are to be taken in case of any disturbance during machining, the dynamic expert scheduler can and will also be utilised for initial scheduling tasks before the manufacturing process has started.

One of the outputs of the Scheduler is a detailed workplan in which the sequences of jobs which have to be performed on the machines have been determined. The Scheduler is have been supported in its task by the Expert System which is fed with relevant data by the Monitoring and Diagnostics System. The resulting workplan will be passed over to the dispatcher which will release the corresponding jobs to appropriate machines on scheduled times and will take care of supplying the necessary tools and information.

The various workstations are supplied with the required information concerning jobs, tasks and equipment by Workstation Control. This system plays a central role in the production control system: it supplies other components with required information in the right format. At last, I/O Interfacing communicates with external storage units such as Material Store, Tool Store etc.

Relevant to this paper are the Scheduler and the Expert System which build up the dynamic expert scheduler. We will now describe both systems in more detail and how they do cooperate.

3. The scheduler

The Scheduler determines the order sequence in a Flexible Manufacturing System. The system is fed by new orders and process plans in a cyclic way, the additional transferring of quick orders, too, is no problem.

The system have to be loaded for about two planning periods, i.e. perhaps two weeks for long-term process-steps or two days for very short-term processing.

The most important basis for the planning algorithm are the set of those process steps of all process plans, which are to be processed within the cell. Usually there will be a due date for the last process step in a process plan, which comes from the higher level planning system. Another important basis of the Scheduler is the set of actual values, which is collected by a monitoring system. Thus, it is possible to dynamically create new schedules in parallel to the running production.

The concept of the Scheduler is to have an on-line, quick-running planning system, which permanently collects the shopfloor data and new orders, and being ready for rescheduling any time.

All decisions about when to reschedule, which dispatching strategies to use in the simulation and which candidate (of several alternatives) have to be taken for the new current workplan, can be made by the operator with simple user interfaces or by the expert system with a complete set of gathered, practically oriented rules.

The final output of the Scheduler is a workplan. It is a time-table for all machines and manual working places. The current workplan additionally contains values about the processing status of the different process steps. The current workplan is the basis for the human or automatic dispatcher. In the case of rescheduling the current workplan will be replaced by a new current workplan.

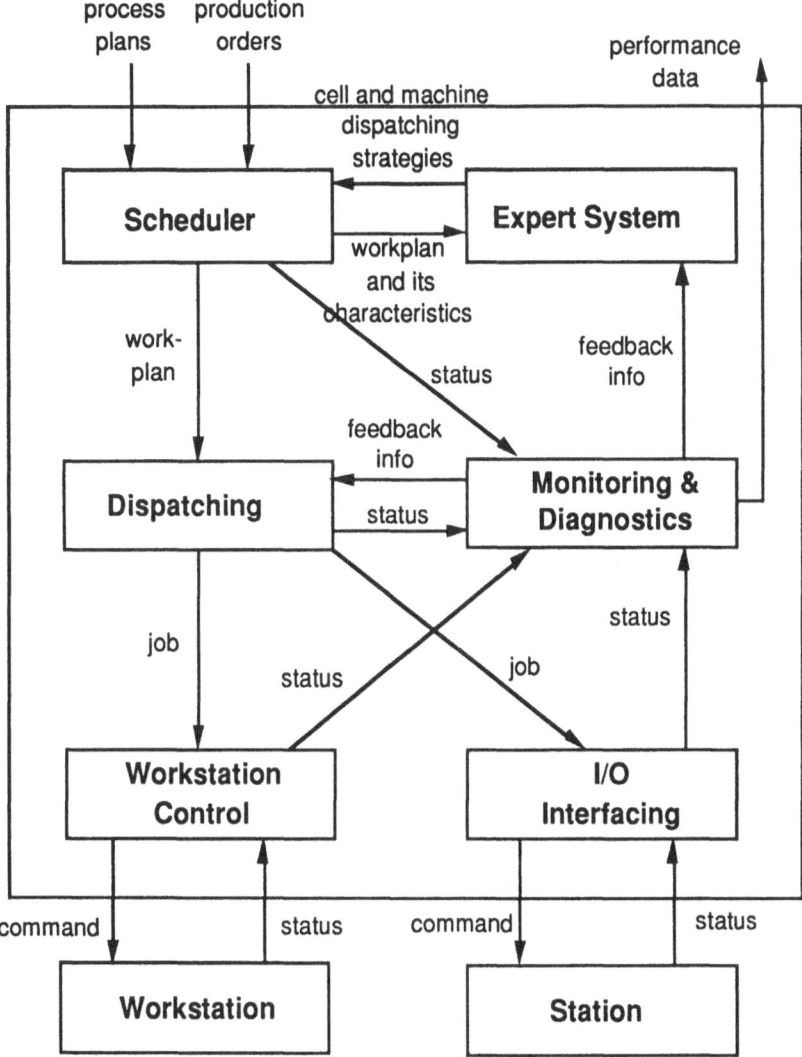

Figure 1. *The components of the Esprit 809 production control system*

The real-time Scheduler consists of seven software packages, which are described in figure 2.

3.1. WORKLOAD BALANCING

Before determining the optimal order sequence, it is calculated, whether it is possible to reach all due dates. Based on the processing times and predefined overheads the workload of all machines in the cell is evaluated. When the workload exceeds a defined upper bound, a balancing algorithm is started. The balancing starts with the machine with the highest deviation from the maximum workload. Iteratively, the other machines are considered, too.

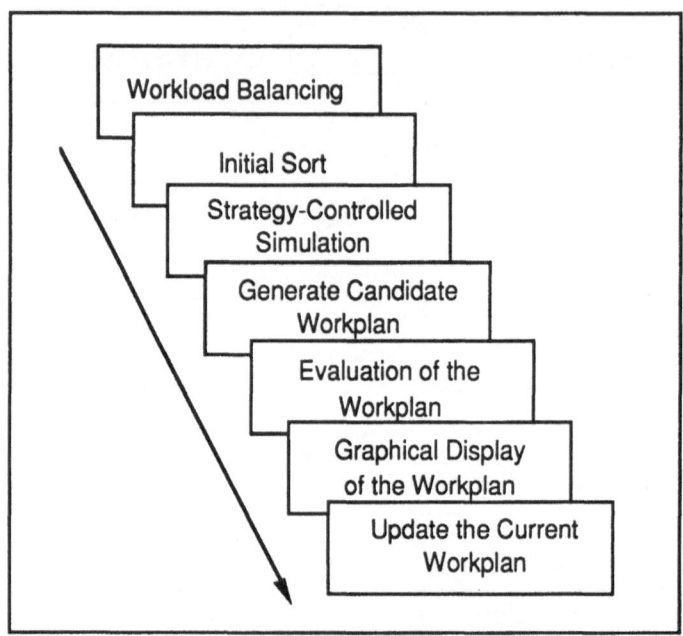

Figure 2. *Structure of the Scheduler*

The workload could only be reduced when there are alternative machines pointed out in the process steps. All alternative steps for an overloaded machine are considered and may be distributed to other machines until its workload is reduced to a second upper bound.
In order to avoid organisational problems, the operator has the last decision on how to distribute the process steps to other machines by defining priority rules.

3.2. INITIAL SORT

The orders which come new into the cell or from one machine to the other are sorted by different dispatching strategies, which are chosen either by the operator or by the expert system. The dispatching strategies are defined in several levels and are active for cell and machine queues. The system can take into account 15 dispatching strategies, e.g. earliest due date, shortest remaining processing time, minimum slack time, machine enhancement, longest imminent operation and others.
The concept of these sorting dispatching strategies is:

- to have simple rules, which can be understood on the floor,
- to have the possibility to very simply add new dispatching strategies , and
- to add dispatching strategies which are dependent of typical features of the specific production and also of technical terms like use of equal materials, tools, fixtures.

This concept shows the main difference to commonly used dispatching strategies on higher levels. This scheduler can react very flexibly to the various situations in different shops or time periods. Beside the time-oriented or utilisation-oriented planning methods there are also technical-oriented principles involved.

3.3. THE STRATEGY-CONTROLLED SIMULATOR

The simulator uses the same strategies as sorting routines. In the automatic simulation run the complete time-dependent flow is calculated. For the simulation run a snapshot of the current situation in the shop, that is in the monitoring system, is kept. This snapshot then is the initial data set for this simulation run and is also used for following runs for comparison purposes. It is evident that the simulation run must work so quickly that the situation in the shop cannot change too much in the meantime. The simulation considers all processing and presetting times and also idle times for transport and others.

3.4. GENERATING CANDIDATE WORKPLAN

From the internal results of the event-driven simulation run, a new (candidate) workplan is derived. The initial values from the old (fixed) workplan are taken and updated to a new time table for all actions in the cell. Beside the pure processing at the machines, there are also auxiliary steps created, which need to be done. That is, for instance, fetching materials, measuring tools and presetting them. By choosing new dispatching strategies new candidate workplans can be generated. The number of alternatives is optional.

3.5. EVALUATION OF THE WORKPLAN

The new (candidate) workplan will be analysed for its quality. This is measured once absolutely (Is the plan good for due date-oriented optimisation?) and next also relatively (Is the plan better than another for e.g. utilisation-oriented optimisation?). The calculated values are the utilisation rates of the machines and the cell, the makespan, the idle times, the mean waiting times in the material flow, the mean throughput time, the tardiness in different priority classes and others. These and additional terms are evaluated in all cases without qualifying the actual meaning of the term in the cell. The values are used in the expert system or can be seen by the operator. Depending on the various goals of a schedule optimisation (only due dates, only utilisations, etc) the calculated terms have a different meaning and importance.

The terms are partly defined in such a way that a comparison is possible with a cell load under quite different circumstances. So, it is possible -in the sense of a control loop- to optimise the workload of a cell in the course of using it.

3.6. GRAPHICAL OUTPUT

In order to have the operator informed, it is necessary to graphically present the cell load. A bar chart for the loading of each machine with the different shop orders or process steps can be shown. It is allowed to have a complete survey on self-defined windows. On demand additional characteristics of the order can be displayed.

3.7. UPDATE OF CURRENT WORKPLAN

After the decision to reschedule has been made -either by the operator or the expert system- it is fed into the Scheduler. A main problem of the Scheduler is the real-time aspect. During the planning time -also when the duration is only a few minutes- something could have happened on the shopfloor. Therefore, in order to come to a new workplan, the candidate workplan cannot be taken itself. The current workplan will be halted for a short while, so the dispatcher has to queue its messages. Then the complete calculations with the chosen dispatching strategy will be repeated now based on the latest actual initial values. The result will be a new workplan, which then is to be taken to make the final comparison with the old current workplan to come to a new one. Now access to the new workplan is permitted. Maybe the dispatcher has immediately some queued new messages and updates the status values. Then the work continues with the new current workplan.

4. The expert system

Job-shop scheduling is a difficult task. The number of possible workplans given a number of jobs and machines is exponentially large, arising from the fact that given n jobs queued at a work station there are n! ways to sequence those jobs and some shop condition at another work station might influence the optimal sequence of jobs at the present work station. Therefore, it is practically impossible to find an acceptable (not necessarily a sub-optimal) workplan that meets the constraints concerning due dates, costs, throughput, idle times of machines algorithmically, by simulation. For this reason, dispatching strategies are used, i.e. some way to select a next job to be processed from a set of jobs awaiting service. Dispatching strategies can be simple or very complex.

At the moment it is common practice that a workplan is generated manually. To accomplish a scheduling task task adequately, quite a lot of experience and judgment is needed. Most of the times some fixed dispatching strategies are used which are known to behave reasonably well under idealised conditions. However, in practice the manufacturing process is disrupted almost all the time by all kinds of disturbances concerning machines (breakdown), tools (breakage, wear), materials (quality) or rush orders. Quite often, such an event has obvious consequences with respect to waiting times, idle times, costs or due dates. However, a decision to reschedule will not easily be taken because the time and cost aspects of manual scheduling are considerable. If these aspects could be diminished by using an automated decision making tool, the overall performance of the manufacturing process could have been improved.

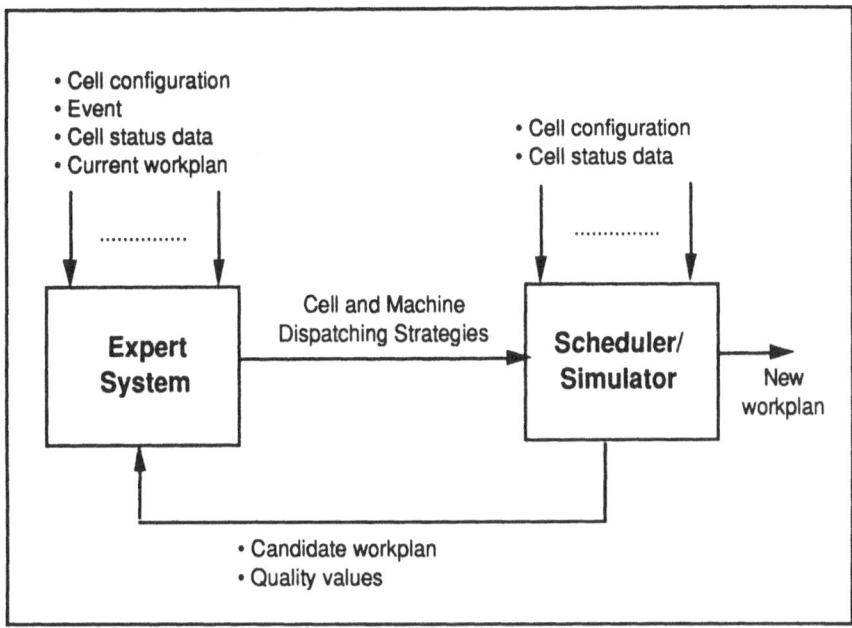

Figure 3. *A schematic overview of the dynamic expert scheduler*

Since the (re-)scheduling process is partly based on heuristics, it is quite a natural choice to apply knowledge-based techniques to it. For this reason, the dynamic expert scheduler contains an expert system for supporting all kind of decision making. Nevertheless, this is not sufficient. Heuristics can give an idea of the necessity to reschedule and if so, which dispatching strategies will have to be used. However a detailed insight into the consequences of a workplan can be achieved by forecasting or simulation. This task is carried out by the Scheduler.

It is expected that automation of the scheduling process using heuristic reasoning (Expert System) and simulation (Scheduler) will improve the manufacturing performance, since the production control system can flexibly deal with disruptions during manufacturing. In figure 3 it is outlined how these two components of the dynamic expert scheduler are interrelated.

Globally speaking, during the manufacturing process the expert system and the scheduler cooperate as follows. After an event has occurred it will be analysed by the expert system taking the data concerning the cell configuration and cell status into account. To be able to make a justified decision on rescheduling the Scheduler is first activated to make a forecast based on the present data and on the data concerning the cell configuration and cell status. In fact, the forecast results in a workplan based on the same dispatching strategies as the current workplan. If the current workplan is still acceptable under the changed conditions the process can continue without rescheduling. Acceptability is decided on a set of so called quality values provided by the Scheduler. These quality values enable the Expert System to judge the quality of the generated workplan with respect to some important characteristics such as due dates, idle times etc. If analysis of a forecast or a candidate workplan leads to the demand to reschedule, the expert system will try to improve it by modifying the dispatching strategies and supplying the resulting set to the scheduler.

The expert system has the following distinct though subsequent tasks, some of which can only successfully be carried out with the help of the Scheduler:

- to analyse events during manufacturing
- to determine whether or not to reschedule
- to determine dispatching strategies for cell and machines
- to analyse (candidate) workplans generated by the scheduler
- to decide on which candidate workplan will be the current workplan

Each of these tasks will be discussed in more depth in the following sections.

For a detailed description of the various parts of the expert system (the tool used, the knowledge base consisting of the heuristic reasoning rules dealing with on-line event handling, strategy selection, workplan evaluation etc) the reader is referred to [Swa 89].

4.1. EVENT ANALYSIS

It is stated before that the emphasis of the dynamic expert scheduler is put on its ability to make the right decisions in case of events during manufacturing so that the process can proceed either without or after rescheduling. This decision largely depends on the kind of event and on the cell and machine status data at the time of occurrence.

In order to demonstrate the applicability of a dynamic expert scheduler, in WP2 only a few events were considered. However, the final system must be able to handle any event that could occur during manufacturing. For this purpose, the number of events needed to be extended and the event handling to be adapted.

There is a great variety of types of events which can be represented hierarchically. All events, coming from various levels, are passed over by the Monitoring and Diagnostics System. The following hierarchy from lower to higher levels can be distinguished:

- shopfloor
- cell (e.g. increasing delay, built-up small delays, jobs getting too late)
- tool/material management
- work preparation
- workload balancing (e.g. workload of machines, bottle-neck problems)
- planning/production control (new orders, high priority orders)

In [Swa 89] an extensive list of all events to be handled by the expert system is enumerated. Here we restrict ourselves to only a few remarks. In the work thusfar emphasis has been put on shopfloor events. Most of them concern tools (tool failure, tool breakage, tool wear, waiting for a new tool), machines (machine breakdown) and resources (resource not available).

Events at cell level concern the jobs and their progress made thusfar. For instance, there could be a slowly increasing delay which would eventually necessitate rescheduling. Other events on this level are jobs coming too late, material not yet available etc.

At this level new (rush) orders can be entered.

4.2. DECISIONS TO RESCHEDULE

Once an event has occurred and analysed by the expert system it has the task to examine the necessity to reschedule. Some events are only of minor importance, for instance an event causing a minor delay. It is evident that these events will most probably not lead to rescheduling. In such a case the expert system can decide -without further calculations (that is without the need of having the Scheduler simulated a forecast under the current conditions)- that rescheduling will not be necessary. However, it is also possible that there is reasonable doubt about the consequences of an event, and then the expert system needs to further examine the consequences. This could only be done by having the Scheduler generated a forecast. The consequences of the disturbance are made explicit and are subject to examination by the expert system. The result may be that the consequences are less serious than earlier assumed and the current dispatching strategies would not need modification. However, if the the analysis shows the inevitability to reschedule, the decision to do so has to be made also based on the characteristics of the earlier rejected workplan.

So, except for obvious minor events, the expert system will start its examination by activating the scheduler to supply a forecast based on the present status. The new workplan with its quality values generated by the scheduler provides all necessary data for the expert system to conclude about the necessity to reschedule.

4.3. DETERMINING CELL AND MACHINE DISPATCHING STRATEGIES

The current workplan is based on a list of dispatching strategies for both cell and machines. There is quite a number of well-known dispatching strategies (in the literature also known as dispatching rules), [Bla] enumerates 34 of them. Thusfar we have taken into account about 15 dispatching strategies, e.g. First In First Out, Earliest Due Date, Longest (Shortest) Remaining Processing Time, Most (Fewest) Remaining Operations, Longest (Shortest) Imminent Operation, etc. The circumstances determine which dispatching strategy is expected to be best for the performance. The corresponding decisions are contained in the knowledge base.

4.4. WORKPLAN ANALYSIS

Once the Scheduler has generated a candidate workplan (as we have earlier indicated, this could also be a forecast based on the current dispatching strategies) it is the expert system's task to analyse its quality: are due dates met. If the analysis shows that the candidate workplan's quality is satisfactory, then the Scheduler is allowed to pass it over to the dispatching module. In case the candidate workplan has been based on a new set of cell and machine dispatching strategies, job sequences could differ quite a lot from those of the current workplan. However, even when the dispatching strategies have not changed, the jobs could and quite often do have different sequences.

What happens when the current workplan appears not to be satisfactory. As said before, the expert system has to generate a new set of strategies. The expert system would do a bad job when new dispatching strategies are selected at random, thus without taking the circumstances into account which have lead to rejection. Then a series of newly generated workplans will in no way tend to converge. A better approach would be that these circumstances influence the selection of the dispatching strategies. Since no algorithmic procedures are available to do this job, the introduction of heuristics is mandatory. Therefore, for this part the knowledge base contains some heuristic rules that will help to make a more intelligent choice of new dispatching strategies.

4.5. DECISIONS TO TURN CANDIDATE WORKPLANS INTO CURRENT WORKPLAN USING COST FUNCTION

It may be that a series of newly generated candidate workplans do not meet the requirements. One after the other may have been rejected for a variety of reasons. For only a few machines and jobs the total number of possible workplans is too large to continue secting new sets of dispatching strategies until a satisfactory workplan is found, if there exists any. Therefore, after a few tries (could be any number) the selection process is stopped leaving the expert system with the task to select the best one out of the number of rejected workplans (again, also including the current workplan). The approach chosen here is to calculate a cost function for each of them and to take the one with minimal costs. Of course, now the heuristics lay in the definition of the cost function and not in the knowledge base. Yet, it is the expert system's task to control this process.

5. Integration of the dynamic expert scheduler in production control

The dynamic expert scheduler organises the order sequencing in a Flexible Manufacturing System (FMS). The definition of a FMS is used here in a more general way than normally used. It is a group of different machines which build up one organisational unit.

The different machines are alternatives for the same process steps or are chosen for other process steps to form a line of consecutive processes. The work of the dynamic expert scheduler is the more effective when there are possible definitions in order sequencing left.

In a medium-sized machine tool company the scheduling problem is realised in two or three levels. The first level is a very rough level, mainly for the schedule of the sales office based on a first level list of parts.

The second level is based on the final list of parts and the defined steps and durations in the process plans. The result of a weekly or even daily *termination run* in a commonly used MRP System is a list of process steps from different orders or process plans sorted by the machines or machine groups, see figure 4. All steps possess calculated due dates.

Normally the dispatcher is left with all the process steps for several machines in a time period of one day to one week.

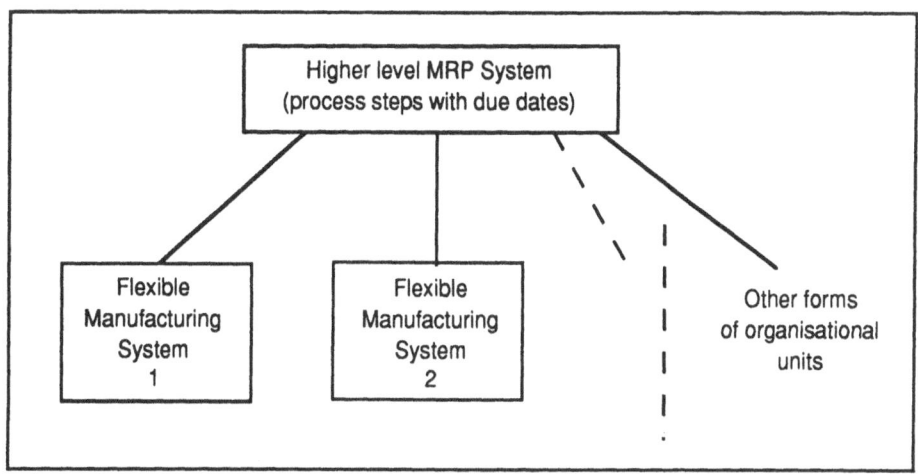

Figure 4. *Configuration of production control*

The dynamic expert scheduler assists the dispatcher on a decentral computer in the cell to organise his work in an optimal and yet flexible way.

The dynamic expert scheduler gets its data (production orders, process plan and process steps) from the MRP system and transfers the information about the processed work back to the MRP system. Within the cell it tries to optimise the sequence order in such a way that all boundaries from the MRP system are fulfilled, but also in a very cost-reducing way by reducing presetting times and optimising reactions on problems in man and machine capacities.

Another aspect, though somewhat out of the context of this paper, still needs to be mentioned. The dynamic expert scheduler has been made a useful and integrated part of the production control system by having it used the central database. All components of the production control system jointly contribute to its complex task by using the central database to or from which data are written and read respectively, see figure 5. For this purpose special database access software has been written by ICL.

From the expert system's point of view some additional software had to be developed allowing it to communicate with the central database: initial cell and machine information, events with their additional information, the Scheduler's workplan and additional quality values etc, are read from and dispatching strategies are sent to the central database. For this purpose the expert system intensively uses Delfi2+'s (the expert system shell used, for more detail concerning Delfi2+ see [Swa 89]) tasking mechanism.

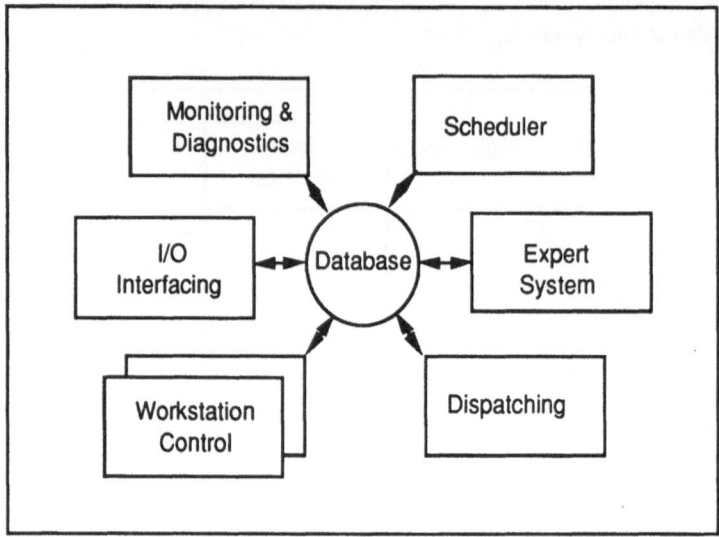

Figure 5. *All components communicate through a central database*

6. Conclusions and results

The dynamic expert scheduler is a powerful instrument for optimising the order sequence in a Flexible Manufacturing System. Other production control systems have a fixed due date- or machine utilisation-oriented strategy and therefore cannot be optimal for all realistic situations in a cell. Contrary to these systems the dynamic expert scheduler emphasises its flexibility of using very different dispatching strategies in different situations. Also new dispatching strategies, which are typical for the special products being manufactured, can be easily applied .

Another advantage of the dynamic expert scheduler is the real-time planning. The cell organisation is very near to the actual manufacturing. That is, all disturbances which occur in the course of the day -short-term breakdowns of machines, breaks of tools or fixtures, missing man capacity for presetting, material not available, damaged parts or parts of wrong quality- touch the cell planning in some way. The dynamic expert scheduler works on the very actual data and is quick enough to make a new plan in a few minutes. On the other hand -contrary to an automatic dispatcher system which makes the decision about the next work on a machine in the last minute- the operator has a plan for the whole period and can follow the plan as long as he wishes.

The cell load can be controlled by the (human) dispatcher in considering the long-term quality factors of the Scheduler. In this way he can optimise the term *sum of processing times/duration of planning period*. Also other values can be followed, e.g. throughput times and utilisation rates.

Another great advantage for the acceptance of the dynamic expert scheduler is that only simple dispatching strategies are combined. The user always knows why the decision has been made to change the plan. Also, he has the influence to make the decision by himself on display.

7. Acknowledgments

The authors wish to acknowledge all who have made this publication possible. Special thanks are contributed to Robert Sever who did most of the Scheduler programming and to Wybo Mentink and Ben van der Pluym who have put much effort into implementing and integrating the expert system.

8. References

[Bla 83]   Blackstone Jr. JH, Phillips DT, Hogg GL. *A state-of-the-art survey of dispatching rules for manufacturing job shop operations.* Int. J. Prod. Res. Vol 20, no. 1, pp 27-45. 1982

[Hei 87]   Heinz-Fischer H, Los R, Poels H, Pohl C, Ringel J, Stienen H. *Expert Supervisory Control of a Flexible Manufacturing System.* Proc. of the 4th Annual Esprit Conference. Brussels, September 28-29, 1987.

[Tie 88]   Tiemersma JJ, Kals HJJ. The Design of a Monitoring and Control System for Small Batch Manufacturing. Proc. of the 20th CIRP International Seminar on Manufacturing Systems. Tbilisi, 1988.

[Tie 89]   Tiemersma JJ, Kals HJJ. *The Design of a Machine Interface for on-line, Real-time Control within Flexible Manufacturing Systems.* (To be published).

[Swa 89]   Swaan Arons H, Mentink W, Pluym B van der. *Knowledge-based On-line Decision Taking in Production Scheduling. (to be published).*

Project no 932

COOPERATING EXPERT SYSTEMS AS CIM MODULES

W. MEYER
Philips Forschungslaboratorium Hamburg GmbH
Vogt-Kölln-Str. 30
D-2000 Hamburg 54
FRG

Abstract

The paper presents a unified treatment and classification of cooperation problems in factory management. Based on a generic CIM model, a coordination algorithm is implemented using qualitative and quantitative symbolic temporal representations.

Key Words: distributed artificial intelligence, problem solving, temporal representation, CIM reference model, production management systems, constraint satisfaction.

1. Origins of Cooperation Problems in Manufacturing

1.1 Complexity: the Need for Coordination

Any task can be broken down into a number of subtasks and operations. Such task disaggregation allows parallel processing and thus translates directly into increased productivity. When different subtasks are performed by different workers the labour itself has become divided. Today, further division of labour is restricted by coordination costs which exceed the productivity gain.

If we treat the specialization of (expert) knowledge in a similar way as an economic resource (Fig. 1) we may formulate the coordination question congruently in three different ways. It is relatively straightforward to explode the task into a thousand subtasks: how difficult and costly is it to assure their proper sequencing, scheduling and interaction - over a period of time? It is quite customary to divide the labour among hundreds of 'incomplete' workers: how difficult and costly is it to maintain their coordination, motivation and performance? We are used to dividing information into billions of tiny bits: how difficult and expensive is it to achieve its requisite integration, record and update?

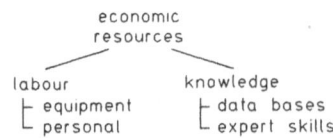

Figure 1.   Knowledge is the most important economic resource today.

The answer to all three questions is that it is progressively more difficult and progressively more costly. As the complexity and costs of integration and coordination become too large, we turn, through necessity, to questions of reintegration.

In this context, the JIT efforts today which aim at flow-oriented site layout, concern the reintegration of physical labour (machines) via flow lines, whereas CIM anticipates the reintegration of special knowledge organized in functional departments by integrated information processing (Fig. 2). Carriers of this integration are product, process and resource models exchanged between the departments. In fact, the amount of shared knowledge determines the degree of cooperation between team tasks. The cooperation costs of Fig. 2, therefore, are communication costs.

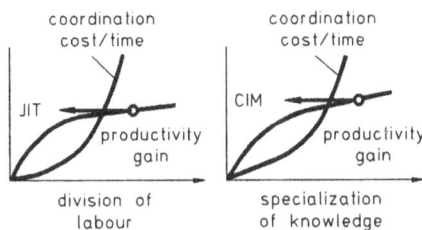

Figure 2.   The two basic approaches to the diversification problem:
             – integration of physical labour (JIT)
             – integration of knowledge contained in functional departments (CIM).

1.2 Complexity Plus Uncertainty: the Need for a Distributed Control system

To cope with task complexity, coordination is required. But the dynamics and uncertainties of the environment afford planning and control. In a multilayer structure the determination of control is split into algorithms which operate on different time scales. Generally speaking, the breakdown of long-to-short term planning decisions indicates levels in the complexity of decisions: with complex decision problems the resolution is sought in a hierarchical approach. A family of decision problems is defined and solutions are generated in a sequential manner, top-down.

Fig. 3 illustrates the general truth that manufacturing complexity and environmental uncertainty afford a distributed hierarchical control system. Complexity and uncertainty are two opposing forces: complexity forcing task distribution ultimately into a heterarchical, functional, horizontal structure; uncertainty pushing in the opposite direction, vertically stretching tasks into a more hierarchical structure. Such a two dimensional control schema is called an organization.

1.3 Managing Complexity and Uncertainty: the Organization

Actions taken by any manager at any level (Fig. 4) are classified as
– planning actions: allocating human, financial, or material resources to activity units oriented towards gols
– control actions: reducing the distance between goals and results of activity units

– coordinating actions: maintaining the integrity and consistency of decisions made by serveral managers engaged in solving a common problem.

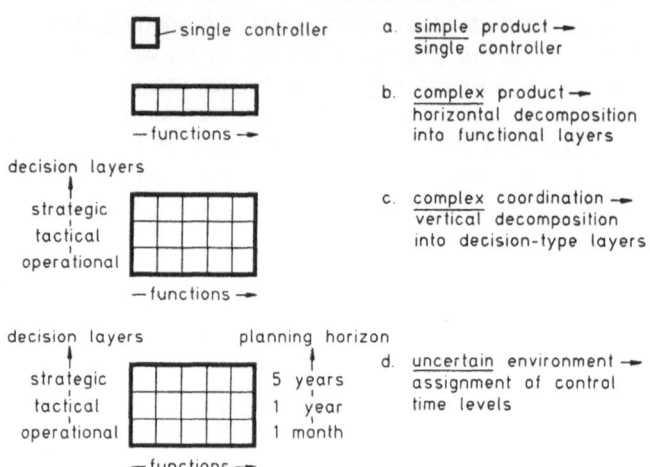

Figure 3. Evolution of the organization as a distributed controller network:
(a) Simple product, simple process, 1 controller.
(b) Product complexity leads to functional decomposition.
(c) Complex coordination task affords specialization of decision making.
(d) Uncertain environment necessitates hierarchical control.

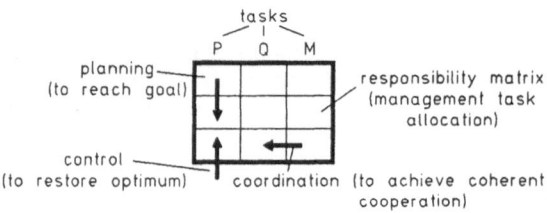

Figure 4. The organization as a distributed management system.
P,Q,M: production, quality, maintenance task.

In Fig. 4 (and in ESPRIT project EP932) we have restricted our scope to the three functional manufacturing areas production, product quality and machine maintenance. Each matrix element of the organizational matrix in Fig. 4 is a manager, either supported or possibly even replaced by computer systems, and performing the basic management activities of planning, control and coordinating (Fig. 5).

```
management
 ├ coordination (of functions and units)
 ├ planning (of action sequence to reach goal)
 └ control (to keep optimum performance)
    ├ monitoring (of plant)
    ├ interpretation (of plant information)
    ├ modelling (of plant state)
    ├ simulation (of plant dynamics)
    ├ diagnosis (of critical events)
    ├ action planning (to restore optimum)
    │  ├ alternative plan conception
    │  ├ alternative plan configuration
    │  ├ alternative plan simulation
    │  └ plan selection
    └ action triggering
```

Figure 5.   Basic management activites.

2. The Organization as a Distributed Problem-solving Network

2.1 Problem Classification

The main factory objective is, of course, to manufacture products through (limited) resources under time and cost constraints. In EP932, we mainly considered resources for production P (labour, workstations, 'blue collar robots') as well as for quality Q and maintenance M (knowledge, experts, 'white collor robots'). The resources are managed by the organization: the organization is a group of individuals and computers whose purpose is to manage tasks in an attempt to achieve a set of goals while observing a set of constraints.

Two classes of problems are prevalent at the factory:

– personalizable problems or task-related problems, which require a high amount of technical expertise and where a single most fitting answer is available (example: diagnosis); and

– collaborative problems or organizational problems, which may involve group consensus and committment (example: management boards), information and opinion exchange (example: conferences), multiple knowledge sources (example: marketing strategy), or balancing of conflicting goals (example: budget placement).

The basic problem in *planning* is, of course, the high data accuracy needed for longterm forecasts as opposed to the actual customer demand and the realtime capabilities of the production facilities. This holds true equally for production, quality and maintenance planning; in fact, the dynamic scheduling problem is even more severe in product and machine repair where duration and cost of repair jobs are difficult to estimate. However, principally the same planning and scheduling procedures can be applied in all three functional areas of manufacturing which only differ in the nature of resources to be managed: workstations, processes and (mainly physical) labour in production, and (mainly human) test and repair experts in maintenance and quality assurance .

In production *control*, main activities as outlined in Fig. 5 are plant monitoring and supervision, interpretation of plant information, modelling and simulation of plant and plans, diagnosis of critical events, and action planning to restore optimum order performance according to the original plan. Main problems here are timeliness, consistancy and relevance of information, as well as the sheer amount of data which must be interpreted in relation to the control problem at hand.

Finally, *cooperation* problems remain the most severe malfunctions detected in the factory. Sufficiently coherent cooperation between the PQM-functions has not yet been achieved (not to mention others like CAD/CAM etc). Coordination between most tasks of production, quality, maintenance is nonexistent, vague, occasional, informal or unstructured. Coordination tools like information networks and databases provide an enormous amount of data, but mostly unstructured and not accessible via user friendly interfaces. Coherent cooperation can be improved by several means (Fig. 6). These are classified as
- top-down goal oriented: goal sharing, decision frames
- bottom-up knowledge oriented: knowledge sharing
- communication oriented: message passing regarding events and plans.

```
coordination by
    ├─ shared knowledge bases (states)
    ├─ communication (events)
    ├─ meta communication (plans)
    ├─ responsibility (goal sharing)
    └─ authority (decision frames)
```

Figure 6.   Techniques and means to improve coherency in cooperation.

Coordination policies determine the right balance between these means thus providing coherency between specialist functions. Coherency depends on the quality of information available for each decision maker, i.e. it depends on
- relevance of information
- timeliness of information
- completeness of information.

Communication policies are trade-offs between these three. Such policies are seldom clearly defined on the shop floor. What could then be expected concerning cooperation policies which are not even regarded as being necessary but just neglected?

The reason for this is the lack of a sound theory of cooperation among intelligent agents. Artificial intelligence or better still: distributed AI may offer new approaches to this problem.

3. AI for Distributed Problem Solving

3.1 Distributed Artificial Intelligence: Definition

Problems addressed by factory management are of different types:
- interpretation problems (example: data reduction)
- constraint satisfaction problems (example: job scheduling)
- optimization problems (example: set-up time minimization).

However, they are all related to decision making and – they are all related to information deficiencies: either too little, too much, too late, etc. We are not astonished about that, as problem solving depends on *understanding* the problem. In fact, problem solving follows the generic cycle of understanding-and-planning (Fig. 7). With other words, problem solving means acquiring and structuring (interpreting and diagnosing) information about the real world and planning actions to change that world.

Problem solving is very much related to decision making: both activities convert information into action. The differences come about when comparing the definitions as coined by decision theory and AI, respectively:

(1) decision making uses a formal, prescriptive framework for making logical choices in the face of uncertainty; and

(2) problem solving is the art of using relevant knowledge in the attainment of desired goals.

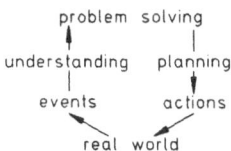

Figure 7.   The problem-solving cycle.

Modern AI techniques are very well adapted to cooperative management styles as favoured by advanced industrial societies. AI techniques like model-based reasoning (Fig. 8) can help to improve integration, cooperation and problem solving within such an environment.The concepts of distributed artificial intelligence will especially support this aim: distributed knowledge sources, distributed control, communication, coordination and organization are all covered by one framework (Fig. 9).

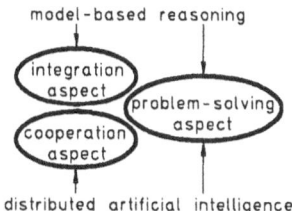

Figure 8.   Organizational aspects covered by AI techniques.

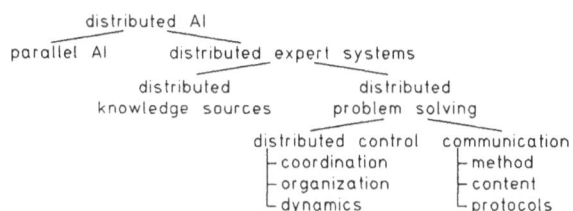

Figure 9.   Distributed artificial intelligence taxonomy.

3.2 Distributed Control

In distributed AI, obtaining global coherence with decentralized control is the main task. Controller nodes must cooperate to exploit and coordinate their answers to interdependent subproblems but must do so with limited interprocessor communication.

The coordination task is intimately linked with the general problem of uncertainty: each cooperating agent (controller) must effectively deal with
- environmental uncertainty: not having an accurate view of the number and location of processors, affectors, sensors, and communication channels
- data uncertainty: not having complete and consistent local data at a node
- control uncertainty: not having a complete accurate model of activities in other nodes.

Research in these three areas draws heavily on the work in knowledge-based systems and organization design. In EP932 we deal with the aspect of control uncertainty: how to achieve and maintain coherent behaviour among autonomous agents (controllers, managers, interest groups). Autonomous systems are located somewhere between fully cooperative and fully antagonistic systems (Table 1). Fully cooperative systems, which are able to tackle nonindependent subproblems, often pay a price in high communication costs. Fully antagonistic systems may not cooperate at all (including blocking each others' goals except where similar goals are being pursued) and often have no communication costs.

Table 1. Cooperation characteristics of distributed systems.

dimension	spectrum of values		
system characteristics	cooperative	autonomous	antagonistic
communication cost	high	medium	low
amount of control low	medium	high	

Most real systems are cooperative to some degree and lie in the middle of the spectrum. In a distributed problem-solving network such as a CIM system, a balance between problem solving and coordination is desirable so that the cost of both is acceptable. Table 1 shows that communication and control are interchangeable to a certain degree. In fact, communication (message passing) is one means to achieve coherent cooperation, and control (authority) is another, as has been indicated in Fig. 6. Coordination policies determine trade-offs between these means. They are seldom clearly defined as no sound theory of cooperation is available yet. Several key ideas of distributed AI, which might form the nucleus of such a theory are
- contract networks
- functionally accurate, cooperative networks (FA/C)
- organizational structuring.

Different organizational structures match different problem situations and performance requirements. It is the task of organization theory (OT) to analyse generic organizational classes to determine their strengths and weaknesses with respect to processing, communicaiton, coherence, and flexibility (for a detailed account, the reader is referred to Fox (1981), for instance). There is great overlapping between organization theory and the evolving discipline of distributed artificial intelligence.

3.3 Communication Versus Shared Knowledge Bases

The nature of communications in a distributed problem solver is probably its most important aspect. With increased coordination and cooperation comes an increasing amount of information – therefore, techniques that keep communication to a minmum, are always desirable. Out of the spectrum of communication methodologies we only deal with the *paradigm* by which communication takes place; for a complete review see Decker 1987.

The *paradigm* by which communication takes place in distributed problem solvers is either shared global memory, message passing, or some combination of the two. These methods have advantages and disadvantages, and many systems use both by providing shared memory communication for processes working locally (or working as a team on one subproblem) and providing message passing for communications between processors or teams of agents.

Much has been written about communication through shared global memory and how it is used in both simple and distributed problem solving. The model used most often is the blackboard model where the shared memory is viewed as a blackboard on which to write messages and find information. A blackboard is usually partitioned into levels which are used for different representations or abstraction layers of the problem at hand. Agents working at a particular layer observe a corresponding level of the blackboard along with the adjacent levels. In this way, data that have been synthesized can be communicated to the higher levels, while higher level goals can be filtered to force to the expectations of lower level agents (or experts) up. The simplicity with which this paradigm represents the classic problem of data-driven versus goal-driven information flow is perhaps the reason that blackboard models are used in one way or another in almost every existant distributed problem-solving system. However, blackboard systems are unfeasible for real distributed AI systems because if only a single blackboard exists, it becomes a bottleneck, and if several exist, the semantics of the situation revert to message passing.

The history behind message passing is also a long one; object-oriented languages implement the message passing paradigm among objects which exist in parallel. From a pragmatic point of view, message passing offers a more abstract means of communication than simple shared memory, and its semantics are well understood. Agents in a message-passing system do not directly affect one another, so they retain the objectlike quality of independent agents. Message passing is easier to program (because it is more abstract) and possibly more efficient (if one can exploit locality of data). Therefore, object-oriented programming is regarded as the most important progress in software engineering throughout the last 15 years (Saunders 1989).

As always, a compromise is the best solution: either one blackboard being implemented with object-oriented languages, or several blackboards being integrated via message passing. These are the approaches we favour in EP932; for further details see Isenberg (1988), Isenberg and Hübner (1989).

4. Specification of the Distributed Controller Network

Throughout the previous sections we have viewed the factory organization (and its subset: the production management system PMS) as a distributed problem-solving network in two different ways:
(1) Problem decomposition (top-down, centralized control): The PMS network is regarded as a system for decomposing a problem and assigning subproblems to its various functions. This view stresses intelligent hierarchical control and the planning aspect.
(2) Communicating problem solvers (bottom-up, decentralized control): The network is viewed as collection of decentralized problem solvers that can communicate. This

second view emphasizes intelligent local control of each individual probelm solver (who decides which local tasks to pursue in order to best contribute to network performance) and the cooperation aspect.

Both organizational modes exist in factories. The first one needs high data accuracy, the second one high coordination efforts to attain coherence. Local control allows locally optimal use of resources but does not guarantee overall optimality, and vice versa. But when problem decomposition and task load are not known ahead of time, one is happy to have a system that runs at all, and optimality is not a central issue. The dynamic and volatile aspects of the factory lead us to value the flexibility in the face of change that local control offers. A production management system, therefore, is a trade-off between solution (1) and (2).

The PMS consists of a number of CIM modules (Fig. 10). The CIM modules (for production, quality, maintenance etc) are generic in that sense as they implement the generic functions of model-based problem solving (compare Fig. 7). For a detailed elaboration see Meyer et al. (1988).

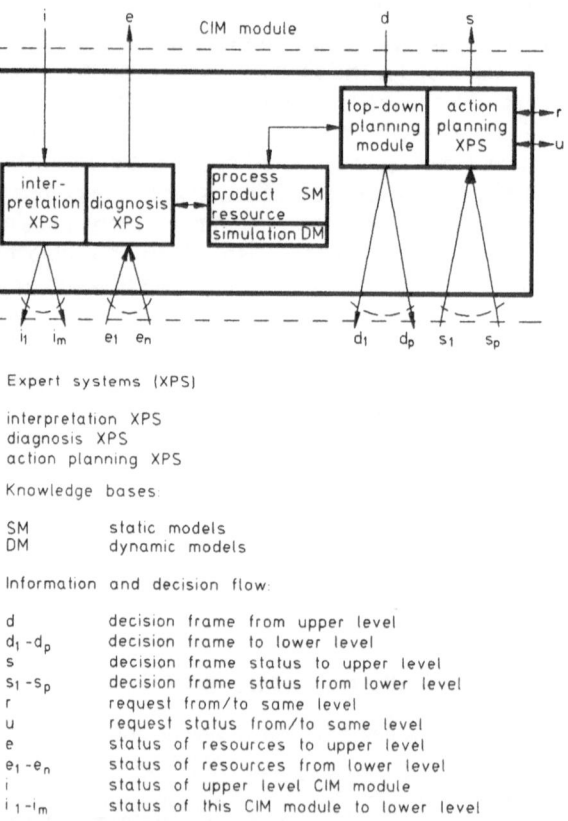

Expert systems (XPS)

interpretation XPS
diagnosis XPS
action planning XPS

Knowledge bases:

SM	static models
DM	dynamic models

Information and decision flow:

d	decision frame from upper level
$d_1 - d_p$	decision frame to lower level
s	decision frame status to upper level
$s_1 - s_p$	decision frame status from lower level
r	request from/to same level
u	request status from/to same level
e	status of resources to upper level
$e_1 - e_n$	status of resources from lower level
i	status of upper level CIM module
$i_1 - i_m$	status of this CIM module to lower level

Figure 10. Generic CIM module architecture

Horizon/Period	Function	Sales Management	Operational Management			Resource Management			Product Design
			Production	Quality	Maintenance	Material	Personnel	Equipment	
COMPANY/FACTORY	4 years / 1 year		strategic plan *qty/prod.family start & end date factory*						
	1 year / 1 month	order forecasts *qty/prod.fam./month*	general master plan *qty/prod.type/month*	quality master plan *products, equipment*	master plan for maintenance *maintenance procedures & availability effort/w.station*	vendor capacities *qty/cpnt/month* orders for common parts *qty/cpnt/month*	forecasted manpower capacities *manpower/qualification/month*	equipment policy, machine, layout *w.station specs., release dates*	development plan *prod.type specs., release date*
SHOP	6 weeks / 1 week	customer orders *qty/prod.type, customer priority*	master plan *qty/prod.type/week*	quality information *quality level/prod.type/week, quality level/w.station/week*	maintenance plan *maint.orders/w.station/week*	short term orders, external subassys *qty/cpnt/week*	allocation of people *manpower/qual./workcell/week*	installation & modification of equipment *installation plan/w.station.date*	request for product modification *prod.type modification, release date*
	3 days / 1 day		released shop orders *qty/assembly, due date*	detailed quality analysis *quality level/prod.type/day, quality le-vel/w.station/day*	released maintenance orders *maint.orders/w.station/shift*	reserved parts & subassys *qty/cpnt/week*	allocation of people to JIT or subassy production *manpower/qual./workcell/shift*	reservation of equipment *w.station/shift*	
WORK-CELL	3 shifts / 1 hour		sequenced production orders	quality data per w.station *quality level/w.station*	machine status & maintenance actions *maint.orders/w.station*	picking list *qty/part*	adjusted allocation *people/qual/w.station*	tooling plan *tools/w.station/shift*	alternative components list *qty/cpnt/shift*

Figure 11. Organization modelling: GRAI information grid.

Horizon /Period	Function	Sales Management	Operational Management			Resource Management			Product Design
			Production	Quality	Maintenance	Material	Personnel	Equipment	
COMPANY/ FACTORY	4 years / 1 year		determine business golas						
	1 year / 1 month	forecast market requirements	balance market requirements & production resources	determine & analyze quality procedures	determine maintenance procedures	define vendor capacities & procure common parts	determine manpower & qualification	define equipment policy & develop new equipment	develop new products
SHOP	6 weeks / 1 week	check customer orders	determine master production schedule	collect & analyze quality information	analyze machine performance & plan maintenance	order short term components & external subassys	plan allocation of people	plan installation & modification of equipment	modify existing products
	3 days / 1 day		release shop production	check & analyze samples	release maintenance orders	reserve parts & subassys	allocate people to JIT or subassys production	reserve equipment	decide on alternative subassys
WORK-CELL	3 shifts / 1 hour		sequence production orders	measure & check product quality	diagnose machine faults & repair machines	pick material from stores	adjust allocation	set up machines	decide on alternative components

Figure 12. Organization modelling: GRAI decision gird.

The CIM modules for the manufacturing functions production, quality, maintenance etc. are arranged in a two dimensional matrix according to Fig. 4. The decomposition criteria for the horizontal levels are temporal: planning horizon and planning update period, as introduced by the GRAI method (Doumeingts 1987). The GRAI factory analysis and design method describes the factory organization in terms of a layered information and decision model sometimes called CIM architecture. The CIM architecture of a typical factory in the electronic mass production sector (car radios) is shown in Fig.s 11 and 12.

Each matrix element in Figs. 11,12 is an activity center whose control activities are described, supported and in some cases even fully conducted by a CIM module according to Fig. 10, be it a human decider or a computer system.

To enable the coordination of these distributed decision activities the CIM modules are connected by two different types of links to a distributed decision network (Fig. 13). These link types are the decision frame and the request link. Decision frames establish the hierarchical relation between CIM modules of different levels of the production control system, whereas request links enable the cooperation of CIM modules of the same hierarchical level.

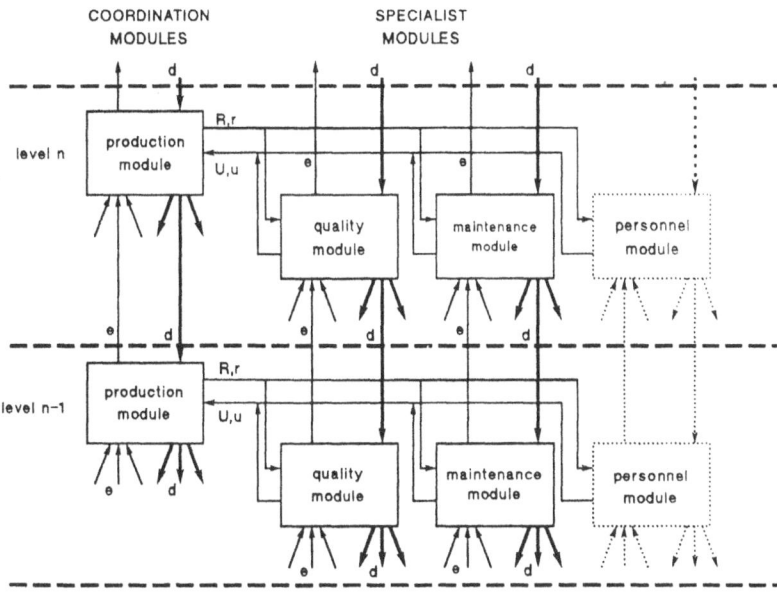

Figure 13. Distributed network of CIM modules, coordinated via decision frames d, requests r and status messages u,e.

Level n in Fig. 13 may refer to the shop level, n-1 to the workcell level, etc. Within one hierarchical level, each CIM module has a certain local autonomy to carry out its own control decisions. However, the decision activities are coordinated by the CIM module of the production planning function. This CIM module is called Coordination Module. The role of the coordination module is to guide the decision making process of the other CIM modules of the same level in that way that their decisions support the overall mission and the overall decision objectives relevant for this PMS level.

The CIM modules guided by the coordination module are called Specialist Modules. The scope of a specialist module is restricted to one aspect of production control (e.g.

material management, maintenance, quality control).

CIM modules of the sales function play a special role in the PMS. Assuming that the production system produces products for a buyer market, market requests can be sent to the coordination module via the sales CIM module. The sales CIM module is the interface function between the customers and the coordination module.

The only requests which should be received by the coordination module are sales-related requests like forecasts or customer orders posed by a sales CIM module. To all other specialist modules the coordination module sends requests. These specialist modules have to fulfill the requests within the decision frame given by their superior specialist module. Otherwise they reject the request sent by the coordination module.

In summarizing, vertical integration among CIM modules is achieved via decision frames, horizontal integration via requests. Both are efficiently represented by temporal nets.

5. Temporal Networks for Cooperative Problem Solving

5.1 Fundamentals: Symbolic Time Interval Logic

It is hard to think of a technical area that does not involve reasoning about time in one way or another. Of course, the passage of time is important only because changes in the physical world are possible. Therefore, a general theory of time must meet two requirements. The first is that it provide a language for describing what is true and what is false over time. The second is that it provide a method to infer change, i.e. to relate action and time.

Causation, action and time have been studied for many years outside and before AI. Previous work has been shortly summarized by Allen (1983) and Shapiro et al. (1988). Important characteristics of any time representation which will prove to be useful in the factory environment are (Fig. 14):

– It should allow signigficant imprecision: much temporal knowledge is strictly relative (e.g., A must be performed before B) and has little relation to absolute dates.
– It should allow uncertainty of information: often, the exact relationship between two times is not known, but some constraints on how they could be related.

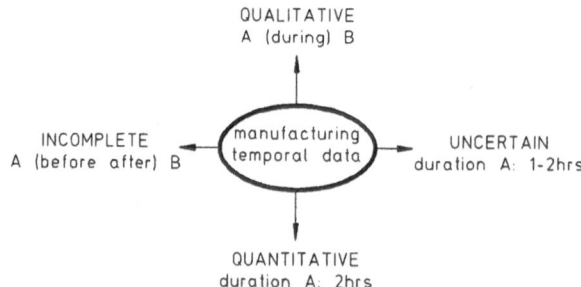

Figure 14.   Temporal information may be uncertain, incomplete, qualitative or quantitative.

– It should allow the consideration of time at all levels of detail: ranging from nanoseconds to years.

– It should allow computationally effective reasoning, i.e. must assure viable implementations.

In EP932, we use a combination of Allen's relational qualitative symbolic logic and Rit's sets of possible occurrences SOPO (Huber and Stephan 1988) which allow the representation and solution of coordination problems caused by qualitative and incomplete temporal data (Fig. 15) and caused by uncertain quantitative temporal data (Fig. 16). Figs. 15 and 16 are taken from a recent elaboration on the topic by Meyer (1990); for further reading the interested experts (factory managers and software engineers) are referred to that book and the paper of Meyer, Bünz and Huber (1989). Basic papers are by Allen (1984) and Rit (1986).

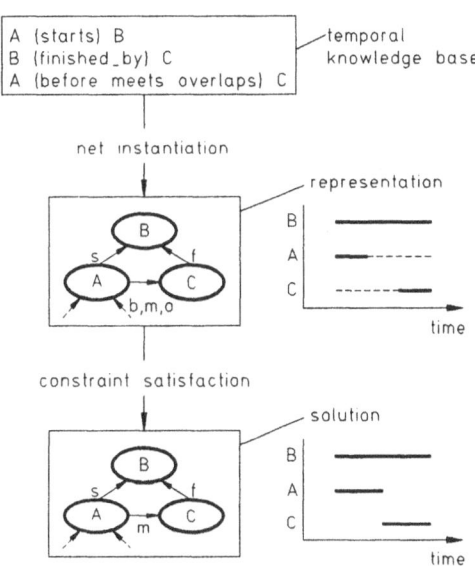

Figure 15.   Qualitative temporal reasoning. A,C may be production orders to be produced within shift B.

5.2 Application

For the representation of plans, events and actions we have adopted the notation of the generalized window. A window (a plan segment) of any time interval is bound by the six independent parameters earliest/latest start, earliest/latest end, minimal/maximal duration. This corroborates with the monthly or weekly plans as they are released for the shop floor by MRP systems like COPICS. The COPICS plan or, more generally, any decision frame is concretized by successive refinement of plan parameters by inserting qualitative (structural, causal and temporal) as well as quantitative constraints (Fig. 17). The two-stage constraint satisfaction process solves the coordination and optimization problems occurring at the shop and workcell level (Table 2) thus providing vertical integration. The time interval representation also provides horizontal integration by considering requests from quality and maintenance departments and others. The concept of SOPOs combined with temporal interval logics, therefore, forms the representational backbone for cooperative problem solving which is always cooperation over *time.*

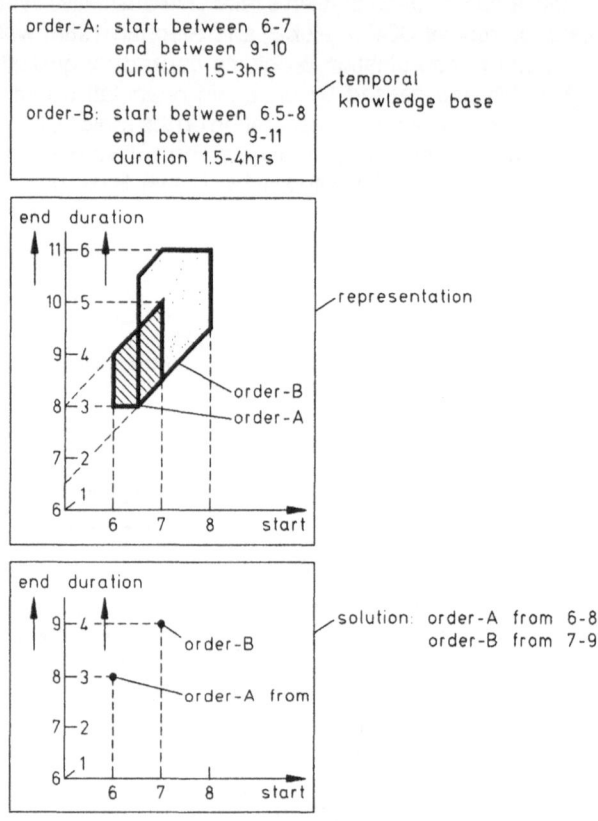

Figure 16.   Quantitative temporal reasoning: Propagation of constraints over SOPOs

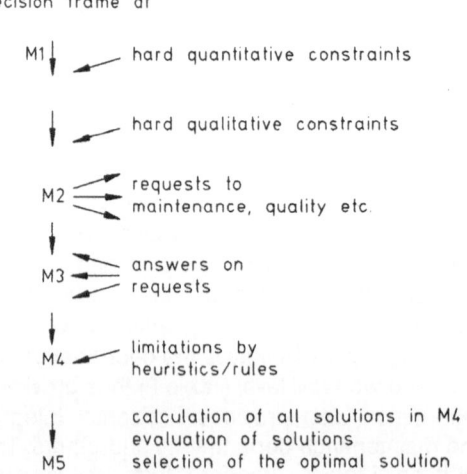

Figure 17.   Incremental constraint insertion.

Table 2. **Decision parameters and constraints determining the organizational tasks at different hierarchical layers.**

organizational	organizational	decision	decision	constraints	result
SHOP	coordination (of several tasks)	1 shift	global goals (flexibility)	qualitative (causal relations)	sequence of tasks
WORKCELL	optimization (of single task)	1 hour	local objectives (cost, time)	quantitative (maintenance times)	schedule of tasks
WORKSTATION	control (of process parameters)	1 minute	performance indicators (overshooting)	mechanical (amplitude)	adjustment of parameters

6. Implementation Example: Horizontal Integration with Temporal Nets

Temporal nets as outlined in section 5 describe the constraints to be considered during the decision making process of a CIM module in a declarative way. According to GRAI (Doumeingts 1987) the decision frame and the requests of CIM modules are the constraints for the decision making. They implicitly describe all elements of the set of feasible solutions. That means that each valid instantiation to be found by the temporal inference engine is a feasible solution of the decision making process. Of course, a feasible solution can be either a good or a bad solution concerning the given decision objectives.

In the following a problem solving procedure is presented which allows the horizontal coordination and the vertical cooperation in a net of distributed CIM modules. The general principles will be illustrated by an example which shows the cooperation betwen a final assembly line FL2, a pre-assembly line FL1 and the maintenance department. We assume a JIT production where the final assembly line pulls products from the pre-assembly line FL1 (Fig. 18).

Figs. 19 to 24 describe this procedure for coordination modules and specialist modules. The main steps of this procedure are:

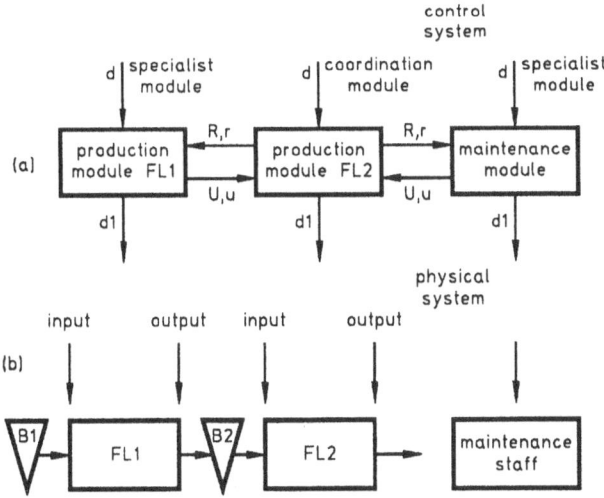

Figure 18. Example: Coordination betwen CIM modules
(a) control system
(b) physical system: B1, B2 buffers FL1, FL2 flow lines (workcells).

704

Step 1:
The coordination module (production planning module "sequence production orders for FL2") receives a temporal net representing the decision frame d (Fig. 19, released orders with quantity, type, start_date, latest_end_date) from the upper level coordination module. It transforms this decision frame d by using its process model (throughput times of FL2) to determine the decision frame d' for the load scheduling. The transformed decision frame d' contains the earliest_start and latest_finished at the loading point of FL2.

d: released orders

```
order  A: qty= 100, type=  A
order  B: qty= 300, type=  B
order  C: qty= 200, type=  C
order  D: qty= 200, type=  D
order  E: qty= 300, type=  E
```

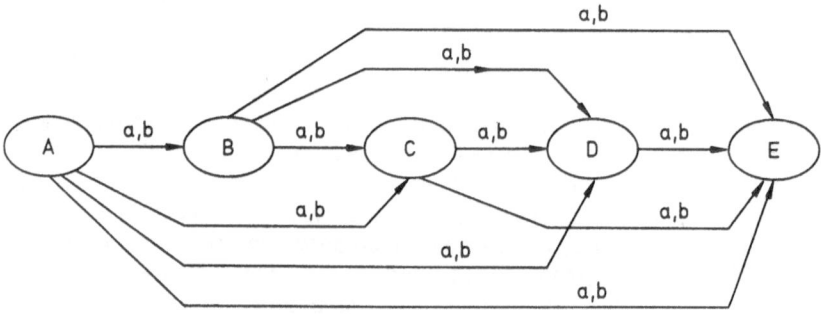

Figure 19:   Decision frame: released production orders.

Thereafter the decision frame d' must be updated. It has to consider the current status of orders (whether they are in process or to be processed). In the example the loading of order E has been finished but some products of order E are still on FL2 and order D is just in the loading phase. Therefore, order E has not to be considered for loading any more. For order D the quantity has to be reduced (Fig. 20).

Step 2:
The coordination module takes this temporal net (its transformed decision frame d') and sends it as the set of admissible requests R to all specialist modules of the same level.

Step 3:
The specialist modules receive this set of admissible requests R. From R they calculate R' which is the set of requests transformed into a notation suitable for the respective module.

In the example (Fig. 21, Fig. 22) the specialist module for FL1 established the relation between the orders for FL2 (e.g. order A') and orders released for FL1 (e.g. order A*). Then it updates its order status and inserts the restrictions U' into the net describing R.

In the same way the maintenance module inserts restrictions (Fig. 23) which reflect the relations between each maintenance order and each production order for FL2.

d': transformed released
orders & order status
at loading point of FL1

order A': qty= 100, type= A
order B': qty= 300, type= B
order C': qty= 200, type= C
order D': qty= 100, type= D

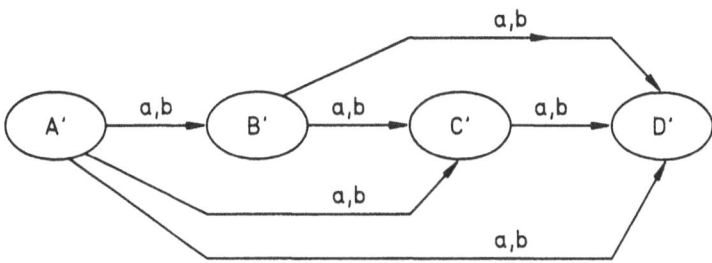

Figure 20. Transformed decision frame d'.

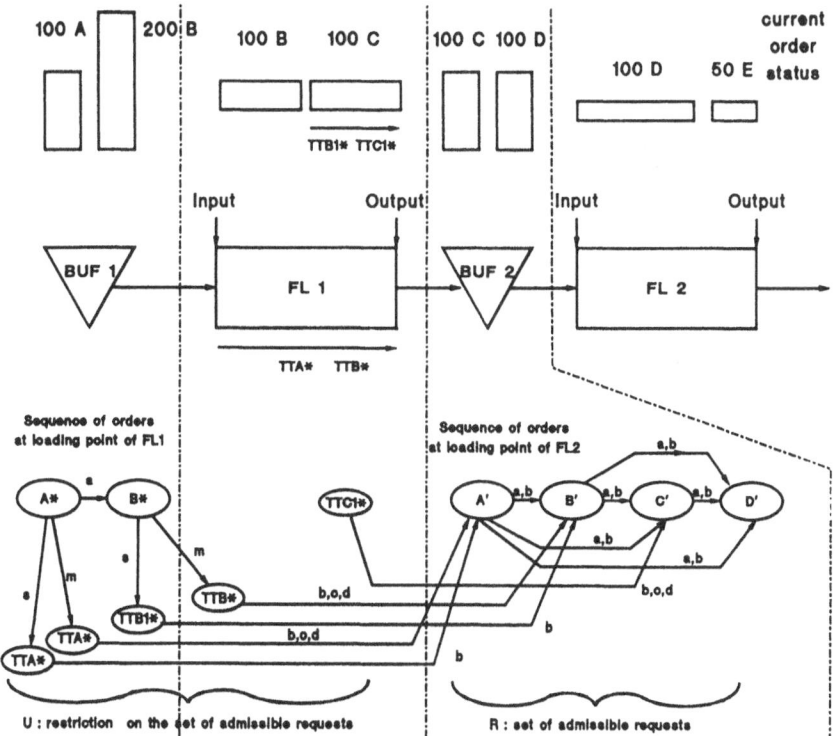

Figure 21. FL1: insertion of restrictions U into the net describing the set of admissible requests R.
TTA* : throughput time order A*
TTB   : throughput time order B
TTB1 : time until the first product of order B* arives at the output of FL1

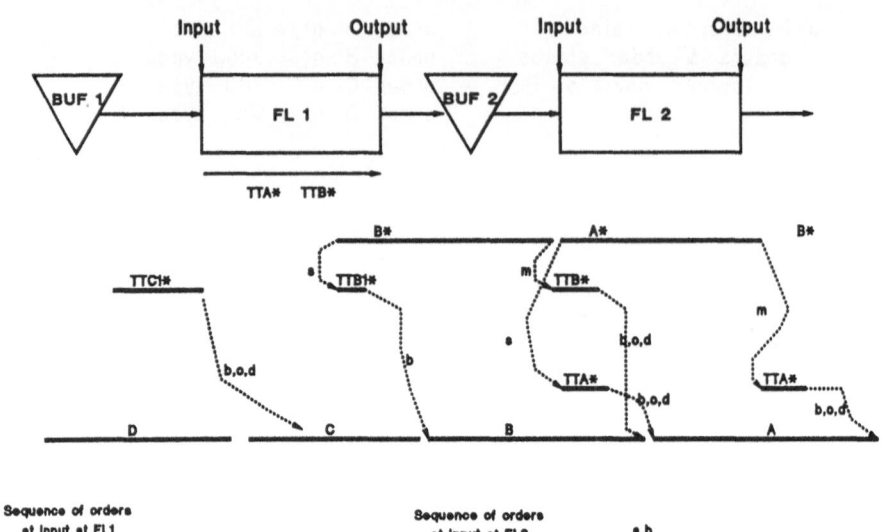

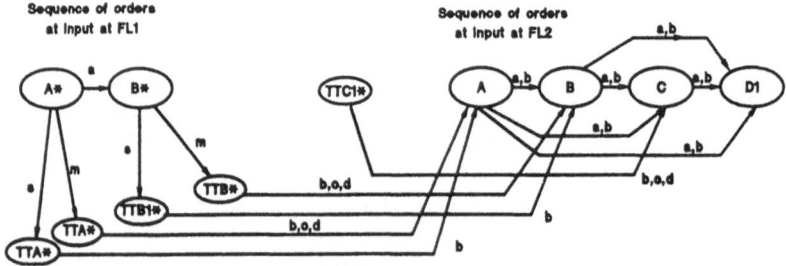

Figure 22. Temporal relations between orders for FL1 and FL2.

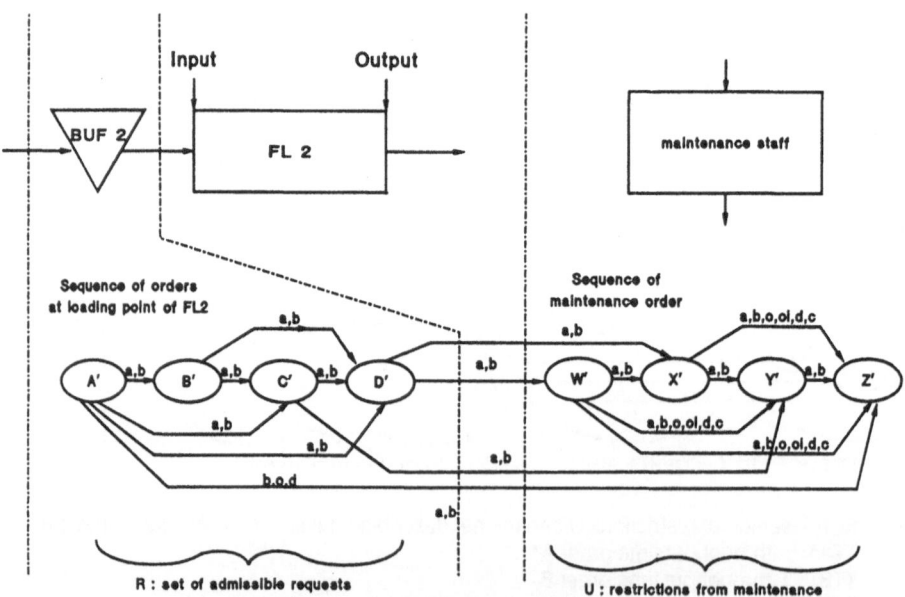

Figure 23. Maintenance: insertion of restrictions U into the net describing the set of admissible requests R.

Step 4:
The specialist modules check whether they can fulfill R' within their local updated decision frames d' considering the current status of resources m'

(e.g. d'   =   released but not finished maintenance orders,
      R'   =   requested maintenance actions,
      m'   =   status of the maintenance resources).

Step 5:
If the specialist module cannot fulfill R' it rejects it and passes a message (set_of_requests_rejected!) to the coordination module and to the corresponding specialist module of the upper level.

    If the specialist module can fulfill R' it determines a temporal net presenting the local restrictions U' which are to be considered for definite request r.
The local restrictions U' reflect the local decision frame d' und the status of resources m'

(e.g. U'   = maintenance_orders + relation_between_maintenance_orders
              + relation_between_maintenance_orders_and_production_orders).

Step 6:
If R' has been rejected by one or more specialist modules the coordination module rejects the decision frame d on its own part and sends a status message (decision_frame_rejected!) to the coordination module of the upper level. The upper level coordination module has then to provide a new decision frame d.

Step 7:
Otherwise the coordination module inserts all restrictions U' receive from the specialist modules into the temporal net describing R (Fig. 24). Then it uses its temporal logic inferencing component to find one element of (R n U). This instantiation of the temporal net should optimize the overall decision objectives of this organizational level.

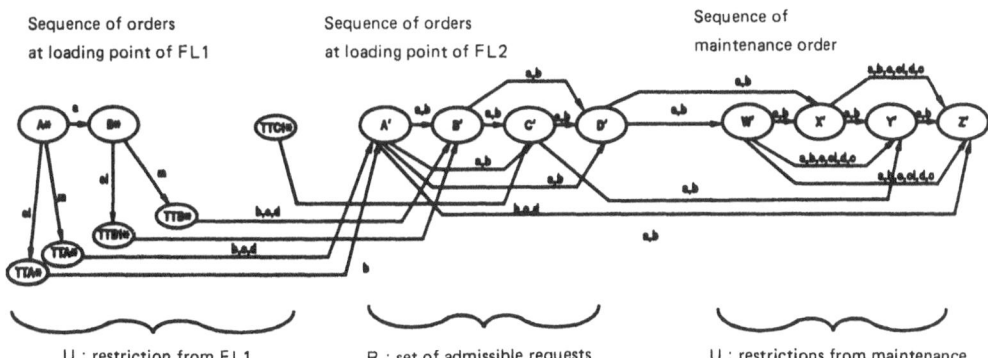

Figure 24.   Temporal net with inserted restrictions from FL1 and maintenace.

Step 8:
If the coordination module cannot find at least one element of (R n U) there is no feasible solution for the decision process on this level. The coordination module sends a status message (decision_frame_rejected;) to the Coordination Module of the upper level which then has to provide a new decision frame d.

Step 9:
If the coordination module has found at least one solution it serves as a decision frame for the lower level and as the definite request for the specialist modules of the same level.

Step 10:
The specialist modules receive the definite requests r and transform them to get their local requirements r'. Considering its local values for r', d' and m' the specialist modules use the still existing freedom to derive control decisions which optimize their local decision objectives.

7. Conclusion

Throughout this paper, the factory organization has been treated as a distributed problem-solving network with a hierarchical control structure. The approaches of the newly evolving software discipline of distributed artificial intelligence have been successfully applied to the class of coordination problems existing in factory management. In EP932, we developed a successive two-stage constraint satisfaction process based on qualitative and quantitative temporal data. The software program developed forms the representational backbone for cooperative problem solving.

8. Acknowledgement

I thank D. Bünz for the preparation of the implementation example chapter 6.

9. References

Allen, J. (1984) 'Towards a general theory of action and time', Artificial Intelligence 23, 123-154.

Decker, K.S. (1987) 'Distributed problem-solving techniques: a survey', IEEE Transact. Systems, Man and Cybernetics, SMC-17, 729-740.

Doumeingts, G. (1987) 'Guiding techniques for manufacturing systems' , in T. Bernold (ed.), Artificial Intelligence in Manufacturing, Elsevier Science Publishers, North Holland, pp. 287-302.

Fox, M.S. (1981) 'An organizational view of distributed systems', IEEE Transact. Syst.ems, Man, and Cybernetics, SMC-11.

Isenberg, R. (1987) 'Comparison of BB1 and KEE for building a production planning expert system', in: Proceedings Intl. Expert Systems Conf., London, pp. 407-427.

Isenberg, R. and Hübner, M. (1989) 'A combined object-oriented and blackboard-based system for the simultaneous optimization of lateness, lead time, utilization and inventory in CIM', in: Proceedings 2nd Intl. Symp. Systems Res., Informatics and Cybernetics, Baden-Baden (FRG).

Meyer, W., Isenberg, R., Hübner, M. (1988) 'Knowledge-based factory supervision: the CIM shell', Intl. J. Computer Integrated Manufacturing 1, 31- 43.

Meyer, W. (1990) Expert systems in factory management: knowledge-based CIM. Ellis Horwood Ltd., Chichester.

Meyer, W., Bünz, D., and Huber, A. (1989) 'Knowledge-based factory supervision: the flow line controller', Intl. J. Computer Integrated Manufacturing 2, to be publ.

Rit, J. (1986) 'Propagation temporal constraints for scheduling', in: Proceedings 5th AAAI, Philadelphia, pp. 383-388.

Sanders, (1989) 'A survey of object-oriented programming languages', J. Object-Oriented Programming 1 (6), 5-11.

Shapiro, S.C. (1987) 'Encyclopedia of Artificial Intelligence I, II' (editor). John Wiley & Sons, New York.

Project no 955

CONFORMANCE TESTING
CNMA PHASE 3 (CCT)

G.R. KNIGHT
SPAG s.a.
Avenue Louise 149
B-1050 Brussels
Belgium

Summary

The prime objective of Phase 3 of Project 955 was the development of conformance testing technology required for products based on the MAP/TOP 3.0 specification. The test tools formed an important part of the testing preparations for the Enterprise Networking Event 88i, staged in Baltimore, Maryland USA as well as having been used within the CNMA project.

The results of the project are now being exploited after further refinement through a distributor network covering Europe, the Americas and Japan.

1. Introduction

The Commission of the European Communities has provided a framework for European collaboration through the European Strategic Programme for Research and Development into Information Technology, ESPRIT.

One of the many projects supported in this way, in the field of Computer Integrated Manufacturing, CIM, is Project 955, which addresses a Communications Network for Manufacturing Applications, or CNMA.

The project started with recognition of the work carried out by the Manufacturing Automation Protocol (MAP) and Technical Office Protocol (TOP) programmes led by General Motors and Boeing respectively from the United States. These programmes use the seven layer model for Open Systems Interconnection (OSI) developed by the International Organisation for Standardisation (ISO).

The first phase of the CNMA project began in January 1986, and resulted in a demonstration in April 1987 at the Hannover Industrial Fair.

The second phase was the project's participation in the Enterprise Networking Event (ENE) 88i as one of the nine sponsored booths. This event built on the work of the MAP/TOP user groups in defining the standards for CIM. CNMA were to remotely demonstrate the capability of intercontinental distributed manufacturing.

CNMA became further involved in ENE by providing a major part of the required MAP/TOP conformance test systems, through the third phase of the project, named CNMA Conformance Testing or CCT.

2. CCT Project

2.1 Initiation

Discussions first took place in the summer of 1987 between the Industrial Technology Institute (ITI) from the USA, the test coordinators for ENE, and various other interested parties. The objective was to determine what test activities could be provided from Europe for the event, which was to be staged in June the following year at Baltimore Maryland, USA.

2.2 CCT Partners

The CNMA partners who financially committed to the project were, BMW, BULL, ICL, Nixdorf, Olivetti and Siemens. Two additional sponsors joined the project; ACERLI a consortium based in Paris, France and SPAG s.a. a consortium of computer and telecommunication companies from Brussels, Belgium.

2.3 Project Responsibilities

In view of the short timescales strict responsibilities had to be assigned:

- Management and co-ordination was provided by Siemens AG located in Karlsruhe, in the Federal Republic of Germany.
- Upper layer test tool development was the responsibility of the Fraunhofer IITB. Also based in Karlsruhe, Fraunhofer is a non-profit making organisation of industrial research and development.
- Test tool development for the lower layer tools was undertaken by The Networking Centre (TNC) based in Hemel Hempstead, in the United Kingdom.
  (Both of these test tool providers had substantial experience of developing protocol conformance test systems.)
- Quality Assurance audits were undertaken by Siemens for the upper layer tools and ACERLI for the lower layer-tools.
- External relationship and marketing was conducted by SPAG s.a.

2.4 Funding

The project, costing over 3 million ECUs, was 50% funded by the Commission of the European Communities.

3.0 Conformance Test Tool Requirements

The requirement for both CNMA and ENE was for test systems that complied with the MAP/TOP 3.0 specification, were of sound quality and would provide comprehensive test coverage.

The total conformance testing platform required tests on each layer of the seven layer reference model.

3.1 CCT Deliverables

CCT upper layer conformance test tools and associated test cases were provided for;

– **Manufacturing Message Specification (MMS)**
The MMS protocol provides the essential link between a central computer and a computer controlling a device, for instance a numerical controller or a robot controller.

– **Directory Services (DS)**
The DS system provides a database of directory information containing the name of every object attached to the network with the corresponding address by which it can be reached.

– **Network Management (NM)**
NM gathers information on the usage of the network, ensures its correct operation and provides reports.

All the CCT conformance testing tools test these protocols over the full 7 layer stack as well as the 3 layer MiniMAP.
The test tools operate over 802.3 and 802.4 networks.
In order to ensure interworking at the application layer, it is necessary to ensure that the Session Layer, the Presentation Layer, the Association Control Service Element (ACSE) and the Remote Operations Service Element (ROSE) are also conformant. Special embedded tests are provided for these other components.

For the lower layers, test tools were provided for;

– **Router**
The router provides an intelligent link between a number of different networks, and is able to route data from one network to another. The test systems provided tests for routers which operate between any combination of 802.4 (Token Bus) and 802.3 (CSMA/CD) networks.

– **ES to IS**
The End System to Intermediate System protocol is the mechanism for communication between an end system, for example a Directory Server, and an intermediate system, for example a Router.
Tests for both the End System and the Intermediate System are provided over 802.3 and 802.4 networks.

– **MAP Bridge**
The MAP bridge provides a link between two 802.4 token bus local area networks.

– **Logical Link Control Class 3**
LLC3 is required in the MiniMAP environment. This is because there are no upper layers (for example Transport class 4 to control sequencing), and therefore MAC with immediate response and LLC Class 3 are required to control the transmission of data.

A more detailed technical definition of the tools is provided a the technical annex attached.

3.2 Hardware and Software Environment

All the tools were designed to run over the same hardware configuration which is a SUN 3/160 workstation with additional couplers depending on whether the Implementation Under Test (IUT) is running over 802.3 or 802.4 local area networks.

3.3 Test Tool Integrity

To ensure the tools provided the necessary coverage expected by the ENE participants, detailed technical documentation in the form of protocol test specifications and protocol test plans were inspected in meetings attended by representatives of the MAP/TOP user group, ITI, the vendors from the sponsored booths and the developers of the tools.

All the tools passed these inspections with minor changes recommended, all of which were incorporated into the test systems before delivery.

3.4 Test Tool Delivery

The next phase was to exercise the tools against an implementor's IUT and this was undertaken both in the USA and also at the development centres' locations. The CNMA project also started to use the test systems at British Aerospace at Salmesbury, Preston U.K.

The tools were then released for ENE on a free-of-charge basis to ITI, the test co-ordinators, who in turn distributed the tools to the appropriate pre-staging areas for conformance testing of products to be undertaken.

4.0 The Success of ENE 88I

ENE was both a marketing and technical success. Over 7000 people attended this very specialised exhibition which connected 59 vendors over a single network built and made serviceable in less than 6 days. The CCT tools were used during the preparations at Baltimore and were actually demonstrated at a booth at the exhibition.

5.0 Project Results

So far, many different IUT's have been tested (which means many more vendors, as some vendors use OEM or bought in implementations) giving confidence in terms of the tools' total capability.

The result is that the SPAG-CCT test tools;

- were developed over a very short timeframe
- have passed inspection by the MAP/TOP Technical Review Committees.
- have been used within ENE and CNMA
- represent a substantial international consensus on the testing requirements for MAP/TOP 3.0

The CCT conformance test tools have also now been recognised by the European MAP Users Group (EMUG) and will be recommended to its members.

6.0 Further Developments

Whilst the CCT tools were developed as testing technology in support of both the CNMA project and the Enterprise Networking Event 88i this is by no means the end of the story.

6.1 Commercial Exploitation

The tools have been further developed so that they can be made available on a commercial basis. They have been productised and integrated into a common environment with complementary test tools from the Corporation for Open Systems International (COS). To be known as the Integrated Tool Set or ITS the tools from both SPAG-CCT and COS will run within the same hardware platform, the same software environment and under a common menu system.

The architecture allows for further tools to be added later. Training, consultancy support and maintenance services will also become available.

6.2 ESPRIT Project 2292

Following on from the success of CCT a new ESPRIT project has just been approved by the Commission of the European Communities. Entitled Project 2292 - Testing Technology - CNMA (TT-CNMA), its prime objective is to support and improve the competitiveness of European CIM product vendors and manufacturing industry in world markets.

7.0 Who Needs Conformance Testing?

7.1 Benefits of conformance testing to vendors

It is not practical or cost effective for a vendor to ensure their product will interwork with that of any other vendor who could also potentially connect to the network. Therefore, vendors will conformance test their products before their release which will significantly increase the probability that their product conforms to the protocol specification to which it has been implemented and will therefore interoperate with other vendors' products. Conformance testing can save vendors money because;

– If conformance testing is introduced at the development stage, the product under development will be constantly checked against an industry standard reference test tool. The product will therefore be conformant before it is passed on to the validation phase. Validation can concentrate on testing the product against its own specification, capability and functionality rather than whether the product conforms to the appropriate International Standards.

– QA managers will use conformance tests to underwrite products' conformance against International Standard Profiles (ISPs) before the products are approved for release.

– Support Managers will use conformance tests to underwrite maintenance release conformance before products are reissued.

– Sales and Marketing managers will find that only conformance tested products are required by the marketplace and that this endorsement will give their products essential credibility.

7.2 Benefits of conformance testing to users

Users will benefit because by calling for independently conformance tested products, they will be significantly increasing the probability of their multivendor networks interworking. Large users will purchase this technology to enable them to integrate their networks and to maintain control by conformance testing new products before adding them to the network as it continues to grow.

8.0 Availability

A worldwide distributor network is being set up by SPAG s.a. who are marketing the tools on behalf of the SPAG-CCT consortium. The Corporation for Open Systems has already been appointed as the distributor for the Americas and others will follow in Europe and Japan.

The tools can be purchased for use internally on a first party licence basis. Test centres are also available where products can be subjected to conformance tests by third parties.

9.0 CCT Consortium

SPAG-CCT is now being established as a consortium so that the results of the CCT project can be exploited. The consortium includes ACERLI, BMW, BULL, ICL, Nixdorf and Siemens and SPAG as funding partners from the CCT project as well as additional members Fraunhofer IITB, The Networking Centre (TNC) and Alcatel-TITN. It will receive funding through sales of the software and will take responsibility for planning and funding future developments based on the existing technology. The consortium has already funded and managed the upgrade of the SPAG-CCT tools to match the specification of the Integrated Tool Set, a joint development with the Corporation for Open Systems International.

New developments will also be undertaken; for example the ESPRIT Project 2292 (TT-CNMA) is a new development work programme.

The consortium has access to all the relevant Intellectual Property Rights necessary to allow forward development of the existing SPAG-CCT software through the inclusion of the original partners. The development partners include all those involved in previous developments and in the case of Alcatel-TITN, the new developer included in the TT-CNMA project.

The consortium will have the legal form " Société en participation" and will be managed by a committee initially being chaired by ICL.

APPENDIX 1

Keywords:

CNMA: Communications Network for Manufacturing Applications

ES to IS: End System to Intermediate System Protocol

ITS: Integrated Tool Set co-developed by SPAG-CCT and COS

LLC3:   Logical Link Control Cass 3

MAP/TOP: Manufacturing Automation Protocol/Technical Office Protocol

MMS: Manufacturing Message Specification protocol

NM: Network Management protocol

OSI: Open Systems Interconnection

Router: Protocol

SPAG-CCT: Consortium comprising ACERLI (F), BMW (GB), BULL (F), ICL (GB), Nixdorf (D), Siemens (D) & SPAG (B) as funding partners from the CCT project as well as additional members Fraunhofer IITB (D), The Networking Centre (GB) and Alcatel-TITN (F).

Appendix 2

Technical Definition of Tools

The CNMA Conformance Test Systems are designed to test protocols based on MAP/TOP 3.0 specifications. These specifications adopt the following ISO protocols:
- For MMS: ISO 9506
- For NM: ISO 9595
- For DS: ISO 9594
- For Router &
-                 ES-IS: ISO 8348 & 8473 & 8648 & 9542
- For LLC3:   IEEE 802.2
- For Bridge: MAP 3.0 Specifications

The CNMA Conformance Test Systems are based on a windowing environment and they provide the operator through menus and mouse facilities with a self-explanatory and straightforward interface.

The product to be tested, also called IUT, is described in the standardised PICS and PIXIT documents. Both documents are available online as simple ASCII files. They are managed by the Test Systems to build a testing environment for the product and to automatically select in test data bases, tests according to product functions.

Product dedicated test suites are ready for execution through one command. After a test campaign, a test report aligned with requirements of the ISO 9646 document, is automatically extracted by the system from traces available for each layer and the conformance of the product can be confirmed.

The Upper Layer Conformance Test Systems (MMS, NM and DS) are generic for ASN.1 based protocols ie: in order to develop a test system for another protocol, it is only necessary to write the grammar and to develop the test cases. Test cases are dynamic: they are parameterized during execution and bound to the vendor specific capabilities.

The components (see figure 1) of the Upper Layer Conformance Test Systems (MMS, NM and DS) are:

- the files (Test Schedules, Test Parameters, Dynamic Test Cases, Log Files, PDU Grammars) containing test cases, parameters, protocol grammars and test execution records;
- the Test Language Controller (TLC) which handles operator commands and initiates associated actions;
- the Test Control Manager (TCM) which executes test cases, maintains synchronisation and produces traces;
- the ASN.1 Tools containing the Compiler translating off line a source PDU grammar into an object PDU grammar, the Encoder/Exception Generator translating parameterized ASN.1 templates into binary encoded PDUs, and the Decoder/Comparer translating binary encoded PDUs into ASN.1 values strings and comparing received PDUs against ASN.1 templates;
- the Layer Access Switch which enables dynamic access to different lower layers;
- the LLCI which provides a standardized interface between the test case and the lower layer (being LLC3) and deals with special requirements of Mini-MAP.

The design of the Lower Layer Conformance Test Systems (Router, Bridge, ES-IS and LLC) is common. The components (see figure 2) are the single Executable Test Case (ETC) process which communicates with an Encoder/Decoder process, the Control Function (TCF) process controlling the ETC process, and the Man Machine Interface (MMI) which provides a user- friendly interface to the Command Interpreter (CI) system.

Several test methods are used depending upon the protocol to be tested. A combination of the Distributed and the Remote test methods is employed for testing MMS, NM and DS User Agent. The Remote test method is used for DS Service Agent. The Transverse test method is used for the relaying and routing tests. The Distributed Single Embedded test method is used for ES-IS in ES testing. The Coordinated Single method is used for LLC.

The CNMA Conformance Test systems are installed as normal devices on the network supporting the product to be tested. Possible types of networks are CSMA/CD (802.3), Token Bus Carrierband (802.4 CB) or Token Bus Broadband (802.4 BB).

The test systems run on SUN 3/160 machines comprising 8MB of memory and a 280MB disk, using standard SUN operating software (SUN OS, SUNLink, SUNOSI) as well as standard Motorola boards depending upon the network used.

CNMA Conformance Test Systems for MMS, NM, DS
System Architecture

Figure 1

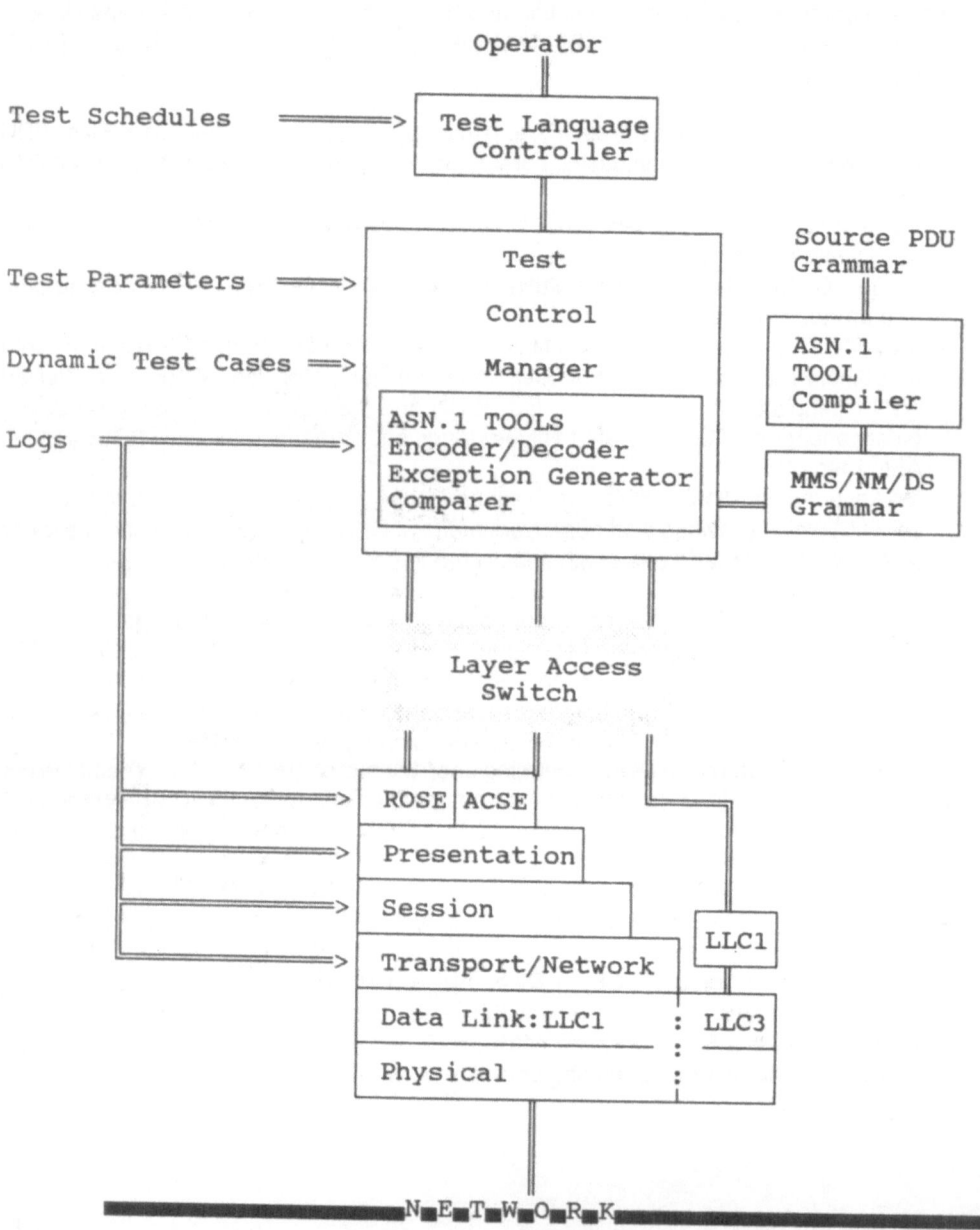

CNMA Conformance Test Systems for Router, Bridge, ES-IS, LLC
System Architecture

Figure 2: Operator

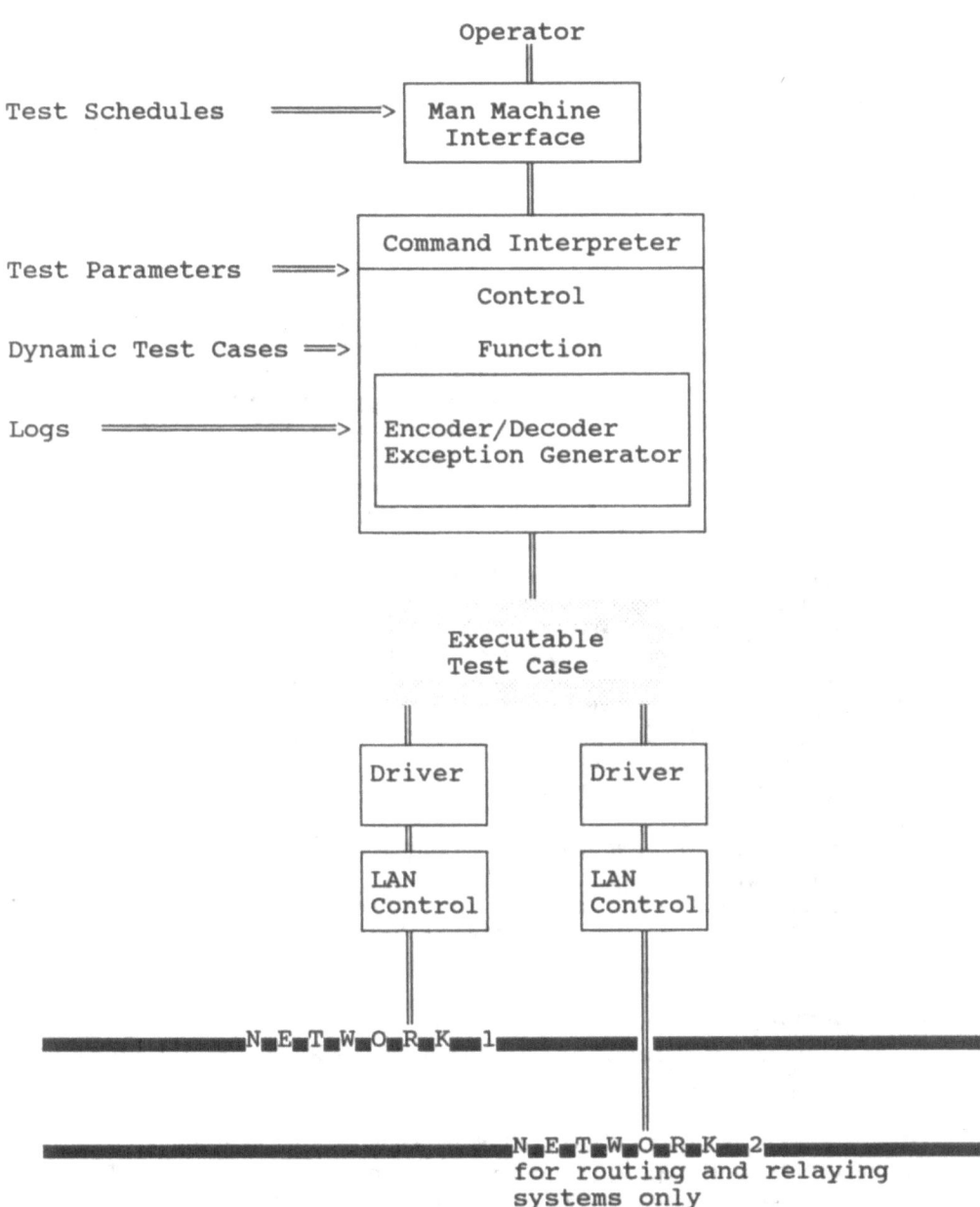

Project no 1062

ACCORD : Developing Concepts Towards Integration of Analysis and Design

J.P. PATUREAU
BERTIN et Cie
Zone Industrielle des Gâtines
BP 3 - 59 rue Pierre Curie
78373 PLAISIR CEDEX - FRANCE

C.W. TROWBRIDGE, C. BRYANT
VECTOR FIELDS Limited
24 Bankside, Kidlington
OXFORD OX5 1JE - ENGLAND

N.H.W. STOBBS, R.G. PARKER, S.J. DENNISS
GEC - Marconi Research
West Hanningfield Road
Great Baddow
CHELMSFORD, ESSEX CM2 8HN - ENGLAND

Abstract

This paper outlines some concepts developed in ACCORD to help integrate analysis and design of systems into a coherent computer aided environment. Such concepts include : computational speed-up through concurrent processing, on-line access to engineering expertise, data sharing across engineering domains.

Four engineering domains, pertaining mostly to electronics systems, were selected for the development and demonstration of the above concepts, viz : electromagnetics (and more particularly device modelling), life cycle costing, reliability prediction, thermal management. This project will end up in a year with the following software packages :
1) a library of mathematical routines optimized for concurrent processing (APPEAL),
2) a novel life cycle costing software facility,
3) an advanced thermal engineering software environment featuring on-line access to specialist expertise,
4) an advanced reliability engineering software facility featuring knowledge-based assistance to the user.

All packages can either be used separately or integrated via a common Database Management System (INGRES). A demonstration on a VAX station will be available at the 1989 Esprit Week.

1. Introduction

ACCORD was set up as a research project with the general aim of improving the current state of the art of computer aided analyses of complex electro-mechanical systems.

Very schematically, the typical design process amounts to tapping into the body of existing relevant knowledge for ideas and information, which are then drafted and analyzed against specified criteria for optimum global performance. This is usually very much an iterative process which may involve several expert teams, each working in isolation with their own computer tools and databases. It is also widely admitted that analysis is becoming very effective in reducing the cost and duration of design, often avoiding costly prototyping and/or experimentation. Yet, building on the ever improving capacities of computer techniques, still larger effectiveness can be contemplated.

Three avenues of research can be explored :

- computational speed-up aiming at truly interactive analysis,
- computerization of engineering expertise, thereby aiding the analyst to work faster, more reliably and more comprehensively,
- integration of a range of design and analysis tools into a single coherent computer environment, allowing for easier management of the design process as well as allowing for improved communication between designers and analysts.

All three avenues of research were, to some extent, considered in ACCORD. However, to keep the project within reasonable limits, the scope was limited to only a few of the analysis techniques required for typical large electro-mechanical systems, namely : reliability and availability prediction presented in section 5, thermal management in section 6 and life cycle costing in section 7.

For the same reason, and although it is the ultimate goal of a fully integrated system integration with CAD tools (for circuit boards, or for mechanical packaging layout and design) was not implemented. Section 4 presents the concepts that were developed to allow easier communication between different engineering domains.

Similarly, the work on computational speed up, described in section 8, was limited to mathematical routines (used to solve differentiel or integral operators) and to a representative range of commercial computers, including FPS and Transputers.

A general discussion of the main issues that formed the background of ACCORD is given in the following section.

Finally this paper ends with some perspective regarding possible exploitation of the results obtained in this project.

2. Main Issues

Some of the main issues arising from an ideal and typical Computer Aided Design environment can be derived from the limitations which are often encountered in current industrial design practices.

a) <u>Centralizing and managing product data</u>.

From its embryo in the mind of a marketing engineer down to the dealer's shelf, a product will have been adressed many times over by many different people using their own language, their own way of describing the product and keeping their own documentation. Yet, all of these people must communicate as efficiently as possible so that each progress step in the life of the product is as fast and as efficient as possible. This calls for a uniquely defined CIM reference model, based on a descriptive language designed to suit the needs of all potential actors. Such a language is already being developed [1], [2] in the sector of integrated circuit manufacturing, and is meeting with a definite commercial success.

Such a language is still to be specified for mechanically based systems. Of course, one tremendous difficulty lies in the extreme diversity of mechanical parts that can be encountered in CIM. This may very well call for a sectorization of the descriptive language to make some kind of stylization possible. In that case, there is some hope that a unique language can cater to several descriptive viewpoints (functional description, various modelling descriptions, various performance descriptions, parts lists, CAD descriptions, ...).

Based on such a language, a computer infrastructure can then be constructed to allow storing and retrieval of product data uniquely defined for all actors. Creating a central product data repository has three oustanding advantages :

– it allows easier integration of design tools which work on the same data,
– it facilitates communication between the people concerned with one product,
– it facilitates filing and archiving of the data in an easy-to-retrieve format essential for future re-use.

Short of such a language, any integration to a CAD system can only be made partial and specific. Because ACCORD was not a proper framework to endeavour the development of a descriptive language the concept of integration was limited to several basic features particularly relevant to the analysis of complex systems. This work is summarized in section 4.

b) Another major issue is **to facilitate the search for existing information, data or expertise.** Traditionally, a design engineer starts with an understanding of the physics attached to a given engineering discipline (say mechanical engineering) and gradually builds up experience form his work on what makes things succeed or fail. Obviously, such experience plays an essential role along his creative paths. Yet, the experience of one person is neccessarily limited in scope and also in time (it is not rare to make twice the same mistake ...). The concept of an on-line access to computerized design information, data and expertise is extremely appealing as it could gradually be built up to contain an entire company's experience and knowhow which could thus be made available to all designers. The amplification factor of everybody's work could thus be very large.

In proportion to the potential benefit, the difficulties to achieve such an environment are also quite significant. Three major problem areas can be identified :

– computer representation of very diverse information must be highly flexible. Certainly the current AI techniques using rules, frames or objects are adequate to encapsulate some aspects of engineering knowhow, like procedures and codes of practice. Such a system (based on the AI toolkit KERIS) is being developed in ACCORD regarding Reliability planning procedures, as described in section 5. Yet, for a significant fraction of engineering knowledge, rules are not well adapted and would be cumbersome to use. This is particularly the case for non-procedural engineering know-how. Fortunately, if one ignores (as premature) the automation of some kind of engineering reasoning, then there is almost no constraint on how to represent a given piece of information. The current work under development in ACCORD will be discussed later in section 6 of this paper.

– the selection mechanisms are particularly important in this context. They must reflect the way that a design engineer would search for information in the context of a given problem ; keywords, associative mechanisms (finding the closest match), high connectivity (finding a useful piece of information from quite different initiating request primitives) are all important aspects.

- enrichment mechanisms must be provided so that the system can be continuously updated with new engineering rules, new experiences, new experimental results ... This is best achieved by the users themselves who can then customize the package to their own specific needs. To this end, a central requirement is the simplicity and generality of the updating interface procedures. Work done along these lines for thermal engineering is briefly presented in section 6.

c) **Processing speed in design analysis** is also a central issue towards interactive engineering. Large mathematical modelling software packages (such as finite element) typically require tens of minutes and sometimes hours of computing, for the size of problems which are being considered in today's designs. In addition to taking advantage of the steadily increasing power available on commercial computers, it is now recognised that the most promising route toward interactive processing is through the extensive use of parallel processing. A well known obstacle on that route is to harness the full capacity of parallelism ; an important aspect of overcoming this obstacle is to properly match the computational algorithms onto the target hardware [3]. Work is being done in ACCORD along those lines and will be presented in section 8.

Finally, another key issue relates to company attitude and organization. The full power of computer based design and analysis will make sense in the long term, only if it can be supported unaminously throughout the company. One often cited example of such a problem area is the maintenance and updating of databases. A database (say about technical component data) is useful only if it is up to date and validated, which requires quite a strong commitment and a corresponding organization. However this issue falls obviously outside the scope of a single Esprit project.

3. ACCORD Aims

ACCORD aims at exploring some of the above issues towards closer integration of analysis and design using state of the art computer aided techniques. Selected areas of design engineering were chosen in accordance to the available expertise within the Consortium.

- reliability engineering particularly regarding the use of generally accepted (and sometimes standardized) methodologies and tools such as Failure Mode and Effect Analysis (FMEA), Reliability prediction of electronic systems, Fault Tree Analysis (FTA) and Reliability Block Diagram Modelling (RBD).
- life cycle costing of complex systems using a novel methodology for cost prediction and analysis.
- heat transfer engineering, as regards the thermal management of system or components where proper heat evacuation is critical to the design performance.
- electromagnetics engineering especially regarding those problems which lead to heavy computational loads (device modelling for example).

This last domain was chosen as a vehicle to demonstrate the potential of optimizing known mathematical techniques for vector processing and for the emergent parallel processing capabilities : this is later refered to as APPEAL (Accord parallel processing engineering application library).

In addition to being developed separately (emphasis on one or another aspect of computer aided design analysis), the first three engineering areas are also used as a test vehicle for integration concepts. Such an integration appears to be quite relevant to the design of sophisticated electronics systems (Telecommunications, Defense).

The following sections will now summarize the progress made so far in all four areas, emphasizing what is going to be the likely outcome of this research when the project ends in about a year.

4. Integrated Environment

For the reasons mentioned earlier, integration was limited to the three engineering domains that were considered in ACCORD.

The purpose was mainly to explore the feasibility and potential interest of several key concepts as described in the following.

Figure 1 outlines the basic elements of the integrated environment that is being developed in ACCORD. Each of the three engineering domains is considered as a separate entity (as it is to be used by different domain specialists in each case). Yet, this entity is very closely linked to the others at two key levels :

- at the top level, through the unique user interface layer (also called Executive level),
- at the bottom level of product data, through a unique database managed by a relational database management system (INGRES in our case).

Each domain is meant to contain all facilities required to conduct fast, efficient and comprehensive analyses in the corresponding domain. Such facilities include :

- knowledge based assistance to planning and analysing,
- a range of tools appropriate for various types of problems,
- databases of relevant information,
- access to domain expertise,
- management tools to label and handle data sets and result sets.

Integration is achieved via direct data exchange through the product database. An example is electronic device temperatures for predicting equipment reliability. Another example is equipment maintenance scenario for calculating life cycle cost.

Integration is also obtained at Executive level where the following functions are available:

- product descriptions in the form of a System Breakdown Structure (SBS), which is a unique and evolutive reference for all integrated domains,
- status log where progress in design is reported and where relevant information is exchanged between and specified to domains.

CAD tools for drafting and modelling were at first considered as being part of one more domain, on the same level as thermal engineering or reliability engineering. This attitude has somehow evolved in the course of the project due to the very extensive work of CAD vendors directed at using CAD tools as a structuring core for wide design environments. Although it is always possible to achieve close integration with specific tools for special purpose applications, it was felt that the only constructive approach which could offer sufficient generality was to develop a descriptive language adequate for the range of products to be considered. This, however, was considered to stretch well beyond the scope of ACCORD and was therefore abandoned.

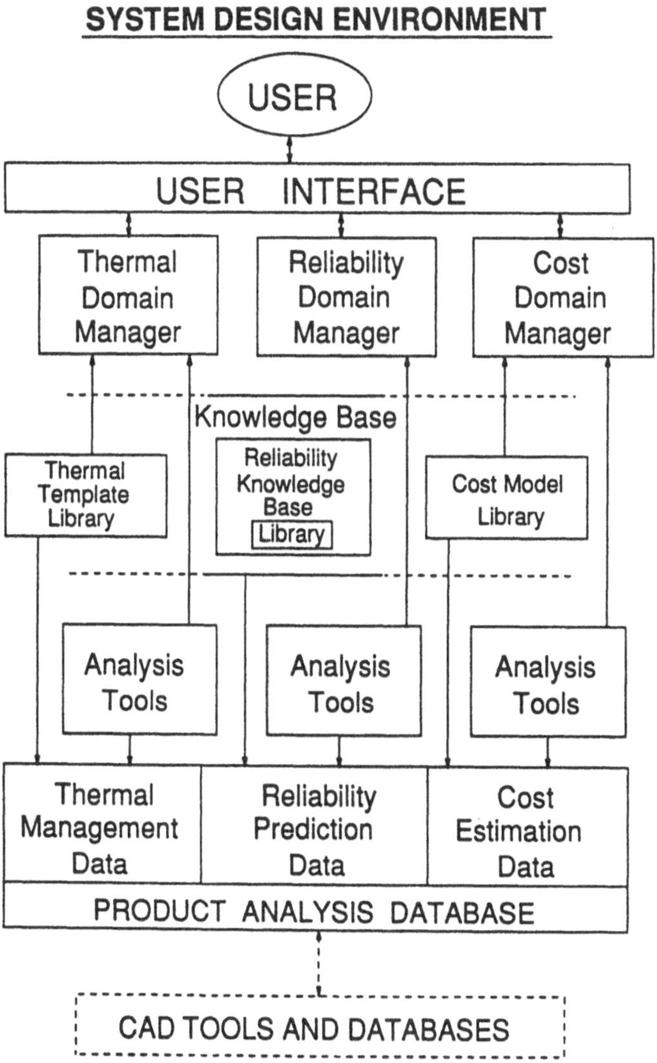

Figure 1

5. Reliability Analysis Domain

5.1. Introduction

The so-called "Reliability Doamin" deals with those analyses which are concerned with the measurement and improvement of Availability, Reliability, Maintainability and other similar characteristics of a system, usually called ARM activities.

There are three main aspects to the work undertaken in this domain : Planning, Analysing and Reporting. The latter being at an early stage in development, this paper will concentrate on the PLAN and ANALYSE functions. The Reliability Plan Function is the first application of a prototype software module by the name of ADVISE (ACCORD Design Verification Intelligent Support Environment). Before considering the individual features encompassed by the Plan function, the ADVISE module, and the philosophy underlying its development will be briefly summarised.

5.2. ADVISE

ADVISE, which has been developed using the AI toolkit KERIS (the roots of which lie in the ESPRIT-funded Portable Common Tool Environment (PCTE) project), is intended to provide the framework for developing software support for a wide range of engineering analysis activities. The central feature of this framework is an Interface Manager which is physically separated from the individual applications.

The physical separation, so far as it may be achieved without diminishing the interface, of the interface manager from the underlying functions, may be seen to be conductive to interface development in several ways. Of a list of perceived benefits identified by Cockton [4], the following have had a strong bearing on ACCORD.

a) Organisational consistency. The adoption of a standard set of interaction proce- dures, which may be applied to arbitrary functions, lends predictability to generic classes of actions, thus reducing the learning time required by the user to become familiar with a new set of functions.

b) Reusability. The generality of the methods adopted in the user interface manager should enable it to be used in conjunction with a wide variety of functions.

c) Personalisation. A standard interface may be developed for a given set of functions. This may then be tailored to the requirements of individual users, without requiring alteration to the functions themselves.

Within ACCORD, the first two of these features have been heavily exploited : from the user's point of view, the software has a consistent 'look and feel' ; from the developer's viewpoint, the ADVISE Interface Manager has enabled user interfaces for a variety of knowledge-based modules to be developed rapidly, and extended easily. The third feature will become significant when the exploitation of ACCORD with regard to a range of classes of users is considered.

The Interface Manager is strongly object-oriented in nature. Object-oriented programm- ing has been looked upon as a promising model for the construction of dialogue managers (e.g. Fischer in [5]), since object based representations provide a modular, coherent and convenient packaging mechanism for encapsulating the information required for a given interface. The representation used in ADVISE centres on a formal description of two key aspects of the interaction process, the decision point and the window. This is supported by various dialogue control mechanisms, aimed largely at both handling menu sequencing and providing some ways around the inherent limitations of menu-based systems.

5.3. PLAN

The Reliability Plan Function encompasses a wide range of features, including facilities for project definition, production of largely complete Reliability Programme Plans (based on standard documents and analytical requirements), task scheduling and advice on remedial actions to be taken when goals are not met. This paper will restrict itself to a single example of this work : a facility for Reliability Modelling.

The Reliability Modelling facility allows the user to model a physical system in terms of an enhanced version of a standard modelling paradigm for Reliability, the Reliability Block Diagram. The system is broken down in a hierarchical fashion, by specifying the number of sub-units (i.e. lower-levels blocks), and the configuration (series or one of various types of parallel redundancy) of each unit, or block, in turn.

This is performed by building and executing commands by clicking the mouse on command selection icons. The units and their relationships are displayed graphically ; each is mouse sensitive, with a view to providing a 'direct manipulation' style of interface. To allow systems to be analysed to arbitrary levels of complexity, a 'zoom' facility is provided.

The graphical display is underpinned by an object-based representation, attributes such as the parent unit, configuration, sub-units and graphical coordinates being encoded in a formal description of each unit in the model. This representation provides the structure for performing numerical analysis ; in addition it allows the modeller to provide input to the module in ADVISE which is concerned with task scheduling.

Once a system is defined, it may be analysed using standard numerical techniques (specifically, an availability analysis may be performed). The user selects a unit with the mouse and enters a value for a chosen Reliability parameter, such as the Mean Time Between Failure (MTBF). Whenever sufficient information is available, the effects of a change are propagated around the system, taking into account the varying configurations, and any user-specified constraints. This process is performed by applying the appropriate equation (according to configuration) to each unit in the system which can be affected by the change, taking into account 'knock-on' effects.

Although a number of tools already exist for calculating the overall availability of a system from that of lower level units, there is less support for working the other way round, to determine the lower level requirements for meeting a particular high level goal (apportioning).

The modeller offers a variety of apportioning strategies. The default strategy is one of equal apportionment (i.e. a value is found such that if all the top-level units in a chosen configuration have that MTBF then the goal MTBF of the configuration will be met ; each top-level unit is then treated as a local configuration, and the process continues level by level to the bottom). The user may, however, over-ride this by either 'fixing' the MTBF of one or more units or applying complexity weightings, to take into account existing knowledge about the units involved. These constraints may be applied in combination, and at any level in the system model.

RELIABILITY DOMAIN
INTEGRATION (DATA FLOW)

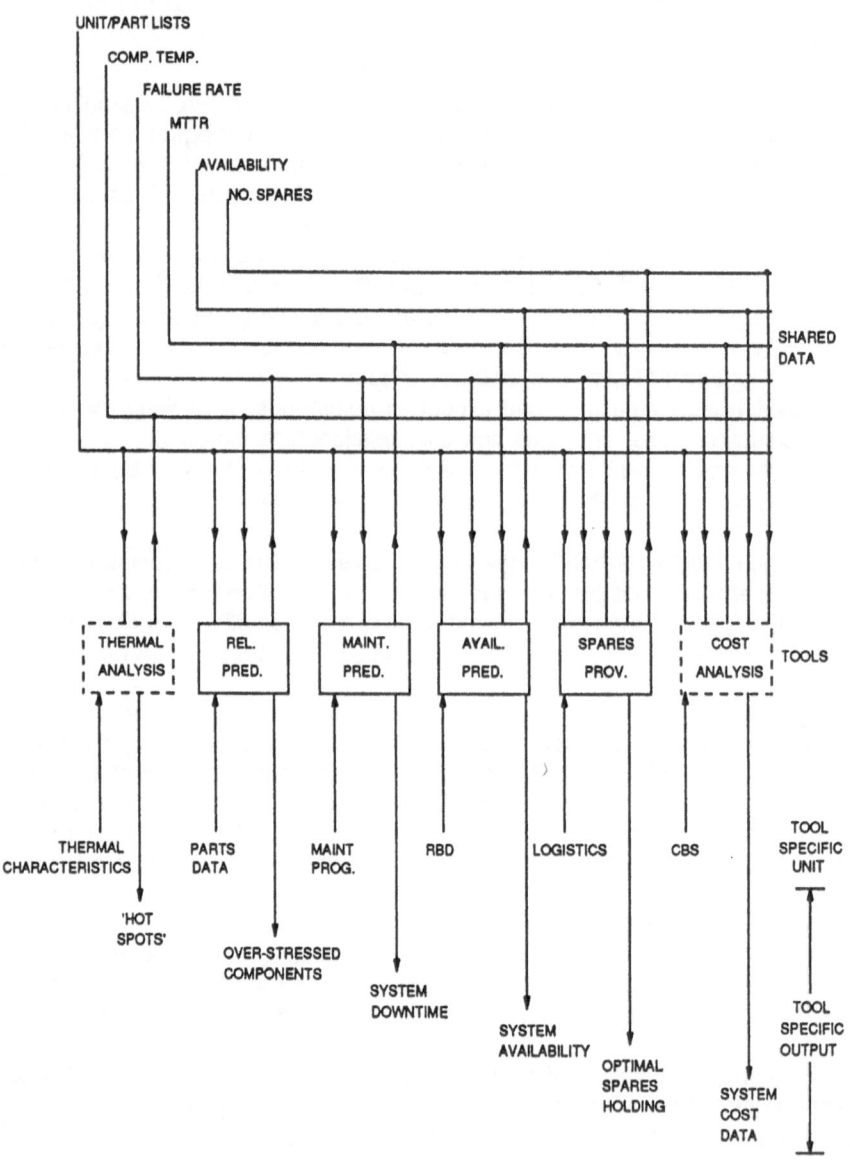

Figure 2

5.4. Analyse

This function contains the tools necessary to perform Logistic Support analysis. These tools are currently run in isolation and the results often arrive too late or not in a suitable format to influence the Design. The objective of this function is to provide an environment where these tools can share common data, and to enhance the user interface to each tool such that they become more efficient ot use and the results become more useful at the design stage. One of the key aspects of the software design is the data sharing via the DBMS INGRES. The principles of this are described below.

5.5. Data Sharing

Figure 2 illustrates the principle of data sharing. The tools shown are a representative selection of the those which could be integrated with ACCORD. The tools in the solid boxes are "Reliability tools". Also shown is the sharing of data with the two other key domains of Thermal and Cost Analysis. As can be seen, each tool has its own "tool specific" data input and output. In addition each tool may use as input, data provided by other tools, or may provide as output, data which other tools may use as input.

Tools may be used in any sequence ; however it becomes readily apparent from the diagram, that tools to the "right hand side" generally rely on tools to the "left" for their data input. This information would be used in planning an scheduling the analyses to ensure the maximum use of common data.

The production of the Data Integration diagram was a very important stage in the evolution of the database design, and the interface between the system and each tool.

Data used as input by a given tool is denoted by an arrow "feeding into" the tool, whilst data provided as output is shown as an arrow "feeding out" from the tool. For example, Maintainability Prediction uses as tool specific input, details of the system maintenance procedures, and combines this with shared data items ; units parts list, and failure rate. It produces a shared data output of Mean Time To Repair (MTTR) and a specific output of System Downtime.

5.6. Conclusions

The Reliability Modelling Facility exemplifies many of the advantages which the features offered by the Reliability Domain may confer on the analyst. The following may be identified as making a substantial contribution to its power :

5.7. Operational Efficiency

The intention of building a graphical interface is to free the user from having to worry about the idiosyncracies of the analysis tool, and allow him or her to concentrate on the system under analysis. For instance, manu of the overheads of data preparation and checking associated with conventional batch-oriented tools mau be avoided ; input errors can be detected and corrected easily on-line. Results are displayed graphically, and all, or a required subset, of the information may be saved at any time during the session. A major benefit of these features is the ease with which explorative analyses may be undertaken.

While the modeller is not yet of product status, it is already becoming an exploitable tool, and has been successively applied to a real sub-system (part of a large communications network) containing slightly over 500 units in various configurations ; starting from scratch, the system was modelled and sets of results for several goals generated in around three hours.

On a broader level, operational efficiency is greatly enhanced through the automation of data sharing between analysis tools.

5.8. REUSE

The formal representation of the models developed by the user provides the basis for developing libraries of sub-models at arbitrary levels, which may then be mixed, matched and tailored to fiture projects as required. At a later stage it is hoped to extend this capability by incorporating domain knowledge about standard subsystems. This could be used, for instance in the apportioning process, in the form of saved values for MTBF and/or system complexity, which could either be proposed to the user as initial estimates or entered automatically, depending on the level of confidence in the informations.

Operational efficiency and reuse may be seen as related aspects of the overall goal of productivity enhancement. While the above details relate largely to the modeller, the drive towards greater productivity, in particular through the pursuit of these two goals, has had a strong bearing on the design of the ADVISE system, and indeed ACCORD as a whole.

6. Thermal Engineering Domain

The thermal engineering software was designed to be used by heat transfer specialists, either as an integral part of a wide analysis environment (referred to as ASSET for ACCORD suite of software engineering tools) aimed at product assurance in general, or as a stand-alone package for thermal management design teams.

The architecture of the package is engineered in such a way as to cater for the basic needs of heat transfer analysis, viz :

- define the product/configuration to be studied and provide means to unambiguously label and manage input and output files for further exploitation (activation of modelling tools, combination of input files, archiving, ...),
- provide access to general purpose modelling tools (like ESATAN for lumped parameter analysis),
- provide access to thermo-mechanical data of usual materials,
- provide access to basic heat transfer knowledge (physical correlations, phenomeno-logical laws, ...) in a form which is directly exploitable for quantitative evaluation,
- provide access to engineering heat transfer knowhow and dedicated software tools (performance of a range of components and/or configurations, experimentally measured performance parameters, semi-empirical laws, field experience data, ...).

The first two functionalities are fairly straightforward, although they involve the development of sophisticated user interfaces (usually multi-windows) and the use of a powerful relational database management system (INGRES running on Vax/VMS was selected for the demonstration case).

Much of the innovation is to be found in the implementation of the last three functionalities. Those amount to developping a selection/activation/enrichment mechanism along with the corresponding interfaces, to operate on a very diverse base of engineering expertise. This expertise takes the computerized form of :

- either inactive files (to be viewed only),
- or active files to be activated (executable code sequences) usually after input of a few appropriate parameters. Each file can be viewed as containing a "piece" of thermal engineering expertise later referred to by the generic term of template, to be presented to the expert user for him to use in his current analysis. The advantages of such a system are to,
- decrease the time spent looking for the "proper heat transfer correlation" or the right value of parameters,
- decrease the number of errors done in numerical evaluation of relevant quantities,
- increase the scope of expertise which can be tapped for each analysis, hopefully much beyond that of the current expert doing the analysis.

The first step towards such a software facility is being taken in ACCORD. Each template is characterized by a number of representative keywords, which can relate to various points of view for example, the area of physics involved, the type of material, the type of product or components, the type of operating environment, ... The keywords are then organised in families and managed by the RDBMS Ingres.

Faced with a given heat transfer problem, the user is then in a position to search for the relevant resident pieces of expertise (templates) via a multi-criteria mechanism. Once selected, the templates must be validated before they can be viewed and/or activated.

Within the timespan of the project, ACCORD will produce the software shell and database (see figure 3) that implements the basic mechanisms and interfaces in the "Template Manager". In addition, a limited number of templates will be incorporated, adressing the thermal management of electronics systems. A representative list of templates is as follows :

- Heat 1 : a lumped parameter model dedicated to the evaluation of chip temperature on a PCB (printed circuit board),
- Thin rectangular plate of variable dimensions with two dimensional uniform network and variable mesh refinement in x and y directions,
- Hollow rectangular box of variable dimensions and wall thickness with two-dimensional uniform network in each wall and variable mesh refinement in each planar direction,
- Thin circular plate of variable diameter with two dimensional radial network of variable mesh refinement in radial and circumferential directions,
- Simple forced flow in narrow rectangular channels with heated rectangular objects attached to one wall,
- Natural flow over heated rectangular objects in a vertical ventialted enclosure,
- Natural flow through narrow rectangular channels freely ventilated at the top, with rectangular heated objects attached to the walls of the channels,
- Natural circulation in a totally enclosed space with randomly distributed, heated rectangular objects,
- Natural flow between totally enclosed parallel plates with heated rectangular objects attached,
- Component thermal resistance tables,
- Fin efficiency for rectangular profile, longitudinal and radial fins,
- Performance data for complex fin configurations,
- Coldwall performance,
- Unitary heatsink ladder calculations,

732

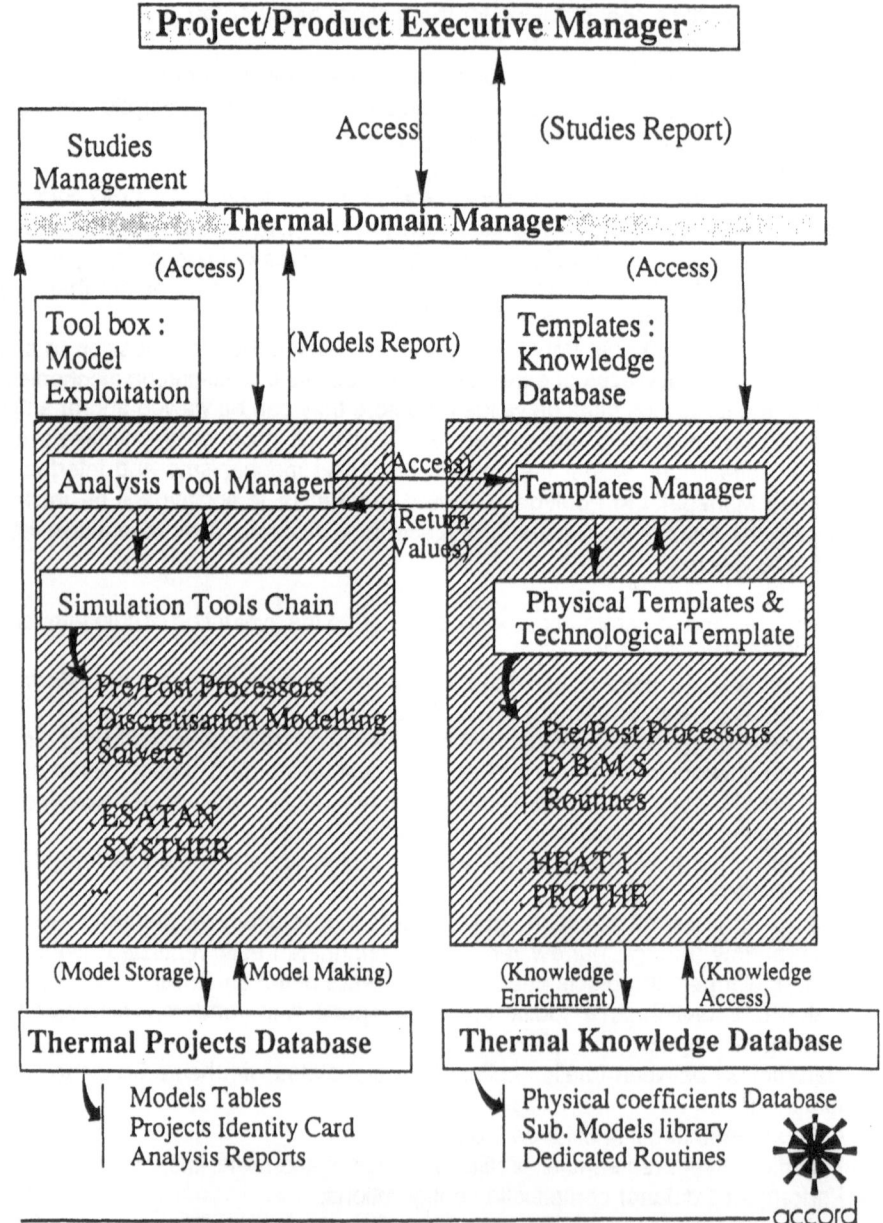

THERMAL ENVIRONMENT ARCHITECTURE BLOCK DIAGRAM

Project/Product Executive Manager

Access   (Studies Report)

Studies Management

Thermal Domain Manager

(Access)   (Access)

Tool box : Model Exploitation   (Models Report)   Templates : Knowledge Database

Analysis Tool Manager   (Access)   Templates Manager
(Return Values)

Simulation Tools Chain   Physical Templates & TechnologicalTemplate

Pre/Post Processors Discretisation Modelling Solvers

ESATAN SYSTHER ...

Pre/Post Processors D.B.M.S Routines

HEAT 1 PROTHE

(Model Storage) (Model Making)   (Knowledge Enrichment) (Knowledge Access)

Thermal Projects Database   **Thermal Knowledge Database**

Models Tables
Projects Identity Card
Analysis Reports

Physical coefficients Database
Sub. Models library
Dedicated Routines

accord

Figure 3

- Heatsink ladder clamps,
- Superposition.
- Natural convection in channels.

Clearly, the power of such a system is in proportion to the comprehensiveness of the template base ; ideally, it can be made to contain a large amount of company expertise provided the effort is made to stylize and characterize that expertise in the form of templates (which are quite flexible indeed and ready to accomodate diverse types of information). The extent to which keywords mechanisms can satisfy the needs for characterizing large amount of diversified engineering knowhow, is not well known. Probably, limitations will appear regarding:

- the limit in organisational complexity,
- the limit in access time for very large bases.

Other techniques can be examined as well, building on this first experience. The concepts of objects (and the rich palette of commercial software tools which support those) might prove closer to a more natural way of structuring engineering expertise. Similarly, if one extrapolates to very large and rich engineering template bases, techniques of pattern recognition could very well prove to be quite useful. In this context, the current promise of neural networks to turn out truly adaptative (and therefore "learning") systems is of particular relevance. The challenge of computerizing expertise in a readily exploitable manner is still open, and is likely to remain so for a quite a few years ...

7. Life Cycle Costing Engineering Domain

The Cost Domain deals with those analyses which are concerned with the cost-effectiveness of systems, usually this is referred to as Life Cycle Costing. In the following, the objectives of the development of the cost domain are given, together with a description of its key features.

7.1. The Objectives

The Cost Domain provides a facility for Life Cycle Costing (LCC) studies. These LCC studies aim to measure the cost-effectiveness of systems ; for example, it is used by :

- customers for competitive tender evaluation,
- suppliers for bid/no bid decisions.

The specification of the cost domain has been completed and it is currently being implemented. The aim of this work is to develop a computerised LCC environment which can support :

- rapid development of LCC analyses,
- an extendible library of cost models,
- interfacing of proprietary LCC tools,
- integration with reliability tools,
- cost sensitivity analysis.

7.2. The Cost Domain Structure

The approach to LCC that has been taken in ACCORD is essentially innovative and has been made possible by the application of Database Management System (DBMS). The DBMS is a key feature which enables data to be readily accessed both within the cost domain and from the other domains. The other key features of the cost domain are the domain manager, the cost model library and the cost prediction and sensitivity analysis tools. The domain manager consists of three main elements known as DEFINE, ANALYSE and REPORT. These are discussed below.

DEFINE

This handles the generation of the LCC model and the storage of the basic system information. The generation of the LCC model consists of :

- selection of Primary Breakdown Structures,
- selection of cost elements and assignment of cost models,
- generation of the LCC model.

The primary breakdown structures identify the classification of cost required with respect to the various phases and activities of the project. The cost model assignment phase requires the analyst to select the cost elements which represent the lowest level nodes on the branches of the hierarchical structure. The analyst then assigns to each node a cost model which will predict the cost for that node. The generation of the LCC model is a system process which verifies the model construction and stores it in the database.

The storage of the system information consists of the analyst specifying the cost information for the system and also the associated accounting information.

ANALYSE

This handles the access to the analysis tools and the tools themselves. Currently the cost prediction tool is being implemented and this is described below. Also there are plans for the development of a sensitivity analysis tool.

REPORT

The report facility enables the analyst to :

i) View system information

ii) Produce documents presenting results for a given tool and specified LCC model.

THE COST MODEL LIBRARY

The concept of an extendible cost model library is a key feature of the cost domain. The cost model library contains cost models which can evaluate costs related to many different types of system. Consequently the analyst can select those cost models which are most appropriate for a given study.

THE COST PREDICTION TOOL

This tool provides the means for the analyst to evaluate the follwoing results :

i) Reference Costs - these are costs which are adjusted to the base year of the analysis.
ii) Reference Cash Flow - this gives a projection of the reference costs onto the project schedule on a year by year basis.
iii) Actualised Cash Flow - this gives the cost on a year by year basis taking account of inflation.
iv) Discounted Cash Flow - this gives the cost on a year by year basis taking account of inflation and discounting.

Prior to the tool's execution the analyst is given the option of modifying the accounting and cost model data.

In conclusion, the cost domain provides cost analysis with a flexible facility for the rapid development of LCC models which are supported by a set of proprietary tools. The key features of this domain are :

i) A flexible modelling facility to construct various hierarchical cost breakdown structures which can be used by multiple tools.
ii) Provision of an extendable set of cost models.
iii) The capability of interfacing to other domains/systems via the DBMS.

8. Computational Speed Up

One objective of ACCORD was to provide an engineering analysis library for concurrent computer architecture which is referred to as APPEAL (ACCORD parallel processing engineering analysis library).

In the first specification of APPEAL two chapters were defined, namely Vector, and Concurrent with the resulting software to be targeted on a number of different architectures. The revised objectives for the vector chapter will focus strongly on the deficiencies in library software by giving priority to algorithms for solving large sparse systems. The machine range will still be broad and encompasses, FPS Vector processors, IBM 3090, Cray XMP etc.

On the other hand "concurrent" systems are a relatively recent commercial option, and the alternative ways of designing a concurrent processing system are legion and thus the market place is gradually filling with systems based on different philosophies. The project has selected the INMOS transputer system as the vehicle for developing routines for the "concurrent" chapter, this product is stable and has sufficient tools to allow software development. The most obvious gains in performance are to be expected in solving the large fully populated systems which arise in physical problems modelled by integral equations, incidentally this approach complements the functionality provided by the vector chapter.

Current Status

a) Vector Chapter

The Vector Chapter of the library has responded to the needs of the market place, and has concentrated on developing well optimised code for techniques that address the solution of a linear matrix equation, where the matrix possesses an irregular sparsity pattern. A range of gradient style methods have been implemented, which between them allow any matrix form to be tackled, together with a new storage [6] scheme that enhances the performance of these very important methods.

To demonstrate the effectiveness of these new implementations two approaches have been adopted : firstly, to replace the solution code in a commercially available package ; secondly, to write two application packages based specifically on these new implementations. These projects are still under way but initial results show the value of using the optimised code.

What can it be used for ?

Many application codes today are based on differential operators approximated by using the Finite Element technique. Such codes have at their heart the solution of large irregularly sparse linear matrix equations. These codes are potential candidate for improvement. The designer is constantly requiring faster results, more robust results, results he can put more confidence in ; and a faster solution time provides flexibility to address these further requirements.

What needs to be done ?

The optimization of gradient techniques has in our preliminary studies shown itself to be a worth while activity but a number of areas need fuller investigation. More work needs to be done in porting the presently available code to other vector machines. Also work is still required to squeeze the best performance from a given algorithm on a given system. These are open ended activities when viewed form the lifetime of the project but both activities are vital to the success of the final products.

b) Concurrent Chapter

Mathematical modelling of continuum problems can be tackled on most occasions by utilising either the differential (viz, the vector chapter) or integral description of the physical system. The integral approach has a number of attractive features associated with it but unfortunately the computations involved are normally prohibitive for very accurate studies. For many years now the work-horse of computational mechanics has been primarily the differential approach. Even this approach is beginning to run out of steam : some engineering and scientific problems require too much computing time to produce meaningful results. One solution to this dilemma is to try and exploit the parallelism within the basic algorithm. This open up a whole new vista which intuitively re-introduces integral techniques. This arises since each discrete point at which the integral equation is evaluated is independant of all other evaluation points.

To examine this hypothesis a boundary integral equation which is specific to linear magnetic problems was implemented using an array of transputers. Each transputer was aware of the geometry of the solution domain and the approximate matrix equation was evenly distributed and formed in situ on each transputer. The optimum topology was found to be that of a ring [7]. With this set up, and for large enough problems, efficiencies of greater than 80 % are routinely available (see figure 4).

What can it be used for ?

As intimated earlier a large class of continuum problems can be posed in an integral form that is suitable for high efficiency implementation on loosely coupled architectures, such as a transputer array. This opens the door for a new generation of integral codes built on the experience of the last decade. As chip technology pushed the speed for an individual device to its limits, "instantaneous" design begins to look a real possibility.

What needs to be done ?

A viable approach has been found for reducing the time and therefore the cost of using integral equation techniques in design. The major remaining task is to demonstrate the effectiveness of the outline approach by utilising greater number of processors and examining more areas of application. By the end of the project both of these goals will have been met.

For dynamic problems it is necessary to obtain both eigen-values and eigen-vectors. An initial study has shown that transputers can also be employed effectively for eigen-value extraction. this work will be intensified.

9. Conclusions and Perspectives

Enhancing design analysis with the help of computers is quite a vast subject indeed. Understandably, the project had to focus on highly selective topics to ensure that concrete results would emerge. This focusing exercise was largely dictated by the expertise available in the Consortium. Nevertheless, it appears that quite useful results will be obtained, demonstrating such basic concepts as multi-disciplinary integration of tools and data, the encapsulation of engineering expertise and the speedup of computational algorithms. The project will produce :

- a library of vectorized and parallelized mathematical routines on FPS and IBM 3090 and transputer hardware, respectively (APPEAL),
- a new software tool (infrastructure) for life cycle costing of systems,
- an integrated reliability analysis package featuring planning facilities (knowledge based), reliability prediction, failure mode and effect analysis and fault tree analysis,
- an integrated heat transfer analysis package featuring advanced interfaces for project management as well as a base of computerized expertise in the form of templates,
- a demonstration of the multidisciplinary integraton of the above packages using one single common database managed by the RDBMS Ingres (on Vax/VMS).

Exploitation of those results can be as broad and as multi-faceted as the subject is: these tools can be tailored and bundled (in terms of scope and interfaces) to adress the needs of currently different classes of potential users. Thus.

738

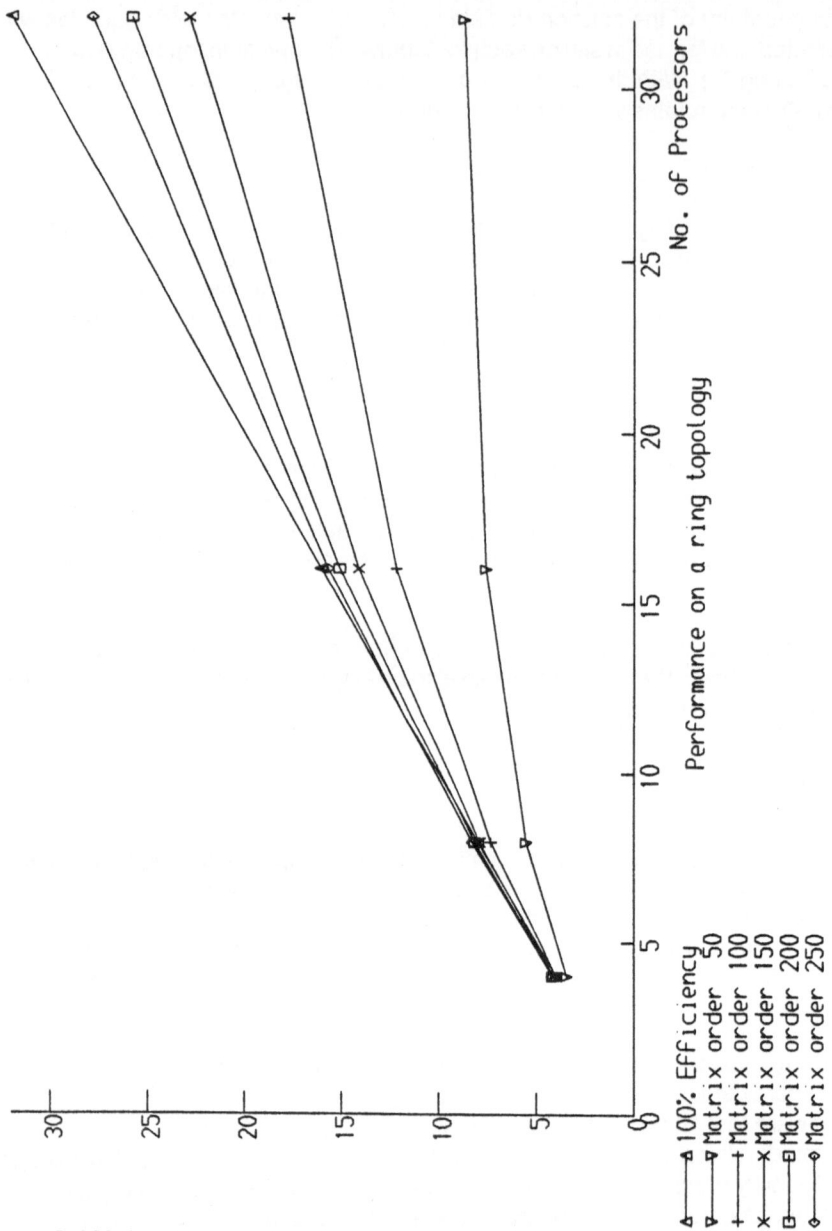

Figure 4

Efficiency comparisons for a range of matrix sizes and ring sizes

- PCB designers (printed circuit boards) could make good use of a reliability prediction tool as well as of a temperature evaluation template. Being usually neither reliability nor heat transfer specialists, they therefore require dedicated tools ; those could be derived from ACCORD and be packaged in the form of add-on optional modules to commercial offers for electronic CAD systems,
- packaging engineers would require similar tools (in nature) as the PCB designers, but of broader scope and focused on mechanics. Thermal templates could also be assembled in the form of an add-on package to commercial CAD systems,
- reliability specialists would make full use of the reliability package as it is being developed in ACCORD. More tools, however, should be added, such as maintainability modelling,
- similarly the needs of heat transfer specialists could be met by the ACCORD facility, provided the range of both templates and general purpose tools, is extended. This will definitely be pursued by several partners,
- the life cycle costing facility can be quite useful to Project Managers who need to evaluate the costs of large projects,
- the multi-domain integrated environment (LCC, reliability, thermal) is also of significant value as a nucleus on which to build a company-wide design analysis facility, although in that respect obviously much more work needs to be accomplished before an effective tool can be obtained,
- finally, APPEAL can be used by package developers to optimise their software tools in the case of selected target hardware platforms.

All of the above routes for exploitation will be given consideration although clearly only a limited number can be actively pursued in the course of this project.

References

[1] Ron Waxman - "The VHSIC Hardware description Language - A Glimpse of the Future"
1986   IEEE   Design & Test

[2] J.D. Nash and L.F. Saunders - "VHDL Critique"
1986   IEEE   Design & Test

[3] Hockney, R.W. and Jesshope, C.R., "Parallel Computers (Architecture, Programming and Algorithms)", Adam Hilger Ltd, Bristol, 1981

[4] Cockton, G., 1988. Interaction Ergonomics, Control and Separation : Open problems in User-interface Management. In Systems, Vol 29 n 4 May.

[5] Fischer, G., 1987. An Object-oriented Construction and Tool Kit for Human-Computer Communication. In Computer Graphics, Vol 21, n 2.

[6] Fitzsimons, C.J., "Performing Matrix Substitution with the storage scheme", ACCORD/WP4/TCD/WD/006/25.11.1988/CF

[7] Bryant, C.F., Roberts, M.H. and Trowbridge, C.W., "Implementing a Boundary Integral Method on a Transputer system", COMPUMAG 1989, to be published.

Project No 1556

SYSTEM BUILDER AREA OF VITAMIN PROJECT DESCRIPTION: MAJOR ACHIEVEMENTS

S.ALLARI (*1), C. RIZZI (*2), M. HAGEMANN (*3) , C. TAHON (*4)

(*1)   TEAM s.r.l. - Via Verbano, 2 - Ispra (VA)- Italy
(*2)   Politecnico di Milano - Dpt. di Meccanica - P.za L. Da Vinci, 32 - 20133 Milano - Italy
(*3)   FHG IITB - Sebastian Kneipp-strasse 12-14 - D 7500 Karlsruhe - 1
(*4)   LGIL Université de Valenciennes - Le Mont Houy - 59326 Valenciennes Cedex

ABSTRACT. This paper describes a prototype for the generation of user interface to CIM applications, in particular scheduling and manufacturing control packages.
The prototype covers the System Builder area of VITAMIN ESPRIT I project n. 1556. Purpose of this toolkit is to demonstrate feasibility of UIMS and System Builder approaches within CIM work area.
It allows the design and the fully automatic implementation of a user interface using graphic and interactive tools. Prototype implementation has been carried out on DEC GPX II workstation and BULL workstation, both using Unix Operating System, and X-Window System.
Toolkit performances have been evaluated through a test benchmark especially aimed at controlling system packages and shop floor monitoring.
Qualifying features of the prototype are portability, introduction of graphics and standards, and easy customization to different application fields.

1 . Introduction

The role of the people on shop floor in manufacturing industry varies from direct action to the supervision role in which their intervention will be necessary outside the automatic mode of production. By this, the man-machine dialogue must be more sophisticated and use more graphic capabilities since graphic information is more powerful and easier to use.
This paper presents a toolkit which has been developed within System Builder area of an ESPRIT I project called VITAMIN (VIsualization sTAndard tools in Manufacturing INdustry)[1]. Project duration is from December 1986 to October 1989. The work is done in co-operation with: Syseca Temps Reel (Prime Contractor), Bull MTS and Université de Valenciennes from France, Mannesmann Kienzle and FhG-IITB from FRG, and Team and Politecnico di Milano from Italy.
The purpose of VITAMIN is the development of two sets of software tools in order to increase the use of graphic systems on shop floor in manufacturing industry:

-   the Active Management Dashboard (AMD) toolkit that has to support management decisions in a wide range of production systems. Syseca ORDO scheduling package has been adopted as reference [2].

- the Active Control Dashboard (ACD) toolkit that provides facilities for the remote control of production process. Mannesmann Kienzle KIBIS package has been adopted as reference [3].

VITAMIN takes into account that in computer integrated manufacturing functions for organization as well as for process control systems must be integrated and available at the workshop place. The developed tools are suited for both and provide several techniques for available graphic presentation and dialogue.
Main goals of the project are the following:

1. *Hardware independence*. The availability of the VITAMIN results at workstations of different manufacturers is made possible using X-Windows System.
2. *Application-independent user interface*. This feature enhances reliability and maintainability of the whole system, since updating the application code has no effect over the Interface code, and vice versa.
3. *Multi-threaded dialogue*. Several interactions can be made by an end user simultaneously and independently from each other. Multiwindowing and multitasking are the basis for an appropriate user interface.
4. *Reduce user interface development time*. Interactive and graphic tools should be developed in order to allow the construction of each part of the user interface.

VITAMIN toolkit design has been based on rapidly evolving UIMS (User Interface Management System) technology [4][5][6].
Several definitions of UIMS can be found in literature. Herein we recall a definition which explains how a UIMS is used [6]: " A UIMS provides a way for a designer to specify the interface in a high-level language. The UIMS then translates that specification into a working interface, managing both the details of the display and its associated input and output and also the interaction with the rest of the program ...".
The above mentioned objectives are reached through the System Builder concept of VITAMIN. In such a context the user interface development consists of two parts: first, a logical model of the user interface is built through a set of interactive graphic tools, in a second step, this logical model is taken as a basis to build the user interface code that controls the man-machine interaction.
Prototype implementation has been carried out on DEC GPX II workstation and BULL workstation, both using Unix Operating System, and X-Window System.

2. VITAMIN Toolkit architecture

Overall system architecture is shown in Fig. 1.
During architecture definition two existing packages have been considered: an application package (e.g.ORDO or KIBIS), and a business graphic tool (application administrator tool).
An External Data Base has been adopted in which application data are mapped and connected with the graphic information.
VITAMIN toolkit generates:

a. the modules to have access to the data generated either by the application package or by the application administrator tool. These data are stored in the standard External Data base;

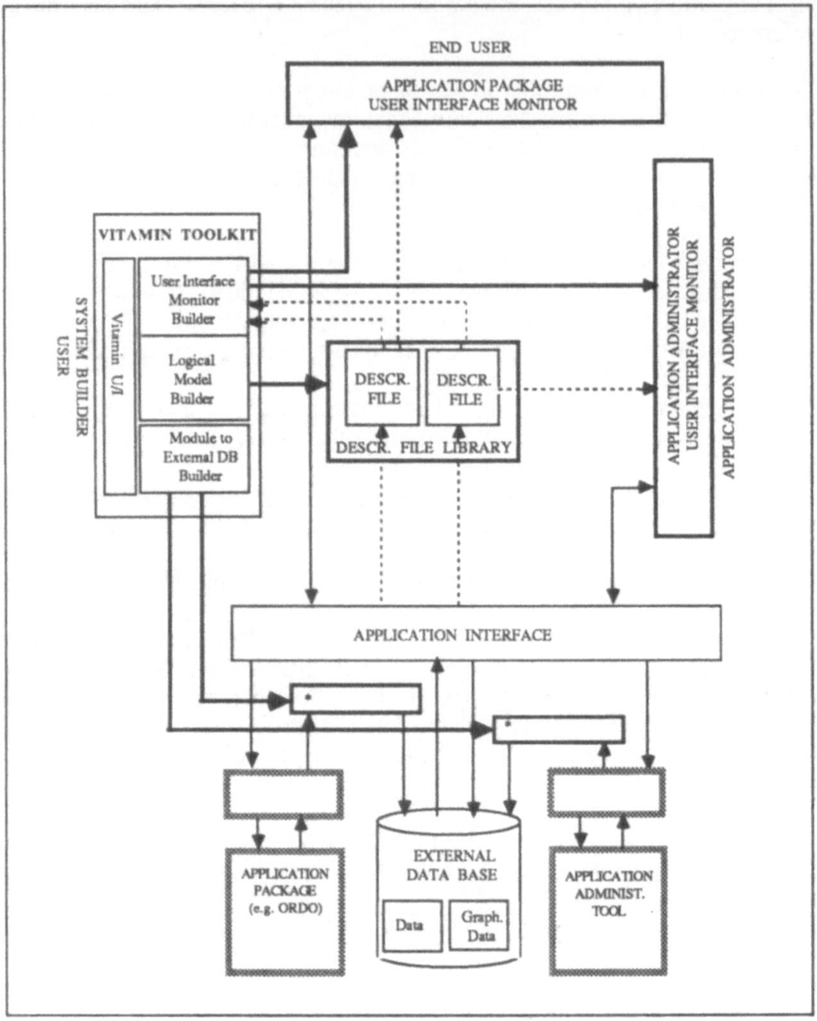

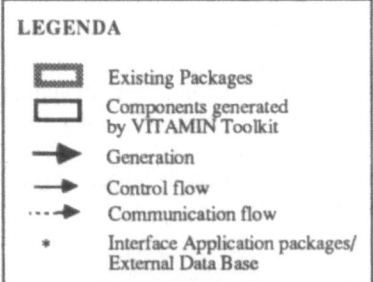

Fig. 1 - Overall system architecture

b. the User Interface Monitor (i.e. the user interface) for the application administrator tool;

c. the User Interface Monitor for the application package.

This type of configuration implies:

a. the *system builder user* : the direct user of the toolkit who is directly involved in user interface design;

b. the *application administrator*: the one using the appropriate user interface to generate the graphic data;

c. the *end-user* : the user of the application package; in the case of a scheduling package he is the manager of the production unit.

VITAMIN toolkit is a UIMS based on the System Builder approach which splits the user interface construction into two phases: user interface logical description and user interface code generation.

Two tools have been considered in order to support these activities (Fig. 2).

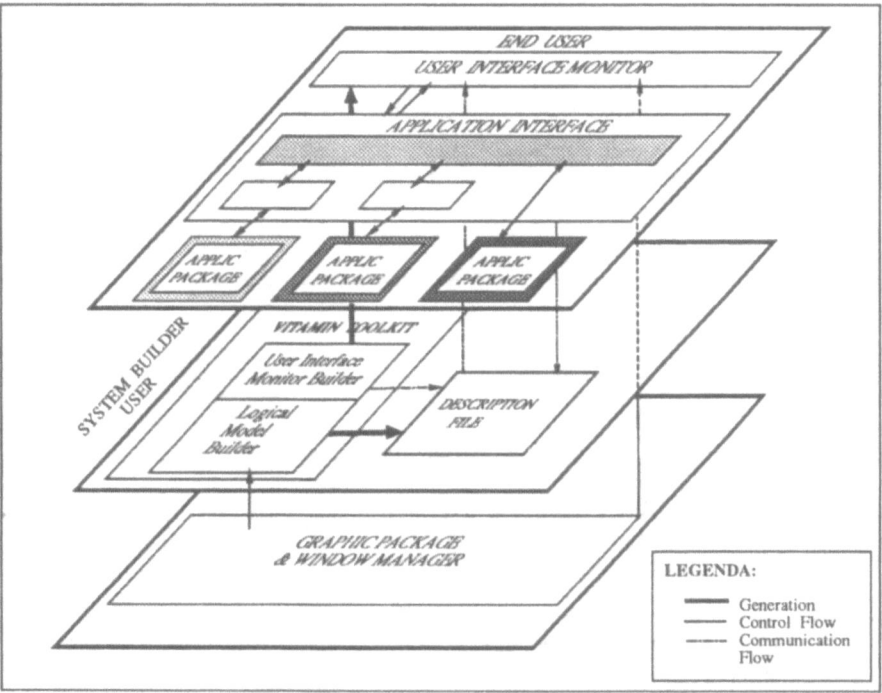

Fig. 2 - Layered System Builder Toolkit

744

The first one is the *Logical Model Builder* which allows the user interface designer (called System Builder User) to define interactively the logical model of the user interface to be constructed and to store it in a file called *Description File* .
The second one is the System *Builder* (User Interface Monitor Builder) [7], a software tool that automatically generates the user interface code (User Interface Monitor) for an application package on the basis of logical description without requiring hand coding.
The *User Interface Monitor* is the program which manages the interaction between the end-user and application modules. Most emphasis is put on the logical specification of the user interface instead of coding, since the System Builder automatically generates the code.

3. Logical Model Builder Tool

At this stage of the project the Logical Model Builder has been completed.
Its structure is based on the Seeheim model which splits a user interface into three components: Presentation Techniques, Dialogue Control, and Application Interface components (Fig. 3)

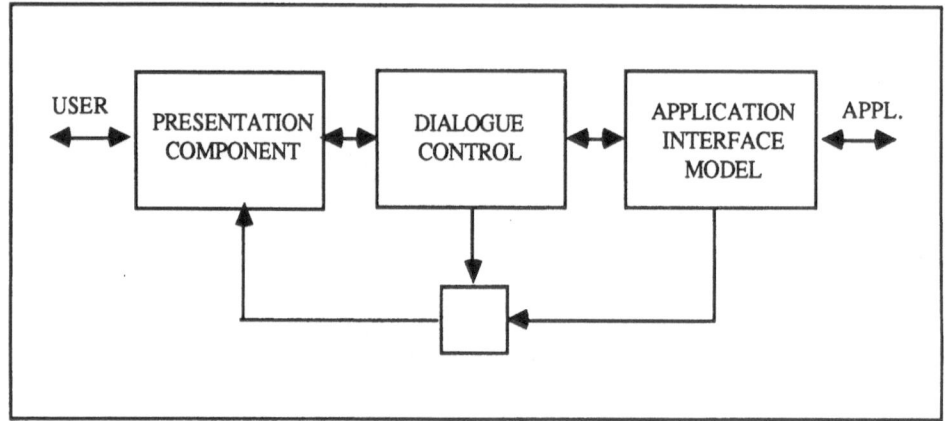

Fig. 3 - Seeheim model

The Presentation Techniques component deals with the physical representation of the user interface (input and output devices, screen layout, interaction and display techniques).
The Dialogue Control component defines the structure of the dialogue between the user and the application and can be considered the kernel of the user interface. It can be viewed as the syntactic level of the user interface.
The Application Interface component defines the interface between the user interface and the application procedures, i.e. the semantics of the application. It is in charge of application routines calls, and in particular in CIM environment, this component must be able to intercept and manage abnormal situation (alarm) coming from the shop floor.
In order to help the user interface designer during user interface definition phase three graphic and interactive modules have been developed: Presentation Techniques, Dialogue Control and Application Interface modules.

Each of them produces a Description File containing a logical model of the corresponding user interface component. This allows the modification of one of the user interface components without affecting the other ones. A common structure for the Description Files and a communication protocol among User Interface Monitor components has been adopted.

This multi-level model allowed a subdivision of workload among partners as follows:

- Presentation Techniques module: FhG-IITB
- Dialogue Control module: Politecnico di Milano and Team
- Application Interface module: Universitè de Valenciennes

3.1   PRESENTATION TECHNIQUES MODULE

Graphic monitors allow the representation of process data adjusted to the needs of the user. In order to enable a comfortable presentation of process states, appropriate tools and methods have to be developed. The task of user interfaces is to present application data to the user in a proper way which has been fixed, for example, by the system designer, and to provide a certain level of independence for the application. Development of a user interface depends not only on the aspects of man-computer interaction, but also on process and control system characteristics.

Especially process requirements make heavy demands on the user interface structure; for example, parallel applications have to be handled and data values have to be displayed continuously.

Methods are offered to the end-user which allow to create his own personal interface according to his needs.

Presentation Techniques descriptions are generated by the system designer. To design them he uses an interactive graphic tool. This tool generates a Description File that is processed through the Logical Model Builder compiler (Fig. 4) and converts it to a form suitable for the User Interface Monitor.

The User Interface Monitor Builder takes this description file and generates from it the target system using a library of object-class-descriptions.

An interesting aspect of our work is the fact that the graphic tool of the Logical Model Builder will be implemented as an application of the User Interface Monitor. That means that the same procedures are used to edit the layout of the screen and to present the layout in the User Interface Monitor.

3.1.1   *Object Oriented Approach.* To achieve the considered goals the paradigms of object oriented programming style have been considered. In addition to some other features, the following items characterize the object-oriented approach:

- abstraction
- encapsulation
- objects
- messages
- polymorphism

A presentation technique object is an entity which is characterized by the actions defined on it. These objects are encapsulations of abstractions, i.e. details of the implementation are not essential to an understanding of its purpose or functionality.

746

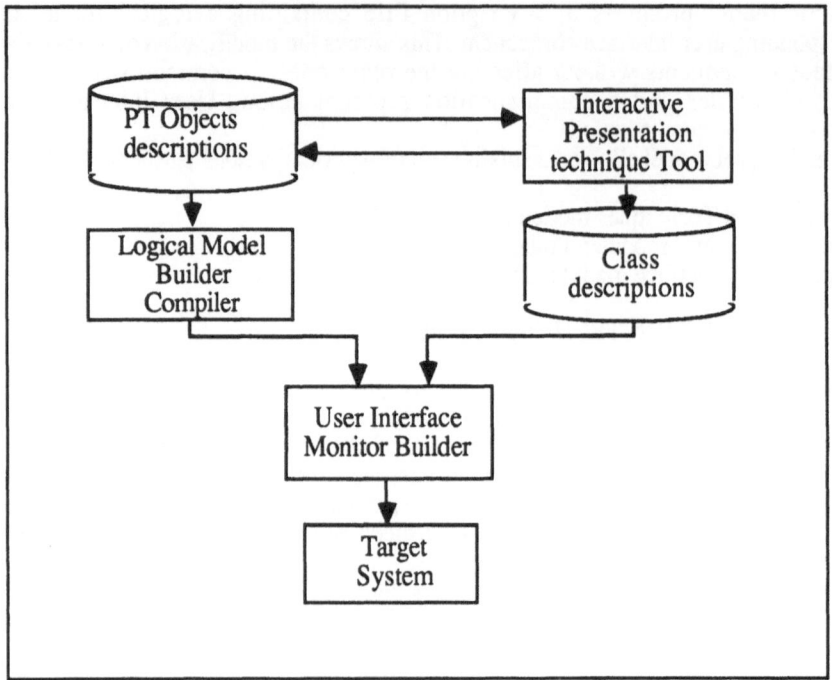

Fig. 4 - The Logical Model Builder (LMB) part of the Presentation Techniques (PT)

Encapsulation is a technique for releasing an abstract-data-type. This includes the data-handling of the concerned structure and the visibility of the functions and procedures that can be used to manipulate them.

The behaviour of an object is characterized by its responses to the various messages it receives.

Messages are data-structure "sent" to objects which interpret these messages by triggering actions.

Polymorphism allows us to send the same type of message to different objects and have each object respond on the basis of the kind of objects it is.

Each object contains information on the graphic aspects and supervises the constraints [8].

For example such constraints make possible the connection between critical values and graphic attributes. This makes the presentation technique especially useful for CIM applications. For instance, it is possible to define graphic equivalences for the presented data, where the representation of the data is changed automatically by the object whenever critical values of the data are exceeded. In addition to this, the behaviour of an object is described by means a transition network. This function can be used by the dialogue designer to integrate syntactical and semantical descriptions of the dialogue.

A presentation object consists of:

- a description of the graphic aspects
- the monitoring description (influence of application data to graphics)
- a buffer for application data
- the description of the behaviour

- the list of generic functions that can be applied to the objects class
- the hierarchy of the object

3.1.2   *Communications with the Dialogue Control.* If the Dialogue Control wants to communicate with Presentation Techniques it has to create a communication line. It can then create a screen-layout.

In order to have access to Presentation Techniques-object the Dialogue Control component at User Interface Monitor level has to call the Presentation Techniques-Access function which returns an internal identifier for the object. This identifier is used in order to send data to the object or to receive data. Whenever the Dialogue Control wants to know if an object was selected by the user it may call a Presentation Techniques_Select function which returns the "id" of the selected object and a selection mode.

3.2   DIALOGUE CONTROL MODULE

This Logical Model Builder module allows the user interface designer to define the logical model of the interaction between the end-user and the application package, i.e. the structure of the dialogue and the commands used by the end-user.

It depends heavily on the model adopted to describe and modelize the dialogue. Different notation [5], [9], [10], and [11] can be found in literature .

Transition Networks are one of the best known notations. They are based on the concept of the user interface state and consist of a set of states and transitions from one state to another. It is possible to distinguish three different types of Transition Networks: Simple Transition Networks (STNs), Recursive Transition Networks (RTNs), and Augmented Transition Networks (ATNs); these last ones are an extension of RTNs and allow the definition of context-sensitive dialogues.

Other notations that have been analysed are Context-free grammar and Event languages [9].

In the evaluation of previous notations the following parameters have been taken into account:

- *descriptive power* that is the range of user interface which can be described by the notation;
- *ease of use* ;
- *ease of learning.*

The last two parameters are mainly considered from the end-user point of view. In fact one of the Vitamin Toolkit goals is to involve more directly in the user interface development people working on the shop floor, who usually have no experience in this field.

The ATN notation has been chosen: this choice is due to the fact that ATN formalism is sufficiently well known and easily understable by non-programmers and the dialogue model can be graphically represented.

An ATN graphic editor has been implemented [12]. It makes available to the user interface designer a complete set of editing functionalities (including object creation, modification, deletion, storage, etc.), and a consistency check facility (based on graph theory) which verifies the syntactic correctness of the ATN. Moreover interactive Dialogue Control module allows expert users to directly create/modify a Description File through the "vi" text editor (Fig. 5).

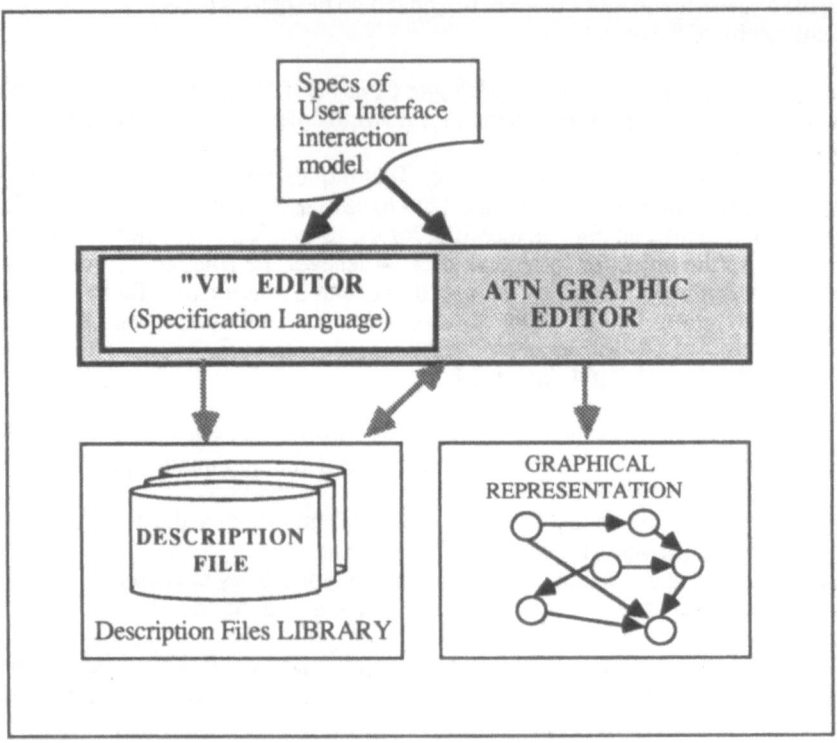

Fig. 5 - Dialogue Control Module

In such a context the user interface interaction model is described by a main network and a set of sub-networks. Each of them consists of:

- set of states (graphically represented by circles);
- set of transitions (graphically represented by directed arcs);
- set of square states (graphically represented by square);
- set of conditional functions (written in 'C' language).

where:
state is a static situation in the dialogue between the end-user and the application package (e.g. when the system is waiting for a user action);
transition describes how the dialogue moves from one state to another;
square state represents a state reachable from any other state;
conditional function is a function attached to each transition which determines if the transition can be performed; this allows the definition of context-sensitive dialogues.
In order to complete the dialogue description and to define the interface with the other two components a set of 'ad hoc' information has been associated to each object.
The information associated to a state refers to the corresponding user interface graphic context, while the information associated to a transition refers to the corresponding rules and actions performed at user and application level. These actions represent the links towards Presentation Techniques and Application Interface components.

3.3   APPLICATION INTERFACE COMPONENT

The Application Interface module architecture is shown in Fig. 6

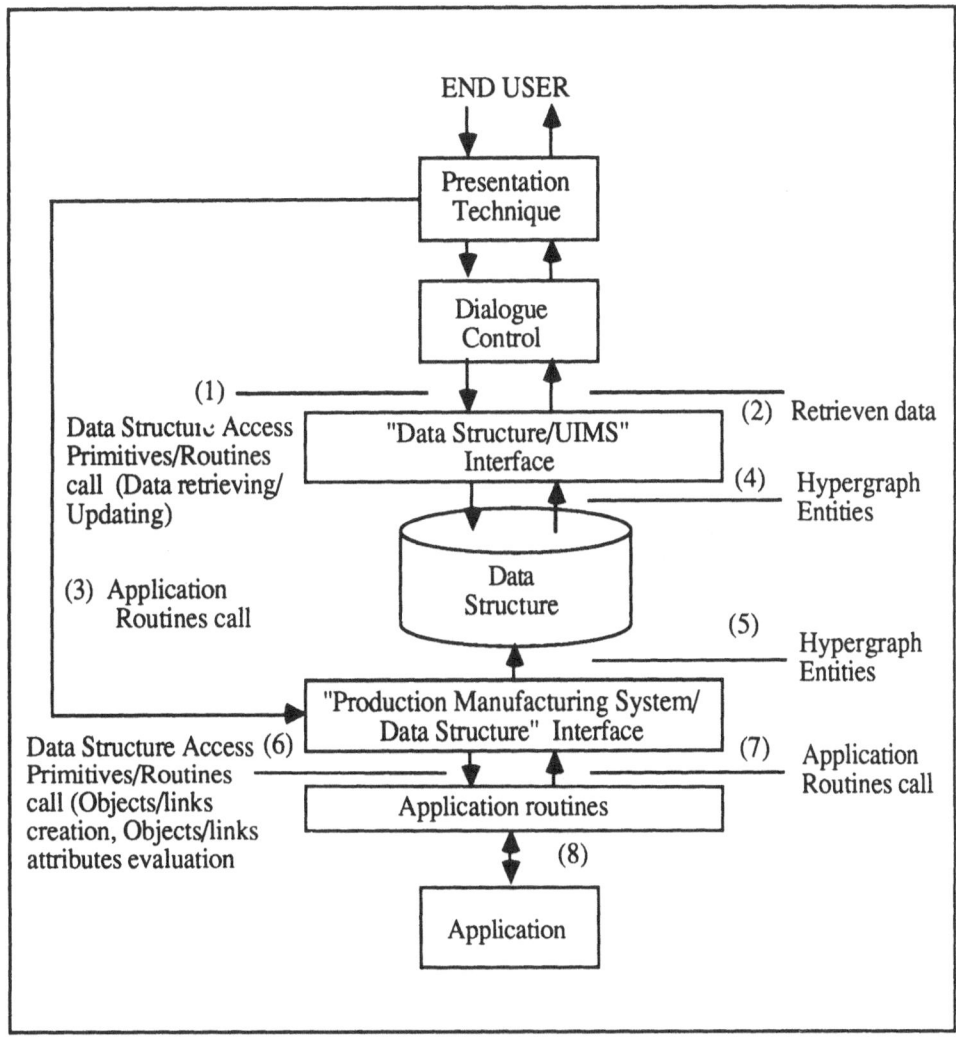

Fig.6 - Application Interface module architecture

This structure is designed to take into account the following objectives:

- management of static data, corresponding to the semantic of the application
- management of dynamic data, representing the current state of the system
- management of data exchange between Dialogue Control and application package

The Application Interface Component is logically divided into three parts:

- Data Structure. It supports the production management and control system. This structure ensures the semantic coherence of the application software package. The logical model of data is derived from the Hypergraph model [13][14].
- Data Structure/UIMS Interface. It concerns updating and consulting functions.
- Production Manufacturing System/Data Structure Interface. It corresponds to the updating functions which are necessary to modify dynamic data.

The following functionalities can be distinguished (Fig. 6):

a) From the application to the Production Manufacturing System data structure (6) (8):

- creation of objects or links at request from the application package (static data management);
- evaluation of objects and links (dynamic data management)

These functions are provided by calls of application routines.

b) From Dialogue Control to the application (3) (7). In case of direct actions from the user to the application package.
c) From Dialogue Control to Data Structure (1): static and dynamic data management
d) From Data Structure to Dialogue Control (2): sending of the data to be displayed, extracted by access routine call, performed by Data Structure /UIMS Interface.

The design of the Application Interface functionalities requires the specification of the static and dynamic data structure and the associated management routines. The used methodology consists in (Fig. 7):

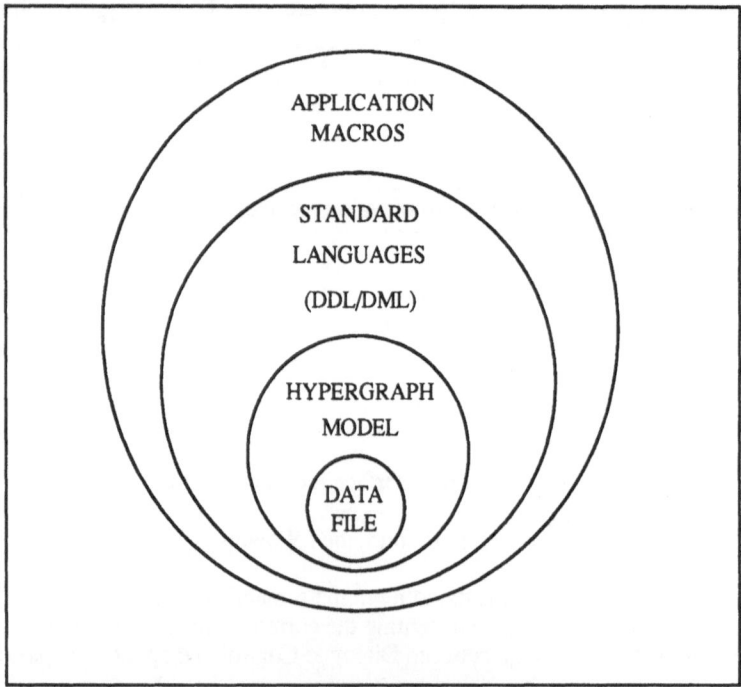

Fig. 7

- static data modelization, using Data Structure standard entities (classes, objects, links, etc.)
- creation of the frame of the static data structure using the Data Description Language (DDL)
- development of 'Macro-Routines' for a specific application using the standard Data Manipulation Language (DML)

For the development of a particular Application Interface the following tools are available:

- the Data Description Language that creates the schema of the data structure.
- the Data Manipulation Language that creates and evaluates Data Structure objects

We can also use these tools in an interactive way:

- a graphic editor developed upon the Data Description Language;
- a graphic editor developed upon the Data Manipulation Language.

These graphic tools provide the Application Interface designer with the possibility to design the Data Structure in an interactive way

4. Benchmark at Logical Model Builder level

In this section a benchmark scenario related to CIM applications for the prototype is presented. The benchmark simulates a manufacturing environment: the prototype has been used to develop and manage the user interface of a monitoring and scheduling package. This test allowed an evaluation of the Logical Model Builder ability to define a logical model of interaction specific to these applications. In such a context the system must allow the management of activities typical of these environments: shop floor data monitoring, and abnormal situations handling, such as alarms and warnings. Purpose of this benchmark is to evaluate the feasibility of adopted approach and integration among the existing modules developed by different VITAMIN partners.

It has been supposed a production system limited to a lathing section (SECT_B) that consists of: lathes D3 and D4, and robot T1 (transportation system). D3 and D4 execute the same working program, while robot T1 loads parts to D3 and D4 and unloads them once they have been machined.

Each element is controlled by a DNC, while a SUPERVISOR (application package) is responsible for system running mode management.

Figure 8 shows a situation in which one of the lathes (D3) breaks down, as it could appear to an ACD operator. The system allows the operator to perform a re-scheduling of the production unit, in order to let robot T1 feed only the working lathe D4, and to stop D3.

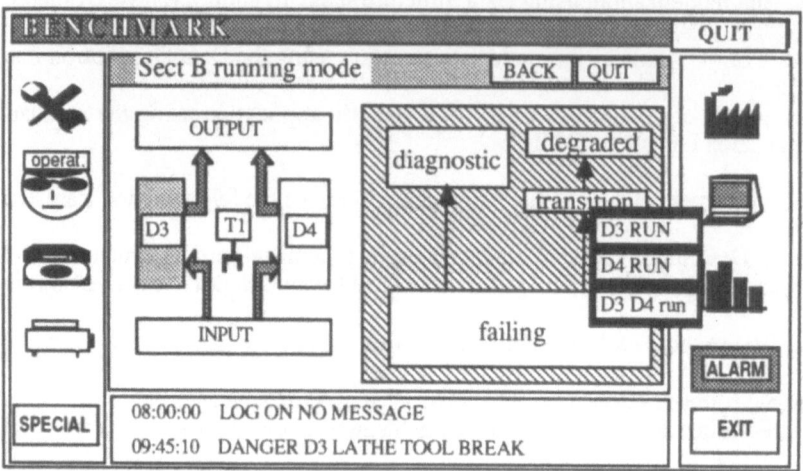

Figure 8 - A shot of benchmark execution controlled by the prototype

As a first step in the definition of the logical model for the user interface, the man-machine interface programmer can "draw" some graphic objects and arrange them on the screen: this scene represents what will actually appear to the ACD operator.
Actions to be performed at this level are:

- Definition of the basic Graphic Objects (e.g. Icons, Menus,.etc.)
- Construction of a whole graphic context, using available Graphic Objects

Fig.8 represents a sample layout that refers to the benchmark.
At Dialogue Control Module level, the user interface programmer can define the syntax of the Dialogue, through the use of a graphic and interactive Graphic Editor. As explained in section 3.2, the Augmented Transition Network formalism was chosen to define Dialogue Control. Figure 9 shows a part of the ATN that is used to control the Alarm Management of the scenario (picture of the actual prototype).
In the Application Interface Module the connection between application programs and variables on one side, and the Dialogue Control Component on the other are set, as well as data structures that allow data exchange.
Reconfiguring the system requires no other effort than the changing of the specifications of the logical model, by means of the tools supported by the prototype: Presentation Techniques, ATN, and Application Interface editors. Neither hand coding nor recompilation or link operation is required, since the User Interface Monitor Builder can directly convert the logical model specifications into an executable runtime User Interface Monitor.
The communications between the User Interface Monitor and the supervisor are performed by message sending (Alarms, Warning and Normal messages).

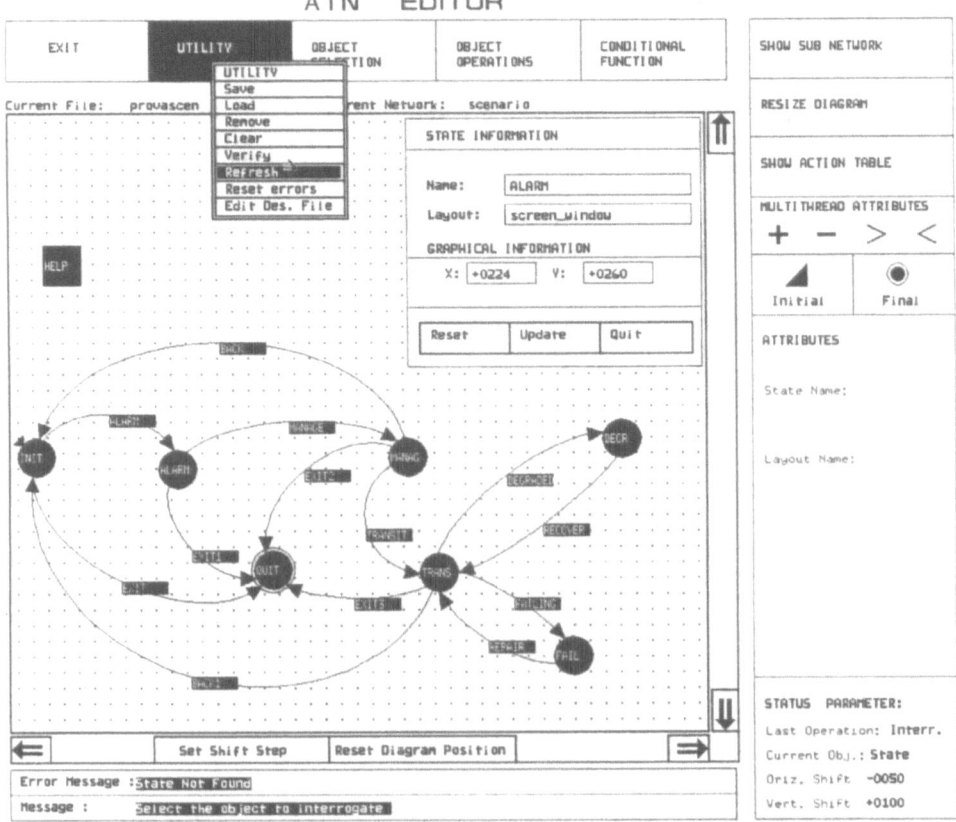

Figure 9 - A Dialogue defined using the ATN editor

5. User Interface Monitor Builder tool

In the final phase of the project the implementation of the User Interface Monitor Builder is being completed. The main tasks of this tool is to build up automatically the source code of the User Interface Monitor, to compile, and to link it with the application modules in order to obtain the executable code.

BULL is the responsible for User Interface Monitor Builder development. Basis of this work is a prototype developed by Politecnico di Milano as study case.

The prototype generates the User Interface Monitor code using the *Skeleton* technique [15] [16]. The skeleton is a predefined standard User Interface Monitor (source code) and it has to be customized with specific code for each different application.

Main structure of Skeleton program which is being developed by BULL is as follows [16]:

- load data structure for user interface control (Dialogue Control loop) from Description Files;

- runtime control of the user interface.

The User Interface Monitor Builder is in charge to complete and to adapt the skeleton for the specific application package.
It generates the lacking source code, retrieving necessary information from the Description files (defined with the Logical Model Builder modules) and produces the User Interface Monitor source code inserting generated code in predefined location into the skeleton.
All these operation are handled automatically by the tool; the user interface designer is not involved in writing code, he has only to deal with logical descriptions at Logical Model builder level.
The complete User Interface Monitor source code is then compiled and linked with application programs.
In the User Interface Monitor Builder implemented by BULL, the code inserted into the skeleton concerns [16]:

- the Presentation techniques Description file;
- the Dialogue Control Description File;
- the application functions to be called.

related to the specific application package.
Results of BULL work is being tested on the same benchmark scenario (proposed by Universitè de Valenciennes) which was already used for the Logical Model Builder part, possibly with minor modifications. This allows an evaluation of integration between Logical Model Builder and User Interface Monitor Builder part.

6. Exploitations of results

Since both industrial and academic partners were involved in the project, it is necessary to take into account the main objectives of both of them in order to evaluate project results.
On a research side, this project revealed to be a fruitful and in-depth experience in the field of Computer Integrated Manufacturing, and the introduction of concepts such as UIMS and System Builder, places it among state of the art developments.
Research institutes involved in the project are strongly interested in continuing development and testing of implemented prototype in order to point out the advantages and the drawbacks of the adopted solutions, and to investigate new applications that could profit from this approach.
VITAMIN Project results have been presented in various national and international conferences, in order to spread qualifying points of the project among research and University institutes, and European industries.
Industrial partners plan to exploit the relevant potentialities offered by the product. The prototype will be used for internal software development, i.e. for office automation and/or control packages.
Moreover, contacts are being established with several companies, both private and public, that showed themselves very interested in the product. Target companies are usually small or medium size (but even bigger ones) mechanical, manufacturing or electronic industries that deals with CIM, supervision, automation & control problems.
An internal testing, possibly with other industrial partners, is also forecasted, in order to verify the system reliability in such critical applications, and to study new industrial applications.

Collaboration between partners coming from different countries, but most of all, showing different viewpoints on project matter, was a stimulating challenge to everyone of us: in fact it represented a checkpoint of our own working methods, and allowed us to deal with different partners' needs.

6. Conclusions

Major achievements of the project can be summarized as follows:

- *Portability.* Development of the toolkit has been performed upon X-Window System. This should allow an easy porting on different and/or distributed hardware. In order to enhance portability the 'C' programming language was adopted.
- *Introduction of graphic and standards in CIM environment.*
- *use of emerging UIMS concepts in industrial environment.* In particular the System builder technique has been adopted for the prototype. This feature will allow an easy customization of the system, in order to fit to different application packages in different industrial environments.

System Builder modularity consented a granular subdivision of work among partners. Each module has been developed as a separate item, since it was sufficient to define links among different modules.

Acknowledgements

The authors would like to thank following people for their valuable co-operation in writing this paper: G. Barzaghi, M. Bordegoni, U. Cugini, R. Nagel, Rorich, R. Soenen, T.A. Shamsi, J. X. Wang, and all other project partners.
A special thank to the reviewers for their helpful comments.

References

[1] D. Morin, U. Cugini,G.A. Mauri, The role of graphics within VITAMIN Project, Graphics in esprit Workshop, Brussels, 31 May and 1 June, 1988.
[2] Syseca Temps Reel, ORDO: Presentation Fonctionelle, Vitamin Document (1986).
[3] Mauri G.A., Machnikowski E., Schmidt P.W., ACD architecture, Vitamin document, V.2 (April 14th, 1988).
[4] Pfaff Ed., G.E., User Interface Management System, (Spring-Verlag, Berlin 1985).
[5] Foley J.D., Models and Tools for Designers of User Computer Interfaces, in Theoretical Foundations of Computer Graphics and CAD, NATO ASI Series, Series F, Vol. 20 (Springer Verlag,1988), pp. 1121-1151.
[6] Lowgren J., History, State and Future of User Interface Management Systems, ACM/SIGCHI, vol. 20, n. 1 (July 1988).
[7] Allari S., Cugini U., Rizzi C., VITAMIN Toolkit Architecture, V.4, VITAMIN Document, (April 22nd 1987).
[8] M. Hagemann, U. Schreiber, R. Nagel, Object oriented Design of User Interface of CIM applications, submitted to Hawaii International Conference on System Science, 1990.

[9] Green M., Report on Dialogue Specification Tools, Computer Graphics Forum 3 (1984) pp. 305-313.

[10] Green M., A survey of three Dialogue models, ACM Transactions on Graphics, 4, 3 (July 1986), pp. 244-275.

[11] Olsen D.R., Pushdown Automata for User Interface Monitor Management, ACM Trans.on Graphics 3,3 (1984), pp. 177-203.

[12] Barzaghi G., Bordegoni M., Rizzi C., Dialogue Control Module: ATN Graphic Editor, Vitamin document, V.2, (December 30th, 1988).

[13] El Yousefi A., Taghboulh S., Soenen R., An Entity Relationship model for Computer Aided Production Management, Universite de Valenciennes 1988.

[14] Boley, Directed Recursive Labolnode Hypergraphs: a new Representation Language", Artificial Intelligence, 9 (1977), pp. 49-85.

[15] Allari S., Sistema per la generazione automatica di interfacce user-friendly, Thesis in Scienze dell'Informazione, Universita' degli Studi di Milano, A.A. 86/87.

[16] Bouchot B., Poncet F., Architecture of UIMB, Version 1, VITAMIN Document (April 25th 1989).

Project no 1561

ADVANCED ACTIVE CONTROL ALGORITHMS FOR INDUSTRIAL ROBOTICS.
OVERVIEW OF PROJECT SACODY.

Jean-Luc FAILLOT
BERTIN & Cie
BP 3
78373 PLAISIR CEDEX
FRANCE

Abstract.

Project SACODY focuses on the development of techniques and tools which are necessary for the implementation of advanced controls for high-speed industrial robots. Specifically, all mechanical manipulators exhibit at high speed coupled dynamic flexions and torsions which limit their performances and can be reduced or suppressed by appropriate control actions.

Issues being adressed in project SACODY deal with the development of modelling, identification and control methods and softwares, constituting a complete set of tools needed to design adapted new control actions.

The performances achieved by new control are tested on laboratory models and will be demonstrated at the end of the project on an industrial robot.

Major outputs of the project are techniques, their implementation into CAD softwares, and their programmation on a robot controller for test on an industrial robot.

This paper outlines the objectives and some of the current achievements of the project.

1. Introduction

Six European organizations :

BERTIN & Cie (France)
KATHOLIEKE UNIVERSITY LEUVEN (Belgium)
AEG AG (Germany)
LEUVEN MEASUREMENT AND SYSTEMS (Belgium)
KUKA SCHWEISSANLAGEN UND ROBOTER GmbH (Germany)
UNIVERSITY COLLEGE DUBLIN (Ireland),

are cooperating in ESPRIT Project 1561, called SACODY.

This project is in the domain of industrial robotics and attempts to solve some of the long-standing problems in the control of fast automated-assembly robots.

Indeed, flexible manufacturing systems demand ever increasing specifications in terms of speed, acceleration and range along with same or improved terminal accuracy. As a result, weight and inertia of mobile parts can no longer be ignored and traditional control must be supplemented by various actions, not only to compensate for inter-axes couplings but also to prevent or damp out unwanted oscillations resulting from excitation of elastic motions. Altogether, this active control problem becomes more complex as the size of the robot increases or as operating speed and terminal accuracy specifications become more drastic.

Project SACODY ambitions to answer partly this challenge in examining how masses in motion can be reduced and how lost structural rigidity can be replaced with on-line control and well-chosen sensors. Hence, lack of rigidity is not really intended but is rather a mechanical deficiency, resulting from weight reduction and other design trade-offs.

Pratical implementation of such advanced controls has many implications, especially in terms of :

- Dynamic modelling of the manipulator structure behaviour,
- Identification of model parameters from experiments,
- Design of innovative control algorithms to provide active damping,
- Selection of sensors, keeping in mind mechanical integration requirements and overall cost limitations, along with associated filtering algorithms,
- Overall control architecture,
- Real-time implementation on a new-generation numerical controller exhibiting increased computational speed and memory size and enlarged inputs/outputs.

Project SACODY is a four-year project, failling into R and D Area 5.4 (CIM-Flexible Automated Assembly) of the ESPRIT (1986) Work Plan.

2. Objectives

The objectives of project SACODY are :

- To understand and solve the methodological problems associated with on-line control of new-generation FMS robots, considered as articulated flexible structures ;
- To develop and test a versatile CAD software serving together to obtain a dynamic model of a robot, to design multivariable controls and to simulate the behaviour on a numerical computer ;
- To develop innovative sensors ensuring accurate trajectory tracking and vibration control as well as measurement of robot performances ;
- To improve upon off-line and on-line identification techniques to satisfy need for more precise models ;
- To upgrade the capabilities of a new-generation numerical controller to make real-time active control possible ;
- To program the control functions on the numerical controller ;
- To install the equiped robot in an automated-assembly cell, for testing and evaluation purpose ;
- To recommend wide-ranging improvements in active control, on-line sensors as well as on mechanical design, with a view to upgrade globally performances of FMS robots expected to work in demanding environments ;
- To demonstrate such improvements on laboratory models exhibiting a number of characteristics representative of that encountered in flexible robots.

Hence, the activities of project SACODY encompass many disciplines, with a high degree of interactivity and cooperation between partners.

3. Main Aspects of SACODY Project Activity

3.1. Development of Control Methods

One of the main aspect of SACODY project is the development and the test of active control of flexible manipulators, i.e. control which takes into account the structural dynamics of the manipulator. The main advantage of such control methods is that the control bandwidth is not limited by the lowest natural frequency of the manipulator. Hence, by using these new control methods, the natural frequencies of the manipulator structure can be lowered without affecting in a negative way the stability, bandwidth and dynamic accuracy of the control system. The weight and stiffness of the manipulator can be reduced, thereby allowing higher operational speeds and accelerations, reducing energy consumption, and maintaining or even improving the overall accuracy. Since the beginning of SACODY, innovative methods have been defined and have been presented on conferences dedicated to robot control (De Schutter et al. 1988).

3.1.1. Proposed control scheme.

Once derived a model representative of the behaviour of a flexible robot, it is possible to design control actions in order to obtain a desired closed loop behaviour of the robot.

To the variables classicaly controlled by kinematic control, i.e. the joint degrees of freedom and their time derivatives, some variables are added in order to represent the elastic behaviour of the robot. The obtained set of controlled variables is regrouped in a vector of generalized joint coordinates.

In order to control globally these generalized coordinates, the classical steps of robot dynamic control, described by diagram figure 1, must be adapted by inclusion of the treatment of flexible behaviour.

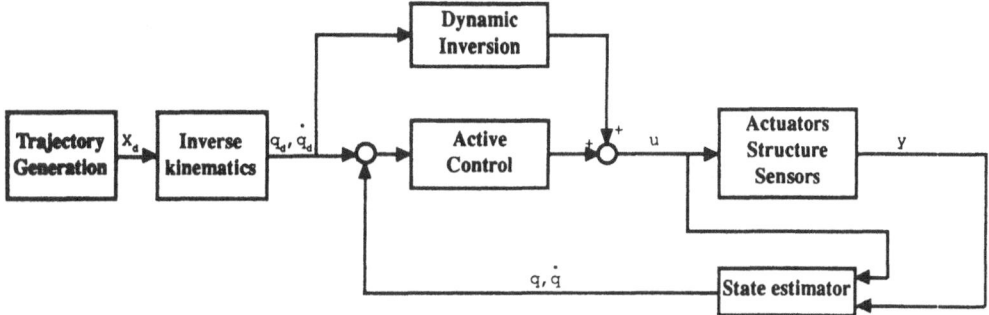

Figure 1 - Dynamic control scheme of robot

- Inverse kinematics :

Static deflection of each link must be taken into account when solving the inverse kinematics problem, i.e. determining the desired joint displacements and their time variations starting from the desired values expressed in a cartesian reference frame.

- Inverse dynamics :

In order to achieve the tracking control, a feedforward action, using inverse dynamics, is computed. This consists of the calculation of the nominal actuator commands at the robot joints as a result of position, velocities and accelerations of the desired robot generalized joint coordinates.

- Tracking control :

For ensuring the precise tracking of the desired joint trajectories and the active damping of vibration modes of the structure, a tracking control is implemented. It is designed on the basis of a low order model (including essential structural modes) ; an observer reconstructs the whole state of the system in minimizing the spillover effects, i.e. disturbing effects of ignored vibration modes on control stability.

The generic form of the model of a flexible robot is :

$$\begin{bmatrix} \Theta'' \\ q'' \end{bmatrix} = \begin{bmatrix} W_{11} & W_{12} \\ W_{12}{}^T & W_{22} \end{bmatrix} \begin{bmatrix} F_\Theta(X) + T \\ F_q(X) \end{bmatrix}$$

where regroups the rigid d.o.f. variables and q the variables representing flexible displacements, moreover

$$X = [\Theta^T \ q^T \ \Theta'^T \ q'^T]^T$$

The tracking control law T is composed of two feedback loops after :

$$T = L_I(X) + W_{11}{}^{-1} L_0(X)$$

The dynamic inversion feedforward $L_I(X)$ which realizes a partial feedback linearization of the rigid modes, so that the closed loop equation of the rigid modes become :

$$\Theta'' = L_0$$

whereas the closed loop equation of the flexible modes, once injected the expression of $L_I(X)$ remain non linear :

$$q'' = f(X) + h(X) L_0(X)$$

The active control loop L_0 which ensures the stability of the rigid modes near the desired motion is, in its simplest form, given by :

$$L_0(X) = \Theta''d - K(X - X_d) = \Theta''d - K dX$$

Assuming an appropriate choice of the full state feedback gain K, the closed loop system can have a good stability. If we linearize the flexible mode equations near the desired state trajectory, the closed loop equation of the system becomes :

$$\Theta'' + K dX = 0$$

$$q'' = f(X_d) + dF(X_d) dX - q''d + h(X_d)(\Theta''d - K dX)$$

which can be summarized by the following matrix equation :

$$X' = A(X_d) dX + B(X_d) K dX + R(X_d)$$

where A, B and R are matrices only depending on the desired state trajectory X_d.

The feedback gain matrix K is adjusted to stabilize with satisfactory dynamics the linear time-varying tracking error system respresented by the matrix pair (A,B).

3 1.2. Experimental validation of the proposed control algorithms.

As a part to the verification process, before extending to the full size robot, considerable effort has been brought on the specification, design and construction of representative breadboard models in order to validate and test experimentally performances of new control laws.

Five laboratory models have been constructed for SACODY, representing increasing complexity in terms of control of flexible manipulators. They exhibit respectively torsional and flexible modes. They have additional distinctive features, one being with electric DC motor and gear reduction and the other being powered by a direct-drive motor. The fifth laboratory model exhibits two planar flexible links. Collectively, these models form a complete collection of testing situations enabling verification of control algorithms against such defects as gear plays and other mechanical backlashes, various unmodelled dynamics, torque saturations, etc.

We illustrate the results obtained by describing briefly the experiments performed on the fourth model, represented in figure 2. Further details on these results are given in (Swevers et al. 1989).

This device is made of a flexural blade, driven by a direct drive motor through a torsional shaft.

Besides the rigid rotational d.o.f which is the functional d.o.f to be controlled, this system exhibits a vibration mode damaging the overall positionning accuracy : Any torque applied by the motor excites this disturbing extra degree of freedom, which, if no specific action is undertaken, damps out with its sole natural damping. The system has been designed so that the flexible behaviour exhibits resonance at 4.9 Hz, the damping ratio being 0.03.

Figure 3 shows the results of the feedback control of this laboratory device. It is to be noticed that, by use of the specific feedback, the first natural frequency of the closed-loop system has been rejected up to 8 Hz, as illustrated in figure 4, where gain Bode plot of closed-loop transfer between payload position and command is compared with that obtained in open loop.

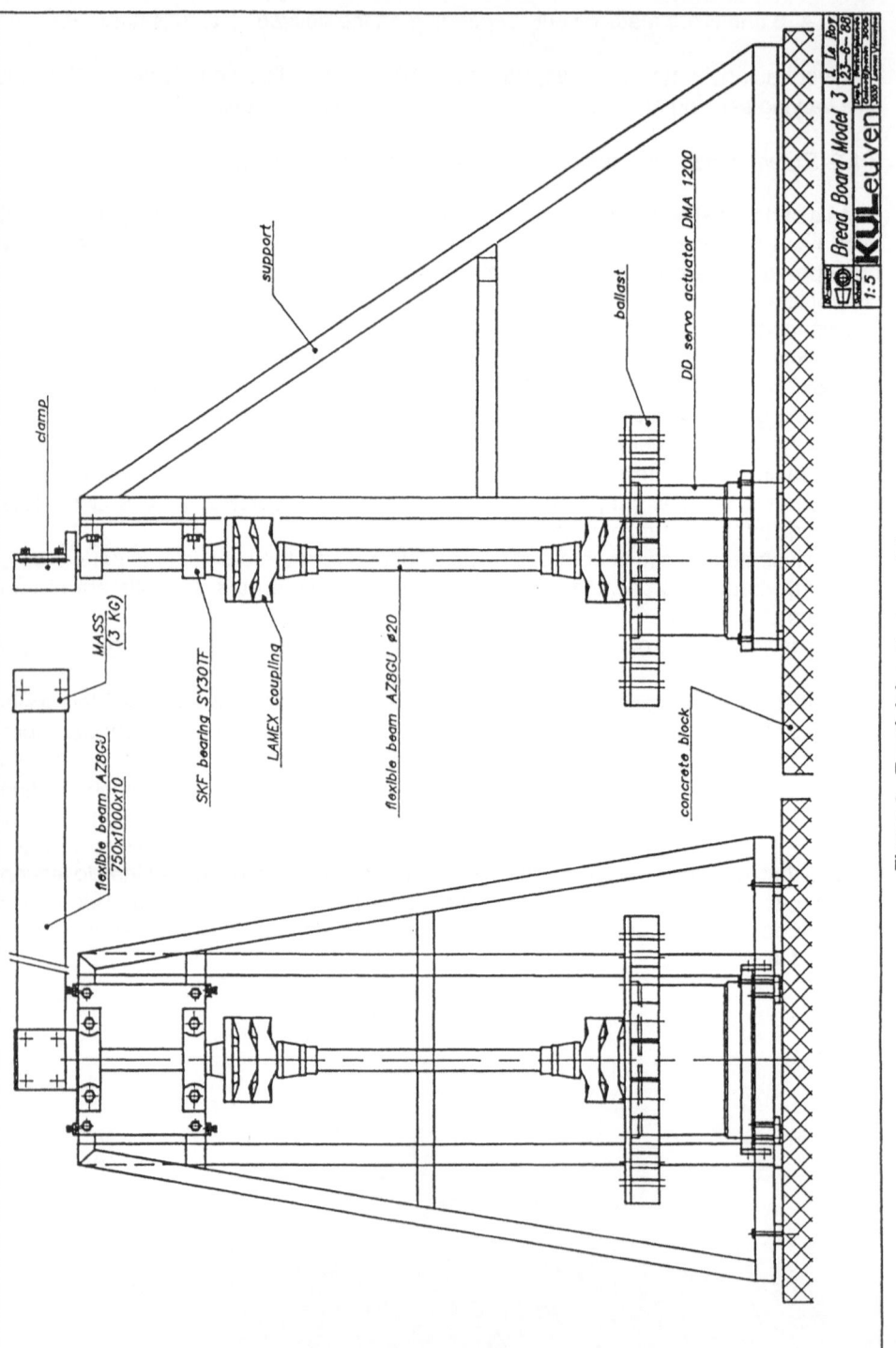

Figure 2 - Fourth laboratory model

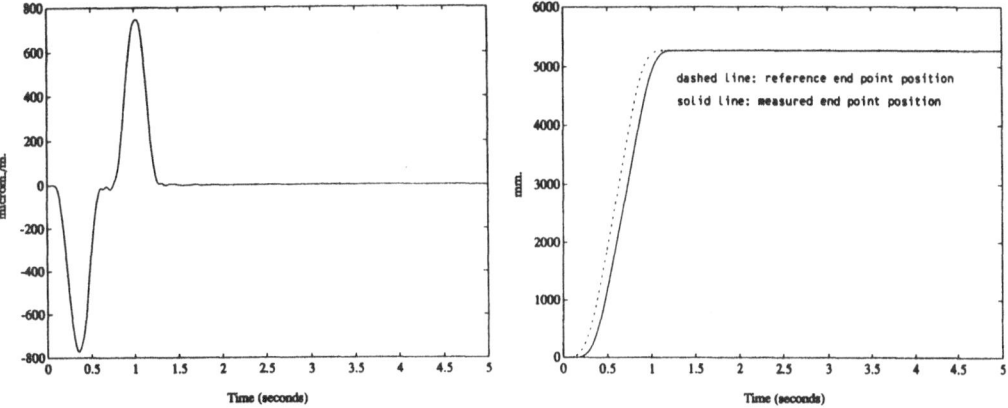

Figure 3 - Result of active control of laboratory model

In the quest for advanced, yet pratical, control algorithms, particular attention is being paid to the computing requirements of the control strategy, since this fully novel control architecture will be implemented on a new digital controller constructed by A E G. Application to the demonstration robot will be preceded by a full computer simulation using the modelling software described in section **3.3.**

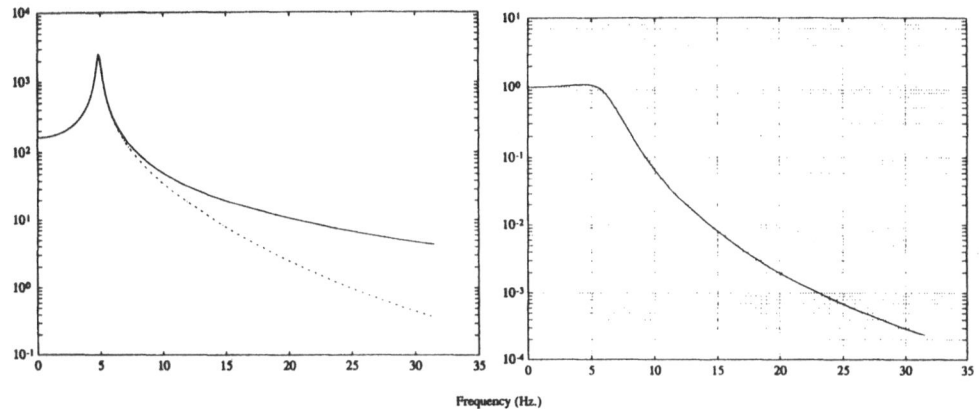

Figure 4 - Open and closed loop transfer functions of laboratory model

3.2. Identification Tools

The assessment of the dynamic robot behaviour is required to provide the necessary information for the optimization of the robot design as well as for the design of a controller which can take these dynamic characteristics into account.

Also for simulation purposes, the experimental identification results can be used to optimize and update the theoretical system model.

We describe in the following section a frequency domain identification technique deriving directly the parameters of a reduced state space model of a flexible robot. From this model, the physical system states as well as a modal system model can be calculated.

3.2.1 *An example of identification method.*

The basis for the described procedure is the state space representation of the input-output relations for a mechanical system:

$$x'(t) = Ax(t) + Bu(t),$$
$$y(t) = Cx(t),$$

Where $x(t)$: state variable (2N elements),

$y(t)$: output signal (No elements),
$u(t)$: input signal (Ni elements),
A   : state transition matrix (2N, 2N),
B   : input matrix (2N, Ni),
C   : output matrix (No, 2N).

From measured frequency response functions (FRFs) $H(s)$, a representative state $X(s)$ and the corresponding output matrix C can be identified using the singular value decomposition of a real symmetric matrix of dimension No. Using the Laplace transform of $h(t)$ and its time-derivative, the following equation for this state can be built.

$$[s^2 I + sA_1 + A_0] X(s) = sB_1 + B_0.$$

From this "measurement" equation, the unknown matrices A_0, A_1, B_0, and B_1 can be identified using a least squares method. The dimension of these unknowns are limited, being either of dimension N_i or N, as compared to N_0 which itself may be considerably larger.

Because of the proper selection of the state $x(t)$, the system matrix A can be constructed from its identified submatrices A_0 and A_1.

Modal parameters are then obtained via eigenvalue decomposition of this matrix. This identification technique feactures the following characteristics :
– The state transition matrix A and input and output matrices B and C are readily available,
– Numerically reliable tools are used : least squares solution and eigenvalue decomposition techniques,
– The number of unknowns is limited, using only a second order linear model with matrix coefficients of dimensions N, the number of modes. This model is valid for a reduced state $x(t)$. Since the output matrix is obtained from a real symmetric matrix, the usual singular value analysis is replaced by an eigenvalue decomposition, drastically reducing computer requirements and load as compared to traditional methods (e.g. ERA),
– A very important problem, namely the dimension of this state space model, is eliminated by the data measurement matrix,
– All measurement data can be analysed simultaneously, reducing the measurement noise, and yielding a consistent model for the system's response.

3.2.2. Integrating sofware into a commercial package.

The integration of this frequency domain direct parameter identification (FDPI) technique into the LMS CADA-X software package for Computer Aided Testing (CAT) was completed.

This package has been set up for general dynamic signal acquisition and analysis, and contains the necessary hard- and sofware to perform for example modal analysis tests of mechanical and mechanical/servo systems. This CAT system also provides general software libraries for modern linear algebra, least squares problem solving, eigenvalue and -vector decomposition, singular value analysis, system theory, etc...

The integration of the identification algorithm in the LMS CADA-X system for computer aided dynamic analysis (CADA) and computer aided testing (CAT) makes communication with other sofware modules straightforward. The measurement data can be retrieved directly from the relational CADA-X project data base, making all necessary information available : used test equipement and circumstances, transducer sentivity and calibration values, the needed frequency response functions, etc... The obtained modal parameters can be stored to the analysis tables of the same data base, and interpreted via a graphical animation module to display the deformation patterns (mode shapes). Other interesting analysis information (such as the singular values, indicating the state space dimension) can also be stored to this data base for later use.

3.3. Dynamic Modelling and Simulation

3.3.1. Modelling a flexible robot.

The design of a control law satisfying the simultaneous goals of fast motions and vibrations control requires the derivation of a model of the robot, describing not only the evolution of the robot configuration, but also interaxes couplings due to inertia forces and vibrations of the robot structure.

Derivation of such a model is known to be a complex task because of non linearities caused by change of robot configuration during motion, and difficulties for modelling accurately elastic behaviour of polyarticulated flexible components. In fact, a systematic tool is necessary to compute equations of motion of such systems.

Within the framework of SACODY, a general purpose computer code developed by BERTIN, called ADAMEUS, has been adapted to the particular case of robots or manipulators with possible flexible links or joints. In its current commercial version, it enables the modelling and closed-loop simulation (i.e. simulation with a feedback control) of any mechanical polyarticulated chain, including actuators and sensors.

ADAMEUS involves an adapted method for obtaining automatically dynamical equations of flexible mechanical systems. It enables a general approach of the problem of flexible systems dynamics for complex bodies behaviours and system topologies (open or closed chains with possible constraints). The difficulty of the coupled rigid-flexible dynamics problem lies in the difference of magnitude order of displacements induced by these two types of behaviour.

The equations are built by use of the principle of virtual work (KANE's method). Rigid body kinematics uses relative coordinates, and a modal synthesis method is associated to deal with the problem of flexibility.

The method used consists of separating the two contributions, and of choosing for the second one a modal representation of the deformation, the modes being the natural modes of vibration or more general deformation modes, as yielded by standard FEM codes such as NASTRAN or SYSTUS.

The advantage of this choice is to minimize the number of variables, in order to run numerical simulations on different computers of reasonable size. ADAMEUS therefore satisfies both criteria of generality (complex configuration topologies, 3D geometries, flexibilities), and efficiency (minimization of the number of variables, automatic elimination of non working degrees of freedom).

3.3.2. Using modelling software for computing control.

Obviously, the robot model is needed to perform all the steps mentionned above. For this reason further developments of ADAMEUS have been undertaken, in order to provide the control designer with all the parameters needed for computing the orders to be applied at the joints.

By inclusion of the resulting new modules, the model yielded by ADAMEUS is used not only for providing open-loop and closed-loop simulations, but also to compute:
- the kinematic inversion yielding the trajectory of joint coordinates while taking static flexibility into account,
- the predictive feedforward resulting from dynamic inversion along the desired trajectory,
- linearized models about prescribed trajectory points, to be handled by control design software to obtain the feedback gains needed by tracking control stage.

All these items are loaded in a control data file, and used during closed-loop simulation for computing the control orders to be applied at the actuators.

The overall architecture of ADAMEUS is given on figure 5.

3.4. Application of Software and Real Time Control Implementation

After state of the art survey and development phases, project SACODY has entered its final steps of integration and pratical application of methods.

This step will end with a demonstration, that will enable to measure the gains in accuracy and operational speeds achieved by new algorithms on an industrial robot. This measurement of new performances will be made by use of benchmark contours and will be completed by a demonstration involving an application process.

In order to prepare this final demonstration, it is necessary to perform all the steps needed to design control, i.e. :
- identification of robot,
- modelling of robot,
- control design and closed loop simulation,
- construction of the robot controller,
- implementation of control on numerical controller,
- integration of additional sensors on the selected robot.

Following paragraphs give examples of the first results obtained by applying developed softwares to the case of an industrial assembly robot.

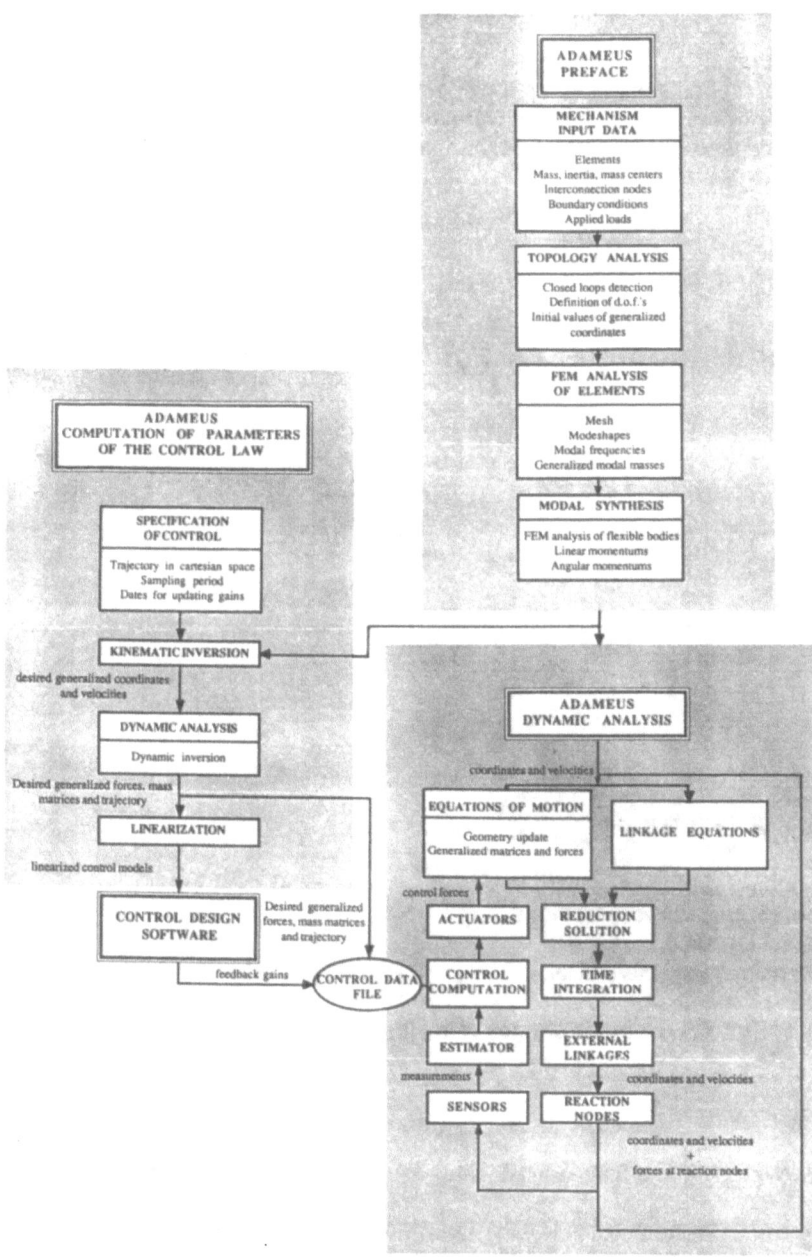

Figure 5 - Architecture of ADAMEUS program

768

3.4.1. Identifying structure of KUKA IR 360.

To illustrate the practical application of the frequency domain parameter identification technique, modal tests have been carried out on the KUKA IR360.

Frequency response functions were measured between 106 measurement locations in 3 directions, and 1 excitation point. The geometrical 3D wire frame model for these nodes is shown in figure 6.

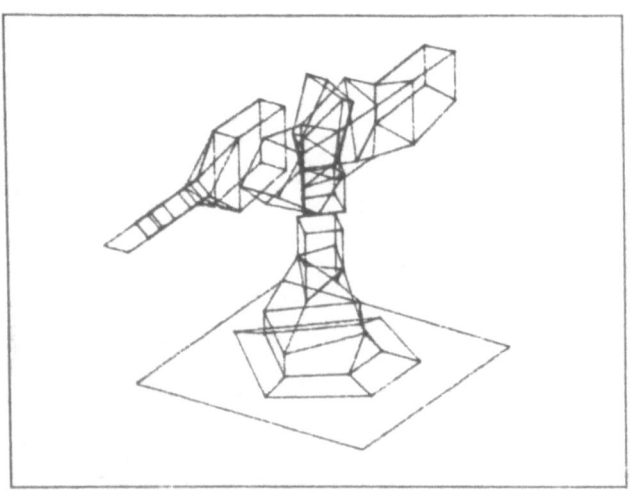

Figure 6 - Geometrical 3D wire of KUKA robot

The modes identified are :

- A bending in the horizontal plane,
- A bending in the vertical plane,
- A skew bending,
- A torsion of the rotary pillar,
- A base translation.

Figure 7 shows the mode shapes of the first two modes, displayed on the 3D wire frame model.

3.4.2. Modelling KUKA IR 361.

KUKA IR 361 is a 6 axis vertically articulated assembly robot with a carrying capacity of 8 kg.

It is made of 3 main links, the third of which carrying the three terminal d.o.f. (wrist).

Gars and belts ensure the transmission of actuation from DC motors to the joints.

In order to describe the behaviour of the robots, including its motors and gears, an ADAMEUS model of the IR 361 has been derived.

It involves 43 elements and 43 grid points (interconnection points), and leads to 41 remaining degrees of freedom.

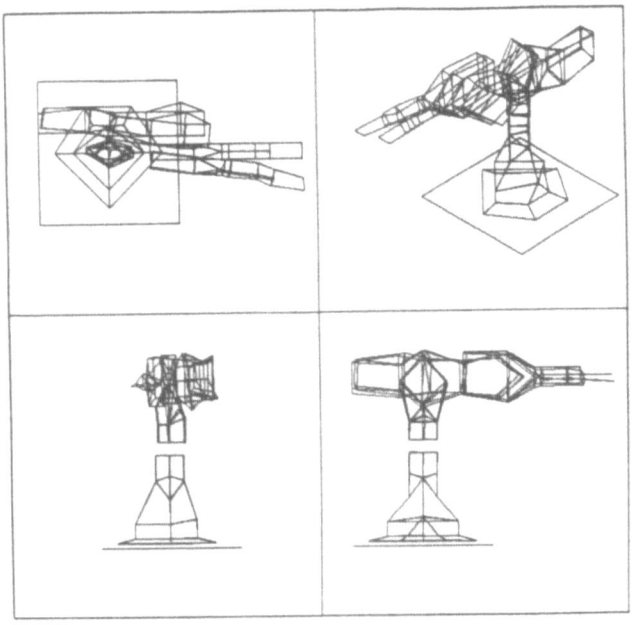

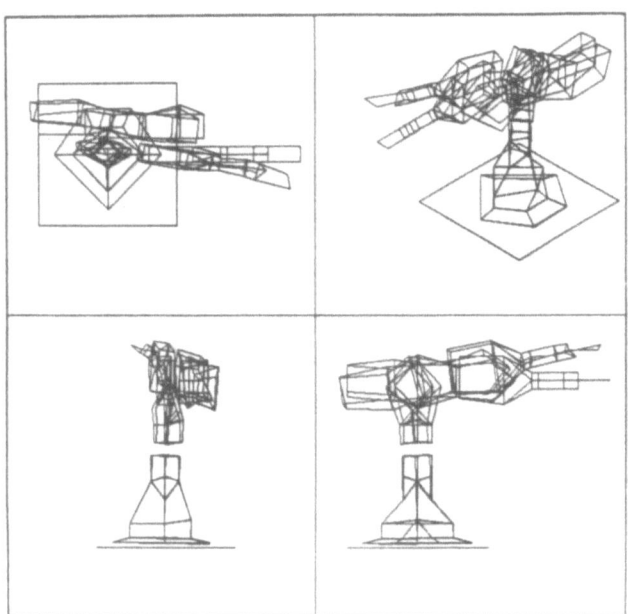

Figure 7 - Mode shapes of mode 1 and 2 of KUKA robot

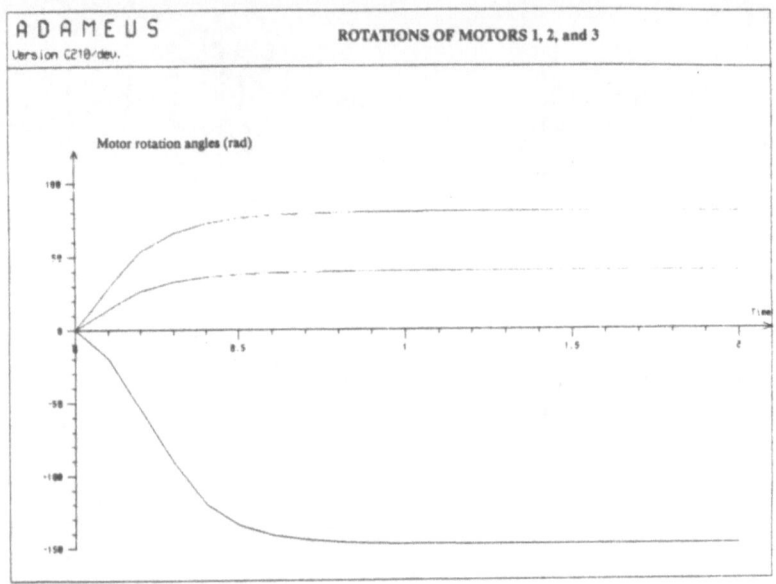

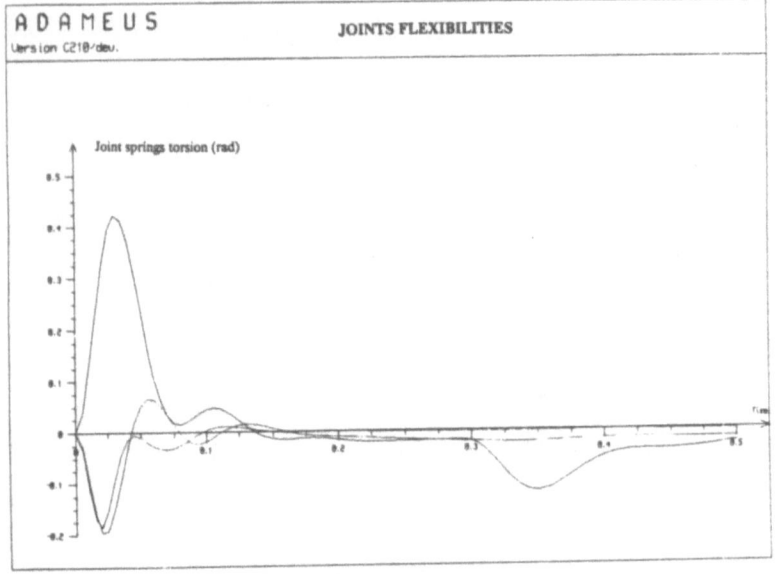

Figure 8 - Simulated motors positions and joints flexibilities of KUKA robot

Study of the identification tests has led to introduce lumped flexibilities (concentrated rotating springs) at the joints, in order to reconstruct belts elasticities.

Introduction of these springs has permitted to reconstruct vibration behaviours at frequencies of 13.3 Hz, 17.8 Hz and 22.2 Hz, to be compared with those identified on the robot in the same configuration respectively 13.3 Hz, 17.2 Hz, 22.1 Hz.

Figure 8 plots results of simulation obtained with the robot driven by classical PID controller.

3.4.3. Control implementation.

The first attempt to identification and modelling of an industrial robot has given acceptable results.

The selection of the best demonstration is currently on the way. Once this demonstration selected, the steps described above will be applied on the robot retained.

Then control law will then be designed and validated by closed-loop simulation. The new servo control algorithms will be programmed on a robot controller and will run for controlling the demonstration robot.

Real time implementation concerning the industrial application of the developed control methods will be performed during the last year of the project.

The objective of this implementation is the demonstration of the applicability and of the efficiency of the proposed methods on pre industrial hardware.

To this aim, an upgraded capabilities robot controller will be available at the begining of 1990. It will be used to control an industrial robot.

The performances obtained will be evaluated by using test contours, and compared with those currently obtained.

A final demonstrator, involving an application process, will finally be delivered at the end of project, to occur in early 1991.

The control action will be designed by applying to this demonstrator the complete set of software methods and developed during the project for identifying, modelling and controlling the robot.

It should enable to improve the absolute accuracy of current industrial robots, by upgrading the precision of its geometrical model used by the controller. Moreover, using new capabilities of software developed within the project will enable to reduce operational cycle times of robot by application of fully novel control mehods.

3.5. Sensor Systems

Besides new algorithms for robot control, new sensor systems are expected to be further fall-outs of SACODY project.

A new measurement system for robot dynamic performance analysis, called RODYM, has for example been developed and constructed in the laboratories of K.U.LEUVEN. It will be used for assessing the gain in performances obtained on the demonstration robot.

The RODYM system consists of a digitizing table, interfaced to a personal computer by an IEEE-bus, and a software package for data acquisition and analysis. X, Y coordinates of a measurement probe relative to the surface of the tablet are acquired by a personal computer. To perform tests, the probe is fixed at the end point of a robot. The tablet is positioned in such a way that the programmed robot movements are in the measurement plan of the board.

A lest squares orthogonal axis transformation is implemented in order to make the base-frame of the robot coincident with the measurement frame. Bias on the acquired error vectors is eliminated by this transformation. The transformation matrix is calculated by minimizing the sum of all deviations between commanded points and acquired points. The achieved (or transformed) points are calculated out the acquired points by using the matrix M.

Different Figures of Merit, ranging from overshoot, setting time to static position accuracy and path accuracy are calculated by a menu driven software package. These figures of merit describe the performance characteristics of the tested robot, as defined in the ANSI/RIA R 15.05 performance standard for industrial robots. The results of a series of tests are reported in a performance data table.

The surface of the digitizer tablet can range up to 1800 x 1200. Two modes of acquisition are provided : point to point mode and stream mode. In the first mode a point is acquired continuously at the maximum sampling frequency. This mode is used for dynamic measurements.

The position of the probe can be acquired up to a distance of 20 mm orthogonal to the surface. When the probe is leaving this region, a signal is sent to the personal computer. This feature is used to synchronise RODYM and the robot under test. RODYM also detects if the robot has reached a commanded point, within a repeatibility zone.

The accuracy of the sensor is improved with a factor of four by a calibration algorithm, up to 0.05 mm ; which is in most cases at least an order of magnitude better than the robot. A high bandwith (200 Hz), portability, a large measurement area (1800 x 1200 mm) and a user friendly menu driven software package, make this system very attractive for robotic applications.

4. Conclusion

Project SACODY is a multi-discipline undertaking which aims at associating advances:

– in dynamic modelling and simulation,
– in theoretical control methods,
– in real-time numerical computer hardware and programming software,

to make practical use of advanced control methods in industrial applications, with a view to compensate for unavoidable mechanical deficiencies of manipulators.

The software packages for identification on the one hand, and for modelling, simulating and control design on the other hand, have been completed and experimentally tested on models of flexible robots.

They will be interfaced and used to design the control to be applied to a 6 axis industrial robot in order to obtain, thanks to a better knowledge of the robot structural behaviour under extreme accelerations, faster motions and, thus, reduction of its operational cycle time

5. Acknowledgements

This paper is a representation of a collective work carried out at each of the six organizations of the consortium. The dedicated work of a large number of people, under a spirit of cooperation, is acknowledged here.

The contractors would like to acknowledge the support of the Commission of the

European Communities, as represented by DG XIII, on Telecommunications, Information Industries and Innovation, and more especially its CIM Group.

6. References

De Schutter, J., Van Brussel, H., Adams, M.,Froment, A., Faillot, J-L., *Control of flexible robots using generalized non linear decoupling* IFAC Symposium on robot control SYROCO, Karlsruhe, October 1988.
Swevers, J., Adams, M., De Schutter, J., Van Brussel, h., Thiclemans, H.
Limitations of linear identification and control techniques for flexible robots with non linear joint friction. International symposium on Experimental Robotics, Montreal, Canada, June 19-21,1989.

Keywords

Light weight robot, Control of flexible structure, Vibration control, Modal identification, Fast assembly robot, Modelling software, Robotic sensor system, Robot control

Project no 1652

METHODS FOR ADVANCED GROUP TECHNOLOGY INTEGRATED WITH CAD\CAM

V. Van de Steen
WTCM
Celestijnenlaan 300 C
3030 Heverlee
Belgium

Abstract.

The key development in this project is the elaboration of an internal representation of parts which allows an automatic achievement of all the advantages of group technology and classification integrated into the different CAD/CAM systems used by the partners in the project. By automating group technology, human interpretation of engineering drawings during the coding process, is excluded. The automatic retrieval of existing similar parts, subassemblies and assemblies will be possible, moreover also the retrieval of appropriate manufacturing information sequences will be in the bounds of possibility. Until now the consortium has worked on two prototype software packages. One working group developed a package for the automatic creation of the computer internal representation of simple parts whereas the other group concentrated itself on the automatic retrieval of assemblies and family parts from a data model, that is based on family information. The two packages were only tested in a limited degree, although those results are satisfactory it is still too soon to draw conclusions.

1. Introduction

1.1. Brief Synopsis of the Project

The MAGIC project aims to improve and facilitate the use of Group Technology methods and tools in small and medium sized factories, by automating these methods and by integrating the tools in existing CAD/CAM systems.

The final results will be a Computer Automated Group Technology (CAGT) software prototype with two main modules for:
- Automatic creation, during the design task on a Cad system, of a computer internal representation (or patterns) of assemblies, subassemblies and parts, including their geometrical, functional and production characteristics.
- Automatic retrieval of:
  1) similar parts based on a rough sketch of a new part on the Cad system, keeping in mind that a picture tells more than a page of literature, or based on user friendly menu inputs
  2) existing assemblies or subassemblies, based on functional or morphological characteristics
  3) appropriate manufacturing information sequences.

1.2. Partners

Industrial partners as well as research centres are involved in the project MAGIC.

The WTCM is the prime contractor of the project. WTCM is a research centre of the Belgian metalworking industry. The department Mechanical Engineering has a research team of 28 engineers, working in the fields of 'CAD/CAM and FMS.

CETIM is a Technical Centre of the Mechanical Industry in France. The Production Engineering department (37 people) directs the research in company management, process planning, machining techniques and group technology.

N.V. Michel Van De Wiele is a medium sized Belgian company (700 employees) manufacturing weaving machines. It has a manufacturing experience of more than 100 years in textile machinery engineering for carpet and velvet weaving machines.

As a Belgium machine tool manufacturer LVD has been involved in mechanical engineering for more than 30 years. The extensive production program covers not only standard machine tools but also special manufacturing systems.

Some new partners are involved in the project 2623. One industrial partner and two software houses will join the consortium.

C.M. MARES is an injection mouldmaker for plastics materials since 1952. The 90% of their turnover is directed at the automotive industry. They are specialized in big tools until 40 tons, with complex surfaces.

Eigner & Partner are German system consultants for the solution of problems in the technical field.

CAP SESA Industry is a French software house that works on a very large range of different fields of Computer Integrated Manufacturing, covering CAD/CAM, Flexible Manufacturing Systems, networks, data management systems,...

2. New Methods and Technical Innovation

The introduction of high technology and advances in computer technology have completely altered the manufacturing environment. As the processes involved have become increasingly complex, they have given rise to a number of new problems. New methods had to be developed to meet these new challenges.

The techniques known at present to store technical information on parts in a way that they can be extracted from a database are the known group technology techniques of coding certain aspects of the parts and storing the code into a database. If one wants to find a part back one has to define the filter (the mapping code) and ask for all parts matching that filter.

Group technology (GT) is considered to be one of the best solutions to solve the problems with respect to a modern manufacturing system, especially the specific problems of small batchsize, multiple-product production systems. A classification and coding system is the basis of successful implementations of a GT concept. By automating group technology human interpretation of engineering drawings, during the coding process, is excluded. The new coding software will avoid degradation of information, misinterpretation and inconsistency between part design and part code.

However, like the technology it depends on, the methodology for problem solving is constantly changing. As the state of the art changes, so does the methodology evolve that describes the technology. Within the project a new approach to solve an old problem has been worked out. The major benefit will be the elimination of the existing implementation barriers for GT applications.

The key development in this project is the elaboration of an internal representation of parts which allows an automatic achievement of all the advantages of GT and classification as mentioned above, integrated into the different CAD/CAM systems used by the partners in the project.

This whole internal part representation is generated during the design of the part on the CAD/CAM system so that no separate coding system has to be maintained. The full integration of the GT principles with the CAD/CAM system will avoid the time consuming coding of parts, the different interpretations of code units by different operators and the possible inconsistency between the part and its code.

One of the most economical applications is the generation of new designs by modifying existing ones.

An optimized research strategy will strongly improve the retrieval speed and quality. The coding of the production oriented part characteristics will allow standardisation and optimization of the operation sequence.

2.1. Design with a Cad System, Using Form Features

A form feature is defined as a number of basic geometrical entities (points, line, arcs, etc.) grouped into logical instance associated with specific production oriented characteristics.

Features represent a higher conceptual level than lines, arcs and text used in existing CAD/CAM systems because they inherently contain more information. Simple entities such as lines, arcs and circles can be grouped logically to denote a functional or manufacturable entity. The typical feature based design systems provide a set of standard engineering shapes, or features, that are ready to use. Designers simply select a generic feature, like a chamfer or fillet, and then enter the necessary values, or parameters, to render the specific feature they have in mind.

The form features always correspond to both simple and repeated functions of the component and, ad to the production, to short sequences of simple manufacturing operations. Thus, the gathering of form features is easy to perform in a company.

In the CAD system, we have set up a catalogue of form features as a set of parametric routines which assign a geometrical feature to the general form of a component. It also contains functional, technical, technological or administrative information. The data (i.e. the values of the parameters) are gathered in a tabulated structure apart from the associated process itself, and which is examined by this process.

When direct coupling of CAD systems and external Relational Database Management Systems (R-DBMS) is possible, the tables will be collected, stored and handled directly by the R-DBMS.

2.2. Automatic Extraction of Miscellaneous Information from Cad Drawings (Fig 1.)

The design department not only designs the geometry of parts but also defines the attached tolerances, the roughness, the materials and their mechanical capabilities. Furthermore, the designer uses the specific functional characteristics of each component. Thus he works with different types of information.

Today's CAD systems allow the storage of several of these types of data but the manipulation capabilities are not easy and powerful enough, except for the geometrical information.

The idea is to collect the different types of information on the CAD system as far as it provides good capabilities and to store them in drafts (in one or several files for the computer), and then to extract the necessary information automatically from the files and store it in relational DBMS. Further information can also be collected, stored and handled by this system.

The related R-DBMS allows miscellaneous operations in collecting, storing, filing, consulting, retrieving, handling and updating information.

The extraction of information is achieved as follows:
- scanning of the CAD geometrical model to get back the form features which have been assigned to the component, starting from the structure of information (forms derived from parametric programs)
- scanning the alphanumerical part of the file to get back the functional and technical properties and the administrative information. All information that is useful for the retrieval of similar parts is extracted.

When a drawing on the CAD/CAM system has been completed, it will be approved for further use. During the process of approval a batch process will be started that scans the design for the GT form features and extracts the input data to the technical database.

Modifying the part on the CAD system ends up with a new approval that automatically updates the database information, by means of the same process. This database will contain information to be examined by research and development (R&D), manufacturing, process planning, ...

When direct coupling of CAD systems and external R-DBMS is possible, the extraction of non geometrical information will no longer be useful since the information will be directly stored in the DBMS.

However the information about the presence of form features within a part is still needed and will be extracted. Moreover, since the R-DBMS is not supposed to manage geometrical data, the adopted principle of a double management by two supplementary systems will still be suitable.

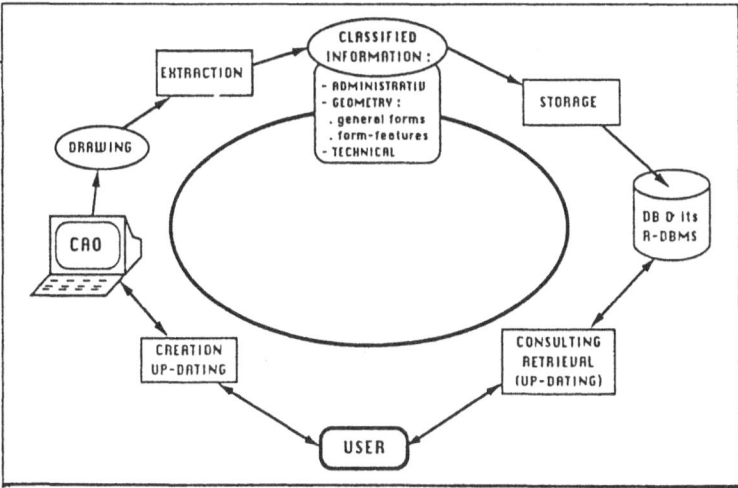

Fig 1. Routing of Information in a Design Office

3. Team-work and Synergy

Since this project 1652 is a one year project the whole of the objectives could certainly not be matched in this development. The consortium decided to elaborate a prototype software package, containing different modules corresponding the different tasks of the project.

Looking at the objectives of the project you can define two different parts in the project. On the one hand you have the automatic creation, during the design on a CAD system of a computer internal representation. And, on the other hand, you have the retrieval of similar parts, existing assemblies and sub-assemblies and appropriate manufacturing operation sequences.

Looking at all the drawings of a company one can notice that a company has simple parts, subassemblies as well as assemblies. It is beyond doubt that the approach is different for simple parts and assemblies.

All the work that has to be done can be divided into four fields. Fig 2. shows the different fields with their links. It is obvious that it was impossible to elaborate for the prototype a software that integrates all the fields.

Therefore, the consortium decided to work into two working groups. One group would work on the automatic creation of the computer internal representation of simple parts whereas the other group would concentrate itself on the automatic retrieval of assemblies and family parts from a data model, that is based on Family information.

For the prototype, the work is divided in two main tasks, one dealing with the group technology idea based on the use of functional form features, the other dealing with the automatic retrieval part.

MVDW and the WTCM were working on the creation of an internal representation. This representation will be extracted from the CAD drawing. The information needed to make this possible will be incorporated into form features that will be used to design.

LVD and CETIM were working on the family data models. In the first phase LVD is mostly interested in a functional approach. That is the reason why they are concentrated themselves first on the retrieval of families of parts.

For the prototype the consortium has limited itself to a representative sample of parts. MVDW decided to concentrate itself to the rotational and simple parts and have chosen 10 parts to be used in the prototype phase. In the first phase LVD is more interested in subassemblies and the functionality of his composed parts. LVD has chosen one family of parts, the pistons.

Although there has been worked into two working groups, the cooperation between these two groups was very good. In the prototype phase, the members of each group worked together very closely and they met each other two times a month. The whole consortium came together monthly to explain the state of the work and to adjust, when necessary, the prototype to the objectives of the other partners.

During the second and third year of the project the prototype will be expanded and the work done for the automatic description of simple data models and family data models will be joined together.

Maybe one can wonder how the approaches of the two working groups can be fitted. Fig 3. tells everything about this matter. Every company has a lot of parts and a lot of these parts can be divided into several families. How those families are defined will be explained later on. Every family contains several parts, subassemblies and assemblies. These can be split into simple parts.

	MVDW-WTCM		LVD-CETIM
	SIMPLE PARTS		SUB-ASSEMBLIES ASSEMBLIES
GENERATION of COMPUTER INTERNAL REPRE- SENTATION	PROTOTYPE MVDW SIMPLE PART MODEL (rotational parts)	ADDING FUNCTIONAL INFORMATION AND RELATIONS BETWEEN DIFFERENT SIMPLE PARTS	2623 topic (FAMILY) FUNCTIONAL DATA MODEL (piston family)
RETRIEVAL	2623 topic		PROTOTYPE LVD

Fig 2. Division of the project in four fields

Until now, MVDW has worked on the 'lower level' of the simple parts. Whereas LVD has worked on the 'higher level' of the part families. During the development of the project LVD will follow path B. They will decent to the level of the simple parts. MVDW, on the other hand will climb up to the level of the families. They will follow path A.

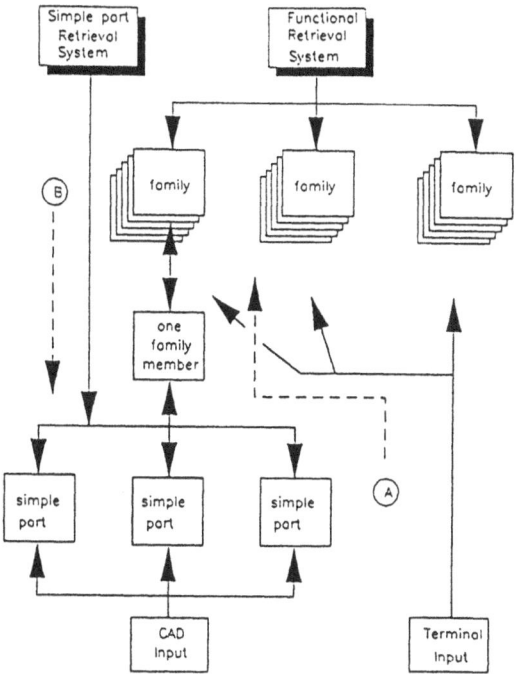

Fig 3. Evolution of the work for the different partners

4. Objectives and Results of the Project by LVD

4.1. Existent situation in the R&D department

The LVD production program covers not only standard machine tools but also special manufacturing systems for the plateworking industry. Designers in the engineering department therefore have to deal with several kinds of engineering and design activities. Fig 4. describes the working method a design engineer usually follows performing his task and at which steps in the design phase computer assistance related to GT should be provided.

Problem description is in most cases based on functional data and cost information on the system that has to be designed. If the job consists in adapting an existing design, the retrieval of the assembly drawing is possible the denomination of the system is a known datum. The adaptation work must be performed by means of the Computer Aided Engineering (CAE) and CAD tools and in addition to that, an advanced retrieval software packages using group technology that enables the designer to find existing parts that he can use.

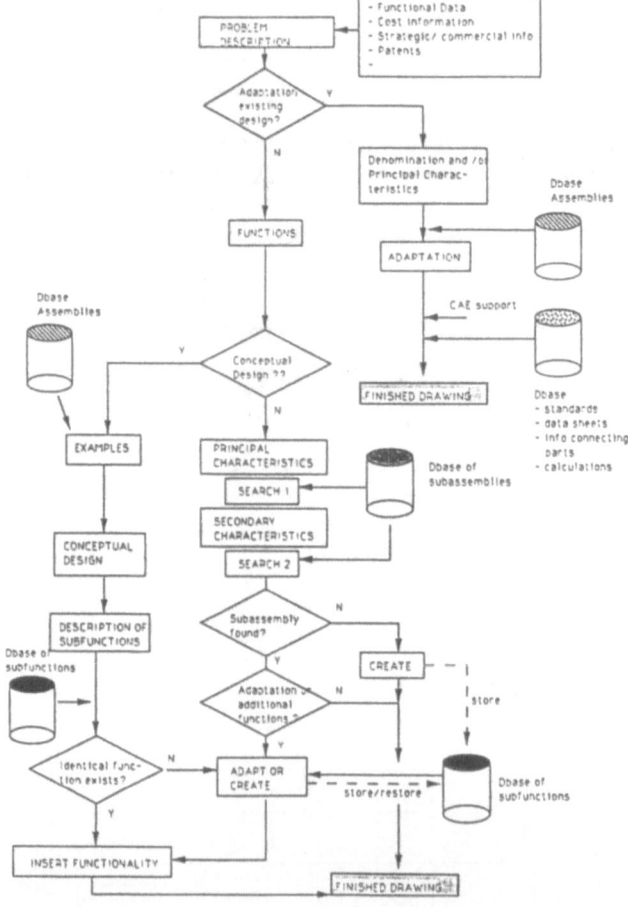

Fig 4. Working method a design engineer follows at LVD

4.2. Industrial Objectives in This Project.

In case that a new system has to be developed at LVD, and there is no need for a new conceptual design, the engineer should try to look in a database for existing subassemblies according to the principal and secondary characteristics of the system that he must create. If no subassembly can be found that performs the required function, it must be created and stored for later use in other designs. If on the contrary a subassembly fits into the design, adaptation can be necessary in some cases.

Every new subassembly or part generated in a design phase must get its internal description provided by the MAGIC tools before storage in the database. The description should be automated as much as possible. If a conceptual design has to be performed, another type of retrieval has to be done. The engineer looks for subfunctions (e.g. linear movement, rotary measurement system, etc) instead of parts or subassemblies. The assemblies performing these functions are stored in a subfunctions database.

4.3. Achieved Results

Based on the mainly functional approach to the part retrieval issue, LVD's task was scheduled as follows:
- Selection of a sample part family.
- Elaboration of the functional properties for the sample part family.
- Determination of the required properties for process planning.
- Elaboration of a prototype that supports the general part classification issue.

4.3.1. Selection of a sample part family.

The first approach to part retrieval was to create a model that could help the user in retrieving of a complete functional entity (e.g. a linear movement system) without even specifying the kind of technique required to realize that functionality (e.g. a hydraulic system, a mechanical system, etc).

In a second phase the first approach was considered to lead us too far from real life, especially within the context of the first year of the project. Therefore, it was decided that the user himself would have to specify the kind of technology that would realize the required functionality (e.g. the user has to specify that he wants a linear movement based on a hydraulic system).

Within this approach, it was decided to select a family of simple parts which are necessary for the realization of the required functionality.

The concrete selection was made as follows:
- functionality: linear movement
- technology: hydraulic or pneumatic system
- part family: piston

An example of the selected part family is shown in figure 5.

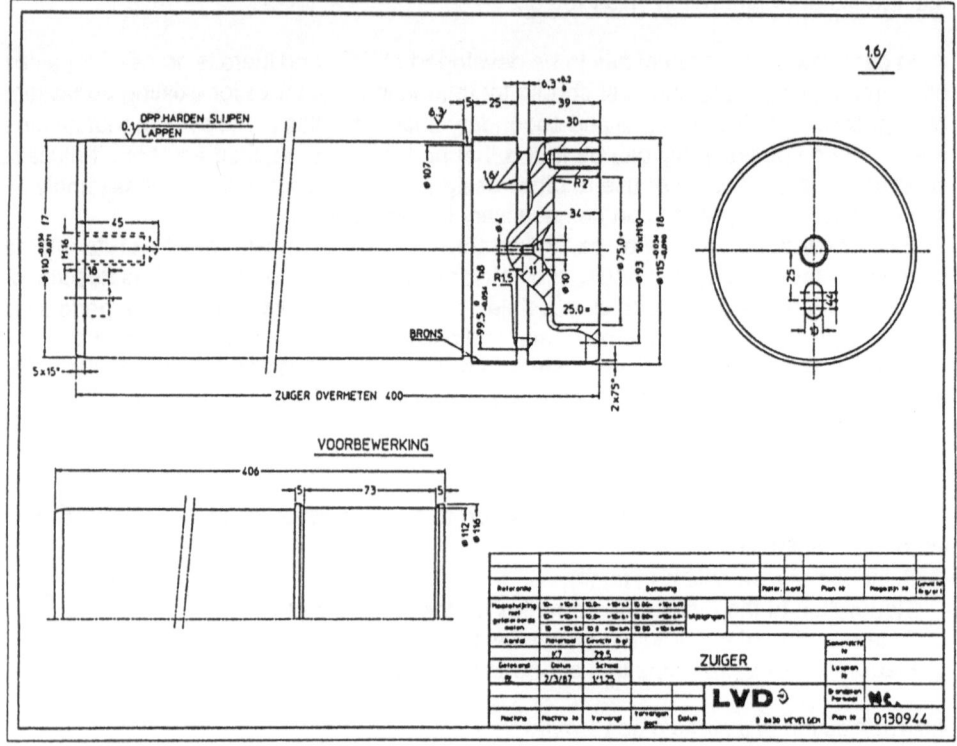

Fig 5. Example of part used for the prototype software at LVD

The intention is to extend the current selection in the upward direction: from part families, over technology to the selection of a complete functionality.

4.3.2. Elaboration of the functional properties.

The first approach to the sample part family is the approach of the designer. He is mainly interested in the functional properties of the part.

The main problem consists in selecting the properties of the part which should be regarded as functional. The study of the piston part family has proved that this is a quite elaborate job. Presenting the complete results of this selection would be outside the scope of this article.

However, we can mention the main categories of criteria which define part functionality. The criteria can be:
– Logical (Yes or No).
– Numerical (with possibly a minimum and maximum value).
– Alphanumerical.
– Selective (a value representing a choice from a list of possibilities).

The analysis of the sample part family has also proved that a certain degree of incompatibility among the criteria can exist: e.g. if one of the logical criteria indicates that a piston does not have a head, it does not make sense to inquire the diameter of the head of the piston, although generally speaking the diameter of the head of the piston may be an important functional property.

The goal is also to store these properties as they are, and not coded based on the family which they belong. This is quite different from the currently existing GT systems. The major advantage of this approach is that the user can set the acceptable range during the query.

4.3.3. Process planning properties.

In a first phase it was agreed that to generate, in a complete automatic way, the required process plan could not be the goal. The approach was chosen to create a set of process plan models, which define the major manufacturing steps for the parts that are covered by that model.

On the other hand, the set of functional properties of a particular family of parts should be extended to also contain the necessary data to select one of the process plan models in an automatic way. Introducing the concept of 'automatic selection' also requires the creation of a set of selection rules, and possible selection values.

4.3.4. Elaboration of a prototype.

The prototype that supports the concepts mentioned above will cover the following topics:
    a. Definition of a part family
    b. Insertion of new parts in a family
    c. Modification of the properties of a certain part
    d. Retrieval of parts based on multiple criteria
    e. Selection of the visible criteria of the retrieved part drawings

5. Objectives and Results of the Project by Michel Van De Wiele

5.1. Existent situation in the R&D department

The administrative tasks in the research and development department are very computerized. For the technical aspect the PRIME/CV CIS MEDUSA CAD/CAM system is being used which has been customised to a large extent to meet the MVDW requirements. Furthermore, calculation programs developed by MVDW and third parties are used for engineering purposes.

As to MVDW the administrative system is connected to the CAD/CAM system for the transfer of administrative data concerning the drawings, such as bill of materials and identification data (part no, drawing no, name,...).

All administrative data is maintained using a relational database. It is possible to query this database using parts of names, drawing numbers, etc. Information can be obtained on where a part is used, price, ... Until now no specific technical information about the parts was stored into a database for retrieval purposes. The only way of gathering information on previous designs is to scan on names and administrative data of the parts and then examine the drawings (or CAD models).

5.2. Industrial objectives in this project.

The main objective for N.V. Michel Van De Wiele in this project is the elaboration of a GT information database that is automatically maintained/updated by drawing and modifying parts on the CAD/CAM system.

Since all mechanical parts in the company will be drawn on the CAD/CAM system, the tools used on this system (macros, ...) should provide the necessary input for the basics of the GT database. During the design phase, at approval time, the GT database shall automatically be updated and filled in by the CAD/CAM system.

In the later stages of the creation process additional information for the different departments involved will be added by a number of additional software modules such as the tools selection software in the manufacturing department.

In the figure 6. the possible use of such a GT information database is illustrated in a system flowchart.

The main departments involved are R&D, manufacturing, process planning,and purchase. The elaborated system shall make it possible to deal with problems of retrieval of identical parts in the view of design, tool usage, machine usage and process plans. Via expert system software tool selection, price evaluation process plan creation and day to day jobscheduling can be made computer assisted in the future, based on the contents of this kernel database.

The objectives during this project 1652 are to define and create the basics of this kernel GT database system. Therefore we have to:
- define the exact contents of such a database,
- elaborate the interface for MVDW in the context of "form features",
- work out the form features into usable tools on the CAD/CAM system.
- build a prototype software for evaluation purposes.

The creation of the retrieval and exploitation tools is foreseen in a next ESPRIT project following this one, as well as the quantification of the results.

5.3. Achieved results by MVDW

Since this project 1652 is a one year project the whole of the objectives could certainly not be matched in this development. Therefore the purpose was to create a prototype system which could extract the GT information of a sample of rotational parts and store the information into a relational Database.

The prototype software to be developed should make it possible to analyse automatically all the data on these drawings and convert them into relevant GT informations in the relational database.

This database could then be examined using SQL-like statements for different retrieval purposes. In a further development the retrieval of parts will be optimized so that user specific language will be used that gets translated by a rules database into these SQL-like statements for querying the database.

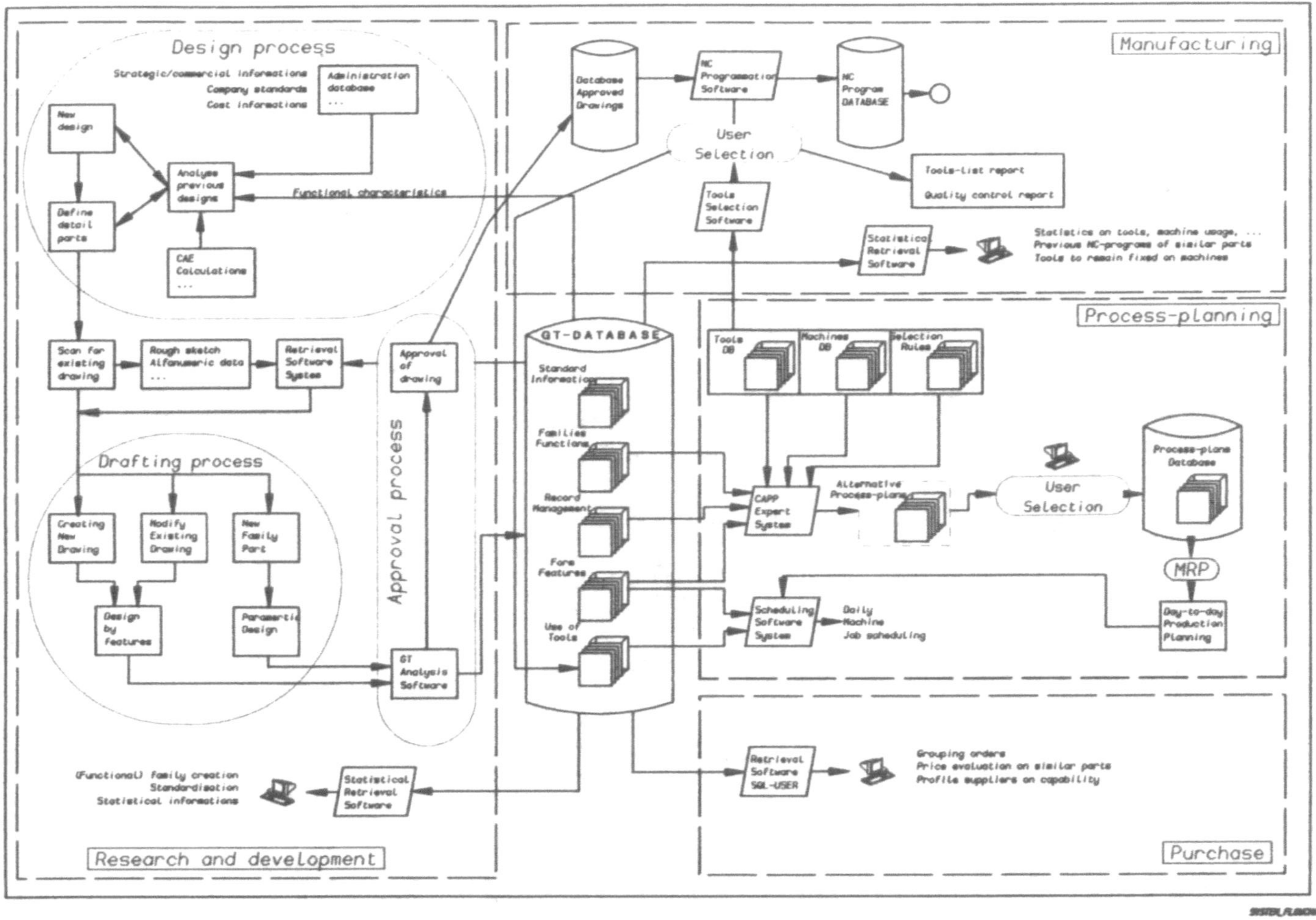

Fig 6. Use of group technology information database

The data that is gathered is divided into 5 categories of data:
- Record Management data
- Material data
- Engineering notes and specification data
- Form features data
- Relationships and tolerances data.

For every category of these data, special tables have been introduced into the relational database.

For this prototype a selection has been made of possible information in the database. The main purpose was to create a prototype software package to evaluate the possibilities. The exact contents of the database have to be seen as experimental and can certainly vary from site to site and will be evaluated through the experience by using the system.

It is worthwhile mentioning that the GT database is not only used as an informational database on the parts, but also acts as a general technical information database on available techniques, methods and materials in the company. As an example of this, all treatments used at MVDW site are stored into one of the standard information relational database tables in the system.

5.3.1. Record Management data.

This information contains further administrative data relevant to the design department such as:
- General data: part no, drawing no, part name, ...
- Drawing data: size, units, pages, scale, ...
- Project data: date drawn, drawn by, status code, project name,

Most administrative data of a part is stored in the relational database of the administrative system at MVDW side. However the design department is more likely to have, next to these part informations, additional information on the level of drawings.

The selection of the contents of the Record Management data for the part in the example on figure 7. are the following:
- Part no   : 3-62-48707
- Drw no   : 62-200583
- Page no   : 1
- Tot.pages   : 1
- Name   : As Nokken Grijperbew. 3V
- ...

The translation into tables and fields in the relational database of the GT system has been defined, grouping these record management data in a more logical way.

787

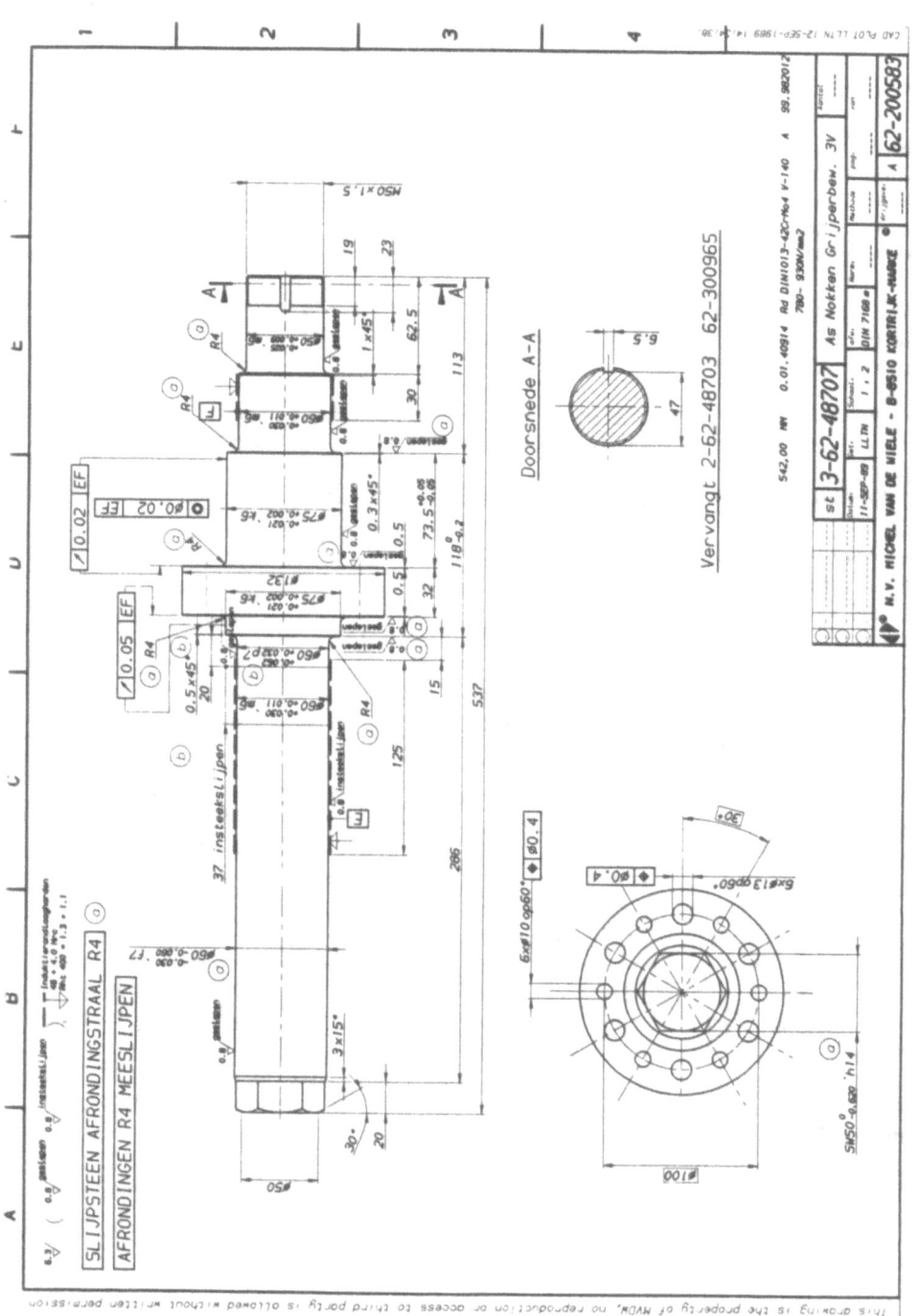

Fig 7. Example of part used for the prototype software at MVDW

5.3.2. Material data contains information of the material being used for the particular part:
Administrative data: Material no, document no, Material name, ... Raw part information: Length, diameter, form,.... technological data : Composition, stress resistance, ...

The contents for the material data in the prototype have been selected as follows:
- Part no   : 3.62.48707
- Material no   : 0.01.40914
- Material name   : Rd DIN1013-42CrMo4 V-140 780- 930N/mm2
- Material spec.sheet   : 99.982012
- Material form   : Round
- ...

Again these data have been stored into relational database tables in a more logical way, where all data on materials are grouped and separate tables indicate what material is used in which part.

5.3.3. Engineering data contains information on the roughness indications on the part drawing and the special treatments indicated such as heat treatments, coatings, finish treatments, ...

For the part in the example the data should look like these:
- Part no   : 3-62-48707
- Part general roughness : 6.3
- Part appearing roughness   : 1.6 0.8
- Number of treatment records   : 1
- Coatings   : 0
- Finish treatments   : 0
- ...

Again these informations have been structured in the GT database into different relational database tables.

5.3.4. Form feature data consists of general data concerning the part and dedicated informations to every single form feature in the drawing.

The General part information consists of a summation of all appearing form features by their name and an indication of the outer dimensions of the part. For a rotational part these are the length and the diameter.

The implementation in the prototype will be a dedicated relational database table for every kind of form feature known to the system. Introducing new form features in the recognition system will end up by introducing new tables in the relational database. In this way we can store different field items for every different form feature type known in the system.

The information structure will contain a first level, indicating the appearance of the form features in the general data tables of the part. To find out more about the form features themselves, the dedicated tables can be examined.

To give an idea about these data for the part in the example some of them are presented hereafter. The summation surely is not complete since the information in the structure is rather elaborated and ends up with quite a lot of database tables.

5.3.4.1. General data:

Part no	: 3-62-48707
Total length	: 537
Total Diameter	: 132
Units	: mm
Number of outer diameter form features	: 8

...

1. Outer hexagonal   OHX
2. Outer Diameter straight   ODS
3. Outer Diameter straight   ODS

...

5.3.4.2.. Examples of form feature details:

Part number	: 3-62-48707
Sequence number	: 1
Form feature name	: OHX
Outer Diameter	: 58
Length	: 20
Length upper tolerance	: 0
Length lower tolerance	: 0
Length DIN code tolerance	: 0
Key	: 50
Key upper tolerance	: 0
Key lower tolerance	: -0.620
Key DIN code	: h14

5.3.5. Relationships and tolerances deal with the indications of place and form tolerances on the sheet.
    For the part in the example an extract of the these data looks like this:

Part no	: 3-62-48707
Tolerance type	: slag
Tolerance sequence number	: 1
Tolerance value	: 0.05
Reference form feature 1	: 2 ODS
Reference form feature 2	: 6 ODS

5.3.6. Description of the prototype system.

The prototype that has been elaborated consists of the following main modules:
– A relational database system to store the different data into database tables.
– A query language to examine the database (SQL)
– An analysis software that analyses the drawing datafiles and extracts the input informations for the GT database
– Interface modules to update the different database tables from the input generated by the analysis software.
    The drawing files are analysed automatically at approval time. This process updates

the database contents. In this way the database information is automatically updated for each modification in the part.

6. Conclusion

After one year in the ESPRIT project MAGIC, the consortium has already achieved practical results. The ideas gathered during the first months, were elaborated and couched in a concrete form. Two working groups have worked on two different work packages.

As explained in previous chapters, two independent parts of a prototype software are elaborated. Due to the short time span it was, until now, neither possible to connect the two prototype software packages nor to test the prototype with a wide range of parts and families. That is also the reason why it was not possible to quantify the results.

For the next two years of the project 2623 MAGIC, it is the task of the consortium, on the one hand, to elaborate the prototype to more parts and more families and, on the other hand, to combine the two prototype software packages to one general system that is able to generate an internal representation of a part or (sub)assembly and to retrieve existing drawings and information.

7. References

Bhadra, A. and Fischer, G. W. (1988) 'A new GT classification approach: a data base with graphical dimensions', ASME Manufacturing Review Vol 1, No 1, 44-49.

Srinivasan, R. and Liu, C.R. (1985) 'Extraction of manufacturing details from geometrical models', Comp. & Indust. Engin. Vol 9, No2, 125-133.

Deitz, D. (1989), 'The power of parametrics', Mechanical Engineering, January, 58-61.

Davidson, H. (1989), 'Variation on a theme', Mechanical Engineering, January, 62-64.

Esfandiar, K. and Melkanoff, M. A., 'Automatic generation of GT codes for rotational parts from CAD drawings, Manufacturing Engineering Program, UCLA, 191-199.

Project no 2617

INDUSTRIAL NETWORK MANAGEMENT
IN CNMA PROJECT

G.SEGARRA
REGIE RENAULT
INFORMATION TECHNOLOGY AND TELECOMMUNICATION DIVISION
Sce 0457
BP103
92109 BOULOGNE BILLANCOURT CEDEX, FRANCE.

Abstract

This paper concentrates on the user's needs and requirements relative to the management of a communication network for manufacturing applications as implemented at the level of manufacturing complexes such as a car assembly plant, a mechanical factory or a car component manufacturing plant. Firstly, the target communication network architecture is presented, the management objectives are listed, the technical choices are discussed and then the user's requirements relative to the network management system are given. At the end, an "ideal" solution is described with the emphasize being put on research aspects undertaken by the ESPRIT CNMA project 2617 such as the building of a network components/objects library, the utilization of knowledge based technics for network trouble shooting or the study of a global, accurate network real time clock for events time stamping.

1. Introduction

The CNMA phase 4 of the ESPRIT 2 project (project EP 2617) has been started at the beginning of december 1988 once the CEC gave its green light to the CNMA consortium proposal to continue its work, started during ESPRIT 1 (project EP 955), on the definition and implementation of a communication network for manufacturing applications. The CNMA consortium has slightly changed with the departure of PSA and BMW and the joining of new partners such as AEROSPATIALE (FRANCE), FIAT with its subsidiary MAGNETI MARELLI (ITALY), RENAULT (FRANCE), ROBOTIKER (SPAIN), with its associate partner the UNIVERSITY OF PORTO (PORTUGAL) and the UNIVERSITY OF STUTTGART (GERMANY).

The other partners remaining in the project from ESPRIT 1 are: AERITALIA (ITALY), BRITISH AEROSPACE (prime contractor from UK), BULL SA (FRANCE), G4S (OLIVETTI from ITALY), GEC (UK), IITB Fraunhofer Institute (GERMANY), NIXDORF (GERMANY), SIEMENS (GERMANY), TITN (FRANCE).

The main effort of the CNMA consortium will be put on the following items of work:
– Network Administration ,
– MMS and MMS conpanion standards ,
– MMS application interface ,
– Identification of other user's needs (ie: FIELDBUS, Critical Time Architecture,... etc)
and evaluation of new technologies such as the fiber optics.

Among these important work items, the Network Management is one aspect we consider as being the most urgent to be completed, particularly for Industrial Complexes where we can not imagine to build a large extended LAN connecting hundreds/thousands computer systems and manufacturing devices being directly involved in the manufacturing process control, without a suitable administration system allowing us to completely master such an important resource. Even for smaller LANs interconnecting manufacturing/process control devices, it is important to know the practical limitation of the installed technology but also the behavior of interconnected stations in order to master such an important shared resource which is an industrial LAN, especially when introducing changes.

All these items of work are currently under standardization within ISO TC 97 (Network Management, Directory Service) and ISO TC 184 (MMS) excepted the MMS application interface which is coming from MAP 3.0 and which is judged very important for users and vendors in order to make the applications portable.

2. The Network Management Problem

2.1. The Target Communication Network Architecture

The target communication network architecture which is currently under development in RENAULT factories is represented on the figure 1. Though this two levels architecture is especially well adapted to Automotive Industry Manufacturing Complexes, its modularity allows us to extract some subsets which are well suited to smaller plants. In fact, a company like RENAULT owns different kinds of factories including small ones, and it is a main goal to select the best technology we think being the most suitable to a given site characteristics.

An example of an architecture and technologies usable in a small factory is given on the figure 2. Such a two levels architecture do not include the connection of low level devices like sensors and actuators to PLCs. In order to include such a capability, it could be necessary to add a third network level. In this case, communication and network management aspects of such a FIELDBUS should be integrated in the global communication / network management system of a factory. Now let's concentrate on an extended LAN architecture for an Industrial Complex, which is becoming the most critical resource to administer.

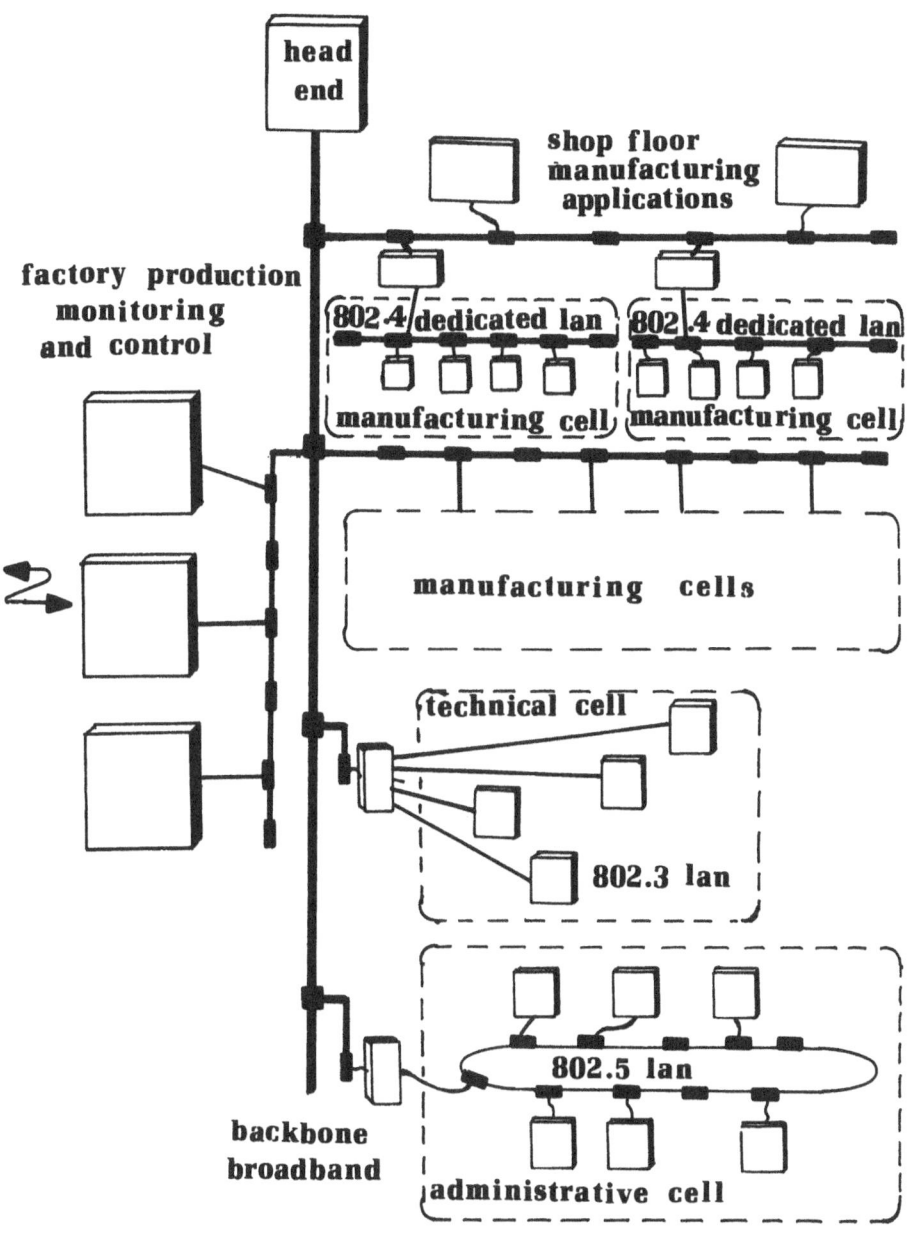

fig. 1.

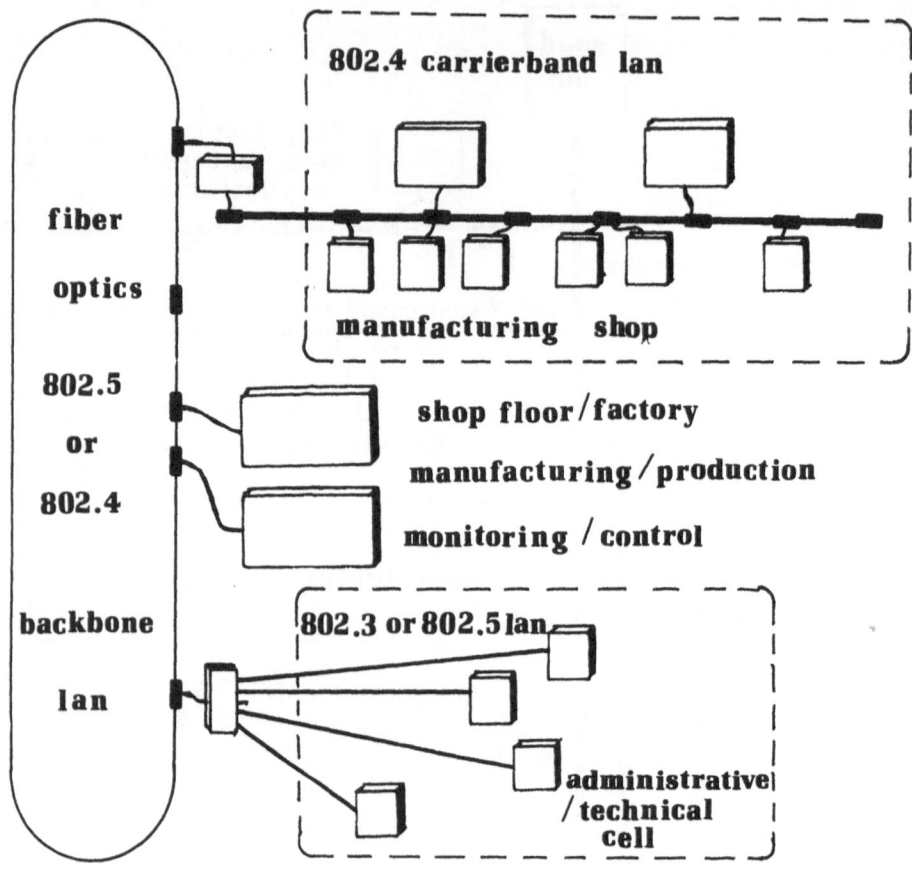

fig. 2.

As already said the factory communication network is a two levels architecture. It's a complete architecture comprising:
– The extended LAN specification,
– The functional profiles specification,
– The application Interfaces,
– The Network Management System.

A/ The extended LAN is a hierarchical assembly of Local Area
   Networks:
   - One BACKBONE LAN which is used to federate N DEDICATED LANs, executing the information transportation between Cells.
   - N DEDICATED LANs which are used for the information transportation within administrative, technical, manufacturing CELLS.
   - The dedicated LANs are connected to the backbone LAN by means of BRIDGES.

B/ Within manufacturing cells and for factory production management (shopfloor and factory CIM levels), the functional profiles which are selected are belonging to the FULL architecture of MAP 3.0 (MMS and FTAM). We consider that MMS and FTAM are frozen standards as they are at an IS (International Standard) status level, and we are expecting that CNMA 4.0 will be upward compatible with MAP 3.0 (excepted for technical bugs correction if any remaining). Some complementary studies will be achieved in order to introduce new profiles (Mini MAP, Connectionless, Connection Oriented multi-peers communication... etc) for critical time and fault tolerant systems.

C/ The RENAULT application interface is based on High Level Services specified by MAP 3.0. Some complementary information have been added to indicate the destination address mode being used (Individual/group) and to give some indication on the requested quality of service level in terms of response time, reliability, security.

D/ The required network management system is described here after.

2.2 Lan Technology Selection

A two levels architecture has been selected because we judged that it was not reasonable to mix several thousands of different equipments having various communication needs (robots, PLCs, vision systems, measurement machines, CNCs, Terminals, cell controllers, area managers, mainframes,... etc) on one local area network and in the other hand, a three levels hierarchy were not necessary and would have increased the complexity, the cost, and decreased the global MTBF. Moreover, with these two levels we keep some freedom to select the best dedicated LAN relatively to the environment constraints, the application requirements but also to the technology evolution.

– For the backbone LAN we have selected the broadband CATV technology which is considered the best technology to satisfy industrial complex requirements:
  - Broadband CATV is well adapted to cover large areas such as encountered in a car factory which is spanning from hundred thousands to several millions of square meters. Its topology and low cost coupling units (taps) allow a capacity cabling.
  - Broadband CATV components are robust and low cost as they are mass produced for cabled cities.
  - Broadband CATV has been used for more than 10 years in USA and EUROPEAN factories, particularly in automotive industry factories.
  - Broadband CATV allows the multiplexing of several channels on a same cable. On a broadband network we can have simultaneously 802.4 token bus LANs, 802.3 CSMA-CD LANs, point to point data links, point to point and broadcast video channels.
  - Broadband CATV is an option of the MAP architecture.
– On broadband we are using preferably the 802.4 token bus LAN as it consumes less bandwidth, is less expensive and imposes less constraints than the 802.3 CSMA-CD. Moreover the 802.4 offers more functionalities which are going to be used for LAN management purpose.
– For manufacturing cells dedicated LANs, the MAP 802.4 carrierband option will be used. But as long as the number of connected equipments remains relatively small (to 3 hundreds), the installed broadband can also be used as a manufacturing cells LAN option. This is due to the current high cost of intermediate systems which have a tendency to wipe out the cost advantage of the carrierband.
– For technical cells dedicated LANs, the 802.3 on twisted pairs will be used.
– For administrative cells dedicated LANs, the 802.5 token ring on twisted pairs will be used.

- Fiber optic could be a backbone or a manufacturing cells LAN alternative provided it offers an equivalent capacity at an equivalent cost when compared to coaxial.

2.3 Network Management Objectives

Being given the network architecture here before described, the network management system objective is to MASTER this important resource which is becoming the nervous system of the factory upon which all a factory production will be quickly depending. By mastering a communication network we understand:

- Having the capability to extend at will the network coverage, the number of connection points and to increase as necessary the traffic relatively to new needs without decreasing the offered quality of service. In summary to have the capability to take the best profit of new technologies and to have the capability to satisfy quickly new communication needs.
- Keeping the control of the offered quality of service in terms of network availability and network performances whatever the possible evolutions. The target is to have a network availability of 100% (The production must never be stopped by a communication network fault) and to get the level of performance adapted to the manufacturing process needs whatever the degree of data and control distribution.

In order for a user to master its industrial communication network, it is necessary to have on the market an integrated management system allowing:
- Network description and configuration management,
- Network performance management,
- Network fault management.

3. The Deduced User's Requirements

3.1 Configuration Management

Network description /configuration is particularly useful when planning some evolution (geographic extension, traffic evolution,...etc) of the network, but also for network trouble shooting in order to locate quickly and precisely a defective component, to isolate and repair it.
It is also important for network tuning in order to maintain a specified quality of service. Network description/configuration comprises the following functions:
- Description of the hardware components of a communication network (cable, taps, repeaters, amplifiers, intermediate systems, end systems,... etc). These generic components description will constitute the network hardware components library.
- Description of the ISO managed objects which are specified in CNMA IG 4.0 (layers, connections, SAPs, connection templates, service entities...etc) and their relationships with hardware/software components. For example for fault removal it is necessary to know the distribution of ISO objects among hardware/software components (definition of the system organization containment tree) such as modems, communication controllers, CPU and main memory... etc. The generic ISO objects description will constitute the ISO objects library.
- Description of the network topology which consist to instantiate generic hardware components and to establish their relationships.
- Installation of software components by local or downline loading them in end systems.

- Configuration of private/ISO managed objects, that is to say setting attribute values.
- Acquiring a global view of the extended local area network state and systems state.
  - For the broadband backbone, this means getting a real time view of the active hardware components status such as amplifiers (amplifiers status monitoring), and remodulators.
  - For each LAN (backbone and dedicated LANs) this means building and up dating the stations live list.

3.2 Performance Management

From a user's point of view we can distinguish two categories of performance which apply to different communication services which are different and sometime difficult to achieve concurrently. These two categories of performances are:
- System/Network throughput which characterizes the capacity of a system/network to transfer a continuous flow of information and is applicable essentially for the transfer of large files. It is expressed in a maximum number of Koctets per second.
- System/Network transmission delay which characterizes the capacity of a system/network to transfer quickly (promptness), in a predefined constrained time period, a small amount of information (messages) whatever the systems /network workload condition. It is expressed in the maximum time it is necessary to transmit a message of the highest priority level and to obtain a response.

Of course from a user's point of view, only the performances obtained at the application level are relevant, it is why it is important to consider all the performances achievable by the chain of subsystems (fig 1 and 2) contributing to the transport of the information, that is to say:
- The cooperating end systems ,
- The crossed intermediate systems ,
- The crossed networks ,

and to characterize these performances under different workload conditions and in particular during peak periods (worst case figures) in order to guaranty a proper operation of the global system even in worst case figures. That is to say to guaranty the service quality level required by the manufacturing process control applications.

3.3 Fault Management

Failure of a system occurs when the delivered service deviates from conditions predetermined in the specification, the latter being an agreed description of the expected service. The objectives of fault management are:

A) Trying as far as possible to avoid a faulty condition by preventive maintenance actions undertaken on a fault prediction basis. Fault prediction can be achieved by a real time analysis of trends, that is to say by monitoring a certain number of variables (network/ systems attributes) and analyzing their evolution in time, in order to detect, according to some expert's rules, trends which would be leading to a fault if some preventive maintenance operations are not urgently undertaken. Such actions will contribute to increase the global system MTBF (Mean Time Between Failures).

B) When a fault can not be avoided, trying to detect and locate precisely the component or subsystem which is at the origin of the fault in order to repair it as quickly as possible.

Such operation will contribute to decrease the MTTR (Mean Time To Repair).

Failures caused by hardware/software component or subsystem faults may have several degrees of criticity (gravity) relatively to the manufacturing process. Consequently, the priority (the effort) must be given to the faults which present the highest degree of criticity. For this purpose it is necessary to classify the various possible faults relatively to their level of criticity. For such purpose we have used a FMECA (Failure Mode Effect and Criticity Analysis) method which demonstrated that the most critical subsystem is the LAN segment connecting manufacturing devices (the manufacturing cells dedicated network in our case).

This method considers three failure mode evaluation criteria which when combined give a value reflecting the level of criticity of a failure.These criteria are:

- The fault severity Index indicates in what extend the manufacturing process is impacted by a given failure mode. The highest value indicates that the failure is immediately stopping the manufacturing process and that several systems are concerned.
- The fault predictability Index indicates in what extend a fault can be predicted. The highest value indicates that the examined failure mode is not predictable.
- The fault periodicity Index indicates the frequency of a given failure. This index is in fact related to the Mean Time Between Failure of the examined component or subsystem.

CNMA will study the utilization of knowledge based technics for:

- Fault predictability on trends analysis in order to achieve preventive maintenance actions.
- Fault detection and diagnosis in order to achieve curative maintenance actions.

The figure 3 represents the management system functions and their relationships being used for fault management.

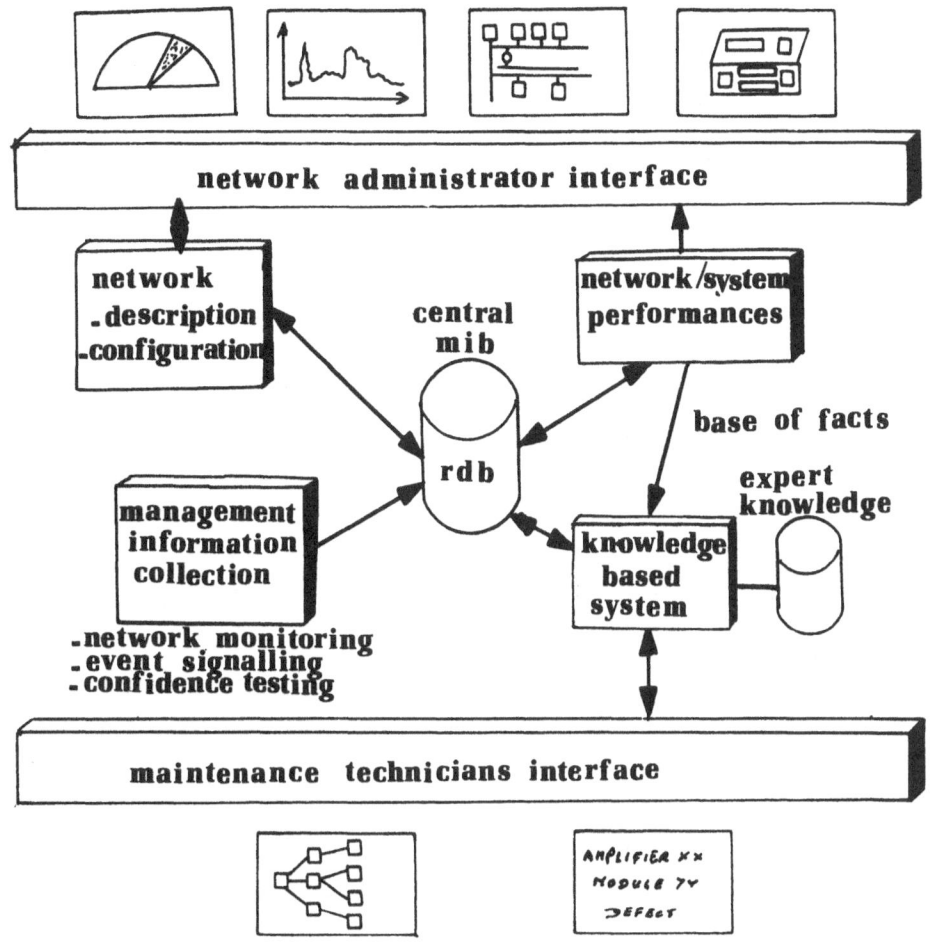

fig. 3

4. The "ideal" Network Management System

4.1 Management System Architecture

The figure 4 represents the "IDEAL" network management system we expect to build from market modules within a 2 years time period. This management system comprises the following modules exchanging management information through CMIS / CMIP ISO services and protocols:

800

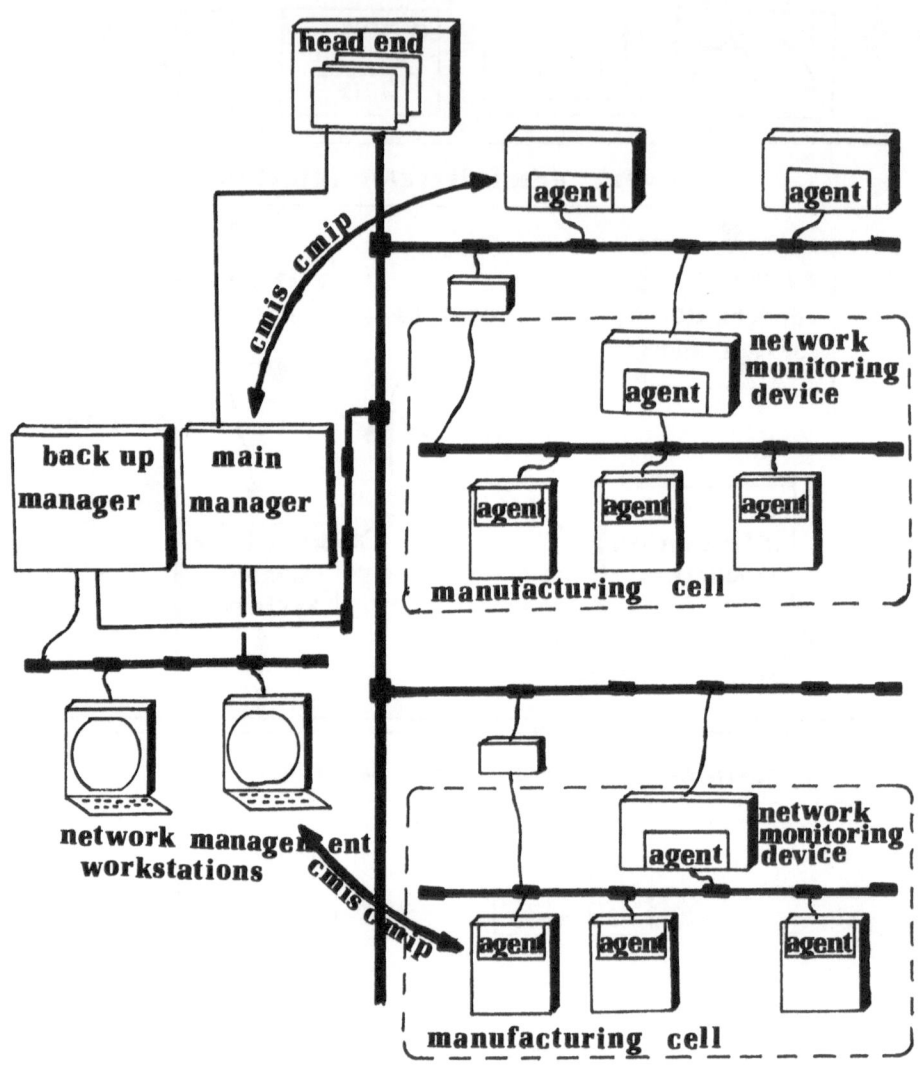

fig. 4

A) Distributed system management agents which under the control of a redundant network manager are setting/reading local system managed objects attributes, are reporting events and are executing actions. System management agents must comply to MAP 3.0 or CNMA 4.0 specifications, supporting a requested minimum set of management services and managed objects under definition.
B) Network monitoring agents either distributed in end systems or centralized in network monitoring devices (at least one per LAN) as represented on the figure 4.

This latter solution is preferred because:

- It allows to simplify the system management agents,
- It is independent of the availability of such monitoring services in commercial systems.
- The network monitoring device could offer Xsphisticated services programmable from the network manager.
- The network monitoring device can be directly connected to the LAN supporting the network manager, this allowing to continue to report information related to the observed LAN even in the event of a failure of this latter one.

C) A redundant manager which executes extended LAN configuration, performance, fault management and interfacing the network administrator and maintenance people.
Such a manager will comprise (fig 5):
- An information collection mechanism ,
- A relational data base allowing to store all the management information being not distributed in agents ,
- The management applications including the knowledge based system,
- The Man/management system interface.

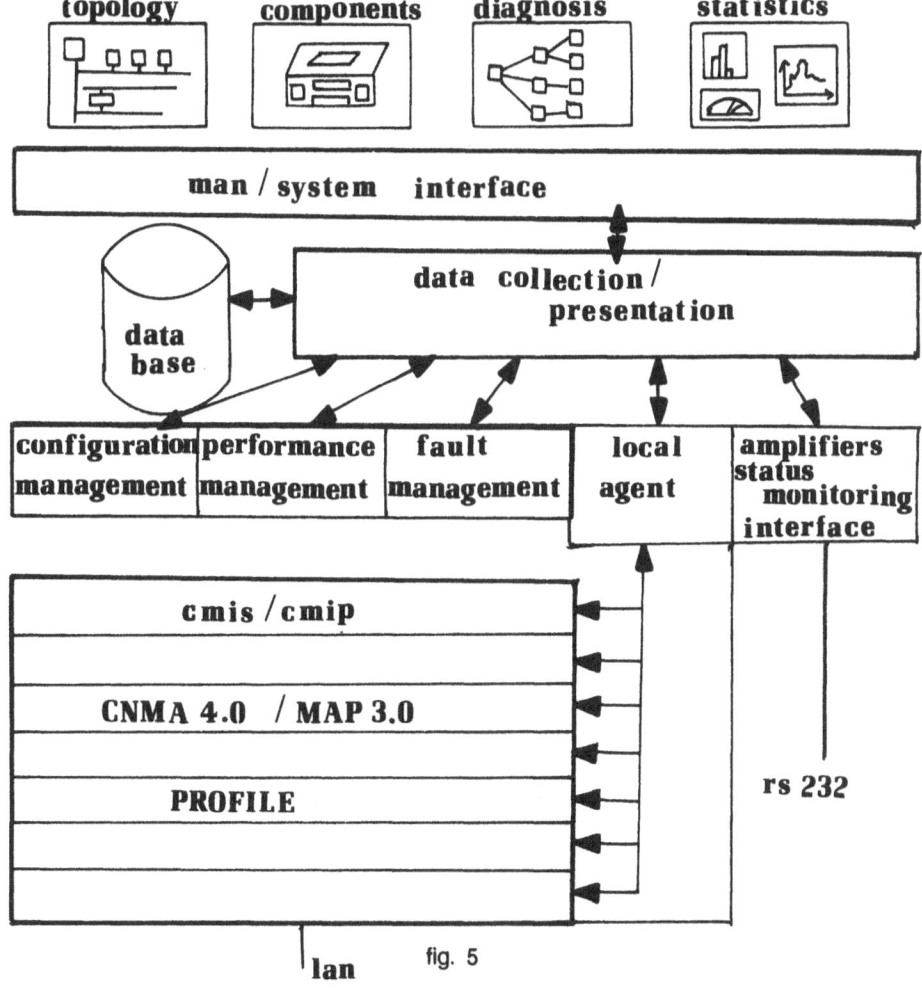

fig. 5

4.2 CNMA Expected Results

The results expected from CNMA are the following:
A) Specification of ISO management services and protocols being used between mana-gers and agents.
B) Specification of ISO managed objects and accessible attributes. After a failure mode effect and criticity analysis it appears that the effort must be done on LANs attributes (PHYSICAL and MAC layers) collecting a maximum number of information allowing the expert system to use them for fault management and performance tuning.
   A and B will constitute the network management part of the CNMA Implementation Guide for phase 4 (IG 4.0).
C) Specification of network management applications including a minimum knowledge based system (KBS) used for fault management.
D) Description of generic hardware / software components necessary to build a network topology and to identify and locate physical components which will have to be replaced when a first maintenance operation is required.
E) Connection to the network manager of private management subsystems such as the broadband amplifiers status monitoring, allowing to get complementary information relatively to the physical network itself.
F) Specification of the Man/System interface allowing to describe, configure an extended LAN, to analyze its performance and get diagnosis information from the knowledge based system.
   C, D, E, F will constitute the IG 4.0 Addendum 1.
   Once the specification phase is finished, a network management system prototype will be implemented by CNMA vendors and be firstly, partly demonstrated at ISW (University of STUTTGART) and completely at the RENAULT pilot site, before being ported to the AEROSPATIALE pilot site.

5. The CNMA Network Management Research Aspects

The four following aspects are considered as being research items:

5.1 Description/standardization of a Network Components Library

– Firstly, for fault removal it is necessary to identify and locate precisely the defective Hardware/Software component which is at the origin of the network failure, in order to replace it by a first level maintenance operation.
– Secondly, all information captured for fault management relate to network/systems managed objects attributes. This collected information is processed to extract some trends (statistics) on network/ systems performances.
– Being given these facts, it is necessary to establish a relationship between managed objects attributes and the hardware/ software components containing them (Network Topology Containment Tree, Systems Organization Containment Tree). Then it becomes important to build a library of generic managed hardware/ software compo-nents describing the characteristics, status and statistics attributes of these compo-nents. In such a case building a network topology will consist to instantiate some generic hardware components (Cable segments, coupling units, power supply mo-dules, amplifiers/ repeaters,... etc) and define their relationships.

4.2 CNMA Expected Results

The results expected from CNMA are the following:
A) Specification of ISO management services and protocols being used between managers and agents.
B) Specification of ISO managed objects and accessible attributes. After a failure mode effect and criticity analysis it appears that the effort must be done on LANs attributes (PHYSICAL and MAC layers) collecting a maximum number of information allowing the expert system to use them for fault management and performance tuning.
A and B will constitute the network management part of the CNMA Implementation Guide for phase 4 (IG 4.0).
C) Specification of network management applications including a minimum knowledge based system (KBS) used for fault management.
D) Description of generic hardware / software components necessary to build a network topology and to identify and locate physical components which will have to be replaced when a first maintenance operation is required.
E) Connection to the network manager of private management subsystems such as the broadband amplifiers status monitoring, allowing to get complementary information relatively to the physical network itself.
F) Specification of the Man/System interface allowing to describe, configure an extended LAN, to analyze its performance and get diagnosis information from the knowledge based system.
C, D, E, F will constitute the IG 4.0 Addendum 1.
Once the specification phase is finished, a network management system prototype will be implemented by CNMA vendors and be firstly, partly demonstrated at ISW (University of STUTTGART) and completely at the RENAULT pilot site, before being ported to the AEROSPATIALE pilot site.

5. The Cnma Network Management Research Aspects

The four following aspects are considered as being research items:

5.1 Description/standardization of a Network Components Library

- Firstly, for fault removal it is necessary to identify and locate precisely the defective Hardware/Software component which is at the origin of the network failure, in order to replace it by a first level maintenance operation.
- Secondly, all information captured for fault management relate to network/systems managed objects attributes. This collected information is processed to extract some trends (statistics) on network/ systems performances.
- Being given these facts, it is necessary to establish a relationship between managed objects attributes and the hardware/ software components containing them (Network Topology Containment Tree, Systems Organization Containment Tree). Then it becomes important to build a library of generic managed hardware/ software components describing the characteristics, status and statistics attributes of these components. In such a case building a network topology will consist to instantiate some generic hardware components (Cable segments, coupling units, power supply modules, amplifiers/ repeaters,... etc) and define their relationships.

5.2 Communication Software Downline Loading

Software downline loading, which can be considered as the action of instantiating communication software components, is necessary:
- After a power failure or system failure in the case where a distributed system do not owns a permanent memory (ie: EPROM or DISK).
- When a new software release has to be installed simultaneously on a certain number of interworking systems.

Such function has been introduced within MAP 3.0 and is based on IEEE 802.1 standard. This item of work is not well advanced in ISO.

5.3 Knowledge Based System For Fault Management

The study of KBS suitability to solve fault prevention and diagnosis problem is an important research item as it could greatly contribute to increase the global MTBF by triggering in due time, preventive maintenance actions and can contribute to limitate the MTTR within a given period of time (ie:less than 1 hour) allowing to reach the 100 % availability goal. In our architecture case this goal will be reached by giving to the cells applications a limited autonomy of 1 hour necessary to repair a defective backbone network component or intermediate/end system. The cells dedicated LANs are considered as being very reliable as they are completely passive.

5.4 Network Global Clock

For performance and fault management it is necessary to time stamp the events which are generated by network management agents. In some circumstances, during traffic peak period or for faults which propagate from a system to an other, it would be useful to keep the time global order of events as they are detected by distributed agents. For this purpose it is necessary that each real time clock of distributed systems indicates the same hour with a high degree of accuracy (ie: 10 to 50 ms). Such feature can be obtained by broadcasting a global clock for example every second on the network, allowing distributed agents to adjust their local real time clock.

At a first glance, such performance could appear very ambitious, however being given the following assumptions it seems possible to reach such performance level.

Assumptions:

A) We have only a two level architecture, that is to say only one level of bridges and the bridges have the capability to serve incoming messages on a priority basis.
B) We use the ISO 8802.4 priority scheme option allowing to guaranty a bounded transmission delay less than 10 to 20 ms whatever the LAN traffic for a maximum number of stations in the order of one hundred per LAN.
C) We broadcast the clock messages using the highest priority level.
D) We use a connectionless architecture similar to the miniMAP allowing to reduce the layers transition time variation. In all cases such connectionless architecture will be necessary to broadcast messages which do not require to be reliable (if we lost one clock pulse, the distributed local real time clocks will be up dated on the next one).

6. State of the Art

A first version of CNMA IG 4.0 is now available. Relatively to network management, this document contains:
- The concepts and an overview of the CNMA service subsets.
- The definition of management communication services.
- The systems management functions specifications.
- The addressing management information definition.
- The managed objects and attributes definition.
- The network management protocol specification.
- A mapping to the lower layers.

The RENAULT requirements for CNMA network management have been produced and will be used to evaluate the CNMA project results.

All the pilots general specifications have been delivered to the CEC. These are relative to:
- ISW pilot which is an ESPRIT pilot 1 showing all vendors products interworking to demonstrate a subset of a CIM system producing some small components customized by means of a CAD workstation integrated in the system.
- AEROSPATIALE pilot showing several cooperating applications such as:
  - Manufacturing cell control and monitoring application, including a handling/transport/storage system.
  - Maintenance management application including network management.
  - shop management application including short term planning leading to lists of jobs delivery to the cell controllers.
- MAGNETI MARELLI pilot showing a flexible manufacturing system associated to an assembly line controlled and monitored by applications using CNMA profiles and used for the production of alternators.
- RENAULT pilot which concentrates on the network management system test, evaluation and demonstration. In order to generate some traffic, a subset of the AEROSPATIALE application and a small video control application will be implemented.

7. Conclusion

From a user point of view, when considering industrial complexes such an automotive factory, it is important to have a global approach in order to facilitate the integration, to optimize the costs and to guaranty a good level of service quality. The administration system is an important aspect of this global approach and is complementing the communication system itself. We believe that only an integrated network management system allowing to obtain a global view of the industrial communication network, from a central point or from a small number of distributed points, is an essential component giving to the network administrators and maintenance agents the capability to achieve and maintain a high level of service quality. Such integrated network management system does not exist today in an opened system environment and we hope that the CNMA project will contribute significantly to the availability in Europe of such a system within a short period of time.

Keywords:

Lan, Management, Iso, Standards, Knowledge Based System, Objects, Fault, Configuration, Performance

ESPRIT Project No 955 and No 2617

COMMUNICATIONS FOR C.I.M. -
A REVIEW OF THE CNMA PROJECT

Tim Simmons
BAeCAM
British Aerospace
The Guild Centre
Preston

The Communications Networks for Manufacturing Applications
project is specifying, implementing, validating, and promoting
emerging communications standards for CIM. This highly
successfully project has commissioned a number of real production
pilot facilities and provided working demonstrations at two
major international exhibitions.

1.0 INTRODUCTION

If European industry is to be competitive in world markets, it must
make good use of Computer Integrated Manufacturing, CIM. For optimum
profitability, a manufacturing facility must be flexible and reliable. It
must be able to produce quality. It must provide accurate, timely
feedback and must respond quickly to management decisions or new
requirements. To be competitive, European industry must ensure that
manufacturing resources are fully integrated.

A wide range of sophisticated computer based devices is now available
for use in or around the shop floor. Most of these use different proprietary languages to communicate. Integration of these devices is expensive
and slow because bespoke interfaces are required. What is needed is a
standard communication language made available for everyone to use. The
development of such a language is a complex, expensive task. To be
successful the standard language must fulfil the needs of a wide range of
users, and it must be adopted by a wide range of vendors. The task of
specifying the language therefore requires input from, and collaboration
between, many different companies and organisations.

The Commission of the European Communities has provided a framework
and funding for such collaboration, through the European Strategic
Programme for Research and Development into Information Technology,
ESPRIT. One of many projects supported in this way is CNMA, which
addresses Communications Network for Manufacturing Applications.

CNMA aims to complement world-wide initiatives in the specification
... implementation ... validation and demonstration of common communication standards. If widely adopted, they will simplify the integration
task and reduce the cost of CIM systems.

806

.0 <u>CNMA EP955</u>

To achieve this, British Aerospace is leading a powerful European consortium which came together in early 1986 with 16 partners, including both users and vendors of computing technology.

It is important to recognise that this European wide combination of experience, skills, and interests, all dedicated to a common goal, has been the major contributing factor towards the success of the CNMA programme.

It must be stressed that CNMA is aiming for the same final profile of communication standards as the American lead MAP and TOP projects. However, this has led some people to question CNMA's purpose : "If CNMA is aiming for the same final profile as the MAP and TOP programmes, why is the CNMA project necessary?" Or, put another way : "Why is this issue so strategic for European industry?"

The MAP and TOP initiatives have done much to further the idea of Open System Interconnection, but the target of open, multi-vendor systems has not yet been realised. CNMA has helped to overcome some of the past limitations in MAPs success.

One limitation was that the take up of MAP among European vendors was far from adequate. The European MAP Users Group (EMUG) has done much to further the acceptance of this technology but is limited in its abilities due mainly to funding constraints.

CNMA is also representing the European users' requirement for multi-vendor interworking down to the level of device controllers. Within a manufacturing cell large American industries often use a single vendor's controllers and thus proprietary communication systems. However, within Europe, many companies do not have the same level of investment, and demand multi-vendor working right down to the level of robots, machine tools, etc.

An earlier limitation of MAP's success was the limited scope for device to device interworking. There were too many options within the standards and specifications and too little conformance testing.

Having noted the main limitations in MAP's European success, we can now appreciate the objectives of the CNMA project. These are :-

a)   specification, implementation, validation, demonstration and promotion of standards and specifications for factory automation applications to ensure the development of standards suitable for European users ;

b)   compatibility with MAP and TOP Specifications, and SPAG and CEN/CENELEC supported standards, in Europe, to ensure that a single international profile is obtained ;

c)   Promotion of European acceptance of standards, to encourage European vendors to adopt them ; and

d)   Encouraging the creation of validation centres, to facilitate the testing of vendors implementations.

In order to achieve these objectives the CNMA Consortium has engaged in the following main activities :

a)   specification of CIM users' requirement ;

b) selection of an unambiguous profile of communications standards to meet the users' requirements ;

c) implementation of the profile on controllers and mini-computers ;

d) development of conformance test tools and conformance testing of the implementations ;

e) validation and demonstration of the implementation in real production facilities ; and

f) promotion of the profile.

Although considerable effort has been expended to establish the standards for industrial LANs, there is still insufficient information on the needs these standards are intended to satisfy. CNMA has the advantage of including major industrial users, and has conducted detailed analyses of the subject. These cover a wide variety of requirements, in both batch manufacturing and the process control industry, including costs, time constraints, reconfiguration, redundancy, reliability, integrity and training.

It was found that reliability is a major concern for most applications, as well as the need for low cost, high integrity installations that allow for redundancy, and which permit reconfiguration as factory layouts change.

Examination of the basic network technologies showed that broadband MAP is not the only solution to users' needs. The results of the studies are contained in reports which are available from the Commission of the European Communities.

3.0 EP 955 PILOTS

The first phase of the project began in January 1986 and resulted in a demonstration in April 1987 at the Hannover Industrial Fair. A real manufacturing cell was established incorporating an advanced machining centre, a tool handling robot and a transport system. It was typical of a manufacturing environment and had the capability of machining real engineering parts.

The control architecture was deliberately made more complex than necessary, so as to be typical of a large-scale implementation, where equipment from many vendors may require integration. It included three vendor's controllers and four vendor's mini-computers. The machine tool was managed by a Computer Numerical Controller or CNC, from GEC. The transporter and tool handling robot were managed by Programmable Logic Controllers from GEC and Siemens. The mini-computers from Bull, Nixdorf, Olivetti and Siemens, provided cell control and operator access functions. These devices were linked by three Local Area Networks and two relay devices. Communications between these devices utilised the CNMA phase one software which included important new work on the automation protocol, Manufacturing Message Specification or MMS. For the eight days of the Fair, manufacturing of parts on the cell was demonstrated to visitors from all over the world.

This demonstration had two major differences from the MAP/TOP facility shown at Autofact in 1985. Firstly, it achieved true interworking in a real production application, rather than "gluing together" of equipment by laboratory techniques. Secondly the CNMA implementation was targeted at compatibility with MAP 3.0 instead of MAP 2.1, and, as such, it involved MMS automation protocols between devices rather than just file transfer or the earlier "Memphis" protocols of MAP V2. This was the first ever demonstration of multi-vendor interworking using MMS, and the only such demonstration before mid 1988.

While exhibitions can be used to demonstrate new communications software, the most thorough validation is provided by using it in real production activities, in a competitive industry, where it must provide all the essential functions with very high reliability. The C.N.M.A. project has validated its communications software in three such environments... at BMW, British Aerospace and Aeritalia.

BMW's position in the highly competitive automotive industry, means that the company must stay at leading edge of technological developments, to produce ever greater efficiency. With the rapid increase in the number of computing devices in its factory, BMW were one of the first companies to recognise the benefits available from adopting standards for communications and this led to their involvement in the CNMA project.

The first CNMA production facility is at BMW's factory in Regensberg, West Germany, where three-series limousine and convertible cars are manufactured. This facility provides a comprehensive maintenance system and is the world's first production application of MMS. It indicates a major industrial company's confidence in the software developed within the CNMA project. The result is that since early 1988, this software has been used in the manufacture of cars which are a status symbol throughout the world.

A consortium of Aerospace companies from six nations in Western Europe, Airbus Industry produces a family of high performance, wide bodied civil aircraft, which are selling very successfully.

CNMA's second production facility at this British Aerospace factory in England, machines 'D' shaped components for the leading edge of the A320 Airbus wing. It also participated in the Enterprise Networking Event in Baltimore and was the Event's only true production facility. CNMA's participation in Enterprise has allowed the projects vast experience in communication software development to be applied to the evolution of the MAP specifications. CNMA's final preparations for Enterprise were witnessed by influential representatives of the world MAP/TOP Federation during a visit to the British Aerospace facility.

The use of a variety of Local Area Networks demonstrates the practicality of selecting OSI upper layers appropriate to the application and OSI lower layers appropriate to the installation.

Communication with the Enterprise Networking Event was achieved by running MAP/TOP 3.0 software over X.25 Wide Area Networks, between Baltimore and British Aerospace. This involvement emphasises how the latest communications standards can link manufacturing environments throughout the world.

CNMA was further involved in Enterprise by providing a major part of the required MAP test system, through the sub-project, named CNMA Conformance Testing or CCT. These test tools, which are necessary to ensure inter-operability of devices, enabled Enterprise to proceed, and a commercial venture, SPAG-CCT now markets these tools world wide (in collaboration with the Corporation for Open Systems (COS)).

CNMA's involvement in Enterprise demonstrates that the same software standards are being adopted on both sides of the Atlantic. This is a major step towards enabling manufacturers to integrate devices from different suppliers, European or American, in a move towards truly Computer Integrated Manufacturing.

In November 1988, at an Aeritalia factory in Turin, CNMA commissioned its third production facility. The CNMA software is used here in the automated production of aircraft wire harnesses. The process is the automated cutting and stamping of wires. An advanced feature of this facility was the ability to receive manufacturing data for wire harness production from the British Aerospace site using a Wide Area Network connection.

This demonstration illustrated the increasingly important requirement for companies in manufacturing industries to be able to transmit manufacturing data to sub-contractors quickly and reliably.

4.0 CNMA PROCEDURE

In order to understand why CNMA was so successful, let us first look at the procedure which was adopted by the partners.

Firstly Vendors and Users came together to identify user needs. Then Vendors (and some Users) produced an agreed and unambiguous method for implementing the standards. This was set out in a document called an Implementation Guide. The Vendors then produced implementation of the standards for their own equipment.

In order to ensure that these implementations were error free and really did conform to the standards, each device was then tested using specially developed conformance test software.

Finally the equipment was integrated into the CNMA pilots, thereby demonstrating the value of the original standards. Because each device had already been rigorously tested, the system integrator could be confident that devices would interwork without a long expensive commissioning process.

5.0 RESULTS OF EP955

By the end of 1988 the CNMA project had successfully demonstrated the use of OSI protocols at two major manufacturing exhibitions and in three production pilots. New OSI conformant products were becoming available from the CNMA vendors. Also the conformance test tools, which had played such a major part in the success of the project, were being marketed world wide by SPAG-CCT.

Partly because of CNMA, the Standards on which the protocols were based had progressed towards stability.

6.0 CNMA EP2617

Because of the success of the CNMA approach, a major new phase to CNMA was begun in late 1988 with the objective of specifying, implementing, validating, and demonstrating the the new standards. This was project EP2617.

The new phase of the project involves important new work on the network management core protocols plus new work on fault management, configuration management and performance management. Other work on companion standards, application interfaces and fieldbus is also under way.

The same operating procedure is being adopted, with the first external release of the Implementation Guide taking place in October 1989, and the final release due in February 1991.

New partners have joined the project. These are Aerospatiale and Renault in France, Magneti Marelli in Italy, Robotiker in Spain, the

Universities of Stuttgart in Germany and Porto in Portugal. Test tools are being developed by the complementary ESPRIT project EP 2292, known as TT-CNMA (Testing Technology for CNMA).

Under the current phase of CNMA, four pilot facilities are being produced. These are :

- one "Pilot 1" demonstrator at University of Stuttgart

- three "Pilot 2" demonstrators. These are real industrial control systems which are being installed at Renault and Aerospatiale in France and Magneti Marelli in Italy.

The ISW Pilot network architecture has been made over-complicated to demonstrate the ability of equipment from a number of vendors to intercommunicate. The cell integrator, ISW, has taken equipment from all the vendors within CNMA, Bull, Siemens, Nixdorf, Olivetti, GEC and Robotiker along with the vendors implementations of the communication protocols. Application specific control software has then been produced by TITN and ISW.

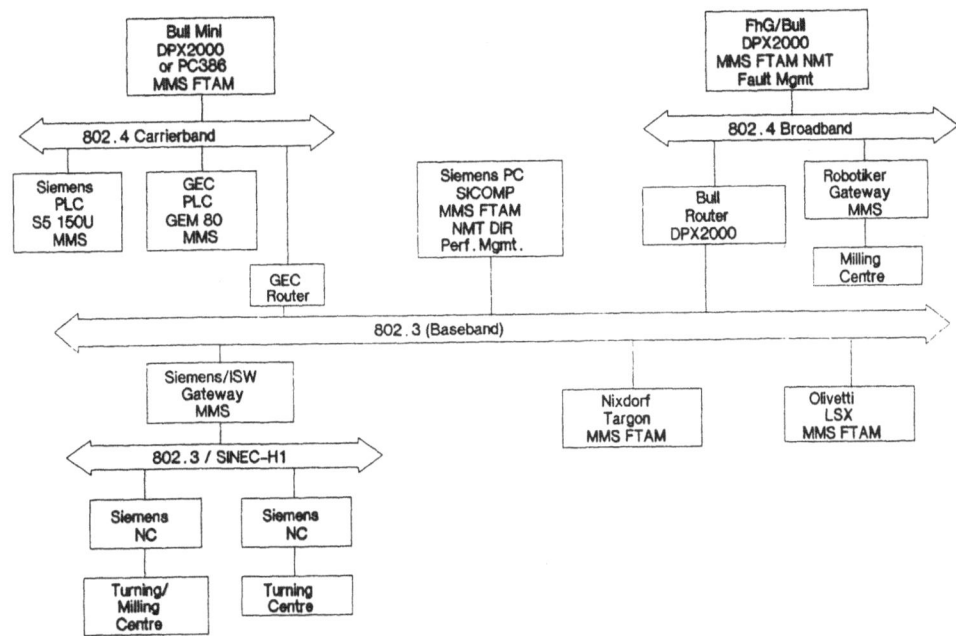

ISW DEMONSTRATOR – COMMUNICATION ARCHITECTURE

Two independent cells operate within the pilot. One cell is designed to manufacture bodies for hydraulic valves. It consists of a turning centre, a boring and milling centre, a linear portal robot and a palette store. Production within this cell will be fully automated.

The other cell consists of a manually loaded 5-axis milling machine. This cell will be used to demonstrate the advantages of LAN communication between CAD and NC systems. The design of free form surfaces will be shown on a CAD system. CAD files will be transferred to a separate mini computer for part program generation. These will then be downloaded to the NC controller which will be started from the cell controller.

This pilot will be open to the public in April 1990 and will be fully operational in July 1990.

The second pilot will be at Renault's factory in Boulogne Billancourt, France. This pilot is to provide a test and demonstration facility for Network Administration systems. In large networks, automated facilities are required to handle faults, tune the network for optimum performance and handle re-configuring of the network. Suitable protocols for this are provided within the Network Management function.

The pilot will consist of a conference room linked to a test laboratory. Equipment for three types of demonstration will be installed in the laboratory. The three demonstrations will be linked together, and to the conference room by Local Area Networks. One of the demonstrations will be a simulation of the Aerospatiale pilot. The second will be a demonstration of the control of video equipment from the control room and the communication of video images via the network. The third will involve injection of simulated faults into the network.

In combination, the pilot will demonstrate the ability of the Network Management function to tune and reconfigure the network and to diagnose and manage fault conditions. This pilot will be operational in October 1990.

The Aerospatiale pilot will control 9 machines which are used for making prototype missile components. The main objectives of the pilot are to improve on delivery times, quality and flexibility.

Four applications will run on the pilot. A Maintenance Management Application will collect status information from the machines, and produce fault diagnosis by expert system. A Shop Management Application will perform short term planning and produce machine shop performance statistics. A Machining Cell Management Application will take job schedules from the Shop Management Application, and will handle all data communication to and from the NC controllers. Finally, a Storage, Handling & Transport Cell Application will control the AGV transporter, the manual load/unload stations, and the storage area. This pilot will be operational in December 1990.

The Magneti Marelli pilot will be installed in the final section of an alternator production line. Three main functions will be performed. A monitoring system will collect data on machine productivity and behaviour, a tracking system will trace all work-in-progress on the shop, and a diagnostic system will collect detailed information on machine states. The pilot will be operational in December 1990.

7.0 CNMA COMMUNICATIONS PROFILE

One of the CNMA project's aims is to specify a communications system for computers and programmable devices.

Communication in CNMA is based on the Open Systems Interconnection basic reference model (OSI/RM), which is defined by ISO in IS7498. It defines a framework for communications, that is the services to be provided to application processes, the breakdown of the communications software into 7 layers and the split of services between the layers. The model is known as the "OSI seven layer model".

Each service requires protocols - specified interactions - between the two systems. Definitions of the services and protocols in each layer are provided in individual standards documents. However, a given service can be provided by a number of different protocol combinations in the lower layers. Hence, an additional document is required to identify the exact protocols used at each layer. Such a selection of protocols is known as as a "profile".

The CNMA Implementation Guides defines the communications profile chosen for use in the project. They represent a considerable amount of work by the participating companies and their publication is one of the primary functions of the project.

MAP V3.0 is also a communications profile based on the seven layer model. However, when MAP V3.0 was specified protocols for layer 7 were not stable and so interim solutions were included.

The main purpose of CNMA is to focus its research into layer 7 issues. CNMA has contributed to the MAP evolution through its research and implementation work in this area, so aiding the definition of a single international profile of communications for use in the manufacturing environment. CNMA has always maintained a close relationship with MAP and other relevant initiatives and also with the standards supported in Europe via such groups as SPAG-CCT, and through the standards bodies such as CEN/CENELEC.

CNMA Implementation Guides devote a chapter to each layer or service.

For layers 1 and 2, CNMA initially uses Local Area Networks (LAN's) and acknowledges the benefit to users of providing a choice of LAN type. This allows a user to choose a type based on : cost; performance; installed base; maintainability; etc. Three options are specified.

MAP initially opted only for the broadband technology and has extended this to cover carrierband, with token bus access, but a recent study in Europe showed that 98% of LAN installations have opted for baseband technology. CNMA's choice of baseband LAN is the same as that supported by TOP.

In CNMA layers 3 to 5 are designed to conform with MAP. These layers are considered to be more stable than layers 6 and 7, and are therefore defined as 'background' for the project. This enables work to be concentrated on the other upper layers.

For layer 6 (the presentation layer), CNMA has specified the use of the kernel functional unit. Layer 7 protocols are covered by a number of chapters. The first defined the Association Control Service Elements known as ACSE, which uses the latest ISO international standard protocol to provide association control.

A further chapter for layer 7 protocols defined the Manufacturing Message Specification - MMS, which is a protocol for passing messages between computers, numerical controllers and programmable logic controllers, robot controllers and transport systems.

Another chapter in the layer 7 group presents FTAM which is used to exchange files between mini-computers.

Another chapter for layer 7 covers network management. This is a protocol which permits a remote device (manager) to access communications related attributes (counters/timers) in other devices (agents). Using this protocol a manager can monitor and influence network performance, configuration etc. This management is achieved by writing to and interrogating the management information base on the agents. This is currently the greatest area of activity within the project.

Another chapter of this series covers directory services, which supports storage and interrogation about named objects (things or people) in order to provide services such as network "white pages" and "yellow pages" type searches.

Further chapters in the IG deal with the specification of the application interfaces for MMS and FTAM. These are to improve application portability.

Implementation Guide version 4.0 is currently being produced and will reference the latest versions of the ISO specifications for all these services.

8.0 CONFORMANCE TESTING

If industry is to reap the benefits of open systems it is essential that conformance tests are established worldwide which provide consistent and uniform results. This is a necessary step towards ensuring interworking of products from different suppliers.

In the first phase of the project the Fraunhofer Institute, an independent testing organisation, obtained and developed the conformance test tools and procedures to verify the CNMA implementations.

During the pre-staging for the Hannover Fair, the test tools were used to great effect, testing the vendors' new implementations. The outstanding success of the demonstration is evidence of the quality of the test tools developed so far.

For the second phase of the project, the sub-project, CNMA Conformance Testing, developed conformance tests for MMS, Network Management, Directory Services, Bridges, Routers, End System/Intermediate System Protocol and Logical Link Control Class 3. These tests were supplied to the ENE staging areas.

Besides the existing CNMA partners, SPAG Services, the Fraunhofer Institute, The Networking Centre and ACERLI are involved. Agreements with ENE Corporation for Open Systems (COS) in the USA have been made to ensure the widest possible acceptance of these conformance tests.

The urgent requirement for test tools for MAP and TOP at ENE '88i provided a rare opportunity for Europe to establish and own a significant test suite by bringing together the necessary resources under the ESPRIT programme.

For the current phase of the project the conformance test tools are being developed by a separate but closely linked ESPRIT project EP2292, known as TT-CNMA.

The MAP 3.0 tool set is now marketed world-wide by SPAG-CCT in collaboration with the American Corporation for Open Systems, COS.

5.0 SUMMARY

In summary, the major achievements of the project can be reviewed as follows :-

a) Three Implementation Guides have been produced defining a communications profile to allow ACSE, MMS, FTAM, Network Management and Directory Services protocols to be exchanged between multi-vendor systems of mini-computers and programmable devices.

b) Multi-vendor control using CNMA communications software has been successfully demonstrated at the 1987 Hannover Fair, the Enterprise Networking Event, and in real production environments at Aeritalia, British Aerospace and BMW. A number of world firsts have also been achieved during these activities.

c)   CNMA is liaising with standards bodies and is having an impact on the final single communications standard, thanks to its experience in implementation and validation of the CNMA profile.

d)   The project has brought many European vendors closer to the final standard.

e)   A comprehensive set of MAP 3.0 test tools are now marketed world wide by SPAG-CCT, a commercial venture launched specifically for the marketing of the conformance test tools.

f)   CNMA vendors such as BULL, GEC and Siemens now market OSI based products. These MAP compatible products include mini-computer interfaces, programmable logic controllers, gateways and bridges. The major impact the project has had on emerging standards is already being built upon with major new work extending into the 1990's. This fact, coupled with the admission of influential new partners, including Renault, Magneti Marelli and Aerospatiale, ensures that ESPRIT CNMA continues to make a major contribution to the lower cost CIM facilities that manufacturing industry so urgently needs.

Current CNMA project members:

Aeritalia, Aerospatiale, Alcatel-TITN, British Aerospace, Bull, Comconsult, Fraunhoffer, General Electric Company, G4S-Ricera, Magneti Manelli, Nixdorf, Renault, Robotiker, Siemens, University of Porto, University of Stuttgart.

KEYWORDS

Communications Networks for Manufacturing Applications
Computer Integrated Manufacturing
MAP/TOP/OSI
Local Area Networks
Production Pilot Demonstrations
Conformance Testing

OFFICE AND BUSINESS SYSTEMS

Multimedia Document Storage, Classification and Retrieval: Integrating the MULTOS System [1]

P. Constantopoulos [2]
E. Lutz [3]
P. Savino [4]
Y. Yeorgaroudakis [2]

ABSTRACT

The development of effective multimedia office document filing systems requires both optical and magnetic storage, combined use of indexing, text and image data access techniques, automatic document classification, and a flexible, intelligent user interface offering rich functionality to non-computer expert users. In this paper we present the latest developments in ESPRIT project MULTOS which led to an integrated prototype system with the above features.

1. Introduction

The information handled in offices often has multimedia form, comprising text, graphs, images or spoken messages. Correspondingly, this reality is faced by computer-based multimedia office information systems. However, due to the complexity introduced by the multiplicity of media, such systems have only recently been developed by virtue of technological advances in powerful, fast, small computer systems, data communications and storage systems.

Our particular concern is the filing and retrieval of multimedia office documents. A *multimedia document* is a structured collection of formatted alphanumeric data, free text, graphics, bitmap images, and audio recordings [Fors84]. Depending on the application, or even the specific document, some components may be missing.

The storage of such documents requires using very high volume storage media and storage strategies that take into account the life cycle of office documents.

Retrieving multimedia documents presents a whole range of issues. Indexing techniques applicable to accessing the formatted data must be combined with text access methods and image retrieval techniques coordinated in a uniform document retrieval strategy. A related issue is that of document classification. Both classification and retrieval must be based on a model for representing multimedia documents and organizing them in collections and hierarchies.

A multimedia document filing system must offer flexible man-machine interaction supporting a wide range of operations and be suitable for non-computer expert users. This requirement is translated into a

---

[1] The partners of MULTOS project are: Battelle - Frankfurt (DM), Institute of Computer Science F.O.R.T.H. (GR), ERIA (SP), I.E.I-C.N.R. (I), Epsilon (GR), Olivetti-DOR Pisa (I), Triumph-Adler (DM)

[2] Institute of Computer Science, Foundation of Research and Technology-Hellas, P.O. Box 1385, Heraklion, Crete, Greece.

[3] TA Triumph-Adler AG, Research/EF, Hundingstr. 11b, D-8500 Nuernberg 80, FR Germany, Phone: +49 911/3327127, Net: ernst%taeva@unido.uucp

[4] Olivetti - DOR, Via Palestro 30, Pisa 56100, Italy.

sophisticated user interface design.

Furthermore, the system ought to be extensible, and feature concurrency control, recovery, security and version control.

ESPRIT project MULTOS has addressed all of the above issues, by designing and implementing in prototype form a multimedia document filing system which features an open architecture, its own document model for classification and retrieval, a combination of magnetic and optical storage, automatic document classification, an intelligent, flexible user interface, and document exchange according to the ODA standard.

A number of previous papers have documented various facets of the MULTOS system [Bert88, Cons86, Cons88, Cons88a, Eiru88]. In this paper, after a brief overview, we present the latest developments integrated into the system, namely the multi-storage file system and the automatic classification server along with the new uniform user interface. The design of the latter has been described in [Cons88a], yet now it is actually running. A sample working session with MULTOS is presented.

2. MULTOS overview.

MULTOS is an open, distributed multimedia document filing system following the client-server paradigm [Svob84]. Clients and servers are independent processes running, possibly, on different computer systems connected in a LAN.

The storage servers perform the data management operations and handle the optical and magnetic storage media. Other servers in MULTOS are: the document printing server that prints documents on a laser printer; the image acquisition server that transfers scanned images to the MULTOS editor; and the automatic classification server that classifies documents according to the MULTOS document hierarchy, in either batch or interactive mode.

All servers are activated by clients which offer a uniform, flexible user interface. Documents can be retrieved by classification attributes as well as textual and pictorial content.

MULTOS runs under UNIX 4.3 BSD, on workstations with a bitmap screen and three-button mouse, connected in an ETHERNET LAN. It has been implemented using the C programming language and the Andrew toolkit which runs on top of the X window system. Only the automatic classification server is implemented in Prolog.

3. Client Processes

A MULTOS Client comprises a number of tools. A *tool* is a set of related operations performing a high level function. A collection of tools performs a *task*. An editing task, for example, may need a retrieval tool to fetch a stored document and an editing tool to modify it.

Tools are distinguished into *basic* when all their operations are implemented using ad hoc procedures, and *composite* when at least one operation is implemented using another tool.

There are seven major composite tools performing *Document Retrieval, Document Handling, Document Input/Output, Document Preparation, Automatic Document Classification, Document Type Administration, and System Administration.*

The client interface of MULTOS supports various kinds of dialogue techniques and levels of verbosity in order to accommodate both expert and novice users, is adjustable to the needs of the individual user and provides constant and incremental feedback as well as on-line help facilities [Cons88]. It has an object-oriented design and is implemented using the Andrew toolkit. The following user-system interaction modes are provided:

- *Simple interaction*: Operations that require keyboard input generate a window, near the area of the screen where the operation is invoked, suitable for getting input from the keyboard.

- *Tool driven interaction*:
  The user can select a tool, tool operation or target object of an operation from a menu with a pointing device.

- *Object driven interaction*: Valid object operations can be selected from a menu associated with a given object. Objects include documents, document types, document collections and document conceptual components.

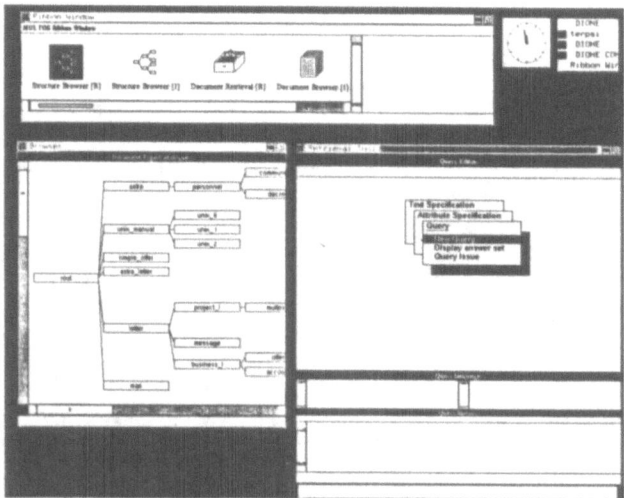

Notice: Multiple instantiations of a tool are possible
Fig. 1. Client subsystem: General layout

Upon logging into MULTOS one gets a ribbon window which is the basic environment where tools run (Fig. 1). It consists of two areas. The first is a ribbon that contains labeled icons (64X64 bit maps). Each icon represents a task or a tool that has been instantiated by the user. The label associated with the icon gives additional information about the function of the task/tool. The second area of the ribbon window is the actual working area. The ribbon may extend on either side of the visible area, thus a scroll bar is attached for viewing. A tool is instantiated by adding its icon to the ribbon. Multiple instances are possible.

Tools are activated by clicking on their icons at the ribbon window. More than one tools can be open simultaneously on the screen, but the user can interact only with one of them, the *foreground* tool, while all the others, if any, are either *inactive* or *background* tools. When the icon of an inactive tool is opened, the related window restores its operating status. If the state of a background tool changes while it is closed, its icon in the ribbon window flashes in order to indicate that a change occured.

In abnormal situations a message is generated in a window next to the mouse position on the screen. All operations are expected to be frozen until the user takes some action, possibly suggested by the message.

During the working session five kinds of pop-up menus are available. Some have a fixed, predefined list of selections while others are dynamic, i.e. the list of allowed selections depends on the tool or object on which the pop-up menu appears. These menus are: the *System Menu*, for general purpose operations like window management; the *Ribbon Menu*, for customizing the appearance and the order of tools on the ribbon; the *Tool Menu*, used to activate/deactivate tools; the *Tool Operation Menu*, used to activate operations

822

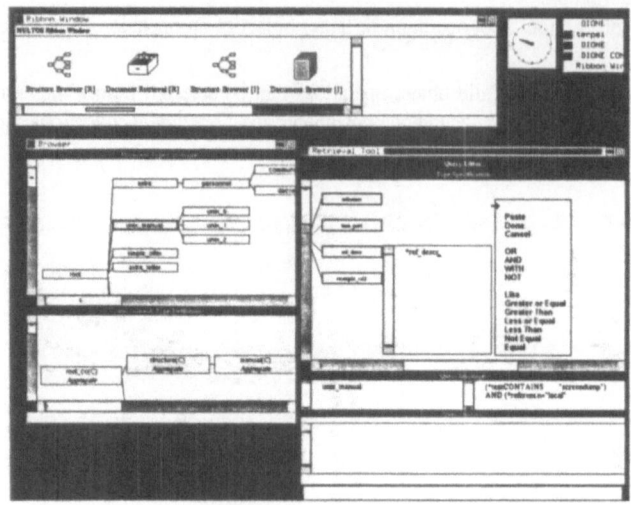

Fig. 2. Client subsystem: Query Formulation

of the foreground tool (depending on the tool status, some selections may be disallowed); and the *Object Operation Menu*, offering the operations associated with the object currently pointed by the mouse.

Pop-up menus in MULTOS features *acceleration*: an action of the pop-up menu can be connected with one of the mouse buttons; when this button is pressed the corresponding action is executed.

3.1. A sample working session.

The function of the MULTOS system is perhaps better understood by following a sample working session. Here we consider a session involving document retrieval.

First the user activates the *retrieval tool* (see Fig 1.) from the ribbon window. Optionally the user may also activate the *structure browser* tool. The document type(s) to be used should be specified in order to restrict the search domain. Document types are either selected from the document type catalogue, displayed at the structure browser tool, or typed directly in the respective area of the retrieval tool. Here, the user has selected the unix_1 type and gets the graphical representation of the hierarchy that represents the structure of the components that define the unix_1 document type. Following the query by example approach [Zloo75], the user may point to the components of the type on the values of which restrictions apply. Components have various data types. For each type different sets of operators are available, defined in [Bert88], and selectable from a dynamic pop-up menu (see Fig 2). The system assumes an AND operator between the components. ORing the components is allowed but must be defined explicitly. As soon as editing some restriction using this graphical interface is complete, the current expression of the query can be observed in textual form in a suitable window, at the bottom right of the editing window. The expert user always has the choice of using the MULTOS query language directly, in parallel with the graphical interface. The sizes of all editing areas can be dynamically redefined by the user. The sequence in which components are selected is unimportant and reformulation of a restriction is always possible.

All documents in MULTOS have at least a textual part. By selecting the text component, restrictions on the textual content of the document can be specified. Queries on textual content may require the presence or absence of a number of words, conjunctively or disjunctively. For editing such queries, a text specification window is opened in the query formulation area where the required words are entered. A pop-up menu will display on request the available operators: AND, OR, NOT, END OF QUERY. The list of operators is extensible. For instance, a LIKE operator is needed to support similarity retrieval. Words entering the query specification may well be part of some document displayed by the document browser (see below). Words or whole parts of text can be transferred using cut and paste operations. Spelling errors, typing effort and having to remember words are thus minimized.

Documents that contain images have an image component. By selecting the image component users may specify restrictions on the image content of the document. Images may belong to one of the *image classes* known to the system which are described by image class attributes and vocabularies of image class objects which also have attributes. Queries on images may refer to their semantic content as well as the relative position of contained objects (structure). The user can query images directly or indirectly, through image captions and in-text references to the image. Image semantic content is specified by image attributes, a pictorial example (the user may either draw the image, or get the image from another document)and associated document components. Image structure is specified by pictorial example. Image specification is accomplished through an image specification canvas where the user can specify the desired image in any of the above described ways.

The above steps may be repeated several times until the desirable query definition has been reached.

Query replies are kept by the *query history*. The query history is by default displayed at the bottom of the query editor (Fig 2.) The query history tool enables inspecting previous queries and their qualifying documents. By pointing on the query number, a window is opened where the query is presented in graphical form. The selected query becomes the current query. The user may inspect the query in this window but without editing permission. To reformulate the selected query, the query editor should be used. By pointing on the number of qualifying documents, a window is opened where icons of the documents are displayed. If a document has been fetched from the server then the corresponding icon is gray.

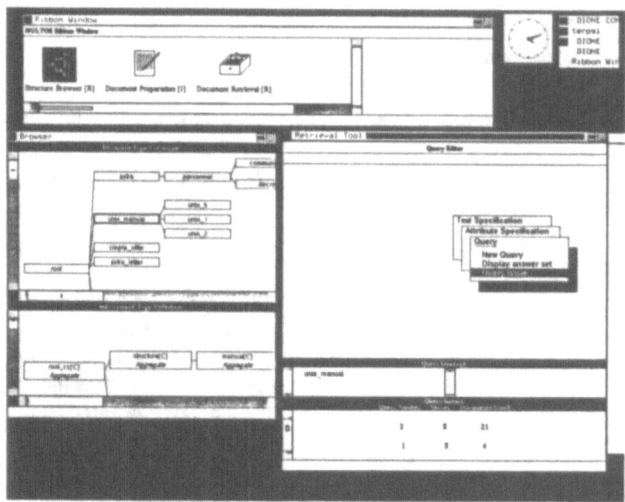

Fig. 3. Client subsystem: Query History

Fetching a document from the server is a time consuming operation to be avoided if not necessary. MULTOS allows the user to quickly inspect the important features of documents belonging to a query answer set in order to decide which of these are interesting. In a first stage, the actual document is not retrieved from the server. Only the characteristic features of the document are retrieved and presented. In the second stage of feature presentation, the user sees more features of documents marked as "interesting"

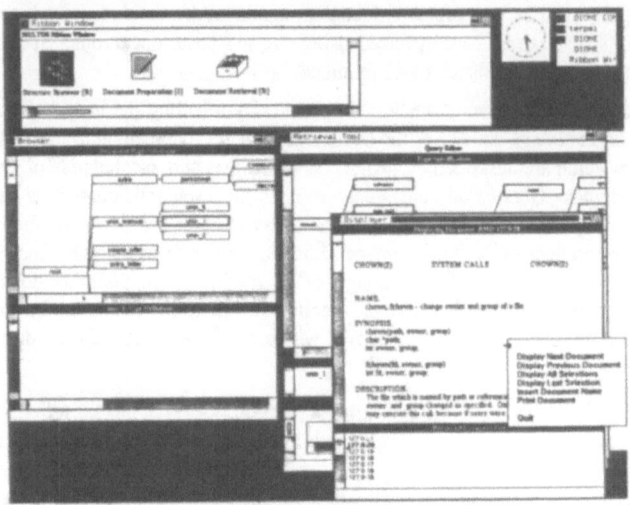

Fig. 4. Client subsystem: Document Browsing

during the first stage. During this stage, documents are fully retrieved from the data base and presented to the user through the document browser.

Four feature views are possible besides full document inspection:

a) *Simple card view*: information about the document such as name, doc_id, size, owner, date of last modification etc.

b) *Type card view*: contains the most significant conceptual components of the type, which the document belongs to.

c) *User card view*: The user can define private *user cards* by selecting some conceptual components and placing them in the card view.

d) *Query card view*: The parts of the document that match a query. Use of the query card requires the extraction of the full document.

In both stages of feature presentation, the user should not only see the presented information but also be able to compare its occurrences in different documents. In order to fit more than one document on the screen of the workstation information may be presented in *miniature* form. A *text miniature* is created in real time by using a smaller point size for the letters. An *image miniature* is created using an image reduction technique, together with the document and stored in the data base. A *voice miniature* consists of a short part, e.g. first three seconds, of the voice recording.

When a document has been selected from the query history, the *document browser* is invoked automatically. The document browser operates on an independent window and allows the user to view documents at various levels of detail. (see Fig. 3 and 4)

The final stage of browsing involves inspecting a single document in its real form. During document browsing the user can move from one page to another (both forward and backward), use a scroll bar to define the approximate desired location, or define the exact page number. Additionally, the user can move to a page that contains a specific word or image or voice recording. Pages are presented in their real size unless they do not fit in the screen. In this case, scrolling inside the page is also provided. Pages may be folded if they exceed the size of the regular pages of the document and can be unfolded by the user. Finally, browsing inside a document can be based on its attribute values. The attribute values of the document are presented in a separate window next to the document presentation window. The user can point to an attribute value and see, in the presentation window, the page that contains it.

4. Document Server Processes

Document server processes support the filing and content based retrieval of Multimedia documents. In particular they offer the following set of functions:

- **Document Management** functions, like access control, security, integrity and version support. These are provided by the Server Controller (see Figure 5), that has also the task of controlling and executing server requests.

- **Type Management** functions,provided by the Type Handler, that allow to *update the type catalogue* by inserting or deleting document types, to *access the type catalogue* and to *retrieve type instances* (Figure 5).

- **Document Collection Management,** for defining new collections, deletion of existing collections, and adding and deleting members to/from a collection.

- **Document Retrieval.** The Query Processor [Bert88] (Figure 5) is able to accept queries in the Multos Query Language and, interacting with the Type Handler, Collection Handler and Storage Subsystem to retrieve the documents that match with the query.

- **Storage of Multimedia documents,** supported by the Document Storage Subsystem (Figure 5). Documents, received in the ODIF format or in the Andrew format are translated into the Multos internal format and the structures used to support the retrieval are created. In particular free text retrieval is obtained by using the signature method based on Superimposed Coding [Chri84], while image retrieval by content is supported by extracting from images information about the objects they contain and the relations among them [Cont88, Cons88b].

Two different servers are provided in Multos, one which allows dynamic document filing (i.e. documents can be deleted and updated) and one which allows static document filing (i.e. modifications on stored documents are not permitted). The dynamic filing service is provided by the **Current Server**; a limited storage capacity is required, so that Magnetic Disks (MD) can be used as storage media. The static filing service, offered by the Multos **Archive Server,** manages public documents whose content is static; the required storage capacity must be very large (as documents are never deleted) but, since documents are static, WORM Optical Disks (OD) can be used as storage media. Table 1 summarizes the characteristics of Current and Archive servers.

The Current and Archive servers have a similar architecture; the differences have been limited at the level of the Multi Storage File System (MSFS). In the Current Server the MSFS corresponds to a standard Unix file system, while in the Archive Server it provides the basic filing functions by integrating Magnetic and Optical Disks and supporting the management of multiple OD volumes. The WORM Optical Disks being removable, enables the design of a Multi Storage File System whereby documents can be inserted in one server and accessed on another. This is possible if the Optical Disk volume is self-contained, i.e. the documents and their access structures used for their access are stored on the same disk volume. In the following more details on the MSFS functionality and architecture are given.

The files managed in the MSFS include *document containers*, *document signatures* and *indices*, used to support retrieval on conceptual component values. These files require different *allocation* and *access* strategies.

Document containers store the document data and their length is highly variable (from 1KB up to several MBytes for documents containing a lot of image and voice components). The document containers are accessed sequentially when the document is read; random access is performed if only one data component (e.g. text) has to be accessed; however a small number of accesses are performed. The data managed in

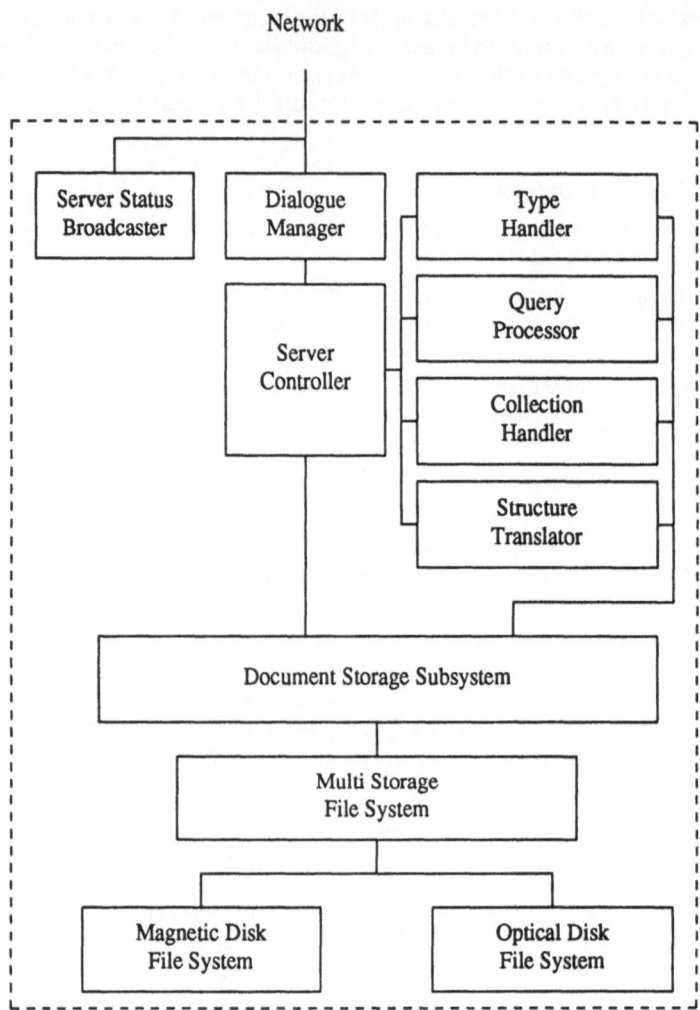

Figure 5. Server Subsystem

	Current	Archive
access rights	public	public
update	yes	no
capacity	50MB - 2GB	> 1GB
storage medium	Magnetic Disk	Optical Disk

Table 1 - Characteristics of document filing servers

the document containers are stable: after their storage they are never modified.

Access to document textual content is obtained by using a signature method based on superimposed coding [Chri84]. The signature file is updated after each document insertion appending the document signature at the end of the signature file. The access to the signature file is sequential.

The access structures for conceptual component values are based on B-trees. They provide a fast access to data and can be easily updated during the insertion of new documents. However they are subject to frequent modification and the access is random.

From these file characteristics, general *file allocation* and *file access* strategies derive.

The allocation of a file may be either *dynamic* or *static*. Dynamic allocation means that the content may be modified and the size is not known at creation time (e.g. signature file, index files); the file is stored on Magnetic Disk. If the file is static its content is written once and the size is known at creation time; the file is stored directly on OD.

The access to a file may be either *sequential* or *random*. If the access is sequential, such as for the signature file, it can be performed directly on OD, while files that need random access (e.g. index files) must be buffered on MD in order to ensure acceptable performance.

access allocation	random	sequential
dynamic	MD	MD
static	MD	OD

Table 2

The MSFS manages virtual volumes that are mapped in part on optical disk and in part on magnetic disk depending on the file characteristics (see table 2 for the mapping strategies). The volume content can be logically subdivided into four segments: *bulk storage segment* used for storing document containers, *signature segment*, *index segment* and a *system segment* (Figure 6) that contains a volume header, the collection catalogue and the type catalogue. The system segment is used for supporting multi-volume queries when a juke-box is provided.

In the MSFS, the optical disk volumes can be in the *hot* or *frozen* state. A hot volume can accomodate new documents while a frozen volume is read only. The strategy followed for the allocation and the access of data is different for hot and frozen volumes.

In a hot volume the bulk storage segments are directly stored on Optical Disk. Document instances are, indeed, static and accessed sequentially. This also enables to recover from magnetic disk failures since all the access structures can be reconstructed. *Signature* and *index segments* are temporarily kept on magnetic storage because they are dynamic and their access is random (for index).

When the volume becomes frozen and new documents cannot be inserted, a reorganization of signature and indices is performed (the signature file is translated from "sequential" to "bit-sliced" organization and a sort on indices is performed) before storing them on the optical disk. This procedure allows to avoid the problem of maintaining dynamic structures (such as B-trees) on WORM devices. When a frozen volume is mounted, the index segment is buffered on magnetic disk while all the other segments are accessed directly on the optical disk.

The presence of multiple volumes requires a strategy for query processing involving off-line volumes too. The MSFS contains, indeed, the access structures that support the query processing during multi-volume query resolution. For all the *loaded* volumes (the *load* operation makes the volume visible inside the system) a set of global indices are maintained: their access enables disregarding volumes that do not contain any document that matches the query.

828

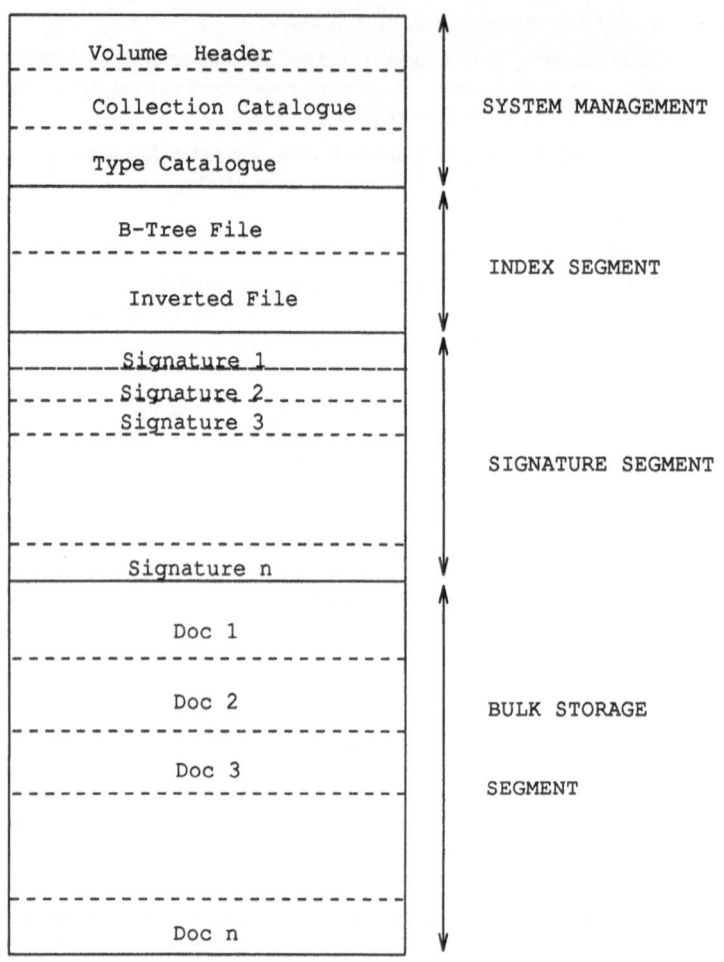

Figure 6. The volume storage organization

To improve access to documents, migration algorithms from frozen to hot volumes can be applied to the most frequently accessed documents [Chri85]; all these algorithms offer the advantage of faster access to these documents at the cost of data duplication on the new hot disk, including the document content the access structures. In Multos a simpler solution is chosen: documents are stored on the hot disk as they arrive to the server and when the disk becomes full it is frozen (all the content of the volume temporarily stored on magnetic disk is moved to the OD) and an empty OD becomes the new hot disk. This implies that the most recently archived documents are stored on the on-line OD. The algorithm is simple and provides fast access to the last stored documents which should be those more frequently accessed.

Thus the volume management has been simplified with the following assumptions:

- Only one **hot** volume is present at a time (insertions/deletions concern this volume),

- Only one volume can be **mounted** at a given moment : it may be either *hot* or *frozen*, and the operation requires the unmounting of the currently mounted volume. It is not possible to mount volumes that have not been loaded previously.

After the mount operation all the information contained on the volume can be accessed.

- A volume not mounted can be **loaded**, so that global information related to it is made available. However, after loading and before mounting, it is impossible to access documents contained in the volume. More than one frozen volume can be loaded at a given time.

The load/unload operation is mainly used to import/export a frozen volume into/from a server. After an unload the volume is no more known inside the server.

The MSFS uses the standard UNIX file system to support the Magnetic Disk File System (MDFS) and the Optical Disk File System (ODFS).

The ODFS has been designed so as to manage a large number of multimedia files with acceptable performance [Losi89]. Some consideration was given to using an ad hoc rather than a a standard file system. Each solution presents advantages and problems.

If an ad hoc file system is used, the advantages include optimal disk space utilization and response time. The disadvantages are that it can only be used for a specific application and that the integration of the file system with existing tools that operate on the magnetic disk is problematic.

If a standard interface is adopted (e.g. Unix), the file system can also be used for other applications [Nels87] and can be easily integrated with the rest of the DSS. This solution, however, does not allow to optimize space allocation and access time to the OD for a specific application. Moreover, part of the facilities offered by the file system are not used.

The data managed in the Multos Storage Subsystem do not have fixed or predefined characteristics, so a general purpose file system is necessary. Also, considering the advantages that derive from the adoption of a standard interface for the ODFS, Unix has been chosen.

The ODFS has been implemented using a Toshiba [TOSH85] and an OPTIMEM 1000 [Opti85, Opti85a] Optical Disk. Initial use of the ODFS has shown that the integration of Optical Disks does not incur significant performance degradation.

We plan to carry out a systematic evaluation of the ODFS performance in order to choose the optimal system parameters, as well as a global evaluation of the complete archive server.

5. The Classification Server

The task of the classification server is the construction of content-oriented representation for documents. Traditional filing and retrieval systems generally use the vector space model based on the coordinate indexing principle for the content-based access to their databases [Fox85]. As long as the application domain was constrained to the bibliographic world the results this method yielded was quite satisfactory. But for use in the office domain this model shows very definite shortcommings [Crof87]. For example, although the vector space model gives good results as far as the document representation is concerned, it cannot be used for data retrieval - a feature that is of quintessential importance in the domain we are looking at here. Moreover, the automatic indexing methods perform best when dealing with a specialized literature and specialized terminology, whereas documents in the office domain tend to deal with less well-defined textual phenomena, using for example a terminology that can change from one environment to the next.

In order to make retrieval procedures more efficient, it is necessary to have an explicit deep structure for the content-oriented access to the documents [Scha80, Rijs86]. Thus, an a priori processing has to be done to classify the documents corresponding to predefined structures. Work done in the area of artificial intelligence has shown new ways for capturing the semantic import of documents for filing and retrieval purposes. Systems like I3R [Crof87] and RUBRIC [Tong83] are using domain knowledge besides their vector-space-representation for the the content of documents. In addition natural-language-based systems seem very promising with this respect [Schan80, Hahn86]. However, these systems only work in very limited domains. The wider their coverage, the less robust they tend to be - which makes them at the moment not suitable candidates for handling the wide range and large quantity of documents that passes through most offices daily.

The MULTOS document model covers the middle ground between the traditional vector-space and the knowledge-based document representation. It embodies a document model of the weak type, which - following the official standards of the ODA (Office Document Architecture [ODA85], integrates the layout, the logical structure as well as the semantic (conceptual) features of the documents it processes [Barb85] providing, moreover, access to documents on the basis of their conceptual structure.

The MULTOS system provides two different classification services for the construction of the conceptual structure. On one hand there is a classification tool recognizing the semantic features of the document automatically, on the other, the user can specify a new conceptual structure using the manual classification tool. In addition manual and automatic classification can be combined, e.g. for reclassification purposes. This section describes the automatic classification server.

5.1. The Document Model

Office documents are showing a wide range of different layout and logical structures (e.g. blocks, tables, etc.). The MULTOS document model describes these structures by using the ODA standard. ODA gives a formal description of the document composition (logical structure) and a formal device-independent description of the document representation (layout structure). In order to make retrieval and distribution procedures more efficient, it is necessary to represent the content of documents. In the MULTOS system this is accomplished by using a conceptual structure definition (CSD) which marks those content parts of documents that are important for office procedures like retrieval or distribution purposes.

Three modeling principles form the basis for the description of the conceptual structure: aggregation, typifying and generalization:

- The documents are described as a tree structure constructed through the aggregation of objects (conceptual components). We distinguish three different kinds of conceptual components. First, there are the complex components; they are built as a sequence or a choice of subcomponents. Next are the basic and the spring components; they are the components that show up as the leaves of the tree structure. The spring components are unspecified and mark positions which can be used for later refinements of the CSD. Finally, there are the basic components; they name those content parts of the document and thus provide a direct link to the document content.
- Each conceptual structure is an instance of a class (or type). Within the concept of the "weak type" typifying means the definition of type properties by collecting similar conceptual structure definitions.
- The spring components can be used to define a refinement rule within the MULTOS type world. New types can only be added to the type structure by specifying spring node(s) of a supertype. Starting off from an unspecified root type and using the described refinement rule a generic type hierarchy can be built.

Figure 7 shows an example of a simple type hierarchy. Subtypes are defined by specifying the spring component root_cc of the type "Document" and specifying the spring component letter_content of the type "Business_l.".

The type hierarchy with the defined CSDs form the basis for the classification. In this framework, classification means the selection of the most suitable type-CSD for the document to be classified. Two different kinds of CSD are distinguished:

- the Type-CSD is the definition of a class representing all properties of the type;
- a constructed document-CSD is an instantiation of a class with links from the basic components to the content parts of the document that are relevant for retrieval.

This document model meets the above mentioned requirements. It integrates the layout, logical and semantic structure of documents by combining the ODA structure and a conceptual structure. It can also be used for data retrieval (the basic components have a link to the document content) and reference retrieval. The type hierarchy and the CSD offer the user very simple retrieval strategies, because formular oriented as well as boolean search can be used.

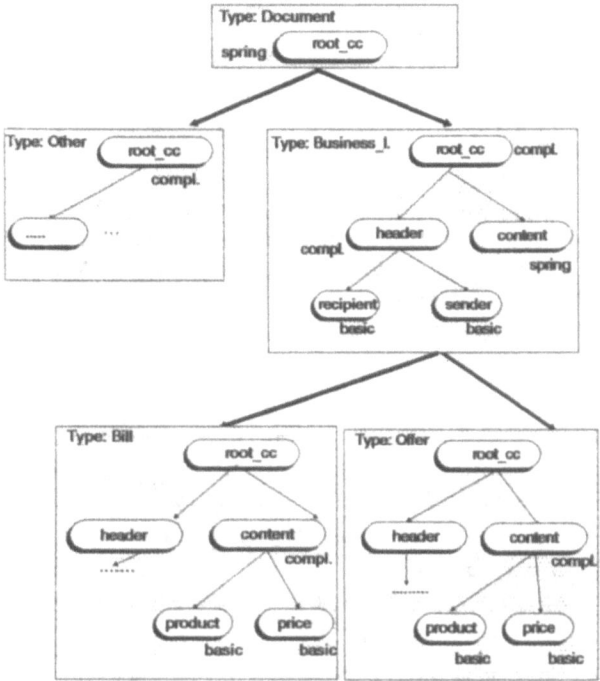

Figure 7: Example for a simple type hierarchy

This monohierarchical model has some limitations, however. It does not allow, for example, to order documents taking into account the many different viewpoints of different users. Also the retrieval procedures based on this type hierarchy and conceptual structure are not sufficient. To overcome these limitations, the MULTOS system provides the possibility of building collections where each user can define his own views and can formulate queries related to this particular view . Moreover, to raise the retrieval quality a full text search based on signatures [Cons86] is integrated in the system.

5.2. The Classification Approach

As already outlined, the vector space model is not sufficient for the office domain. At the same time, systems with a richer functionality such as natural language systems cannot yet handle the large variety of office documents. The MULTOS classification system tries to combine the most effective characteristics of both approaches. It uses a type- oriented, top-down approach, using the types and their conceptual structure definitions as tools for the abstract description of the semantic structure of the document. The classification system then tries to prove if there are any text segments that contain the semantic structure of a document of a special type. The MULTOS system thus does not engage in full text understanding but picks out special content parts in a goal-directed fashion, and then uses these parts to identify concepts.

In this approach, three different kinds of knowledge are used. There is structural knowledge about documents, represented by the type hierarchy and the conceptual structure definition. It was described more fully earlier on. But when dealing with office documents, it is often also possible to identify semantic units by looking at the layout or logical form of the document or of parts of the document (price-tables, adress-blocks, etc.). Concepts often occur in specific parts of this layout structure that are fairly easy to specify and identify (e.g. the names and prices of products in the same line, sentence, ...). One can thus pick out relevant content parts by searching for special keywords and then try to identify concepts within these document parts using deeper knowledge. The keywords and the relations between keywords and concepts and the relations between concepts and concepts are stored as background knowledge about the domain of application.

5.2.1. Realization of the goal-directed approach

The basis for the classification procedure is the predefined type hierarchy with the related conceptual structure definitions. For the identification of the relevant content parts of documents, a set of predicates is added to each conceptual component. The predicates are formulated in an abstract description language called CDL ("Content Description Language") and relies on the types of knowledge described in the previous paragraph.

The CDL forms the basis for the realization of the goal- directed approach by using an abstract description of how to find concepts that are relevant for a conceptual component. The CDL allows content description on many different levels. It provides predicates on the logical level and on the level of layout and syntax as well as predicates referring to semantic and linguistic relations. The following predicates are integrated in the CDL:

- predicates that describe layout and logical structures (e.g. tables, lines);
- position operators for the layout and logical structures (e.g. after, first);
- keyword- and token-match predicates;
- semantic relations pertaining to background knowledge (e.g. instance_of, part_of relations);
- grammatical relations (e.g. possessive, temporal);
- references to other conceptual components (variables);
- position operators for keywords, relations and references (e.g. after, between);
- combination of predicates that use set and logic operators;

As a result, MULTOS has three different levels at which to describe the structure of a document. The type-hierarchy represents the first level. It defines all types of the application world. Starting off from a common root type, the classification process traverses top-down through the type hierarchy to find the best fitting type - where "best fit" is defined as a function of the appropriateness of leaf nodes as well as other nodes in the hierarchy for the classification of the document.

The second level, the CSD, describes the conceptual structure of each type of the type hierarchy. Here a top- down strategy through the CSD-tree is used as well. For each conceptual component of the Conceptual Structure Definition the relevant content part will be located in the document. Because a generalisation relation has been defined within the type hierarchy, the already identified components of the super type will be inherited by the subtypes. Consequently, only the new components of the Conceptual Structure Definition of the subtypes must be proved. If the tree node is a complex component, the identified new relevant content is used as search domain for the subcomponents. This means that one can stepwise constrain the search domains to find appropiate content parts. The relevant content identified as a basic component will be linked to the component provided for retrieval operations or conditions for invoking office procedures such as the distribution of information.

The CDL constitutes the third level of document description. The classification system derives knowledge from the CDL definition of a conceptual component on how to find relevant content parts. In other words, the CDL represents the rules of the classification. Using this knowledge, the system is able to decide whether or not the documents contain the specified conceptual components of a particular type.

Figure 8 shows a very simple example of the description of document types on these three different levels. The first level is the document of type offer, probably a subtype of the document of type business_letter sharing the components recipient and sender. The Conceptual Structure Definition describes the components (like sender, product, price, ...) which are typical for an offer. The related CDL predicates represent a description of how the conceptual components usually occur in a document of this type (like tendered products and prices in tables).

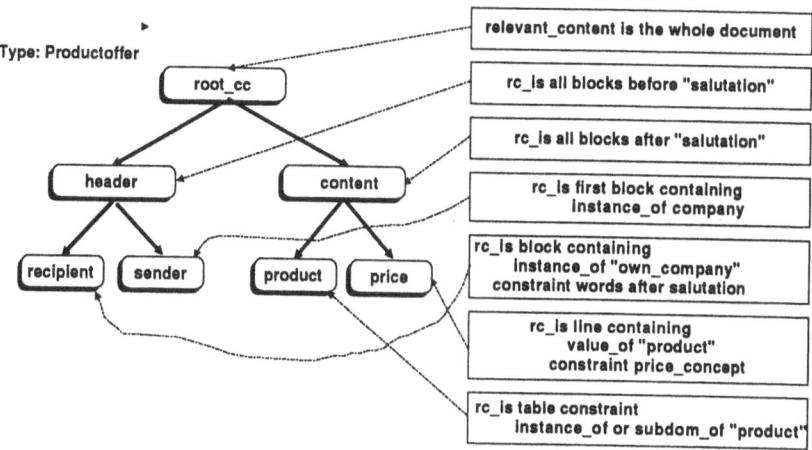

Figure 8: Example-CSD with CDL sentences

5.2.2. The Classification system

With the above mentioned conditions the configuration of the classification system is predefined. The system has to use the three different kinds of knowledge - the structural knowledge, the representation of the application world and the layout and logical structure of the current document. Since the classification system in its current state is an experimental system, its architecture was kept strictly modular: this allows for a flexible change of representations or of formats when necessary. The following figure shows the architecture of the current classification system implemented in PROLOG.

Figure 9 demonstrates the intended modularization. The application world, some knowledge on text analysis (like recoginition of prices, dates,etc.) and the various possible modes of access are represented by independent modules. The only predicates defined in the classification handler are predicates like "give_instances", "give_relation", etc.; they are independent from the knowledge representation. Consequently, the representation can be changed easily. The current classification system works with structured inheritance network similar to the KL-ONE system [Brac85].

The layout and logical description of documents are generated by a document preprocessor. The preprocessor recognizes all CDL-defined layout and logical structures which is represented by an intermediate document representation. This representation subdivides the document structure as follows:

Layout objects:

      pages as an aggregation of blocks, composed of lines

Logical objects:

      chapters as an aggregation of paragraphs, composed of sentences with sentence parts

Other content portions, important for the identification of relevant content parts (tables, quotations, brackets, etc.) are also identified and represented by additional structures.

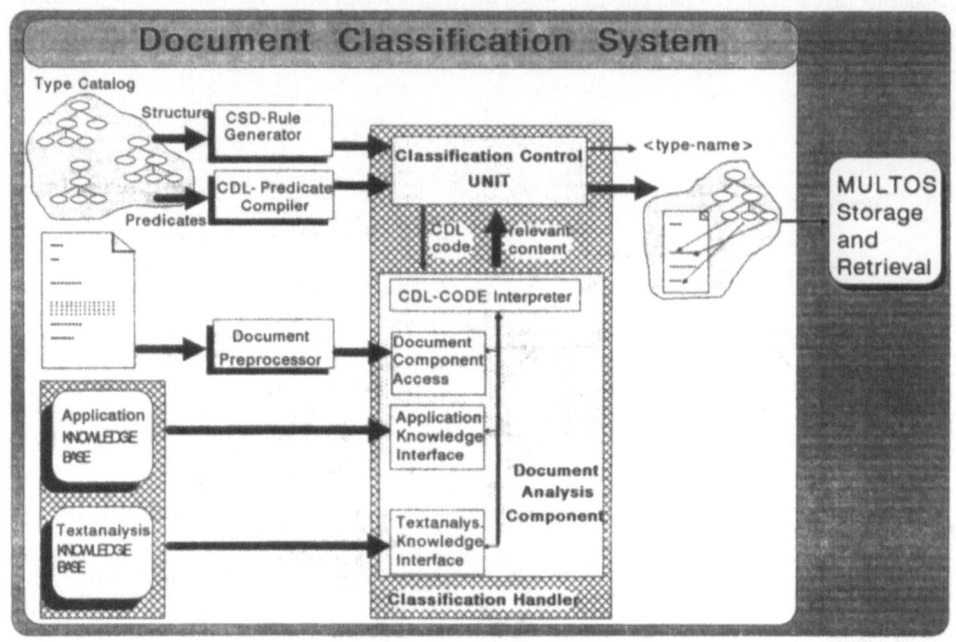

Figure 9: The Architecture of the Classification System

The internal document representation is word-oriented. A specialized component inside the preprocessor recognizes the tokens of the document. A morphological analysis is carried out as well as the detection of descriptors known to the system, because the CDL predicates and the objects of the knowledge refers to normalized forms. The morphological expert maps the token of the document onto the dictionary to explore the normalized forms for each token (Schott algorithm for the german language [Scho78]), Lovins algorithm for the english language [Lovi68].

The results this preprocessing yields is the internal document representation, which provides a direct access to the descriptors (keywords) and logical/layout structure of the document.

Since type, CSD and CDL definitions must be very flexible and easy to change, a special procedure was developed for the access to the structural knowledge. The type definitions are transformed automatically into a PROLOG predicate representation by the "Rule Generator". The CDL is compiled into a horn clause program which contains a set of predefined, independent predicates representing the classification rules. Using this method, the updating of types is a really independent procedure which has no influence on the way documents are classified. In a similar way, the set of CDL features are also dynamically extendable. Furthermore the independent CDL predicates offer the possibility of optimizing the interpreting sequence, thus avoiding performance problems.

The classification handler is the kernel of the system. Using the compiled structural knowledge, the Classification Control Unit controls the top-down traversal through type and related Conceptual Structure Definition definitions. It manages moreover the proof procedures over the classification rules and stores intermediate results.

To prove whether the document contains a specific conceptual component, the control unit delivers the compiled CDL predicate set to the CDL-Code-Interpreter (CCI). The optimization component of the interpreter chooses a predicate of the set and distributes it to the subcomponents to prove the related rules. The interpretation procedure for the CDL predicates has a defined final state, but the proof sequence itself is arbitrary. The result of the proof procedure will be a fail if the document does not contain the concept asked for; else it will produce the relevant content part(s) of the document containing the concepts searched for.

6. Conclusions

The storage, automatic classification and retrieval of multimedia office documents with a combination of magnetic and optical disks by non-computer expert users through a flexible man-machine interface are provided by the MULTOS multimedia document prototype filing system. Actually, the development reported here are the major enhancement features by the second (and last) MULTOS prototype. A first prototype, intended only to demonstrate the concept and the basics of the architecture, was demonstrated in ESPRIT Conference 1987.

Further work in the project includes experimental installation under real working conditions and evaluation of the system.

References

[Barb85].

F. Barbic and F. Rabitti, "The Type Concept in Office Document Retrieval," *Proc. 11th Conference on Very Large Data Bases*, Stockholm, 1985.

[Bert88].

E. Bertino, S. Gibbs, and F. Rabitti, "Query Processing in a Multimedia Document System," *ACM Transactions on Office Information Systems*, vol. 6, no. 1, pp. 1-41, 1988.

[Brachman].

J. Brac85 and J. G. Schmolze, "An Overview of the KL-ONE Knowledge Representation System," *Cognitive Science*, vol. 9, 2, 1985.

[Chri84].

S. Christodoulakis and C. Faloutsos, "Signature Files: An Access Method for Documents and its Analytical Performance Evaluation," *ACM Transactions on Office Information Systems*, vol. 2, pp. 267-288, 1984.

[Chri85].

S. Christodoulakis, "Issues in the Architecture of a Document Archiver using Optical Disk Technology," *Proc. ACM SIGMOD*, pp. 34-50, Austin, Tex., May 1985.

[Cons86].

P, Constantopoulos, et. al.,, "Office Document Retrieval In Multos," *Procs of Esprit Technical Week '86*, September 1986.

[Cons88].

Constantopoulos P., Theodoridou M., and Yeorgaroudakis Y., "Flexible Query Formulation for Multimedia Document Retrieval," *Proc. First European Conference on Information Technology for Organisational Systems, Eurinfo '88.*, Athens, 1988.

[Cons88a].

Constantopoulos P., Yeorgaroudakis Y., Matteucci A., and Savino P., "Client design for the MULTOS multimedia office filing system," *ESPRIT Conference '88*, Brussels, 1988.

[Cons88b].

Constantopoulos P., Orphanoudakis S., and Petrakis E., "An Approach to Multimedia Document Retrieval on the Basis of Pictorial Content," *RCC/CIS/TR/1988/011*, Technical Report, Institute of Computer Science, 1988.

[Cont88].

P. Conti and A. Martelli, "Document Management Tools - Retrieval Effectiveness," *Internal Report*, Olivetti DOR/Pisa, 1988.

[Crof87].

W. B. Croft, "A Framework for Office Document Retrieval," *Proc. IEEE Office Automation Symposium*, Computer Society Press, Gaithersbourg, 1987.

[Eiru88].

H. Eirund and K. Kreplin, "Knowledge Based Document Classification Supporting Integrated Document Handling," *Proc. ACM Conference on Office Information Systems*, Palo Alto, March, 23-15, 1988.

[Fors84].

H.C. Forsdick, et. al.,, "Initial experience with multimedia documents in Diamond," *Bull. IEEE Comput. Soc. Tech. Comm. Database Eng.*, vol. 7, no. 3, pp. 25-42, Sept. 1984.

[Fox85].

E.A. Fox, "Composite Document Extended Retrieval - An Overview," *Proc. 8th ACM SIGIR Conf. on Research and Development in Information Retrieval*, Montreal, 1985.

[Hahn86].

U. Hahn and U. Reimer, "TOPIC Essentials," *Proc. of the 11th International Conference on Computational Linguistics, COLING86*, pp. 497-503, Bonn, 1986.

[Losi89].

M. Losi and P. Savino, "The use of Optical Storage," , North-Holland, 1989. (to appear).

[Lovi68].

J.B. Lovins, "Development of a Stemming Algorithm," in , vol. 11, pp. 22-31, 1968.

[Nels87].

T. Nelson, "Implementing a UNIX File System on WORM Devices," *UniForum Conference*, Washington, D.C., 1987.

[ODA85].

ODA, *Office Document Architecture*, ECMA-101, 1985.

[Opti85].

Optimem, *Optimem 1000 - Interface Manual*, 1985.

[Opti85a].

Optimem, "Optimem 1000 - Optical Disk Drive - OEM Manual," , 1985.

[Scha80].

R. Schank, J. Kolodner, and G. DeJong, "Conceptual Information Retrieval," *Research Report #190*, Yale University, Department of Computer Science, New Haven Connecticut, December 1980.

[Scho78].

G. Schott, *Automatische Deflexion deutscher Woerter unter Verwendung eines Minimalwoerterbuches*, 2, pp. 62-77, 1978.

[Svob84].

Svobodova S., *File Servers for Network Based Distributed Systems*, pp. pp 353-398, , Dec. 1984.

[Tong83].

R. M. Tong, D. G. Shapiro, and B. P. McCune, "A Comparison of Uncertainty Calculi in an Expert System for Information Retrieval," *Proc. IJCAI-83*, Karlsruhe, 1983.

[TOSH85].

TOSHIBA, *WORM - Drive Reference Manual*, 1985.

[Zloo75].

M.M. Zloof, "Query-by-Example," *Proc. AFIPS Conf.*, pp. 431-438, NCC, 1975.

Project no 237

AN INSTRUMENTATION SYSTEM FOR CSA AND SIMILAR ARCHITECTURES

G. Horne
MARI Applied Technologies Ltd, U.K.
Old Town Hall
Gateshead
Great Britain

Abstract

This paper describes the Communications System Architecture (CSA) Instrumentation Machine (IM) developed within the CSA project. Although produced primarily for CSA the IM is intended to be a more general machine which can be configured by the architect or designer to demonstrate behaviour of other similar architectures.

The internal model of the system being instrumented is built up in the form of a hierarchical graph from a sequence of events emanating from the system. The modeller has control over the effects of these events upon the model being constructed, defining its structure in terms of nodes, representing objects, and arcs representing for example communications between them, or other more abstract relationships. Modellers then graphically display parts of the system according to their interests.

The use of the IM within, and impact on, CSA are explored giving some indications of the requirements to be met by other architectures before they too could be instrumented. Possible development paths are explored.

1. Introduction

The IM was developed to demonstrate the architecture and behaviour of CSA. The appendix provides background information about the CSA architecture which it is not appropriate to present in the main text. See also the companion paper in this volume describing CSA in general.

The IM was developed along with CSA, an approach few other systems apart from Jade [1] have taken, as it was recognised that distributed operating systems are complex in terms of their usage, error states and internal data structures. It was perceived as useful for two groups of users: system programmers seeking to build and debug CSA components and secondly, people who are unfamiliar with CSA concepts, who find graphical display an aid to their understanding.

In common with other object-oriented architectures CSA uses message passing between system objects to direct its processing [2]. These messages are of a predefined CSA format and it is these which are used by the IM to build up the picture of system behaviour presented graphically to the user.

Subsequent sections of this paper describe the evolution of the IM, the model used to represent the system, the instrumentation method and examples from our experience of the impact of instrumentation on the instrumented ("donor") system.

2. Evolution of the IM from Particular to General System

The IM has been developed in two stages. An initial prototype was constructed with limited functionality. This worked by receiving copies of all inter-object messages and using them to update a screen representation of the donor system behaviour. It was possible to run the system live or store messages in a database and replay them later and in both cases to filter messages so that only those of interest affected the screen [3].

This system was deficient in several respects [4]:

- It was CSA specific; intelligence about CSA messages was built in. Major effort would be expended if the format of messages changed
- Presentation was rigidly defined and not under the control of the user
- It revealed nothing about the behaviour inside an object or the relationship between Local Object Machines (LOM) making up a Domain Object Machine (DOM). Further programming effort would be needed to accomodate these.
- Hardware and software limitations meant that further development on the prototype platform was impractical.

The current version of the IM took these limitations into account. A strategy was defined to remove all intelligence about CSA and the aesthetics of the visual display from it, though the same basic functions were to be retained. Once all intelligence about CSA is removed from the IM then a configuration excercise needs to be performed to make it useful. There seems no reason why this cannot be done for any system, not just CSA, provided the basic data needed can be identified and provided. Obviously in our case configuration for CSA has been carried out. It is worth pointing out that configuration consists only in reading data from text files so there is little effort in reconfiguring for another system. Only new files are needed and must be read.

The increased functionality required was accomodated by implementing the current version in C under Unix with a graphical interface in X11 or NeWS. This has also produced a system which is more portable than the original.

By taking this route the final IM has been developed in parallel with the DOM as the exact details of DOM implementation were not relevant to it. All that must be done is to ensure that all the data needed to make a meaningful representation of DOM behaviour is available.

3. The IM Model of the Donor System [5]

The internal model chosen to represent the donor system was a hierarchical graph of nodes connected by directed arcs. This was chosen as it can encompass a wide range of user applied semantics. Nodes and arcs exist as basic proformas to which the user can add data so that the user view of a node as objects or atomic particles, arcs as messages or electromagnetic waves, for example, can all be supported. Its hierarchical nature permits partial views and successive focusing on finer or coarser levels of detail. The graph is also a familiar computing paradigm.

The graph model used represents the interaction between system objects: it shows when they come into being, how they affect and are affected by other objects and their disappearance. Other approaches to distributed system monitoring have concentrated much more on the internal data structures of objects (eg MANDIS monitoring group working with the Amoeba system [6, 7]) about which the CSA IM can say little except by further subdivision into nodes and arcs.

The graph holds information with each node or arc but details of how to depict them, in terms of colour, shape, size, labelling etc., is not held here but is provided by the user during configuration.

Diagram 1 shows a possible graph which might be built up during the instrumentation of an object-oriented architecture. Successive generations of nodes from the root represent finer-grained views. In CSA terms the root node represents the DOM, the first generation LOMs, second generation objects on LOMs and the third generation processes making up objects.

4. Instrumentation Method

4.1. Personality Packages [8]

Before any instrumentation session can be run the user must supply the configuration data to the IM so that it can make sense of the messages it receives and paint the screen appropriately. As the user may be interested in highly conditional tracing of events within the donor system, eg for aberrant behaviour, illegal access, deadlocks etc., and because there is a limit to what can be assimilated it must be possible to restrict the interactions which end up being displayed. Two types of restriction can be applied: what messages affect the graph and what parts of the graph are displayed.

There are four types of data the IM needs:

– how to find the data needed from an incoming message
– criteria to decide whether a particular message contains relevant information
– how to react to the relevant messages, ie how to update the graph
– how to display particular screen objects

All this data forms a Personality Package (PP). Within one PP there can be many sets of instructions about the visual display, known as *templates*. In a session one PP is used, but the availability of several templates allows the user to view each of those parts of the graph which are of interest.

A special language has been provided for users in which they can write PPs. These would usually be written, in our case, by someone familiar with the workings of CSA though not necessarily of the IM. The rules can be expressed in terms of the high level concepts of donor systems rather than the nodes or arcs used by the IM.

The next three sections describe in more detail the sequence of events from the arrival of a message to its effect becoming visible on the screen whilst running in real time mode. Diagram 2 depicts this as a successive filtering of the large amount of data coming in from messages to leave only the desired information on the screen.

4.2. Messages

All messages passed in the donor system are copied to the IM. The messages are tested against the selection criteria defined by the user, called here the Message Rules, by looking at fields whose type and location has also been defined by the user. The message is regarded simply as a record where the user is effectively asking for a member to be examined. In the case of CSA, the message format used provides a lot of built in information about the location and size of fields but other formats will need more

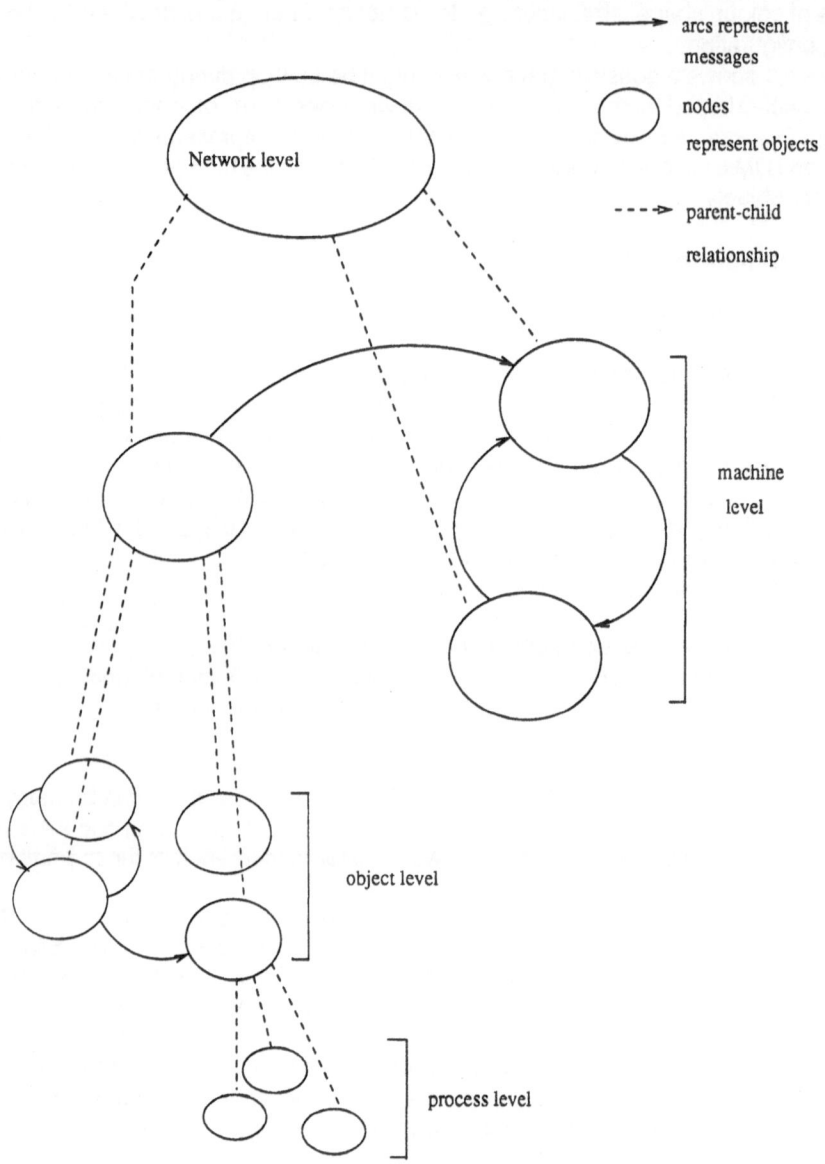

Diagram 1: An Example of a Graph which might be Generated during
Instrumentation of an Object-oriented Architecture

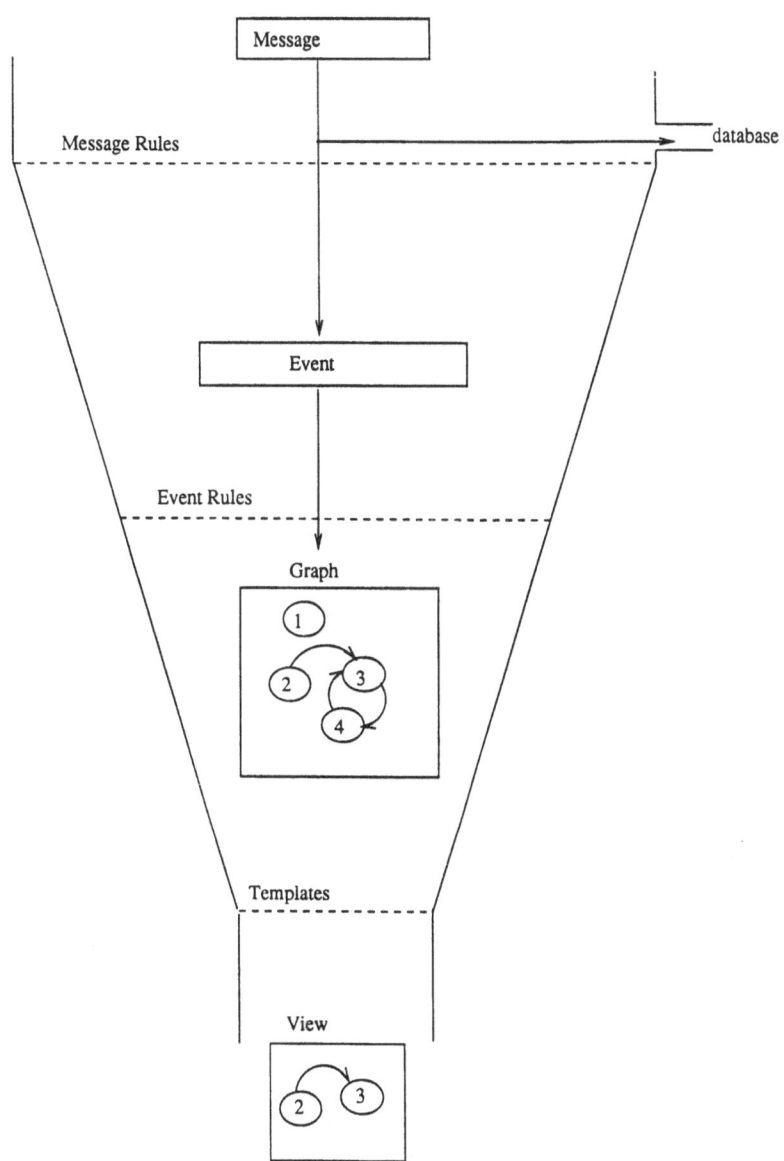

Diagram 2: Conceptual Overview of IM Operation for a Real-Time Session

information to be provided by the instrumenter to make them recognisable. The ability to view a message in this way is the key to making the IM general.

If the message is recognised, i.e. the specified fields match the required criteria, an *event* is created. This is a concatenation of several pieces of data from the message. It represents the occurence of something in which the user has declared an interest through the definition of the message rules.

The IM can operate in several modes. In real time mode, where messages are received directly from the instrumented system and manipulated in real time, a recording option is available where all events produced are copied to a database. This means that they can be reused later and given a different interpretation at the stages described below.

4.3. Events

Once an event has been recognised its effect on the graph has to be determined. The user can define rules to determine this for any particular type of event. This may be the creation, alteration or deletion of a node or arc, or storage or retrieval of information from the IM's memory.

4.4. Views

If an event causes a change in the graph then the screen representation could also potentially be changed. The PP defines which arcs or nodes appear on the screen and their appearance. These rules are referred to as templates in the diagram. The user is free at any time to change template, ie to alter the level of granularity of the view or to display alternative views at the same level. This is possible because the whole graph is stored internally.

5. Data and Message Collection from the Donor System

The next two sections describe the practicalities involved in collecting messages from the instrumented systems and how the need to provide data for instrumentation affected our donor system, CSA.

5.1. Message Collection

It was a requirement of the IM from the beginning that it should have a minimal effect on the performance and no effect on the behaviour of the donor. The IM is therefore placed on a separate computer, the prototype version on a PC, and the current version on a Sun 3. Copies of messages are sent to an Instrumentation Server (IS) which is resident on the LOM. This is known to the low level parts of CSA as a destination for message copies. The LOM is perfectly capable of operating without it and copies to it messages which are to be sent anyway, so the impact of instrumentation is minimal.

The IS is connected to the IM via a reliable network whose responsibility it is, as usual, to ensure that messages are transferred in the correct order and without errors. In the case of CSA there is no universal clock so messages are ordered only by their arrival times. Any timestamping which is required must be done by the user. This could be added to the functions of the IS, as the IS also has to be provided by the user as it is donor system specific.

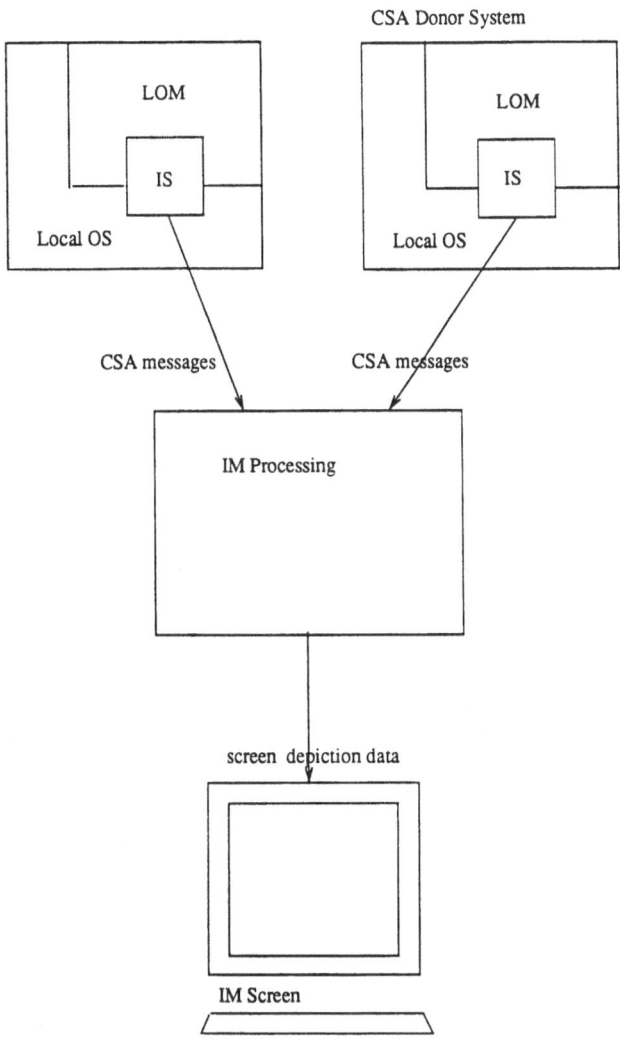

Diagram 3: Instrumentation Configuration for CSA

5.2. Data Collection

A distinction will be made here between message collection and data collection. The previous section referred to the process of ensuring messages from the donor system reach the IM, called message collection. Data collection refers to the more important task of ensuring that information of interest is in the messages in the first place, either naturally or by design.

Before a data collection strategy can be defined for the donor system the user must decide how the graph is to be mapped on to it, ie what nodes will represent and what arcs will represent. Once this is known it can be determined what traps have to be placed in the donor to detect events of significance.

In the prototype version copies of inter-object messages were sent every time the low level message **Send** routine was called. This routine simply transfers messages between the process groups used to implement objects. This was found to be adequate here as the significant events: creation, alteration and deletion of objects, were all detectable by reading messages. This is because inter-object messages must identify their source, destination, the nature of the communication and any parameters needed. This is a case of naturally available data since such messages are present whether or not the IM is in use.

The final version has a wider scope as described in section 2. It seeks to demonstrate the behaviour inside an object, the processes making up the process group in our case, and over the sets of objects residing on several LOMs [9, 10]. This has the effect of increasing the set of significant events we want to see. Process creation and deletion and their intercommunications, messages between LOMs are the obvious ones.

Inside an object the processes making it up communicate via message queues but do not pass CSA messages down these queues. This means that when processes are created or destroyed or send data to each other, nothing can be learnt by using the trap placed in \fBSend\fP except by later inference. This is a case where data collection has to occur "by design", special purpose traps have to be added to send CSA messages giving the data we need when these events occur. This does have the undesirable effect of altering the performance of CSA more than the prototype version did, but it is inevitable that there should be some cost associated with a more detailed display.

On the Domain level CSA messages will be used as in a LOM so the required data should be obtainable without instrumentation-only messages being added.

In a more general case the determination of instrumentation points (to monitor the signifiant events to be represented as node or arc creation or deletion at various levels of abstraction) may be strongly implementation dependent.

6. Plans for Future Development

The IM emerging from the second development phase can instrument any donor which can be modelled as a hierarchical graph. This then has applications to other object-oriented operating systems. It can be also be applied to other systems if they can be easily modelled as graphs, which many computer systems can.

In order to instrument an architecture similar to CSA, the required displays must be designed, and a minimal set of traps implemented to deliver the required information to the Instrumentation Machine. A Personality Package must then be written, which takes the information given, and transforms it into the required displays.

Another possible avenue of development is to alter the role of the IM from its current one of a passive display tool to one of a system or network management tool where it could influence the behaviour of the donor.

7. Conclusion

The IM has been presented as a general monitoring system to capture information about the interactions occuring in object-oriented or other similar systems. It can be used as a demonstration tool for those unfamiliar with a system or a passive debugging tool for those working on it. We have given some indication of the type of events which need to be instrumented and examples of how and where data to do this has been found in CSA. The development path following from the work has been outlined.

8. Acknowledgements

The work described in this paper has been undertaken within a collaborative project comprising representatives of MARI Applied Technologies, Plessey Research and Technology, Philips Forschlungslaboratorium Hamburg and Synergie Informatique et Developpment. The author of this paper acknowledges the support of her employer in the project and production of this paper. She also acknowledges the help of her colleagues in producing this paper.

Appendix - Communications System Architecture

The Communications System Architecture is an architecture which is being designed and implemented to solve the problems caused by heterogeneous computing systems.

It brings together a set of machines into a domain such that the domain then presents a uniform interface to all users and applications. The resources of all machines are treated as a single shared resource where issues of location, hardware, operating systems and communication are transparently handled.

The architecture is object-oriented, with an object being the building block for applications and the provider of kernel operations and system services. The object is conventionally considered to consist of a set of information and associated operations allowed on the data. All the usual object composition mechanisms are available: inheritance, sharing, export and import.

Within the domain objects communicate via messages, which will consist of an invocation of one of the object's operations and any necessary parameters.

The domain consists of a hierarchy of object machines, the most basic of which is the Local Object Machine or LOM. This provides the first level of abstraction from the local operating system. It provides a machine independent interface which offers the services usually provided by the native operating system and kernel functions of object and invocation management. These two are conveniently split into two layers; the Operating System Kernel Emulator (OSKE) and Object Interface Provider (OIP).

The OSKE provides the operating system services such as processes, memory and message passing. Objects are modelled as process groups each with a pair of mailboxes for messages.

The OIP supports the kernel functions of object and invocation management and a set of primitives for the construction of applications as a set of objects.

LOMs are brought together to form the domain object machine or DOM. In order for this to happen communication facilities are needed between LOMs to provide a location

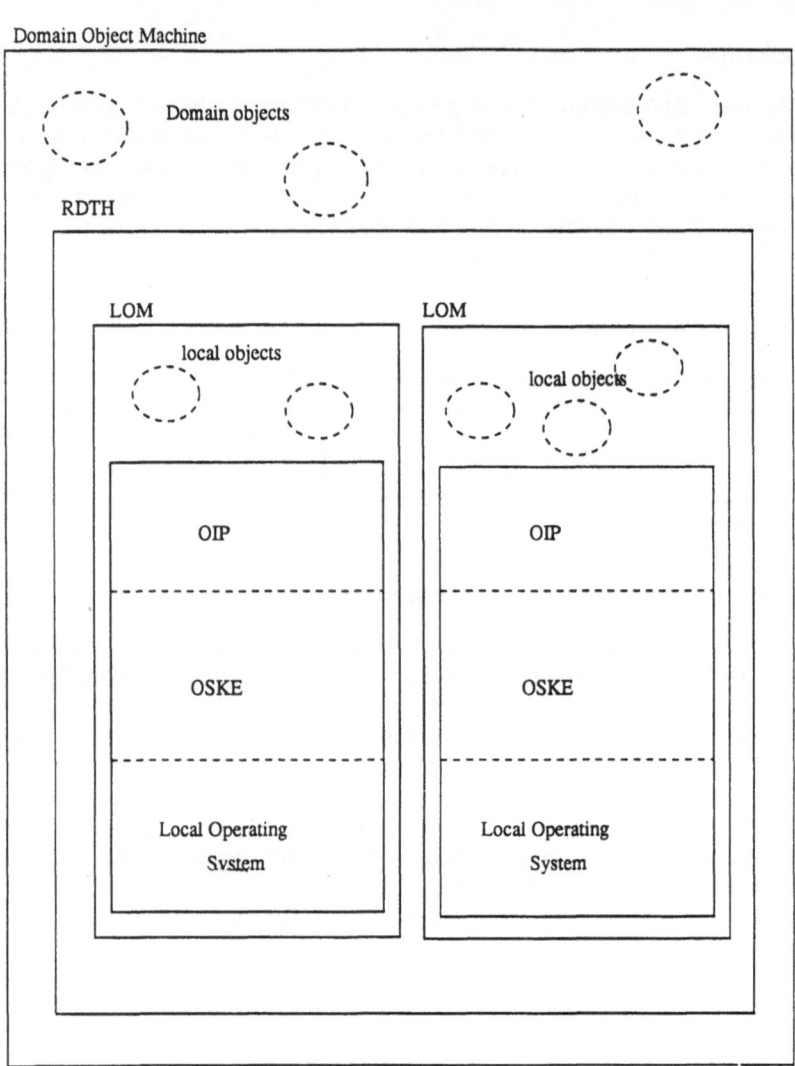

Domain Object Machine

Domain objects

RDTH

LOM

local objects

OIP

OSKE

Local Operating
System

LOM

local objects

OIP

OSKE

Local Operating
System

Diagram 4: A Domain made up of two LOMs showing objects

at a local and domain level

transparent interface for distributed applications and domain objects. These are provided by a domain object, the Communications Manager and a local object, the Reliable Data Transfer Handler which works over an abstract network.

The diagram gives an overview of the architecture.

Glossary

CSA - Communications System Architecture
an object-oriented architecture developed for distributed office systems. It provides a uniform environment for applications, abstracted from the details of the underlying architecture.

DOM - Domain Object Machine
A set of LOMs interconnected via an abstract network to provide a distributed machine kernel. It provides applications with the view of a single resource sharing system abstract from distribution.

IM - Instrumentation Machine
Machine which receives messages from an instrumented system and uses them and user defined rules to graphically represents aspects of the system behaviour.

IS - Instrumentation Server
Component added to an instrumented system to forward all messages to the IM and to which all messages must be copied.

LOM - Local Object Machine
The main purpose of the LOM is to abstract from the local operating system. It provides a similar set of operations as the operating system and the kernel operations of object and invocation management.

PP - Personality Package
the set of data required to configure the instrumentation machine. It can be described in EBNF as:

$$< peronalityPackage > ::= < messageRules > < Eventrules > < Template > +$$

where these are contained in several text files. Also refers to a program written in the configuration language which when processed produces the above files.

References

1. J. Joyce et al, (May 1985), "Monitoring Distributed Systems", Project Jade Research Report.

2. J.P. Behr et al, (1988) "Development of a Prototype CSA Object Machine and the Support for Applications", *Esprit Project 4.3.1/237*, Esprit Technical Week 1988 Conference Proceedings

3. "Design of Instrumentation and Monitoring for Fine-Grained Machine", *Esprit Project 4.3.1/237*, project deliverable, (1987)

4. R.A. Martin, (1988) "A General Instrumentation Machine", *Esprit Project 4.3.1/237/M100*, MARI, Newcastle Upon Tyne.

5. R.A. Martin, (1988) "Instrumentation System High Level Design", *Esprit Project 4.3.1/237/M102*, MARI, Newcastle Upon Tyne.

6. J.M. Bacon et al, (1988) "MANDIS: Architectural Basis of Management".

7. D. Holden et al, (1988) "An Approach to Monitoring in Distributed Systems", SIGDSM seminar paper.

8. G. Horne, (1989) "Instrumentation Machine - Template Editor Design", *Esprit Project 4.3.1/237/M114*, MARI, Newcastle Upon Tyne.

9. R.A. Martin, (1989) "The Instrumentation of Intra-Object Communication", *Esprit Project 4.3.1/237/M129*, MARI, Newcastle Upon Tyne.

10. R.A. Martin, (1989) "Instrumentation of the CSA Domain", *Esprit Project 4.3.1/237/M132*, MARI, Newcastle Upon Tyne.

Project no 237

DEVELOPMENT OF A PROTOTYPE DISTRIBUTED SYSTEM BASED ON THE COMMUNICATION SYSTEMS ARCHITECTURE (CSA)

M.J.WILSON
Plessey Research and Technology
U.K.

P.SAMPSON
Mari
Newcastle, U.K.

G. DURAND
Synergie Informatique
et Developpement
Paris

R.STECHER
Philips Forschungslaboratorium
Hamburg

Abstract

Within the office environment there is a trend towards multi-vendor systems with computing power distributed throughout. By its nature, the information and input/output resources are distributed on different nodes and accessed by many users from a variety of terminals. It is this complex scenario which motivated the project to investigate how such systems can be constructed and managed in a consistent manner. The project was convinced that major benefits could be realised if a consistent platform could be provided to application system irrespective of whether the underlying resources were single system/single vendor or distributed over multi-vendor systems. Our approach was to define an architecture (Communication Systems Architecture) that incorporated a number of strategic interfaces, observing standards where appropriate. An object oriented philosophy was adopted. From this architecture, the project constructed a number of platforms (Local CSA Machines) based on single end systems and moved on to implement a prototype platform based on a number of computing nodes linked by both local and wide-area networks (CSA Domain Machine). This paper outlines the architecture, describes the implementation of a Local CSA System, the design of the CSA Domain Machine and concludes by considering the relevance of the architecture to other environments and in which areas standards may be influenced.

1. The CSA Architecture

CSA addresses the problems of providing distributed applications in the office environment. Such "office systems" are typified by a number of distributed resources and computing nodes which are employed by a collection of users and maintain information which is held in a number of different nodes. This collection of users may be dispersed over a large geographical area and members may use resources which are physically remote from themselves. Each member's use of the resources at his disposal may be grossly variable with time regardless of their relative physical locations. The means of data transmission used to connect the various users and their resources may differ wildly, and the reliability of the connections may also vary, as may the costs involved with making and maintaining them.

It is our contention that no single scheme can be devised which will suit such variable requirements and that only a model which abstracts from the particular organisation of any group and its resources will be suitably adaptable to be employed in a general or strategic sense, or will be able to cope with changes in the office over a period of time.

To cope with such changes it is essential that the architecture adopted interfaces to both computing and communication resources that are manufacturer independent and are designed to best serve the requirements of the office system users.

Office system "users" can be divided into many interest groups. The <u>application writer</u> is concerned with the development of applications and application services by using the underlying resources. The <u>system administrator</u> may need to duplicate application services and information. He may need to move these and input/output resources to minimise networking costs and reduce the possibility of system bottlenecks. Security against unauthorised access to the office system may be demanded. The <u>system configurator</u> is concerned with fast addition of new resources and the use of more powerful computing and communication resources to satisfy the evolving needs of the system's users. <u>System users</u> require a uniform access to the system services but may wish to personalise the way they interact with the system. These interfaces are depicted in Figure 1..

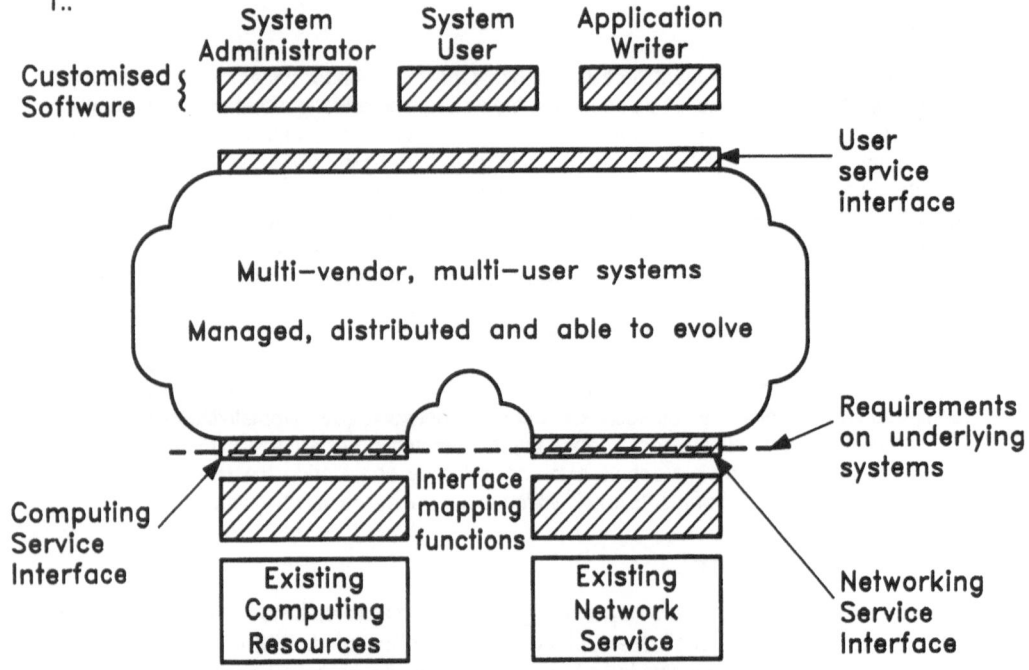

Figure 1: The Interfaces at the boundary of the CSA Architecture

1.1. The User Service Interface

CSA has adopted an object-oriented approach to application construction and system design. The user perceives the application as a set of interacting objects. These may in fact be distributed over a network. The application writer needs to specify object types and to be supported by a mechanism that allows new types to be introduced to the system. Type modification at this interface also provides greater potential for in-service update and enhancement. The system user requires a mechanism to create new object instances and invoke operations of an object to perform a particular service.

The system administrator is supported at the user interface by primitives that allow objects to be moved or copied. Thus the two main functions supported by CSA are object management and object interaction.

1.2. Architecture Overview

The CSA Architecture [1], [2], describes an object model as a means for structuring application systems. It offers great potential for distribution and parallelism along with the many other benefits in the software development cycle[6]. From the users viewpoint a CSA object is seen in the classical sense as having a data part and an accompanying set of operations (the executor part). The data cannot be accessed from outside the object but only through invocation of a particular operation.

The CSA System also associates a controller with each object. This controller layer is responsible for ensuring the consistency of the executor part in dealing with multiple invocations and synchronising invocation patterns between objects. Thus a CSA object can be represented as shown in Figure 2..

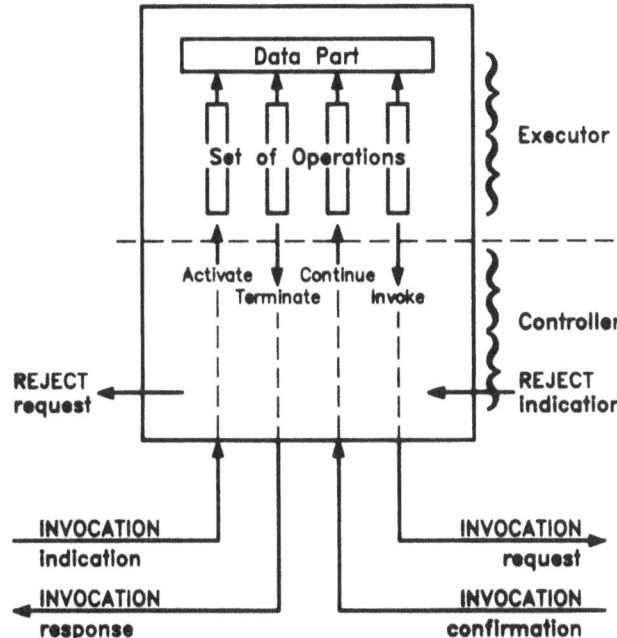

Figure 2: The CSA Object Model

CSA Objects are supported by a platform known as a CSA Object Machine. It provides the users a set of primitives that enables objects to be managed and to interact. An object machine is designed to solve a particular problem and to hide this from the user level. In particular, applications can be written separately from the problem addressed by the supporting object machine.

The CSA Architecture covers three problems relevant to the office environment. These are:

- The problem of using different manufacturers equipment (heterogeneity).
- The problem of users and resources being geographically separated (distribution).
- The problems of communicating to other office systems possibly using different architectures (openness).

Thus three object machines are prescribed, each solving one of the above problems. However, the CSA Architecture goes one step further in proposing that machines can be constructed as a hierarchy[3]. Having provided an object machine to solve the heterogeneity problem (the CSA Local Machine) the application interface can be used to construct a second level object machine (the CSA Domain Machine) to solve the distribution problem. Finally, this second level is used to construct the third machine to solve the problems associated with external communications.

An application may run on any of the three levels since the interface provided by each is the same. Figure 3. shows the resulting model.

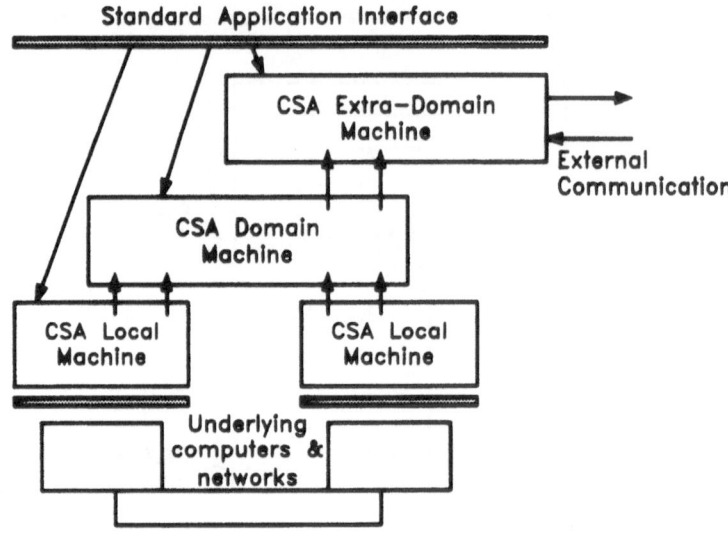

Figure 3: Hierarchy of CSA Object Machines

2. Implementation Issues of the CSA Local Machine

To solve the problem of using different manufacturers equipment, the CSA Local Object Machine (LOM) was introduced. Thus, the task of the LOM is to provide a homogeneous interface for different underlying local systems such as work-stations, personal computers and mini-computers.

The interface provided by the LOM supports the object model (figure 2). The users of the LOM's interface are objects. Therefore the LOM has to map the objects and their interactions onto concepts supported by the underlying operating systems or possibly directly onto hardware.

2.1. Lom Architecture

Within the LOM an intermediate interface was defined [5], the *Operating System Kernel Interface*, that provides the necessary concepts to best support the functions of object management and interaction. This interface is used to construct the *Object Interface Provider* layer which contains components that implement these functions. The *Operating System Kernel Interface* is mapped onto the facilities of the underlying operating system by a set of emulation routines collectively known as the Operating System Kernel Emulator (OSKE). The introduction of this internal interface assists in the portability of the whole OIP to different manufacturers equipment and provides a set of requirements against which existing operating systems can be evaluated. Figure 4. illustrates the architecture of the CSA Local Machine.

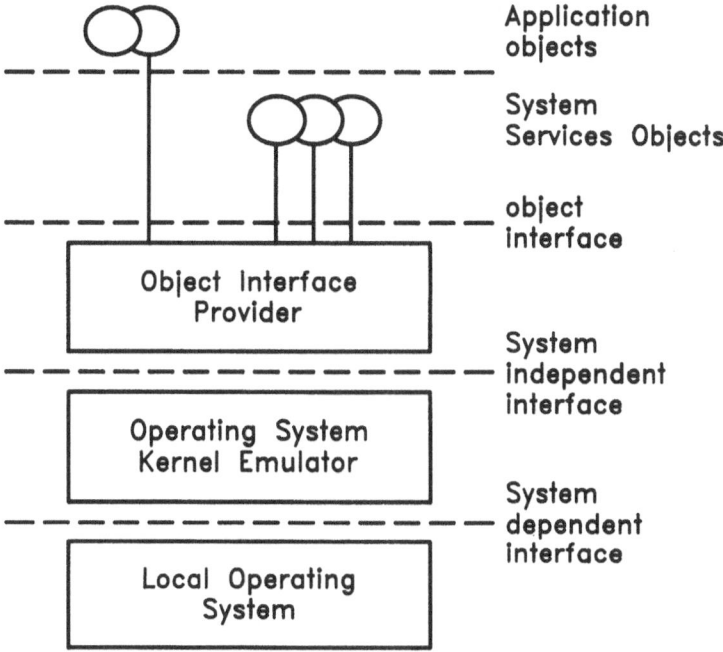

Figure 4: Architecture of the Local Object Machine.

OSKEs have been implemented on top of both UNIX systems and MS-DOS systems. These operating systems were selected not because of their suitability but because they are popular and provide many existing tools and applications.

2.2. The Operating System Kernel Interface

The OSKE interface has been defined as a set of primitives that give access to the underlying resources. In the prototype implementation these have been written as a set of Modula-2[4] procedures which are exported to the Object Interface Provider.

The concepts incorporated into this interface are:
- process group
- process
- mail box

- message
- location

A *process group* is an entity for structuring processes. It may contain any number of processes and it defines a boundary between processes within the group and processes in other groups. Data associated with the process group may be shared by all the processes within that process group. Furthermore a process group serves as the only access point for sending messages. This construct provides an ideal mechanism for the construction of CSA objects.

Procedures have been implemented that allow the dynamic creation and deletion of process groups. A user of the OSKE may specify certain characteristics that specialise a process group during its creation. This includes information about the capacity of the mailboxes attached to the group, the data requirements of the process group, and information that allows an initialisation process (the *boot* process) to be automatically created within the new group. This boot process customises the particular process group.

A *process* is an executing entity and is contained within a single process group. A process never crosses its process group's boundary nor does it communicate with processes in other process groups. However, within a process group different styles of interaction between processes are allowed.

New processes can be created dynamically. There is no requirement for a process to be able to create a process within another process group. This supports the object view that the way it is implemented is hidden from the outside world. The process created will execute the program indicated by a parameter. It may also access initialisation data which is stored in the location given by a second parameter.

Two *mailboxes* are attached to each process group in order to receive messages sent by processes of other (or the same) process groups. A *message* is a location with a structured content and serves as a container for information to be sent from a process into a mail box of a specified process group. A *location* is a piece of contiguous memory that can be used to store any data, programs, or messages. It can be qualified as foreground, virtual, or background. This technique assists the portability of the software which uses the OSKE Interface. Figure 5. summarises the listed concepts.

Procedures are provided that enable communication between process groups. A Send procedure places a Message into a specified mail box of the given process group. A Receive procedure gets a Message out of the specified mail box. Send and Receive are blocking i.e. Send, blocks the requesting process if the mail box is full, Receive blocks if it is empty.

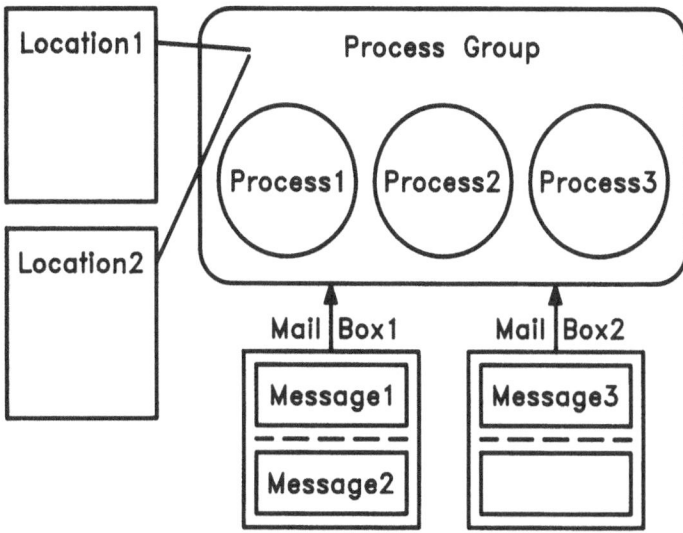

Figure 5. Concepts at the OSKE Interface

In the current prototype implementation, all the process groups (and the processes they contain) are incorporated into a single UNIX process. This enables a more lightweight process model to be adopted and a higher granularity of object to be investigated. It does however mean that the OSKE has to implement a scheduling policy. Scheduling in the OSKE is carried out in two steps. The first step decides which process groups contain processes that are ready to run and schedules one of them according to some priority. In the second step a ready process within the scheduled process group is selected to become active. The scheduling policy attempts to complete all processes within a scheduled group (i.e. until they are waiting or deleted) before the next group is scheduled.

There are other OSKE procedures that allocate, deallocate and resize locations, and others that deal with timers.

2.3. Object Interface

The part of the LOM kernel which implements the mapping from the object world to the OSKE concepts is named the *Object Interface Provider* (OIP). The concepts of the OSKE are used by the OIP to provide an object-oriented interface to the applications. A number of implementation decisions have been taken on how such a mapping is best achieved.

Each object within the CSA prototype is implemented as a process group. An operation of the object will be executed on the receipt of an invocation. Each execution of an operation is mapped onto a new process within the corresponding process group. An invocation of an operation is mapped onto a message and is sent to one of the mail boxes of the target process group. This mail box is called the *activation mail box*. A reply to an invocation is mapped onto a message that is sent to the second mailbox of the process group, termed the *reply mail box*. Finally locations map an objects state space onto real memory and store the code of the operations.

As well as implementing the CSA Objects, process groups are also used to implement the entities within the OIP which are responsible for the management of objects and for handling interaction between objects. Messages are used to carry requests to the OIP and also to support I/O in the CSA Local Machine.

2.3.1. Object Management.

Object Management is concerned with how objects are specified by the application writer through some language and how these specifications can be analysed by the system and mapped onto the underlying resources to produce a set of objects that are capable of reacting to a well defined set of invocations.

As with many object-oriented systems CSA recognises two levels in the management of objects[10]. The first level is concerned with the specification of an object type. The application writer will provide information about the operations that the type supports, how the new type is implemented (the code for the operations and the structure of the state space), and data that enables the object controller to manage incoming invocations (access rights, concurrency information and parameter types). Modula-2 was chosen as an implementation language for an object's operations. This was extended to support the constructs defined by the CSA Architecture. Concurrency control over the execution of the operations of a single object are expressed by Open Predicate Path Expressions[7] extended to support a number of features of CSA. The ability to introduce new object types to the system in a dynamic manner is implemented by a system component called the Type Manager and is implemented as a single OSKE process group.

The second level of object management defines how instances of a particular type can be dynamically created and destroyed. Each internal representation of an object type is contained in a system component called an Instance Manager. Thus one Instance Manager represents one type in the system. Because of the dynamic nature of Instance Managers they are not implemented as a part of the CSA Local Machine kernel but as special system objects (termed system services) that reside on the Local Machine. Thus they are invoked like any other object. An Instance Manager is able to dynamically create a new object of the type that it represents. The newly created object will be allocated a name given as a parameter and can be invoked by referring to the name.

2.3.2. Invocation Management.

Two objects interact by means of invocation. The OIP provides an invocation transfer service to the applications. This service is provided by two OIP components termed the Identification Manager and Communications Manager. The Identification Manager solves the problem of naming and identifying objects whereas the forwarding of invocations to the identified objects is carried out by the Communication Manager. These components are implemented as normal procedures which can be called by processes within the process groups.

The Identification Manager maps an object name given by a user (*given-name*), to an OIP internal and unique object identifier called a *designator*. It is implemented as a look-up table. Many given-names may map onto one designator but not vice versa.

The Communication Manager forwards an invocation or a reply message from one object to another. For this task it maintains a mapping table from designators to addresses (in fact an address at the OIP level is equivalent to the process group identifier at the OSKE level). This mapping is a one-to-one mapping because in the LOM there exists exactly one representation (process group) per object. The mapping may be changed however, because the representing process group may be moved from foreground to background with a resulting change to its process group identifier. A designator is bound to exactly one object through the life time of the system.

Figure 2 outlined the system view of an object consisting of two parts - the object Controller and the object Executor. The Controller part comprises that functionality of an object representation which controls the flow of invocations between the objects. The processes that implement the controller functions share the same process group as those implementing the operations of the executer.

The Controller part of an object representation is implemented by five processes. The Access Controller receives messages from the activation mail box of its process group and checks whether the requested operation is known within the current object. It also checks if the parameters sent with that invocation are of the correct type. This information is included in the internal representation of the type description. Moreover the Access Controller checks whether the requesting object has the access right to invoke that operation. The access rights are stored with the invoked object in an access right list.

If the checks of the Access Controller are positive then the invocation request is moved on to the Concurrency Controller. The Concurrency Controller obeys the concurrency constraints given by the application programmer. Dependent on the result, an invocation may proceed immediately or at a later point in time to the next controller.

The Activation and Reply Controller sets up a new process for the execution of the invoked operation within the Executor part of the object. It provides the new process with the parameters given in the invocation. When the executor process terminates, the Activation and Reply Controller collects the results of the operation and sends a reply to the requesting object using the Communication Manager of the OIP.

If an executor process wishes to invoke an operation of another object it uses the Invocation State Controller to mediate that invocation. The Invocation State Controller uses the Identification Manager to resolve the name given by the executor process and uses the Communication Manager to forward the invocation message. Moreover it observes the reply mail box of the process group in order to receive incoming reply messages.

The Executor part of the object representation is currently able to carry out two kinds of operations. The first kind are interpreted operations which are carried out by an M2-Code interpreter. M2-Code is an intermediate code originally designed for Modula-2 program compilation. The second kind of operations are compiled operations which are linked with the LOM software prior to system boot time. Both modes can be supported in a single object instance.

Compiled operations are faster but have the disadvantage in our current implementation that they can not be introduced into the system during run time. Tools that enable dynamic linking of compiled operations are not currently available but would greatly assist future work in this area.

3. The Domain Object Machine (DOM)

Local Object Machines have been defined and implemented, thus the problem of heterogeneity of hardware and native operating systems is no longer a concern. The Domain Object Machine (DOM) may now concentrate upon the aspects of the architecture which remove concerns over distribution.

The sort of distribution elimination talked about here is not merely removing the necessity for users to explicitly log in to another machine, for example by establishing a PSS connection. Nor is it only to remove awareness of remote system addresses such as might be used in unix mail, after all, that can already be hidden quite easily with mail aliases.

A *complete* elimination of all knowledge of physical machines (in our case represented by local 'abstract object machines') is the ultimate aim. Naturally, knowledge of *location* in some sense may still be allowed, but only at the object's given name level. One would remain free to specify, for example, the use of a 'Philips Research Hamburg LaserPrinter' should one so wish. Clearly, in the unlikely event that Philips Research were to be relocated to somewhere else then the name of the object would become misleading. However, the choice of the object's name has no bearing on the matter of how the domain machine positions objects. The name is up to the creator of the object and the domain managers will certainly not attempt to read the name in order to deduce where the object representation should be placed.

Each LOM, considered in isolation, provides a set of services that are available to the creators of the DOM. These include:

- Dynamic creation of object types. (Instance Managers).
- Dynamic creation of objects. (Instances of those types).
- Various principles of object composition. (Importation, inheritance, sharing).
- A means by which one object may invoke an operation on another,either synchronously or asynchronously with a reply, or asynchronously without a reply. (Invocation).

It is conceivable that one might wish to stop there. However, the real point of the architecture is to solve the problem of connecting many machines, as described in Section 1.

Since we believe that object-oriented methods are highly suitable for application development, and also that we have successfully demonstrated this with the earlier LOM, we would like to see exactly the same principles carried up into the domain level. Accordingly we adopt, for the composition of a DOM, a structure similar to that which we used for a LOM.

3.1. Domain Managers

The DOM therefore employs the same concepts of object management and invocation management as does the LOM, plus some others to support functionality specific to a distributed system. The machine is an environment supporting domain applications by using domain objects interacting via the mechanisms of domain invocation. As far as possible, this is constructed out of local application objects inhabiting the LOMs which comprise it. Thus:

- The domain type manager turns descriptions of object types into instance managers for those types.
- A domain instance manager creates instances of domain objects.
- Domain object type descriptions are specified in the same language as are the local type's and have the same concepts of importation, inheritance and sharing.

Invocations between domain objects are supported by a domain identification manager and a domain communications manager.

It is stressed that there is no necessity to construct a DOM in exactly the same way as was the LOM. It would be possible to use a different type description language for example, or to support invocations with different mechanisms. We are merely adopting a principle of self similar, recursive architectures.

To actually create 'concrete' realisations of domain objects (as local objects), the DOM employs a location manager in the same way that the LOM employs a (local) location manager to create object representations which may be executed' on a real machine.

Domains, or distributed systems in general, have problems which local machines do not. Obviously, some sense of the composition of the DOM in terms of LOMs and resources provided, communicability and network management in general, has to be found within the system itself. All this has to be done so that the actual distributivity is hidden here, otherwise the DOM as a whole will fail to abstract from distribution - the principal reason for its existence. Most of this abstraction is performed by the domain managers themselves, but there remain certain resources (usually, but not exclusively hardware in nature) tied to specific LOMs. We can go a long way to solving this problem by registration of such resources (and their associations with LOMs) as named objects. A domain resource manager supplies this functionality, and only the domain location manager need derive LOM information from it.

A domain transaction manager is introduced into the architecture at this point because 'real people' will be using the DOM in an essentially unpredictable way. A transaction manager is not felt to be necessary at the LOM level since the software resident upon a LOM is under the control of systems programmers and is not subject to misuse by possibly naive, and certainly unregulated, application writers. This is not to say that there is no concept of transactions within a LOM but merely that any such concepts are built into the LOM directly.

3.1.1. Domain Manager Interface.

The managers together, that is the domain

- Type Manager
- Location Manager
- Resource Manager
- Transaction Manager
- Identification Manager
- Communications Managerform what we call the **"domain kernel"** .

We choose not to include domain instance managers within this kernel since we regard these as (service) objects which, whilst having a well known' set of operations, can better be described along with other domain objects.

3.1.2. Object Management.

The interfaces for object management within a domain are a little different to those within a local environment. More application support is required and we still have to solve the problem (despite abstraction from distribution) of ensuring that objects can be placed *conveniently* within the domain, either for the purposes of load balancing or to ensure that special hardware requirements are met. Note that this is still fairly low level management - application writers will be unable to access domain managers directly, except for the Type

Manager which can be accessed just as if it were an object.

The domain type manager provides domain applications with the ability to create object types dynamically in much the same way as the local Type Manager. A new feature here (and in the interfaces to the other domain managers) is the concept of **resource requirement**. Instance managers themselves, which are objects of essentially one object type, do not *as a class* have any special resource requirements. However, the application writer may wish to ensure that a particular instance manager be created if and only if certain other instance managers already exist, notably the Instance Managers which represent types that the type being created might inherit, import or share. This is only an example of resource needs - no application writer will be forced *by the architecture* to do this, although of course a particular tool development support environment may easily be written at the application level to enforce such techniques. Operations are also provided by the Domain Type Manager for the deletion of object types and to enable object type migration and duplication.

Domain Instance managers are functionally very similar to Local Instance Managers. They support operations that allow the dynamic creation and deletion of instances of the object type that they represent. Additional operations are provided at the Domain level that allow objects to be copied and renamed.

As one might imagine, the semantics of these operations are largely identical to those of the Type Manager, although instances of object types are now created and then manipulated.

The Domain Location Manager is concerned with how Domain Objects are realised in terms of the underlying resources i.e. Local Objects and ultimately process groups, processes and locations. It provides operations that allow these realisations to be created, deleted, copied or moved. The create operation returns a designator for the realisation which is unique within the context of the Domain machine and is used for further references to that object realisation. It also supports the common requirements of both type and instance managers as well as maintaining realisation positioning at the LOM level.

A Resource Manager is introduced at the Domain level and acts as a directory of 'special' objects. Objects may be registered as LOM dependent resources at the time they are created. Only those objects which *are* in fact LOM dependent (such as those which take advantage of particular hardware on particular LOMs) will be so registered. This registration can still be performed without specifying the LOM directly as an operational parameter since the operation will in actuality be performed on the relevant LOM anyway. The Resource Manager also provides a facility for obtaining information on specific resource and will be used solely by the Location Manager both for the purposes of load balancing as well as in attempts to position objects depending on the 'special objects' (which is what resource requirements actually are) appropriately.

Finally, a Transaction Manager is introduced at the Domain level and provides a transaction service to the Domain Objects. The creation of a transaction is performed by creating a transient (and identifiable by the transaction controller within each object) transaction handler. Transactions may be nested, hence an extant transaction may be supplied to the transaction manager in the CreateTransaction operation. When a transaction has been completed, the corresponding transaction handler is destroyed as a result of the DeleteTransaction.

3.1.2. Invocation Management.

There is little further to say about the interfaces exported by domain identification and communications managers as their purpose remains the same as their counterparts in a LOM.

The domain identification manager provides (and maintains) a total surjective function which relates the given names (the name given to the object by its - usually human - creator) of domain objects into the domain designators of those objects. Neither the given name nor the designator give any clues as to the whereabouts (i.e. upon which LOM) of the objects identified. Domain designators are unique throughout the domain, not re-usable, and last precisely as long as the domain object so designated.

The domain communications manager provides (and maintains) a total bijective function which relates the domain designators of domain objects and the domain addresses of their realisations. The domain addresses do contain information about the LOMs upon which the realisations are resident. They are composed of the LOM address and the given name of a local object which realises the domain object. The communications manager, as for the LOM, additionally supports mechanisms for forwarding invocations and replies from one object to another.

Location transparency is achieved by burying location information in the Communication manager. Nothing above it will ever see a domain address. It is also the case that only a very few components - largely object controllers - within the domain will ever see even designators. Application writers will certainly never see them.

3.1.3. Domain Manager Construction.

There are a number of ways in which the services to be provided by the domain's managers could be realised. However, there is an overriding practicality which must be taken into account if we wish to permit inter-domain object invocations. This practicality demands that the domain invocation managers (identification and communications) must have a presence on each LOM in the domain, otherwise domain invocations will be unnecessarily difficult.

We choose to realise these two managers, then, as a set of local objects, one on each LOM, which provide a service access point to other realisations of entities which wish to interact via domain invocation. Each such local object will, to its users, implement the interfaces appropriate to the manager. Each set of objects has its own (private) interfaces which will allow the manager as a whole to work. It should by now be clear that it is possible for one domain object to access these public interfaces of domain invocation managers by a local invocation between two *local* objects which actually realise (wholly or partially) the domain object and the domain manager in question. Communications between peer components of the same domain manager may of course be performed with the help of the Reliable Data Transfer Handler present on each LOM (again with a local invocation).

As for the other domain managers (type, location, resource and transaction), no such service access point concept is required since the domain communications manager can transmit messages (on behalf of the invoker or the service requester) to the desired manager irrespective of where it might be. However, since the domain is intended to be a dynamic environment, it is in fact convenient to implement each domain manager as a set of co-operating local objects, one per LOM. In this way we avoid some of the problems associated with having a single LOM upon which a manager resides (overloading, bottlenecking and unreliability).

There is a 'middle' path in which only *some* LOMs are used in the realisation of a manager, but the general solution in which each manager is realised on (possibly unequal) subsets of all LOMs is, we believe, unnecessarily complicated since knowledge about the distribution of the manager components would have to be maintained in some other manager (subject to the same problem!). This realisation of the domain managers unconnected with invocation management is by choice and not by necessity.

3.2. Domain Objects

As on the local abstract object machine, domain objects have an executer and a controller. The controller is however extended with the addition of a *transaction controller*. Object descriptions are written in the same language as are local object descriptions, although this is not by necessity but by choice. There are a number of ways in which such a domain object could be realised.

One extreme is to map each controller process and each executor process as a distinct local object. The idea of a state space which is accessible by each executor thread would however make this an unattractive proposition. Another means, one which avoids the disruption of a state space concept, would be to realise a domain object as two local objects, one for the executor and one for the controller.

The easiest choice however, and the one which we have made, is to realise one domain object as one local object (in the sense that a composite object consisting of one importing object and one imported object, for example, consists of two objects). A composite object, which may be pictured as a tree of objects related by importation and inheritance, may be realised within the domain in one of several ways.

The easiest choice to make is to simply take over the method employed by a LOM. This means that such an object is instantiated in such a way as to position all of the objects on one LOM. This has certain advantages in that the LOM's implementation of the representation of a composite object would allow local invocations as the method of communication between objects within the same tree. However, this would be an optimisation and should not be considered as an architectural feature.

There are also drawbacks to consider. There is no guarantee that the instance managers required for all components of a particular composite object will be present on the relevant LOM since any given object type may in principle be employed by many different composite object types. This is not a problem as such since it is possible for an instance manager on one LOM to realise an object on another (through the domain location manager which has a component on each LOM).

More seriously, perhaps, is that it may be impossible to satisfy a resource requirement for a given composite object on one LOM. We might answer this point by saying 'so be it'. If an object needs to share an array processor (which is present upon only one LOM) and also a flatbed plotter (supported on a different LOM) then it would be impossible to satisfy the resource requirement and the object would thus not be created. On the other hand, this type of problem is not restricted to composite objects. In cases such as these, it would be preferable to leave the two 'special' objects out of the resource requirement, or to choose only one of them. The use of the thereby 'remote' object or objects would still of course be possible, since we have domain invocation, but a little slower.

4. Conclusions

The CSA project has throughout its 5 year program attempted to maintain a distinction between architectural, design and implementation issues. The architecture described was original and in order to prove its applicability it was essential that the prototype system conformed. This has led to a disciplined approach to implementation particularly where efficiency conflicts with the structure defined by the architecture. However, this approach has enable certain aspects of the architecture to be modified based on the practical experience gained in prototype implementation. A stronger and much more mature architecture has resulted.

The approach, the interfaces and protocols in the Architecture have been and are continuing to be influential in a number of areas:

- Many of the principles are shared with the Advanced Network Systems Architecture[8] and are thus being fed into the Esprit 2 project on Integrated Systems Architecture.
- Experience gained in developing invocation and distribution protocols is influencing the ISO working group on Open Distributed Processing.
- Development of an interface that encompasses the computing requirements of managed object-oriented systems is promoting interest in the work on Intelligent Networks and telecommunication service management. Use of different manufacturers equipment is a problem in these areas.

There are an increasing number of areas that are now recognising that distribution is a key problem in their work and that an architecture based on the concepts described in CSA is fundamental.

In addition to developing CSA prototype systems for the Local and Domain Machines, the project has developed a number of applications. These were designed in an object-oriented manner independently from the models prescribed in CSA. It is stressed that the project is concerned with object-oriented systems and not object-oriented programming. This approach enabled the project to assess the facilities provided by the CSA application service interface in a more objective manner.

The object-oriented approach to building application systems has proved very powerful despite the lack of available tools in this area. Dynamic update and linking of object types enables objects to be modified quickly and improves application development types. The ability to do this for compiled as well as interpreted operations would be advantageous. Some tools have been produced on the project for testing object types in isolation and reporting object diagnostics. Even these simple tools have been of great benefit to the development of the application objects.@partext = Finally it was realised that in order to demonstrate the features of the architecture some sort of monitoring system was required[9]. It is not sufficient to demonstrate solely using the applications. This deficiency has been noted in many system demonstrations in the past and the CSA project was determined to avoid this. Indeed it was decided to develop a general monitoring system for object oriented approaches. As well as enhancing the CSA demonstration it is envisaged that this could be used by similar projects.

5. Trademark Notices

UNIX is a registered trademark of AT&T.MSDOS is a registered trademark of Microsoft Corporation.

6. References

1. Esprit Project 4.3.1/237, Strategic Architecture Design Document, October 1986

2. Esprit Project 4.3.1/237, CSA Strategic Architecture, March 1986

3. J-P. Behr, U. Killat, R. Kraemer, CSA: A Hierarchical Object Oriented Architecture for Distributed

Office Systems, Informatik Fachberichte Nr. 160, Springer Verlag, 1987. Proceedings of "Communication in Distributed System".

4. N. Wirth, Programming in Modula-2, Springer Verlag, 1985

5. R.H. Cunningham, C. Durel, R. Kraemer, J.B. Wright, Prototype Design of a Local Abstract Object Machine, September 1987. ESPRIT Technicak Week 1987.

6. J. Jeffcoate, K. Hales, V. Downes., Object Oriented Systems: the Commercial Benefits, Ovum Limited, 1989.

7. M.R. Headington, A.E. Holdehoeft, Open Predicate Path Expressions and their Implementation in Highly Parallel Computing Environments, Proceedings of the Conference on Parallel Processing 1985.

8. A.J. Herbert, et al., The ANSA Reference Manual, Advanced Network Systems Architecture, Cambridge, March 1989

9. G. Horne, An Instrumentation System for CSA and Similar Architectures, Esprit Conference 1989

10. Yasuhiko Yokote, Fumio Teraoka, Mario Tokoro., A Reflective Architecture for an Object-Oriented Distributed Operating System, Proceedings of the Third European Conference on Object-Oriented Programming, Cambridge University Press, July 1989

Project No. 285

THE OSSAD METHODOLOGY

E. BESLMÜLLER [1]
IOT
Kästlenstr. 32
D-8000 München 82

D. W. CONRATH [1]
CETMA
Boulevard des Camus
F-13540 Puyricard

ABSTRACT. The OSSAD Methodology aims at providing a means for conducting office support systems analysis and design projects. The Methodology includes: data collection instruments and procedures, means to model both organizational and technical support systems, methods to analyze the present situation and proposed alternatives, ways of generating such alternatives, the conversion of a chosen alternative to technical and organizational specifications, issues of implementation and how they can be handled, and the conduct of periodic post implementation "audits" to ensure that the system is accomplishing its objectives. The approach chosen for the development of the Methodology included five applications. While the Methodology is already exploited and sold more elaborated data collection instruments and software tools (based on OSSAD's modeling techniques) would represent an essential improvement of the Methodology and its market potential.

---

[1] Further contact persons: G. de Petra, IPACRI, Via a Rava 124, I-00142 Roma – V. De Antonellis, Universita di Milano, Dip. di Scienze Dell' Informazione, Via Moretto da Brescia 9, I-20133 Milano

1. INTRODUCTION

Computer-based systems are being used to support an increasingly broad spectrum of office work, and there is little doubt that their use will continue to grow. While early systems were usually developed for a single purpose, current technologies (e.g. the micro-computer and local area networks) support a variety of tasks, and the trend is to increase the level of integration across not only tasks but people as well. The consequence is that the nature of office work is changing and will continue to change, and to accomplish this work effectively an office (an identifiable set of people organized to process data and information in the pursuit of interdependent goals) depends on the appropriate integration of the organizational and technical support systems. The resulting need is to ensure that this is done in a well planned, comprehensive and beneficial fashion. That is the goal of the Office Support System Analysis and Design (OSSAD) Methodology. The means to achieve this goal are contained in a document, the OSSAD Manual [1] (see fig. 1). On behalf of ESPRIT complementary work has been done, for instance, in the FAOR project [2] with the key feature of benefits analysis techniques and the TODOS project [3] focussing on providing a set of (computer-based) tools to improve the design of Office Information Systems.

The OSSAD Methodology has been developed on the assumption that those who use it are not experts in the design and implementation of integrated organizational / technical support systems. We do assume, however, that persons applying it have some knowledge of both the capabilities of computer-based information technology and organizational structure. As a consequence the target audience of the OSSAD Manual is composed of those who work in the areas of information systems, office automation, organization and methods, organizational design and the like. Therefore, the Manual concentrates on what the Metholdogy is and how to apply it, and little is written about office support technology per se, the details of organizational design and the concepts which lie behind information systems. These are assumed to be known or obtained from other sources.

A vital work stream of the OSSAD project was concerned with the application of intermediate results (see fig. 1). This, we felt, is essential, yet the major shortcoming of virtually all the current approaches to the stated primary goal of the OSSAD Methodology is that they have not been properly evaluated. To our knowledge, no follow-up studies have been conducted to determine whether the design results were effective or the procedures (methods) used appropriate. Without such feedback, however, one has no more than the product of intuition, and the risk is too great to rely only on that. The execution of the field studies is described in a report entitled "Field Test Report" [4].

This paper is structured as follows. First, we outline the ultimate output of project 285, the OSSAD Manual, and present the coverage of the OSSAD

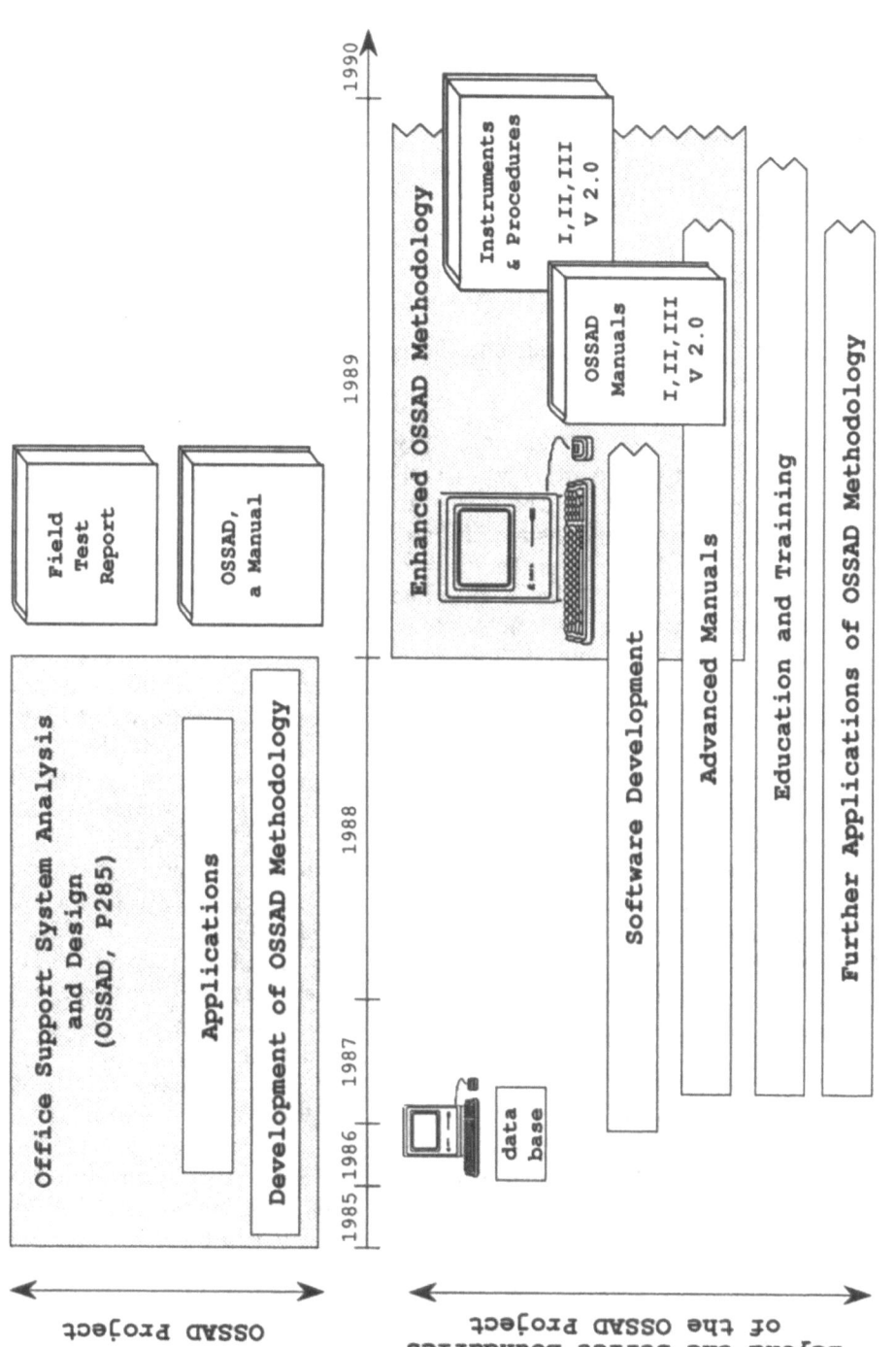

Figure 1: Past and Future Aspects of the OSSAD Methodology

Methodology with respect to the development life cycle of office support systems. Then we describe briefly the five applications of the Methodology during the course of the project. This is followed by reporting on the significance of the results of the OSSAD project in general. We conclude with a presentation of exploitation issues currently undertaken by some of the partners of the OSSAD consortium.

2. RESULTS OF OSSAD

2.1 Office Support System Analysis and Design, a Manual

As should be clear from the opening paragraphs, the OSSAD Methodology is devoted to more than just analysis and design. It is concerned with the entire process - from the initial contact with someone interested in changing the existing support systems, to their design, implementation and ex post evaluation. As a consequence, such things as planning, data collection and analysis, creativity and the management of change are discussed. The result is a methodology which provides a number of options and allows one to generate methods that are specific to the circumstance in which they are going to be applied. The Methodology includes: data collection instruments and procedures, means to model both organizational and technical-support systems, methods to analyze the present situation and proposed alternatives, ways of generating such alternatives, the conversion of a chosen alternative to technical and organizational specifications, issues of implementation and how they can be handled, and the conduct of periodic post implementation "audits" to ensure that the system is accomplishing its objectives.

Figure 2 displays the five basic Functions that have to be accomplished if one is to undertake a complete OSSAD Study: DEFINE PROJECT, ANALYZE SITUATION, DESIGN SYSTEM, IMPLEMENT CHANGES and MONITOR SYSTEM PERFORMANCE (see [1]). In addition, it shows how they are linked to one another. On the surface our approach resembles the life cycle of a system as described by many researchers and practitioners. Where we differ from the others is how we suggest that one respond to the various aspects of a system's life cycle. That is the purpose of the OSSAD Manual - discussing and describing each of the Functions in detail. In what follows we present a brief outline of important issues.

DEFINE PROJECT involves establishing the basis upon which a particular study will be undertaken. While most approaches say nothing about this step, since it provides the framework for everything which follows we feel that its role is too important to ignore. Participants do not naturally do all of the things that are required to ensure a successful study, thus we wish to call attention to those which are important and/or which might be overlooked. For example, in addition to objectives, performance criteria

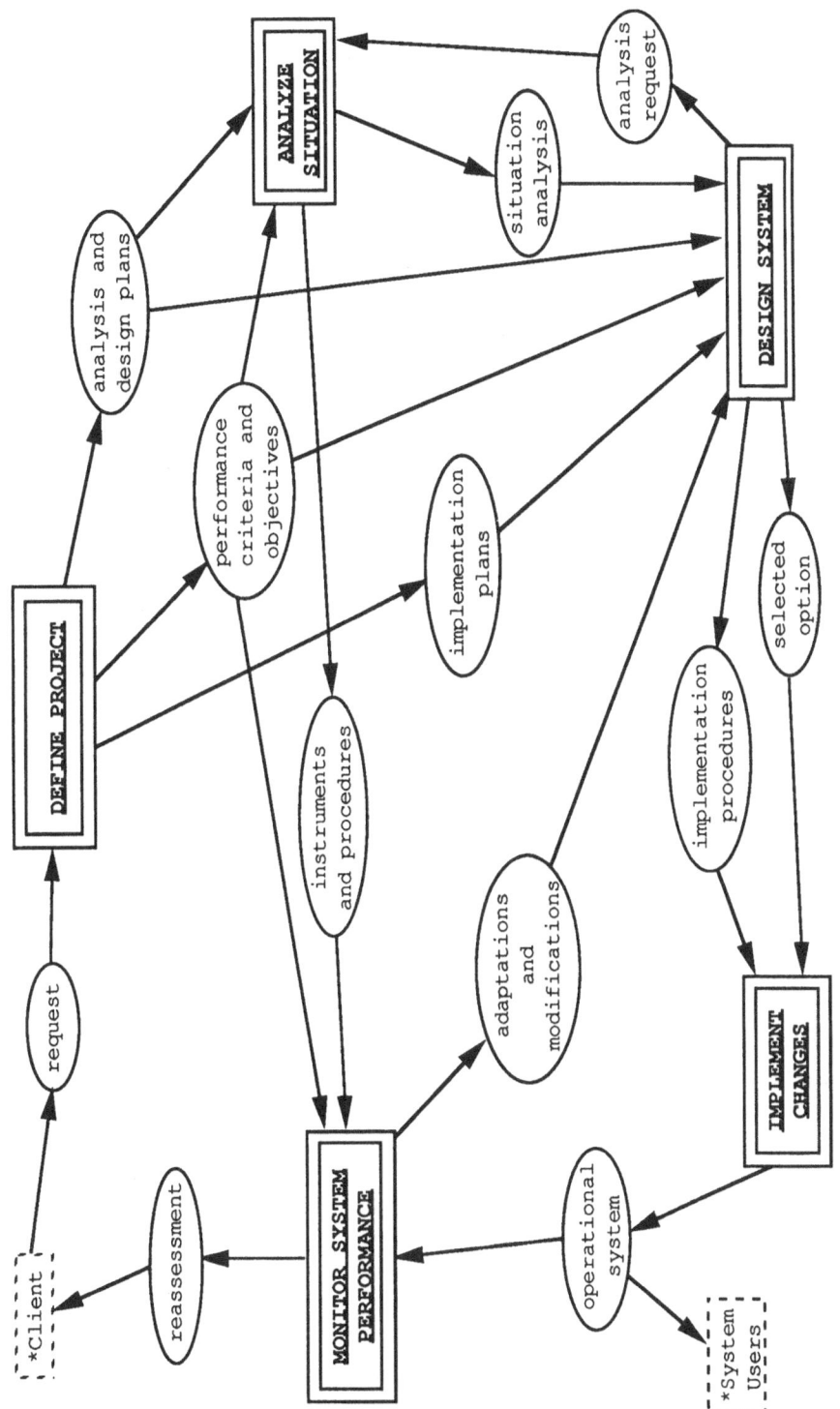

Figure 2: OSSAD Methodology, an Overall View

should be delineated at the very beginning. If they are not, there is little basis for determining whether or not the entire analysis, design and implementation exercise has been a success. Also at this stage, preliminary consideration has to be given to the approach one will take to design, since it will have major implications regarding how the study will proceed. Furthermore, problems of implementation have to be considered from the very start, both because of the early information dissemination requirements and because they may affect system design.

ANALYZE SITUATION concerns the collection, processing and presentation of the data needed to describe the organization and its environment for diagnostic purposes. This does not depart greatly from the system's life cycle approach. One major exception is that this Function provides both the base-line data and the instruments and procedures that are essential elements to the effective evaluation and selection of options and the monitoring of system performance after implementation. In fact we emphasize issues of data collection instruments and procedures, something which the vast majority of analysis and design methodologies treat lightly, if at all. Another exception, which is not obvious from the figure, is that the analysis is not conducted in a linear fashion. Feedback and revision is the rule, and if a prototyping approach to system design is taken, analysis and design are linked so closely that they are almost merged into one.

Our approach to DESIGN SYSTEM, like analysis, is not a radical departure from existing procedures. One searches for alternative organizational and technical systems that will effectively respond to the identified problems and improve performance in general. Much of where we do differ is not obvious from figure 2, as it lies in the details of execution. Briefly we consider a number of different approaches to design, from prototyping and pilot testing to "speculation", the choice of which depends on the circumstances (budget, breadth, deadlines, complexity,...) of the study. The figure does display, however, two feedback loops (analysis request to ANALYZE SITUATION, and adaptations and modifications from MONITOR SYSTEM PERFORMANCE), which provide evidence on a global scale of our allegiance to the principles of Experimentation and Iteration. We also show that implementation procedures are a distinct output of DESIGN SYSTEM, and not just something which is of concern after the fact.

No matter how ideal a selected option might be, poor handling of IMPLEMENT CHANGES can negate all of its benefits and may even make the situation worse than it was before. Appropriate procedures depend upon several things. For one, they should take into consideration what is known about organizational change. There is a substantial literature on the subject and it should not be ignored. For another, as we have already noted, implementation issues should be raised at the very beginning (at the first meeting) and should be recognized during the execution of each of the Functions which follow. Motivating users to accept a new system is not a one shot process, and without motivated users successful system implementation is problematic. This is the primary reason we have stated user Participation as a principle, for active and meaningful participation is perhaps the most effective source of motivation.

No new or revised system will be used exactly as intended, nor will the effects be precisely as forecast, as the complexity of today's integrated organizational-technical systems makes it virtually impossible to discern all of their nuances before a system becomes operational. Thus there is a need to develop what one might call an audit procedure to MONITOR SYSTEM PERFORMANCE, looking at system use and its users to determine its economic, organizational and human consequences. This should be done for several reasons. First, an effective means of monitoring system performance would enable one to detect and resolve problems before they reach crisis proportions (the most common basis for launching most system reassessments). Second, one would like to identify the aspects of a system which work well and those which need to be changed. Note that the concept of working well is not restricted to the technical system, but should include its effects on individual users (motivation, job satisfaction, turnover, etc.) and on organizational structure (authority, direction and control). Third, a system which is appropriate for today may be inappropriate tomorrow. No support system can remain fixed for long and be as effective as it was intended to be when designed. Thus there needs to be an informed basis for adaptation. Fourth, since there will be a need for a system reassessment at some time in the future, feedback on the existing development process is needed if one is to improve that process. An organization should try to find out what worked and what did not, recognizing that the overall process is situational, so that it can build on the positive aspects and revise the negative ones during the next application of the OSSAD Methodology.

2.2 Modeling

Modeling provides what might be called the backbone of the OSSAD Methodology, as it is the process whereby we represent the various systems and their components (see also [5, 8 - 10]). We use the term model in a very general sense, as a means of describing structure - components and their relationships. The purpose of our modeling techniques is to enable and enhance communciation among those who must understand the workings of the systems being represented. Communication is critical to the success of any organized endeavor, and this includes the application of the OSSAD Methodology.

The Abstract Model is normative in the sense that it indicates what needs be done to accomplish the goals and objectives - the raison d'être - of the organization. It covers what might be called "organizational imperatives", those elements which are essential to the business independently of how they are executed. This model also identifies the boundary between what is to be and is not to be considered as pertinent to the system under study. The Abstract Model, in effect, defines the extensiveness of the analysis and design.

Descriptive Models are used to describe both the existing office and proposed alternative configurations (called "options"). A Descriptive Model indicates how the organization does or can accomplish the' Activities identified in the Abstract Model. Basically the Descriptive Model provides a description of behavior. To do this it focuses on the dynamic aspects of the office, not only in terms of resource utilization and the conversion of inputs into outputs, but also in terms of the rules governing this behavior (e.g. job assignment, control, priority, synchronization and the like). There are a number of basic concepts which characterize the Descriptive Model. They fall into two general classes: one concerns the organization of people and the other focuses on the work to be done.

3. MAJOR ACHIEVEMENT: FIVE FIELD STUDIES

In what follows we describe the five case studies conducted during the evolution of the OSSAD Methodoloy (see also [4 - 13]). While these field studies were undertaken to evaluate the concepts, instruments and procedures developed by the OSSAD project at that time, the intention of outlining these case studies here is to show the variety of circumstances in which the Methodology has been applied and could be applied in future office support systems analysis and design projects.

CASE I was carried out in a regional savings-bank and aimed at the accomplishment of two strategic objectives. First, more autonomy was to be given to local branches. In order to accomplish this, issues of organizational structure, departmentation, allocation of authority and control had to be redefined to ensure the desired performance of the local agencies. Second, the goal of enlarging the uses of computers in the front offices was defined. Due to various reasons options for local data processing at the branches was to be investigated. For example, the break down of the telecommunication network, interlinking the branch offices and the central data processing facility caused serious problems and had to be minimized by analyzing and reflecting the needs of the tellers. During this study these goals were narrowed to the problem of handling unpaid checks and card-debits which were rejected by the existing computer-based system. For design a prototyping approach was chosen and implemented, resulting in the definition of a new office support system for the branch offices.

Case II was also carried out in the bank business. It delt with a most important economic core-activity of the bank: the corporate loan business. Faced with a decline of margins in an environment of strong competition and the necessity to improve services for clients and to reduce costs, the bank decided to "streamline credit procedures". The analysis of the actual situation covered issues such as "reduction of processing time", "reduction of correction loops in the process of granting loans" and "reviewing the functionality of forms used". Proposals for alternative accomplishment of office work were elaborated. In doing this, organizational, technical and human measures were envisaged to fit, on the one hand, a clients needs and, on the other, the accomplishment of the goal to reduce the costs of providing bank products and services.

Case III took also place in the banking business and was concerned with developing strategies for the improvement of the actual counter-business. This had to be seen in a broader context as the savings-bank investigated is in a process of gradual transformation to achieve a more effective relation with the market. Like Case I, the application of the OSSAD Methodology focused on branch operations, elaborationg strategies for technological support of the tellers, for improvements and adaption of the organizational structure and the working environment. A combination of two options for the alternative accomplishment of the tellers' work was implemented. One of these options is concerned with removing all those teller transactions that do not directly involve the customers and to assign these transactions to a new role to increase the service time available to customers.

Case IV was concerned with the application of the OSSAD Methodology in another branch of business, namely a sales department of a paper machine company. In this case, the goal was to implement new technologies (teleservices) according to the requirements of the sales department reflecting their worldwide communication needs.

Case V was carried out in an educational branch of a large organization. The subject here was to investigate options for improvements of the nationwide procedure of "application for seminars". This included the definition of a solution-mix containing technical, organizational and human aspects in order to improve capacity planning, the use of the capacity by participants and the reflection of the political goals of that organization. A major aspect in conducting this case study was the fact that active participation of employees in redesigning their work is a must for this organization. Therefore, valuable experience was gathered on the comprehensiveness of the OSSAD Methodology for the end users of office support systems.

4. SIGNIFICANCE OF RESULT AND ACHIEVEMENT

During the evolution of the OSSAD Methodology we experienced growing evidence that vendors and purchasers of computer-based office systems are trying to enhance their know-how with respect to the implementation /

introduction of their computer-based systems. One reason for this is that both user organizations and vendors have had a lot of negative experience with office automation products and, therefore, see the necessity for changing office systems in a well-planned and well-organized way. The strategy of vendors might be characterized as follows. To the extent that hardware and software is implemented and used beneficially, the market potential of their products will grow. A precondition to accomplish this is that a vendor or purchaser provides also "orgware". As this is an essential aspect of the OSSAD Methodology our result are significant for vendors and purchasers.

User organizations of office technology also recognize this evolution, but to a certain extent they are more concerned with organizational and human aspects in redesigning office support systems. While the support given by vendors for introducing office technology often regards the organizational structure and the personal situation as given and, more or less, stable, user organizations tend to see the most urgent need for support of redesigning an office support system in reflecting organizational and human aspects. As far as the OSSAD team is able to determine there is a shift in the need of support in user organizations from the "analysis task" to the "design task" of office support systems projects. Here too the OSSAD Methodology can provide a significant product.

5. EXPLOITATION ISSUES: BEYOND THE STRICT BOUNDARIES OF THE OSSAD PROJECT

Figure 1 displays some activities currently undertaken by members of the OSSAD consortium, though different members put different emphasis on these activities.

Already during the development of the OSSAD Methodology a partner built a data base system for supporting a case study. It can be seen as a system containing an "Abstract Model" of the banking business. According to the "description schemas" developed by OSSAD, the functions of a bank (e.g. approving a loan, establishing an account for a private client), their subdivision, and their relationships to other functions are stored. An intention of this data base is to provide a banking business oriented approach for future projects. In other words, in future applications of the OSSAD Methodology in the banking area the determination of "what should be done in an office" under consideration will be merely a matter of checking the functions stored in the database against the functions to be investigated in a particular office. The benefit of using this data base system was mainly seen in speeding up the application of the OSSAD Methodology.

Another prototype was developed by another team and aimed at the graphical representation of the "Abstract Model". During the field studies it was realized that drawing up the graphic representation of the office under investigation was quite labour intensive and the completeness of the data was not as desired. The major cause was seen in the fact that during an interview where the "Abstract Model" was developed, both the interviewee and the interviewer needed support and hints to ensure that all those conditions and interrelationships of functions, as defined in the OSSAD Methodology, could be easily fulfilled.

A further prototype was developed. Its intention was to show how data collection and data evaluation could be supported by a computer-based tool. The framework for the development of this software consisted of the data gathering and analysis as it was done in one field site. It is concerned with investigations of the throughput time of office work procedures, the automatic reconstruction of these, providing suggestions for changes of the organizational structure, and estimations of benefits if a given system were implemented. For example, the impact of changing the authority of credit clerks on the throughput time of the processing of a corporate loan request can be estimated.

The development of these software prototypes was not only a matter of investigating the feasibility of computer-based tools for the OSSAD Methodology, a major reason was concerned with people interested in the OSSAD approach. As a result of many talks with potential clients it turned out that many representatives expected the existence of computer-based tools for analyzing and designing office support systems. In this way, the prototypes developed have been used several times for presentations of the OSSAD Methodology.

A second activity undertaken by some national teams is concerned with providing enhanced instruments and procedures. The framework for this activity is the OSSAD Manual. Note, there is a close relation between the software development and the "instruments & procedures" as shown in fig. 1. It is intended to integrate these components into an enhanced OSSAD Methodology, given time and budget constraints.

In addition, the Methodology and the case studies performed are being used in educational programs. This ranges from courses at universities to particular seminars for interested parties offered by those organizations of the OSSAD consortium engaged in consulting.

Another activity of exploitation is concerned with "simply" applying the OSSAD Methodology in other projects which are, like the field studies, also concerned with the analysis and design of office work in the banking business. On the basis of the experience gained in the banking sector some partners plan to intensify their consultation for the banking area in general.

Another exploitation issue (not shown in fig. 1), which is actually taking place, is related to using the OSSAD framework for other areas or subareas of business. For example, some results and experiences from banking have been transferred, where appropriate, to provide a framework for analyzing organizational aspects of the production sector and the implementation of

CAx systems. While this is not concerned with the development of technical components, many organizational problems in implementing these systems are similar to those found in offices.

6. CONCLUSION

The experience gained in the case studies provide evidence that the OSSAD Methodology has, at least, two characteristics reflecting the needs and requirements of practice. First, the inclusion of organizational and human aspects as well as technical ones in office analysis and design projects is a convincing strength. Second, the design component, which reflects also these aspects, and its close relationship to analysis has been of great significance in launching and conducting OSSAD projects.

Nevertheless, two areas for future work have been recognized where computer-based tools would improve the performance of the Methodology. First, a tool for the configuration of data collection instruments would speed up not only the drawing up of such instruments but also data evaluation in light of options for alternative accomplishments of office work under consideration. Second, it seems feasible to develop a tool based on the modeling techniques provided by OSSAD. This could support an office designer in simulating and evaluating different design alternatives (taking account of organizational and technical dimensions). In addition, such a tool could be utilized for an advise giving facility supporting the process of configuring data collection instruments.

ACKNOWLEDGEMENT

We should like to thank all colleagues within the OSSAD team and are very grateful for the cooperative discussions with the Commission and the experts during the evolution of the OSSAD Methodology.

REFERENCES

[1]   Conrath, D.W. and Dumas, Ph. (eds.) "Office Support System Analysis and Design, a Manual", Institut für Organisationsforschung und Technologieanwendung (IOT), München, 1989.

[2]   Wolfram, G. and Pulst, E. "The FAOR - TODOS intersection – methodology and tools for analysis and design of office information systems," in ESPRIT '88, Part 2, the Commission of the European Communities (eds.), North-Holland, Amsterdam, 1988, 1120-1139.

[3]   Rames J.R., Rolland, C. and Pernici, B. "Office logical design in TODOS,"

in ESPRIT '88, Part 2, the Commission of the European Communities (eds.), North-Holland, Amsterdam, 1988, 1109-1119.

[4] Baron, R. and Beslmüller, E. (eds.) "Field Test Report, Report on OSSAD Applications in Finnish, French, German and Italian Field Sites", Institut für Organisationsforschung und Technologieanwen- dung (IOT), München, 1989.

[5] Beslmüller, E. "Office modeling based on Petri nets," in ESPRIT '88, Part 2, the Commission of the European Communities (eds.), North-Holland, Amsterdam, 1988, 977-987.

[6] Beslmüller, E., Conrath, D.W., Caserta, S. and Dumas, Ph. "OSSAD Methodology: Results of the analysis phase," in ESPRIT '87, Part 2, the Commission of the European Communities (eds.), North- Holland, Amsterdam, 1987, 1077-1090.

[7] Beslmüller, E., Conrath, D.W. and Simone, C. "Bridging the Gap between Users and Vendors of Office Support Systems," in ESPRIT '85, Part 2, the Commission of the European Communities (eds.), North-Holland, Amsterdam, 1986, 1025-1032.

[8] Conrath, D.W., De Antonellis, V., Simone, C., "A Comprehensive Approach to Modeling Office Organization and Support Technology", in Proceedings IFIP Working Conference on 'Office Information Systems: The Design Process", Linz, Austria, August 1988.

[9] De Antonellis, V., Bodem H. and Coccia, A. "Relevance of Models for Office Support Systems Analysis and Design," in ESPRIT '85, Part 2, the Commission of the European Communities (eds.), North-Holland, Amsterdam, 1986.

[10] De Antonellis, V. and Simone, C. "Evaluation Criteria for the Analysis of Office Representation based on Petri Nets" in ESPRIT '87: Achievements and Impacts, the Commission of the European Communities (eds.), North-Holland, Amsterdam, 1987.

[11] Dumas, Ph., de Petra, G. and Charbonnel, G. "Toward a methodology for office analysis: introduction and field study," in ESPRIT '86, Part 2, the Commission of the European Communities (eds.), North-Holland, Amsterdam, 1986.

[12] Sorg, St. "Organisationsanalysen als Instrumentarium zur Ein- führung von Bürokommunikation," presentation at 3. Europäischer Kongress für Büroanalyse + Informations-Management, IOT-Schriften- reihe Nr. 29, IOT, München, 1986.

[13] Sorg, St. and Weber, F. "Erneuerung der Bürokommunikation in Banken - Anforderungen an Methoden für Büroanalyse und -planung, dargestellt an einem Fallbeispiel aus dem Kreditbereich," in Geldinstitute, Heft 2, and Heft 4, 1988.

Project no 295

The Treatment of Office Documents: Bridging the Gap Between Paper and Computer

U. Boes
AEG Electrocom GmbH
Bücklestr. 1-5
7750 Konstanz
West-Germany

G. Maslin, P. Wright
Plessey Research
Roke Manor, Romsey
Hampshire, SO51 0ZN
United Kingdom

H. Keil
Philips GmbH
Forschungslaboratorium
Vogt-Kölln-Str. 30
2000 Hamburg 54
West-Germany

G. Fogaroli
Ing. C. Olivetti & C. S.p.A
Via G. Jervis,77
10015 Ivrea
Italy

Abstract

This paper describes the work performed in ESPRIT Project 295, "The Paper Interface". The project deals with the interaction between the electronic and paper forms of documents. It covers the input from handwritten or printed documents, the manipulation of the electronic version of the document, and the output to paper. The input can be from a document scanner or from a digitising tablet in real time. Techniques have been developed in order to recognise both the textual and graphical parts of the input document so as to enable their subsequent manipulation by computer. Editing and printing facilities have also been developed. The paper highlights the Final Technology Demonstrator which is a prototype office workstation with the "Paper Interface" components integrated into it.

1. Introduction

For centuries paper has been the major means of storing and manipulating information. Since the advent of computer technology and its establishment in everyday life, information has begun to be created, manipulated and exchanged electronically. However, paper remains the prevalent medium for information processing and will remain so for at least the next decade [1]. Consequently, both paper and electronic data processing will continue to exist side by side. Although printed output from computers is well established and OCR techniques have become widespread, other areas such as handwritten input are still in their infancy. The paper interface project aims to provide a full range of tools to deal with the interaction between the paper and electronic worlds.

The five year project that ended in the summer of 1989, splits naturally into four main areas:

FROM-PAPER: This deals with the scanning and analysis of composed paper documents, which may contain multicolour photographical areas, geometric graphics and text.

WITH-PAPER: This deals with direct handwritten or handsketched input onto a tablet or electronic paper.

TO-PAPER: This area considers the rendition of electronically prepared documents onto paper.

SYSTEM ASPECTS: This encompasses all interface components by providing a means for composing and manipulating documents. A special emphasis is given to standards; in particular, the Office Document Architecture (ODA) Standard[2] is promoted.

The project culminated in an integrated demonstrator which covered these four areas. The final demonstrator, which is described in this paper, shows the applicability of the project results in the office. By way of example, a specific scenario, that of the preparation of an estate agent's document, has been chosen. The use of Paper Interface techniques in order to create this document using the final demonstrator, is described in detail.

Four major companies collaborated effectively in the project and a high level of success was achieved. Despite the changes that occur naturally during a project of this duration, all goals have been achieved. The following chapters give a synopsis of the final demontrator and the technology involved.

2. The Final Technology Demonstrator

Although the four aspects of the project have been developed independently, these are brought together in the final integrated demonstrator, shown in figure 1.

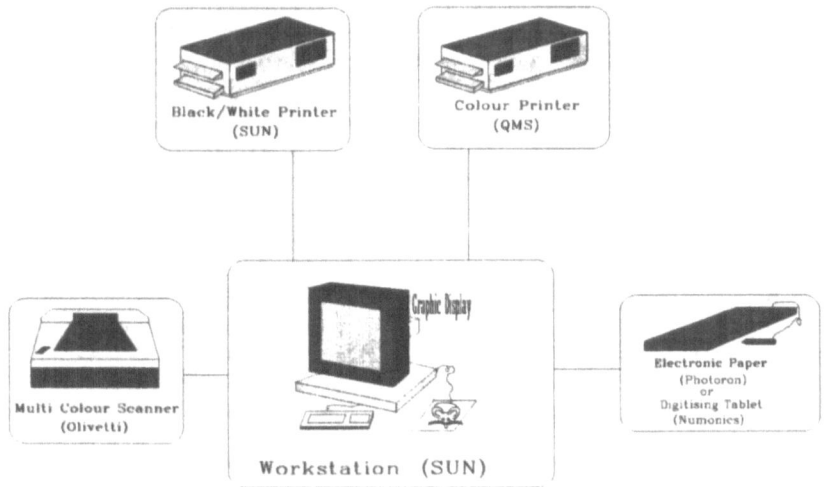

Figure 1: Configuration of the Final Integrated Demonstrator

Most of the processing is performed by a central workstation, and various peripheral devices carry out the I/O functions. Early in the project a SUN machine was chosen as the central workstation due to its widespread use in the technical area. However, porting to European workstations is now taking place. There are four main peripheral devices attached to the workstation. These comprise a Multi Colour Scanner, a digitising tablet or electronic paper, a black and white printer and a colour printer.

The Multi-Colour Scanner, developed in this project, allows scanning of binary, grey-level and colour images up to a resolution of 400 dots per inch and from 1 to 24 bits per pixel. Both flat bed (up to A4) and passthrough (up to A3) are supported. The scanner is connected to the workstation by means of a SCSI interface allowing a speed of up to 1.5

MByte/sec. An A4 page scanned at a resolution of 300 dots per inch and one bit per pixel for instance, can in theory be transmitted from the scanner to the workstation in approximately one second. A colour image on the other hand, coded with eight bits per colour and eight bits per pixel takes 24 times as long. However, the practical maximum scanning speed is approximately 6 documents per minute. An automatic sheet feeder has also been developed.

Direct handwritten input is handled either by a digitising tablet or electronic paper. The digitising tablet is an input-only device which encodes the strokes of handwritten input. Electronic paper, on the other hand combines this function with that of a flat screen visual display. Electronic paper by combining both input and output functions gives much greater scope for natural editing, and allows the manipulation of text and graphics in a way that rivals, if not surpasses, existing keyboard and mouse systems. The electronic paper also enables alteration of existing electronic documents regardless of their original source.

The black and white printer is a standard 300 dots per inch laser printer working to the PostScript standard. The colour printer is a 300 dots per inch thermal transfer device. It also works to the PostScript standard thus ensuring interchangeability.

The functional centre of the final demonstrator is the editor and document manipulation system, which runs on the central work station. The system encompasses only that functionality which is necessary for the Demonstrator. It is not intended to rival commercial editors and those being developed by other ESPRIT projects. The editor is based on the ODA[2] standard and uses NeWS (Network extensible Window System) for the user interface. Also incorporated in the central workstation are the from-paper and with-paper recognisers.

The information exchange between the components of the system occurs via ODIF[2], CGM[3], and plain ASCII. For transmission of images, however, various standards which are available today have been considered. The most promising one appears to be the TIFF standard[4], but currently only true bitmap transfer has been implemented.

The bitmap image from the scanner is transferred to the from-paper recogniser which initially separates the document into three content categories: geometric graphics, raster images and text. Different techniques are used for the recognition or handling of these content categories. Geometric graphics is recognised and encoded as polylines; raster images are stored as pure bitmaps with no further processing; text is recognised by a sophisticated technique resulting in an ASCII character code. Additionally the layout of the document is analysed and the results are stored in ODIF.

The WITH-PAPER recogniser receives encoded handwriting or handsketched strokes from the digitising tablet or electronic paper. These are analysed and compared with a database of handwritten characters or sketched shapes as appropriate. This enables handwriting to be recognised and converted to ASCII characters, and rough hand sketches to be recognised and converted to draughtsman-like quality. Both unconnected and cursive handwriting can be handled by the WITH PAPER recogniser.

The FROM-PAPER and the WITH-PAPER recognisers both utilise post processing techniques to improve their performance. These include lexical look-up, and both syntactic and semantic analysis.

The printing activities concentrated on processing and imaging of colour and binary documents. Algorithms have been developed for rendition of colour, grey-level and binary images. Standardized physical and logical interfaces are used and transformations between them have been implemented.

3. Editing and User Interface Aspects

For a user of the "Paper Interface" system, the document editor plays a central role because it acts as a platform for the integration of the peripheral components. A major point for this integration is concerned with the representation of documents. For this purpose, the ODA standard was utilized.

The structural model of ODA describes two views of a document, namely a layout view and a logical view. The layout view associates the content of a document with physical regions on an output medium, whereas the logical view relates the content to objects of a logical meaning. These objects are used to form a specific layout structure and a specific logical structure. In the case of editing, the logical structure is of relevance. It is this structure that is created and manipulated by the user. This is true because the user usually thinks of a document in logical terms; the layout of the information is of secondary importance.

The document editor described here is built according to the client-server model. It consists of a client process which is implemented in C++, and a server process which was realized on top of the window system NeWS. Both processes are programmed in an object-oriented style. New classes comprising data and methods were designed for the ODA-specific constituents of a document. As NeWS is an extensible window system, part of the overall functionality can be realized in the server. Thereby the following criteria are to be considered:
- Computations should be performed by the client.
- The amount of communication between client and server should be kept minimal.
- The interface between client and server should be simple.

The advantage resulting from this approach is a clear and well-designed software architecture. The client on the one hand can be focused on processing of document structures (editing, formatting, and printing), while on the other hand all display and user interface activities are performed by the server.

Structure and content of the document as well as the results of editing operations have to be visualized on the workstation display. Documents composed of text, geometric graphics, and raster images are rendered in WYSIWYG (What You See Is What You Get) form, thus there is no difference between screen representation and printer output.

Special attention was paid to the integration of the "Paper Interface" subsystems, i.e. the multi-colour scanner, the electronic paper and the recognition units. In addition to the traditional keyboard and mouse, these devices can be used to fill in text, geometric graphics, or raster images into basic logical objects. This feature has been called "pouring" which is originally a concept from the publishing industry.

Besides the functionality of the document editor, the user interface is of major importance for the acceptability of the overall system. In figure 2, a document loaded into the document editor is shown.

Three areas are to be distinguished, namely an "image" area, a "logical" area, and a "document control" area. In the "image" area, the current page of the document is displayed in WYSIWYG form. The specific logical structure of the whole document is shown in the "logical area". Basically two representations are appropriate, namely a table of contents, or a tree structure. Both views allow the sequential order of logical objects and their hierarchical relationships to be depicted. However, the view as a table of contents seemed to be more informative to the novice user, therefore this one was preferred. Each logical object is represented as a rectangular box. The sequential order of objects starts

882

with the logical root of the document at the top of the "logical" area, followed by the immediately subordinate object in the vertical direction and so on. Hierarchical relationships among objects are marked by indenting subordinate objects. The string within the box designates the user-visible name of that object. In order to make a distinction between

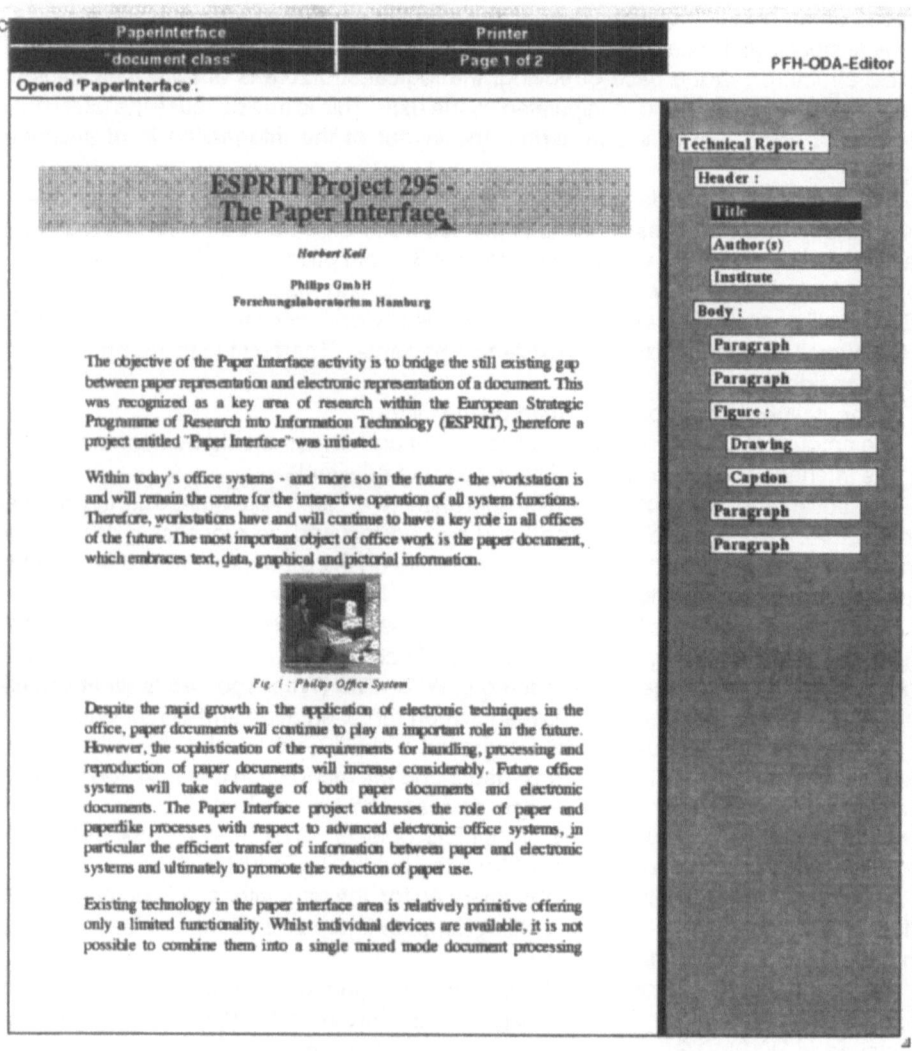

Figure 2: Desktop of the Demonstrator's Editor

composite and basic logical objects, the user-visible names of the composite ones are displayed with a colon at the end. Finally there exists a "document control" area. This area is divided into subareas which serve to display the name of the document, the name of the underlying document class, the printing service, the number of the page currently shown in the "image" area, and system messages to inform the user of what is going on.

4. Recognition - A Key Element for the Paper Interface

Written input is provided either by scanning or by direct input from a digitising tablet. The initial stages of these two processes are quite different and therefore yield very different data, and hence different data structures. The scanning process produces a digitised raster image of the document, whereas a digitising tablet (or electronic paper) produces a series of coordinates recording the travel of the stylus over the tablet. These initial differences have a knock-on effect throughout the subsequent stages of the recognition processes. Consequently, different algorithms have been developed for the FROM-PAPER and WITH-PAPER recognisers. These are therefore described separately below.

4.1 The FROM-PAPER Recogniser

The FROM-PAPER recognition system, or document analysis system, has to process and read bitmap images of documents which have previously been scanned. The documents may consist of several content categories which are processed separately. The objective of automatically reading documents of any kind will be met by a recognition system which uses Artificial Intelligence techniques. A general purpose analysis system was considered[5], but due to its very generality it is unsuited to practical requirements. A more practical approach was therefore pursued leading to a less complex analysis system[6] which is a sub-set of the original concept. This system consists of a control module, a data base and the recognition algorithms as shown in figure 3.

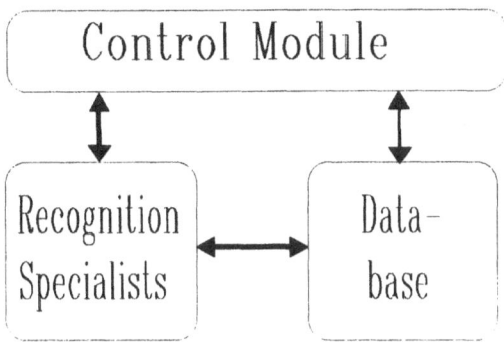

Figure 3: The Document Analysis System

The recognition algorithms are usually known as specialists since each performs a specific task. The specialists are invoked by the control module and access the data-base in order to read, alter, manipulate and write the data. The specialists perform the following functions:
image preprocessing,
connectivity analysis,
discrimination of content categories,
processing of graphics,
layout analysis,

character recognition,

contextual analysis.

The data-base contains all intermediate and final results. It is constructed hierarchically, and thus corresponds to the layout of documents. This is also consistent with ODA. The database is object-oriented and designed to be very flexible, so that objects can be deleted and created, thereby re-ordering the hierarchy. It has been designed in order to fulfil the needs of the recognition processes.

The most important component of the document analysis system is the control module. There are various approaches to the development of a control module. The most well known uses a so-called "blackboard architecture" which stems from HEARSAY-II[7]. The blackboard architecture allows dynamic allocation of specialists. In practical systems, however, the recognition algorithms still run in a predetermined sequence. Unfortunately, pure sequential processing is not sufficient to deal with all problems occurring in document analysis, such as disturbances, dark margins due to copying, handwritten correction marks or even merged characters. Therefore, in order to achieve dynamic allocation of the specialists a new algorithm has been proposed, and tested for the case of merged characters.

This approach assumes that the recognition process can be modelled by a tree structure. Using the uniform cost algorithm[8], an optimal path through this tree is found by invoking specialists at every node and evaluating the costs inferred from the recognition results. Each node represents an object to which a hypothetical value, some attributes and an estimated quality are assigned. An object might be a character, a graphical object or a composed text object. The value and the attributes can be changed again if it is necessary. The branches of the tree are represented by operations to separate or merge objects. The first results obtained using this control structure have been very promising, and therefore a more general application is being planned for the future

One of the most important specialists in the document analysis system is the single character classifier. Also here, tree searching methods are applied successfully. The classifier uses a decision theoretic approach to assign the meaning of a character to its image. This approach led to a polynomial classifier, which can recognise both machine- and hand-printed characters adequately. However, this "one shot" classifier has the disadvantage that the error rate depends on the number of classes to be classified; the greater the number of classes, the higher the error rate. Furthermore, the total mathematical framework has to be computed numerically for each character. Consequently, research work has been carried out in order to obtain a structured classification system which has better performance and is more efficient than the older version The result was a fully hierarchical classifier system. Each node of the classifier tree has a single polynomial classifier. Different strategies have been developed to select the desired branches. This classifier is flexible, and highly efficient, yielding a higher throughput recognition system.

The classifier examines only single characters and issues a list of possible character alternatives which have to be reduced by contextual analysis. Here, different methods have been investigated. Some standard contextual analysis procedures have been applied like the evalution of the geometrical relations of characters in one line. The most favourable approach seems to be the applications of dictionaries for correction of unrecognised text.

The functionality of the document analysis system and the necessary recognition performance will be achieved by a software system independent of the special type of hardware. This software can be implemented either on dedicated hardware, general purpose hardware, or it may be a pure software recognition system. In this demonstration configuration, text recognition is performed with a speed of up to 50 characters a second. For machinewritten text, the error rates are less than 1%; the error rate of human input via keyboard in comparison is approximately 3%. The recognition of handprinted text turned out to be more difficult, the error rate amounts to more than 3%. This is due to the greater variations that usually occur in handwriting. Any future research will therefore address this problem, with emphasis on contextual analysis techniques.

4.2 The WITH-PAPER Recogniser

The WITH-PAPER recogniser consists of two separate parts, a handwritten-script recogniser and a sketched-graphics recogniser. The basic techniques employed in these two systems are, however, essentially the same. For instance, the process involved in recognising a letter 'O' is very similar to that involved in recognising a circle. Sketched-graphics recognition, since it involves a smaller number of distinct recognisable shapes, can therefore be regarded as a subset of handwritten-text recognition. The description that follows therefore concentrates on handwritten-text aspects.

The development of the WITH-PAPER handwritten-text recogniser has progressed through a number of distinct stages:

1. Recognition of a single user's unconnected script with a limited character set (a-z).
2. Extension of the initial system to provide user independence.
3. Expansion of the character set from lower case only, to include upper case characters, numerals etc.
4. Development of recognition algorithms to cater for connected or cursive hand-writing.

This final stage, the recognition of cursive handwriting, is a natural and accepted process for a human but is extremely difficult for a machine. This phase of the project was planned to be investigatory only and there was no intention of developing a working cursive recogniser. However, the degree of success that has been achieved with cursive script is such that a working prototype has in fact been developed.

The cursive recognition system uses a digitising tablet and stylus to encode pen position while writing. The resulting stream of X,Y coordinates is then further encoded using a technique based on the Freeman encoding mechanism used for encoding curves. In this instance, the path of the character as it is being written is quantised into of one eight allowable vector directions. The travel associated with each vector direction is also accumulated in order to produce a final normalised encoding of the complete vector path following the original pen trace. Each vector in the path has a percentage travel measure associated with it. In order to simplify later stages in the recognition process and also to perform a degree of normalisation on the input characters, character slant and baseline drift are removed at this stage. The next stage, which is unique to cursive recognition, is to identify and remove all possible ligatures which join adjacent characters. The vectors which now remain represent the letters, or parts of letters, of the word. By systematically reinserting the ligatures which have been removed, and passing the resulting vector strings through a character recogniser, it is possible to construct a letter net which defines the possible structure of the word. Figure 4 shows a letter net for the cursive word "and".Individual character recognition is performed by comparison of the received vector string with strings stored in a database. Matching is performed initially by vector sequence and then by vector size. This results in a ranked list of alternatives, the top six of which are then used to build the letter net. For clarity, only one alternative is shown at each node in figure 4.

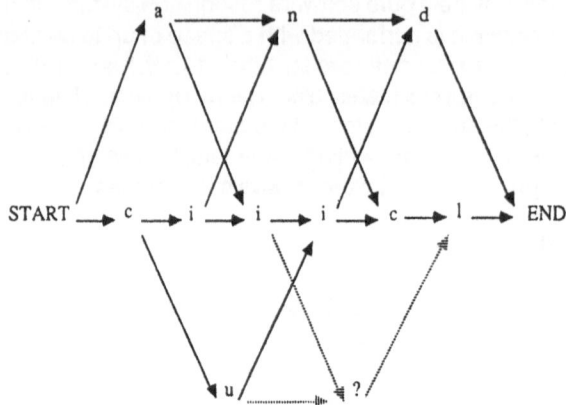

Figure 4: Letter Net for the Cursive Word "and"

Each path through the letter net constitutes a possible interpretation of the input cursive word. Each of these possible words is then passed to a lexical post processor which performs a high speed dictionary look up function. This eliminates all nonsense words and the resulting legitimate words, of which there are usually only a small number, are ranked according to the aggregate ranking of the individual characters. Results for both the unconnected and cursive recognisers have been very good. Error rates of 5% and less have been achieved for the unconnected script recogniser and the system is able to cope with a wide range of writing styles. The cursive recogniser on the other hand has to be trained to a specific user and at present can only recognise the lower case alphabet. Recognition performance
is very dependent upon the size of the dictionary tree used but provided the correct word is included in the dictionary then a word recognition at the 95% level can be achieved. Further work needs to be done in optimising dictionary size and in dealing with misrecognised words. For instance, if one letter is misrecognised and therefore not included in the letter net then the whole word will be unrecognised. The word "misrecogniti?n" for example, would be rejected completely. There is clearly scope for the use of wild cards or a similar technique.

Apart from the development of the lexical post-processor, methods for the syntatic and semantic analysis of sentences and paragraphs have been developed. As a simple example "line" and "graphics" have a stronger intercorrelation than "me" and "graphics". This information can then be used to resolve ambiguities Another area which has been investigated, is the development of a natural editing tool which can fit around the recogniser and permit the user to manipulate the text rapidly using the pen as a selection and pointing mechanism.

5. A Prototype Application

The preceding sections have described the technical aspects of the Final Demonstrator. This section is devoted to an application of this demonstrator in the office environment, and an example application will be described. This application can use to advantage the functions offered by the demonstrator.

An office workstation of the type represented by the Final Demonstrator could be applied in many different scenarios. The example application chosen for the Final Demonstrator is that of creating house sale particulars in the office of an estate agent. This application was chosen because it contains geometric graphics, raster graphics and text, and is assembled using data from a wide variety of sources, some of which would be handwritten. Large numbers of this type of document are produced by estate agents and the use of Paper Interface techniques would be beneficial in this process.

A typical document of this type is shown in figure 5. It includes a standard header which gives the company name, address, logo etc. This could be a colour raster file drawn from disk or it could be part of the preprinted stationery. The document also includes a colour photograph of the house. This is normally a separate photograph which is glued onto the document itself. The use of a scanned image and colour printer would eliminate this process. Estate agents'particulars sometimes also include further graphics in the form of a site plan, or a sketch showing room layouts. The document will also include descriptive text giving information about the location, the size and appearance of each room etc.

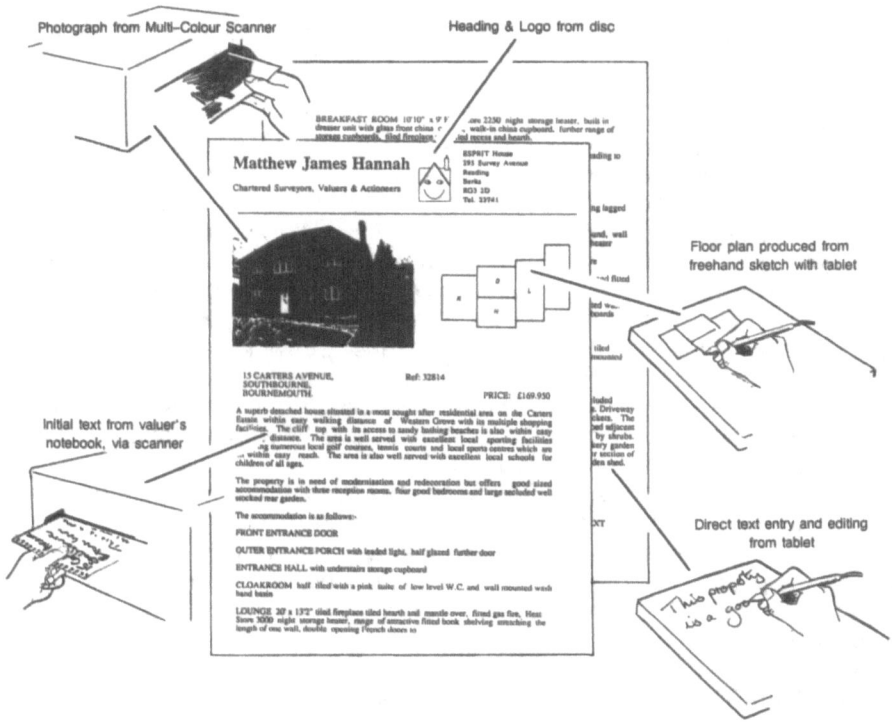

Figure 5: Producing an Estate Agent's Particulars by means of the Final Technology Demonstrator

When an estate agent first receives instructions to sell a property he will make written notes regarding any special requirements of the vendor. He, or his surveyor, will then visit the property, photograph it and make further handwritten notes or sketches. He will next contact the local authority in order to ascertain the rateable value of the property. All this information then has to be integrated into an attractive finished document in order to attract buyers to the property. Figure 5 also shows how these parts come together to form the finished document by using the various component parts of the Paper Interface Demonstrator.

6. Conclusions

The Final Technology Demonstrator described in this paper has been successfully built and a high level of performance achieved. The office environment requires a sophisticated user interface in order to have an effective man machine communication and Paper Interface techniques will form a significant part of this interface in the future. The editor, which is central to most office systems, plays a crucial role since it is the user's direct link with the other parts of the system such as the scanner, OCR, printer etc. While the mouse and keyboard are familiar tools for controlling an editor the use of a tablet or electronic paper is much less familiar. These two devices, however, have potentially much more power for editing. Electronic paper in particular, enables the user to interact with the machine in a totally familiar and natural way. When combined with script recognition it integrates all functions into a single device.

The basic techniques developed during the project in the fields of FROM-PAPER and WITH-PAPER recognition have proved highly effective. These enable handwritten text to be input into the machine, either by scanning a paper image or by writing directly on a digitising tablet. The use of post processing by lexical look up in order to improve the recognition rate has proved highly beneficial and in conjunction with the technique for ligature extraction has enabled cursive script recognition to be achieved. The use of syntactic and semantic post processing still requires further work, but the initial techniques developed so far have shown the considerable potential of this approach. In order for recognition machines to match the performance of the human reader, when dealing with hard to read documents, a high level of syntactic and semantic post processing will be required. The demonstrator has shown, however, that for co-operative users with reasonably neat and regular handwriting, an acceptable level of performance can be achieved using only character and lexical methods.

In the FROM-PAPER area a recognition system has been developed which can directly be applied in OCR and document recognition products. The recognition system is designed to be highly portable and applicable in a variety of products, ranging from high end to low cost systems. The advanced hardware technology of today enables the efficient implementation of the OCR techniques described in this paper. Thus a fast OCR system with high performance can be established. Specific recognition modules could be provided which would have application in areas such as publishing, banking etc.

The WITH-PAPER recogniser has demonstrated the feasibility of real time recognition of cursive script and sketched graphics. This opens the door to several markets where keyboards are not particularly suited e.g. in a vehicle, on a building site, on a train etc. Particular users who might benefit could include emergency services, salesmen, architects, surveyors etc. In the office, the technology could be used to provide a portable memo-pad, a means for unobtrusive communication in meetings, or a method of rapid graphics entry. Also, the area of form filling would seem to provide almost limitless opportunities (for example, home shopping, point-of-sale etc.).

The TO-PAPER part paid special attention to the growing importance of page description languages, in particular PostScript. Investigation resulted in the design and implementaion of an ODIF-to-PostScript conversion program. With regard to the widespread use of PostScript in the printer area, the availability of such a conversion is essential for today's office systems.

References

[1] Lahr, R.J. (1984) 'The Non-death of Paper', Journal d'Electronique et de Microtechnique.

[2] ISO - International Standard 8613, (March 1988) Information Processing-Text and Office Systems-Office Document Architecture (ODA) and Interchange Format.

[3] ISO - International Standard 8632, (November 1985) Information Processing Systems-Computer Graphics Metafile (CGM) for the Storage and Transfer of Picture Description.

[4] Andrews, N. und Frey, S. (1988) 'TIFF: Der neue Graphikstandard', Microsoft System Journal, Juli, August, pp.19 - 23.

[5] Bayer, Th. (1988) 'Dokumentinterpretation und Analysestrategie in einem Frame-System', Proceedings 10th. DAGM Symposium, Zürich, pp. 284-290.

[6] Angele, J., Boes, U. and Schultes, N. (1987) 'Moderne Datenerfassung: Automatisches Lesen von Dokumenten', Informatik-Fachberichte 156, 17th. Annual Conference of the GI.

[7] Hayes-Roth, B. (1985) 'A Blackboard Architecture for Control', Artificial Intelligence Journal 26, pp. 251-321.

[8] Pearl, J. (1984) Heuristics: Intelligent Search Strategies For Computer Problem Solving, Addison Wesley Publishing Company.

Project no 385

HUMAN FACTORS IN INFORMATION TECHNOLOGY - RESULTS FROM A LARGE COOPERATIVE EUROPEAN RESEARCH PROGRAMME

H.-J. BULLINGER, K.-P. FÄHNRICH AND J. ZIEGLER
Fraunhofer-Institut für Arbeitswirtschaft und Organisation
(IAO)
Nobelstraße 12
D-7000 Stuttgart
W. Germany

M.D. GALER
HUSAT Research Centre
The Elms", Elms Grove
Loughborough, Leic. LE11 1RG
United Kingdom

Abstract

The project 'Human Factors in Information Technology' (HUFIT) is an extensive project of cooperation in the European ESPRIT programme, involving eleven companies and research institutes in eight European countries. The project has two major objectives: it aims at improving the design of IT products by increasing the awareness of Human Factors issues and by providing methods and tools for a user-oriented design. The second major objective is to further develop user interface techniques, especially for multimedia and multimodal interfaces, and to provide tools for prototyping and implementation of these interfaces.

This paper presents the areas of research in HUFIT. The focus is on the presentation of the major outcomes of the project: the HUFIT Toolset and the software tools INTUIT, DIAMANT and MULTEX.

1. Introduction

Human Factors (HF) and user interface design have become increasingly important for the development of Information Technology (IT) products. This issue has been taken up in the European HUFIT project, which aims at providing knowledge, methods, and tools for the development of products, which are more closely matched to the tasks, requirements, aspirations, and characteristics of users, and which provide effective, usable, and flexible user interfaces.

The preparatory work for HUFIT began in 1982 and 1983 by a study on "Ergonomics in Information Technology (IT) in Europe - A review" (Shackel 1984). A complementary study, comparing Human Factors knowledge in Europe and world wide, was performed in Germany (Fähnrich 1985).

The reports analysed major shortcomings in research and its application in Europe. Parallel to this study, members of the European IT industry were forming a consortium for a large multinational project in information technology ergonomics. The Fraunhofer-Institut für Arbeitswirtschaft und Organisation (Institute for Industrial Engineering) IAO, Stuttgart, took on the main management of this work. The project began in 1984. The industrial project partners are Bull, ICL, Olivetti, Philips, and Siemens along with their respective software ergonomics laboratories and external academic experts acting as consultants. Also represented in the project are Münster University, West Germany, The University College Cork, Ireland, the Piraeus Graduate School of Industrial Studies, Greece, and the University of Minho, Portugal. The scientific and technical content of this sizeable research programme is being coordinated by the IAO and HUSAT Research Center, Loughborough, UK. HUFIT takes on an integrating and co-ordinating function for the area 'Human Factors and IT products' within ESPRIT (the European Strategic Programme for Research and Development in Information Technology).

2. Areas of research in HUFIT

The programme is made up of three project areas, each area with different but complementary objectives.

2.1. Area A: From Conception to Use - An Integrated Hf Contribution to Product Design and Development

The initial proposition, supported by preliminary research, was that human factors knowledge should be introduced as early as possible into the IT product design process. Therefore, this project area is concerned with IT products from the moment of their conception, right through the design and development process, their installation and use. Its particular contribution is to the development of an integrated human factors input to this whole process. Specific areas of work are:

- analysis of the development processes for IT/products in different companies;
- development of an integrated toolset for analysis, design and evaluation;
- development of a HF support environment for designers;
- gathering human factors knowledge in a public database.

2.2. Area B: Advanced Human-computer Interfaces

The second area of interest of the project is concerned with the interaction of the user with the system, involving theoretical and empirical analyses of advanced techniques of human computer interaction.

Interaction techniques, such as direct graphic manipulation and speech input/output, were empirically analysed in order to develop guidelines for their use in interface design. A major research issue in the project is the development of multimedia and multimodal interfaces integrating different input/output media and corresponding interaction techniques. The project area is developing User Interface Management System (UIMS) components for rapid prototyping and implementation of such interfaces. Major activities are:

- modelling of human computer action for analysis of learning requirements, efficiency etc.;
- empirical studies and development of design guidelines for different types of interfaces;
- tools for design and implementation of integrated direct manipulation and multimedia

user interfaces;
– implementation of demonstrator systems.

2.3. Area C: Transfer of Human Factors Knowledge

The third area is concerned with the transfer of HF knowledge, i.e. a broad dissemination of HF knowledge in the European IT community. This part of the project aims at raising the level of awareness, knowledge, and practice of human factors in the European IT industry. This is achieved by the development of practical human factors tools for designers in IT supplier industries. In order to support human factors consideration at both design team and managerial levels seminars and workshops are provided. In addition, human factors consultancy is offered to assist companies to tailor the tools precisely to their product life cycle. It is also offered to companies wishing to set up and operate Human Factors laboratories.

3. Specific Outcomes of the HUFIT Project

Specific outcomes of the HUFIT project to be discussed in this paper are:

– The HUFIT Toolset.
  Human factors tools and techniques are being developed for use by designers throughout the stages of product design.
– INTUIT.
  A design support system to deliver human factors methods and guidance integrated in a structure analysis software engineering approach.
– DIAMANT.
  The Rapid Prototyping Toolset.
– MULTEX.
  Tools and demonstrators for multimedia integrated interfaces.

An overview on research results gained in the past may be found in Fähnrich, Ziegler (1987). In addition, the following tools and methods have been developed to support the design process of IT products:

- methodology for the identificaton of user and task characteristics (Jeffroy, 1988)
- the human factors evaluation toolset (Novara et al., 1987),
- the documentation tool set (Allison et al., 1987),
- the computer user satisfaction inventory (Kirakowski, 1987), and
- IT support for managerial tasks (Laios et al., 1988).

In the following, the HUFIT Toolset and the design tools INTUIT, DIAMANT, and MULTEX will be described in more detail.

3.1. The HUFIT Toolset

A user-centred design approach has been taken in the development of human factors tools for use in product design. The users of the human factors tools are the actors in the design process in Marketing, Product Planning, Development, Quality Assurance, and so on. Hence the tools developed in the project must not only fit the requirements of these users within the constraints of the product life cycle but must also be usable by people who do not necessarily have human factors expertise.

The HUFIT toolset is being developed in close collaboration with the users. It currently comprises sets of human factors tools which address particular issues in user centered product design.

3.1.1. The User Requirements Toolset.

The design of successful IT products relies upon a good understanding of the target users and their tasks. A set of tools called the User Requirements Toolset has been developed by HUSAT to help planners and designers of IT products effectively to use information about users and their tasks. The User Requirements Toolset provides a structured method for looking at user and task information and generating product requirements. The User Requirements Toolset comprises these core tools:

- User Mapping
- User Characteristics
- Task Characteristics
- User Requirements Summary

The four core tools are paper based methods for use during the Planning stages of the product life-cycle. An on-line version, the User Requirements Program is also available on Apple Macintosh^{TM} using Hypercard^{TM}. However it is anticipated that companies will wish to produce their own versions to be compatible with other on-line design tools in use in their companies.

User Mapping - identifies all user groups for the planned product and examines the benefits and costs for those groups of using the product. It is intended to be used very early in the Planning stage when it can help to decide whether or not a product idea is viable and begin to target design and marketing activities more accurately.

Later in the Planning stage of the process the following pair of tools is used:

User Characteristics - enables designers to describe the characteristics of the user groups and derive the product requirements resulting from them.

Task Characteristics - enables designers to describe the task characteristics of the user groups and to derive the product requirements resulting from them.

The requirements generated by these tools are collated in the **User Requirements Summary** which arranges it into a format more suitable for designers.

The on-line version of the tools called the User Requirements Program is intended for use alongside the systems analysis and task analysis methods used by the designers and are complementary to them. The tools are intended for use by planners and designers without any previous human factors training or experience. They present a process to follow, with supporting explanation and examples. The output of the Program is of use to marketeers, designers and evaluators of the product at various stages in the design process.

The demonstration version of the tool was developed on a Macintosh^{TM} using Hyper-card^{TM}. Working versions will be implemented using database software. It is expected that companies will wish to produce their own versions to be compatible with other on-line design tools.

The Program follows a form filling pattern to retain the benefits of structure perceived in the paper version of the tools. Each title for a field to be filled by the user is supported by HELP which contains an explanation of the term and examples to assist in its completion.

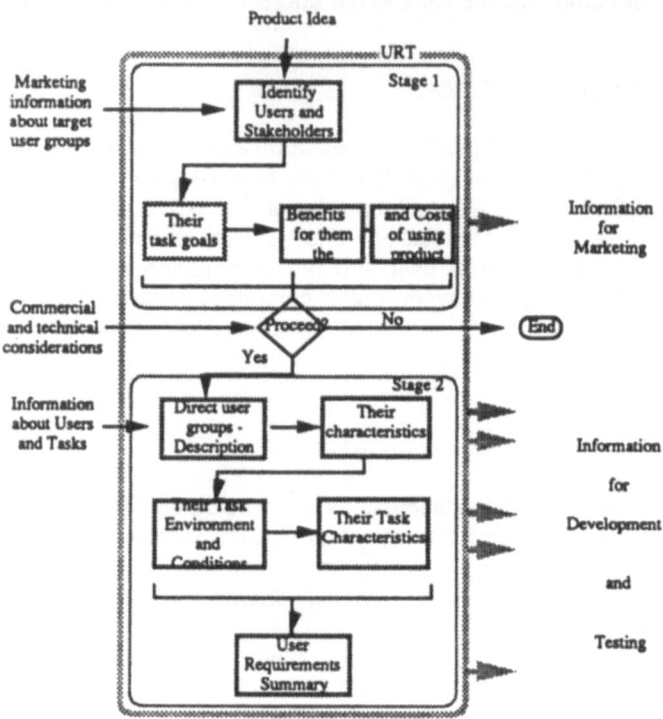

Figure 1. User Requirements Core Tools

3.1.2. Specification and Design Tools.

Having completed the User Requirements Specification phase, the design team must then attempt to amalgamate these requirements with the proposed technical specification and, on approval, move to the design phase. In the design phase they will particularly address the user interface design issues which will have the greatest bearing on the system's usability.

The Specification and Design toolset comprises the Functionality Matrix,m the User-Computer Interface Design Tools and the on-line version Dialogue Adviser. The Functionality Matrix supports user and task issues in the development of an enhanced functional specification by presenting complex human factors, business and technical data in an easily understandable matrix format presented in both paper and spread sheet forms.

Having identified a range of appropriate product functions, the software desinger has to build a usable interface to allow the user to access them. The User-Computer Interface Design Tool offers a step-by-step approach to interface design in a concise format.

An on-line version of the User Computer Interface Design Tool, called Dialogue Adviser, has been implemented in Hypercard[TM] on the Macintosh[TM]. This comprises two parts:

an interactive checklist where the designer specifies the sorts of activities the user will be carrying out with the system and a detailed reference section where the designer can obtain advice about constructing and implementing the various forms of dialogue style. In hypertext fashion the designer checks boxes and then presses a button to jump directly to the appropriate design advice. A facility is incorporated for compiling the selected sections of advice into a brief report which can then printed for use off-line. Planned versions will also include animated examples of dialogue components such as pop-up menus, dialog boxes, forms and iconic buttons. An example screen from a prototype version of Dialogue Adviser is shown below:

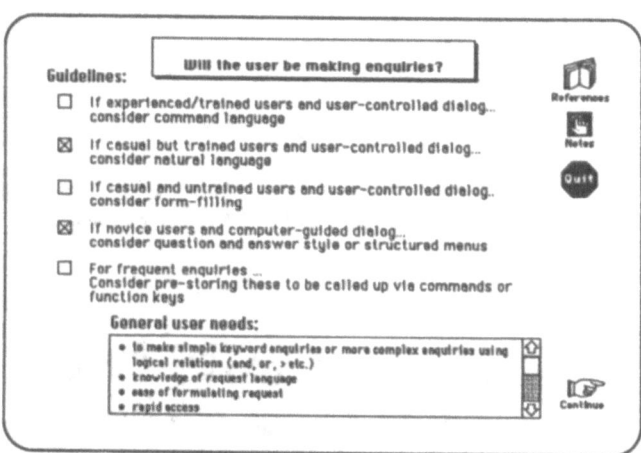

Figure 2. Example screen from Dialogue Adviser prototype

The HUFIT Toolset provides an integrated human factors input to the design of IT products by enabling designers themselves to take a user-centred approach. The Toolset has been well received in Europe as a major contribution to the design of usable IT products.

3.2. INTUIT

The investigation of design processes used by IT companies showed that the classic forms of knowledge transfer, such as publications, handbooks, design guidelines, have an only limited radius of influence on industrial practice. Another approach is to create a computer-based tool allowing integration with other software development tools available to the IT companies. This idea forms a further part of the HUFIT project, whereby a support system for product designers is to be developed. It consists of four main parts:

- a software engineering environment supporting structured analysis and graphical system specification at different levels;
- a knowledge-based system that has acess to several HF data and information sources;
- simulation and evaluation modules;
- links to a prototyping component.

These modules are integrated in the HUFIT product INTUIT (Russell et al. 1988, Pettitt 1989).

INTUIT is a software development tool which uses knowledge engineering techniques and encapsulates expert human factors knowledge to assist a design team. It provides a CASE tool that promotes good human factors practice and delivers contextually relevant design guidance. INTUIT is a tool for user-centred design in the construction of usable application software and in particular in the building of user interfaces (currently restricted to conventional alphanumerical interfaces). Both the display formats and dialogue components of interfaces can be produced with assistance from the system. The technology being used is frame and object based and in consequence offers the considerable benefits of inheritance of attributes and properties as the user steps through the design process.

INTUIT could be used as the basis for other aspects of application design. It could, for instance, be used to perform other activities such as data base design or as a basis for rapid-prototyping. It is intended to provide in-line support for the development process to allow human factors expertise to be applied to the development process directly. INTUIT is intended to operate proactively using human factors expertise to generate usable systems.

Figure 3 shows the logical processes carried out and the basic components within INTUIT, which is based on the Structured System Analysis and Design Method (SSADM).

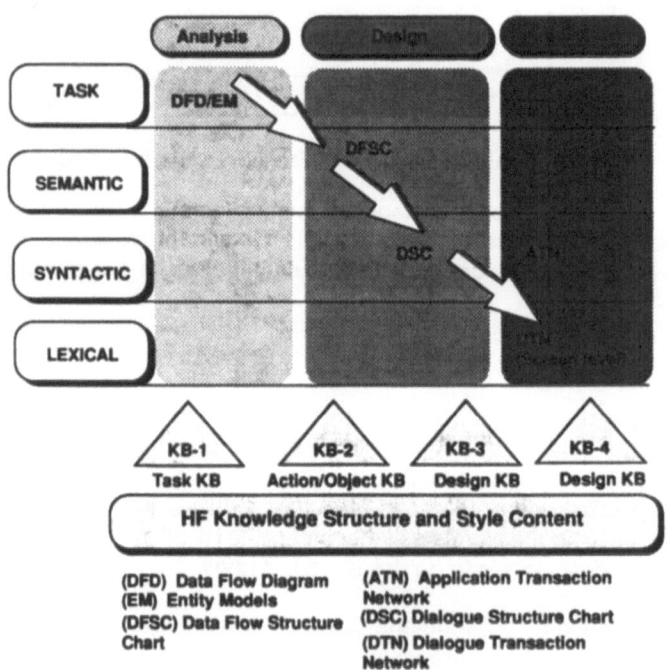

Figure 3. Logical processes and basic components within INTUIT

SSADM is a method widely used within the UK and adopted as a standard by the UK government. This standard approach to the analysis and design of computer based information systems provides *structural standards*, which define the structure of a development project in terms of explicitly defined tasks and outputs. It also provides *procedural standards*, which define the set of proven techniques and tools, plus detailed rules and guidelines on when and how to use them. Finally, SSADM provides *documentation standards* which provide the means of recording the products of a development activity at a detailed level.

The system consists of four main functional subsystems. These provide a set of design constructs, the Human Factors Knowledge Base (KB), the referencer, and the user interface.

The basic mode of interaction is by direct manipulation of the application objects presented. INTUIT possesses a powerful graphical *user interface*. The interface uses multiple windows in which both text and graphics (particularly diagram techniques) can be displayed and manipulated using a keyboard and mouse.

One of the windows is used for the *referencer* of guidelines and examples. INTUIT offers a contextually determined selection of advice materials from which a user may choose. The selected guideline or example is then displayed in a special window. Examples that demonstrate dynamic aspects of the interface may be interacted with by the user.

The objects in the *design constructs* subsystem represent the design information required by the development process. These objects are similar to the objects one finds in a data dictionary. They are defined by a rich frame-based schema that describes not only their structure but also the operations that can be applied to them.

The *human factors knowledge* subsystem is founded upon a taxonomy of interaction components. These knowledge bases may be regarded as libraries of components that can be used in the design session. Currently these are dialogue task actions, and interface actions and objects.

3.3. DIAMANT

The aim of this tool is to enable and to support prototyping and rapid development of direct manipulation user interfaces for experimental, evaluation, and implementation purposes. The tool's support for prototyping range from quick interface demonstrators to fully functional system implementations. The tool has been developed on the basis of an extensive survey of User Interface Management System (UIMS) and dialogue modelling approaches.

UIMS provide an architecture and components for defining and implementing user interfaces. They aim at a separation between user interface and application which is realised to a different extent in the different approaches. This separation allows to build specific productivity tools for implementing the interface components and to modify and port them more easily. Moreover, interface implementation becomes feasible for persons who are not programmers but, for example, HF or application experts.

Two versions of a UIMS called DIAMANT (Dialogue Management Tool), have been developed.

DIAMANT I is implemented on a PC and allows to rapidly develop applications with direct manipulation interfaces. The tool handles windows, icons, buttons, menues and fields. After designing the different information windows containing text, graphics, icons, buttons, and data fields, the designer is provided with a comfortable way of specifying the dialogue. The underlying mechanism is an event-response model. User events (user input) and the corresponding system actions (like opening windows, actions on button presses) can easily be specified in a tabular form. A number of different commercial applications have already been developed with this tool.

In DIAMANT I, all information visible to the user has to be specified prior to using the system. DIAMANT II (Koller et al. 1988, Trefz & Ziegler 1989) has much more powerful mechanisms for handling dynamic objects (like the varying number icons in a directory window). A User Interface Description Language (UIDL) has been implemented, which consists of object-oriented event handlers which can communicate with each other. The UIDL of DIAMANT II is an object-oriented programming language, which was designed to be well suited for the implementation of dialogue managers based on an event driven model. In such a model, each action in the user interface is considered as a direct response to a recognized event. Event driven models are well suited to describe user interfaces, which require no complex command parsing but allow multiple command threads or even parallel input.

The tool is well suited to distributed systems, for example for managing the communication between an interface running on one machine and a data base running on another. High-level graphical definition methods are currently being developed which facilitate the design and modification of the interface. Standard solutions for good interface design can be embedded in this system. A prototype version of DIAMANT II, running under UNIX and X Windows on SUN, has been completed. Figure 4 illustrates the architecture of DIAMANT II.

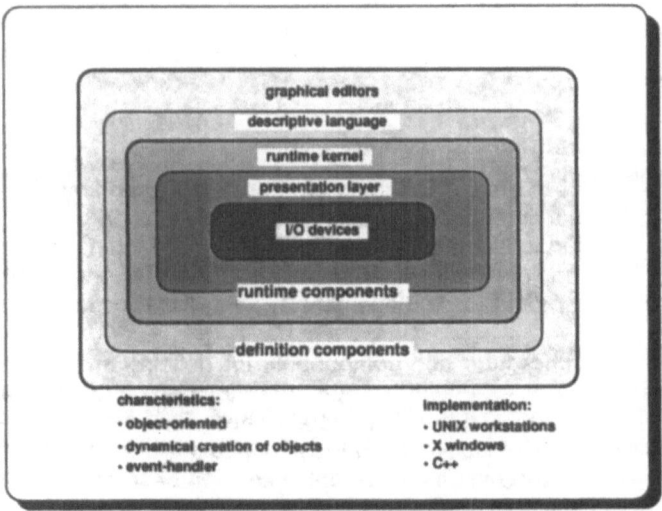

Figure 4. Architecture of DIAMANT II

The basic architecture of DIAMANT II is very flexible, so that it is quite easy to extend the system by providing new toolkit classes. The UIDL is well suited for combining the predefined toolkit classes and for defining their communication with other components.

Future work will address the issues of introducing manager components for new media, like video and animation, and graphical tools sitting on top of the UIDL, as well as facilitating the design and modification of relevant, often-used user interface techniques, such as desktop modelling.

3.4. MULTEX - A Multimedia Expert System

Future user interfaces will draw heavily on the availability of multimedia presentation of information and on the integration of different interaction techniques connected to these different media. From a human factors point of view, it is important to understand what these media can contribute in the context of the users task and how the user can handle the different types of information. From a technical point of view, tools are needed to produce multimedia material and to adapt the system to the particular application and user needs. The aim of the project MAITRE (multimedia, adaptable interfaces and tools) is therefore to provide guidelines and software tools for the development of multimodal integrated interfaces combining different forms of information presentation (text, graphics, images, animation, video, speech) and different forms of dialogue techniques.

There are numerous application fields which can benefit from the usage of multimedia interfaces, like education, training, and many others. There is already a number of activities in the field of multimedia documents. As a field of investigation diagnostic expert systems have been chosen where the user can get advice on how to detect fault conditions and how to repair the target system. The system MULTEX (multimedia expert system) demonstrates these capabilities for the example of bicycle repair (Koller 1988). The system asks for states of the bicycle and gives instructions to the user either on how to check the state of the device or on how to repair a detected fault. The system follows a strategy to minimize the overhead of work during maintenace using heuristic rules.

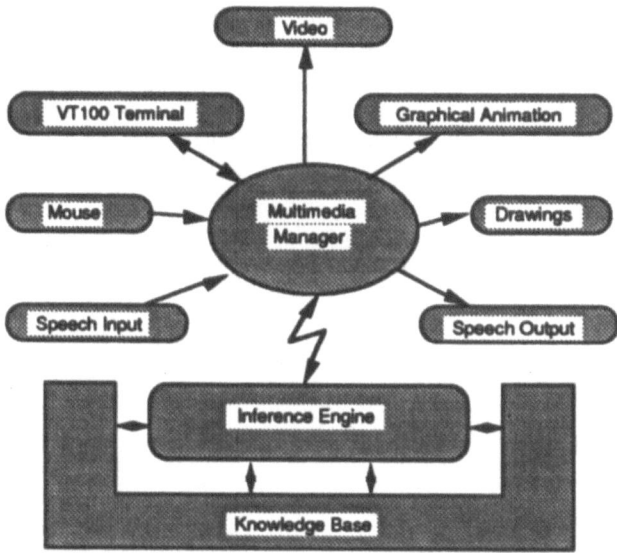

Figure 5. Architecture of MULTEX

Once the system is started, the first question will be asked, such as "Turn the front wheel and check whether the rear light is on". This question is printed on the screen, as well as spoken by the system (synthetic speech). Possible answers will be displayed by soft buttons. The user may answer by speaking, by clicking a soft button, by typing, or by pressing a function key. The system repeats the answer by voice medium. During a session, the system may give the instruction "Unmount the rear light reflector" and starts an animation sequence demonstrating how to do it. Longer sequences of repair actions are shown by video which can better convey complete work procedures but are less focused than e.g. animated graphics.

Although this simple example is used so far for demonstration purposes, the system is not restricted to the diagnosis of bicycles. It is well suited for most kinds of industrial machine diagnosis. In order to use it for other products, it is necessary to produce a new knowledge base, new video sequences, and new animation sequences.

On the basis of this demonstrator a collection of guidelines and appropriate software components are developed which allow the designer to choose appropriate combinations of media and interaction techniques suited for the particular problem. The development of the software tool is based on the dialogue management system DIAMANT II and will include the numerous findings on specific interaction techniques like speech interaction, direct manipulation, dialogue structuring etc. from other activities in HUFIT.

4. Summary and Outlook

HUFIT is an approach to influence all relevant stages and issues of IT product design from a Human Factors perspective. This approach comprises activities to strengthen user-oriented aspects in the complete design process as well as methods and tools which can be used for the design and implementation of user interfaces. Some of these tools have already influenced the design of industrial products or have become products themselves.

Figure 6 shows some outcomes of development which have already lead to products.

The project runs until the end of 1989. In the final project phase, most of the activities are directed towards an completion of the tools and a stronger integration of the results.

One major mechanism for this integration is the design support system INTUIT which will incorporate relevant methods and tools developed in the project.

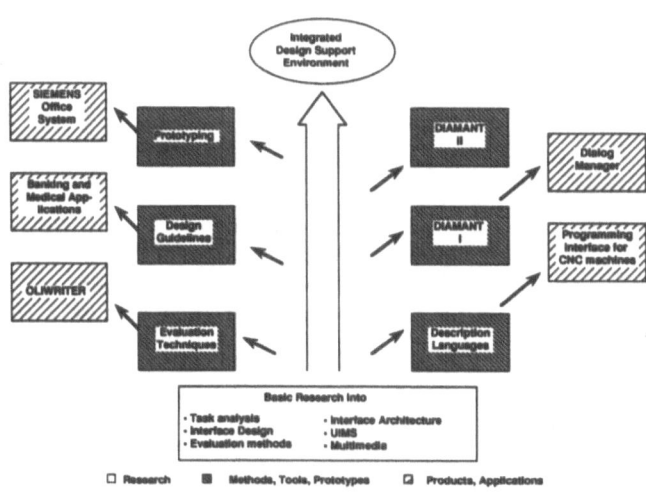

Figure 6. Products and Applications developed from HUFIT results.

5. Acknowledgements

The authors would like to thank Astrid Beck and Petra Walderich for their help in preparing this paper.

6. References

Fähnrich, K.-P. (1985): European Human-Factors Laboratory in Information Technology. In: Bullinger H.-J. (Editor), Proc. "Human Factors in Manufacturing", IFS Publications UK.

Fähnrich, K.-P. Ziegler, J. (1987): HUFIT (Human Factors in Information Technology). In: Salvendy, G. (Editor), Proc. "Cognitive Engineering in the Design of Human - Computer Interaction and Expert Systems", Amsterdam.

Jeffroy, F. (1988): Course of Action - A Theoretical and Methodological Framework for Analysis of Users and Tasks Characteristics, ESPRIT Project 385, Deliverable A4.3/4, HUFIT/5-BULL-07/88.

Kirakowski, J. (1986): A Flexible Interface to a Real Task and iots Reception by Naive and Non-Naive Users, ESPRIT Project 385, Deliverable A6.1a, HUFIT/HFRG-1-9/86.

Koller, F., Trefz, B., Ziegler, J. (1988): Integrated Interfaces and their Architecture, ESPRIT Project 385, Working Paper B3.4/B5.2, HUFIT-23-IAO-11/88.

Laios, L., Lioukas, S., Giannacourou, M. (1988): Towards a Taxonomy of Managerial Planning Tasks for SME's. In: Bullinger, H.J., Protonotarious, E., Bouwhuis, D., Reim, F. (Eds.): EURINFO 88, 1st European Conference on IT for Organisational Systems, Athens. North Holland, 410-415.

Novara, F., Bertaggia, N., Dillon, A., Bonner, J. (1987): The Evaluation of Products Using the Usability Methodology with Proposals for Product Development, ESPRIT Project 385, Working Paper A5.3a, HUFIT/04-OLI-11/87.

Pettitt, P. (1989): INTUIT - A knowledge and structured design approach to user-centred application design In:Commission of the European Communities, Directorate-General TELECOMMUNICA-TIONMS, INFORMATION INDUSTRIES and INNOVATION (Ed.): ESPRIT '89, Proceedings of the 6th Annual ESPRIT Conference, Brussels. North Holland,

Phillips, K.E. (1987): The 1987 Computer Human Factors Classification and Thesaurus and the Collation of Computer Human Factors Litearature, ESPRIT Project 385, Working Paper A3.1d/A3.2c HUFIT/12-HUS-11/87.

Russell, A.J., Elder, S.A., Pettitt, P., Coney A. (1988): Functional Specification for INTUIT, ESPRIT Project 385, Working Paper, HUFIT/2-ICL-03/88.

Shackel, B. (1984): Ergonomics in Information Technology in Europe - A Review. HUSAT Memo No. 309. Report for the Commission of the European Communities.

Trefz, B., Ziegler, J. (1984): DIAMANT - Ein User Interface Management System für grafische Benutzerschnittstellen. In: Maaß, S., Oberquelle, H.: Software-Ergonomie '89. Stuttgart: Teubner.

Project no 385

INTUIT, a knowledge and structured design approach to user-centred design.

Phil Pettitt.
International Computers Limited,
Lovelace Road,
Bracknell, RG12 4SN
United Kingdom,

Abstract

This paper describes INTUIT, a prototype of an integrated human factors tool for software designers. Human factors expertise has historically been made available to the designer as either guidelines or in terms of product evaluations. Neither of these forms of delivery are ideal. INTUIT adopts an alternative approach of providing a CASE tool that promotes good human factors practices and delivers contextually relevant design guidance. The emphasis is upon providing human factors expertise proactively and in-line with the design process.

1. Introduction.

INTUIT is a software development tool. Using knowledge engineering techniques, it encapsulates expert human factors knowledge to assist a design team. It is being developed as part of HUFIT (ESPRIT project 385) which is looking at improving the human factors of systems by providing appropriate methodologies and tools to designers. [1] provides a fuller description of the HUFIT project as a whole and the work of our partners from Bull (France), Siemens (Germany), Olivetti (Italy), Fraunhofer IAO (Germany) and HUSAT (UK). ICL's major contribution is in the development of INTUIT, the implicit provision of human factors expertise in tools to be directly used by a software designer. [2]

INTUIT is a tool for user-centred design. The driving concern of the tool is in the construction of usable application software. Its main emphasis is upon the building of user interfaces. However, the interaction aspects of application design are not completely separable from the rest of the design process.

- INTUIT will produce user interface designs. Both the display formats and dialogue management components of interface design can be produced with computer assistance from the system.

- INTUIT could be used as the basis for other aspects of application design. As interface and other design issues are not completely separable, the system will have to cater for design information that is not solely relevant to user interaction. Therefore, there exists an opportunity to perform other activities such as data base design.

INTUIT is intended to provide **in-line** support for the development process. Human factors expertise should be applied to the development process directly. A designer should exploit it as a natural part of the job otherwise it is too easily ignored or perceived as mere overhead to the work in hand.

INTUIT is intended to operate **proactively**. Human factors expertise should be used to **generate** usable systems. A reactive system for evaluation has a role but emphasis on this aspect merely leads to rework and wasted effort in the development process.

HUSAT's work on human factors tooling complements the INTUIT approach. Their tools concentrate upon the "softer" or earlier stages of development where paper based tools are more appropriate but some of this information can later be transferred to INTUIT to contribute to the design process.

2. Requirements.

An INTUIT-like tool could support a wide range of tasks within the software development life cycle. Human factors knowledge needs to be applied both to high and low level constructs. Ideally, the entire product life cycle needs to be catered for. However, to achieve results in a timely fashion, the requirements for INTUIT have been carefully scoped.

2.1. User Role.

INTUIT is particularly dedicated to designers. Support for this group of users provides a high degree of leverage in supplying improved usability in information products. This group also presents a reasonably well understood task to which support tools can be applied. INTUIT could later be diversified to support the other groups of users.

A software designer may use structured methods or prototyping. These are both modern and acceptable ways in which a designer may function. Hopefully, the designer is not stuck back in the "hack and wack" era and it is improbable that formal, mathematical methods are in use. At each of the latter extremes there may be little recognition of human factors expertise anyway.

- Structured methods provide the tools and techniques for transforming a concept into a product. The spirit of such methods is to impose order and mechanisability to the development process. In so doing, they provide the opportunity to attach human factors expertise directly to the design constructs and transformations used by such methods.

- Prototyping provides the feedback for evaluation of the designs constructed. In particular, interface evaluation is a popular aspect of prototyping. This style of working does not presuppose the use of structured methods but is an invaluable adjunct if exploited correctly to evaluate evolving designs.

2.2. System Role.

As a CASE tool, the system needs to take account of the technology relevant to software development. This technology consists of both the notations (how the design is documented) and techniques for moving from one stage in the development process to the next. Only a subset of all those available can be realistically catered for. This subset must be adequate for both demonstrating the use of INTUIT and its relevance to current software development practice.

The system provides human factors support for interface design by a process of step wise refinement. It provides a window onto the software development process from a user interaction perspective. The initial representation analyses the task whilst the other three loosely correspond to the semantic, syntactic and lexical levels of interface design. A designer would be expected to transform the initial task analysis into some high level representation of a dialogue which in succeeding steps would be translated into low level screen formats and key strokes. The process adopted is based upon Jackson's system design techniques as embodied in the SSADM methodology for ICL QUICKBUILD implementation [3,4,5]. INTUIT's support is integrated into this design process.

– The task analysis is represented as data flow diagrams. These are augmented by the use of entity/relationship models to represent the application objects/data of the task domain. One of the first design operations performed upon the task analysis is to identify the user roles and events implied by the data flow diagrams.

– The semantic level of interaction description is represented as data flow structures. This provides an abstract level of dialogue organisation in terms of the data flowing between the user and the computer system. The data flow structure can be derived from the data flow diagrams of the task analysis by examination of the data flow associated with the application's events and represented using structure charts.

– The deep syntactic level of interaction description is represented as dialogue structures. The information in the semantic level data flow structures is re-grouped to provide a dialogue structure that meets the requirements of chosen interaction styles and usability. The dialogue structure is primarily represented using structure charts although a corresponding generalised transition network can also need to be generated for subsequent refinements.

– The lexical level of interaction description is represented as screen formats. Information about the structure of information represented in preceding stages is translated into choices of interface objects. The surface syntactic level of interaction description is also required, it will constitute an elaboration of the generalised transition networks derived in the previous phase. The transition networks used to describe the details of dialogue structure are based upon the notation developed by Siemens in their work within HUFIT on dialogue design [6].

For rapid evaluation of evolving designs a prototyping facility is provided. Once a dialogue design has been refined to the lexical or surface syntactic level it can be animated on the INTUIT workbench. These animations make direct use of specific interaction knowledge and encodings of house style. The prototyping activity can thus be supported by a structured methodology in which design information is properly captured and exposed to human factors guidelines.

Interaction styles knowledge is filtered to ensure that it is relevant to the current design task. Interaction styles knowledge will only be relevant to particular types and quantities of application objects/data. Certain kinds of user operation also effect interaction relevance. Part of the task of INTUIT is to ensure that only the knowledge relevant to the designer's current problems is provided, or at least that the designer is not bothered with that which is irrelevant.

– User characteristics may influence the interaction styles relevant to a design. Particular groups of users have particular skills and work habits that affect how they might interact with an application. Thus, a significant global filter on the interaction choices available to the designer will be those user characteristics. INTUIT filters out interaction style information that is inappropriate to the characteristics of the target users.

- Target systems may limit the interaction possibilities. It is inappropriate to support WIMP style interactions on glass teletypes. More interestingly, choices about target systems will often imply the adoption of particular house styles. INTUIT helps to focus the interaction styles knowledge base by use of target system information if available.

INTUIT adopts the underlying design process as its operating metaphor. Given the complexity of the task, a user will need to possess a clear model of how the system operates. This model needs to be based upon users' existing knowledge. Therefore, the user can operate according to the disciplines of a structured methodology which is the one common body of knowledge that can be assumed for the intended user base.

2.3. System Tasks.

The system supports the drafting of diagrams. In this respect, INTUIT will operate as a computer aided drafting tool. Only diagrams relevant to INTUIT's interface design process are supported.

The system offers direction to design choices. Where possible, a designer will be able to make design choices by selecting from lists. In the simpler cases this will amount to offering a menu of choices. However, for more complex situations, such as selecting a characterisation for a dialogue task, a more sophisticated browsing mechanism will need to be supplied. Note that the direction given will depend upon the local design context.

The system enables interface designs to be prototyped. Automatic screen generation and interface animation allow a designer to evaluate proposed designs and evolve them rapidly.

The system provides guidelines and examples. This facility augments the implicit provision of human factors expertise offered the methods listed above. INTUIT will provide this explicit advice to justify its implicit methods and enable its users to learn and improve their own human factors knowledge. This advice will be accessible via the design constructs presented by the drafting facilities. The guidelines and examples presented in any situation will be automatically chosen to be instructive and relevant to that context.

3. Architecture.

INTUIT consists of three main functional subsystems. These provide a set of design constructs, the human factors knowledge base and the user interface. Each of these subsystems is further partitioned and the communication between subsystems is far from simple. However, this division reveals the purpose of each of the components. See diagram below.

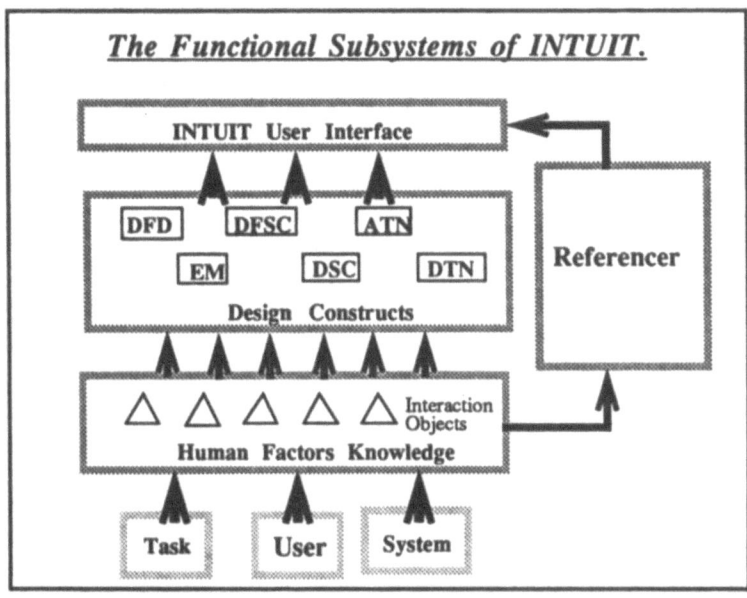

3.1. User Interface Subsystem.

INTUIT possesses a powerful, graphical user interface. The intended users perform a set of complex tasks requiring high bandwidth communications with the system. Such complex interactions can only be supported in an effective manner using a graphical interface. The interface uses multiple windows in which both text and graphics can be displayed and manipulated using a keyboard and mouse.

Application objects are directly manipulable. Due to the nature of the software design task, a large number of complex application objects are involved. In such circumstances, adequate usability is achieved by presenting users with an environment in which there are no intermediaries between them and the application objects about which they are thinking. This amounts to ensuring that application objects are displayed to users in meaningful ways and that the relevant actions are immediately and clearly accessible. For instance, using icons and problem-oriented (software design) diagrams upon which actions can be invoked via pointing and other locator-oriented operations.

The user interface supports a set of views onto the underlying design constructs and human factors knowledge. The diagrams of a design are displayed in graphical form in the main work area window. Associated windows are used to access human factors knowledge in such forms as textual guidelines. Note that the designer need not access the taxonomy of human factors knowledge directly, rather this retrieval is performed implicitly by INTUIT via the displayed design constructs. *See following diagram*

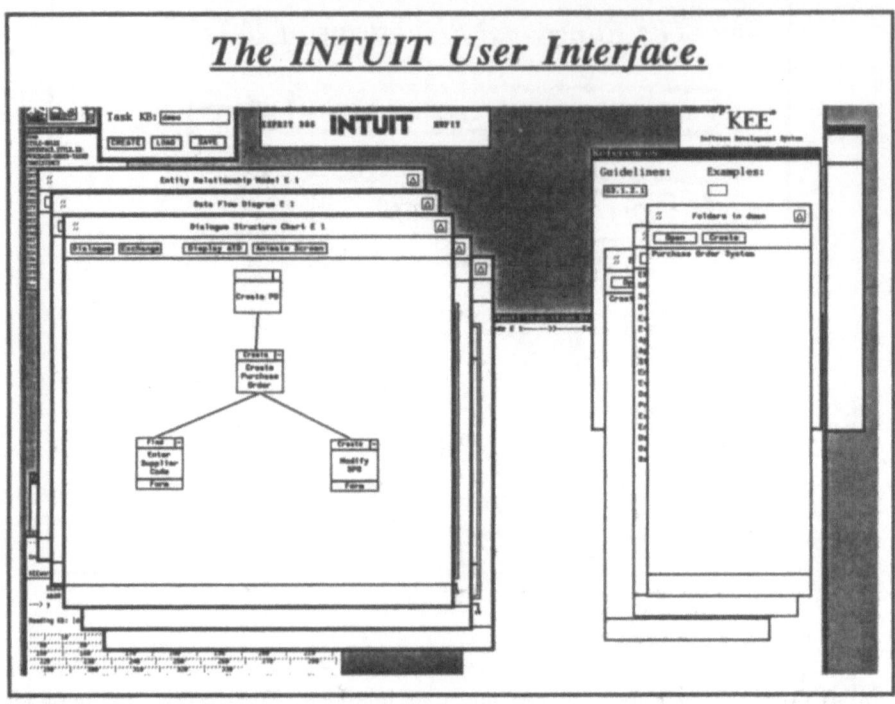

A separate window is used for the <u>referencer</u> of guidelines and examples. INTUIT offers a contextually determined selection of advice materials from which a user may choose. The selected guideline or example is then displayed in a special window. Examples that demonstrate dynamic aspects of the interface may be interacted with by the user. The guidelines appear in a window that supports a rudimentary form of hypertext, the user is able to point to references within the guideline shown and navigate to other advice. The user is thus free to browse through all the advice related to the issue under consideration. The **referencer** module providing access to guidelines and examples was constructed by Fraunhofer IAO.

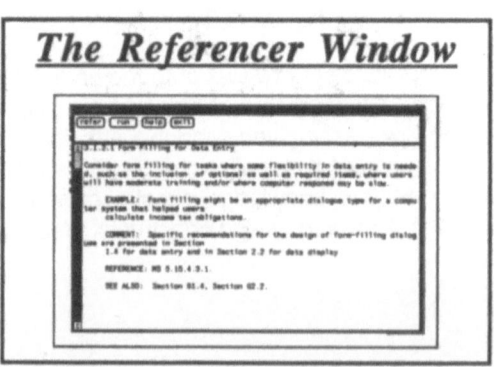

3.2. Design Constructs Subsystem.

The objects in the design constructs subsystem represent the design information required by the development process. These objects are similar to the objects one finds in a data dictionary. They are defined by a rich frame-based schema that describes not only their structure but also the operations that can be applied to them. Inheritance is used extensively to reuse existing definitions and explicitly show similarities. This information consists of the views a user can take of the design and the elements that represent its details. The following diagram shows an example of how constructs are defined

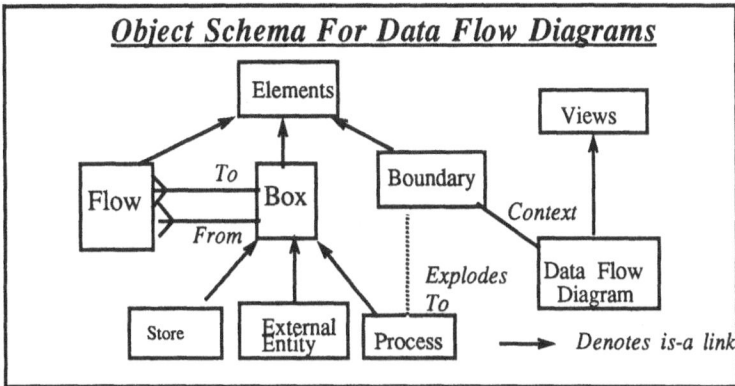

The design constructs subsystem supports transformations. Some of the work a designer would be required to do with INTUIT can be reduced to clerical chores that should be automated. The priority for such automation is in packaging **human factors expertise** to support designers rather than more general engineering tools for tasks such as data base generation. An example of such automation is in the area of **UIMS** where a high level dialogue structure can be used to generate a user interface under the control of a knowledge base encoding a specific interface style.

3.3. Human Factors Knowledge Subsystem.

The human factors knowledge subsystem is founded upon a taxonomy of interaction components. These components include both the interface objects that a user would manipulate and the interface actions that constitute those manipulations. This taxonomy includes both very generic forms of component such as ones that describe selection actions in general terms and increasingly specific descriptions such as about selection from a menu on a particular target system using a particular style. The interaction components taxonomy is partitioned into a collection of knowledge bases which enables its content to be managed in an effective manner as it modularises the knowledge into distinct chunks.

– A generic interaction components knowledge base lies at the root of the taxonomy. This knowledge base defines both the general structure of the interaction components and how they are organised according to human factor principles. All other interaction component knowledge bases will be based upon this root.

- Subsidiary interaction components knowledge bases will package distinct interface styles. The knowledge about each style will be related back to the generic knowledge in the root knowledge base.
- Dialects of interface styles will be accommodated by further knowledge bases. These will allow variants according to target system limitations to be accommodated. It should also enable a rich structure of inter-relations between interface styles to be represented.

Human factors expertise is attached to this taxonomy which acts a framework. Currently, this expertise consists of guidelines and examples that have been produced by the team and extracted from the *Mitre Guidelines Manual* [7]. It is hoped that further guidelines and examples will be added to the system. Other forms of delivery of expertise could also be used; for example, rule bases could be incorporated at the contextually relevant points in the taxonomy to aid a designer in making choices about how to proceed with the interface design process. Using this framework, the human factors expertise formulated by other partners in the HUFIT project will be introduced.

Interface design can be guided by an orderly migration through the taxonomy. Ideally, design proceeds from general and abstract to more specific. Using the taxonomy, a designer will be directed to make initial, general decisions about dialogues that are to be constructed. Introducing further constraints about the user, the task and the target system will imply movement through the taxonomy of interface components. Eventually, the designer will be located at a set of consistent points in the taxonomy that correspond to a realisable interface design and that lie within the knowledge base for one interface style. Thereby, the design process is directed by the human factors issues appropriate to the choice of task, user and target system.

3.4. Design Filters.

INTUIT uses three main design filters. These filters ensure that the design enforces the constraints defined by the task, user and target system characteristics. These filters are shown at the bottom of the preceding diagram that describes the functional subsystems of INTUIT. Currently, there is little substance to these filters, INTUIT really only caters for one group of users and targets in terms of the interface styles it provides. However, this should be a major area for medium term development.

- Task Characteristics are described in a taxonomy of dialogue tasks. These are associated with the pertinent interaction components for supporting such tasks. This task taxonomy currently only contains relatively simple dialogue tasks. It is anticipated that this taxonomy will grow to include more complex tasks. That which exists at the moment merely provides a base that can be used to bootstrap a more comprehensive knowledge base of task characteristics.
- User characteristics are not explicitly represented in the current version of INTUIT. A simple classification scheme will be adopted in the short term which at least provides a coarse filter on appropriate interface styles. A more sophisticated filter for user characteristics will depend upon the deliverables from other parts of HUFIT.
- Target system characteristics will be described relative to the UIMS that they support. It seems that the most appropriate way of using target system information is as a filter on the interface styles that can be effectively supported. This in turn depends upon the

availability of user interface delivery systems for those styles on the targets described.

3.5. Development Route.

INTUIT is being developed using KEE™, a hybrid AI toolkit. It provides frame-based representation facilities together with powerful rule based programming. Traditional procedural programming can be provided using the underlying COMMON LISP in which KEE is embedded. KEE also offers advanced expert system facilities such as truth maintenance and hypothetical worlds. [8]

INTUIT is being developed on SUN™ workstations. These combine the ability to host KEE with a powerful graphical interface in a UNIX environment. The minimum configuration for development requires a SUN™ 3/60 with 12Mbytes of memory and 130 Mbytes of filing system.

3.6. Object Structure.

The architecture is object-oriented. It is implemented using the KEE toolkit in which objects (called units in KEE) provide the basic building blocks. The system consists of a large number of intercommunicating objects. These objects are grouped, according to purpose, into the functional subsystems presented earlier. The behaviour of the objects is further constrained by an object schema that describes how the objects relate to each other. This object schema elaborates some of the detail in the arrows shown on the diagram of the functional subsystems. *See diagram following*

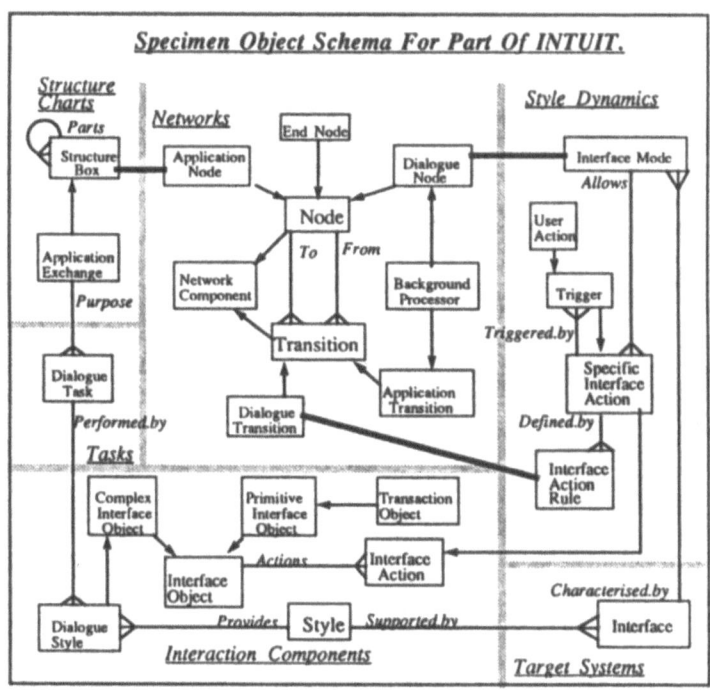

4. Design Principles.

The design is conditioned by a set of underlying principles. These principles shape the form INTUIT takes and carry the basic messages of its technical positioning. These messages are re-iterated here for emphasis.

The support provided for the development process is **in-line** and **pro-active**. Human factors expertise has historically been made available to the designer as guidelines; however, a designer needs to exploit them as a natural part of the job otherwise the guidance is too easily ignored or the effort of referring to it perceived as mere overhead to the work in hand. Similarly, ergonomic evaluation services are offered but these suffer from occurring too late in the design process to significantly affect a product. Neither of these forms of delivery are ideal. INTUIT adopts the alternative approach of providing a CASE tool that promotes good human factors practices and delivers contextually relevant design guidance to the development process directly.

4.1. User-Centred.

INTUIT is a user-centred design tool. In designing INTUIT the emphasis has been on the user, what the user does and how the user interacts with the system. This emphasis is exhibited in two forms, INTUIT is intended for user-centred design and is devised by user-centred design.

- A user-centred approach to interface design is adopted. The structuring of dialogues and formatting of screens are both driven by the needs of the user and the task according to human factors considerations. Such issues as the structuring of physical data in the implementation will always take a subsidiary role in directing the design.

- The CASE tool is designed to meet the needs of its intended users. Human factors support will only be included in INTUIT if it will be effective for application designers and acceptable to them. The emphasis is upon **proactive** and **in-line** tools that enhance the productivity of designers. Too often, human factors support has been offered in forms that are not useful to their intended users.

4.2. Interface Styles.

Knowledge is organised around interaction styles. A designer using INTUIT will want to investigate possibilities for organising interactions in the dialogue; the step-wise refinement process described above implies a process of choosing and elaborating on interaction styles to develop a dialogue design. Other human factors knowledge may be relevant but it is accessed according to interaction style considerations.

- Generic interaction style knowledge is required in defining dialogue structures. The deep syntactic level of interaction design depends upon generic knowledge about such issues as how to make selections; *for example, should it be a menu.* Very specific knowledge about particular proprietary interfaces may not be relevant and can certainly be obscure at this point in the design process.

- Specific interaction style knowledge is required in defining the surface syntactic level of dialogues. The generic interaction knowledge is specialised to provide further support at the screen format and "keystroke" level. Thus, the interaction style knowledge is organised so as to help the designer in the step-wise refinement of dialogues.

Interface standards are introduced as part of the specific interaction style knowledge base. Such standards might be proprietary or industry standards or company house styles. *An example of such a standard is the Digital Research GEM interface style.* A useful contribution of INTUIT will be to lead designers automatically towards adopted interface standards. This becomes a natural part of the design process as the human factors knowledge, that directs the design process, encompasses such standards at its most specific levels of knowledge.

4.3. UIMS.

User interface management systems (UIMS) are packaged as interface styles. Each realisable interface style loaded into INTUIT is to provide prototyping facilities. The results produced from using INTUIT should also enable working systems to be generated. Human factors expertise is thus packaged in the form of high level interface constructs that relieve the application designer of the details of interface design. This UIMS technology is smoothly integrated within the overall interface design process presented by INTUIT so that the designer is directed towards a rational choice of constructs within the UIMS by characteristics of the application and under the control of human factors expertise.

5. In Conclusion.

The last year of the project has been spent on integrating the results from our collaborators where this material is relevant to a CASE tool approach. The HUFIT project has constructed a coherent set of methodologies and tools, of which INTUIT is only one instance. In its present form the system is essentially a *CASE shell* and the exciting work is to considerably enhance the knowledge bases with our collaborators' insights into the human aspects of system design.

– Work by Bull MTS on task analysis has been integrated. Their situations of action model which allows tasks to be structured so that design rules can be associated with relevant task elements [9,10]. INTUIT exploits the model by allowing designers to browse a library of tasks from applications that have been developed previously; attached to these tasks are guidelines based upon detailed human factors investigation.

– Dialogue design guidelines from Siemens have been incorporated. Detailed studies have, amongst other things [6], resulted in an important set of guidelines that can now be accessed from INTUIT.

– Some of the non-functional information collected by HUSAT's tools for user and task analysis have also been incorporated into INTUIT. The HUSAT tools tend to complement INTUIT in that whereas the latter deals with the traditional *hard* data of software engineering they concentrate upon the difficult *soft* aspects that influence design, such as user and organisation derived system constraints [13,14]. However, we have started to integrate the information collected by their tools so that these non-functional requirements cannot be ignored during the design process.

- Fraunhofer IAO have concentrated upon direct manipulation techniques and UIMS [15,16]. In contrast, INTUIT is based upon methodologies for traditional character-based systems. To integrate INTUIT with the IAO work would require the evolution of design methodologies towards the object oriented techniques implied by modern interface technology. ICL and IAO have already started this investigation into the convergence of design methodology and direct manipulation interface technology.

6. Acknowledgements.

INTUIT was developed by a team at ICL consisting of myself together with Fred Russell, Simon Elder, Stuart Thompson, Andrew Coney and Steve Green. The referencer was built by Bernhard Trefz of IAO. Grateful acknowledgement is also made of our partners in Bull, HUSAT, IAO, Olivetti and Siemens.

7. References.

1 Bullinger H-J., Fahnrich K-P., Ziegler J. (1989) 'Human Factors In Information Technology - Results from a Large Cooperative European Research Programme.' In: Proceeding of the ESPRIT Conference '89. Kluwer Academic Publishers, Dordrecht.

2 Russell A.J., Elder S., Aldred W., Thompson S. (1987) 'Functional Description of INTUIT.' HUFIT deliverable A2.2/3.

3 SSADM Reference Manual, Version 3. (1986) CCTA.

4 ICL QUICKBUILD User Guide. (1985) ICL.

5 SSADM Implementation Guide (QUICKBUILD). (1987) CCTA.

6 Mittermaier E., Haubner P., (1987) 'A Method of Describing Dialogues.' HUFIT deliverable 07/-SIE-5/87, 1987.

7 Smith S.L., Mosier J.N., (1986) Guidelines for Designing User Interface Software. MITRE Corporation.

8 KEE™ Reference Manual, Version 3.00. (1986) IntelliCorp®.

9 Mazoyer B. (1986) 'The Analysis of Tasks and Users for the Design of an I.T. Product.' HUFIT deliverable A4.4a.

10 Jeffroy F. (1987) 'Maitrise de l'exploitation d'un systeme micro-informatique par des utilisateurs non-informaticiens. Analyse ergonomique et processus cognitif.' Doctoral these, Universite de Paris.

11 Novara F., Allamanno N., Bertaggia N., Howey K., Fox S., Ophert W. (1986) 'Methodology for Assessing Usability Of Products.' HUFIT working paper A5.2.

12 Dunker J., Melchior E-M., Bosser T. (1987) 'Empirical Methods for the Validation of Models of User Knowledge.' HUFIT working paper B7.1a.

13 Allison G., Dowd D., Galer M.D., Maguire M., Herring V. and Taylor B. (1987) 'Development of Human Factors Tools to fit the Design Process." HUFIT working paper A1.3 HUFIT/13-HUS-11/87

14 Taylor B. (1989) 'HUFIT - User Requirements Toolset.' In: Megaw E.D. (Ed) Contemporary Ergonomics, Taylor & Francis, London.

15 Trefz B., Hoppe H.U. (1987) 'Interface Prototyping - Comparison of Methods and Description of the Interaction Cell Environment (ICE).' HUFIT report B3.2b.

16 Dannnenberg M., Ziegler J. (1987) 'A Dialogue Management Tool for Prototyping and Development of Interactive Systems.' HUFIT working paper B3.4a.

Project no 612/1593

MODELLING AND SIMULATION OF THE VISUAL CHARACTERISTICS OF FLAT PANEL DISPLAYS TECHNOLOGIES UNDER OFFICE WORK CONDITIONS. (DISSIM)

I. PLACENCIA PORRERO	J.P. LEVIS	J.DUPREZ	D.BOSMAN	G.SPENKELINK
Océ-Nederland B.V.	*Myfra*	*Barco Ind.*	*Twente Univ.*	*Twente Univ.*
P.O. Box 101	*A.Briand 83*	*T. Sevenlaan 106*	*P.O.Box 217*	*P.O.Box 217*
Venlo	*Paris*	*Kortrijk*	*Enschdede*	*Enschdede*
The Netherlands	*France*	*Belgium*	*The Netherlands*	*The Netherlands*

ABSTRACT. The aim of the DISSIM project is to identify user requirements for flat panel display developments. Those requirements will be identified from ergonomic experiments, running on a real time interactive flat panel display simulator. The experiments are designed in the context of office work conditions. As these may be too time consuming and costly in the design phase, in addition a "quick look' vision model (a kind of standard observer in software) is developed to estimate the appearance of such display designs, and e.g. the perception of certain fonts in relation to the display parameters.

1. Introduction

The visual characteristics of flat panel display technologies differ a lot from the ones of the well known cathode ray tube (CRT). The picture elements of the CRT have got a gaussian distribution due to the emission of light where the electron beam hits the phosphor. Flat panel displays have spatially fixed addressable electrodes overlapping and producing a grid of rectangular display elements. Between the electrodes, there is an active layer sandwiched that emits or modulates light.

These differences in the visual characteristics are responsible for the differences in the performance of the user, when carrying out the same task with various displays.

To be able to determine optimal characteristics, it is necessary to experiment with the display properties, varying them in a flexible way, while performing a certain task. For this reason a flat panel display simulator is built which permits, to realise ergonomic experiments with various simulated display technologies under office work conditions.[5]

In order to characterise a certain display it is necessary to model three different display aspects:

- Light distribution of a display element (Point Spread Function)
- Switching characteristics of the active layer (rise and decay time)
- Color aspects of the display element.

In the simulator, models of the different technologies calculate a description of these three visual aspects when a certain display definition is given by its engineering parameters. This permits the visualization of the simulated flat panel display on a high resolution monitor while being able to run interactive applications in real time.

The real time aspect is one of the most important characteristics of this simulator. This permits to have a user running any office application built under UNIX with a simulated display. At the same time it is possible to monitor the development of the experiments and change some of the parameters during operation.

The Simulator has two different interfaces to facilitate the use of the system to two types of users, the display engineers and the ergonomic experts:

-For the first one, engineering display parameters are given as input for the technology models like for instance materials, sizes, and electrical characteristics.

-For the second one, the Point Spread functions, Rise and Decay time, and Color aspects can be interactively changed acting directly over the visual aspect of the simulated display. This last interface permits to change easily parameters like:

-Pixel size, shape and separation.

-Luminance ,contrast, color, reflections.

-Temporal aspects.

The DISSIM consortium consists of the following partners and main workpackages:

- Océ-Nederland B.V. (NL)

Displaymodel development , implementation and projectmanagement

- Twente University (NL)

Ergonomic experiments,

Vision modelling,displaycover modelling and simulator control software

- Cimsa-Sintra S.A. (F)

Hostcomputer and image processor

Participation in EL modelling

- Myfra S.A (F)

Very high speed hardware: convolutions, switching characteristic processing up to 20 Gigaops/sec

- Barco Ind (B)

High performance digital controlled monitor

- GEC Research (UK)

Participation in LCD modelling.

The simulator machine is based on a Unix computer, modified with very complex dedicated hardware. A general description of the system architechture is given, taken care, not to forget the fuctional description. Obviously, special attention is given to the monitors that permit the visualization of the simulated images.

Furthermore, the software structure and the interfacing aspects are described. The software is built in a modular way, this permits the development of the system in parallel and gives the possibility of using simplified hardware. In addition, a summary of the possibilities and functionality of the simulator is included. However, the description of the physical models of the different flat panel display tecnologies is quite condensed. The emphasis is again placed in thè functional description and not in the physical bases.

In order to evaluate the visual quality of the simulated displays a model of the perception of the human eye is implmented. With this model it is possible to show numerically and graphically the differences between two simulated display technologies. However, this does not mean that we have found an objective measurement of display quality, we have to see these models as a tool to help to the evaluation of these devices.

Finallly, ergonomic experiments are realized with two purposes, the first one is the validation of the simulator machine, the second one is to characterize the optical properties of the display technologies. The validation not only permits to determine the accuracy of the machine, but provides with guidelines for the interpretation of the second ergonomic experiments.

With the use of the DISSIM display simulator and the present tools we are identifying user requirements for flat panel displays. Furthermore,, transmiting these requirements to display manufacturers can mean an improvment for the characteristics of visual displays.

2. General Simulator Description: Hardware and Software

The simulator is built on a Unix host computer and a graphic processor from Cimsa-Sintra. They are modified with very high speed dedicated hardware that permits the realization of simulations in real time. The system is covered with a software layer that converts it in a dedicated machine. The idea is to have a turnkey system that can be used for simulating flat panel displays and realizing ergonomic experiments without having to deal with the architecture of the system.

2.1. SOFTWARE

The software of the system is divided in two main parts. The first one is the software control, that is the part that is in charge of the monitoring and control of the realtime simulations. The second part contains the software models of the different flat panel display technologies, together with the vision modelling part.

2.1.1. *The control Software.* The control software can be split in three parts. The first one is the user interface and control of the system, the second is the interface with the system hardware and the last one is a simplified interface for non display expert users.

The control software provides the user machine interface and consist of a set of programs available to the user. These programs provide the user the tools to set up a simulation according to his own needs. To realize the appropriate actions required by the user this software should call the necessary routines from the other software layers. For this function a menu driven command interpreter is chosen. It basically realises three functions:

a. Initialization and Off-line Operations
the user has the possibility of describing a new model, to use some utilities like back up or data base operations and at last to go to the next fuction to run a simulation

b. set up simulation and start up application program
the user has to specify a certain number of parameters before a simulation session can start: the choice of the model, the font to be used, static or dynamic simulation, the monitor choice between color and b/w, what reflection image to show, monitor in landscape or portrait mode and the choice of the application program.

c. runtime changing of parameters
this function is introduced in the system to provide a simplified interface for ergonomic experiments. Visual parameters of flat panel displays can be changed without having to calculate them from the display engineering parameters. This means that during a simulation session it is possible to change parameters like: viewing angle, display element size, display element shape, separation between elements, contrast, luminance, color, reflection images, rise and decay times, the fonts used, the line lenght, interline distance and scrolling.

The interface with the hardware is called the cover layer because it has the property of hiding the complex hardware from the operator of the machine. Also by changing this cover layer we can use other basic systems without having to modify the rest of the high level software.

The function of the cover layer is to interface the hardware with the high level software. It is built in the form of a library where all functions available are present. Here first, there are functions to configure the hardware, there are also functions to handle bitmaps of synthetic images and recorded reflection images, functions to load and read tables, like color look up tables and luminance controller tables. Furthermore there are also functions to load and write convolution

918

kernels and of course, functions for the handling and definition of fonts are also present.
A part of this cover layer, has been placed in the hardware itself and there are high level cover
layer routines that are communicating with the low level ones.

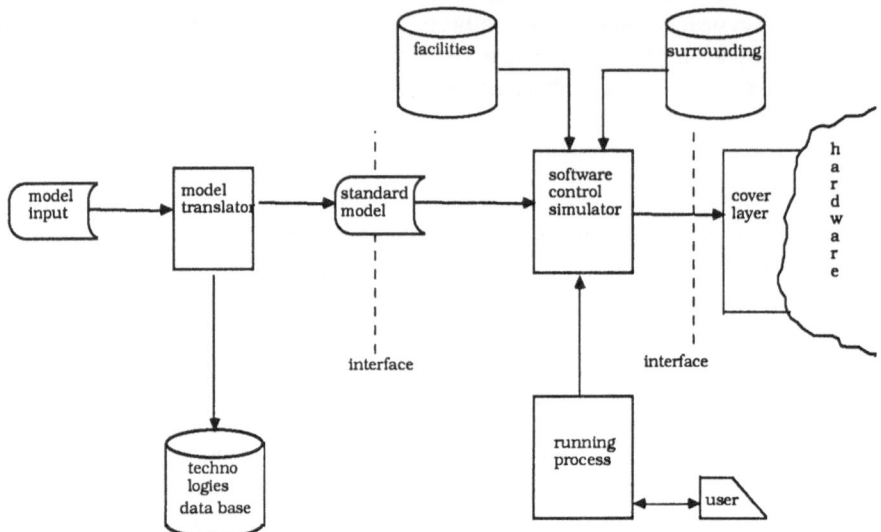

Fig. 2.1. **software structure**

2.1.2. *Display Modelling Software.* A very important part of the software of the simulator machine is what we have called the model translator. Here the description of a certain display
technology is given by its engineering parameters (materials, sizes, electrical characteristics) and
the spatial and temporal visual parameters of that particular display are calculated. These operations are in advance of a simulation session.[6]

The model translator has two main parts, one the database the second the calculation
modules. In the data base the materials and all their characteristics are stored, but also different
input and output models. The data base is provided of an interface that permits the search and
easy store of all this elements. The calculation modules are separated for each technology although some of them are treating similar phenomena. The reason for this, is a modular construction of the system that permits to work independently with each technology and add new ones if
wished. This modular approach has also been applied to the complexity of the models and to the
images shown in the display simulator monitor.
From all the existing flat panel display technologies we have chosen to implement the model of
the electroluminiscence and liquid crystals. Both are in our opinion, the most promising technologies for office applications.

An electroluminiscence (EL) display is basically built up of a phoshor layer sandwiched
between two dielectrics that are contained into two layers of patterned electrodes. These layers
are transparent and are deposited on a glass plate. When a voltage is applied to the panel the
phosphor layer emits light.
The EL model is first calculating the electrical behaviour of the sandwich, the light-voltage
relation.
The dielectric layers are considered in the model as perfect capacitors, the phosphor also behaves

like a capacitor below a threshold voltage, above threshold, it behaves like a resistor with a power compsumption proportional to the emitted light in the device.

Then, the optical behaviour of the panel is calculated, how the emitted light is transformated until it is coming out of the panel, how it is reflected in the back electrode and the interference effects that are occurring. It is also possible to calculate the reflection of the ambient light at the EL panel. This provides not only the amount of light but also the spectral distribution.

Once the light output is calculated and taking into account the geometrical shape of the pixels, the calculation of the point spread functions of the device is done. From the light-voltage curve it is also possible to obtain the temporal characteristics of the phosphors, which is quick enough and comparable to the CRT one.

The last part calculated is the total voltage distribution accross the panel. Because of the lenght of the electrodes and their a high resistivity, there is a non uniform distribution of the voltage in the panel. When addressed at one end, there is a time delay and a voltage drop at the other end of the electrode. This can cause a difference in light output at different areas of the panel.

Besides the point spread function, the light output, color and temporal effects, some other geometrical parameters are also transmitted to the output, like the shape and sizes of the pixels but also the size and shape of the panel and its resolution.

The basic construction of a liquid crystal panel is similar to the EL one. The main difference is that instead of the phosphor emitting light, here, a liquid crystal layer is modulating the light that is passing through the panel.

The input and output parameters required, are also the same as for the previous model, only that for the liquid crystal the temporal effects become more important.

A LCD display can work in three different modes: the transmissive, the reflective and the transflective mode. In the first case, the LCD has a back light, which emits the light that goes through the panel. In this case we presume that the image that is seen by the user is formed by the back light and the LCD panel structure. In the reflective mode, the ambient light that goes through the LCD configuration, will be reflected at the back of the panel, will go again through the LCD structure and will get to the user. In this mode the ambient light is needed by the LCD to form an image. In this second case the LC layer is projecting a shadow on the back plane that is disturbing the displayed image. The transflective mode is the combination of boths.

The first part that is calculated is the behaviour of the LC material when a voltage is applied. The molecules of the LC tend to orientate in the direction of the electric field but, the viscous forces, present in the liquid, are opposing to this movement. By minimising the Helmoth free energy the orientation of the molecules is calculated for the on and off state.

The second calculation is concerning the dynamic behaviour of the LC material. The relation between the orientation of the molecules and the time, when changing from the off to the on voltages and viceversa is calculated. These dynamic characteristics are the base for the temporal behaviour of the LC panel.

The final calculation is the optical response of the LC display. By giving the position and orientation of the molecules in the liquid crystal layer and the wavelength dependent refractive indices of the the liquid crystal the propagation of the light through is obtained. This process has to be repeated for each wavelenght in order to obtain the colorimetric data.

Finally the consideration of the geometrical characteristic permits the obtention of the point spread function of the display. This together with the temporal characteristics are providing the complete characterisation of the visual aspects of the display.

Both models have been applied to existing devices and the outputs have been verified with the mentioned displays, the results are encouraging similar.

2.2. HARDWARE

2.2.1. *General configuration.* The Simulator system is based on VME/UNIX V.2 computer with a 68020 processor, interfaced with a graphic processor Triade 60 of Cimsa-Sintra. In the computer and graphic processor are present the part of the software that is responsible of the general control and communication of the system. Besides this hardware is taking care of some special functions of the simulator. Basically these functions are two. First the generation and compilation of images: fonts and some graphics. Second the handling and storage of the reflection images.[4]

To generate these characters first the basic display elements have to be defined. They are formed by a matrix of pixels of the memory. Using the display elements as basic unit, different sets of fonts can be designed, apart from shapes, they can vary in sizes, number of gray levels and colors. When the images are generated they are stored in a bitmapped memory of 905* 1280 words of 8 bits.

The configuration of the memory is corresponding with the one of the monitor. In fact the real size of the memory of the system is larger and this fact is used for the simulation of the scrolling function. Changing the starting address of the image sent to the monitor.

The reflected imags are stored also in a bitmap memory of the same size that the generated image being 905*1280pixels of 8 bit deep. These images have been previously recorded with video equipment and are stored on tape or hard disk. They are downloaded in the system for a simulation session. They correspond to different display technologies, viewing angles and display covers.

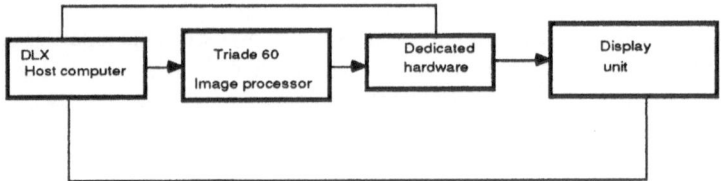

Fig. 2.2 General configuration

2.2.2. *The dedicated hardware.* This hardware is calculating in realtime three main functions of a simulated display: the rise and decay times, the point spread function and the color characteristics.

-Top level design Luminance controller and color look up table.

The luminance controller is the part of the system that calculates the switching characteristics of the simulated flat panel display. This means that it takes care that the on and off luminances that are previously calculated in the models are shown in the monitor at the right time.

Basically it is a matrix of 905*1280 state machines, one for every pixel. The switching characteristics are stored in a look up table in advance of simulation runs. Each luminance value is at the same time the address to the next luminance value in the table. Every frame this value is at the same time send to the monitor and stored in a state memory. In this way, in the next frame the state memory is read and the cyclus is repeated. This means that the sample frequency of the luminance curves is the 80Hz refresh rate of the monitor. In fact it is a table, selectable and configurable during simulation in the frame retrace time. The basic idea behind a configurable table is that dependent on the simulation mode a choice has to be made of how many gray/color levels will be used:

In the case of a bilevel simulation, only one bit/pixel is needed, in this case the table size is 512 byte and due to the amount of available memory (64k*8) it is possible to have 128 tables avail-

able in this part of the machine, which is useful in the case of head movement experiments.
In the case of grey/color level simulation, 8bits/pixel can be used. The table size will increase to 64kbyte. This implies that only one table can be resident in the machine.
Between those two extremes any possibility can be configured.

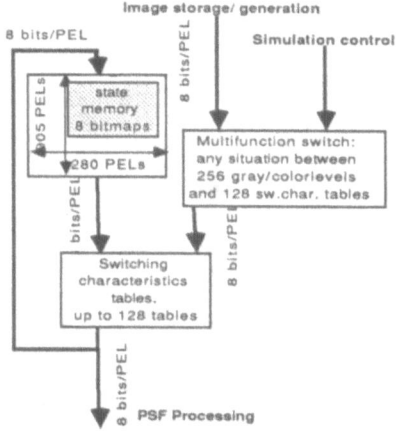

Fig 2.3   **Switching characteristic processing**

The rôle of the color look up table is to mix the data coming from the convolver (generated image) with the data from the reflected image, in order to provide the monitor with three 8-bit words, corresponding to the three fundamental colors.
The Color look up table has also to generate the line and frame synchro signals and the blanking signal.
From the Triade 60, the Color look up table receives the reflected image as eight 8-bit pixels every 72ns. Normally, this image is a large number of clock periods late in comparison with the reflected one. This delay corresponds to all the pipeline registers located in the luminance controller and the convolver. Then, also the simulated image, 8 bits per pixel coming from the convolver are added together at the correct time in the color look up table. This implies a very large size table of 64k* 24bits, but gives the possibility of choosing 256 colors out a palette of over 16 million colors.

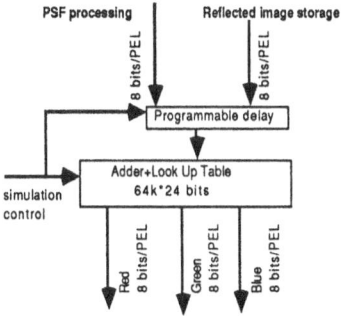

Fig 2.4   Colormapping; adding reflected and generated image

-Top level design convolver

The convolver is used to calculate the luminance profile (point spread function) of a certain display element and the influence that it has on the neighbouring elements. It performs a convolution of the input image with a maximum kernel size of 9*9 pixels. This will be done in real time with a throughput of 9ns/pixel. Because about 20 gigaoperations per second are needed, that implies the necessity to use eight parallel processes.

The convolver is build up from a family of Marconi components. Input is coming with 8bits/pixel from the switching characteristics processing part.

In order to be able to simulate changes in viewing angle and different pixel shapes within one simulation session, we have the possibility of storing up to 16 different convolver kernels in the hardware. This gives the possibility of switching tables every frame within 0.7ms.

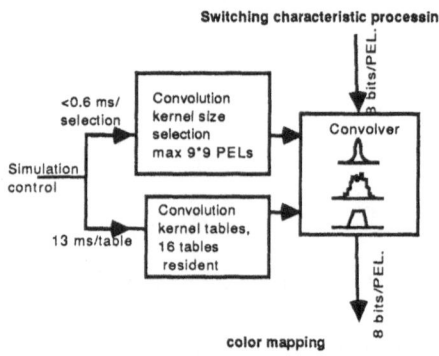

Fig 2.5   **Point spreadfunction processing**

2.2.3. *The Monitors.* The display system system can be divided in three main parts :
    a- the control rack
    b- the B&W monitor

The reason for using two different monitors is mainly due to the insufficient resolution of the available colour picture tubes and the disturbance of the shadow mask in some simulations.The control rack contains all circuitry common to both monitors. All the normal adjustments and table loading can be done by data entry either from a terminal or from a host computer.

    a-Control Rack

    Linebuffer.: The video input signals are coming from the Myfra dedicated hardware. They are in digital format (3*8 bit + sync and clock, 9nsec). Gamma correction is provided by means of 3 (256x8)look-up tables. The pixel clock is converted from 9 ns to 7 ns.This timing conversion is needed for correct operation of the monitors. The pixel period of the video output is 143 MHz (7ns/pixel period).The digital video signals are converted to analog signals by means of three 8-bit D/A converters (300 MHz). The background signal is derived from a second D/A converter. The analog output signal is the sum of video and background signals. The reference input of the video D/A is derived from a third D/A converter. This allows digital control of the video amplitude(gain control). The reference input of the second and third D/A converter is a dynamic waveform generated by the shading correction circuitry (see paragraph 2.2)

    Shading correction:The luminance across the screen of a conventional CRT-display is not constant (shading). Variations up to 20% are normal for B&W CRT's. Color CRT's can vary by 50%. The variation is not exactly the same for the three primary colors, which causes color shift over the screen area. It is possible to correct this shading error by modulating the incoming video signal with the appropriate waveform. In the DISSIM-display system, this waveform is derived from 8 points on the horizontal axis and 5 points on the vertical axis (40 points total). A two-dimensional interpolation between those points produces a smooth waveform. The correct values for each measuring point are obtained from a calibration procedure. During this procedure, a luminance probe is placed on the screen (40 points in sequence). A control loop adjusts the correction value for that particular point until the desired luminance is reached.

Built-In Test: The control rack has a built-in videogenerator. This allows the user to verify correct operation of the monitor, independent of the incoming videosignal. The BIT is also used to generate the pattern for the shading calibration.

Microprocessor System: The μP-system receives commands from a hand-held terminal, the host computer or the luminance probe. After receiving a command, the μP takes the necessary actions to execute the desired functions.The most important functions are :

- contrast and background (independent for the three primary colors of the color monitor)
- gamma (either an exponent for the built-in power-law, or a complete table from the host)
- calibrate shading
- landscape/portrait orientation
- built-in test

Monitors: Most important specifications common to both monitors :

- addressed resolution : 1280 x 905 pixels
- horizontal frequency : 82.67 kHz (12.61 μs line period)
- horizontal video blanking : 2.6 us
- vertical frequency : 83.676 Hz
- vertical video blanking : 1 ms
- video bandwidth : 180 MHz (2.5 ns fall time)

The screen luminance at zero video input (black level) is stabilised by a feedback circuit (Automatic Kinescope Biasing).

b- B&W monitor and color monitor. The B&W CRT has a useful screen diagonal of 38 cm and a deflection angle of 70 deg. The color CRT has a useful screen diagonal of 43 cm and a deflection angle of 90º deg.. The tube face is of the 'flat&square' type.Summary of measured results:

B&W monitor	Color Monitor	
Spotwidth (FWHM) : 0.17mm vert, 0.24mm hor	- Dot trio pitch : 0.26 mm	
Luminance (at max. gain) : 100 cd/m2	- Spotwidth (FWHM) : approx. 0.31mm	
Shading correction accuracy : +/- 10%	- Luminance :	red : 20.5 cd/m2
White point : 6164 K (x=0.318, y=0.345)		green : 65.5 cd/m2
Decay time (10%) : 4 ms		blue : 14.75 cd/m2
Diffuse screen reflectance : 19.6%	- Color	red x=0.607 y=0.342
Gamma : 2.3		green x=0.312 y=0.586
Background luminance range : 0.1 - 5 cd/m2		bluex=0.151 y=0.066
	- Diffuse screen reflectance : 10.5 %	
	- Gamma : approx 2.2	

The aim of the project was to realise a display tool which could be characterized and specified for as many characteristics as possible.The display system realised for the DISSIM project is entirely digitally controllable . All sensitive parameters are corrected and/or stabilised. The whole display system is fully characterised for both B/W and color monitors. So, from the display development's point of view, all goals which were set forward were obtained.

3. The vision model

In the DISSIM project the subjective quality of the image can be rated in real time by ergonomic experiments: improvements to obtain better acceptance can be brought about by changing design parameters in the simulator software. As ergonomic experiments still are time consuming, it was deemed adviseable to also develop an off-line visual model providing the ability to obtain a provisional indication of the visual characteristics of a proposed design, including display artifacts, by coupling the display design software to a 'visual model', a kind of standard observer, also defined in software. Of course, being a model, its judgement is incomplete, it can only provide answers to the most pregnant defects but as such is already valuable. It produces a prediction of how characters and symbols displayed on the screen are perceived, just as the technology model predicts the radiation distribution at the screen. An example of the latter is that the CRT response to every addressed pel is a Gaussian radiance distribution; these responses overlap when the pel pitch is smaller than the 'dot size', adding locally to produce higher luminances than the peak magnitude of a single response. The result is that the radiation distribution over the character addressed area is non-uniform, a familiar phenomenon in poorly designed displays, which can be very annoying when viewed for prolonged periods. However, the calculated radiation distribution is not exactly what we see, because the early visual processing in the eye modifies the image (often to our advantage as can be inferred from the actions associated with some visual characteristics described below). Cognitive aspects are not considered because most of the display design flaws are thought to be identifiable in the result of early visual processing; moreover, cognition rather is a function of the displayed image itself and should be improved in the software defining that image.

The literature on visual performance under office work conditions is rather poor, particularly in the engineering sense: most data is valid under threshold conditions whereas display engineering models are required to have predictive power over a range of suprathreshold conditions. It was considered that the now mature image processing techniques, in combination with available physical descriptions of early visual processing and the abundance of threshold data, could lead to a workable engineering model of early visual processing by the eye. Image processing is based on calculations of the effects of point spreading (optical,neural) on every picture element (pixel,or pel) of the image. In display simulation the display element (del) is made up of many pels of the CRT screen, therefore we prefer to speak about dels instead of pels in the visual model.

The eye is a marvel with many features of which only a few need modeling in the context of display visual-system engineering. For instance, the image is scanned by angular jumps (saccades), with still periods in between (fixations) during which small regions of the total image are taken in, buffered and then combined in the brain. The finer the detail (e.g.small print), the smaller the saccades; the more difficult the perception, the longer the duration of fixations. Therefore the model is limited to the foveal region of the retina, the region of about 2 degrees around the visual axis, and the total image is built-up through convolution of the processing results in this region over the entire image as displayed. Each del is assumed to be smaller than the foveal region so that only small disc vision data need to be considered; involving the optical and the neural point spread functions of the eye and the detection threshold associated with photon and neural noise. In its present form the model can be applied to monochrome and bicolour displays having differing foreground and background colours.

Early visual processing is responsible for, among others, the following phenomena:

a) attenuation of spatial high frequencies, rounding sharp direction changes

b) a relation for threshold sensitivity between luminance L and target (dot) area A (Ricco, small A: LA=C, Piper, intermediate A: L√A=C, large A: no area dependance, Ogle, small A: threshold increases with blur)

c) inhibitory lateral action, strong stimulation of a channel reduces response of neighbouring channels

d) a stimulus dependent threshold which produces a non-linear luminance to brightness function e) several temporal effects

f) adaptation

From a) to d) can be deduced that in the static perception of a character displayed in e.g. ELtechnology, the following can be observed:

A) the square display element, when seen under a small solid angle, will seem round instead of square. As a consequence the separation gaps widen at the intersections increasing their visibility but this effect is offset by the smearing which decreases their contrast with the dels;

B) when all dels are driven at the same level, a del with many neighbours (eg the upper del of the vertical bar of a T) is seen at reduced brightness, therefore will seem smaller than end dels;

C) the inhibitory lateral action produces Mach bands around the character features, with non-zero background. These are less conspicuous at the inside of characters where Mach bands of both sides overlap.

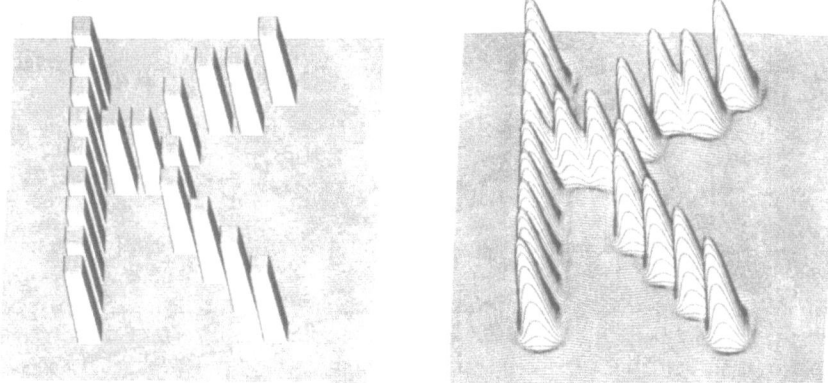

Fig.3.1.a)Original EL radiation pattern and b) response of vision model

Because small (round or square) dels are the natural input primitives to the eye, it was decided to base the analysis of the vision model on contrast sensitivity data of discs, rather than on contrast sensitivity for Line spread Functions or (periodic) gratings; using image processing methods to calculate the effects of point spreading (optical, neural) on every del in the image.[2] The detector model obtained represents ideal energy detection under the presence of noise . Display engineering models must have predictive power over a range of (supra threshold) conditions, whereas contrast sensitivity data represent but one point on the function involved. Because the vision model is largely based on mathematical descriptions of the physics involved in processing of the image, it is postulated that estimation of the parameters of the functions of the vision components engaged in early visual processing, in combination with the abundance of threshold data, would yield a workable engineering model.

3.1. THE MODELING APROACH

In the context of display visual-system engineering the model can be kept fairly simple; e.g. the main optical processes in combination with an ideal detector, the processing area in the retina limited to the foveola. Image formation resulting from a sequence of fixations is mimicked by convolving the processing area over the entire image. Small disc vision data deserve special attention because small areas are stronger affected by the optical and neural point spread functions of the eye and its detection threshold associated with photon and neural noise.

The target image Ew (x,y) is blurred and distorted by the "defects" which are the object of modeling. to form Er(x.y) which denotes the spatial illuminance distribution at the retina; the pair (x.y) the distances from the centre determined by the intersection with the visual axis. In the transmission of the image the lens system (1) spreads the light rays by aberration; and by diffraction due to the finite size of the pupil (2). Throughout the body of the eye scatter occurs causing diffusion, but scatter "particles" located remote from the retina only bend the light rays to regions outside the fovea, thus contributing to the overall illumination of the retina (veiling glare): scatter close to the receptors is mainly instrumental to blurring.

Just in front of the receptors the mess of nerve tissue and bloodvessels (3) is so dense that it can be the major cause of image blurring by diffusion. Thus the image actually projected onto the retina Er(x,y) can be conceived of as the convolution of the 'ideal' image Ew (x,y) with three point spread functions, associated with aberration: PSF1(S,V), with diffraction: PSF2(S,V) and with scatter PSF3(S,V); S and V signify the distances from the center of the PSF. The PSF3(S,V) is a composite function with a very slender center and a wide skirt, because the scattering layer is inhomogeneous: a fraction ϕ of the luminous flux passes undisturbed. Further it is assumed that 50% of the incident light is reflected (fundus photographs) leaving for the scattered fraction $(1-\phi)/2$.

Because the order of the factors involved in convolutions is not important, one may write

$$E (x,y)= PSF1(S,V)*PSF2(S,V)*PSF3(S,V)*E (x,y)= PSFop (S,V)*E (x.y) \quad (2.1)$$

where * stands for convolution and PSFop (S,V) for the total transmission point spread function of the eye optics.

At the retina the image Er(x,y) is sampled by the receptors and subsequently, convolved with two neural point spread functions associated with two different pooling nets of receptors: considering a receptor located at (xo,yo) as the center, a slender foreground pool PSF4 (S,V) and a collocating, surrounding, a much wider background pool PSF4b(S,V). The result of all this spreading is that the center magnitudes of the both total PSFp(S,V) = PSFop(S,V)*PSF4S,V) have decreased. Within the retinal network, local illuminance differences are 'measured' by comparing the two outputs Epa(x,y) and Epb (=input flux weighed by neural P~F volume):

$$Epa(x.y) = PSFop(S,V)* PSF4a(\sim.S,V)* Er(x.y) \quad (2.2a)$$

$$Epb(x.y) = PSFop(S,V)* PSF4b(S,V)*Er(x.y) \quad (2.2b)$$

$$AEp(x,y) = [Fa(Epa)- Fb(Epb)] \quad (2.2c)$$

The functions F(Ep) represent the non-linear behaviour of the detector, including effects of noise due to the stochastic nature of light (photon counting receptors) and of neural noise, together responsible for the detection threshold. Its effect is to reduce blur or, when overcompensating, to cause the so called Mach bands. The shape of the horizontal cross sections of the neural PSFs can be adapted to local need (e.g. oblong), their volumes must be approximately equal to ensure zero output for zero input contrast (uniform image). The combined action of the PSFs can be associated with the notion of receptive field. The function F(Ep) can be determined as follows. The detector is assumed of discriminate with 50% probability between to magnitudes of the retinal illumination differing by the standard deviation of the total noise given by the RMS sum pt of the photon noise at the retina pl and the neural noise pn (classic detection theory). The photon noise amplitude is proportional to the square root of the mean value (Poisson process), the neural noise is assumed to be independent of the illumination. The contrast threshold at the input side of the eye optical PSF thus is found by multiplication of the threshold at the detector with the PSF acting on the signal:

$$Cvth = \Delta Lth/L = PSFop(S,V).pt$$

For each luminance level the adaptation of the eye is determined.
Then the perception of one Just Noticeable Difference in brightness is calculated and gives an idea of the quality of the images.
With these data the static model can be established; the results of comparison of model data for discs[1] with experimental data of Blackwells show good agreement, see figure 1.

luminance	size of disc target in minutes of arc				
cd.m-2	.6	9.7	18.2	55.2	121
3400	.3	0.0	-0.3	-0.1	-0.1
340	-.3	0.1	-0.3	0.0	0.0
34	-.1	0.1	0.1	0.3	0.1
3	.5	0.3	0.4	0.9	1.1

Fig.3.2. Contrast sensitivity differences ACS between model and Blackwell data.

This is a first step towards the establishment of a reproducible standard observer capable to objectively rate image quality, or estimate the allowable upper threshold for unwanted artefacts such as the shape of dels, jaggedness, contrast of separation lines in flat panel technologies and defects like mottle, shading and crosstalk.
    The vision model is designed to calculate a response to an entire image, but in a 512*512*8 format. Its principle is to process the input data by convolution with a non stationary PSF, which is continuously determined depending on the local luminances on the local image. If, however the input image luminances have a restricted distribution as it is the case in office systems, one may relax the requirements, allowing to use an invariant PSF.
The results of the vision model processing can be made visible in two ways. Either the low end simulator is used with its image processing capabilities and hardcopy recorder; or the calculations are carried out by a GP machine an then plotted as either numerical tables where each number represents del luminance, or cross sections through the image providing a luminance distribution. Also it is possible to obtain statistical data about the luminance distribution like mean, standard deviation and histogram.

4. THE ERGONOMIC WORK IN ESPRIT DISSIM

4.1. .INTRODUCTION

The ergonomic contribution to the DISSIM project was identified as one of the key factors: the usability of the simulations for research and display development ultimately depends on the quality as it is perceived by human observers. Of course this quality is determined by the combined hard- and software of the machinery. It is very difficult however, if not impossible, to predict display quality in advance from its hardware and software specifications. So here lies the first task for the ergonomists: the empirical estimation of the quality of simulated displays. During its lifetime, display manufacturers and other interested parties may use the simulator for designing and evaluating displays or some aspects thereof. In order to provide future simulator users with the tools to investigate the human factors, instruments and procedures have been developed that allow gathering objective (performance) and subjective data for these goals. This is the second task. The third part of the ergonomic work pertains to a preliminary investigation of three display technologies: ELD, backlit supertwisted LCD and non-backlit supertwisted LCD. These investigations partly coincide with simulator quality estimation and tool development. The work has concentrated on display applications in traditional office environments. The main tasks that are performed in such an environment pertain to word processing, data base/spreadsheet manipulations and form filling. Graphical man-machine interfaces have been ignored since the simulator was not designed for fast graphic operation.

4.2. INSTRUMENTS AND PROCEDURES

First of all an experimentation office was built in which the investigations take place. The room has been designed for minimum reflections and colouring of the ambient light. The illumination level is variable between zero and higher than 750 lux, which is well beyond the range that is normally found in offices. Secondly a number of experimental tasks and procedures have been developed. The experimentation instruments are designed specially for the simulator and fit its capabilities. The instruments are used both for performance data collection (in experiments) and as a reference for the subjective technique. Thirdly, instruments have been developed or selected for the subjective evaluation of displays and their simulated counterparts. Human observers are well able to judge displays (see eg. [1]).

The experiments that have been defined are:

- a threshold experiment, finding the threshold detection time for alphanumerics
- a number verification task in which the presence of a target character has to be verified in short alphanumeric strings
- a search task in which occurrences of a target character have to be located in text, quasitext or nonsense text
- a display matching task in which observers can interactively match a simulated display to the actual sample
- an optimization task in which observers can redefine a number of display parameters in order to define their own, subjectively optimal display.

The most difficult parts of generating the simulations themselves are the definition of a font and the mapping of the spatial characteristics of existing displays onto the simulator. Although in a way the font is independent of the display (one can define all kinds of fonts on a particular display), there are intricate relations between the display's resolution (display element size and separation) and the appearance of a dot matrix font. The 'best' font cannot be specified in terms of the dot matrix structure, but only in terms of the final appearance (such as height width ratio, relative spacings and line thicknesses). In designing a font according to these specifications and mapping it onto the displays's resolution, compromises have to be made. Compromises also have to be made when mapping the display element size and shape of an existing display on the simulator. The most pregnant aspect to map has shown to be the gap between display elements. These gaps usually are of a very small size. In order to simulate a gap at least one simulator pixel has to be used. Depending on the ratio of actual pixel and gap the display elements (simulated pixels) can become very large when an almost perfect match between simulation and actual situation is wanted. Therefore, in order to maintain character subtenses that compare with normal display work (about 20'), the viewing distance of the simulator display has to be increased. The simulations in the experiments are all sub-optimal mapping solutions and the viewing distances have been adjusted.

The subjective evaluation element is investigated by two instruments: the Task Load indeX (TLX) and the Display Evaluation Scale (DES). Supplementarily a new scale is in development: the Task Evaluation Scale (TES). The first two instruments serve a different goal: the TLX measures the experienced task load while the DES measures the subjective impression that a display invokes. Although both instruments are applied in connection with a working (experimental) task on a display the TLX results are task specific, while the DES results are not. Therefore the TLX can also be used to evaluate the tasks that are carried out by experimental subjects. The Task Load Index or TLX [2] was developed at the NASA Ames Research Centre. It is a multi-dimensional rating procedure providing an overall workload estimate based on individually weighted ratings on six sub- scales. The reasons for selecting the TLX as the instrument to evaluate subjective workload were the following:

- There is no doubt that workload is of a multi-dimensional nature. The TLX accounts for this.

- The TLX uses individually weighed scores and thus accounts for individual differences in the experience of workload.

- The rating procedure is fast.
- The method provides insight in the sources of workload.
- The ratings can be obtained retrospectively.
- The method has been tested and validated in a variety of settings.

The TLX scale consists of six subscales, rating for mental demand, physical demand, temporal demand, estimated performance, invested effort and experienced frustration. In addition to obtaining scores on these subscales the items are ranked according to the importance in contributing to overall workload with a pairwise comparison task. An overall workload score is then calculated as the sum of the subscale ratings * importance weights. There was one problem with the application of the TLX in our experiments: it is written in the English language while we used Dutch experimental subjects. Therefore difficulties with the interpretation of the scale descriptors might arise. It was decided that a Dutch translation, although not validated, would be preferable to the original version. The Dutch TLX was implemented on a PC for automatic administration. Instead of using paper and pencil, the respondents perform the ratings on a display by positioning a cursor on a fixed length line with the left and right cursor keys. The pairwise comparison task is also computerized. The translated TLX was put to the test in an experiment on a number of different actual displays [3]. It appeared that the TLX discriminated better between the displays than an overall workload score that was obtained directly. The TLX workload score was also in good agreement with the performance results. Furthermore, low correlations were found between

930

the individual weights and the task load index (a high weight for a scale did not automatically correspond with a high rating on the same scale) which indicated that the subjects were well aware of the difference between the rating and pairwise comparison task. Lastly, no subjects indicated having trouble comprehending the meaning of the scale items (as provided by the item descriptions). In a second experiment that was carried out both on actual displays and in simulations the TLX produced very good results. These are discussed in the results section.

The Display Evaluation Scale (or DES) was specially developed by us to measure the subjective quality of displays. The scale consists of a number of items that together provide an almost complete description of a display from the ergonomic point of view. The DES items cover the following display characteristics: brightness, contrast, fore- and background colour, sharpness of the image, response speed, flicker, reflections and font. The DES is also used in a computerized version. For reasons of compatibility the actual scoring procedure is exactly the same as for the TLX. Assuming that the DES items provide a more or less complete description of a display, an overall appreciation score may be obtained by following the same procedure as the TLX. The usability of such a procedure will be investigated by including a pairwise comparison procedure in the second version of the DES and calculating a weighted overall score. This score can be compared with a directly obtained appreciation score that is also obtained with the DES.

The development of the Task Evaluation Scale, abbreviated as TES, is not complete, but then it is primarily intended for a global evaluation of the experimental tasks. The reason that the tasks are evaluated is that non-discriminating tasks or otherwise unexpected performance results may be identified and interpreted from this scale. Furthermore, this scale provides a good opportunity to have the subjects directly assess the experienced workload. The implementation and rating procedure are identical to the DES.

4.3. SOME RESULTS

A number verification task (as mentioned in section b) was carried out on three actual displays

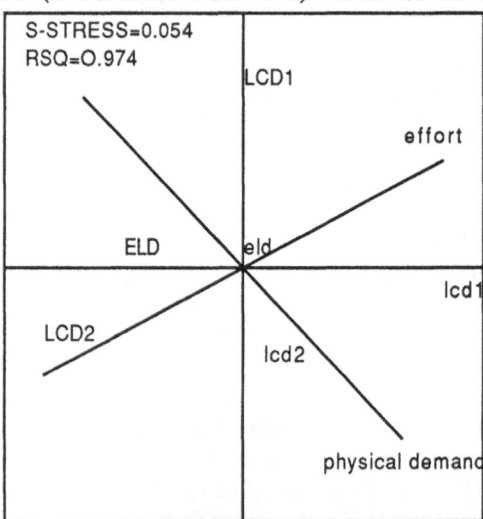

Fig 4.1 -D MDS solution for displays, based on TLX data. Lowercase letters indicate actual display, uppercase their simulated counterparts. eld= electroluminescence, lcd1= non backlit LC, lcd2= backlit LC.

and their simulated counterparts by sixteen subjects. The displays were an electroluminescence display, a non backlit LCD and a backlit LCD. The simulations were carried out on a simulator prototype. The data obtained with the TLX and with the DES were subjected to a multidimensional scaling procedure.

The data allowed a reliable representation of the six displays (three actual and three simulated) in two dimensions in either case. Figure 1 shows the found solution for the TLX, figure 2 depicts the result of the scaling based on the DES items.The 'vectors' in both figures are regression lines obtained by regressing the TLX and DES item scores over the MDS solution. Only significant regressions are shown. The item descriptions are printed near the end corresponding with a higher score.

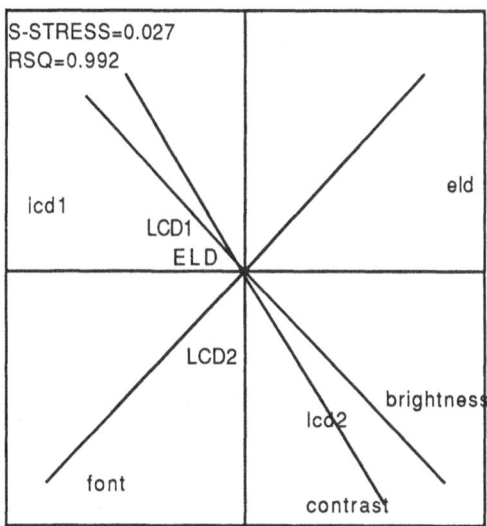

Fig 4. 2-D MDS solution for displays, based on DES items. Legend as for fig. 1.

As can be seen in figure 1, the simulated displays are all represented in the left half of the plot, while the actual displays all lie in the right half. Thus there must be a systematic difference between both categories in the sense that one or several of the TLX ratings differentiate between simulated and actual displays. It can also be seen that the subjectively experienced physical demand highly discriminates between actual and simulated displays: the perpendicular projection of the displays on the regression line draws the actual displays far apart from their simulated counterparts. Furthermore, the item discriminates well between the three actual displays, but not between the three simulated displays. This can be explained by the fact that the simulations were all carried out with the same instrumentation, including the subject's response keyboard, whereas the instrumentation for the actual displays differed from display to display. These differences may have led to differences in experienced physical demand, eg. the simulation display (which was the same CRT in all three simulations) may have imposed a lower physical demand on the subjects, just as the response keyboard may have. The lesson to be drawn is that 1) physical demand may vary between display/keyboard combinations and 2) a close match between the simulated and actual images alone is not sufficient: the physical environment must also be simulated with high fidelity. With respect to the other significant dimension, the invested effort, it can be remarked that it discriminates between both actual and simulated display technologies, but that actual and simulated display do differ from each other. This latter is especially true for the LCDs. The invested effort is consistently rated lower for the simulated display than for the corresponding actual display. Figure 2, based on the DES, provides a very clear picture of experienced differences between the six displays. The actual displays are drawn apart in space, while the simulated displays are much closer to each other. The significance hereof is that although the simulated displays are rated differently, those differences are much smaller than the differences between the actual displays. Furthermore, the spatial pattern between the simulated displays is not strongly deviating from that between the actual displays. It would seem therefore, that a number of display technology characteristics are maintained in the simulations, but that the differences between these simulated displays are smaller than between the corresponding actual displays. This result is not surprising, since the simulations did not include a number of relevant characteristics, such as response speed and reflections of the cover, and since colour and luminance could not be simulated 100% correct. The regression results can also be seen in fig. 2. Contrast and brightness share the same direction, while font is about orthogonally related to these. The two dimensions can be treated as independent from each other. This implies that if we are interested in investigating the luminance characteristics of a display we can use a standard font and vice versa and that if we want to optimize a display we have to pay attention to these two factors at least.

4.4. CONCLUSIONS

The prototype of the simulator, which provides only static simulations and in contrast to the 'high end' simulator cannot simulate response speed and reflections, has a display resolution of only 512 * 512 and cannot handle the full luminance range, was used for developing the experiments and subjective methods. Therefore the simulator prototype still lacked validity: although the simulations exhibited a number of display technology related characteristics the differences between three actual displays could not be fully simulated. Two aspects of simulation, the mapping of the pixel dimensions of actual displays and the mapping of a font onto the simulator have been the most difficult parts and resulted in relatively large display elements. In order to compensate for these large display elements, the viewing distance had to be increased, in some cases up to over 2 meters.   The employed subjective methods are sensitive to differences in the presented images and can be used both for investigating display technology dependent characteristics and for simulator validation.   It can be expected that the high end simulator, with (according to the specifications) its dynamic properties, the capability to simulate reflections and the better simulation display resolution, improved stability and luminance range, will produce better discriminating results and will therefore be very suitable for the investigation of flat panel display characteristics.

5. CONCLUSIONS

The Dissim project finishes at the end of this year, the simulator will be then available for third parties. Furthermore, the simulator will be used in the Esprit II project ADOT to research the use of color in flat panel displays. In this way, we have achieved one of the most important objectives of DISSIM "to increase the cooperation between users and display manufacturers".

6. ACKNOWLEDGEMENTS

This paper has been written with the contribution of all the partners of the DISSIM project. Specially, I would like to mention Mr. Offermans and Mr. Huntjens from Océ-Nederland for their modelling work, and Mr. van Huijstee from Twente University for his help during the integration phase. Finally, I would like to thank Mr. van der Meulen from Océ-Nederland B.V. for his help even after he left the project.

7. REFERENCES

1) Bakker W.and Bosman D., Engineering, how we see a display. SID 1988 Digest of technical papers,lecture 22.5

2) Koenderink, J.J., et al. Models of the retinal signal processing at high luminances, Kybernetik 6, pp 227-237, (1970)

3) Levis J.P. Final report: Convolver, luminance controller and color look up table, Myfra S.A.

4) Meulen A.E., Architecture, requirements and specifications display simulator. Esprit Technical week 1987,(North-Holland, Amsterdam)

5) Meulen A.E. and Placencia Porrero I., Requirements,Specification, and Architechture of a Real Time Display Simulator. SID 1988 Digest of technical papers,lecture 23.2

6) Placencia Porrero I., Display modelling and its software implementation. Esprit Technical week 1987,(North-Holland, Amsterdam)

Project no 834

Comandos Integration System (CIS)

S. Copelli, W. Cossutta, D. De Sano, G. Oldano, Q. Zhong
Applied Research Group
Via Pio La Torre, 14
20090 VIMODRONE (MILANO)
ITALY

Abstract

The COMANDOS Integration System (CIS) is one of the technical streams of the CO-MANDOS project. The main goals of CIS are:

– the definition of an integration methodology based on the O-O paradigm and the development of all the techniques needed to support the methodology
– the definition of tools which will support the development of O-O interfaces on top of pre-existing application systems

With respect to the COMANDOS objectives, CIS aims to allow the COMANDOS applications to access pre-existing application systems that cannot be re-implemented on COMANDOS machines. The architecture of CIS, however, has not been conceived only for COMANDOS machines, therefore it can be used as a stand-alone Generalized Integration System.

CIS was developed with the objective to overcome, as much as possible, the limitations of the traditional mapping approach in order to allow the integration of a wide spectrum of database and non-database applications. Two new original mapping techniques has been defined: Operational and Protocol-Emulation Mapping.

One of the possible kind of applications that can be integrated using CIS are the on-line telematic services, in this area some exploitations of CIS approach are already on-going. In fact, products giving the possibility of accessing large set of telematic services with an homogeneous, easy to learn and manage end-user interface are strongly requested on the European market.

The CIS system has proved to be a good tool for the development of such kind of products; some industrial and pre-competitive research projects, based on CIS, has been already started in A.R.G. in this area, as explained below.

1. Introduction

The COMANDOS Integration System is one of the technical streams of the COMANDOS project. The primary objective of COMANDOS is to set up a flexible, reliable and easy-to-use environment for the development and management of distributed applications in the office context. One of the peculiar characteristic of the project is the attempt to define the overall environment under the Object-Oriented paradigm; both the architecture and the tools are modelled in terms of objects. A new operating system and a new programming language was designed in order to cope with the needs of both generic applications and database applications in a distributed office environment.

With respect to the COMANDOS objectives, the goal of CIS is to allow the COMANDOS applications to access pre-existing application systems that cannot be re-implemented on the COMANDOS machines.

The architecture of CIS, however, has not been conceived only for COMANDOS machines. Therefore CIS can be used as a stand-alone tool and it can be seen as a Generalized Integration System giving to the application programmers the ability of interfacing a wide spectrum of pre-existing application environments based on several heterogeneous Information Management Systems.

CIS shares with COMANDOS project the Object-Oriented approach and, in particular, the CIS Data Model is a subset of the COMANDOS data model.

Note that the term "Information Management System" is used instead of "Database Management System" because CIS aims to integrate not only Database Systems but also File Based environments and Information Services in general, such as Public Data Banks.

The main goals of CIS can be summarized as follows:

- the definition of an integration methodology based on the O-O paradigm and the development of all the techniques needed to support the methodology

- the definition of tools which will support the development of O-O interfaces on top of pre-existing application systems

The main functionality provided by CIS is an Uniform Interface to Heterogeneous Information Systems. At this level of interface a global application has a view of different and independent information systems which can be accessed through the same interface.

Another important aspect of the system is the architectural one. The CIS infrastructure must be accommodated in a great variety of heterogeneous application environments, therefore flexibility in the allocation of CIS components is a major requirement. As an example, when a Data Bank is integrated, it is generally not possible to physically allocate any software component on the host machines.

2. Problem Statement

In order to Integrate heterogeneous systems it is necessary to have an environment and tools which support the integration process.

The capabilities of integration tools can be measured at two different levels:

a. The degree of integration provided. Two different levels can be defined:

1. the uniform access level (e.g. Multidatabase approach);
2. the globally integrated level (e.g. Transparent Global Schema approach).

Uniform access means to provide the applications with a single data model and data manipulation language for interacting with Information Systems characterized by differences in either the data model and the data manipulation languages.

Integrated access means to provide, in addition to a uniform data model and language, a semantically integrated description of different sets of data (global view), possibly managed by different systems, that however refer to the same sets of entities.

b. The degree of heterogeneity of the systems to be integrated, from the viewpoint of:

1. hardware and software environments;
2. data models used by the systems (e.g. relational, Codasyl, etc., or no data model at all);
3. degree of access to existing systems.

This latter aspect is a relevant issue since it affects the degree of control that the system integrator has on the system that manages the data to be integrated.

In certain cases the system integrator has the control of the data management system and therefore has the availability of all the functionalities provided by the system and can access directly the data. This situation is typical of the application environment located on the user workstation.

In other cases the data management system is completely hidden and therefore the data can be only accessed through a pre-existing application interface. In this case the system integrator has no control on the underlying data management system.

The on-line services belong to the last class. They are located remotely with respect to the user workstation and they can be manipulated only through a pre-existing user interface based, in general, on a TTY oriented protocol. No direct manipulation of data can be implemented.

3. Existing Approaches

Several generalized systems has been developed in past years specifically for the integration of DBMS's systems. However recent trends in software technology are showing the need for providing systems and tools for the integration of a wider spectrum of applications.

In fact, even a perfect database management system cannot be the sole repository for data in a complex computing system. There will be always other data storage subsystems as well, because either these other facilities predate the database system or their functions are so specialized that it is difficult or unwise to incorporate them into a generalized database.

The area of integration of non-database applications is still a research issue. Most of the projects currently start in this area are based on the definition of ad hoc solutions for integrating specific applications instead of generalized integration systems. In the following two of these projects are reviewed.

3.1. Generalized DBMS Integration

A first limit of these projects is on the degree of heterogeneity of the systems to be integrated.

In fact they are focused on the integration of DBMSs only, where the system integrator has the control of the systems to integrate.

The main reason of this drawback can be found in the kind of technique used for mapping between the common global data model and the local systems, called Schema Mapping.

Using this approach structural mappings between the data elements of the various pre-existing Local Schemata and the data elements of the common data model should be defined by the designer of the Integration environment.

Once the mapping is defined, the distributed DBMS supports the translation of operations issued by the integrated applications in terms of operations on the data schema of local databases. The mapping is defined only once and all integrated applications use this definition.

The Schema Mapping technique is very powerful from the application programmers' point of view. However, it has a limited applicability because there are many situations in which this technique does not work properly, e.g. when the semantics of data is deeply dependent on the way in which applications manipulate data and is only partially expressed by the Schema, or the Schema does not exist at all (e.g. applications built on a file system).

Another drawback of the Schema Mapping approach is that it does not provide any support for making available at global level operations on data that local applications may implement. Thus distributed applications cannot be generated by reusing partially pre-existing local application programs.

3.2. Specific Applications Integration

The importance of integration has led several organizations to build their own integrated applications using ad hoc solutions, without defining generalized tools. We recall here two projects: [ELLI88] and [DGIX86] which can be considered in some sense internal to the European Community.

The solutions applied by these projects are interesting, because they have been used in a real-life context; however they are specific to an application environment and not general.

In particular, [ELLI88], a System for Trans-National Accounting, uses a functional abstraction mechanism (instead of a traditional data mapping mechanism) which is in the same direction of the Object-Oriented approach of CIS.

Under the framework of DGIX [DGIX86], two internal integration project have been started by EEC: UAS (User Agent Service) and IDS (Information Dissemination Service).

UAS addresses the problem of connecting to a set of independent local systems and IDS aims at providing a uniform access by means of a unified query language. IDS requires a complete restructuring of all the local Data Bases to allow their integration. This strategy could not be followed when the systems to be integrated belong to different organizations and the system integrator has no complete control on them (e.g. Public Data Banks).

4. CIS: an O-O Approach to Integration

CIS was developed with the objective to overcome, as much as possible, the limitations of the Schema Mapping approach in order to allow the integration of a wider spectrum of non-database applications.

A description of the data model, the mapping techniques and the architecture of CIS is given in the following.

4.1. The Data Model

The data model used in CIS as common global model is the result of merging the concepts of Object-Oriented Databases together with some principles of Object-Oriented Programming Languages.

An Object-Oriented database consists of objects that represent real world objects, such that the relationship between the former and the latter is one to one. The relationships between real world objects are modelled as relationships between the corresponding database objects; such relationships are commonly called properties.

In order to make manageable large collections of objects stored in the database and supporting query facilities the concept of Class is defined. Classes are special objects that do not represent any real world object: they rather abstract sets of objects having the same properties.

The notion of class has both an intentional and an extensional aspect. Intentionally, a class gives the "type" of an object in terms of a set of properties possessed by the object. Extensionally, a class denotes the set of objects that at a certain time belong to the class. The extensional aspect is fundamental for defining the meaning of queries over an O-O database.

The notion of a Database Class as defined above has no direct counterpart in Object Oriented Programming Languages; however it can be seen as the result of merging the notion of (intentional) "Class" with the notion of (extensional) "Collection". Therefore, the generic operations on abstract classes (defined later on) are derived from the operations on "Classes" and on "Collections" of Object Oriented Programming Languages.

In summary, the abstract data model of CIS has been obtained by:

a. Defining the notion of Class in the Database sense, i.e. both intentional and extensional,
b. Defining as generic operations on Classes the typical Object-Oriented Programming Language operations on Classes and Collections.

A database object consists of some private data and a set of operations (methods), inextricably bound to it, that can access that data. Operations on objects are executed by sending to the object a message telling what to do. Upon receiving the message a piece of code, implementing the operation invoked, is executed by the receiver object.

Objects are identified by an *object-identifier*. This provides a handle by which an object can be manipulated by a message expressions, the sole legal operations on object identifiers. Object identifiers must uniquely identify an object instance inside all the instances that may coexist in the system at run-time.

Object-identifiers are created by CIS at run-time, when an object is extracted from the local application environment and brought in main memory. An object-identifier is valid only during a working session of the integrated application with a particular local environment and it cannot be use in subsequent sessions.

Message expressions are embedded in the Application Programming Language (the C-language for the first prototype of CIS) by means of some special constructs and primitives allowing: classes and objects management, distributed environment configuration and query request definition.

Classes, have an abstract interface (called "abstract class") and one or more implementations (called "implementation class(es)").

The abstract class defines the behaviour of an object, i.e. all the operations to which the object can respond to and all the properties of the object that can be accessed by another object.

The implementation class provides data structures and procedures that implement the abstract class operations.

In order to take advantage of the existence of indexes or query facilities supplied by the underlying applications (e.g. an indexed file or the query language of a data bank) the notion of "key" is included in the data model.

Only attribute properties can be defined as key. A set of relational operators are defined for each kind of basic type domain, so that simple predicates can be defined on key attributes. Then the boolean operator AND, OR and NOT can be used in order to express predicates involving several key attributes inside the same class. The query predicates defined using these mechanisms can be used by means of specific operations defined on abstract classes.

Two levels of operations on abstract objects are provided (fig. 1):

1. <u>Object-at-a-time Operations</u>. They can be issued on a single class or object instance at a time; they return, as result, the identifier of an object or the value of some attribute properties. They can be subdivide in two sub-categories:

<u>Generic Operations</u>; they must be defined on all abstract classes or explicitly dropped. These operations are derived from those defined on "Classes" and "Collections" in O-O Languages (e.g. First, Next, Remove, ReadAttributeValue, ...) <u>User-Defined Operations</u>; They can be defined in addition to the generic ones for specific classes. They main purpose of these operations is to allow the exporting of operations implemented by pre-existing application programs at integrated system level.

2. <u>Associative Operations</u> (Query Language). They give access to the query support of CIS. The query support is implemented on top of the Object-at-a-time interface supplied by the CIS O-O data model; thus it is completely independent from the underlying local applications. Queries are executed interactively interpreting a query-program generated using object-a-time operations.

The Query Language is based on a first-order logic language, called calculus query language. This kind of language allows to express in a more clean way the query structures. However, other more friendly languages (such as SQL-like ones) could be defined, in the future, on top of the calculus-query language.

4.2. CIS Mapping Techniques

The separation between abstract classes and implementation classes allows the definition of new integration techniques able to solve some of the problems that cannot be addressed with the Schema Mapping approach.

In particular, two new techniques have been defined by CIS called: **Operational Mapping** and **Protocol Emulation Mapping**. The first technique can be used when data managed by the pre-existing application are directly accessible and the lack of a data model prevents the use of a Schema Mapping approach. It consists of writing the implementation of the object operations explicitly as local programs working directly on the Information System to which the implementation class is associated.

If the data are not directly accessible (e.g. Public Data Banks), the *Protocol Emulation* technique can be used. It consists of interfacing the application managing the data through its interface as it is typically defined in the application's user manual. In this case the abstract object oriented operations are implemented through an interaction with the

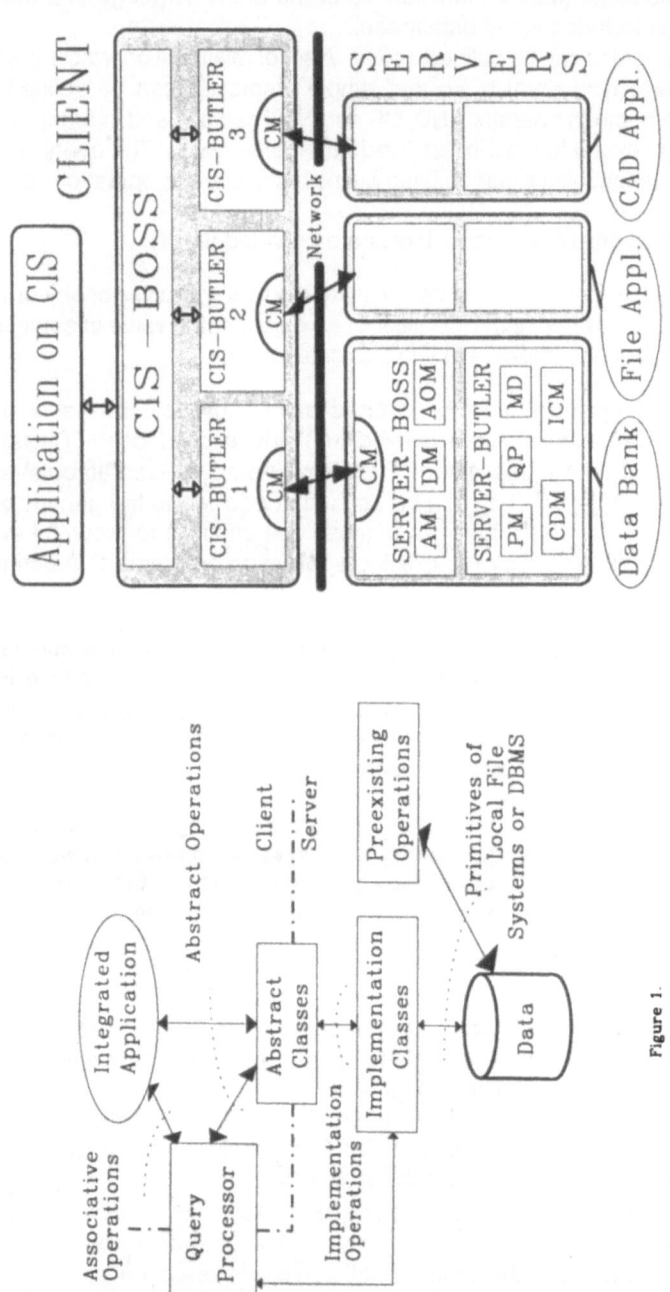

Figure 2.

Figure 1.

local application. Of course, this technique may impose some constraints on the abstract object oriented view, since not all the abstract operations might be implemented.

The possibility of having multiple implementation classes corresponding to the same abstract class provides a powerful support for dealing with heterogeneity (both at hardware and software levels). In fact, given an abstract class, there may be several implementation classes, one for each system storing data that must be accessed through the abstract class. In this way a uniform interface in terms of abstract classes can be provided on top of heterogeneous local environments.

Another aspect where the CIS approach appears relevant with respect to the problem of integration concerns the issue of software reusability. Because it is possible to make available at O-O schema level those operations, implemented by local pre-existing application programs, that may be of interest in the implementation of distributed applications. This objective is achieved associating at the abstract class level the abstract specification (operation name plus the definition of input/output parameters) of the operations implemented by local applications.

5. The CIS Architecture

CIS supports the development of integration applications accessing data managed by heterogeneous, independent applications located on different nodes of a computer network. Thus the local applications may run on different hardware and operating systems with, in general, different data representations and storage management.

CIS models this situation with the notion of *logical-node*. Two types of nodes are defined: *Clients* and *Servers*. There is no relationship between CIS logical-nodes and the physical nodes of the network. Several logical-nodes may be located on the same physical node, viceversa a single logical-node could be implemented on top of a distributed system.

A Server node is the abstraction of (or part of) a pre-existing application environment. The role of a CIS Server is to implement the mapping from the local environment to the Object-Oriented interface based on the CIS data model and to manage the active objects in main memory. A Client node is the abstraction of an integration application based on CIS accessing the Object-Oriented data view supplied by one or more servers.

Servers cannot communicate with each other; moreover an O-O schema must be implemented by a single Server. Thus the only relationships between Servers are implemented by Clients. This structure guarantees a high degree of site autonomy and keeps very low the degree of connection between the pre-existing application environments.

At design level no assumption is made on how Server units are implemented, because they depend on the pre-existing heterogeneous application environments on top of which they work. In the first prototype of CIS, each Server is implemented as an independent process running on the same host of the interfaced application.

The Client node is made by two modules:

- a programming library that supplies the O-O schema manipulation primitives in a procedure call style, embedded in the application programming language (C has been used in the first prototype) and manages the communication with the Servers.

- the integration application code written by a CIS user (application programmer) and calling the library primitives

A refinement of the internal architecture of CIS functional units is given in figure 2. With respect to the CIS_CLIENT we have:

- One *CIS_BOSS*, this module is in charge of support for the interface with the *CIS_USER* and manages the CIS_BUTLER modules.

- Zero, one or many *CIS_BUTLER* modules. There is one of this module for each CIS_SERVER that is accessed in parallel by the Client. This module performs CIS operations in terms of parameter type checking, parsing of predicates (keyed operations) and of queries (associative operations). Then prepares the messages for the CIS_SERVER related to it, according to the CLIENT_SERVER_PROTOCOL.

- one *CIS-Communication-Manager (CCM)* for each CIS_BUTLER. This module has a symmetrical functional entity allocated on the CIS_SERVER and it is in charge of accessing the network services. The Client CCM and its counter part on the Server communicate by means of a lower level protocol called CCM_Protocol and they provide to the higher level of Client and Server an interface called CCM_Interface.

The CIS_SERVER has a similar BOSS/BUTLERS architecture:

- *SERVER_BOSS*, there is a only one instance of this module inside a SERVER. It is composed by three sub-modules:
  - *Allocator Manager*, that is in charge of allocating and deallocating main memory needed to store objects.
  - *Active Object Manager (AOM)*. This module is in charge of activating (i.e. bring in main memory), deactivating (i.e put back in the underlying application environment) and managing (when active) all the CIS-Object.
  - *Dictionary Manager (DM)*; it manages the Object-Oriented Dictionary of the server, containing all the information needed to validate and process the Client requests.

- *SERVER_BUTLER*, there is one instance of this module for each CLIENT connected to the SERVER at the same time. This module can be subdivided into the following sub-modules:
  - *Message Dispatcher*, that has the role of managing the synchronous communication with the CIS_CLIENT BUTLER and of dispatching the messages inside the SERVER.
  - *Query Processor (QP)*. This module processes the queries previously parsed by the CLIENT_BUTLER. The following activities are performed by the QP: type checking, query decomposition and translation into query programs using the abstract operation interface (this operation are invoked by means of the Message Dispatcher module)
  - *Implementation Class Module (ICM)*. This module is strictly dependent from the pre-existing application interfaced. It implements the O-O abstract operations in term of the primitives of the local environment. All the mapping operations are performed by this module.
  - *Predicate Manager (PM)*, that is responsible for checking the key_operation predicates against the CIS schema
  - *Connection/Disconnection Module (CDM)*, that is in charge of implementing the operations of connection to, and disconnection from the underlying application environment.

6. CIS Prototype Implementation

The prototype implementation of CIS has been carried out taking into account the following objectives:

– demonstrating the feasibility of the Object-Oriented integration techniques defined within the COMANDOS project

– verifying the CIS approach in a distributed (network) environment characterized by heterogeneous hosts, Operating Systems and applications.

The feasibility demonstration depends on the selection of a set of pre-existing applications, to be integrated, suitable for stressing the modelling features of the new O-O integration techniques and, in particular, the operation oriented mapping mechanisms. Taking into account the above considerations, three kind of applications has been selected:

1. A file-based application using both indexed and sequential access files.
2. A graphic application implemented on a CAD management package running on MS-DOS.
3. A public data-bank, with a TTY-oriented interface protocol accessed through telephone line. The bank selected is an Italian public bank supplied by the "Corte di Cassazione" storing data concerning laws of the Italian Government, sentences passed by lawcourts, press news concerning legal aspects, etc..

The test of the prototype in a heterogeneous network environment allows to verify two other important aspects of CIS approach related to distribution problems:

– the Client/Server Architecture with respect to the partition of OIS functions between the two logical modules and the high level communication protocol defined between them.

– the portability on different Operating Systems of the CIS Software and, in particular, of the Server module that, usually, resides on the same host as the pre-existing application to be integrated.

6.1. Development Environment

The Prototype 1 has been developed on a mixed MS-DOS and UNIX (XENIX) network environment. All the MS-DOS workstations are 80386 or 80286 IBM compatible machines, the XENIX multi-user system is an INTEL 320. All these systems are connected by means of a local area network. Some VT-100 terminals are directly connected to the serial ports of the XENIX machine. One of the serial ports of the XENIX system is connected, through an Hayes (automatic dialling) modem, to the switched telephone line, in order to access remote data banks.

The local area network is the OpenNet™ by INTEL. The net software emulates, on MS-DOS sites, the IBM MS-NET™ network services and it is, at low level, fully compatible with NETBIOS system calls interface. Moreover the XENIX system is seen by the MS-DOS workstations as a file server.

The Prototype 1 has been developed in C language. Under MS-DOS, the MicroSoft-C, version 5.1, development environment has been used. On XENIX system, the standard C compiler and its related development tools has been adopted. The user interface of the demonstrator applications has been implemented under the MicroSoft-WINDOWS 2.03 environment on MS-DOS workstations.

6.2. The Demonstrators

Three main goals are addressed by the final demonstrators of CIS prototype:

- giving the feeling that the three heterogeneous application environments, described in the previous section, can be seen, through CIS, using the same common O-O model. Moreover, they can be browsed, updated and queried using the same manipulation operations, i.e. those supplied by CIS.

- showing the integration capabilities of CIS system producing, by means of a CIS application, some documents containing data coming from all the three applications interfaced.

- demonstrating how CIS make more easy the development of end-user applications implementing high level user interfaces for accessing public telematic services.

The last objective has been considered because a short term industrial exploitation of CIS results has been advised in this direction. In fact, there are many requests coming from the market, both in Italy and in Europe, of products that give the possibility:

- to access large set of telematic services with an homogeneous, easy to learn and manage user interface

- to export very easily data collected from the remote services inside the user local application environment (e.g. a local information system, a document management system, etc.)

Three demonstrator applications have been developed, one for each of the above outlined objectives:

6.2.1. A CIS Interactive Interpreter.

This demonstrator allows the user to manipulate, interactively, an application environment interfaced through CIS.

The Interpreter pilots the user in the definition of the parameters of the selected operation by means of a set of dialog boxes (data windows). Before invoking the operation it checks that all the parameter values are correct, then, it calls the CIS client operation and returns the results to the user. If an error occurs, an alert message is displayed and the user can decide to recover and continue or to stop the execution.

Also the query processor can be invoked from the Interpreter interface. The queries can be stored on and recalled from mass storage. The query results are shown in an "edit" window and they can be browsed and/or saved.

This application meets very well the objective of giving the flavour that a set of heterogeneous application environments can be seen through the same data model and manipulated using the same operations. Moreover, the Interpreter is a very useful debugging tool for the developers of the Implementation Class Module on CIS-SERVER, because, without any additional programming, it allows to call all the O-O manipulation primitives implemented by a Server.

6.2.2. A Sample Integration Application.

This application produces a printable document containing information retrieved from all the three application environments interfaced by the CIS Prototype: the file-based application, the graphic application and the data bank.
This application and in particular its source code, demonstrates how, using CIS interface, it is easy to retrieve data from heterogeneous environments without taking care of their original sources.

6.2.3. A High Level User Interface for Accessing the Data Bank of "Corte di Cassazione".

CIS is used here as an internal tool that encapsulates the data bank native interface supplying an O-O interface. Thus the end-user interface manager is completely decoupled from the data bank native protocol. Changes on the native protocol (e.g. changes to command syntax) will affect, in general, only the CIS-Server (implementation class modules) and not the User-Interface Manager.

7. Conclusion and Future Exploitations.

The prototype implementation has pointed out that the CIS approach has good performances with respect to the integration of a wide range of pre-existing application environments in a real-life context.
As said above one of the most relevant areas with respect to industrial exploitations, is the integration of public telematic services. In this area A.R.G. is currently working at both industrial level and pre-competitive research level.
With respect to the industrial environment, CIS based product are under development or planned in a short term period, in particular:

– A product , based on CIS, is sponsored by the Local Government of "Regione Toscana", one of the Italian regions. This system will be an enhanced version of the prototype for accessing the data bank of "Corte di Cassazione" developed inside COMANDOS. The integration of other telematic services available on the computer network of "Regione Toscana" is also planned: a second data bank and the internal electronic mail service.

– Another product for accessing public data banks specifically oriented to the professionals' office environment (lawyers, notaries, etc.) is currently planned.
With respect to pre-competitive research, the data-model and architecture of CIS are under evaluation inside two ESPRIT II project:

– TOOTSI - a project concerned with the development of high level interfaces for accessing public data banks. CIS is considered as a mean for encapsulating the native protocols of the services supplying a uniform manipulation interface to the higher level of the TOOTSI architecture.

– ITHACA - the project aims to design and implement an integrated environment for the development of distributed application. In this project CIS is considered, as a first step for the development of an integration tool working at application level in a network environment.

With respect to the continuation of COMANDOS I and COMANDOS II, CIS will be the basis for the development of the data management support and, in particular, of the query processing facilities of the COMANDOS language.

REFERENCES

[ARG87] A.R.G., "Architecture of the COMANDOS Integration Systems", Chapter 4 of COMANDOS Global Architecture Report T2.1/D2, September 1987.

[BANJ87] Banerjee J., et Al., "Data Model Issues for Object-Oriented Applications", ACM TOOIS, Vol5, N.1, 1987.

[BATO85] D.S. Batory, W.Kim, "Modeling Concepts for VLSI CAD Objects" ACM TODS, Vol.10, N.3,1985.

[BATI86] C. Batini, M. Lenzerini, S. B. Navate, "A Comparative Analysis of Methodologies for Data Base Schema Integration", ACM Comp. Surv., Vol. 18, n. 4, dec. 86.

[BREI86] Breibart Y.J., Olson P.L. Thompson G.R., "Database Integration in a Distributed Heterogeneous Database System", Proc. Second Int. IEEE Conference on Data Engineering, Los Angeles, 1986.

[DASE86] "DASE Model", ECMA TC32-TG2 Working paper /86/89, October 1986.

[DAYA82] Dayal U., and Hwang H.Y., "View Definition and Generalization for Database Integration in MULTIBASE: A System for Heterogeneous Distributed Databases", Sixth Berkeley Conference on Distributed Data Management and Computer Networks, 1982.

[DEMU87] Demurjian S.A., and Hsiao D.K., "The Multilingual Database System", Proc. Third IEEE Conference on Data Engineering, Los Angeles, 1987.

[DGIX86] DGIX - CEC, "Guidelines for an Informatics Architecture", November 1986.

[ECMA85] ECMA-TR/32, "OSI Directory Access Service and Protocol", Dec. 1985.

[ECMA86] ECMA-TC/32-TG5, "Framework for Distributed Office Applications", Dec. 1986.

[ELLI88] Ellinghaus, Hallman, Holtkamp, Kreplin, "A Multidatabase System for Transnational Accounting", Proc. EDBT Conference, Venice 1988.

[FERR83] Ferrier A., and Stangret C., "Heterogeneity in the Distributed Database Management System SIRIUS-DELTA", Proc. Eighth Int. VLDB Conference, Mexico City, 1983.

[GLIG85] Gligor V.D., Popescu-Zeletin R., "Concurrency Control Issues in Distributed Heterogeneous Database Management Systems", in "Distributed Data Sharing Systems", Schreiber F.A. and Litwin W. editors, North Holland 1985.

PROJECT 870 TALON

TESTING & ANALYSIS OF LOCAL AREA OPTICAL NETWORKS

by

George Georgiou and Keith Ralphs

Cossor Electronics Ltd
Harlow, UK.

&

J.Lund Nielsen

NKT Elektronik
Denmark

ABSTRACT. The project, Testing and Analysis of Local-area Optical
Networks, or TALON, sits within Section 3, Office Systems, R & D
Area 3.2, Communication Systems of the ESPRIT Workplan. It is a
B-type project totalling some 400 KECUs spread over three years.
The TALON team comprises two partners, Cossor Electronics in the UK
and NKT Elektronik of Denmark.

1   INTRODUCTION AND OBJECTIVES

Optical local area networks will make a major contribution to
Information Technology in the future, as recognized by the ESPRIT
initiative. The overall objective of the TALON project is to
formulate a test philosophy for the installation and maintenance of
these systems, and where necessary, to develop new techniques for
optical test and analysis. In this way it is hoped to enhance system
reliability and reduce operating costs.

In essence, the project was devised out of a belief that existing
techniques and methods for testing optical fibre systems could not
automatically be applied to optical fibre LANs. This is largely due
to the generally more complex structures of optical LANs, the environment in which they tend to be installed, the shorter distances
involved, and the effects that failures have on systems. The project
was structured essentially in two phases. The first comprised a
study of the whole subject of optical LAN testing. The second was
to build and evaluate a demonstrator based on a novel test equipment
identified during the study phase.

947

2   TEST PHILOSOPHY STUDY

This study had the objective of developing a mode of thinking in respect of optical LAN testability. As the test philosophy was developed, two approaches for investigation emerged. Firstly, that of employing techniques such as fault tolerance and self-diagnosis at the design stage of an optical LAN, and secondly, that of stand-alone test equipment.

By the end of the test philosophy study, it was decided to concentrate on test equipment rather than fault tolerant techniques which are intimately associated with practically every aspect of network design. A thorough investigation would have been beyond the scope of the project.

The emphasis of the study was toward the LAN community, and more particularly to those whose job it is to install and maintain the network.

Three areas of test equipment need can be identified:

i)   Installation/Commission
ii)   Routine Maintenance
iii) Fault-finding repair/replace.

i)   Installation/Commissioning

This will normally be carried out by a contractor, not the end user. The fibre must be fully tested to ensure it will function correctly and has sufficient margin for ageing etc. The tests carried out will be:

- Fault Location (Backscatter) - to ensure there are no breaks or excessive bend losses.

- Attenuation - to ensure that the absolute loss does not exceed specified limits.

- Bandwidth - a check on the quality of the fibre and its transmission capability.

- Power - to ensure enough transmitted signal from the system's optical transmitter is received at the far end of the cable.

ii)   Maintenance

The customer or his appointed contractor would be responsible for maintenance and will need to check any tests against the original installation data. In most cases the equipment used will be the same as the original installation equipment.

This equipment may have been handed over to the customer by the installation contractor.

iii) Fault Finding/Repair and Recommissioning

Initially most optical LANs will be used to carry important information, which justifies their cost. Thus the rapid repair of faults will be vital. The topology of systems will mean some cables will be more important that others, with major data highways being vital to the overall operation of the system. Customers will dictate their fault finding policy upon the potential loss in revenue of system/cable failure. Some customers can afford to contract repair services to companies with a guaranteed turn around time, although many may have to provide their own capability.

Test Equipment must identify the fault condition to allow a clear repair action. In many cases substitution and temporary measures will be taken.

When correctly installed, optical LANs are reliable. It is therefore quite likely that any test equipment could be bought and rarely used. This will affect the price that customers are prepared to pay. The factors of cost and wide range of potential parameters for measurement have led us to investigate the feasibility of a multifunction tester.

3   MULTIFUNCTION TESTER

To meet the needs of LAN users, a novel multifunction Test Equipment to measure faults, attenuation, power and bandwidth was devised.

At present, 4 instruments are required to measure each individual parameter. Each instrument has expensive optical transmitters or receivers. Tests must be performed using each piece of equipment. This increases time and requires manual correlation and interpretation of each test.

The TALON equipment will reduce time and cost by using one instrument, and one set of optical components to measure all parameters. Two concepts for TALON equipment are shown in figure 1. Concept A will be achieved in the the project timescale. This novel instrument comprises a reflectometer with an additional remote receiver to measure attenuation and bandwidth parameters.

Concept B is the eventual aim, which will give a complete test with access to only one end of the fibre. In this case the reflectometer receiver will perform the function of the remote receiver described above. However, it is not yet clear if the information received from backscatter will be sufficient to perform the other functions (total attenuation and bandwidth) satisfactorily.

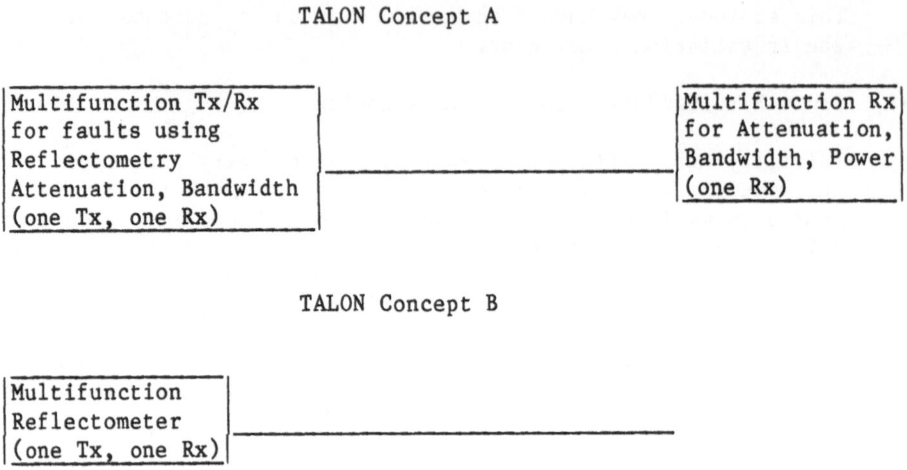

TALON Concept A

Multifunction Tx/Rx for faults using Reflectometry Attenuation, Bandwidth (one Tx, one Rx)

Multifunction Rx for Attenuation, Bandwidth, Power (one Rx)

TALON Concept B

Multifunction Reflectometer (one Tx, one Rx)

Figure 1   System Concepts for Multi-function Tester

The type of reflectometer employed was selected to facilitate measurements other than fault location. The technique chosen is known as Optical Frequency Domain Reflectometry.

4   OFDR (Ref 1)

High resolution fault location is essential in an environment where an error of one metre could mean wasting time in the wrong room or floor. Optical frequency domain reflectometry (OFDR) had been identifed in the earlier stages of the TALON project as being the most promising of new fault location techniques.

Compared with currently used techniques such as optical time domain reflectometry the operation of OFDR affords design trade-offs favourable to achieving high resolution. In addition the signal format can also be used to measure bandwidth.

The OFDR technique selected uses a modulated CW light source. The modulating frequency is varied in steps and the amplitude and phase of the reflected signal is measured at each frequency. Eventually amplitude and phase measurements are made at all the modulating frequencies. These frequency domain measurements are transformed to the time domain by using an inverse Fourier transform. This gives a similar trace of fibre characteristics to the OTDR technique. The upper and lower modulating frequencies used are set by resolution and maximum fibre length, respectively.

Previous investigations of OFDR have focussed largely on long distance, low resolution applications. The main emphasis of the project was to explore applications to shorter distances and higher resolution performance.

OFDR permits the use of a laser that is significantly cheaper than the high power pulse lasers employed in OTDR fault location of comparable resolution.

Box B can determine the bandwidth of the fibre by measuring how the amplitude of the received signals falls off as the modulating frequency is increased. Similarly by comparing the amplitudes of the received signals with a measurement made via a short test lead, the attenuation of the link can be determined. The receiver section of Box B can also be used for power measurements on a functioning system.

A block diagram of Box A is shown in Figure 2.

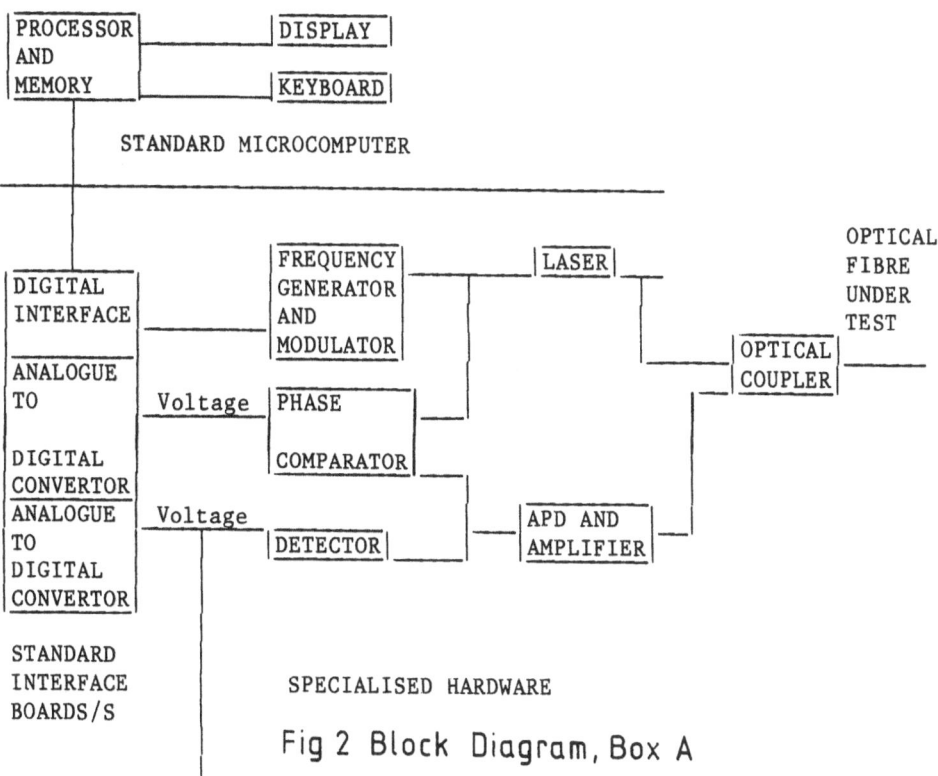

Fig 2 Block Diagram, Box A

A photograph of the demonstrator hardware is shown in Figure 3.

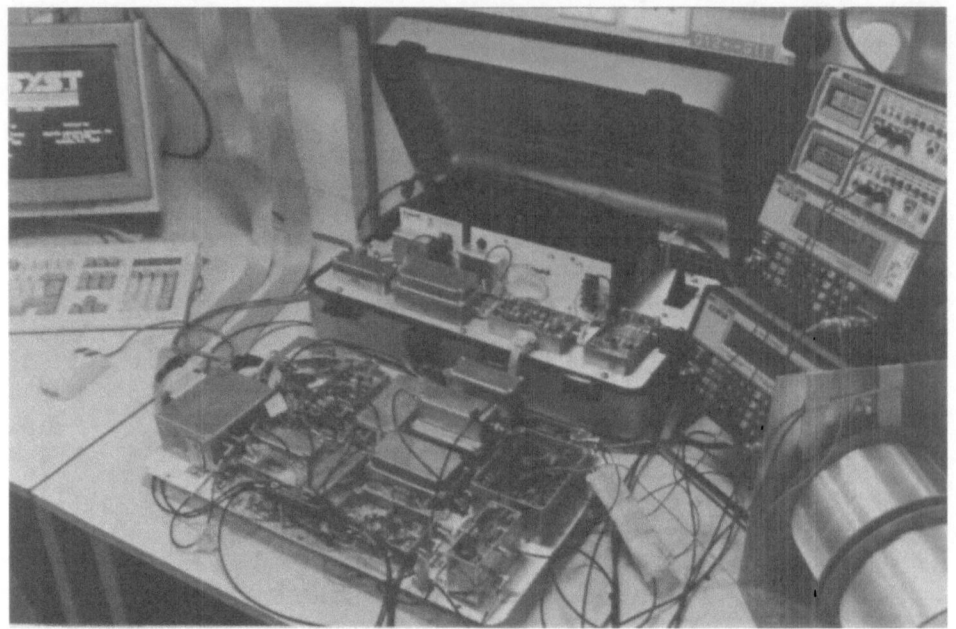

Fig 3 Demonstrator Hardware

5 RESULTS

The Evaluation Phase has been completed, successfully demonstrating (to varying degrees) all the measurement functions expected from the unit. Also an insight had been gained into the fundamental mechanisms and limiting factors of the OFDR method which affect the quality of the fibre traces obtained.

Presented here are a selection of results from OFDR tests carried out as described in Section 5 showing measurement of the various fibre parameters at progressive stages during the evaluation phase. Each set of results show both frequency domain plots plus the resulting inverse transformed time domain plot.

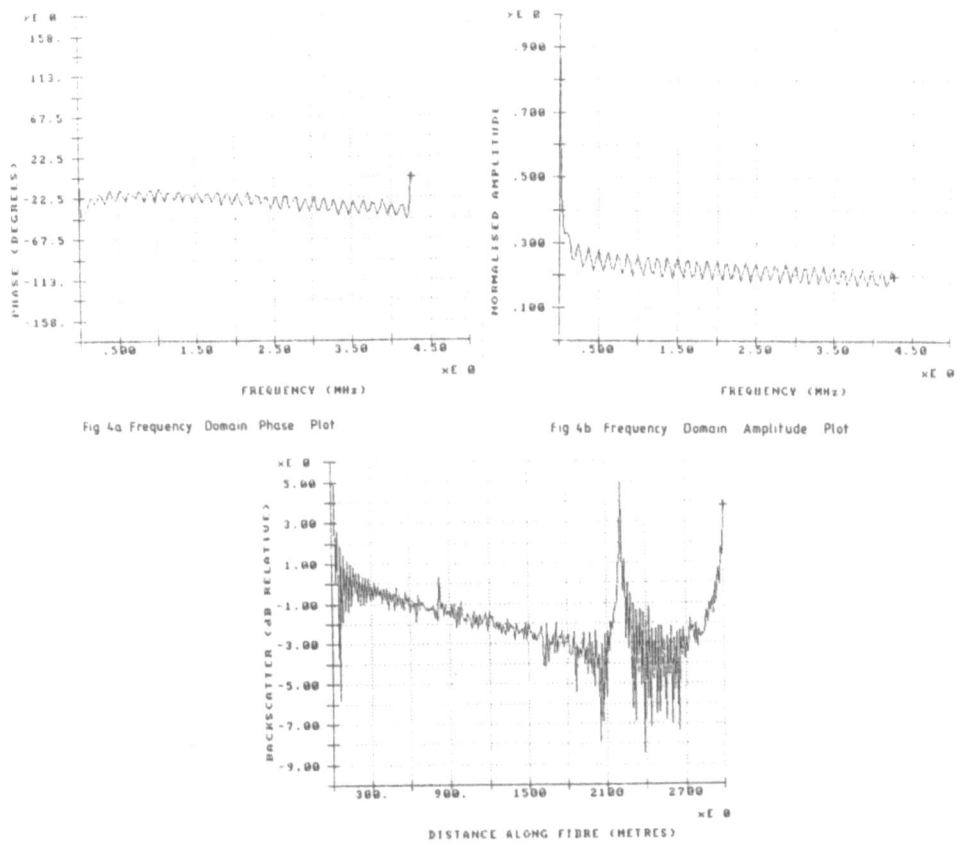

Fig 4a Frequency Domain Phase Plot

Fig 4b Frequency Domain Amplitude Plot

Fig 4c Time Domain Fibre Trace

Figs 4a, 4b and 4c: Early results on a 2.2Km fibre with frequency domain plots showing the characterstic "long fibre" shapes with periodicity due to the strong fibre end reflection superimposed.

954

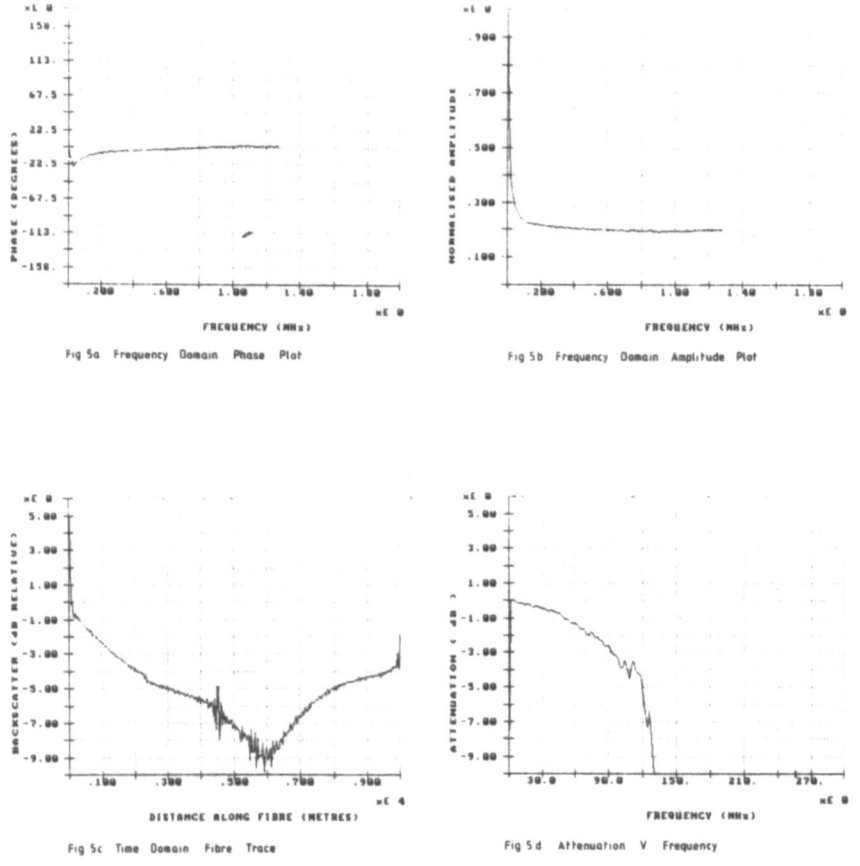

Fig 5a Frequency Domain Phase Plot

Fig 5b Frequency Domain Amplitude Plot

Fig 5c Time Domain Fibre Trace

Fig 5d Attenuation V Frequency

Figs 5a, 5b, 5c and 5d: Results from two spliced 2.2Km fibres
including a Bandwidth Profile and showing splice loss on the time
domain trace. Notice the disappearance of periodic variations from
the frequency domain traces, because of the lower amplitude fibre and
reflection, leaving just the "long fibre" shapes.

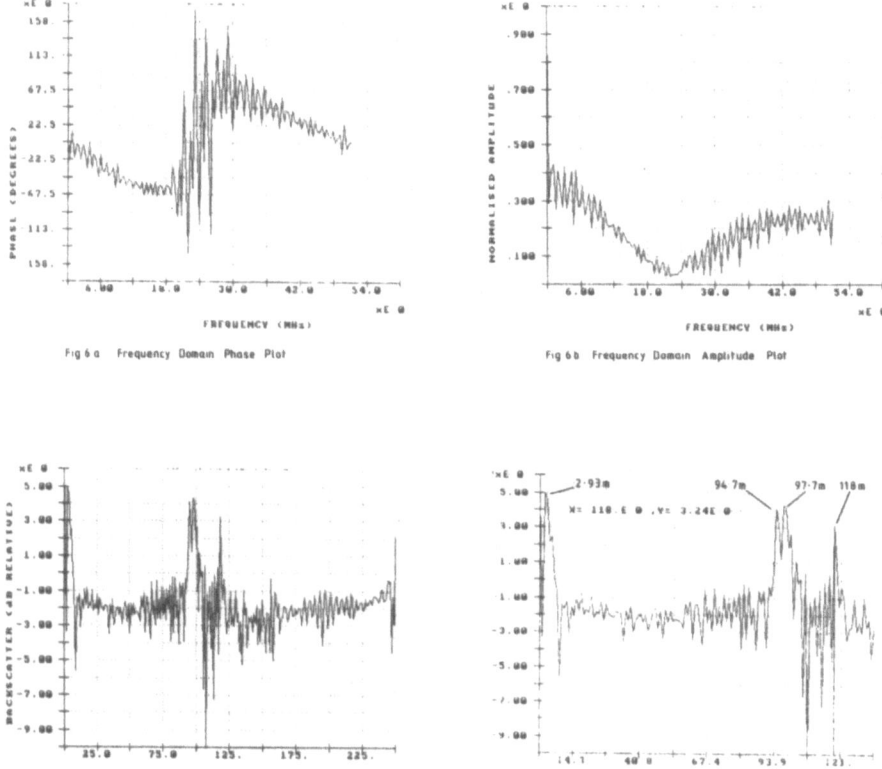

Fig 6a Frequency Domain Phase Plot

Fig 6b Frequency Domain Amplitude Plot

Fig 6c Time Domain Fibre Trace

Fig 6d Expanded Trace

Figs 6a, 6b, 6c and 6d: Results from a composite fibre comprising a
92m fibre, a 3m patch lead and a 20m fibre, connected throughout
using Stratos Connectors. The frequency domain traces now display
multiple periodicities due to the multiple strong reflections.

In essence, these results and others obtained in the course of the
project indicate that the OFDR technique is suitable for the purposes
of a Multi-function Tester. The areas for further study have also
been identified, in particular the reduction of strong periodic
signals.

6   CONCLUSIONS

A test philosophy for optical LANs has been developed. Coupled with an analysis of commercial consideration, a novel test equipment has been defined.

A demonstrator has been designed, built and tested, based around the OFDR technique. Attenuation, bandwidth and backscatter fault location have been successfully demonstrated. The project is now completed and has provided a technical basis on which further equipment development may now proceed.

Reference 1

H.Ghafoori-Shiraz and T.Okoshi — "Fault Location in Optical Fibres Using Optical Frequency Domain Reflectometry".
Journal of Lightwave Technology — Vol LT-4 No.3, March 1986.

ON THE USE OF LOTOS TO SUPPORT THE DESIGN OF
A CONNECTION ORIENTED INTERNETTING PROTOCOL

L.FERREIRA PIRES
University of Twente
PO Box 217
7500 AE Enschede
the Netherlands

Abstract

The design of systems with the help of Formal Description Techniques has moved from the theory books to the reality. In this paper I present an Internetting Protocol for Connection Oriented Traffic which has been designed and specified in LOTOS. First the functional requirements of the protocol were selected, following two approaches: the service support analysis, and the scenarios analysis. The actual selection of features is determined by the relevant functionality identified in both approaches. Further, the functional requirements are specified, and the specification is then applied as a starting point for the development methodology that ends up in an implementation. In order to comply to these objectives, the specification has to be designed according to certain principles, which are discussed in this paper. Therefore, the relevance of this paper is to provide a set of ideas to guide the development of what we call a System Architecture, in other words, the first formal specification of the system functional requirements. The importance of language tools in the verification of specifications is also addressed.

1. Introduction

In this paper the design of a Connection Oriented Internetting Protocol is described. The specific characteristic of this work is the use of the Formal Description Technique LOTOS, in order to specify the protocol functional requirements. The use of LOTOS brings some benefits, e.g. it allows unambiguous and precise descriptions. Another benefit is the possibility of specification verification using tools, allowing the designer to check whether the desired behaviour is really specified. Some more benefits are related to formal support of implementation oriented steps.

In PANGLOSS a methodology is being developed, which is based on a cyclic approach. One of the aspects of the cyclic approach is that the complete functionality of the system is not specified from the beginning, but only a manageable and representative part of it. The specification must be made 'open-ended', so that in subsequent working cycles more functionality can be included. This approach allows rapid evaluation of the system feasibility and prototyping [12, 13].

Therefore, the protocol which is described in this paper does not contain all features one could expect. Only the features which are considered relevant because they are mandatory for the protocol operation or because they cause impact on the implementation are included.

The specification that is discussed here is the first step in the implementation design methodology and it is often called an *architecture*. The main characteristics of an architecture is its high level of abstraction and detachment from irrelevant implementation details. It follows that an architecture must allow a family of implementations, which are consequences of different implementation decisions.

In section 2 the architectural background required for the specification development is presented. In section 3 the selection of features for the protocol design is discussed. In section 4 the design is described, in terms of the selected features and their specification. In section 5 possibilities and results of the specification verification are presented. Conclusions are presented in section 6.

2. Architectural Background

The purpose of this section is to find the architectural delimitations for an internetting protocol. Therefore the problem is first introduced, and the architectural approaches available for its solution are worked out.

Suppose we have an environment with systems like LAN's, X.25, ISDN, etc which are expected to be integrated for the purpose of information exchange. Considering the difference in the nature of these systems, an architectural framework needs to be defined. Figure 1 shows an example of interconnection of systems of different technologies.

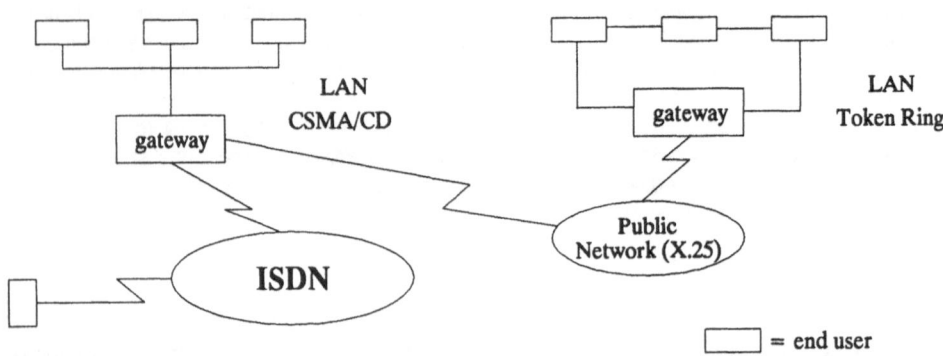

Figure 1. Interconnection of different networks.

The initial question to be answered is: in which of the OSI Model layers are we going to interconnect our systems? The answer to this question takes into account the definition of the function for each layer given in the OSI Model and the degree of generality desired for the interconnection system: interconnection is considered to be done at the Network Layer (layer 3).

Our decision is also based on the so called 'Internal Organization of the Network Layer'[1], an ISO standard which presents some concepts, definitions, approaches and scenarios for interconnection of different sub-networks using the Network Layer. Protocol roles or functions are identified in this ISO standard, which reflect the functions to be performed by the whole layer. Figure 2 depicts the protocol functions and their position with respect to the Network Layer.

From the definition of protocol roles or functions we define sub-layering in the Network Layer. Figure 3 depicts this architecture, considering the existence of an intermediate system.

The 3A protocols provide access to the sub-network, and therefore are sub-network dependent protocols. Therefore the 3A service can also be called Sub-Network Service. The 3B protocols (de)-enhance the 3A service to a certain harmonized level, providing to 3C entities a sub-network independent service. The 3C protocols have to perform relaying and routing functions, and provision of the Network Service at end systems.

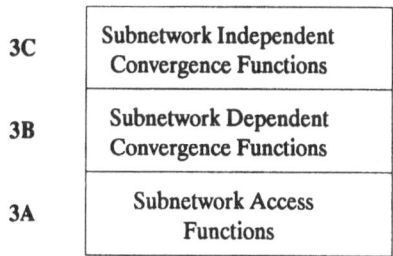

Figure 2. Internal Organization of the Network Layer.

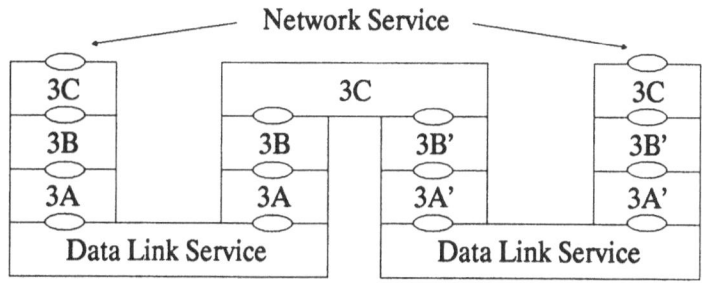

Figure 3. Network Layer sub-layering.

The service definitions are left to the system designers, which have to consider that there is a complexity balance between the service definitions and the protocols. For example, a very complex 3B service may lead to a very simple 3C protocol, while a very simple 3B service may lead to a very complex 3C protocol. On the other hand, a very complex 3B service may require a complex 3B protocol to provide it.

Following these principles, it is considered that the protocol for the lower layers of the OSI Model are already available, and we concentrate the research on solutions for the interconnection problem, which will occur in fact at the 3C sub-layer. For this reason an Internetting Protocol (3C) has been developed, and a 3C entity, that is considered embedded in an intermediate system, has been formally specified in LOTOS. This 3C entity is called a *gateway*.

The gateway uses the 3B service, which will be called throughout this paper the sub-network service, keeping in mind that in fact it represents the sub-network service (3A service) enhanced to a certain sub-network independent level. Figure 4 shows how the internetting protocol and the gateway are positioned.

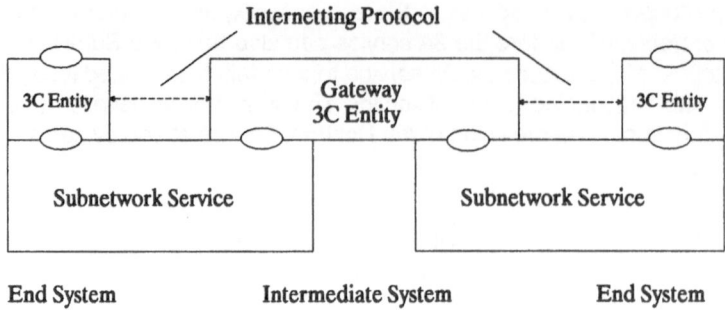

Figure 4. Position of the Internetting Protocol

3. Selection of Features

Since our protocol is now delimited, the selection of features can start. There are two main ways to identify and select the features for a protocol definition: the analysis of the service to be provided and the underlying (available) service, and the scenarios analysis.

In the first approach, the protocol is expected to fill a gap of functions between the services. In the second approach the existing protocols are taken into account, and their conformance to the requirements can be evaluated, as well as their weakness and lack of completeness.

The actual selection takes into account functions which are identified using both approaches, and are paramount for the protocol working, or have a big impact on the implementation structure.

3.1. Identification of Features using Service Definitions

The selection of features starts with assumptions about the underlying service, or the sub-network (enhanced) service. The sub-network (SN) service is supposed to be Connection Oriented, similar to the Data Link Service [2] or the Network Service [3].

From this initial assumption we can conclude that our protocol must be able to establish and release sub-network connections and receive and send data over these connections. It means that handling the sub-network service primitives by interacting at the SN service boundary are identified functions which are selected from the beginning, because if they are neglected communication becomes impossible.

Considering that the Internetting Protocol must provide the Network Service to the service users at the end systems, some more functions must be identified. These functions can be classified according to the local phases of connections they relate most: connection establishment, data transfer or connection release.

3.1.1. Connection Establishment

During this phase, negotiation of options, negotiation of Quality of service and routing are the most important functions to be performed by the protocol entities. These functions have an impact on the establishment of SN connections.

SN connections are some of the resources to be allocated in order to support the Network Connection. The SN connections establishment can support the Network connection establishment in at least three different ways: for example, the SN connections can be started from the beginning of the end and intermediate systems' operation, and allocated to the Network Connections as soon as they are required, or they can be dynamically started when they are needed. Considering the second option, each system can wait for a Connection Request Network Protocol Data Unit (NPDU), start the next SN connection and only then forward the NPDU, or the Connection Request NPDU can be sent as data in the SN connection establishment primitives in a fast connection procedure.

Of course the choice among these SN connection handling methods has to do with the supposed underlying SN service definition and performance requirements of the system. Considering that resources such as underlying connections have some cost and are scarce, dynamic allocation of connections seems more appropriate. The choice for starting up SN connections before conveying the Connection Request NPDU results in a discipline that can be applied to SN services that may or may not support data as SN connect primitive parameters, so it is more general.

More functions can be identified, such as:
- re-try of refused SN connections, via alternative routes;
- splitting of a Network Connection over SN connections to accomplish the required Quality of Service;
- multiplexing of Network Connections on a single SN connection.

3.1.2. Data Transfer

Functions more related to this phase can be:
- explicit or implicit connection identification;
- data relay or data forwarding;
- local or global flow control;
- data transfer reset;
- data transfer re-sequencing;
- segmentation and reassembling;
- expedited data transfer;
- data acknowledge;
- error check and error correction;
- priority and protection (Quality of Service provision).
Connection identification and data relay are considered the essential features.

3.1.3.Connection Release

Since it was decided that connections are going to be dynamically started, the release of Network Connections means that the SN connections must also be released. Alternatively, the gateway could wait for some time, until a new connection establishment to the same destination and the same quality of service values is requested (re-usage of underlying connections).

In the case of Network connection release because of forced SN connection release, the gateways could try to continue supporting the Network connection, by starting alternative connections. Nevertheless, a discipline for connection re-start would be necessary, to avoid the undesirable effect of multiple connections wrongly started.

3.2. Identification of Features Using Scenarios

In this section two scenarios are presented and briefly discussed. The scenarios follow from the application of the architectural framework on existing systems.

3.2.1. Scenario #1

It consists of assuming the X.25 Packet Level Protocol (PLP) as an internetting protocol, acting on top of a Data Link Protocol, either LLC2, LAP-B, or HDLC (Circuit Switched operation or not) and it is depicted in figure 5.

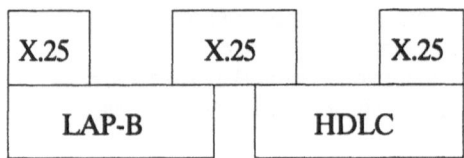

Figure 5 - Scenario #1 for selection of features.

This scenario is supported by OSI standards [4] and [5], which are the ISO 8208 protocol and the use of X.25 PLP to provide the Network Service respectively. ISO 8208 is the ISO version of X.25 PLP, very similar to the definition found in the CCITT X.25 [11], but extended with a DTE/DTE communication description.

Nevertheless, the X.25 PLP informal specification presents some problems when we try to apply the structuring proposed in the section 2, mainly due to its original development goals of being an access protocol, or an interface.

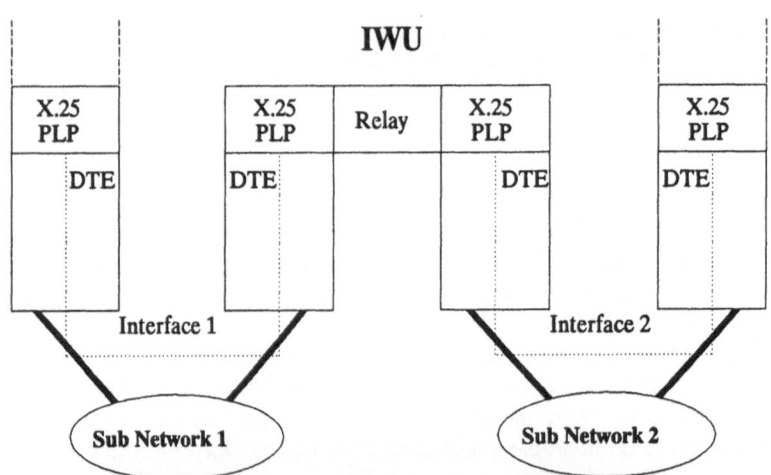

Figure 6 - Model for Intermediate Systems (ISO TR 10029).

Since we consider that protocols provide a service, by using an underlying service and executing a set of functions, we also expect in the protocol description to find references not only to the execution of the functions and provision of the services, but also to the underlying service usage discipline, preferably in terms of the interactions at the underlying service boundary (service primitives). Unfortunately that is not the case in X.25 PLP (or in ISO 8208), which provides a description of some protocol mechanisms, PDU formats, but no underlying service interactions.

The conclusion follows that if ISO 8208 is expected to be used as an internetting protocol, as indicated in [6], the description of the underlying service interactions has to be added. Furthermore, the description of the intermediate system given in [7] is a very preliminary one, because it defines the behaviour of an Interwork Unit (gateway) only in terms of packets relay between two DTE/DTE interfaces. There are no references to global routing and global addressing problems in this standard, and these seem to be the major challenges. Figure 6 shows the model used in this description.

3.2.2. Scenario #2

Another possible scenario would be to have X.25 PLP playing the role of a (enhanced) sub-network protocol (3B), and a generic Internetting Protocol (IP) on top of it, as it is depicted in figure 7.

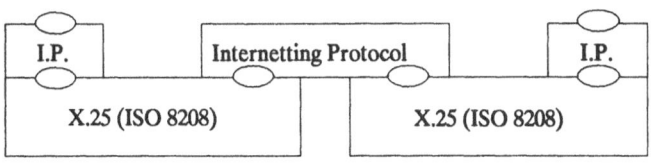

Figure 7 - Scenario #2 for selection of features.

This is also the most intuitive scenario, which comes from studying the IONL and considering X.25 PLP as a subnetwork protocol, either to public or private (sub)-networks.

In this scenario, the 3B service can be the Network Service [3], provided by X.25 PLP using [5], and then the internetting protocol must define the sequences of primitives and functions like routing, negotiation of QoS, negotiation of options, etc.

3.3. Selected Functions

In our work, we try to develop a protocol that does not deviate too much from the X.25 PLP definition, only filling the gaps for the interactions at the sub-network service boundary, in such a way that both scenarios could in principle be supported. Therefore we have selected the following functions:
- SN service primitive handling;
- dynamic connection establishment;
- negotiation of QoS;
- negotiation of options (expedited data and data acknowledge);
- data relay;

- SN service (local) flow control;
- wrong PDU types detection and disconnect;
- disconnect handling.

4. Protocol Design

In this section the elements of the protocol design are presented. Initially the criteria for the protocol specification are presented and then the specification structure is described, pointing out the general guide-lines, which can be generalized to the specification of any system.

4.1. Specification Development Criteria

The specification development criteria is very much attached to the objectives of the specification. These criteria influence the specification structure, forcing us to use a suitable specification style.

In our case, since the specification of gateway functional requirements is considered to be at the highest level of abstraction, no structure should be introduced in the specification that could be misinterpreted as the structure of its implementation. However, specifications have to be structured in some way, since they are supposed to be understood by human beings.

Furthermore, we want to describe independent requirements independently, keeping a considerable separation of concerns. The specification has to accept more (or different) functionality, without major rearrangements. The specification is also expected to be easy to verify using the LOTOS tools [8, 14] and the adopted style must support eventual specification development exercised by a designers' team.

Therefore we use the constraint oriented style [9] in the specification, where constraints are specified and composed in parallel (synchronized) at the systems' interaction points (gates in LOTOS).

Misuse of specification styles for system specifications at the highest level of abstraction may disturb the design objectives, due to the fact that structure in terms of LOTOS hidden gates introduced in specifications may be difficult to get rid of in later design phases.

4.2. Specification Structure

The gateway specification presents a single LOTOS gate, which represents the whole sub-network service boundary of the gateway. The sub-network service access points are distinguished by their addresses, which are represented as parameters at the sub-network boundary interactions. According to this representation, a service access point, which is by definition the interaction point for the service primitives to occur, is modelled by a LOTOS gate followed by a service access point address.

A connection end point identifier, which identifies a specific connection end point at a service access point, is also represented as an interaction parameter.

The interactions at the sub-network boundary are the sub-network service primitives, which are also represented in the specification as parameters of the interaction (event). Therefore, the general gate structure for interactions can be represented in LOTOS as:

sn ?snaddr: SNAddr ?sncei: SNCEI ?snsp: SNSp

where
- *SNAddr* represents the sort of sub-network access points addresses;
- *SNCEI* represents the sort of sub-network connection end-point identifiers;
- *SNSp* represents the sort of sub-network service primitives.

Using the constraint oriented style, we could express constraints of different concerns in different language elements. By doing this, we generate levels of concern in the specification structure. The highest level presents three processes, one to represent that interactions can only take place at connection end points (CEPs) and service access points (SAPs) which are defined (*SNIdentification*), another to handle the CEP usage, avoiding more than one connection per CEP at the same time (*SNAcceptance*) and another one to represent the network connections support (*NCsSupport*). These processes are composed in parallel, synchronized at the interaction gate *sn*, since they represent constraints that are applied in every interaction.

Intuitively, the synchronized composition of processes means a logical "and" of the constraints described in the processes on the interactions. Figure 8 shows this organization.

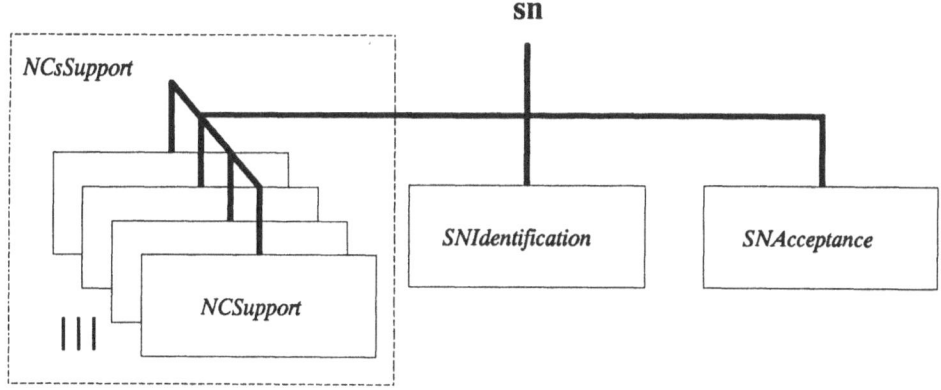

Figure 8 - High Level structure of the specification.

The *NCsSupport* process is specified as a set of independent network connections support, which are expressed with the help of the pure interleaving (| | |) LOTOS operator. Each network connection support is represented by the *NCSupport* process. The interleaving can be interpreted as a logical "or" applied to the interleaved processes, meaning that only one of the interleaved processes participate in one interaction.

The *NCSupport* process handles sub-network primitive interactions and protocol functions. Therefore, the *NCSupport* process is structured in terms of constraints which are local to the sub-network access point attachments, and constraints which are remote, because they relate both sub-network access points. The constraints at the two sub-network service access points are in principle independent, and are interleaved. Figure 9 shows the organization of the *NCSupport* process.

The local constraints define the handling of sub-network primitives in terms of allowed primitive sequences and parameters. The remote constraints, described in the *EndToEnd* process, relate events at both connection end points, and are structured as a synchronized composition of processes which represent the following concerns: network connection

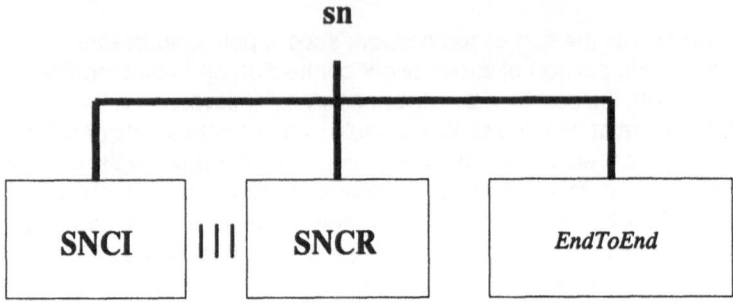

Figure 9 - Structure of the remote constraints.

establishment (*NCEstab*), data transfer (*DataTransfer*), network connection release hand-ling (*NCDiscHandling*) and network connection release monitoring (*NCDiscMonitor*). Figure 10 shows how the remote constraints are structured.

The structuring of the remote constraints closely resembles the local phases of a

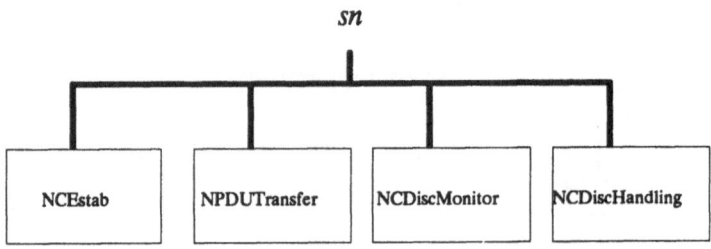

Figure 10 - Structure of the NCSupport process.

connection (connection establishment, data transfer and connection release), with the only exception that there are two processes to express the connection release phase. The reason for this duplication of processes comes from the fact that connection release can either be requested by the gateway because of some error situation, or externally, through Disconnect NPDU exchange.

The processes at this level are again sub-structured as a composition of constraints of different concerns and every time a process with an acceptable degree of difficulty and that represents a concern which is considered hard to sub-structure again is found, it is specified in terms of the sequence of events (monolithic style).

The application of the separation of concerns criteria in the structuring of specifications may lead the specification designer to difficult situations. An example of such a situation is the specification of routing, which can be divided in two parts: the determination of the service access points for the next hop sub-network connection, and the determination of the connection end point at the local service access point. The second part is determined by two constraints, namely the constraint on the service access point address determined by the first part, and the global constraint on the availability of connection end points at the service access points. The specification must allow a family of implementations, so one could think about an implementation mechanism that tries to find service access points which still have connection end points available. We could end up in an internal deadlock, however, if the routing at the network connection basis provides a set of service access points for reaching the next hop, and no connection end point is available on these service access points.

Actually, there must be a communication between the routing process in the network connection support (*NCSupport* process) and the global constraint on connection end points (*SNAcceptance* process), which we do not want to describe in this abstraction level, because it must be subject to design. Therefore, we modelled this internal communication by a LOTOS internal event, which literally "solves the deadlock". It is interesting to stress the flexibility of our solution to the problem, although it does not sound very elegant. Implementations that have this explicit communication, or implementations that solve this problem via a timeout in the routing process comply equally to our specification.

5. Specification Verification

The verification of LOTOS specifications using tools is the only way the designer can be sure that a big specification shows the desired behaviour. Since some of the objectives of specifying using formal description techniques such as LOTOS is to avoid ambiguity and to express precisely the required behaviour, the use of tools plays an important role in the specification design.

Motivated by the need to verify our specification against our desired behaviour, we submitted our specification to tools. The available tools are a syntax checker, a static semantics checker, a simulator [14] and a syntax oriented editor with built-in static semantic checking [8]. The syntax and static semantic checkers just verify if the specification complies with the rules defined in [10] for syntax and semantic requirements. They are equivalent to the compilation and linking requirements in programming languages.

The simulator translates a specification to a transition model, which can be simulated in an interaction by interaction basis, and allows the designer to verify the behaviour of the specification against his (her) expectations. Our tools allow symbolic or non-symbolic simulation, giving the flexibility to verify specifications even if they are incomplete. The symbolic simulation is very often required, mainly in case the details are irrelevant at the specification abstraction level, and are not fully described.

In the case of the *Gateway Architecture*, we developed a complete specification, in the sense that its abstract data type definitions can be completely evaluated. Therefore the specification can be simulated either symbolically or non-symbolically. The simulator allows one to get as results the list of allowed interactions at each "state", the trace to reach a certain state or the specification behaviour tree. It follows part of the trace of the *NCEstab* process error-free behaviour:

```
START : ncestab(rt1)
  sn ?$0.1:snaddr ?$0.2:sncei ?$0.3:cosnsp                                    (1)
  [(iscosnindication($0.3) or issnconnresp($0.3)) = true]
  sn !sn1 !snc1 !conind                                                        (2)
   sn ?$2.1:snaddr ?$2.2:sncei ?$2.3:cosnsp
   [(iscosnindication($2.3) or issnconnresp($2.3)) = true]
   sn !sn1 !snc1 !conresp                                                      (3)
    i(exit) !conresp
      sn ?$5.1:snaddr ?$5.2:sncei ?$5.3:cosnsp
      [(iscosnindication($5.3) or issnconnresp($5.3)) = true]
      sn !sn1 !snc1 !crpdu                                                     (4)
       i(exit) !cr1
        i(exit) !cr1
         i(exit) !cr1
          i  { $4$src:snaddr, $3$dst:snaddr }
          [(route($4$src, $3$dst) belongsto rt1]
          i  { sn5, sn6 }                                                      (5)
            sn ?$12.1:snaddr ?$12.2:sncei ?$12.3:cosnsp { $7$snqos2:snqos }
            [(issnconnreq($12.3) implies
                (sn5 iscallingof $12.3) and sn6 iscalledof $12.3))) = true]
            [(((issndareq($12.3) and iscrq(npdu(userdata($12.3))))
                implies (op1 areoptionsof npdu(userdata($12.3)))) = true]
            [(issnconnreq($12.3) implies ($7$snqos2 isqosof $12.3)) = true]
            sn !sn5 !snc1 !conreq  { q2 }                                      (6)
```

The trace consists of a list of LOTOS events optionally followed by guards. The events in the trace represent the interactions that have occurred in the simulation, and the guards, which appear between square brackets, represent the conditions for the events to occur. The event sequence is indicated with indentation. For example, event (1) represents a symbolic execution, while event (2) represents an instance of event (1) with pre-defined actual values. Event (3) occurs after (1) (or (2)). Internal choices on values are indicated between brackets and internal events are represented by *i*, following the LOTOS conventions.

Therefore, the trace shows that in the *NCEstab* process the execution of a Connection Indication Service Primitive (SP) (2) can be followed by a Connection Response SP (3). After this a Connect Request NPDU may be received (4), and there is a non-deterministic choice on the next route (pair of SN SAP addresses) through which the destination Network Entity can be reached (5). Then a new SN connection is started to build this route, through the execution of a Connection Request SP (6).

Our verification discipline was to start verifying the low level constraints using simulation techniques as the one mentioned before, and then to move through the hierarchy verifying each level, until we reached the complete specification, or the highest level. The constraint oriented style proved to support this discipline.

6. Conclusions

Using formal specifications to support system design proved to be useful in at least two ways:
- to make us sure that our protocols do not present misbehaviour or gaps ("bugs");
- and make us sure that everybody (acquainted to the formal description technique) will have the same understanding of our specifications.

Following the guide-lines of the PANGLOSS Method, the specification was developed to be a high level specification (*Gateway Architecture*), which shall allow a large class of implementations to comply to it. In our *Gateway Architecture* only architectural functions and observable behaviour are addressed. Implementation structure or internal interfaces should be considered in further decompositions of this architecture, following criteria derived from the design goals (performance, modularity, failure resilience, etc.). In this respect, attempts to develop implementations from the specification proved that it attends the requirements of the design goals.

The specification style adopted in a specification development plays a dominant role. The constraint oriented style was used in this specification development. This style supported functionality extension without major rearrangements in the specification, and it allowed the development of a verification strategy, which gives a lot of confidence on the contents of the specification.

The use of tools in order to check the specification were revealed to be very important, since we found and corrected a considerable amount of problems in the specification during the testing phase. The tools have also a certain instructive effect on the user, in the sense that more about the details of the formal description technique can be learned while using them. After some time using the tools, the designer starts to make fewer mistakes, and the testing procedure becomes less time consuming.

Furthermore, a slightly different version of the specification discussed in this paper is being used in the development of a Stream Oriented Internetting Protocol for switching systems. Again the constraint oriented style supported the modification of the concerns in which the traffic is expected to be different, causing relatively small rearrangement efforts.

Still the protocol is not completely specified, because we did not specify (yet) the behaviour of the end systems to support it. Any way, most of our specification effort can be directly used in the development of the end systems' specification.

7. References

[1] ISO/IS 8648 - Data Communications - Open Systems Interconnection - Internal Organization of the Network Layer

[2] ISO/DIS 8886.2 - Data Communications - Open Systems Interconnection - Data Link Service Definition

[3] ISO/DIS 8348 - Data Communications - Network Service Definition

[4] ISO/DIS 8208 - Data Communications - Information Processing Systems - X.25 Packet Level Protocol for Data Terminal Equipment

[5] ISO/DIS 8878 - Data Communications - Use of X.25 to provide the OSI Connection-mode Network Service

[6] ISO/DIS 8880 - Information Processing Systems - Protocol combinations to provide and support the OSI network service

[7] ISO/IEC TR 10029 - Information technology - Telecommunications and information exchange between systems - Operation of an X.25 interworking unit

[8] van Eijk, P. - 'LOTOS Tools based on the Cornell Synthesizer Generator' in: Proceedings of IFIP WG 6.1 Ninth International Symposium on protocol specification, testing and verification, North Holland 1989

[9] Vissers, C.A. et al. - 'On the use of specification styles in the design of distributed systems', presented at TAPSOFT '89 March 13-17, Barcelona

970

[10] ISO/IS 8807 - Open Systems Interconnection - LOTOS - A formal description technique based on the temporal ordering of observational behaviour

[11] CCITT Recommendation X.25, Interface between data terminal equipment (DTE) and data circuit terminating equipment (DCE) for terminals operating in the packet mode and connected to public data networks by dedicated circuit

[12] Bogaards, K et al. - 'The PANGLOSS Method', Proceedings ESPRIT Technical Week 1988, Brussels, November 1988, North-Holland

[13] Bogaards, K - ' LOTOS supported system development', Proceedings of the 1st International Conference on Formal Description Techniques FORTE 88, Stirling UK, September 1988, North-Holland

[14] van Eijk, P. - 'Software Tools for the Specification Language LOTOS', Ph.D. Thesis, Twente University of Technology, Enschede, Netherlands, 1988.

Key Words:

LOTOS, PANGLOSS, internetting, internetting protocol, formal description techniques, formal specifications, system design, specification development

SYSTEMATIC DESIGN OF TELECOMMUNICATION SYSTEMS USING FORMAL DESCRIPTION TECHNIQUES

JEROEN SCHOT
University of Twente
P.O. Box 217
7500 AE Enschede
the Netherlands
earn: schot@henut5

Abstract

In Esprit Project #890 - PANGLOSS, we developed a methodology for the design of concurrent and distributed systems, based on the use of a formal language (LOTOS). The major benefits of the method are its systematic approach, that allows the designer to ignore implementation details in the first place, and concentrate on the sole functionality of the system required, and the fact that the initial design is based on a small but relevant subset of all requested features, which makes it manageable. Since formal languages show an ultimate precision, the first steps on the design trajectory can be immediately assessed on their correctness and feasibility with respect to the required system, this is also termed 'rapid prototyping' and is a very profitable technique. In this paper I described the main characteristics of this design methodology, as well as the issues encountered during application of this method on the design of a telecommunication switching system. As the design of a switching system still requires tremendous R & D efforts, the industrial benefit of a concise and parsimonious design methodology is expected to be considerable. In this respect, the results obtained in the PANGLOSS project are promising.

1. Introduction to the PANGLOSS Method

The method that is discussed in this paper (see also [4]) follows from the Esprit project PANGLOSS, which is a joint effort of the SEMA Group (UK), 7-technologies (DK), Twente University (NL), and Liège University (BG), as main contractors. The project is nearly finished, and its objectives were:

- the development of a systematic design and development method for distributed and concurrent systems, and

- to prove this method by designing a high performance networking gateway according to the principles of OSI.

Unlike most existing methodologies, the language in which requirements are expressed and which is used in our approach contains formal semantics. In this case we favoured LOTOS - Language of Temporal Ordering Specification [1], which became an ISO International Standard in 1988. The major reasons that advocate the use of a formal language are that it enables us to:

– express unambiguously and precisely the required behaviour of the system at different abstraction levels,

– verify the equivalence relations that must exist between various descriptions,

– automatically generate test suites for conformance testing,

– rapidly prototype the system to assess its capabilities at an early stage.

It is evident that drawbacks can also be identified, such as the need to learn the language and to practice in using it, but it is our firm belief that this investment is regained many times over during employment of the formal description technique (FDT) in the design process. This opinion is reinforced by the progressions observed in the development of software tools that support the usage of FDTs, such as syntax editors, syntax & static semantic checkers, and simulators [2]. Apart from the use of LOTOS, the design methodology is characterized by its application of Step-Wise Refinement and a Cyclic Approach.

1.1. Step-wise Refinement

In the development process we distinguish several descriptions that differ in abstraction level:

user requirements - the initial, informal definition of the system functions and requirements;

the top-level architecture - the first formal specification being a completely implementation independent description;

intermediate design descriptions - various descriptions in which implementation concerns are increasingly incorporated. One of them, termed the *Reference Architecture*, has a special significance in our project, as will be explained later;

the implementation - the final design, being a complete blueprint for the realization of the system.

As a consequence, our first step in the design process is to formalize the requirements of the system as much as possible. This formalization step transforms the user requirements into the (top-level) architecture. In practice we can express functional requirements perfectly in LOTOS, but for the description of non-functional requirements such as timing and sizing aspects, we have to invoke conventional formalisms. Thus, the top-level architecture, as well as all other intermediate descriptions, consist of two parts: a LOTOS specification of the functional requirements, and an annotation with all other requirements. The obtainment of the top-level architecture bears great benefit to the user and its developers, since the functional behaviour of the top-level architecture can be easily assessed using simulation tools for LOTOS, and the extent to which the design complies to the user's and developers' expectations can be evaluated. The definition of a formalized architecture also provides us with a concise starting-point for the rest of the development trajectory.

After a suitable top-level architecture is established, a certain number of transformation steps are taken, in which implementation concerns that are applicable on that level of abstraction are incrementally incorporated. The consistency of the intermediate designs with the top-level design can be checked after each transformation step. In practice, we try to avoid the need for (extensive) verification of the behaviour of the transformed specifications with respect to the specification it is derived from. This is achieved if the decomposition is based on generalized transformation rules that preserve correctness, also called CPTs (correctness preserving transformations). Eventually, such transformations can be supported by software tools; first results obtained in this area are found in [7].

Finally, a design is obtained that is easily mapped on hard- and software components available, thereby constituting the implementation of the system. Figure 1 illustrates this step-wise refinement process.

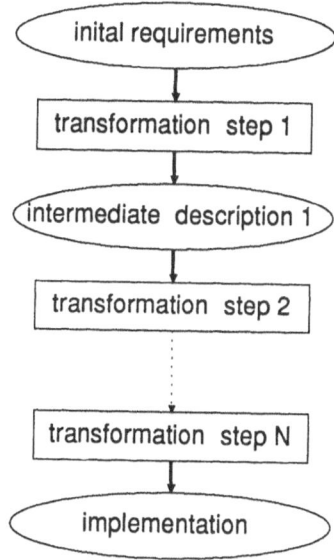

figure 1 - step-wise refinement process

1.2. Cyclic Approach

In general, a designer will base her/his design decisions and implementation choices on a small set of well chosen features of the system, since it will be difficult to maintain an overview of all possible details in an unstructured way. Therefore the method is based on a cyclic approach in which in each cycle the complete or part of the implementation trajectory is traversed based on a limited functionality, and in each next cycle the functionality is increased until in the last cycle the full functionality, defined in the user requirements is supported.

Which functions are selected for the first cycle, is subject to criteria of which the most important is the impact that a function will have on the structure and organization of the system. We term such functions *key features*. The proper selection of key features to be considered firstly should lead to formal descriptions that are apt to inclusion of new functions in subsequent cycles.

1.3. Work Division

The step-wise refinement method defines a development trajectory, on whichdecisions of different concerns are separated, e.g. in order to make them manageable. In particular, the decisions to be taken during the earliest steps of the design process are more related to architectural aspects of the system (i.e. the requirements), whereas decisions taken in later steps are more related to specific implementation details. However, choices that are made at the beginning of the development process, are paramount to the success of the design, since the structure that is introduced at description level i of the process will be reflected in all i+1,..,N descriptions, where description N corresponds to the implementation of the system. Therefore, there is a demand for verification of the suitability of the design with respect to its (time) performance, soon after few transformation steps have been taken.

This threefold characterization of activities performed in the design process, which comes quite naturally, has brought us to the definition of three tasks, in which the people contributing to the development process can be grouped, according to their skills and expertise. This applies particularly in the case of our Esprit project, in which people with large variation in background attempt to cooperate. The three disciplines are:

– *Architecture Task*, concerned with the high-level design of the system (the first steps in the design process);

– *Implementation Task*, committed to the mapping of abstract functions onto implementation elements (e.g. hard- and software);

– *Performance Analysis Task*, in charge of assessing the performance of the intermediate design, before transformation to the implementation is performed.

For the cooperation of the three task groups, a proper interface has to be defined. We selected a formal intermediate design as interface, and termed it the *reference architecture*. It is a structured refinement of the top-level architecture, containing the coarse outlines of the organization of the system. Hence it permits an early analysis of its appropriateness with respect to the user requirements. For the implementation task, it serves as a starting point for further transformation.

The use of an intermediate description as interface between the three tasks should not prevent them from working in parallel. In practice it has been confirmed that they can work concurrently, i.e. the architecture group can increment their design while the implementation group implements the previous reference architecture and so on. Only during the first cycle, the implementation task has either to wait for the first reference architecture as input for their transformations, or they have to anticipate it.

1.4. Iteration

In the previous sections, the design methodology has been displayed as a strict top-down process, with the user requirements as input and an implementation as output. However, this is too idealistic a view, since the exact consequences of the design decisions taken in early transformation steps can not be completely predicted. Therefore, results obtained at a later stage are fed back to earlier descriptions, or even to the initial user requirements in case they appeared unimplementable. In fact, this process can be regarded as 'tree search', in which a negative result leads to 'backtracking' in the tree. This is illustrated in figure 2.

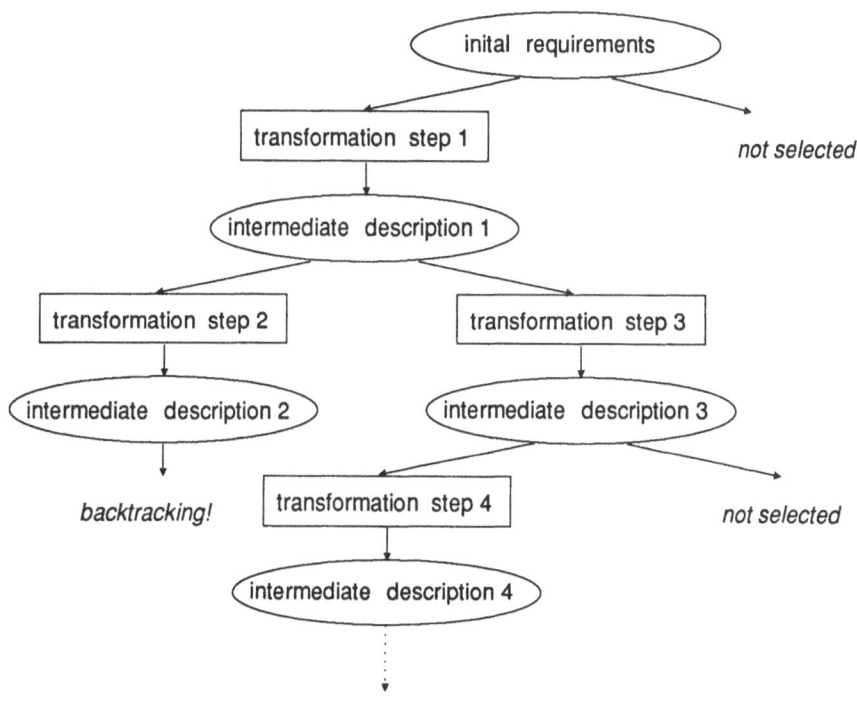

figure 2 - tree search (iterative design)

1.5. Aggregate Design Methodology

In figure 3 the final design method is sketched, as a composition of step-wise refinement, cyclic approach and iteration. Also task responsibilities are indicated. In brief: starting with the user requirements, we make a selection of *key features*, that are formalized in LOTOS leading to a *top-level architecture*, which is then transformed to the implementation via intermediate steps. Feedback from description i to description 0,1,...,i-1 is always enabled. After the complete path is traversed, more user requirements are considered, until the complete set is implemented.

1.6. Specification Styles

In the step-wise transformation process from user requirements to implementation we identified several intermediate descriptions, each expressing particular concerns. Accordingly, they should be written using the appropriate specification style. For the initial description, we select a style that shows no structure in order to avoid implementation decisions to be taken in this phase. Such a style is the *constraint oriented style*, in which the required behaviour of the system is described as the conjunction and interleaving of

976

a set of constraints, logically grouped on basis of architectural criteria such as orthogonality, generality, and separation of concerns.

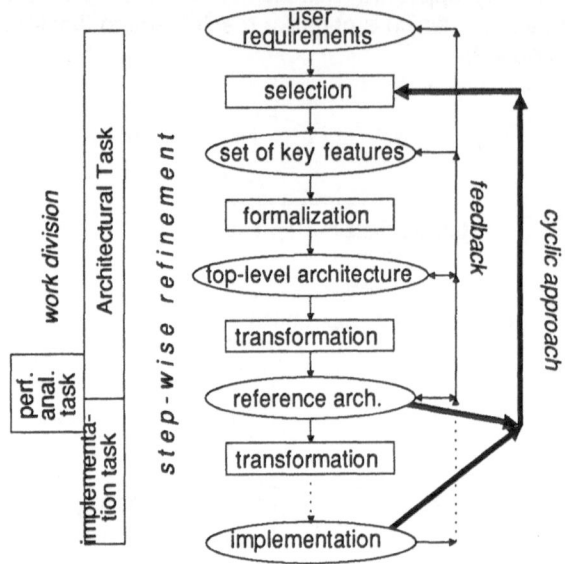

figure 3 - aggregate design methodology

For the descriptions that are more implementation oriented, i.e. contain a structure that should be incorporated in the final implementation, we use a *resource oriented style*, in which separate components are identified. Note that the components together show a behaviour which is equivalent to that of the initial description, and that the components themselves are again represented as black boxes (i.e. not showing structure) and therefore specified using again the constrain oriented style. More information on specification styles can be found in [3].

A major characteristic of the PANGLOSS method is the fact that it employs the formal description techniques LOTOS for the specifications of the functional behaviour of the system, at various stages in the design process. LOTOS was designed to be useful for a large class of applications, and its generality indeed permits us to describe systems at different levels of abstractions. As a result of this, LOTOS is easily misused, and interpretations of different people over a single specification often tend to diverge. An example of a language element of LOTOS that is easily employed to express many objectives is the internal event i. Hence, adherence to a certain specification discipline is cardinal.

With LOTOS, we are able to describe the temporal ordering of events (= interactions of the system with its environment) the system should allow, and parameter value sets associated with these events. When such an external specification is implemented, these events should have their implementation as well. This implies that events are abstractions from actions that occur on the system level: this is the semantics of the abstract events used in the specification and ought therefore be presented at specification level.

When specifications are transformed into specifications with a lower abstraction level, also called an implementation of that specification, we require that the behaviour of the implementation is equivalent with the behaviour of the specification it is derived from.

However, there are numerous equivalence relations defined (see also [9]), and omitting a statement about which equivalence relation valid implementations of the specification should satisfy, renders the specification meaningless (hence useless in the design process).

In summary, we have to realize continually that the usage of a formal method only makes sense in the context of our model (= the limited perception of the world in terms of entity and event definitions), which therefore has to be defined precisely and accepted by the group of people working with the specifications.

2. Application of the Method

In order to demonstrate the relevance of the design methodology developed and briefly described above, we applied it onto several realistic design questions [5], the one discussed in the rest of this paper is the design of a switching system for stream oriented traffic. An example of stream oriented traffic is (video)telephony traffic. Other applications of the method that are performed in our project are the design and implementation of network protocols for connectionless and connection-oriented traffic [8], and network management.

There are at least two reasons that justify the application of the design method to the development of a switching system. Firstly, the design of a modern switching system that supports a tenable number of connections and incorporates advanced features that are becoming more commonly accepted (i.e. so called supplementary services), has proved to be a formidable task which claims large amounts of resources of manpower and the like. Secondly, the interconnection problem contains a large amount of intrinsic parallelism, since many connections are to be supported simultaneously. In order to fully exploit the parallelism that is inherent to the problem in some form of parallel system architecture, the transformation from the requirements to the implementation should preserve concurrency as much as possible. This is an important characteristic of the design method presented here, mainly to the benefit of the formal description language that is used. Where more conventional approaches abstain from formal techniques, it turned out that confidence in the correctness of parallel solutions decreased due to the fact that the human mind seems less suited to maintain an overview of many events occurring concurrently. This can be made plausible, if we consider the number of states as a function of the number of independent events, which is an exponential function (the combinatoric state space explosion problem).

We will describe the design of the system according to the different description levels that are identified in the methodology, such as the user requirements, top-level architecture and reference architecture. The trajectory from reference architecture to final implementation in not covered in this paper, since it is dealt with in other papers submitted to this conference. Moreover, similar issues and techniques as described here apply to that part of the design process.

2.1. User Requirements and Top-level Architecture

The objective of this section is to explain how we came to the definition of a top-level architecture for a switching system, that can serve as input for the transformation process towards the implementation. As explained in the previous section, this process consists of two phases, selection of key features and formalization of key features.

In the first cycle, we only select those features that are cardinal for the functioning of the system. In our case, we selected *basic interconnection* as the key feature for the first cycle. This is the capability to arbitrarily interconnect end users connected to the network, supporting a single quality of service (i.e. 64 KBit/sec. full duplex). Notice the correspondence with CCITT's concept of *bearer service*. It is supposed (for the time being) that routing information (information about the logical organization of the network) is to be held within the system.

The second step is to come to a top-level architecture, that exactly describes the functionality required from our switching system. The concept of architecture is extensively used in our method and will therefore first be defined for a common understanding.

2.1.1. The Concept of Architecture.

We consider the architecture of a system as the **external behaviour** (or external functionality) of that system. Hence we regard the system as a **black box**, or a **process**, with which the environment (e.g. the user) can interact. In order to be able to interact with that process, **gates** are defined at which the interactions take place.

Since the system is only regarded externally, no information about the internal structure of the system (i.e. how it realizes the functions that it performs) is revealed. Such a behaviour expression can be given by an extensional formal description in LOTOS. It describes which interactions with the system can occur, which parameters and values are related to those interactions, and the sequence in which those interactions can happen. The specification of the architecture refrains from defining structure internally in the system, since such structure would suggest implementation decisions which are difficult to ignore. Thus, the architecture is defined in terms of externally visible interactions, their relationship being specified in terms of constraints defined in a logic-oriented style. Hence it exactly defines the total class of implementations that are potentially solutions to our problem, without favouring any particular implementation.

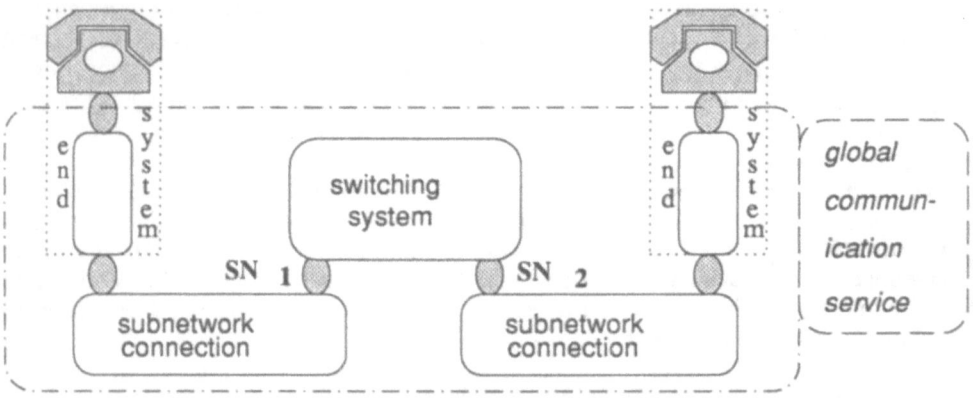

figure 4 - top-level architecture showing 2 of N subnetworks and end systems

2.1.2. An Architecture for the Switching System.

In figure 4, the setting of the switching system as an embedded system in a larger whole, is depicted. In this picture we see that the switching system is modelled as a black box, that interacts through gates SNi (which can be considered as a single gate with parameter i, where i = 1,...,N and N equals the number of subnetworks attached) with its environment, in this case the attachment to the rest of the communication network which it forms part of. At this gate, all information related to subnetwork connections, as well as the data that is transferred over these connections, 'arrives'. Remember that the subnetwork connections are used to build the global or network connections, according to the OSI Reference Model for Open Systems Interconnection [6]. See also section 2 of [8] for an architectural discussion on modelling of interconnection functions.

In this figure, it can be seen that the global communication service is provided by the composition of (in this case 2) subnetwork services, and protocol entities at the end system and at the switching system. Since these services and protocol entities are formally described as LOTOS processes, we can proof the correctness of the functional behaviour of the switching system (in a given network configuration) as follows: the sequential composition of the processes out of the network configuration should be an implementation of the global communication service specification. Accordingly, a suitable implementation equivalence relation should be selected. Following a top-down approach, we can develop the specification of the switching system by decomposing the global communication service into protocol entities and underlying services.

In this way, correctness of the switching system is implied by correctness of the steps taken during the decomposition process.

A first significant observation is, that two different types of information related to a subnetwork connection (between end system and switching system), namely the **status** or **control** information of that connection, and the **user data** (e.g. speech) that is transparently transferred over that connection, are not treated separately at this level of abstraction. This is essential, since those two types of information are architecturally related to each other, hence they should not be considered independently yet. This is opposed to more implementation oriented descriptions, that separate these two information streams on the basis of their characteristics, or on the basis of implementation structures that facilitate this communication.

What we can do at this level, is classify information on the basis of the subnetwork connection endpoint which it is utilizing, and come to a second parameterisation of the gate SN: SN i,j, where i stands for the identifier of the subnetwork it interacts with, and j stands for the subnetwork connection identifier interactions are associated with (OSI: connection endpoint identifier).

At the gate SN i,j the following primitives may be executed: ConnReq, ConnInd, ConnResp, ConnConf, DaReq, DaInd, DisReq, and DisInd. Associated to each primitive, a set of parameters is defined (in order to exchange data between user (end system's or switching system's entity) and provider (the subnetwork)). Furthermore, the architecture is determined by the temporal ordering of the execution of primitives allowed, together with the parameter value possibilities. It is outside the scope of this paper to discuss the architecture which has been specified in our project in further detail. However, I would like to add that we have tested the specification obtained after this step, by symbolic execution of the specification.

2.2. Transformation Towards the Reference Architecture

Our second step in the design process is to come to a more structured specification, the reference architecture, that reflects implementation characteristics. This is achieved by decomposition of the single process (the top-level architecture) into a number of processes, each containing a subset of the complete functionality and therefore easier to implement.

The composition of these processes should yield a behaviour that is equivalent to the behaviour which the unstructured process it is derived from showed. This property can either be proved formally, or checked with software tools that disclose this behaviour from the specification of the processes. Our preference is to decompose on the basis of CPTs (see end of section 1.1), and coerce correctness by inheritance of that property, which turned out to be more time efficient. After these steps a reference architecture is obtained, that leaves the implementers with sufficient freedom to transform further towards an appropriate implementation, but also gives them firm guide-lines about the direction in which they will be able to find one. Their constraint is that the interactions which are defined between the processes of the defined structure, must all have their corresponding events in the implementation.

2.2.1. Structuring Principles.

The criteria along which a single process is transformed into multiple processes, can arise from architectural as well as implementational concerns. From the architectural viewpoint, we separate out those functions from another that emerge to be mutually independent or orthogonal, in order to manage the complexity of the process. From the implementational viewpoint, we substructure a process on the basis of implementation issues, e.g. the implementability of a certain function, given the current status of technology.

Examples of implementational limiting factors are communication bandwidth and processing power. This brings us to define two types of functions that represent abstractions from two types of implementation elements, viz. communication functions and transformation functions. The corresponding implementation elements are interconnection structures and processors, respectively. A communication function can be described as a process that contains many input and output gates, and is capable of transporting information transparently from input to output gates according to a predestined scheme. A transformation function is characterized by the fact that it possesses only few interaction gates, but is capable of performing operations on data, according to some predestined function.

In other contexts (e.g. database systems), one could identify storage as a potential implementation bottle-neck, and accordingly identify a memory function as building brick for decomposition purposes. In our case of the switching system, we did not envisage storage of information as an potential difficulty, at least we considered it of minor significance in comparison to communication and transformation.

2.2.2. Transformation of the Top-Level Architecture for the Switching System.

When we recall the architecture of the switching system, as described in section 2.1.2, and apply the structuring criteria as discussed above onto it, we can make the following observations.

Within the top-level architecture of the switching system, we can distinguish two major functions related to switching, namely a function that is able to forward the stream of user data from one point to another, and a function that is concerned with the control of connections (i.e. set-up and release), and that governs the former function. Notice that their characteristics roughly correspond to the definition of communication and transformation function as we presented in the previous subsection.

The interactions between the switching system and its environment are organised on a subnetwork connection basis, and no discrimination between the control data related to the subnetwork connection and the global connection it supports (also called: signalling data), and the user data that is transferred over these connections, is made in the top-level architecture. In system descriptions, which are more implementation oriented, we often see that those two information flows between end user and switching system are separated, and even make use of distinct underlying subnetwork connections. Also from an architectural point of view, those two information streams can be separated, since their characteristics are quite different: the user data shows the property that it is a continuous stream of bits or octects, while the control data is of a bursty nature. This results in a substructuring of the SN i,j gate into two gates, CNTRL i,j and DATA i,j, where at the CNTRL gate the control information exchanged between end system's and switching systems entities, plus the primitives related to the establishment and release of subnetwork connections are found, and at the DATA gate the user data stream is located.

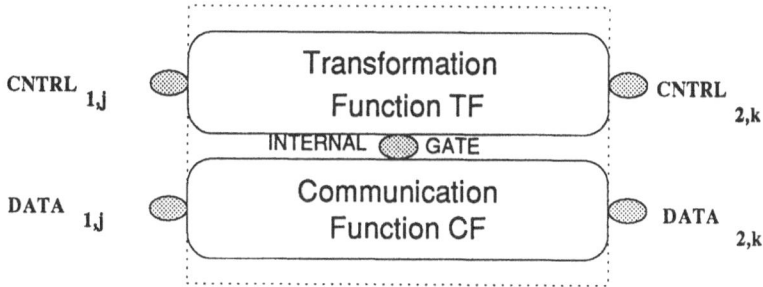

figure 5 - first decompostition structure, with 2 subnetwork access points

We now come to a first decomposition of the top-level architecture into an intermediate design, as is illustrated in figure 5. In this figure we see that the transformation function TF is interacting with all kinds of control information related to global as well as subnetwork connections, and that it, in accordance with the protocol defined for the interconnection of end users, will set the communication function CF in such a way that actual transport of stream oriented user data from calling to called user and vice versa can take place. However, we previously defined a transformation function as a function that contains only few communication gates. The TF in figure 5 does not conform to this definition; it has too many gates. Therefore we have to substructure this function again, and decompose it into a communication function that accounts for the transport of the control data coming from all CNTRL gates to the single internal gate of the transformation function, see figure 6. This CF' has a connection-less functionality, as opposed to the CF which is connection oriented. It has to schedule the primitives, that are executed at a large number of gates, onto the single gate with which the transformation function is equipped. Although the service provided by this communication function is characterized as connectionless, the ordering of primitives related to a specific subnetwork connection (SNi,j) should be preserved. This requirement is fulfilled by the constraint that the maximum number of primitives queued in this CF is limited to one (again with respect to a specific subnetwork connection).

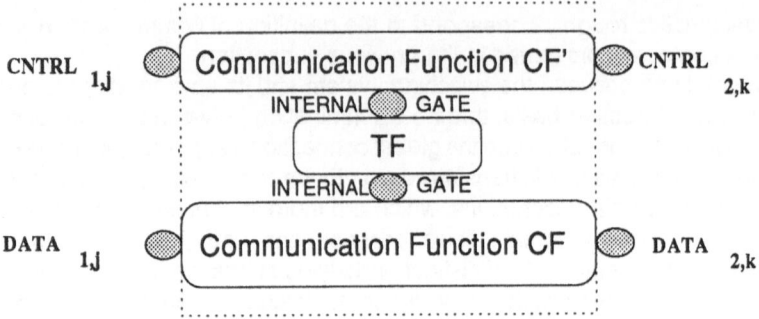

figure 6 - second decomposition structure, with 2 subnetwork access points

2.2.3. Further Refinement.

The transformation function TF in the structured specification of figure 6 is still quite complex, due to the fact that it should manage a large number of connections concurrently. If we take a detailed look at the procedure it has to follow for the support of an end-to-end connection, we see the following. First it is notified that a local connection from calling user to the switching system is requested, second information is received that contains the request for a global connection, next it has to establish a second local connection to the next switching system or end user in order to build the global connection. Hence this transformation function is concerned with network as well as with subnetwork handling. Therefore, an alternative transformation of the first decomposition structure (figure 5) is presented.

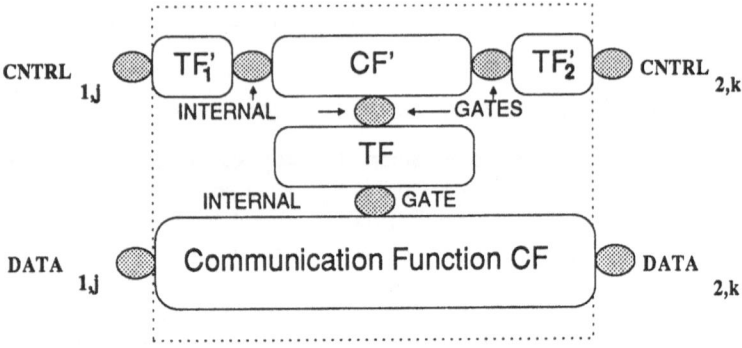

figure 7 - third decomposition structure, with 2 subnetwork access points and 2 TF's

In this structure (figure 7), those functions which are related to subnetwork handling are distributed over N transformation functions (TF') (where N equals the number of subnetworks attached), located at the periphery (the subnetwork boundary). The network handling is still centralized in one transformation function (TF). As a positive side-effect of this further refinement we found that the requirements for the CF' in this decomposition are much lower in terms of bandwidth and delay, than in case of decomposition no. 2.

As is explained in section 1.4 of this paper, this 3rd transformation step can be regarded as a result of backtracking in the tree of possible transformation steps, due to early performance analysis. However, a complete rewriting of the specification was not required; many parts could be reused. In fact, decomposition no. 2 can be seen as one extreme of the range of possible decomposition structures; all its processing functions are centralized. The other extreme of this range is the structure in which all processing functions are distributed over the peripheral TF's; for figure 7 this would imply TF to have a null functionality. In our development path, neither of the extremes is favoured, and the reason why structure 3 is preferred over structure 2 has been explained in the previous. With respect to the fully distributed structure we remark the following: it shows as a major drawback that implementation of systems management functions becomes more complicated, since status information about connections is scattered over the distributed processes.

The architecture that we have selected now (figure 7) corresponds to the reference architecture of the first cycle of the project (i.e. for that part which is concerned with interconnection of stream oriented traffic). It was passed to the performance analysis task that mapped it onto a queueing network model, and to the implementation task that tried to develop implementations for the processes defined in the reference architecture's specification.

2.3. The Second Cycle

In the next cycle, we selected more functionality for the definition of the top-level architecture. Then the transformation of the top-level architecture to the reference architecture was repeated for this new architecture (which is not a complete re-do of the steps performed in the first cycle, to the benefit of the constraint oriented specification style we obeyed).

Functions that additionally were selected are quality of service negotiation and resource handling. Since no protocols for the realization of these functions are standardized, we had to define the appropriate mechanisms ourselves. Next we formalized them in LOTOS, and after testing, this resulted in a second version of the top-level architecture of the switching system. To this version, we added more structure to the communication function of the reference architecture that is concerned with the switching of the data streams (see section 2.2.2). It has been decomposed in more communication functions and a transformation function, which can be justified with two reasons. In the implementation task it appeared that the CF of the first reference architecture had to be mapped onto several smaller interconnection structures, in order to obtain a switching structure that is capable of handling so much communication. Moreover, in general the capacity of the interconnection structure (the central switch) of a switching system is (much) smaller that the potential number of connections, representing the scarcity of (communication) resources in the system and the improbability of all possible connections being requested simultaneously. A further structuring of the central communication function hence is a realistic design step, regarded from an architectural as well as from an implementational point of view.

3. Conclusion, Evaluation

The use of FDTs in the development of a concurrent system requires a lot of extra effort during the first steps of the design, since new skills have to be acquired, and the fact that FDTs require uncompromising precision from the specifier. However, this extra effort is regained in multiplicity during the rest of the design process, because allocation of system functions can be arranged in various pilot decompositions, and their benefits can be assessed without obligation to transform them into real implementations. Thus convergence towards a proper implementation can be triggered off at an early stage.

The example application of the PANGLOSS method onto the design of a switching system, has shown the industrial significance of the method, bearing in mind the complexity of such a system. It has proved that in case we are concerned with development and definition of highly concurrent systems, the use of formal description techniques (language and software tools) is an outstanding addition to existing approaches.

4. References

[1] ISO: International Standard 8807 - Information Processing Systems - Open Systems Interconnection - LOTOS - A formal description technique based on temporal ordering of observational behaviour' Genève, 1988

[2] Eijk van P.H.J.(ed.), Vissers C.A.(ed.), Diaz M.(ed.), 'The Formal Description Technique LOTOS', Results of the ESPRIT/SEDOS project, North-Holland

[3] C.A. Vissers et al. - 'On the use of specification styles in the design of distributed systems' - presented at TAPSOFT "89 March 13-17, Barcelona, advanced seminar on foundations of innovative software development

[4] Bogaards, K. Pires, L. Pras, A. Schot, J. 'The PANGLOSS method', Proceedings Esprit Technical Week 1988, Brussels November 1988, North-Holland

[5] Bogaards, K. 'LOTOS supported system development', Proceedings of 1st International Conference on Formal description Techniques FORTE 88, Stirling UK September 1988, North-Holland

[6] ISO: International Standard 7498 - Information Processing Systems - Open Systems Interconnection - Basic Reference Model - Genève, 1984

[7] Eijk, P.H.J. van (1989) 'Tools for LOTOS Specification Style Transformation', submitted to FORTE "89, Vancouver; also published as internal report 'Memoranda Informatica 89-35', June 1989, Department of Informatics, University of Twente, Enschede, the Netherlands

[8] Pires, L.F. 'On the use of LOTOS to support the design of a Connection Oriented Internetting Protocol', Proceedings Esprit Conference 1989, Kluwer Academic Publishers, Dordrecht, the Netherlands

[9] Brinksma, E. Scollo, G. and Steenbergen, C. 'LOTOS Specifications, their Implementations and their Tests', in Proceedings of the 6th International Workshop on Protocol Specification, Testing and Verification - 1986

BMS: A Knowledge Based Budget Management System

Chao CHEN, Jacques BICARD-MANDEL, Michel TUENI
Advanced Studies Department, BULL MTS
Rue Ampère, Massy Cedex , France
Tel: 64 47 84 76
Electronic Mail: tueni@masbull.gtmy.bull.fr

Abstract

This paper describes the Budget Management System (BMS) which is a knowledge based system for budget management. The BMS has been designed with two major goals: on one hand to allow the creation and modification of a hierarchical organization model for a specific budget, and on the other hand to provide intelligent assistance to office workers while operating in a real budget setting. The knowledge domain is naturally modelled by means of constraints and rules. The constraint formalism postulates numeric relations among budget items with the purpose of transforming the incoherent numeric situations into explicit symbolic pointers to guide the budget management. It allows communication between symbolic reasoning and data processing. The constraints are loosely coupled with a satisfaction mechanism. To satisfy violated constraints, the user can either activate the reasoning mechanism by choosing the corresponding knowledge base or decide to alter directly the budget model. The system can be considered as an expert system shell dedicated to the representation of constraints. In addition, a generic budget model with an appropriate Man Machine Interface is provided to allow end-users to interact with the system by direct graphical manipulations.

1. Introduction

Budget preparation in large organizations is a most complex task because of the uncertainty that is embedded in selecting important items on which money should be spent, and the amount of money that should not be exceeded. In addition to the quantitative analysis tasks, the budget preparation process includes numerous unstructured tasks with justification, criticism, approval and distribution of the budget hypothesis before final consolidation.

The raison d'être for drawing up a budget is the management of financial resources and the manner in which these resources are to be utilized by various sub-organizational units. We can divide budget management into two phases: budget elaboration and budget tracking. During the elaboration phase, the manager establishes a general forecasting on the basis of the tasks to be fulfilled, taking into account the organization's internal and external influences and his past experience. Then, he draws up an estimated budget. This is a simplified scenario when the manager is in the position to make all the forecast. A more complex scenario is one when the process of budget preparation is distributed among various people. Here coordination is vital, otherwise the result could be catastrophic. The elaboration phase is now distributed among the various expert. Each expert is given a rough estimation in accordance with the tasks that have to be fulfilled. With this as benchmark, each expert then draws from his/her own experience to formulate alternatives within the confines of the global strategies. The outputs of the experts' elaboration phase are then given back to the manager for budget consolidation. Here many iterations may be necessary before a consolidation can be drawn up. The budget tracking phase is a monitoring process whereby the manager keeps track of the evolution of the budget. This phase consists of periodically ascertaining the discrepancies between the allocated resources and the consumed resources. The discrepancies may be wide enough to revise the budget. If this is the case, then the budget goes through the same cycle as before, only this time attention has to be focused on the items causing the discrepancies.

Software such as DBMSs and Spreadsheets have greatly increased the productivity of management information systems in finance. Most budget preparation tasks are now relying on these tools. Although they are efficient in data processing, they lack the following features:

- explicit description of the budget organization with its internal and external influences.
- representation of experts' knowledge in budget management.
- advice for budget allocation and reorganization.
- collaborative mechanism to support decisions in a decentralized environment.

AI and expert systems technologies (object oriented models, constraints, production rules, intelligent interface etc.) might be used to solve these problems. Recent years have seen the emergence of the integration of knowledge-based systems into existing office products, encouraging the investigation of a much larger spectrum of office applications. Syntel [1], a financial risk assessment system, which applies functional programming techniques to integrate databases and expert systems; CLP [2], a constraint logical language developed for an option trading application; and FAME [3], a financial marketing advisory system using an object oriented approach, were developed in the field of finance. Few expert systems have tackled the budget management field.

An exception is the FRM (Financial Resource Management) project [4]which integrates rules and constraint formalisms in the spreadsheet paradigm. This system introduces a constraint formalism and takes the financial resource management as a constrained planning problem. The constraint formalism is defined in terms of logical relations among database values, and incorporates a satisfaction mechanism which can be triggered to activate the corrective database modification. If simple corrective actions are not available, the system invokes a planner, which uses production rules to satisfy violated constraints. Constraints are a natural way to represent some domain situations, and can be used as a problem solving method to limit the search space or guide the planning. The constraint formalism shares some similarities with production rules. They have two parts: a condition part to express a logical relationship in the fact domain, and an action part to indicate the actions to take. But they are conceptually different. For a rule, if the conditions are all satisfied, then the actions are started. But for a constraint, if one of the conditions is not satisfied (that means this constraint is violated), then the actions are started and all these actions have only one pragmatic goal, namely, to satisfy the violated condition (constraint). After the satisfaction actions, the constraint may still not be satisfied. Some relaxation mechanism may be used to relax the rigidity of the constraint. Although this kind of purist approach could be used to investigate the general problem of planning under constraints, it may prove to be less practical in solving real problems in budget management. Moreover the FRM spreadsheet style interface is not explicit enough to represent the hierarchical organization model for the budget management.

Other systems have been developed with a constraint oriented approach to solve real life problems like operations research (scheduling, layout, warehouse location etc.). A new generation of tools, such as CHIP [5] combining a declarative aspect and the efficiency of constraint solving techniques has appeared. CHIP, developed on top of Prolog is expressive, efficient and flexible but too strongly devoted to operations research. This type of system emphasizes the numeric aspect of the solution itself and not the actions to be executed to change the status of the constraints in order to reach the solution.

ISIS system [6] describes the structure constraints for the job shop scheduling application. This very generic constraint formalism is used to guide the beam search process at different levels, and some constraints could be relaxed according to their importance and utility. This constraint approach to communicate the information between higher and lower levels is more flexible than the traditional hierarchical search mechanism where the lower level search space is more restricted by the higher level search process. Logic expressions [7] are used to define constraints allowing systematic deduction to produce a set of logic propositions that satisfy all the constraints. But the deduction process is not explicit in terms of domain knowledge.

Traditional spreadsheets and DBMSs are efficient but lack the capability to perform qualitative analysis and reasoning. Constraint satisfaction or relaxation is a suitable approach for solving problems with conflicting goals, but the few expert systems that do incorporate constraint formalism are not specialized in the budget management. FRM, whose domain of application includes budget management, extends the spreadsheet paradigm with planning under constraint formalism. However, its constraint management is somewhat rigid and lack a highly manipulative budget model interface that will allow various hierarchical representation of the budget to be shown.

In this paper, we propose a budget management system which assists organizations to prepare, modify and track a budget. The system is composed of a model that reflects the organization structure generally represented in tree form, a budget-oriented constraint layer and a rule system to represent the domain expertise to solve the constraint violation situations.

We will describe in the following sections the model components for the organization structure, the constraints and the solution generation mechanisms based on constraints satisfaction. Finally, some examples will illustrate the system operations.

2. A generic formalism for budget representation

Financial experts recently have adopted spreadsheet packages to prepare their budget: these packages are used mainly to set up costs and to establish relations between the different budget cells. However, the spreadsheet model lacks a tool to describe the global organization, and the entities involved in the process of preparing the budget independently of their relationship. BMS tries to reflect these elements by defining a model.

The model is divided in three basic components *Budget Entity*, *Constraint* and *Income*:
- *BudgetEntity* represents a organizational unit which will play a role in budget funding. *BudgetEntities* might be linked together according to rules defined by the organization. In general these rules are hierarchical. When one entity reports to another, we say that they are linked with a *part-of* type of link. For example, in fig.1, a BudgetEntity instance called budgetEntity1 has two sub-budgets: BudgetEntity2 and BudgetEntity3. The budget model hierarchy is thus formed.
- *Constraint* defines a relation which might exist between any two entities. The link between two entities having a constraint relation is called *has-constraint*. In fig.1, BudgetEntity1 has a constraint instance constraint1,which is related to BudgetEntity4.

– *Income* represents the income expected to be assigned to a specific entity. The link is a *has-income* type of link.

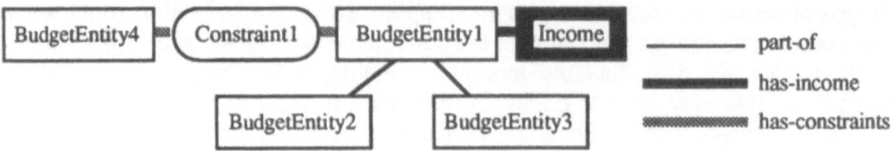

Figure 1. Semantic network for budget model

Using an object oriented model, experts explicitly build a semantic network where nodes are budget-oriented entities (budget-entities, incomes, constraint) and links have a budget-oriented semantics (part-of, has-income, has-constraint) as introduced in Figure 1.

Each type of link has its own methods for regulating information through the network. Costs are aggregated through the *part-of* links which permit the value of a global budget to be computed, taking into account its constituent components. Two association links model specific budget relations. The *has-income* link is useful to calculate the balance of a budget. Finally, the *has-constraint* association is involved in the constraint propagation mechanism: should BudgetEntity1 value be changed, the system must check BudgetEntity4 value, and furthermore activate other constraints attached to BudgetEntity4 which might trigger modifications in the rest of the taxonomy.

However, the semantic network is hidden from the financial expert. He only handles a natural budget representation made of "part-of" and "has-income" links through a tree-like graphical interface (figure 2).

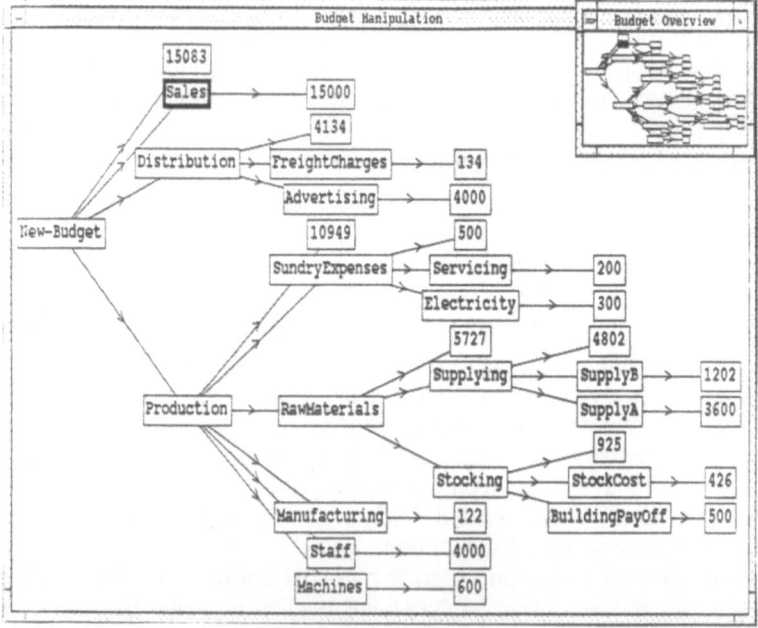

Figure 2. A sample budget

2.1. Budget Entity

Three major information elements constitute the budget entities: expenses , income and balance. During the preparation phase, evaluation is processed to determine the expenses according to the value of expected incomes. On the other hand, specific incomes could be attached to some budget entities. The balance is introduced also in order to represent the relation between income and expenses.

To support reasoning, we had to deal with symbolic knowledge. Two basic attributes are considered to handle this aspect: impact and category.

- Category classifies BudgetEntities into classes from which knowledge bases used in the constraint satisfaction procedure are instantiated.
- Impact represents the importance of the BudgetEntity. To state that some entities are more important than others, an attribute called impact is introduced. The greater it is, the more important the entity is. This kind of weight will allow to sort entities within the same category; in case of budget reduction, the component with lowest impact will be reduced first in order not to affect the global budget too strongly.

The different characteristics of a standard Budget Entity are represented as follows:

Attribute	Meaning of the attribute	Type
Expenses	Value of the expenses	Numeric value
Income	Value of the incomes	Numeric value
Balance	Difference between Income and Expenses	Numeric value
Part-of	List of the sub-components	part-of link
Has-Constraint	List of associated constraints	has-constraint link
Has-income	Name of the associated Income object	has-income link
Impact	Importance of the entity	Symbolic reasoning
Category	Class of the entity	Symbolic reasoning

Figure 3. The standard characteristics of a Budget Entity

2.2. The Constraint Formalism

After meeting with several budget managers, we discovered that constraint formalism was very appropriate to model the domain knowledge. All the budget entities can take different values during the budget elaboration, but they must always satisfy particular relations called constraints. Constraints guarantee the overall coherence of the model. They are often expressed with relations between several budgets entities. For example, purchases should not exceed their allocated incomes.

In addition to the mathematical equation, constraint has attributes: tolerance and priority

- Tolerance is a way to express laxness, which means that the relation between entities should be considered as true when the arithmetic relation is not strictly satisfied, let's say at 10 %. To provide more flexibility, it is interesting to allow some tolerance in the constraint verification mechanism. The end-user can set up tolerance as he wants.

- Priority expresses that some relations are more important than others and should be verified and satisfied first. It is divided into two parts: Test Priority and Satisfaction Priority.

The constraint with higher Test Priority will be tested first. This attribute can also be used to relax some momentarily uninteresting constraints: using Test Priority as a strength indicator, the system is able to detect constraints with low strength and to deal only with those higher than a fixed threshold. This mechanism could avoid conflicts between contradictory constraints by relaxation of some of them. On the other hand, some minor constraints might be relaxed and their satisfaction delayed after the resolution of major problems.

The violated constraint with highest Satisfaction Priority will be satisfied first. The plan generated after the satisfaction of the most important constraint will alter the budget and lower constraints might be verified at the same time. This mechanism largely improves the efficiency of the global satisfaction process.

We have introduced a constraint as an explicit numeric relation among budget items, with a real budget semantics. Moreover, the financial expert can easily handle constraints with a graphical interface and build constraints freely on any entity and independently of its attributes.

There are four advantages for the constraint paradigm to be introduced in the budget model:

1) The constraint formalism is efficient to model other kinds of relations than hierarchical, especially external influences.
2) It explicitly helps the budget formulation by the constraint violation warning mechanism and provides hints for interactive corrective actions
3) It offers intelligent solutions through attached corrective actions or through the activation of the corresponding knowledge base.
4) It provides transparent reasoning by tracing the propagation of the violated constraints. The constraint formalism can be considered as a starting point of an explanation mechanism in budget management.

3. The Constraint Management Cycle

Now that we have explained the basic model and the constraint mechanism, we describe the Constraint Management Cycle. The budget management has functionalities for coherence maintenance similar to the commercial spreadsheet software packages. For example, the modification of the value of a sub-budget activates the change of the value of the global budget, like what happens in spreadsheet where the modification of one cell triggers the alteration of others. The difference lies in the processing of the coherence maintenance, which is performed through a cycle where the scheduling process is activated to control and execute specific actions to verify and satisfy the constraints invoked by the modification.

3.1. The Schedulers

All the constraints bound to the updated entities are forwarded to the agenda of the Test Scheduler. These constraints may have become violated because one or more of their parameters have been changed, and their status should be checked in order to detect numeric inconsistency in the budget.

When the user decides to control the coherence of his budget model, the system will verify these constraints. According to the priority of verification, the constraints are filtered: the violated ones are transmitted to the agenda of the Satisfaction Scheduler, those which are verified within the tolerance rate are stored in the agenda of the Tolerance Scheduler, and the verified constraints are dropped. At the next step of the cycle, only the constraints on the satisfaction agenda will be processed, while the tolerance agenda gives only further explanations about the model coherence. Secondly, according to the Satisfaction Priority, the most important constraint in the Satisfaction Scheduler will be satisfied. Using the corresponding knowledge base, the inference engine will generate a plan that alters the budget model. Some constraints may have their parameters modified. These constraints are then stored in the Test Scheduler and are subject to the same verification process as that described above. This cycle (cf. figure 4) will end when the Satisfaction Scheduler agenda gets empty.

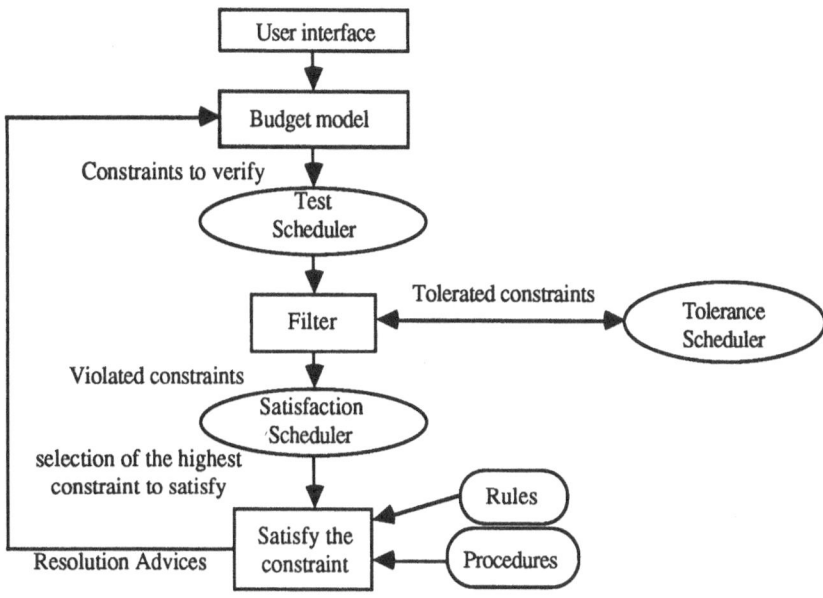

Figure 4. BMS Constraints solving cycle

3.2. The Inference Engine

To generate a solution which satisfies a given violated constraint, the system has to compute the current goal of the planning mechanism. Once the main goal has been generated, the system loads the corresponding knowledge base (rules and computing procedures) to divide the goal among the different categories and generate the solution by means of the KOOL [8] inference engine. Since each assignment of an attribute can trigger the inference engine, we decided to use this generic goal entity as engine starter. Consequently all the rules are instantiated from this object.

In the BMS context, the KOOL engine can be described as follows :
- Assign or update an attribute of the generic goal object
- The system chooses among all the loaded rules those which test the most recently updated attributes

- It tests all these rules. A rule is said to be true if all its premises are true. The system stores an instance of the rules said to be true in a conflict set. To instantiate these rules from the budget model, category and impact can be used in the precondition part.
- It starts the rule with the stronger interest. If two rules happen to have the same interest, it starts the rule that refers to the most recently updated object. This characteristic allows a data driven behavior. The action part of the rule modifies the budget model and the generic current goal: new rules will be processed and new constraints will be filtered in the next BMS constraints cycle.
- Premise new check - Before starting the chosen rule, the system checks that this rule is still true. Indeed, if the attributes in its premises have been modified between the moment it was put in the conflict set and the moment it is sent out, its premises are newly checked. If the rule is false, it is abandoned and the system chooses another rule.

The sequence will stop when the conflict set is empty. It is worth noting that the knowledge engineer has to be aware of the importance of the interest attribute of the rules because it is the major procedure in the inference engine.

The whole resolution process has clearly been divided into two different steps. The first step consists in the detection of numeric inconsistency in the budget model. Constraints can be considered as a sort of deep knowledge, which models very precise points in the budget setting. During the second step, rules can be considered as heuristic knowledge: once a constraint has been selected to be satisfied and the goal has been formulated, rules express the means to reach the goal. They are completely dependent on the budget setting (another situation would require other rules) whereas a coherent budget setting is subordinated to the constraint definitions.

4. Example

In this section, a concrete sample budget (cf. figure 5) with a working scenario is outlined.

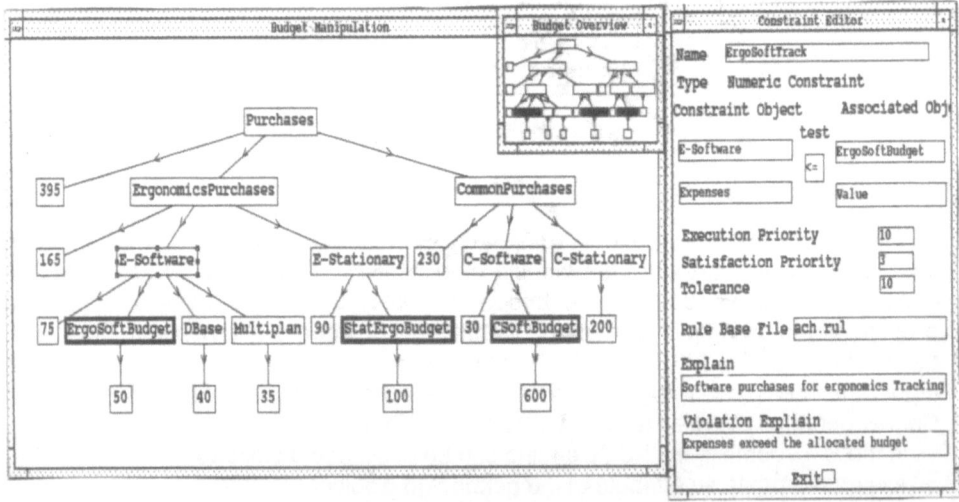

Figure 5. Purchases Budget

A constraint called ErgoSoftTrack[1] (cf. window Constraint Editor in Figure 5) which requires the expenses for the "E-Software"[2] Purchases budget not to exceed the value of the allocated resources ("ErgoSoftBudget" income object) has been constructed using the constraint editor; this constraint is violated.

The user then activates the reasoning process of the adequate rule set. The system loads the corresponding rules. The goal to achieve is to find another sub-budget where the reserve (balance) is high enough to take over the last purchases.

In order to achieve this goal, we have written the following rules:

Rule	R1
Interest	1000
If	the purchases in software for the ergonomics section "E-Software"[3] exceed the allocated budget "ErgoSoftBudget"
Then	find the latest purchase

Rule	R2
Interest	500
If	the sub-budget purchases in stationary for the ergonomics section "E-Software" can support this purchase
Then	charge "Stationary_E" with the last purchase

Rule	R3
Interest	300
If	the sub-budget purchases in stationary for the ergonomics section "ErgoSoftBudget" cannot support this purchase
	the sub-budget purchases in software for the other sections "CSoftBudget" can support this purchase
Then	charge "Software_CP" with the last purchase

The result is shown on the figure 6. R1 and R3 have been triggered and R2 is false. The software "Dbase" is suggested to be charged to the "C_Software" sub-budget, instead of "E_Software".

---

1   The constraint ErgoSoftTrack is a numeric relation between the budget for Purchases of software for the ergonomics section represented by the "E_Software" object and the allocated income for these purchases designed by the "ErgoSoftBudget" object
2   This object is selected in the Budget Manipulation window of Figure 5 and framed with spots
3   In the following rules, " " refers to objects described in Figure 5

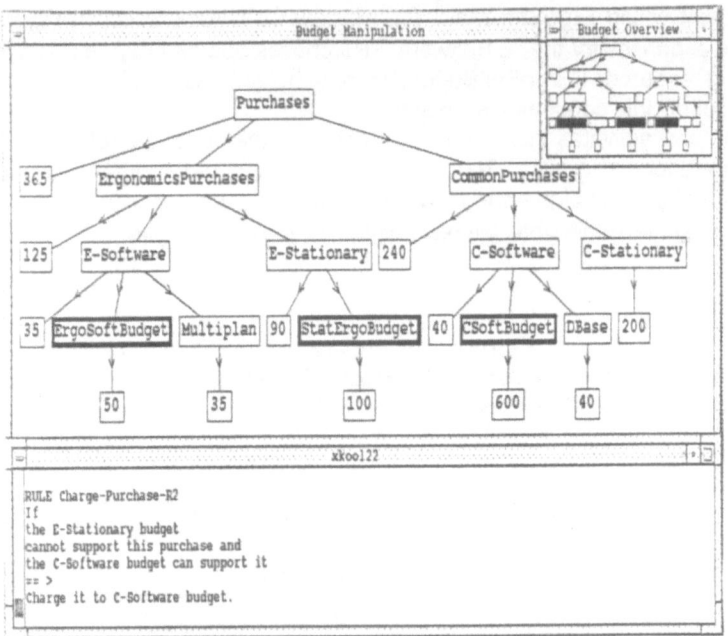

Figure 6. The new budget after reorganization

We have intentionally introduced a quite simple example to explain the satisfaction mechanism. In fact, real budget setting consists of a succession of this type of reasoning. One modification in the model might trigger many constraint violations. Consequently, a lot of knowledge bases might be activated, some of them might even be contradictory. The knowledge engineer has to see to the problem of knowledge base consistency. Moreover, rules are specific to a given situation: another organization with other strategies will use another set of rules.

A second example will emphasize the constraint propagation mechanism. The tree-like hierarchy is given by Figure 2. Let us assume that the marketing department forecasts that the company will be able to sell 1000 items during the next period. In order to compute the Production and Distribution budgets, the following constraints have been defined. The production section will be asked to manufacture 20 % more items than the forecasted sales because of misproduction (CProd). To produce, the section needs machines, each machine can process 500 items (CMachines), the number of items manufactured on those machines is equal to the number of item produced in order to calculate the manufacturing cost (CManufac), one new item is made of 3 items A (CSuppA), and a half item B (CSuppB). All these supplies must be stored (CStock1, CStock2, CMaterials to compute the number of units for the RawMaterials, StockCost, Stocking budgets). All the sold items must be distributed and carried (CDistrib, CFreight). The management of these constraints is summed up in Figure 7.

Let us assume that the production manager wants to know what would happen if the number of sales is changed into 1500. Constraints CProd, CFreight, CDistrib will be stored in the Test Scheduler and then in the Satisfaction Scheduler because one of their parameters has been changed and consequently these constraints became violated. CFreight is selected because of its higher priority. Rules to compute the number of carried items are started, with the same set of rules the system is able to compute the number of distributed items (both are equal). During the next constraints resolution cycle, CProd will be selected (CFreight has just been satisfied, CDistrib vanished because it has been satisfied during the same cycle as CFreight). The number of items to produce will be computed and so on. Every constraint will be satisfied according to its priority. The number of A and B supplies will be computed when the number of items to produce will be known, the number of supplies when A and B are known etc.

Constraint	Priority	Relations	Introduction in the Test Scheduler
CFreight	110	FreightCharges Number = Sales Number	When Sales Number is changed
CDistrib	109	Distribution Number = Sales Number	
CProd	100	Production Number = 1.20 * Sales Number	
CManufac	80	Manufacturing Number = Production Number	When Production Number has been computed
CMachines	70	Machines Number = (1 / 500) * Production Number	
CSuppA	50	SupplyA Number = 3 * Production Number	
CSuppB	49	SupplyB Number = 0.5 * Production Number	
CStock1	40	StockCost Number = Supplying Number	When Supply Number has been computed
CStock2	30	Stocking Number = Supplying Number	
CMaterials	10	RawMaterials Number = Supplying Number	

Figure 7. Constraints for the Production Budget

Figure 8 shows the reallocation for the Production and Distribution budgets.

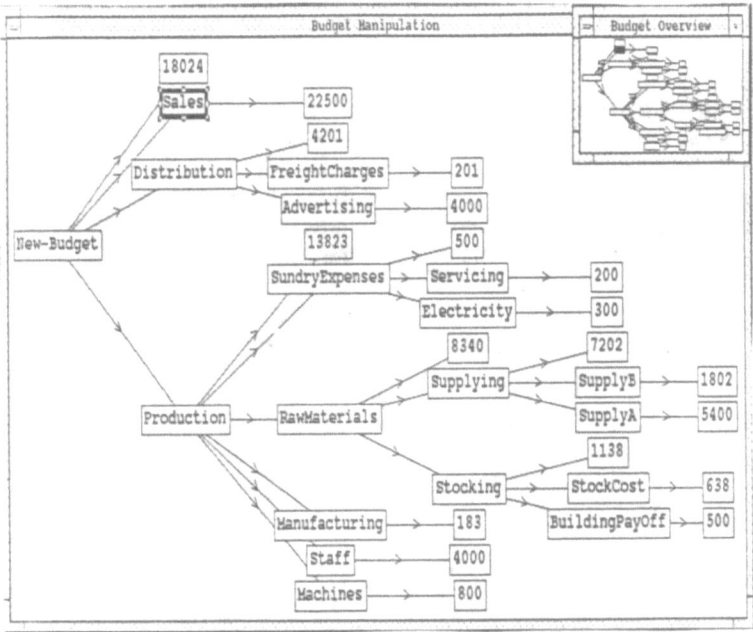

Figure 8. Reallocation for the Production budget

This example shows a propagation mechanism similar to spreadsheet software. Nevertheless, the constraint formalism is used as a means of explanation of budget reallocation and the smart graphical interface allows explicit representation of the model modifications. Moreover production rules can be used for the satisfaction of each violated constraint, just as in example 1.

5. System architecture

The BMS architecture, as depicted in figure 9, is composed of a User Interface, a Organization Model, a Knowledge Interface, Rules and Constraints, Schedulers and a Constraint Management Cycle. The budget model can be constructed by the user through an interactive interface. The constraints and rules can be edited by the domain experts by means of the Knowledge Interface. The advice obtained by the Constraint Management Cycle may modify the budget model.

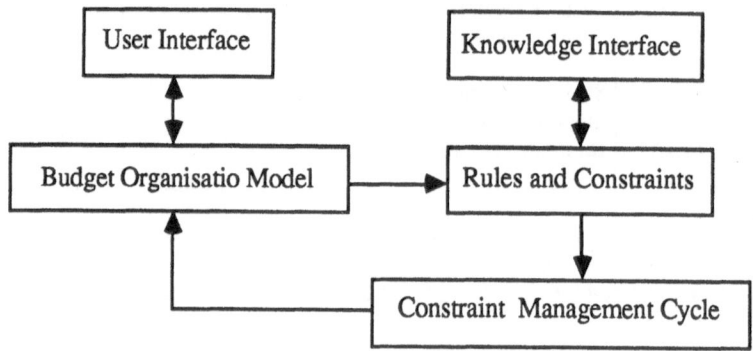

Figure 9. The BMS architecture

BMS is implemented on top of KOOL, which is an object oriented language with a rule package developed by BULL-CEDIAG. The tree-like hierarchical budget organization interface has been constructed with GSE, a Graphical Structure Editor [9], and the form-filling and menu interface has been developed with the FORMS package [10]. These two packages are major high level tools of an advanced MMI model built on top of X11. The code has been developed with Le_lisp under Unix (figure 10).

Budget Manager System		
Budget Model	**Constraints Layer**	**Satisfaction Cycle**
KOOL	GSE	FORMS
Le_Lisp		
UNIX + X11		

Figure 10. The software hierarchy

6. Final remarks and conclusions

BMS is a knowledge based system which provides assistance to the users involved in budget preparation and budget tracking. It consists mainly of a constraint mechanism and a solution generator built on top of Kool, an oriented object language and inference engine. It can be considered as a dedicated shell for budget management expert systems.

The constraints represent the information exchange between the world of symbolic reasoning and that of data processing. The goal is only to transform the incoherent numeric situation into explicit symbolic hints to guide the budget management. Our constraint formalism is limited to represent numeric inconsistency and is able to activate associated knowledge bases built on the rule formalism. This is adapted to the budget management domain where explicit constraint representation is needed to guide remedy process. This constraint layer is loosely coupled with the satisfaction mechanism. The user can either activate the reasoning of the relevant knowledge base or make decision to alter interactively the data to satisfy the violated constraints. The constraints may still be violated after the satisfactory mechanism, nevertheless the violation degree is reduced. The tree-like graphic interface explicitly represent the organizational hierarchy. The graphical tools help the domain user to set constraints.

In order to validate concepts, we have exploited BMS in field trial to construct a budget preparation for our division. The discussions with domain experts clarified the problems and the desirable design criteria for the assistance part. This was very crucial in the understanding and the design of the suitable formalisms to represent the budget management expertise.

Budget planning is a complex process entailing the assessment of many pieces of information. Budget description is not restricted, such as in the BMS, to expenses and incomes forecasting. Internal and external data are used to compute financial concepts, which are evaluative indices showing if the budget meets the initial objectives of the company.

The creation of budget samples is not a straightforward, linear process as economic data can be aggregated in several ways. Such data can be analyzed and consolidated according to:

- the organizational and functional structure of the company.
- the product and project division of the company.
- the chart of account according to the inflow and outflow of the company.
- time analysis. Here, information can be distributed over several time periods. Auto-matic conversion is required to transfer data from one time scale to another (month, quarter, semester, year).

Finally, each organization will have to rely on its own heuristics for budget manipulation. What BMS can do here is to reduce the order of difficulty for formulating these rules (or heuristics) by providing some general purpose formalisms. These formalisms constitute the basic building blocks for rules, but the manner in which these blocks are assembled is organization specific.

Some others issues have been pointed out during the experiments on budget:

- Budget elaboration is a cooperative work. Experts at the top level of the company express some strategies and goals which are in turn transmitted to the different departments.
- Because firms make many decisions involving immediate outlays and long-run benefits, it is desirable to bring time value into the analysis of budget in a sensible fashion.

Nevertheless, each field of the budget has its own forecasting period. For example, treasury situations must be planned day after day because of the uncertainty of the financial market, staff administration needs month or quarter to pace up with the working planification, a capital budgeting decision is characterized by costs and benefits that are spread out over several time periods. This last point requires some elaborated financial concepts such as the time value of money, discount rate and internal rate of return, paybacks periods etc. to evaluate alternatives correctly [11].

The principal procedure in BMS is the constraint violation warning and satisfaction mechanism which provides the feedback information and interactively assists the budget construction. A constraints manipulator system, derived from the BMS constraint layer might also be efficient in other financial fields such as bank credits, insurances, company diagnostics and could open the way for most generic constraints programming systems.

Acknowledgments

We thank Najah Naffah, director of the Advanced Studies Department for his support and for his providing numerous suggestions during the course of this research. We also acknowledge Louis Sauter, leader of the COCOS Esprit project, for many useful discussions. We are grateful to Florence Hermelin for her contribution and her help to understand the budget reality. Gabriel Jureidini, who developed the GSE package, and Michel Texier who designed the FORMS package also provided constructive ideas for the BMS graphical interface. We thank aswell James ANG for his comments on an early draft of this paper.

References

[1] Duda R.O., Hart P.E., Reboh R., Reiter J., Risch T. "Syntel, Using a functional language for financial risk assessment" IEEE Expert, Fall 1987

[2] Lassez C., McAloon K., Yap R. "Constraint Logic Programming and Option Trading" IEEE Expert, Fall 1987

[3] Mays E. Apte C. "Organizing knowledge in a complex financial domain" IEEE Expert, Fall 1987

[4] Gelman "FRM, Financial Resource Management" Internal report, Stanford University 1987

[5] Aggoun A., Dincbas M., Herold A., Simonis H., Van Hentenryck P. "The CHIP System" Technical Report, European Computer-Industry Research Center GmbH München

[6] Mark S. Fox "Constraint-Directed Search: A Case Study of Job-Shop Scheduling" Pitman, London, 1987

[7] "A hybrid structured object and constraint representation language" AAAI-86 (986-990)

[8] "Kool Manual" Bull Cedlag, Sept. 1987

[9] Jureidini G. "GSE Manual" DEA Bull MTS, March 1987

[10] Texier M. "FORMS Manual" DEA Bull MTS, Sept. 1987

[11] Bierman H. Jr., Smidt "Financial Management for Decision Making" Macmillan Publishing Company New-York 1986

Keywords

Office automation, Expert system, Management

Project No. 1024/2374

ODA/ODIF - THE STANDARD SOLUTION TO DOCUMENT INTERCHANGE

J. NELSON
International Computers Ltd.
Lovelace Road
Bracknell
Berks. RG12 4SN
England.

C. BATHE Nixdorf, Germany
I. CAMPBELL-GRANT ICL, United Kingdom
K. FISCHER IBM, Germany
P. KIRSTEIN University College London, UK
E. KOETHER Siemens, Germany
G. KROENERT Siemens, Germany
O. MOREL Bull, France
A. PASSARO SSS, Italy
R. PENNELL ICL, United Kingdom
F. REYNAUD TITN, France
M. SCHOONWATER Océ, The Netherlands
G. TRAVERS BT, United Kingdom

Abstract

Extended communication facilities now mean that documents often require to be exchanged between document processing systems with different hardware and software and incompatible data formats. More importantly the recipient needs to be able to manipulate and process the document in a way that conforms to the originator's intentions. This is only possible if the document structure, and the rules by which the document has been composed, are transmitted with the document itself. Office Document Architecture (ODA) is an international standard which provides the framework to enable the user to do just this.

In supporting the adoption and future developments of ODA the PODA (Piloting the ODA) Project, which is funded within the ESPRIT II Programme, has the twin objectives of accelerating the development of ODA-based products and advancing the application of the ODA Standard. The standard is now well established with the standards bodies and product announcements are imminent.

1. Introduction

With the advent of powerful word processing and electronic publishing systems the "electronic document" has become an important basis for information exchange. Today's extended communication facilities mean that sender and recipient will often have different document processing systems (different hardware and software with incompatible data formats). In such cases document exchange is restricted to the content-only representation of the document e.g. facsimile or ASCII encoded form. Facsimile merely allows the recipient to produce an image of the transmitted document as created by the originator. The image may be displayed on a video screen, or printed to hard copy. In order that the recipient may edit, reformat or process the document in any way an understanding of the structure of the document is required and, as yet, no facility exists to allow the automatic derivation of document structure from content. It is necessary therefore to transmit the structural information, together with the rules for editing and formatting the document, with the document itself. ODA (Office Document Architecture) is an international standard which has been designed to fulfil that function.

ODA permits multi-media documents to be exchanged between conforming computer systems in such a way that the received document can be presented in the originator's format but, more importantly, it can be edited or reformatted by the recipient.

The Standard does not define rules for structuring documents but provides the framework for the originator to do so. OSI standards such as X.400 and FTAM have made possible the exchange of data between proprietary systems. ODA, which sits at level 7 of the OSI seven layer model but above such data interchange standards as X.400 and FTAM, permits the transfer of information in a way that allows it to be manipulated and processed by the recipient.

Work began on the ODA Standard in 1981. Since then it has been supported by a number of projects, most significantly the HERODE, PODA-1 and INCA ESPRIT I Projects. This support took the form of validating and promoting the development of the Standard through early prototyping, the results being fed through the appropriate standards bodies. Wide exposure for the developing Standard was achieved through public demonstrations. The Standard now has international acceptance. ODA workshops are now firmly established in Japan, the USA and Europe.

PODA-2 (Piloting of the Office Document Architecture - 2) is an ESPRIT II Project that started at the beginning of 1989 in a smooth continuation from PODA-1. This paper seeks to review the current status of ODA and its proposed developments and to explain how the PODA-2 Project is achieving its twin objectives of accelerating the adoption of ODA and supporting its continued development.

2. ODA - the Standard

2.1. What is ODA?

ODA is an international standard which permits multi-media documents, i.e. electronic documents which contain any mix of text, raster and geometric graphics, to be interchanged between heterogeneous computer systems.

ODA describes both the logical and the layout structure of the document in addition to its content. The logical structure shows how the document content relates to such objects as paragraphs, headers and diagrams. The layout structure describes the physical layout of the document. As well as the logical and layout structures specific to the document ODA includes generic logical and layout structures which define the intentions of the originator and drive the layout process respectively. The generic structure holds the rules for generating and amending the specific logical and layout structures.

Aspects of both the structuring information and the content in a document can appear many times throughout the document. ODA embraces the concept of "classes" i.e. within the document objects (or elements) in the structure which have the same characteristics are classified into groups. An object class is a specification of the set of characteristics that are common to its members. The set of object classes in the document is in fact the generic structure. The concept of object classes simplifies the creation of the document, improves document transmission efficiency and preserves the integrity of the document structure after editing.

As stated above, the document content can be of different types. Those currently covered by the standard are text, raster and geometric graphics. Each content type has its own defined content architecture which specifies content structures, coding schemes etc. An important feature of content architectures is their independence. This enables new content architectures to be easily included.

ODA employs the structures above to represent the document in three forms. The formatted form allows the document to be printed or displayed conformant to the originator's intentions. The processable form allows the document to be edited and/or reformatted conformant to the originator's intentions. The formatted processable form is a combination of the two previous forms. It allows the recipient to image the document and if he wishes he can edit and/or reformat it as well.

In a practical situation a document generated on any document processing system is converted into the ODA-encoded form, Office Document Interchange Format (ODIF), prior to transmission to the receiving system where the ODIF encoded document is decoded into the internal format of the receiving system. The sending and receiving systems need not be ODA compatible with the exception of the ODIF converters which require to be specifically designed for the systems they serve.

2.2. Current Status of ODA

ODA was first proposed by the European Computer Manufacturers' Association (ECMA) in 1982 and published as ECMA 101 in September'85. It was modified and extended by the International Standards Association (ISO) into DIS (Draft International Standard) 8613 in April '86. It was finally approved as IS (International Standard) 8613 by ISO in August '88 and as the T.410 Series by the CCITT in November '88. IS 8613 and T.410 are essentially the same.

Because ODA is intended to be an all embracing standard for office documentation, Document Application Profiles (DAP's) have been defined to permit the implementation of interworking document handling systems at different levels of complexity. Each DAP is the specification of a combination of features that are defined in various parts of IS 8613, (Campbell-Grant and Krnert 1989).

The main organisations engaged in the development of DAP's are CCITT, EWOS (European Workshop for Open Systems), NIST-WOS (National Institute for Science and Technology Workshop) and AO-WOS (Asian and Oceanic Workshop for Open Systems).

PAGODA (Profile Alignment Group for ODA), which draws its membership from these three regional workshops and CCITT, is responsible for aligning the DAP's. Where possible, common DAP's are being developed.

Four DAP's are currently being developed for international use. Q111 restricts use to character text only between commonly available PC Word Processing packages. Q112 supports character text, raster and geometric graphics between word processing workstations with basic graphics capability. Q113 supports character text, raster and geometric graphics between desktop publishing systems. Q121 is intended to support the interchange of simple character text messages in messaging systems.

Although the Q-Profiles are formally those developed, and now approved by, EWOS they are recognised internationally. Q111 and Q112 (but not Q121) are upwards compatible. Thus a system transmitting a document which conforms to the Q111 Profile will be understood by both a Q112 and a Q113 system.

The DAP's used by the PODA Project Partners at the CeBIT '87 and '88 Exhibitions (see Section 3) played a major part in the definition of the Q111 and Q112 DAP's.

The way in which future developments of ODA will be supported by the PODA-2 Project is now described in Section 3.

3. The PODA-2 Project

The PODA-2 Project - ESPRIT Project No. 2374 - started at the beginning of 1989 as a smooth continuation to the ESPRIT I funded PODA Project and has a planned duration of 2,5 years.

In supporting the further adoption and future developments of ODA the PODA-2 Project has the twin objectives of accelerating the development of ODA-based products and advancing the application of the ODA Standard.

Key elements in the approach to the Project are multi-vendor interworking as a methodology for progress, the development of standards, the transfer of the technology, and, where appropriate, collaboration with related activities.

The partners in the project are:-
- British Telecom, UK   (Partner)
- *Bull, France   (Partner)
- IBM, Germany   (Partner)
- *ICL, UK   (Prime Contractor)
- Nixdorf, Germany   (Partner)
- Océ, The Netherlands   (Partner)
- *Olivetti, Italy   (Partner)
- *Siemens, Germany   (Partner)
- *TITN, France   (Partner)
- University College London (Associate Partner/ICL)

* Founder members of PODA-1.

Nixdorf, Océ and University College London joined PODA-1 after it had started. Nixdorf and University College London were also members of the INCA Project.

The PODA-2 Project consists of two main activities viz. ODA Interworking and ODA Advancement.

3.1. ODA Interworking

The objective of this activity is to accelerate the development of European products which can interwork using ODA.

3.1.1. Interworking Demonstrations

The public demonstration of interworking between heterogeneous state-of-the-art document processing systems has proved to be a most effective means of promoting ODA with potential users and suppliers. Following successful demonstrations at the Hannover CeBIT exhibitions of 1987 and 1988 as part of PODA-1, the partners in PODA-2 (except for TITN) participated in the CeBIT '89 Exhibition demonstrating multi-media document interchange between proprietary systems using ODIF. The National Computing Centre, UK provided a demonstration of conformance testing. Each of the partners supplied software to convert between their internal document formats and ODIF. Figure 1 shows the demonstration configuration which had a wide variety of proprietary systems interconnected via X.400 as the interchange carrier. Network communication was via public X.25. Ethernet and Floppy disc were used by some partners as alternative interchange media to X.400. The toolkit developed for the demonstration is described in Section 3.1.2.

The Document Application Profile used by the partners at CeBIT'89, the H'89 Profile, is a subset of Q112 which was defined by EWOS in December '88. The H'89 Profile allows the interchange of documents containing any mix of character text, raster and geometric graphics in formatted, processable or formatted processable form.

The demonstrations were well received by visitors from 27 different countries. A contacts database of 300 inquirers was compiled of users, suppliers, etc. interested in further information on products, standards, toolkit and ODA in general.

Following CeBIT'89 the partners have agreed that public demonstrations in the near future must concentrate on a thorough implementation of Q112 rather than extended features. Using ODA for reliable, practical interchange of documents between heterogeneous, non-ODA based systems will require incompatibility reporting, fallback procedures and perhaps user involvement. All of this needs rigorous investigation.

There is also a requirement to develop a portable demonstration kit to allow the Project to respond rapidly and cost-effectively to requests for demonstrations anywhere in the world. Such a facility should permit any number of partners to participate either locally or remotely.

Additionally the partners intend to use the developing ODA technology in a permanent network for demonstrations and intra-project communication. Currently a permanent X.400 network is being used for text-only messaging. Some partners are even now able to send ODIF documents via the network. As the project progresses the X.400 network will be used to exchange mixed-mode, ODIF documents between all of the partners.

3.1.2. ODA Toolkit

To support interworking in a multivendor environment between proprietary document processing systems not originally based on the ODA technology a common toolkit is being developed. The toolkit has the broader objective of generally reducing the cost and improving the quality of converters for effective interworking between proprietary document processing systems. The toolkit will be available to organisations outside the Project to facilitate their use of ODA in interworking.

Figure 2 shows the document conversion software architecture for a proprietary document processor. Each proprietary document processor requires its own document internaliser and externaliser to convert documents between its internal format and a stored ODA (SODA) representation. The document internalisers and externalisers require to be produced specially for the document processor they serve and conformant to the appropriate DAP. The other components of the architecture form a common base for ODA interchange and, together with test utilities for screen and printer output from SODA files, comprise the tools within the toolkit. These major tools are as follows:-

(a) The ODA Storage Manager which provides random access to the constituents of the SODA representation. The ODA Storage Manager and SODA are being developed as a portable tool.

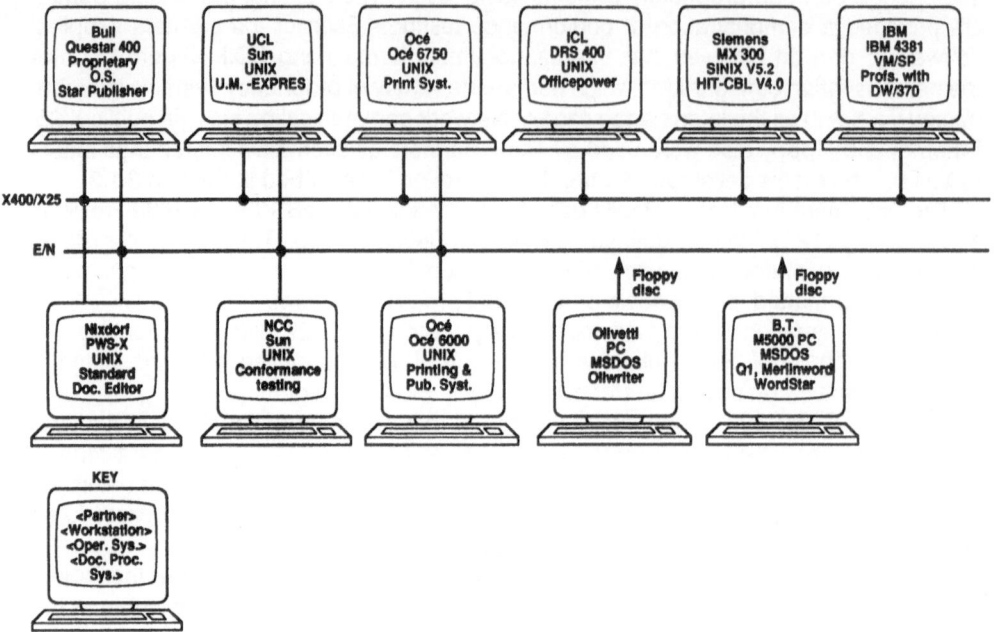

Figure 1 CeBIT '89 Demonstration Configuration

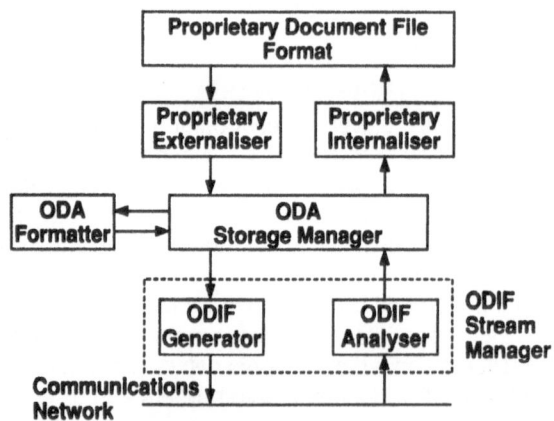

Figure 2 Document conversion software architecture

(b) The ODIF Stream Manager which generates and analyses the ODIF stream. Unlike SODA, ODIF is a serial stream format specifically designed for document interchange over communications networks.

(c) The ODA Formatter which generates the formatted processable form of the document from the processable form.

3.1.3. Technology Transfer and Liaison

Effective technology transfer is crucial to the successful adoption and promotion of ODA. This is being achieved through:

(a) Liaison primarily with ESPRIT projects and European Institutions. In addition, there is a specific PODA task dedicated to liaison with the US-based EXPRES Project, one of whose goals is the promotion and advancement of ODA in the United States, through prototypes and pilot systems.

(b) The generation, collation and dissemination of information on ODA to as wide an audience as possible.

(c) The education of potential users and suppliers to the opportunities and benefits of ODA. Public demonstrations such as CeBIT'89 are obviously a very effective means of achieving this. Negotiations have begun with EurOSInet to explore the possibility of the collaborative promotion of ODA. EurOSInet is a European Marketing Association which promotes and demonstrates the practical application of Open Systems in a multi-vendor environment.

(d) Increased awareness and understanding of ODA in the appropriate standards bodies. PODA Project Management has proposed a formal liaison directly to TC29, the Technical Committee within ECMA which is responsible for the development of the ODA standard. Input to other standards bodies such as ISO and CCITT will normally be via ECMA, after refinement and formal endorsement in TC29.

3.2. ODA Advancement

The objective of this activity is to enhance the ODA Standard and thereby to extend the capability and integration of future ODA products. This is being achieved through the development of an ODA architecture to satisfy the needs of a comprehensive range of office systems applications.

In each of a number of application areas user requirements are first specified. The results of this analysis, together with results from other relevant projects and developments in associated standards, are used to define the extended ODA architectural model and to contribute to the extension of the ODA standard through the standards bodies. In a number of cases early software prototyping is appropriate to validate the applications techniques and this will be included in public demonstrations in the latter stages of the ODA Interworking activity. The application areas being addressed are as follows;

3.2.1. Document Storage and Access

In particular, five interrelated application aspects of document storage and access are being investigated.

3.2.1.1. Partial Documents. Partial documents are defined as components of a "compound document" which may or may not be physical distributed and which may contain references to information stored in a number of locations. The structures of compound documents to satisfy various operational factors, such as efficient references to external objects, are being defined.

3.2.1.2. Retrieval of ODA Documents by Content. ODA does not specify how to search documents by content and, in most cases, existing search engines work to their own storage architectures. To meet performance demands the retrieval of ODA documents by content may therefore need special self-identifying storage formats which will have wide system implications.

3.2.1.3. Hypermedia. In a hypermedia system windows on the user's screen bear a one-to-one relationship with objects in a multi-media document database. Users can construct links between these objects, i.e. permanent cross references, to enable browsing through the documents in the database. The links may be internal to a document or between documents.
The objective is to define the new structures required by ODA to support hypermedia applications for both partial and multi-documents.

3.2.1.4. Distributed Document Editing. This work is concerned with the management of the multiple editing process whether sequential or simultaneous and involves local and remote access to whole and distributed partial documents.
In addition to generating extensions to the ODA document architecture model, distributed editing will require the definition of new communication services.

3.2.1.5. Communication Requirements. There is a need to develop communication services and protocols that allow for the exchange of and access to ODA documents. Existing and evolving standards such as CCITT's DTAM (Document Transfer, Access and Manipulation) will be reviewed in this context. In addition, other document storage and access applications will require a directory service and in this respect the suitability of such standards as CCITT's X.500 will be evaluated.
This work item will involve liaison between the two areas of standardisation involved namely OSI protocols and document architecture.

3.2.2. Computable Data in Documents

IS 8613 does not contain rules and mechanisms for the inclusion of data in ODA structures. Data is here defined as structured information of type integer, boolean, real or character string and includes records or arrays comprising these datatypes. High priority is being given to enhance ODA to allow such data to be incorporated into the document in human comprehensible form such as text, tables (e.g. spreadsheets) or business graphics. This will facilitate interworking between data and document processing applications.
Extensions to the ODA Architectural model need to be developed which take into account the relationships between data in a document and the content architectures. Enhancements to the logical structure to incorporate rules for handling data need to be specified.
Related to the incorporation of data into the ODA model is automatic content generation

which is the automatic process of extracting and presenting organisational information contained within a document. Such items of information include table of contents, cross references and indexes.

First papers on user requirements have been submitted to ISO (SC18 SWG on ODA Directions and Extensions).

3.2.3. Document Editing

Work on document editing is in four disparate areas, the results of which will be of significance in supporting the implementation of an ODA document editor. The four areas are as follows;

3.2.3.1. Document Structure Editing. Previous work on ODA editors has been based on the approach of using the generic structure of the document to control the editing function. This is called Document Class Controlled Editing and releases the user from many tasks but incurs the penalty of an "unfriendly" user interface for certain applications. Document Class Controlled Editing is well suited for applications like form filling or writing formal reports but for applications such as the production of marketing papers or advertisements it is felt to be too restrictive for editing the logical structure. Document Class Supported Editing attempts to cater for both types of application. It combines the freedom of structure editing with the support offered by the document class concept and enables the experienced user to employ "short cuts" to build up the logical structure in a convenient manner. Layout revision and automatic layout generation facilities are to be built into the editing function.

3.2.3.2. Graphics in Documents. Graphics content architectures, both geometric and raster, are defined in the ODA standard. This work is concentrating on developing two editors, one for geometric and one for raster graphics, which satisfy the standard and which can be integrated into an ODA-based, multi-media editor. The work is a continuation of the graphics task in PODA-1 in which two such stand-alone editors were produced. Colour will be introduced into both editors conforming with the ODA colour model defined in the addendum to IS 8613.

3.2.3.3. Annotation, Revision Control and Management. Facilities are being developed to provide for;
(a) annotations, such as comments, to be added to the document,
(b) the control of revisions such as insertions, deletions and changes e.g. their identification within the content and structure,
(c) the management of the complete revision history of the document.

3.2.3.4. Editing Primitives. The goal of this work item is to specify a set of editing function primitives which can be used as building blocks in a document processing system. Primitives will be developed to manipulate text and graphics separately and an assessment will be made of the functionality required to manipulate integrated multimedia documents.

The availability to the project of a prototype editor based on the ODA technology enables early evaluation of intended enhancements to the ODA standard, especially those which are related to the logical and layout structure. This class driven editor, the Standard Document Editor, supports document class specification and structure editing.

3.2.4. Document Printing and Display

The demand for increased functionality and quality in printing makes this an important topic. Standard Page Description Language (SPDL) is an emerging ISO standard which refines the ODA Formatted Form. It describes, in a device-independent manner, documents comprising text and graphical material for presentation on paper or other media. The suitability of SPDL will be evaluated for use between document processing systems, which generate ODA Formatted Form, and printing services. Converters to work between ODA, SPDL and other PDL's, such as Postscript, are being written as part of this process. Emerging ECMA standards for Document Print Applications and Print Access Protocol are also being evaluated and enhanced.

3.2.5. ODA/SGML Interworking

Standard Generalised Markup Language (SGML), which is the document exchange standard that has emanated from the world of publishing, is of increasing significance as the office and publishing markets converge. Consequently, SGML applications are being evaluated to assess which features could be beneficially mapped into the ODA model and vice versa. Requirements for tools to transform between the two structures are being defined.

4. Other Future ODA Developements

Topics not yet covered in the PODA Project include:-
(a) security features in documents,
(b) user document classes i.e. document classes defined for particular applications,
(c) relationship with Electronic Data Interchange (EDI) which is the documentation standard of the commercial world,
(d) audio content and
(e) moving images.
In addition it is vital that the ODA standard be evaluated and refined via pilot projects with a range of potential ODA users.

5. Conclusion

There is clearly a world wide commitment to the use of ODA for document interchange. The standard is well established with the standards bodies and ODA product announcements are imminent. This paper has described both the current status and projected developments of ODA and how the ESPRIT funded project PODA-2 is, through its activities, supporting both the adoption and advancement of the Standard.

6. Bibliography

(1) I.R. Campbell-Grant, "Introducing ODA", ICL Technical Journal, Nov.'87, vol. 5, Issue 4, pp. 729-742.

(2) "Office Document Architecture (ODA) and Interchange Format", ISO IS 8613, '88.

(3) I.R. Campbell-Grant and P.J. Robinson, "An Introduction to ISO DIS 8613, Office Document Architecture, and its Application to Computer Graphics", Computer and Graphics, '87, vol. 11, No. 4, pp. 325-341.

(4) E. Koether et al., "ODA - From Theory to Real Life", ETW '88.

(5) I.R. Campbell-Grant and G. Krönert, "First Applications of the ODA Standard", IFIP Paper on ODA, '89.

(6) Technical Annex, PODA-2, '88.

(7) P.J. Robinson and S.M. Strasen, "Standard Page Description Language", Computer Communications, April'89, vol. 12, No. 2.

7. Summary

ODA is an International Standard (IS 8613) which permits multi-media documents to be exchanged between conforming, computer systems, such that the received document can be presented, printed or stored with the same layout as the originator or, more importantly, it can be edited or reformatted by the recipient. The Standard currently supports document content types of any mix of text, raster and geometric graphics. Document Application Profiles (DAP's) have been defined to permit the implementation of interworking document handling systems at different levels of complexity.

The PODA-2 Project is funded within the ESPRIT II Programme - Project 2374. It started at the beginning of 1989, following on from the PODA-1 and INCA projects which were part of the ESPRIT I Programme. PODA-2 has 10 partners, including many of Europe's leading IT companies.

In supporting the adoption and future developments of ODA the PODA-2 Project has two major objectives.

7.1. Acceleration of the Development of ODA-based Products

This is being achieved through;
(a) the public demonstration of ODA interworking (including the results of the ODA advancement work) between heterogeneous systems,
(b) the production of tools to give European companies a common technology base on which to develop products,
(c) the validation and alignment of DAP's.

7.2. Advancement of the Application of the ODA Standard

This is being achieved through the development of an ODA architecture to satisfy the needs of selected office systems applications. These include document storage and access, editing, printing and display and the incorporation of data in documents. User requirements are being defined for each application and, where appropriate, extensions to the Standard are promoted through the relevant standards bodies. The relationship with other standards, such as SGML (Standard Generalised Markup Language) and SPDL (Standard Page Description Language) are being established.

Collaboration with such organisations as EurOSInet (currently being explored), with the US-based EXPRES Project and with other relevant ESPRIT projects is regarded as important to the successful promotion of ODA.

The partners are using the developing ODA technology in a permanent network for demonstrations and for intra- project communication.

Keywords: Communications, Document Architecture, Document Interchange, Document Application Profile, ODA, ODIF, Office Systems, OSI, Multi-Media Document, PODA

Project no 1059

Dynamically Adaptable Multi-service Switch (DAMS)

Reinhard Pohlit, Manfred Bleichrodt
Telenorma
Mainzer Landstr. 128-146
6000 Frankfurt-1, Germany

Abstract

Currently, most voice and data services are provided by separate networks which use either circuit or packet switched techniques. The DAMS project is aimed to establish an integrated solution for the business environment. A distributed system architecture on the basis of a LAN/MAN was selected. It consists of a standardized optical backbone ring system connecting several subsystems which provide terminal and network access (**figure 1**). The system, where possible, is based on standardized protocols. The subsystems are specified to operate as stand-alone systems. They are subdivided into the following modular blocks:
- Port Unit (to adapt to various terminals and external networks)
- Local Switch Unit (to provide internal switching functions)
- Backbone Unit (to connect the subsystem to the backbone ring)
- Control Unit (for local system control)

In order to determine the best solution for the backbone ring standard, a performance analysis was carried out to investigate the two alternative proposed standards: FDDI-II and IEEE 802.6 (MST). The results showed that:
- FDDI-II has a similar performance as IEEE 802.6 (MST)
- certain management strategies improve the overall efficiency of bandwidth usage
- no general problems arise by using ATM as a transmission technique, however further investigations are necessary

Investigations into the available technology and the support of these standards by semiconductor manufacturers showed that there was more commitment to FDDI-II. For this reason it is expected that, for the first solution, the transmission scheme will be based on the emerging FDDI-II standard. It supports circuit switched (CS) and packet switched (PS) traffic types, with dynamic bandwidth allocation. Since asynchronous transfer mode (ATM) is expected to be a final integration mechanism, a strategy towards a long term solution based on ATM was elaborated. This includes an overall ATM system design and a migration strategy from a first hybrid solution (CS + PS) to ATM. Three possible signalling standards (CCITT No7, DPNSS & ISDN LAPD) have been considered suitable to be adapted to the DAMS requirements. A subset of CCITT No7 is preferred to be implemented on the FDDI-II backbone network.

1. Introduction

At present, the vast majority of voice and data services, both public and private, are handled by separate networks which employ circuit switched or packet switched techniques. The main objective of the DAMS project is to address the problems associated with the integration of these services within the business environment, and finally to provide a suitable solution. This solution must not only be capable of supporting the services in use at present, but must also be capable of supporting future services which, although

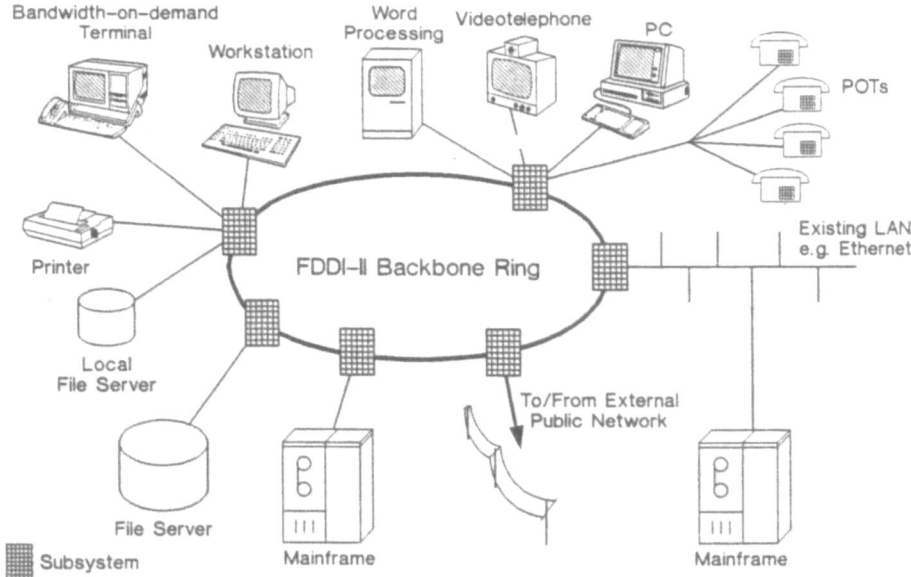

Fig. 1 – DAMS Connecection Domain

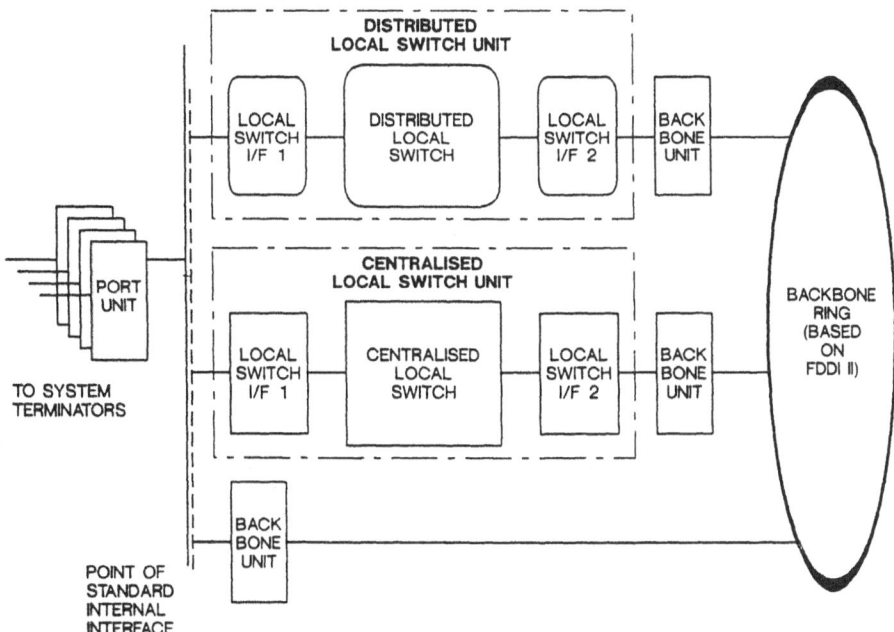

Fig. 2 – DAMS Architecture

not yet specified, are expected to require higher bit rates.

The project was predicted to take a total of 5 years and was divided into two consecutive stages in line with the European ESPRIT programme:

- Stage 1   -Initial study and specification (2 years)
- Stage 2   -realization and validation of the concepts outlined in stage 1 (3 years)

This contribution summarizes and presents the results of the first stage.
The work carried out by the partners **Plessey (UK), JS Telecom (F) and Telenorma (FRG)** was complemented by the efforts of two subcontractors:

a marketing subcontractor, engaged to assess the potential and requirements of a system with dynamic bandwidth allocation.
a university subcontractor, employed in order to ascertain the suitability and performance of the currently proposed fibre transmission standards which support the integration of voice and data services.

2. General Structure OF DAMS

2.1 Architecture

The final DAMS architecture is shown in **figure 2** and demonstrates the relationship between the major system elements. Key to the architecture is the concept of a service independent interface called the Standard Internal Interface (SII) which allows Port Units to interface functionally to the Local Switch Unit or the Backbone Unit. The architecture incorporates subsystems, connected to the so called Backbone Ring (BBR). Each subsystem, which incorporates a Local Switch Unit, may be stand alone and any number of subsystems may be interconnected by means of one or more Backbone Rings, thus providing a variable and flexible system capable of meeting a wide range of traffic, service and topological requirements. DAMS may thus be implemented as a small stand-alone subsystem or as a number of subsystems spread across a large physical area interconnected by one or more Backbone Rings.

2.2 The DAMS Subsystem

The Local Switch Unit

The Local Switch Unit is made up of two functionally separate elements: the Local Switch (LS) and the Local Switch Interfaces (LSI). The Local Switch provides the actual connections whereas the two Local Switch Interface variants provide interfacing between the Local Switch and the Port Unit using the SII, and between the Local Switch and the Backbone Unit.

The Port Unit

The Port Unit provides the interface between the Standard Internal Interface and the system terminators. A system terminator is any network or terminal to which DAMS interfaces. The Port Unit interfaces to the Backbone Unit or the Local Switch Interface via the Standard Internal Interface.

The Backbone Unit

The Backbone Unit provides an interface between the Standard Internal Interface or the LSU and the Backbone Ring. It has to seperate traffic flow and to adopt different speeds of LSU and Backbone Ring. It also includes the hardware and software necessary to implement the low layer protocols.

The purpose of the Backbone Ring is to provide a transmission path which will link together individual subsystems and to provide, by direct access, higher bandwidth services.

The Control Functions

Control facilities have two functions: the first is that of connection control and the second, higher layer, is that of call control.

Connection control involves the establishment of paths of required bandwidths between isochronous interfaces (circuit switching) and the passing of messages between non isochronous interfaces. These mechanisms are an inherent part of the subsystem design.

Call control on the other hand deals with the higher layer functions such as digit analysis and translation, call routing, customer and network signalling protocols such as DPNSS1, CCITT No 7, ISDN protocol LAPD and the provision of supplementary facilities. The detailed aspects of call control were outside the scope of this phase of the project. However general considerations of the information transport system which includes the interaction concept of all the control entities within the DAMS system have been carried out with the following results:

– control function communications use a uniform protocol from layer 2b upwards
– control messages use the same physical medium as user information
– control entities at each location within DAMS can communicate with each other
– control messages use the same channel as user packet data

2.3 Technical Solutions

Two different technical solutions for the described general system structure have been investigated.

The first was considered under short term realization aspects and is based on hybrid operation techniques. Hybrid operation means the system wide coexistence of synchronous channel switching and asynchronous packet switching.

The LSU supports both circuit switched and packet type connections in an integrated fashion. The Standard Internal Interface has elements corresponding to each connection type: the isochronous and non isochronous parts respectively. The functionality of the Standard Internal Interface is the same regardless of whether it is a Backbone Unit or any type of Port Unit.

The Backbone Ring is implemented by means of an FDDI-II based optical fibre ring. The diagram shown in **figure 1** shows only a very basic system configuration which could be enhanced by the addition of further parallel rings, the direct linking of Backbone Rings or the linking of Backbone Rings by a high speed ring.

The second solution was regarded as a long term solution, based on the system wide asynchronous transfer mode (ATM).

3. Hybrid System Solution

3.1 Local Switch Unit

Hybrid Local Switch Bus (LS Bus) Structure

All entities for the Local Switch are located in one cabinet.

This LSU uses a parallel bus in order to reduce the transmission rate and hence the level of technology required. Some 64 Mb/s could be reduced efficiently to a bit rate where standard components can be used.

Concerning the transmission of PS traffic and CS traffic a common transmission medium is used for both traffic types (hybrid bus). This hybrid solution allows the total bandwidth to be used flexibly as well as having increased reliability due to the fact that the number of wires, connectors, output drivers etc. are reduced when compared with the separated version. The lower number of lines in the hybrid solution will be advantageous for VLSI implementation because of the reduced number of pins needed on the VLSI device.

Access Methods for the Hybrid Bus

As a hybrid bus carries CS and PS traffic it is necessary to separate logically both traffic types from each other. For synchronous CS traffic transmission, a central unit, which is connected to the hybrid bus, provides a 125us cycle which is divided into slots. When a request for CS-transmission is received, the central unit assigns one or more slots to the stations which form the connection, using the Circuit Packet Access (CPA) signal. This signal, running on a separate line, indicates whether the actual slot on the hybrid bus is used for CS-traffic or not. The unused slots (no CS-traffic assignment) can be used for PS traffic transmission.

Under the premises of high throughput, acceptable complexity, little redundancy and possible independence of access procedure capabilities (PS traffic) and traffic load, appropriate access protocols (e.g. IEEE 802.4 token bus, IEEE 802.3 CSMA/CD) have been investigated. A decision was drawn towards an access procedure using one separated access line (Request Line) with a specific bus medium access procedure. At the same time, a similar access procedure was proposed by the IEEE P1396 (Hybrid Communication Bus Standard) [12]-[14].

Local Switch Control

The Local Switch Control (LSC) performs tasks which are directly concerned with the function of the switch. The LSC uses the LS bus to communicate with other control entities. These entities are the Subsystem Control Unit and entities which control the Local Switch Interfaces (LSI).

The LSU will allow plugging in or disconnecting printed circuit boards containing an LSI without disturbing the operation of the Local switch.

The LSC has to assign certain slots of the transmission frame to requested connections and release them on demand. It has to enable the involved LSIs to identify the slots which are allocated to certain connections, e.g. via access control functions or via a special protocol (e.g. via the Bus Control Device).

Local Switch

The Local Switch (LS) is the hardware configuration for the physical transport of the data. It consists of the LS bus and the associated Bus Control Device (BCD). The LS bus consists of parallel data lines and a number of control lines. The control lines are the internal clock and frame synchronization and Circuit Packet Access (CPA) to separate the two traffic types on the data bus.

The PS traffic access is controlled by a Request Line as described above which is set and observed by the LSIs.

Local Switch Interface

The Local Switch Interface (LSI) has to manage the information transfer between the LS bus and the Port Unit or the Backbone Unit. It consists of the Medium Access Controllers for CS and PS traffic (CS/MAC and P/MAC). The P/MAC has to manage the access to the LS bus via the Request Line. The CS/MAC detects the CS time slots on the LS bus destined for its connected Port Unit or Backbone Unit.

3.2 Backbone Unit

Hybrid Backbone Ring Protocols

Different hybrid protocols were investigated under the following points of view: technical criteria, performance and standardization [15]. As a final favourite FDDI-II was selected [1]-[4],[9]-[11]. The transmission capacity of 100Mb/s includes a fixed cycle overhead. This leaves a usable capacity of 99.072Mb/s which is divided into 16 wide band channels (WBCs) of 6.144Mb/s each and a dedicated packet data group (PDG) of 0.768Mb/s. WBCs are able to carry either circuit switched or packet switched information.

Performance Analysis of FDDI-II and MAN/MST (former IEEE 802.6 proposed standard)

The comparison [15] of both systems revealed very little difference (**figure 3**). Only for high packet loads and 75% CS traffic a minor decrease in performance for the Multiplexed Slotted and Token/Flexible Bandwidth Access (MST/FBA) protocol was shown [5]-[8]. This is due to the fact that the PS timed token protocol is identical, that the slot size was chosen to be identical to the size of a WBC and that an increase of CS slots within the MST/FBA protocol causes an increase in slot-overhead whereas the overhead in FDDI-II is constant. An investigation concerning the residual PS capacity versus CS capacity implied that the information throughput would become worse if the size of the FBA slots was less than 2Mb/s because of the slot-overhead. However, if one single slot is used, as in the original MST proposal, the residual PS capacity would be higher or equal to the FDDI-II system depending on the utilization of reserved FDDI-II wide band channels.

Bandwidth Management

The investigations dealing with bandwidth management strategies for CS traffic within one WBC of FDDI-II covered several kinds of grouping, hunting, allocation and reservation strategies [15].

The results show the blocking probabilities and throughput for three service classes

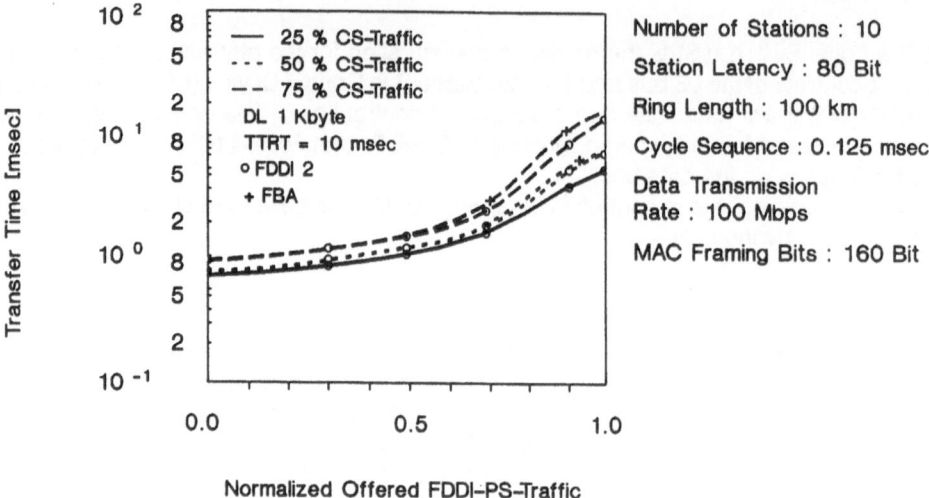

Number of Stations : 10

Station Latency : 80 Bit

Ring Length : 100 km

Cycle Sequence : 0.125 msec

Data Transmission
Rate : 100 Mbps

MAC Framing Bits : 160 Bit

Fig.3 Mean Transfer Time vs Offered PS Traffic

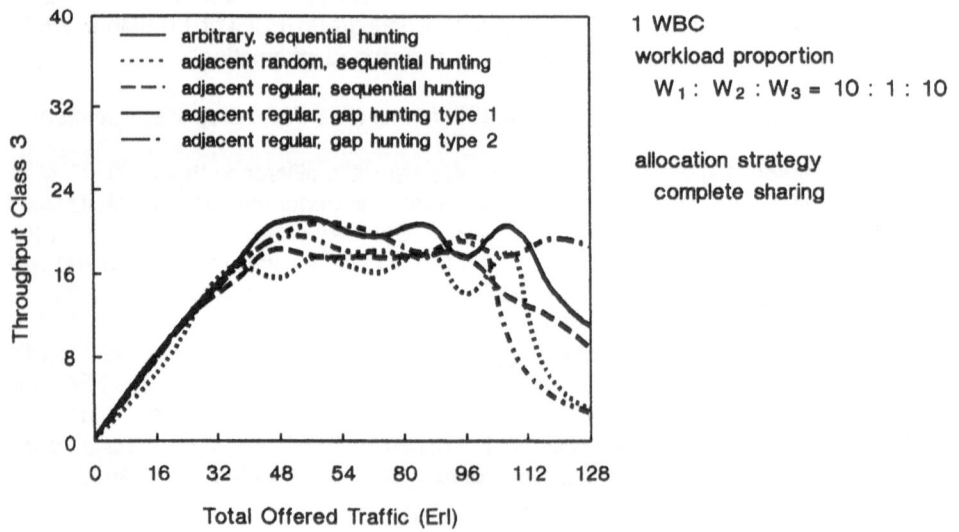

1 WBC

workload proportion
$W_1 : W_2 : W_3 = 10 : 1 : 10$

allocation strategy
complete sharing

Fig. 4 Throughput vs Total Offered Traffic

requiring an integer multiple of 64 kb/s (multiples 1, 6 and 32 corresponding to 64kb/s, 384kb/s and 2.048Mb/s). As an example these results are reproduced in **figure 4** for service class three.

If the total bandwidth of one WBC is available to each service class (allocation strategy: complete sharing) arbitrary grouping and sequential hunting yield the best throughput behaviour. Arbitrary grouping allows a multi-channel connection to use every free channel disregarding its position. Sequential hunting searches for free channels sequentially from the beginning.

Based on these grouping and hunting methods, allocation strategies were investigated which improve the performance by applying bandwidth restrictions. The most effective throughput was given by the sum limitation strategy. Connection requests are rejected if the total number of channels occupied exceeds a predefined limit. This strategy allows the blocking probabilities to be equalized by choosing the same limit for all service classes. Another allocation strategy was proposed which uses bandwidth restrictions for each service class, called class limitation strategy. Although a small deterioration in throughput has to be accepted, this method is very flexible because the limits could vary depending on the actual load situation.

Reservations of WBCs for multi-channel connections did not improve the performance significantly but had the effect of severely decreasing the performance of single-channel connections as long as up to four WBCs were reserved. Therefore, reservation schemes were not considered to be an adequate solution.

3.3 Signalling System

General Signalling Requirements

In order to avoid the definition of a new call control signalling protocol, signalling systems as CCITT No 7, ISDN LAPD protocol and DPNSS were investigated. The first two ones seem to be more appropriate to the DAMS system, because of their widely defined standards and their international acceptance.

In both protocols enhancements are required to solve the DAMS specific problems.

A convergence function should be required which can be called the Logical Link Services (LLS) entity, resolving the significant incompatibilities between LAPD type entities and LLC II (802.2) entities as layer 2 protocols to provide services to the layer 3 (Q931) entity.

ISDN D Protocol Enhancements

CEPT is already in charge of defining the required protocols for layers 1,2,3 used in the ISDN primary rate access for interface between PBX's and Local Exchanges.

However, to be used in an appropriate manner within the DAMS system, some other enhancements are needed as messages and procedures for different flexible bandwidth allocation schemes.

CCITT No7 Enhancements

Obviously, as in the Q931 utilization, new messages and procedures must be defined for assuming flexible bandwith allocation schemes. One advantage of this standard for our purpose is that there could be the possibility of developing the concept of channel selection and reservation being done at a Network Control Point, which may be a distinct third station and this concept converges with the trend of the FDDI-II standard.

4. ATM System Solution

The aim of the investigations was to propose a universal communication system based entirely on the technique of ATM.

This DAMS system is divided into two functional parts:

1) service independent transport system (TS) and
2) service dedicated elements (Port Adapters and corresponding software)

If services change, some modification of Port Adapters (PAs) has to be made. If the demand (bandwidth/delay) on the transport system increases or technology reducing costs become available then only the transport system has to be changed. In **figure 5** the boundary between the TS and the PAs is shown by circles.

4.1 Transport System

The TS consists of a subsystem internal ATM-bus and the external Backbone Ring (BBR). Since the defined packet structure is the same for both the ATM-bus and the Backbone, the Backbone Adapter (BBA) does not need to modify anything of the packet contents nor of the header.

For the subsystem internal medium, a centrally controlled parallel bus (ATM-bus) is proposed. An Access Control and Clock Unit (ACCU) provides slot access to the ATM-bus and information for bandwith control.

The different Port Adapters have to convert the corresponding standards to the uniform ATM format of the system.

4.2 Packet Structure

The ATM packet format is a key element of the transport system. The packet structure has to take into account the advantages and drawbacks of both synchronous time division techniques (e.g. 64 kb/s) and packet techniques (e.g. X.25).

ATM packet definition has to be:

– suitable for all sections of the transport system
– fast routing/processing capability
– supporting security aspects as far as reasonable
– various possibilities for source and destination relations
– supporting message and connection oriented transmission

A proposed ATM packet (**figure 6**) consists generally of n words of m bits each. The first word is the header. The word width corresponds to the width of the ATM bus. The information field contains any data of the application whereas the header is divided into defined fields. For example the address field may represent the physical final address structured into: SS defining the subsystem number, PA for Port Adapter number and P selecting a port of a Port Adapter.

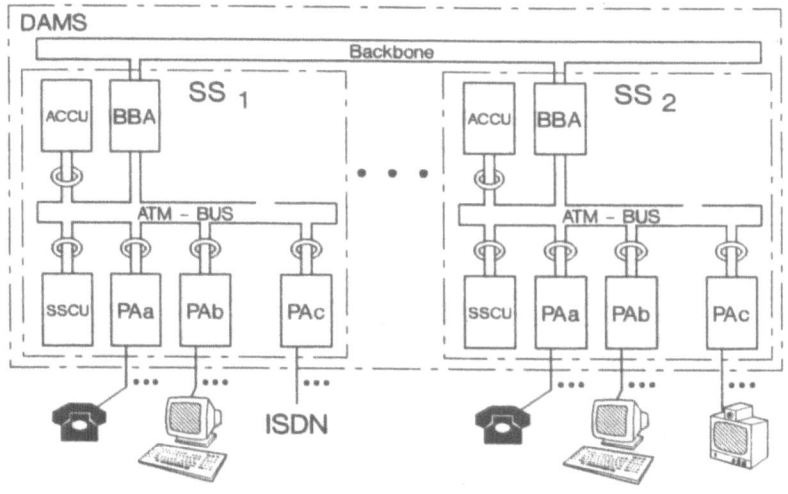

SS   : Subsystem   SSCU : Subsystem Control Unit
BBA : Backbone Adapter   PA   : Port Adapter
ACCU : Access Control and Clock Unit

Fig. 5 ATDM – System Concept

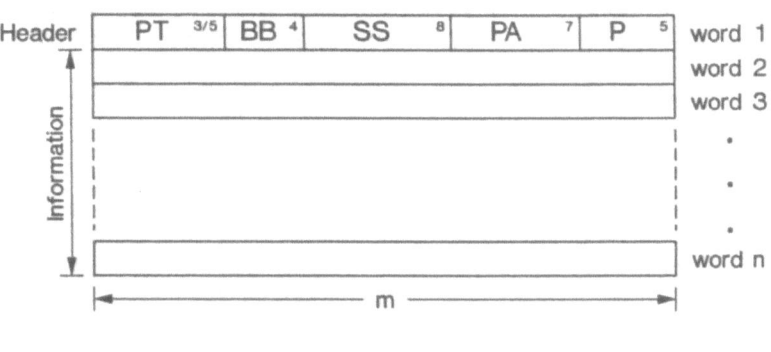

PT – Protection (PC, CRC) Address Spaces :
BB – Backbone Number   – to all
SS – Subsystem Number   – physical address
PA – Port Adapter Number – controller address
P   – Port Number   – logical channel (broadcasting)

Fig. 6 ATDM – Packet Structure

4.3 Allocation of Bandwidth

A connection implies a defined maximum bit rate (BW). Each Port Adapter has to guarantee the given bandwidth limitation for a connection. The required bandwidth is requested from the Subsystem Control Unit (SSCU). SSCU knows the current situation of all relevant system elements in the subsystem of the source, the subsystem interconnection (all Backbones) and of about the same subsystem elements of the destination.

To provide this information, the SSCUs will be cyclically informed about all current traffic situations by the ACCUs and BBAs.

4.4 Basic Performance Investigations

Basic performance investigations of the ATM technique focus on the delay behaviour of packets for different slot sizes and different packet/header ratio assuming one hierarchy with a transmission rate of 140 Mb/s.

The asynchronous transfer mode (ATM) structure of the transmission medium consists of equal-sized time slots and was regarded as a clocked queuing system. Incoming bitstreams or messages are packetized and provided with a header containing at least the address of the receiver.

For considered packet sizes the average waiting times were very small (<0.05 ms) for a basic load of less than 90% (**figure 7**), however, increasing proportionally to the packet size as long as the ratio packet/header is kept constant. The probability for a packet exceeding 0.1 ms waiting time is very low even for a packet of 64 bytes and 90% basic load. Due to this fact, the packetization time was identified to have the major impact on the transfer time of a packet. As an example, a packet of 64 bytes requires 8ms packetization time if voice of 64 kb/s is considered.

The investigations showed that it is possible to meet real-time requirements with ATM techniques without special priority mechanisms using an information field length of 64 byte.

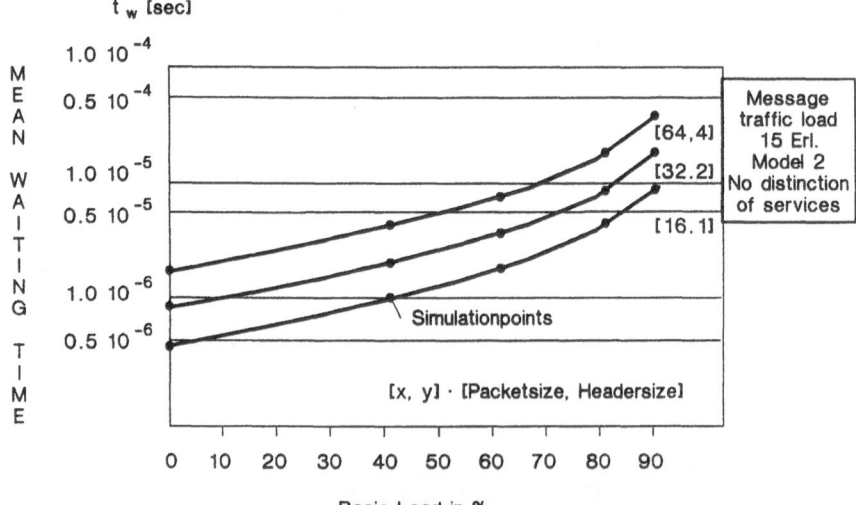

Fig. 7 Mean Waiting Time as a Function of Basic Load
for Three Different Packet Sizes

5. Evolution Of DAMS from the Hybrid Approach to an ATM Approach

A strategy has to be provided to migrate progressively from a hybrid to an ATM solution (as dictated by the market), without replacing the whole communication system. ATM environment like terminals and/or networks need to be adapted to DAMS via new Port Units. The general approach is to use as far as possible the DAMS system internal infrastructure like backplane, Local Switch, Backbone Ring and control entities.

One solution would be to offer complete ATM LSU and BBU functions to these new terminations. This solution has to provide, from the beginning of the design and development phase, some basic hardware provisions. One of them is at the backplane level to reserve an additional free bus for ATM communication.

All new ATM boards will be able to access the unused bus while still being connected to the other common busses for control or I/O. Control unit boards can be kept identical but an interworking unit may have to be developed to allow interoperability between the hybrid and the ATM world.

Further hardware solutions have been developed for the hybrid Local Switch Unit. This defines a common bus for the packet, circuit and ATM data.

A possible realization of implementing ATM into the FDDI-II standard (and hence migrating the backbone standard towards ATM) has been evaluated. The main idea is to reserve a one or more wideband channels for ATM transmission.

6. Conclusions

The marketing subcontractor has indicated a requirement for a DAMS type system in the near future. They foresee a market window between the current generation of PBX and the future IBCN compatible private exchange. There is a warning that there will be stiff competition from Japan and the US since this market has traditionally commanded high revenues and is also the driving force for the introduction of new services and facilities. A failure in this sector may have negative consequences on the telecommunications industry within the common European market envisaged after 1992.

Technical studies carried out within the project have shown that FDDI 2 offers a sufficient performance and flexibility for a short term solution of office communication systems. Under the proposed system architecture it allows a cost effective design for narrowband voice communication.

The ATM technique with its most advanced ability of service integration, may be applied in a long term system design. From the short term towards the long term solution migratory steps are required and technically feasable.

The results of the project encouraged to step into the second stage and demonstrate the feasibility of the specified DAMS system by a prototype model. Beyond this an intensiv participation in standardization process is necessary.

7. References

[1] FDDI Token Ring Media Access Control, Draft Proposal ANSI X3T9.5, (1986), Rev. 10

[2] FDDI Token Ring Physical Layer Medium Dependent, Draft Proposal ANSI X3T9.5, (1986), Rev. 5

[3] FDDI Token Ring Physical Layer Protocol, Draft Proposal ANSI X3T9.5, (1986), Rev. 12

[4] FDDI Token Station Management, Draft Proposal ANSI X3T9.5, (1986), Rev. 2

[5] Metropolitan Area Network (MAN), Multiplexed Slotted and Token, Medium Access Control, Draft of Proposed IEEE Standard 802.6 (1986), Rev. A

[6] Metropolitan Area Network (MAN), Multiplexed Slotted and Token, Physical Layer, Draft of Proposed IEEE Standard 802.6 (1986), Rev. A

[7] Metropolitan Area Network (MAN), Isochronous Management Working Paper, Proposed IEEE Standard 802.6 (1986), Rev. B

[8] Metropolitan Area Network (MAN), Station Management, Draft of Proposed IEEE Standard 802.6 (1986), Rev. B

[9] FDDI-II Working Paper, Feb. 1987

[10] FDDI-II Working Paper, Draft Proposal ANSI X3T9.5, April 1987

[11] FDDI-II Working Paper, Internal Working Document, ANSI X3T9.5, June 1987

[12] A Proposed Standard Hybrid Communications Bus, IEEE P1396, Oct. 1986

[13] Hybrid Communications Backplane Bus Standard, IEEE P1396, March 1987, Version D1.01

[14] A 155 Mbit/s Communications Bus, IEEE P1396, Nov. 1987

[15] Final Report for Esprit 1 Project 1059, Dynamically Adaptable Multi-service Switch (DAMS), July 1988

Project no 2054

UCOL: EVOLUTION OF THE SYSTEM CONCEPT DURING THE REALIZATION PHASE

D. CAPOLUPO, A. FIORETTI, S. FORCESI, E. NERI
Alcatel Face Research Centre
Via Nicaragua 10
00040 Pomezia
Italy

Abstract

The UCOL project intends to demonstrate experimentally the performance and the flexi-bility, in the context of LANs, of a coherent multichannel system; it is to be stressed that this project is unique, in that it adopts an all-round approach to the development of a system aiming to provide integrated support of narrowband and broadband services (data, voice and video) directed at the needs of specific market segments. This means that the project addresses both technology issues (specific of the physical layer) and the development of the necessary higher layers, for management and resource allocation for example, thereby insuring that the mutual interdependencies are taken into account. The basic feasibility of the key building blocks of the network concept under development has already been demonstrated.

1. Introduction

The UCOL project follows a feasibility study investigating the promising characteristics of coherent optical techniques, such as high bandwidth, improved receiver sensitivity and fine selectivity, and FDM multichannel capability.

The result of the preliminary study is a project for an Ultrawideband Coherent Optical Local Area Network, that will demonstrate experimentally the performance and the flexibility of a coherent multichannel system and allow a validation of coherent optical technology and the related network architecture conceived during the feasibility study. It is important to point out that the system will be compliant with OSI recommendations and that it is actually the first candidate to be an ultrawideband network standard.

Many applications can be foreseen for this network, including broadband video-tele-phony and video-text, teleports and MANs. First potential application areas are large industry sites, high technology and business organizations, universities and hospitals.

In the following section a description of the fundamental solutions adopted for the design of both the physical and the network layers will be given, together with a description of the expected performance of both the future system and the demonstrator.

2. System Requirements

Key aspects of developing a coherent network have been studied within the feasibility phase of the project, taking into account the optical hardware characteristics as well as the operability implications at network level. In the following sections, the description of these aspects will be divided into physical section and network section.

UCOL is a network interconnecting a large number of network interfaces (i.e. transmitter/receiver pairs) to a central passive star by means of single mode fibers operating at a wavelength of 1550 nm. Two configurations are foreseen: the first gives priority to the number of network interfaces and allows the connection of 512 network interfaces located at a distance from the central star not exceeding 10 km, the second allows longer connections (up to 25 km) and 128 network interfaces. These configurations make UCOL suitable for large industrial plant applications as well as metropolitan optical islands.

All the interfaces are capable of transmitting over a set of 25 FDM digital channels, frequency locked to a common reference source, each one offering a TDM access mode. This concept allows maximum flexibility, since it makes possible transmission of different information flows (ranging from fraction of Mb/s up to 155 Mb/s). A maximum of 64 network interfaces can operate on each channel. The network can therefore handle from a simple telephone call up to digital TV.

The network system gain, that is the difference between the available transmitted power and the receiver sensitivity, is 45 dB for a bit error rate of 10^{-9}; like other networks based on a star topology, most of the losses are introduced by power splitting inside the central star, but transmission, splices and aging losses also need to be considered. The modulation scheme is the Differential Phase Shift Keying (DPSK) with a gross transmission rate of 200 Mb/s; this scheme offers good detection sensitivity and makes TDMA implementation easier. The receiver is based on a polarization diversity scheme since this prevents problems related to the state of polarization of signals coming from different sources; the optical front-end of this receiver will be based on both channel waveguide and fiberoptical realization.

The network architecture is conceived with the aim of making available to the network users the enormous bandwidth inherently provided by coherent optical transmission systems.

A set of core network functions is developed internally to each UCOL node, in such a way that a wide spectrum of user applications can be easily built on it. Besides asynchronous bursty data services, traditionally available on present LANs, UCOL provides isochronous services with low end-to-end delay requirements (voice) and high speed (about 30 Mbit/sec) connectionless data services. To exploit better UCOL multiservice capabilities, distributed management functions responsible for the balanced distribution of the traffic among the various channels are provided.

3. Optical Hardware

A diagram of UCOL hardware is given in Figure 1. The system is composed of a number of Laser Subsystems and the central star. A Laser Subsystem is composed of a Master Laser Block, a Comb Generator and Divider, and several Stations each one consisting of one or more network interfaces; the maximum number of network interfaces (transmitter and receiver pairs) per Laser Subsystem is 16.

Within each Laser Subsystem a comb of optical carriers is derived by phase modulation from the Master Laser which is shared over the Laser Subsystem stations. All the Master Lasers are locked to one of them, working as common reference. Each transmitter and receiver pair in the station is fed with a set of reference harmonics. In each pair the appropriate harmonic is selected first, then, by means of fine frequency shifting units, the desired channel carrier and local oscillator are obtained: the carrier thus generated is then phase modulated by means of a phase modulator. The Master Laser Block is composed of a tunable external cavity laser, in order to obtain the necessary narrow linewidth required

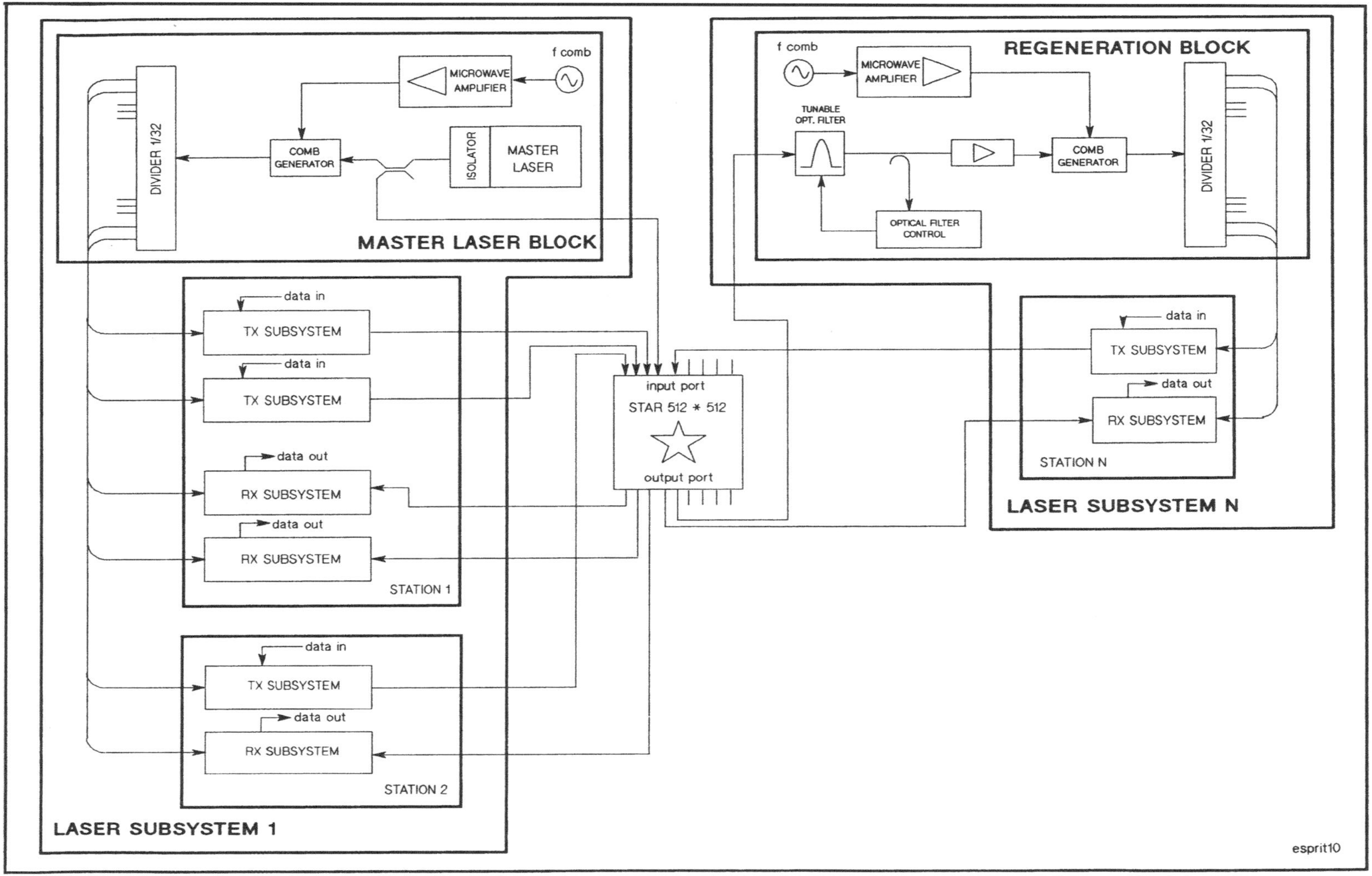

Figure 1 – **UCOL Block Diagram**

from this modulation scheme, whose frequency is locked to the reference line by means of a control circuitry when it operates as slave. The Comb Generator, connected to the Master Laser output, is realized with a traveling wave phase modulator driven by a microwave signal (whose frequency corresponds to the comb line spacing) and a Divider to feed the transmitter and receiver pairs. The transmitter is composed of an optical PLL, whose function is to effect the frequency shifting and to narrow the line of the DFB laser operating as VCO. It also includes a LiNbO3 data phase modulator and an output switch to disable the output. The receiver uses the same type of OPLL used in the transmitter for the local oscillator, followed by the polarization diversity optical front-end. In the receiver the incoming signal is combined with the L.O. by means of a 3 dB coupler, divided in two orthogonal polarization components by means of two TE/TM splitters and then detected by four photodiodes. The two IF signals, each corresponding to a polarization component, are fed to a double arm DPSK demodulator.

All the network interfaces are connected to the central star by means of a dedicated fiber. The central star acts as frequency multiplexer and on each output all carriers are present.

The current phase of work focuses on some of the present technological limitations of the optical components. The linewidth of DFB lasers used as VCO in the OPLL represents the main issue, as it affects the OPLL circuitry complexity and the adjacent channel. Improved spectral characteristics of lasers could open up a greater number of available optical channels and increased receiver sensitivity, due to a reduction in interference. This last point, together with a higher transmitter output power, would increase the system gain with a consequent increase in central star size and, therefore, in network dimensions.

4. Network Architecture

To satisfy the growing needs of high performance communication networks there is a common desire among network providers and manufacturers to define a set of protocols, interfaces and network architectures that will allow the planning the next-generation of communication systems.

The major requirement of such networks is "flexibility", that is the capability to support different types of services and to facilitate the implementation of future and unknown services. As previously mentioned, UCOL relies on a star topology which allows passive interconnection of stations which are not integral parts of the transmission medium, thereby reducing demands on station reliability as, although coherent detection offers improved receiver sensitivity and an extended dynamic range, its adoption implies scheduling and management tasks further complicated by the the multi-channel nature of the operating environment.

It is however important to point out here that the integration of any number of services and the flexibility in bandwidth allocation with agile mixing of high and low bandwidth users requires an efficient multiplexing scheme which, as explained in [1], is implemented in UCOL by taking advantage of the fact that the only actual "broadcast" section of a star topology is the centre of the distribution hub.

Efficient handling of real time packets has led to the recognition of the need to operate a TDMA scheme on each optical frequency.

Frequency domain switching between channels is therefore exploited essentially to obtain inexpensive adaptive network reconfiguration without hardware modifications.

In particular the project will show that the tunability of optical frequencies makes the network flexible and open to future evolution because of the capability to self-reconfigure without physical changes.

This last aspect of the system demonstration is important because it shows one of the main advantages of the multichannel system over both the baseband approach and another competing technique: Wavelength Division Multiplexing [2], [3] and [4]. Although in principle this could offer an alternative solution, in practice is severely limited in a flexible multichannel environment by the rigid channeling scheme and the wide channel separation.

The project will therefore have to demonstrate that the bandwidth and the optical frequency tunability offered by coherent optics can be conveniently exploited and used in an OSI context to achieve a universal network capable of connecting pre-existing networks and supporting wideband communication. Multichannel networks can therefore be considered as a natural evolution of already existing high speed LANs and not as new systems not open to communication with pre-existing networks.

The latter point has led to the consideration within this project of the results of the rapid progress made in the development of ATM (Asynchronous Transfer Mode) techniques, particularly in public broadband communication.

From the technical standpoint ATM emerges as a universal transfer technique with no special functions tailored to particular services, and is therefore a relevant reference point for a system such as UCOL which is aiming at providing a transparent communication world.

Currently, heavy standardization activities are led in parallel by CCITT and IEEE in order to address the critical common broadband aspects of MANs and B-ISDN. The points on which agreement has been reached are:

- synchronous transfer in labelled fixed cells,
- use of an identical cell format for all services,
- provision for cell-based isochronous traffic,
- provision for cell-based connection-oriented non-isochronous traffic.

It is expected that the final standard of the IEEE 802.6 MAN will be available before the end 1989 and CCITT is targeting 1990 for the B-ISDN user network interface.

UCOL can carry out a primary role in the currently emerging scenario where private networks are interconnected, in a city-wide area, by means of MANs and in wider areas by the public B-ISDN, provided it allows easy interworking with the above networks.

These considerations have led to a review of the networking aspects of UCOL which is paving the way to the adoption of ATM techniques with the specific aim of insuring that the objectives of the project are met, via:

- to define a network architecture independent of the specific services;
- easy interworking with public and local networks that will be based on ATM techniques, thereby minimizing the gateway functions;
- to avoid non-open solutions.

The most apparent impact of the introduction of ATM techniques in UCOL is related to the way in which information is carried through the network. The information flow is segmented into a number of fixed-length cells. Each cell is made up of a header field for control purposes and an information field containing user data. In contrast with the synchronous time-division multiplexing (STM) mode where calls are identified by the position of the time slot within a frame, ATM techniques associate calls with virtual channel identifiers (VCI) contained in the header field of the cell.

The asynchronous nature of ATM is meant to be a non deterministic ordering of different information streams in an ATM multiplexed channel. There are no implications for the transmission systems used to carry ATM information. In UCOL the transport mechanism along the fiber is a framed TDM, therefore the ATM cells are inserted in time slots within the frame structure.

The first classification of services supported by the UCOL network, based on the way the information is generated by different sources, can be represented by the following two types of traffic [5]:

– Continuous bit stream oriented services (CBO), such as voice, video, etc.
– Bursty oriented services, such as data, variable rate video codec, etc.

CBO services generate information units at regular intervals and require a reserved amount of network capacity during information exchange. CBO traffic is connection oriented and usually presents time transparent requirements (low delay and low delay jitter).

Bursty services generate variable length packets with variable time intervals between two consecutive packets. Bursty traffic can tolerate greater delays and can be both connection oriented and connectionless.

Owing to the different nature of the information sources, different interface functions must be performed to adapt the external traffic to a common internal transfer mode. The CCITT I.121 recommendation defines a layered protocol architecture (Figure 2), where the adaptation layer is responsible for making the ATM layer independent of the type of service. Since the ATM layer protocol data unit (PDU) has a fixed size, the adaptation layer must perform segmentation and reassembly functions and must guarantee the correct ordering of service data units (SDU) provided by the higher layer. It goes without saying that future services and terminals will not need adaptation functions because they will be conceived according to the user-network interface (unfortunately not yet standardized).

The mapping of the adaptation layer and the ATM layer in the UCOL node is presented in Figure 3, where three types of services are offered:

1. connection-mode isochronous (CBO/CO),
2. connection-mode non-isochronous (VBR or Bursty/CO),
3. connectionless non isochronous (Bursty/CL).

In the adaptation layer the assembling/disassembling functions are outlined, while the ATM layer has been assigned the basic function of handling the identifiers of connections (VCI). Further functionalities must be defined (service priority, error detection, clock recovery, etc.) and shared between the two layers. The proposed architecture is mapped in the MAC sublayer, the lower part of the Data Link layer.

The connection-oriented isochronous services require establishing a connection before the actual information exchange starts. A VCI value must be assigned to the call and this value, as cell header, will unambiguously identify each information field of this connection carried through the UCOL network. Furthermore the requested bandwidth must be guaranteed throughout the connection period. The signalling protocol is a version of the Q.931 recommendation (Figure 4) enhanced with functionalities such as handling of a multichannel network, handling the call parameters according to the ATM characteristics, signalling for distributive services, etc.

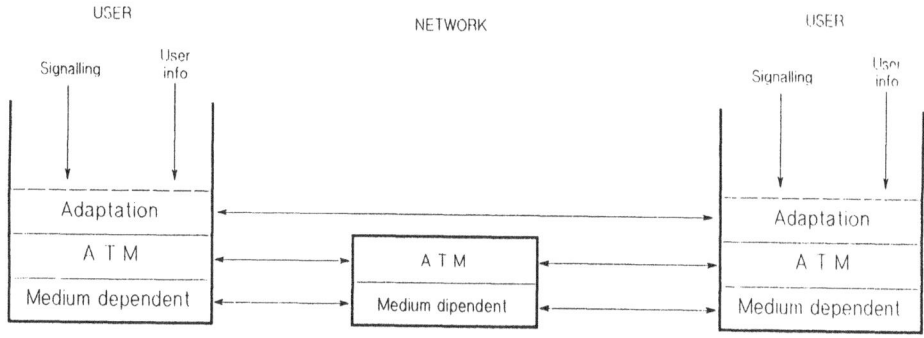

Figure 2 – Protocol Model

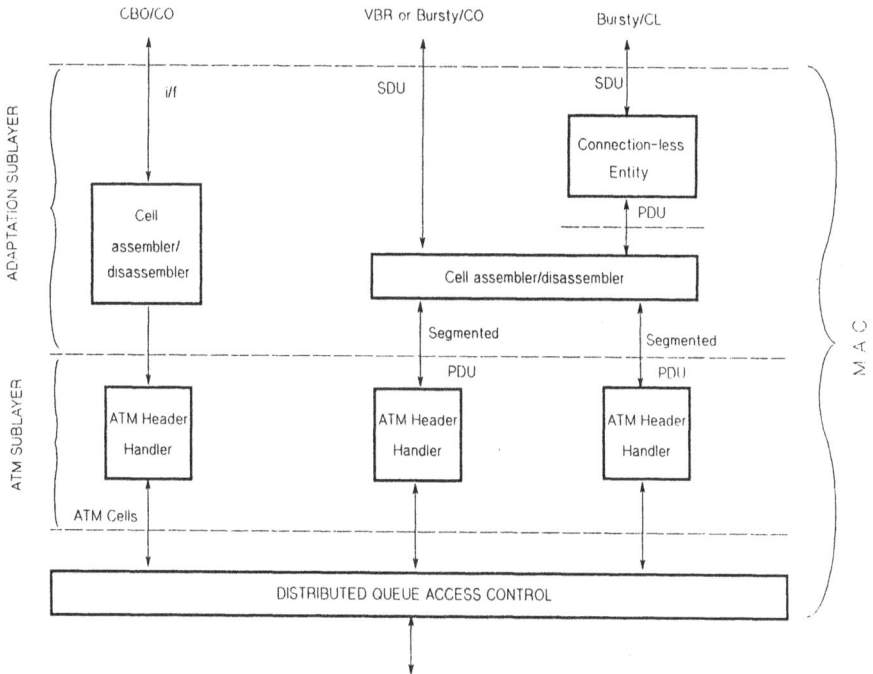

Figure 3 – ATM-based UCOL Functional Structure

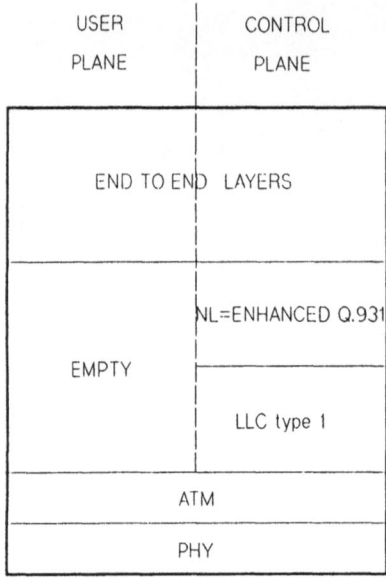

Figure 4 – Connection-oriented Isochronous
Service Profile

CLASS 4 TRANSPORT PROTOCOL	
INTERNET PROTOCOL (IP) ISO DIS 8473	
LLC TYPE 1	
ATM	
PHY	

Figure 5 – Connection-oriented Data Service
profile

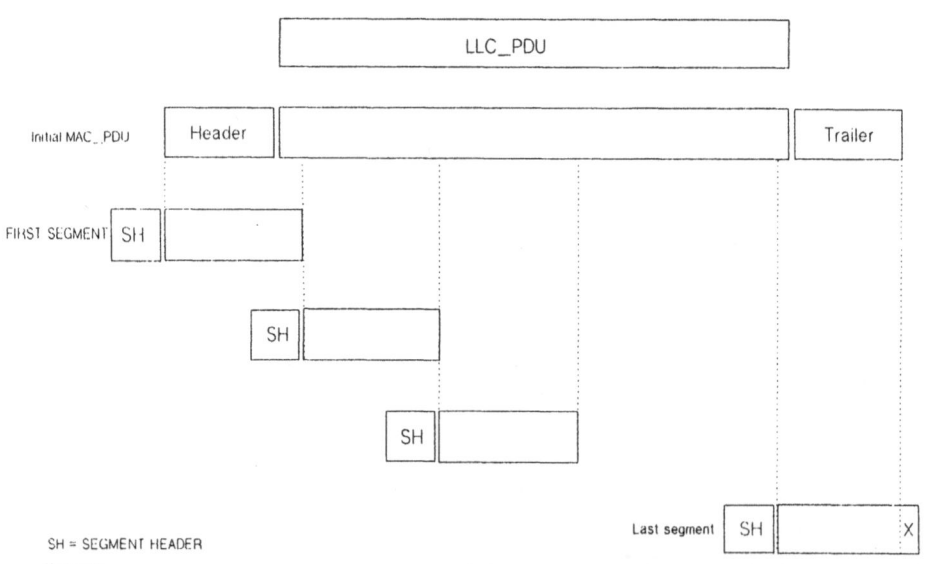

SH = SEGMENT HEADER
X = spare

Figure 6 – Initial MAC_PDU Disassembling

The connection-oriented non-isochronous services basically refer to data communications, such as LAN interconnection, workstation networking, etc. A possible functional profile as stated by ISO according to the OSI reference model is shown in Figure 5.

The connectionless non-isochronous services need specific consideration. The maximum packet length handled by the LLC IEEE 802.2 protocols is about 9000 bytes and since the information field of the ATM cell is far shorter (at present 64 bytes seem the most promising length), the problem of segmenting the original LLC protocol data unit (LLC_pdu) and provide a robust mechanism to univocally reassembly the packet at the destination side arises. One possible solution is to emulate a connection for the duration of the packet transmission, assigning the same VCI to all the segments of a single packet. Figure 6 shows the two phases of the CL service handling:

– building of the MAC_pdu (adding source and destination addresses)
– MAC_pdu segmentation (segment overhead provides the reassembling mechanism).

The lower functional block of the layered structure of Figure 3 is the Distributed Queue Access Control. Its aim is to guarantee ordered and collision-free access to the passive star centre, which is the only network resource shared among all the nodes. A 1-msec framed structure is depicted in Figure 6. The basic principles of the frame format shown are:

– a synchronization event is transmitted by one of the stations (FRAME_ALIGN).
– Each station knows in advance where to allocate its blocks of data. This condition is matched reserving a fraction of the frame (QUEUE STATUS FIELD) for each station to notify all others of the status of 5 queues of different priorities. The highest priority queue is reserved for synchronous (SY) traffic, the other queues are assigned to different asynchronous (ASY) services. A simple deterministic priority-based scheduling algorithm solves access competition.
– Each station knows its distance from the star centre and anticipates the transmission of an amount of time equal to the propagation delay along the fiber to the star (a guard time, GT, between data coming from different stations is inserted to compensate for possible measurement imprecisions).

5. Conclusions

The final goal of the UCOL project is the realization of a demonstrator that will show all the essential characteristics of the whole network. The optical hardware of the demonstrator will be composed of the following subsystems:

– 16*16 central star
– two Laser Subsystems
– three stations equipped with 8 transmitter/receiver pairs with a connection length up to 10 km;

With this reduced hardware it is possible to demonstrate the functionality of each block, together with the overall system operation. In order to clarify further the nature of the demonstration it is necessary to point out that the network layer will be interfaced directly to the external world, in order to show its ability to provide connection oriented and datagram services with appropriate quality of service parameters.

FRAME FORMAT

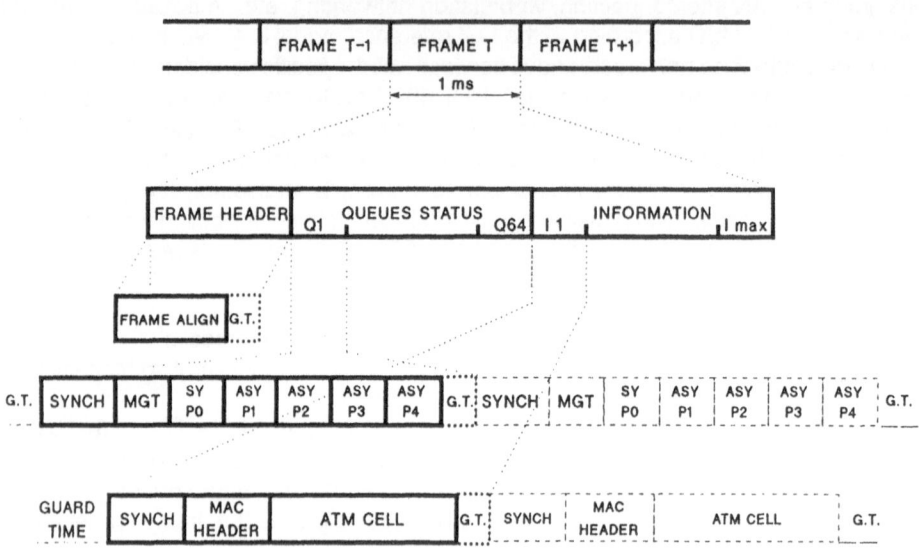

Figure 7 – Frame Format On Optical Carrier

In particular the synchronization method, the frequency allocation capability, the FDM transmission and the receiver sensitivity in a multichannel environment will be demonstrated, together with their impact on performance, flexibility and OSI compatibility. The demonstrator will therefore be used to show that UCOL is a transparent network capable of connecting pre-existing networks and of supporting wideband communication.

References

1. Fioretti, A., Rocchini, C.A., Wilkinson, P.J. and Haylett, A.J. 'A new protocol for multiservice integration over a high speed fibre optic LAN based on star topology' IFIP WG 6.4 Workshop, Aachen, Feb. 16-17, 1987

2. Yasui, T. and Goto, H., 'Overview of optical switching technologies', IEEE Communication Magazine, 25, pp. 10-15, 1987

3. Midwinter, J.E., 'Photonic switching technology, component characteristics versus network requirements', IEEE Journal Ligthwave Technology, Special Issue on the 1988 OFC/OFS Conferences, Vol. JLT-6, pp. 1512-1519, 1988

4. Dewilde, C., Mondor, D.R., Wagner, B. and Huang, N.C., 'Integrated switch and cross-connect systems as a flexible transport network', IEEE Globecom 88, Florida, pp. 675-677, 1988.

5. Byrne, W.R., Papanicolaou, A. and Ransom, M.N., 'World-wide standardization of broadband ISDN', International Journal of Digital and Analog CabledSystems, Vol 1, pp. 181-192

Keywords

Local Area Network (LAN), Coherent Optics, Metropolitan Area Network (MAN), Broadband ISDN (B-ISDN), Asynchronous transfer mode (ATM), Passive star, Differential phase shift keying (DPSK), FDM transmission, TDMA

Project no 2105

MULTIWORKS - A MULTIMEDIA INTEGRATED WORKSTATION

G.FOGAROLI
Olivetti DOR
Via Jervis
10015 IVREA
Italy

L.SAUTER
Bull MTS
7, Rue Ampere
F 91343 MASSY CEDEX
France

Abstract

This paper presents MULTIWORKS, a large Technology Integration Project aiming at the development of a low cost integrated multimedia workstation for the office of the future. In this project full use is made of high-performance European RISC technology and design on silicon techniques in order to produce significant improvements in processor architecture and speed of execution. The software also uses European technology for operating system, AI environment and applications. It is UNIX based and will adhere to international standards. The combination of new features with advances in user interface design produce a significant increase in the usability of the system over products presently on the market. A complete description of the structure of the project is given, with details on the most significant innovative choices at the basis of this development.

1. Introduction

MULTIWORKS is a low cost integrated multimedia workstation for the office of the future. It is based on European Technology and complies with international standards.

Full use is made of high-performance European RISC Technology and design on silicon techniques in order to produce significant improvements in processor architecture and speed of execution.

The low-cost of MULTIWORKS is essential to the project and will influence the extent to which certain features will be integrated by access to state-of-the-art VLSI technology.

MULTIWORKS software also uses European Technology for operating system, AI environment and applications. It is UNIX based and will adhere to international standards.

The objective of the software architecture is to provide good real-time support for all hardware devices and to provide an object oriented platform for multimedia documents and hypertext information systems. Since the key applications deal with documents in the widest sense, a multi-media editor, Optical Character Recognition, script recognition and support for 3-D and hypertext are also developed.

The combination of these new features with advances in user interface design produce a significant increase in user-friendliness and thus usability of the system over products presently on the market. MULTIWORKS is therefore an ideal system to increase the productivity of the office worker in the service sector, and since this sector is now the largest employer in Western democracies, such productivity increase addresses one of the largest market available.

The project is planned to cover 4 years and to spin out significant results into the companies throughout the duration of the project and not only at the end.

The Companies contributing to MULTIWORKS are a good mix of large and small companies, with a very wide technical expertise: they are in a position to guarantee the successful accomplishment of the ambitious workplan and the capacity to turn the results into winning products on the market.

The Consortium is organized in:
- Main Partners:
    Olivetti, prime contractor
    Bull
    AEG Electrocom
    ICL/STL
- Associated Partners:
    SGS Thomson
    Acorn Computers
    TA Triumph Adler
    Philips Components
    Philips Kommunications Industries
    Chorus Sistemes
    Harlequin

2. Structure of the Project

MULTIWORKS is a very large and complex project, therefore it has been divided into groups of hierarchical working units: each of them has been assigned for coordination responsability to a well defined partner.

The working units have been structured according to the Commission definition and consist of Workpackages, Tasks and Activities. The following is a description of the various workpackages.

2.1. Workpackage on requirements, specifications and standards.

This workpackage deals with analysis of Market opportunities, general specifications of architecture, interfaces, etc. and with necessity to specify or adhere to standards.

Since a fully commercial approach is not practicable for a pre-competitive research project, we have started from a technology base rather than a market requirement. However some way must be found to develop the market view in parallel with the technology.

Many factors will affect how a low cost multi-media workstation would sell in the early-1990's. The level of technology, for instance CPU power, store size, scanner resolution, the ability to recognise characters in scanned text, the object oriented programming paradigm etc., are all clearly important. However, the level and types of services on networks, the international standards pertaining to this class of product and most of all the expectations of the potential customer are clearly just as crucial to the success of this programme.

Increasingly, the workstation in addition to its powerful dedicated capabilities, is becoming a window onto a whole range of services, both commercial and public. Access to corporate mainframe services, and to public multi-media information services such as videotex, fax, telex, are likely to remain or become important.

This workpackage then is concerned primarily with establishing the needs of the markets, as if this were a normal commercial development.

2.2. Hardware architecture and CPU development

The aim of this workpackage is to design and produce the building blocks required to construct a state-of-the-art microprocessor system which can compete in a world market against this background of rapid technological advance. The work requires expertise in system architecture and VLSI design, and knowledge of and access to leading edge semiconductor processing.

The particular focus of this development is to achieve hardware capable of supporting state-of-the-art workstation performance in a very cost effective manner.

The last years have seen the introduction of several new 32 bit microprocessors. These have included a number of first generation RISC designs, some of which have emerged from companies new to the microprocessor business and some from the established semi-conductor suppliers. Current forecasts suggest that the processing power of 32 bit microprocessors will double almost every year for the next few years and RISC architectures are likely to lead this advance because of their shorter design cycles.

At present all but one of the RISC microprocessors are targeted at the high-end workstation and mini-computer markets. This may be concluded from characteristics such as expensive packaging, high power consumption, requirement for external cache built from static memory, etc, all of which make the construction of a low cost workstation difficult. The one available RISC microprocessor which is targeted at personal computer and low cost workstation applications and which is at the base of MULTIWORKS, is actually capable of delivering a performance similar to the high-end CPUs at a much lower cost level. The 'low cost' target does not appear, therefore, to conflict directly with achieving 'high performance' in the MULTIWORKS VLSI CPU.

This workpackage consists of different tasks, strictly related to each other:

2.2.1. Hardware Architecture and Chip Set Definition

This includes the use and some developments of a 2nd generation high performance RISC CPU chip set.

2.2.2. Peripheral Subsystem Definition

We have defined three classes of peripherals:
- I/O peripherals, consisting of all devices needed by the user to interact with the workstation, that is to introduce data and to get the information resulting from the workstation processes represented and possibly reproduced.
- Communication peripherals, for connecting the workstation in both local and long distance networks.
- Mass storage peripherals, with special attention to new archiving media, since the working memory will be mainly implemented in RAM.

The peripheral subsystem needs to efficiently support real time high bandwith information exchange, like voice, video, text and images. A standard bus which supports these requirements has therefore been chosen. However the project has identified also key areas for cost reduction through VLSI integration, avoiding the cost overhead of the standard bus by close integration with the CPU system of some peripherals, as the scanner and the printer.

2.3. I/O Peripherals

The wide range of advanced functionalities which are required for the workstation has shown the necessity of a complete set of input/output pheripherals, like keyboard and display, electronic paper, printers and scanners, Optical Character Recognition, voice and video interfaces. Let's describe some of the most significant ones, that most clearly sets MULTIWORKS apart from traditional workstations.

2.3.1. Electronic Paper

A flat panel display is covered by a transparent graphic tablet which can be written on as if it were paper. This "electronic paper" is particularly appropriate for the office, since it lies flat on the desk top surface and occupies the same physical space as a sheet of paper; it can also be used as a friendly input of handwritten data (text and graphics) as well as commands and can be therefore considered as an alternative or complementary I/O device to the traditional CRT/keyboards. On line script recognition and free hand editor techniques are also being integrated in order to fully exploit the electronic paper features.

2.3.2. Optical Character Recognition

This module has to read text parts of documents which are delivered to the recognizer in a bitmap representation. The bitmap of a document may originate from an input device like the scanner or via a telecommunication link. The image may also be manipulated or selected interactively. The recognition module will assist the input and the processing of data and will recognize all kinds of writing like machine printing, bookprint and handprinting in a fast and reliable way. Despite the desired high performance it will be a low cost system.

To meet the challenge of offering a low cost and high performance recognition system, existing character recognition algorithms, the quality of which is unsurpassed, are improved and implemented using up to date hardware design techniques. The aim is to develop a very high integrated, yet flexible application specific circuit.

2.3.3. Voice Interface

Voice is a real time information to be handled in the Multimedia Integrated Workstation. Voice related services can be divided into three classes:
– Voice communication.
  The transmission channel capability needs voice signal compression. Algorithms for very high compression are being defined and implemented.
– Man-machine dialogue related services.
  Voice is the most direct way for human communication: for this reason future workstations need to emphasize these aspects of input/output interface. Real time voice synthesis and recognition are then strategic to this project.

As a first step the recognition of a limited number of words and a voice generated from pre-stored elements can support this level of communication.

As a longer term perspective, we will consider continuous speech recognition with wider vocabulary and a synthesis that must support automatic voice generation from text.
- Voice manipulation.

In this class we include voice message store and retrieval, automatic answering, voice annotations to multimedia documents, etc.

2.3.4. Video

Video camera input is considered very important for MULTIWORKS, where the emphasis lies on multimedia: this adds new interesting features to the general information to be communicated or stored in the mass memory devices. The output from the camera will also be displayed on a window of the window environment, through a special hardware and software video controller

2.4. Communication Peripherals

The Workstation has to provide the user with information in different forms, data, voice, image and video, operating as part of a local or extended area network rather than as a stand alone equipment.

MULTIWORKS is intended to comply with these needs and therefore is considering the following existing or emerging networks:
- Telephone Network
- ISDN (Integrated Services Digital Network)
- FDDI-2 (Fibre Distributed Data Interface) Local Area Network, as the emerging standard among future high speed LANs

2.4.1. Telephone network

An integrated telephone line controller is considered in the hardware configuration.

This telephone line can be used for multipurpose connection, that is for voice, data and fax communication. In fact, although the future of communication will see the large diffusion of ISDN network, it is necessary to consider in MULTIWORKS conventional telephone network, that will supply the great majority of communication links for quite a few years in the future.

Appropriate line interface, for both voice and data (modem) will be provided, in order to allow for the implementation of related functions, that will be realised also by using the high processing power of the CPU and the other peripherals like the display/tablet for telephone enhanced features and the integrated scanner printer for facsimile transmission.

2.4.2. ISDN

ISDN will become the ideal network to provide a wide connection between equipment with multimedia capability.

The ISDN moduke of MULTIWORKS will make available to the computational section of the machine a transparent network access at defined speed, by implementing the transceiver functions as well as protocol functions or the establishment and de-establishment of the connection.

Moreover, the B-Channel must be partitioned and assigned to the multimedia sources, according to an appropriate frame format and speed.

The specific objective of this task is to define the system architecture and partitioning of the ISDN module and to implement multimedia dedicated functions on silicon with VLSI technology.

2.4.3. FDDI 2 Local Area Network

This work covers the integration into the MULTIWORKS workstation of an interface to an FDDI-2 high speed integrated services LAN subsystem. FDDI has in fact become generally accepted as the "next generation" open LAN standard to replace current Ethernet and token ring offerings.

The realisation includes the following steps:
- Liason with FDDI standard committees.
- FDDI-1 chipset and board level product evaluations.
- FDDI-2 subsystem design, including full specification definition and hardware and software integration and testing.

2.5. Mass Storage Peripherals

This workpackage deals with the integration of all the major classes of mass storage peripheral technology.

2.5.1. Winchester hard disk and floppy disk drives.

The hard disk being integrated has a 3.5 in. form factor, very low power consumption, high capacity (starting from 80 MB), high transfer rate (10 Mb/sec).

The floppy disk drive has a standard 3,5 in. format and the maximum capacity available at this moment (2 MB). In this case the availability of a data interchange standard is important and during the project development will be closely monitored.

2.5.2. Optical disks

Optical read-only media (CD-ROM), write once media (CD-WO/WORM) and erasable/re-writable optical media will be made available in MULTIWORKS. This means the selection of hardware (controllers and drives), the development of software to adapt these hardware components to the operating system, and of the File Management Systems.

This task consists of different activities:
- CD-ROM and WORM optical disks
  The integration of these devices presents some problems due to their "non rewritability". In order to overcome this limitation it is necessary to develop a SW device driver and filing system with unique features.
- Rewritable optical disks.
  This type of technology is still in development stage.
  First prototypes will be based on the Magneto-Optical technology. Other basic technologies like phase-change and dye polymer are still behind in development and will not be available before some years and therefore are not considered in the present project.

Rewritable optical disks combine the erasability/ rewriteability of magnetic storage with very high capacity, offering new opportunities for storage of images and voice. Being the media removable, the compliance of recorded data to standards is an important issue.

2.5.2. Semiconductor Memory Cards

It is a type of storage which is based on the use of RAM chips plus a battery back-up mounted in a housing having approximately the dimensions of a credit card, in order to obtain a non volatile fast access time memory, with a capacity in the range of 1 MB at a reasonable cost. It can be used in addition or as replacement of FD for I/O of personal or general purpose data.

In MULTIWORKS we will integrate all types of cards selected in accordance with the requirements of the workstation and will follow and promote the related standard activities.

2.5.3. Digital Audio Tape (DAT)

This type of memory device presents challenging aspects offered by its high performance and modest cost at the same time. Being tape life and reliability media parameters not yet stabilized, we are dedicating particular attention to their evaluation, to the implementation of redundancy mechanisms and to the consequent impact on the global device performance.

This activity will be soon followed by the integration of a first model of device. The integration activity will then be continued on an optimised model and will be specifically oriented to partially compensate the intrinsic long access time by the implementation of appropriate software procedures. Other activities will be dedicated to the interface selection and to the standards for media interchangeability.

2.6. Software Overview and Objectives

The MULTIWORKS project will produce a hardware platform with extremely wide capabilities at a cost which should ensure the possibility of wide market penetration. However the effective exploitation of this capability is largely dependent on the availability of appropriate software. The cost-effective implementation of such software is in turn dependent on the availability of adequate tools to support the software applications developer.

The primary requirement on the MULTIWORKS software is to provide the programmer with effective access to the full range of available facilities within a suitably structured and supportive environment. This requirement covers three major areas: multimedia document handling, development support for complex applications (particularly knowledge based systems) which fully exploit the workstation, and management of distributed functionality.

The structure of the project allows for each of these areas to be explored independently during the initial phase, but envisages that these will be brought together in a general purpose application development environment.

The MULTIWORKS software activity is defined in the following workpackages.

2.7. Operating System, Basic Languages and Protocols

The objective of this workpackage is to provide the workstation with all basic software necessary to support higher level developments, i.e., the operating system, compilers and network protocols. They will be implemented on the successive hardware prototypes so that they can be used by other workpackages as soon as possible.

The workpackage is based on existing technologies that will be enhanced and completed to meet the project objectives and fit with the hardware architecture and distributed operating environment.

2.7.1. Operating System

Sitting in between the integration of new hardware products and the integration of new office application support tools, the operating system is one of the key elements for ensuring good integration of the project results.

The CHORUS operating system, used in MULTIWORKS, is a state-of-the-art distributed system incorporating the most successful concepts, proven from several research projects: it has a message-passing kernel, distributed virtual memory, threads, network addressing and transparent file naming.

In addition to the operating system, there will be support for optical memory (file service on CD-ROM, CD-WO/WORM and erasable/rewritable), for X.400 services (taking full advantage of the multimedia devices) as well as support for the languages, basic libraries and tools necessary for the other developments (in particular the multimedia libraries and the interactive environment).

2.8. Multimedia Libraries

The objective of this workpackage is to create a set of libraries allowing to deal with various media. Extensions to the standard user environment will be made when required. The MULTIWORKS Interactive Environment will allow applications to have a unified object-oriented interface to the libraries.

A number of tasks are included in this workpackage:

2.8.1. 3D Object Manipulation

This task consists of interdependent modules which will generate a complete platform for producing high quality 3D moving objects, with an ergonomy oriented towards non computer specialists.

2.8.2. Postscript, as a full integration of the PostScript language and imaging system.

2.8.3. Document Recognition Software

The document recognition functionality is an essential component of the workstation. Character recognition requires to deal with a broad range of text fonts, print and paper qualities and formatting possibilities.

Starting from the current recognition technology, improved recognition methods are implemented.

2.8.4. Script Recognition

This task can be partitioned into two subtasks, the adaptation, extension and modification of script recognition algorithms, and the interface between script recognition and the user and between script recognition and application programs.

2.8.5. Voice

This task consists in the integration of well known algorithms from the speech domain. Its aim is to provide libraries of routines controlling voice functions in the workstation. These cover two separate aspects : voice capture, storage and annotation of documents, and speech recognition and synthesis.

2.8.6. Video

This task will allow to open a video window with moving pictures originating from a local or remote device.

2.9. MULTIWORKS Interactive Environment (MIE)

The objective of this workpackage is to produce an extensible interactive programming environment based on an object-oriented programming paradigm. This is considered important for the development of the new multimedia applications that will emerge to exploit the MULTIWORKS hardware.

The object layer is actually an extensible interface, containing initially a set of interactive object classes similar to those in conventional window manager plus a set of ODA-compatible document objects for displaying and interactively editing document parts. The interface also allows the transmission of instructions, procedure definitions and object descriptions from the application environment to the interactive environment.

A prototype of the MIE will be produced as early as possible to enable experimentation. Two prototype applications (a knowledge engineering application and a multimedia editor) will be developed to validate the MIE design, and to provide requirements for further work.

2.9.1. Language environment and object management tools for MIE

This task will provide the facilities needed to support communication between the Application Environment and the Interactive environment. The task also includes the implementation of a version of CLOS and the LispWorks environment. Finally, it will provide a set of tools for the management of the object-oriented interactive code and a set of diagnostic and debugging tools.

2.9.2. Object oriented document representation (OODR)

The aim of this task is to investigate and define an object-oriented internal document representation for MULTIWORKS. A type model for object definition and management will be defined, supporting the document type model defined by the applications. A full ODA-compatible document representation with multimedia extensions will be built in Common Lisp + CLOS.

2.9.3. MULTIWORKS Interactive Environment and User Interface Toolkit.

The MIE provides the consistent, clear, rapidly understandable view of the system facilities which the user needs to build, maintain and interact with his applications.

2.9.4. Persistent Hypermedia Object Store

A persistent object store will be developed in the second phase of the project to support the hypermedia applications.

2.10. Applications

The main objectives of this workpackage are to define a limited but representative set of application programs, that allow the creation, the storage and retrieval of compound documents containing complex and linked objects. The following applications have been envisaged:

2.10.1. Multimedia Editor

This editor will be based on the background of an existing multimedia editor. It will be extended to cover additional media such as 3D animations, voice and video. As the MIE becomes available, the editor will migrate to the rich, object-oriented interactive environment.

2.10.2. Storage and Retrieval

This task provides the storage and retrieval functions for two important application areas, multimedia document management and hypermedia management. The work carried out in this workpackage is based on the Esprit MULTOS project (Multi Media Office Server), that provides content oriented filing and retrieval capabilities for large amounts of multimedia documents.

2.10.3. Hypermedia System

The hypermedia system will handle a network of nodes containing multimedia data connected together by links. It will heavily rely on the the work described above. A significant effort will be devoted to the user interface issues.

A major feature of the system will be the development of a specific programming language: MultiTalk. This language will aim to overcome the limitations of existing hypermedia languages. A tool for providing support to idea elaboration during office document preparation will be implemented using the hypermedia system and MultiTalk.

2.10.4. Knowledge Engineering Environment

A knowledge engineering environment will be developed to facilitate the construction of knowledge based application systems. It will exploit the multimedia features of the workstation. It will assist knowledge-engineers and/or end-users by providing a problem specific abstract view on the application. To support these more advanced multimedia interaction techniques, conceptually more powerful and efficient reasoning techniques will be implemented.

2.10.5. Computer Supported Cooperative Work (CSCW)

This task aims at the development of the basic environment needed to support teamwork activities on the MIW workstation. It focusses on the scenario where groups of authors cooperate to produce documents through real-time interaction.

The task will develop tools for group coordination and a basic support for computer conferencing extending the operating system and user interface of the MULTIWORKS workstation.

ESPRIT HOME SYSTEMS PROJECT, A STATUS REPORT

R.van Dootingh
Philips International - CE
P.O.Box 218
Building SWA 8
5600 MD Eindhoven
Netherlands

Abstract

This paper summarises the status of the ESPRIT project 2431 on standardisation of Home Systems as per nov 1989.

1. Project Scope and Objectives

The primary objective of the project is to propose a comprehensive standard for integrated electronic systems for use within the home. Application areas are broadly defined as Home Control, (Tele)Communications, and Audio/Video distribution. As such, the proposed standard covers a vast spectrum of products which may have a functionality within the home, or in relation to external services such as telephone, TV and utilities. The proposed standard will include aspects related to transmission media, communication protocols and interfaces, and application command language. The standard will provide a basis for compatibility of consumer products of various kinds, serving a variety of applications, thus providing opportunities for bringing new products onto the market for multi-brand and multi-application environments. The prime area of application of the standard is the single family home, though extensions to cover service premises and small businesses will be provided.

2. Current Status

The project commenced in January 1989, will be of two years duration, and hence is about halfway through at the time of this seminar.

2.1 Requirements and Architecture

The work in our project started with a careful consideration of requirements which are to be met, and of the input conditions which are to be taken into account. The first quarter of 1989 thus was devoted to:

- definition of user requirements, both generic requirements which we feel are to be met by any home system, and requirements specific to particular application areas
- evaluation of documents from the Eureka project 84 on Integrated Home Systems.

Within the project the group GLOBAL has embarked on devising application scenarios, anticipating the way people in developed industrial countries want to live in the not too distant future. Analysing these scenarios, requirements can be formulated for the Home System in terms of

- what the network should offer, and what applications are to be supported
- scope and size of the network, number and classes of units, expandability and modularity
- performance in number and class of concurrent channels, response times, security of access, and future proofing
- user interface, ease of use, input and output devices, user means for controlling and accessing the system
- network installation, both in the physical sense and in the sense of initialisation, and management and control of (changes in) the network configuration
- links to external networks and services, such as telephone, radio and TV broadcast on cable, terrestrial and satellite transmission, and to various company specific networks which are to be recognised as sub_buses to the HS
- safety, reliability, life

We completed this initial part of the project with the issue of a User Requirements Specification in April 1989. This document will be the reference against which we will measure the proposed standard; the report has also been sent to the CEC as the first project deliverable.

Subsequently, during the second quarter of this year, we studied and defined an overall architecture for the Home System, taking into account the requirements and the standardisation frameworks as provided by ISO/IEC JTC1/SC83 and by CENELEC TC105. The architecture being layered in the spirit of the OSI reference model, we have defined the services to be provided by each of the layers.

In the next phase of this part of the project we are currently preparing the detailed technical specifications for the protocols for the upper communication layers and associated command language, and for the network management layer.

2.2 Media and Interfaces

Our input conditions were largely set by the preceding work of the Eureka project 84 on Integrated Home Systems. This project completed its work in February 1989 by delivering draft media specifications, and protocol and command language specifications. Also demonstrators were handed over for enhancement. Although the Eureka results are owned by the member companies of the Eureka Consortium, these results were openly made available to the ESPRIT HS project to further the work. Our MEDIA and INTERFACES group did most of the evaluation. Most of the media specifications were found to need manageable extensions in order to provide the communication services required to support the application requirements of ESPRIT HS. Significant additional work was found to be required to integrate the various specifications into a common architectural framework, and to provide a network management definition which we consider an essential basis for user friendliness. We have issued reports to the CEC on the evaluation of the Eureka results, and on the definition of ESPRIT media.

As it stands today, we have almost completed the specification of media and media specific protocols for twisted pair, coaxial cable and mains wiring. We are also rather optimistic in completing similar specifications for infra red and radio shortly. This leaves only optical fibre for draft completion at project end, not surprisingly so because there was no preceding Eureka work on this subject.

At the same time, the network layer and protocol is being defined in a joint effort of GLOBAL and MEDIA, defining among other things the very important issues of addressing and routing of messages, and of functionality of gateways of various kinds.

2.3 Applications

In the APPLICATIONS group we continued work on the Eureka demonstrators to meet the extended ESPRIT objectives, and are on the verge of providing access to these demonstrators for a selected audience. These demonstrators are both a test set up on which validity of parts of the proposed standard will be proven, and pilot implementations of typical applications which can be expected on the basis of the standard. Future demonstrators, due at the project end, will provide similar proof of validity and model implementation of applications on a more integrated basis.

This group is also actively involved in the definition of gateways, particularly to other non_HS networks, and to external networks and services.

It is evidently impossible to define the complete spectrum of applications. However we have analysed what we believe to be a representative set of applications. In addition, a validation concept has been defined, describing how we are going to validate the standard, using demonstrators as target systems on which to run test scenarios.

2.4 Installation

Other areas were less well prepared or completely missing from the preceding Eureka project. A notable example is the preparation of installation requirements, recommended practices, etc to arrive at an integrated approach to energy (mains voltage) distribution with control and signal distribution when it comes to installing a network in a home. Within the ESPRIT HS project a dedicated INSTALLATION group addresses this difficult and complex area.

The domains of power distribution and of data communications have various aspects in common which require a common approach:
- The Home System network and its attached devices need reliable power for proper operation: a modern approach to providing reliable power may very well utilise advanced data communication techniques for signalling and control of circuit breakers, outlets, etc.
- Techniques and standards of how to install a power distribution system in a home are likely to change once switches and controls are recognized as devices on the signalling and control network, instead of devices which directly switch power on and off.
- The mains wiring will be one of the data transmission media in the home system for any application. Other wired media are expected to be pre-installed, integral with the mains wiring, for both cost and ease-of-installation reason and because the mains outlets and the data outlets need to be close together in most cases.

Electrical and logical topology are different for the various media, yet the physical topology should allow for integrated installation, both for pre wiring newly built houses, and for retrofit in existing houses.

The merger of installation rules of a mains installation with those for twisted pair and coaxial cabling, meeting functional requirements as well as regulations which are different in various countries, has turned out to be a major problem within the project. Our document on Installation Concept Definition is dated September 1989, and we are currently working on detailed specifications.

3. Expected Results

Since the project is still in full swing, it is too early to report on final results and achievements. However the advances of the project are expected to be highly visible in both the state of the art, and in terms of industrial activities leading to introduction of new consumer goods on the market.

3.1 Relation to Current and Anticipated State of the Art

The proposed standard will provide compatibility across various products on the basis of an integral approach to data communication for a number of media and for a number of application areas. The level of integration exceeds that currently available in similar standards, such as HBS and CEBus, by virtue of a wider variety of media and by virtue of extensive facilities for inter working of different application areas.

The proposed HS standard also addresses the difficult problem of physical installation, integral with the power distribution system in a house, which we believe is a novel approach badly needed to make Home Systems viable and affordable.

3.2 industrial impact

The list of companies participating in this project includes the leading European companies in the areas of entertainment and audio/video, household appliances, installations, security systems, and telecommunications. In most cases these companies have, in the past, provided their own company specific and application specific network. The end user, i.e. the consumer who "invests" in his house, thus is confronted with the need to duplicate costs of his infrastructure, with complex installations, and with incompatibility. The member companies of this project have joined in the pre-competitive standardisation effort of this project to alleviate this situation, and thus to open up new market opportunities for Home Systems. The name of the game is compatibility: in a prewired house, it must be possible to gradually hook on a wide variety of consumer goods with the minimum of additional costs, and with a minimum of installation and initialisation problems.

A glimpse of what is ahead can be obtained from the demonstrators which are about to be presented. These demonstrators may well be seen as model implementations, resulting from the advanced developments of the various contributing partners.

INFORMATION EXCHANGE SYSTEM

Y-NET: OSI Services for Researchers in EC Programmes

G. Autier
BULL
Route de Versailles 68
F - 78430 Louveciennes

Abstract

Y-NET is an initiative by a group of manufacturers in the ESPRIT framework supported by the IES, the infrastructure part of ESPRIT, to provide OSI services for researchers. It primarily addresses participants of ESPRIT and other programmes and activities of the European Community but will be open for other researchers as well. It will be embedded into the overall OSI communications to be aimed at within the COSINE framework, and by this will allow information exchange between a wide range of users, both from academic and industrial research. Y-NET will concentrate on support for researchers from industry with particular emphasis on small and medium sized enterprises (SMEs).

The Y-NET configuration will be based on OSI developments from various ESPRIT projects. The configuration will comprise service points in each of the countries of the Community commonly managed at European level. Users will be enabled to access, at a national basis, the service points for international communications through X.25 PAD connections or through other simple telecommunication connections. It will start with X.400, include FTAM subsequently and will be extended to other OSI services as soon as stable products are available (directories, structured document transfer, etc.). At the carrier level, it will use international public X.25 and the Commission sponsored RARE/COSINE international X.25 backbone (IXI).

CACTUS: A RETROSPECTIVE VIEW

J. SARAS, A. LANCEROS, J. BERROCAL
A. AZCORRA, J. SEOANE, J. RIERA
Dpto: Ingenieria Telematica
ETSI Telecomunicacion
Universidad Politecnica de Madrid
28040 MADRID, SPAIN

J.DELGADO, F.JORDAN, M.MEDINA
Dpto. Arquitectura de Computadores
ETSE Telecommunicación
Universidad Politecnica de Catalunya
BARCELONA, SPAIN

This paper outlines some of the experience gained from the development of the CACTUS project. The paper focuses on four more or less independent subjects:
– The use of Formal Description Techniques to develop higher layer protocols;
– Problems and experiences found when implementing the P7 protocol;
– General reflections about ASN.1;
– Comments on the underlying OSI tower used in X.400.
A brief description of CACTUS goals and practical achievements is also included.

1. Introduction

This paper reports on the experience gained from the development of the CACTUS project. It is relevant both to the implemented standards (holes, errors, pitfalls within the definition of P7, ASN.1, etc) and to difficulties found in their implementation.

The conclusions are divided into four main categories, relating to:

– the development methodology used;
– the communication infrastructure - RTS, Session, PC connection layers;
– ASN.1 encoding and decoding;
– the P7 protocol.

A full section is dedicated to each subject.

A short description of the CACTUS project can be found in the next section although a better reference is [1].

2. Brief CACTUS Project description

The CACTUS project (Carlos Addition for Clustered Terminal User Agents) is an extension of the CARLOS project which implements the CCITT X.400 series of recommendations in a form suited to medium-sized private organisations.

The CACTUS project provides components which enable existing users of personal computers to access the X.400 world via the telephone network on asynchronous lines.

Initially the connection to the OSI environment is via the Session service from CARLOS.

A CACTUS system may be connected to the Public Messaging Transfer Services and directly to other Private Messaging Systems, thereby saving the cost of using the public service. To do this CACTUS supports the P1 and P2 protocols to other Message Transfer Agents (MTAs) and Messaging Systems. The CACTUS MTA is essentially an endpoint and consequently does not act as a relaying transfer agent.

CACTUS components, called the Mailbox Service Agent (in ECMA terminology, Message Store in ISO/CCITT), Mailbox Client Agent (or Message Store User in ISO/CCITT) and Remote Operations Server have been implemented to enable the user to be remote from the CACTUS Box and permit mail to be delivered in the user's absence. CACTUS intercepts the ECMA version of the P7 protocol standard [2] (upon which the CCITT 88 message store access version is based).

The user can utilize the PC for other functions and at intervals will log into the Mailbox Server which has been holding any incoming messages. These are then transferred to the PC's disk for perusal by the user. The user can also transfer messages prepared on the PC (probably using a standard PC word processing package) into the Mailbox System for onward routing via the MTA.

CACTUS software was successfully tested against SPAG. In Spain, three pilot installations have been started (two in Madrid and one in Barcelona), to build an experimental network. This network is connected to the Spanish Academic Network via another X.400 node, running the well-known DFN version of the EAN software (some interworking tests were made with it). Nodes in the CACTUS network are accessible by remote PC users via asynchronous lines, or through the PSTN. It is planned to expand the means of access (e.g. running P7 over a LAN).

A more detailed project description can be found in [3] [4]

3. Development Methodology Used

Before going into other technical details, a few words should be said about the utility of using Formal Descriptions Techniques (FDTs) for protocol development.

ESTELLE and LOTOS [5] [6] were used to specify all the protocols developed for CACTUS (one was specified in LOTOS, the remainder in ESTELLE). As far as is known, this was the first attempt to mix a protocol developed from a LOTOS specification into an ESTELLE environment [7]. Due to the different types of synchronisation used in each language (infinite queues in ESTELLE and rendezvous in LOTOS), some care had to be taken when connecting both worlds. Surprisingly, only one bug was found when both worlds were integrated. Due to the multitasking kernel that the LOTOS compiler generates to control parallellism in LOTOS the whole software derived from LOTOS was isolated inside a UNIX task. This could lead to inefficient implementation if one interleaves layers specified in LOTOS and in ESTELLE (because separate tasks must be used to run each implementation, resulting in an implementation with a different task for each layer).

The ESTELLE and LOTOS compilers (both developed inside the department, the latter within another ESPRIT project called SEDOS) were used to translate the specifications into C code. On average, 50% of the final code was obtained in this way (less in the higher layers of the OSI stack, more in the lower).

An ASN.1 toolset (see [1] and following section on ASN.1) was used to encode and decode all Protocol Data Units (PDUs). Its use helped to avoid much of the otherwise repetitive and error-prone work required for various levels (especially when developing Application protocols, since the PDUs are rather complicated).

It was noticed during the testing phase that although some typical errors were found (bits wrongly set, etc.), it was rare that the error was due to the omission of processing some event in a state (which was not the case for other systems against which we tested, even those supposed to be fully conformant to X.400). It is thought that this was due to the use of FDT's, since their use and method forces the specification of answers to all possible events inside any state. For example, if ESTELLE is used to specify the opening of an RTS connection when there is an outstanding open Connection Request, it is necessary to specify an answer to any (and all) possible incoming interactions (those specified inside the ESTELLE channel to the Session layer). This would mean that a response would be required if a DATA Transfer Indication arrived (although unexpected).

Typical implementations first codify normal behaviour (that without exceptions) and once they have a skeleton, commence adding further functionality to consider exceptional situations. This obviates most possibilities liable to be experienced by implementations where unexpected events are not considered, or are discarded (possibly leading to deadlock). Although an ordered design could also be followed, FDT's force the use of a structured design method (much in the same way that high level languages discourage the use of "GO TO" and the like).

Finally, it appears that the higher in the OSI architecture a protocol is, the simpler are its open procedures, as specified in the standard; thus apparently decreasing the importance of using ESTELLE and LOTOS, which were designed mainly to specify behaviours. On the other hand, the higher a protocol is, the more complicated its PDUs are, which stresses the importance of using some kind of encoding/decoding tool to help in the task. Consequently, within CACTUS, where several kinds of protocols are involved (low and high), the combination of tools mentioned in this section proved invaluable during the development process, reducing the development time and increasing its robustness.

4. RTS, Session and Below

The following stack: RTS, Session, Transport Class 0 and X.25 - are used in X.400 to provide a reliable data transfer service to the Message Handling application. It is enough to count the number of service primitives of the RTS and Session services to realize that the RTS protocol is not straightforward. While the RTS service is easy to use and understand, the Session service leads both to errors in its use and to misunderstandings among different RTS protocol entities. Hence any Application Service element built directly on the Session Service will have a complex task.

One of the problems encountered during the CACTUS certification tests was related to the session connection identifier in the connection establishment phase. This is a parameter whose value is chosen by the connection initiator in accordance with internal rules. The responding entity must not change the chosen value, since if this were done both entities would have different identifiers and the connection could not be resumed in the case of failure. The result is that some implementations changed this value, encoding it as an ASN.1 type. This does not prevent basic interworking, but any recovery mechanism done by the RTS would fail.

This is only an example of interoperability problems, but it would be possible to enumerate others (synchronization points, functional units, etc.). The conclusion is that the session *service* is highly complicated to use and may provoke problems when interworking between open systems. This is one of the reasons why entities like the RTS are highly desirable. Instead of having a complex set of tools to structure a dialogue (the session service), RTS offers one kind of dialogue appropriate to reliable transfer APDU's.

The same justifies the complexity of RTS itself.

During the implementation of the RTS many errors and 'holes' in the 1984 version became apparent. Contributions were made to the X.400 series Implementors' Guide on some of the bugs and clarifications.

5. ASN.1 Encoding/Decoding

Within the project, a general purpose library to encode and decode ASN.1 data structures was built. This library interprets a set of C tables obtained via an ASN.1 compiler developed in parallel (see [1]). The library was used to encode and decode all protocols' data units that are used inside each protocol (P7, P1, P2, ROS, etc) and that are defined using ASN.1. This gave us the opportunity to concentrate efforts into only one development and to reconsider deeply how ASN.1 performs the encoding and decoding. These reflections are exposed in the following paragraphs.

As ASN.1 recognises, the tagging mechanism is a detail of the basic encoding rules of ASN.1 embedded inside the abstract notation one. In other words, it is a detail of how to encode/decode values of some type inside the own-type definition. It is our belief that, at least, the context specific tags could be suppressed from ASN.1 and generated automatically when encoding/decoding. Alternatively, some kind of parser could generate automatically all context specific tags necessary in a ASN.1 module where this feature had not been used. Its output would be a standard ASN.1 module with the minimum tags needed.

For example, the following ASN.1 type without any tagging:

```
AType::= SEQUENCE {
        f1 INTEGER OPTIONAL,
        f2 INTEGER OPTIONAL,
        f3 BOOLEAN,
        f4 INTEGER }
```

would be transformed into the following (solving the inconsistencies):

```
AType::= SEQUENCE {
        f1 [0] IMPLICIT INTEGER OPTIONAL,
        f2 [1] IMPLICIT INTEGER OPTIONAL,
        f3 BOOLEAN,
        f4 INTEGER }
```

The above feature would only require that all new elements in a set or sequence be added at the end (in order to respect the automatic numbering). In addition, an automatic tool could only generate the minimun tags required to solve ambiguities (in opposition to present practise to number every field in a sequence/set to avoid problems). The former would mean an increase in efficiency.

Although the parser idea was introduced to respect the present definition of ASN.1 (and allow the reuse of all developed ASN.1 software), the numbering rule can be included completely inside the encoding rules.

The same ideas could also be applied to the application and private wide tagging. In this case, the code generated in each use of an application/private wide tag may be different (if tagging is necessary to resolve ambiguity), but this will be hidden inside the encoding/decoding routines (and it should be):

OldApplicationTagType ::= INTEGER

AType ::= SET {
        OldApplicationTaggedType,
        INTEGER }

AnotherType ::= SEQUENCE {
        INTEGER OPTIONAL,
        OldApplicationTagType OPTIONAL}

after automatically tagged would result in:

OldApplicationTagType ::= INTEGER

AType ::= SET {
        [0] IMPLICIT OldApplicationTaggedType,
        [1] INTEGER }

AnotherType ::= SEQUENCE {
        [0] INTEGER OPTIONAL,
        [1] OldApplicationTaggedType OPTIONAL}

When decoding the above ASN.1 data, one would like to know which is the value of the "OldApplicationTaggedType" field in both types (Atype and AnotherType) without taking care if in the first case its tag code is 80H and in the second is 81H. If tags numbers were assigned application wide, we would achieve to keep the same code in both places (say 40H). From the abstract syntax point of view however, this is irrelevant.

Another confusing feature, from our point of view, is the existence of two kinds of lengths. In ASN.1, there are two ways to encode the length of a field: via a count of the whole number of octets (definite form) or via a field terminator (indefinite form). It is our belief that from an implementation point of view only one kind of length should exist: *the indefinite form.* When encoding it is clear that, in general, it will not be possible to know in advance the total length of a field (think of the total length of a PDU before encoding it) and some kind of backtracking will be necessary (an unwanted requirement). But, even when decoding, the only advantage of definite form lengths is to copy fields (without interpreting). If one has to decode a field with a definite form length, it is worse because in addition to following the internal structure, as when decoding an indefinite form length field, a count has to be maintained to see how many octets still rest.

Finally, when tagging explicitly in ASN.1, a pair of code/length fields is added to the old pair. It is clear that the length field is redundant and it could be suppressed if universal types followed a different encoding schema.

6. The P7 protocol

6.1. Problems with ECMA P7

The ECMA work on P7 was completed and adopted as an ECMA standard [2] in June

1987. The reason for the invention of the P7 protocol was mainly the obvious drawbacks [1, 8] of the 1984 P3 protocol [9]. In the P3 concept, a 24 hour availability of the UA is implied because the MTA "forces" messages to be delivered to the UA in the sequence they arrive at the MTA. The only way the UA can turn off the flow of messages is to use the optional service *hold for delivery*, which works like a water tap. Furthermore, a message delivered to the UA is "trapped" in that UA system; this causes problems for a mobile user, who may want to access his mailbox from terminals at different locations (like office and home), or from a "portable" terminal that he carries around.

The P7 approach set out to alleviate these problems. By delivering the messages to a separate entity, the Mailbox Server (MBS), the requirements to have 24 hour availability at the UA disappeared. The P7 protocol is always initiated from the Mailbox Client (MBC) - on behalf of the user - and the messages are retrieved by the user rather than being forced on him. Examples of other features offered by the P7 protocol are listing messages to select in which order to retrieve them and the possibility of leaving messages in the MBS for further retrieval later. The latter is useful if you sit with your portable PC or "laptop" in a hotel room on the other side of the globe and have just used up your last free floppy.

Due to the P3 shortcomings described above and the advantages [1] introduced by P7, the **CACTUS** project decided in 1987 to intercept the ECMA P7 protocol standard [2] which has been recently ratified in the 1988 version of MHS/MOTIS [10]. In the development phase, some minor bugs (errors in the P7 abstract syntax) were identified and fixed without a big effort. Nevertheless, we detected other problems related to the design of the P7 protocol and which probably will remain forever unsolved in the ECMA version - luckily, most of them have been solved in the 1988 ISO/CCITT P7. Of these problems the following are of interest:

(1) The abstract syntax of the ECMA P7 protocol is *cumbersome*. The number of operations (23!) is excessive. Besides, there are no more than 6 distinct operations. All of the 23 would be grouped into dispatch, analyse, list, fetch, delete, register. The adopted solution of specifying different operations (only with slight differences in its definition) for each mailbox object causes a very heavy protocol. As most of them are mandatory, or become mandatory if you want to provide the functionality of the specified object, this provokes severe criticisms. As a result of our implementation, we can say that the compilation (using our ASN.1 compiler + the C compiler) of the P7 abstract syntax produces an object of 80 Kbytes (only for the ASN.1 tables without any encoding/decoding function!). As an example, we show 3 of the 4!! different *delete* operations defined in the ECMA P7 standard:

```
deleteMessages OPERATION
ARGUMENT ::= SET {
        messages [0] CHOICE {
        range [0] IMPLICIT Range,
                numbers [1] IMPLICIT SET OF SequenceNumber}}
RESULT NullResult
ERRORS {
        serviceNotAvailable, accessControlViolation,
invalidMSSequenceNumber, invalidRange, newMessagesPresent}
::= 107
```

```
deleteInlogEntries OPERATION
ARGUMENT ::= SET {
        cutoff [0] IMPLICIT Cutoff}
RESULT NullResult

ERRORS {
        serviceNotAvailable, accessControlViolation,
        noEntries, mSMessagesNotYetDeleted}
::= 110

deleteACLogEntry OPERATION
ARGUMENT ::= SET {
        requestedEntry [0] IMPLICIT SequenceNumber}
RESULT NullResult
ERRORS {
        serviceNotAvailable, accessControlViolation,
        invalidDispatchSequenceNumber}
::= 116
```

In the ISO/CCITT P7 protocol this problem has been solved by defining a general *delete* operation. It is defined in ASN.1 as:

```
Delete ::= ABSTRACT-OPERATION
ARGUMENT DeleteArgument
RESULT DeleteResult
ERRORS {

        DeleteError, InvalidParametersError, RangeError,
        SecurityError, SequenceNumberError, ServiceError}
::=116

InformationBase ::= INTEGER {
        stored-messages (0),
        inlog (1),
        outlog (2)} (0..ub-information-bases)

DeleteArgument ::= SET {
        information-base-type [0] InformationBase DEFAULT stored-message,
        item CHOICE }
          selector [1] Selector,
          sequence-numbers [2]] SET SIZE (1..ub-message) OF SequenceNumber}

DeleteResult ::= NULL
```

(2) The idea of providing different operations to dispatch or fetch a message in the General Mailbox Services (dispatch-Message, fetch-Message) and IPM-specific Mailbox Services (dispatch-IPMessage, issue-IPM-Notification, fetch IPMessage) is not a good technical solution. It only increases the size of the P7 abstract syntax and confuses the user about which operation (dispatch-???, fetch-???) to use for submitting/fetching

a message (thought this may be hidden by the user interface). A better solution would be to provide generic dispatch and fetch operations and to include in it - as optional fields - the parameters related to the IPM-specific Mailbox services. This is the approach adopted by ISO and CCITT in their '88 version of the P7 protocol.

(3) The implementation in the ECMA P7 of *auto-correlation* services - logging facilities for automatic correlation of dispatched messages and any returns (notifications and replies) relating to these messages - as another object - *Auto-correlation Log* object - creates unnecessary complexity to the implementation. If the information related to the submission of a message is fully registered in the mail box object named *Outlog*, it seems more reasonable to store there the auto-correlation information, using for that some attributes of the outlog entry. The fact that auto-correlation logging is requested on a per-message basis reinforces our assumption. In this way, the extra operation *readACLogEntry* only defined for the *Auto-correlation Log* object and the rest of operations defined for this object would be removed. The solution of including auto-correlation facilities as outlog attributes would help to provide a uniform access to all the logs and eliminate the overhead associated with different access for each log object. An identical solution has been adopted in the 88 version of P7.

4) Different ways of structuring the information in the different mailbox objects, i.e. different methods of storing and retrieving the same or related information. The mailbox object named *Message Store (MS)* stores information for incoming messages; this information is called the MS Message. While the other mailbox objects (Inlog, Outlog, Auto correlation Log) only store *entries* - an entry is a list of attributes - the MS Messages consist of three elements : the MS Entry, the MS Envelope and the MS Content. Each of these elements can be individually fetched but an MS Message is deleted from the MS as a whole. This special structure of the MS means that special operations for handling it are needed. These operations *(listMessages, deleteMessages, fetchMess-age)* can not be used with the other mailbox objects, where only general operations are used *(list-Entries, delete-Entries)*. It seems more logical that all the information stored in mail box objects is collected in entries, each entry being composed of attributes. The only requirement to satisfy is that both the content and the envelope are stored as attributes of the MS Entry. This solution would provide a coherent structure for all the mailbox objects (ms, inlog, outlog etc.).

(5) Poor facilities for auto-forwarding messages. When an MBC (on behalf of a user) instructs its MBS to automatically forward the incoming messages, it can only select the recipient of the auto-forwarded message. All the messages are auto-forwarded, the user is not allowed to specify a filter to indicate that only the messages satisfying the filter criterion should be auto-forwarded. Besides, the user is not allowed to specify several recipients to whom the auto-forwarded messages should be sent.

6.2. Migration to ISO/CCITT P7

The latest achievement in Message Handling base standards is the joint MHS/MOTIS texts [10] produced by CCITT and ISO in late 1987 and early 1988. These texts form the basis of the 1988 "Blue Book" X.400 series of Recommendations in CCITT and the MOTIS International Standards in ISO/IEC.

An important 1988 MHS extension is the Message Store (MS) which is an intermediary between the User Agent (UA) and the Message Transfer System (MTS). The primary function of the MS is to accept delivery of messages on behalf of a single MHS end-user, and to retain them for subsequent retrieval by the end-user's UA. The MS also provides indirect message-submission and message-administration services to the UA, in effect, via "pass-through" to the MTS. This enables the MS to provide additional functionality compared to submission directly to the MTA; such as forwarding of messages residing in the MS and logging facilities.

ISO/CCITT have based the Message Store service and the MS Access Protocol (P7) on the ECMA work. Thus, most of the changes or additions [1] are only in the form and not in the model. Other changes affect the model, e.g. the inclusion of the MS Content and MS Envelope as attributes of the MS Entry and the mechanism created to allow a "complex" delivered message to be split up into several entries - the result is that a Stored Message entry can have three different possible roles: main-entry, parent-entry and child-entry.

The MS offers its services via three ports: Indirect Submission Port, Retrieval Port and Administration Port. It is mainly the *Retrieval* port services - Summarize, List, Fetch, Delete, Register and Alert - that are specific to the MS. For providing the other port services the MS consumes the services offered by the MTS Delivery Port, Submission Port, and Administration Port. In CACTUS, a similar breakdown has been followed in the specification of the Mailbox Server (MBS) [11]; the MBS has been split into two modules:

(a) the *Mailbox Access Entity (MBA)* [12] which provides the kind of services offered by the Retrieval and Indirect-Submission Ports; and
(b) the *Mailbox Delivery Entity (MBD)* [13] which provides the kind of services offered by the Delivery, Administration and Submission Ports.

By 1989 it could be expected that public services of X.400 would be committing to support the *Message Store Access Protocol (P7)* or the *Message Transfer System Access Protocol (P3)*, and that terminal and workstation vendors would be developing the capability to attach these services. Equally, the Office Automation vendors are likely to adopt P7 for LAN based office systems with distributed processing capability. We can expect LAN based mail hosts which allow P7 access from workstations on the LAN. A question of interest, is whether it makes sense to use a combination of P3 and P7 in an implementation or not. Our suggestion is that normally, if you implement the P7 protocol, it is overkill to also implement the P3 protocol. But there is no rule without exceptions. The exception is where the MTA is in a public domain and the delivery from the MTA is made over the P3 protocol into a private domain. In this case it may still be of interest for the private domain to store the delivered messages in an MS and allow access over the P7 protocol to that MTS.

Although P7-based products and services will not begin to appear in force until 1991, the CACTUS decision to "intercept" the ECMA P7 protocol was advantageous. The migration of the CACTUS P7 (based on ECMA work) implementation to the ISO/CCITT P7 may be done without much effort [1] and in a short time. This is because the CACTUS project has been following the standardization process of the P7 protocol - and participating in ISO meetings - and because of this the structure adopted in the ISO/CCITT P7 was followed in the internal design of the CACTUS MS. The main work to be done (in the migration process) will consist of the encoding and decoding of the P7 operations (P7 PDU's_ because there is a great difference between the original abstract syntax defined by ECMA and the new one standardized by ISO and CCITT. Nevertheless, as CACTUS has developed a very powerful ASN.1 library, we hope this process will not need much work and only we need to change the contents of some modules specialized in encoding/decoding PDU's. Also, as the design of the message data base - managed by the Message Store - was made bearing in mind the new structure of the ISO/CCITT protocol, only slight modifications will be needed to support the new data structures defined in the 88 version of P7. The main work to be done is the inclusion of the Application Control Service Element (ACSE) and the new Remote Operation Service (ROSE) with the old Mailbox Server (MBS) and Mailbox Client (MBC) ESTELLE entities. As both entities have been specified in ESTELLE, it is easy to change the specification (the generation of C code is made automatically from the specification) to include a new supporting ASE - ACSE - and to expand the ROS of ECMA to the new ROSE standardized by ISO and CCITT.

Due to the previous considerations and because MHS standards have reached a high level of stability, this migration should be made as soon as possible to place the CACTUS products in an advantageous situation in the market place.

6.3. Suggested extensions to ISO/CCITT P7

In the MHS/MOTIS standards a comprehensive set of facilities for the IPMS user are provided, intended to satisfy a broad range of service requirements. However, in the case of the stand-alone UA participating in MHS activity by means of the Message Store (MS), a number of substantial service problems are not addressed. The MS does not store submitted messages, it only delivers messages. This is to avoid the MS becoming a generalized filing system which would, no doubt, need to have a lot more features than the current MS solution.

Most of the problems arise from the limited scope envisaged for the MS in the present standard, which regards it as a temporary "in-tray", rather than a permanent repository of messages. There is a proposal [14] about extending the MS and currently ISO/SC18/WG4 is working on it. The goal is to extend the functionality of the MS itself to that of a long-term message depository. This will imply some kind of interworking between the UA, the MS and the Document Filing and Retrieval (DFR).

The danger of the current position is that the MS is *not sufficiently rich in facilities* to be an attractive base for the development of stand-alone UA's. Vendors may opt instead to co-locate their UA and MTA products and offer access to remote workstations by means of a windowing interface. This may be entirely satisfactory for users of high bandwidth LANs, but is less attractive for those utilising slower network technologies. The lack of a functionally adequate MS limits the options available for devising MHS topologies to suit differing requirements.

We agree with the conclusions reached in [14] and we supported it in the last ISO/SC18/WG4 meeting (May 1989) in Stockholm. We recommend the addition of the following facilities as extensions to the ISO/CCITT Message Store (MS):

(1) The ability to store a copy of a message upon its successful submission. This is a common feature of existing messaging facilities.

(2) The ability to store messages in multiple message *folders*. At present the stored-messages information base is a solitary, flat, sequential store of messages, with no provision for the permanent grouping of messages. The current MS model could be extended to allow the user to create named information bases to operate as additional message folders. Alternatively, a message could be considered to be a member of some folder

by virtue of its possession of the appropriate folder attribute.

(3) The ability to suspend the activity of message composition and to store a *draft message*, with the intention of completing and dispatching it on a subsequent occasion. One of the arguments originally advanced in support of the concept of the MS - that messages should reside on open systems rather than be locked away on workstation discs - applies equally to draft messages. It should be possible to begin composition of a draft message on your office workstation and later complete and dispatch it using your personal computer at home.

(4) The ability to automatically remove *(auto-purge)* expired and obsolete messages from the MS. It would be very useful to allow the UA to attach an expiry time to a message; this facility provides a simple method for the management of messages whose useful lifetime is limited. When the user first inspects a message, his UA might allow him to select one of a number of expiry time periods (one day, one week, one month) to be attached to the message.

6.4. Porting P7 into a PC

One of the main ideas of the CACTUS project is to be able to access the X.400 world from a PC. To do that, it has been necessary to implement the P7 protocol in a PC. More exactly, the Mailbox Server (MBS) part is in the CACTUS box side and the Mailbox Client (MBC) is in the PC.

The CACTUS PC software may be divided into two main parts:

- 1. The User Interface (UI), independent of protocol;
- 2. Communication dependent parts: MBC and lower layers.

The UI provides the human user with an easy-to-use and powerful environment that allows him to make use of all CACTUS features. This UI has been designed to be easily interchangeable with other distributed applications (see [15]), different from X.400 messaging. In addition to the Man-Machine Interface, the UI has a local database and a Local Name Directory.

The MBC and the lower layers (ROS, SRTS AND DLL) were developed in a UNIX environment and then ported to the PC.

With regard to the P7 protocol integration into a PC, we can distinguish two main points: First, the limitation of the PC MS-DOS Operating System resources and second, the complexity of both the P7 protocol itself and the needed lower layers.

One big drawback of PC's is the limitation of resources, i.e., memory shortage (640kBytes), no parallellism in execution etc. Thus the main problems found when porting software developed in a UNIX environment were:

- 1. Size of modules
- 2. Parallel execution
- 3. Interprocess communication.

The most important problem is perhaps the size (although its importance will decrease with time). The MBC is that part of the software in charge or performing the access to the

Message Store, by using the services given by the Remote Operations Service (ROS), the Simple Reliable Transfer Service (SRTS) and the Data Link Layer (DLL). Therefore all these entities have been integrated inside the PC, resulting in a large amount of code.

On the other hand, problems related to modules' execution (MBC, SRTS, ROS and DLL were designed as ESTELLE modules and this implies parallellism) have been easily solved because the software development has been made from an ESTELLE specification (and therefore communication and parallellism follow strict rules). As a reference information, the code size of the full ported software results in about 300 kBytes, or about half of the PC's standard memory.

The complexity of the P7 protocol has been discussed before, but this has a strong relationship to the developed software. For instance, the great number of operations defined in P7 results in a more complex software interface to use these services. However, we have tried to design an interface between the UI and the MBC as general and simple as possible.

One may notice that the software related to the P7 part in the PC may be a little complicated perhaps, because the final code was obtained via compilation of the specification in ESTELLE and LOTOS. But this is the penalty we have to pay to obtain fine and safe software derived directly from the specification.

Again, with respect to P7 problems, in the design of the User Interface we have tried to solve some of the restrictions of the P7 protocol. For example, users can create folders by using a command that permits the attachment of a "keyword" to a message or a group of messages. In this way it is possible to retrieve messages from the PC local database, grouped by folders, or keywords.

The use of a local database in the PC also helps to solve other P7 problems: sent messages are locally stored, you can handle several draft messages locally stored etc. From the users' point of view, all these features are given by the CACTUS system, the human user does not need to know about P7, he only interacts with the UI.

7. Conclusions

Although the main goal of the CACTUS project was to build a prototype, some general knowledge can be extrapolated from the experience.

The use of a development methodology based on FDTs proved to be very useful. In particular, the mixture of FDTs designed to specify behaviours (ESTELLE and LOTOS) and ASN.1, designed to specify syntaxes, permitted proper modelling of all features at any layer (taking into account the peculiarities of each layer).

Some reflections on how ASN.1 specifies the encoding/decoding of PDUs have been outlined. Some modifications are proposed to the standard in order to simplify it and make it more efficient.

The P7 protocol has been analysed at length. Some comments, errors and simplifications of the P7 protocol (specially of the ECMA version [2] on which the project based its implementation) have been shown. Some of the difficulties experienced in its implementation have been also identified (of probably general scope). Finally, some future extensions were proposed.

Acknowledgements

This work has been partly supported by the Commission of the European Communities under the ESPRIT programme, in the project CACTUS (ESPRIT-718).

References

Some CACTUS documents are not available outside the consortium. For information, contact one of the authors.

[1] "CACTUS: Opening X.400 to the low PC world", J.Saras et al, ESPRIT Conference Week 1988.

[2] ECMA, "Mailbox Service Description and Mailbox Access Protocol Specification" Standard ECMA-122, July 1987.

[3] CACTUS, "CACTUS Functional Specification", March 1987

[4] CACTUS, "CACTUS Architectural Specification", September 1987

[5] ISO, "Estelle - A Formal Description Technique Based on an Extended State Transition Model", ISO DIS 9074, July 2, 1987.

[6] ISO, "LOTOS - A Formal Description Technique Based on the Temporal Ordering of Observational Behaviour", ISO DIS 8807, July, 1986.

[7] "The SRTS experience: Using LOTOS from Requirements Capture to Implementation", Arturo Azcorra, DIT internal paper (to be published).

[8] Manros, C., "The X.400 Blue Book Companion", Technology Appraisals, 1989.

[9] CCITT, "Recommendation X.411 - Message Handling Systems: Message Transfer Layer", Fascicle VIII.7, Red Book, Malaga-Torremolinos, October 1984.

[10] CCITT/ISO, "Message Handling Systems (MHS)/Message Oriented Text Interchange System (MOTIS)", CCITT Recommendations X.400-X.420 / ISO DIS 10021 Parts 1-7, April 1988.

[11] CACTUS, "MBS Formal Specification - Part 0: Internal Refinement", September 1988.

[12] CACTUS, "MBS Formal Specification - Part 1: Mailbox Delivery (MBD)", September 1988.

[13] CACTUS, "MBS Formal Specification - Part 2: Mailbox Access (MBA)", September 1988.

[14] Shaw, S., "Required Extension to P7", ISO/IEC JTC 1/SC 18/WG 4 N1006, Expert Contribution, January 1989.

Project no 719

Testing the OSI Directory

S.E. Kille
P. Barker
A.D. Turland

Department of Computer Science
University College London
Gower Street
WC1E 6BT
UK

ABSTRACT. The OSI Directory has recently been standardised, and a number of implementations have been undertaken in the commercial and research world [CCI88]. It is clearly important to test any such implementation in a systematic manner. In future, conformance testing will meet many of these requirements, but this is not available for early implementations.

THORN is an Esprit project (719/720), which has implemented the OSI Directory [SA88], and is experimenting with its usage through a Large Scale Pilot Exercise [Kil88a]. THORN recognised the importance of testing, and so contained a substantial task to tackle this area. This paper reports on the results of this task. There are a number of aspects of testing the OSI Directory which have not been dealt with in previous work on testing of OSI Applications. This particularly relates to the distributed nature of the OSI Directory.

KEYWORDS: OSI, Directory, Testing, Conformance, Distributed System, THORN, ASN.1.

1 Introduction

The first part of the paper introduces the OSI Directory, and describes the the general OSI Conformance testing model [ISO88c]. A model for OSI Directory Testing is then developed in terms of the OSI Testing model.

This testing model is related to the Abstract Service Description of the Directory, and its refinement into Directory System Agents (DSAs). This enables testing of the overall behaviour of the OSI Directory and testing of the behaviour of individual components to correctly provide this overall behaviour. The advantages and disadvantages of this approach are then discussed.

The approaches taken to implement this model for the THORN project are then described, covering the tools used to generate, manipulate, and apply tests. It is shown how useful results were achieved without the levels of complexity and effort required for a full conformance testing suite. The problems in designing both the tools and the test suite are discussed. This is then illustrated by description of a simple test.

Finally, experience in use of the testing system is described. The THORN testing system was developed independently from the rest of the THORN Directory, which was very beneficial from two viewpoints. From within the project, a high level of confidence in many aspects of operation were achieved before putting the system to real use. Secondly, as the testing system is general and independent of THORN, it can be used to test other X.500 systems. The advantages and disadvantages of the approaches taken are analysed.

2 Testing Model

2.1 The OSI Directory

The OSI Directory is specified in the joint CCITT X.500 recommendations / ISO 9594 [CCI88] [ISO88a]. In essence, the Directory Service is a database, to which a User is given remote access using OSI. The database has certain special characteristics, which have led to it not following well known systems of database structure (e.g the relational model):

1. The Directory is intended to be capable of becoming large, and highly distributed. This is a fundamental criterion. The other characteristics can be viewed as restrictions which facilitate this goal, but are still acceptable to users of the Directory. A global relation database might be useful — unfortunately it could not be implemented at the current state of the art.

2. The data will be structured in an hierarchical manner.

3. Read access will dominate over update access

4. Following an update, temporary inconsistencies in the database are acceptable.

The Directory Service is intended for use by communication oriented applications, but also has applications in a much wider environment. This will be made clear by later examples.

It is useful to consider the Directory Service according to the Object Model. The Directory can be viewed as an object, which provides a number of services. A user accesses the directory by use of a *Directory User Agent* (DUA), which can be considered to represent the user from the viewpoint of the Directory. There is a one to one correspondence between the User and DUA. The DUA accesses the directory over an OSI Association, by use of the *Directory Access Protocol* (DAP). The DAP uses the Remote Operations Service (ROS), which allows the Directory Service to be specified cleanly as a set of operations which are functionally like procedure calls [ISO88d].

The Directory has an associated database, which is termed the *Directory Information Base* (DIB). The DAP operations allow the user to read, search, and modify the information in the DIB. From the standpoint of most applications which use the Directory Service, the basic model is sufficient. However, the Directory has internal structures and can be *refined* into a collection *Directory System Agents* (DSAs) which provide the Directory Service in a distributed manner. This structure is shown in Figure 1.

Each DSA has an associated database, which represents some portion of the DIB. The DSAs have rules for distributed operation, and will behave collectively to provide the overall service defined for the directory. The specification of this distributed operation is complex, and is not pursued in detail here. It is worth noting two basic modes of operation:

Chaining Where a DSA passes a query onto a further DSA, and then returns the answer to the DUA.

Referral Where a DSA returns a pointer to the DUA, indicating "ask the same question to that DSA".

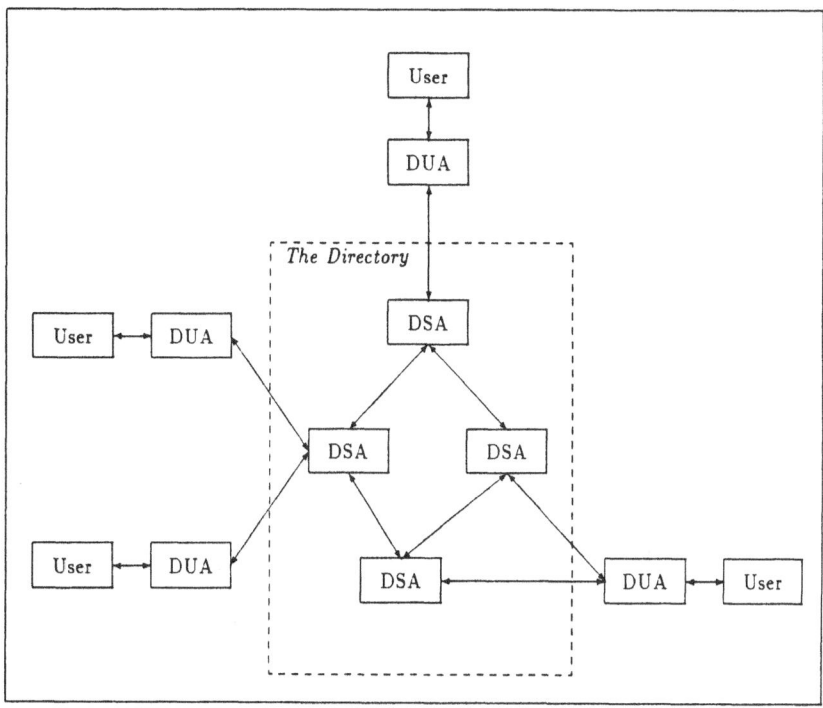

Figure 1: Distributed Directory Model

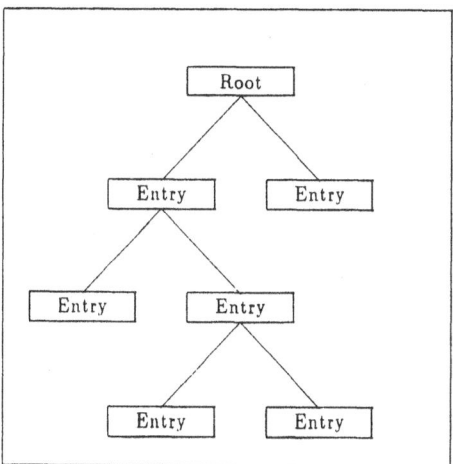

Figure 2: Overall Structure of Directory Information Tree

In the first version of the standard, only an hierarchically structured DIB is considered. When this is referred to in a manner which relies on its hierarchical nature, it is called the *Directory Information Tree* (DIT). The DIT is composed of entries which usually correspond to real world objects. For example, an entry hanging off the "root" might correspond to a Country, an entry below that to an organisation, and a "leaf" entry correspond to a person. This equivalence is important, as it is the mechanism whereby objects are related to the Directory Service. Once this relationship is established, it becomes possible to refer to the object in the context of the Directory Service. Each entry has a *Relative Distinguished Name*, which uniquely identifies it relative to its parent. Thus one can construct for each entry a *Distinguished Name*, as a sequence of Relative Distinguished Names, which uniquely identifies the entry. The Distinguished Name is in fact the only form of name defined in the Directory Service, and so is the manner in which an object is named.

2.2 OSI Conformance Testing

The work on OSI conformance testing is well developed [ISO88c]. This standard has substantial terminology, with carefully defined meanings. During the THORN work, we found that much of the vocabulary of testing has normal English meaning, and is frequently open to mis-interpretation. Therefore, we have been increasingly careful to use the terms of the standard, and to be careful about the meaning. Whilst initially offputting, this leads to a better understanding due to the improved precision. In this paper, where a standard term such as *test case* is used, it is emphasised in italics. A few terms which we use extensively are now defined informally.

IUT Implementation Under Test — the implementation which is being tested.

SUT System Under Test — an instantiation of the IUT.

PICS Protocol Implementation Conformance Statement — a definition of the capabilities of an IUT in terms of an OSI standard.

PIXIT Protocol Implementation Extra Information for Testing — Extra information on an IUT, which is useful but beyond the scope of the PICS

test purpose A description of some specific objective of testing

test case A generic, abstract, or executable test case, which realises a test purpose.

2.3 Testing the Directory

There are many ways in which a complex system system such as an OSI Directory Implementation is tested. In addition to the work described in this paper, the following types of testing were done on THORN, and would be expected for other implementations:

Component Testing The THORN implementation is built from a number of distinct modules (e.g., User Interfaces, Protocol Engine, Database). Each of the system components has its own testing facilities to ensure that it corresponds to the defined interfaces and functionality.

Installation time Tests There are some basic tests which are made when the system is installed, to see that it is available and performing simple operations. This basic type of confidence test does not require sophisticated tools.

Testing by Usage For large distributed systems of this nature, there is no substitute for operation in a "realistic" environment. One part of testing is to configure a non-trivial system and use it. The Large Scale Pilot Exercise [Kil88a], has been a significant help here.

Interoperability Testing No form of internal testing (conformance or otherwise) can ensure interoperability. Conformance testing does not per se guarantee interoperability. For this reason, interoperability testing has been seen as an important part of the THORN project.

In many cases, OSI conformance testing in conjunction with analysis of the *PICS* of two implementations can lead to non-interoperability being deducible. How far this goes, will depend on how carefully the conformance sections of the standard has been written.

Given these forms of testing, the role of the testing work described in this paper can now be explained. Attempting to implement full conformance testing with the resources available was not practical. Use of an external X.500 conformance testing laboratory could not be relied on in the timeframe of the project — development of conformance testing facilities takes many years.

The main intention is to provide systematic testing of the overall system. This has much of the *flavour* of conformance testing. The major difference is not in the testing approach or tools, but in the level of coverage provided by the *test suites* developed. In principle the tools developed would be a useful basis for conformance testing. Systematic testing is important for the following reasons:

- Manual testing may lead to believing what the system says, even though behaviour is subtly wrong.

- Boundary conditions may be exercised.

- Once tests are developed, re-application is relatively easy.

- For a large system, ad hoc testing would lead to some areas not being exercised. Whilst the approach taken to systematic testing is no less ad hoc in itself, it does provide a basis for structuring testing effort.

A related function is to provide regression testing, which can be provided with the same facilities. This is important for two reasons:

- To check that developments do not have undesirable side effects on previously working aspects.

- To verify that the system has been ported or installed correctly in a new environment.

The intention of the work was to focus on directory related aspects of testing. It was assumed that OSI layer services (Presentation, Session and below) and OSI Application Service Elements (ACSE and ROSE) used by the directory would be tested by other means. These aspects are also less interesting, as much other work has already been done on the testing of these components, for example in the CTS-WAN programme [Dwy88]. In practice, much confidence testing of these layers was achieved by the testing procedures. The general approach was to use *multi-layer testing*. This minimises the need to bind service interfaces into the testing components, and thus avoid systematic errors.

First, the distinct aspects of testing a directory are considered. Most layer services can be tested entirely in terms of protocol and services provided and used. The basic approach of OSI Conformance Testing relies on this, and *Real Effects* are excluded. It is interesting to note that an

FTAM implementation can conform in a non-trivial manner, without being able to transfer files. The state of the Directory Information Base (DIB) is a *Real Effect*. It clearly does not make sense to test a Directory without considering this! The service semantics of changes to the DIB being permanent are clearly an important aspect of the Directory, which needs to be reflected in the testing approach.

The second distinct aspect of the directory is its distributed behaviour, and the distributed operations of the directory. Each DSA contains some part of the DIB, and does not conform merely by following the protocol and the directly associated services. It must behave correctly in the light of the *knowledge* which specifies the overall configuration of the directory, and how the DSAs cooperate to provide the overall service.

The object oriented view of the directory is a useful basis for considering the testing approach. The top level view gives two components: the Directory and the DUA. It is appropriate to test both of these. The original plan proposed DUA testing, although this was not implemented in the end — enhancement of other aspects was given a higher priority. The approach to DUA testing was to use a *distributed test method*. This meant that the tester would replace the directory, and the *Abstract Service Primitives* would be communicated to the DUA manually by a user following a testing script. Unlike the other tests described, this would not have been suitable for automation to provide simple regression testing.

Testing the directory is done through the Directory Abstract Service by a *remote test method*. An alternative approach would have been to use a *local test method* in conjunction with a *lower tester*. The former was chosen, as it decreases interface dependence, and does not require any changes to the DSA or tight coupling with testing tool components. It also allows for the tests to be used in two distinct ways:

1. Where the Directory is provided by a single DSA (a recognised OSI Directory Conformance Class), the tests can be used directly to verify conformance of that DSA, as the tests are considered to relate to the Directory Abstract Service, as well as the Directory Access Protocol. A given test is relative to a DIB which is the current state contained in the DSA.

2. The same test suite can be used to test a distributed directory, to verify that it provides the Directory Abstract Service. If distributed operation is performed only by use of Chaining, then this can be done transparently — the *System Under Test* (SUT) should behave externally in the same manner as a single DSA. If DSA Referral is used, then the *tester* must be able to follow these referrals, and similarly the distributed nature of the testing must be reflected in the *test cases*.

The DSA Abstract Service and Directory System Protocol are tested in a manner analogous to the Directory Abstract Service. Here, the issues of distribution are inherent to all but the most basic tests.

Two *test suites* were developed, one each for the Directory Service and DSA Abstract Service. Sets of related *test cases* were collected together in *test groups* to facilitate management of the overall suite. Each test needs to be applied relative to a fixed DIB state, and so there is always an optional *preamble* to load the requisite initial state. To provide for regression testing, it is desirable for tests to be idempotent. For this reason, each test has a *postamble* to ensure that the final state is the same as the initial one. This facilitates *test groups* to use the same base state.

3 Design and Implementation

This section describes the testing tools which were implemented, and also a number of interesting features of the design which would have been implemented given more time. The key feature is the design of the *test case*. We chose to use an *executable test case* as the basic specification mechanism. The overhead of using a notation such as TTCN (Tree and Tabular Combined Notation) [ISO89], and then manually translating to an executable form would have been to high. The *test case* is an ASN.1 encoded file, which contains all of the information needed to apply and analyse the *test case*.

Each *test case* contains *PICS* (Protocol Implementation Conformance Statement) information on features required. This can then be matched against the *PICS* of the *IUT* (Implementation Under Test), in order to determine whether the *test case* is appropriate. The *PICS* is not yet standardised, and so we defined information as needed. The major component is a specification of which Attribute Types and Object Classes are required. When the *test case* is applied by the test application tool (called TAPP), the *PICS* and *PIXIT* (Protocol Implementation eXtra Information for Testing) information about the *IUT* is supplied.

Each test consists of a set of OPDUs (Operation Protocol Data Units) which are supplied to ROSE (Remote Operations Service Element) [ISO88d], or ACSE (Association Control Service Element) [ISO88b]. The rules of ROSE, ACSE, and ASN.1 basic encoding rules are built into the test applier and will always be followed, as these are not the subject of testing[1]. The OPDUs come in groups, with information on relative timing of the groups:

- Bind, to define a new association within the test.

- Unbind, to release a specified association.

- Operations, giving the operation type, invoke id, and which association is used.

In principal, this approach is entirely symmetrical. However, we have only implemented the case where TAPP is the initiator of associations and the invoker of operations. This is the most common usage, and the most important situation to test. Each of these groups has the following information:

- An OPDU to be sent by the test applier.

- One or more valid OPDUs which may be returned.

- Error conditions which are conformant, but would lead to an *inconclusive verdict*. For example, returning a service error "DSA Busy".

ROSE OPDUs which are sent by TAPP are simply treated as raw ASN.1. This allows for both legal and illegal OPDUs to be sent. This is important for testing that plausible but illegal OPDUs are dealt with correctly. The OPDUs used for the analysis of information returned by the Directory are more interesting. These need to be analysed at a high level. It is useful to consider why various lower levels of equivalence are not useful.

- An exact byte equivalence is not useful, as ASN.1 may be encoded in more than one way.

---

[1] This is not strictly true, as the *test case* allows specification of some variance in the ASN.1 encoding

- A basic ASN.1 abstract equivalence is also not useful, as the directory applies additional rules. For example, some string attributes should be compared in a case independent manner.

- A directory abstract syntax equivalence is not sufficient, as some aspect s are non-deterministic. For example, if there is a size limit, it is not specified which entries are returned.

This implies that the test analysis needs to be done on the basis of significant knowledge as to how the directory works. This "semantic checking" built into TAPP has proved to be most useful.

There is clearly a need to be able to generate tests in a straightforward manner. An initial command driven interface was not successful, due to the excessively large number of parameters needed. Therefore, a procedure was derived to map between the ASN.1 and text form. This text form can be manipulated easily by use of a normal editor. It was also hoped to be able to automatically develop a range of tests by use of a reference directory (e.g., generate a given test, varying parameters over given ranges), but this was not done.

Recording the behaviour during the application of a test is also an important element of the testing process. A test format for this record was chosen with the degree of verbosity selectable at run time. TAPP produces a "dti" (directory test information) file corresponding to the test file which it applies. At its most verbose the "dti" file records a structured ASCII representation of the PDUs sent and received interleaved with statements recording the progress of the application of the test. Such a record of the application of a test contains information which can be useful in assessing the circumstances of a test failure and may provide some indications of the reasons for failure. It was straightforward to introduce time-stamping into the account of test application providing an initial basis for monitoring the performance of the *IUT*.

A final component of testing was a mechanism to bulk load data into the directory. Whilst also a very useful tool for operating a directory, this sort of tool is critical for testing. There is a need to load in a reference database prior to tests, in a clean and efficient manner. Text formats were developed for all DIB information, and in particular for a range of attribute syntaxes, which were also used for the *test cases*. The ability to represent commonly used attributes in a convenient fashion proved very useful. This applied to both standard attributes and frequently used system specific attributes such as Access Control Lists.

A *PICS* for the *IUT* must be generated. Each *test case* also includes a *PICS* and these are used within the test applier to judge the applicability of a test to the given *IUT*. (There is no point in sending ModifyRDN operation tests to an X.500 implementation which does not support ModifyRDN operations). A tool to support the creation of an ASN.1 encoded form of a project defined *PICS* structure is included in the testing tools. The *IUT PIXIT* is subsumed into the tailoring/invocation of the test applier, as information such as the presentation address of the *IUT* is easiest to deal with dynamically.

Generating tests is accomplished by editing a text format specification of the test and then encoding it. *Test cases* are gathered into *test groups*, which in turn form entire *test suites*. As one of a number of *test groups*, it is sensible to include one which tests the ability of the directory to respond to erroneous (as opposed to illegal) requests with the appropriate error.

4 Example

This section attempts to highlight some of the elements of testing an implementation using the THORN testing tools. The steps which have to be taken in order to test an implementation are

```
testId:test201N;
testPurpose:modify the rdn of a non-leaf (DSP ModifyRDN);
objectInTestMode:no;
asn1Style:cen-cenelec;
bulkLoadFile:;
```

Figure 3: Test — basic parameters

briefly described by use of a simple example. In practice, many tests are much larger and more complex than this one.

The example considered is that of attempting to perform a ModifyRDN operation on a non-leaf entry in the DIT. The construction of the test illustrates the importance of developing the DIT held by the *IUT* at the same time as developing the tests. In order to test the attempt to ModifyRDN operation on a non-leaf entry it is necessary to know that the entry referred to exists in the DIT and is in fact a non-leaf entry. With this in mind the availability of suitable data and the loading of the *IUT* with this data is assumed.

An example text format test is given in Figures 3–11. This would be represented in a single file, but is broken into separate figures for clarity here. This text format would be used to generate the ASN.1 encoded test. The test proceeds as follows:

Figure 3 A testId is assigned to the test. This test id is useful when structuring individual *test cases* into *test groups* and *test suites*. The purpose of the test is described. Several parameters are set for TAPP, such as the style of ASN1 encoding is to be used.

Figure 4 The test-case *PICS* is described. In this example the object classes and attributes which need to be supported for this test to be applicable have been left undefined.

Figure 5 Before a directory operation event can be applied a directory bind event must have occurred, in this case the test is intended as a test of the Directory Service protocol and this is reflected in the context and syntaxes specified. This generates a DSP association with handle 1.

Figure 6 The main point of the test is the ChainedModifyRDN operation specification. The DIT held by the *IUT* has been constructed so that the named directory entry is a non-leaf entry. Specification of the ChainedModifyRDN event involves a number of aspects. The chaining arguments are shown in this figure.

Figure 7 The operation itself is described with just three parameters: the old name, the new name, and a single flag.

Figure 8 A list of service controls, common to all operations is specified.

Figure 9 Then, a few fields indicate how the operation is controlled. In particular, a set of errors which are conformant, but would lead to an *inconclusive verdict* are specified.

Figure 10 This figure shows the outcome expected for the test — an error should be returned.

Figure 11 The final event is to unbind from the directory.

```
testObj:directory;
setObjClass:;
setAttType:;
setAttSynt:;
supportRead:yes;
supportComp:yes;
supportAban:yes;
supportList:yes;
supportSearch:yes;
supportAddEntry:yes;
supportRemEntry:yes;
supportModEntry:yes;
supportModRDN:yes;
secSimple:yes;
secStrong:no;
secExternal:no;
accessControl:none;
```

Figure 4: Test: PICS parameters

```
initBind:;
handle:1;
context:directory directorySystemAC;
abstractSyntaxes:directory directorySystemAS;
callingAdd:;
credName:;
credPw:;
versions:1988;
```

Figure 5: Test: DSP Bind

```
initOp:;
handle:1;
invokeID:102;
dspModifyRDNOp:;
originator:c=gb@o=Test Org@ou=Test OrgUnit@cn=Test Person;
targetObject:c=gb@o=Test Org;
nameResPhase:Not started;
nextRDNToBeRes:0;
aliasdereferenced:no;
aliasedRDNs:;
returncrossrefs:no;
referencetype:Superior;
```

Figure 6: Test: Directory System Protocol Parameters

```
objName:c=gb@o=Test Org;
newRDN:o=Test Organ;
deleteRDN:no;
```

Figure 7: Test: The ModifyRDN Operation

```
preferChaining:yes;
chainingProhib:no;
localScope:no;
dontUseCopy:no;
dontDerefAliases:no;
priority:Medium;
timeLimit:;
sizeLimit:;
scopeOfRef:None;
requestorDN:;
nameResPhase:Not started;
nextRDNToBeRes:0;
aliasedRDNs:;
```

Figure 8: Test: Service Controls

```
maxDelay:0;
serviceError:no;
securityError:no;
partialOutcome:no;
```

Figure 9: Test: Operation Analysis Parameters

```
opError:;
updateError:;
problem:NotOnNonLeaf;
```

Figure 10: Test: Expected Outcome

```
initUnbind:;
handle:1;
```

Figure 11: Test: Unbind

Having generated one such specification similar and related variations on the theme can be quickly generated by editing the initial *test case*. For example, a systematic approach to testing might suggest that every test where it is possible for an alias to be appear in a directory name should be tested both with and without an alias in each relevant name. The systematic approach is easily supported by a text format such as the above and suitable structuring of the DIT generated to support the tests.

5 Experience

An early decision was to keep the coupling between the testing tools and the rest of the THORN implementation as weak as possible. This proved to be a good decision, as integrations of new releases of most components did not affect the testing tools. This is particularly important for regression testing, which also helped to eliminate systematic errors. In many aspects, the testing tools therefore had to reimplement many parts of the directory. To prevent the testing component becoming larger than the rest of the project, many parts of the testing tools were derived from QUIPU, which is an openly available implementation of the OSI Directory from UCL originally developed as a part of the Esprit INCA project [Kil88b]. The interface was done at the ROS/ACSE level (the so called THORN IROS interface). This is a relatively simple interface compared to those above and below it (Session and DUA procedural interface), and so minimised the coupling.

The tests and testing tools were used throughout the integration phase of the development, which surprised and pleased us! We had expected that the major use would only be at the final stages of checking out. The separation of the testing tools from other aspects of the implementation was a major reason for this. It was able to expose many errors in the implementation. Interestingly the tools were also able to expose problems in the implementation from which the tools were derived, due to the semantic checking. This feature, which was viewed as not so important initially, proved crucial for the success of the tools.

The boundary between testing and debugging can be seen to concern the amount of information made available about the circumstances of failure: in retrospect a greater emphasis on generating such information would have provided greater utility at little extra cost. Designing the presentation of an account of test application should be an important part of the overall design.

Generating *test cases*, and designing large test suites is hard work. The importance of first-class editing tools is high, due to the complexity of even the simplest test, such as the example given in the previous section. Once provided, the text form allowed for simple manipulation of tests to operate on a range of related problems. We generated several hundred tests for a range of problems. This covered all operations and basic error conditions in both the Directory Abstract Service, DSA Abstract Service, and the associated protocols (DAP and DSP). Whilst use of these tests suites did not guarantee conformance, it gave confidence that a wide part of the specifications were implemented correctly. Tests were also developed for THORN specific aspects, such as Access Control, Schema, and Distributed Operation management. It was important that the tools could be used for general and system specific purposes. Tests were also developed for illegal OPDUs, which facilitated "bullet proofing" the DSAs. There was also a range of tests utilising the capabilities for multiple associations, which showed that the DSA could operate correctly in non-trivial conditions.

Experience with the implementation of testing tools and the generation of *test cases* strongly emphasises the importance of the relationships between service and protocol on one hand and *Real Effects* on the other. The relationships between the data and knowledge held by DSAs and the behaviour of DSAs in processing an operation can be subtle and complex. A combination of approaches could be studied in future, including the generation of expected results (used for semantic checking) from a DIT description and the operations specified in a *test case* and the generation of a suitable DIT from the operations and expected results specified in a *test suite*.

The one aspect which was not tested as much as would have been liked were some of the aspects related to distributed operations. These aspects are currently highly system dependent, and so specification of tests is hard. The multiple association and attribute manipulation facilities allowed some tests to be developed, and it is clear that more complex tests can be developed by use of the existing tools. Further work could usefully concentrate on this aspect.

6   Conclusions

We have described how a set of testing tools and tests was designed and built for use with the THORN directory, and describes usage to test Directory Abstract Service, DSA Abstract Service, the two protocols, and distributed operations. This has shown some of the issues that will arise when performing testing of a directory system, and suggests a some approaches which might be taken for conformance testing systems. The use of executable *test cases* which allow multiple associations to be managed is a useful basis for testing such distributed systems. We believe that this work has greatly facilitated the progress and quality of the THORN implementation.

References

[CCI88] The directory - overview of concepts, models and services, December 1988. CCITT X.500 Series Recommendations.

[Dwy88] D. Dwyer. Conformance testing services — wide area network. In *International Open Systems 88*. Online, April 1988.

[ISO88a] ISO. The directory - overview of concepts, models and services, April 1988. ISO DIS 9594.

[ISO88b] Information Processing Systems — Open Systems Interconnection — Service Definition for the Association Control Service Element, April 1988. ISO/IEC Standard 8649/CCITT Recommendation X.217.

[ISO88c] OSI conformance testing methodology and framework, July 1988. ISO/IEC JTC 1/SC21.

[ISO88d] Remote operations: Model, notation and service definition, December 1988. ISO/CCITT.

[ISO89] The tree and tabular combined notation (TTCN), February 1989. DP 9646-3 from Sydney, ISO/IEC JTC 1/SC21.

[Kil88a] S.E. Kille. Experience with the use of THORN. In *Esprit Conference Week*. North Holland Publishing, November 1988.

[Kil88b] S.E. Kille. The QUIPU directory service. In *IFIP WG 6.5 Conference on Message Handling Systems and Distributed Applications*. North Holland Publishing, October 1988.

[SA88] F. Sirovich and M. Antonellini. The THORN X.500 distributed directory environment. In *Esprit Conference Week*, November 1988.

The EUREKA COSINE Project - Status Report

Mr. Andrew J. HILL and Mr. Nicholas K. NEWMAN
Commission of the European Communities, Directorate General
Telecommunications, Information Industries and Innovation
Project Officer and Secretary, COSINE Policy Group
200, rue de la Loi, B-1049 Brussels - Belgium

Abstract

The aims, progress and current status of the EUREKA COSINE project are described. COSINE means "Cooperation for Open Systems Interconnection Networking in Europe" and is a "EUREKA project to provide a computer communications infrastructure for European researchers, and thereby also for EUREKA and other programmes".

Following a Specification Phase carried out by the RARE (Réseaux associés pour la recherche européenne) association of users and providers of computer networks for researchers, the project is now entering its Implementation Phase. A COSINE Project Management Unit (CPMU) is being established for the Implementation Phase and the first implementation activities have started.

The paper describes the technical areas addressed by COSINE, the services which it will provide, and the progress so far. It gives an overview of the principles by which the project will operate during the Implementation Phase, with particular reference to the project's federative approach.

1. Introduction

COSINE means "Cooperation for Open Systems Interconnection Networking in Europe". It is a EUREKA project being carried out by all of the countries participating in the EUREKA programme, and the Commission of the European Communities (termed 'the Commission' in the rest of this paper), to establish an open pan-European computer communications infrastructure, conforming to international technical standards. It aims to satisfy the diverse requirements of industrial, academic and state-run research establishments.

EUREKA is a major programme for industrial development projects in Europe, supported by the European Economic Community (EEC) countries, the European Free Trade Association (EFTA) countries, Turkey and the Commission of the European Communities.

The birth of COSINE was largely dictated by the ever increasing amount of international collaboration in research, and the difficulties of computer communications at international level.

The initial services envisaged for COSINE are those for which Open Systems Interconnection (OSI) standards and products are available, or which are so urgent and important that they need to be set up even before the relevant standards have stabilised. These are:

- interactive access to remote computers and data-bases;
- message handling and group communications systems and services;
- File Transfer, Access and Management (FTAM) mechanisms and services;

- directories and name-servers;
- network management.

In addition, because many of the interconnections of national packet-switched networks have proven inadequate in quality and bandwidth, COSINE is to establish a pilot pan-European X.25 Backbone infrastructure, in cooperation with the telecommunications administrations.

The first steps towards the implementation of these services are now under way, in the context of a Memorandum of Understanding between the Commission (acting on behalf of COSINE) and RARE. As relevant standards become stable, further services will be added, such as:

- computer job transfer and management (JTM) mechanisms;
- virtual terminal access;
- graphics transfer;

Surrounding and supporting all these services, there is a need to ensure adequate confidentiality and integrity of data, and also accounting and charging mechanisms.

COSINE will create a mature environment within which open services can be made available and can evolve in autonomous fashion, responding to the needs of the users with well-supported OSI-based products. It is planned that government intervention will cease once the COSINE services are established, at which time it will no longer exist as a project, but will be self-supporting, with financing coming from its users.

2. European Community Activities

Europe has been highly active in the area of open computer communications since the late nineteen-sixties and early seventies. For example, the PTTs, with encouragement from the Commission, set up EURONET [1], a Europe-wide packet-switched network used mainly for remote interrogation of data bases of largely scientific and technical information. Today, EURONET has been replaced by the inter-connected national packet-switched networks, but the related Commission-sponsored information service, the DIANE system, now numbers some 870 data bases on over 80 host machines throughout Europe, with a master index of data-bases and general information being kept on the Commission's "European Communities Host Organisation" or ECHO machine in Luxembourg.

Following a few isolated actions and smaller programmes, international collaboration in research in Europe was greatly increased from 1983 through the European Community's ESPRIT (European Strategic Programme for Research and Development in Information Technology) programme.

Projects in such programmes are carried out by teams in several countries, often involving industry-academic cooperation, and sometimes including countries outside the twelve member states of the European Community itself.

When ESPRIT was started, plans were also made to establish an Information Exchange System (IES) [2] to cater for the computer communication needs of widely dispersed research teams. Because it is impossible and inappropriate to impose any given proprietary communications architecture on all ESPRIT participants, who use practically every type of computing equipment available, the Commission's policy is to recommend the use of OSI standards.

3. RARE

The RARE [3] (Réseaux Associés pour la Recherche Européenne) association of users and providers of computer networks for researchers was founded following the first European "Networkshop" hosted by the Commission in Luxembourg in May 1985.

RARE aims to ensure a modern, open computer communications environment and services for researchers throughout Europe, conforming to OSI standards for the same reasons of dispersion and heterogeneity of makes and models of computer as in the ESPRIT IES.

RARE also aims to use publicly available data-conveyance networks, and to act as a forum for the exchange of information, experience and mutual help. RARE reflects the current research networking situation in Europe. In most countries, there are one or more national networks. International communication is being provided through networks such as EARN or EUNET, or operated through the high energy physics community for example. All countries contributing to COSINE are members of RARE, with a special status for the Commission in recognition of its catalytic role in European computer networking. Turkey also intends to join. In addition, there are international members (mainly user organisations and networks), associated national members in other parts of the world, and liaison members from related fields or bodies.

4. EUREKA COSINE

The EUREKA programme covers a broad spectrum of topics, including biotechnology, the environment, information technology, robotics, materials science, transport etc... Projects are proposed, generally by international consortia with 2 or more members, and if accepted, they are given the "EUREKA label".

COSINE was accepted at the second EUREKA Ministerial Conference in Hannover in November 1985. Its aims were defined as the creation of a pan-European environment for computer communications for researchers, conforming to OSI standards, using publicly available data-conveyance networks, and using supported industrial products, thus leading to a "market pull" effect.

RARE proposed a project plan to the newly constituted COSINE Policy Group in June 1986. This was followed by the first COSINE Workshop, hosted by the Commission in Brussels in November 1986, during which RARE presented their plans to a wider audience and the initial steps were taken towards the management, financing and organisation of COSINE. One of the first steps was the creation of the COSINE Policy Group's Bureau, consisting of its Chairman, two Vice-Chairmen, Secretary and Project Officer, with the Secretariat and Project Officer being provided by the Commission. Figure 1 indicates the relationships of the organisations involved in the COSINE management.

5. The Phases of COSINE

COSINE is being developed with a phased approach. During its Preparatory Phase, the contractual and funding aspects were refined, and a precise workplan was created for the Specification Phase.

The Specification Phase started in July 1987. RARE produced, either through its own membership or through sub-contracts with outside firms, over forty technical and organisational reportscontaining COSINE specifications. From this material, eight deliverable reports were produced [4]. All these reports are in the public domain. In addition, a series of specially formed task forces defined the management, organisation and sub-projects of the subsequent Implementation Phase.

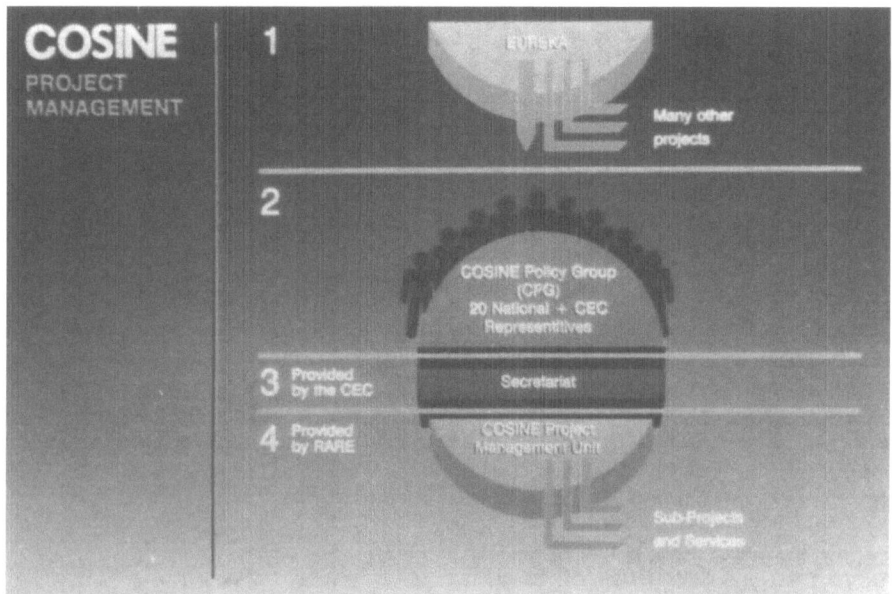

Figure 1. Organisations involved in the management of COSINE

Also during the Specification Phase, national plans for the creation or further development of OSI networks, or for transition to conformity with OSI, were produced for each of the COSINE countries by the COSINE Policy Group members or their delegates. The production of these national plans, which is co-ordinated by the Commission of the European Co}munities' PrOJect Officer allocated to COSINE, has already resulted in a vastly increased awareness of the issues involved, and a cohesion right across Europe is beginning to appear. Figure 2 illustrates some of the national networks which are planned to be linked to the COSINE service.

After a short Transition Phase to make organisational preparations, the project is now entering the three-year Implementation Phase. Under a Memorandum of Understanding, the Commission (acting on behalf of the COSINE Policy Group) and RARE have decided to take the first implementation steps, namely the establishment of the CPMU and the creation of detailed workplans for the most urgent implementation activities. Provisions have been made to ensure that certain urgent actions such as the setting up of a pilot X.25 Backbone infrastructure can be undertaken directly, while other projects are being further defined.

6. Federative Approach

One of the most important principles of COSINE is its federative approach. Europe's computer communications depend on a large majority

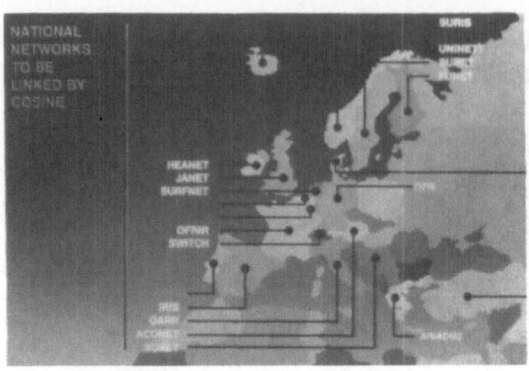

Figure 2. National Networks

of computer centres and isolated users communicating in the same standardised way, with equipment and software procured according to common specifications. This can only be achieved by a high degree of national and international cohesion and harmonisation. It is because of this that a federative approach is being taken. Indeed, the whole idea of OSI, i.e. of distributed networking, points to such a federative approach. Of course, coordination at an international level is also needed on a permanent basis to bring about such a harmonised networking environment.

Therefore, there are three levels of activity:

– international level;
– national level, within the general international framework;
– local level, with computer centre managers, procurement agencies, research departments and individual users.

It is only through a coordinated approach throughout these three levels that the intended "market pull" effect will be felt, as only then will potential suppliers perceive a relatively uniform market, both in space (elimination of unnecessary variations between countries or regions, and hence different versions of essentially the same products) and in time (transition towards OSI undertaken more or less in step throughout the community).

Apart from setting up international services such as directory and information services, one of the major tasks of the Implementation Phase will be the creation of common procurement specifications useable in all COSINE countries, and hopefully elsewhere. Alignment of these specifications with those emerging in the area of government administration, or in company areas different from R&D such as design or manufacturing, is essential and will be sought.

7. Results of Specification Phase

The overall conclusions from the Specification Phase are that international OSI standards, as further clarified by functional standards, are sufficient as a basis for the services foreseen over the coming years.

The major points from the deliverable reports are as follows:

Scope of COSINE

The COSINE Community is estimated at 500,000 industrial, academic and public sector researchers. They are served today by many different suppliers and operating systems. The basic services required are electronic mail, file transfer, remote terminal access and remote job entry. There is a large pent-up demand for services, and traffic is growing rapidly.

There are considerable differences between countries at the detailed level, both in number of researchers as a proportion of the population and in approaches to the management and funding of research centres.

Public Services

National public packet-switched data networks have been installed, or are in an advanced stage of installation, in nearly all COSINE countries. However, they generally conform to X.25 (1980), with a wide spread of options, and need to be upgraded to X.25 (1984). Also, the reliability and bandwidth of the international links need to be improved and monitored.

While national ISDN services are now being set up, they are not yet harmonised across Europe, so they are unlikely to be used by the international research community for some time. Several telecommunications administrations are setting up X.400 messaging services, but there is as yet no coherent naming strategy across Europe.

Protocol Profiles

The service requirements were evaluated, in terms of the European Development Norms, for:

- remote terminal access based on X3, X.28 and X.29 (XXX),
- file transfer based on ISO FTAM, where the currently available European Development Norm with Limited File Management and Recovery and Restart is adequate for an initial service, and:
- electronic mail based on X.400 (84), with an upgrade to X.400 (88) as soon as possible, and attention to user interfaces.

Future Services: These are:

- full screen services, for which it is recommended that pilot projects be defined on the basis of the ISO Virtual Terminal Protocol (Basic Class) with its second draft addendum;
- job transfer and manipulation, for which it is recommended that FTAM should be used pending the availability of adequate JTM standards;
- higher speed services, where a need has been identified for a 2 Mbit/sec. pan-European network by the end of 1989, with a 34 Mbit/sec. network to follow about three years later.

Although standards for directory systems are only now stabilising, these and network management are discussed under "Operational Aspects" below.

Migration Strategies

Great care is required with migration of any service to use of a new generation of system, in order to disturb the user and the service itself as little as possible, and in the case of computer communications it is a particularly complex matter. Considerable emphasis is placed on it in the COSINE Specification Phase.

Two main cases can be distinguished:

- transition from a proprietary implementation or interim standard to OSI;
- transition from one generation of an OSI standard to another, as from X.25 (1980) to X.25 (1984), from X.400 (1984) to X.400 (1988) etc.

In either case, it is often necessary to set up a new service in parallel with the older one, and move user installations across to the new one in a controlled manner, while maintaining converters between the two versions of the service so as to ensure continued connectivity.

Operational Aspects

While the existing functional standards cover actual services, there are few if any in the area of operational support. Specific recommendations are made concerning:

- user support and the establishment of an International Information Service;
- directories, naming and addressing, which involve other communities and service providers, and where a pilot directory service, probably based on the ESPRIT IES project THORN (THe Obviously Required Name Server), is proposed;
- conformance and operability, where most of the services required by COSINE can be met by the CTS (Conformance Testing Service) [5] Programme of the Commission;
- security and authentication;
- accounting and charging, where accounting is required not only as a basis for charging but also as the basis for capacity planning, and the detection of (often involuntary) misuse of a network; charging strategies will vary according to national and local policies, ranging from "the end-user pays for every byte" to government subvention at the national level. These issues become extremely complex when data transfer across gateways and converters is involved;
- management of services, where facilities managers need information, training, resources and procurement specifications, and need to apply the OSI Management standards;
- protocol converters, which are essential to the implementation of the migration strategies outlined above;
- network matters and LAN/WAN interworking.

8. The Implementation Phase

During the Implementation Phase, a number of operational tasks will be carried out, services mounted, and projects undertaken, at all three levels of the federative structure outlined in paragraph 6 above. In order to perform the detailed management of these activities, a CPMU is now being established by RARE.

The main tasks of the CPMU are:

T1:   to maintain, and update as required, the workplan;
T2&T3: to define and manage projects and services;
T4:   to maintain, and extend where necessary, the COSINE specifications;
T5:   to make sure the international dimension of COSINE is fully reflected at nati
           level, through harmonisation, information, documentation, training, liai-
           son, and promotion activities;
T6:   to encourage and aid the transition of existing networks to the use of OSI st
T7:   to manage the human, financial and other resources of the CPMU.

The projects envisaged for the Implementation Phase are:

P1:   pilot gateway services to the USA, for FTAM and remote access to computin
           vices, prior to transition to operational services;
P2:   pilot international directory, support and information services;
P3:   pilot activities to support international user groups (e.g. mathematicians, hig
           ergy physicists);
P4:   pilot transitions to use of OSI by existing networks or user groups;
P5:   development of tools and techniques to aid the transition to OSI, such as pr
           converters, diagnostic packages and accounting mechanisms;
P6:   pilot demonstrations of interworking of multi-vendor products conforming to
           SINE specifications;
P7:   pilot procurement exercises based on COSINE specifications;
P8:   development and proving of security mechanisms;
P9:   investigations into future facilities such as use of OSI over basic and primary
           ISDN, full screen terminal services, etc..;
P10:   LAN-WAN interworking.

The Services intended to be provided during the Implementation Phase are as follows:

S1:   provision of a pilot X.25 Backbone infrastructure;
S2:   message handling services, based on use of X.400;
S3:   information services, including the provision and updating of an OSI user ha
           up-to-date catalogues of OSI products, and COSINE-related news;
S4:   gateways to North America, in particular for file, message and news transfer;
S5:   international directories;
S6:   security key management.

An initial core set of projects and services has been identified for early implementation, comprising S1, S2, P2, P3, P4, P8, the CPMU and part of P1 (pilot gateway services to N. America for FTAM).

In addition, it is intended to carry out market research to determine how COSINE may give the greatest benefit to its users. Further tasks, projects and services may be added as the project progresses, in the framework of an annual rolling plan. So as to have the maximum effect, a number of operational targets are being defined, concerning the penetration of COSINE within the user community, the degree of progress of the underlying infrastructure etc..

9. Conclusion

It is now (August 1989) just over three years since the first meeting of what was to become the COSINE Policy Group. During that time, a great deal has been accomplished in research networking.

Since that time, RARE has been built up and has completed the technical work of the COSINE Specification Phase, thereby validating the work in the standards field against actual user requirements, and creating a pan-European consensus on requirements for research networking.

Further, twenty countries and the Commission of the European Communities have come together at government level to map out a strategy for computer communications for researchers, and have begun to apply it, thereby leading to enhanced cooperation between and within widely dispersed research teams in industry and academia. A pilot pan-European X.25 Backbone infrastructure is to be established to support the COSINE services.

Contacts have been made with further industrial users, particularly those represented in user organisations or participating in EUREKA or ESPRIT projects. The European IT supply industry has been approached and is defining the way it can contribute.

Formal liaison has been established with EWOS (the European Workshop on Open Systems), and with North American networking organisations through the CCIRN (the Coordinating Committee for Inter-continental Research Networking).

With the establishment of the CPMU by RARE, the first activities for the three-year Implementation Phase of COSINE are now under way.

A federative structure has been defined for COSINE, which creates an overall framework for the creation of OSI-based networks, or their transition to use of OSI, in all member countries of COSINE, with an aggregate budget an order of magnitude greater than the budget needed at the international level (several tens of millions of ECUs), thus having a maximum "gearing effect". COSINE will make a major contribution not only to the efficiency of European research, but also to open communications and vendor-independence for many other user sectors.

10. Ackowledgements

The authors wish to thank their colleagues, who made many helpful comments during the preparation of this paper; the RARE project team who pulled together the results of the Specification Phase; the Chairman of the COSINE Policy Group, Dr. Peter Tindemans, for his drive and enthusiasm throughout the project; and the members of the COSINE Policy Group's Bureau and Task Forces, who did so much for the preparation of the COSINE Implementation Phase.

References

[1] P.T.F. Kelly: Interconnection of national packet-switched data networks in Europe and their impact on new services, ICCC, London 1982 (North Holland).

[2] M.S. Audoux, H.J. Helms, C.J. Horn, N.K. Newman, J.F. Renuart: European Computer Communications Policies in the Research Area and the ESPRIT Information Exchange System, ICCC, Munich 1986 (North Holland).

[3] RARE Information Bulletin, 16 November 1988, RIP (88)5, RARE Secretariat, P.O. Box 41882, NL-1009DB Amsterdam.

[4] RARE: COSINE, EUREKA Project 8, Specification Phase, RARE, Amsterdam 1988.

[5] Fact sheets on Conformance Testing Services published by the Commission DG XIII E4 in 1988.

BASIC RESEARCH

TOWARDS AN INTEGRATED THEORY FOR DESIGN, PRODUCTION AND PRODUCTION MANAGEMENT OF COMPLEX, ONE OF A KIND PRODUCTS IN THE FACTORY OF THE FUTURE

J.C. Wortmann
Eindhoven, University of Technology
Department of Industrial Engineering
Pav. D16
P.O. Box 513
5600 MB Eindhoven, The Netherlands
Telephone: (31)40-472290
Telefax: (31)40-451275

Summary

The ultimate goal of the contracted research is to obtain a designer's workbench for the development of CIM in production systems. Such a designer's workbench requires a method to describe operations in a production system. These operations include product design, tendering, logistics, quality control, etc.

There are many methods for describing the operations in production systems, but these methods are all fragmented in nature. In fact, they are based on a set of fragmented theoretical notions.

For several reasons, the research is restricted to production systems providing the market with one-of-a-kind products. However, a generalization of results to other production systems is strived at.

The action consists of 5 work packages, apart from project management. In the first package, current theory for describing operations in production systems is examined, ordered, and described. Four real world production systems will be analyzed. At the time of writing of this status report, about 50% of workpackage 1 has been realized. A comprehensive questionnaire for describing production systems has been compiled. A perspective on synthesizing various viewpoints is emerging. There is clearly a need for a typology of one-of-a-kind production. A first version of such a typology is presented in the paper.

1. Introduction

The present approaches to CIM may deserve respect from a pragmatic point of view, but these approaches are not based on a thorough theoretic foundation. Presently, such a thorough foundation for CIM does not exist, because current theory is highly fragmented.

The fragmentation of theory is twofold in nature. Firstly, there is a fragmentation of production phases. Design, process planning, logistics and manufacturing have all become "computer-aided". CIM denotes the effort to cross the boundaries between these phases. The question, whether the boundaries are always appropriate, remains to be investigated.

Secondly, there is a fragmentation in views on CIM. A communication-oriented view differs from an organizational view. These views are rooted in different theoretical frameworks. It is currently not clear whether these views overlap or contradict, nor whether all important aspects of reality are covered. The major part of available theory is based on questionable assumptions such as:

- complete and reliable information is available before design or production operations take place
- CIM-components are installed once and for always
- boundaries between such operations as product design, process planning, manufacturing, and logistics are fixed and for all companies identical.

The research to be performed here tries to unify these fragmented theoretical parts. The successive production phases can be described in a unified way, if it is acknowledged that:

- each phase reduces uncertainty about the final production output
- each phase may include design activities, process planning activities, logistics activities, etc.

In order to emphasize this unified nature of our approach to various production phases, we will use the term "operations" for such phases. The integration and coordination of different phases is called "operations management". In our project, different views will be unified by multidisciplinary teams comparing these views in order to reveal contradictions, overlap, complementary insights or unexplored aspects of reality.

The effort to unify and expand the existing body of knowledge is carried out in the first two work packages of the project. The first work package describes the relevant views for operations management. Each view is applied to all operations of at least two different CIM-systems. Presumably four CIM-systems are described in this way. This results in a coherent description of current theory. The second package synthesizes the different views to a unified approach. This leads to a systematic way to describe existing or future production systems. This systematic way of describing is called "the conceptual model". It is to be expected, that the conceptual model, although unified, will be multidisciplinary in nature.

Although the ultimate aim is to develop a general theory for any type of production (or even for any type of operation), it seems too ambitious to develop and test such a theory within two and a half years. Therefore, the project proposal is restricted to the production of "complex, one of a kind products".

The third workpackage of the project focusses on formalization of the conceptual model. In this workpackage, languages and specification methods for the conceptual model are investigated. If suitable, existing specification methods, languages and tools will be adapted, and brought together into a workbench. Unfortunately, the budget does not allow for the development of new languages or tools in the workbench. Consequently, it is foreseen that we are forced to restrict ourselves to a subset of operations at this point. Workpackage 4 should provide guidelines and examples of the way in which to employ the utilities in the workbench for the design and description of production systems.

The last workpackage of the project comprises a demonstration of the capabilities of the workbench for simulated practical applications. For example, it should be possible to investigate a gradually changing implementation of CIM in a simulated production system, with consequences for such different aspects as cash flow, required personnel, stock levels, flexibility, production capabilities etc.

Currently, the project is at the end of its fourth month. Therefore, we are in the middle of workpackage 1. We will describe the status of this workpackage in section 3. Before describing this status, however, it is worthwhile to position the project in the current and anticipated state of the art. This will be the subject of section 3. Our work in workpackage 1 revealed, that there is a need for a typology of production systems for "one-of-a-kind" products. A first attempt to develop such a typology is given in section 4.

2. State of the Art

As indicated in section 1, the current state of the art is fragmented. In the preliminary work for this project, seven different views on CIM have been identified, viz.:

– Operations Research and Cybernetic
– Human centered
– Organizational
– Functional
– Communications and linguistic
– Databases
– Linguistical

Each of these approaches, and many other approaches, contribute to our under-standing of CIM. However, because they are not unified, each approach also contributes to the confusion. Of course, others have signalled this unfortunate state of affairs, e.g. the AMICE-project (ESPRIT nr. 688, Emond [1988]). Although four views are defined in this project, only two views (the function view and the information view) are described in some detail up to now. Nevertheless, this project provides valuable input for the research discussed here.

The **operations research** and/or **cybernetic view** of production systems which is often called **logistics** control, considers a production system principally in terms of well-defined tasks to be done and in terms of required resources for these tasks (materials, capacities, tools, ...). Traditionally, much effort within this view is spent on deterministic optimization in scheduling. Currently, much research effort is spent on the application of AI in this field. However, the notion of uncertainty has been largely neglected both in theory and in applications.

Surprisingly enough, much knowledge for dealing with uncertainty is available within the OR and cybernetics discipline, but this has hardly been applied to the production area. We feel that our research can contribute significantly at this point, especially if we concentrate on event-driven production situations, where emphasis is no longer on planning and scheduling, but more on feedback and hierarchical control.

The current state of the art, is described e.g. in Browne et al. [1988]. In Bertrand et al. [1989], the theory of logistics control is extended towards one-of-a-kind production.

The **human-centered view** considers a production system in terms of the positions which humans can fulfill within it. We feel that the distinction between capacity and capability is essential for describing future production systems. However, additional research is required to operationalise the concept of capability in a more formal way. Our proposal aims at contributing to this operationalization.

The human view is challenging the world of automation and tayloristic work organisation where the human factor is reduces to specific functions and subject of replacement by machines. One main objective in the latter world is, to develop architectures and technologies in order to replace the human in the production process. But the success and the applicability of this philosophy is limited to well defined production processes. This means: to all production processes where the embedded know-how can be made explicit. But a considerable part of production know-how cannot be made explicit and cannot be described in formal rules. According to the human centered view there is a "hidden situation" in every firm beneath and alongside the formal organisation of the firm, a situation which is largely unrecognized by the management or by external system experts. The conventional automation approach tries to overcome this situation by developing more sophisticated methods and techniques maintaining the fundamental objective to become independent from humans. In contrast, the human view tries to use both human capabilities and machine capabilities in an optimal way. This seems to be promising, especially in one-of-a-kind production. In this world the procedures are usually ill defined, there are simple organization structures, also ill-defined division of labor and low automation especially in the key-areas. The production situation is usually defined by uncertanties and the need for real-time acting and decision-making outside of the formal decision procedure. In this approach, many socio-technical ideas are applicable, such as employed in ESPRIT-project no. 1199, Human-centered CIM systems (see e.g. Murphy [1988]). Furthermore, attention will be paid to group-decision making and multi-criteria decision making.

The **organizational view** considers the production system as consisting of a number of departments connected together in order to fulfill the mission of production. The organizational view studies issues such as line/staff responsibilities, the game of budget control, responsibility accounting, and organizational strategy.

Current models for production systems in the organizational view are based on:

- small-scale production with low investment levels and high craftmanship
- large-scale production with medium investment levels and considerable direct labor with less skill
- dedicated automated factories such as chemical plants.

We feel that the models within the organizational view should be formulated for production systems which supply complex one-of-a-kind products to the market. Of course, the work in this field should be based on classical writers such as Galbraith [1973] and Mintzberg [1983]. The results of Esprit project 418 (see Roboam et al. [1988]) will be used.

The **functional view** studies each of the manufacturing functions in detail. These functions are e.g. tendering, product design, process planning, manufacturing, logistics, quality control, etc. Currently, these functions are treated as production phases, and their interfaces are subject to standardization. This approach is questionable. It leads to the implicit assumption that design can only start after tendering has been completed, that work preparation can only start after design has been completed etc. In future practice, production functions, such as design or logistics may occur in each production phase. Therefore, we feel that a reformulation of the functional view is necessary in a way which enables us to decouple production phases and functions. Such a reformulation would naturally include various forms of comakership. For this view, the work done in AMICE (Esprit 688, cf. Emond [1988]) may provide valuable input.

The **database view** considers a production system as a (possibly distributed) data base. The structure of this data base should correspond to the structure of the production system in reality. Each decision in reality is reflected by a change in the content of the data base.

The data-oriented view is in principle a suitable basis for application software development. However, two problems with the data-oriented view concern us here. The first problem lies in the fact that a change in the structure of the data base is not easily translated into a corresponding change in application software. A second problem in the data-oriented view is the fact that communication can only be modelled with great difficulty within this approach. We feel that object-oriented application development environments can help to overcome these two problems. Our task during this research is to investigate the possibilities of object-oriented methods for production system modelling.

Another option with the data-oriented view is to give the allowed operations a better mathematical foundation. We will investigate to which extent modern array theory can simplify the specification of allowed operations to a data base. Again, we will restrict ourselves to production systems which supply complex one-of-a-kind products to the market.

Finally, our intension is to investigate modular decomposition of the conceptual data model. Such a decomposition may provide a key for a gradual change of CIM-systems (cf. Pels and Wortmann [1985]). In the data-oriented view we will also use the result of Esprit pilot project 5.1/34 (cf. Yeomans et al. [1985]).

The **communication view** considers a production system primarily as a system where messages are exchanged between nodes in a network. The communication view studies protocols for the exchange of messages as well as the speed, capacity, and security of data transmission. Much has been achieved recently due to the OSI Reference Model and various recommendations which conform to it, such as MAP, TOP and ESPRIT 955: CNMA (cf. Goulding [1988]). Another inmportant source is, again, Yeomans et al. [1985].

The semantic aspects of communication (the meaning of the messages) has almost completely been left unspecified in OSI. It is assumed to be dealt with by the applications layer (OSI layer 7). We believe that different types of communication should be distinguished according to their semantics. If messages can be "typed" according to their semantics, the development of applications is simplified. This requires theoretical investigation of the meaning of messages in production systems. We are going to make a first study of such possibilities.

The pragmatic aspect of communication refers to the question what a receiver of a message is expected to do with this message. We believe that the ability to model the pragmatic aspect of communication provides many possibilities for intelligent support of such communication. If we were able to model the pragmatic aspect of communication, we would improve considerably our understanding of the production system. The semantic and pragmatic aspects of communication are called together: the linguistic aspect. Current applications of linguistics to the field of production are hardly existing. Therefore, we think that our work in this view will be explorative in nature first.

3. Status of the Project

As has been mentioned earlier, the project has just started. Each one of the seven partners provisionally takes care of one of the above seven views. In principle, such a view represents a particular branche of scientific knowledge. Our aim in the first workpackage is to obtain a precise description of these views. However, these descriptions are by no means goals in themselves. Rather, they serve as raw material for a synthesis.

For pragmatic reasons, the description of the views has taken the form of a **questionnaire**. In this way, a local research team is enabled to describe a particular factory according to all views. The questionnaires recently have become available and their usability seems to be good. The resulting descriptions of real-life factories are almost complete. These descriptions will give the research teams feedback on the quality of their questionnaires, and on the clarity of their viewpoints.

The results thus far lead to several observations:

1. **Some views are overlapping, and some views are missing**.
   Especially the communications view and the linguistics view are intertwined. Therefore, we decided to combine these two views into one view. Presumably, an "economic" view has to be added. The quality control view is still being discussed.

2. **The description of each view may be different in focus**.
   The description of a particular view (say, the data-oriented view) could be focussed on detailed rules about the symbols by which the view is to be described (e.g. a data-structure diagram). The view could also focus on detailed rules on the meaning of key-concepts (e.g. what is an entity type). It could focus on detailed rules on how to proceed. Obviously, the seven views described in this project are different in their focus.

3. **Views can be normative**
   A view is normative, if it assigns a predicate such as "good", "well structured", "consistent", etc. to a particular description of an existing or perceived production system. For example, a conceptual data model could be required to be in fourth normal form. The view becomes even more normative if it is always required to have product-data, process-data, and resources data in the conceptual model. If a view is normative, this is not necessarily an unfortunate state of affairs. After all, each engineering discipline consists mainly of prescriptions of this kind!

It will be clear, that these points and other ones have to be settled, before the project can proceed to workpackage 2.

4. Typology for One-of-a Kind Production

The term "one-of-a-kind" refers to a wide range of different organizations. Ii is hazardous to set up a typology, because these organizations are as different as species in biology, and they are changing substantially during their lifetime as well. However, a typology is a neccessity if we want to proceed in this field.

The basic question in a typology for one-of-a-kind production is, of course, the question **which** activities, procedures, resources, and messages are influenced by the unique product, and which are not. This basic question can become more pronounced, if it is replaced by the following two questions:

A. which **activities** in the primary process are customer-order driven?
B. which **investments** (in e.g. product-design, resources, procedures, or supporting activities) are customer-order independent?

In Fig. 1, these two questions are depicted as two dimensions of a matrix, representing different types of production. Several remarks must be made here: Firstly, question A yields the well-known distinction between make-to-stock, assemble-to-order, make-to-order, engineer-to-order. Although this distinction is often used, it is highly simplifying. Many production systems are producing some products for stock, and other products customer-order driven. In our typology, we accept this simplifiction currently. Secondly, it is possible that some upstream activity, e.g. purchasing, is customer-order driven, whereas a downstream-activity, e.g. transportation to the customer, is not customer-order driven (e.g. because transportation to this paricular customer is performed anyhow at regular intervals). We will neglect this complication and assume, that whenever an activity is customer-order driven, all downstream activities are also customer-order driven. Thirdly, we have replaced the expression "one-of-a kind" by the expression "customer-order driven". This equalization is not always justified, but seems to be acceptable for the time being. Fourthly, whereas the horizontal dimension in Fig. 1 is quite common, the vertical dimension is less well-known. The vertical dimension, however, explains many differences encountered in practice between production systems which are positioned at an equal point in the horizontal dimension. This wil be demonstrated already in Fig. 1, and will be elaborated upon below. Fifthly, both dimensions in Fig. 1 suggest discrete values, whereas in reality there is a continuum.

Despite of these shortcomings, Fig. 1 reveals some interesting differences, even if it is simplfied further, as is done in Fig. 2. Here, the vertical dimension is dichotomized into two types, **product-oriented production**, and **capability-oriented production**. The term "product-oriented" is meant metaphorically: a well-specified type of service (e.g. car-washing) is also called a "product". Generally speaking, a product-oriented company operates on the market by offering predefined products. It has made investments which allow the company to deliver these products cheaper, faster, or better than without such investments. However, the customer is restricted in his choice of products by standard-characteristics which have been predefined. For this reason, a service which is the same for all customers is seen as a standard product, whereas a servive which is negotiated, is not. A sauna, a restaurant, or a museum, are product-oriented; a doctor, a lawyer, or an artist, are not. A maintenance shop for cars or airoplanes offering standard preventive maintenance sevice at fixed prices is typically operating in a product-oriented mode. If a shop specializes in repair after crashes, it operates in a capability-oriented mode. Capability-oriented companies invest pimarily in skills. Whenever they invest in machinery, these investment are of a general-purpose nature. Sales and distribution channels are short. Top management's involvement is focussed on customers, personnel, technological innovation. In contradistinction, product-oriented companies' top do not focus on customers, but on marketing, not on personnel, but on systems, not on technological innovation, but on product innovation. For simplicity, we have dichotomized in Fig.2 also the horizontal axis, leading to four categories of one-of-a kind production, generically denoted as **the customization shop, the engineering company, the fill-in-the-forms company, and the jobber.**

Question A → Question B ↓	The customer-order drives (part of the) product design and the remainder of production and distribution	The customer-order drives component-manufacturing and purchasing assembly and distribution	The customer-order drives assembly and distribution	The customer-order drives distribution only
Investments are made largely independent of the customer orders ←→ **product oriented**	Engineer-to-order e.g. customer-modified professional equipment	Make-to-order e.g. standard professional equipment manufacturing	Assemble-to-order e.g. truck manufacturing large computers	Make-to-stock e.g. furniture
capability oriented → Investments are as much as possible delayed until the customer-order contract is signed	Engineer-to-order e.g. civil engineering construction company	Jobber	Assemble-to-order e.g. building construction company	Repetitive subcontracting of customer-specified parts e.g. caroutlets

Fig. 1

	engineer-to-order	manufacture-to-order
product-oriented	customization shop	fill-in-the-forms company
capability-oriented	engineering company	jobber

Fig.2. Simplified typology for one-of-a-kind production

We will shortly review the differences between these types of companies from the seven viewpoints mentioned earlier in this paper.

The **economic** viewpoint has been touched upon already in the above discussion. Our hypothesis is, that capability-oriented companies will require less capital investments than product-oriented companies, and they will take longer time to depreciate their assets (because these assets are general purpose). The cost-accounting system of a capability-oriented company will be strongly based on individual projects; product-oriented production requires a focus on the profitability of each product-line and each market segment.

The **functional** viewpoint describes the transformations of information or physical products in the primary process and in supporting processes. In a sense, Figs. 1 and 2 are based on the funcional view, because both dimensions are based on questions which are key-issues in the functional view. First, consider the horizontal dimension. Not surprisingly, we expect to find more customer-order driven design activities when moving from right to left. Second, consider the vertical dimension. We expect to find more customer-order independent activities, when moving from the bottom (capability-oriented production) to the top (product oriented production). This is certainly the case for product-design, but it holds also for the development of comakership relations, for marketing activities, distribution relationships, etc. Activities related to improvement of the primary process will be observed in both types of companies. However, in the product-oriented companies these activities are related to particular product-lines, whereas in capability-oriented situations they are not. Third, customer-order driven work-preparation acitvities are important for all situations except for the fill-in-the-forms company.

The **cybernetic** viewpoint is predominantly determined by the amount of uncertainty. Uncertainty should be distinguished into market uncertainty (what has to be delivered, and when?) and process uncertainty (how is this job to be done?). Market uncertainty is less for product-oriented than for capability-oriented production. Process uncertainty is less for manufacture-to-order than for engineer-to-order, because the job to be done is more clear upon order entry. Therefore, the fill-in-the-forms company has relatively few uncertainty. Production control will resemble MRP II and elements of repetitive manufacturing control techniques will be used. The jobber has more market-uncertainty; therefore, order-acceptance is a key element here. Work-preparation is required for each order, and gives some opportunity to handle changing market requirements. The customization shop has some elements of the fill-in-the forms company and the engineering company. Therefore it is the most complex control situation. The engineering company has the most uncertainty, but it has also the best means to design in an optimal way for production; therefore, it will not need better systems, but more communication and in-process flexibility.

The **communication** viewpoint describes regular communication pattrns on a formal or informal basis. In th capability-oriented company we expect much more informal communication. An interesting point is the communication between engineering and manufacturing. In the engineering company, this should be frequent and informal. In the fill-in-the forms company, this will be less frequent and much more formal. In the customization shop there is a distinction between the customer-order independent engineering and the customer-order driven engineering. The latter part resembles the engineering company, whereas the former part resembles the fill-in-the-forms company.

The **organizational** viewpoint describes the structure of authority, responsibility, and decision making. In make-to-order companies, this will be simpler than in engineer-to-order companies. In product-oriented companies, organizational structures based on different staff-professions and based on different product-lines should be expected. Hierarchy is emphasized; authority is based on organizational positions. In capability-oriented companies, an organizational structure based on projects and professionalism in the primary process (as opposed to staff-professions) is to be expected, with authority based on skill or role, rather than on organizational position.

The **human** viewpoint refers to management style, task-allocation, and personnel management. In the capability-oriented company we expect to find much attention for professionalism and skill, together with a customer-order orientation. Tasks are given to groups in stead of individuals, and communication between disciplines is important. On-the-job training, informal procedures, low turnover, and flexible working hours should belong to the culture. In the product-oriented companies more attention is given to systems than to humans, even in the customer-order independent design projects.

The **data-oriented** viewpoint describes the objects and interconnections which are important in information systems for support of the primary process, the management processes, or supportive processes. In the **fill-in-the-forms** production situation, customer-order independent data play a major role. Bills-of-material, routings, drawings and instructions are all available before the customer order is entered. Furthermore, these data are available in precise form: they should be complete, consitent, and up-to-date. Planning, plant monitoring, quality control and costing, all rely heavily on the existence of accurate data independent of the customer order. This does not mean that the customer order is an irrelevant object, but that its role is modest, and that information on the customer order is generated by these customer-order independent data sets. A typical example of such a generated data-set is automated process-planning. With the **jobber**, the situation is reversed. Here, there is hardly any product- or process-information in the primary process which is not related to customer orders. The description of resources, capbilities, and capacities may be independent of customer orders, but product information is not. In cases where the jobber uses a group-technology strategy, a kind of **reference routing** or **preferred routing** may exist, serving as a tool for work preparation to fit the work into predefined capabilities. These routings are not necessarily complete, consistent or up-to-date: human intelligence is needed, and perhaps supported by computer-aided process planning. In the **engineering company**, customer-order independent information exists, but it is not necessarily in complete, consistent or up-to-date form. These data concern primarily the nature of projects, in the form of **reference project structures**, or **preferred project structures**. Archive information on past projects may serve the same goal. Again, these data serve as an aid for the professional, not as a basis for his replacement. In the **customization shop**, finally, we expect to find similar reference data sets on **products**, in addition to the data sets mentioned above for the engineering company and for the fill-in-the forms company. Again, this type of shop seems to be the most complex situation of Fig. 2.

References

[1989] Bertrand, J.W.M., Wijngaard, J. and Wortmann, J.C. *Production control - A structural and design oriented approach*, Elsevier, Amsterdam

[1988] Browne, J., Harhen, J. and Shivnan, J. *Production Management Systems, A CIM perspective*, Addison-Wesley, Berkshire

[1988] Emond, J.C. "CIM-OSA: Key concepts overview and DEMONSTRATION", in: *Esprit '88 - Putting the technology to use*, part 2, pp. 1509 - 1527, North-Holland

[1973] Galbraith, J. *Designing Complex Organizations*, Addison-Wesley, Reading, Mass

[1988] Goulding, M. "Communications for CIM - an overview of the CNMA project", in *Esprit '88 - Putting the technology to use*, part 2, pp. 1589 - 1599

[1983] Mintzberg, H. *Structuring in Fives: Designing Effective Organizations*, Prentice-Hall, Englewood Cliffs, N.J.

[1988] Murphy, S. "Human-centered CIM systems", in: *Esprit '88 - Putting the Technology to use*, part 2, pp. 1615 - 1629, North-Holland

[1985] Pels, H.J. and Wortmann, J.C. "Decomposition of Information Systems for Production Management", *Computers in Industry*, vol. 6, no. 6, pp. 435 - 452

[1988] Roboam, M., Doumeingts, G., Zanettin, M., and Kiesewetter, S. "Study of manufacturing systems: Need of integrated methodology", in: *Esprit '88 - Putting the Technology to use*, part 2, pp. 1442 - 1461, North-Holland

[1985] Yeomans, R.W., Choudry, A. and Ten Hagen, P.J.W. *Design rules for a CIM-system*, North-Holland, Amsterdam 4.

Keywords

production theory, factory-of-the-future, CIM, one-of-a-kind production, typology.

THE DYNAMIC INTERPRETATION OF NATURAL LANGUAGE*
Report on ESPRIT Basic Research Action BR 3175 (DYANA)

Ewan Klein and Marc Moens
Centre for Cognitive Science
University of Edinburgh
2, Buccleuch Place, Edinburgh EH8 9LW, UK
eurokom: ewan klein
email: klein@epistemi.ed.ac.uk
     moens@epistemi.ed.ac.uk

1. Introduction

The Basic Research Action DYANA ("Dynamic Interpretation of Natural Language") is concerned with foundational research towards the development of an integrated computational model of language interpretation, covering the spectrum from speech to reasoning. The programme of work focuses on the following themes in natural language understanding:

• Dynamic interpretation

• Partial information

If one strives after a theory of interpretation which is both mathematically precise, and at the same time takes into account the fact that interpretation is a cognitive process, then obviously one wants the resulting theory not only to say something about the outcome of this process, but also to provide some insight into the nature of the process itself. As far as formal theories of phonology, syntax and semantics are concerned, this *dynamic* point of view has only been developed in a very sketchy and incomplete way: up to very recently, a large part of the field has almost completely neglected procedural aspects of interpretation.

Similar remarks holds for the theme of partial information, which is what is being manipulated in the process of interpretation. It is extracted, modified, transduced at various levels of representation—phonological, syntactic, semantic and pragmatic. These levels can be completely characterized by the kind of information involved and the way it is acted upon. Again, this may be too obvious an observation to make, but it should be remarked that up until a few years ago, the essential role of information had been underrated rather severely. This

---

* This paper is condensed from the project proposal, and as such represents the work of many authors. We would particularly like to thank the following for their contributions: Steve Isard, Chris Mellish, Michael Morreau, Mike Reape, Mark Steedman, Frank Veltman and Henk Zeevat.

neglect is, of course, connected with the neglect of the dynamic aspect of interpretation. What the process of interpretation amounts to is a change in the information state of the interpreter. How exactly the information state is changed depends on which information is to be interpreted. One might very well equate the meaning of this information with the change in information state that it induces. In doing so, meaning — and here we think of meaning at all levels of interpretation — becomes a dynamic notion; it is to be regarded as a function from information states to information states. It is easy to see now where the partiality comes in: both the domain and the range of such a function will be states of *partial* information — since complete information states hardly need an update.

Partial information arises in natural language processing, however, not only as a result of the dynamics of the context, but also as a result of the dynamics of the interpretation process itself. Problems of ambiguity are encountered at each stage of the interpretation process, and the result is that complete and reliable information cannot pass between the levels in a predefined way. In such a situation, it is desirable, if possible, to maintain a system that is 'opportunistic', yet monotonic. That is, as much information as possible should be communicated between levels, but not at the expense of passing results which later have to be retracted. In this way, each processing stage can have available as much information as possible to resolve its own uncertainties, but pathological global backtracking behaviour can be avoided. An incremental parsing system of this kind relies crucially on the ability to summarise states of partial knowledge and pass such information on to other places that can make use of it. A fundamental operation in a system of this sort is the merging of partial descriptions in such a way that complementary information is pooled, consequences of the merging are apparent and contradictory elements are identified. Whether we are dealing with information of a phonological, syntactic, or semantic nature, for instance, at a suitably abstract level similar operations are required and similar criteria apply.

In the past few years, people in a number of subfields of theoretical and computational linguistics, cognitive science, and formal semantics have been independently pursuing the study of partial information and the dynamics of interpretation within their own domains. For instance, within formal semantics there is considerable interest in partial logics and partial models, and in syntax the advent of unification-based grammar formalisms has also led to detailed accounts of information flow. The idea of semantic interpretation as a dynamic process of integrating new information into an existing knowledge representation has also been a key idea in much recent work on semantics, with influences coming from a number of sources: discourse representation theory, the semantics of computer programmes, and certain approaches in computational semantics.

In what follows we will illustrate the above, rather general, observations with a more concrete sketch of the role that information plays in disambiguating phonetic input within a model of language understanding.

2. An exemplary problem

Recent research has led to progress on a number of fronts towards the plausible and attractive picture of language comprehension, whereby successive portions of an utterance contribute to a gradual accretion of meaning as soon as they are encountered. If we assume for the moment that the input is spoken language rather than written, then the hearer has the task of identifying words from sounds, and also of fitting the words together into syntactic constituents. According to the hypothesis of *incremental interpretation*, candidate words may emerge before all of their sounds have been heard, and may be semantically evaluated more or less as they are proposed, rather than being strung together into a complete syntactic representation of the sentence which is subsequently subjected to semantic analysis.

In this style of analysis, the status of mappings between completed entities at distinct levels of representation remains broadly as it has been in standard models of grammatical organization. However, dynamic interpretation at all levels adds an extra dimension to the interfaces between them, and makes it more difficult to study the processing at a given level without regard to its neighbours.

To see some of the issues which arise with interacting levels of representation, consider the sentence

(1)   The bat slept.

We assume that (1) will occur in a discourse context—perhaps in response to the query *What did your pets do today?* – and the addressee must decide whether and how to update a mental model by incorporating the information content of the response, (1). As part of this process, the utterance will be assigned a phonological analysis (2a), a syntactic parse (2b), and a semantic representation (2c) —precise details are immaterial at this point:[1]

(2)   a.   /dh@batslept/

        b. $[_S[_{NP}[Det N][_{VP}[V]]]$

        c. $\exists!x[bat'(x) \wedge PAST(sleep'(x))]$

Consider the tasks which must be addressed by a processing device – human or mechanical – in arriving at (2c). A first problem is to recover a sequence of lexical items from the speech signal. If something like the form (2a) could be produced as an intermediate stage, the initial /dh@/ would have to be identified as an occurrence of the definite determiner *the* and subsequent portions as the noun *bat* and the past tense of the verb "sleep" respectively. Second, the words have to be fitted together into grammatical phrases; that is, the determiner and noun are combined to form a noun phrase, and this is combined with the intransitive verb to form a sentence. Third, the meanings of the words have to be combined to form a semantic representation for the whole sentence; that is, the definite description is represented by $\exists!x[bat'(x)...]$ (read 'there is a unique *x* such that *x* is a bat'), and the verb contributes both the predicate sleep' and the tense specification *PAST*; and the argument of the verb has to be bound by the interpretation of the subject noun phrase.

Early levels of representation will be subject to massive ambiguity. A given portion of speech wave will normally support several phonetic interpretations. In English, for example, the same physical signal can be heard as /rek@naizspiich/—"recognize speech"—or /rek@naisbiich/—"wreck a nice beach"—according to context. At the next level, even an unambiguous phoneme string of any length will be compatible with a large number of lexical hypotheses. The phonemes of the English word *associated*, for instance could also be segmented as *a sew sea eight had*, among numerous other possibilities, if syntactic and sense

---

[1] We use the symbol sequence 'dh' to indicate a voiced dental fricative, while '@' indicates a schwa.

constraints are not taken into account. (The *had* could arise via the contracted form *eight'd*, as in *There were only twenty students in the class to begin with, and eight'd dropped out before the exam*.)

In our example *The bat slept*, two additional analyses of the same utterance would be

(i)   The bat's leapt

(ii)   The bats leapt.

A further consideration is the lexical ambiguity between *bat* as 'noctural flying quadruped' and 'implement for striking a ball'.

There are grounds for believing that humans make phonetic, lexical, syntactic and semantic hypotheses very early on in the course of processing sentences, possibly on a word-by-word basis. Thus it is probable that lexical and syntactic knowledge will enable a choice of the phonetic string /dh@bats/ over /v@batz/ in casual speech, and discourse context will allow a selection among *the bat, the bats* and *the bat's* almost as soon as the sounds have been heard.

Consider briefly the kind of information that would play a part in eliminating the unwanted interpretations (i) and (ii), noted above. (i) is only grammatical if *bat's* is analysed as a possessive determiner whose head noun is taken to be a null anaphor; the value of the anaphor has to be sought in the discourse context; cf. *The bird's young fell out of the nest, but the bat's leapt*. Similarly, the short definite descriptions *the bat/the bats* carry a presupposition that some individual, or set of individuals, is familiar from the prior context. (This observation is not adequately captured by the kind of logical structure illustrated in (2c)—we return to this point shortly.) Presumably, one of these resolutions will be more plausible than the other. The lexical ambiguity between *bat* = 'flying quadruped' *vs*. 'implement' can be largely resolved by the sense-semantics of both verbs *sleep* and *leap* since they carry a presupposition that the subject is animate. Finally, other features of discourse coherence may indicate that the verb *slept* is to be preferred over *leapt*.

One might imagine carrying out the interpretation task in the following manner. First the speech signal is segmented into all possible sequences of phoneme hypotheses. Each of these sequences is in turn segmented into every possible set of substrings which match possible surface realizations of words in a given lexicon. The surviving word sequences can then be given a more detailed syntactic analysis, relative to some grammar, and the structures which result will be used to induce a compositional semantic representation. The induction proceeds bottom-up on the parse trees, associating appropriate representations with the lexical nodes in (2b), and using these to build complex representations for the dominating nodes until a representation for the whole sentence is achieved. By virtue of the semantic interpretation, it ought to be possible to filter out yet more of the analyses which were proposed earlier in the parse process.

The recognition regime we have just described has a pyramidal structure which can be summarised as follows: multiple hypotheses about the input sentence are proposed at the lowest level of the grammar, and these are progressively filtered out at higher levels until—hopefully —only one complete analysis remains. Such a regime fits fairly well with a model of grammar that has distinct rule components associated with distinct levels of representation, and where complete structures generated at one level are mapped serially into the structures generated at some other level. The architecture has one strong factor in its favour. Since many potential parses can be discarded at early stages of the recognition procedure, it is computationally advantageous to postpone detailed syntactic and semantic analysis of hypotheses which will only be thrown away. However, as applied to speech understanding tasks, the same parsing architecture suffers from the considerable (perhaps fatal) drawback that there is a combinatorial explosion of analyses for utterances of any significant length.

The architecture is also extremely implausible as a model of human sentence processing, both because it is known that interpretations at all levels are produced before sentences are complete, and because, if taken literally, it does not permit higher levels to compensate when lower level information is lost due to noise or error. In the case of function words like *the*, which are often pronounced in a reduced and indistinct form, it is not at all obvious that reliable phonetic hypotheses can always be generated bottom-up from the speech wave, and it may often be reasonable to arrive at such hypotheses top-down from the syntactic or discourse level. Such considerations can perhaps be viewed as phonetic level analogues to the 'null anaphor' case mentioned above.

The considerations adduced above argue for a theory of grammar and an interpretation regime whose joint effect is to assign lexical interpretations to partial utterances and semantic representations to sub-sentential phrases, and to allow evaluation of these partial representations relative to a discourse context. An appropriate model of grammar is one which associates rich descriptions with individual words, characterising them simultaneously at the phonological, syntactic and semantic levels, and where semantic representations are constructed as a side-effect of combining words to make phrases. Moreover, the semantic interpretation of words and phrases should be couched in terms of their contribution to a discourse, rather than an isolated sentence. Recall, for example, that the logical representation of the definite article proposed in (2c) was wholly inadequate for capturing the anaphoric nature of most definite noun phrases. Even more important than the properties of individual levels of the grammar are the *interfaces* between those levels. It is well known that there are mismatches between phonological, syntactic and semantic structure, yet the relation is far from arbitrary. So far, we lack a theoretical model which is descriptively adequate, formally well-defined, explanatorily powerful, and able to support an incremental interpretation procedure.

3. Objectives

3.1 RESEARCH OBJECTIVES

The research on DYANA is being carried out on an interdisciplinary basis, drawing on concepts from artificial intelligence, theoretical and computational linguistics, logic, and cognitive psychology. The following are some of the key research questions to be explored:

- What declarative theory of grammar (extending from phonetics to discourse) would best support incremental interpretation and top-down information flow?

- How can existing unification-based grammar formalisms be developed so as to increase their ability to express high-level generalisations, while retaining a clear semantics and a computational interpretation?

- How is the informational content of an utterance dynamically integrated into the current discourse context?

- What formal models can adequately capture the defeasibility and non-monotonic character of human reasoning?

- To what extent is it possible to synthesize the formal methods used for modelling partial information in speech, grammar, semantics and reasoning?

The work programme is divided into three interdependent themes:

(1)   Grammar Development, Speech and Prosody

(2)   Meaning, Discourse and Reasoning

(3)   Logic and Computation.

In **Grammar**, the emphasis is on formal models, and on abstract specification and parametric variation of linguistic structure. The formal model that we have taken as our starting point is the family of categorial grammars. These have attractive computational properties and provide a theoretically satisfying architecture for the interaction of syntactic, semantic, computational and phonological information. Abstract specification adds to the formal model by systematically characterising the properties of a language without any particular theoretical bias or concern for the computational effects of this specification. Parameterisation is a device to make the specifications as simple as possible, by allowing the expression of generalisations over natural languages or families of natural languages. In order to develop this approach, descriptive work is being carried out on identifying the parameters of variation in the Romance and Germanic language families.

In the area of **Speech**, there are two main goals. The first is to incorporate intonational structure into the representation of utterances, a task that requires a considerable amount of empirical study and modelling of intonation recognition before it can be carried out. The second goal is to investigate the interplay of intelligibility, predictability and discourse context in speech recognition and production. Both of these efforts will feed into an integrative task which attempts to construct a theory of declarative, constraint-based phonology within the framework of a unification-based model of grammar.

In the **Meaning, Discourse and Reasoning** theme, there are several related subthemes oriented around the intertwined topics of meaning, reasoning and discourse.

In **Meaning**, the emphasis is on further exploration of the family of dynamic and partial models of meaning that have been emerging in the past decade. The aim here is to arrive at a fuller development of the perspectives in question in the hope of eventually achieving a unification of these different perspectives. In the shorter term, we are developing more extended theories of meaning that exhibit the advantages of the new models; advantages such as computational tractability, analysis of information, and integration of discourse factors that were once classified as pragmatic factors. In contrast to earlier approaches, the proof theory of the semantic representations will play a prominent part in the investigation, so that the relationship between these theories and reasoning can be investigated directly.

In addition, the theme is investigating the basis of human reasoning. Much of this work is grounded in a thorough study of nonmonotonic logic, and it has two main aspects. First, it provides a basis for our accounts of nonmonotonic and conditional reasoning and of defeasibility in lexical concept combination. Second, it provides a basis for the fundamental problem of constructing a default interpretation of various aspects of linguistic structure in the face of only partial information.

The **Logic and Computation** theme is directed towards the development of formal mathematical and computationally tractable models of natural language and in providing computational support for grammar development and for the Action in general.An important question that is being addressed is how powerful a representation language one needs to express the generalisations of interest to linguists. In this context we are investigating formalisms that have feature-value pairs as their basic building blocks and unification as their basic operation to evaluate their usefulness for specifying information about linguistic objects. This work is proceeding along three axes.

First, we are attempting to derive an abstract characterisation of the kinds of description spaces that can be captured by standard unification systems—i.e. systems based on feature structures that take the form of trees, directed acyclic graphs and arbitrary directed graphs. We are considering , amongst other things, the behaviour of primitive descriptions, the kinds of sets of incompatible elements that can arise and various kinds of local finiteness that can exist in a description space. This enterprise will give us a formal understanding of what a description space must be like in order for these representational techniques to be appropriate.

Second, these formal results are used to investigate the linguistic arguments that have been proposed for various unification formalisms by determining whether they really do involve descriptions that do not have the required properties for a simpler mode of representation.

Finally, although formal results may indicate that, in principle, a certain representation language is adequate for a given problem, in practice that language may be computationally unattractive because of the size of the descriptions that arise. We therefore aim to isolate a set of unification extensions which yield a formalism that is not only representationally adequate but can also be implemented efficiently for the examples considered.

3.2 EXPLOITATION

The research themes that constitute the DYANA project aim to meet some of the challenges that lie ahead on our way to computer systems that use spoken or written human language. Apart from contributions in the form of technical reports, work will also be carried out towards the development of prototype implementations of some of the language processing techniques studied. Some of this work will also be used in other ESPRIT funded projects, such as the SUNDIAL project.

4. Participating teams

The work on DYANA is being carried out by the following teams:

University of Edinburgh:
> Centre for Cognitive Science, Department of Artificial Intelligence, Department of Linguistics, Centre for Speech Technology Research, Department of Computer Science

University of Amsterdam:
> Instituut voor Taal, Logica en Informatie

University of Tübingen:
Seminar für Natürlich-sprachliche Systeme

University of Stuttgart:
Institut für Maschinelle Sprachverarbeitung

The Project Manager is Dr Ewan Klein at the Centre for Cognitive Science, University of Edinburgh.

Esprit Basic Research Action: Sprint Project

SPRINT: Speech Processing and Recognition using Integrated Neuro-computing Techniques

Abstract

The objectives of the SPRINT project are to explore the particularities of neural networks (non-linearity, self-organization, parallelism, ...) to extend the capability profile of an automatic speech recognition system in important directions, in terms of adaptation to new environments (speaker, channels, ...), and noise reduction. This is done through a scientific and analytical approach, with the aim to design and evaluate the performance of neural networks at various stages of the speech processing chain, with the goal to solve the problems mentioned above. The analytical approach should allow to better understand the behaviour of neural networks, when applied to speech, to appreciate their usefulness, and in the future to efficiently implement and use them in speech recognition devices, to tackle the speech variabilities (inter-speakers, noise).

1. Introduction

Whatever the speech processing consists of, the main problems remain constant, i.e. noise interference, speaker insensitivity, large vocabulary. The performance of a system, usually characterized as a sole number reflecting the decoding rate, can be better specified as a capability profile, that is the capacity of a system to perform in the presence of noise, its multi-speaker mode, its vocabulary size, etc...

On the other hand, the efficiency of neural networks in various areas of pattern recognition has been demonstrated experimentally, for applications such as character recognition, vision, ..., excluding speech recognition, for which a lot of basic work still remains.

The aim of the work carried out by this project will is to examine, both theoretically and experimentally, whether connectionist techniques can be used to improve the current performance of automatic speech recognition systems (ASR) by expanding the capability profile in different main directions, in particular towards speaker independence and noise insensitivity.

The problem will be divided into individual tasks. On each task, research will be performed on the best use of the connectionist approach (which means some original research on unknown behaviour of neural networks), but with the global aim of improving the capability profile.

2. The Connectionist Approach

If doing research with the aim of improving the capability of speech recognition systems has obvious goals, many elements advocate the choice of exploring the connectionist approach. Obviously, for the general problem of pattern processing, the interest in using neural networks is that they have certain types of nice properties, namely their non-linearity, their ability to generalize and the automatic nature of their interaction and self-organization, and their parallelism (in that many hypothesis are tested at once). Apart from that, other reasons, more related to speech, are of great importance.

First of all, speech presents a great variability, as well as an evident noise corrupted pattern production. The capability of connectionism to perform well on these characteristics is now well accepted.

Next, if connectionism is old , its recent growth in research may put it in a position of quite new theory in the field of pattern recognition. This implies that the internal behaviour of this model is not well understood in many areas, especially in ASR. Still focusing on speech processing, most of the research done up to now leaves aside important problems, and has mainly concentrated on the use of neural nets for static word recognition. This is obviously far from sufficient, and lot of work still has to be done.

Last, but not least, an historical view on the techniques used for ASR highlights their different status: Dynamic Time Warping is now at a standard industrial level, Hidden Markov Models are now over prototype realization and are already available on the market, and if connectionist products can be developed, their application to speech is still in the laboratory. It is quite safe to say that the efforts put into thisapproach now will be rewarded in the future.

3. Tasks

Instead of trying to solve the speech recognition problem as a whole, the project is viewed as a set of tasks, each addressing a part of the speech recognition process. Speech is divided into levels of representation, and in this framework, each task consists of going from one level to another. For each of them, neural networks will be designed, and results compared to standard techniques.

In each task, neural networks will be investigated, including the choice of network architecture and of training method, and experiments, bearing in mind the general problem mentioned above, of improving the capability profile. Comparison with standard existing methods will be performed. The emphasis will be on research on the time dependency of neural networks.

The tasks are :

- Task 1: signal to parameter
- Task 2: parameter to parameter
- Task 3: parameter to phonetic
- Task 4: phonetic to sub-lexical
- Task 5: parameter to lexical

The tasks can be considered as relatively independent, although results achieved in one task (e.g. signal to parameter) would be used in the others (e.g. parameter to lexical).

By this analytical approach, we hope to first have a better and a deeper understanding of the behaviour of neural networks, especially when applied to a temporal phenomenon, such as speech, second to appreciate how and where neural networks are useful in speech recognition, third to achieve interesting results in speech separation from noise, and speaker adaptation.

4. Future impact

This research is oriented towards a fundamental problem, which is the use of connectionist techniques in the field of speech recognition. By using this approach, we actually have three distinct issues in mind. Of course, we want to have software that recognizes speech in the presence of noise, speaker independent or speaker adapted, and, if possible, with a better accuracy.

But it has seemed important to us to use the connectionist approach, not only for their proven capability to achieve certain tasks in noisy conditions, but for the rigorous underlying mathematical formulation. Markov models were a first step towards this, and let us recall that their success is mainly due to the underlying philosophy, that is the Shannon paradigm. And the connectionist networks are a first approach based on two powerful theories: non-linear filtering, and dynamic systems. Exploring this, applied to one of the most difficult problems of pattern processing, seems very promising.

Then, in this research, we have left room for an important problem which is not directly related to speech, but very important for a better understanding of neural networks: time dependency. This will help the connectionist approach to escape from its static framework, and should be useful in other areas than speech.

5. Conclusion

This project has started in April 1989, and first results will be obtained 15 months later. The future of the project can now be foreseen as going in two different directions:

- on a pure research basis, there is still plenty more to do. Let us quote, for instance, source separation (separation of the speech signal from other sources around), wavelet representation for speech coding (using wavelet-like neural nets), decomposition in independent components (expected to be more meaningful than the correlated ones).

- but the results obtained within SPRINT could be used in more industrial projects, having as a precise goal the integration into highly parallel machines, to respect the real-time constraint.

Those two possibilities will be the guarantee that basic research is important to achieve, and that its industrialization is not an impossible goal. It must be quoted that this project is carried out by different groups, coming from industry, from research centers, from universities, and that we took care that specialists in speech processing and specialists in neural networks would work on common tasks. This synergy will allow an inter-disciplinary research to be carried out and should insure a high quality project.

The partners involved in SPRINT are:

CSI: Cap Sesa Innovation , France (Prime contractor)
Khalid Choukri
Serge Soudoplatoff

ENST: Ecole Nationale Supérieure des Télécommunications, France
Gérard Chollet
Jean Pierre Tubach

IRIAC: Institut de recherche en Intelligence Artificielle et connexionisme, France
Francoise Fogelman
Patrick Gallinari

RSRE: Royal Signals and Radar Establishment, England
Roger Moore
Andrew Varga

SEL : Standard Elektric Lorenz, Germany
Heidi Hackbarth
Manfred Immendorfer

UPM: Universidad politecnica de Madrid, Spain
Thierry Michaux

Key words:

Speech Processing and Recognition, Neural Techniques, speaker insensitivity, speaker adaptation, multi-speaker mode, time dependency, noise reduction, network architecture, self-organization

DIALOGUE AND DISCOURSE*
Report on EBRA Working Group 3351 (DANDI)

Ewan Klein and Marc Moens
Centre for Cognitive Science
University of Edinburgh
2 Buccleuch Place, Edinburgh EH8 9LW, UK
eurokom: Ewan Klein;
email: klein@epistemi.ed.ac.uk
        moens@epistemi.ed.ac.uk

1. Introduction

The DANDI Working Group, supported by the ESPRIT Basic Research Actions, is a collaborative framework involving five consortia with significantly overlapping research interests in the area of Dialogue and Discourse. The main aim of the Working Group is to increase the interaction and scientific cooperation between the European research centres involved in DANDI, by means of visits within and between the consortia, through the exchange of research results and through the organisation of special interest group meetings and workshops.

The overarching research interest of the Working Group is to investigate how the flow of information in dialogue and discourse is structured and controlled. This involves the study of how information encapsulated in the speaker's message is integrated in a coherent way in the listener's expanding knowledge. It is modelled as a process of continuous information growth, where all participants in the discourse try to maintain a partial yet consistent model of the world and of the dialogue they are participating in.

2. Information

A central idea in the classical philosophy of language has been that one knows the meaning of a sentence if one knows the conditions under which it is true. The technical paradigm embodying this view is Tarski's extensional truth definition. Its format is a recursive, compositional specification of the following notion:

an expression ϕ is true in model $\mathcal{M}$ under the variable assignment g.

---

* Many thanks to Henk Zeevat for helpful comments.

Here, $\mathcal{M}$ is a model of the real world, or some other situation under discussion, often construed in a timeless fashion (e.g., as some frozen 'snap-shot' of the universe). The assignment g fixes the values of the variables occurring in ϕ: different assignments may give different truth values. In Tarski's machinery, complex sentences ϕ have truth conditions which are constructed systematically out of those for the components of ϕ, thus following Frege's Principle of Compositionality.

A richer picture arose in the intensional truth definitions of Kripke and Montague in the sixties. There sentences are evaluated relative to different 'indices', each of which represents one possible world or history in a universe $\mathcal{M}$ containing all relevant courses of events. The meaning of a sentence ϕ may then be identified with the set of worlds at which ϕ is true. In other words, the propositions expressed by sentences can be taken to be sets of possible worlds. This is the influential proposal known as 'possible world semantics'.

With the introduction of further parameters representing entities like addressee, location, and speech time, possible worlds semantics reached a considerable sophistication. Yet it is still beset by one fundamental problem. As we have seen, classical possible worlds theory treats propositions as sets of possible worlds. This means that two logically equivalent sentences (like the tautologies in (1) and (2)) are taken to represent exactly the same proposition, since they are true in exactly the same possible worlds:

(1)   White is white and black is black

(2)   Either John is at the party or he isn't

Yet, we have the feeling that they say quite different things, that their information content is distinct. There have been a number of attempts to modify the classical possible worlds framework to avoid this problem. In the early eighties, various theories were put forward (e.g. Situation Semantics and Discourse Representation Theory) advocating a major shift in focus, treating information not as a mere additional parameter in the truth-conditional scheme, but rather as a central notion of semantics. As a result, the meaning of a sentence is conceived of as a mapping from partial information states to partial information states.

3. Research Themes

How information is encoded in and extracted from discourse is the central concern around which the research interests of the Working Group are centered. Within this general area, several different foci of interest can be discerned, some of which we will briefly discuss here. Note that this discussion is based on the contributions of the 50 or so researchers in this Working Group; it is unfortunately impossible to associate the names of these researchers with every idea mentioned.

3.1. QUESTIONS AND ANSWERS

The semantics of questions and answers is an important topic of research in the Working Group for two reasons. One is that questions and answers constitute an important part of natural language dialogue; a semantics for questions and a theory of what constitutes an appropriate answer to a given question is of central importance to a theory of discourse. Another reason for studying questions is that the way the listener integrates newly conveyed information with discourse-specific old information, with knowledge about the domain of discourse, and with general background knowledge, can be viewed as a question and answer process—a process which has been called "cognitive questioning" (Seuren). The assumption is that the same structuring principles underlie these implicit questions as well as explicit linguistic questions: in both cases, a valuation is sought for a particular parameter. The specification of this value is the answer—the implicit cognitive answer or the explicit linguistic one.

3.2. FOCUS, TOPIC, COMMENT AND PRESUPPOSITION

The interesting point about cognitive questioning is that speakers apparently anticipate certain forms of cognitive questioning their listeners are likely to engage in. This is reflected in the topic-comment structure of the utterances. Although most people are convinced that notions like topic, comment and focus play an important role in natural language discourse, work in this area has often been quite intuitive and pre-theoretical.

The question-answer perspective seems to provide better tools in this respect. It is commonly accepted that notions like topic can often be assigned only in context. The key idea is that what the speaker puts in topic position is assumed by the speaker to be a question that might arise in the listener's mind; the comment is the answer to this question. Depending on the context in which it occurs, in the sentence

(3)   It was John who left

the *it*-clause can be construed as the answer to a cognitive question like *Who left?* In other words, the speaker assumes that the listener already knows that *someone left* (the topic of the discourse) and adds the comment that it was John.

There is clearly a considerable overlap between such an account of topic/comment distinctions and any theory of presupposition: (3) pragmatically as well as semantically presupposes (4), which was identified as the topic of the discourse:

(4)   Someone left

The study of presupposition in an integrated theory of language and information also constitutes an important area of research in DANDI. We will return to this strand of research in section 3.4.

3.3. INTONATION

The given/new or topic/comment structure of utterances is also influenced by intonation and accentuation. Consider Ladd's example:

(5)   A: Why don't you make some FRENCH TOAST?

        B: There's nothing to make French toast OUT of.

In B's utterance, the lack of accent on *French toast* is an indication that it is already known material. Rather than making speech processing easier through clear articulation, deaccenting signals that these words do not need to be newly decoded at all. Compare this to (6), where the word involved *is* new:

(6)   A: Why don't you make some FRENCH TOAST?

       B: There's nothing to make the BATTER out of.
            *There's nothing to make the batter OUT of.

A proper understanding of the role played by intonation in indicating what sort of dialogue 'move' the speaker is making will have to develop in conjunction with more thorough and precise theories of dialogue and questioning. Consider

(7)   A: Are you going home now?

       B: Yes

If B's reply is spoken with a low fall, the impression is that he takes A's question as a natural one, whose motivation he can probably understand and does not find worrying. By putting a low rise at the end, he would convey that he wonders why he is being asked, and is perhaps slightly concerned about what A is going to follow on with. It is these pragmatic functions of intonation that the DANDI Working Group is especially concerned with.

3.4. TEXT COHERENCE

The basic assumption made in interpreting an expression in context is that the text in which it occurs is coherent. Sometimes such coherence relations are marked explicitly, by means of connectives like *and* or *but*, which specify how the information in the clause is to be integrated in the knowledge structure built so far; at the same time, the current information state may influence the interpretation of the element. Another coherence device is anaphora -- the use of elements which require an appropriate antecedent to be recovered for their interpretation. Both devices are the topic of research by teams involved in DANDI .

But also in the absence of such devices will a reader assume that the text she is processing is coherent, and she will continually attempt to accommodate the clauses she is processing in the model of the discourse she is constructing. This means that a natural language semantics not only has to allow for the introduction of new discourse referents and the interpretation of the following text with respect to them, but also has to allow for the possibility of adding presuppositions and assumptions to the discourse model if they are necessary for interpreting a part of the text such that it is coherent with its context. Recently, van der Sandt has made promising proposals for integrating presuppositions within the general anaphoric framework of DRT. The presuppositional part of (3), for example, was identified before as in (4) and is repeated here as the first clause of (8); the propositional content of (3) can be paraphrased as the second clause in (8):

(8)   Someone$_i$ left and it$_i$ was John

It is intuitively clear that the *it* in (8) behaves just like a pronoun, yet it specifies the same entity as the one introduced in the discourse by the presuppositional part of (3).

Finally, apart from strategies for creating coherent representations of the discourse one is processing, one also has to provide strategies of repair if the interpretation of an expression requires a previously introduced referent that, in fact, has not been introduced. With respect to these strategies of repair or accommodation there have to be flexible limits, which, when stretched too much, result in unacceptable texts. These repair strategies are also an active area of investigation among DANDI teams.

3.5. LEXICAL SEMANTICS

A final strand of research that wil be pursued by several of the participating teams is the context dependency of lexical meaning. One problem that will be studied is how words can be stored in such a way that they can interact with the information emanating fom preceding discourse. A specification of this is of crucial importance not just for a better understanding of the processes underlying the incremental construction of a discourse model but also for the processes involved in the generation of language. The accurate encoding of a semantic message into language depends on the efficient choice of the right words from a vast lexicon. In this production process, just as in the interpretative process, the emerging meaning of the discourse has to be taken into account, as well as the meaning associated with each item in the lexicon.

4. Objectives

One important theoretical thread that runs through all these research foci is the development of discourse models that are psychologically motivated. This involves a study of the mental models of dialogue participants and of the inference abilities they bring to bear on these models. A study will be made of what logics can be used to formalise this process of model building.

The results of this work can feed into the formulation of a more plausible computational model of dialogue phenomena. This computational model in turn will lead to the formulation of guidelines for the development of more efficient natural language front-ends. The specification of the architecture of person-machine dialogue systems will be advanced greatly by a deeper understanding of how the exchange of information is achieved in person-to-person dialogue.

5. Participating Consortia

There are five consortia participating in this Working Group, comprising in total some 20 European teams:

Lexical Semantics and Discourse Representation:
> University of Tilburg, Max-Planck Institute for Psycholinguistics (Nijmegen), University of Nijmegen, Institute für Deutsche Sprache (Mannheim);

Cognitive Pragmatics:
> University of Milano, Ecole Polytechnique de Paris, University of Cambridge;

Linguistic Pragmatics:
> University of Antwerp, University of Edinburgh, The Linguistics Institute of Ireland, Universität des Saarlandes (Saarbrücken);

Incorporation of Semantics into Computational Linguistics:

UMIST (Manchester), Belgian Institute for Management, University of Oslo, Istituto Dalle Molle (Geneva), Copenhagen School of Economics, University of Liège;

Questions and Discourse:
University of Essex, University of Nijmegen.

The activities of the Working Group DANDI commenced on 1st May 1989, and will continue for 30 months. The Coordinating Contractor for the Working Group is the Basic Research Actions project 3175 DYANA (Dynamic Interpretation of Natural Language); contact person is Dr Ewan Klein, at the University of Edinburgh, Centre for Cognitive Science.

INDEX OF AUTHORS

INDEX OF PROJECT NUMBERS

INDEX OF ACRONYMS

Acronym	Prj. num	page
SACODY	1561	757
SEDOS DEMO	1265	528
SKIDS	1560	569
SMART	1609	582
SPRINT	3228	1108
TALON	870	947
THORN	719	1065
TOOLUSE	510	375
UCOL	2054	1023
VITAMIN	1556	740
WSI	824	126

INDEX OF KEYWORDS